T0073872

The Badgers of Wytham Woods

The badgers and the wood are filled with fascination and beauty, and so too is the science that grows from the quest to understand them. Reproduced with permission from an original painting by Lucy Grossmith.

The Badgers of Wytham Woods

A Model for Behaviour, Ecology, and Evolution

David W. Macdonald

and

Chris Newman

OXFORD
UNIVERSITY PRESS

OXFORD
UNIVERSITY PRESS

Great Clarendon Street, Oxford, OX2 6DP,
United Kingdom

Oxford University Press is a department of the University of Oxford.
It furthers the University's objective of excellence in research, scholarship,
and education by publishing worldwide. Oxford is a registered trade mark of
Oxford University Press in the UK and in certain other countries

Impression: 1

Published in the United States of America by Oxford University Press
198 Madison Avenue, New York, NY 10016, United States of America

British Library Cataloguing in Publication Data
Data available

Library of Congress Control Number: 2022935318

ISBN 978–0–19–284536–8

DOI: 10.1093/oso/9780192845368.001.0001

Printed and bound by
CPI Group (UK) Ltd, Croydon, CR0 4YY

For Hans Kruuk and the great tradition of naturalist scientists.

Preface

The origins and purpose of this book, and the project that it recounts, are explained in the Prologue. Here, therefore, our task is to explain enough about our approach to serve as a sort of users' guide, and to enable us to thank the many people on whose giant shoulders we have stood while trying to glimpse the horizon beyond which lie truths about badgers and nature more widely.

First, what sort of book is this? We will have failed if the answer is not vibrantly clear to readers a couple of pages hence. This is a scientific book, and we expect it to be judged by the high standards of science; we hope it is also enjoyable and accessible to a readership far beyond professional scholars, indeed by anybody with a fascination for nature. Second, what sort of team are we? Chris began as David's doctoral student at Oxford, and grew into his right hand—a hand without which the Wytham badger project would not have survived. In this sense, as a team should be, we need each other, and so it has been as we have written the results of our 31-year partnership. This shared journey by two people caused us a grammatical hiccup: mostly we have a shared story to tell, and the pronoun 'we' captures the reality, but on other occasions we want to write something about each other. The solution is that we sometimes refer to each other by our first names; for example, as Chris observed or as David recalled. Third, our modus operandi deserves brief mention: we have not had the financial or logistical resources, far less the mental capacity, to keep all strands of this project running at full steam for the nearly 50 years about which we write; rather, we have dipped into particular topics for a few intense years, before largely turning to another. Happily, this means we've covered a lot of topics; sadly it means we cannot always tie everything together simultaneously.

How have we been able to amass the information synthesized in this book? This has been made possible only through the generosity of our sponsors. We thank them all, but foremost amongst them, for more than 20 years, was the Peoples' Trust for Endangered Species.

The then Chairman, Sir John Beddington, and David were thrown together by many shared interests and opinions, all of which converged on badgers. Without John's support, and that successively of Valerie Keeble and Jill Nelson, the Wytham badger project would not have survived. We thank them deeply.

This project has been enriched by the intelligence and hard work of many graduate students. We warmly acknowledge them as the engine of our discoveries. They have gone on to distinguished careers, and we delight in their varied successes—several of which are mentioned in the book. In approximate chronological order they are: Heribert Hofer, Jack da Silva, Rosie Woodroffe, Paul Stewart, Dominic Johnson, Xavier Domingo-Roura, Frank Tuyttens, Chris Newman, Christina Buesching, Sandra Baker, Hannah Dugdale, Inigo Montes (Coventry), Pierre Nouvellet (Sussex), Yung Wa (Simon) Sin, Geetha Annavi, Michael Noonan, Kirstin Bilham, Nadine Sugianto, Tanesha Allen, Julius Bright Ross, and Ming-Shan Tsai. In addition, and no less valued, a large number of other members and associates of the WildCRU have contributed in major ways, of whom an incomplete list includes: Greg Albery, James Bunyan, Ruth Cox, Stephen Ellwood, Rebecca Fell, Becci Foster, Kerry Kilshaw, Bart Harmsen, Graeme McLaren, Rubina Mian, Philip Riordan, David Slater, Sophie Stafford, Katrina Service, Tom Simmons, Josephine Wong, and Nobuyuki Yamaguchi. Those whom we have omitted will hopefully take pity on the disarray of our records and deterioration of our memories, and trust that we valued them nonetheless. It is no slight to the others, and their achievements and contributions to mention especially Dr Christina Buesching with special warmth and gratitude: from her start as a doctoral student, Christina toiled for the project for 25 years of dedicated work, participating with Trojan effort in almost every catch-up as an indispensable member of the core team.

This is a book about badgers, but most especially about the badgers of Wytham Woods. The woods, and our capacity to function there, depend on the skill and

David Macdonald

Chris Newman

goodwill of the woodland staff, and we take great pleasure in what approaches a lifetime of friendships with Nigel Fisher, Kevin Crawford, Nick Ewart, and Neil Havercroft. We thank them for their tolerance, support, and interest.

In addition to our research collaborators, many others have supported us; we thank them all and single out particularly Robin and Karrie Langdon for their voluntary devotion to database management.

The preparation of the book has been a weighty task, and sometimes caused us to stray beyond our core expertise. In keeping us from straying too far onto intellectually thin ice, we thank those who read hunks, and in some cases all, of the text: Greg Albery, Adi Barocas, Jan Bayley, Malcolm Bennett, Kirstin Bilham, Stephan Bodini, Julius Bright Ross, Dawn Burnham, Liz Campbell, Chris Cheeseman, Poul Christensen, Christl Donnelly, Julian Drewe, Linda Griffiths, Gordon Harkiss, Andy Gosler, Paul Johnson, Jan Kamler, Keith Kirby, Annie Kerriage, Tom Langholm, Sil van Lieshout, Gus Mills, Pierre Nouvellet, Christopher O'Kane, Beth Preston, Mark Stanley Price, Chris Ruiz, Pavel Stopka, Simon Sin, Margarida Santos Reis, Nadine Sugianto, Ming-shan Tsai, Frank Tuyttens, Neil Watt, Andrew Wood, and Tristram Wyatt. Nobody has supported the rigors of the writing process, and its endless drafts, more loyally, tolerantly and perceptively than Dawn Burnham. We are also grateful for the assiduous grammatical precision of Stephan Bodini in reining in our brutality to the English language, and the tireless loyalty of Christopher O'Kane in always responding to endless pleas for help. In the same vein we are grateful for the publishing prowess of Charlie Bath and Ian Sherman at Oxford University Press—David has known Ian since he began his career at Oxford University Press, they have conspired together often, and it seems fitting that our book should be amongst those that bookend that distinguished publishing career.

As our thanks broaden, we think especially of Tom and Dafna Kaplan—although badgers missed the boat by about 35 million years in failing to capture the Kaplan hearts that beat for wild felids, nonetheless the Recanati-Kaplan Centre, home to the WildCRU at Tubney House, has nurtured the badger project along with the diversity of projects that flourish therein thanks to the foundation laid by Tom and Dafna. Badgers may not be cats, but, in their role as a model for behaviour and evolution, they have a contribution to make. Further general thanks goes to our ever-supportive Oxford College, Lady Margaret Hall (LMH). It is a wonderful continuity that as we wrote this book Chris was supported as the H.N. Southern Fellow in Conservation at LMH: Mick Southern, latterly David's doctoral supervisor when Hans Kruuk left Oxford, had been a pioneer ecologist in Wytham in the days of Charles Elton—his widow, Kitty (née Paviour-Smith) died in 2017 bequeathing to David at LMH the funds that fuelled Chris' Fellowship. Without Mick, complete with pipe and leather elbow pads, we'd all know less about Wytham's ecology, and without Kitty's memorial bequest, we could not have written this book.

David W. Macdonald and Chris Newman
Lady Margaret Hall and WildCRU,
Department of Zoology,
University of Oxford, UK

Contents

Prologue   xi
Maps of Wytham Badger Setts   xiii
Foreword   xvi

1   Setting the Scene: Births and Beginnings   1

2   It's Tough at the Bottom   22

3   Apprenticeships for Badger Society   30

4   Setts, Society, and Super-groups: The Geology of Social Behaviour   44

5   The Sum of the Parts: knowing One's Place in Badger Society   67

6   Social Odours: The Perfume of Society   92

7   Sex: How and Why, and with Whom?   115

8   Social Behaviour in an Uncooperative Society   144

9   Who Goes There: Friend or Foe?   164

10 The Ecological Foundations to Badger Group Living   191

11 The Economics of Survival: Population Size, and Crashing Through the
    Ceiling   221

12 Weather: Actuarial Insights   248

13 Weather: Badgers Adapt or Die   259

14 The Game of Life   272

15 In Sickness and in Health   305

16 The Story of Badgers and TB: Perturbation and Beyond   337

17 Genetic Mate Choice—Quality Matters   385

18 Senescence, Telomeres, and Life history Trade-offs **413**

19 Of the Same Stripe, or Not—Exceptions That Prove Rules **440**

References 479
Index 550

Prologue

Anything appears greatest to him that never knew a greater.

Lucretius,[1] *De Rerum Naturae* (Book VI)

Beginnings are tricky to identify, perhaps impossible, because each has another standing behind it, but you've got to start somewhere. We have two beginnings to deal with here: our badger story; and the story of the badger. As to the beginnings of badgerologists, David credits his Glaswegian father, doctor of medicine and scratch golfer, teaching him, aged 10, to use plaster of Paris (intended to splint broken limbs) to make casts of badger and fox paw prints in the sandy bunkers of St George's Hill Golf Club (although, considering the influences of fate or family culture, the journey to Chapter 16—on the topic of tuberculosis—may have been written in the tealeaves when his pioneering great aunt Elizabeth, one of the first women doctors in Glasgow, wrote her prize-winning MD thesis *On the Value of the Tuberculo-opsonic Index in Diagnosis*[2]).

Chris, for his part, with his fascination for the natural world inspired by his great aunt Mabel Walker, recalls introducing himself from his crutches at York Sixth Form College having dozed off while perched precariously—too precariously as it turned out— in a tree while trying to catch a glimpse of badgers on Haxby Common. Chris, however, was still looking forward to his third birthday when David first encountered the badgers of Wytham Woods: in the autumn of 1972, as a doctoral student who at first considered lumbering badgers a bit of a nuisance in his single-minded pursuit of exquisitely graceful foxes. David determined to make himself into a naturalist scientist, to which end he sought out the great natural historians of the day to teach him badger-craft: Ernest Neal, author of *The Badger* (1958), and film-maker Eric Ashby with his artificial sett in the New Forest (it was the cohabiting foxes that first lured him there). Above all he worshipped at the feet of Hans Kruuk, his doctoral supervisor—of whom more in a moment—and standing behind him, Niko Tinbergen, Oxford's Nobel Laureate pioneer of ethology, whose 1964 masterpiece *Social Behaviour in Animals* had been his bedside companion in draughty boarding school dormitories. It is a perspective on the historical span of our study that David first began naturalistic stalkings in Wytham more than a decade before DNA-fingerprinting was invented, while as this book went to press we published the first full genome of one of our study badgers[3]—an achievement now approaching commonplace but that would then have been dismissed as science fiction.

In the way of academic genealogies, Chris, more motivated by curiosity about how the natural world works than by naturalist observations (that broken foot had dampened his enthusiasm for hours of treeborne vigil), became David's research assistant in 1991, became his doctoral student in 1994 and later in this book the lineage consisting of Niko, Hans, David, Chris

---

[1] Lucretius, *c.*99–55 BCE, the Latin poet, wrote, as translated by Cyril Bailey of Balliol College, 'A little river seems to him, who has never seen a larger river, a mighty stream; and so with other things – a tree, a man – anything appears greatest to him that never knew a greater'. To this translation of *On the Nature of Things*, we might add to the list—a tree, a man—the badgers of Wytham Woods, as we do our best in this book to follow Lucretius' exhortation to think out of the box, and to share with readers our own surprise at these extraordinary models of behaviour, ecology and evolution.

[2] University of Glasgow (n.d.). Surgeon Elizabeth Fraser. https://www.universitystory.gla.ac.uk/ww1-biography/?id=834 (accessed 29 March 2022).

[3] Chris Newman, Ming-shan Tsai, Christina D. Buesching, Peter W.H. Holland, David W. Macdonald, University of Oxford and Wytham Woods Genome Acquisition Lab, Wellcome Sanger Institute Tree of Life programme, Wellcome Sanger Institute Scientific Operations: DNA Pipelines collective, Tree of Life Core Informatics collective, Darwin Tree of Life Consortium (2022) The genome sequence of the European badger, *Meles meles* (Linnaeus, 1758). Wellcome Open Research.

will stretch on to include Chris's students. But what of Hans Kruuk, whose influence infuses this book (and who published his own book, *Badgers*, in 1989)? Briefly, Hans left the Serengeti and its spotted hyaenas to join Tinbergen's Animal Behaviour Research Group in Oxford where, in October 1972, David became one of his first three graduate students.

Hans pioneered the study of badgers in Wytham Woods, and was the first to appreciate the paradox that badgers there live in social groups (Hans described these as occupying exclusive, defended territories, ringed with border latrines), but, unlike most group-living carnivores (or indeed most group-living vertebrates), appeared to show little evidence of cooperation. Hence our opening question, inherited from Hans: why do badgers live in groups? From what perspective do we frame this question, and the cascade of others that follows? Our perspective is that of biological function—why are things the way they are—expressed first by asking what is the adaptive significance of badger behaviour, most particularly their social behaviour; how this reflects their life history and physiology; and how it can be interpreted as an emergent property of their ecological circumstances, framed by the constraints of their mustelid phylogeny and, in turn, their mammalian ancestry.

That first, catalysing question is about adaptive significance, and its tendrils extend from sociology deep into ecology and far beyond. It launched us into an exploration of how a species adapts through natural selection optimally to fit, and to thrive in, its environment. So while, in the British colloquialism, this book

reveals what makes badgers 'tick', its higher purpose is to offer general insight into evolutionary ecology through the prism of the badgers of Wytham Woods as models for behaviour, ecology and evolution.

That is one beginning for our story of badgers, but where to begin the badger's story, and thus the book? This is a puzzle of the chicken and egg variety: lives cycle, generations pass, but we have to break into that cycle somewhere, and we have chosen not the egg but the hatching: birth. In the first chapter we deliver the infant badger—pausing briefly for a first, taxonomic, glance at the contextual question: what is a badger? In the next two chapters we follow the badger cub's progress towards joining the society that it, like us, will need to understand. Chapters 4–9 provide answers to questions that are largely about badger behaviour and become the building blocks of an appreciation of badger society, which is then, in Chapter 10, set within the framework of badger ecology. Next, we leap a stratum of ecological organization and devote four chapters to populations, the emergent entity of assembled individuals, examining how life histories are influenced by social and environmental circumstances, ageing, and attempts to find optimal fitness strategies while subject to the vicissitudes of ever-changing conditions. And then we dive into detail, through genetics, immunology, endocrinology, and pathogens, ending the Seventh Age of the badger's life in senescence, until we emerge, in Chapter 19, equipped to evaluate variation at two levels: within and between species, to deliver a deep answer to the question, what is a badger?

David (left) and Chris with an anaesthetized badger during one of the 130 catch-ups, yielding 11,488 capture records and the life histories of the 1,823 individual badgers monitored from cradle to grave that are the foundation of this book.

Maps of Wytham Badger Setts

Readers of *The Badgers of Wytham Woods* will come to know some of the badgers personally, and will refer often to the setts in which they live. The woods are dotted with over 1000 badger holes, but only about 70 of them have names and it is principally 23 of these more major setts that feature prominently in our story. The aerial photo shows the landscape, the borders of the wood, and, as white dots, those setts that feature by name in the text. On the facing page each of these setts is given an abbreviation and here we provide the key to the names these codes abbreviate. Readers who wish to follow the personal stories of badgers and their social groups in detail can find on these maps the sett names referred to in the text. This map is a sliding amalgam across time: some named setts are no longer there (e.g. BP, RCO), the fortunes of others have waxed or waned—sometimes prominent, sometimes in cobwebbed disuse—and yet others are more or less closely linked (e.g. those on the outskirts of main setts such as HC, P, TC, and HH). Nonetheless, we have monitored individual badgers in every one of these named setts, and for readers eager to delve deeply into their story the map will offer spatial reference.

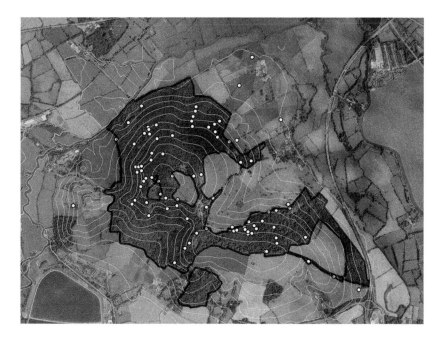

An aerial photo of Wytham Woods and surrounding landscape, depicting contours, and with white dots marking major badger setts.

Key: DO, Ditch outlier; FB, Firebreak; GO, Thames, Thames outlier; Great Oak; GOO, Great Oak Outlier; GOA, Great Oak Annex; GOB, Great Oak border outlier; Chalet, CH; CHA, Chalet Annex; CHO, Chalet Outlier; HC, Hill Copse; HCTop, Hill Copse top outlier; HCWO, Hill Copse western outlier; HC Flat, Hill Copse flat outlier; HCEO, Hill Copse eastern outlier; KH, Keeper's Hut; Oblivian, Oblivian outlier; GAH, Great Ash Hill; GAHSH, Great Ash Hill Side holes; GAHB, Great Ash Hill bottom outlier; GAHT, Great Ash Hill top outlier; B Grid, Ingrid-B outlier; GW, Great Wood; Clearing outlier; HH, Holly Hill; HHNSH, Holly Hill northern side holes; HHSSH, Holly Hill southern side holes; TC, Thorny Croft; TCU, Thorny Croft upper outlier; TCL, Thorny Croft lower outlier; SH, SHO2, Sunday's Hill outlier #2; BBO, OL64, Outlier #64; NCP, New Common Piece; OCP, Old Common Piece; OL1, Outlier #1; HS, High Seat; GATES, Gates Sett; OL1A, Outlier #1 Annex; TR, Terrace Outlier; MT, Mount Sett; AGH, Andy Gosler's Holes; HSBC, HSBC (bank) outlier; CC, Chunky Cap; RC, Radbrook Common; RCO, Radbrook Common outlier; US,

Upper Seeds outlier; CLO, BB, Brogden's Belt; Brogden's Belt outlier; LS, Lower Seeds; RAT-P, Rat Pens; NC, BP Burkett's Plantation (no longer there); Sunday's Hill, SH; SHO1, Sunday's Hill outlier #1; Nealing's Copse; LR, Lawson's Ridge; P, Pasticks; PFO, Pasticks flat outlier; POOH, Pasticks Outlier outlier holes; PO, Pasticks Outlier outlier holes; PLO, Pasticks Lower outlier; PUO, Pasticks upper outlier; PMO, Pasticks middle outlier; PO1, Pasticks Outlier #1; WSV, Wormscat Views; PUO2, Pasticks Upper outlier #2; JH, Jew's Harp; JHA, Jew's Harp Annex; JHO, Jew's Harp Outlier; SW, Singing Way; SWO, Singing Way outlier; PV, Park View; M2A, Marley #2 annex; M3B, Marley #3b; M2, Marley #2; M3A, Marley #3a; JKBr, Jack Bracken; JKBrOL, Jack Bracken outlier; MM, Marley Main; UF, Upper Follies; BL, Botley Lodge; McBr, MacBraken; GC, Gipsy Camp.

And, on the periphery of the woods: At Northfield Far: RD, Road outlier; MF, Mick's Field; FSTN, Field Station; OB, Old Barn. At Tilbury Farm: HEC, Hill End; TFC, Tilbury Farm Clumps; Higgins, Higgins Copse.

Outline of Wytham Woods, and superimposed contours, as depicted on previous aerial photo, showing the abbreviated names of major badger setts mentioned in the text.

To appreciate better the circumstances of the setts, they are depicted (top) in the context of land use (at about 1992; there have been some changes to land use since, for example, the pattern of grassland and cropped, with somewhat more grassland in recent years to the east of the woods), and (bottom) soil types. To place these maps, and setts, in the context of social groups and sup-groups, readers should refer forward to Maps 9.1 and 9.2 (pages 173 and 174).

Sources: The aerial photograph is credited @2021 Microsoft Corporation Earthstar Geographies SIO. The contours are by Ordnance Survey and subject to the terms at www.ordnancesurvey.co.uk/opendata/ licence. © Crown copyright and database right 2021. The soil map comes from Cranfield University (https://cranfield.blueskymapshop.com/maps/os-openmap?x=446822.922&y=208030.9835&z=4&w= 4000&h=4000&f=&p=[]&m=).

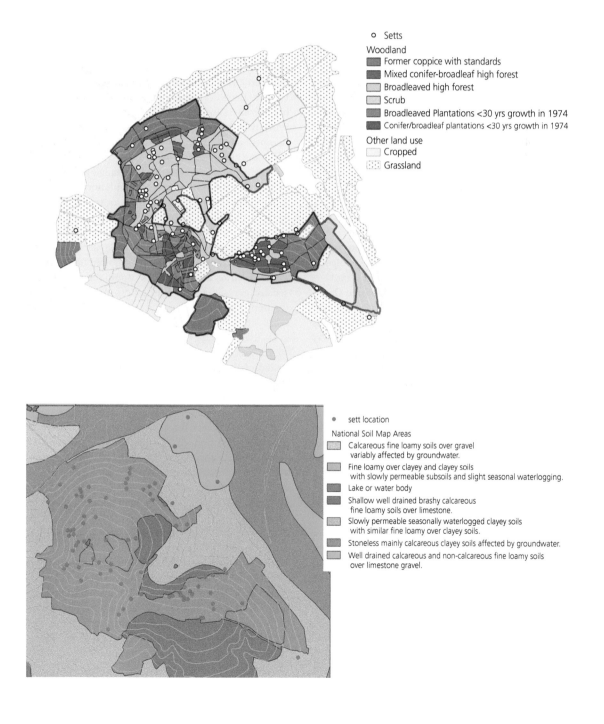

Setts

Woodland
- Former coppice with standards
- Mixed conifer-broadleaf high forest
- Broadleaved high forest
- Scrub
- Broadleaved Plantations <30 yrs growth in 1974
- Conifer/broadleaf plantations <30 yrs growth in 1974

Other land use
- Cropped
- Grassland

sett location

National Soil Map Areas
- Calcareous fine loamy soils over gravel variably affected by groundwater.
- Fine loamy over clayey and clayey soils with slowly permeable subsoils and slight seasonal waterlogging.
- Lake or water body
- Shallow well drained brashy calcareous fine loamy soils over limestone.
- Slowly permeable seasonally waterlogged clayey soils with similar fine loamy over clayey soils.
- Stoneless mainly calcareous clayey soils affected by groundwater.
- Well drained calcareous and non-calcareous fine loamy soils over limestone gravel.

Foreword

I expect I am not the only person whose first childhood encounter with badgers in literature was through reading the Rupert Bear Annual, in which one of Rupert's closest and most cheerful chums, Bill Badger, appears dressed in a tight blue jacket, striped trousers, a red or yellow waistcoat, a white shirt and a black bow tie. A few years later, the avuncular badger of Wind in the Willows cemented my anthropomorphic childhood view that badgers are jolly decent chaps that wear smart clothes and drive cars. I was too young and naïve to know better. But myths about badgers have a much longer history and are more widespread.

In his 1646 treatise *Pseudodoxa Epidemica or Enquiries into very many received tenents and commonly presumed truths*, otherwise known as *Vulgar Errors*, the medical doctor and polymath Sir Thomas Browne dedicates Chapter 5 of book 3 to The Brock or Badger.

Browne's methodology is to apply the three Determinators of Truth: Authority, Sense and Reason. Unfortunately for Browne, David Macdonald's and Chris Newman's monumental monograph on *The Badgers of Wytham Woods* was not yet available as a Determinator of Truth. But nevertheless, he reached the correct conclusion.

In the middle of the 17th century it was widely believed, even by those who had observed them at close quarters, that badgers have shorter legs on the left side than on the right, to allow them to walk comfortably on sloping hillsides. Brown concludes that "on indifferent enquiry, I cannot discover this difference", that it is "generally repugnant to the course of Nature" (animals are bilaterally symmetrical) and that "The Monstrosity is ill contrived". In short, it's not true.

In the same book, Browne also debunks many other animal myths, for instance that beavers bite off their own testicles to escape hunters, that elephants have no joints, that salamanders can endure fire and that bears lick their misshapen new-born cubs into bear-like form.

It is just possible that one of the sources of evidence that Sir Thomas Browne used in reaching his conclusion about the badger is the beautiful early 16th century fresco in the great cloister of the Abbey of Monteoliveto Maggiore by Giovanni Antonio Bazzi (known as Il Sodoma). The relevant scene is of the first miracle of St Benedict, but also includes a self-portrait of the artist. While St Benedict humbly performs his miracle (repairing a broken sieve) in the background, Il Sodoma struts proudly in the foreground wearing fine robes, with his two pet badgers at his feet. The badgers show no sign of having asymmetrical legs.

Macdonald's and Newman's detailed and comprehensive studies of the badger population of Wytham Woods near Oxford leave no room for errors or myths. In nearly 500 pages, enhanced by colour photographs, diagrams, and maps, they show how approaching 50 years of research, using a wide array of ingenious techniques on a single population, has revealed detailed truths about badgers' population dynamics, social organisation, communication, mating and reproduction, disease, and death. It will stand as a reference work for many decades to come.

Wytham Woods is home to some of the longest-running ecological studies anywhere in the world, including the great tit population study that celebrated its 75th anniversary in April 2022. Paradoxically, as it has become increasingly difficult to obtain funding to support long-term ecological research because funders want quick answers, the value of these long-term studies has become more and more apparent. Long run data sets, properly stewarded and analysed, have allowed ecologists to detect subtle signals of the impacts of climate change as well as the effects of changes in land use and habitats. The Wytham badger study is no exception.

In recent decades badgers have been in the news as villain or victim, depending on whose view you accept, in the long-running (and still running) debate about bovine tuberculosis. Although Macdonald's and Newman's book is emphatically not a book about badgers and TB, Chapter 16 gives the reader a very thorough and enlightening history of the debate and shows how important it is to know the details of badger behaviour and ecology if one is to understand what is going on and develop appropriate policies.

The central questions in the controversy are (a) Can badgers transmit TB to cattle and (b) if so, is killing badgers an effective control strategy for bovine TB? When, in 1996, I was asked by the then Government to review the evidence and make recommendations on these questions, David Macdonald was a key witness to the enquiry. He, along with others who were studying badger ecology and behaviour, showed how any successful analysis of the problem depended on understanding the details of how badgers behave. His finding that culling badgers can disrupt the relatively stable, group territorial, population structure and cause increased movement of badgers across the landscape, the so-called perturbation effect, was particularly relevant. The randomised badger culling trials that resulted from my report revealed an unexpected and surprising result. While culling badgers appeared to reduce TB infections of cattle in the cull area, it unexpectedly resulted in an increase in infections in the penumbra. This may be explained at least in part by the perturbation effect.

Those of us who chose a career in scientific research are mainly motivated by curiosity. Macdonald and Newman are clearly in this mould. Their passion is to document and understand the life of the badger.

As one famous scientist put it, doing curiosity-driven research is a bit like a predator stalking its prey. The prey are the secrets of Nature that the scientist is trying to unlock, and requirements of the predator are persistence, cunning and imagination, attributes well displayed by Macdonald and Newman. But unlocking Nature's secrets through curiosity-led can also lead to direct, tangible, and often unanticipated, benefits to society. In the case of bovine TB and badgers, for a variety of reasons, the policy for controlling TB has sadly not made the best use of the unlocked secrets of Nature. But in the decades ahead the breadth and depth of studies such as that of the *Badgers of Wytham Woods* will not only fascinate but also help to provide answers to the most intractable problem that humanity faces today: to avoid destroying the planet on which we live and live in harmony with the millions of other species that share the planet with us.

Professor Lord John Krebs, of Wytham.

CHAPTER 1

Setting the Scene

Births and Beginnings

When one tugs at a single thing in nature, he finds it attached to the rest of the world.

John Muir

Like grave robbers, four of us, clutching long sections of pole, a metallic briefcase, and a radio-tracking receiver, shuffled suspiciously through the moonlight of the crisp January night. For some weeks previously our ingenious field assistant, James Bunyan, had been visiting this badger sett daily, his antenna sweeping the ground like an archaeologist's metal detector in search of ancient treasure. James' reward was not historic booty, but the mesmerizing bleeps emitted from the radio-collar worn by Badger F773—a 4-year-old female snoozing at some unknown subterranean depth beneath his feet (years later we would be able to locate her descendants accurately, using magnetic resonance tracking; Chapter 8). Guided by his map of clustered day fixes at 'Clearing Outlier' (a sett at the centre of Wytham Woods, see map, p. xiv), we waited until nightfall, when the badgers would be out, and began to drill down, adding section after section to our 25-mm diameter auger until, at around 2 m, resistance ceased. Tentatively, we unwound the cable, attached to our (at the time high-tech) infrared (IR) pinhole camera (VideoProbe® XL PRO™). Eureka! This upside-down periscope revealed a large subterranean chamber, lined promisingly with bedding material. Emboldened by this success, we bored out our narrow earthy aperture to 40 mm and inserted a robust plastic tube into which we set a miniature camera equipped with built-in IR light emitting diodes[1] along with a microphone and thermometer. We stoppered the tube's top to prevent any chilly draughts (mindful of the sett microclimate, which turns out to be important (see Chapter 10; Kaneko et al. 2010; Tsunoda et al. 2018). Later that night we implanted three more voyeuristic cameras at Clearing Outlier, one in a tunnel, another in a second chamber, and the last one in a cavernous void ($c.1.5 \times$ 2 m) that these badgers had excavated amongst blocks of limestone rag jutting into the sandstone strata (see Coombes and Viles 2015).

We had timed our intrusion right, in that we (and our scent also, apparently) were long gone before the subsequent video revealed that at around 5 a.m. a badger returned to the chamber. It was a female, in badger vocab a sow (males are boars,[2] porcine vocabulary that reveals the evolutionarily wayward, but naturalistically forgivable, rustic pigeon-holing of badgers). This sow was not F773, our original intended target, but F695, a 6-year-old female, born in 1998, whom we had ultrasound scanned the week before: we knew she was pregnant.[3] Insofar as most badgers look the same, how could we identify these individuals? Each year since 1987 it has been our practice to intercept as many badgers as possible for quarterly roll calls and health checks by luring them into cage traps with peanuts. On first capture each individual was bestowed a unique tattoo for lifelong identification (we refer to badgers by their tattoo number and sex, e.g. F695); also, for short-term identification (the tattoos can be read only on anaesthetized badgers), some were given recognizable fur-clip marks.

[1] 37CHRIR PAL, 480TVL CCD, 3.6 mm lens, RFConcepts Ltd.

[2] The species that was, until recently, referred to as the 'Eurasian badger' was originally mistakenly classified as a type of bear, *Ursus meles* ('honey bear'), by Linnaeus (*Systema Naturae*, 1758). Four years later, Brisson correctly reassigned badgers to the *Mustelidae* (*Regnum animale* in classes IX, 1762).

[3] Under special licence from English Nature, as Natural England was then known.

The Badgers of Wytham Woods. David W. Macdonald and Chris Newman, Oxford University Press.
© David W. Macdonald and Chris Newman (2022). DOI: 10.1093/oso/9780192845368.003.0001

(a)

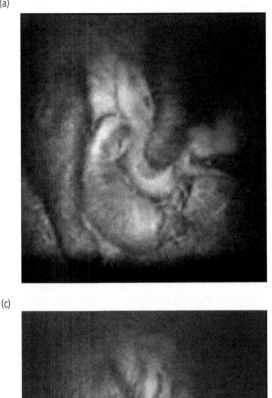

(b)

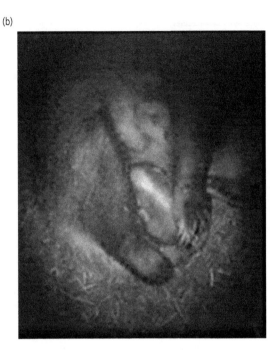

(c)

Photo 1.1 Photos from the infrared video cameras placed at the Clearing Outlier sett, inside the chamber in which Briar Badger (F695, pictured here) whelped her cubs.

Our luck held. F695 (more affectionately known to us as Briar Badger) could have whelped in any one of the many chambers at Clearing Outlier, but 2 weeks later, on 5 March 2004, James excitedly saw two new-born cubs in the video chamber (Photo 1.1) that later turned out to be females (who became known as F945 and F960).[4]

That was in 2004, 17 years ago, longer than the lifetime of the two longest-lived badgers, F699 and M32, ever to pad through the Dog's Mercury of Wytham's woodland floor (Chapter 11). But even in 2004 we knew quite a lot about the Who's Who of the badgers of Wytham Woods—we'd already been voyeurs into their

[4] Readers already apprehensive about these unmemorable numerical names, do not despair: for the most part you don't need to memorize them—although they will help the

aficionado make links between the characters of different chapters—and where it matters we'll set out family trees to make plain who relates to whom; further, some particular characters are known by nicknames, which often began with B until our imaginations ran dry at Boadicea Badger.

Photo 1.2 The core badger team, from left to right, Chris Newman, Christina Buesching, and David Macdonald, discussing what might be occurring beneath them.

private lives for 18 years, which amounts to at least four to five badger generations. Only around 20% of badgers born in Wytham survive to be 5 years, or older. Indeed, we recorded 51 badgers born in that year (although using the enhanced estimation procedure we will introduce in Chapter 11, this was in reality more likely nearer 61); and while 7 of them survived till their 10th birthday, the oldest amongst this cohort, F945 and F959, died in 2017, making them 13 years old.

And so we embark upon our journey into the lives of badgers, the scene set in Wytham Woods (Photo 1.2), Oxfordshire, using our 30 years of womb-to-tomb analyses detailing over 1823 badger life histories; personal lives that weave the warp and weft of this society, which we set out to document. Indeed, that fabric was subsequently enriched by our genetic pedigree, based, in due course, on freezers full of blood samples stretching from 1987 to 2020, and also by the DNA 'fingerprinting' of 1350 individuals. Although we were never able to determine the parentage of F695, we do know a lot about her family tree, and, as in all populations, and the societies that structure them, it is the individual, and the links between individuals, that constitute the working parts of evolution. So, while readers interested only in the generality may quail at the dynastic detail, it is worth making the effort to get a grasp of those details, since therein lies the route to getting inside the badger's skin—a treasure trove of information, illustrated for F695 and her cubs, F945 and F960, in Box 1.1.

There were 21 so-called main setts[5] in the woods at this time, each associated with a group of badgers; in

addition, some main setts had a constellation of outlier setts. These 21 ranges, which we termed at the time 'territories' (however, we were to reform our views; see Chapter 9), together encircled 69 clusters of holes (setts) each with more than 5 (and some with as many as 80) entrances.

As for the soap opera in which these badgers are entangled, these are just a few strands of the tapestry of private lives that comprise our 'data'—and as those data are the foundation on which this book is written it is worth illustrating the detail we have assembled. While F927 succumbed before her first birthday, the lives of her two half-sisters, F776 and F798, born to Female 730, were to unfold through our database for the next decade. The family trees were a revelation, to which we'll return fully in Chapter 17, but they also included tantalizing gaps: we were never able to determine the parentage of F695, mother of those two new-born cubs recorded by our spy camera; the evidence pointed to F773 being her sister, making F945 and F960 nieces to F773, and F776 and F798 cousins to F945 and F960.

F945 was a lifelong resident at Clearing Outlier (aside from also visiting the Holly Hill sett in 2011, and even the Hill Copse sett (a step further west than the Chalet sett) in 2012). She produced a male cub, M1141, in 2007, fathered by Male M893, who lived at Great Ash Hill, the adjacent sett *c*.250 m away within this same range (see map, p. xiv). She too was last seen aged 9 in 2013, at home in Clearing Outlier 9 years after that first grainy image in their birth chamber. Their mother, F695, disappeared from Clearing Outlier during the year when the sisters were born, whereas their father, M507, was actually last caught, aged 8, at the Hill Copse sett/range in 2003, and thus his breeding success in 2004, described here, was assigned posthumously. Given that badgers have delayed implantation stretching their pregnancies out to a full year (see Chapter 7), M507 could well have been long dead (i.e. after fertilizing F695 in 2003) by the time his brood was born in 2004. We return to the question of how long

---

[5] In an earlier naturalist vocabulary it was helpful to distinguish between main setts and outliers. The distinction was always a bit blurry in terms of sett architecture, and even more so in terms of sociological implications, with the result that at Wytham the vocabulary has come to seem less acceptable as our understanding of setts deepened. Nonetheless, as a

shorthand, acknowledging at the extremes that a massive, multi-generational earthwork represents a different investment to one or two recently excavated holes, we use the vocabulary where it is helpful. A main sett is well-established, relatively large, and used for breeding; outlier sets are smaller, used less often, and may not be bred in. Devils in the detail include the fact that in Wytham a sett, such as Clearing Outlier, began as a small subsidiary, earning the 'outlier' label, only to become, some years later, a conspicuously main sett with 17 entrance tunnels. And so some outliers become main setts, some group ranges have multiple main setts (seemingly with equal standing), and some main setts in one group range can be smaller than an outlier used by a neighbouring group.

Box 1.1 Genealogies—the deeper story

Combining trapping records with genetic pedigree, what more can we glean about the intricacies of this badger group and the fate of these two neonatal protagonists as the years rolled by?

These infant females F945 and F960 were revealed as daughters of Male M507 (Figure 1.1), who sired them when he was 8 years old and co-resident at Clearing Outlier sett. In the previous year (2003) he'd sired F927, also born at Clearing Outlier as the single assigned daughter of F773. A serial polygamist, M507 could count amongst his lifetime progeny F776 and F798, born to F730 in 2000 at the adjacent Chalet sett, in which he resided at the time (indeed, M507 was to make his home in no fewer than six setts in three ranges throughout his life (Chalet, Clearing Outlier, Great Ash Hill, Great Wood, Hill Copse, and Lower Seeds; see map, p. xiv), but, as it happened, he never sired a surviving son—his first mate, F730, produced F798 and F776, siblings; his second mate, F773, singleton F927).

These are the sorts of details availably only because our study is very long term, and they are essential to a deep understanding of badger biology.

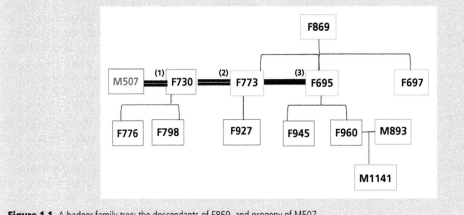

Figure 1.1 A badger family tree: the descendants of F869, and progeny of M507.

---

badgers live in Chapter 11. Remember, each year since 1987 we lured as many badgers as possible into cage traps baited with peanuts, which is how we kept tabs on M507 and all his companions. In the case of M507, we caught and examined him 30 times throughout his life.

To introduce better the actors on the stage, let us start at the beginning: when are badgers born? The answer varies, of course, with latitude, given the European badgers' distribution from Norway to Iran (Ibis et al. 2015; see Johnson et al. 2002a; Chapter 19). This is linked to the severity of winter weather, with births occurring from early January in southern Spain, and peaks in late January in south-west France, to the latter half of March in Sweden, and as late as early April in the former Soviet Union (see Neal and Cheeseman 1996; Revilla et al. 1999; Zhou et al. 2017, all of whom document birth dates for related species). In southern England 76% of births occur between mid-January and mid-March, with a clear mid-February peak (Neal and Cheeseman 1996). However, birth dates also vary from year to year in any one place, as in our case in Wytham Woods. We know this because, with the aid of a borrowed portable ultrasound scanning machine transported from vet clinic to woodland, we could see tantalizing glimpses of badger embryos from their early stages of development in their sedated mothers— as pulsating greyscale images. This is the very technology with which an earlier generation of human parents were familiar (more modern babies are viewed via high-resolution 3D colour Doppler imaging). Dunmartin et al. (1989) had published the lengths, postmortem, of badger foetuses from studies elsewhere at various stages of development, enabling us to calculate a regression equation that related each embryo's length and growth rate, measured on the scanner, to the number of days since it had implanted. Knowing that badger gestation is normally 53 days, we could therefore anticipate each embryo's birthday, to an accuracy of ±2.6–3.2 days (Woodroffe 1995).

The timing of our woodland obstetrics was delicate. The short days of winter trigger implantation of embryos (Chapter 7), and we aimed to measure them around the start of the second trimester; sooner, and the germinal vesicles would be too small to detect; later, and we would risk upsetting the gravid mother. Christmas Day was generally, if inconveniently, optimal, leading us to Yuletide woodland antics that stretched the remarkable dedication of Julie Saxton, the vet who, year after year, operated the borrowed scanner (see Macdonald and Newman 2002). Later, armed with a genetic pedigree, we could prove which of these scanned females went on to give birth successfully (demonstrating, in passing, that our intervention had not affected the pregnancies). We learned that, for healthy, chubby females in Wytham, embryonic implantation typically commences around 21 December (our earliest record was 17 December), but for their scrawnier counterparts it was as late as 31 December (11 January was the record (Woodroffe 1995)). So, while many births are in mid-February (revealed when trails of bedding often litter badger paths leading to setts), the extremes result in estimates of birthdays from *c*.8 February to *c*.5 March. What determines the all-important plumpness of pregnant females? The answer lies at the intersection of prevailing weather conditions, foraging success, and thermoregulatory costs—but more of that anon (Chapters 11–14; see Noonan et al. 2015c; Macdonald et al. 2015e).

So, now aware of the general birth patterns of badgers, we return to James and his matronly supervision of our subterranean maternity ward: both cubs were around 110 mm long at birth (i.e. appearing slightly longer than the length of their mother's hind foot), and probably weighed around 100 g (i.e. about the same as a pet golden hamster), with sparse, silky, grey fur and, even on these grainy images, they sported badger facial stripes. As for maternal care during these first, entirely subterranean, 6 weeks, the mother was with her cubs for 56% of the time for which they were in view and, importantly, during those weeks no other badgers entered their natal chamber, whether the mother was present or not.

The nursery chamber was flying saucer-shaped (hemi-ellipsoidal), about 1 m across, with a single access tunnel. Its roof was about 2 m below the surface, and 2 m from the nearest sett entrance. The floor was layered with fine dry grass and straw. Of the time the mother was in view, we recorded that she spent 86% of it in nursing, and while she was present the cubs spent 80% of the time lying still, tucked amidst her legs. She spent 12% of her time grooming them, and 1% grooming herself. Those episodes of suckling lasted on average 9 min 35 s (±18:24 SD),[6] and generally ended with a bout of cub grooming lasting, on average, 1 min 22 s (±2:00 SD). These measurements were made over 6 weeks, but we could detect no difference in the mother's behaviour throughout this time. We watched her leave the chamber on 80 occasions, on 63% of which she returned within 5 min, often with fresh bedding. As we shall see in Chapter 7, her longer absences were not only to feed, but potentially also to mate, initiating the next breeding cycle: badgers exhibit postpartum oestrus and delayed implantation (Chapter 7; Yamaguchi et al. 2006; Sugianto et al. 2020, 2021).

In their book, Neal and Cheeseman (1996) suggest that prior to giving birth sows select larger than average chambers situated close to an entrance to provide better ventilation and to isolate and defend the litter from interlopers. We lowered thermometers through the chimneys into the breeding chamber and recorded a temperature range of 16.8–17.6°C, consistently warmer than the 4°C and 6°C means for outside air in February and March, respectively (in Chapter 19 we'll develop the idea that one function of setts may be to provide a temperature-controlled environment during harsh weather). The cubs often lay on top of bedding for prolonged periods, but on five occasions we saw the mother pull a blanket of bedding over them as she left; more often than not, however, they clambered back on top of the bedding as soon as she was gone.

By the time the cubs were 10 days old, and still effectively immobile, their mother began to move them around, one at a time, between at least three chambers. We were in luck as one of these was the other chamber we had instrumented with a surveillance camera. This

---

[6] Readers familiar with descriptive statistics will know that it is a helpful snippet of information to know, when speaking of an average or mean, how wide is the spread of values, variance, amongst the numbers contributing to that mean. This is commonly expressed by the notation +/− and one of two measures of variation known as standard deviation (SD) and standard error (SE). Readers familiar with statistics will appreciate the distinction between SD and SE, but for readers unaware of the niceties the details don't matter, but what does matter can be seen at a glance: if the measure of variation is small relative to the mean, then most elements of the sample have roughly the same value, whereas if the variation is large relative to the mean, the underlying data include a wide spread of values. That is really all readers less familiar with statistics need to know to glean quick insight from these +/− measures of variation given throughout the book.

was similar to, but smaller than, the one in which they had been born; that is, about 0.7 m in diameter, two thirds filled with coarse dry straw, and with a solitary, very narrow entrance, impinged by a large root.

In infrared we could see, but in the total darkness of the den, Briar Badger F695 operated purely by touch and smell, often giving her the appearance of ineptitude, to the point of brutality, with her fragile cubs. She would roll onto them, clumsily bash them against the edges of the chamber entrances, or drop them roughly before pawing around to relocate them with her heavily clawed feet. To move cubs, F695 used her forepaws to manipulate them into a position in which she could pick them up by the scruff of the neck, before shuffling backwards out of the chamber. Between 4 and 7 min later she returned for the second cub. Once, we saw her enter the new chamber backwards with one cub and then forwards with the other. Why would she move her cubs? We will come back to that question, and the itchy topic of parasites in Chapter 8 (Cox et al. 1999; Stewart and Macdonald 2003; Albery et al. 2020a).

Nonetheless, F695 was also diligent: during those first 6 weeks before their eyes were fully open (in contrast, puppies' eyes open at around 2 weeks), the cubs spent 97% of their time in close contact with her while she was present. She stimulated them to defecate and urinate by licking their bottoms, then consumed the waste, keeping the chamber hygienic. When she was absent, the cubs snuggled side by side and were motionless for 83% of the time (they lay apart for 11% of the time, interacted actively for 4%, and clambered around separately for 2%). Thus passed 6 weeks, at the end of which their eyes were open, they were around 300 mm long, appearing to weigh around 1 kg (i.e. guinea pig-sized), and began to stray from their chambers, thereby enabling them, for the first time, to meet other badgers. Insofar as part of our story centres on social behaviour, these first interactions are the start.

Before describing the cubs' first encounters with their social group, an aside is required on the remarkable magnitude of badger earthworks: at Wytham the largest sett is called Great Oak, with over 80 holes stretched over 80 m of bank, accrued over centuries (see p. xiv). Impressed by these huge setts, we could scarcely avoid the presumption that burrows, especially natal burrows, were likely an important element of group living in badgers, and perhaps also across the Carnivora. Guided by our former Canadian graduate student (and international wrestler) Mike Noonan (now at the University of British Columbia, via a stint at the Smithsonian Institute), we reviewed every study we could find (Noonan et al. 2015a). This led us to coin the 'Fossorial Benefits Hypothesis' (FBH; Noonan et al. 2015c;[7] Macdonald and Newman 2017), which proposes that fossoriality correlates positively with the extent of offspring altriciality (prematurity). In short, the extreme altriciality seen in badgers is associated with a life in burrows that provides protective, womb-like conditions during early neonatal development (Case 1978a, 1978b; Wolff 1997). Any naturalist seeing the expanse of setts like Great Oak would guess that these labyrinths are somehow important in badger evolution, and the FBH turns out to be a first step towards an understanding of that importance, as will become clear—after we have assembled many other elements in the argument—in Chapter 19.

Returning to the socialization of cubs F945 and F960 deep underground at Clearing Outlier, the first evidence of socialization came when a non-breeding co-resident, 8-year-old F869, who was present somewhat irregularly, entered the cubs' chamber and groomed them both vigorously for approximately 40 s. Plausibly, she may have been their grandmother (i.e. she was mother to F697, who was likely sister to F695). The cubs' mother, Briar F695, left the chamber at 20:12:49 and likely grandmother F869 arrived just over 47 min later; their mother returned at 05:15:52. Three days later, when their mother left at 02:05 the cubs followed her a minute later, returning separately about 40 min later. Then, at 03:40:59 their grandmother arrived again and groomed them both for about a minute and a half. Both cubs then followed her when she left the chamber, but returned alone a few minutes later. Their mother returned to the cubs at 05:08:12. We never saw another adult in the nest chamber with the mother.

Do badger cubs benefit significantly from the care of non-breeding adults, as for example the cubs of some canids assuredly do (e.g. Macdonald and Moehlman 1982)? For years, biologists watching non-breeding individuals of many species of mammal and bird interacting with youngsters dubbed them 'helpers', too often in the absence of evidence that they were, in fact, performing a helpful function (see Woodroffe and Vincent 1994). The question of whether non-breeders were good, bad, or indifferent for badger infants in Wytham has kept recurring—the answer is to be found in Chapter 8.

When the cubs were nearly 8 weeks old, one was seen resting underground with two other adults but outside of the natal chamber; one adult was a male, likely M507, their visiting father, who at the time was

---

[7] From Latin *fossor*, meaning 'digger', i.e. an animal adapted to digging that lives primarily, but not solely, underground.

resident at Great Ash Hill within the same range. However, again to appreciate both the detail of our data, and the untidiness of badger behaviour, it is important to point out that generally, for that 3-year period, M507 was engaged in a game of musical chairs between Great Ash Hill and a further two setts: Clearing Outlier and Great Wood (later it will become relevant to know that he moved to Hill Copse and then Lower Seeds). After 2 h the adults left, seemingly paying no attention to the cubs (see Fell et al. 2006). Three days later, one of the cubs shared the chamber with likely the same adult male for several hours, during which time both the cub and the adults came and went several times. We watched, waiting—it turned out fruitlessly—for the exuberant greetings and effusive sociality that typifies, for example, the behaviour of fox cubs living nearby. In contrast, these badger cubs and the associated adults seemed detachedly oblivious of each other.

During the ensuing months we followed these sisters, by then known as F945 and F960,[8] into prime adulthood (during which they became notorious for disrupting our long-term study of Wytham's mouse and vole populations; see Buesching et al. 2011): this study involved a geometric grid of Longworth live traps, each the potentially comfortable temporary residence of a rodent lured within by a handout of grain. As youngsters, F945 and F960 developed the knack of prising open the traps that contained tasty grain bait (there was never evidence that they also extracted rodents). This ruinous intervention in our study reached a peak when the sisters began following our field assistants around the grid as they set the rodent traps each evening, an enthusiasm we were compelled eventually to curb with 400 m of electric fencing to protect the traps. During her life F960 left no known progeny, but remained in the Clearing Outlier sett until 2011 (aside from a brief spell in the Great Ash Hill sett in 2007, which also constituted part of the same range). Thereafter, forcing us to acknowledge yet more untidy detail, she wandered both westward to the adjacent Chalet sett and eastward to the adjacent Holly Hill sett in 2010, being trapped again at the Chalet sett, then the Chalet Outlier sett, in 2011. In 2012 she returned to Holly Hill, living there into 2013, although in the autumn of that year she was caught for the final time, back at home at Clearing Outlier, suffering from a heart murmur, aged 9—a decent innings for a badger, as we shall see when we tackle questions of badger demography and senescence, respectively, in Chapters 11 and 18.

Although we followed sisters F945 and F960 from that first ultrasonic encounter in their mother's womb to their dotage, this complete history is not typical for badger cubs because the intensive underground surveillance is not routine. So it's tricky to estimate what proportion of cubs dies along its journey from birth to emergence from the sett (i.e. this being the point at which we can first see them), to May, this being our first opportunity to catch them and administer a numerical tattoo against which to record their life histories in our database. In a general review, Roper (2010) estimated that about 20% of cubs die during this subterranean phase, but to answer this question for Wytham we followed two lines of evidence. First, from a sample of pregnant females examined during our winter ultrasound scanning (Photos 1.3 and 1.4) we knew the typical number of embryos each was carrying, despite the practical difficulty imposed by the fact that many badgers are torpid at this time, with the result that we catch fewer of them than during summer (i.e. 20.89% ± 6.6 SD in winter, of those later realized to have been alive, vs 50.5% ± 9.3 SD in summer, with an overall annual trapping efficiency of 81.3% ± 6.1 SD; Noonan et al. 2014, 2015c; see Chapter 11). This number of embryos averaged around 2.10 (SE ±0.14) (similar to the estimate from post-mortem studies of 2.7 (Anderson and Trewhella 1985) or 2.9 (by Neal and Cheeseman 1996)).

Second, we noted the proportion of females caught in the spring that had lactated, as compelling, but not infallible, evidence that they had given birth to cubs. This involved measuring maternal teat size using a length × diameter gauge (similar to those used to measure fingers for ring sizes) (Dugdale et al. 2011a). In 2010 we did this for the 136 mothers that, at the time, our genetic studies had confirmed as having produced cubs in that year (see Chapter 17). Females that had just ceased nursing cubs in our spring catch-ups had teats measuring >10 mm diameter by 10 mm long (and their teats remained distended to a diameter greater than 5 mm (SE = 0.332) and 2 mm long (SE = 0.16) until late August), whereas unproductive females had teats of less than 3 × 3 mm. On average 37.67% (SE = 2.02) of females aged 2 years or more (therefore, sexually mature) showed signs of lactation each year.

Crucially, therefore, dividing annual cub cohort sizes by our number of lactating females yields mean litter sizes ranging from just 1.62 ± 0.62 SD (1987–2000; Macdonald and Newman 2002) to more recently 1.46 ± 0.65 (Macdonald et al. 2009; Annavi et al. 2014a).

---

[8] We gave each badger a unique number, tattooed into its left groin.

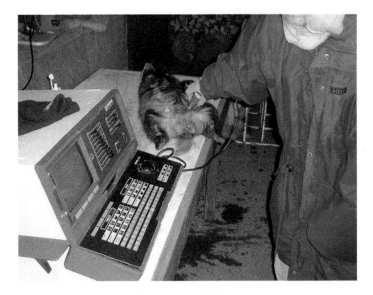

Photo 1.3 Ultrasound scanning an anesthetized badger to check for pregnancy.

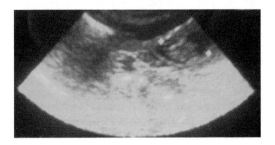

Photo 1.4 An ultrasound image from a pregnant badger.

Compared with ultrasound-derived litter sizes (and assuming that all lactating females had been pregnant), this implies a neonatal pre-emergence cub mortality rate of between 23 and 30.5% (more on this topic in Chapter 11). For comparison, in Woodchester Park, a similar high-density population of badgers about 50 miles to the west of Wytham, in Gloucestershire, Cheeseman et al. (1988a) estimated 38% pre-emergence mortality, while Mallinson et al. (1992) estimated 35% mortality using post-mortem data.[9]

What causes the death of neonatal cubs? Some doubtless succumb to developmental malformations and congenital disorders; indeed, of those few tiny neonatal skeletons we have found, around one third exhibited cranial or skeletal deformities, especially involving the pelvis, suggesting hip dysplasia. Others surely suffer a miscellany of viruses, of which candidates discussed in Chapter 15 include parvovirus, canine infectious hepatitis, and canine distemper; although Newman and Byrne (2017) review evidence that badgers generally are quite resistant to distemper, with Delahay and Frölich (2000) finding 0 of 468 badgers sero-positive in a British study. Duarte et al. (2013) detected parvovirus in all three sick badgers examined in Portugal, while Barlow et al. (2012) reported parvovirus enteritis in five sick cubs in the UK, all of which suffered severe diarrhoea and died. So, while the short answer is that—owing to lack of access to neonatal cubs—we do not know in detail what causes their deaths while they are nestled underground, a very strong clue lies in what kills them post-emergence. At this juncture, the emphatic answer is that there is a highly pathological unicellular coccidian parasite

---

[9] We do not know the extent to which females lose entire or just partial litters, although the fully distended teats evidencing lactation would not be apparent if entire litters were lost early during cub development.

at work: *Eimeria melis*. We have studied this intestinal parasite intensively amongst those cubs surviving long enough to be caught in spring (late April to early May) and thus to provide a faecal sample (Anwar et al. 2000; Newman et al. 2001). These cubs, gravely afflicted by the parasite, have swollen pot bellies, and produce copious, watery faeces, which, as it turns out, are composed substantially of egested parasite eggs (oocysts)—a subject to which we will return in Chapter 2.

In some years the size of the cub cohort falls well below the long-term population average of 45 (or 52.5 from the enhanced Minimum Number Alive method described in Chapter 11). For example, in 2011 we counted only 13 cubs in our May/June trapping (of ultimately just 24 born that year, the smallest cohort across our dataset; Chapter 11) and, intriguingly, this coincided with unusually high antioxidant capacity amongst these youngsters (Bilham et al. 2013). Antioxidants are compounds produced by cells that inhibit tissue damage by quenching/neutralizing what are pleasingly, and somehow evocative of the Sixties, called 'free radicals'. Free radicals are the dangerous by-products of cellular respiration—the burning of oxygen to power the badgers' lives. So a possible interpretation, to which we return in Chapter 15, is that disease weeds out the cubs with less developed immune and antioxidant defences that are less able to cope with juvenile parasitoses, leaving only the strong (see also Bilham et al. 2018), a process termed 'selective mortality' (Ricklefs and Scheuerlein 2001). Clearly too,

if mothers die while their litters are dependent on lactation, their cubs are doomed. For example, in 1996 four females from a single social group on the periphery of our study site (Botley Lodge) were all run over on local roads within a 2-week period during the critical April nursing season. Subsequently, and for the only year in our records for that sett, we caught no cubs from that group in that year.

A sociologically interesting—in terms of how the behaviour evolved—cause of cub mortality in group-living mammals is infanticide (Photo 1.5), executed either by disenfranchised or cuckolded males (Agrell et al. 1998), or by females in order to extinguish the reproductive fitness of their rivals (Wolff and Peterson 1998). This remains a tantalizing topic in badger biology. In 2002, we found dead cubs at three separate setts in Wytham. Dramatically, at one sett (Thorny Croft) a pair of cubs, approximately 4 weeks of age, had each been killed by a simple canine puncture to the cranium with no other mutilation. However, nearby we found an adult badger's head attached to its spinal column with all flesh torn away and the appendicular skeleton scattered in the undergrowth, along with bite-sized patches of skin and flesh. The damage was such that we could not even decide the sex of the dismembered adult, and could only ponder its involvement in the demise of the cubs. Possibly it was the mother killed defending her cubs, or the aggressor, killed by the mother. The brutality of the scene inclined us to think that several badgers had joined this fray. Infanticide still simmers in the midst of badger society; some

Photo 1.5 A gruesome scene—an adult badger that was dismembered in a potential infanticide attack. The identity and sex of the badger and its involvement in the incident remain unknown.

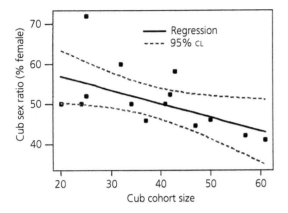

Figure 1.2 The annual badger cub cohort size as a predictor of the corresponding cub sex ratio between 1987 and 2001. Solid black line shows the regression line, and dashed line shows the 95% confidence limits (from Dugdale et al. 2003, with permission).

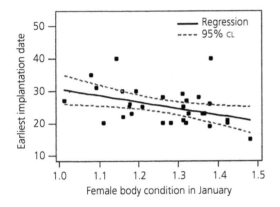

Figure 1.3 The body condition index of each pregnant female in January as a predictor of the earliest implantation date (number of days after 1 December) of that female in the years 1993–1995, 1997–1999, and 2001. Solid black line shows the regression line, and dashed line shows the 95% confidence limits (from Dugdale et al. 2003, with permission).

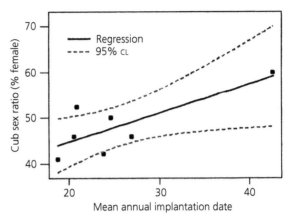

Figure 1.4 The mean annual implantation date (number of days after 1 December) of all pregnant female badgers as revealed by ultrasound scanning, as a predictor of the badger cub sex ratio in the years 1993–1995, 1997–1999, and 2001. Solid black line shows the regression line, and dashed line shows the 95% confidence limits (from Dugdale et al. 2003, with permission).

sort of process is at work here, but quantifying exactly what it might be has been just beyond our reach.

An intriguing intricacy in mammalian reproductive systems is the manipulation of sex ratio (Charnov et al. 1981). Now widely supported by empirical data, the theory, originally conceived by R.A. Fisher (1930), was energized by Trivers and Willard's (1973) proposal that for mammals, generally, it requires greater maternal investment to raise male offspring of high reproductive quality than to raise females. Indeed, for Wytham's badgers, over the period 1987–2001, we found (Figure 1.2) that in the years with the largest

cub cohorts—when we deduce conditions were most favourable—sex ratio was biased towards males by as much as 60:40, from a long-term mean of 50:50 (mean female % = 50.95 ± 7.80; 1987–2001) (Macdonald and Newman 2002; Dugdale et al. 2003; see Chapter 11). Conversely, in years when cub cohorts were small, on average only 43% of cubs were male. Exactly why this occurs, and how, will be the topic of Chapter 14, but a relevant fact is that fatter females implant earlier (Figure 1.3), and the earlier they implant the more male-biased the resulting cub cohort sex ratio (Figure 1.4): of cubs implanted before Christmas 56% are male, while of those implanted after Christmas, only 43% are male.

An alternative hypothesis to explain variation in sex ratios lies in the local resource competition hypothesis (Clarke and Fredin 1978; Silk 1983). This proposes that in years when female body condition is poor, it will be advantageous for females to lower competition for local resources by producing offspring consisting chiefly of the dispersing sex, that is, males; however, we found no evidence to support this in Wytham (Dugdale et al. 2003) (the topic of dispersal is introduced in Chapter 3). In either case, the reverberations of ecological circumstances on the lives and life histories of individuals are far reaching: see Chapters 10 and 14. For wider context, another denizen of Wytham illustrates the point: roe deer, *Capreolus capreolus*. Analysing cull records for these deer, David and Paul Johnson, WildCRU's statistical wizard, found that the minority of pregnancies were singletons, but when singletons occurred they were heavily biased towards males, and all the more so for roe deer mothers in better condition (Macdonald and Johnson 2008). Although roe does producing a single fawn are generally lightweights, they nonetheless have the opportunity to heap investment on their 'only child' (roes, especially heavier ones, mostly give birth to twins), and the greatest maternal investment can come from females in the best condition, who can then anticipate the greatest rewards (measured in grandchildren) from producing strapping sons. As regards the majority of roe mothers, who give birth to twins, across the population they have an equal sex ratio, but for any individual mother there is a highly significant tendency to have a 'pigeon pair': one fawn of each sex. In coin-tossing terms, these mothers throw a head and a tail much more often than the average gambler. Further, this excess of mixed-sex twins was greater when the average condition of does was high. Why? David and Paul suggested that mixed pairs of fawns offer mothers the greatest opportunity to optimize maternal investment during lactation, when conditions will be unpredictable, and that this is increasingly the case when average condition is poor. The obvious question, and one particularly germane to badgers, is how are these controls over sex allocation achieved? The reason the question about roe deer is particularly relevant to badgers is that roe deer, like badgers, display delayed implantation (Chapter 6), and during the delay up to 60% of blastocysts may be jettisoned—providing the opportunity for mothers to favour the fortunate 40% depending on her circumstances. Is it really possible for mother roe deer, and badgers, to manipulate which embryo prospers—to better maximize their reproductive investment? A quick jump, mentally and geographically, to Brazil adds to the growing chorus of evidence that the answer is yes. In parallel with the badger study, David was researching the behavioural ecology of capybaras, giant South American rodents. On the island of Marajó (it looks like a stopper in the mouth of the Amazon), the ranchers harvested capybaras in preparation for Holy Week (the Catholic Church considers them as fish because of their aquatic habits, but that's another story), and this allowed David and his student, Jose Roberto Moreiro, to examine the wombs of the slaughtered females. The average litter was 4.4 viable young and 0.9 dead embryos, and they found that 18% of implanted embryos are lost during pregnancy. Male embryos were heavier than their sisters—larger litters were female-biased whereas smaller ones had a preponderance of males (amongst capybaras,

successful males secure a preponderance of matings). The embryos were not randomly positioned in the uteri—they tended to be arranged next to an individual of the opposite sex. Indeed, females in better condition tended to have more dead embryos in their uteri and those that survived were predominantly males. Roberto and David concluded that females in better condition were weeding out daughters in their uteri, in order to take advantage of the opportunity to invest in highly competitive sons (Macdonald and Tattershall 1996). Exactly what mechanism manipulates sex ratio in badgers (or roe deer) is unknown, but the capybaras illustrate the possible sophistication of sex ratio allocation.

The question we posed next was: might sett size, or type, have a bearing on either litter size or survival, or perhaps on both? The lexicon of setts, which can be large or small, has traditionally been to classify them as either main or outliers—the main setts being the sociological hub of the badger group (see footnote 4; Thornton 1988). It seems that where you are born matters: of 29 cubs born within a single focal badger group range (Hill Copse) between 1987 and 2004, we recaptured 15 out of 18 cubs (83.3%) as healthy yearlings, born at the main sett, whereas only 3 out of 11 cubs (27.3%) survived at the 5 outliers distributed across this range (Kaneko et al. 2010) (Figure 1.5).

What might account for the poor prospects of cubs born in outlier setts? One clue might be in the architecture of these den systems. In Wytham outliers tend to have shallower tunnels, dug in inferior clay soils (Macdonald et al. 2004d; Tsunoda et al. 2018), and are generally cooler with less stable deep interior temperatures during the cub-rearing period of February to April (Kaneko et al. 2010). Examining the thermal properties of 11 setts through 2016, we found that although the weights of 18 adults captured consistently in those setts were not associated with interior temperature, the weights of 9 cubs born in spring 2017 (aged *c*.3.5 months when caught and weighed) were heavier (up to 3.8 kg) in setts that had remained warmer over the interceding winter (vs minimum of 1.8 kg in the coolest sett; Tsunoda et al. 2018)—this result will be worth recalling when, in Chapter 19, we consider how social thermoregulation—back to the hot water bottles idea—may have impacted the badgers' tendency to cohabit.

But is quality the key factor determining cubs' chances of survival? If so, is it the quality, on the one hand, of the sett, or on the other, of the individuals that reside there? Adult badgers assigned primary residency at outliers tend, on average, to be 7% lighter than

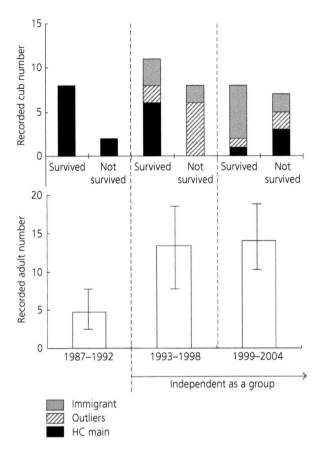

Figure 1.5 Cub survivorship (upper pane) and adult numbers (lower pane) in the Hill Copse range between 1987 and 2004. Cubs are categorized by where they were born—in the main Hill Copse sett (solid black), in an outlier sett (diagonal lines), or in another sett not belonging to the Hill Copse cluster (solid grey) (from Kaneko et al. 2010, with permission).

the residents of the main sett (Kaneko et al. 2010). Both the quality of the accommodation and of the residents may impact the cubs' prospects. For, in the words of the brutally unegalitarian Biblical aphorism: 'For to every one who has will more be given, and he will have abundance; but from him who has not, even what he has will be taken away' (Matthew 25:29[10]); that is, the poorest females may be pushed to the periphery of badger

society where they give birth in the least advantageous setts and in turn bear smaller litters and/or suffer greatest infant mortality.

Despite new technologies for tracking badgers (Noonan et al. 2015b; Ellwood et al. 2017; Bright Ross et al. 2020), these first few subterranean weeks of life remain the most inscrutable period of their lives, of which the only knowledge we possess derives largely from inference and anecdote. From this point onwards, however, as badger cubs mature to emerge above ground and become large enough to trap and track, our knowledge of their biology grows more comprehensive. And so we now proceed to examine how badger cubs become integrated into their natal groups and broader society, and the challenges that face them in a hostile world.

The transition to adulthood is a crucial time for cubs (a 'sensitive period' in development; Bateson 1979). Cub recruitment is a major driver of population dynamics (Chapter 11), and much affected by the

[10] Featuring in the Gospels of Matthew, Mark, and Luke, and based on the Parable of the Talents (a unit of currency: 80 lb of silver), this concept is not only recurrent in ecology (and this book) but also an established concept in economic theory—the Matthew Effect of Accumulated Advantage. The term was coined in 1968 by sociologists Robert Merton and his wife Harriet Zuckerman, formalizing the adage that 'as the rich get richer, the poor get poorer'. This does indeed reflect any tendency in human society to brutalize the poorest, perhaps a trait in some contexts of badgers too, but the Reverend Professor Andrew Gosler points out to us that it is a misunderstanding of the original text.

Photo 1.6 Taking measurements from an anaesthetized badger in the field lab.

clemency of each year's weather and thus its food supply (Chapters 10, 12, and 13), with potential to influence the ultimate lifespan written into its DNA (Chapter 17). To probe further, we recruited Nadine Adrianna Sugianto, whose talent as a veterinary student from Indonesia won her a prestigious government scholarship; as part of her thesis research she elucidated cub development and the endocrinological basis of puberty.

As is common with mammals (Badyaev 2002; Isaac 2005), on average, adult male badgers are $c.3.4\%$ longer than females (average 704.1 mm, $n = 1368$, vs 680.8 mm, $n = 1385$) with 5% wider heads (Photo 1.6) (zygomatic arch; Sugianto et al. 2019b). These differences between the sexes are real, but small, and there is visible overlap in size between them. Although they are sexually dimorphic, badgers buck the mammalian allometric[11] trend (termed Rensch's rule[12]) by being less sexually dimorphic than expected for their body size (the explanation for this phenomenon lies in their feeding ecology, and will emerge as important to understanding badger society not only in the wider context of musteloid, but also of more general

carnivore evolution; see Chapter 19; Noonan et al. 2016). Unexpectedly, slight though their sexual dimorphism might be, attaining this size differential places males and females under different developmental stresses, in terms of both rate and duration of growth. We used our database to extract records of body length and head (zygomatic arch) width for 501 weanlings, of which 364 made it to adulthood (1995–2016). Furthermore, in 2016 Nadine took a host of other measures from 48 weanlings of which 40 made it to adulthood (Sugianto et al. 2019b).[13]

Up to age 3–4 months both male and female weanlings had similar body lengths, but from 6 months onwards axial (head and spine) skeletal growth curves differed significantly (Figure 1.6), so that male body length grew on average 2.57 mm/month more than females'. This rate disparity continued until age $c.11$ months, after which it decreased to 1.48 mm/month, and reduced further as both sexes slowly reached their final size. The outcome is that females reached 99% of their end body length at 16 months of age, 2 months earlier than males. Limb (appendicular) metrics ceased to grow at a younger age than body length and zygomatic arch width, but exhibited similar dimorphic trends. Only tail length was an exception, which showed no sex pattern but high interindividual variation (which, combined with pelage variation, can enable observers to recognize individual badgers; Dixon 2003).

This divergence from $c.11$ months onwards coincides with the onset of puberty in badgers (Sugianto et al. 2019c; see Chapter 2), when sex steroids induce the release of sex-specific growth hormones (Badyaev 2002). Relating to these growth trajectories, and to the aforementioned Trivers and Willard hypothesis (1973) on sex ratio variability, maternal investment is typically higher in male than female offspring (Clutton-Brock et al. 1985; badgers, Dugdale et al. 2003).

Due to the way bodies change in shape with size—so-called scale allometries (a subject discussed in detail in Chapter 14, Carbone et al. 1999; McClune et al. 2015) and rapid early growth, the relative metabolic rate of cubs is higher than in adults; therefore, any exposure to inclement weather (Chapter 12), which exacerbates the impact of the aforementioned coccidial gut parasite *E. melis* (causing coccidiosis; Chapter 2),

---

[11] Allometry refers to the disproportionate rate at which things happen with respect to body size. Examples might be growth of a part or parts of an organism as the organism changes in size, or the size of the home range travelled by a species relative to its body mass. These disproportionalities are very important in biology, and crop up throughout this book in diverse contexts.

[12] It becomes important to delve into Rensch's rule in Chapter 19, but to whet readers' appetites it postulates that in comparisons across closely related species, male body size relative to female size increases with the average size of the species.

[13] Those measures were axial (head–body length, to the nearest 5 mm; zygomatic arch width, mm and appendicular (averaged left/right, mm, pastern length, radius–ulna, humerus, tibia–patella, patella–femur, and tail length—all made using callipers) (Sugianto et al. 2019b).

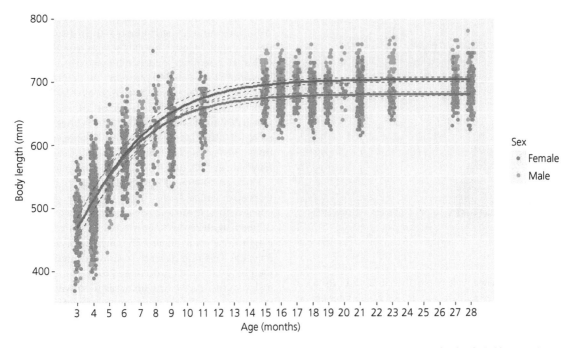

Figure 1.6 Head–body length growth curves in males and females aged 3–28 months. Females are shown in red and males in blue. Raw data are shown as points (from Sugianto et al. (2019b), with permission).

can impair growth, causing smaller final body size (e.g. Mowat and Heard 2006) or a lowering of their probability of survival. Interestingly—and this is a topic to which we return in Chapter 14 as we examine the relative drivers of sex-specific demographic contributions amongst adults—male cubs prove sensitive primarily to social factors, growing more slowly towards a smaller ultimate size in social groups with more adult males present (Sugianto et al. 2019b). In contrast, female final size is predominantly affected by weather conditions and associated food availability (Leberg and Smith 1993; Post et al. 1999; Leblanc, M. et al. 2001). This greater susceptibility of male offspring to social factors is likely due to the lengthier nature of their growth period, which facilitates compensation for shorter-term factors such as adverse weather, but provides no such buffering against long-term adverse social conditions (Festa-Bianchet et al. 1994; Leblanc, M. et al. 2001); whereas females lack the ability to compensate for short-term adverse weather effects, due to their shorter growth periods, but are less affected by longer-term effects (Sugianto et al. 2019b). Similar patterns have been reported for other mammal species: in big horn sheep (*Ovis canadensis*; Leblanc, M. et al. 2001) and white-tailed deer (*Odocoileus virginianus*; Leberg and Smith 1993) population density correlates negatively with body mass in both sexes, but the effect is significantly greater in males than in females, resulting

in reduced sexual size dimorphism at higher population density.

Recently, sexual selection has provided the dominant explanation for why males tend to be larger than females amongst mammals (Mueller, 1990; Karubian and Swaddle 2001; but see Lammers et al. 2001). From this perspective (and there are other factors, such as niche partitioning, Chapter 19), male-biased sexual dimorphism arises mainly because larger males are assumed to have an advantage over smaller rivals in competing for, and attracting, mates (Hirotani 1994; Howard et al. 1998; Wong and Candolin 2005). But because our Wytham badgers are highly promiscuous (Dugdale et al. 2007), a fact that forms the basis of Chapter 7; and since, as we reveal in Chapter 5, they don't have a clear dominance hierarchy, even when it comes to mating (Macdonald et al. 2002b; Yamaguchi et al. 2006; Dugdale et al. 2011b; Chapter 7), there may be less competition between males than might have been expected. Furthermore, small males may be favoured when food is a limiting factor because, in absolute terms, they require less food (Blanckenhorn et al. 1995; Bright Ross et al. 2020). On the female side of the equation, and as we will detail in Chapter 7, we have seen every adult female in a group engaged in mating; on average, however, only 45.2% of them are ultimately assigned young (Annavi et al. 2014b), prompting questions about reproductive failure, perhaps due to social suppression, and infanticide (Woodroffe and Macdonald 1993).

This seems like a recipe for intense competition between females, perhaps putting a premium on larger body size (Noonan et al. 2016). This combination of factors resonates with the 'Ghiselin–Reiss small-male hypothesis' for sexual size dimorphism evolution (Blanckenhorn et al. 1995). That is, in mating systems dominated by scramble competition, male breeding success is largely determined by opportunity, and not by dominating pugilism. Amidst all the rarefied nuances of mate selection, could it actually be that all that matters to badgers is which males a female bumps into, or is she choosey on the basis of genetic traits?

That question will be addressed in Chapter 17, where we demonstrate that encounter rate amongst badgers provides the most parsimonious explanation of mate choice rather than any directed selection for genetic traits (Annavi et al. 2014b). So with topics ripe for deep dives accumulating, we reach the point of badger cubs emerging from their setts and entering our database. The remainder of their precarious journey to adulthood is the subject of Chapter 2. First, now that you have been present at the birth of badgers in Wytham Woods, you need to appreciate their evolutionary birth. To enable you to appreciate the badgers of Wytham Woods as models for behaviour, ecology and evolution, and to set the scene for the 17 chapters that lie between us and the denouement of this story in Chapter 19, we need to answer the question, what is a badger?

What is a badger?

Pausing now at our second preparatory pit-stop to fine-tune information needed for the journey to intra- and inter-specific comparisons, we ask: what is a badger?

Tricked by the English language we might look up Wikipedia to learn that nine species share the moniker badger, their distributions mapped in Figure 1.7.

Key:

- Gold = honey badger (*Mellivora capensis*).
- Red = American badger (*Taxidea taxus*).
- Teal = European badger (*Meles meles*).
- Dark green = Asian badger (*Meles leucurus*).
- Lime green = Japanese badger (*Meles anakuma*).
- Blue = Chinese ferret-badger (*Melogale moschata*).
- Indigo = Burmese ferret-badger (*Melogale personata*).
- Azure = Javan ferret-badger (*Melogale orientalis*).
- Purple = Bornean ferret-badger (*Melogale everetti*).

However, all that glitters is not gold, and students of convergent evolution will be familiar with the trap into which the koala's resemblance to a bear may lead one to fall. In fact, the koala is no more a bear than the red panda is kin to the giant panda (which is a bear). When it comes to the badgers of Wytham Woods, one thing is sure, badgers are not bears, contrary to the binomial *Ursus meles* ('honey bear') Carl Linnaeus first assigned them in 1758 (*Systema Naturae*). It was Mathurin Jacques Brisson who, 4 years later, correctly re-assigned badgers to the *Mustelidae* (*Regnum animale* in classes IX, 1762). And this is also the Carnivore family to which all of the colloquial 'badgers' also belong, their relationships schematized in Figure 1.8 from Koepfli et al. (2008).

How is it possible to diagnose a badger ancestor from the fossil record? As hinted by French anatomist Georges Cuvier's memorable quip 'Show me your teeth and I'll tell you who you are', the answer lies largely in their dentition. Anatomically, members of the *Melinae* lineage are united by a broad posterior cingulum on premolar 4 (P^4),[14] the enlargement of molar 1 (M^1) and of the talonid (latin for heel) on M^1; also the loss of the supraorbital (above eye) ramus (branch) of the medial meningeal artery (Flynn et al. 1988) and the loss of the suprameatal fossa (for the anatomical enthusiast there's more[15]).

Although bonded colloquially by the surname 'badger' the last common ancestor of the Melinae that bonded Chinese ferret badgers, genus *Melogale*, with contemporary *Meles* and *Arctonyx* species, lived 10–11 million years ago; all retain the marten-like original musteloid feature of four premolar teeth in each jaw. Despite their unmistakably badger-like appearance, the American badger, genus *Taxidea*, and the Afro-Asian honey badger, or ratel, *Mellivora capensis*, have only ancient links to *Meles*. Indeed, both are so distinct that Koepfli et al. (2008) argue that they merit their own subfamilies, *Taxidiinae* and *Mellivorinae* (Figure 1.9). The honey badger is, in fact, a weasel in badger's clothing, with its ancestry within the *Mustelinae* from which it diverged around 13–12 million years ago

---

[14] The cingulum refers to a convex protuberance at the cervical third of the anatomic crown on the lingual or palatal aspects of the incisors and canines; the suprameatal fossa is a small triangular depression on the inner side of the temporal skull bone.

[15] They share the canoid (dog)-type inflated auditory bulla (inner ear) with an inflated hypotympanic sinus posterior to the promontorium (a hollow skull prominence, formed by the projection outward of the first turn of the cochlea of the inner ear); this latter feature is also present in the *Mustelinae* sister subfamily. Nevertheless, these cranial features are not unique to badgers, occurring in multiple ancestral badger-like digging ecomorphs, such as *Trochotherium*, and *Trochitis*, as well as amongst analogous putative procyonids (raccoon family), such as *Stromeriella* and *Zodiolestes*.

Figure 1.7 Map of the distribution of badger species.

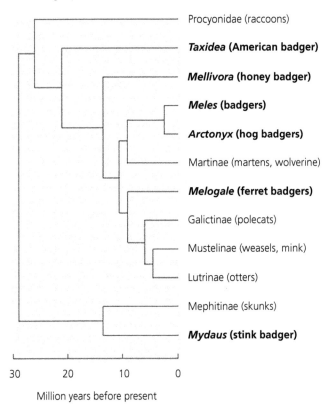

Figure 1.8 Phylogenetic relationships of badgers (bold) and related carnivores (after Koepfli et al. 2008; Nadler et al. 2011).

(Mya) (Vanderhaar et al. 2003; Koepfli et al. 2008). *Taxidea* may descend from the Asian *Ferinestrix*—a large and more carnivorous distant meline ancestor that immigrated to North America, likely via the Beringian land bridge, no later than the Pliocene

5–2.5 Mya (Wolsan and Sotnikova 2013). It may be a skunk disguised as a badger in that, unlike Old World *Melinae*, modern *Taxidea* and skunks (*Mephitinae*) have a fossil-distinguishing auditory epitympanic sinus in the squamosal-mastoid region of the skull. Similarly,

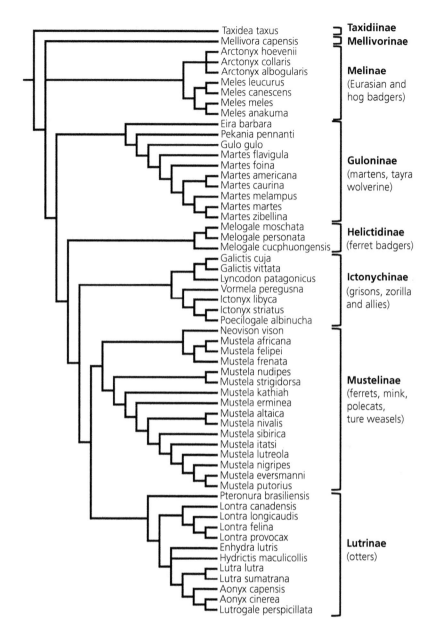

Figure 1.9 Phylogenetic tree showing the pattern of relationships among all extant genera and 54 of the 63 proposed, or recognized, species within the family *Mustelidae*. Bracketed clades or lineages define the eight recognized subfamilies. The following six species are not depicted in the tree due to absence of formal phylogenetic studies (but see Nyakatura and Bininda-Emonds 2012): *Melogale everetti, Melogale orientalis* (*Melinae*); *Martes gwatkinsii* (*Guloninae*); *Mustela lutreolina, Mustela subpalmata* (*Mustelinae*), and *Aonyx congicus* (*Lutrinae*) (from Koepfli et al. 2017, with permission).

Asian stink badgers, genus *Mydaus*, are actually Old World skunks (subfamily *Mephitinae*), diverging from the rest of the mustelids at least 30 Mya (Dragoo and Honeycutt 1997). They are endemic to Southeast Asia (Long and Killingley 1983), the Indonesian stink badger, or teledu, *Mydaus javanensis* (Desmarest 1820), and the Palawan, or Calamian stink badger, *Mydaus marchei* (Huet 1887), being closely related, with speciation arising from geographic isolation on the Palawan and Calamian Islands (Zhou et al. 2017).

More broadly amongst the Musteloidea, the relatively rapid rate of divergence within their family tree

(Bryant et al. 1993, Dragoo and Honeycutt 1997, Marmi et al. 2004), and the exceptional ecological diversity this brings with it—that is, from swimmers to climbers to diggers (Macdonald et al. 2017)—have provided taxonomists and phylogeneticists with much to debate (Sato et al. 2012; Koepfli et al. 2017).

Now that we have the vocabulary straight—a biological badger is a member of the subfamily Melinae and thus of the genera *Meles* or *Arctonyx*—let us consider a different sort of answer to the question of what constitutes a badger: what is the essence of being a badger? The roots of the answer take us back 34 million years to the time when Antarctica drifted to the South Pole—a geological moment of global cooling, causing a '*Grande Coupure*' ('great break') in ecosystem continuity. Everything has consequences, and in this case it was that atmospheric CO_2 reduced, disadvantaging plants using C3 photosynthesis (dicotyledons, notably trees) but favouring monocots—grasses—using C4 photosynthetic pathways (see Macdonald and Newman 2017). Why is this relevant to the evolution of badgers, or their ancestors? Because it resulted in the forest ecosystems that abounded in the Middle to Late Eocene (47.8–33.9 Mya) becoming wooded grasslands in the Early Oligocene (33.9–28.1 Mya). Some 60% of terrestrial mammals went extinct, to be replaced by new species adapted to the avant garde grassland biome.

By the mid-Oligocene (*c.*28 Mya) fleet-footed predators pursued large ungulate grazers out on the open plains; however, smaller rodent grazers exploiting this new bonanza needed a different strategy to seek refuge from both predators and climatic extremes: they burrowed. It is here that the evolutionary origins of the proto-mustelids, and amongst them the badgers, lie, with evolution along two trajectories.

To catch the burrowing rodents, one evolutionary option was for weasel-type slender ecomorphs to squeeze into their burrows, much as extant black-footed ferrets hunt prairie dogs (see Eads et al. 2010). Alternatively, robustly muscular forearms can rip open rodent burrows: this was the badger option, as exemplified by extant American badgers excavating ground squirrels (Weir et al. 2017). The epitome of the early excavators was *Megalictis*, the giant weasel, weighing in at 60–100 kg.

Reconstructing the badgers' fossil lineage is hindered by missing links, especially from the Miocene, so there are several competing phylogenies, but all agree on a clear progression to omnivory, with shearing carnassials reduced in favour of tubercular, grinding, molar teeth (see Ginsburg and Morales 2000).

Midway along this road, a mustelid progenitor, *Rhodanictis* from the Middle Miocene (MN5[16]), combined various marten-like characteristics with more badger-like teeth. Further along the line, *Promeles* emerges as a definitive badger ancestor in the European Miocene strata (from a division termed the Turolian age, 9.0–5.3 Mya (see Ginsburg and Morales 2000).[17] At roughly the same time, other ancestors of actors relevant to our plot emerge, each with its own dental quirks; for example, *Taxodon*[18] and, in Asia, *Melodon*, which survived until the Lower Pliocene[19] (3.6–2.5 Mya), to be superseded by new genera *Palaemeles* and *Arctomeles*, both with the distinctive defined lingual cingulum on M^1 that unifies true badgers. In summary, by the Pliocene, in Europe, the two genera closest to the heart of our comparisons were extant: *Meles* and *Arctonyx* (descended from *Arctomeles* ;Ginsburg and Morales 2000), the hog badgers of subcaudal pouch fame in Chapter 6, to which we return later.

Inspection of Figure 1.9 reveals four modern species of the genus *Meles*—how did that come about? Two of the earliest fossil *Meles* are *M. polaki* and *M. maraghaus*, from the Lower Pliocene of Iraq, but none is found in European strata until the Upper Pleistocene (mid-Villafranchian, 2 Mya) emergence of *M. thorali* near St-Vallier, near Lyons in France (Kurtén 1968). This is where modern molecular techniques have brought hindsight with which to reinterpret the fossil evidence (see Box 1.2). The outcome is that western badgers evolved through *M. thorali* to the modern European badger, *M. meles*; and *M. canescens* the eastern lineage evolved to modern Asian or white-tailed badger, *M. leucurus*, and the Japanese badger, *M. anakuma*[20]

---

[16] The MN zonation (from Mammal Neogene) is a 16-zone system used to correlate mammal-bearing fossil localities in the Miocene and Pliocene epochs of Europe.

[17] Promeles has a clear and individualized metaconulus on M^1, and a broader, more elongate talonid on M_1, with a separated hypoconulide (Ginsburg and Morales 2000).

[18] From slightly earlier in the Upper Miocene (MN5-6) *Taxodon* was similar to *Promeles*, except for a lingual wall on the talonid of M_1, which divides into multiple small cusps.

[19] In biostratigraphy, the younger rocks generally occur higher up the order, due to superposition; thus 'Upper' is synonymous with 'Later' in palaeontological nomenclature.

[20] *Meles thorali* has dental features that mix those of two contemporary species: *M. meles* (Pm1 not reduced, Pm2 long with two roots) and *M. leucurus*, whose fossils had split by the Early Pleistocene, *c.*2.6 Mya (Abramov and Puzachenko 2005). Another similar species, *M. chiai*, occurs in the mid-Villafranchian (3.5–1.0 Mya; overlapping the end of the Pliocene and the beginning of the Pleistocene) in China, characterized by the absence of Pm1 and M1, with a well-expressed external notch characteristic of modern *M. leucurus*.

Box 1.2 Badger phylogeography

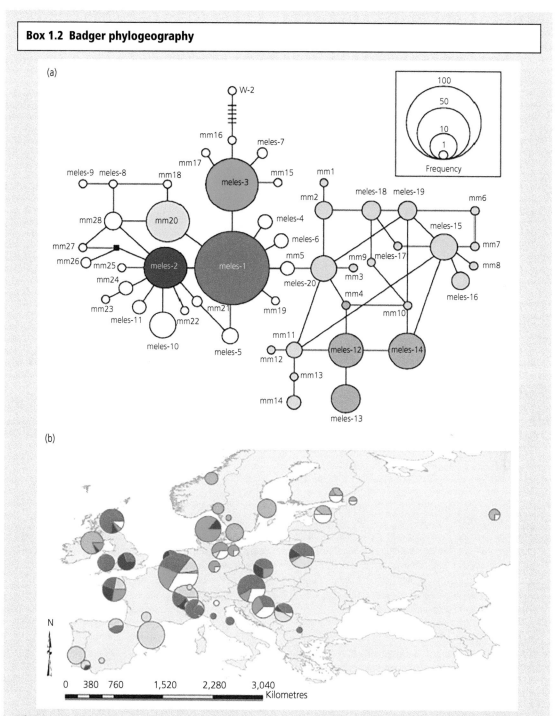

Figure 1.10 Median-joining network (a) and geographic distribution (b) of 49 badger mitochondrial control region haplotypes. The same colours represent the same groups of haplotypes in both figures. Missing haplotypes are indicated by a small black square in the network. Horizontal bars represent mutational steps when greater than one. The sizes of the symbols in the network and the map are representative of haplotype frequency and sample sizes, respectively (from Frantz et al. 2014, with permission).

Box 1.2 *Continued*

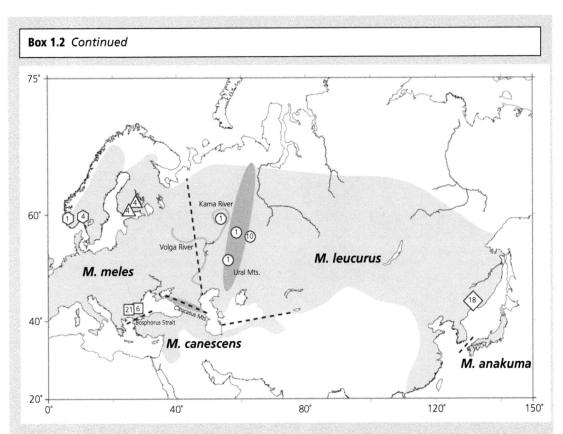

Figure 1.11 Sampling locations in the distribution (grey area) of Eurasian badgers, *Meles*. The symbols and numerals correspond to sampling locations (Bulgaria, within squares; Finland, within triangles; Norway, within hexagons; Ural, within circles; far east Russia, within a diamond) and numbers of examined samples, respectively. Broken lines indicate the hypothetical geographical boundaries. The Ural Mountains are shown in dark grey (from Kinoshita et al. 2017, with permission).

How did the contemporary distribution of badgers come about? Their phylogeography, as the geography of genealogy is called, took shape as the ice melted at the end of the last (Weichselian) glaciation (115,000 to 11,700 years before present (ybp)). The suspicion that all the badgers in western Europe had expanded from a single refugium (Pope et al. 2006) was elaborated when Frantz et al. (2014) considered eastern European sites too, and found 49 *Meles meles* mitochondrial control (mtDNA) region haplotypes[21] depicted in Figure 1.10. Core haplotypes formed a star-like pattern with the meles 1–3 cluster distributed throughout eastern and central Europe, the Balkans, and the UK,

along with a newly identified haplotype, mm20, found in the Balkans, Poland, and Estonia. A different cluster of haplotypes—mm1–4, mm6–14, and meles 15–20—was largely restricted to the Iberian Peninsula. Sweden and Norway were home almost exclusively to a Scandinavian cluster of haplotypes meles 12–14, which was more closely related to the Iberian than to the eastern haplotype clusters. This suggests northern colonization initially over the North Sea Doggerland land-bridge after the Younger Dryas. Some Scandinavian haplotypes also occurred in Finland, Estonia, and central Russia, as well as—surprisingly—in Ireland, along with mm4.

[21] A haplotype is a group of genes within an organism that was inherited together from a single parent, as per 'haploid' gametes with only one set of chromosomes.

Box 1.2 *Continued*

From the haplotype map Frantz et al. (2014) deduce at least two glacial refugia source populations, one Iberian and one 'Eastern'.

Curiously, however, a similar mitochondrial DNA cluster analysis performed by O'Meara et al. (2012) grouped the Irish population with Scandinavia and Spain, whereas the majority of British haplotypes grouped with those from central Europe.

Looking at Ireland in more detail, Chris worked with a large group of authors, led by Adrian Allen, on a comparative phylogeographic analysis that showed that admixture[22] played a role in the composition of badgers in Ireland, with badgers from north-eastern and south-eastern counties being more genetically similar to contemporary British populations—transported there by people *c.*600–700 (confidence interval (CI) 100–2600) years ago, likely for baiting as sport (Allen et al. 2020).

Those animals found furthest from the refugium populations typically exhibited reduced genetic diversity, especially if founder populations were small. This probably explains why in Wytham Woods (Chapter 17), as in other badger populations on the western and northern fringes of Europe, badgers have low microsatellite diversity, especially compared with central and south-eastern populations (alternatively, a bottleneck caused by population contraction following disease, habitat degradation, or earlier historic persecution is also possible).

What of the relationship between *M. meles* and *M. leucurus*? Nowadays they bump into each other along a narrow contact zone around the Volga and Kama Rivers of Russia (Figure 1.11). There, 17/71 badgers had mixed hybrid genotype (Kinoshita et al. 2019).

[22] A genetic admixture occurs when previously separated genetic lineages mix.

(from the Japanese meaning hole-bear). By the way, *M. canescens* isn't generally accorded a common name, but since canescent is an English word, derived from the Latin, meaning growing white or hoary, we henceforth call it the hoary badger. Mitochondrial DNA evidence indicates that the hoary and European badgers diverged between 2.37 and 0.45 Mya, with the Japanese and Asian badgers diverging between 1.09 and 0.21 Mya (Marmi et al. 2006).

The ancestral *Meles* species probably had a wide Palaearctic distribution towards the Late Pliocene, and Early-Pleistocene Europe was inhabited by badgers very similar to modern species: *M. iberica* reached the Iberian Peninsula before the beginning of the glacial–interglacial cycles in the Northern Hemisphere (*c.*2.6 Mya) (Madurell-Malapeira et al. 2011a), along with *M.*

dimitrius from the Early Pleistocene of Greece, and *M. hollitzeri* from Austria and Germany: these species were at least similar to, and probably conspecific with, *M. thorali*.

Readers are now equipped at two levels for the chapters that lie ahead, introduced not only to the badgers of Wytham Woods, our models of behaviour, ecology and evolution, but also to badgers generically; framed within the context of musteloid evolution. That context will be essential to the final assembly of our argument in Chapter 19, but meanwhile, in each of the intervening 17 chapters, hold it in mind as the frame within which all the adaptations revealed in these chapters can be interpreted. Meanwhile, we return to Chapter 2 and the perils faced by badger cubs in Wytham Woods.

CHAPTER 2

It's Tough at the Bottom

> We all create the person we become by our choices as we go through life. In a real sense, by the time we are adults, we are the sum of the choices we have made.
>
> **Eleanor Roosevelt**

The third of May 2001 was memorable not only because of the abundance of cubs that were to spill from Pasticks sett at dusk, but also because the teacher accompanying the children that Chris had positioned to watch them announced that, more than a decade earlier, she too had been led by Chris as a schoolgirl on the same excursion. Their reflections on the longevity (as it seemed two decades ago) of our project were interrupted as a stripy pate tentatively poked from a nearby hole. The head disappeared into the tunnel, as if to sound the 'all clear', and immediately a pair of cubs erupted from the sett (Photo 2.1) and started to tussle, kekkering[1] (see Chapter 5) as one grabbed its sibling's tail. A moment later two adults emerged (Photo 2.2), of whom one, a lactating female, was an elder stateswoman of the group, Badger 'H' (more formally, F298); the other M553. Next came a smaller, slower, and more ponderous, pot-bellied cub, nicknamed Bloat and later to be known as M814, son of M553 and, probably, Badger H, to whom he clung close while the others played rumbustiously. That pot belly was a sign of gastro-intestinal infection, boding badly for Bloat, although with hindsight we know that despite his sickly start he survived until at least 2006.

A moment later a commotion at another hole heralded the emergence of two more cubs, Females 779 and 819, who ploughed into the throng, chasing, pouncing, and making their distinctive kekkering calls, closely followed by their mother, F400. As the confusion of cubs tumbled into the vegetation, they numbered at least five and likely nine, along with nine identifiable adults.

How did this profusion of cubs fare? It was ominous that just 2 weeks later, during our post-weaning spring trap-up, of the likely nine, we caught only five cubs from the Pasticks sett (F779, F816, F818, and M823 and Bloat (M814)). A hundred metres to the west, at Pasticks Outlier, we caught two more (F819 and M821). Cubs F779 and F819 were both later assigned to Female 400 (although with different fathers, M300 and M367, respectively; exactly how we know this will become clear when we detail the badgers' genetic pedigree in Chapter 17). Cub F816 was assigned to F507 (and fathered by M533), but we could not with any degree of accuracy assign parents to the three remaining cubs (see Figure 2.1).

Before we perfected the DNA parentage tests, there was another way to deduce the mothers of at least some genetically anonymous cubs. To illustrate the detective work that underlies our generalizations, clues to the identity of the mothers of those unassigned cubs (F818, M821, and M823) lay in the teats of adult females at Pasticks and nearby Pasticks Outlier. Previously, we had calibrated how swollen teats indicate recent lactation (Dugdale et al. 2011a). Aside from F400 and F507 (to whom we had already assigned cubs), five other females had clearly lactated that spring (i.e. F434A, 324, F298(H), F301, and 'Beta Badger'), whereas the teats of another three were ambiguous (F350, F417, F65W—we ruled out F763 because she was a yearling). That ambiguity can arise either when an older female has teats still distended from nursing in previous years, or because a mother's cubs of that year died early in their infancy. In this case we judged that the distension of F417's teats was carried over from nursing her assigned cubs of the previous year, leaving F350 and F65W as the

---

[1] The term kekkering probably derives from the German word *Keckern* used to describe the vocalizations of foxes, reminiscent of a football rattle, and in English often rendered as gekkering. Although keck means cheeky in German, the origin is almost certainly onomatopoeic.

The Badgers of Wytham Woods. David W. Macdonald and Chris Newman, Oxford University Press.
© David W. Macdonald and Chris Newman (2022). DOI: 10.1093/oso/9780192845368.003.0002

Photo 2.1 Adult and cub interacting outside the sett.

Photo 2.2 Badgers interacting outside the sett.

strongest candidate mothers. As part of the process of elimination, F221 (first caught as an adult in 1991, and now very skinny—something that will turn out to be important in Chapter 14), F323 (aged 8), and F713 (a yearling) had tiny nipples that had not lactated.

Why are we wading through these personal intricacies? To illustrate the whittling down of cub numbers that continues after their emergence, and to illustrate also the inferential soup that allows us to piece together personal histories. These two types of information allow us to answer different sorts of question: by analogy, the cohort sizes of cubs is analogous to primary school class sizes, and in Chapter 11 we will analyse how these impact the population's demography. On the other hand, the fates of the individual cubs in the cohort (or, analogously, of children in the class) inform questions regarding how individual characteristics influence success or failure. For now, our principal purpose is to report how cohort sizes varied from year to year, and why, but just for illustration of the underlying detail, of the Pasticks badgers in 2001 at least seven, and possibly as many as ten, females were initial candidates as mothers to the nine cubs we'd probably seen earlier in May. The genetics assigned two to mother F400 (Figure 2.2); likewise two to mother F301 (whom she'd been seen to suckle in May). In June we caught seven (comprising a maximum of six litters because F779 and F819 were sisters, born to F400), and by August there were only three: of the seven cubs we had tattooed, four (F779, F818, M821, and M823) were never seen again beyond their first summer). In short, there was a high rate of cub mortality following emergence (Figure 2.3).

Although heavy infant mortality is not unusual amongst mammals, it is also not universal. In a comparative analysis across mammal species, Sibly et al. (1997) found that in 21 of 27 cases, the juvenile mortality rate differed significantly from the young adult mortality rate; in 18 of these the rate for juveniles was higher. Amongst the badgers of Wytham Wood, as each cohort of cubs is born it is whittled down both in the neonatal pre-emergence phase of life, and thereafter throughout juvenile development. The question we next pose is therefore: what proportion of cubs survives to be recruited as adults? And then we reveal a first, and very important, answer to the question of what causes them to die (the full actuarial analysis awaits the reader in Chapter 11, and in Chapter 12 we discover more about their causes of death).

Figure 2.1 Genealogy of some individual badgers mentioned in the text, showing that M553 sired M814 and F816 by two different females.

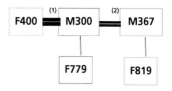

Figure 2.2 Genealogies of some individual badgers mentioned in the text, showing that F400 produced cubs F779 and F819 in a single litter sired by two different males.

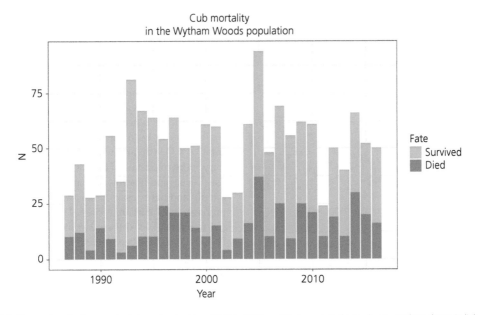

Figure 2.3 The number of cubs emerging into each cohort from 1987 to 2016, and their survival, those who survived are shown in light blue, and those who died by their first birthday in dark blue, based on Bright Ross et al. 2020, with permission.

In 2002 we published our first life table analysis (based on the 366 cubs whose life histories we had followed), giving first year survival (Lx) values of 75% for 1987–1996 (69 and 80% first-year survival for emerging female and male cubs, respectively, tagged post-emergence) (Macdonald and Newman 2002). Subsequently, we refined the estimate using data from 1987–2001 and more sophisticated statistics to 67% (±3% standard error (SE)) cub survival (Macdonald et al. 2009). Over the years, there were bad periods; for example, 2012–2014 when cub first year survival rate fell to 43.1% (Bilham et al. 2018). Of course, as more years of data come in, mortality rates are subject to continuous revision, as we establish who was actually still alive. Overall, and with yet more data (1987–2015) and even more sophisticated statistics that incorporate the probability of still being alive, but undetected in the population, the average was 65% (Chapter 11). Our most complete analysis of cub mortality of the badgers of Wytham Wood was published in 2020 (Bright Ross et al. 2020). It reveals a more than fourfold annual variation in the cub cohort—lowest in 2011, highest in 2006—and similar inter-annual variation in the percentage of post-emergence mortality (Figure 2.3). What kills them? We turn first to infant diseases.

Infant diseases

So, about a half of Wytham's badger cubs typically succumbs before maturity, but to what? The most revealing clue is that characteristic pot belly displayed by lethargic cub Bloat at Pasticks. As soon as we saw the unfortunately named Bloat we suspected gastrointestinal infection, but that year, as every year, clinical investigation of that crop of cubs had to wait until late May. Then, at c.3 months old, it becomes safe to catch and sedate cubs, prior to which they're too small and delicate and it is therefore essential not to separate them from their mothers.[2] As an aside, being caught young clearly affects a cub's attitude to research (Photos 2.3 and 2.4): those caught as diffident youngsters in spring, often sleeping while awaiting their turn for sedation, tend to be compliant when subsequently recaptured. Those caught for the first time as juveniles later in the summer can be feisty and aggressive—a trait they may adopt lifelong thereafter. Furthermore,

---

[2] Through continuous refinements to our anaesthesia (McLaren et al. 2005; Thornton et al. 2005) and handling procedures (Sun et al. 2015), we can nowadays safely capture cubs as small as 1.5 kg, with a body length of c.350 mm (Sugianto et al. 2019b; processing takes only 5–8 min so we do not need to induce deep anaesthesia; see Davison et al. 2007).

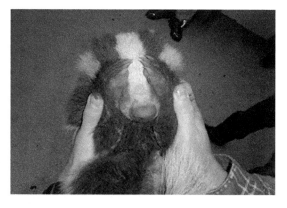

Photo 2.3 Badger cubs have different reactions to being captured depending on their age at first capture, some are very relaxed, and others very anxious or aggressive.

Photo 2.4 A badger cub entering the dataset for the first time.

our personality dossiers on how each individual reacts to capture revealed that badgers first caught when young (probably during a sensitive period; Serpell and Jagoe 1995) tended to be more accepting of, and less anxious during, future recaptures for the rest of their lives, compared with individuals caught originally as adults (Sun et al. 2015).

Our exploration of cubs' parasites began under the genial guidance of the late expert parasitologist Professor Ali Anwar (who had a formative role in characterizing malaria), and culminated in Chris' doctoral thesis (submitted in 2000, and based in major part on the products of warm, soapy enemas). Once faeces had been produced by badgers and collected, the process involved washing the contents of carefully labelled pots of pickled faeces through a 2-mm-gauge sieve to remove large fragments of semi-digested food. To separate the faecal suspension into a clear liquid supernatant, the murky residual fluid was decanted into test tubes and centrifuged twice at 2000 rpm for 3 min. That liquid was discarded, and a firm pellet of fine material forced out of suspension by centrifugal force (see Newman et al. 2001). This pellet was then weighed, macerated in a saturated salt solution, then topped off to form a positive meniscus. Over the next 20 min, buoyant parasites float up to adhere to a cover slip placed over the test tube (termed the 'salt floatation technique'; Parker and Duszynski 1986; Lloyd and Smith 1997; Newman et al. 2001[3]). Dropping this cover slip onto a slide and popping it under a microscope revealed that samples from those pot-bellied cubs were teeming with what looked like little double-yolked eggs—termed 'oocysts', each *c.*20 μm in size (they're fiddly to work with; about three or four of them standing in a line would be the thickness of a cigarette paper).

Ali confirmed these as 'coccidia', a type of unicellular protozoan parasite (belonging to the apicomplexan class Conoidasida). Specifically this was *Eimeria melis* (Figure 2.4), a coccidian species unique to badgers from the genus *Meles*. In the spring and early summer every single cub carried coccidia and the most heavily infected cubs carried a lot: more than 20,000 oocysts per gram of faeces (which would, if they all held hands, represent a string of protozoa almost half a metre long coiled into every gram of faeces, or, to continue with the cigarette paper metric, represent a wadge well over 5000 papers thick—the point being the oocysts didn't leave much room for anything else).

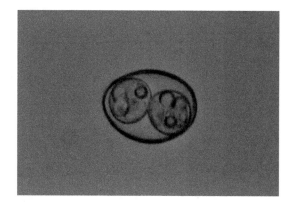

Figure 2.4 A microscope image of *Eimeria melis* (20 ± 0.18 × 15.7 ± 0.02 μm).

---

[3] Latterly, we replaced this technique with the cuvette method.

In the vocabulary of parasitology 'prevalence' defines the number of infected individuals expressed as a percentage of the number of individuals sampled (%), whereas 'intensity' defines the number of units of infection (here oocysts) recorded per gram. Mean intensity is thus the average of individual intensities across the sampled population. Note that the absence of coccidial oocysts in a faecal sample does not guarantee absence of infection, which could be present at undetectable levels (as typically occurs in sheep and goats; Pellérdy 1974). So what does coccidiosis do to the innards of a badger? Piecing together our lines of evidence, many cubs look sickly, with swollen abdomens; a proportion becoming lethargic and disoriented, while many die. Clinically, under sedation, these cubs exhibited flaccid skin due to dehydration and often produced copious, soft, dark faeces, riddled with E. melis, associated with diarrhoeal enteritis. In captivity, these symptoms lead to chronic wasting, morbidity, and, ultimately, death (Rewell 1948; Ratcliffe 1974; Neal 1977; Lindsay et al. 1997). Once ingested, oocysts release 'sporozoites' that invade the intestinal lining cells and set up a reproductive cycle of infection in neighbouring cells. The output of this cycle is ultimately more oocysts,[4] which pass with the faeces. Sporulated oocysts can then survive for up to a year in the right environment, where the warm moist conditions of a natal chamber are perfect for transmission between cohorts.

Already faced with the burdens of growth and development in an often nutrient-limited environment (Chapter 12), and competing with litter mates—factors that can kill even parasite-free cubs—it is no wonder that a proportion of cubs cannot contend with the additional toll this disease imposes. It may be more surprising that any survive. So why don't all cubs die? Early in life, a cub's only line of defence against coccidiosis is through cell-mediated immunity (innate

pro-inflammatory responses; Yun et al. 2000). It is not until they are around 8 months old, in October, that they start to 'acquire immunity' and produce antibodies as their B-lymphocytes learn to recognize the pathogen (see Chapter 15)—a slow process due to the complex life cycle and intricate host immune response to Eimeria (Reeg et al. 2005). Badgers that clear this hurdle generally stop shedding oocysts, but they may start to do so again if stressed by pregnancy or lactation.

The experiences of infected badgers were similar to those of the gallant human volunteers participating in clinical trials with the closely related coccidian Isospora belli: diarrhoea, steatorrhea (the excretion of abnormal quantities of fat with the faeces due to reduced absorption in the intestine), headache, fever, malaise, abdominal pain, vomiting, dehydration, and weight loss (Lindsay et al. 1997). Chae et al. (1998) graphically describe the symptoms exhibited by piglets afflicted by the porcine variant of Eimeria, which endure intestinal epithelial sloughing and villious atrophy (villi being the tiny finger-like projections lining the intestines). It seems wondrous that any cubs survive E. melis, but in order to discover which cubs do, we undertook two systematic studies: first sampling 402 individual faecal samples collected from 159 badgers through 1993–1995 (Newman et al. 2001); then a further 445 faecal samples collected from 259 individuals between 1996 and 1997 (Anwar et al. 2000).

Of 14 cubs sampled in the spring of 1993, oocyst counts averaged 7359 oocysts g^{-1} faeces (standard deviation (SD) 15,537), of which, of 10 cubs caught in November, 1 still had over 1000; although the remainder had none. Similarly, in spring 1994, oocyst counts averaged an astonishing 10,802 oocysts g^{-1} faeces (SD 2873, $n = 2$), or including July ($n = 4$), 5101 oocysts g^{-1} faeces (SD 4928), but by November, of 7 cubs caught only 1 still had over 1000, while 6 had functionally nil. Confirming this trend, cub oocyst counts averaged 5972 oocysts g^{-1} faeces (SD 2707) in spring 1995 ($n = 25$), with all 10 cubs caught in November still shedding E. melis, but averaging just 293 oocysts g^{-1} faeces (SD 134) (Figure 2.5).

In contrast, adults (sampled in spring 1993, $n = 15$, and 1994, $n = 10$) averaged just 1.2 oocysts g^{-1} faeces for males (18.7% prevalence), and nil for females (i.e. below the threshold of detectability). Similarly, over our 1996–1997 dataset, cubs had a significantly higher intensity of E. melis infection (mean 369.8 oocysts g^{-1} faeces, peaking at 16,157 oocysts g^{-1} in one cub caught in May) compared with adults (mean 18.1 oocysts g^{-1} faeces, peaking at 570 oocysts g^{-1}).

---

[4] When the sporulated oocyst is ingested by a susceptible animal, the sporozoites escape from the oocyst, invade the intestinal mucosa or epithelial cells in other locations, and develop intracellularly into multinucleate schizonts (also called meronts). Each nucleus develops into an infective body called a merozoite; merozoites enter new cells and repeat the process. After a variable number of asexual generations, merozoites develop into either macrogametocytes (females) or microgametocytes (males). These produce a single macrogamete or a number of microgametes in a host cell. After being fertilized by a microgamete, the macrogamete develops into an oocyst. The oocysts have resistant walls and are discharged unsporulated in the faeces. Oocysts do not survive well at temperatures below ~30°C or above 40°C; within this temperature range, oocysts may survive ≥1 year.

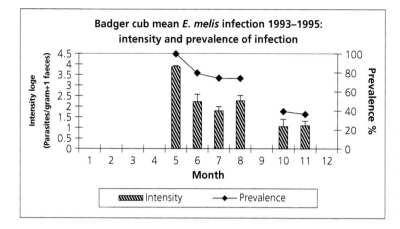

Figure 2.5 The bars show intensity of *Eimeria melis* parasite infection (measured as parasites per gram of faeces) in each month of the year between 1992 and 1995, with standard error. The points and lines show the percentage prevalence of *Eimeria melis* in each month (from Newman et al. 2001, with permission).

As a rule, the younger the cubs were when caught and sampled (the earliest at *c.*12 weeks old), the higher their prevalence and intensity of infection; the unwelcome absolute record going to F322, caught on 21 May 1995 with 53,033 oocysts g^{-1} faeces. Surprisingly, this female survived, living to be at least 5 years, having one cub in 1995 and two (by different fathers) in 1997. Moreover, the consequences of coccidiosis are brutally self-reinforcing: as a cub is weakened by infection so it becomes more susceptible to further infection.[5]

How do neonatal cubs become infected with *E. melis* in the first instance? *Eimeria* species have direct life cycles (i.e. they are monoxenous), transmitted via the faecal–oral route, and so at first we thought cubs were infected via their mothers or other infectious adults. However, the low prevalence and intensity of coccidia being shed by adults suggests otherwise. For comparison, Aramini et al. (1998) detected antibodies to *Toxoplasma gondii* (a similar apicomplexan parasite) in 92% (11/12) of wild cougars, indicating they were infected, but only 8.3% (1/12) actively produced infectious oocysts. Possibly, adult badgers could pick up persistent viable oocysts while sniffling at latrines, and, although unaffected themselves due to their antibodies, transport them back to susceptible cubs. More plausibly, neonatal cubs might contract coccidiosis from the sett itself, with the burrow soil acting as a fomite (i.e.

a substrate likely to carry infection; see also Courtenay et al. 2006—*Mycobacterium bovis* can persist viably in burrow soil, which is relevant to Chapter 16). To test this, we sampled soil from 2 m deep inside badger setts (aided by an old baked bean can attached to a broom handle) and found oocysts, evidencing soil contagion. Oocysts can remain viable for over a year (Robertson et al. 1992; Dubey 1998), so infection could be passed from one cohort of cubs to the next, a process termed vertical transmission (as opposed to horizontal transmission between contemporaries).

Variation between individuals in the intensity of infection (and thus the rate at which immunity develops; Chapter 15) became greater as cubs matured. Of 27 cubs sampled in November (1993–1995), 15 were completely free of infection—mirroring adults—10 had *E. melis* levels of 100–500, and just 2 still had intensities characteristic of cubs, in the range of 3000 oocysts g^{-1} faeces. This trajectory of infection implies that, prior to emergence, infectious intensities amongst neonatal cubs are likely extremely severe, leading us to think that coccidiosis contributes substantially to pre-emergence mortality, at a time when, unfortunately, we are unable to sample them (Chapter 1).

To explore the consequences that varying degrees of intensity of coccidial infection have for badger survival, we split cubs born in 1993–1995 into two categories: those with a greater, and those with a lesser parasite burden than the average for their year (Newman et al. 2001). We compared each cub to the mean seasonal value of its own year, because in a year with a good food supply even heavily infected cubs might survive, whereas in a hard year (notably the drought of 1995) a lower absolute severity of infection could exert a greater relative effect. Only 57% of the more intensely infected cubs survived to their first birthday, whereas

---

[5] This is consistent with severe coccidiosis as recorded for the juveniles of other species, ranging from domestic dogs infected with *Cryptosporidium parvum* (Lloyd and Smith 1997) and *Isospora canis* and *I. burrowsi/I. ohioensis*, also infecting wolves (*Canis lupus*), through to wapiti (*Cervus elaphus*) infected with *E. wapiti* and *E. zurenii* (Foreyt and Largerquist 1994), and red squirrels (*Sciurus vulgaris*) infected with *E. sciurorum*, *E. andrewsi*, and *C. parvum*.

72% of less infected cubs survived to maturity. This highly statistically significant effect still approached significance by the badgers' second birthday.

To put these statistical generalities into the context of the lives of individual badgers, consider some of the cubs born in May of 1995, a hot dry spring/summer. On 19 May we caught female cubs F472 and F476 at Radbrook Common sett—they were probably littermates—and they were afflicted with 10,254 and 9982 oocysts g^{-1} faeces, respectively. We never saw either of them again. In contrast, a week later, on 24 May, we caught their contemporary, F487, at the Chalet sett; this animal had only 3652 oocysts g^{-1} faeces. She lived to be 10 years old, changing her group of residence three times (Chalet, Hill Copse, and Great Wood), and was assigned three litters, totalling five cubs surviving to adulthood over her lifetime.

To complement the information on individual infections, delivered courtesy of enemas, we also tackled the overall and seasonal prevalence of infection in faeces collected from latrines throughout 1993 (ten samples per month, at ten social groups; Newman 2000) (latrines, scent marking sites, are the topic of Chapter 9). From May through August, E. melis oocysts increased in latrine faeces, which mirrored the increasing use of these latrines by maturing cubs. Cubs begin using sett latrines in their first weeks above ground, and as summer advances they graduate to other latrines throughout and at the borders of their natal range.

There was a minor peak in E. melis counts at latrines in January, before the cubs were born. We suspect this reflects immune-suppression of pregnant females (Woodroffe and Macdonald 1995a), due to their increased levels of oestrogens, progesterone, and corticosteroids (Lloyd 1983; Alexander and Stimson 1988; Brabin and Brabin, 1992; Newman et al. 2001; van Lieshout et al. 2020).

For those that survive coccidiosis, does this cubhood infection have consequences into adulthood? Yes; we have even found recently that this insult to their development can actually be written into their very DNA (see Chapter 18). At the level of population demography, coccidiosis results in the 'selective disappearance' of heavily infected cubs, consequently the surviving fraction comprises those more robust individuals: only better-quality individuals survive to grow old (Hämäläinen et al. 2014; Campbell et al. 2017; Chapter 18). Once again drawing on personal lives, consider males growing up at Brogden's Belt sett where, interestingly, average body lengths tend to be large, at c.725 mm. On 18 May 1995 (the drought year

with the most prominent effect) male cub 471 had coccidia levels well below the annual average, with just 1541 oocysts g^{-1} faeces, and grew to a gargantuan (for Wytham's high-density population) 755 mm. In comparison, his full brother, M479, suffered a much worse infection (3998 oocysts g^{-1} faeces), well above the average for the year, and grew to the relatively modest adult size of just 708 mm. Taking a population perspective, and again splitting cubs born from 1993–1995 according to whether they suffered above or below average annual rates of infection, the more heavily infected surviving cubs grew up stunted, with a 7.5% body-length disadvantage (head–body lengths of up to 50 mm shorter; Newman et al. 2001). Splitting this tendency, which was statistically significant overall, by year, male cubs born in the drought summer of 1995 (when it happened that we did this work) and suffering heavy parasitosis were highly significantly stunted compared with their less infected contemporaries, whereas the effect was not as significant in the less challenging summers of 1993 and 1994. Although the same trend was apparent for females, it achieved only marginal significance by age 2 years in the 1995 cohort.

What of the greater impairment of body-length development in males compared with females? Adult male badgers in Wytham are typically c.3% longer than females (c.22 mm and generally up to 20% heavier, c.2 kg; see Noonan et al. 2016). This arises because males grow more rapidly, as discussed in Chapter 1, attaining this 3% length advantage by the August of their first year (Sugianto et al. 2019b). The physical demands of rapid growth, while simultaneously burdened by coccidiosis, may ultimately exert a greater toll on body length of males than it does on females (see Figure 1.6). Furthermore, while body length is, for us, a ready reckoner of the impact of infant illness, in all likelihood other less easily measured aspects of physiology may be compromised by developmental coccidiosis, from physical development to immune function (see Chapter 15).

One such trait plays out in the improbable theatre of the roof of the badgers' mouths. While scoring the tooth wear of sedated badgers, our curiosity was piqued by the variously symmetrical, melanistic (black) blotches on the otherwise pink roofs of their mouths (akin in appearance to Rorschach inkblots in a psychological perception test). Might then the (a)symmetry of these palate markings arise as the product of developmental stress, perhaps even in direct response to coccidiosis? The symmetry of various traits certainly does respond to developmental insult (Van Dongen

(a)   (b)   (c)
Template   Typical marking   Superposition
                                                   of all marks

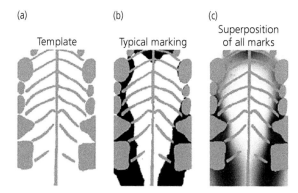

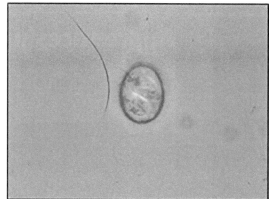

Figure 2.6 Digitized badger palate image: (a) template used for drawings, grey areas show the teeth/alveoli and maxillary-palatine ridges; (b) palate with representative markings, black areas show melanin deposits; and (c) superposition of all 971 palate images, darker pixilation indicated the most frequently marked (maculated) area. From (c), some regions are maculated in up to 94% of all instances among the 971 badger palates recorded; notably, maculation is disposed towards medial-alveolar deposition (especially around the incisors and molars). Maculation size is highly variable (range = 0–75%) (from Nouvellet et al. 2011, with permission).

Figure 2.7 A microscope image of *Isospora melis* (32.8 ± 0.34 × 26.9 ± 0.19 μm).

et al. 1999), including avian tail feathers (Møller and Höglund 1991), the whisker spots on a lion's muzzle (Packer and Pusey 1993), and skulls of African wild dogs (Edwards et al. 2013). Van Valen (1962) was amongst the first to appreciate that the visible asymmetry that arises from an inability to maintain trait canalization under environmental stress, such as childhood disease, can later provide a criterion for mate selection (see Chapter 17). Selecting a mate whose asymmetry reveals deeper imperfections entails the risk of offspring inheriting that propensity for infirmity. We therefore posed two questions: Was asymmetry in the palatal maculation a consequence of earlier serious coccidial infection? Was it a cryptic proxy of qualities involved in mate selection?

Equipped with a blueprint outline of a badger's palate, between 1994 and 1997 we dutifully sketched the maculation of 323 individuals, amounting to 971 records, including repeats (225 were drawn on multiple occasions, averaging 3 ± 1.4, up to 10 times). Quantifying and analysing these irregular shapes (Figure 2.6) fell to the mathematical expertise of our former doctoral student Pierre Nouvellet, now Reader in Evolution, Behaviour and Ecology at the University of Sussex. Our suspicions grew with the discovery that although expanse of the maculated area was quite heritable, its symmetry was not, suggesting

a developmental interaction. Comparing the palatal markings of badgers that, as cubs, had suffered the highest and lowest intensities of coccidial infection, revealed that greater infection was associated with greater asymmetry (Nouvellet et al. 2011). Indeed, what of the infamous badger F322, highlighted on p. 27, with her 50,000+ oocysts g⁻¹ faeces? Well, as it turns out, her mouth was almost entirely pigmented down her right palate, but free of all but a few minor gum-line blotches on the left.

The faecal samples revealed oddly few parasites, other than coccidia, but amongst them were two nematode (roundworm) species *Strongyloides* sp. (present as both egg and larval stages) and, very rarely, the egg of a *Trichuris* sp. There was also a second coccidian, *Isospora melis* (Figure 2.7), present at very low incidence in around 35% of badgers, including adults (Newman et al. 2001), for which we could detect neither morbidity nor age-specific pattern of infection. We return to the topic of how these coccidia interact with the rigours of growth—a development that will be seen, in Chapter 15, to challenge the badgers' immune system. The key discovery we highlight to carry forward from this chapter is that infant mortality is high in Wytham's badgers, and that the development of survivors may be impacted, to varying degrees, by the lasting effects of juvenile coccidiosis. But given these rites of passage, that subset of individuals that does make it to adulthood is already selected for toughness. For at least some of them, it will turn out that Friedrich Nietzsche's (1888) aphorism holds true 'Out of life's school of war—what does not kill me makes me stronger'.

CHAPTER 3

Apprenticeships for Badger Society

If there is no struggle, there is no progress.

Frederick Douglass

Remember sickly male cub Bloat (M814), with his coccidial pot belly? In Chapter 2 we learnt that he cheated the Grim Reaper, whereas most members of that 2001 cohort born into the Pasticks sett apparently succumbed. What hurdles next awaited Bloat? Had he been growing up in the society of any normal mustelid (a member of any of the 54 or so species that live in the intra-sexual territories[1] that characterize all but 7 species in the family; those 7 being at least occasional exceptions), his fate at maturity would have been predetermined: dispersal (the same end to adolescence awaits all pubertal males (and most females) in most of the 6339 extant mammal species; Chapter 19; Greenwood 1980).

But for Wytham's badgers, the career options are different: sometimes they disperse but, then again, sometimes they do not. What proportion disperses? A first answer awaits us towards the end of this chapter. And for those that don't disperse, the alternative scenario is graduation into philopatry, which etymologically means love of fatherland but biologically amongst mammals is more often about residence in the mother's or parents' land. As a term in ecology, philopatry was defined by Peter Waser and Thomas Jones (1983) as the continued close spatial association with siblings and parents—a situation that may arise when group life brings explicit benefits (e.g. Stacey and Lignon 1991; Marino et al. 2012), or where striking out alone brings costs (Wolf 1994). The denouement of these benefits and costs, and their sociological consequences, modelled by Macdonald and Carr (1989), is the topic of Chapter 10, but for now the key fact is that in the unusual case of the badgers of Wytham Woods the choice is not between homely biding, on the one hand, and lonesome pioneering, on the other; rather, badgers that disperse are neither involved in commandeering unoccupied territory, nor yet do they usurp expelled residents; instead, they join a different group. So, irrespective of whether they disperse or not, they will pass through an apprenticeship to group living. Badger cubs that survive the journey described in Chapter 2 are confronted with two questions about the future: what apprenticeship prepares them for membership of the society into which they were born; and will they opt for philopatry or disperse to a new group? This chapter tackles these two questions in sequence.

Apprenticeships

Again, Bloat the Badger offers insight, remembering that while he clung close to his putative mother 'H Badger' (named after her fur-clip), she, for her part, largely ignored him; in fact, she seemed more interested in grooming her sister, named 'Beta Badger' (another alphabetical fur-clip). That hint of indifference was further evidenced later in the summer of 2001 when

---

[1] The prevalent spatial arrangement amongst most members of the order Carnivora, and indeed many mammals, involves males living what we term semi-detached lives from the females whose home ranges they partly share. The result is two distinct sets of tessellating territories, the larger ones of males overlapping the smaller ones of females. Imagine the bathroom tiles in your shower, ideally hexagonally shaped, forming a neat mosaic—those are the females' territories. Then imagine a similar mosaic, but with each platelet two or three times bigger, superimposed on this—it could be a polygonal pattern on a transparent shower curtain, so you can see how the two systems overlap—these shower curtain polygons are the territories of males, and each overlaps two or three female tiles. Each sex behaves territorially to members of its own sex, excluding them, hence the system is referred to as intra-sexual territoriality; while the two sexes are clearly in contact, and partly cohabit, they do so in a semi-detached way.

The Badgers of Wytham Woods. David W. Macdonald and Chris Newman, Oxford University Press.
© David W. Macdonald and Chris Newman (2022). DOI: 10.1093/oso/9780192845368.003.0003

we arranged for our Earthwatch[2] team (Silvertown et al. 2013) to maintain an evening vigil at the Pasticks sett. Watching a clique of adult badgers, while four cubs were engrossed in scratching under a rotting log, one poor community scientist—who happened to be wracked with hayfever at the time—writhed in the attempt not to sneeze. She failed, and the tranquil woodland scene erupted with an explosive 'Achoo!' The adult badgers dived for the safety of their burrows, making no attempt whatsoever to round up the youngsters, call them, or in any way encourage them to safety. The four cubs stared, seemingly more bewildered by the exit of these adults than by the sudden sneeze; yet more remarkable, they made no attempt to run for cover, but nonchalantly stood their ground. Glancing at one another with what an anthropomorphic ethologist might have construed as a shrug, they returned to excavating their treasure beneath the log.

While there's danger in reading too much into one (in fact, far from isolated) observation, the absence of any attempt by the adults to shepherd or summon the youngsters to safety is a commonplace phenomenon in badgers; something that would be shocking to any naturalist accustomed to watching most social carnivores, such as African wild dogs (*Lycaon pictus*) or meerkats (*Suricata suricatta*), or even lions (*Panthera leo*), amongst all of whom a flurry of adults fizzes with excitement as it dotes on, and privileges, youngsters. In jarring contrast to the sociality of canids, mongooses, and even famously haughty cats, an odd thing about badgers is that as cubs become mobile and interact with the general throng of adults they are mostly treated with insouciance. An alternative way of looking at this behaviour would perhaps be that in forming this impression of parental, indeed adult, indifference, we were just placing too much emphasis on immediate protective behaviours; that is, perhaps cubs were getting effusive parental, and perhaps even alloparental, care in other contexts? After all, in most species of carnivore as conspicuously gregarious as badgers, not only are parents dotingly assiduous, but the benefits of being brought up in that pack, band, or pride are axiomatic (e.g. Macdonald and Moehlman 1982; Jennions and Macdonald 1994; Macdonald et al. 2004b). Indeed, WildCRU's work on both wild dogs and meerkats has contributed to evidence

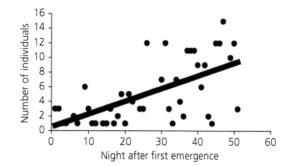

Figure 3.1 As cubs mature they associate with more members of their social group. Relationship between the age of cubs (night after first emergence) and the total number of individual cubs interacted with each night (from Fell et al. 2006, with permission).

that cooperative care by alloparents (i.e. care bestowed by adults other than the pup's parents) can deliver higher juvenile survival (Doolan and Macdonald 1999; Angulo et al. 2013).

The question of whether alloparents are helpful to young badgers has such great evolutionary importance to understanding the species' social system that we will devote a large part of Chapter 8 to its answer. But, to set the scene, what is the tenor of the cubs' social experiences as they mature? Cub social circles expand progressively (Figure 3.1) from their mother to include, first, other lactating females; then successively to incorporate adult males, subadult males, subadult females, and lastly, non-lactating adult females (Fell et al. 2006).

Two behavioural yardsticks of social relationships are sniffing and grooming (Table 3.1). Cubs sniff, and are sniffed by, their mother less often as they mature, but with this gradual severing of the apron strings all of the other adults increasingly sniff them; while it is more and more often the cubs that initiate these interactions. At first, only their mother grooms cubs; then adults of both sexes groom them (Figure 3.2). Cubs also initiate an ever-greater proportion of these grooming bouts with their mothers and other adult females, but rarely initiate grooming with males. Interactions with subadult and adult males commonly involved playfighting, progressively instigated ever more by the cubs (Fell et al. 2006; Dugdale et al. 2010). Overall, within social groups, we found no relationship between an adult's genetic relatedness to a cub and the time it spent in proximity to that cub after weaning (Dugdale et al. 2010), which occurs between 12 and 16 weeks of age; in other words, from mid-May to mid-June. How then, after weaning, does integration progress?

---

[2] Earthwatch is a charity dedicated to citizen scientists. David was their Chairman between 2010 and 2013, and many teams were active in Wytham, where Chris and Christina fruitfully coordinated their small mammal surveys for many years.

Table 3.1 Summary of the main changes in cub behaviour with increasing maturity.

Changes in behaviour with increasing age	F	R-coefficient	P-value
Cub emergence			
Emerged earlier from sett	30.78		***
Cub integration			
Increased grooming	45.60		***
Interacted with a greater no. of adults	25.88		***
Reduced time spent alone	5.21		*
Increased time with Adult males		0.195	**
Adult females		0.207	**
Subadult females		0.322	***
Subadult males		0.128	
Reduced time spent with mother and other lactating females		0.412	***
Cub–adults social interactions			
Followed adults and subadult females less but subadult males more often	4.00		**
Increased scent theft	7.30		**
Increased close contact with badgers of all classes except lactating females	5.24		***
Increased playfighting with adult males and subadults but decreased with adult females and mother	2.93		*
Initiated more playfights	12.85		***
Groomed less with mother but more with badgers in other classes	2.32		*
Initiated more grooming with adult females and mother but less with other badger classes	2.98		*
Reduced sniffing with mother but increased with all other classes	3.65		**
Initiated more sniffing with adults and mother but less with subadults	3.79		**
Overall			
Increased interaction with all classes except lactating females	4.55		***

The *P*-values are shown in a simplified format.
* < 0.05, ** < 0.01, *** < 0.001.
F and R-coefficient values represent effect sizes from the statistical model.

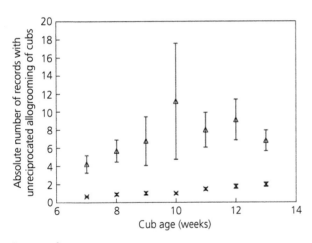

Figure 3.2 Mean absolute number of 'cub and group member records' in which group members allogroomed cubs without reciprocation, against cub age in weeks. Group members were classified as breeding females (triangles) or other members of the group (crosses). Error bars display standard errors. The graphs display the mean predicted values from generalized linear mixed model, which controlled for repeated measures of the individuals nested within a social group as a random effect, and the following fixed effects: the number of times and individual was seen on screen with cubs; observation time; group size; the number of cubs in the group; and social-group identity (from Dugdale et al. 2010, with permission).

Integration

The beguilingly simple question of how integration progresses (Table 3.1) is initially difficult because the cubs are underground, and remains difficult because although we can recognize adults by sight or on infrared videos from designs clipped into their guard hairs (Stewart and Macdonald 1997), we never clip youngsters until they are 9 months old in order to avoid the danger of them becoming chilled after the guard hair trim. Nonetheless, a first answer came from the observations of intern Rebecca Fell, who videoed 18 cubs attributed to 9 mothers (via lactation and association) from their first emergence in mid-April through the springs of 2005 and 2006 at the Chalet/Chalet Outlier sett and the Firebreak sett. At first, these cubs emerged only at the dead of night, around 01:00 to 02:00, and remained close to the sett entrance in the company of an adult—likely their mother (Fell et al. 2006). By late May, then aged 3 months, they began to emerge with adults around dusk (20:00). So, how did the process of integration progress? Anticipating the detail that will become evident in Chapter 6, social cohesion amongst badger groups is maintained largely by shared scent, particularly, in the close social context; this being a type of perfume produced by a pair of subcaudal glands (unique to meline badgers) that lies within a pocket between their tail and anus. Adult badgers anoint each other with these fatty secretions, and scent marks are smeared preferentially onto prime-age breeding males and females (Buesching et al. 2003), greasing the wheels of badger society. However, for cubs aspiring to become cogs in that society, the fact that they do not produce subcaudal secretion until aged c.4–6 months is problematic (Buesching et al. 2002a; Sugianto et al. 2019c). Even then, in May/June, male youngsters produce scarcely measurable traces, even less than females' smears, which averaged a whiff of only 0.04 ± 0.09 ml ($n = 72$). Why? Because the secretion consists largely of fat, which these growing youngsters allocate preferentially to growth (Allen 2020). It is only at c.11 months old, when approaching their first breeding season, that the subcaudal glands of both sexes become more productive (0.38 ± 0.36 ml in males, $n = 41$; and 0.19 ± 0.13 ml in females, $n = 47$). They only reach full odorous productivity at 2 years old (males, 1.17 ± 0.58 ml; females, 0.52 ± 0.41 ml)—the age when most badgers start their reproductive lives (Chapters 7 and 11).

So, infant badgers produce next to none of this socially bonding scent and, in accordance with the aforementioned indifference they display towards them, adult badgers do not scentmark the scentless youngsters. It was Christina Buesching who first noticed that cubs crawl under the bellies of adult group members (a behaviour readily confused with suckling), squeezing out between their hind legs, thus, in effect, forcing the adult to scent mark their backs—a behaviour we term 'scent theft' (Fell et al. 2006). Christina has been a central member of the badger project, participating tirelessly in every single catchup since she arrived to start her doctorate in 1996 since when she has become an international authority on scent communication (see Buesching 2019) and led our explorations of this topic. Christina and Rebecca observed that cubs engage in this scent theft with increasing frequency as spring turns to summer, 'stealing', in particular, from subadult females, then subadult males, diminishingly so from their mothers. Anointing themselves thus serves as an olfactory flag of the cubs' social affiliations as they insinuate themselves into the group's scentscape. And why do they steal scent (or perhaps are allowed to borrow it) from particular categories of individual in order, starting

first with subadult females, graduating to subadult males, and then on to adults? Perhaps they begin with the social classes that are 'soft touches', least likely to bite them, and closest to their own unchallenging status.

Talking of olfactory flags, badger cubs also have a visual one: careful scrutiny of the new-born cubs photographed through our periscope in Chapter 1 reveals that they are born with a stripey facial mask (Pearce 2011). Even by the time they might encounter a marauder in the dimly lit entrance to a burrow, and surely by the time they first emerge, these stripes herald the stridency of adults. For adult badgers are uncommonly fierce (Macdonald and Newman 2017), something they advertise to would-be assailants with their salient aposematic (Greek for 'keep-away sign') facial stripes (Pocock 1908; Newman et al. 2005). These strident markings are very evident in low-light, nocturnal conditions; larger predators quickly come to associate them with an unexpectedly vicious retaliation, whether that comes in the form of the bite of a badger, or the anal gland repulsion of a skunk (see Buesching and Stankowich 2017). And so the idea arose that it might serve cubs well to mimic adults in order to bluff aggressors—badging them as combatants of a formidable stripe. Moreover, the long guard hairs that badger cubs fluff up, or pilo-erect, when threatened, making them appear nearer to the daunting size of an adult, reinforce this youthful fakery.

It seems then that in the contexts of both scent and visual signalling, badger cubs embody the aphorism 'Fake it till you make it', which suggests that by imitating, in these cases clan membership and ferocity, respectively, the cubs can realize those qualities to achieve the results enjoyed by adults (respectively, acceptance and intimidation). In contrast to wide-eyed, snub-nosed puppies, or Mickey mice, there is nothing neotenic[3] about the faces of badger cubs; that is, the 'cute' features Konrad Lorenz (1943) conceptualized as a 'baby schema' (Kindchenschema) that elicit caregiving responses at the endocrinological level (Glocker et al. 2009; Kortschal 2012). Just as we speculate that cubs defer the synthesis of their own fatty scents to invest first in growth, while plagiarizing adult scent, so too we speculate that there is a syndrome of adaptation whereby the tender rewards of a kindchenschema are outweighed by those of mimicking adult ferocity. And while the contemporary lives of the badgers of Wytham Woods are comfortably predator-free, it was not always thus, and nor is it today for many of their continental brethren (Chapter 19). Indeed, in Chapter 19 we will

---

[3] Neoteny is the retention of juvenile features.

argue that both precociously stripey faces and enormous communal setts are heirlooms of a past bedevilled by predators.

Puberty and maturation: one pattern

Puberty is awkward, heralding the fading distinction between cub and adult at sexual maturity, marked by the development of secondary sexual characteristics (Adkins-Regan 2013), the first occurrence of ovulation/oestrus in females and the onset of spermatogenesis in males (Evans and O'Doherty 2001). Initially we documented these changes from the outside. As cubs' hormones developed, so too did their genitals. Around three quarters of male cubs exhibited scrotal (i.e. fully descended) testes for the first time at the age of 5–6 months. This, however, may not signal precocious reproductive capacity (it may just be an aspect of growth where the testes are released outside the body cavity) as the testosterone levels were still very low at 5–6 months (Sugianto et al. 2019c). These young males experience their first mating season when they are 11 months old, in January, at which point only around 41% had fully descended (although still small) testes. By the following spring, when they were 15 months old, the majority (94.8%) had fully descended testes and, thereafter, the ascent and decent of their testes followed the adult seasonal pattern through their second year, by which time their testes attained full adult volume, or the size of an ample cashew nut[4] (Figure 3.3). Male carnivores have a penis bone, or 'bacullum', which seemingly functions to prolong matings (helpful in the context of sperm competition[5]), induce ovulation (see Chapter 7), and to 'tie' in some species (not, however, in badgers). Bacullum length grows steadily, reaching mean adult size (86.03 mm) at an age of 23–24 months.

For females, we documented the manner in which the shape of vulvas changed with age; that is, being initially closed and flat, but steadily opening to what we termed a 'rose bud' condition as a possible indicator of receptivity.

Watching the external manifestations of maturation was clearly a blunt instrument with which to monitor the endocrinological tumult of adolescence. Documenting the underlying hormonal changes

was the doctoral topic of our student Nadine Adrianna Sugianto—whose veterinary background was introduced in Chapter 2—who undertook enzyme immunoassays of 119 blood samples from 7 males and 63 from 17 females at various stages of their journey to sexual maturity, selected from across our 1995–2016 dataset. To appreciate Nadine's discoveries, first check Box 3.1 for a primer on puberty.

Linking our external observations to internal hormonal changes, we inspected 1174 records (repeat records of the progress of multiple individuals) for females aged <28 months and noticed that their vulvas (Photo 3.1) were swollen to different extents. Mindful of the sexual swellings of chimpanzees, we wondered whether this was directly related to receptivity.

Nadine tested the assumption that a pink and swollen vulva with mucosal secretion indicated oestrus. She analysed sex steroid hormones from plasma samples collected from sexually mature adults during the winter mating season, the spring post-weaning period, the summer minor mating peak, and autumn reproductive quiescence, to compare oestrone ($n = 143$) and oestradiol ($n = 36$) levels in females with the condition of their external genitalia (categorized as swollen, intermediate, normal). More about these female hormones in Chapter 7, but the point here is that sex steroid levels exhibited seasonal patterns, but vulval swelling was an imperfect indicator of hormonal receptivity in badgers:[6] there was only an approximate match between both oesterone and oestrodiol titres and the extent of vulval swelling (Figure 3.5) (Sugianto et al. 2018).

Nonetheless a small proportion of maturing females exhibited this sign of sexual maturity by their first mating season (4.4% swollen, 20% intermediate), and all followed the adult pattern during their second year (Sugianto et al. 2018, 2019c). Genital condition was more indicative of breeding status in males (over 1136 observations): those with descended testes (vs intermediate or ascended) of near full adult volume had higher adult-like testosterone titres (Figure 3.6), although this was only during the spring mating season (Figure 3.7).

In Wytham, both sexes commenced endocrinological puberty[7] at *c*.11 months (Sugianto et al. 2019c).

---

[4] Winter, 6650.82 mm^3; spring, 5776.31 mm^3.

[5] Promiscuous mating can lead to ejaculates from two or more rival males being present in the female genital tract simultaneously. Sperm competition is a form of post-copulatory sexual selection whereby sperm physically compete (race) to fertilize a single ovum (see Chapter 7), and thus the bigger the testes, the more swimmers the male can commit to each race, and potentially to multiple races.

[6] Season was the only significant predictor of oestrone levels, while neither vulva condition nor the interaction term of vulva condition/season was significant.

[7] We defined puberty as the sudden peak (approaching adult levels) of hormone levels at the age of 11 months. Before that, the hormone levels were consistently lower than adult levels. The hormone levels started to follow the seasonal adult patterns thereafter.

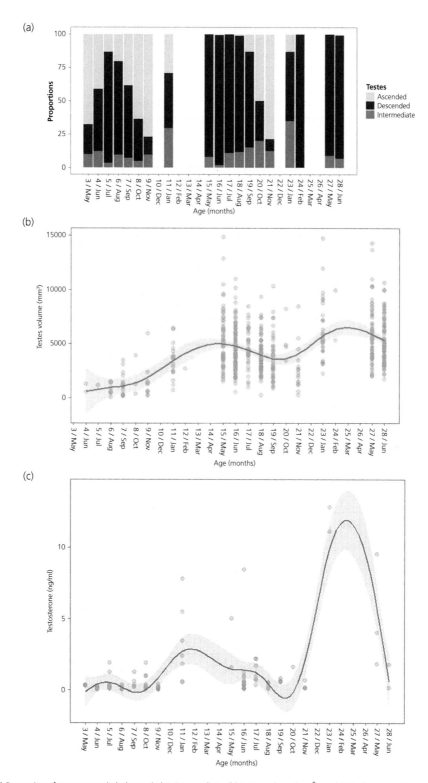

Figure 3.3 (a) Proportion of testes ascended, descended or intermediate; (b) testes volume (mm³); and (c) testosterone titres, in males aged 3–28 months (from Sugianto et al. 2019c, with permission).

Thereafter, yearling hormone levels followed the seasonality of adulthood (Figure 3.8), but at lower levels. Some yearlings (3 of 7 males sampled; 9 of 17 females sampled) had hormone levels suggesting they could be capable of breeding, but most reached full sexual maturity during their second mating season (22–28 months). In fact, our genetic pedigree (Chapter 17) revealed that only 22 females and 19 males bred as yearlings amongst the 1650 life histories we monitored between 1987 and 2015. Due to delayed implantation, this means they were assigned cubs at age 2, requiring them to have mated aged 1. This is just 2.5%, suggesting that early reproduction is a bold, perhaps premature, move (see also Dugdale et al. 2011c); a pace of life strategy we explore further in Chapter 14.

Serendipitously, the first two yearling breeders we identified, M64 and F59, mated with each other. Both were born at the Great Oak sett in 1987, and the union of these adolescent sweethearts gave rise to no fewer than four progeny (F156, M159, F160, and F183), born

---

Box 3.1 Puberty

Endocrinologically, puberty involves the activation of the hypothalamic–pituitary–gonadal (HPG) axis (Figure 3.4). This triggers the episodic release of gonadotropin-releasing hormones (GnRH) by the hypothalamus, which in turn activates the anterior pituitary gland to secrete luteinizing hormone (LH) and follicle-stimulating hormone (FSH), which instigate the generation of gametes and release of sex steroids (Plant 2006). In females, FSH stimulates the ovarian follicle(s), causing an ovum or several ova to grow, and also triggers the production of follicular oestrogen.

This rise in oestrogen then causes the pituitary gland to cease production of FSH and instead to increase LH production. In turn, elevated LH levels cause ova to be released from the ovary, resulting in ovulation (Adkins-Regan 2013). In males, LH stimulates testosterone production from the interstitial cells of the testes (Leydig cells), and FSH stimulates testicular growth and promotes the production of an androgen-binding protein by the Sertoli cells, which later are a component of the testicular tubules necessary for sustaining maturing sperm cells (Adkins-Regan 2013). Consequently, oestrogen and testosterone levels are low throughout the pre-pubertal period, but increase immediately prior, during, and after puberty, until they reach adult concentrations (Fitzgerald and Butler 1982; Beehner et al. 2009).

Image credit: Nguyen K. (2018) https://commons. wikimedia.org/wiki/File:Anatomy_of_the_ovaries.jpg, CC BY-SA 3.0; Gilbert Scott F. (2006) https://es.wikipedia. org/wiki/Archivo:T%C3%BAbulo_seminifero.png, CC BY 3.0.

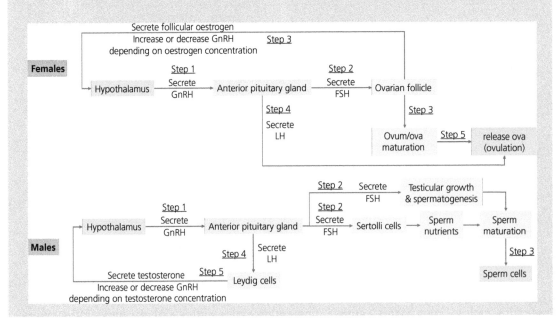

Box 3.1 *Continued*

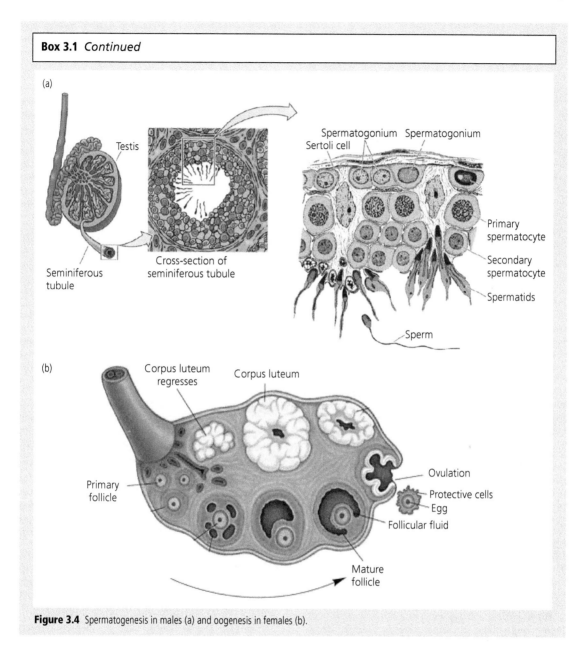

Figure 3.4 Spermatogenesis in males (a) and oogenesis in females (b).

at Great Oak, in 1989, implying that this couple mated in 1988 as yearlings. Perhaps by living fast they died young, because actually M64 was never caught again after 1987, suggesting his reproductive success was posthumous to the 1988 February breeding season, while young widow F59 was not caught beyond 1989, leaving her four young as orphans (see Pace of Life, Chapter 14).

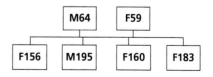

Just as choristers' voices break at different times, so as the young badgers grew, individuals in the same cohort varied in the development of secondary

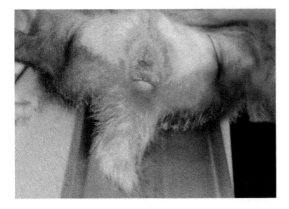

Photo 3.1 The vulva of a badger (anterior to the 'smile' of the subcaudal gland).

sexual characteristics (Sugianto et al. 2018, 2019c). Was this evidence of heterochrony (individuals differing in the timing of key pubescent stages)? It was. Male cubs divided into two distinct endocrinological phenotypes at age 11 months, with three early developers (of seven tested) with higher testosterone (HT, 5.58 ± 2.19 ng ml^{-1}; see Figure 3.9) reaching sexual maturity (adult testosterone profiles) at this age (first mating season), and four late developers with low testosterone (LT, 1.15 ± 1.14 ng ml^{-1}) not reaching sexual maturity testosterone levels until 22–26 months (i.e. during their second mating season). This hormonal divergence (and concomitantly different testes development) amongst adolescent males was reflected in distinct growth curve trajectories (technically, different phenotypes; see Chapter 1), with faster growth linked to more precocious endocrinological maturity: HT males achieved 95% of their mature length aged 11 months, whereas LT males took until 14 months, although body length converged between types at *c*.19–20 months. Early developers also underwent more advanced testes development (Figure 3.10), although, interestingly, these phenotypes also produced divergent subcaudal gland volumes by age 11 months, with precocious males at 0.33 ± 0.32 ml, and laggards at 0.14 ± 0.08 ml (indicative, but not statistically significant, differences) (Chapter 6). Social effects were also evident. Male cubs born into larger social groups tended to follow the late developer phenotype, suggesting a density-dependent constraint on their rate of development (Chapter 14); a similar group-size effect on male development exists in relation to patterns of sexual size dimorphism (discussed in Chapter 1; Sugianto et al. 2019b). Indeed, there may be a biogeographic effect of density, with Ahnlund (1980) reporting that in a

low-density Swedish badger population the majority of females were sexually mature at 1 year old. Similar negative correlations between population density and individual growth rate are familiar in carnivores, for example, the northern fur seal (*Callorhinus ursinus*), polar bears (*Ursus maritimus*), and American black bears (*Ursus americanus*), and amongst female baboons (*Papio cynocephalus*), in which first menstruation occurs earlier in smaller groups where individuals experience less social stress and competition.

Which of these two badger stratagems yielded the greatest reward in reproductive success? Revisiting our genetic pedigree (see Chapter 17), we sought to compare the reproductive success of LT and HT males, expecting the latter to get off to a racing start. On the contrary, and mindful of our small sample size, whatever its consequences for sociological advancement, precociousness brought no immediate reproductive advantage insofar as none of this sub-sample of seven males was assigned cubs in its first breeding season, irrespective of LT/HT status.

Amongst 17 adolescent females tested, endocrinological phenotypes were also apparent, diverging at age 15–18 months into those 9 with above-average, high oesterone (HO, 86.06 ± 12.72 ng ml^{-1}) and those 8 below average, with low oesterone (LO, 38.31 ± 15.32 ng ml^{-1}); this divergence occurred considerably later than for testosterone in males. In contrast to males, we found no predictive relationship between endocrinological phenotypes and growth curve trajectories, extent of vulva swelling, or subcaudal gland volume, aged 15–18 months. In terms of reproductive performance, in this small sample neither phenotype appeared to be hormonally capable of ovulating until its second breeding season, although 22 females aged 2 were assigned cubs in our pedigree implying they mated aged 1 year.

What of the sociology of female hormones? Group-size effects, somewhat depressing oestrone levels, and genital (vulva) development were much weaker in females than in males, but nevertheless there was an indicative trend. As for the question of why males might respond to density effects more strongly than females, a possible answer might be that the age at which puberty occurs is not just a pre-programmed genetic switch; rather, it is modulated either by intrinsic and extrinsic/environmental factors, or perhaps by both, as explained in the context of sex ratio theory in Chapter 1 and explored further in Chapter 14, in which we see that the fact that males tend to grow more quickly may make them more vulnerable to resource limitation and social competition—this possibility would fit with the observed delay in puberty

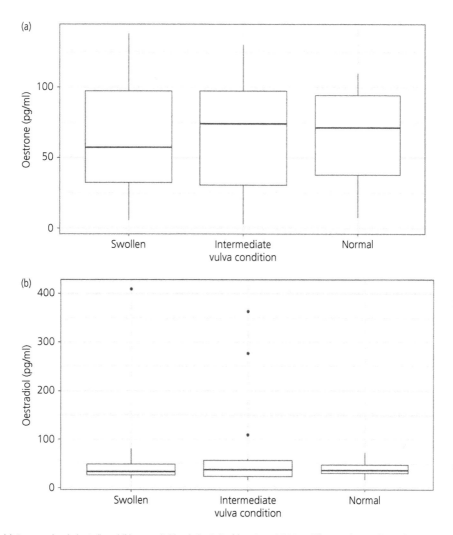

Figure 3.5 (a) Oestrone levels (pg/ml) and (b) oestradiol levels (pg/ml) of females exhibiting different vulva conditions (swollen, intermediate, and normal) (from Sugianto et al. 2018, with permission).

in larger social groups (LeBlanc, P.J. et al. 2001; Festa-Bianchet et al. 1994).

In summary, as youngsters face the transition to adulthood, and the decision to go or to stay, the general rule, not surprisingly, is that individuals growing up without good nutrition, subject to disease (coccidiosis; Chapter 2), or with social stress generally reach puberty later (Plaistow et al. 2004; Alberts 2012; Palombit et al. 2012). What then?

Dispersal and philopatry

To stay or to go—this dilemma between dispersal and philopatry is familiar to all mammals. To remind you of the general answer offered at the opening of this

chapter, at maturity most young mammals disperse (Greenwood 1980), a phenomenon typical of almost all mustelids (Powell 1979; Macdonald and Newman 2017). But not badgers. Later we will delve deeply into dispersal amongst the badgers of Wytham Woods, but, to set the scene, of the 5255 locations at which 267 individuals were (re-)captured between 1987 and 2003, we found that only 19.1% (51 individuals) dispersed to another group during their lifetime (Macdonald et al. 2008); 28 of these were males and 23 females. Indeed about a third of individuals (96/267) were never trapped outside of the social group into which they had been born. This was the situation for the badgers of Wytham Woods, and in Chapter 10 we develop an ecological argument for why it is so. In

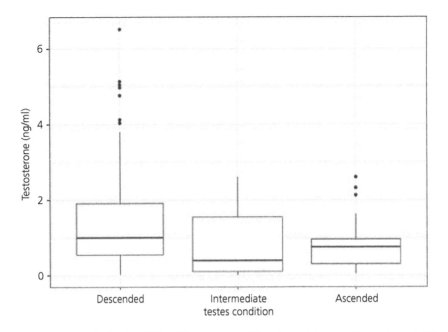

Figure 3.6 Testosterone levels (ng/ml) of males exhibiting different testes conditions (descended, intermediate, and ascended) (from Sugianto et al. 2018, with permission).

many other places, where ecological circumstances are different, young badgers at puberty are compelled to disperse—indeed are likely expelled by the occupying mother—if they add an unsustainable burden to the territory's food supply: intra-specific variation that is the topic of Chapter 19.

Where circumstances allow philopatry, what are the options for mating, and with whom? Indeed, in the calculus of whether to stay or go, traded off against celibacy or genetic purging, an important factor is that if two closely related individuals breed, this inbreeding can result in the pairing of recessive alleles, and lead to genetic defects in the progeny (Pusey 1987; Perrin and Mazalov 1999). So disadvantageous is inbreeding that psychological barriers develop such that animals (including humans) raised in close proximity (such as, but not exclusively, siblings, parents/offspring) become desensitized to sexual attraction,[8] a phenomenon first proposed by Finnish anthropologist Edvard Westermarck in his book *The History of Human Marriage* (1891) as an explanation for the incest taboo.[9] Most adolescent mustelids avoid this

propinquity by moving on (Powell 1979; Macdonald and Newman 2017), but not the badgers of Wytham Woods. This is a reminder of one of the central themes of this book: phylogenetic baggage. The question to hold in mind is, to what extent has the badger's contemporary behaviour—and how this, in turn, compares with that of other species of carnivore—been affected by its mustelid phylogeny? More on this when we unveil the mating system of Wytham's badgers through observations of their behaviour (in Chapter 7), and its genetic outcomes (in Chapter 17) in the wider context of inter-specific variation of musteloid social systems and beyond (Chapter 19).

For now we turn to the question of why, as they mature, some badgers within our Wytham population disperse and some do not. The argument develops from the ancestral system of intra-specific territoriality. Later we will find it a convenient shorthand to refer to this type of relationship between mated males and females as semi-detached. In brief, the territories

---

[8] Although for emerging contrary evidence, see de Boer et al. (2021).

[9] 'The idea that boys want to sleep with their mothers strikes most men as the silliest thing they have ever heard.

Obviously, it did not seem so to Freud, who wrote that as a boy he once had an erotic reaction to watching his mother dressing. But Freud had a wet-nurse, and may not have experienced the early intimacy that would have tipped off his perceptual system that Mrs. Freud was his mother. The Westermarck theory has *out-Freuded* Freud' (Steven Pinker, *How the Mind Works*).

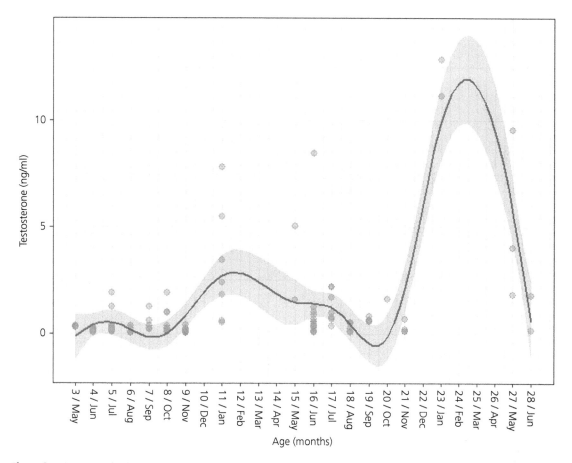

Figure 3.7 Testosterone levels (ng/ml) in males aged 3–28 months (from Sugianto et al. 2019, with permission).

of males and females, in this semi-detached relationship, are sexually dimorphic (as often are the body sizes of the males and females), with the much larger territories of males each overlapping several smaller territories of females. This system is fully explored through the lens of the badgers of Wytham Woods in Chapter 19, but suffice it here to note, as our work on American mink revealed, that even the overlap of male and female territories adds a heavy energetic cost to cohabitation (Yamaguchi et al., 2003): for predatory mustelids there just aren't sufficient resources available to enable adolescents to stay at home, causing them necessarily to be solitary (Powell 1979). We have formulated this chapter to ask how pubertal badgers decide whether to bide philopatrically or to disperse, but of that 19.1% of Wytham's badgers that dispersed, only a minority did so as they reached adulthood. For instance, between 1987 and 1995, badger M1 was

caught 27 times, but always within a group of ranges associated with Brogden's Belt, where he was born; with occasional dalliances, however, to the neighbouring Sunday's Hill sett—all his 13 progeny, bar M238 who was born at adjacent group Nealings Copse in 1991, were born at Sunday's Hill.[10] In contrast, M93 (an individual memorable for his cataracts and the development, as he aged, of a strange lump on his forehead) totted up 38 lifetime captures between 1988 and 1999, oscillating between the (adjacent) Pasticks and Jew's Harp setts, except when, in 1989, in some sort of pubertal rite of passage, he explored, briefly, living at the Marley Main sett (1 capture) to the west; and also the Sunday's Hill sett (3 captures). All his offspring were

[10] All born at SH: M193 (90), F199 (90), M212 (90), M233 (91), M285 (92), M340 (93), M341 (93), M342 (93), M419 (94), M437 (94), and F468 (95).

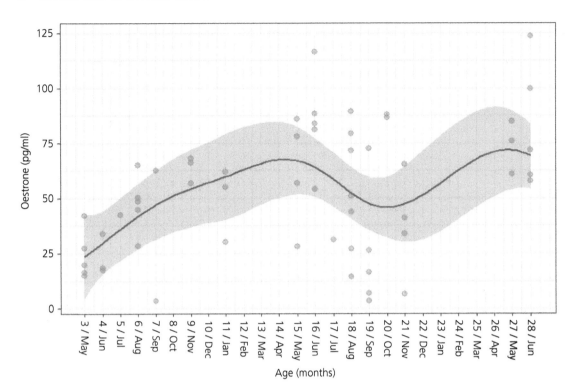

Figure 3.8 Seasonal variation in oestrone levels (pg/ml) in females aged 3–28 months (from Sugianto et al. 2019, with permission).

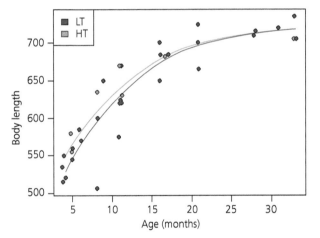

Figure 3.9 Body length growth curve of HT (higher-testosterone; orange circles) and LT (lower-testosterone; blue circles) groups, age: 4–33 months (from Sugianto et al. 2019, with permission).

also born at his main residences (M243 and M269 at Jew's Harp in 1991, M300 at Pasticks in 1992, and F589 at Jew's Harp in 1997).

If, then, dispersal at 1 or even 2 years old is rare, what of the lifetime picture? Of those badgers surviving to age class 1 (i.e. to the calendar year after their birth) for which we had sufficient data to record potential dispersal, only c.15% dispersed (at rates similar for males and females of all age classes). For those badgers surviving to age 5, the rate, or probability that an individual had dispersed, was 30%. On the other hand, a third of the individuals we recaptured (96/267) were never found even to visit setts beyond the group range in which they had been born.

We began in anticipation of a fork in the road for adolescent badgers at sexual maturity. In Chapter 19 it will

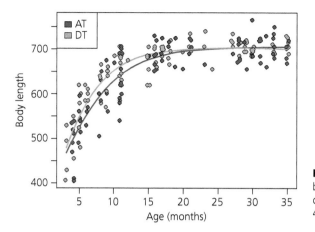

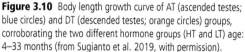

Figure 3.10 Body length growth curve of AT (ascended testes; blue circles) and DT (descended testes; orange circles) groups, corroborating the two different hormone groups (HT and LT) age: 4–33 months (from Sugianto et al. 2019, with permission).

become clear that just such a fork looms in the lives of some populations of European badgers, in some other species of meline badgers, and in some of the pot pourri of species colloquially called 'badger'. However, for the badgers of Wytham Woods, this is not an option at puberty when they are 1 year old, and rarely so even at sexual maturity when they are 2 years old. Dis-

persal, when it happens, is a feature of middle age, born, perhaps, of frustration, and perhaps also linked to a reproductive advantage associated with our discovery that dispersing males tended to move to larger groups and to groups with a preponderance of females (Macdonald et al. 2008). Next we will explore badger groups.

CHAPTER 4

Setts, Society, and Super-groups

The Geology of Social Behaviour

Nonconformists travel as a rule in bunches. You rarely find a nonconformist who goes it alone. And woe to him inside a nonconformist clique who does not conform with nonconformity.

Eric Hoffer

Assembling glimpses, and jumping to conclusions, is risky. In human society, the '7/11 rule' has it that we make 11 decisions[1] about the people we meet in the first 7 s of encounter—decisions derived from assumptions that, unsurprisingly, often prove wrong (Polly and Britton 2015). Snatched, if exhilarating, fragments of voyeurism into the private lives of shy, nocturnal badgers may also 'give the wrong impression'. That said, in the early days, glimpses were all we had to add insight and vibrancy to the staccato pulsing of radio-tracking fixes (even when the latter existed in the thousands) and to such actuarial insights as may be gleaned through the flicker book of trapping records. This difficult aspect of detective work is made worse insofar as prominent (i.e. trappable/observable) members of society might not be representative of the less observed throng (e.g. Tinnesand et al. 2015). Kruuk's (1978a) heroic efforts between March 1972 and 1975 to equip 12 Wytham badgers[2] with homemade, harness-mounted VHF transmitters, which operated on average for 3.5 months, were pioneering, but was this a representative dozen? Certainly, they were uniquely influential in the interpretation that the home ranges of badgers affiliated to different social groups were exclusive, and that intruders would be repelled by fierce attack (see, for example, Roper 2010). Without Kruuk's inspiration there would be no Wytham badger study, but it seems that this early dozen either hid successfully some of the habits that, we have subsequently established, result in 50% of cubs having extra-group fathers (Chapter 17), or indicate that times have changed. Almost 50 years later, have circumstances changed and what nowadays do we think is going on?

Remember, in badger terms, 50 years is around 10 generations. By extension, the human equivalent would involve the passing of about 250 years (Devine n.d.); that is, looking back to a time in which David's clansmen were about to fight the Battle of Culloden against the English monarchy, in which, having proved victorious, the latter would soon find itself limbering up for the American War of Independence and the Napoleonic Wars. Society can change a lot in ten generations—as can the factors that impact it (e.g. smallpox was rampant back then, but Covid-19 did not exist). Indeed, to push the parallel, in this span of human history feudal aristocracy has fallen and the human population increased nearly tenfold, while amongst Wytham's badgers the population was estimated at 58 in 1969, less than a fifth of its 1996 peak, and the number of social groups has increased from 12 to 23 (and the number of main setts from 13 to 63; see map, p. xiv). And thinking laterally, today the badgers of Wytham Woods live at a population density of 10 badgers km^{-2}, which is 20-fold the current national average of $c.0.5$ badgers km^{-2} (Judge 2014, 2017). So, things have surely changed for the badgers

---

[1] They are the following: (i) education level; (ii) economic level; (iii) perceived credibility, believability, competence, and honesty; (iv) trustworthiness; (v) level of sophistication; (vi) sex role identification; (vii) level of success; (viii) political background; (ix) religious background; (xx) ethnic background; and (xi) social/professional/sexual desirability.

[2] Actually, with huge effort, 23 were collared—and helping Hans to effect these captures while seeing the master at work—was one of the most formative experiences of David's career, as a 21-year-old; imagine, then, the frustration on finding that only 12 yielded sufficient data.

The Badgers of Wytham Woods. David W. Macdonald and Chris Newman, Oxford University Press.
© David W. Macdonald and Chris Newman (2022). DOI: 10.1093/oso/9780192845368.003.0004

of Wytham Woods. In Chapters 11 and 12 we'll answer the question of what drove those changes, but hold in mind that they did not stem from a low base, but rather from an unusually dense to an exceptionally dense population; European badger circumstances elsewhere are, in many ways, different, providing us with a wonderful opportunity to solve the algebra of intra-specific variation—how things differ from place to place (Chapter 19).

In arriving at what we nickname the Tidy Territorial Model of badger society, Hans Kruuk was not only impressed by the c.300 m spacing between setts dug in the wooded band of diggable sand (this drew him to a different conclusion to that reached by Ernest Neal in his 1958 classic; Neal believing that badgers do not defend a specific area against other badgers). He also saw fights—five of them, of which four 'were on the range boundary while one started near a sett which was 300 m from the boundary, and finished on the boundary'. Hans' beautifully observed field notes of these encounters (Kruuk, 1986) fitted closely with his observations of group territoriality in spotted hyaenas (Kruuk, 1972) leading him to summarize (Kruuk 1978a), 'When badgers met within a range, they appeared to recognize each other, either as individuals or as fellow members of the same clan; if another badger did not originate from the same sett, aggression followed immediately', and 'the defense of clan ranges against neighbours was confirmed by direct observation of aggression between neighbours on the range boundary'. It was, after all, miraculous that Hans could make these nocturnal observations at all, made possible only by the generous gift of Niko Tinbergen's Nobel Prize money to buy the first 'Hot Eye' (Chapter 10), but even so requiring a talent for fieldcraft, and dedication to fieldwork—qualities that were rare at the time, and even rarer now. But with more evidence, and increasing numbers of categories of that evidence, does the conclusion of aggressive, exclusive territoriality hold up today? Indeed, was it right at the time? And how generalizable might these observations be of the badgers of Wytham Woods?

We first felt we had the evidence to reappraise the exclusivity of badger group ranges c.30 years after Kruuk's (1978a) pioneering description of badger society. By that time, our database documented 5255 capture events accrued from our seasonal trapping regime between 1987 and 2003, involving 267 different individual badgers (Macdonald et al. 2008). This stocktaking revealed that 863 (16.4%) captures involved individuals caught at a social group differing from that of the previous trapping occasion; however, theirs was

a temporary visit, evidenced by the fact that they subsequently returned to the group to which their trapping history suggested they were mainly affiliated. That is, these visitors were making sorties to other setts—not dispersing.

Kruuk (1978a) had attributed a spurious bait-marking return—that is, food fed to badgers containing harmless coloured plastic bait chips in one territory turned up in their faeces, but was deposited in a different territory—as a badger 'visiting' another group's territory[3] (we explain bait marking thoroughly in Chapter 9). Our routine trapping had revealed that such visits, as evidenced by badgers getting caught where they conventionally ought not to be (Photo 4.1), were, by the late 1980s, far from occasional: overall, at any given time, we encountered 19.8% of the population at the sett of a social group other than their principal, or natal, affiliation. These visits were most common in autumn (when 17.1% of recaptured badgers were caught at what we originally thought of as the 'wrong' sett) and least common in winter (10.9%). This tendency for making visits did not differ between the sexes. There was, however, strong evidence that a badger's movement history affected the probability of its accumulating scars; adjusting for age and number of captures, badgers that had a history of movement between social groups were significantly more likely to be scarred.

Of course, our seasonal cage-trapping regime has provided only occasional snapshots of the social flux—at best three or four badger placements each year—that together compiled a sort of flicker book image of their affiliations. Flicker book or not, these observations shook the foundations of our previously tidy assumptions about badger society. For example, while about a third (96/267) of individuals were never trapped outside of their natal social group and remained totally philopatric (as mentioned in Chapter 3), 39.9% of badgers were captured in (but just visiting) the setts of at least two groups during their lifetimes (16.0% in three groups, 5.2% in four groups, and 2.9% in more than four groups), with males tending to visit more groups than did females. In Chapter 3 we reported that permanent dispersal[4] was rare, with just 51 (19.1%)

---

[3] Similarly, although bait-marked faeces are often deposited deep into neighbouring 'territories', any patterns of latrine use incongruent with imposed borders are generally disregarded as 'erroneous' data (see Delahay et al. 2000).

[4] A badger was flagged as 'resident' within a social group if the two most recent captures—as well as at least one of two previous captures before that—were made within the same social group.

Photo 4.1 David returning a badger to its home sett after routine sampling.

badgers (28 male, 23 female) satisfying our definition; that is, 80.9% of badgers exhibited lifetime group residential fidelity. Indeed, in a society seemingly so open to free movement, the question arises as to the benefit, or necessity, for such a decisive commitment to shift affiliation (see below). When permanent dispersal did occur it was generally unadventurous; that is, the animal moved no further than a neighbouring group, where the average 530 m (standard deviation (SD) = 393 m, standard error (SE) = 55 m) dispersal distance approximates the average distance between adjacent group main setts in Wytham (Figure 4.1).

So our observations, some seven generations on from the badgers Hans Kruuk observed, were different. Was this a genuine change in circumstances and social system, or did it, perhaps, expose some deeper understanding of a social system that was always there, but only revealed by subsequent years of detail?

The insights of naturalist scientists tend to advance in quantum leaps, often springing from a new technology. In the beginning there was stealth and fieldcraft,

aided only by binoculars and a notebook. The study of Wytham's badgers surged forward with developments such as VHF radio-tracking, night-vision equipment, and then DNA fingerprinting, heralding the advent of a burgeoning range of high-tech arrows in our methodological quiver (e.g. from Amlaner and Macdonald 1980 to Bright Ross et al. 2020). In this vein, one such recent addition that was able to revolutionize the fine granularity with which we could elucidate badger society, especially interactions between groups, was active radio frequency identification (aRFID). Essentially these are the security tags used in high street clothes stores to protect against shop-lifting. We wanted to know how frequently badgers within and between groups came into close proximity, both at their setts and their boundary latrines—aRFID could tell us.

The adaptation of this technique enabling us to track many individual badgers simultaneously, and with less disturbance, was the brainchild of technical whizzes Stephen Ellwood and Andrew Markham (Ellwood et al. 2017). Their deft soldering of motherboards produced badger-collar-borne aRFID tags able to communicate with static, wirelessly networked aRFID detector base stations. These were placed either at setts or at latrines shared between neighbouring groups (established from bait marking; see the following paragraphs), to record badger locations within a 31.5 m radius (90% within 27.9 m, 80% within 22.5 m).

The reality that each aspect of the badgers' biology is linked to all of the other facets challenges any one linear route through this book, but here, to make sense of our narrative, it is now necessary briefly to introduce border latrines (they are the detailed topic of Chapter 9). These shared defecation sites ring-fence that division of the populations' range co-opted primarily by badgers resident at the heart of that area, and were long thought to signal the borders of aggressively defended territories, functioning as 'Keep out!' signposts.

In parallel, we must also touch here briefly on a complementary, but less technical method from the antiquity of badger surveying: bait marking. In 1973, David, the apprentice, followed Hans Kruuk around the wood with a bucket stickily filled with peanuts and thin strips of coloured polythene (3–5 mm long × 2–3 cm wide; later superseded by alkathene chips, about the size of rice grains) drenched in syrup. Adapting an idea that Swede Peravid Skoog had applied to badgers that had been fed marked herring baits (1970), Hans put a couple of scoops of this tacky mix at each badger sett, thus feeding the occupants of each sett a different colour of bead/mix. Resident badgers will ingest this confection and defecate undigested bait into the latrines they use.

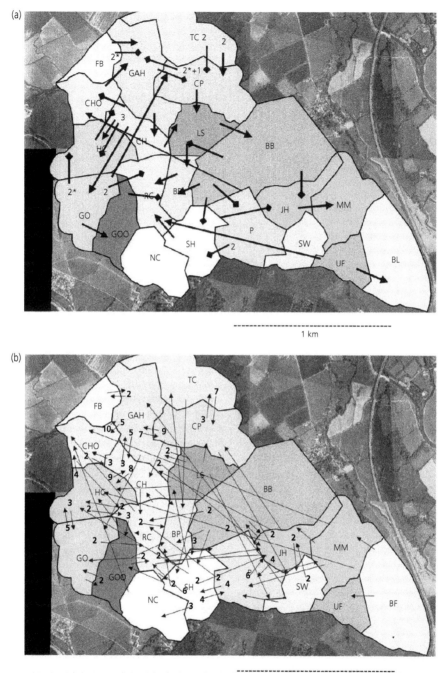

Figure 4.1 (a) Dispersal events. Males = triangles; females = diamonds. An asterisk (*) denotes approximately simultaneous events. Range boundaries as determined in 1997. (b) Non-dispersal intergroup movements for the period 1996–1998 (from Macdonald et al. 2008, with permission).

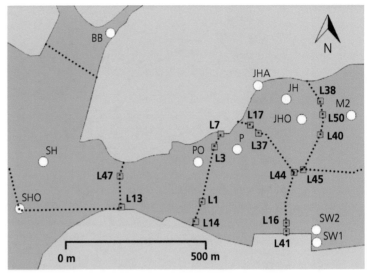

Figure 4.2 Study area map. Dark green = wooded; light green = agricultural land; white circles = sett base stations (BB, SH, SHO, PO, P, JH, JHA, JHO, M2, and SW1 and 2); blue squares = base stations at latrines (Ln) shared by two or more social groups (none found outside woodland); dotted lines are notional woodland territorial borders separating our seven *a priori* defined notional social groups (SH = SH + SHO; BB; PO; P; JH = JH + JHO + JHA; M2; SW is a dispersed sett requiring two base stations, SW1 and SW2) (from Ellwood et al. 2017, with permission).

These latrines can then be surveyed and plotted onto a map, and then coloured droppings can be tied back to their sett of origin (i.e. the hub). Spokes are then linked by a polygonal rim to denote the particular zone primarily 'owned' (or utilized) by each resident group (see Delahay et al. 2000). These 'rims', or borders, butt up against those of juxtaposed neighbours, resulting in a tessellating odorous faecal/mosaic. Moreover, neither spoke nor rim was purely an invention of the procedure; in reality each existed visibly on the ground as a series of interlinked, well-trodden paths, with latrines characteristically marking every junction. It was these tiles of garish colours that were interpreted as territories, according to the definition offered by William Burt as 'an area defended' (Burt 1943) (we will return to more sophisticated definitions of this phenomenon in Chapter 10). Henceforth, we use the adjective 'boundary' in a solely geometric sense, to describe the latrines that bisect the hinterland between main setts, and not in a functional or interpretative sense that implies defended exclusivity (although the latter is exactly the implication of the terms 'border' or 'territorial latrines' that pervade the literature on badger society).

First, what did we learn from the aRFID monitors about visits to boundary latrines? These aRFID-tagged badgers were remarkably enthusiastic about toiletries. The perfunctory activities of defecation and urination could have been much more swiftly accomplished than the $c.12$ h week^{-1} each badger committed to loitering at latrines. Experiments reported in Chapter 9 cause us to interpret each latrine as, metaphorically, a newspaper, poster-board, or, nowadays, a Facebook (or should one perhaps say 'Scentbook'!) page of scent signals, so

we infer that badgers took this extra time to check the messages, suggesting something more along the lines of a reading party rather than the defence of a border outpost, in which they appeared deliberately to seek encounters with neighbours at latrines. Before raising an eyebrow at this seemingly extravagant use of time, remember that the average American checks his or her mobile phone 96 times a day (once every 10 min) and one sample of British teenagers revealed the fact that the latter spent $c.60$ h a week on social media, five times more than the average badger spends at latrines (as this chapter progresses it will become increasingly clear that what is not communicated is a simple 'keep out' signal; the detail, however, of what is communicated at boundary latrines is the subject of Chapter 9).

So, what sort of match, if any, existed between our cartographic representation of territory geometries (and the location of latrines) and the question of how these latrines and main setts were used, and by whom, as revealed by our aRFID stations? Over a 13-week autumnal study period in 2009 we collared 32 badgers initially—approximately half of 50–60 adults typically resident in the 7 focal groups involved in the Pasticks area of Wytham[5] (see Figure 4.2; and map, p. xiv), which eroded to a sample of 18 due to gradual collar loss and device failure. To analyse social networks we coded detections, post hoc, into time intervals (Grolemund and Wickham 2011) and compared all detections

[5] The seven focal groups (which encompass ten main setts) are: Brogden's Belt (BB), Sunday's Hill (SH), Pasticks (P), Pasticks Outlier (PO), Jew's Harp (JH), Singing Way (SW), and Marley 2 (M2). For further social groups see Map 9.1, page 173.

iteratively per base station, per night, to identify when two badgers were in the same vicinity (termed 'dyadic overlaps' in space and time) with this 'co-location' (which may not be the same as thing as a meeting, but might sometimes approximate it)—the latter measured in seconds.

Of 1834.1 h of detections (n = 161,333 events) over this 13-week period, 56.1% (males 454.1 h, females 575 h) were made at sett base stations, and 43.9% at border latrines (males 297.6 h, females 507.3 h). Forty-three per cent of these detections (785.1 h) involved badgers co-locating with another collared individual (likely also co-locating with many uncollared individuals). These badgers meet up frequently, but with whom: cohabitant or outsider?

For those of us brought up with the tidy territoriality mindset, the results were astounding. When we tabulated the co-locations, badgers were actually in the company of members of other groups for 16.1% of their co-location time (0.16 h dyad^{-1} week^{-1}, SE = 0.02), and with cohabitant group members during the remaining 83.9% of their time in company. More amazing yet, co-location rates between members of different social groups were similar at boundary latrines and at main setts, implying core exchange visits, not just border rendezvous: that is, those visits detected when during routine cage-trapping we caught individuals at the 'wrong' sett on 16.4% of occasions no longer seemed out of the ordinary; rather, they started to seem like everyday occurrences. According to aRFID co-locations, during liaisons at latrines and setts, there were roughly equal rates of associations between the sexes. Within the same social group, more time was inevitably spent co-locating at setts (0.67 h dyad^{-1} week^{-1}, SE= 0.08; 68.9%) than at latrines (0.15 h dyad^{-1} week^{-1}, SE= 0.02; 15.0%) due to implicit co-residency (Figures 4.3 and 4.4).

These results were completely at odds with the Tidy Territoriality Model. What pattern underlay the affiliations of the tagged badgers? We sought a quantitative answer through social contact network (SCN) analysis (Farine and Whitehead 2015). This characterizes the interaction between actors in a network, termed 'nodes' (individuals or groups) by linking their extent of connectivity with ties, termed 'edges'. This involved the conversion of weekly dyadic co-location data into network graphs, in which duration of co-location defined 'edges'; that is, lines connecting 'nodes' (in this example, badgers) in the network, such that the thickness of the edge equated to the dyadic strength of association between each pair of badgers. Anticipating that the social dynamic at setts might differ from that at latrines, we calculated three networks:

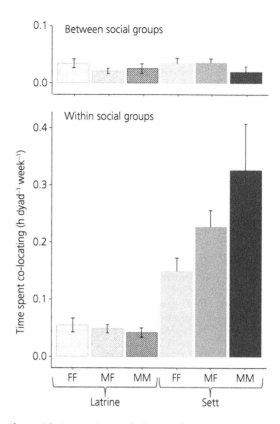

Figure 4.3 Average time, per dyad, per week, spent co-locating, per combination of gender–dyad group (male–male, female–female, male–female), site type (sett or latrine), and dyadic social group relationship (intra-, or intersocial group) (from Ellwood et al. 2017, with permission).

one each for setts, also for latrines, and then the combination of both. Using the amusingly titled 'Fast and Greedy' algorithm (Clauset et al. 2004) to estimate 'community strength', we were able to compare the picture of groupings emerging from the three aRFID networks with that which we had developed from our routine trapping records.

The networks that emerged from these analyses were complicated, but even so they must still represent a simplification of reality, remembering that we had instrumented only about half of the badgers resident in this cluster of setts. First, and considering weekly snapshots of social dynamics, the tagged badgers were arranged into a smaller number of more populous communities based on latrine interactions, rather than as communities revealed by interactions at setts (Figure 4.5).

The sett-based SCN communities corresponded more closely with social group memberships apparent from our cage-trapping records (Figure 4.5a). Combining encounters at setts and those at latrines into a single

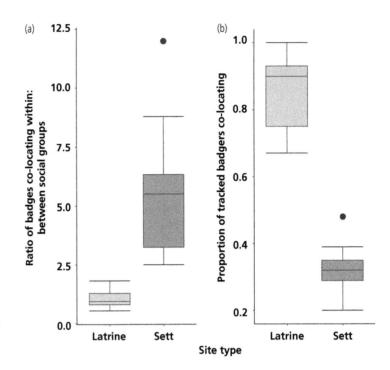

Figure 4.4 Box–whisker plots showing co-locations at setts and latrines: (a) ratio of badgers co-locating within social groups; and (b) proportion of badgers co-locating between social groups (from Ellwood et al. 2017, with permission).

network produced a comparable number of similarly composed communities to those arising solely from the examination of co-location at setts, due to a higher rate of at-sett encounters weighting the overall picture, but with much greater inter-group connectivity than would have been evident from a study of setts alone.

The original Tidy Territoriality Model of the badgers of Wytham Woods was thus starting to crumble; interactions (and visits) between at least some group neighbours were commonplace, rather than the risky lifetime-defining moment that we had assumed a shift in residency to be. We began to wonder whether Kruuk (1978a) had been victim to 'confirmation bias' (Loehle 1987; Nickerson 1998) fitting badgers into the paradigm of tidily tessellating hyaena territories (Kruuk 1966, 1976), or was it rather the case that the badgers' behaviour had changed during the intervening years, or perhaps a bit of both? Keen to avoid Einstein's satirical pitfall, 'If the facts don't fit the theory, change the facts', we were mindful that the very untidy pattern of aRFID associations reinforced our mounting evidence of inter-group trapping visits (Macdonald et al. 2008). The conclusion that the society of the badgers of Wytham Woods in the twenty-first century was much less segregated than it had appeared to be in the 1970s was also congruent with other observations, reported in Chapter 5 of neighbours amicably sharing a meal at experimental feeding sites at territorial borders (Macdonald et al. 2002b), and similarly tolerant

encounters seen while testing food repellents (Baker et al. 2005). Moreover, the absence of even a single bite wound (which we would have spotted at recapture) over the weeks of the aRFID study amongst these inter-group protagonists was at odds with our expectation of blood-letting border clashes anticipated from the model of antagonistically enforced territoriality. These are all clues that will inform, in Chapter 9, our exploration of what it means to be a neighbour in badger society.

Building on this revelation of inter-group contacts, we next pondered whether the liaisons were the prerogative of a few highly connected individuals with particular neighbourly ties (perhaps some grandmothers popping round to catch up with a diaspora of welcoming grandchildren). Not so. They were all at it. At setts, in week 1, 15 individuals (out of 31 collared; 48%) were involved in inter-group contacts, decreasing to 5 in week 4 (out of 25; 20%). An even greater proportion of individuals engaged in between-group assignations at latrines, where between 8 (out of 12, in week 13; 67%) and 24 (out of 24, in week 4; 100%) individuals were revealed to be simultaneously in the vicinity of their neighbours—although these groups were turning out to be so jumbled that the word neighbour was starting to feel imprecise. It remains useful in terms of primary sett residence; nowadays, however, as far as Wytham is concerned, 'neighbour' has become ambiguous to the point of confusion in the context of

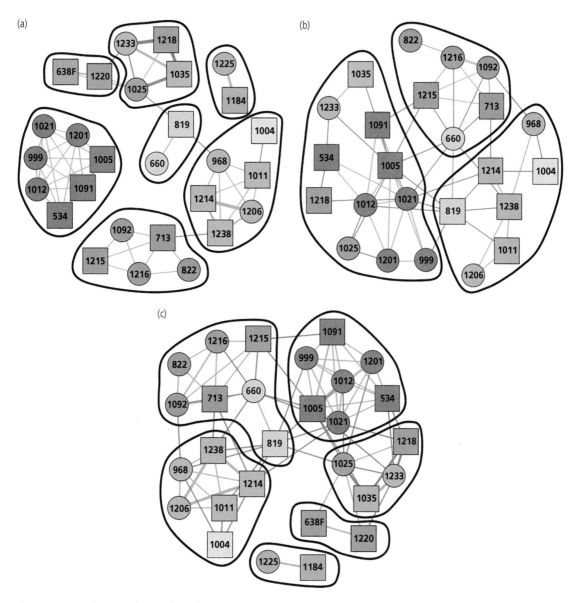

Figure 4.5 Network patterns from week 2 co-locations at: (a) setts; (b) latrines; and (c) setts and latrines combined ('All'). Node numbers, shapes, and colours depict badger IDs, gender (square = female, circle = male), and social group affiliation from trapping (light blue = BP, dark green = SH, light green = BB, red = PO, gold = P, fawn = M2, orange = JH, dark blue = W), respectively. Edge thickness is proportional to co-location duration. Edge colour indicates intra- (green) versus intersocial group (red) co-locations. Black borders indicate network 'community' estimates (from Fast-Greedy algorithm) (from Ellwood et al. 2017, with permission).

spatial organization. One male (M660) and one female (F819) stood out as being completely itinerant (across weeks) between adjacent social groups—possibly the well-linked type of 'influencer' individual that can tie societies together (or, more sinisterly, spread diseases; see Chapter 15). Revealingly, on the topic of sampling, both of these networking individuals were only ever 'trapped' at Pasticks sett, indicating how our flicker

book of inter-group visits intercepted, of course, only a fraction of visits reflecting where they typically slept, but not limiting their theatre of nightly activity. F819 was born at Pasticks Outlier where she was first caught as a cub in June 2001, but was brought up at Pasticks, to judge by her capture in that location through August and November of that year; she remained in Pasticks until we last caught her in November 2007. Over her lifetime F819 was assigned three cubs, all of which she bore at Pasticks sett: M1089, to father M822 (from adjacent Singing Way), in 2006; and M968 and M1019, to father M862 (from Chalet, some three social groups distant), in 2004 and 2005, respectively.

Although 25 years had passed without our giving it much attention, these larger communities around latrines forcefully brought to mind an earlier suspicion that groups of badgers might be assembled into larger collectives: a.k.a. super-groups. By analogy, nowadays Chris lives in Nova Scotia, in which several native American 'bands' are all members of the Mi'kmaq First Nation. This suspicion that at high population densities affiliated groups of badgers shared memberships of super-groups had grown from our first sortie into genetic affiliations (in a different population of badgers, in Gloucestershire). We will return to those genetic insights, using allozyme techniques that ornithologist Peter Evans had developed for starlings, in Chapter 17 (Evans et al. 1989). Suffice it to say that if two entirely different methodologies—that is, genetic and social clusters—both suggested the same phenomenon, however unusual it might be, it could not so easily be dismissed.

What selective advantages might cause super-groups to arise in Wytham's ultra-dense badger population? Propinquity—social 'closeness'—consanguinity, territorial familiarity, and the benefits of local knowledge, are all candidates, as is physical geography. Two obvious possibilities, with potentially very different sociological implications and ecological frameworks, arise: on the one hand, that some neighbouring groups coalesce into a super-group; on the other, that a formerly united group fissions into its components.

The ontogeny of super-groups: internal geometry

The case of Pasticks main sett and its outlier hints at the mechanism whereby super-groups arise. Pasticks Outlier did not exist until it was excavated in 2001, in an area which was, so to speak, in the backyard of Pasticks main. But the story goes back further. In 1974 the now monumental Pasticks sett, with its 20–30 entrances, was merely an outlier of the Jew's Harp sett/group (sett number 4 on Hans Kruuk's map), lying 100 m to the north-east. Today Jew's Harp has a clear-cut boundary of border latrines within a range that encompasses four outlier setts, each forming a social sub-hub. By 2003, 2 years after it was first excavated, Pasticks Outlier was assigned five (three male, two

female) residents, whereas Pasticks main was home to ten (four male, six female). Their fortunes were to reverse, however, so that in 2005 Pasticks Outlier had nine (four male, five female) residents to Pasticks' seven (three male, four female). Subsequently, the outlier (the name had already become an anachronism) has consistently accommodated more occupants than the main sett. Over the years, our bait marking[6] (2005, 2009, 2011) revealed an incipient 'latrine-marked border' halfway between these two setts, sometimes extending north and south into, and then fading out in, the farmland beyond the thin isthmus of woodland occupied by these setts. The attempt to partition two groups that coexisted not much more than 100 m apart, via the use of latrines (if that is indeed the function of the latrines—the topic of Chapter 9), may stretch one's credulity as to the degree of precision with which it is reasonable to expect badgers to partition resources and their society; today, however, 19 years after its excavation, Pasticks Outlier's latrines define the western edge of this conjoined group, with Pasticks main latrines defining the eastern flank, although the border between the two continues to be ambiguous—from which we deduce that partitioning on such a small scale is not viable. Imagine, in contrast, that circumstances had caused Pasticks to spawn an outlier 1 km away, as might typify a more conventional, lower-density badger population; in that situation, there would have been scope for a robust break between the two.

Continuing the theme of fission, in 2017 one of the Jew's Harp outliers budded off to become 'Park View' to the east. Two hundred metres further east lies the Marley # 2 sett (with its boundary, well marked by latrines). And 100 m south of Pasticks, the Singing Way sett (encompassing two clusters of outliers) completes the constellation of sett ranges between which the aRFID tracking (and indeed our routine trapping) shows that badgers make frequent visits, in a game of meline musical chairs that defines what we have come to recognize as the Pasticks super-group.

(For readers with a fascination for natural history and, particularly, with this process by which variously complete fissioning creates super-groups, the complete compendium may be found in Box 4.1.) In short, Pasticks is, nowadays, one of 64 principal named

---

[6] As mentioned earlier, bait marking is the technique in which marked food that has been fed to badgers in one territory is found in faeces, sometimes in the same territory but on other occasions in a different territory; thereby providing insight as to where the individual went subsequent to eating the bait. This information is highly relevant to our discussion, in Chapter 9, of relationships between neighbouring groups.

Box 4.1 Fissioning to create super-groups

Following the model of the Pasticks super-group, similar case histories on the other side of the wood offer further insight into the ontogeny of super-groups and their internal geometry. It was 1972 when David first visited Great Ash, which Hans Kruuk estimated, in 1974, to house 10 resident badgers within its 26 holes (at the time Great Ash was the principal outlier in a territory named Great Wood). Aided by bait marking, radio-tracking, and naturalistic stealth, Hans plotted the boundary ringing Great Ash with well-trodden paths and latrines, and estimated it at 50 ha (Figure 4.6).

Forty-six years later, this range remains identical, differing only insofar as it is partially cleaved from the triangle of its eastern tip, where an outlier was dug in 1997. Nowadays, that new outlier is 23 years old and is known as Firebreak sett, neatly partitioned by boundary latrines from Great Ash Hill, but nonetheless still regularly sharing members (see map, p. xx). Great Ash Hill (marked sett number 13 on Hans Kruuk's map) became the prominent main sett through the 1990s to 2015, with foxes utilizing Great Wood in the intervening years. The original Great Ash Hill sett that

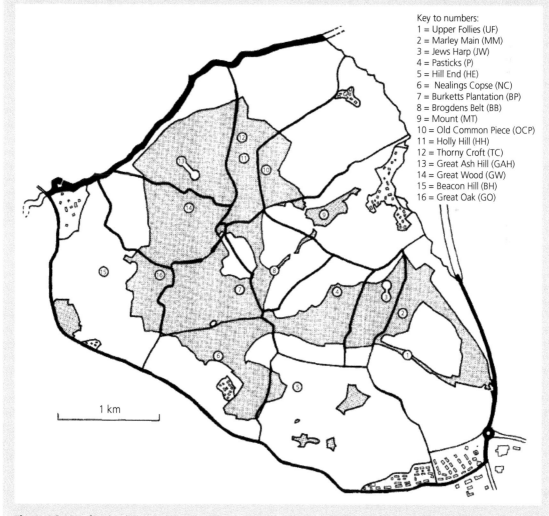

Key to numbers:
1 = Upper Follies (UF)
2 = Marley Main (MM)
3 = Jews Harp (JW)
4 = Pasticks (P)
5 = Hill End (HE)
6 = Nealings Copse (NC)
7 = Burketts Plantation (BP)
8 = Brogdens Belt (BB)
9 = Mount (MT)
10 = Old Common Piece (OCP)
11 = Holly Hill (HH)
12 = Thorny Croft (TC)
13 = Great Ash Hill (GAH)
14 = Great Wood (GW)
15 = Beacon Hill (BH)
16 = Great Oak (GO)

1 km

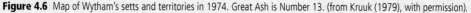

Figure 4.6 Map of Wytham's setts and territories in 1974. Great Ash is Number 13. (from Kruuk (1979), with permission).

Box 4.1 *Continued*

David visited in 1972 became less used through the 2000s, when the badgers moved 10 m down slope to excavate 13 new holes, and then two more a further 10 m downhill (the period of hole proliferation is the remarkable story of Chapter 10, during which time another two well-used outlying holes were dug uphill of the original sett, adjacent to the woodland track).

More broadly, in 1994, a small outlier noted by Kruuk (OL57) grew in prominence on the boundary between Great Ash Hill/Great Wood and the Chalet group range lying to its south-west. Its occupants expanded their range, which soon encompassed *c.* 20 ha formerly occupied by the Chalet group. We mistakenly deduced that Outlier 57 was socially budding off from the Chalet group, and named it Chalet Outlier. We were wrong, with the result that the name Chalet Outlier is a misnomer, which persists to this day. It became clear that the badgers at Chalet Outlier actually hailed from Great Ash Hill. Nonetheless, they were engaged in a cleavage that resulted in the establishment of a separate group of badgers, typically comprising 6 resident adults occupying, nowadays, 20 holes in an area in which, in 1974, there had been only 1.

Chalet Outlier (née OL57), now a free-standing group range, was not the only outlier of Great Ash Hill. Others to the east of the main sett gained prominence from 2005. The first 'Ingrid-B' (the B in deference to nearby Grid B of our small mammal survey grid [Buesching et al. 2010]) was some 30 m due east of Great Ash Hill sett, along a well-worn path, comprising at maximum three active holes (with towering yellow sandy spoil heaps). It remains an outlier. Another, *c.* 150 m east of Great Wood sett, is Clearing Outlier, with which our story opened as we sunk our periscope to spy those new-born cubs (Chapter 1). In 2005 Clearing Outlier started with the excavation of 2 holes, but by 2010 this had grown to 14 holes (nowadays more than 20);

however, while it dominates the south-east quadrant of that super-group range, it has not completely fissioned from its progenitor the fourth Great Ash Hill outlier, named Thames Outlier (due to its proximity to the river), which lies *c.* 300 m north, and comprises three holes deep in an ancient hazel coppice amongst hawthorn: Thames Outlier has never had residents and is seemingly used as a bothy en route to riverside farm pasture. The rise to prominence of these outliers, still within the curtilage of the Great Ash Hill range, has had consequences for the occupancy of the original main sett, which, between 2007–2015, was occupied by *c.* 20 resident adults. Subsequently, while these memberships are difficult to measure, because of the to-ing and fro-ing between setts typical of modern badgers, our best estimate in 2018 was that occupancy at Great Ash Hill has fallen to eight or nine residents, while Clearing Outlier and Firebreak both average five.

Together, but distinct from Chalet Outlier, the setts named Great Ash Hill, Clearing Outlier, and Firebreak comprise a sociological cluster. Their members were linked, but if this was a process of fission they were partly fissured, sharing the same curtilage but with contours of affiliation bonding the individuals based in the outliers (readers of a biological bent might think of cell division at mitotic anaphase with chromatids segregating either side of the metaphasic equator). Badgers in the Great Ash Hill cluster quite often visit (in the sense reported earlier) the badgers in the Great Wood, Chalet, and Chalet Outlier groups—this, in spite of the clear-cut boundary latrines between them. In marked contrast, these badgers seldom visit those residents in the Thorny Croft or Holly Hill group ranges to its east. In the same vein, visits within the Pasticks super-group are common, but they seldom venture into Brogden's Belt, Upper Follies, or Marley Main country. If they do, it is rarely to visit, and generally to disperse permanently.

---

setts (and numerous small outliers with ≤3 holes) that make Wytham a Dutch cheese of burrows in 2022. They distil to 23 distinct (latrine-bound) group ranges, encircled by paths punctuated by boundary latrines. These, as revealed by the flux of visits between them, assemble into *c.* ten super-groups, each comprising two or three group ranges.

The super-groups mimic almost exactly Kruuk's progenitor group ranges. Consider this, together with the case histories of fission related in Box 4.1 and, QED, the ontogeny of super-groups arises from the

fission of progenitor groups and not from the coalescence of previous neighbours. With this mechanism revealed we can see a progression from the traditional model of badger society, the Tidy Territorial Model centred around a main sett with occasional outliers, through partial distancing of social hubs within the curtilage of the group range—a partial cleaving that sometimes nudges towards fission with the establishment of ephemeral latrine borders (as between Pasticks and Pasticks Outlier during well over a decade), to culminate in complete fission—where space allows;

and in Wytham, in which location high density often precludes any full cleavage, forcing this complex and tantalizing sociological intermediary to persist. All the while, as we will elaborate in Chapter 11, badger numbers have been increasing and food resources have been burgeoning (Chapter 12) during the period in which climate ameliorated.

A torrent of questions flows from this, but most will be better tackled when we have set out some more stepping stones to the answers. In the meantime, one shrieks for immediate response: does the aggression that Hans Kruuk concluded was typical of his badger clans, the progenitor of social groups nowadays fissioned into the seemingly amiable components of super-groups, still prevail at super-group borders? A clue may lie in our analysis of the history of bite wounding amongst Wytham's badgers, a topic of Chapter 5. Meanwhile, having answered the first question of this section, revealing that super-groups arise by the cleaving of a progenitor group, while remaining nonetheless quasi-united, we turn now to the second question: what dictates the spatial geometry of borders of super-groups?

Super-groups: external borders and geological background

If an understanding of super-group borders in Wytham is the goal at which we are aiming, a helpful route can be found in the 1933 thinking of German Geographer Walter Christaller, who developed central place theory to understand the dispersion of human settlements. In a society in which it is important to maintain equal provision of goods and services (and making a number of assumptions variously not upheld by different populations of badgers), Christaller noticed that in those days the largest human settlement in a hierarchy was generally linked to three sub-settlements—he called this the K = 3 system (and it arose from the so-called marketing principle). The parallels between human settlements and the spatial arrangements of other mammalian societies may be approximate, but some of the analogies are powerful (e.g. the average distance of a wildebeest hunt determines the spacing of spotted hyaena territories, while the distance between frontier rail settlements in the United States was 12 miles, the distance of track a gang could maintain). It is therefore revealing, although not exact, to consider parallels between Christeller patterning of human settlements and badger settlements. Given that, as the membership of a 1970s group of badgers increased, each individual surely sought to maintain approximately equal access

to resources (the good and services embodied by food and safe shelter), so it may be noteworthy that the modal number of groups in a super-group is three (and in Chapters 9 and 10 it will become clear that the number of cliques in each group is similarly small). It will become crucial to understand what enables the badgers in each group (or super-group or clique) to maintain approximately equal access to resources (a fact that is an extremely unusual occurrence in mustelid societies, as illustrated by the overwhelming predominance of the intra-sexual solitary occupancy territorial system illustrated in Wytham and surroundings by the weasel and American mink (Macdonald et al. 2004b, 2015d); the final answer for the badgers of Wytham Woods will emerge in Chapter 10. An early step to that answer, closely aligned to central place theory, is the question: where do badgers select to dig their setts?

No student of badgers would doubt that setts are an important resource (see Rogers et al. 2003; Kaneko et al. 2010; Noonan et al. 2015a) and that excavating them is a big investment (Roper 1992; Stewart et al. 1999; Macdonald et al. 2004a; Rosalino et al. 2005b; Loureiro et al. 2007). Sliding precariously between the trees that vertiginously cloak the steep bank, one slithers to a ledge composed of a massive pile of sand excavated from Great Wood sett. This sett comprises in excess of 40 holes, extending into the vegetation over 30 m in both directions. Generations of holes interlaced with paths transecting spoil heaps terrace the entire topology of the bank. This sett has been active since David first visited Wytham in 1972. Indeed, it is probably *ye olde sett of the Great Oak* recorded in a deed in 1538, when King Henry VIII dissolved the Benedictine Monastery of Abingdon, and the Estates there held.[7] The deed defines one boundary with reference to this venerable badger sett, as part of the transfer of these lands to Lord Williams of Henley and his successors Lord Norreys' family, thence to the Beartie family.[8] By the time the Bearties became the Earls of Abingdon from

---

[7] To appreciate the sense of continuity: badgers were, therefore, already sniffing the air as they emerged at dusk from the sanctuary of this marvellous sett in Wytham when, with the 1534 Act of Supremacy, King Henry established himself as the Supreme Head of the Church of England, separating England from the Church of Rome, and enabling the king to sell off the monasteries to fund his military campaigns.

[8] Sir Henry Norreys, Usher of the Black Rod, was executed in 1536 for his involvement in the downfall of Anne Boleyn. His son and namesake was restored in blood, created Baron Norris in 1572 and subsequently acquired the Oxfordshire manor of Rycote, near Thame, and, through his marriage into the Williams family, the estate of Wytham. His grandson was created Earl of Berkshire in 1621, but the earldom became extinct on his suicide by crossbow the following year.

1682, the badgers were still in residence, and remain so, 13 English monarchs later (Grayson and Jones 1955).

Evidently, then, nobody should underestimate the persistent investment badgers make in constructing and then maintaining their setts—keep this in mind because it turns out to be pivotal to the denouement of our story in Chapter 19. The 1130 set entrances in Wytham represent, at a loose approximation, the better part of 1000 tonnes of spoil![9] In Wytham, setts are neither evenly nor randomly dispersed. What criteria do badgers use to decide where to dig them? In 1978 Hans Kruuk reported that in Wytham a narrow belt of calcareous grit underlain by a sandy stratum occupies only 14% of the estate, but contained all the main setts and 89% of all setts, 'i.e. woodlands are preferred within the calcareous grit and sandzone, and grit sand is selected within the woodlands' (Kruuk, 1978a). The attraction of the sand is that it is easy to dig and well drained; therein, setts are generally situated on convex and moderately inclined north-west-facing slopes, at 65–165 metres above sea level (masl) (Macdonald et al. 2004d).

Although badger tunnels typically do not extend deeper than 2 m (Kaneko et al. 2010; Noonan et al. 2015a), they can be very extensive, and involve a great deal of excavation. Coombes and Viles (2015) estimated that Wytham's 64 principal setts at which we monitor the occupants each represents consistently the displacement of around 304–601 ± 72 m³ of soil. Who does that digging? Some 60–90% of sett excavation was undertaken by just 20% of badgers, mostly males, who may thereby encourage receptive breeding females to move in (Stewart et al. 1999) (an aspect of cooperation explored in Chapter 8). There are consequences of sett location. For example, badgers residing in setts dug into clay soils are, on average, 7% lighter in weight (Macdonald et al. 2004). Why? Perhaps clay setts are

located in ranges that provide less food, or perhaps the shallower tunnels typical of clay are cooler, causing their inmates to burn more energy (Kaneko et al. 2010; Tsunoda et al. 2018). Alternatively, perhaps the occupants of these less complex, damp clay setts are somehow socially peripheral individuals.

What then of the geology that underlies the social organization of Wytham's badgers? As evidenced by the coral limestone caps of its twinned hills, 175 million years ago, in the Jurassic era, Wytham was a tropical coral reef off the coast of the Welsh mountains (Savill et al. 2010; see map on p. 57). Beneath this limestone lies the sandy strata—the palaeo-beach. Lower still, along the Thames River valley, are deeper, former sea, clays (Oxford clay/Denchworth series). Wytham's badgers exploit this diggable ('friable') band of sand (Cotteswold Sandstone/Frilford soil series) halfway up the slope to excavate their setts; furthermore, this system is faulted to create three repeats of the sandy strata as the northern hill slumps into the river valley, where low-lying clays are too damp for sett construction, and prone to winter flooding, but which harbour copious earthworms. So we see that accessible outcrops of sand in woodland determine first where badgers dig and, consequently, the dispersion of potential social hubs from which they set out to forage. We speculate that back in the early 1970s, the setts that Hans Kruuk mapped were, on the basis of first mover advantage, in the best locations (with the easiest digging and the most advantageous slope and aspect), and subsequent diggers have made the best of the options remaining to them. Where sandy outcrops are extensive, badgers can dig subsidiary setts, or outliers; these can become the nuclei for new groups if fission occurs. Occasionally badgers try in the limestone rag/Shelborne soil series above the sandy strata, but these rocky, crumbly setts rarely develop or persist. In contrast, where there are no diggable outcrops, typically where terrain dips below the topography of the sandy strata outcrop, sett digging is a poor prospect (Figure 4.7).

This geology plays some part in segregating clusters of setts into super-groups, although this factor is combined with the way food resources are dispersed (Chapter 10) and also the way that family trees have branched (Chapter 17). Consider the Pasticks conurbation: to the east the terrain drops down off the plateau. Here the slope is flat and does not transect the sandy strata which lies at 1–3 m below the surface, so no diggable soil surfaces for 250 m until the sandy outcrop that accommodates the four sett groups which together comprise the Marley Main super-group. Similarly, walking south from the Pasticks super-group takes us through a drop in terrain of c.50 m and through a slim

---

The barony passed through the female line to Bridget Norreys, who in 1648 married as her second husband Montagu Beartie, second Earl of Lindsey (d. 1666). His son James Bertie (1653–1699) was created Earl of Abingdon in 1682. His great-great-grandson, Montagu Beartie, succeeded to the earldom at the age of 15 in 1799, and married in 1807 a daughter of General Thomas Gage, the scapegoat of British humiliation at the hands of the American colonists. This takes us back about the same duration of time, in human generations, that is equivalent, in badger generations, to the start of our story (see p. 44).

[9] Even considering just the 64 principal setts at which we routinely catch the badgers of Wytham Woods, and estimating a mid-value of 450 m³ excavated per sett, and assuming that the density of sand is 1.6 tonnes m⁻³, those setts alone might represent 720 tonnes of burrow excavation.

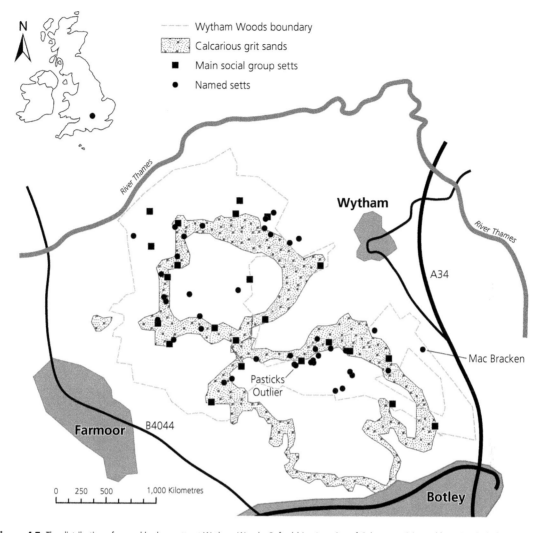

Figure 4.7 The distribution of named badger setts at Wytham Woods, Oxfordshire. Location of Calcareous Grit sand layer in which the majority of setts are dug is also shown, digitized from Kruuk (1978a). Reconstructed setts (MB and PO) are indicated (from Coombes and Viles 2015, with permission).

isthmus of woodland to arrive at the Upper Follies constellation. North-west of Pasticks there is a 500-m wide grazed pasture area, dipping some 60 m elevation into a valley—the latter bisected by a spring-fed stream where limestone hits clay with the sandstone elapsing away in this succession. The Pasticks super-group holds the south-eastern side of the valley, while the five sett groups of the Brogden's Belt constellation occupy the north-west, with a clear line of latrines peppering the valley floor: here it's easy to visualize an isopleth of food depletion (aligning with slope contours) as badgers fan out from their respective groups on either side of the valley—an idea we'll develop in Chapter 9. Notice, finally, and as a reminder that

geology is influential but is not the sole explanation for the phenomena we've been discussing, that there are no main setts of fortress dimensions in the south-west band of sand—this is heavily ploughed arable land with no protective woodland. There are, however, some small setts, Hill End, Clumps, Cottage, in which we caught badgers for our vaccination campaign (Chapter 16).

Modern communications enabled us to draft this chapter while Chris sat in his home in Nova Scotia, and David in his in Scotland. Chris conceived of a super-group Federation of badgers, partitioned into Provinces that splinter into Municipalities, while David envisioned the Clan Donald super-group, cleaved into neighbouring septs (the Macdonalds of Clanranald, the Isles, Glencoe, Glengarry), with further tensions between families and their clients partly fissioning within. Either way, Christaller's K = 3 might have been at work but, crucially, not dispersed in the homogeneous landscape that the geographer had envisioned with its hexagonal plots; rather, the badgers of Wytham Woods populate a world described by layers of heterogeneity. Some of that heterogeneity is, on the time scale of badger generations, entirely spatial—such as the geology that partly dictates the geometry of diggable sites and, equally important, that of the undiggable interstices that locate some of the hard edges to super-groups. But overlain on that geological heterogeneity are the dispersions of foods (variously mapped onto habitats), whose availabilities are both spatially and temporally heterogeneous, sometimes on a time scale of seasons, sometimes of nights, and often of hours. In Chapter 10 we will refer to this spatio-temporal patchiness in Wytham Woods as an RDH landscape—with reference to the resource dispersion hypothesis that we argue explains how, at the intersection of ecology, behaviour, and blood ties, the badgers partition the woods into super-groups, groups, and cliques, much like those Macdonald clansmen.

In much the same way that geographers analyse human settlements, the foregoing summary provokes a host of important questions. Why, for example, do some groups cleave completely, some only partly? Why don't today's super-groups, descended as each is from a Kruukian group of old, simply operate as a cohesive band? Indeed, why don't several super-groups coalesce into a Wytham herd of badgers? We arrive at the answers for the badgers of Wytham Woods in Chapter 10, and more broadly for badgers elsewhere, and their more distant kin, in Chapter 19. But to reach that point we need better to understand what it means to be a neighbour for a modern Wytham badger (Chapter 9), and before that to appreciate cooperation amongst them (Chapter 8), all of which necessitates appreciating how their groups are structured (Chapter 5), how they communicate (Chapter 6), and how they reproduce (Chapter 7). First, however—and anticipating the considerable importance of setts and the enormous task of constructing them—a brief detour to introduce the sett dispersion hypothesis, first articulated by Patrick Doncaster and Rosie Woodroffe, is required.

Sett dispersion hypothesis

Both Patrick and Rosie had been doctoral students of the WildCRU (in Patrick's case in the ancestral Oxford

Foxlot[10]) and both have gone on to distinguished professorial careers. They postulated that suitable sett sites might be a limiting factor shaping the distribution of main setts and thus the territorial partition of Wytham Woods, which we later termed the sett dispersion hypothesis (see also Blackwell and Macdonald 2000).

Testing the hypothesis in Wytham enabled us to refute it as a limitation in that site (Macdonald et al. 2004). Rather, it became clear that badgers could, and did, very readily dig new setts, and lots of them (some consequences of this are important to our discussion of badger demography in Chapter 11). Specifically, in 1974 Hans Kruuk counted 14 badger setts[11] and 56 outliers (<5 holes). But when we began our trapping regime in 1987, major setts (>5 entrances) numbered 23, with 67 outliers. Until 2019, when the fieldwork ended, we trapped at 63 major setts, and have mapped 216 outliers (defined minimally as a 'badger hole'). Concomitantly, by 1993 the number of social groups had inflated to 20 (Macdonald and Newman 2002)—in the mid-1990s it seemed that the increase in setts paralleled the increase in badger numbers (Chapter 11), and continued to do so until there were 23 by 2002 (Macdonald et al. 2004); both sets and groups settling in that location around this new equilibrium (Figure 4.8a, 1978; 4.8b, currently).

Although there was no shortage of diggable sites, even at Wytham excavating a modest modern retreat might bring different benefits to occupying a decades-old underground fortress (and where wolves, or even predatory people, abound, a deep labyrinth might have particular value, and offer different insulation from the chill of winter or swelter of summer). Setts are a major resource, best shared, and better yet inherited (Chapter 19); better a room in a shared palace than a private squat in a ghetto. Further, the intersection of geology and physical environmental conditions with food dispersion, foraging distances (Chapter 9), and the stresses of crowding (Chapters 10 and 11) create different outcomes in different circumstances, and so, for example, sett sites did turn out to be a limiting factor for

---

[10] Between the submission of his thesis in 1975, and creation of the WildCRU in 1986, David assembled a dedicated group of fieldworkers, all eccentric, most using radio-tracking, many researching the behaviour of one or other species of fox—hence they became known widely as the Oxford Foxlot. Two early Wytham badger students were amongst them: Heribert Hofer and Jack da Silva.

[11] One of which, the farmland sett Hill End, fell off our regime of monitoring due to being in a restricted access children's education centre, until we began bovine tuberculosis (bTB) vaccinating in 2018 (Chapter 16).

(a)

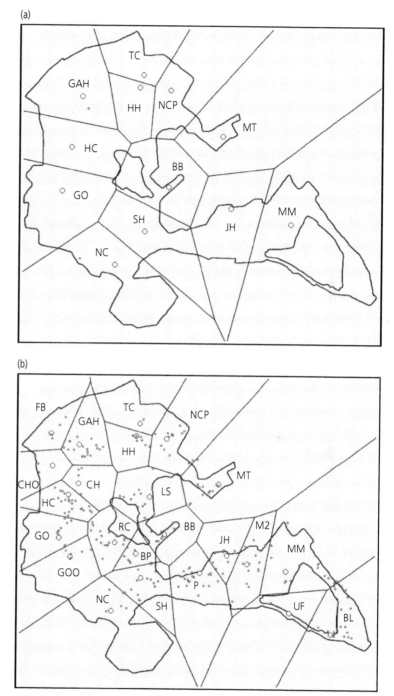

(b)

Figure 4.8 Setts (dots) and ranges (dividing lines) (modified from Macdonald et al. 2004). (a) 1978, and (b) currently (from Macdonald et al. 2015, with permission).

badgers in south-west Portugal's Mediterranean cork oak woodlands (Chapter 19).

Individuals in space

The spatial arrangements of the badgers of Wytham Woods are now resolved into a series of layers, that is,

the clans, septs, and cliques, but what is known of how individuals' movements overlap?

Delving into history the first answer takes the shape of Hans Kruuk's 1978a map (see Figure 4.9) of two males and two females who shared ranges of c.50 ha each at the Jew's Harp sett. Two males from the neighbouring sett (Platform) also overlapped closely with each other, but not at all with their Jew's Harp neighbours on one side or their neighbours, that is, at the Marley Main sett, on the other. These ranges, shared by multiple females and multiple males immediately—and with evolutionary significance that we will explore in Chapter 19—break the ancestral mustelid mould of semi-detached male and female relationships; each inhabiting single occupancy intra-sexual territories, those of males much larger than, and thus overlapping several of those of females (we use American mink, nearby on the River Thames, as the architype of this ancestral system, e.g. Macdonald et al. 2015d). The evolutionarily avant garde arrangements of these 1970s badgers of Wytham Woods prompted the question of how their relationships matched Charles Darwin's (1871) generalization that 'The law of battle for the

possession of the female appears to prevail throughout the whole great class of mammals'. These multi-male, multi-female arrangements fitted, indeed defined, the Tidy Territoriality Model (Kruuk 1978a). The broadly congruent territories of multiple males and multiple females became less tidy, and more interesting with respect to our findings regarding visits, when Kruuk described the relationship between badgers at the Marley Main sett and Botley Lodge sett—the females did not overlap with their female neighbours but, to quote Kruuk, 'However, two males appeared to belong to both Marley and Botley setts, spending the day in either of them and having nocturnal ranges completely incorporating the ranges of the females of both setts (combined male and female range 107 ha)'. In those days there were only 90 badgers, including cubs, in

Figure 4.9 Range of the Jew's Harp clan (l), the bachelor males (2), and the Marley females (pecked, 3), Botley females (dotted, 4), and Marley/Botley males (solid, 3/4). Determined by radio-tracking (from Kruuk 1978a, with permission).

the entire woods. Of course, as we turn our attention to the evolutionary ontogeny of badger groups, culminating in Chapter 19, the congruent ranges of two males overlapping the separate ranges of two females could be interpreted as the male infilling of the semi-detached arrangements of conventionally solitary mustelids (Powell 1979; Macdonald and Newman 2017). There will be more variations on this theme—intra- and inter-specifically—later, but for now we note that even in the 1970s there was evidence of visits between main setts, and the coalescence of the ranges around some centrally placed and distinct bait-marked setts into affiliated super-groups, at least unified by being overlapped by the movements of some males.

More than a decade later, when the badgers of Wytham Woods had doubled in number to an estimated 188 individuals (132 adults, 56 cubs; enhanced minimum number alive, see Chapter 11) Rosie Woodroffe and David did some further VHF radio-tracking of badgers on foot by night (Woodroffe and Macdonald 1995b). Comparing 1990 and 1991 they found that overlap between tracked female group members varied between months and years, with minima each September, and least overlap in 1991 at

12.4% and almost fourfold more the previous year (46.3%) (they noted that female group members that successfully reared cubs had the least overlapping home ranges).

Perhaps the best, and certainly the most recent, insight into the overlaps in movement between members of groups and their neighbours came from a second burst of aRFID tracking—between 20 August and 5 November 2014—that took us back to Great Ash Hill, Firebreak, and Chalet Outlier groups. The base stations are mapped on Figure 4.10. Remember that in these group ranges we had placed RFID base stations at both main and outlier setts (using these terms purely to distinguish main setts with ten or more entrance holes, from outlying setts with two or three). We instrumented all nine setts (three main, and six outlying; Figure 4.10) in the 49-ha study (sub-) area occupied by the three focal groups with aRFID-detecting base stations (for details, see Noonan et al. 2015b). Base stations broadcast signals at a frequency of one fix per minute, over a detection range of 15–20 m. Simultaneously, we used a different system, magnetic inductance, to track the badgers underground, where we recorded proximity amongst the tagged badgers from 20 August 2014 until the final collar batteries expired on 5 November

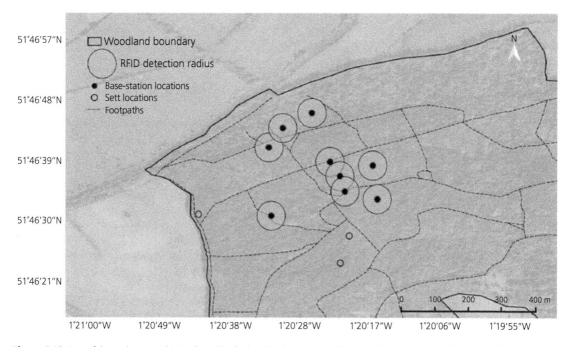

Figure 4.10 Map of the study area within Wytham Woods, depicting the locations of active radio frequency identification (aRFID) base stations at monitored sett sites, as well as unmonitored, nearby sett sites. Base stations are shown as dots, unfilled circles are sett locations, and the large circles show RFID detection radius.

2014, generating 176,788 relocations across all animals. So, mindful that the ranges we plotted reflect only movements between base station-equipped sites, and do not represent each badger's complete home range, to what extent do badgers of each sex partition their shared ranges?

The aRFID yielded 103,797 relocations for 15 adult badgers. These six males and nine females, tracked over four autumnal months occupied home ranges averaging just 7.06 ± 0.15 ha (but remember these were constructed around just the base stations, and did not represent the full home ranges). The task of analysing these data fell to Mike Noonan, who found that the movements of female badgers from neighbouring groups overlapped rather little, on average by 4.0% (±5.1%)—remarkably similar to the 1970s result—and that such overlap as *did* occur was driven primarily by F1333, a female in oestrus. While we will reserve most intra-specific comparisons for Chapter 19, in light of this rather meagre overlap between neighbours at Wytham we note that in populations of about one-tenth the density, in Ireland, O'Mahony (2015) found that inter-group contacts were negligible, at just 0.35% of all contacts recorded, while Böhm et al. (2009) reported a contact rate frequency with neighbours of just 0.02%. The base station home ranges of males were not statistically larger than those of females, averaging 8.67 ±

2.33 ha, although males tended to visit multiple setts in neighbouring groups, and males from one group averaged an overlap of 12.7 ± 11.1% with females of the neighbouring group. The visits males made, to-ing and fro-ing to neighbouring groups, resulted in each male, on average overlapping with males from neighbouring groups by 23.9 ± 16.1%. Interestingly, males with higher body condition indices (BCI) tended to visit more setts than males with lower BCIs, whereas females tended to visit a fixed number of setts irrespective of BCI.

And so (reminiscent of most mustelids), it is males that seem to wander (and philander), in line with the fundamental expectation that in addition to food and shelter, they are principally motivated to configure their homes ranges to ensure access to females. In contrast, assured of male interest, females are principally motivated to stake a claim on food resources to sustain pregnancy and lactation (although in the case of badgers, all those visits might suggest females may also have been keeping an eye on potential mates; see Chapter 17) (see Box 4.2).

The aRFID finding corroborated our trapping data: Macdonald et al. (2008) reported that males were captured at approximately twice as many social groups as were females (body weight provided no additional explanatory value in models including sex). Indeed,

---

Box 4.2 Private lives

M1434 had been born at Chalet Outlier in 2012, but a year later was caught visiting both Great Wood and Ditch Outlier (Great Wood falls within the curtilage of the Great Ash Hill range, whereas Ditch outlier is within that of Chalet Outlier). The following year, 2014, he was caught successively in Great Ash Hill, Great Wood, and then Chalet Annex before returning to Great Wood. It was August to November in 2014 that we followed him using aRFID, so the intensive schedule of visits can be seen within the snapshots revealed by our routine trapping. During 2014 M1434 sired his only known offspring, F1610, born at Chalet in 2015 (as a yearling, in 2016, F1610 moved to New Common Piece sett (NCP), but was never caught again). After the aRFID tracking, we continued to intercept M1434 during our routine trapping in 2015 but that year we encountered him only in the Great Ash Hill range (at both Great Ash Hill sett and Great Wood sett). He began his last year, 2016, at Great Ash Hill, but then moved to Clearing Outlier during which transition we know only that he acquired a torn ear.

A second focal female, F1187 was first caught as a 2 year old in 2008 at Ingrid-B, the outlier 40 m from the Great Ash Hill, and we caught her twice more that year at Great Ash Hill. In 2009 she was caught twice at Great Ash Hill then once in the neighbouring Firebreak sett. 2010 saw her at Thames Outlier, a more distant outpost contested between Firebreak and Great Ash Hill, lying almost exactly on the latrine-marked range boundary. Throughout 2011, 2012, and 2013 we caught F1187 seven times, always at Great Ash Hill, which is where we fitted her aRFID tag in August 2014. She was still at Great Ash Hill in 2015. Over these years, in 2012 she gave birth to M1442 (we caught him, as a yearling, at Firebreak, then again in 2013 at Firebreak, in 2014 at Chalet, and finally in 2015 at both Ditch Outlier and Firebreak). In 2015, F1187 gave birth at Great Ash Hill to two cubs, F1608 and F1627; however, both died as cubs after taking up residence at Firebreak.

Box 4.2 *Continued*

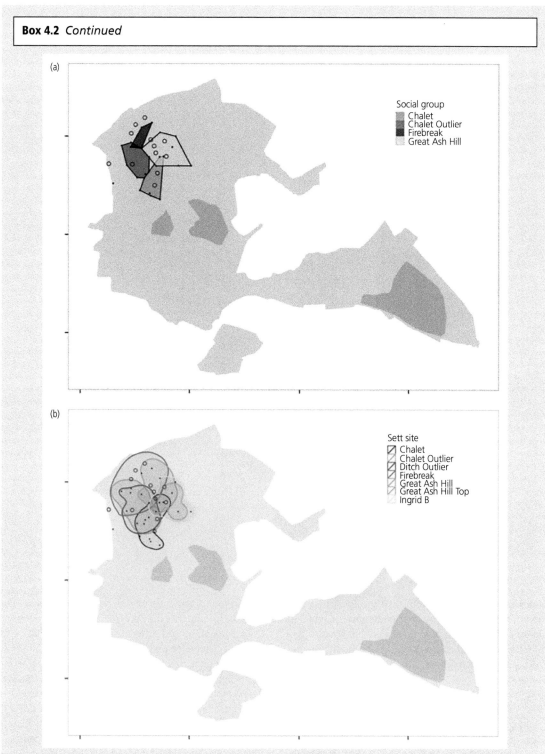

Figure 4.11 Map of the study site depicting: (a) minimum-area convex polygons (MCPs); and (b) kernel utilization distributions drawn around bait-marked latrine sites. Hollow circles represent sett sites within the surveyed area, and solid circles the location of latrines. Latrines containing unmarked faeces were excluded from analyses.

Box 4.2 *Continued*

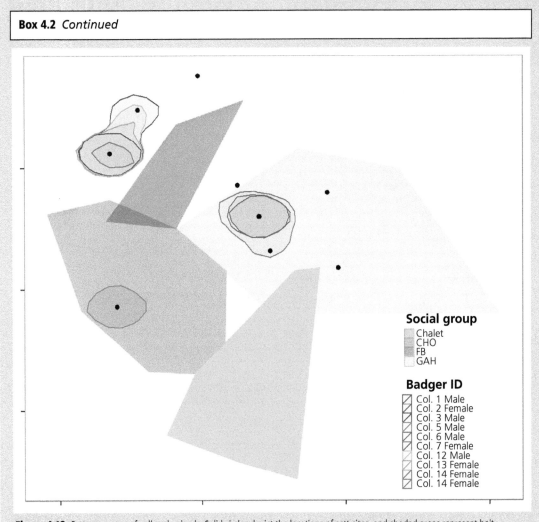

Social group

- Chalet
- CHO
- FB
- GAH

Badger ID

- Col. 1 Male
- Col. 2 Female
- Col. 3 Male
- Col. 5 Male
- Col. 6 Male
- Col. 7 Female
- Col. 12 Male
- Col. 13 Female
- Col. 14 Female
- Col. 14 Female

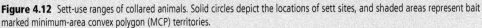

Figure 4.12 Sett-use ranges of collared animals. Solid circles depict the locations of sett sites, and shaded areas represent bait marked minimum-area convex polygon (MCP) territories.

males tended particularly to visit groups with large memberships and with a preponderance of females. On the other hand, trapping revealed that females predominantly visited groups with fewer resident females (and, one might suppose, less breeding competition) (Table 4.1).

At the same time as we tracked these aRFID badgers, Mike Noonan deployed magnetic inductance tracking at the setts across the same three-group sett. Mike logged a total of 103,797 (re-)visits to setts over the ensuing 552 'badger days' Until 5 November. In Chapter 9 we will document what they did

underground, and with whom, but here we answer the question of whether they entered each other's setts.

The magnetic inductance data revealed that both males and females predominantly occupied the main sett on which their spatial group's minimum convex polygon (MCP) was centred (Figure 4.12). Indeed, despite the numerous comings and goings of individuals of both sexes to the vicinity of setts in ranges adjoining their own, only one female entered a sett affiliated to a group MCP to which she was not assigned residency (this accounted for just a 3.9 ± 9.6% overlap in inter-group sett occupation across all individuals).

Table 4.1 Number of social groups with which individual badgers had been associated by each age class (an asterisk (*) denotes significantly different male–female comparison). So, for males in their first year 18 were caught at two different social groups, in their next year 5 were caught at three social groups. (from Macdonald et al. 2008, with permission).

	Male: no. social group					Female: no. social group						Test: males versus females
Age	1	2	3	4	5	1	2	3	4	5	6	
0	286	18	0	0	0	287	8	0	0	0	0	*
1	153	44	5	0	0	154	18	1	0	0	0	*
2	88	45	16	0	0	100	36	8	1	0	0	
3	50	41	12	2	0	65	31	15	1	0	0	
4	32	27	13	2	1	49	29	12	2	0	0	
5	21	18	13	1	3	36	24	14	2	0	0	
6	12	18	12	4	2	26	22	15	0	1	1	
7	9	14	10	2	0	18	17	14	1	0	1	
8	5	10	9	4	1	9	15	10	2	0	1	

There was no significant difference between sexes in the frequency with which they entered main—and outlying—setts. For females, the tendency to use outlying setts was greater for individuals with higher BCIs. For males, however, it was the opposite: those with the lowest BCIs tended to use outliers. An obvious sociological hypothesis would be that higher-quality (fat) females had the capacity to set up independent homes, distancing themselves from some domineering matriarch in the main sett (an alternative might be that the fattest females in autumn were those that had not bred that spring, and were in that sense lower quality). Regarding males, similar speculation might propose that the best-quality (fattest) males occupied the prime property of the main sett, whereas scrawny underlings made themselves scarce in the outliers (an alternative might be that the most testosterone-rich males were exhausted and worn away by autumn, when they associated themselves with the corpulent females they had inseminated). The remarkable outcomes of the intersection of fat, the weather, and sociology is a topic of Chapter 14.

Polygons and contacts

Remember, from earlier, that the aRFID ranges of males and females are different: the females cohabiting in the main sett used the same shared range but exhibited minimal overlap with females from neighbouring groups, whereas males overlapped the ranges of multiple females from numerous groups. How did these movements relate to the boundary latrines that appeared to ring the three study groups? To answer this, while we were tracking these ten badgers above and below ground, Mike Noonan also mobilized our bait-marking study. We found 157 bait-marked faecal deposits at 27 latrine sites, with 1 to >40 pits at each site (the full story on latrines is detailed in Chapter 9). The tradition with badger bait-marking studies, ever since the days of Peravid Skoog in 1970 has been to link latrines used by individuals originating from each spatial group by drawing an MCP around them. Following tradition, we did this too, and true to form revealed four non-overlapping, contiguous 'territories' (Figure 4.11a). Less traditionally, we also applied a kernel estimation approach separately to both the main and outlier setts, factoring in both the location and number of faecal deposits. These kernels are depicted as an amoeboid contour around the area marked by badgers that had eaten coloured bait at each sett, whether main or outlier. These avant garde bait-marked group ranges no longer matched the group MCPs, and didn't even look like exclusive territories either (Figure 11b).[12] Indeed, two setts classified

[12] Very few latrines occurred in the northernmost section of the survey area, so those boundaries were not resolved as robustly, although all data points were included in the analyses.

as 'outliers' were encompassed within areas marked by individuals that had eaten bait at four different main setts—raising the thought that they functioned as bothies used by travellers from different places.

In short, although MCPs drawn around bait-marked latrines often have the semblance of exclusive territories, striving to put aside inductive inference (*sensu* Popper 2012) and to treat the data without preconception, the badgers' movement patterns were neither constrained by, nor congruent with, the latrine boundaries. Half of the animals collared here made frequent visits to setts outside of their MCP latrine-based group range, and all collared badgers made at least occasional visits to neighbouring latrine-based group ranges. The patterns revealed by detailed study of these 12 aRFID-tagged individuals match closely the patterns we have described from our sett-based routine trapping: to develop the metaphor, we could now see that the full cine-film matched the flicker book of occasional stills (Macdonald et al. 2008).

There is much more to be said about latrines and the messages conveyed at such sites (Chapter 9); also the selective pressures that determine why badger ranges, and the groups that occupy them, are neither smaller nor larger (Chapter 10). Meanwhile, hold in mind *modus tollens*, the term in formal logic that asserts that where one factor is true, so must be its contra-positive: that is, if the latrine-based MCP looks like a border, but does not, in fact, restrict visits between setts either side of that border (which is what borders do), then the MCP does not, in reality, serve as a border, despite appearances. We must seek to discern alternative functions for these encircling latrines.

CHAPTER 5

The Sum of the Parts

Knowing One's Place in Badger Society

a cultivation of the self within a wider community.

Seneca (65 CE)

The spatial characteristics of Wytham's badgers distil to a fascinating layering of subgroups, groups, and super-groups, ultimately within the population. Like an onion's rings, these structures, from individual (Photo 5.1) to population, were the meat of Chapter 4, in which we described how a process of more or less complete fission involved the progenitor groups, now embodied as super-groups, cleaving into a larger number of groups, themselves partly fissioned into subgroups. Each annulus in this socio-spatial taxonomy emerges from the interactions of individuals that comprise the membership of badger society. It is in the flux of social interactions between these actors, sniffing, grooming, and squabbling, that we will find the answers to the question of how their society is structured, and thereby clues to its adaptive significance: how it works. Shortly, we will look for signs of hierarchy within badger groups. In other species, hierarchies may be revealed through glimpsed clues in ephemeral moments, a flick of the ears, an averted gaze, and in this chapter we will look for such signs (following which—owing to the fact that badgers are so essentially osmic—we devote all of Chapter 6 to their scents). Fleeting moments that betray sociological structure may leave enduring signs—a suppurating wound (see the text that follows), a mating secured (the topic of Chapter 7), and ultimately the conception of a youngster (Chapter 17). Do these interactions reveal a cooperative heart to badger society? Are badger groups, like those of most social carnivores, the product of selective pressures fostering cooperation? We answer these questions in Chapter 8. First, we ask: what is the composition of badger groups?

Photo 5.1 David and Chris getting to know an anaesthetized badger: at every encounter more than 30 aspects of each individual were recorded.

Group composition

The average size of a badger group depends on how you calculate it. If you take it to be, on average, the size of the total population in a year, divided by the total number of groups, the answer is 10. This is because we have a mode of 23 groups (although for bovine tuberculosis (bTB) vaccination, Chapter 16, we have added 3 more peripheral farmland setts whose existence became evident during 2018–2019), and approximately 230 badgers (see Chapter 11), neatly giving an average group size of 10. Approaching the question in a different way, if we sum the annual residencies at all groups between 1987 and 2019, this would yield a 33-year average of 5.46 adults (standard deviation (SD) ± 3.79) and 1.35 cubs (SD ± 1.76, right-skewed).

These averages vary across space and time, partly due to the aforementioned flux of individuals shuffling

The Badgers of Wytham Woods. David W. Macdonald and Chris Newman, Oxford University Press.
© David W. Macdonald and Chris Newman (2022). DOI: 10.1093/oso/9780192845368.003.0005

between groups and between setts within groups.[1] However, in terms of an individual badger's experience, its sense of community, crowding, or perhaps hostility with co-residents, then it is the occupancy of each sett, as trapped—each hub community—that is perhaps a better indicator of an individual's experience. This calculus yields 33-year averages of 5.06 (± 3.82) and 1.6 cubs (± 1.93) over all setts that had at least one resident per year.

These averages smooth out details, such as the fact that our trapping efficiency varies with conditions (Chapter 12; Noonan et al. 2015) and with whether sett residents were accustomed to capture from their cubhood (Chapter 12; Sun et al. 2015), and can differ substantially between groups.

Individual groups vary in size about these averages due to various selective processes, along with the vicissitude of random events, notably traffic accidents, which also contribute to periods of sex bias skew, although, overall this is remarkably close to parity at 1.02 males per female. For example the average skew was towards males in the Common Piece group over the period 2006–2016 (mean 8.73 males (range 7–13); mean 7 females (range 2–10); producing cubs every year, mean 4.8 (range 2–10)), whereas the skew over the same years in the Jew's Harp group was towards females (mean 4.82 males (range 2–8); mean 7.82 females (range 3–11); failing to produce cubs in 2006, 2007, 2011, and 2012, despite 3, 5, 7, and 9 females being respectively resident—mean cubs 2, max 6).

Similarly group size varies across space and time. Some groups, such as Marley # 2, are consistently small (2006–2016: mean 1.6 males (range 1–3); mean 2.2 females (range 1–4); with cub numbers ranging up to 3 (in 2007 and 2014), but reduced to an average of just 1.1 by a dearth of cubs over 2010–2013, again despite 3 females being resident in those years). Others have more variable histories, for instance Great Oak, where over the same decade an average group size of 3.5 conceals variation from 9 residents in 2007 to just 1 and 2 females in 2011 and 2012 (at which point no males lived there full time), through to 2 then 1 resident male in 2015 and 2016 (with no females assigned there full time). Further, from 2011 to 2014 at Great Oak we recorded not a single surviving cub during the annual May/June trappings (this dragging down the average annual number of cubs from a maximum of 3 to a mean

of just 1.1). A further variation is that cubs were born at Great Oak in 2015 and 2016 (2 and 1, respectively), but soon thereafter these mothers moved to spend the majority of their year elsewhere.

So how does this social mélange operate? How are social bonds and, indeed, hostilities mediated, reinforced, and maybe occasionally tested? How does a badger know where it stands in its society?

These questions matter because the answers may reveal which individuals (and thus which categories of individual) enjoy greatest fitness or, most directly, greatest reproductive success.

Some badgers were surely successful in the reproductive stakes, but was their skewed share of reproductive success a reflection of social supremacy or merely some anarchic throw of the dice? Consider M10 at Pasticks sett. M10, known affectionately as Lecher B, arrived at Pasticks in 1990, from across the super-group divide (Chapter 3) from the Brogden's Belt group; his age was unknown as he was already mature when first caught at the inception of our regime in 1987— but we estimate he was at least 8 years old in 1992. A brief introduction establishes him as a reproductive success. In 1990 F19 bore his cubs; in 1991 F217 and F140 bore his cubs, as they, together with F221, did in 1992. Then, in 1993 Lecher B sired cubs with F221 and F140 again. Lecher B continued to sire cubs with Pasticks resident females, year after year until 1996. In total, he sired 11 cubs. That makes him interesting, because it's the second largest tally in our entire 33-year dataset—the story of the record-holder, Satyr B, lies in Box 5.1 (and that of Casanova B is worth a look in Box 17.1).

But Lecher B was not an island; his reproductive successes were in the context and company of other adult males, and the mothers of his cubs were not the only females in the group. While Lecher B and his consorts were stacking up descendants, amongst their companions others were failing on these direct fitness stakes. What distinguished the winners and losers? With an average of 20 groups per year studied over 33 years, the synthesis in this book comes from over 660 group-year histories: more detail than any reader could bear. However, the averages and trends that we distil, and the population processes they exemplify (in Chapters 11–14), are emergent properties of individual lives, so for readers intrigued by how personalities assemble into populations in Box 5.1 we delve deeper, as an exemplar, into Lecher B's life at Pasticks sett in 1995. That year the cast of characters in the Pasticks group numbered ten adult females alongside five or six resident males.

---

[1] In this case, group size is estimated each year according to the residency rules set out in Chapter 3 (hence the sum of group memberships adds up to the total number of badgers estimated alive in that year).

Box 5.1 Lecher B's dynasty at Pasticks sett

Of the five or six males resident in Pasticks in 1995, three (of which one was a cub) and five of the ten females (of which one also was a cub) were assigned no offspring in 1995. Indeed none of these males (M239, M242, and M302) and females (F240, F323, F324, F327, and F469) that failed to reproduce in 1995 left any grand-descendants. They were members of the community, but do not feature further in this dynastic story. Meanwhile, Lecher B clocked up 11 offspring, and 10 grand-offspring; the females, by contrast—for example, F217 and F221—accumulated lifetime tallies of 5 offspring each, and 7 and 1 grand-offspring, respectively (a population-wide analysis of the characteristics of the most successful reproducers awaits in Chapter 14).

F217 and F221 were, as mentioned earlier, probably sisters. Observations of allogrooming and allomarking patterns identified F217 and F221 as the more revered matriarchs of this society, in that they received attention from other, generally generationally younger, females, but gave little in return. Indeed, their greatest tolerance was for one another. They had arrived together at the Pasticks group in 1991 from origins unknown (a rare occurrence in our detailed records), with F217 already pregnant with triplets (M232, M239, and M242). F217 again had cubs in 1992 (F297 and F305), fathered, as stated, by Pasticks resident, Lecher B, who himself had arrived at Pasticks in 1990, across the super-group divide, from the Brogden's Belt group. He was already mature when we first met him at the inception of our cage-trapping in 1987, likely making him at least 8 years old in 1992.

After 1992, F217 ceased to breed successfully but remember, from earlier, that F221 produced cub M302, fathered by Lecher B in 1992, and cubs MXXX, F323, and F327 in 1993, one of which, MXXX, was again fathered by Lecher B (the other was sired by M4, a similarly ancient founding father, although steadfastly resident in the neighbouring Brogden's Belt super-group throughout his life). In 1993, F221 produced two more cubs, M466 and F469, this time fathered by M243 (an inexperienced 2-year-old, born in 1991 at the adjacent Jew's Harp sett within the Pasticks super-group). F221 bore no cubs in 1994, but then no other female did either: no cubs were caught at the Pasticks sett that year. The patterns assembled from life stories of this sort emerge in Chapter 14.

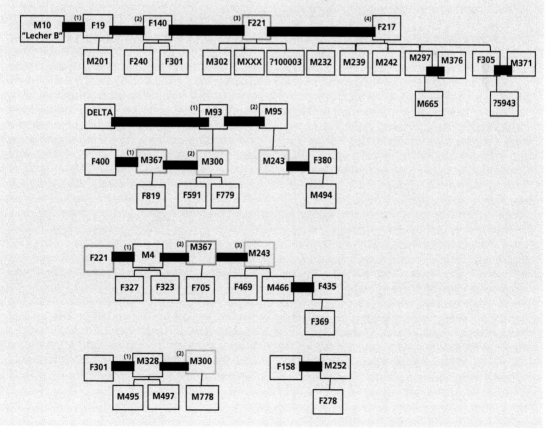

Box 5.1 *Continued*

This brings us back to 1995 and the ten adult females alongside five or six males then resident at Pasticks sett. By this point F217's daughters F297 and F305 were 3 year olds, continuing group philopatry, alongside F221's 2-year-old daughters, F323 and F327. In addition, female F240, a 4 year old (another immigrant daughter of Lecher B, from a mating with F140, born at Brogden's Belt), was part of the matriarchy, along with her full sister, F301, but born a year later in 1992. F301 evaded capture at Pasticks in 1995, but nonetheless was assigned two male cubs, M495 and M497, born at the adjacent Pasticks Outlier, which F301 frequented off and on throughout her life. Interestingly this pair was fathered by a 4-year-old male, M238, resident at the Nealing's Copse group/super-group, separated from the Pasticks group by the entirety of the Brogden's Belt super-group (see map, p. xiv (also 174)). F324 completed this ensemble—a 2 year old, born at Pasticks to unassigned parents, but potentially another daughter of F217 (Table 5.2).

The joker in this 1995 pack was female F380, who had first been caught at Pasticks in the previous summer, already as an adult, and thus likely another immigrant female. Although, from her toothwear, F380 was likely no more than 3 years old, for a female she was quite big (*c.* 720 mm long) and heavy, weighing 8.5 kg when she arrived in July 1994 (badgers being lightest in summer), peaking at 12 kg that October, and still clocking 11.3 kg by January 1995 when we caught her for ultrasound scanning—she was pregnant. She went on to bear a male cub, M494, that spring, fathered by M367, her age contemporary, by then residing at Pasticks, but hailing from the neighbouring Jew's Harp group (within the Pasticks super-group), from whence F380 likely also originated. When she arrived at Pasticks, F380 kept well away from established successful breeder and matriarch F221 (a rather small individual), who also bred again that year, giving birth to M466 and F469, fathered by M243, a 4-year-old, resident in Jew's Harp. By now, F221 was mother to five members of the group (her matriarchal line included still-resident daughters F323 and F327). While F380 kept her distance from F221, she was conspicuously hostile to F221's putative sister F217, and missed no opportunity to tussle and brawl with her (although never to cause wounds serious enough for us to detect). In January 1995, F217 had been pregnant (as detected by ultrasound), but was assigned no surviving cubs. Whether they were aborted, or died neonatally, we cannot say, but she was not seen suckling that year, despite her teats being suspiciously distended. By May 1995 F217's weight had slipped from 10.8 kg to 7.3 kg, decreasing through 6.3 kg in July, to an emaciated 5.5 kg

in August, at which point we noted she was 'riddled with lice'. Although F221 continued to spend time with her sister, F217 had become a pariah within the group; whether she was voluntarily reclusive or ostracized we could not tell. We never caught F217 again, but her daughters F297 and F305 both bore her grandchildren (M665 and F594, respectively) in 1997; indeed M665 was fathered by M367, from Jew's Harp, the former mate of bullying female F380. F380 in turn stayed on at Pasticks until 2002, but was finally caught as an elder at the Sunday's Hill sett (a component of the neighbouring Brogden's Belt super-group) in 2003; however, she was assigned no further cubs. In contrast F221 also lived on until 2002, remaining at the Pasticks group throughout, being assigned her final cub (F705) in 1999, once again fathered by the Lothario M367 (although his 4 offspring, including F819 born at Pasticks Outlier to F400 in 2001, could not compete in number with the 11 offspring assigned to Lecher B, or the 13 assigned to Satyr B (a.k.a. M4).

Intriguingly, and we will follow this up in Chapter 17 where we delve into genetic relationships, in 1995 only M243 (also born in 1991) enjoyed some resident mating success in Pasticks super-group, although he was actually resident at Jew's Harp from where he hailed originally, where his mother (Delta) and father (M95) both lived. He was a visitor to Pasticks sett in 1995, although clearly he'd met F221 in the previous year, given his having fathered her 1995 litter (see Table 5.1). Subsequently he also went on to produce two other cubs (M565 and M592) with F417, a Jew's Harp resident, in 1996 and 1997. Aside from M243, the other paternities went to extra-super-group M328. Part of the reason for this general outbreeding may well have been inbreeding avoidance (see Chapter 17), with young (born in 1991) resident males M232 and M242 being sons of F217, brothers to F297 and F305 (fathered by Lecher B), and cousins to almost every female present. M302 (born 1992) was a son of female F221 in her mating with Lecher B.

Another factor favouring breeding beyond the group might be that the resident males were also all quite young—4 years old or less. M300 (son of Delta and M93—mentioned in Chapter 3 with regard to the memorable lump he developed on his head in old age), was born at Jew's Harp in 1992 but emigrated to Pasticks as a yearling, where he fathered two litters with F400, of the Pasticks parish; one such litter resulting in female cub F591 in 1997 (she died before maturity) and female cub F779 in 2000 (who also died before maturity); and to a litter with F301, also of the Pasticks parish (another daughter of roving M10 from one of his repeat breedings with F140), producing male cub M778

Box 5.1 *Continued*

Table 5.1 The reproductive pairings of Lecher B (M10) and co-residents at the Pasticks social group in 1995, with offspring and great-offspring.

Year of birth	Sett of birth	Pasticks resident 1995	Offspring	Year offspring born	Other parent	Grand-offspring
<1990	?P	F217	M232	1991	M10	M379 M462 M464 F558 M572
			M239	1991	M10	*
			M242	1991	M10	*
			F297	1992	M10	M665
			F305	1992	M10	?5943
<1991	?P	F221	M302	1992	M10	*
			F327	1993	M4	*
			?100,003	1993	M10	*
			F323	1993	M4	*
			M466	1995	M243	F639
			F469	1996	M243	*
			F705	1999	M367	*
1991	P	F240				*
1992	P	F297	M665	1997	M367	*
1992	P	F305	?5943		M371	*
1993	P	F323				*
1993	P	F324				*
1993	P	F327				*
1995	P	F469				*
<1990	HC	M252	F278	1991	F158	F522 M572 M884 M896 M934
1991	P	M239				*
1991	P	M242				*
1992	P	M302				*
1995	P	M466	F369	1998	F435	F639
<1987	BB	M10	M201	1990	F19	*
			F240	1991	F140	*
			M232	1991	F217	M462 F464 F558 M572
			M239	1991	F217	*
			M242	1991	F217	*
			F301	1992	F140	M497 F526 M533 M778
			F297	1992	F217	M665
			F305	1992	F217	M543
			M302	1992	F221	*
			MXXX	1993	F221	*
			?100,003	1993	F221	*

? = probably.
M = male.
F = female.
Asterisk = zero.

Box 5.1 *Continued*

(who at least made it to age 4). Male M367 wraps up the tale of these brethren, but only born in 1993, at Jew's Harp (parentage unassignable; he was yet to reach his breeding prime (see Chapters 2 and 18)). Nevertheless, he was the chosen young sire of female bully F380's male cub, M494. Much bullied, he was clearly a subordinate member of the Pasticks clan, although in an Oedipal twist, he did ultimately go on to mate with the much older female F221 to produce female cub F705 in 1999, and mate with F297 (daughter of F217) to father male cub M665 (Pasticks) in 1997; he bred as well with F400 (a former partner also of M300, mentioned earlier) to produce female F819, in 2001 (Pasticks).

Table 5.2 The parenthoods of cubs born in Pasticks sett in 1995, showing the father's group at time of conception.

Cub	Mother	Father	Father's group	Father's super-group
M466	F221[a]	M243	Jew's Harp	Pasticks
F469	F221	M243	Jew's Harp	Pasticks
M494	F380[b]	M243	Jew's Harp	Pasticks
M495	F301[c]	M328	Nealing's Copse	Nealing's Copse
M497	F301	M328	Nealing's Copse	Nealing's Copse

[a] F221 born elsewhere perhaps 1988 (presumed beyond wood).
[b] F380 born Brogden's Belt super-group 1991.
[c] F301 born Pasticks 1992.

The births in Pasticks (Box 5.1), built around Lecher B's life, add to the questions that piled up in Chapter 4 relating to the sociological and, ultimately, genetic relationships between groups and super-groups. In this exemplar, Table 5.2 shows that in 1995 two of three Pasticks sett mothers had originally immigrated there as 3-year olds (at least one, and probably both, from a different super-group); both fathers were, at the time of their offspring's conception, from a different group and one was from a different super-group. Neither father was, at the time of conceiving his cubs (in spring 1994), resident in the groups to which he had been born. What, then, is the mating system of Wytham's badgers, and how does it map onto their spatial arrangements? The ultimate answer will come through genetic analyses (detailed in Chapter 17), but the scene is set through some sociological insights here.

Making sense of badger interactions

We began this chapter asking whether there is structure to badger groups. For example, are males dominant to females, does youth defer to age, or vice versa? Such structure, for example some form of hierarchy, might provide general insight into the functional significance of badger groups and, perhaps, even a start to answering the question of which characteristics launch individuals towards greater reproductive success. Some males, such as Lecher B, and some females, such as Dowager B (aka F221), left more than the average (indeed more than 2 SD more than the population average) number of descendants (this inequality, called reproductive skew, is common in animals, particularly

males for which a minority of high achievers may widely outperform the also-rans). Does rank play a role in distinguishing these reproductive stars from the reproductive flops?

Empirically, hierarchies exist in the societies of almost every social carnivore (Creel 2005) and primate (Majolo et al. 2012), and reproductive success is, depending on species, generally associated with high rank, most notably amongst males. Nonetheless, Majolo et al. 2012 conclude that studies of the fitness-related benefits to dominant individuals produce contradictory results, with some showing strong positive association between rank and fitness, others weak associations, and some even revealing negative associations. Theoretically, insofar as reproductive success in Wytham varies 13-fold amongst male badgers and 8-fold amongst females (Chapters 14 and 17), then in evolutionary terms this differential is something for which it is worth competing—of course this reproductive differential is not just about the opportunity to conceive (expect impacts of the latent ability to gestate, lactate, and wean offspring, as determined by the availability of resources (food; Chapter 12), by the genetic advantage of the offspring (Chapter 17), and by disease (Chapters 1, 2, and 15), all complicatedly shaken and stirred in the context of Pace of Life tactics (Chapter 14)). However, to win a race, you have to enter; high status may help to get you off the blocks, so we now ask how rank manifests amongst the badgers of Wytham Woods.

When it comes to a contest over prized resources, size often matters—at least big disparities do. Certainly throwing one's weight around, combining physical size advantages with aggression is associated with dominance in various group-living species, such as gorillas (Wright et al. 2019), but brawn is not always the chief explanatory variable (Jennings et al. 2011; Holekamp et al. 2015). Understanding what determines the position of individuals in (linear) dominance hierarchies is central to the evolution of social behaviour (Hausfater et al. 1982; Goessmann et al. 2000; Wittemyer and Getz 2007), where this can have a profound influence on access to food and other resources in ways that ultimately affect health and reproductive success (Clutton-Brock et al. 1984; Alberts et al. 2003; Archie et al. 2012; Sapolsky 2005).

Franz et al. (2015) lay out two alternative explanatory hypotheses: the 'prior attributes hypothesis' reflects our preceding discourse on the crude effects of size, fighting ability, bold personality, and background social attributes (e.g. family background) on dominance, positing that hierarchies will emerge purely based on dyad-level (pairwise contest) differences (Chase and Sietz 2011). In contrast, the 'social dynamics hypothesis' posits effects beyond the dyad level, based on more complex social processes; that is, winners tend to get better at winning and losers more resigned to losing in a succession of encounters (Chase et al. 1994; Rutte et al. 2006; Hsu et al. 2006). Moreover, success breeds confidence, while failure breeds defeatism, and so a win against one adversary also affects the chances of winning against others. Consequently hubris can outweigh prior attributes in the dynamics within a social group. Of course, these hypotheses are not mutually exclusive: an individual's prowess may lead to successes that enhance its confidence, which position it in society to grow physically fitter, and so forth.

Mindful of the soap opera at Pasticks (Box 5.1), what clues might explain winners and losers in contests between individual badgers? Obviously size. Then, remembering how the young females at Pasticks appeared to seek to appease F380 both by grooming her (a common diagnostic of primate rank, e.g. Kaburu and Newton-Fisher 2015) and by scent marking her (similarly a sign of rank in various carnivores, travelling up the hierarchy in female farm cats (Macdonald et al. 2010b) and down it in male meerkats (Macdonald and Doolan 1997)). Indeed, both allogrooming and allomarking flowed centripetally towards F217 and F221. Perhaps they were the 'dominant females' of the group, and F380 a challenger to their rank. Were there subtle signals to be read? For example, amongst primates looking away is a tell-tale sign of subordination (Harrod et al. 2020), as is, amongst canids, flicking of the ears or lashing of the tail (e.g. Macdonald 1980c).

Hierarchies

Pecking orders are no longer the prerogative of hens, although it was to describe their social dynamics that Thorleif Schjelderup-Ebbe's seminal 1921 paper first formalized the idea of social stratification, or hierarchy. The idea is so intuitively familiar to humans, not just from nobility to serfs but in the flow of deference in countless everyday interactions, that it is widely assumed in other animal societies, and indeed has been documented in many (e.g. Chase 1986; Creel and Macdonald 1995; Creel 2005). Throughout the natural world, competition occurs when resources are limited, and there is a benefit to having more, or better, access to them. This competition can be variously personal: at one extreme an impersonal scramble to a first-come-first-served outcome; and at the other, to a face-to-face contest and confrontation for privileged access (Milinski and Parker 1991). Such contests are expensive and, if the outcome is predictable, wasteful. Therefore, when individuals compete repeatedly, as they do within mammalian social groups, they tend to eschew escalated fights and settle the conflict according to precedent: the result is a dominance hierarchy.

That is why—in spite of the fact that our goal throughout this book is to interpret their behaviour *de novo* without squeezing it uncomfortably into some pre-existing model—considering that Wytham's badgers live in groups, our first expectation was that they would be arranged hierarchically. Further, experienced naturalist badger-watchers readily speak of dominant boars and sows; and, indeed, the expectation that carnivore, indeed mammalian, societies would have at least despots—even if they do not exhibit a tidily linear or stratified pecking order—is so widespread that mention of 'dominant' boars, or sows, permeates mainstream badger books (e.g. Kruuk 1989; Neal and Cheeseman 1996; Roper 2010). Now, in primate societies, where dominance has traditionally been most rigorously studied, it is commonplace for the costs and benefits of group living to be distributed unevenly amongst individuals, and conspicuous that certain individuals exert disproportionate, often bullying, influence over others (Rowell 1974; Hawley 1999). Are badger groups organized hierarchically?

We answered this question for the badgers of Wytham Woods by mining the thousands of hours of infrared video we had amassed. Selecting 3 years (1995, 2004, and 2005) for which we had good libraries of videos during the cub rearing and mating seasons (1 February to 31 May), at two social groups—that is, Sunday's Hill and Pasticks (1995) and Pasticks and Pasticks

Outlier (in 2004–2005)—we had 11,230 h of observation on 319 calendar nights, involving respectively 9, 14, 7, 7, 9, and 5 resident badgers. From these we extracted 659 instances of directed aggression (a bite, nip, or charge)—surely enough to reveal consistent

dominance relationships if they existed. For the most part, they didn't (Hewitt et al. 2009).

The detailed results of our unsuccessful search for clear-cut structure in badger social groups are given in Box 5.2. In the absence of generalizations about badger

---

Box 5.2 The quest for hierarchy—aggression

Following the methodology of de Vries et al. (2006), we focused on two revealing properties of dominance hierarchies: (i) steepness, from how great a superior height does one individual look down on another, based on a cardinal rank measure that quantifies the rank distances between them based upon their relative likelihood of winning dominance encounters? (ii) Linearity, how reliably does A dominate B, B dominate C, C dominate D, and on to the hapless individual that, like Groucho Marx, cowers beneath the bottom rung of the social ladder? This is an ordinal measure describing the consistency of the direction of dyadic interactions (Vervaecke et al. 2007). Large rank distances produce steep and despotic hierarchies, whereas small, shallow rank differences characterize more egalitarian societies. However, across our six social-group years of study that we published in 2009, no single attribute of dominance emerged. The contest asymmetries, that is, who won or lost, were largely context, or contest, specific. They did not plot out to give a linear dominance hierarchy based, for example, on size. Something else was determining particular outcomes—perhaps previous knowledge of the opponent's demeanour and strengths, or perhaps hunger for the reward.

Confusingly, against that generality that no consistently straightforward hierarchical structure was detectable either to an observer or statistically, there were glimpses of structure. For example, in this early analysis, in some social-group years, badgers were more likely to receive unreciprocated allogrooming (this might seem like an apology) from badgers to whom they directed more aggression (this was, however, statistically significant only for Pasticks badgers in 1995 and Pasticks Outlier in 2004). Our curiosity piqued by this result, we would later come back to it with a suite of more modern analytical techniques (see Box 5.3). The most compelling result from our paper (Hewitt et al. 2009) was that the flow of aggression was not conspicuously structured (Figure 5.1). What of the flow of more amicable interactions? Was there an asymmetry in the flow of grooming, creating a 'giver' and a 'receiver'? Such asymmetry could suggest either that one badger ranks higher than the other, or that the giver is soliciting favour, because they want something—perhaps in the context of a male ingratiating himself with a potential

mating partner (Stopka and Macdonald 1999). In an earlier study of licking in farm cats the same question led to the answer no—although some pairs of cats groomed more than did other pairs, the flow of licking within each pair was approximately symmetrical (Macdonald et al. 1987).

However, in that same study of farm cats the result was different for the flow of rubbing (allomarking), which was strongly asymmetrical within pairs of cats, and moved centripetally; for example, from adult daughters to their mother. For cats, then, the flow of marking therefore appeared to be diagnostic of rank; what of badgers? Stewart and Macdonald (2003) extracted from the same videos 838 instances of unreciprocated allogrooming and 1476 of sequential allomarking (topics fully developed in Chapters 8 and 6, respectively). Our tireless intern Stacey Hewitt sought patterns in these instances (Hewitt et al. 2009). We also asked whether genetic relatedness structured badger groups, perhaps through affinities of hostilities between closer relatives. This was not consistently the case. Genetic relatedness was not statistically related to directed aggression across social-group years, and the flux or asymmetries in allomarking or grooming correlated with genetic relatedness in these three groups over those 2 years wasn't either.

One, not at all intuitive, statistical result was that dominance hierarchies based on aggression were significantly steeper than random in five out of six social-group years (all except Pasticks in 1995); there was also a suggestion of asymmetric (one individual outranks another) relationships in three social-group years where the statistics indicated a significant degree of linearity (Sunday's Hill 1995, Pasticks 2004, Pasticks 2005). In these cases there was a hint that sex was sometimes related to dominance: in two of those three social-group years with significant linearity (Sunday's Hill 1995 and Pasticks 2004) there was an effect of sex on dominance rank according to aggression: females, especially older, breeding females (but never yearlings) outranking males, although rank was not related to age, or the likelihood of displaying a bite wound (see the text that follows), or of an individual's breeding status. Another hint that female dominance was sometimes involved came from the observation that those groups with the highest female

Box 5.2 *Continued*

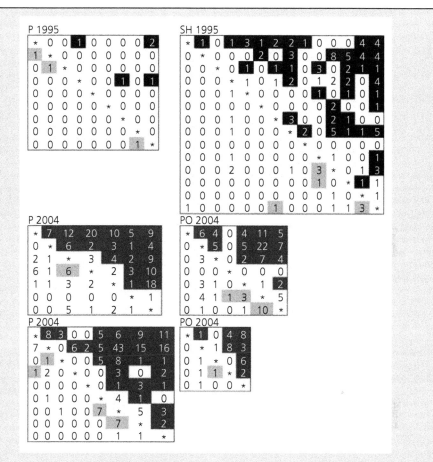

Figure 5.1 Matrices of the number of directed aggression events observed between dyads in each of the six social-group years. Rows show individuals that received directed aggression and columns individuals that initiated the directed aggression. Cells in each matrix are colour coded to show individuals in each dyad that directed the most aggression in that dyad and had the highest (black cells) or lowest (grey cells) DS rank. White cells indicate dyads in which no directed aggression or an equal number of directed aggression events were observed, or they indicate an individual in a dyad that directed the least aggression (from Hewitt et al. 2009, with permission).

skew in sex ratio (Pasticks 2005) and the lowest (Sunday's Hill 1995) had significantly linear hierarchies, whereas the 2 years in which there was no trace of linear hierarchies (Pasticks 1995 and Pasticks Outlier 2004) had intermediate sex ratios (Hewitt et al. 2009) (Table 5.3).

If all this sounds a bit like trying to grasp a mirage, that's exactly what it was like: shadows of hierarchy here and there, but no strident pattern discernible across groups or years. That mirage-like quality—something might be happening, but we cannot tell what it is—would be unexpected to anybody who has watched, even casually, an interacting

throng of any other species of social carnivore. Badgers, at least those in Wytham Woods, are conspicuously different to, for example, Ethiopian wolves, banded mongooses, or giant otters (comparators chosen because all form, like badgers, multi-male and multi-female societies—and in the case of giant otters, are taxonomically second cousins to badgers (Chapter 19)). Amongst badgers no dominant pair, sprinkling urine, and standing tall, pulls conspicuous rank over cowering (reproductively suppressed) subordinates, whose rank is instantly obvious to a human onlooker. We mention an onlooker deliberately, to highlight that the

---

Box 5.2 *Continued*

possibility that an on-sniffer might perceive things differently. Regarding allomarking, albeit in just 1 year, Pasticks 2005, badgers more frequently allomarked individuals to whom they directed more aggression—a counter-intuitive result from which we deduce no more than the fact that it merits further exploration—which is the topic of Chapter 6.

There, we will investigate the possibility that allomarks are smeared up (as in farm cats) or down (as in meerkats) their social ladders. Before that, inspired by the hints that the flow of allogrooming might mirror that of aggression, and equipped with a new generation of analytical tools, we dug deeper: did that change everything? See Box 5.3.

Table 5.3 Results of the dominance hierarchy linearity and steepness analyses based on the flow of aggression (from Hewitt et al. 2009, with permission).

	1995		2004		2005	
	P	SH	P	PO	P	PO[a]
Total observation time (h)	1383	1242	2444	798	3872	1491
Directed aggression events observed	8	122	164	118	198	37
Number of individuals	9	14	7	7	9	5
Linearity index (h' or h)[b]	0.27	0.39	0.93	0.61	0.68	—
Directional consistency index	1	0.85	0.70	0.74	0.85	0.84
Number of circular triads observed	29.25	79	1	6.75	11.5	1.75
Number of circular triads expected	21	91	8.75	8.75	21	2.5
P_{right}[c]	0.61	**0.040**	**0.007**	0.20	**0.020**	—
Steepness of dominance hierarchy	0.05	0.21	0.63	0.43	0.46	0.51
P	0.41	**0.0001**	**0.0002**	**0.005**	**0.0001**	**0.035**

Significant results ($P < 0.05$) determined by permutation tests are shown in bold.
[a] Social-group year contained too few individuals for testing the linearity of the hierarchy.
[b] The improved linearity index h' was used where there were unknown relationships between dyads (SH 1995, P 1995, P 2005, and PO 2004), otherwise Landau's h index was used (P 2004).
[c] P_{right} tests whether a hierarchy is significantly linear. It provides the probability that a randomly generated dominance matrix has a degree of linearity $\geq$ the linearity in the observed dominance matrix.

---

social structures, there was no shortage of special cases. Indeed, for every potential predictor of personal affinity or hostility there were undeniably cases in which each predictor seemed to offer a plausible explanation of a relationship. Blood did appear to be thicker than water in the sisterly alliance between F221 and F217. Heavyweight muscle power seemed to afford F380 considerable sway over her flimsier companions, all variously related to F221 and F217. Amongst the seemingly deferential underlings, asymmetries in the upward flow, as evidenced by grooming and also allomarking, surely signalled their subordination to F380 (only two of the adults resident in Pasticks when F380 arrived seemed to form amiable bonds with her: F240 (a daughter from F140's lineage) and M300 (himself an earlier immigrant from the neighbouring Jew's Harp group)). However, as described in Box 5.2, our intensive quest for structure across the three groups over 3

years, revealed that there were exceptions, and plenty of them, to every rule. Sometimes clear-cut relationships between two badgers were vivid for a season or a year, for example, conspicuous dislike and even dominance, but these were invariably ephemeral. Similarly, there were some groups, in some years, in which signs of dominance were statistically detectable and in some cases some or all of the group members could be sorted into a linear hierarchy.

In those cases, the generality was that females, especially older, breeding females (but never yearlings), outranked males, although rank was not statistically related either to age or to an individual's breeding status. Finally, turning from individual interactions to the arena in which such interactions occur—the group—we asked whether the steepness or linearity of hierarchies varied with group size. Again, there was no obvious pattern: steepness was greatest in Pasticks

Box 5.3 The quest for hierarchy—allogrooming

Badgers grooming each other are concerned with more than their coiffure. While the motivational roots of grooming might lie in hygiene (Freeland 1976; more of this in Chapter 8) even a glancing familiarity with pet behaviour, from budgies to borzois, makes plain that it may also have a social function (Sparks 1967), perhaps nurturing social bonds (e.g. Miyazawa et al. 2020), appeasing dominant individuals (Baker and Aureli 2000; Schweinfurth et al. 2017), or, by way of an apology, conciliation after aggressive interactions (de Waal 1984). From vervet monkeys, *Chlorocebus pygerythrus* (Seyfarth and Cheney 1984), to domestic cattle, *Bos taurus* (Šárová et al. 2016) allogrooming is strongly correlated with dominance rank. One interpretative framework is the biological trade model, which posits that allogrooming might be traded by subordinates for rank-related commodities from individuals loftier on the social ladder (Noë and Hammerstein 1994, 1995). Within that framework we would expect that whether or not an individual returned the favour when groomed by another would depend on their relative dominance rank, which might in turn be revealed by the flow of aggression between them (Barrett and Henzi 2006). This trade model describes well the transactions amongst diverse social mammals from Norway rats, *Rattus norvegicus* (Schweinfurth et al. 2017), to male chimpanzees, *Pan troglodytes* (Kaburu and Newton-Fisher 2015), with meerkats providing a carnivore example (Kutsukake and Clutton-Brock 2010). If obsequious badgers trade grooming for greater tolerance, we might expect a positive correlation between an individual's eagerness to invest in grooming another, and the likelihood of that individual directing aggression towards the supplicant as payment for increased tolerance (Henazi and Barrett 1999). In that sense, the grooming would not strictly be unreciprocated, but repaid in a different currency (e.g. David and his Czech postdoc, Pavel Stopka, now Professor of Evolutionary Biology at Charles University, Prague, found that male wood mice that groomed females were rewarded with the opportunity to sniff her genitals; Stopka and Macdonald 1999). Another possibility amongst the badgers was that allogrooming was distributed on the basis that blood is thicker than water—Japanese macaques, *Macaca fuscata*, for example, preferentially groom their close kin (Mehlman and Chapais 1988).

Convinced that allogrooming was a promising candidate to indicate social status, we returned, 25 years after the first observations had been made, to the actor-received matrices originally analysed by Hewitt et al. (2009). This time we were led by Catherine Nadin, academic granddaughter of the badger project, at the time studying with Hannah Dugdale at Leeds University and nowadays researching bioacoustics of humpback whale super-groups (see Findley et al. 2017

Table 5.4 Social-group year compositions by sex, excluding cubs, which were not included in any analyses.

	1995		2004		2005	
	SH	P	P	PO	P	PO
Females	4	7	4	4	7	4[a]
Males	10	4	3	3	2	2
Total number of individuals	14	11	7	7	9	6
Sex ratio (proportion of females)	0.29	0.64	0.57	0.57	0.78	0.67

SH = Sunday's Hill sett.
P = Pasticks sett.
PO = Pasticks outlier sett.
[a] One badger was excluded from all analyses as it led to structural zeros (i.e. they were not observed with at least one other individual and so there was at least one dyad where allogrooming could not be measured)

and Pirotta et al. 2021 for comparison with badgers in Chapters 4 and 10) off the coast of South Africa. Cat led us back to the two attributes of hierarchy introduced in Box 5.2: linearity (directional consistency of who does what to whom) and steepness (the disparity between them). In Stacey Hewitt's day our focus had been on the flow of aggression, but now we turned to who groomed whom (Table 5.4).

Despite the alluring hypotheses and Cat's mastery of the newest, sophisticated techniques[2]—we optimized the linear ordering of each social-group year matrix using the hieroglyphically named 'improved I&SI method' (Schmid and de Vries 2013)—this 2021 analysis, like its 2009 predecessor, once again discerned only weak hierarchies,[3] with marginal linearity, shallow steepness, and variation from group to group and year to year (Nadin et al. 2021). The new techniques affirmed a positive correlation between the flow of grooming and that of aggression[4]—suggesting a background murmur of traded appeasement, but no correlation with relatedness (Table 5.5).

[2] We used DomiCalc version 14 May 2013 (Schmid and de Vries 2013), which measures the linearity degree of a hierarchy via an unbiased estimate of Landau's (1951) linearity index *h*.
[3] We weighted actor–receiver matrices of allogrooming by the total number of bouts in which both members of each dyad were seen on camera together. A bout was a period of badger activity on camera, terminating when the last badger left the screen and there was no further activity for 1 min (Hewitt et al. 2009). This accounted for differences in the amount of time that each dyadic pair was seen on camera, and thus had the opportunity to be observed interacting. The whole matrix was then multiplied by 100 and rounded to whole numbers.
[4] Directed aggression was defined as an actor initiating aggression (bite, nip, or charge) at a receiver who did not reciprocate the aggression. Directed aggression ended when the dyad moved at least two body lengths apart for at least 20 s (Hewitt et al. 2009).

Box 5.3 *Continued*

Table 5.5 Kendall's row-wise correlation results, testing whether individuals show: (i) higher levels of unreciprocated grooming towards individuals that direct more aggression at them; (ii) lower levels of unreciprocated grooming towards more related individuals; (iii) higher levels of reciprocated grooming towards more related individuals; and (iv) higher levels of unreciprocated grooming towards individuals that reciprocate allogrooming more.

	Year	1995		2004		2005	
	Social group	SH	P	P	PO	P	PO
Unreciprocated allogrooming and directed aggression	Kr	65[b]	28[b]	−22	27	0	12
	n	14	11	7	7	9	5
	P_{right}	0.109	0.051	0.926	**0.045**	0.514	0.075
Unreciprocated allogrooming and relatedness	Kr	17	11	21	8	−20	9
	n	14	10[a]	7	7	9	5
	P_{left}	0.613	0.686	0.906	0.745	0.215	0.867
Reciprocated allogrooming and relatedness	Kr	25	−1	13	0	17	3
	n	14	10[a]	7	7	9	5
	P_{right}	0.367	0.517	0.202	0.521	0.246	0.400
Unreciprocated allogrooming and reciprocated allogrooming	Kr	124	54	56	38	106	20
	n	14	11	7	7	9	5
	P_{right}	**0.020**	**0.008**	**<0.001**	**0.003**	**<0.001**	**0.008**

n, number of badgers; *P*-values in bold are significant at $P < 0.05$.
[a] One badger was not genotyped.
[b] Results differ from Hewitt et al. (2009), as they excluded two badgers from P 1995 due to structural zeros in their hierarchy analysis, which in this analysis we have included in our row-wise correlation, and re-analysis of Sunday's Hill 1995 led to the inclusion of three more unreciprocated allogrooming events.
SH = Sunday's Hill sett.
P = Pasticks sett.
PO = Pasticks outlier sett.

Badgers that stay together groom together: reciprocal and unreciprocal allogrooming were positively correlated in all six social-group years (Table 5.6). But any linearity in their relationships was only weakly detectable, and evidence of steepness, although statistically detectable in four out of six social-group years, was only shallow. We did find a statistically significant unreciprocated allogrooming linear hierarchy in one social-group year (again, it was Pasticks in 2005)(none had a significant non-linear hierarchy either).

Life is a succession of special cases, and so we analysed the grooming interactions of the Pasticks badgers in 2005 (Figure 5.2) from every possible angle (aided by the I&SI method) to reveal that whatever was determining each individual's rank in that exceptional social-group year, it was not its sex. Nor was sex-related steepness (as revealed by a technique called—no relation—David's scores). Similarly, there was a significant correlation (Kr value) for unreciprocated allogrooming and directed aggression matrices in only one social-group year (Pasticks Outlier in 2004) (although the special case, Pasticks 2005, missed significance by only

a whisker), although something was happening as further statistical delving dredged out an overall significant link between delivering unreciprocated grooming and being the butt of aggression.

Almost inevitably, as the steepness of the hierarchy increases, the outcome of dyadic encounters become more predictable, and too linearity become more likely (Sánchez-Tójar et al. 2018). The dramatis personae of Pasticks badgers in 2005 collectively illustrate an exception to the overall rule in the six social-group years we analysed: amongst the badgers of Wytham Woods it is fair to state that there are clearly defined relationships between some badgers, but a step too far to say these are framed by more than a weak social structure. This will not be the last time that individuality appears to shrug off conformity in the lives of the badgers of Wytham Woods, and in Chapter 14 we will start to see at least part of that individuality as phenotypic strategizing.

We conclude that the glimpses of hierarchical structure amongst the badgers are context dependent and could result from individuals having different motivations to offer or solicit allogrooming. What might affect those motivations? Thoughts of grooming lead quickly to fleas. It seems likely

Box 5.3 *Continued*

that individuals suffering higher cutaneous irritation will initiate allogrooming (in the hope that it will be reciprocated to the disadvantage of their fleas) more than individuals with a lower 'itch burden'—a topic that emerges as unexpectedly relevant to cooperation amongst badgers when we scratch beneath the surface of reciprocal grooming in Chapter 8 (Willadsen 1980). This raises the contrarian possibility that unsolicited allogrooming may be annoying and be rebuked.

Also, badgers blighted with greater cutaneous irritation may more often attempt unsolicited allogrooming, provoking an irritated response. Alternatively, directed aggression may elicit conciliatory allogrooming.

Overall, badger societies appear protosocial, with some indicators of social structure within allogrooming interactions, but overall it is likely context dependent.

Table 5.6 Linearity and steepness of unreciprocated allogrooming hierarchies for all six social-group years, with significant results in bold. The *h* value represents the unbiased estimate of Landau's (1951) linearity index (de Vries 1995). Improved linearity test (de Vries 1995) right-tailed *P*-values < 0.05 indicate a significantly linear hierarchy, whereas, left-tailed *P* < 0.05 indicate a significantly non-linear hierarchy.

	1995		2004		2005	
	SH	P	P	PO	P	PO
Number of individuals	14	11	7	7	9	5
Number of pairs	91	55	21	21	36	10
Unreciprocated allogrooming events (weighted by number of bouts)	360	189	259	230	852	167
Linearity index (*ho*)	0.248	0.275	0.571	0.784	0.700	0.950
P_{right}	0.303	0.428	0.194	0.067	**0.006**	0.121
P_{left}	0.697	0.572	0.806	0.961	0.994	0.879
Steepness of hierarchy	0.209	0.143	0.302	0.379	0.488	0.529
P_{right}	**0.001**	**<0.001**	0.284	**0.034**	**0.002**	0.105

H = Sunday's Hill sett.
P = Pasticks sett.
PO = Pasticks outlier sett.

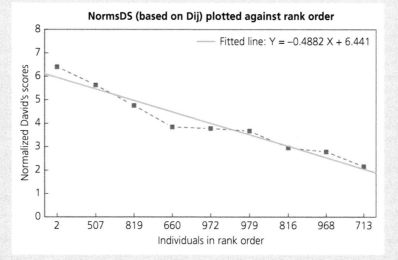

Figure 5.2 Normalized David's scores plotted against rank order (calculated from unreciprocated allogrooming hierarchies) for the individuals of Pasticks 2005 that displayed a significantly linear, steep unreciprocated allogrooming hierarchy (from Nadin et al., (in review)).

Box 5.3 *Continued*

To further emphasize the mirage-like quality of structure in badger groups, amalgamating Stacey Hewitt et al's (2009) analysis of the steepness of badger dominance hierarchies, based on directed aggression and Cat Nadin's *identification of* two social-group years in which directed aggression was correlated with unreciprocated allogrooming were instances when there was either no significantly steep direct aggression hierarchy (Pasticks in 1995), or had the fourth

shallowest hierarchy (Pasticks Outlier 2004) out of the five social-group years with a significantly steep hierarchy. Where do all these indecisive results leave us? Nadin et al. (subm) conclude, 'Overall, we provide a new technique for analysing allogrooming structures and demonstrate rudimentary indicators of a social structure in allogrooming badgers, reaffirming the protosocial nature of high-density group-living badgers'.

---

in 2004 with group size 7, shallowest at Pasticks in 1995 with group size 9, and intermediate at Sunday's Hill 1995 with group size 14.

In short, it would be wrong to say that dominance and subordination—according to the foregoing yardsticks—do not sometimes characterize the relationships between some badgers, but then again wrong also to say that this does not sometimes shake out into a linear hierarchy. On the other hand, in striking contrast to most social carnivores, none of the measures thus far described arranges badger groups into stable dominance hierarchies, or even into predictable substructures based on sex or age. This conclusion raises two possibilities. First, while significant social structuring of badger groups may not be apparent to us, it may be completely obvious to the badgers, most probably because we have measured the wrong attributes (we explore this in Chapter 6). Second, although badger relationships are not consistently amicable or egalitarian, perhaps their groups are largely free of hierarchy.

Insofar as anarchy means the absence of authority, but not necessarily the absence of structure (or, for that matter, the inevitability of chaos), we might contrast a hierarchical hypothesis of social dynamics within badger groups with an anarchic hypothesis (without pushing the analogy with political theory too far, Ward (1966) associates anarchy with groups that are (i) voluntary, (ii) functional, (iii) temporary, and (iv) small—attributes to be borne in mind as we describe badger society). We will return later to some logical consequences of that possibility, especially in the context of the observed skew in reproductive success observed amongst, particularly, male badgers, but also females. First, however, lest we have been looking at the detail through the wrong lens, we will turn to two additional diagnostic candidates of social structure: first, jostlings over food; and second the infliction of wounds.

The motto of the British Special Air Services is 'who dares wins', which one might adapt to who cares, dares. In the context of hierarchies, amongst a litter of fox cubs David had noticed how the smallest and weakest, a vixen named Wide Eyes, cared so much about even a scrap of food that she attacked her larger, nonchalant sisters ferociously; the latter were disinclined to resist. If a measure of dominance was acquisition or defence of food, the runty Wide Eyes was at the top of the hierarchy (Macdonald 1980c, 1987). Such nuances make hierarchies difficult to interpret amongst carnivores, as amongst people, which brings us to Little B—a badger who will be increasingly relevant as our story ages (Chapter 18). Reminiscent of Wide Eyes the fox, Little B exemplified pugnacious disregard for the implied status of size. Her mature length was in the region of just 625 mm (versus average 680.8 mm), and she rarely topped the scales above 7 kg (typically, Wytham females average between 7.57 ± 1.16 kg in June and 9.41 ± 1.86 kg in November) (weights from Sugianto et al. 2019c). Nonetheless, it was Little B who cared enough to dare, and routinely drove larger badgers away from the peanuts with which Chris provisioned them outside the Chalet during the summer of 1996. As her pile of nuts diminished Little B would charge other badgers, sometimes as many as ten in sequence, making sure she always had the best of what nuts remained on offer. The diminutive but short-tempered Little B did rather well: every year she had visibly swollen teats, indicating that she had given birth to cubs (although, and unexceptionally, over her 5-year lifespan our pedigree studies detected not one survivor amongst them).

Reminded of Drew's (1993, p. 289) definition of dominance ('an attribute of the pattern of repeated, agonistic interaction between two individuals, characterised by a consistent outcome in favour of the same dyad member with a default "yielding" response of its opponent rather than escalation'), then surely Little B was consistently dominant in these feeding bouts. To quote Shakespeare's *Midsummer Night's Dream* (Act 3, Scene 2): 'Though she be but little, she is fierce!' And so we arrived at Paul Stewart's Manna from Heaven experiment.

The Manna From Heaven apparatus involved a 'pot on a pulley' that enabled now long-since fledged doctoral student, Paul Stewart, who became one of the most noted natural history film-makers of a generation, to pull strings from a tree hide to deliver *c*.125 g of peanuts to target badger dyads at three social groups. When a particular badger arrived, the pot, containing the *hors d'oeuvre* could be lowered to it. Did stable dominance relationships consistently result in some individuals monopolizing the pot, or perhaps conceding it only to some individuals but not to others? Generally not. Encounters over the dinner table were low key, and generally a matter of first come first served, irrespective of identity. From his treetop perch Paul watched 362 encounters between 12 badgers at the Sunday's Hill sett, 80 involving 4 badgers at the Burkett's Plantation sett and 302 involving 9 badgers at the Radbrook Common sett (the individuals were identifiable from fur-clips; Macdonald et al. 2002a). In total, the sample offered the chance to diagnose priority of access between every member of each group (50 pairwise relationships in the combined Sunday's Hill and Burkett's Plantation community, and 20 more amongst the badgers of Radbrook Common). However, any ranking that existed over access to the pot was no more obvious than it had been in our observations around the setts. No individual badger consistently monopolized the feeder, and none was routinely deferred to either. An individual that took precedence over another on one occasion was often in the reverse role on the next occasion. Age, a good barometer of dominance in, for example, brown rats and spotted hyaenas, had no effect on the odds of monopolizing the peanut pot. Weight and body length, both obvious candidates as predictors of rank, did have a statistically significant effect on monopolizing the pot, but this effect was overwhelmed by prior ownership. As for particular relationships, consider M319 and M340—both Sunday's Hill residents at the time. They were not close relatives,[5] and they were sociologically equivalent (both aged 2 years, 8 kg, and *c*.720 mm long), so one might think them prime candidates to be vying for status. However, either might monopolize the pot,

depending on which arrived first: it seems that, for most badgers, the adage that possession is nine-tenths of the law is apt. As around the sett, some structured personal relationships were apparent. For example, both these young bucks would routinely defer to F327 (a female contemporary, weighing less than 7 kg at 685 mm long and not closely related to either of them).

We had wondered whether the seemingly unstructured society of badgers around the sett would be revealed to be hierarchical when food was involved. It was not. Of course, the experiment was far from perfect. It took account neither of how hungry each badger was when it arrived, nor of the tenor of its previous encounters with badgers it had met over the pot; we even explored whether there might be a cadre of shy, low-ranking badgers that never approached the pot (there wasn't). However, these observations broadly confirmed the conclusion that although there were surely clusters of consistent relationships between some badgers, these tended to be more ephemeral than emphatic. For many badgers and much of the time, it was difficult to dismiss the anarchic null hypothesis. However, one further line of evidence remained, concerning wounds.

Fighting, biting, and social stress

Most of the aggression we saw, whether 'live' or by video, amounted to little more than an irritable grunt or hunch-backed bristling and barging, and, curiously, over the past 20 years it has been mostly sweetness and light amongst the badgers of Wytham Woods, but this was not always so. The previous decade and more was very different. More curious yet, in those earlier years the badgers existed at lower density, so at first sight, and counter-intuitively, it seems that biting and fighting have diminished at higher population density. What evidence leads to this conclusion? Bite wounds. Whether fresh or taking the form of scars, bite wounds provide material evidence that an individual has been the butt of hostility from another. The sex, age, characteristics, and circumstances of badgers bitten by others might reveal patterns in the social flux amongst them. In deciphering such patterns two families of pressures, sociological and ecological, will be interwoven here, and will be untangled at the level of the individual in Chapter 10 with reverberations through to the population in Chapter 11; the latter revealing 1996 as a turning point, during which the badgers of Wytham Woods reconfigured their socio-spatial arrangements. But for now, we ask which badgers get bitten, and why?

[5] M319 (son of F277 and M222) and M340 (son of F286 and M1).

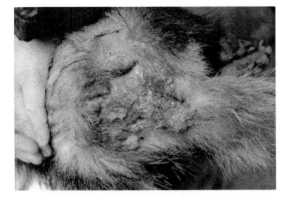

Photo 5.2 A badger bite wound to the rump.

Photo 5.3 The same wound as in Photo 5.2, starting to heal after 2 months.

(a)

(b)

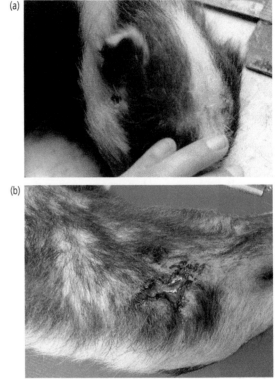

Photo 5.4 Canine puncture wounds to the head.

Badgers are renowned for their devastating lock-jaw bite, generating 207 Newtons of force (Christiansen and Wroe 2007)—about seven-fold the 30 or so Newtons reputedly sufficient to crush the leg of a 'wolf-sized' dog (Wroe et al. 2005).[6] Clearly, they have considerable potential to damage each other. Indeed Pocock (1911) proposed that badgers' characteristic black and white facial stripes may have evolved as an aposematic warning of this disproportionate capacity to put up a fight—a trait we (Newman et al. 2005) have found across a variety of small but fierce musteloid

species (Macdonald and Newman 2017) (a topic mentioned in Chapter 3 and developed in Chapters 10 and 19; badger vocalizations are introduced in Box 5.4).

How do badgers get bitten? Badgers fight with their teeth, circling, with each combatant trying to keep its head away from the other, so strikes are directed against the opponent's hindquarters. As the badgers whirl, one contestant typically gains advantage by sinking its teeth into the opponent's rump, manoeuvring swiftly to prevent the bite being reciprocated. Ultimately the bitten individual will pull away, at the least with nasty perforations, but sometimes with hideous lumps of flesh torn away (Photo 5.2).

We examine these wounds at routine captures. Fresh wounds are often purulent with secondary infection, maggots, and necrotic tissue. Less dramatic but probably more dangerous were those canine punctures to the head (Photo 5.4) that became infected. Eventually the wounds heal (Photo 5.3), leaving scars, and each time we examine a badger we note the details of scars and wounds (collectively 'injuries'), recording severity, freshness, and location.

---

[6] Badgers, as a type, have a reputation for ferocity, captured by the 2007 *BBC News* report that residents of Basra believed the British army was secretly releasing honey badgers as a 'Man eating weapon to subdue the populace'. More plausibly, it's worth noting that the badger's distant cousin, the comparably toothed and muscled wolverine (*Gulo gulo*), is recorded to kill reindeer, or fight an 80-kg wolf to exhaustion (Andrén et al 2011).

Box 5.4 Badger voices

Intern Josephine Wong joined us in the late 1990s. At the time, guided by Paul Stewart, we had two setts, Pasticks and Sunday's Hill, under intensive infrared video surveillance (Stewart et al. 1997). We decided that Jo should devote her time to compiling the first, and still only, vocal ethogram of these badgers—an effort that not only resulted in a benchmark publication on this topic (Wong et al. 1999), but also launched Jo into a research career on bat communication.

Suspending inertan hemispherical pressure zone (PZM: frequency response 20–18,000 Hz) microphones over these setts (shielded with overhead plastic boards from the otherwise deafening noise on playback of military DC10s flying in to Brize Norton) we analysed sound recordings as audio-spectrograms, examining four prime features of call structure: duration, fundamental band frequency, units per call, and inter-call interval (Wong et al. 1999). The keen ear of the sonogram machine listened intently to 445 adult and 149 separate cub vocalizations and, using agglomerative hierarchical cluster analyses (after Newton-Fischer et al. 1993), we sorted them into 10 and 11 distinct calls, respectively (Figure 5.3). Their sounds appear to fit two categories, agonistic or affiliative (but not warning). Some rural eavesdroppers mention badger 'screams'—Chris heard these often in 1993–1994 as Chalet sett budded off and became independent from Hill Copse (the Dutch cheese of bolt holes appearing between them in consequence).

We are pleased to have documented the badger's vocal repertoire (Table 5.7), but do not dissent from Kruuk's initial judgement: badgers don't have that much to say for themselves.

Aggressive/defensive calls include:

- *Growl*—low-pitched rumble, sustained and coarse, long duration, often accompanying snarls and hisses.
- *Hiss*—sibilant 'cat-like' sound, often with growls and snarls. Mostly adults (cubs were only ever noted to hiss at human observers).

These were both used as the primary warning, or defence, calls, produced during inter- or intra-specific competition (see Macdonald et al. 2004c), most commonly over food, but also in response to unfamiliar individuals. In addition, their repertoire included:

- *Snarl*—noisy, harmonic, sibilant, short call, often given in combination with growls, barks, and hisses. Indicative of a higher intensity or escalation of aggression, used in the 'threat-attack' context. Unlike simple growls, snarls betray the intention to physical confrontation.
- *Kecker*—frequently heard rapid series of staccato sounds. Generally reinforcing or accompanying the repertoire of aggressive sounds.

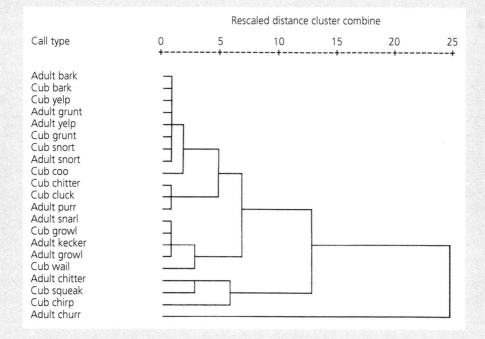

Figure 5.3 A dendrogram showing the relative distance between call types (produced by hierarchical cluster analysis using the mean values of call types) of European badger (from Wong et al 1999, with permission).

Box 5.4 *Continued*

Table 5.7 Summary of the structural characteristics of vocalizations of adult and cub European badger. Values for average fundamental frequency bandwidth, duration, units per call, and inter-call interval are the means ± SD (from Wong et al. 1999, with permission).

Vocali-zation	Type of sound	Average funda-mental frequency bandwidth (Hz)	Duration (s)	Units per cell	Inter-call interval (s)	Given alone or in series	Auditory quality	n
Bark (adults)	Noisy		0.11 ± 0.03	1.00 ± 0.00	0.52 ± 0.42	In series	Moderate-pitched, harsh and gruff	3
Bark (cubs)	Noisy		0.11 ± 0.02	1.00 ± 0.00	1.08 ± 2.39	In series	Moderate-pitched, harsh and gruff	13
Chirp (cubs)	Harmonic	980 ± 211	0.12 ± 0.03	1.00 ± 0.00	1.69 ± 2.65	In series	Moderate-pitched and abrupt	17
Chitter (adults)	Harmonic	498 ± 139	0.48 ± 0.21	4.18 ± 0.75	1.15 ± 0.95	Mostly in series	High-pitched, rapid querulous chatter	20
Chitter (cubs)	Noisy		0.74 ± 0.28	5.00 ± 0.00	0.89 ± 0.67	Mostly in series	High-pitched, rapid querulous chatter	19
Churr (adults male)	Noisy		0.98 ± 0.52	13.8 ± 2.50	1.49 ± 2.07	Mostly in series	Low-pitched, deep vibrant purrs	81
Cluck (cubs)	Noisy		0.69 ± 0.45	4.79 ± 1.32	0.41 ± 0.42	Mostly in series	Moderate-pitched staccatosounds	4
Coo (cubs)	Harmonic	214	0.06	1.00 ± 0.00	Isolated	Alone	Low-pitched, soft, dove-like sound	1
Growl (adults)	Harmonic and noisy	32 ± 5	1.59 ± 0.87	1.00 ± 0.00	1.62 ± 1.87	Alone or in series	Low-pitched rum-bling, sustained and coarse	88
Growl (cubs)	Harmonic and noisy	46 ± 20	1.33 ± 0.36	1.00 ± 0.00	Isolated	Alone	Low-pitched rum-bling, sustained and coarse	2
Grunt (adults)	Noisy		0.18 ± 0.06	1.50 ± 0.71	Isolated	Alone	Low-pitched and blunt	2
Grunt (cubs)	Noisy		0.50 ± 0.55	1.33 ± 0.52	0.59 ± 0.36	Alone or in series	Low-pitched and blunt	6
Hiss (adults)					Isolated	Alone	Sharp, sibilant	7
Hiss (cubs)							Sharp, sibilant	
Kecker (adults)	Harmonic	170 ± 73	1.19 ± 0.44	1.00 ± 0.00	0.73 ± 0.66	Mostly in series	Variable pitch, rapid series of staccato sounds	33
Purr (adults females)	Noisy		0.78 ± 0.52	5.50 ± 1.74	0.46 ± 0.52	Mostly in series	Similar to churr but softer and lower in intensity	51
Snarl (adults)	Harmonic and noisy	61 ± 12	1.15 ± 0.51	1.00 ± 0.00	1.78 ± 1.58	Mostly alone	Moderate-pitched, sibilant growls	31
Snarl (cubs)							Sibilant growls	

Box 5.4 *Continued*

Attention calls include:

- *Bark*—moderate-pitched, harsh, loud; usually in a series of two to four. Typically given in the contexts of surprise, warning, defence, and play, accompanied by jumping back and piloerection.
- *Chitter*—high-pitched, short, querulous chatter; harmonic in adults, noisy in cubs; most often given in series. Used in the context of pain and fear and the inferred context of anxiety and frustration; often uttered when a female is being harassed by an amorous male.
- *Squeak*—cubs only; high-pitched and shrill, short call, usually accompanying yelps and chitters. Has a similar context to the chitter in adults, but also expressed during play.
- *Snort*—low-pitched, noisy, nasal, short duration. Associated with startle and surprise, similar to the bark. Does not elicit a response from individuals in the vicinity, so not a warning call per se, but may startle potential threateners.
- *Yelp*—high-pitched, abrupt, sharp, noisy; single, double, or triple units; mostly cubs. Voicing fear or pain, actual or anticipated.
- *Wail*—cubs only; perpetual cry of the harmonic type. Used to signal infant distress, usually if a cub is left above ground or isolated from its mother.

    Contact and contact-seeking calls include:

- *Chirp*—cubs only; moderate-pitched, soft bird-like, in series with yelps, chitters, and clucks.
- *Coo*—soft, gentle 'dove-like' sound; only in cubs, very quiet and hard to detect.
- *Cluck*—noisy, moderate-pitched, staccato sounds, reminiscent of a duck's quack; in series or combined with similar calls.
- *Grunt*—short, low-pitched, noisy, blunt sound. Usually given singly by adults but in series by cubs.

These are all used during greeting, grooming, foraging, and play, and particularly associated with general close-contact affiliative situations.

- *Churr*—insistent, deep, throaty purrs, with an oily quality; noisy; usually given in series. Used only by males in the context of sexual arousal (equivalent of the female purr), signalling mating interest. Could possibly function in mate-choice selection where call and rate could provide a measure of male physical quality.
- *Purr*—female purr similar to churr, but softer and less intense; often followed by a 'click' sound; average of six units per call. Used to lure or encourage cubs; the use of the click adds reinforcement.

The ontogeny of calls seems firmly linked to social development. Gradations and transitions in these sounds characterize cub calls, with a higher and wider frequency range, usually associated with social play (as per Morton's (1977) motivation-structural rules (Box 6.1)).

   This ethogram is a great start, but it is clear that while we have described the badgers' vocabulary, but not their conversation. Clearly assigning descriptive terms to these sounds is subjective. Other authors also describe 'squeals, screams, trills, wheezes, whickers, whines, yarls' (Simms 1957; Paget and Middleton 1974a; Ferris 1986, 1988; Kruuk 1989; Christian 1993; Neal and Cheeseman 1996). While these studies made verbal descriptions based on the sounds they heard, we used quantitative analysis of audio-spectrograms.

   As suggested by Cohen and Fox (1976), and reiterated by Schassburger (1993), the only modest complexity of badger acoustic communication may reflect the not very sophisticated stage of their social development relative to the classically social Carnivora (Chapter 19); the idea that complex social lives are reflected in complicated brains—the 'social brain' intelligence hypothesis (Dunbar 1998)—may initially have relied too heavily on encephalization traits amongst canids, and neglected extinct species (Holekamp 2007; Finarelli and Flynn 2009; see Newman and Buesching 2018).

The severity of the wounds and their frequency of occurrence both tell interesting stories and, remarkably, both have changed over the past 30 years or so. But to begin at the beginning, we began logging bite wounds in 1987, and as the years passed we increased the detail we recorded to include the specific sites (head/neck, rump, back, underside) of fresh injuries (which we refer to as 'wounds' in Macdonald et al. 2004c) and also of older scars. Although we occasionally found injuries from other sources (e.g. entanglement in wire, glass cuts), most injuries were clearly identifiable as bite wounds. Wounds

Table 5.8 Observed values for male and female wounding and scarring levels aggregating data from 1990–1999 (from Macdonald et al. 2004c, with permission).

	% Females	% Males	No. of females	No. of males	χ_1^2	P
Wounds	17.7	24.9	237	261	3.8	0.051
Severe Wounds	9.3	14.9	237	261	0.84	0.36
Head Wounds	12.3	13.0	195	216	0.04	0.84
Rump Wounds	8.7	21.3	195	216	12.5	<0.01
Scars	39.2	43.3	237	261	0.84	0.360
Severe scars	20.3	31.4	237	261	8.02	<0.01
Head scars	18.5	19.0	195	216	0.02	0.889
Rump scars	34.4	43.5	195	216	3.61	0.057

Percentages indicate individuals recorded with a bite wound in each category at least once during the study period. Severe wounds are a subset of wounds in general, which were also rated as minor and moderate; the same applies to scars (from Macdonald et al. 2004c, with permission).

were defined as fresh bites (open/haemorrhaging, infected/suppurating, or lightly scabbed) (Table 5.8). Sometimes entire patches of skin were pulled away, with severe subcutaneous trauma to soft tissues. The wounded badgers could be visibly disabled, and could develop severe secondary infection. Scars were defined as areas of healed epithelialized granular tissue indicative of previous wounds. We classified wounds and, separately, scars as minor to severe (on a 0–6 scale). Wounds always become scars; however, only the largest wounds produce scars visible through the fur (Macdonald et al. 2002a).

So, what did these wounds tell us about badger society and the flux of animosity through it? First, and frustratingly, insofar as we see the consequence rather than the act, we know the identity (and details) of the bitten, but not of the biter, and by the same token we cannot classify the wounds into those inflicted between individuals from different groups or super-group.

During the period 1990–1999, 24.9% of 261 male badgers presented severe wounds (of these, 21.3% of wounds were on rumps; 13% on the head and neck), versus just 17.7% of 237 females (of these 8.7% were on rumps; 12.3% on the head and neck) (Macdonald et al. 2004c) (Table 5.7). So, females bore fewer wounds than males, and of their wounds a smaller percentage was on their rumps.

Fresh bites were most prevalent amongst badgers in very poor condition (Delahay et al. 2006b), perhaps suggesting hunger and competition for food as a risk factor (see Chapter 11). These conflicts left scars on 43.5% of males, and 34.4% of females, particularly on their rumps. As badgers aged, they accumulated more scars. What were the characteristics of the badgers that ended up most scarred? Of course, a badger receiving a vitally placed neck wound may not survive to grow old, but the oldest badgers were the most

scarred. Other than age, it was the heftiest amongst them that were most scarred: lifetime scar acquisition correlated positively with (seasonally adjusted) badger weight. Wounding and scarring levels correlated with volume of subcaudal gland secretion, which is more copious in older, heavier animals of breeding status. Tthe next question is therefore how scent, and particularly subcaudal gland scent, functions in communication within badger groups? The answer awaits in the next chapter, Chapter 6.

These figures all concern adults. To get a wider perspective, and a much larger sample, we combined our data from Wytham with those our team had gathered between 1995 and 1999 at our study site at North Nibley while researching bTB (see Chapter 16), and with yet more gathered in Woodchester Park by our colleagues in the Department for Environment, Food and Rural Affairs (Defra) (Delahay et al. 2006b). This enormous dataset amounted to 4312 badger examinations. Wounds inflicted on cubs were categorically different: although recorded on only 1% of male and 2% of female cubs, 53% of bite wounds to cubs were inflicted on the head, drawing to mind the gruesome episode we described in Chapter 1 (p. 9). It seems likely that many bitten cubs would not survive to be recorded (and might add to the growing suspicion of reproductive conflict amongst adults, and associated infanticide).

So much for the first act of the play, from 1990–1999. Interesting as these figures are for that window of time, much more interesting is the way in which they changed. For the past 20 years we've almost never seen a bite wound! Even within the 10 years of that first study, things were changing. Goodwill amongst badgers appeared to dwindle between 1990 and 1999. In 1990, for example, 7% of the badgers we examined had bite wounds. A decade later that percentage had nearly trebled, to 20% (Figure 5.4). What had changed?

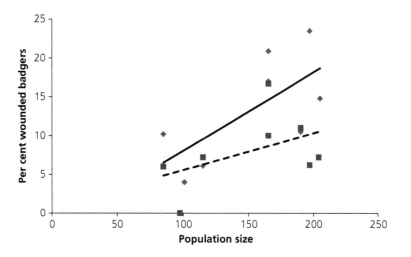

Figure 5.4 Percentage of wounded males (diamonds) and females (squares) plotted against population size for each of the 10 years (1990–1999) (from Macdonald et al. 2004c, with permission).

The full answer awaits in Chapter 11, with causal factors identified in Chapters 12 and 13, but an important clue lies in the fact that over the 1990s the population of badgers residing in Wytham Woods had doubled, to c.200 (Macdonald et al. 2004c). In fact, it continued to rise and we anticipated Armageddon. Confounding our expectations, bite wounding plummeted to less than 1%. Insofar as the pattern of bite wounding might reveal something of badger social structure (of course we wondered whether the lowest ranking individuals might be most bitten, the highest ranking the least—but recall the nebulous nature of rank described earlier—or perhaps the most aggressive neighbours, or most flagrant trespassers, might be most bitten). But whatever that pattern might reveal about badger society, it would appear that that society, as reflected by the frequency of bite wounds, changed (as bite wounding increased), and then changed once more, this time in more radical fashion (i.e. as the evidence of biting almost disappeared) as the 1990s progressed.

During the years of increasing wounding, until 2002, at any one time the rates of male and female wounding (and the ratio of bitten males to females) were higher within larger social groups. We deduce that bite wounding at least partly reflects hostility within groups, and that it is related to the headcount of males in a group. That said, there is also evidence of a between-group component, insofar as both wounds in general, and severe wounds in particular, were more frequent for males (but not females) in proportion to the number of badgers living in adjoining group ranges (Macdonald et al. 2004; Dugdale et al. 2007).

Insofar as wounding is a barometer of social tensions, one would expect it to vary between seasons and circumstances. However, at Wytham we could detect no such seasonality in wounding, whereas at Woodchester Park there was a peak in wounding during the late winter mating season (Delahay et al. 2006b). Similarly, Warren Cresswell et al. (1992), used comprehensive data collected from road kills throughout the year to identify peaks in wounding levels in male, but not female, badgers during the spring, and concluded that they were associated with reproductive activity (Keeble 1999) (in Chapter 7 we will document surprising equanimity amongst the badgers of Wytham Woods during sexual encounters). Cresswell also pointed out that insofar as wounding was associated with competing for something, it did not peak at seasons when food was thought to be in shortest supply. The point is, insofar as hostility leading to bite wounds varies in some places (but not Wytham) from season to season, and place to place, and, even more interestingly, time to time in the same place (most strikingly in Wytham), we can deduce that the social dynamics leading to these wounds also vary on these dimensions.

A further clue is that in both Wytham and Woodchester Park fresh bites were most prevalent amongst badgers of both sexes in very poor condition (Delahay et al. 2006b). Remembering the ferocity of Little B, perhaps increased desperation for food is the common denominator of greater rates of bite wounding. However, when we come, in Chapter 10, to consider badger foraging behaviour, it will become plain that they generally forage alone for tiny items not worth fighting over. Therefore, it seems more probable that biting is more about social discord than ecological shortage. Being bitten was not the prerogative of one sex, any particular age group, or yet of one reproductive category or even of season; something changed, however, c.20 years ago to make the badgers of Wytham Woods much more cordial. The framing of that remarkable change is one of the themes of Chapter 10. Meanwhile, to complete the context of bite wounding, in order to be able to bite each other badgers must be together; so the question arises: when are they together? When they are underground, which is where we join them next.

Underground privacy

A feature of badgers' central place sociality is that when active they are generally alone, in contrast to the tight-knit groups of, say, coatis or banded mongooses. We see them in company around the sett, but only for a tiny fraction of each day. For the most part, when badgers are together it is in privacy, underground. Perhaps this is the venue for socializing (for comparison, David remembers watching Kalahari meerkats, just visible inside a short burrow during a midday siesta during which every single individual's body was entwined in a communal sleep heap). What do badgers do underground? In the context of this chapter and the quest to describe the social dynamics amongst group members, might the locations of bedchambers, or the sleeping companions occupying them, reveal a new side to badger sociality?

It has been widely assumed that badgers underground are 'inactive' by day, each individual snuggled, for the most part on its own, in a single chamber (e.g. Roper et al. 2001; Kowalczyk et al. 2003a). Sitting, forlorn, on a badger sett listening to the monotonous chirp in the headphones of a VHF radio-transmitter[7] on the badger below certainly induced sleep in the biologist, who generally imagined that that was what the badger was doing, more comfortably below (e.g. Biggins 2012; Weber et al. 2013). Into this somnolent scene sliced the cutting edge of technology, in the form of Zimbabwean computer scientist Andrew Markham.

Andrew, who has pioneered inventions ranging from miniature microphones to mesenteric temperature probes (both of which WildCRU has applied to lions (respectively, Wijers et al. 2018, 2020; Trethowan

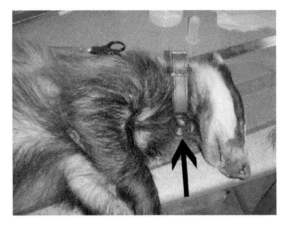

Photo 5.5 A sedated adult badger wearing an assembled magnetic induction tracking collar. Arrow indicates the tag and battery (from Noonan et al. 2015b, with permission).

et al. 2017)), has also pioneered the machine learning algorithms (Dhir et al. 2017) necessary to cope with the 'big data' that these gadgets produce. Faced with a badger sett, Andrew rather characteristically stroked his chin. Quite a lot of chin stroking later, we received our first (indeed, *the* first) magnetic induction (MI) transmitters, which we introduced briefly in Chapter 4. Not only did these transmitters, worn on leather collars (Photo 5.5), enable us to locate the badgers underground, in 3D, to the nearest 10 cm, they also turned the badgers into research assistants—each carrying the mapping technology through the tunnels whose labyrinths we could thus plot—a task that might have taken a man on a digger days of destruction (Roper 1992) (see Figures 5.5 and 5.6).

Mike Noonan (introduced more fully in Chapter 1) led our pilot study (Noonan et al. 2015b). Four MI tracked badgers (three males, one female) spent, on average 12.3 ± 1.69 h in the sett each day—emphasizing that most of our research for the preceding 30 years had been blind to about half of the badgers' time. Surprisingly, we detected no significant difference between the time they spent underground by day (6.4 ± 1.0 h) and by night (5.8 ± 0.88 h). Interestingly, none ever spent more than 2.2 ± 3.1 consecutive hours away from the sett; each returned to it an average of 2.3 ± 1.45 times throughout the night, with resting periods during these visits lasting a bit more than three-quarters of an hour (mean of 0.77 ± 0.91 h). Why did they keep coming back home? Perhaps to conserve energy, for a rest, or even to warm up, or perhaps deliberately to share information on foraging (although probably not) (see Ward and

---

[7] Even the newest Global Positioning System (GPS) technology didn't help because the signals can't penetrate the soil.

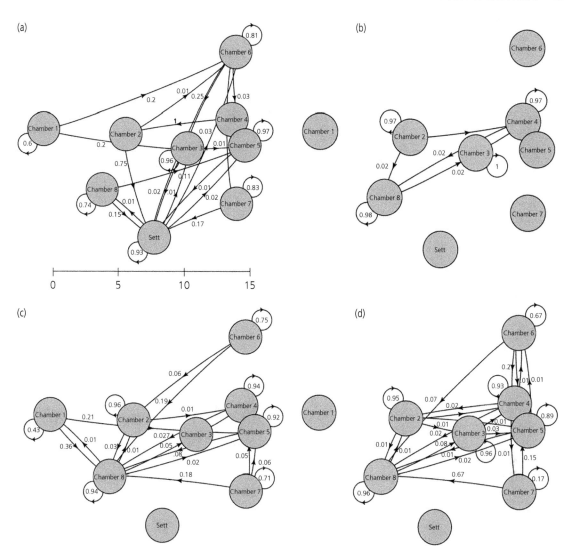

Figure 5.5 Markov chains representing probabilities of transition between chambers for badgers at different time points: (a) 1100; (b) 1300; (c) 1400; and (d) 2202 at a resolution of one fix per 10 min. Only transitions between chambers within the sett and tunnels are presented. The position of each chamber depicts the actual architecture of the sett. Scale is in metres (from Noonan et al. 2015b, with permission).

Zahavi 1973) or maybe even on the whereabouts of the wolves whose presence, we increasingly suspected, had preoccupied badgers' minds in formative times.

Badgers returning for a nocturnal visit used the sett differently to the way in which they used it by day. During the night, apparently returning for a nap, badgers tended to use chambers located near the outer circumference of the sett, whereas for longer diurnal rests they used more central, deeper chambers. Badgers used 1.55 ± 0.61 chambers day^{-1} ($n = 49$, range 1–3) and, although each badger used a personal subset

of the available chambers, each used more than one chamber on nearly half of the days. Did individuals have a special, favourite chamber? Not for long. Their preferences were mercurial; each typically favoured a chamber for 2.7 ± 1.7 consecutive days ($n = 16$, range 1–7 days), before moving to a different chamber. All of this shifting of favoured chambers every few days, and considerable insomniac movements during every day, contrasts radically with previous impressions. For example, Kowalczyk et al. (2004) used VHF tracking in Białowieża Primeval Forest, Poland, to report

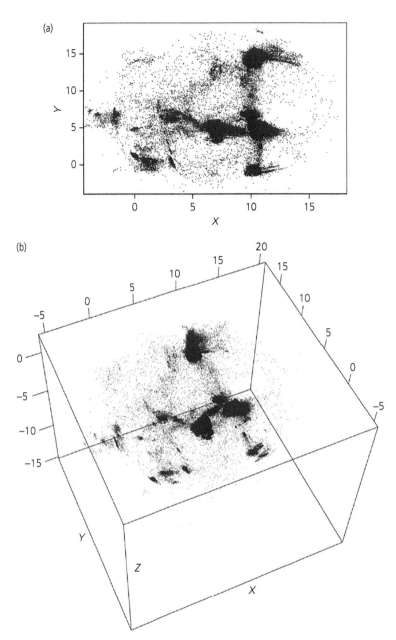

Figure 5.6 X, Y, Z coordinates of underground positional fixes, for all instrumented badgers, over the course of the study period presented in (a) 2D and (b) 3D. Axes are in metres (from Noonan et al. 2015b, with permission).

just two instances of diurnal badger movement over a 4-year study. Had we sampled only at 12 a.m. each day, we would have recorded just 4 chamber changes over the 2 weeks, not the 110 we actually logged while sampling at 10-min intervals; let alone the 802 we saw sampling at 3-s intervals. In short, our old ways of doing things could have missed 99.5%

of underground movements. Furthermore, the daily 'day-fix' entirely missed a remarkable aspect of badger behaviour: far from indulging in an imagined deep sleep for daylight hours on end, individual badgers napped frequently and occasionally shuffled between chambers. We wonder whether fleas contributed to their restlessness (Chapter 8).

Our pilot study of magnetic inductance tracking was a revelation, but it was based on only four individuals sharing their sett with at least another eight untracked companions. We achieved only a glimpse of the sociology of sleeping associations, but even that glimpse confirmed that these badgers (three males and a female) were daytime associates, spending around half of their time in the same sector of the sett; occasionally in the same chamber and often engaging in a sort of moving that was akin to musical chairs as they swapped between chambers. The kaleidoscope of juxtapositions of these four instrumented adult badgers left, returned to, and moved around within the sett autonomously. However, our underground studies reveal ephemeral subterranean associations but not according to any pattern that contradicts the above-ground impression we had gleaned of a rather anarchic society.

From flagrant despotism to cryptic competition

Two facts lead to an important deduction. First, while the members of badger groups surely know each other well, and doubtless have varied affiliations and hostilities, we have yet to identify any recurrent, clear-cut structure permeating their interactions, or even signs of friendships (there were some clear-cut animosities) (we put aside for a moment the fact that we may, so far, have been looking in the wrong place, for such structure, which is why we devote Chapter 6 to scent, and Chapter 7 to mating). Second, insofar as a fundamental tenet of ecology is that organisms compete for limited resources, and that an equally fundamental tenet of natural selection is that success in such competition will be reflected in fitness—basically reproductive success—theory leads us to expect to see winners and losers amongst the members of badger groups. And so we do. Remember Lecher B (M10) and his 11 descendants, and F221 and her 5. Some societies are despotic: consider the dominant chimpanzees who enjoy (doubtless at a cost to their immune system, Chapter 15) the best access to food, the greatest deference from subordinates, and the most fecund reproduction (Wittig and Boeach 2003). But considering their natural history, what, within badger groups, would competition secure for the dominants or deny the subordinates? In Chapter 10 we describe the diet of Wytham's badgers, but already it is clear that badgers forage alone for small titbits—not a resource that can be monopolized by even the mightiest individual. In

Chapter 7 we describe the mating behaviour of badgers (and in Chapter 17 the genetic outcomes of that behaviour), but already it is clear that more than one male within a group secures paternities; further, that some of these are secured by males beyond the group, and indeed beyond the super-group. Might it be that battling for mates, much like battling for food, in the badgers' reality is nugatory work? Worse, considering the appalling biting power of the deeply articulated mandible with its bulging masseter muscles, might the likely costs of fighting for a monopoly of mates, meals, even nest chambers (of which there appear to be plenty to go around) stack up poorly against the likely rewards (Gallagher et al. 1979)? Badgers seem eager to avoid aggression (Enquist and Leimar 1987, 1990). Indeed, ponder for a moment the raw maths of the neighbourhood sociology of the badgers of Wytham Woods, and the task entailed in imposing, even remembering, dominance. Each group of badgers can have up to five conterminous neighbouring groups, each with an average group size of ten. That amounts to over 50 badgers with whom each individual has personal dealings, and the emerging evidence of super-groups is that these dealings may not distil readily into a simple anonymized matrix of them and us. Furthermore, we might wonder how rapidly this kaleidoscope of identities changes—in Chapter 11 we will detail the mortalities that cause about one in four of the actors on this stage to be replaced each year. From all these considerations the question arises: is it worthwhile for individual badgers to invest heavily in the physical domination of other members of their group or even, more radical yet, their neighbours? Military strategists, or politicians, pick their battles wisely, and one interpretation of the results in this chapter is that if you are a competitive badger, you just can't win.

However, some badgers manifestly do win—Satyr (M4) and Lecher B (M10) and F221 as examples—and, in the evolutionary stakes, being a winner is existentially preferable to being a loser (analysis of >30 years of data reveals that 33.1% of females that reach age 3 (146 of 441) never have any sign of reproduction—a reproductive success of zero; Bright Ross et al. 2020). So what, then, is the deduction that arises from the two salient facts: the absence of consistent winners or stable hierarchy in the daily flux of badger interactions; and the presence of clear-cut winners in reproductive skew? If competition is not happening vividly, through tooth and claw, it must surely be happening cryptically (Chapter 17), in the cosier surroundings of the females' wombs.

CHAPTER 6

Social Odours

The Perfume of Society

A wonderful fact to reflect upon, that every creature is constituted to be that profound secret and mystery to every other.

Charles Dickens, *A Tale of Two Cities*

Try as we might to interpret interactions amongst the members of Wytham's badger groups, we *saw* little evidence of the types of social structure that typify other group-living carnivores (Chapter 5). Perhaps not seeing the evidence was the point: while humans *look* for patterns, badgers are deeply olfactory, so perhaps their communication of social structure lay beyond our perception. To paraphrase G.K. Chesterton's lamenting song of the dog Quoodle (1896): 'They haven't got no noses, and goodness only knowses the Noselessness of Man'.[1] Therefore, continuing the quest to characterize, and thence to understand the adaptive significance of, badger groups, we turn now to scent. Badgers are the antithesis of noseless. Of course, other carnivores also have exquisite olfactory capabilities as well, but for badgers, reliance on scent seems to dominate their other senses. Also, of course, badgers have sharp ears and rely too on vocal and visual signals, see Boxes 6.1 and 5.4 (Newman and Buesching 2018).

But in this chapter our priority is the continuing quest to describe the social relationships, and the signals that reveal them, within badger groups (we turn to relationships between groups in Chapter 9). In Chapter 5 we identified allomarking as the most promising diagnostic of structure, if not of hierarchy. Specifically, here we ask what can be learnt about badger society by understanding olfactory communication within their social groups.

Decoding the communication of these olfactory supremoes is a challenge because human nose-blindness thwarts our hunger for knowledge of

what Kenneth Grahame, in *The Wind in the Willows* (1908), called the mysterious fairy calls from out of the void, for which 'we have only the word smell, to include the whole range of delicate thrills which murmur in the nose of the animal night and day, summoning, warning, inciting, repelling'. Whereas communication by scent, involving 'chemoreception' (of a latent signal, where the sender need no longer be present), is evolutionarily ancient, the science investigating chemical signals (semiochemistry) is young; only recently has technology, such as gas chromatography and mass spectrometry, come to the aid of human noselessness (Buesching et al. 2016b).

For olfactory context, remember that badgers are accustomed to a dim, cluttered, nocturnal realm, where moonless nights are 100 billion times darker than bright sunlight (Kirk 2006), and subterranean tunnels can be absolutely dark. In this domain, olfaction—communication through scent and sense of smell—is key. Furthermore, badgers do not see colour, because despite having three visual pigments in their retinal photoreceptor cells,[2] as do most other mustelids and carnivorans generally, these are not trichromatic, as in our eyes; rather they see in sepia tones (Kitchener et al. 2017). Nevertheless, badgers have evolved to use, as best they can, what light there is and, to enhance light receptivity, they have a reflective *tapetum lucidum* behind their retina (Ollivier et al. 2004), and relatively larger and more curved eyeballs than other mustelids,

---

[1] Although modern discoveries suggest that humans are not as talentless with regard to scent as the Quoodle supposed; see McGann 2017.

[2] For domestic ferrets these include a rod opsin, with maximum absorbance (λ_{max}) of 505 nm (green), a short (S) wavelength cone opsin with λ_{max} of $c.430$ nm (violet), and a medium–long (M/L) wavelength cone opsin with λ_{max} of 558 nm (yellow) (Calderone and Jacobs 2003).

The Badgers of Wytham Woods. David W. Macdonald and Chris Newman, Oxford University Press.
© David W. Macdonald and Chris Newman (2022). DOI: 10.1093/oso/9780192845368.003.0006

Box 6.1 Senses

The acuity of the carnivore sense of smell is hard to comprehend, but necessary to bear in mind when trying to put oneself inside a badger's skin. One famous tracking dog can reputedly detect an orca scat floating on a choppy sea from a mile away, an astonishing feat dwarfed by grizzly bears smelling elk carcasses under water from far over the horizon, not to mention polar bears smelling seals through ice (polar bears, by the way, have, relatively speaking, the largest olfactory bulbs amongst mammalian carnivores, the sizes of which correlate with the bears' home range size). Almost nothing is documented of the neuroanatomy of badger olfaction, but a reasonable approximation can be gleaned from domestic dogs: bloodhounds have 300 million olfactory sensory neurons in their nose (dachshunds[3] have only 125 million, which sounds paltry by comparison, until you remember that humans have only 5 million), a brain dominated by an olfactory cortex, and a vomeronasal organ (Jacobson's organ) that also contains olfactory epithelium. According to Padodara and Jacob (2014) a grizzly bear's nose outperforms a bloodhound's sevenfold; and can detect prey 18 miles away. Where badgers rank amongst olfactory Olympians is unknown, but it seems a safe bet that they are far, probably very far, beyond the bloodhound league. Certainly, they too have a vomeronasal organ (VNO), a mammalian device for detecting heavier, non-volatile molecules (Lüps and Wandeler 1993; Kelliher et al. 2001)—literally allowing them to 'taste the air'. The badger's VNO has not been studied, but it's a fair guess that it's similar to a ferret's, which consists of a paired cartilaginous capsule of $c.\,1.6\,cm^2$ that lies at the base of the nasal septum. Signals from the VNO are processed in a discrete, but tiny (0.15% by volume), part of the olfactory bulb of the anterior brain. When the VNO is at work, mammals engage in flehmen (a sort of nose-wiffling grimace, as though the animal were poised, anticipating a sneeze, such as you might see in your horse or cat), a facial expression we also see performed by badgers (Tinnesand et al. 2015).

Irrespective of modality, powers of perception relate to (see Buesching and Newman 2018):

1) The evolved adaptiveness of the receiver's sensory systems to the cue (e.g. Endler 1992). For example, receptor olfactory sensitivity to biologically important pheromones is often extreme (Wyatt 2010).
2) The receiver's 'psychological landscape', which determines detectability, discriminability, and memorability of the cue after it reaches the receiver's sensory organs (Guilford and Dawkins 1991), allowing, for instance, conspecifics to recognize each other's vocalizations or scent marks even after periodic absence (e.g. Buesching et al. 2002a).

3) The efficiency of the cue's transmission through the environment (e.g. Endler 1992); for instance, whether over land, water, or through subterranean tunnels—subdivided into habitat types, such as the different properties of dense forest, or open farmland.
4) The detectability of the cue, which is largely dependent on the physical characteristics of the cue's modality (visual, acoustic, olfactory, or tactile) (Table 6.1), as well as the environment and/or media—such as the effects of temperature, wind speed /noise, humidity, etc. (reviewed in Buesching and Stankowich 2017).

Putting other badger senses in context, nobody has studied audition in badgers, but their first cousins, ferrets, have an upper limit at around 44 kHz at 60 dB SPL (sound pressure level), with lowest thresholds in the region of 8–12 kHz (Kelly et al. 1986). Similarly, American mink can hear up to 40 kHz (Powell and Zielinski 1989), while the hearing range of the least weasel (*Mustela nivalis*), which have a keen interest in hearing mice squeak, extends from 51 to 60.5 kHz for intensities of 60 dB SPL, with a region of best hearing extending from 1 to 16 kHz (Heffner and Heffner 1985). Interestingly, Heffner and Heffner (1987) consider that the acuity of sound localization by mustelids is typically lower than that for larger carnivores, linked to inter-aural distance

Table 6.1 Differences in signal quality between visual, acoustic, and olfactory signals.

Visual/acoustic signals	Olfactory/chemical signals
• Directed: indicates exact location of sender	• Not directed: only gradient will lead to scent mark
• Immediate communication	• Delayed communication
• Short-lived signal	• Long-term signal
• Long-range signal	• Short-range signal
• Sender has sole control over time of communication, although the receiver can sometimes provoke them, it cannot seek out signals	• Time of communication is dependent on receiver, which has to seek out scent mark

---

[3] *Dachs* is German for badger, for which these dogs were bred to hunt.

Box 6.1 *Continued*

(i.e. their ears are close together), reducing perception of phase intensities (Newman and Buesching 2018). For comparison, at 60 dB SPL (re 20 µPa) red foxes perceive pure tones between 51 and 48 kHz, spanning 9.84 octaves with a single peak sensitivity of −15 dB at 4 kHz, whereas the domestic cat can hear through the range of 55–79 kHz, being most acute through 500 to 32 kHz.

Badger vocalizations are described in Box 5.4, but when it comes to what they hear, in the context of communication rather than predation, or indeed, in terms of the main theme of this chapter, what they smell (the latter is also one of the themes of Chapter 9), it may be worth recalling

Eugene Morton's (1977) motivation-structural rules. You, or your dog, may use a different lexicon, but you both attend instantly to a gruff warning as opposed to a falsetto whine. Morton, in the context of auditory communications, highlighted the convergent use of harsh, low-frequency sounds by hostile animals and more pure tone-like, high-frequency sounds by fearful or appeasing animals (by way of analogy, we note that car mechanics tune exhaust notes to maximize the 'psychoacoustic' effect on listeners [Schirmacher 2002]). We wonder whether this relationship between sound structure and function might also apply to social odours.

---

with high rod/cone ratios. But this comes at the cost of acuity (Kitchener et al. 2017). While visual and auditory signals reveal a lot about badger interpersonal relations, they offer little to enable us to reject the sociological randomness described under what we might call the anarchic null hypothesis of Chapter 5. Perhaps the ebb and flow of these interpersonal relationships will be strikingly revealed, like invisible ink rendered visible on the page, through olfactory signals? After all they include such promising candidates as faeces, and associated anal sac secretions, urine, and the badgers' unique subcaudal gland—to which we turn now. Later, in Chapter 9, we will turn to relationships between neighbouring groups, and pose in particular the question: what is the communicative function of boundary latrines, redolent as they are with the odours of faeces, urine, and anal gland?

Badgers at the sett frequently back up to another denizen of their underground realm, with the result that they briefly, if awkwardly, sit upon a compatriot (the allomarking mentioned in Chapter 5: allo-, by the way, is from the Greek *allos* for other, therefore allomarking is marking another); likewise, cubs crawl under adults (as described in Chapter 2), forcing the same effect of being sat upon. More rarely, but theatrically, two badgers will, on occasion, even reverse towards each other in a synchronized mutual attempt to sit, resulting in a 'bum kiss', as Hans Kruuk indelicately termed it (Kruuk et al. 1984). Occasionally, too, they will make the same brief squatting bob, for no more than a second (far too quickly to urinate or defecate), and touch their bottoms on the ground. So, what is it that they so assiduously impress upon the

substrate and one another? It is the secretion of the subcaudal gland. But what is that?

A unique gland

Some might consider this adornment unappealing. However, to those intrigued by the intoxicating influence of its fragrance on animal behaviour, the badgers' unique subcaudal gland, a greasy cavity just above the anus and beneath the tail (Photo 6.1), is spellbindingly interesting. And nobody has ever been more spellbound, or contributed more to understanding of this gland and all that wafts from it, than our team member Christina Buesching (Photo 6.2), her expertise in olfaction aided by her early career in the endocrinology of lemurs. Only badgers in the genus *Meles*, and their Asiatic kin, hog badgers, genus *Arctonyx*[4] (Zhou et al. 2017), have this specialized gland. It has no homologue amongst other mammals (Brown and Macdonald 1985) (the positionally similar, and even more exuberant, glands of hyaenas are, entirely differently, derived from anal gland anatomy; Burgener et al. 2009). Presumably it was *Arctomeles*, the common Pliocene ancestor of both genera, *Meles* and *Arctonyx* (Chapter 1), that first gained selective advantage from evolving a subcaudal gland (towards the end of this chapter we will take a retrospective look at what that advantage might have been) (Zhou et al. 2017).

Since the days of Pocock (1920) and Ernest Neal (1958) badger anatomists have known of the subcaudal gland. Both sexes have this gland, which consists of a

---

[4] *Arctonyx alboguralis, A. collaris,* and *A. hoevenii.*

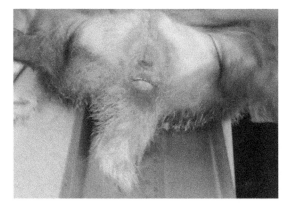

Photo 6.1 Underside of the tail region with subcaudal gland between anus and vulva and reminiscent of a 'smile'.

Photo 6.2 Christina Buesching, 25-year stalwart of the badger team and leading authority on their scents.

blind-ending pouch lined with several layers of sebaceous and apocrine gland cells (Stubbe 1971), partly divided by a median septum and opening into a 20–80-mm-wide horizontal slit—giving the glum appearance of an inverted smile (Photo 6.1). These are pockets in which Christina Buesching has delved for the past 25 years, during which she has pioneered our attempts to discover what the aromas mean. We know what these scents are: mainly long-chained, unsaturated fatty acids, proteins, and water (Gorman et al. 1984; Buesching et al. 2002), but knowing the chemical composition of Chanel No. 5 tells you nothing of what the wearer experiences when it is deployed, what it feels like to smell it, or, for that matter, what messages the perfume might convey.

The social function of allomarking behaviour

Who marks whom? Does the answer allow us to deduce the functions of the odour, and thereby to

Table 6.2 Summary of badgers each year, and the numbers of nights they were filmed. Details of sett observations and number of (sub)adult badgers living at each sett (M = males, F = females) (adapted from Buesching et al. 2003, with permission).

Sett	No. of cameras	No. of film nights	Badgers 1994	Badgers 1995	Badgers 1996
Sett 1: Pasticks	4	516	7 M, 9 F	8 M, 13 F	8 M, 14 F
Sett 2: Sunday's Hill	2	495	3 M, 4 F	7 M, 4 F	12 M, 6 F
Total	6	1011	10 M, 13 F	15 M, 17 F	20 M, 20 F

diagnose, any further than did the evidence adduced in Chapter 5, the social structure within badger groups? We turned again to our mountainous video library accumulated at Pasticks and Sunday's Hill setts between November 1994 and April 1996, and scrutinized the videos for the three types of subcaudal marking described by Kruuk et al. (1984): (i) substrate or 'object' marking; (ii) social or sequential allomarking, in which one badger marks the body of another individual; and (iii) mutual allomarking during which two badgers press their backsides, and hence the openings of their subcaudal pouches, together (the aforementioned bum kissing).

Table 6.2 summarizes actors on this stage each year; also the numbers of nights on which we documented their theatre. Details of sett observations and number of (sub)adult badgers living at each sett (M= males, F= females). These are the same badgers whose dynasty we detailed in Box 5.1, so the two females receiving the allomarking attention were F217 and F221, and the top male was Lecher B, hero of Chapter 5.

Over the 18-month span of the study period,[5] led by Christina Buesching we documented 3021 incidences of allomarking behaviour amongst 40 adult badgers, with some individuals participating across years (cubs were excluded) (Buesching et al. 2003). Ninety-five per cent involved sequential allomarking (one badger marks another; 2866 instances), but on 5% (155) occasions we saw bum kissing (mutual allomarking). Considering, first, sequential allomarking, there was a seasonal peak during the late winter/early spring season of conceptions (Figure 6.1). What was the social flux of these allomarks?

---

[5] Owing to technical problems, no data were collected in August and September.

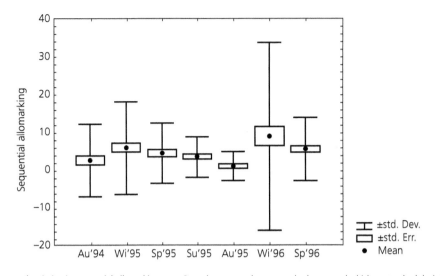

Figure 6.1 Seasonal variation in sequential allomarking rates. Dots show mean, boxes standard error, and whiskers standard deviation. Au = Autumn, Wi = Winter, Sp = Spring, and Su = Summer from Buesching et al. 2003, with permission).

Overall, males allomarked sequentially more than did females (they also bum kissed more); specifically they marked females seven times more often than they marked fellow males (for relative marking dynamics, see Figure 6.3a and b). In contrast, females showed no preference in the sex of individuals they allomarked, distributing their smears in proportion to prevailing adult sex ratio in the group. Adult females did, however, allomark other adult females four times more frequently than they marked subadult females; yearling females marked adult females five times as often as did subadult females. Similar to the flux of allomarks between females, male subadults marked mature males twice as often (three times for yearlings) as adult males marked one another. In comparison, in the flux of bum kissing, we could detect no preferences: badgers engaged in this occasional behaviour with any and all partners in proportion to their level of involvement with that group member. Overall, inter-individual variability in allomarking frequencies was much higher in winter (mating season) and spring (cub-rearing season) than it was in summer and autumn (Figure 6.2).

Are there perhaps finer-grained differences in allomarking fluxes associated with reproductive status? Yes. Non-reproducing females tended predominantly to mark reproducing females and males (Figure 6.4). Similarly, in spring, non-lactating females sequentially allomarked other females significantly more often than did lactating individuals. These patterns, with respect to reproductive status in the flux of marking by

females, were not matched by any discernible pattern amongst males. As a crude measure of male reproductive status, we asked whether the (palpated) degree of testes descent had any bearing on the flow of male sequential allomarking patterns? The answer turned out to be no. But later we were to discover how crude a measure this was of circulating testosterone (see Chapter 5).

So, are some individuals favoured or disfavoured in the deployment of allomarks? Yes. Within the overall categorical patterns of flux, sequential allomarking revealed social cliques within badger groups. Notwithstanding the asymmetries of sex, age, and reproductive status, some clusters of badgers allomarked amongst themselves more than expected. Furthermore, those cliques of badgers that allomarked more sequentially also allomarked more mutually (albeit still at about a twentieth the rate).

Make no mistake: these results, first presented in Buesching et al. (2003), are stunning. In the quest for seemingly elusive structure within badger groups, the strawman of the anarchic null hypothesis has been dealt a death blow. Whatever it is that badgers are saying to each other, they are systematically saying it more to certain classes of individual and, over and above that, to certain individuals. While this is not random, it neither suggests a holistically integrated group, nor does it offer much hint of the adaptive significance of the grouping (see Chapter 8). But let us take stock. Acknowledging that different classes of individual may be deploying their scent signals for different

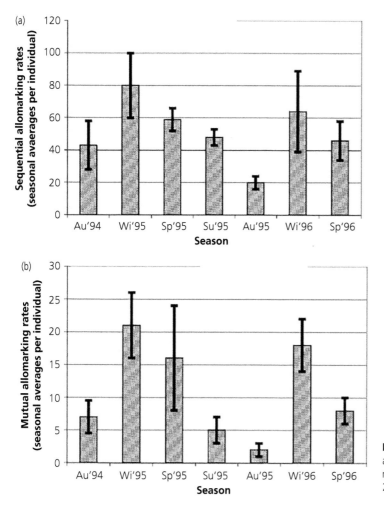

Figure 6.2 Seasonal variation in (a) sequential allomarking rates, and (b) mutual allomarking rates (per individual) (from Buesching et al. 2003, with permission).

reasons, it is noteworthy that males convey a message to (or at least post it on) female group members more than to other males; further, that this is not merely a mild favouritism: it's sevenfold. It may not be unexpected that males take a special interest in females, but it is a considerable insight that while females deploy their subcaudal marks equally amongst both sexes, of those they bestow on females four times the aliquot of odour (to judge by the frequencies of marking) is smeared on their adult contemporaries as on the youth. There is an emerging sense of sisterhood, perhaps matriarchal hierarchies (remembering how F221 and F217 were the source of all breeding success in the Pasticks group for many years, from Chapter 5), although the question of how blood ties weave through badger groups will await genealogical dissection in Chapter 17. Furthermore, amongst these adult females there is a predictable flow of allomarks: two closely

linked indicators of reproductive status, reproductive and lactating females, were both in heavy net receipt of subcaudal smears from non-breeding females.

These emerging social structures prompt the question, how linear is the link between channel width (marking rate) and the importance (indeed, the tenor) of the relationship? Does the fact that juvenile females allomark adult females at fivefold the rate they allomark their own generation indicate that this relationship is five-sevenths as important as the average relationship between an adult male and adult female and 20% more important to the subadult female than is the average relationship between adult females?

Indeed, how much should we make of the difference (50%) in the greater rate at which juvenile males mark yearling males as opposed to adult males (thrice versus twice the rate at which they mark their contemporaries) remembering that meanwhile juvenile females

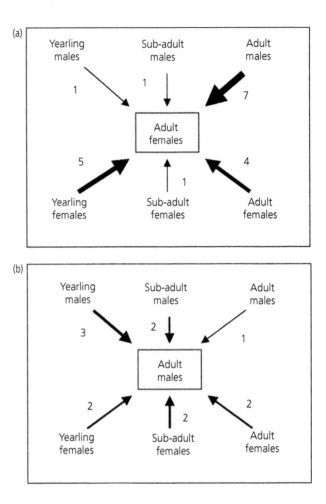

Figure 6.3 Sequential allomarking of (a) adult females by other sex/age classes (relative rates are given next to the arrows), and (b) adult males (from Buesching et al. 2003, with permission).

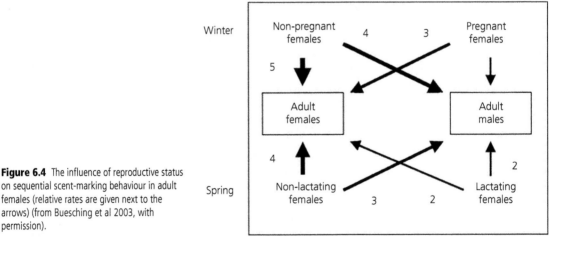

Figure 6.4 The influence of reproductive status on sequential scent-marking behaviour in adult females (relative rates are given next to the arrows) (from Buesching et al 2003, with permission).

are marking adult females at five times that rate? It might, for example, be particularly important for juvenile males to attend to, ingratiate themselves with, or otherwise integrate with the generation of males just one step above them.

Some answers become clearer in the text that follows, but a preliminary conclusion is: allomarking flows in predictable ways within badger groups, and insofar as females are net recipients from males, the mature from the immature, and the reproductive from the non-reproductive, the hypothesis emerges that the flow of marks is towards important individuals. Such a flow was strikingly reminiscent of the arrangements David described amongst farm cats, for which he coined the term centripetal society. Does that parallel shed light on our quest to understand the dynamics of badger groups? We will briefly tackle that question now, but keep firmly in mind that the flux in allomarks is a coarse measure—each mark may contain myriad messages (the topic of our chemical analyses in the text that follows), and each class of individual may be signalling with a different blend of purposes; furthermore, they may be discerning different emphases from each message (the topic of our field experiments that follow the chemical analyses). A further crucial result from Buesching et al. (2003) was the following: something that had been no more than a hint during our analyses of hierarchy in Chapter 5, now became—with more data and Christina's expertise with social odours—much clearer: that is, the flow of allomarking between individuals within each group correlated significantly with the flow of allogrooming. That is, badgers are selective in which companions they anoint and groom, and their selection is broadly the same in both cases, further emphasizing an emerging structure.

What of farm cats? In 1987 David began his study of cat sociology, aided by an Oxford undergraduate, Peter Apps, who, for the duration of the fieldwork, squeezed his lanky form and laconic wit into a caravan on a farm owned by Maurice Tibbles, an internationally renowned wildlife cameraman. Peter is doubly relevant to our story as he became a world authority on semiochemistry, most recently developing and testing innovative theories on the territorial urine marking of African wild dogs that will prove relevant to our analysis of badger neighbourly relations in Chapter 9. The early cat study led not only to a BBC TV 50-min documentary (*The Curious Cat*), but also to a brutal parody of David in *Private Eye* magazine and a succession of doctoral students whose work converged on the centripetal model of cat behaviour (Macdonald et al 1987,

2000b). In a nutshell, farm cats, living largely independent of people, form colonies ranging in size up to >50 members. At first glance, this jumble of cats seems structureless but, with persistence, three highly non-random attributes become apparent. Of these, the most striking is rubbing. Rubbing involves a cat rubbing its chin and lips (both well endowed with sebaceous scent glands), along with its saliva, onto the face of a companion. This does not happen randomly, either in terms of the individuals so anointed or the direction of flow of the anointing: younger cats allomark older ones, socially peripheral ones allomark socially central ones, and the flow is heavily asymmetrical, largely non-reciprocal (think back to the unreciprocated allogrooming in Box 5.3) and, in short, centripetal. That is, marks flow towards important members of farm cat society, and in the case of cats that means up matrilines. There are corollaries to social centrality amongst the sisterhood (in this case it is genetic) of farm cats, including greater reproductive success.

Of the other two non-random attributes of farm cat society, the determining one is spatial proximity—who sits with, or near to, whom. Broadly, propinquity reflects matrilineal relationships. Cats sit closer to their kin, and it is amongst those cliques that rubbing flows. So too does licking (i.e. grooming), but in a different way: it flows reciprocally but with different channel widths. A particular pair of cats may sit close, groom each other a lot, and in egalitarian fashion; the flow, however, of rubbing will be a one-way street.

Insofar as this description has elements in common with our unfolding account of subcaudal gland deployment within badger groups, is there more to learn from this parallel? First, why are farm cats living together in the first place? The answer is strongly ecological: they have access to a highly clumped food source. In Chapter 10 we will answer the question of how their food supply facilitates badger society. Second, cats carry with them the phylogenetic baggage of their ancestry. So, just as farm cats have peri-oral and cheek glands that they smear onto each other (as house cats do onto their owners), and just as young male farm cats bump heads and entwine tails, so too do young male lions. Recently we have explored the social networks of lion prides, documented the flow of rubbing, grooming, and propinquity amongst them (Mbizah et al. 2020), and, in the context of subgroup fission, linked the social system to the pattern of their food availability (Mbizah et al. 2019). In Chapter 19 we will answer the question of how the issues of intra- and interspecific variation in behavioural ecology frame the badger paradigm.

These questions of who allomarks whom, remind David of his habituated groups of meerkats (members of the mongoose family, *Herpestidae*) in the Kalahari desert. Meerkats have succulent anal glands opening into an anal pouch that when shut is reminiscent of the badger's subcaudal 'smile'. In the case of meerkats, allomarking flows down from dominants (the opposite of badgers) and dominants also do most marking of the environment. As David meandered around the desert with a foraging meerkat mob, in moments of collective excitement the dominants would smear their everted anal glands around his ankles. Did they consider him a harmless subordinate amongst them, or, insofar as they also clambered up to use his shoulders as a look-out point, did they treat him as a conveniently mobile, comfortingly ever-present, tree? Either way, he considered it a privilege to be daubed with this membership badge (but, eager not to cause offence, was careful to wash it off before visiting a neighbouring band).

Substrate

Exactly the same subcaudal gland secretion that is used to anoint companions with perfume is also used to smear onto the environment. So, having dissected the flow of allomarking, Christina and David revisited the same selection of video tapes to document the details of substrate marking with subcaudal secretion (Buesching and Macdonald 2004). During the 14 months of observation they documented 442 instances of object marking (189 at Pasticks sett and 253 at Sunday's Hill). How did the badgers' tendency to mark their companions align with that to mark the ground? A very close parallel: seasonal and sex-related differences in allomarking activity were mirrored by the seasonal differences in object-marking behaviour. Indeed, they corresponded closely to the seasonal and sex-specific differences in volume of subcaudal secretion each badger had stored in its pouch (having controlled for sex and season). For both sexes, subcaudal scent-marking activity varied significantly with season, with marking rates being highest during the conception season in winter (ten observed object marks per night in February, Figure 6.5) and lowest in late summer (one object mark per night observed). In mammals it is commonly the case that high rates of scent marking are associated with the breeding season (Brown and Macdonald 1985), and this is generally so for musteloid carnivores (Buesching and Stankowich 2017), as it now proves to be for subcaudal gland marking behaviour by Wytham's badgers. While there was no sex-related difference in scent-marking activity in winter, and autumn, during the cub-rearing season in spring females marked more than did males.

How did badgers deploy their substrate marks? Typically, a badger scent marking the ground, or an object, with its subcaudal gland would simply squat, briefly, mid-stride. This peremptory, fleeting behaviour is what we saw on 82.3% of occasions; generally the dipping of its rump was not associated with any change in the carriage of its head that might betray a visual cue to intensive sniffing. However, in 13.7% of instances the badger sniffed the ground before deciding where to mark.

We were confident in distinguishing the posture of badgers' subcaudal marking (rump hits the ground) and those urinating (crouching, but with rump suspended off the ground). Sometimes both happened: on 34 occasions badgers combined subcaudal marking with urinating on their setts, and on 29% of those occasions the scents were deposited at specially dug pits (23 instances; 9 by males, 14 by females). Four times, twice by males and twice by females, we saw a badger cock its leg against a tree. Of the remaining seven instances (two by males, five by females) when we saw badgers urinating they simply squatted with hind feet set apart while walking. During these observations we saw badgers defecate at the sett on five occasions, invariably in specially dug pits.

Only 13% of marks were smeared on objects that we judged to be visually conspicuous (8% on branches and roots on the ground, 3% on rock fragments, and 2% on leaves). The majority (87%) was deposited seemingly without reference to any landmark; that is, the deposit was made wherever the badger happened to be at the time—generally on flat, bare ground (although 3% were pressed against the base of vertical tree trunks). This general inattention to visually conspicuous marking sites is notable in its contrast to the habits of some other carnivores, notably wild canids, which position many of their scent marks on visually conspicuous objects (e.g. Macdonald 1979a; Sillero-Zubiri and Macdonald 1998).

Did the badgers mark the substrate at a particular, perhaps strategic, time or under a special set of circumstances? Our observations were heavily biased to the vicinity of the sett. There, 79.8% of substrate marks were made when a forager returned to the sett, with the remaining 19.9% being made by badgers emerging from underground. The badgers sharing a sett tend to synchronize their first emergence of the evening. Indeed, all badgers known to be resident in these setts emerged within a 30-min interval on any given night (between 16:30 and 19:00, depending on

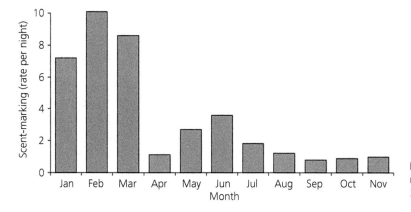

Figure 6.5 Seasonal variation in object marking (from Buesching and Macdonald 2004, with permission).

season) on 90% of nights. On average, of all the sub-caudal marks seen to be deposited on any given night, 56.3% were made within 1 h of the first badger's emergence. A further 21.9% of marks were deposited as badgers returned to their sett around dawn (between 04:30 and 06:00). The remaining subcaudal scent marks (21.8%) were made during nocturnal return visits to the sett (see also Noonan et al. 2015b), predominantly (60% of these 96 events) around 02:00 (these comings and goings, and early hour siestas, are documented in Chapter 5).

Of course, with our restricted field of view (c.7 m diameter) badgers often meandered out of view having scent marked—this occurred in two-thirds (68.7%) of instances. Whatever these badgers did next, they did not go down the tunnel closest to where they had marked, but of those 138 instances where they stayed in our field of view, on 118 occasions (85.5%) the badger that had marked subsequently entered the tunnel closest to the mark it had recently made, perhaps signifying some sense of tenure, or as a navigational aide memoire (we suspect reproductive females, for example, may stake a claim to a preferred natal chamber below). A minority (4.6%) also marked subcaudally a second time while still in view on the video, usually within a metre of the original mark (80 ± 57 cm, $n = 20$). We never saw a badger emerge from the sett, mark, and then retreat underground.

Males with fully descended testes marked more than males with ascended testes (although remember that testicular descent turns out to be only a crude reflection of circulating testosterone; Chapter 5). Oestrous females scent marked at higher rates than the average female subcaudal marking rate for winter (although vulva swelling also turns out to be a coarse barometer of hormonal state; Chapter 5). However, the rate of subcaudal marking varied neither with pregnancy,

nor lactation. For both sexes, subadults marked less than adults. Cubs were observed to mark subcaudally only four times during the study (twice in July, once in September, once in December).

How do badgers react to each other's subcaudal gland marking around the sett?

On the videos we could pinpoint the exact location of each subcaudal scent mark, and so we could tell if the same spot was marked again, by the same individual or a different one. Some 37.6% of subcaudal marks were over-marked by other individuals, at a rate that decreased exponentially with the age of the original mark (Figures 6.6 and 6.7) (Buesching and

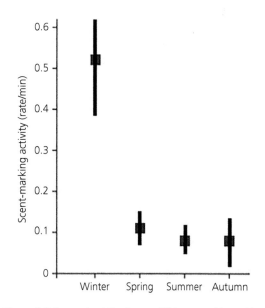

Figure 6.6 Seasonal variation (mean ± SE) in over-marking activity (from Buesching and Macdonald 2004, with permission).

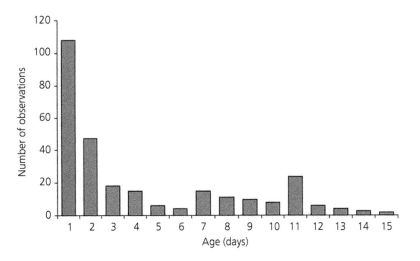

Figure 6.7 Decrease in over-marking activity with the age of the existing scent mark (from Buesching and Macdonald 2004, with permission).

Macdonald 2004). Who did the over-marking? Of over-marks, 20.4% were made by the original marker;[6] 79.7% by a different badger. On average, individuals reinforced their own scent marks after 1.5 ± 2.34 days; 57.6% of those that did so, added their over-mark on the same night as the original mark, while 78.8% did so within 24 h. Was the time course of over-marking by the original marker the same as that by different markers? No. On average, different over-markers acted several days later; male over-markers after 4.1 ± 6.3 days and female over-markers after 3.5 ± 5.09 days. In the case of different over-markers, 44.4% occurred during the same night, and 61.1% within the 24 h. Were there differences between the sexes and seasons in tendency to over-mark? Yes, both. Males over-marked only during winter and spring, with females over-marking at similar rates during all seasons. The ratio of over-marking to scent marking in new locations remained the same in both seasons. However, over-marking was more of a female prerogative than a male one: the proportion of scent marks placed at new sites rather than as over-marks was significantly higher in males than in females. Nonetheless, the two sexes responded similarly to marks made by either sex: the proportions of over-marks made by males and females did not differ depending on whether the original mark had been made by a male or a female.

In summary, badgers around setts in Wytham engage in object marking with their subcaudal glands at rates that vary seasonally and individually in parallel to their tendency to allomark each other (with one important exception): there are times for more marking, and times for less, the time for more being late winter and spring (remembering that births occur in early February and immediately after are followed by the next round of conceptions). The aforementioned exception to the parallels between allo- and substrate marking was the higher rate of substrate marking by females around the sett during the cub-rearing season. It also appears that, at least around the sett, the right place to mark is wherever the badger happens to be when the urge takes it, so substrate marks, a bit like radio-tracking fixes, reveal where the badger has been. In summary, the right moment to mark appears to be when the animal emerges for a night's activity or on returning home—the latter either temporarily for a nap, or at the end of the night. Finally, badgers responded to their own subcaudal marks by topping them up, although at about one-fifth of the rate at which other individuals over-marked them, and on a slightly more immediate time scale. Either way, it seems that both the original marker and, even more so, other individuals (and particularly females) have an interest in reinforcing a mark.

What do these patterns tell us about the motivations for, or intended recipients of, the signals? How might the answers vary between allomarking and substrate marking? Some answers may become clear as we ask whether the components of subcaudal secretion differ between social classes or even individuals; in reply to

---

[6] Of course, when we speak of the original mark, that refers to the first mark of which we are aware during the sequence of videos—there could have been historical marks, but this possibility recedes as the days of observation pass.

which question a first and superficial clue lies in its colour.

The colour of scent

So much for the flux of allomarking and the deployment of substrate marks. What of the content of the message? A first step was to look at the secretion, which, considering this was through the lens of a different modality—sight—proved remarkably informative. We use a round-ended stainless steel spatula gently to scoop secretion from the pouch of sedated badgers (Photo 6.3). Bringing precision to this rather earthy pastime, we attempt to remove the entire volume of secretion, which ranges from 0.0 to 3.5 ml (mean 2.3 ml ± 1.12) (Buesching et al. 2002). Within this range of volumes, are consistent differences discernible between the sexes and age classes? Our recent graduate student, Tanesha Allen, got to grips with this question almost 20 years after Christina first led us into these pouches. Tanesha used subcaudal gland samples and concomitant badger biometric data collected between January 2010 and November 2017 (n = 1840 adults) to calculate that the subcaudal pouches of males consistently yielded around 1.0 ml more secretion than those of females (that is much more than predicted by their 5% size dimorphism, and female secretion volume matches males' only when they are lactating). In winter, the volume of secretion we excavated from the pouch correlated positively with the body condition of both sexes, as it did in summer but only amongst males. This might suggest that producing the secretion is a

Photo 6.3 Collecting subcaudal gland secretion using a rounded stainless steel spatula.

cost most readily borne by the chubbiest badgers, or that the plumpest badgers had most reason to do so. For now, we note that older adults (>5 years) yielded less secretion ($c.1.9$ ml) than did younger adults ($c.2.3$ ml) (one of many declines in performance upon which we elaborate in Chapter 18).

Having estimated the volume of each scoop, we immediately froze the secretion at –70°C in glass vials (plastic vials emit volatiles—i.e. they smell—which can contaminate subsequent chemical analyses). Cubs produce only minute traces of secretion during their first 4 months (Chapter 2), which we collect using a small wad of sterile (surgical quality) cotton wool held with a pair of stainless steel tweezers, wiped carefully through the subcaudal pouch, then they too were deep frozen.

We have examined secretions from almost every capture of every badger (n = 11,687), and (in adults and cubs mature enough to have it; see Chapter 2) it varies conspicuously from white, through cream, beige, to dark brown: differences obvious to the human eye that likely belie at least some differences to the badger's nose. We matched each sample to the Borger colour chart (designed to code the naturalistic colours used in fishing fly tying), which we then categorized on a 6-point scale from light to dark. Through the fuzzy lens of secretion colour we could, at least hazily, see some elements of the smell. From an early subsample of 975 aliquots of subcaudal secretion, collected under Christina Buesching's watchful eye between May 1996 and November 1998, we answered the question of whether the colour of secretion varied with the donor's sex—it did (Buesching et al. 2002).

Male badgers produce paler secretions (tending to beige/white), whereas the subcaudal secretions of females are darker (brown to chocolate brown, Borger Number 64; one can scarcely miss the coincidence with Chanel Number 64). There is seasonal variation in the shade of these secretions, which, for both sexes, are palest in spring (Figure 6.9). Males produced the greatest volume during the winter conception season, whereas females produced their greatest volume in the spring, only after pregnancy and when they are nursing and, one might speculate, have cubs to protect (Figure 6.8).

Was the colour of a badger's subcaudal gland secretion linked to its physical attributes? Yes: not only did the biggest females (greater head–body length) have, in comparison with other females, the greatest volume of secretion in their subcaudal glands in spring and summer, but their secretion was significantly the palest.

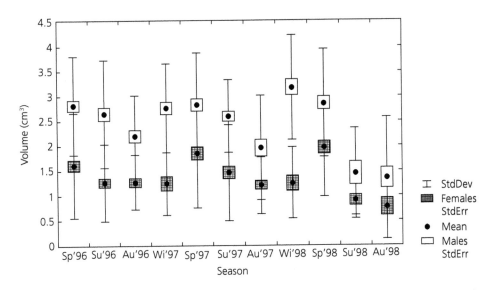

Figure 6.8 Sex-related differences in the volume of subcaudal gland secretions of badgers at different times of the year (from Buesching et al. 2002c, with permission).

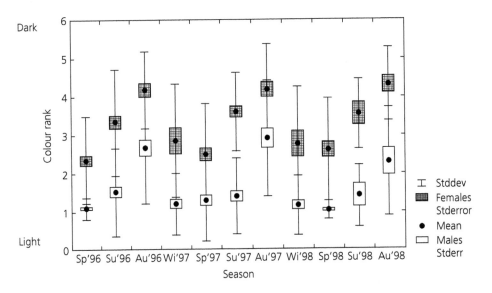

Figure 6.9 Sex-related differences in colour of subcaudal gland secretions throughout the year (from Buesching et al. 2002c, with permission).

Amongst males, the result was similar, but seasonally inverted: the longest males also produced the palest (and most voluminous) secretion, but in their case this was significantly so in all seasons except spring. It's a tentative first guess, but perhaps the spring cub-rearing season was the most socially controversial, and thus demanding of communication, for females, whereas, by contrast, it was the least so for males, after the post-partum peak in mating

activity had subsided? But what did the paler shade of secretion signify—are there distinct male and female attributes to secretion, or just differing proportions of similar constituents? This was a question awaiting chemical analysis (see the text that follows).

A final question regarding secretion colour was whether or not it aligned with any aspect of reproductive status. Amongst males, our early analyses revealed that, in January, those with descended testes produced significantly more copious secretions than did males with ascended testes, and, as with the bigger males, it was paler (Buesching and Macdonald 2003, 2004). However, this became harder to interpret when we discovered that neither secretion colour nor volume related to testosterone titres (see also Buesching et al. 2009). Neither pregnancy in January, nor lactation in spring had any effect on either secretion volume or colour: circulating progesterone titres were, however, significantly negatively correlated with secretion volume, but had no bearing on secretion colour (Buesching et al. 2002a). More of that later.

Although the hues of subcaudal gland secretion constituted an admittedly cloudy lens through which to explore olfaction, it would be surprising if such hues did not at least partly reflect the siren signals within. So, we can add to our store of clues that not only did males, females, old, young, portly and thin, fecund or not, differ in their deployment of scents; they differed also in the colour of the scents they deployed and thus, we might suspect, in some of the messages conveyed. And what of those paler secretions characteristic of both the largest males and the largest females? A whiter shade of pale was memorably inscrutable to the listener to the Miller's Tale in Procol Harem's enduring 1960s lyrics: 'And so it was that later, As the miller told his tale, That her face, at first just ghostly, Turned a whiter shade of pale.' Struggling to interpret the message, the lyrics continue: 'And although my eyes were open, They might have just as well've been closed.' Ours too! So, to find an answer, we turn to semiochemistry.

Semiochemistry

We could, therefore, see marked variation in the colour of subcaudal gland secretion, but frankly it all smelt pretty similar to us—a bit like stale, caked washing powder. The reverse was likely the badgers' experience: what matters to them, above all, is the smell. What chemical signals were communicated by these secretions? This question had prompted Martyn Gorman and his colleagues in Aberdeen, including Hans Kruuk, who had by then moved to the Scottish town from Oxford, to do an early gas chromatographic analysis of badger subcaudal gland secretion collected between 1976 and 1978 (Gorman et al. 1984; Kruuk et al. 1984). They found evidence that the mainly long-chained, unsaturated fatty acids, proteins, and water comprised subcaudal gland secretion—explaining its 'mayonnaise-like' consistency. Peaks in types of these components encoded both individual identity and group membership. However, they found no evidence

that the scent coded any other characteristics of the marking individual (e.g. sex, age). Gorman et al.'s preliminary conclusion was that the primary function of subcaudal scent marking was tied to territory 'ownership' and group defence (in Chapters 9 and 10 we revisit the concepts of territoriality and group defence insofar as they apply nowadays to the badgers of Wytham Woods).

With the passage of time, and the invention of increasingly sophisticated technology, we've revisited the question of what information is encoded in subcaudal secretion and, first by deduction (later, by experiment), decoded its communicative function in badger society. The new technology, a combination of gas chromatography with mass spectrometry, was at the heart of Christina Buesching's doctoral thesis (Buesching et al. 2002a, 2002b) and later, in his thesis, Mike Noonan worked with Christina to introduce new statistical techniques to detect nuanced patterns (Noonan et al. 2018b). We found that 66 badger subcaudal gland samples contained 110 different components, of which 21 were present in every profile. Although the secretion's composition varied over seasons and between sexes, no single peak diagnosed the producer's sex, and there was no systematic variation in components according to season. The composition of the secretion was, however, influenced by the badger's age, body condition, and reproductive status.

Importantly, subcaudal profiles were highly individual-specific, consisting of unique combinations of 23–58 different peaks (average over all samples 33.4 ± 8.05), and the relative percentage area of all peaks (i.e. the amount of each fragrance present) varied substantially between individuals. For those less familiar with chemical analysis the statement that concentrations of up to 58 compounds can characterize a badger's individually recognizable profile might make it sound as if one badger, in identifying another, is accomplishing a remarkable computation. There is a neurological sense in which that is correct, but we imagine that what the badger experiences is not a mind-exploding computation but rather an holistic chemical image, in much the same way that humans perceive the features of a person's face (features that could doubtless be measured by a morphometrist according to 58 dimensions). Later, in Chapter 10, we will estimate that an average badger might know personally more than 50 individuals, so it's interesting to reflect on what proportion of that number it is reasonable to expect them to remember. Although the cognitive capabilities of mustelids and canids may differ (that would be an interesting topic for study, see Chapter 19), it may be relevant that a border collie recently demonstrated that it could associate names, spoken in English, with more than 100 different items (Pilley and Reid 2010).

A particular feature of scent is its potential persistence, or 'latency' (Box 6.1). A posture or facial expression is gone in the blink of an eye, whereas not only may a scent endure for hours (like the smell of fresh bread in a kitchen, giving away what was baked), or even weeks, it may transform with time, adding a new dimension to the message. Thus, in contrast to instantaneous visual and acoustic signals, olfactory signals persist in the environment and are often used for delayed communication, in which the receiver usually encounters scent marks only after the sender (i.e. the marker) has left the area.

The stability of the signal is thus important—there are circumstances when it might be advantageous for the signal to be durable and unchanging, but others where it might be informative (whether to the advantage or disadvantage of the signaller) to signal the passage of time through changing nuances in the message; perhaps the signal is fresh and the signaller is around to meet, or if it's old, the greater the likelihood that the signaller might have left (even died). Does the make-up of subcaudal scent change as time passes? By comparing subsamples allowed to age naturally with that sample's original composition, we discovered that the answer was yes (Buesching et al. 2002b). Nevertheless, it looks like those changes affect the nuance, not the substance, of the message. We collected samples from the same individual 3 days apart, and found that they varied little at the outset, while over 40% of all shared peaks showed less intra- than inter-individual variability over the course of 1 year. Badgers thus appear to have a unique individual-specific subcaudal scent signature, and once written, the autograph remains unique, even as the ink fades. By analogy, you might consider that your friend, or even your enemy, looks pretty much the same from year to year, even if his hair greys or beard lengthens or suntan fades.

Our equipment is sensitive, so the question arises as to whether it is better, more nuanced in its perceptions, than the badger's equipment? Probably not. Badgers, like other carnivores (Kitchener et al. 2017) inhale air through their noses, warming and moistening it in transit across the maxillo turbinate bones, en route to the olfactory receptors carpeting the naso- and ethmo turbinate bones (Van Valkenburgh et al. 2011; Green et al. 2012). Of course, we do that too, but judging from relative anatomy and nasal membrane sizes

badgers' sense of smell is likely (although this is difficult to imagine) 700–800 times better than ours (in comparison, a wolf can smell a deer, wind permitting, from a couple of miles away).

What, in summary, can we deduce that badgers can likely read from one another's subcaudal gland secretion? Well, as a start, the signals are there, if the badgers can register them, to discern its sex, age, body condition, reproductive status, individual identity, and the time since the message was left. All of these, however, can also be gleaned from urine and anal glands, but next we introduce one additional message encoded in these secretions and for which there may be no redundancy: group identity.

This message has a special context in the evolutionary history of the subcaudal gland's competencies. All six of the aforementioned dimensions potentially signalled by contemporary subcaudal secretions would, it seems likely, have been advantageous to an ancestral badger such as *Meles thorali*, living at low population density. Such an ancestor, like some populations of contemporary badgers, could have posted individually signed memos to friend or foe on their olfactory notice board. Nowadays, adapted to agricultural landscapes, badger populations can reach higher density and form large groups, as at Wytham (see Chapter 10); elsewhere, also, in lowland Britain, Ireland, and beyond (a wider context that is the topic of Chapter 19). Under these ecologically modern circumstances might subcaudal gland secretions convey additional information; that is, would they constitute a badge of group membership? And so we investigate next what turns out to be a story of bacterial body odour.

The club tie

Understanding of the badger's membership badge—reminiscent of a club tie or team shirt—has its roots in an inspirational partnership that David was fortunate to join in the mid-1970s with pioneering chemist, Eric Albone (who did much of his science while a school teacher in Bristol; e.g. Albone 1984), and a bacteriologist called Georges Ware. Their collaboration began in the acrid confines of red fox anal sacs (e.g. Albone et al. 1978), but it is in the much more capacious circumstances of badger subcaudal glands that the sociological implications bloom. To understand the power of the idea it is necessary to appreciate how subcaudal scent is generated. The apocrine sweat glands lining the walls of the pouch secrete primary products and they act as a medium on which bacteria grow, producing secondary products: it is these that characterize the perceived odour (akin to the action of bacteria on sweaty sportswear). The products of different communities of bacterial species smell recognizably different (greater bacterial populations of *Corynebacterium jeikeium*, for example, may be found, and detected by scent, more in the armpits of men than of women; Bratt and Dayan 2011). Badgers that share a microbial community should develop some degree of a shared smell.

Revisiting a sample of 46 profiles drawn from badgers simultaneously resident at 7 different groups, we analysed the relative area (% area) that every peak contributed to the overall profile area (Buesching et al. 2002a). Martyn Gorman and his colleagues had been right in their insight that groups had distinct odorous membership badges—our more sophisticated apparatus confirmed that group members have more similar profiles than do badgers from different groups, with profiles clustering by their Euclidean distance (similarity) to one another (Figure 6.10).

So, the importance of bum kissing is, as Kruuk anticipated, clear: mutual allomarking facilitates the exchange of pouch bacteria between group members at least in proportion to propinquity—a fact that mirrors group membership, generating distinct group-specific scent components (Sin et al. 2012a). It is easy to imagine that even badgers allomarking on the rumps of companions may cross-infect their bacteria when several of them mark the same rump, and it's a small behavioural step, but very much more efficient, to insert the microbes at source beneath the tail. The obvious question is how often would it to be necessary to swap bacteria to ensure that the membership badge remained up to date? Perhaps there is a trade-off between the uniformity of odour necessary to bestow a group scent, and the maintenance of some individual odour to the pouch secretions (Buesching et al. 2003, 2016b; Buesching and Macdonald 2004) (Figure 6.11). One might even speculate as to whether, if the group structure mirrors feline centripetal hierarchy, badgers closer to the hub signal their social centrality by greater conformity with the group mean.

Ecological communities change, and doubtless this includes those living in the badger's subcaudal pouch. If the function of bum kissing is to ensure that the symphony of subcaudal scents remains in harmony throughout the group, the question arises as to how quickly bacterial communities drift into individualistic discord? A clue may lie in anecdotes from badger rescue centres which report that after a few weeks in captivity, and thus enforced abstinence from bum kissing, released badgers are not so readily accepted

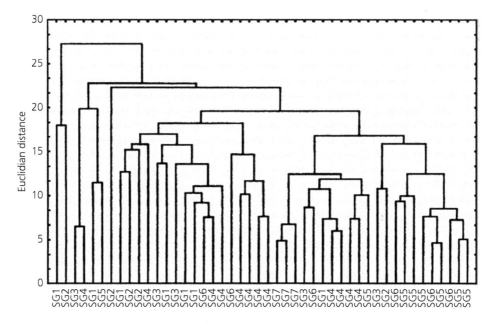

Figure 6.10 Dendrogram of cluster analysis for 46 badgers from 7 different social groups (SG = social group). The horizontal axis shows a tendency for samples from groups (SG1–7) to cluster along its length in relation to scent profile similarity (from Buesching et al. 2002a, with permission).

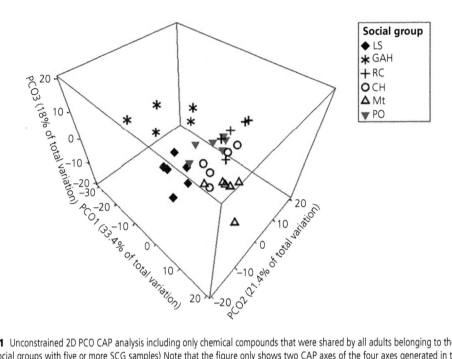

Figure 6.11 Unconstrained 2D PCO CAP analysis including only chemical compounds that were shared by all adults belonging to the same social group (six social groups with five or more SCG samples) Note that the figure only shows two CAP axes of the four axes generated in the model (from Buesching et al. 2016b, with permission).

back into their original social group (Buesching et al. 2016b).

In wider context, Buesching et al. (2016b) note that shared group odours have been implicated in social acceptance (e.g. Hurst et al. 1993) and group cohesion (e.g. peccaries; Buyers 1985). Olfactory advertisement of group membership must be reliable yet flexible. Eric Albone's proposition that the generation of shared group odours relies predominantly on the assimilation of microbial communities amongst group members has also been demonstrated for the anal gland secretions of spotted hyaenas (Theis et al. 2012, 2013).

Which bacteria?

Clearly bacteria in the subcaudal pouch play a crucial role as parfumiers, but which bacteria are involved? Building on the original insight of Albone et al. (1977), and mindful that microorganisms are, at least partially, responsible for odours spanning the entirety of the human anatomy, from armpits to breath (Shelley et al. 1953; Tonzetich et al. 1967, 1971; Fredrich et al. 2013), Christina led us in a deeper investigation of the 'fermentation hypothesis' (Buesching et al. 2016b). Our former graduate student Yung Wa (a.k.a. Simon) Sin (now, following a stint at Harvard, running

his own lab in the University of Hong Kong) had introduced us to the dauntingly polysyllabic technique of terminal restriction fragment length polymorphism analysis (TRF) of the PCR-amplified 16S rRNA gene (Sin et al. 2012a). This revealed that the subcaudal pouches of Wytham's badgers were home to 56 bacterial species (when bacteriologists use the term species they mean 'operational taxonomic units') belonging to four phyla (Actinobacteria, Firmicutes, Proteobacteria, and Bacteroidetes). It is a sobering reminder to mammalogists accustomed to considering badgers and bower birds as very different that they, along with fish, frogs, and amphioxus are all members of just one phylum, Chordata. So the four types of denizen of the badger's backside might all look rather similar to the undiscerning human eye, but they are as evolutionarily distinct as we are from starfish. A total of 76% of bacterial species resident in the subcaudal glands of Wytham's badgers belonged to the phylum Actinobacteria. Not only did the bacterial communities differ between social groups of badgers, they also change with the badger's age: cubs have considerably more diverse microbial communities than do adults, dominated by Firmicutes, and less so by Actinobacteria. Remembering that cubs 'steal' scent from adults (Chapter 3; Fell et al. 2006), it seems that acquisition of a 'mature bacterial community' is an ontogenetic process, requiring social integration and pouch exchange with adults.

More on signatures

Armed with Simon's TRF data, Christina aligned them with her gas chromatography–mass spectrometry for subcaudal secretions from 66 adults belonging to 6 different social groups (Buesching et al. 2016b). The question was, are variations in the bacterial species' composition of each badger's glandular ecosystem reflected in the chemical mixture of each individual's secretion? Simon had discovered 50 TRFs, and Christina had documented 125 different chemical compounds, but amongst all that diversity it turned out that the relative abundance of a subset of four TRFs best explained the structure of the chemical matrix. A simple interpretation of the fermentation hypothesis might have predicted that each badger group was as well defined by its bacterial community as by the semiochemicals they synthesized.

Simplicity fails: although semiochemical profiles were group-specific, microbial profiles were not. Another factor must, therefore, intervene between the microbes and the odours. One possibility could be that the substrate, on which the bacteria work, the primary secretions of the gland, is different and group-specific. That could be plausible if group members were more genetically similar to each other than to their neighbours and the chemistry of the primary secretions inheritable—this possibility can be evaluated in the context of our genetic findings in Chapter 17. An alternative, but not exclusive, possibility is that the commonality that drives the group effect lies in the wider circumstances of incubation, also perhaps the particular conditions of the particular sett (or perhaps even locally specific diets)—this possibility can be evaluated in the context of our findings on spatial organization within and between groups in Chapter 10. Anyway, a person surprised at the subtle bouquets distinguishing the bacterial products of badger subcaudal glands might, by the same token, question a sommelier on the variety of wines that arise from the actions of yeast on different kinds of grapes.

The power to discern individual chemicals in scent secretions (and then to identify the bacteria that produce them) seems almost magical, but our interpretations are complicated by several factors. First, the biological relevance of each chemical component of the subcaudal gland secretions is unknown. Our analyses concentrated on 21 peaks detectable to a greater or lesser extent in every profile, but other, possibly much smaller, peaks might be far more important to badgers. Second, in complex chemical mixtures the ratio between a small number of components might be more important than the relative abundance of each individual compound (White and Chambers 1989).

Heavy molecules

One problem with the analytical methods underpinning most of our semiochemical analyses is that they detect only volatile components soluble in hexane. This is only part, although likely an important part, of the odorous picture: it doesn't detect (is anosmic to) heavy, non-volatile molecules. Could badgers be reading these heavyweight messages? Yes, because they have a functional vomero-nasal organ (Box 6.1), which is what mammals use to taste very long-chained, non-volatile molecules—to judge by the quivering snout and clacketing jaw of a male dog as it licks the urine of a receptive bitch, these molecules may be thrillingly informative (Lüps and Wandeler 1993).

Costs of social media

To judge by the amount of it they do (perhaps 200 times each night) badgers really value subcaudal gland marking. Depositing the scent is not arduous, but is it expensive to make? Certainly it's not free and has to be specially customized, in contrast to faeces and urine (which we consider in detail in Chapter 9).

How quickly can it be replenished? Recapturing badgers whose subcaudal gland we had recently emptied revealed that between 80 and 120% of the amount removed was replenished within 2 days. How much does this cost the badger? The answer is relevant to Gosling and Roberts' (2001a) proposal that scent marking generally provides a cheat-proof signal to competitors and potential mates. They had in mind specifically male mammals showing off their capacity to synthesize an inexhaustible supply of perfumes (analogous to the famously energetic bellowing of red deer stags or the roaring of lions) to signal their 'resource holding power' (RHP). This RHP signal can inform adversaries on the risks of escalation in relation in the assessment of potential costs and benefits to picking a fight. The decision to risk a challenge might be informed via three criteria, in which receivers may: (i) detect intrinsic properties of scent marks (e.g. concentrations of volatile components honestly signalling quality); (ii) associate past encounter outcomes with the individual producing that scent; and (iii) remember the smell of marks encountered recently and match this smell with potential opponents whom they meet subsequently[7] (Gosling and Roberts 2001a) (we return to RHP in Chapter 9).

Against these criteria, is the badger's subcaudal gland secretion a convincing candidate as an honest signal? Which brings us back to our opening question, how expensive is it to produce? Remember, males use up their secretion faster than do females; they produce more of it and mark at higher frequency than do females (Buesching and Macdonald 2004; Buesching et al. 2003; see also similar findings by Silwa 1996 on aardwolves' (*Proteles cristatus*) anal gland marking[8]). Presumably then they must replenish it more quickly through biosynthetic processes?

Tanesha Allen, led the enquiry into how costly it is to produce, and deploy, subcaudal gland secretion. Does the decision to produce it face each badger with an awkward 'trade-off' of investing in secretion versus other essentials? Related to this is the possibility that subcaudal gland secretion signals a 'handicap', demonstrating that an individual is fit and strong enough to thrive despite investing in expensive perfume, much as a peacock flaunts its wildly exorbitant tail (Zahavi 1975; Gadagkar 2003). The 'quality' of glandular scent marks is thus predicted: (i) to correlate directly with the 'quality' (i.e. prowess) of the marker—a concept known as 'index' (Smith and Harper 1995); and (ii) to allow conspecifics to infer information about the signaller's fitness. These compelling arguments have resulted in the widespread presumption that scent marks function as honest (i.e. 'unfakable') fitness advertisement signals (Gosling and Roberts 2001a; Coombes et al. 2018). It was against this background that Tanesha used bomb calorimetry (mostly used in dietary analyses; Baer et al. 2016) to determine the energy content of subcaudal gland secretion (in MJ kg^{-1}; and for those schooled in Imperial measures it takes 4.184 kJ to equal a kilocalorie, generally, and confusingly, referred to as 'a calorie'). Using a 6100 Parr bomb calorimeter,[9] a standardized amount of the sample's fresh/wet weight was freeze-dried (measuring its original water content) and placed into a 'bomb'—a thick, sealed, stainless steel container. This bomb creates a high-pressure oxygen

environment within a larger, insulated container filled with water. An electrical charge travelling through an ignition wire then initiates the combustion process. Chemical bonds within the sample are broken through oxidation, releasing energy that raises the surrounding water temperature. This 'heat of combustion' or 'calorific value'—is then compared to the heat of combustion from the same amount (dry weight) of benzoic acid or another material with known calorific value (Jessup 1970).

Working with samples from 51 males[10] and 31 females[11] collected between January 2010 and November 2017 from known individual badgers, the volume of subcaudal gland secretion in this subsample ranged from 0.2 to 5 ml (median = 1.20 ± 0.95 ml) for males and from 0.1 to 3.5 ml (median = 0.60 ± 0.76 ml) for females. What was the calorific value of secretion? It ranged from 3.7 to 25.9 kJ ml^{-1} in males, or 1–8.38 calories (median 6.5 ± 2.8 kJ), and around 5.0–27.4 kJ ml^{-1}, or 1.2–6.6 calories in females (median 6.8 ± 5.7 kJ ml^{-1}).[12]

Is that a lot? For comparison pure fat contains around 9 calories per gram, where 1 g = 1.04 ml, and mayonnaise (which, unappetisingly, looks and feels much like subcaudal secretion) contains *c*.6.8 calories per gram. Were badgers of different sexes, age classes, or condition producing particularly expensive secretion? Generally not, except that females in spring and summer produced significantly less rich secretion, which we might think results from dilution to allow the marks to be spread further (Figure 6.12).

So, how much energy are Wytham's badgers burning in their enthusiasm for subcaudal gland marking? Multiplying calorific value by the volume we extracted from the subcaudal pouch of each individual revealed that on average male pouches contained between 2.2 and 32 kJ of secretion, or around 0.5 to 7.7 calories (median 7.0 ± 7.0 kJ), whereas for females this ranged from 0.8 to 25 kJ, or 0.2 to 5.9 calories (median 4.0 ± 5.8 kJ). Statistically, this resulted in total pouch secretion energy content being significantly higher in males than females, although with wide inter-individual variation. Did this vary with season? Yes, in males, but not females, pouch energy content varied with season, being highest in spring (Figure 6.13), although neither body condition nor reproductive life stage affected pouch energy content in either sex.

Just how much of a peacock's tail is a badger fluttering each time it dips down to smear subcaudal

---

[7] Game theoretical analysis shows how territorial intruders may switch from using intrinsic properties of marks to scent matching when making decision about whether to remain in a territory (Gosling and Roberts 2001a).

[8] Schulte et al. (1995) report comparable sex-specific colour differences in the anal gland secretions of North American beavers (*Castor canadensis*), which provides the basis for sexing beavers, due to their genitalia being concealed.

[9] Parr Instrument Company, Illinois, USA.

[10] Spring = 21, summer = 8, autumn = 17, and winter = 5.

[11] Spring = 13, summer = 5, autumn = 15; no winter female samples were substantial enough for analysis.

[12] Extrapolating from dry and water content.

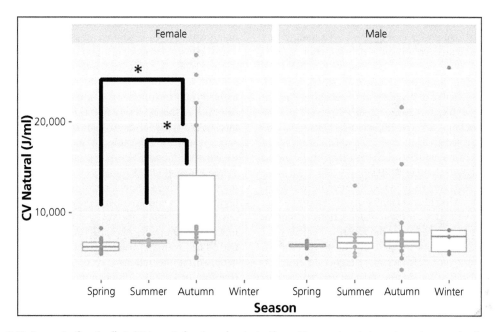

Figure 6.12 Season significantly affected CV$_{natural}$ in females as females had lower CV$_{natural}$ values during spring and summer than in autumn. Male CV$_{natural}$ values were not significantly affected by season (from Allen et al. in press, with permission. Statistically significant comparisons indicated by *).

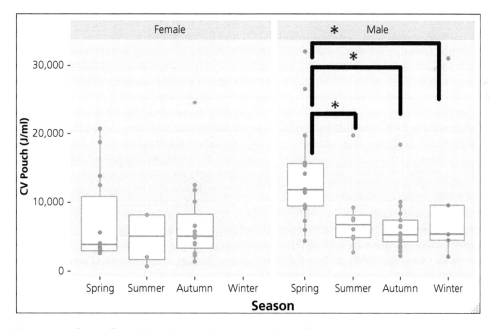

Figure 6.13 Season significantly affected CV$_{pouch}$ in males with males having higher CV$_{pouch}$ values in spring than in other seasons. Female CV$_{pouch}$ values were not significantly affected by season (from Allen et al, in press, with permission).

secretion on a companion or the ground? Drilling down to answer the question, what is the unit cost of each scent mark, Tanesha first multiplied the energy content of secretion (kJ ml^{-1}) by the approximate spreading surface of the pouch (opening (smile)) size (median 56 mm males, 55 mm females) × width (c.3 mm) × smear thickness (0.1 mm) to estimate that the energy content of both male and female scent marks was in the broad range of 0.06 to 0.4 kJ. Of course, this is a rough estimate, and, equally, there is no guarantee that smaller smears are deployed by badgers with lower total pouch energy. Perhaps a badger that has manufactured calorifically rich secretion spends it thriftily, or generously, or vice versa. Nonetheless, there is order of magnitude consistency between Tanesha's estimate of a pouchful of secretory content (males, enough secretion to deposit between 11 and 282 marks (median 71.65 ± 51.19); females, between 7 and 292 marks (median 34.01 ± 56.15) and Christina's field observations that badgers each make about 200 marks each night; Buesching et al. 2003).

In total we recorded 197 videos featuring 162 identifiable responders (13 adult males, 15 adult females, and 1 female yearling). Could the badgers detect chemical differences associated with the calorific content (and sex) of the secretions? Apparently. Of secretion from females (although not from males), higher-energy samples were sniffed for significantly longer than lower-energy secretion (although we don't know if this time was devoted out of greater interest in the richer secretion, or owing to difficulty in spotting the difference in two closely matched secretions). Neither the sex of the donor, nor the energy content of the secretion, had any relationship to the tendency to the time spent sniffing or the 11 instances of over-marking and 49 proximity markings.

With so much information packed into subcaudal gland secretion, one might ask what else would the average badger like to communicate? The answer may lie in a sweaty T-shirt.

From T-shirts to why subcaudal glands evolved

What additional benefit might accrue to badgers to communicate through their subcaudal gland secretions (a question that could be asked of any source of social odours)? Two candidate answers are: suitability as a mate on genetic or health grounds. A linked question, is why did subcaudal glands evolve in the first place?

The answers may lie in a sixth century BCE Sanskrit medical text by Sushruta Sambita, who observed 'By the sense of smell we can recognise the peculiar perspiration of many diseases, which has an important bearing on their identification'. This clinician's insight exploded into evolutionary theory when in 1982, Hamilton and Zuk proposed their landmark hypothesis that animals should inspect a potential mate's urine and faecal odour to select for parasite-free/resistant status. Freedom from disease is one criterion on which to select a mate, genetic compatibility

is another, and while urine and faecal odours may be diagnostic, so too, as Sambita observed, is sweat.

Is there serious evidence that odours, and we have a particular reason to be interested in sweat, can communicate details of health and/or genetics that assist in successful choice of a mate? Yes, plenty. For example, a Russian study found that the smell of axillary (armpit) sweat from men infected with gonorrhoea was distinguishable by, and aversively 'putrid' to, women sniffers (Moshkin et al. 2012). In the same vein, Olsson et al. (2014) triggered an innate immune response in men, mimicking ill-health (this involved an injection of endotoxin, lipopolysaccharide), and 4 h later their sweat became aversive to experimental sniffers. Beyond humans, mandrills (*Mandrillus sphinx*—an Old World primate) use olfaction (smelling faeces) to avoid social interactions with parasitized conspecifics (Poirotte et al. 2017). To hammer home the point—although this time, we're straying beyond sweat—lab mice can discern all this immune-genetic major histocompatibility complex (MHC) information (and more) from urine (Penn and Potts, 1998a, 1998b). Female mice can detect, and avoid, from the smell of their urine, males infected with mouse lice, *Polyplax serrata* (Kavaliers et al. (2003), and male mice avoid females infected with the parasite *Trichinella spiralis* (Edwards and Barnard 1987).

Thence to T-shirts. In a seminal experiment in 1995 Swiss zoologist, Claus Wedekind gave 44 men clean T-shirts to wear for 2 nights, which were then presented to 49 women to sniff and score for 'intensity', 'pleasantness', and 'sexiness' (Wedekind et al. 1995). The women (at least, those of them not on the contraceptive pill) preferred men whose immune systems differed from their own with respect to the MHC (a group of genes that code for proteins found on the surfaces of cells, and which help the immune system recognize foreign substances, explained in Chapter 15).[13]

It thus emerges that MHC genes play a role in odour-based individual recognition and mate choice (although not in avoiding diseased partners), but also in avoidance of the dangers of inbreeding. When sniffing out the MHC genes of a potential mate, each sex (especially choosy females) seeks to achieve a better blend of disease-resisting immunological genes in its offspring, either by acquiring 'good genes' (absolute criteria) or 'complementary genes' (self-referential

---

[13] For completeness, the T-shirt findings, while fascinating, may not entirely withstand the test of time (Havlíček et al., 2020).

criteria) from that match (e.g. Rymešová et al. 2017; Chapter 17).

Against this background, Chris and Christina advanced the general proposition that chemosensory cues evolved not only to enable mating partners to select the best immune gene repertoires for their offspring, but also to avoid partners from which it was adaptive to 'stay at a healthy distance' (Newman and Buesching 2019). Postponing, for a moment, whether there is any evidence of either genetic or health status being communicated in the badger's subcaudal gland secretion, we ask first, how does this line of thought entwine with the question of why subcaudal glands evolved? Further, why have we emphasized particularly the medium of sweat?

For context, remember that other carnivores appear to achieve the selection of immune-appropriate mates, and the avoidance of sickly ones, without subcaudal glands. Why, then, did badgers (*Meles* and *Arctonyx*), and only badgers, find it adaptive to evolve the subcaudal gland? The answer is unlikely to have its origins in the currently useful, bacterially branded, membership badge. We do not know when in the badger lineage the ancestral, solitary intra-sexual territoriality was swapped for group living—we can't tell from their fossils which side of this divide *Meles thorali* fell (Chapter 1), but it may have been the solitary ancestor that had no more need of a group badge than do most solitary mustelids today. Remember, however, that the glands lining the subcaudal pocket are apocrine glands; that is, sweat glands. Remember also that it is bacterial products of sweat, seemingly more than the sweat itself, that demonstrably convey immune and health profiles in other species. Consider then, where might sweat and bacteria meet in close confinement? Human armpits aside, the area beneath the tail of an ancestral badger might be a good candidate. This might be just the sort of anatomical cranny to be afflicted by something akin to 'intertrigo', a medical condition in which bacteria build up in folds of unwashed skin—a condition also associated with a musty (think *Mustelidae*) smell (Neri et al. 2015). The subcaudal gland probably started as a deep fold of skin, breeding bacteria in the anogenital region, and evolved into an involute pocket, lined with apocrine and eccrine cells.

What evolutionary process might lead from an informative signal born of cutaneous sweat to the development of an involuted subcaudal gland? Well, if bacteria nourished on badger sebum and sweat produce odorous products that provide cues to immuno-competence, and if immuno-competence and associated health have heritable survival value, they will be under positive sexual selection. More descendants will be left by the individuals with the most informative sweat: how useful to have, like a canary singing healthily in a miner's cage, a perfume dispenser puffing out signs of immuno-competence and good health.

But why an ancestral badger, but not an ancestral canid, felid, or even polecat, wolverine, or weasel? After all, none of them has a subcaudal gland. Is it that selection for optimal immunity is more pressing in badgers than it is in other carnivores? Or perhaps the cosy cavity beneath the meline tail is somehow a better starting point? Badgers, and to judge by their skeletons, their immediate ancestors, burrow (Macdonald and Newman 2017; Chapter 19). Burrows bring many benefits (Noonan et al. 2015a), but, being warm and humid, they are also ideal for breeding ecto- (e.g. fleas; Chapter 8) and endo-parasites (e.g. coccidia; Chapter 2). In burrows bacteria proliferate, be these *Mycobacterium* causing tuberculosis in badgers (see Chapter 16), or, sometimes, those infecting the bite wounds (Chapter 5), so ferociously inflicted by badgers on one another (e.g. *Eikenella*, *Staphylococcus*, *Streptococcus*, *Pasteurella*, and *Corynebacterium* species) (Newman and Byrne 2017).

By now it will be clear that we are speculating wildly, if enjoyably. But perhaps both hypotheses as to 'why badgers?' have merit. Theirs may be a lineage whose fossorial lifestyle made fighting infection pivotal to survival, and whose robust, corpulent, musclimorph (as Macdonald and Newman 2017 call it) physique made their nether regions a particularly sweaty bacterial hothouse. Remember as well, that once the subcaudal gland got going, it was soon possible for it to communicate almost everything a badger might want known about itself (and plenty that it did not want known)—a veritable Facebook of solipsism and disclosure.

On reflection

So, where does this leave us? Hopefully these reflections on the roles of subcaudal gland secretion reveal the intricacy of olfactory communication—those fairy calls from the void—and are therefore interesting in themselves; after all, the subcaudal glands of Wytham's badgers may now be the most intensely studied scent gland of any mammalian species. But what of our original purpose in this chapter? Have these explorations of subcaudal gland scent enabled us better to describe social relationships amongst badgers? Emphatically yes, although predictably answering one set of questions leads to us facing a different, yet more difficult set.

Postponing for a moment the role of adult males within social groups, the flux of allomarking amongst females and youngsters reveals a clear-cut social structure that was apparent only as a shadowy mirage in our studies of hierarchy in Chapter 5. Strands of fleeting deference permeate badger groups, evidenced by the loosely centripetal directionality of smeared allomarks—an olfactory doffing of the cap.

This deference flows, like the braided rivulets of a delta, back towards mature, especially breeding, females. Although this directionality might, technically, imply hierarchy, it is so undemonstrative in comparison withthe writhing, urinating, tail-lashing, whimpering subservience on the bottom rung of a canid (or primate) social ladder, that the word 'deference' seems a more appropriate choice than hierarchy. While this olfactory doffing of the cap may be fleeting, and not eccentrically obsequious, it exists and is clear and re-emphasized many times a night, and it flows inwards, centripetally into a matriline. It is also compartmentalized into social cliques, making it hard to avoid the suspicion that it is based on matrilines (this inkling is even harder to deny when the comparison with farm cats comes to mind). As we shift from seeking hierarchy to describing structure, it is tempting to think of the flow of allomarking as indicative less of subordinance and more of fealty, even perhaps friendship.

In contrast, is the torrent of male allomarks daubed onto female badgers a sign of possessive guarding? Male mammals generally scent mark more prolifically than females, and they commonly do so in ways associated with securing and commandeering mates (Brown and Macdonald 1985). Alternatively, is the flow of marks from males to females a further expression of the centripetality that puts reproductive, senior females at the nucleus of badger groups? If so, this would be eerily reminiscent of spotted hyaenas, one of the very few mammalian species whose societies are characterized by female dominance. Both, of course, could be true—adult female badgers might inspire in males both deference and desire, but in both cases Chapter 5 suggests that neither message is strident when translated into social structuring.

So much for the flow of messages, what of the content? We have demonstrated seven dimensions to the information encoded in subcaudal gland secretion, and speculated on two more (genetics and health)—a holistic compendium of content, but when posted on the messageboard, for which readerships is this narrative intended? Is the answer the same for allo- and substrate marks? Remember, these messages endure, so the recipient wears the signal that can resonate both in its (often her) own nose, the nose of the signaller, and those of other readers. Each might glean some of the same and some different things. If the flow hints at deference, is the purpose to signal allegiance or ownership?

CHAPTER 7

Sex

How and Why, and with Whom?

So the kaleidoscope of clues—horrific wounds, amiable encounters with neighbours, and the occasional hint of hierarchy—with their beguiling whiff of anarchy were, through the medium of the subcaudal gland, starting to coalesce into a discernible, loosely centripetal, social dynamic within Wytham's badger groups. But what is the adaptive significance of this dynamic? Natural selection had, of course, been up to something, but what? We move now closer to the sharp end of the evolutionary jostle for fitness: mating. This is the fulcrum about which all societies pivot—imperative as it is as a gateway to reproductive success. All mammalian species, even, counterintuitively, solitary ones, are variously social, as memorably captured in the title of Paul Leyhausen's insightful 1965 paper: 'The communal behaviour of solitary carnivores' (later, Mikael Sandell (1989a, p. 164) defined solitary behaviour in carnivores as 'no collaboration between members in feeding, defence of territory, offspring-rearing, and mating even in cases where ranges overlapped', and in Chapter 8, we interrogate the badgers of Wytham Woods against each of those criteria).

Even the maintenance of standoffishness requires social interaction, as surely does any variant of courtship. At the nub of the adaptive significance of social systems is the question: who mates with whom and with what consequences for reproductive success and, ultimately, fitness? A first step to competitive success might be securing food at minimal cost (we travel the road to food security in Chapter 10), and thereby surviving 'to fight another day'. How that fight is best fought can vary between and within species. Much of the answer turns on how individuals invest energy into reproduction, which we tackle through pace of life theory (Chapter 14). En route, and more earthily, we now ask of the badgers: do all individuals mate, do they all leave descendants, and, if not, which factors determine winners and losers?

For empirical behavioural ecologists (voyeuristic natural historians) like us, fascinated by the individual lives, loves, and adaptations of our subjects, there is nothing more captivating than the relationships, coalitions, and intrigues of animal society. Our thesis is that the drama of these lives plays out on a stage set by ecology, defined in terms of biotic and abiotic factors that determine the distribution and abundance of individuals, the topic of Chapters 10, 11 et seq. (Macdonald 1983; Andrewartha and Birch 1986). Remember that two inheritances circumscribe the ecology of badgers: they are mammals and they are mustelids (Chapter 1). In the later chapters of this book these ancient constraints will increasingly be revealed to set the course that the badgers of Wytham Woods navigate amidst the modern ecological rapids on their evolutionary journey. As mustelids, their ancestral sociology involved semi-detachment between males and females—crucially often living at low population density and occupying intra-sexual territories (Powell 1979). It will be Chapter 19 before the opportunities, and the burdens, of that ancestry become fully clear, although in this chapter one heirloom of low population density will determine much that plays out on the stage of sexual relations.

In answer to the question of 'what is a badger?', we described, in Chapter 1, their evolution during the colder, less productive days of European glacial advances over the past four million years. Therefore, the rich food abundance and dispersion now characteristic of southern England's (and Ireland's) agricultural landscape and climate (Chapters 10–12), and the associated formation of social groups that can comprise, *in extremis*, 25 animals (Chapter 5) is a highly unusual circumstance for this species. It seems likely that under these new, and still localized, high-density conditions, their mating system is expressed under circumstances different to those that moulded it. In the following

The Badgers of Wytham Woods. David W. Macdonald and Chris Newman, Oxford University Press.
© David W. Macdonald and Chris Newman (2022). DOI: 10.1093/oso/9780192845368.003.0007

pages, as you read what we have seen of the sex lives of the badgers of Wytham Woods, bear in mind that this is the response of an erstwhile solitary mating physiology and behaviour, adapted to a paucity of mating opportunities and partners, which has therefore been compelled to change by contemporary high-density conditions. As a result, all will not be as it may, at first, appear.

In October 1994, under the inspiration of Paul Stewart, we began to record endless hours of infrared video (already much scrutinized in Chapters 5 and 6) at the main setts of two large social groups (Sunday's Hill and Pasticks). Neither we, nor anybody else, had anticipated how unusual, indeed Rabelaisian, the mating system of Wytham's badgers would be in contrast to their mustelid kin, and indeed mammals in general. Paul videoed at the two setts from October 1994 until April 1996. Ten years later, long-awaited success with our genetic studies rekindled the topic (Chapter 17), and, this time led by then doctoral student Hannah Dugdale, we repeated the videoing at the same two setts between 1 February and 31 May in 2004 and 2005. Hannah then analysed these data together with those Paul had gathered between the same dates in 1995. In the account that follows we combine descriptions from the full 1994–1996 dataset, together with the statistical analyses of the three February-to-May windows, dates selected to coincide with peaks in mating activity we had observed in the first study, and also to coincide with what was at the time believed—really an untested supposition—to be the first of two annual oestrous periods (post-partum in February–March, the second oestrus being expected in August; Dugdale et al. 2011b).

Our raw material aggregates to 634 nights (taped from 18:00 until 06:00) of badger behaviour during 1994–1996 (329 nights at Sunday's Hill and 305 at Pasticks) resulting in 7608 h of video to watch (a dedicated assistant, Sophie Stafford, helped us with this, an experience that prepared her for a career as Editor of *BBC Wildlife* magazine). The second dataset, 1995, 2005, and 2006 (incorporating 4 months of the first) generated a further 11,230 h of video. It takes about three times as long to analyse a video as to record it. According to Malcolm Gladwell's book *Outliers* (2008), it took 10,000 h of practice to transform talent into genius in Tiger Woods, Bill Gates, and Mozart,[1] so our, *c*.20,000 h of transcribing videos should give us an edge in badgerology! The nuggets panned from this dross were Paul's copulatory documentations on 69 video nights

(46 calendar nights), and Hannah's on 319 calendar nights.

So what did we see? A lot of 'mounting', the frequency and duration of which we recorded. The first dataset revealed 332 mountings, the second 198 mountings (Table 7.1); of the latter there were 89 mountings (on 50 calendar nights) during which both mounting partners were identifiable, from clip marks in their fur (Stewart and Macdonald 1997). This term, mounting, is not mere euphemism: it highlights an important distinction from 'mating' insofar as we could not reliably detect actual intromission or ejaculation (on some occasions, there was no observable thrusting; a further 59 failed mountings did not involve genital contact). It became clear that there were periods, with a peak in February and March (Figure 7.1), when a given female engaged in intense sexual activity and received prolonged mountings; we called these periods 'behavioural oestrus', but could not be certain this correlated with physiological oestrus for those females involved (subsequently we have discovered that vulval swellings that can indicate oestrus in other mammals are only loosely related to hormonal receptivity—i.e. oestrone or oestradiol levels—of badgers; Chapter 5; Sugianto et al. 2018).

What does a typical badger mating encounter involve? To give you a sense of the richness of these observations, we will detail one, lasting 8 min, on a crisp evening in March 1995, involving five badgers. The story is intricate, so we'll give them nicknames, all beginning with B, to ease your navigation through it. Three males, Bob, Billy, and Bruce, and two females, Betty and Bella, are lurking outside their sett. This vignette of the five Bs is typical of what we saw and so will save a lot of description when we list some generalizations (in the text that follows). The clock starts at 0. Four of these badgers are in close proximity; the fifth, Bella, more than a metre away, staring into the darkness. Of the four, three are grooming. Specifically, Bob and Billy are both attentively grooming Betty; Bruce is snuffling the ground. Without preamble, Bob mounts Betty, and in that same instant Billy turns to self-grooming; picking at his left chest. Betty doesn't look up, or show any discernible interest in her partner. Riding further up Betty's back, Bob takes hold of the scruff of her neck with his incisors, and, gaining purchase, slides his forelegs into her hunched hips. Bella has wandered off, but Billy and Bruce stand 50 cm apart and, seemingly nonchalant, self-groom. Rocking his pelvis from side to side, Bob wriggles to get a better grasp on Betty, likely trying to manoeuvre his intromission, as Betty wraps her tail sideways. Patient,

---

[1] Sadly, for the dogged amongst us, more recent research demotes the importance of practice relative to aptitude.

Table 7.1 Summary of females mounted and male mounting. The names on the top row represent the clip-marks of males observed mounting the females named in the column.

Female	Mounts	No. males	Mounts per male	Unmarked	M	Sandy	Slash	Smiley	Submarine	Equals	U	Less	Archer	Rewind	VI	Omega	Twin blob	Blob	Target	Ti	Horse	Hurdle
Sergeant	80	4	20.00	19	29	11																
Unmarked	76	13	5.85	9	11		18	12	10	1	2	2	4				2	2	2			1
Phone	57	3	19.00	13	18	26																
H	33	3	11.00	11	2	20																
X	31	3	10.33	11	4	16																
Saddle	15	5	3.00	5	4	4								1	1							
3 slash	9	4	2.25			3	2				4	2										
A	8	3	2.67	4				1			2				1				1			
J	5	5	1.00	1										1	1					1	1	
Ni	5	4	1.25	1	1	2								1						1	1	
Much	4	2	2.00							2						2						
Cross	3	2	1.50	2		1																
Trident	6	2	3.00		5	1																
SUM	332	53.0	82.8	76	74	84	20	13	10	3	6	4	4	3	2	2	2	2	3	1	1	1
MEAN	25.5	4.1	6.4	7.6	9.3	9.3	10.0	6.5	10.0	1.5	3.0	2.0	4.0	1.0	1.0	2.0	2.0	2.0	1.5	1.0	1.0	1.0
Number of female mates				10	8	9	2	2	1	2	2	2	1	3	2	1	1	1	2	1	1	1

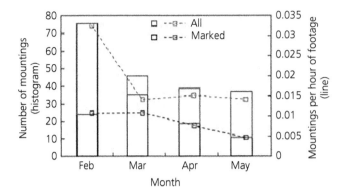

Figure 7.1 Histogram of the number of mountings observed, by all (including unmarked) and by marked badger pairs, per month (1 February–31 May in 1995, 2004, and 2005). The number of mountings, per hour of footage observed, is shown as a line (from Dugdale et al. 2011b, with permission).

Billy grooms the neck of mounted Bob. Bruce resumes snuffling for food in the near vicinity. The scene is eerily quiet, aside from occasional 'churr' sounds from Betty (Charlton et al. 2020; see Box 5.4). After 1 min 5 s of Bob's rhythmic thrusting, at a rate of no more than one lunge per second, Billy starts to nibble at Betty's ears. Billy 'side mounts' Betty, now also with a hold on her neck, and starts thrusting on her left hip. Betty stands relatively still, placid to this attention, without sign of coercion or concern. Bruce grooms her belly. Side-mounted Billy hops off Betty and starts reciprocally allogrooming the neck of Bruce—1 min 27 s have now elapsed since mating commenced. Bob's thrusting appears to have downgraded to rhythmic pushing, as Betty stumbles forward and corrects her stance by rebracing her front legs. Her head is aloft and her neck held up quite stiffly; the scene remains silent. Billy and Bruce are now fully engaged in grooming one another's right shoulders, necks intertwined in embrace, and completely absorbed, seemingly oblivious to the mating pair. After 1 min 42 s Billy and Bruce stop their mutual grooming; Bruce returns to his itchy belly, and Billy starts once more to grab at Betty's neck. Bob thrusts on, now making rotational pelvic lunges and occasionally correcting his footing. Billy now emerges from behind Betty to stand in front of her, almost facing her, but still holding her neck while trying half-heartedly to pull her away from Bob—Billy, however, makes no attempt to directly insinuate himself into the proceedings. Bob suddenly starts to thrust rapidly, around four times per second. Billy then sniffs, perhaps nips, at Betty's front paws. The whole mating ensemble wobbles and totters somewhat, but Betty remains placid. Two minutes have passed. Billy returns

to the far side of Betty, as though to resume thrusting on her hip, whereupon Bruce, who has seemingly dealt with the flea on his underside, comes between Billy and Betty and half mounts Billy, rearing up over his back and shoulders, but then sliding back off to the nearside. Billy shrugs off this advance, and again goes around the front of Betty and pulls at her neck, over her left shoulder. Bob remains firmly in place and reverts to less energetic rhythmic thrusting, while, at the same time, visibly pulling himself onto Betty with his forelegs latched into her hips. Billy leaves, to the left; Bruce falls out of camera view to the right. Bob and Betty continue copulating. Three minutes have passed. At 3 min 35 s occasional glimpses make clear that Bruce is still nearby. After 4 min, the pair has turned right around as Bob pulls at Betty causing her gradually to pivot round in a circle of tiny front-foot steps. After 4 min 24 s Billy returns, approaches the now rotated couple from behind, puts a paw on mounted Bob's hip, takes a bite-hold of his neck, puts his forelegs around Bob's chest, in a sort of Heinrich manoeuvre, and starts to pull, using his back legs and arched back to try to lever Bob off. This threesome of stacked badgers continues for a further 30 s, accompanied by some guttural rumbling growls from the males, until at around 5 min Bob turns to bite back at Billy, who has been tugging at his neck in what appears to be more dissatisfaction than ferocity. With a sudden move, Billy finally levers Bob off the still-nonchalant Betty. Bob gives Billy a bit of a shove; Billy shakes himself down. Bob sniffs Betty's genital area, with his nose wedged under her conspicuously raised tail, mounts again, and resumes thrusting. Billy tries briefly to sniff between the bodies of the mounted pair. Thereafter, around 6 min, Billy resumes the nipping and pulling at Betty's neck, while Bob continues slow pelvic thrusts. At 6 min 47 s Bob looks over his left shoulder, as Bella trots into view. Briefly distracted, Bob almost loses his grip as Billy takes the opportunity to give Betty's neck a firm yank away from Bob; Bob, however, recovers his position. Billy breaks off his advances and goes to greet Bella, silently, with a brief mutual neck groom. A moment later (around 7 min) Bruce returns to the scene and, running around the back of the mating pair, grabs Bella by the scruff and works his way from mounting her off-side hip towards intromission, accepted by her raised tail. Billy, now standing between two copulating couples, briefly grooms Bella's neck, then leaves. Bruce and Bella mate for around 40 s, then Bella turns her head over her left shoulder and nips at Bruce, emitting a harsh 'kek'. He dismounts, and leaves in the same direction as Billy. Bella begins nipping the left flank and underside of still-mounted Bob with more 'keks', forcing him to let go of Betty's neck. Bob had been mating Betty for 8 min when Bella barges him off her. Bob backs up three steps, turns, and trots off. Betty and Bella groom each other's shoulders for around 40 s, before trotting off together.

The key point of this, and other, mating scrums is the lack of animosity amongst the several participants: it is common to see more than one male mate with more than one female, with mutual grooming amongst them even while mating goes on. Theoretically, when females engage in promiscuous and repeated mountings it is evolutionarily intriguing, because matings are expected to be costly and therefore females are expected to be choosy.

Generalizing from the vignette of the five Bs, the vital statistics of 198 within-group mountings were summarized by Dugdale et al. (2011b). The mean mount duration was 230 ± 95 s (median = 34 s). We categorized the mountings into three classes: short (<1 min, range 1–58 s, mean = 20 ± 3 s, N =127), medium ($1 \leq t < 5$ min, mean = 141 ± 21 s, N = 43), and long duration (≥ 5 min, maximum = 82 min, mean = 23 ± 9 min, N = 28). Eighty-nine of the mounts involved badgers we could recognize individually, and their duration averaged 235 ± 191 s. Sixty-two of these mounts were short (mean = 22 ± 4 s), 16 medium (mean = 130 ± 28 s), and 11 long duration (mean = 27 ± 17 min). In this sample of known couples, and considering only long-duration mountings, the females were seen mounting 0–2 nights previously, and the males 0–15 nights previously. In this sample, two different males mounted three females for long durations, at intervals of anything between 6 s and 2 days. The mean time between two males mounting the same female on the same night was 14 ± 28 min (range 0–53 min, median = 3 min, N = 5).

Although the bacchanalian episodes of multiple males mating were vivid in our thoughts, individual females each mated with only one male on 50 (72%) of our first sample of 69 video nights; however, on the remaining 19 nights individual females were seen to mate with more than one male—polyandrously (and remember that during all these nights we saw only what occurred within the limited field of view of our surveillance cameras). Indeed, all but one marked female were seen to mate with more than one male during a single night, or on two consecutive nights, at some point during the first study; the remaining exception was a female seen to mate with a second male after an interval of only 2 days. On nights when a female was mounted by just 1 male, she was mounted on average 2.1 times (standard deviation (SD) = 2.9), whereas when mounted by more than 1 male, females received a mean of 10.8 (SD = 7.6) mounts; or 4.4 (SD = 3.1) mounts by each male involved. In the second dataset, of 16 females that were observed being mounted, 12 were mounted more than once (Dugdale et al. 2011b). Over how long a period of time did their receptivity stretch? Seven were mounted in more than one 4-week period in the same year. Seven females were observed being mounted with an interval of between 4 and 25 days; 5 of these females at intervals of more than 12 days. Later it will be relevant to ask how long badger oestrus lasts.

Male mating behaviour

We have recorded badgers mounting in every month of the year. However, as winter advances and the chill nights of early spring begin to shorten as February turns to March there appears to be a sexual frisson in the air; even amongst juveniles and cubs there is play-mounting. As the winter advances, adult males engage in sexual horseplay, mounting youngsters and each other, as they increasingly turn to more concerted attempts to mount females. Such approaches begin with the male emitting a characteristic churring mating call (Charlton et al. 2020; Box 5.4). If there is no receptive female above ground, males churr into the entrances of the sett (Charlton et al. 2020). If no female emerges in response, a churring male may enter the tunnel. Some males, especially those involved most frequently in mounting, intersperse churring with digging and scuffing the ground.

Females that emerge from their burrow while a male is churring might ignore, greet, or even attack him. Males typically respond submissively to this female aggression, but ultimately attempt to seize the scruff of her neck and mount her. Allogrooming, normally the main, if fleeting, social grease of badger interactions around the sett (Chapters 5 and 8), is less in evidence between male and female in this pre-mounting flirtation (Stewart and Macdonald 2003; presumably male badgers have priorities above itchiness at that moment), although subcaudal allomarking and substrate marking are much in evidence (Chapter 6).

In the theory of mate selection, an important factor is whether females are coerced or willing participants. The vignette of the five Bs was typical: we saw that badger neck skin is sufficiently loose that a female could, at will, turn on her side to break free, or bite the male, and hence terminate the mating. We saw no evidence that males were using brute force to coerce unwilling females. Incidentally, and relevant in the context of sperm competition, although badgers do have a penis bone, or baculum (Rosie Woodroffe celebrated her doctorate by making one into a tie-pin for David), we never saw anything resembling a copulatory tie, as seen in, for example, canids and pleasingly dubbed by Grandage (1972) 'a paradox in flexible rigidity' (e.g. Asa and Valdespino 1988).

As with the vignette of the five Bs, participating males are strikingly amicable with each other. Indeed, aggression, or any hint of competition between male suitors to a female, was rare.[2] On the contrary, as per the five Bs: males often groom each other immediately after mating with the same female, several males cluster around a receptive female and, while one mounts her, the others remain close by, often in physical contact; in addition, one of the waiting males may groom the copulating female or the male mounting her, and may attempt to 'join in' with the mating, mounting the copulating male. When one male dismounts, his place may be taken, within seconds, by another. Against this remarkably lackadaisical, even orgiastic, tolerance amongst males courting the same female, the most grievous signs of impatience involved one male actively pushing another off the mounted females or

biting the mounted male's paws. These generalities of male mounting behaviour, although observably most intense during February and March, continue through every month of the year.

Female mating behaviour

Our impression from the occasional moments of excitement amongst the drudgery of the thousands of hours of video is that females call the shots, accepting placidly any male attempting to mate with them, but able to terminate mounting whenever they choose. The generality, building on the vignette of the five Bs, is that a libidinous female stays close to a chosen male, nudging him while trying to get her head under his front leg, as if to somehow pull him onto her back, and then lifts her tail whenever the male initiates mounting. Females even sometimes mount males momentarily.[3] Sometimes a female detectably favoured one of several available males by the inclination of her head, by moving towards him, or by leaving with him. Sometimes a male seemed to resist and reject a female's advances and sometimes, also, a jilted female attacked a favoured female, although female–female interactions are also generally amicable amongst the mating throng.

So much for the generality of male and female mating behaviour, but what determines winners and losers amongst them in the quest for matings and, much further downstream, reproductive success and fitness? One reason we sought so assiduously evidence of hierarchy in Chapters 5 and 6 is that the generality amongst polygynous mammalian species, such as the fallow deer living alongside Wytham's badgers, is that males compete intensely and those achieving higher rank secure more matings (e.g. McElligot et al. 2001). Amongst the badgers, what determines which males secure most matings? As a preamble to our observations of badger sex we alerted readers that all was not as it appears to be, and mentioned that European badgers evolved under conditions of low population density; a circumstance that bequeathed to them an adaptation with crucial consequences for the interpretation of the mating scrums—that piece of phylogenetic baggage is called induced ovulation.

Induced ovulation

For species living at densities at which a female is assured that sufficient males will compete for her to ensure that the winners are of high quality,

---

[2] Although the generality was amicable behaviour amongst courting males, in the subsample analysed by Hannah on 11 occasions (amongst 29 mounting events), a second male was in view. On eight of these occasions she detected some aggression from the mounted male towards the other, and on four occasions vice versa. In six of these episodes the recipient retaliated, and in eight of them the males also engaged in allogrooming.

[3] The behaviour of 'bulling' in cows is a sign of oestrus.

spontaneous ovulation is adaptive (Conaway 1971; Larivière and Ferguson 2003). In contrast, induced ovulation appears to be an adaptation to low population density in which females risk failing to encounter any male; still less one of a good quality.

Induced ovulation, as the name suggests, is the case when ovulation does not occur spontaneously, but is triggered or induced to happen by some form of encounter with a male. It is a phenomenon that has arisen independently in diverse mammalian taxa—squirrels, mole-rats, koalas, and camels amongst them. The nature of the inducing trigger varies. For example, in the case of rabbits, ovulation is induced by the mere presence of the male. Amongst Bactrian camels *c.*1.0 ml of semen is required in the vagina—it is ovulation-inducing factor (OIF) in the seminal plasma that stimulates luteinizing hormone (LH) release to induce the ovulation. In alpaca—South American relatives of camels—prolonged copulation, causing abrasion and inflammation of the uterus, may enhance absorption of OIF (Adams and Ratto 2012). It will become important to interpreting badger behaviour to know the interval between induction and ovulation, and a first clue is that in the case of camels 66% ovulate within 36 h and the rest by 48 h (Chen et al. 1985).

Of Carnivora, species with induced ovulation exhibit a less strictly seasonal reproductive pattern, potentially because mates are not constrained to meet during the short time window of a spontaneous ovulation (Heldstab et al. 2018). Amongst bears, as rabbits, only the presence of a male is needed (Himelright 2014), but perhaps most can be learnt from domestic cats, which illustrate the felid situation. During intromission, the penis causes distension of the posterior vagina and induces release of gonadotropin-releasing hormone (GnRH) from the hypothalamus via neuroendocrine reflexes. A surge of LH occurs within minutes of mating, and ovulation follows 1–2 days later. More is better: with multiple matings, the LH surge is greater and lasts longer than when only one mating occurs (Shille et al. 1983).[4]

What can be said of badgers? The likely summary is that after intromission the LH surge process (hormonally conserved amongst the phylogenetic baggage of all mammalian taxa) takes at least 24 h, and perhaps 48 h or more to cause ovulation. With multiple matings the LH surge is greater, causing more ova to ovulate (superfecundity) (Amstislavsky and Ternovkaya 2000;

Larivière and Ferguson 2002). As vividly apparent, modern female badgers in Wytham Woods face no risk of failing to find a male when they need one. However, it was not always thus, and it is not necessarily that way nowadays, either, for their contemporary continental cousins across broader parts of their range (Chapter 19), which is probably why induced ovulation constitutes part of their phylogenetic baggage.

The circumstances of the American marten, an ecologically and thus sociologically traditional mustelid, with low-density, diffuse populations, lengthen the odds of being in the right place at the right time to mate, risking that a female ovulates wastefully. Worse, as exemplified by ferrets (domesticated polecats, *Mustela furo*), if jills fail to mate, prolonged high levels of oestrogens cause the cervix to remain partially dilated, allowing uterine bacterial infections that can develop into pyometra and fatal toxaemia. One might expect all these factors to encourage the ancestral female mustelid to be enthusiastic about mating, and that is exactly what is revealed, at least for female American mink, in remarkable experiments by Thom et al. (2004b) reported in Chapter 17.

Induced ovulation[5] is a solution: mating becomes the trigger for ovulation rather than vice versa (Bakker and Baum 2000): any time (albeit within a window of opportunity, oestrus) becomes the right time to achieve conception. More precisely, any time about 36 h after the first mating, because it takes a while for the hormonal cascade linking copulation to ovulation to kick in (including using LH to induce ovulation in human fertility treatments; Wallach et al. 1995).

In the olden days, when ancestral mustelids evolved induced ovulation, their low population density and solitary (semi-detached, intra-sexual territorial) lifestyles likely resulted in the greatest risk to a female spontaneous ovulator that an ovulation would be

---

[4] As it happens, domestic cats are not obligately induced ovulators and can ovulate spontaneously too (a mix termed a reflex ovulator).

[5] Vertebrate reproduction is controlled by GnRH, but spontaneous and induced ovulation involve different ways by which the mediobasal hypothalamic (MBH) release of GnRH leading to the preovulatory secretion of LH is regulated by the anterior pituitary gland (Bakker and Baum 2000). In spontaneous ovulators steroids secreted by the maturing ovarian follicle cause GnRH to be released by the hypothalamic median eminence in pulses. This stimulates a preovulatory LH surge. In induced ovulators, the preovulatory release of GnRH from nerve terminals in the median eminence, and the resultant preovulatory LH surge, are induced by genital somatosensory stimuli during intromission that activate noradrenergic neurons in the midbrain and brainstem. The median eminence is integral to the hypophyseal portal system, which connects the hypothalamus with the pituitary gland and produces regulatory hormones.

wasted because no male was available—induced ovulation mitigated that risk because it guaranteed a male had to be present to trigger the ovulation. However, under the contemporary circumstances of the badgers of Wytham Woods, females face different quandaries: if numerous males are to hand, and some are more desirable than others as fathers to their cubs, and yet at least the first 36 h of mating are not only (i) essential to induce ovulation, but also (ii) generally not going to lead to conception (depending on the lifespan of sperm), how should the sexes play their cards? The answer is clearly an extremely complicated conditional game, made vastly more complicated because what we already know of badger cliques, groups, and super-groups means that as many as a couple of dozen would-be fathers are in the pool of candidates, each one of whom is strategizing on which of a couple of dozen females to concentrate, and in which order.

If this is a difficult problem for both female and male badgers, it is a worse problem for researchers. The badgers, at least, may have an olfactory inkling of where in this process a particular female is situated. They may know whether she was first mated more than 36 h ago, and thus whether any mating now is potentially procreative or merely fun. In contrast, the researchers have much less idea: we cannot be sure that the first mating we see a female experience is her first such mating, and so we cannot know whether a given mating is the real thing (and is being treated by the participants as such). When, in a moment, we come to ask whether there are particular features of males that females prefer in a mate, we cannot know whether the same answer applies to inductive or procreative matings, and we cannot discount the possibility that both females and males treat these very differently: clearly both sexes could play a game of optimality odds to be in the right place, with the right partner, at the right time. The outcome might look promiscuous, even orgiastic.

Of course, when we see a female mate for the first time we can at least deduce that any mating more than about 36 but less than 120 h later is almost sure to be procreative, but sadly that sort of sieving of our observations decimates our sample sizes. Further, not only is the first 36–48 h of mating tantamount to practice, a further problem is knowing if and when a given female enters into subsequent cycles of induced ovulation: the female population arrives in the follicle-stimulating hormone (FSH) window fairly synchronously, within which an LH surge will induce ovulation, but this window probably closes after 4–5 days—however, we have repeatedly seen particular females being mounted in successive months—that is, at intervals far longer than the few days for which that FSH window remains open. The badgers may know, but the researcher has no idea of whether subsequent matings represent a further procreative opportunity or fulfil some other function of non-conceptive matings (we return to the question of 'when is oestrus?' in Box 7.3).

What is to be done? Although we cannot know which of the mountings we documented were potentially procreative, we categorized their durations carefully into short, medium, and long. We assume that long mountings are more likely to be procreative, and so we analyse those three durations separately. Importantly, although doubtless a proportion of the mountings we recorded was solely inductive, out of the whole sample, including many long ones, it stretches credibility that a proportion was not procreative. From the foregoing descriptions we can say confidently that there was not some subset of mountings that was behaviourally different—on the contrary, the generality of the badgers' behaviour was consistent in the promiscuous amiability we describe. It was not the case, for example, that some significant minority of mountings was associated with ferocious defence of the female, betraying that these were the only occasions that really mattered. Therefore, although the uncertainties make our analytical laser unfocused (frustratingly so, considering the remarkable evolutionary interest of these male and female tactics) we nonetheless present our analyses of the short, medium, and long mountings, knowingly ignorant of which were inductive or procreative; but nonetheless confident that a proportion (maybe 3 out of every 5 days) was procreative, and even more confident that if there are qualities associated with male success due to female choice, they will be most likely to be revealed in the long-duration mountings.

So, there's a final interpretative conundrum to have in mind as we consider the results in the text that follows. Times change, and behaviour adapts: the advent of the contraceptive pill, disarticulating sex from reproduction for people living at high densities in the 1960s, had sociological consequences. Is the contemporary sexual behaviour of the badgers of Wytham Woods best interpreted as finely honed adaptation or pathological anachronism? Is induced ovulation, as a piece of phylogenetic baggage, now nuancedly incorporated into a new complex of behaviour that fits intricately with modern high-density circumstances? Or has evolution failed to catch up with the changing times, so that in terms of adaptation the sexual behaviour of the badgers of Wytham Woods is inappropriate—a mismatch[6]—like a piece of clothing that no longer fits, leaving the badgers to make the best of a bad job? Two considerations cause us to interpret the badgers' behaviour as if it is fit for purpose and well adapted. First, evolution can work very fast, over just a few generations, and while the lowland agricultural landscape of western Europe is new, even at its extreme it's been emerging over several hundred generations of badgers. Second, quitting on the quest for adaptation encourages sloppy thinking, insofar as if you don't look for adaptive significance you certainly won't find it. Nonetheless, the wider question of whether the badgers of Wytham Woods are really up to speed with the selective pressures imposed by their extraordinary modern population densities is an important one, upon which we will elaborate in Chapter 10.

Before turning to our detailed results, it is necessary to put in place one more feature of badger reproductive biology, which relates to the duration of pregnancy. Intriguingly, in a comparative analysis of seasonality in the breeding of Carnivora, Heldstab et al. (2018) found that, size allometries aside, some species can shorten their gestation period to squeeze reproduction into a short time window of optimal environmental conditions. However, others lengthen their gestation periods in order to bridge long winters, which they achieve not by decelerating intra-uterine growth but by delaying implantation. The latter is a crucial aspect of badger reproduction.

Delayed implantation

When contemplating mating badgers, perhaps in late February or early March, bear in mind that the fruits of their coupling will not be born until almost a year later. This is because of delayed implantation (DI), a.k.a. 'embryonic diapause'. The enigma of DI occurs in 34 species of mustelid (Mead et al. 1989; Sandell 1990), accounting for almost half of the 54 carnivore species exhibiting this phenomenon (and in *c.*100 mammal species overall, spread between eight different orders, from armadillos to sea lions). Mustelids, remember, are one of four families comprising the superfamily *Musteloidea* (Chapter 1), amongst which, curiously, DI is absent from the procyonids (but tentatively suggested for the red panda; Roberts and

Kessler 1979). Obligate[7] DI disarticulates the mating and the birthing season, allowing up to an entire year between the two in post-partum breeders, rather than the 30–60-day gestation typical of musteloids. The generally accepted adaptive significance of this disarticulation is that it avoids the need to mate during the depths of winter for cold-climate species, which would otherwise be needed to achieve spring parturition dates, and thus also allows offspring the full summer to mature before harsher winter conditions return (Mead 1993).

However, great variation in the length of the delay suggests there's more to its adaptive significance than inconvenient seasonality (Thom et al. 2004a). The American mink, *Neovison vison*, that we studied on the River Thames just outside Wytham Woods, has a short interval delay in implantation of only a few days (Yamaguchi et al. 2006) and only if the females are mated early in the season. This short delay can be explained, in conjunction with superfoetation (SF), as a mate assurance mechanism. SF is the process whereby females have further oestrous cycles after conceiving blastocysts, potentially increasing their litter size and/or their choice of which zygotes ultimately to advance. SF permits additional oestrous cycles during pregnancy (or at least during DI,[8] in the badgers' case) that release multiple ova with possible non-synchronous fertilization; thus allowing the simultaneous occurrence of more than one stage of developing offspring in a single mother (Yamaguchi et al. 2006; Mainguy et al. 2009a; Roellig et al. 2011; Sugianto et al. 2021b). The combination of DI and SF is discussed in Box 7.3.

Other mustelids, including western spotted skunk (*Spilogale gracilis*), American badgers (*Martes* spp.), fisher (*Pekania pennanti*), wolverine, etc., have 'long-interval' delays lasting months (Mead 1980), and for the European badger it lasts *c.*10.5 months. There are also curious pairings of ecologically and phylogenetically similar mustelids with and without DI: the

---

[6] Mismatch is the term used by Daniel Liebermann (e.g. in his book *Exercised*) to describe the way in which modern human lives are out of kilter with our evolutionary adaptations, often with medically damaging consequences.

[7] Facultative or lactational DI is known to occur in some rodents, insectivores, and marsupials. If a female copulates while still lactating for a previous litter, the suckling stimulus will cause the embryos to enter into diapause.

[8] In pregnancies without DI, SF is achieved via a second oestrus immediately prior to parturition of the foetuses that developed from the first ovulation event (e.g. brown hare, *Lepus europaeus*, Caillol et al. 1991; North African gundi, *Ctenodactylus gundi*, Gouat 1985; casiragua, *Proechimys semispinosus*, Weir 1973). Therefore, although the second ovulation occurs prior to parturition, each set of ova develops separately in essentially two parallel pregnancies leading to two distinct parturitions (Yamaguchi et al. 2006).

North American river otter (*Lontra canadensis*, with DI) and Eurasian river otter (*Lutra lutra*, no DI); the stoat (*Mustela erminea*, DI) and weasel (*M. nivalis*, no DI); and the western (*Spilogale gracilis*, DI) and eastern (*S. putorius*, no DI) spotted skunk (see Thom et al. 2004a).

As we consider the mating behaviour of female badgers, does the patchy expression of this trait provide an interpretative backcloth? In the quest for generalizations, Lindenfors et al. (2003) suggest relative body size is a driver (weasels (no DI) versus larger stoats (with DI)); Sandell (1984) argues for high selection pressure for early breeding in stoats and for a high potential population growth rate in the weasel. Mead (1989) had rejected this size-based explanation because body mass of female mustelids does not differ consistently between species with and without a delay. Instead, Mead postulated that DI is characteristic of species living at low population density, in large home ranges; features often associated with mustelids living in seasonally unproductive environments (i.e. seasonality, temperature, snow, latitude, and primary productivity were all inter-correlated predictors of the trait). This cap fits for the American marten (*M. americana*).

Winners and losers in the mating game

Informed by this brief primer in badger reproductive basics, we return to the question of what determines winners and losers amongst Wytham's badgers in the quest for mounts, and to what extent mounting, in the context of induced ovulation, might be linked to reproductive success and fitness. More broadly as we tackle this as a question about function, keep in mind Tinbergen's taxonomy[9] of the four ways in which any question can be considered, and ask: why do badgers mate? The question may seem fatuous, in that the answer appears self-evident, either to an evolutionary biologist (the pursuit of fitness) or any functioning adult (the pursuit of pleasure), but as you read be on the alert that, as it turns out for Wytham's badgers, far from all sex leads to conceptions, a fair amount of it never has any possibility of doing so.

[9] In 1963 Niko Tinbergen formulated four questions about animal behaviour: (i) Function: why does the behaviour improve the individual's fitness? (ii) Phylogeny: how did the behaviour evolve through natural selection over time? (iii) Causation: what stimuli cause the behaviour? (iv) Ontogeny: how did the behaviour develop during the individual's lifetime?

Variance in mounting: predictors of success

Over the 89 mounting events (at setts) documented by Dugdale et al. (2011b) for which both individuals were identified, males and females were observed mounting promiscuously both within a season and even during the course of the same night. Although all identifiable males and females were caught *in flagrante* on our cameras, individuals certainly differed in the frequency with which we managed to sample their mountings. Amongst males, the number of times each individual was seen to mount varied between 1 and 84, and the number of different females each was seen to mate with varied between 1 and 8. On 14 occasions individual males were observed to mate with more than 1 female in a single night.

Some sort of selection was at work. Within groups, some males were observed mounting more than would be expected if all males were mounting equally, at random,[10] but statistically this was true only in two and four social-group years, respectively (out of the total of six group years analysed)—reminding us of transient patterns of hierarchy from Chapter 5. However, the patterns were sufficiently indistinct that, statistically, equal sharing of mounting amongst group members could not be ruled out for males in five social-group years and females in three social-group years.

Males

What features characterized frequent mounters? Diagnosing the answer for males can be tackled by disentangling two aspects of the question: (i) who mounted most often; and (ii) which male mounted the same female repeatedly? For the first question, at a first pass it seemed that males with the lowest tally of mounts had a higher body condition index (between May and August following post-partum oestrus) than those that mounted most (or vice versa, insofar as we imply no causality). However, there are two complications. First, this statistical test included an effect of implied hierarchical rank, and six of the males in the sample came from group years with no detectable hierarchy (Chapter 5). When the analysis was repeated omitting rank, the effect of body condition on mounting frequency disappeared. Second, we were able to measure body condition only between May and August, that is, at least 2 months after the mountings were observed

[10] Controlling for the number of activity bouts in which individuals were observed on screen, group size, and overall levels of group activity.

(and thus 8 months or so before the cubs are born), so deducing a male's condition at the time of mounting, and its implications, is difficult. Nonetheless, hold in mind, body condition might be relevant to which males father most cubs (as distinct from mount most females), and it turns out that although being in good condition a few months after mating in the late winter of one year is associated with siring more cubs that emerge during the spring of the following year, this association is highly nuanced (Chapter 14).

So, what were the attributes of males that mounted the same female repeatedly (although without exhibiting a fidelitous attachment)? Was it, perhaps, that a female consented more easily and frequently to courtship from a 'desirable' male, his quality assessed from physical indicators such as being heavy, large, or in the best body condition? No, none of these applied. However, some *je ne sais quoi* was at work for the Casanovas: for those females mounted by several males in one night, there was a strong correlation between the number of times a female transferred to be mounted by a specific male in a copulation sequence, and his overall mounting success (in this context measured by the frequency with which he mounted all females), but, once more, not by his body condition, body weight, or body length.

Females appeared, thus, to favour high frequency mounters ('For whosoever hath, to him shall be given, and he shall have more abundance', Matthew 13:12 and 25:29), but this characteristic also failed to correlate with weight, length, or condition—our crude candidate measures of male quality. Of course, all the inductive matings might be not merely a trigger, but also an assessment. The 1965 pop song by Savoy Brown advised 'You got to separate, The *truth* from the lies, You got to *taste* and try, Before you buy'. Perhaps then it was only during the very last mating of a sequence when the female was selective. If so, we cannot tell which mating was the last. But more probably, during the 2 or 3 days of procreative oestrus there may have been several matings with several males (to judge by the sexual activity we observed during behavioural oestrus), so the point stands that it seems unlikely that a crude visible measure of male quality escaped our detection (of course, a female influenced by a male's character, odour, or even ability to tell good jokes would be undetectable to us). It is also possible that none of this matters and that the selection of fathers is all achieved cryptically, much later, and deep within the females' bodies—a possibility we explore in Chapter 17, where we will ask whether less crude measures of quality might reside in the right parental blend of genes, especially those shaping the immune system.

The difficulty of disentangling the detail of badger romances under the blurring muddiness of nocturnal field conditions is illustrated by our frustration at not having the precision offered by naturalistic experimentation. The point is made vividly by the highly relevant experiments on wild rats conducted by David and former student, post-doc, and lifelong friend, Manuel Berdoy. The astonishing fact is that despite thousands of publications annually based on lab rats, the behaviour of these animals in the wild was almost unknown—this blinding imbalance motivated the WildCRU's research programme, which included observations of wild rats living naturally in a tennis court-sized enclosure at Wytham (Macdonald et al. 2015a). Manuel and David watched mating chases, where a string of males raced after a female, and several of them mated with her (once one male was mating, the others generally didn't interfere); the situation was close enough to our observations on badgers to be relevant here, and they asked all the same questions. But with the rats they could answer them experimentally. Within the enclosure they built a mating arena consisting of four chambers. A female rat was placed in one, and from eight candidate males, three were chosen to occupy each of the others during each replicate of the experiment. The female had access to all three male chambers through a narrow hole, but because of their greater size, the males could not exit their individual compartments. Prevailing wisdom had been that females only mated multiply because they had no choice, and that all the decisions were made by the males.

However, given the choice, 7 of 8 test females chose to mate with all three males, and the eighth mated with 2 of them (further, these female were not coy, on average they visited each male's chamber 62 times during the period of their peak oestrus). Manuel and David wondered whether females would prefer to see males displaying their prowess in direct competition, making it easier for them to select the dominant one (Berdoy et al. 1995a). So, in some chambers they put males alone, whereas in others they put two males— it made no difference to the female's choices (it is true that where two males were together, the dominant one monopolized the matings, but each female nonetheless made a point of visiting single males and mating with them as well). Nonetheless, the females were choosy: all the rats lived in the mating arena for several days before the female was due to come into oestrus (monitored by vaginal swabs), and the female exhibited distinct friendships, spending more time with some males than others before her oestrus. Then, when she came into oestrus, while she mated with them all, she mated most with those with whom she had bonded beforehand. It turned out that males behaved in two different ways as a female approached oestrus: less preferred males (those with whom she spent less time beforehand, and which had less access to her) always took every opportunity to mate with the female, before, during, and after her oestrus. However, preferred companions often opted not to mate with her, even pushing away her solicitations, until c.2 h before she ovulated, whereupon these preferred males increased their rate of mating to about fivefold that of the less preferred males, peaking at about 25 mating chases per hour (Berdoy and Macdonald 2005). How we wished for the facility to disentangle our observations on badgers with similar precision—and how greatly success in behavioural ecological research depends on choosing the right study species under the right circumstances for each question—but making the best of our circumstances we asked whether, in our other observations of the nightly relationships within badger groups, there were clues as to what strange attractors might explain any aspect of mounting success.

Of aggression, there was no relationship between the number of times males directed aggression at females and their observed mounting success with these females. Of allogrooming, there was a statistical hint that males that allogroomed females more achieved a greater number of mountings with them. This was another of those tantalizing Scarlet Pimpernel effects—now you see it, now you don't—females were mounted more often by males that allogroomed them more often in two, but only two, of the six social-group years analysed by Dugdale et al. (2011b).

That brings to mind those social cliques revealed in Chapter 6, defined by the flows of subcaudal allomarking—what did this have to tell us about mating partner preferences? There was no relationship between the number of times that females were observed to be mounted by males and the number of times those males allomarked them. Looking for clues further afield, a distantly related social musteloid, the ring-tailed coati, *Nasua nasua*, provides one; with this species, simple familiarity and propinquity appear to help a male's troth with a female[11] (Hirsch and

Maldonado 2011). Again, was some measure of propinquity (membership of the same clique, subgroup, a shared bedchamber) a predictor of who mated with whom, and how often? We don't know. However, we do know from our genealogical studies (Chapter 17) that in the sample of observations analysed by Dugdale et al. (2011b) there was no relationship between mounting frequency and relatedness of mounting pairs.

Not for the last time (and Chapter 14 will bring this to a crescendo) we felt the tension between the quest for generalization, subsets of data, and individual case histories of particular badgers at particular times under particular circumstances. The overall analysis, published in 2011, over a specified set of months during 3 years (1995, 2004, and 2005; Figure 7.2) suggested that neither body size nor condition was a good predictor of male mounting frequency, whereas poring over only the 1995 data, although over a rather longer set of months, we had concluded that the number of mounts made by each male did correlate significantly with his body condition score. This intuitive impression was fuelled by the fact that during that period two males, one called M (M367) another called Sandy (M300) who motored between the Pasticks setts, were both powerfully built, bulky, individuals and they, between them, were responsible for 51% of all the mountings we recorded in that period. A rather typical badger conclusion, at the interface of generality and individuality, is that while big, brawny, fat males can monopolize mountings, they don't always do so.

Females

As for males, while all marked females were seen to be mounted, they varied individually in the number of mountings they received; also in the total number and diversity of males by which they were mounted (Table 7.2). Again, whatever it may be that makes a particular female attractive or receptive—as indicated by the number of mountings she received—it was not the individual's size, condition, or weight. Presumably a female is also motivated to have her ovulation induced.

Remember, we classified mounts as short (<1 min), medium (1–5 min), and long (>5 min), with the overall average just under 4 min (230 ± 95 s (see Dugdale et al. 2011: median = 34 s, N = 1982]). These distinctions will turn out to matter, as will the exact timing (i.e. the dates) of conception (see the text that follows). For example, at the Pasticks sett, in May 1996

[11] In social psychology, propinquity is one of the main factors leading to interpersonal attraction. It refers to the physical or psychological proximity between people. Propinquity can mean physical proximity, a kinship between people, or a similarity in nature between things.

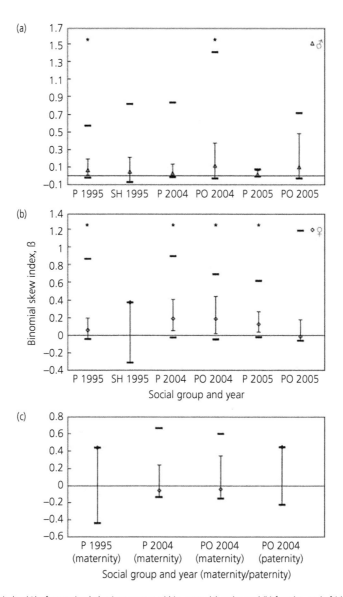

Figure 7.2 Binomial skew index (β) of mounting behaviour among within-group (a) males and (b) females, and of (c) parentage. β is >0 when mounting/parentage is distributed among fewer individuals than expected at random, showing reproductive skew. Error bars display the two-tailed 95% confidence intervals. * indicates that β is significantly greater than zero (one-tailed test; horizontal line indicates β = 0), suggesting some skew away from equal sharing. Data were collected in two neighbouring groups each year: P = Pasticks; PO = Pasticks Outlier; SH = Sunday's Hill. Solid black dashes indicate the minimum (equal sharing) and maximum (monopolization by one individual) possible values of β within each group (from Dugdale et al. 2011b, with permission).

Paul Stewart documented 17 females being mounted, and 10 of them received medium-duration mountings (in May 1995). The other seven, seen to receive only short mountings included three yearlings, and the two oldest females in the group (both born in 1990). Nearby, at the Sunday's Hill sett, 3 of 10 females were observed in long-duration mounts. Other shorter mountings included a yearling that was repeatedly, but briefly, mounted, an exceptionally small adult (6.6 kg/630 mm), and two others, both of which subsequently had cubs. Importantly, for any given female in a given year, all prolonged (i.e. >5 min) mountings occurred within either a single night, or 2–3 consecutive nights. Clearly this raises the question we posed earlier: exactly when is oestrus? Remember, the literature answered that it occurred once post-partum, in late February/March, and then again in autumn (but see the text that follows).

What of the duration of male mountings? Seven of 10 males at the Pasticks sett were observed in medium-duration mountings (in May 1995). Of the three mature males seen to engage in mountings that were only of short duration, two were large males, Lecher B (M10) (see Box 7.2) and Greek (M243), that were notable as being immigrants (Lecher B was based at Brogden's Belt; M243 was always assigned to Jew's Harp; we also knew, however, both from videos and trapping records, that he often visited Pasticks). Despite the short duration of their mountings, they were often seen in the vicinity of receptive females. The third male seen to mount only briefly, Line (M326) (a 2 year old), was notable only insofar as he had recently moulted.

Similarly, of 15 males resident at the Sunday's Hill sett, 4 had imperfect clip marks so the data were unusable; all 11, however, of the clearly recognizable males were observed in medium-duration mountings, of which, however, over half involved just two males, Slash/M1 and Submarine/M340. The obvious question is how did all this relate to the parentage of cubs?

Sex and reproduction—or not

We saw every female mate, but adult female badgers do not produce cubs every year (see Chapter 11). Amongst the females we had seen mounted most, five showed no sign of distended teats the following spring (teat size is an indicator of having nursed cubs; Dugdale et al. 2011). Subsequently, our genetic pedigree confirmed that no cubs had been recruited from these five females that year (acknowledging that cubs which died before spring do not enter the database, this would nonetheless cause a degree of teat distension depending on how long suckling cubs survived). So, which of this cast of multiply mating badgers was assigned cubs by genetic tests? Some readers, like only one of us, will be happily riveted by the personal details of badgers that lead to the answer to this question. For those, read Box 7.1. Other readers, feeling like Methuselah that a thousand years pass as they read, will be reminded, less happily, of schoolroom Old Testament memories ('And Adam begat Seth; and Seth begat Enos, and Enos begat Kenan, and Kenan begat Mahalaleel, and Mahalaleel begat Jared, and Jared begat Henoch, and Henoch begat Methuselah, and Methuselah begat Lamech', Chronicles 1:1). For those readers who find the detail more dreary than delightful, skip Box 7.1 to arrive at the punchline in the next paragraph.

Methuselah's genealogical journey is complete. Amongst this bewildering detail of personalities and intimacy, what is the bottom line: which badgers in the mating race of 1995 arrived at the finishing line with cubs in 1996? None. Of all 332 mounting episodes we recorded at Pasticks and Sunday's Hill in our first marathon of mating videos, observed across 46 separate calendar nights, not a single one produced living offspring that survived to enter our database, assigned by pedigree. While this is a sobering reminder of the challenge, in the wild, of delivering reproductive success, happily we can glean more conclusions by amalgamating data from our second marathon of video

---

Box 7.1 Methuselah's chronicles

The timing of the mountings we watched in early 1995 should be associated, thanks to delayed implantation (DI), with cubs born in 1996. Of all the 13 females, only F301 produced cubs in 1996 (F526, who died in infancy, and M533) at Pasticks. These were fathered by M371. F301 had previously had two cubs (M495 and M497; Figure 7.3) in 1995, sired by M238. None of the Sunday's Hill resident females were assigned cubs in 1996. Indeed, four of these Sunday's Hill females (F323, F2?0, F223, and F327) were never assigned cubs—a lifetime reproductive success of zero.

F134 had a lifetime assignment of five cubs (Figure 7.4), all fathered by M1 (across 4 years); the last of these, however, F468, was born in 1995, so her matings in 1995, described earlier, bore no (surviving) fruit. F305 did bear a cub (F594) sired by M317, although this was in 1997. F461 had a single lifetime cub in 1995 (F1022) fathered by M873. Finally, F400 went on to be a relatively prolific mother, assigned four cubs: cub F324 was fathered by M222 in 1993; M300 fathered her two cubs, F591 in 1997 and F779 in 2000; and M367 fathered her last cub, F819, in 2001.

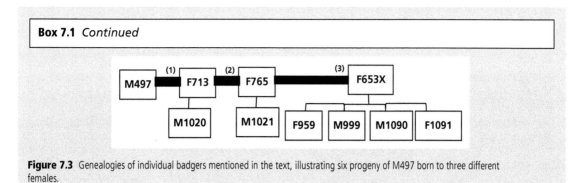

Figure 7.3 Genealogies of individual badgers mentioned in the text, illustrating six progeny of M497 born to three different females.

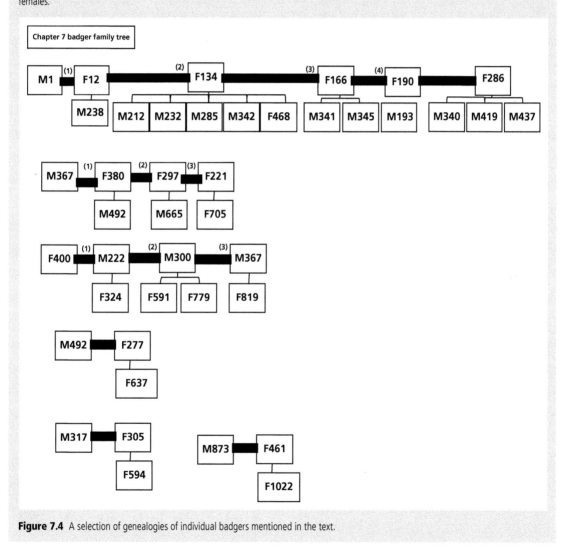

Figure 7.4 A selection of genealogies of individual badgers mentioned in the text.

Box 7.1 *Continued*

So much for the reproductive success of those multi-mated females. What of the males?

Of the 19-strong male cast only 3 were assigned young in 1996, sired during a 1995 mating: M232 (Figure 7.5) produced cubs F558 and F507X with mother F399 (an off-camera mating as F399 hailed from Brogden's Belt from the neighbouring super-group); M343 sired two cubs, one definitely (F566) and the other probably (F529) assigned to mother F134 (resident at Sunday's Hill, but not amongst those we saw to be mounted). M411 fathered cub F532 (which died in her natal year) with mother F284 (with Marley 3a as natal group). Indeed, of these 19 males, 8 were assigned no cubs throughout their lifespans. In contrast, M1

was a prodigious breeder, leaving 12 offspring (all of which bar 1 were sons), to 5 different dams,[12] of which the last was assigned in 1994, and was thus not a product of the numerous matings we recorded in 1995.

Of those males mating in 1995, to father cubs in 1996, only M343's reproductive success was entirely limited to that year. M232 had also fathered a prior cub with F140 in 1993 (M379) and a pair with F277 as mother in 1995 (M462, F464). Similarly, M411 did better with his reproductive outcomes after 1996, producing three cubs in 1999; two of which (M709 M750) were born to F351, the other (M711) born to F353 (Figure 7.6).

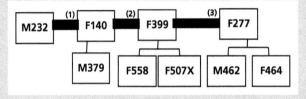

Figure 7.5 Genealogies of individual badgers mentioned in the text, showing five cubs sired by M232 with three females.

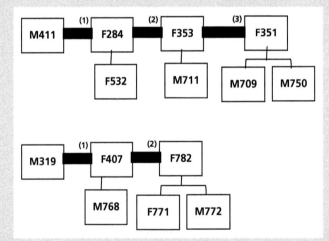

Figure 7.6 The mates and offspring of Males 411 and 319, as discussed in the text.

[12] F12 = M238 (1991); F134 = M212 (1990) M233 (1991) M285(1992) M342 (1993) F468 (1995); F166 = M341 (1993) M345 (1993); F190 = M193 (1990); and F286 = M340 (1993) M419 (1994) M437 (1994).

Box 7.2 More chronicles

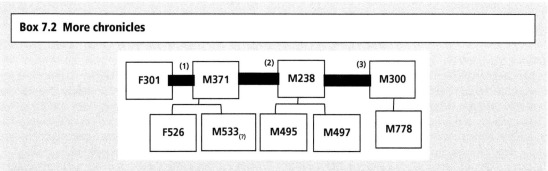

Figure 7.7 Genealogies of individual badgers mentioned in the text, depicting the offspring of F301 sired by three different males.

Recall that M and Sandy, M367 and M300, together accounted for 51% of all mountings in the 1995 conception season. Neither of these males was assigned cubs in 1996 (although both did sire cubs subsequently). A sceptic might wonder correctly whether the apparent failure of all those 332 mountings was a delusion because we misassigned paternity. That seems unlikely, because although the mountings we saw led to no assigned cubs, other matings did so.

Only F301's cubs (Figure 7.7) (F526 and M535) were born at Pasticks that year, and only M343's cubs with F134 (Figure 7.8) (F529 and F566) were born at Sunday's Hill. Actually, four cubs born to two females in these linked groups is not an atypical result. Further, as we repeatedly emphasize, our cameras recorded only a small, carefully positioned field of view, so our sample of mountings represents only a small proportion of the total. While F301's 1996 cubs were not fathered by a cast member, and the actual father, M371, was not caught in 1995, he *was* caught both in earlier and later years at Jew's Harp: probably that's where he was based, as the boy next door, while we were watching the numerous matings of F301 at Pasticks sett.

Similarly, Pasticks resident M232's 1996 cubs were born to mother F399, who was residing in the Brogden's Belt group to the north-west. This liaison was therefore not merely between groups, but across the super-group divide. Even more extreme, in terms of gallivanting, M411 (a permanent Sunday's Hill resident) conceived cubs with F284 in 1995, a remarkable fact given that she lived half the woodland away at the Holly Hill/Common piece super-group:

their liaison stretched over a kilometre, and several group ranges.

F134 was the 1996 mother of M343's cubs, although she resided at Sunday's Hill. Interestingly, although M343 was caught regularly around Sunday's Hill throughout his life (born at Sunday's Hill in 1993, returned there in 1996 and 1997 after his capture at Brogden's Belt in 1995, then moved to Radbrook Common (next sett to west) in late 1997 and 1998, although caught once (of four times) also at Pasticks Outlier in 1998 (next sett east of Sunday's Hill) not being caught thereafter), in 1995 he was caught at the neighbouring Burkett's Plantation sett (*c.*150 m west). He appears to have aged into something of a vagabond (11 of his 15 lifetime captures were at Sunday's Hill). Furthermore, at the time of his fruitful mating in 1995 with F134, she was 5 years his senior, he being only 2 in 1995.

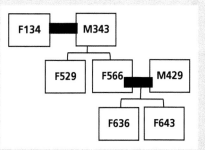

Figure 7.8 Genealogy of badgers mentioned in the text, depicting the granddaughters, F636 and F643, of M343 and F134.

analysis of matings in 2004–2005 (and thus births in 2005–2006) (Dugdale et al. 2011). Overall, this revealed that males assigned more cubs in a year had a higher index of body condition in the months (May to August) following the post-partum mating season in which the cubs were conceived than did males that were assigned fewer cubs, controlling for age, social group, year, and repeated measures on an individual.

The relationship is both important and perplexing. It is important because it's the only generality we could detect on reproductive success, and it held when restricting the data from all 453 badgers involved to 188 badgers of known age. However, the number of mounting events did not correlate with the number of offspring to which an individual was assigned parentage of the following year. We did not observe any of the 15 assigned parent pairs (of the cubs born the following year) mounting together. It is also perplexing because it is the exact opposite of the, albeit statistically tentative, first answer to the question of what characteristics were associated with making the greatest number of mountings. We arrive at a point where a statistical hint that the thinnest males mounted most frequently turns into the reality that the fattest badgers sire most offspring. We'll return to this puzzle shortly, but first, for those strong-hearted enough to bear another bout of Methuselah's ancestry, remember males M and Sandy (M367 and M300) at Pasticks and Sunday's Hill, who monopolized 51% of mountings in early 1995 (Box 7.2; for the faint-hearted, skip the box).

Thus far, this chapter has been concerned with the sexual encounters amongst group members. Sometimes, however, a newcomer would arrive in our field of view, cause a copulating male to pause, and possibly even to dismount (several times) to chase away the newcomer; this created interludes during which the female might wander off, or a different male might mount her while the original male squabbled with the introducer. Our suspicion was that some of these interlopers were visitors from neighbouring groups. On one occasion, we saw an identifiable extra-group male arrive at a sett and mount a resident female, at a moment when no other male was present on screen (Dugdale et al. 2011b).

When is oestrus?

For each female badger we saw periods of 2 or 3 nights during which sexual activity was intense, and we called these periods behavioural oestrus. Oestrus is the period during which mammals are generally receptive to sexual activity, coinciding with the presence of ripened ova that are ready to ovulate, either spontaneously or, in the case of badgers, when induced by intromission. How does oestrus map onto the timing of inductive and procreational mountings, and onto the observed period of behavioural oestrus? Further, insofar as we recorded high levels of mountings each month between February and July (Figure 7.1), and at least some mounting in every month of the year, we ask how many oestrous periods do the badgers of Wytham Woods experience annually, how synchronized are those periods between individuals, and, as seems likely, are they mounting, and having non-conceptive sex outside of oestrus, and, if so, why? Nadine Sugianto, introduced in Chapter 1, devoted her doctorate to solving our store of endocrinological questions.

During the 25 years that had passed since our observations of the vignette of the five Bs, guided by the literature, we had assumed that the badgers of Wytham Woods had at least two oestrous periods a year—one post-partum in early spring (late February, early March), and another in autumn (Neal and Harbison 1958). In line with this general belief, as recently as 2015 spring and autumn oestruses were confirmed by Corner et al. (2015) in the Republic of Ireland, and this broadly fits the findings of Katrina Service, an earlier post-doc on our project, when she and her colleagues studied hormones in the urine of urban badgers in Bristol (Hutchings et al. 2002). In that lower-density population (4.4–7.5 badgers km^{-2}) they found female badgers cycling through a series of up to five oestrous cycles, at intervals of about 28 days, from late winter through to autumn. A key point is that these observations mean that during the period of DI, between post-partum oestrus in February/March, and the start of embryonic development in December, badgers were entering additional periods of oestrus during which they could add *in utero* to their increasing litter of blastocysts. Two non-exclusive hypotheses offered functional explanations of the bi- or multi-annual oestrus.[13] First, fertility assurance, the autumnal oestrus providing a second bite of the cherry for any female not

[13] The vocabulary of oestrus is: polyoestrous, oestrous cycles throughout the year (cattle, pigs, mice, and rats); seasonally polyoestrous, multiple oestrous cycles only during certain periods of the year (horses, sheep, goats, deer, and cats); monoestrous, one breeding season per year (bears, foxes, and wolves).

conceiving in spring (see Yamaguchi et al. 2006).[14] Second, in his book, *The Velvet Claw*, David had suggested that the second oestrus provided what might be a sociologically crucial opportunity for the fatherhood of a litter to catch up with changes in male membership of the group (Macdonald 1992).

Both hypotheses struggle to accommodate emerging knowledge of the badgers of Wytham Woods: surrounded by amorous males, it is apparent that females would be unlikely to need a second bite of the cherry, and the sociology of selecting the best male is more complicated considering the frequent exchanges of

males between groups documented in Chapter 3, the broad participation in mating scrums, and the fact that superfecund badgers conceive an excess of blastocysts, only to implant a small proportion, implying cryptic selection for quality and immuno-compatibility. Furthermore, the need to explain the second oestrus was turned on its head when Nadine found that it does not, at least nowadays, generally exist in Wytham (Sugianto et al. 2019b, 2021b). On the contrary, in the Wytham population as a whole there is now only one, early spring post-partum hormonal oestrus annually. This prompts a swarm of new questions, including why the number of oestruses in European badgers per annum varies from place to place, whether it has changed from two to one in recent years in Wytham, and, if there is only one oestrus a year in Wytham, what the badgers

---

[14] Unfertilized raccoons, or those losing their litters soon after parturition, may have a second oestrus around 4 months later in the year (Hirsch and Gompper 2017).

Box 7.3 The interaction of delayed implantation and superfoetation

Badgers are reproductively unusual, being one of only two species of eutherian mammals, both mustelids, to combine delayed implantation with superfoetation (DI × SF): the other is the American mink (*Neovison vison*; Yamaguchi et al. 2004); however, the two species do it differently. In mink, DI is of short duration (average 27.5 days) and SF occurs during a single additional oestrus. Badgers, uniquely, have one of the longest known durations of DI, up to 330 days (Canivenc and Bonnin 1981; Mead 1981) spanning from the post-partum oestrus around mid-February, after which SF can occur during one or several additional oestrous cycles (Canivenc and Bonnin 1981; Corner et al. 2015; Roellig et al. 2011), until blastocyst implantation is triggered by short-day photoperiod in winter (Bonnin et al. 1978). True gestation then lasts 47–51 days (Canivenc and Bonnin 1981). This DI × SF mating system, with extended female receptivity, is also reflected in male circulating plasma testosterone titres (Buesching et al. 2009; Sugianto et al. 2018, 2019b, 2020) and spermatozoa levels (Page et al. 1994), which peak around March, coinciding with the main post-partum mating season, and then decline to a minimum in October or November when testes ascend into the body cavity (Sugianto et al. 2018; 2019b).

From post-mortem data, supported by progesterone assays, Corner et al. (2015) reported the presence of secondary and tertiary follicles throughout the year in badgers

culled from a population at relatively low density (1.9 badgers km^{-2}) in the Republic of Ireland;[15] blastocyst turnover indicating that mating was taking place throughout the period of DI. This corroborates Hutchings et al. (2002), who, based on sex hormone measurements from urine, reported a series of up to five oestrous cycles at about monthly intervals between late winter/spring and autumn, at moderate population density (4.4–7.5 badgers km^{-2}). Two other studies, at higher population densities (although half that of contemporary Wytham) of 25 badgers km^{-2} (in south-west England, Mallinson et al. 1992; Page et al. 1994; and Woodchester Park, Carpenter et al. 2005), both revealed the main post-partum peak and secondary autumnal peak that we had incorrectly assumed was the general rule at Wytham. Mallinson et al. (1992) reported that c.40% of females had conceived blastocysts by the end of March, increasing to 80–90% by the end of April.

---

[15] Secondary and tertiary follicles are the later phases of follicular development (cluster of cells containing an ovum that simultaneously develops within it); tertiary being the final form before ovulation. The presence of both of these follicles all year round indicates that females within this population do indeed have the ability to reproduce throughout the period of DI due to SF as indicated by the blastocyst turnover during this period.

Box 7.3 *Continued*

Nadine used enzyme immunoassays to measure population-level seasonal variation of circulating sex steroids for 97 females in a cross-section of badgers. She assayed oestrone and oestradiol. Both oestrone and oestradiol are oestrogens. Oestrone is secreted by badgers to maintain a suitable uterine environment for the survival of pre-implanted blastocysts during DI, and extends the lifespan of the corpus luteum (which prevents the blastocysts from being aborted and allows implantation by producing high progesterone in winter), while oestradiol functions mainly to regulate reproductive development and mating activity in females.

Oestradiol was highest in spring in non-parous females, decreased in summer, and remained low during the following seasons, evidencing control of spring oestrus.[16] Remember that oestradiol is the oestrogen that regulates mating behaviour and associated changes in females. Oestradiol was high only during spring (and was very low in other seasons)—this is the evidence that amongst the badgers of Wytham Woods, females are monoestrous; that is, they have only one oestrous period each year.

Oestrone (Figure 7.9) was consistently higher than oestradiol (Figure 7.10); it was elevated in spring, lowest during summer, peaked in autumn, and remained elevated

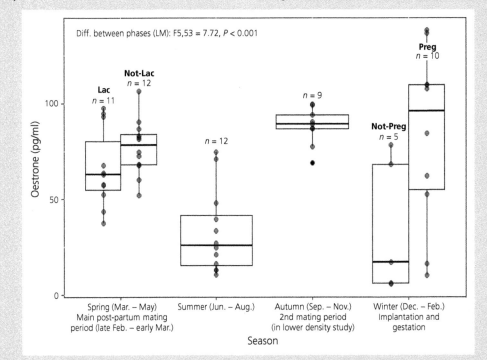

Figure 7.9 Variation in oestrone levels with seasons and reproductive stage. The main mating period in this population occurs in early spring (late February to early March) during the post-partum period, hence during this period some females may be lactating (Lac) and some not (Not-Lac). Summer for badgers tends to be the period of lowest food abundance. For lower density populations, the second mating period occurs in late summer to early autumn (August–September). Afterwards there is a period of reproductive quiescence (inactivity) throughout late autumn (October–November). The delayed implantation period itself spans from spring to early winter (December). In winter, implantation of the blastocysts accumulated throughout the delayed implantation period occurs and gestation commences (hence two categories of females, not pregnant (Not Preg) and pregnant (Preg); Sugianto et al. 2021b).

[16] This is consistent with measures of female oestrogen levels during DI in female black bears (*Ursus americanus*; 7–8 months DI; Garshelis and Hellgren 1994; Tsubota et al. 1987), where oestradiol levels were also high only during oestrus and low during the period of DI. Similarly, in spotted skunks (*Spilogale putorius*; 6–7 months DI; Mead and Eik-Nes 1969), oestradiol levels were only detectable during oestrus, late stage of pre-implantation, and post-implantation.

Box 7.3 *Continued*

for pregnant females in winter, indicating that oestrone sustains pre-implanted blastocysts throughout DI.[17] Progesterone was low throughout (Figure 7.11), and increased significantly only during winter pregnancy, associated with

implantation and luteal development, triggered by a short-day photoperiod (Bonnin et al. 1978; Yamaguchi et al. 2006). We concluded that there is a beautiful balance, such that the low levels of progesterone secreted by the corpus

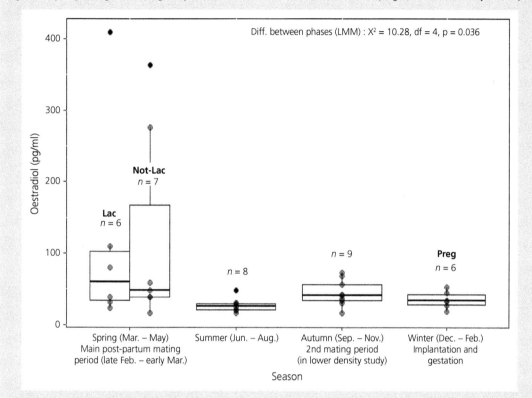

Figure 7.10 Variation in oestradiol with season and reproductive stage. The main mating period in this population occurs in early spring (late February to early March) during the post-partum period, hence during this period some females may be lactating (Lac) and some not (Not-Lac). Summer for badgers tends to be a period of lowest food abundance. For lower-density populations, the second mating period occurs in late summer to early autumn (August–September). Afterwards there is a period of reproductive quiescence (inactivity) throughout late autumn (October–November). The delayed implantation period itself spans from spring to early winter (December). In winter, implantation of the blastocysts accumulated throughout the delayed implantation period occurs and gestation commences (hence two categories of females, not pregnant (Not Preg) and pregnant (Preg); Sugianto et al. 2021b).

[17] The main function of oestrone is to sustain pre-implanted blastocysts during DI. This matches our findings that oestrone is high during spring (after mating season) to autumn (which is the known period of DI). It also accords with our findings that oestrone is high in winter only for pregnant females, insofar as non-pregnant females will have no blastocysts. Summer is a period of nutritional stress during which some females probably reabsorb their blastocysts, causing the population oestrone level to drop during this season.

Box 7.3 *Continued*

luteum during embryonic diapause were insufficient to inhibit further ovulation, but sufficient to maintain comfortable uterine conditions for the pre-implanted blastocysts (without the fatal trauma of menstrual shedding, which would have swept away those pre-implanted blastocysts).[18] We analysed these hormone data at the population

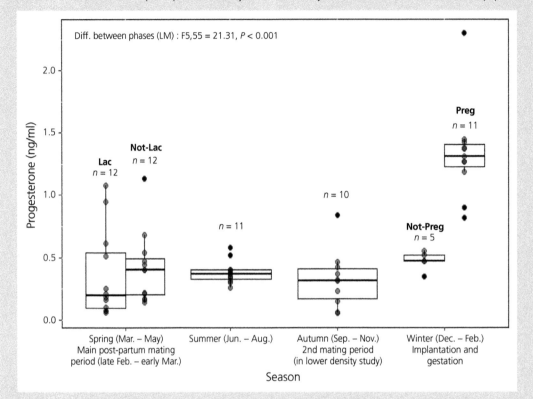

Figure 7.11 Variation in progesterone with season and reproductive stage. The main mating period in this population occurs in early spring (late February to early March) during the post-partum period, hence during this period some females may be lactating (Lac) and some not (Not-Lac). Summer for badgers tends to be a period of lowest food abundance. For lower density populations, the second mating period occurs in late summer to early autumn (August–September). Afterwards there is a period of reproductive quiescence (inactivity) throughout late autumn (October–November). The delayed implantation period itself spans from spring to early winter (December). In winter, implantation of the blastocysts accumulated throughout the delayed implantation period occurs and gestation commences (hence two categories of females, not pregnant (Not Preg) and pregnant (Preg); based on Sugianto et al. 2021b).

[18] Badgers can undergo SF because the corpus luteum (the tissue formed in the ovaries to accompany each ovum that is ovulated and which survives only if the ovum is fertilized) and which functions to produce progesterone, produces only a small amount of progesterone (due to it being underdeveloped during the DI period), delicately balanced such that it neither triggers pregnancy nor inhibits further ovulation. This balance keeps the female badger's reproduction in a state of suspended development, with all options open.

Box 7.3 *Continued*

level—questing for generalizations—so this conclusion does not preclude the possibility of some individuals being alluring, and even conceiving outside of the main peak of oestrus.[19]

We could detect no significant difference in sex-steroid levels between females that lactated and those that did not. The significance of this is that it means that females were able to conceive while lactating, shortly after parturition.

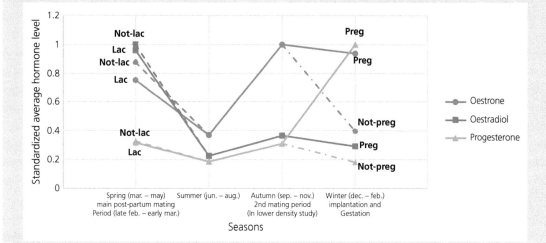

Figure 7.12 Variation in standardized measures of oestrone, oestradiol, and progesterone with season and reproductive stage. The main mating period in this population occurs in early spring (late February to early March) during the post-partum period, hence during this period some females may be lactating (Lac) and some not (Not-Lac). Summer for badgers tends to be a period of lowest food abundance. For lower density populations, the second mating period occurs in late summer to early autumn (August–September). Afterwards there is a period of reproductive quiescence (inactivity) throughout late autumn (October–November). The delayed implantation period itself spans from spring to early winter (December). In winter, implantation of the blastocysts accumulated throughout the delayed implantation period occurs and gestation commences (hence two categories of females, not pregnant (Not Preg) and pregnant (Preg); based on Sugianto et al. 2021b).

[19] We speculate that although one corpus luteum (for one blastocyst) may not produce enough progesterone to prevent further ovulation (as it is underdeveloped), the presence in the badger's womb of multiple corpora lutea from multiple blastocysts accumulated during the mating period may produce a sufficient shower of progesterone to stop the body from cycling but insufficient to prompt implantation (which eventually occurs in winter by which time the corpora lutea are fully developed, triggered by LH secretion due to the change in photoperiod). Perhaps individuals that did not accumulate this 'threshold' of blastocysts may still cycle (as their progesterone levels are not high enough to prevent it) and may therefore conceive outside the main mating period. Thus Nadine Sugianto led us to the idea that in higher population densities where most females are likely to fulfil this 'quota' of fertilized eggs during the main mating period most of the population may then stop cycling, whereas in lower-density populations, where mate availability may be low, the blastocyst 'quota' may less often be achieved and thus individuals may cycle and mate all year long. In Wytham, this results in a generalization, at the population level, that most females opt for one oestrus, which generally supplies all the blastocysts they need, whereas some continue to cycle sufficiently to produce at least a strong whiff of sexual interest, and perhaps even leave the door open for additional conceptions in autumn or throughout the year.

Box 7.3 *Continued*

This is unusual in that in most mammal species lactation inhibits oestrus, because high levels of prolactin cause a decrease in GnRH, thereby inhibiting ovarian folliculogenesis (causing low levels of oestrogens) and inhibiting also the pulse of LH secretion required for ovulation and subsequent formation of corpora lutea (Jameson et al. 2013). Finally, Nadine's work suggests that females with no blastocysts from a first round of mating may then have additional cycles—indeed, and importantly—even females with insufficient blastocysts (and attendant corpora lutea) to generate sufficient progesterone to put a full stop of subsequent ovulations could continue to be at least alluring and possibly fertile (a speculative mechanism we elaborate in footnote 19). The purist might consider this outwith the definition of SF, which, strictly speaking, involves a female adding to her tally of unimplanted blastocysts with subsequent secondary cycles, which is a different thing to starting the mating–pregnancy cycle again.

In summary, and for the purposes of visualization, the hormone levels depicted in Figures 7.9–7.11 can be amalgamated to a standardized median score (ranging between 0 and 1) for all three hormones. This is depicted in Figure 7.12 to show the fluctuations relative to one another within the same period.

---

are doing mating over several months and, at least in some cases, during every month? This takes us into the realm of ART, or alternative reproductive tactics (Taborsky et al. 2008), in which, despite a historical focus on males (reviewed in Kokko and Rankin 2006; Taborsky and Brockmann 2010), females too can adapt their reproductive behaviour and/or physiology to situations of high or low mate availability (Kokko and Mappes 2005).

Generally, once fertilized, females, as the limiting sex (Chapter 11), are more likely to experience too much, rather than too little, sexual attention from males, particularly at high population densities, and are thus likely selected for reproductive strategies to discourage costly attention from males (Kokko and Rankin 2006; Wolff and Macdonald 2004). In mammals, this is especially important if additional oestrous cycles (e.g. Buchanan 1966) or superfluous matings (e.g. presenting risks of contracting sexually transmittable diseases; Thrall et al. 2000) have the potential to be harmful to the female or to her reproductive output (Chapman et al. 2003).

DI and SF evolved for different reasons (Yamaguchi et al. 2006), but in combination they create a pre-adaptation (or exaptation; Gould and Vrba 1982) such that the DI × SF combinations provide the European badger with what is essentially a unique physiological mechanism adapting its reproductive biology to a wide variety of ecological, and thus sociological, circumstances.

The earthily phlegmatic term 'convenience polyandry' was coined by Thornhill and Alcock (1983) (but see Lee and Hays 2004; Rowe 1992) to describe one outcome for which females may opt when balancing the gains (in fecundity) and risks

(of mortality) associated with multiple matings. That balance may be relevant to two puzzles occasioned by the behaviour we have observed amongst the badgers of Wytham Woods, and differences between their behaviour and that of badgers elsewhere. One puzzle is why females accept mountings with such apparent nonchalance (remembering some mountings may not be matings, and some matings may be inductive while others will be procreative). The second puzzle is why females in Wytham, on a population level, have only one oestrus, and those in some other places have two, or more. With respect to both puzzles, it is noteworthy that the balance of gains and risks of multiple mating differs when population density is higher or lower (Kokko and Rankin 2006). In low-density populations, female badgers in oestrus may face a greater risk from failing to encounter sufficient males than from encountering too many. This is the general circumstance leading to females in low-density populations remating more willingly than those in high-density populations (Härdling and Kaitala 2005). Indeed, Corner et al.'s (2015) suspicion that the breeding rate in their low-density Irish population may be limited by failure to conceive illustrates the reality of this risk for badgers. Under these lonely conditions the DI × SF combination enables a female badger to secure multiple matings when it pays her to do so, thereby ensuring fertilization, and also provides her with opportunities for cryptic mate choice through selective implantation of only the 'most suitable' blastocysts (Andersson and Simmons 2006; Chapter 17). Conversely, at very high densities, males are on hand to the point of nuisance and, with their (super-) abundance, fertilization is assured; then the net consequence of additional oestrous cycles may be weighed heavily by increases in such life-threatening risks as endometriosis and sexually transmitted diseases (e.g. genital herpes: Kent et al. 2018; Tsai et al. 2020; Chapter 15). Further exacerbating the already dangerous corollaries of superfluous matings by, for example, risking wounding and discord through repulsing suitors, or provoking competition between them, may simply not be worth it. It was this balance of considerations, in which resisting harassment may be more costly than accepting matings, that led Annavi et al. (2014a, 2014b) to suggest that the observed easy come, easy go, approach during mating scrums was an expression of convenience polyandry. This may explain not only why, for females (and consequently for males), it is not worth fighting over matings during one oestrus, but also why, for most females, one oestrus a year is enough. But answering both these questions immediately flushes out another, potentially even more interesting question: why do the badgers of Wytham Woods persist in sexual activity outside the main oestrous period? We explore that question later, and yet again it will lead us to the recurrent issue of whether the behaviour of the badgers of Wytham Woods should be interpreted as nuanced adaptation or wayward anachronism.

Considering counter-factuals—'what if' questions—often aids the quest to diagnose adaptive significance, although these types of questions sometimes engender the risk of arriving at Just-So answers. Myriad temptations to speculate arise from this story of badger oestrus. If, for example, the badgers of Wytham Woods can afford to drop one of two oestruses, why drop the autumnal one, retaining the 10–11-month DI, rather than going straight for a conventional gestation? Much about this topic emphasizes the anchors of phylogeny—the Irish joke, in the tradition of the wise fool, answers the question of how to get somewhere with the response, 'I wouldn't start from here'. But in adapting to their new, extreme, circumstances, the badgers of Wytham Woods must start from 'here', where they have arrived from a several-million-year history of DI and other adaptations to low population density, fossoriality, omnivory, and the rest (see Chapter 19).

What is the adaptive significance of the single annual (population level) oestrus in Wytham? The answer lies at the intersection of three reproductive phenomena: the timing of the single oestrus, induced ovulation, and DI, all in the context of the badgers' evolutionary history and phylogenetic baggage (Chapter 19). Why does the number of annual oestruses differ between badger populations? The answer seems to relate to population density in ways explicable by the badgers' baggage of mustelid phylogeny. So, the badgers studied long ago by Ernest Neal (Neal and Harbison 1958), with their spring and autumn biannual oestruses, likely lived at a population density less than half that of their Gloucestershire descendants today, and those, also with two oestrous periods annually, studied in 2015 by Corner et al. lived at 1.9 badgers km^{-2}. In 2017 when we analysed their endocrinology, Wytham's badgers were living at 20 times Corner's density, $c.40$ badgers km^{-2}. Even Hutchings et al.'s Bristolean badgers were living at about one-tenth the density of those at Wytham (4.4–7.5 badgers km^{-2}). We do not know the oestrous schedule of those living in Poland at 0.04 badgers km^{-2} (Griffiths and Thomas 1993), but we would predict they work hard to achieve fertility assurance.

Sperm

Considering the promiscuous matings described earlier, we sought to study badger sperm and the competition between them. An amusing anecdote compensated for our early failure. A local, much respected, vet, remembered fondly for his eccentricity, was the regional specialist in artificial insemination. Using his standard electro-ejaculation approach, he set about delicately inserting a rectal probe into a sedated badger, to stimulate its prostate. Accustomed to ready success with pigs, sheep, and goats, the vet grew conspicuously perplexed at the badger's lack of performance—and feared we did of his. Eager to demonstrate that his equipment was not at fault, he whipped the probe from the subject badger's anus, and inserted it directly into his mouth to test for voltage on his tongue. Misinterpreting our wide-eyed expressions, he cheerily assured us not to worry, explaining that the voltage delivered only a tingle to his tongue! Years later, we were to make more headway with badger sperm through the efforts of Ming-shan Tsai, our Taiwanese doctoral student, although this time extracted post-mortem from badgers culled in Ireland.

Functional significance—why?

A clipped summary, then, of observations of mating amongst the badgers of Wytham Woods is that all adults of both sexes mate, each with more than 1, and up to at least 13 partners (and an average of about 4). For context, there are parallels amongst badgers' musteloid, inductively ovulating, distant kin: several male raccoons can mate with a single female (and a male may mate with several females) resulting in 88% of litters having multiple paternity (Nielsen and Nielsen 2007). Similarly several male coatis mate with each female, although, in contrast to badgers, amidst much aggression (Hirsch and Gompper 2017). As a model of traditional mustelid behaviour we settled upon American mink: they too illustrate multiple paternity (Yamaguchi et al. 2004) and, as we detail in Chapter 17, when given the chance, their females are eagerly promiscuous (Thom et al. 2004b). What evolutionary explanations might underlie these observations? The basics for 'standard' spontaneously ovulating mammals are that because males produce many thousands of small gametes[20] they are expected to mate with as many females as possible, whereas females are predicted to be choosier because they produce much fewer and also larger gametes and often invest more in parental care (Trivers 1972) (although other factors such as sex-specific mortality rates may alter this; Kokko and Jennions 2008).

Mindful that the phylogenetic baggage carried by badgers, and shared with mustelids with more traditional lifestyles (we opt particularly for comparisons with American mink that lived on the Thames beside Wytham, Chapter 17) was in part also shared by other mammals, David turned to representatives of the most ancient lineage of mammals to scuttle (in tiny 30-m diameter ranges) in Wytham beneath the badger's feet: the common shrew, *Sorex areneus*. Although they last shared a common ancestor some 90 million years ago, some of the problems that ancestor faced might be shared by badgers and shrews today, or even by the shrew-like morganucodontids from which all mammals descend more than 200 million years ago (Mya). David and his student Paula Stockley, whose distinguished career has led her to be Professor of Evolutionary Ecology at Liverpool University, asked common shrews exactly the same questions we were putting to badgers about their mating habitats. There were interesting similarities in the answers, and they put our badger research in a wider mammalian context. Young male shrews fell into two types: some matured early; some late (see Chapter 2 for the badger parallel). The early sexual maturers expanded their youthful 30-m diameter ranges to about 70-m diameters, and thereby

overlapped the ranges of several females (remember the semi-detached intra-sexual territory of ancestral badgers and contemporary mink), whereas late developers stayed in adolescent-sized ranges but ranged far and wide (200 m or more) in search of females. Pioneering DNA studies revealed not only that most litters had three fathers (one had six) but also that the early maturers inseminated more females and sired more young, and did so in direct proportion to the number of female ranges they overlapped in their semi-detached way. However, the late developers had an ace up their sleeves, or actually somewhere cosier: their sperm counts were higher than those of the early developers, and the higher their sperm counts, the more young they sired (clawing back some, but not all, of the advantage they had lost by their developmental late arrival). Paula and David noticed that many of the mating shrews were close relatives (think badgers again) and so they asked whether these incestuous liaisons were problematic. They were: the offspring of sibling matings were smaller at weaning and less likely to survive until adulthood—remember that finding when we come to consider parenthood in badgers in Chapter 17 (Stockley et al. 1993, 1994, 1996). Paula and David concluded that one advantage of multiple paternity could be to reduce the proportion of embryos likely to be inbred. These tiny shrews illustrate that the questions we now ponder for badgers are widely relevant amongst mammals.

What opportunities does a female mammal have for selecting between candidate mates? Remembering that the quality of the males making inductive intromissions may have little consequence, once ovulation is triggered (induced) the female may have an interest in, but no control over, male quality during a critical window for fertilization. Darwin (1859, p. 175) anticipated the shift in competitive arena when in the *Origin* he wrote 'All Nature . . . is at war. . . The struggle very often falls on the egg & seed'. Sperm competition provides another, albeit cryptic, selection process as the ejaculates from a multitude of males race to the female's awaiting ova, with selection for the fastest swimmer against a tide of immuno-incompatibility, and the chances of winning being enhanced by the number of 'tickets' each male enters into the 'raffle' (Parker 1990). It may therefore be relevant that (see earlier) females mounted by several males in one night were mounted by each of them more than twice as often as were those mounted by only one male. However, the relatively small testis size of European badgers suggests that sperm competition has not been paramount in their evolution, and, surprisingly, may not be today.

[20] As a point of comparison, normal canine spermatozoa are 6.8 μm in length, of which the midpiece is 1.1 μm and the tail is 5.0 μm. Dog semen ranges in volume from 1 to 30 ml per ejaculate and contains 300 million to 2 billion sperm, of which more than 70% are progressively motile and morphologically normal.

Across mammals, the general rule is that testis size is relatively larger in species in which competition between males for females plays out largely *in utero*. Badgers have quite small testes compared with a 10 kg dog. This is unexpected considering the general rule that higher levels of multiple paternity are associated with larger relative testes size, supporting the hypothesis across multiple mammalian taxa that relative testes size reflects the intensity of male post-copulatory intra-sexual competition. If we infer that the size of their testes suggests that badgers are likely not subject to much post-copulatory sperm competition, why would this be so? This is a question that we might well ask, especially considering that the mating scrums and multiple male mating we observed might have been thought exactly the circumstance that would intensify sperm competition. Three answers come to mind. The first is that the theory, while generally supported, simply may not apply to badgers—exceptions prove rules and in this case the capybaras studied by David and his student Jose Roberto (known as Beto) Moreira investigated a species that bucked the testis size trend. Female capybaras swim round in a lagoon while a gang of males try to mate with her—exactly the circumstances where sperm competition might be expected. Amongst the males a dominant will, as swiftly as their rather laborious swimming allows, divide his attention between mounting the female (she gets dunked beneath his weight) and shepherding away half a dozen or more subordinate males. Beto and David expected that capybaras would have large testes, and dominants would have the largest of all. As with the badgers, the expectation was not fulfilled (Moreira et al. 1997a, 1997b). Capybaras testes were small and, moreover, under the microscope they were packed with the highest proportion of interstitial tissue known amongst mammals, rendering even smaller the volume of spermatogenic cells. These interstitial cells are the factory that produces testosterone. It seemed the dominant male capybaras were the product of different evolutionary investment decisions. Male capybaras have a unique, bulbous scent gland on their snout, the morrillo (aptly, Spanish for hillock), which glistens with sebaceous goo, and dominants have the most bulbous and glistening morrillos. Testis size correlated with nose-gland size, but the extra testicular bulk is made up of yet more interstitial tissue producing androgens—these dominants are awash with testosterone, and one might speculate that they need it to motivate the endless shepherding of subordinates. It seems as if dominant male capybaras make their investment in androgen-dependent signals and

prowess, not sperm production. In contrast, subordinate capybaras have smaller testes, but a greater proportion of spermatogenic tissue, and concentrate their efforts on sneak matings (which collectively outnumber matings by the dominant, despite his getting the largest share). Perhaps badgers, like capybaras, don't follow the general rule, but why not? The second possible answer as to why badgers do not have large testes to produce flotillas of sperm might be that despite the apparent promiscuity, and thanks to induced ovulation, the female may have precise control over which male inseminates her during the critical ovulatory window. That seems unlikely from our observations. The third, more likely, answer is that the selection is made later (Chapter 17), later even than fertilization, somewhat exempting the males from sperm competition or any other vote on the outcome. DI provides a leisurely opportunity—10.5 months—for the female gambits such as the selective implantation and death of blastocysts (we mentioned this for roe deer and capybaras in Chapter 1 in the context of badger sex ratio manipulation). This mechanism would vest all control with the females, and render large, spermiforous testes irrelevant baubles in the mating game. In line with this speculation, Page et al. (1994) found unimplanted blastocysts in female badgers for 11 months of the year (and in over 80% of adult females in July–November), yet in January and February only 32% of adult females had implanted embryos: implantation was thus a case of 'some or none', and of those that implanted 36% of foetuses died. Males find themselves surrounded by females prepared, indeed seemingly eager, to mate. By the end of the season they have likely mated all candidate females in their social sphere, both within group and neighbours. Insofar as they therefore consider themselves candidate fathers for any cubs born to these females, this results in the sort of paternity confusion that probably reduces the risk of infanticide by males (Agrell et al. 1998; Yamaguchi et al. 2006).

So why might a female badger mate on multiple occasions without apparently caring about the quality of her partner? Not least, as we will explore in Chapter 15, promiscuity risks sexually transmitted infections, which can compromise fertility in badgers, as in other species. Clearly, the first few days of mating are inductive not procreative. Further, badgers are superfecund: persistent mating can cause multiple ova to be released sequentially, each potentially fathered by a different male, conceiving additional zygotes and then developing to the 'blastocyst' stage (200–300 cells) while earlier blastocysts remain suspended *in utero* (therefore not forming a placenta) during DI.

This, more generally, was the question addressed by Jerry Wolf and David in 2004 in a wider review of the 133 mammalian species amongst which female promiscuity had, at the time, been recorded to involve multi-male mating (MMM) (Wolff and Macdonald 2004). They concluded that the evidence of anti-infanticidal benefits to MMM was sufficient to explain the evolution of this behaviour without recourse to other benefits, and thus the occurrence of MMM even amongst those species without induced ovulation. Infanticide is a widespread feature of mammalian societies, predominantly perpetrated by males, but important amongst females in the context of reproductive suppression (see Hrdy 1979; Ebensperger 1998), as we described amongst Wytham's badgers in Chapter 1. The adaptive link between infanticide and MMM is that the latter may be a mechanism to confuse paternity— the male recalls he mated that female—and thereby reduces the risk of the former; an explanation that assumes males cannot identify their own progeny (Orr 2012). Why might males be infanticidal? The answer lies in the Machiavellian imperative (Byrne and Whiten 1989): all cubs use resources, taking these resources away from the infanticidal individual's own progeny, or those of his lineage, with no return in inclusive fitness (Hamilton 1964; see also Taylor 1992). Consider the farm cats with which we drew a parallel earlier: Herodotus observed, in the late 400s BCE, that tomcats killing kittens brought the bereaved mothers swiftly into oestrus and thus available to bear the killer's progeny—Herodotus was right: nearly two thousand years later David witnessed an interloping male cat kill most of the kittens in a communal litter (Macdonald et al. 1987, 2010a). In less sociable species, thinking of the badgers' past, infanticide by bears (Steyaert et al. 2013) or leopards (Balme and Hunter 2013) may simply remove future competitors for the killer or his cubs.

There are more than a dozen other hypotheses to explain MMM. In addition to paternity confusion, three might apply to badgers: first, promotion of sperm competition (perhaps the alternative to getting dangerously entangled in the aggression between competing suitors) (Møller and Birkhead 1989) (we have already seen that the badgers' rather small testes do not emphasize this mechanism); second, rectifying a mistake by devaluing the previous male's sperm (Walker 1980; McKinney 1985); and third, promoting genetic diversity (Williams 1975) and avoiding genetic incompatibility (Zeh and Zeh 1996). The devaluation hypothesis predicts that a female that has been mounted by a second male should not allow the first male to mount

again. At Wytham one female mounted by two males (both for long durations) on one night was then mated again the following night by the first male. And as for good genes, we have already mentioned sufficient evidence of multiple paternity amongst Wytham's badgers that this phenomenon deserves further exploration, which it will get in Chapter 17.

We have emphasized things we do not know, and particularly the fact that we cannot tell whether a given mounting, even intromission, is merely inductive or potentially procreative, and we don't know, either, how tightly synchronized are the single annual oestruses of Wytham's contemporary badgers—perhaps some of those periods of late sexual activity in, say, May or even June were individuals entering oestrus particularly late, or even having another go if things had gone wrong the first time, or topping up on an inadequate quota of blastocysts. These are things we do not know, but whatever the answers, one thing is inescapably clear: many mountings, involving many males and females, occurred across such a span of time that, if our interpretation of the endocrinology is right, there is a possibility, at the individual level, of such mountings leading to conceptions for females that have not fulfilled their blastocyst 'quota'. The question remains, does non-conceptive, extra-oestral sex occur, and what is its function in the affairs of the badgers of Wytham Woods?

Non-conceptive sex

Non-conceptive, extra-oestral sex is unusual, recorded in mammals so far only amongst some cetaceans (especially dolphins; Mann et al. 2000) and three species of anthropoids: chimpanzees, bonobos, and humans (Furuichi et al. 2014). There's more to this than procreation, but is the badgers' membership of this elite a matter of convergence or coincidence, homology or analogy, or just an evolutionary mistake? An answer might lie in which characteristics are shared by club members, and which others are not.

Badgers appear to qualify for membership of the non-conceptive mating club insofar as they all have fission/fusion societies, longish pregnancies (albeit none of the others achieves these by DIs), and polygynandrous mating. Badgers differ from the other members of this sexually non-conformist group in that the others have large neocortices and are conspicuously intelligent, features with which even we cannot flatter badgers. There is some heterogeneity over testis:body weight—it is high amongst dolphins (Murphy et al. 2005) and chimpanzees, huge amongst bonobos (Wrangham 1993), modest amongst humans, but relatively small amongst badgers. Large testes are associated with sperm competition (notwithstanding some controversy; Brown et al. 1995), for which there is evidence in cetaceans (Brownell and Rails 1986), chimpanzees and bonobos (Møller 1988; Harcourt 1997), as well as humans (Baker 2014), but for badgers, in spite of the fact that the circumstances seem right, evidence is absent. Female bonobos and chimpanzees exhibit perineal swellings, but only bonobos mate outside the swelling phase; badgers clearly engage in mounting outside of oestrus (although their seasonally swollen vulvas turn out to be only loosely related either to oestrus or to mating; Sugianto et al. 2018), while oestrus in humans is somewhat cryptic. Of the features that Furuichi et al. (2014) link to non-conceptive sex the high social status of females might resonate with badgers (Chapter 5), as might the existence of social cliques and alliances (Chapter 6), and within- and between-group tension (and paternity confusion), although none of these seems particularly defining. A feature shared by the other non-conformists is a post-reproductive lifespan, but is this a feature of badgers? We tackle this question in Chapter 18.

According to Richard Wrangham (1993), there are three functions, beyond conception, to sex in bonobos and chimpanzees: paternity confusion (as likely for badgers), practice (hard to dispute for badgers insofar as induced ovulation demands practice before any ova are shed), and exchange for favours. Female bonobos trade sexual favours for access to food—trading sex, however, is not the sole prerogative of clever anthropoids: female wood mice trade the opportunity to sniff their genitals for grooming (Stopka and Macdonald 1999), but bonobos make such trades outside oestrus (De Waal 1995a). According to Wrangham, only bonobos use sex purely for communication about social relationships, and he concludes that their hypersexuality and 'communication sex' are closely linked to the evolution of female–female

alliances, with origins in relaxed feeding competition. There is a shadow of badger circumstances in De Waal's (1995b) allusion to the merits of making love not war[21] for a species equipped to inflict grievous physical damage, cohabiting in close proximity and needful of a conciliatory mechanism. Recall the badger' low-density ancestry, which probably involved an obligate propensity to mate with little scope for discrimination, then parachute that behavioural blueprint into very modern ecological circumstances that facilitate high-density cohabitation, not unlike the circumstances of a tenement block explored for humans by Baker (2014).

This brings us to the possibility that extra-oestral sex in badgers is, in evolutionary terms, a mistake (by which we mean, not adaptive). If the ancestral selective pressure for induced ovulation for females was the risk of failing, at low population density, to coincide with a male, and if the same pressure (and the advantageous opportunity for pre-implantation intra-uterine selection of the fittest blastocycts) favoured the evolution of SF during additional recurrent oestruses at intervals throughout the period of DI, and if the costs of rebuffing suitors and getting caught in the cross-fire of their battles has led to convenience polyandry and remarkable tolerance by both sexes in mating scrums, then perhaps it was advantageous for males and females alike to engage sexually, in an opportunistic fashion, when there was even just a whiff of oestrous odours. Our endocrinological findings reveal that in Wytham, unusually and perhaps uniquely, at a population level there are no secondary oestrous periods during DI, but the endocrinology, and perhaps the signals and expectation of these erstwhile opportunities for SF lie so close below the surface that they occasionally manifest by mistake or, as we speculate, as a top-up. Or perhaps the badgers of Wytham Woods have unpacked their rucksack of phylogenetic baggage to assemble a modernized socio-sexual adaptive syndrome that awaits understanding.

[21] Radical activists Penelope and Franklin Rosemont and Tor Faegre helped to popularize the phrase by printing thousands of 'Make Love, Not War' buttons at the Solidarity Bookshop in Chicago, Illinois, and distributing them at the Mother's Day Peace March in 1965. They were the first to print the slogan.

CHAPTER 8

Social Behaviour in an Uncooperative Society

Nonconformity is the highest evolutionary attainment of social animals.

Aldo Leopold 1920

In Wytham, as in other parts of the UK and elsewhere, badgers live in groups, but why? Indeed, why do animals, generally speaking, live in groups? Beyond the axiomatic answer that individuals live gregariously when the likely fitness benefits of doing so exceed those of living alone (Krause et al. 2002), and remembering the firm caution (Chapter 4) that living alone does not imply asociality, the general answer is that fitness benefits to group membership arise from two broad categories of selective pressure: ecological (the topic of Chapter 10) and sociological (which we explore here, and which can be broadly categorized either as reproductive or non-reproductive).

When hunting, a pack of African wild dogs may gain a marginal, if diminishing, per capita advantage from each additional member (up to, but not beyond, a maximum) (Creel and Creel 2002), and beneath a critical size they are in a downhill spiral (the group Allee effect, which causes packs below a certain size to struggle to raise pups; Augulo et al. 2013). So, how do they get started? Across the Animal kingdom the general answer is by recruiting offspring. Another factor is division of labour, which amongst eusocial[1] insects involves dramatic differentiation of individual roles (Page and Erber 2002; Ulrich et al. 2018), whereas amongst mammals the division may be less specialized but nonetheless fulfils the aphorism that many hands make light work (e.g. Lima

---

[1] Eusocial species: any colonial animal species that lives in multi-generational family groups in which the vast majority of individuals cooperate to aid relatively few (or even a single) reproductive group members. Many examples are insects, but a mammalian example is the naked mole rat.

1995). In Chapter 19 we offer an interpretation of intra-specific variation in badger society (further interpreted in the context of inter-specific variation in wider carnivore societies), but, in anticipation of that, keep in mind the fact that members of different taxa begin this evolutionary journey towards sociality from different starting points: the starting point for all *Canidae* is a socially monogamous pair (Macdonald et al. 2019), whereas that for all *Mustelidae* is semi-detached sociality of intra-sexual territoriality with mother–offspring nuclei (see Powell 1979; Johnson et al. 2000; Macdonald and Newman 2017). And for phylogenetic context, the mustelid model is characteristic of most carnivore families on both sides of the 60 million-year-old divide between the feliform and caniform lineages, illustrated for example by both *Felidae* (cat branch) and *Ursidae* (dog branch) (Macdonald 1996).

From these different starting points the canids and mustelids launch down their converging paths to groups of jackals and coyotes comprising mature offspring cohabiting with both parents and groups of wolverines and Japanese badgers cohabiting with their mothers (Eisenberg 2014; Kaneko et al. 2014) (by the way, the matrilineal arrangements are similar for felids, Macdonald and Loveridge 2010; and ursids, Dahle and Swenson 2003). En route to the synthesis of Chapter 19, for now the obvious question is: if adult offspring are to be recruited to the workforce of the 'family firm', or at least not to be shown the door at sexual maturity, how can there be space (resources) to accommodate them in the family home? Even the partial overlap of adult male and female territories in the classic mustelid system has implications for resource consumption (Yamaguchi and Macdonald 2003 calculate that both female and male American mink have

The Badgers of Wytham Woods. David W. Macdonald and Chris Newman, Oxford University Press.
© David W. Macdonald and Chris Newman (2022). DOI: 10.1093/oso/9780192845368.003.0008

to expand their territories to accommodate the drain imposed on each by cohabiting with the other). After all, the convention amongst solitary species of mustelid is a general exodus of all adolescent youngsters (occasionally, as per the matrilineal predisposition, with a male bias) (Lodé 2001; Blundell et al. 2002b; Macdonald et al. 2008) with the general consequence of inbreeding avoidance (Pusey 1987; Perrin and Mazalov 2000).

A factor influencing the success of dispersal is habitat saturation—if there are few unoccupied territories, the odds of acquiring one will be poor, favouring philopatry. Macdonald and Carr (1989) emphasize how a youngster's decision on whether to disperse or to bide hinges on the balance of odds of securing an unoccupied territory (an ecological factor) and the marginal gain (to all concerned) of swelling the membership of the family group (largely a behavioural factor measured by the rewards of cooperation offset by the costs of cohabitation). This balance is a fundamental ecological consideration across taxa—scarcity of breeding sites is a key factor determining cooperative breeding in birds (Koenig et al. 1992; Cockburn 1998)—and formalized as the delayed-dispersal threshold model (Bowman et al. 2002): the result is kin-structured groups even in the absence of any mechanism for kin discrimination. However, the glaringly obvious problem in a youngster's evaluation of whether to go or stay (and the parents' evaluation of whether to tolerate or expel) is the capacity of the natal territory to feed more mouths. Those evaluations are affected if larger groups cooperate in ways that bring marginal gains to all concerned (the nub of Macdonald and Carr's (1989) argument), but however helpful the stay-at-home offspring may be, the question remains of how to secure the real estate to accommodate them. Remember, those female American mink have to expand their territories even to compensate for the drain of cohabiting with a male, and to add to this burden the accommodation also of even one adult daughter would likely necessitate an approximately *pro rata* expansion. This dilemma can be solved in only two ways, as argued by Kruuk and Macdonald (1985): one is expansionism—the acquisition of more territory to accommodate a larger group, a strategy selected for if the rewards of cooperation outweigh the costs of expansion (accountancy affected by the metaphorical rent—helpfulness—the recruits offer (the pay-to-stay hypothesis; Komdeur 1996; Kokko and Johnstone 1999)). The second solution lies in ecological circumstances that shift the balance of costs and benefits of tolerating cohabitation by making it approximately cost-free and without any requirement for expanding the real estate. This occurs under conditions described by the resource dispersion hypothesis (reviewed by Macdonald and Johnson 2015a), and is the topic of Chapter 10. Here, in this chapter, the key point is that the costs and benefits of cohabitation (versus dispersal) are potentially affected by potential rewards of cooperation, which is why, when we consider Wytham's badgers, we ask whether there are any.

Humans, of course, have a deep intuitive, and doubtless evolutionarily based, appreciation of cooperation, with much of our philosophy, religion, and social justice guided by the principle of ethical consistency or the 'Golden Rule': a unilateral moral commitment to the well-being of another without the expectation of anything in return (Vucetich et al. 2021), expressed biblically as 'do as you would be done by'.[2] This maxim of altruism is crucially different to the maxim of reciprocity *'do ut es'* ('I give so that you will give in return') (Spooner 1914). Two remarkable evolutionary biologists rode the wave of these principles into evolutionary ecology. The first, Bill Hamilton (1964) rationalized that what might be gained in return was the fitness of a blood relative (inclusive fitness); the second, Bob Trivers (1971) deduced that selection could favour helping even unrelated recipients at a fitness cost to the donor if that favour is likely to be returned. These scholarly quantum leaps revealed answers to the question: through which mechanisms might cooperation evolve? The three salient answers were clearly apparent by the time Lee Alan Dugatkin published his 1997 book: kinship, reciprocity, and by-product mutualism. Through any or all of these mechanisms, what forms of cooperation are candidates as potential functional advantages for badgers forming groups?

Several can be dismissed quickly. There is no evidence that the badgers of Wytham Woods, or indeed any European badgers, hunt prey cooperatively, or jointly defend captured prey. In Chapter 4 we found oddly contrarian evidence in Wytham that individuals generally do not aggressively defend territories alone, far less collectively, and further paradoxes of this ilk lie ahead in Chapter 9. Badgers do give and respond to warning signals (Wong et al. 1999; Box 5.4), but make no effort to warn, protect, or round up cubs (Chapter 2). They react to the ghosts (or at least the ghostly sounds) of predators past (and present, in the

---

[2] A very important element of the purpose of conservation, and of associated issues of social justice, is whether the intrinsic value that earns all people consideration within the Golden Rule also applies when one considers the behaviour of other species. Vucetich et al. (2019) argue that it does, in which case the humans who constitute the readers of this book and also the badgers that are its topic would be embraced by the contour of concern that acknowledges that both have interests and intrinsic value.

form of people) (Chapter 19; Clinchy et al. 2016), but there is little evidence of coordinated vigilance as, for example, in meerkat bands (and certainly not when away from the sett, when they are generally alone). An interesting possibility, associated with sharing a den (Chapter 4), is that they share information or body heat. The shared information is often thought about in terms of the whereabouts of ephemeral food patches that they visit, much like bees (Frisch et al. 1967) from their centrally placed home (Ward and Zahavi 1973) (although in Chapter 19 we raise the possibility that badgers might exchange information on the whereabouts of wolves). This is a promising idea, and we have several times sought to test it (looking at the directions from which radio-tracked individuals returned to the sett, and those in which they and others then set forth), but we could neither refute nor support the hypothesis. What's left? We'll explore three candidate behavioural facets of cooperation: care of young, digging (which may lead to protection in a burrow and, through the hot-water-bottle effect, warmth), and hunting (albeit of an unexpected quarry).

Cooperative care

We have described the unexpected indifference of adults to cubs (Chapter 3). It is unexpected because a career spent probing carnivore sociology more or less spans the professional generation from the earliest observations of alloparental care (e.g. Macdonald 1979c; Moehlman 1979) to the rather complete documentation of variants of this phenomenon from communal, alloparental nursing, provisioning, and even adoption, and a syndrome of linked phenomena such as reproductive suppression (Macdonald et al. 2004b, 2019). And it is rife amongst social carnivores, amongst which the word indifference would be the last thing to come to mind as adults interact with youngsters. Cooperative care by alloparents (i.e. care bestowed by adults other than the cubs' parents) can deliver survival benefits, and involve conspicuous assiduousness with far-reaching sociological and demographic consequences amongst a wide array of carnivores, including wild dogs, meerkats, Ethiopian wolves, and giant otters (Doolan and Macdonald 1999; Sillero-Zubiri et al. 2004; Angulo et al. 2013; Groenendijk et al. 2015). So, do badgers engage in cooperative care of their cubs?

Consider first, how do badger cubs get on with adults once they are weaned but still diminutive? A simple question not so easily answered because we generally see cubs only above ground (although, when we did film them underground, Chapter 1, they had no adult contact beyond that with their mother until they were around 6 weeks old). A first answer came from the observations of intern Rebecca Fell, who videoed 18 cubs attributed to 9 mothers (via lactation and association) from their first emergence in mid-April through the springs of 2005 and 2006 at the Chalet/Chalet Outlier group and the Firebreak group (see maps, pages xiv and 173). At first, these cubs emerged only at the dead of night, around 01:00 to 02:00, and remained close to the sett entrance in the company of an adult—likely their mother (Fell et al. 2006). By late May, then aged 3 months, they began to emerge with adults around dusk (20:00). But what of alloparental care? If it occurred, we didn't see it.

During her doctoral research Hannah Dugdale (now a professor of evolutionary biology in Holland) recorded 186 occasions when an adult, male or female, carried a young cub, suggesting some interest in them whether helpful or not. A total of 72% of carrying events involved breeding females, sometimes definitely the mother; on 10% of occasions it was a male, generally during play (Dugdale et al. 2010). On one occasion we saw a mother and a subadult male tussle over one of four newly emerged cubs, yanking it in different directions (Fell et al. 2006). This was horribly reminiscent of a tussle amongst bush dogs, *Speothos venaticus*, where a 'tug of love' killed the pup (Macdonald 1996; and remember the dead cubs and adult from the violent scene reported in Chapter 2). The badgers' behaviour smacked more of abduction than helpfulness. Anyway, cubs busying themselves outside sett entrances were alone for *c*.58% of the time (a seemingly dilatory level of care compared with, for example, wild dogs; Courchamp et al. 2002); they were with a single adult group member (including their mother) around 20% of the time, and two or more adults for the remaining 22%. We began to doubt that adults offered meaningful support to youngsters; supervision was neither obligatory nor, apparently, beneficial (Woodroffe and Macdonald 2000; Dugdale et al. 2010).

Insofar as it is commonplace amongst communally breeding mammals, from foxes to capybaras (Macdonald 1983; Herrera and Macdonald 1989), we had expected to see badger sows suckling each other's cubs. But, while the opportunity was there, insofar as several litters of cubs were sometimes raised in one sett (mean candidate mothers 5.6 (range 5.2, 6.0); fathers 5.8 (range 5.4, 6.2)), allosuckling was never rigorously established (Dugdale et al. 2010). That is not to say it never happened. Consider Female 819, at the Pasticks group, who in 2005 had one genetically assigned cub (M1019, a male, fathered by M864, as had been her single cub in 2004) and appeared to suckle four cubs on four occasions in April.[3] Nonetheless, at most we've seen allosuckling only a handful times in more than 30 years. Unless it was happening only underground, this form of cooperation is rudimentary in badger society.[4]

The proof of the pudding lies in the eating—did the presence of potential allomothers/parents translate into improved circumstances, and ultimately survival, of badger cubs? Indeed, contrary to the helpfulness expectation, the presence of non-breeding adults in badger groups is associated with cubs gaining weight more slowly than might have been expected (Woodroffe and Macdonald 2000), perhaps implying that adults compete with cubs for available resources. Similarly, Dugdale et al. (2010) found that the number of other members of the group (excluding mothers) correlated negatively with long-term fitness. Using the 18 years of data between 1988 and 2005, we found that neither litter size nor the probability of a mother surviving to the next year was related to either the number of within-group mothers or the number of other members of the group (excluding mothers) when we analysed social-group years in which all cubs were assigned a mother or in which at least one mother was assigned. Digging deeper, and mindful of grandmother effects in other species (e.g. Mech et al. 1991; see Sugianto et al. 2020), we asked whether the number of mothers in a cub's group was related to the probability of that cub breeding or its lifetime breeding success—it wasn't. On the contrary, and just the opposite of what we might have expected for a social carnivore, the number of other members of the group (excluding mothers) within a social-group year had a negative relationship with both the probability of a cub breeding and indeed its lifetime breeding success.

In concluding that allomaternal care is at most an occasional occurrence amongst Wytham's badgers,

and consequently not a major cooperative advantage of group living, we ask why don't badgers do it more? This is not a question of the fatuous variety of why don't pigs have wings. On the contrary, not only is alloparental care widespread in group-living birds and mammals, it is close to ubiquitous in social carnivores, and although there are rather few social mustelids to compare (but see Chapter 19), the phenomenon is documented in the superfamily *Musteloidea* (most notably in otters and coatis). Badgers therefore appear to be an outlier, an exception that merits explanation. That explanation might lie in the badger's physique, dubbed 'musclimorph' by Macdonald and Newman (2017). Badgers are squat, robust, and, seasonally, very fat—in short they are unathletic. In a comparative analysis of mammals, Heldstab et al. (2017) found a negative association between the occurrence of alloparental behaviour, such as provisioning, and annual variation in body mass. They reason that tending youngsters is energetically expensive, and that stockpiling that energy as fat would be debilitating for a mother in an agile, athletic species—it is therefore better, under those circumstances, that a different individual (generally kin) shares the burden. However, for habitually fat species—of which badgers are a corpulent and unnimble epitome—there is little impediment to the mother storing her own reserves (although fatness can have thermoregulatory costs for badgers in old age, Chapter 14; Bright Ross et al. 2020). Interestingly, the negative relationship revealed between annual body mass variation and alloparental provisioning does not apply to allonursing, the suggested explanation being that allonursing involves no additional influx of energy but merely a redistribution of maternal help across different mothers. Heldstab et al.'s argument (focused on the demands for variation in fat stores) prompts us to suggest a related but different explanation for the, at most, low level of alloparental care in badgers. Having abundant on-board reserves—fat—as badgers do, diminishes a mother's need for help, although even a chubby mother would presumably like some sucker to do her work for her. But badger babies are few, and their food (when abundant) is easily caught by even an idiot infant (and whether or not abundant, their foods are generally not readily transported by adults back to the den, and musteloids can't regurgitate; Chapter 19). This combination of circumstances likely renders helpers not very helpful, so the pay-offs to them are unlikely to outweigh the costs: there are better investments to be made.

---

[3] Of 23 occasions when cubs of 8–13 weeks old were seen to suckle, the mean duration was between 1 and 2 min. On the only two occasions on which breeding females were seen suckling more cubs than their assigned litter size, the duration was short (Dugdale et al. 2010).

[4] Because we genotyped cubs in June, but watched them nursing from April (observing seeming mother–offspring associations), we are alert to the possibility that some of the suckling we saw before genetic assignment could have been allomaternal. While possible, for this to be commonplace would require a curious lack of observations of cubs moving between females, and remarkable cessation of this behaviour in June.

Cooperative digging

Digging in the sisyphean shifting sands of the Kalahari, a line of burrowing meerkats will pass the spoil backwards as they excavate communal dens (Macdonald 1992), so might badgers benefit from cooperating in digging their sometimes enormous setts? Like icebergs, badger setts can be much more extensive beneath the surface than is apparent from above (Macdonald et al. 2017; Chapters 1 and 19). The important fact (and we make much of it in Chapters 3 and 11) that there are plenty of sites suitable for setts in Wytham doesn't diminish the reality that digging them is a major undertaking, and a task that intuitively therefore seems best made light work of by many paws. Major badger setts in Wytham Woods can have more than 50 entrances, stretching 100 m along a bank, and involve hundreds of metres of underground tunnels, representing dozens of tonnes of earth moved (Macdonald et al. 2015e).

We might have expected this kind of civil engineering to benefit from cooperative effort. Indeed, in his 1198 account of the Topography of Ireland (*Topographia Hibernica*), commissioned by King Henry II, Friar Giraldus Cambrensis remarked that:

some [badgers] are born to serve by nature. Lying on their backs, they pile on their bellies soil that has been dug by others. Then clutching it with their four feet, and holding a piece of wood across their mouths, they are dragged out of the holes with their burdens by others who pull backwards while holding on here and there to the wood with their teeth. Anyone that sees them is astonished.

We have sadly not been amongst those privileged to be thus astonished; what we have seen, by contrast, appeared much more individualistic than strategic: one badger will often kick spoil from its excavations directly into an opposing hole where another of its group is busily at work digging in the opposite direction (the multiple entrances to Clearing Outlier, described in Chapter 4, illustrate this vividly, where excavation of one generally blocks another).

Documenting such digging on infrared video revealed what might be a division of labour, but certainly not a fair one. Between 1994 and 1995 we watched two large groups of badgers, Pasticks and Sunday's Hill;[5] the former comprising 24 identifiable

(clip-marked) badgers and the latter numbering 20 identifiable individuals (Stewart et al. 1999). Just 20% of adults and yearlings were responsible for 60–90% of observed digging and bedding collection.

Amongst diggers there were signs of self-interest (or perhaps happenstance), insofar as the more times an individual was seen to use a particular sett entrance, the more likely it was to engage in sett maintenance at that location. Most of that minority of individuals that we saw digging assiduously were males, and predominantly big ones. Of course, we wondered whether these super-diggers were amorous males, perhaps constructing a nuptial suite attractive to females, akin to the bower of birds of that ilk (Borgia 1995). An anecdote may be relevant: as chance would have it, in the spring of 1998 all but one (F399) of the prime females in the Botley Lodge group, on the periphery of the woods, were run over by cars. Soon afterwards we observed a massive expansion of digging effort by the bereaved males (M411, M587, and M687). Whether or not this sprucing up of the Botley Lodge sett was a contributory factor, by the following year two immigrant females had moved in (F457 from Great Oak Outlier and a previously unmarked one from outside our population); they joined F399 and a philopatric female (F558) that in the meantime had matured to reproductive age. Three cubs were born at Botley Lodge in 2000 (F783, F784, and M785) and both F399 and her daughter F558 lactated, so that, by the end of that year, 2 years after the traffic deaths, the group comprised four breeding-age females and three mature males.

Why do we speculate that the digging activity could have attracted the incoming females? That question goes back to the fundamental one of why badgers live in communal setts at all. More on that big question in Chapters 10 and 19 but two factors to bear in mind are: (i) that bigger and more elaborate setts have stabler interior environmental conditions, from which more young are weaned (a result achieved in the absence of wolves or predatory people, so there's more to it than inter-specific protection; Chapter 2)— it might be worth considering a hot-water bottle effect or, more technically, what Hans IJzerman (et al. 2015) called social thermoregulation (see also Campbell et al. 2018; Chapter 19); and (ii) that fleas can build up in underground chambers (Butler and Roper 1996; Albery et al. 2020), and so, to dodge fleas, having more underground options (i.e. in the form of sleeping chambers) to swap between, and perhaps even to leave fallow, in response to ectoparasites could

---

[5] One comprising 20 identifiable (clip-marked) badgers (12 adults, 6 yearlings, and 2 cubs) and living with 3 unidentified badgers (2 yearlings or adults, 1 cub); the other group comprised 24 identifiable individuals (19 adults, 1 yearling, and 4 cubs, plus 5 unidentified adults/yearlings).

make a sett, and therefore its male occupants, attractive (disuse of a chamber can slow or even terminate the development of immature fleas; Reichman and Smith 1990). This brings us to the broader topic of how cooperative behaviour might alleviate each badger's itchiness.

You scratch my back and I'll scratch yours

Badgers itch. The relevance of this fact to cooperation amongst them, and its evolution, will become clear, but first we will introduce the source of the itch: fleas. Badgers have their own species of flea, *Paraceras melis*, and, at least in Wytham, plenty of them. Badger fleas are a pleasing chestnut brown colour and big enough (over 1 mm) to be easily seen (and, as we can attest, leave an extremely itchy bite). Having chased them around badger bodies we know that they can jump at least 10 cm. To digress, this hopping mechanism, which involves pumping the insects' blood into their legs, which then spring up like piston-driven pogo sticks, was discovered by the remarkable naturalist Miriam Rothschild, daughter of the banking dynasty, who, already beyond her three score years and ten, bedecked in sail-like black and purple chiffon scarves would flounce flamboyantly through Oxford's Zoology Department where she swapped stories with David of fox cubs that they were hand-rearing in the 1970s (Rothschild et al. 1975). Anyway, all this jumping by fleas is a problem for the badgers, as it makes it a difficult task to catch them. The question relevant to this chapter is whether badgers cooperate in hunting fleas. More prosaically, we will go on to ask: can asymmetries in soliciting itch relief be explained by idiopathic sensitivity to flea bites that differs between individuals?

First, some fundamental theory links communal living with avoiding parasites. When Hamilton (1971) proposed the 'selfish herd' effect he was thinking primarily about reducing predation risk, but noted that the same principle might apply to avoiding parasites. Solitary individuals suffer higher ectoparasite burdens if they lack conspecifics either to absorb collateral damage from the local ectoparasite population or to remove ectoparasites by allogrooming. By grouping, therefore, animals may reduce their individual risk of exposure to parasites (Mooring and Hart 1992). Regarding badgers and fleas, an amalgam of field observations sets the scene in Box 8.1.

Box 8.1 Cooperative flea hunting

Female Beatrix sits on her rump, almost reclined onto her back and scratches her chest with both forepaws, contorting awkwardly to nip at her chest with her incisors. Female Betsy enters, approaches Beatrix and sniffs her left hind foot. Immediately, Beatrix folds forward (a movement akin to a 'sit-up') and grooms Betsy's left shoulder intensely—having seen hundreds of such encounters, we interpret this as an invitation and, as 10 s elapse, one to which Betsy is slow to respond. Beatrix, having been maintaining her awkward sit-up position, rolls backwards onto her back and resumes pawing at her belly. Seven seconds pass before Betsy accedes to the invitation and grooms along Beatrix's left flank and belly as Beatrix reclines. Immediately, Beatrix reciprocates, half sitting to groom Betsy over her left shoulder along the back of her neck and between her shoulder blades. After 30 s Beatrix appears to feel less urgency, Betsy flops over onto her left side with Beatrix cuddled up with her belly to Betsy's back, spooning her; Beatrix rests her right hind foot on Betsy's right hip. Beatrix then leans over Betsy and, with renewed urgency, intently grooms her chest and belly. Betsy stretches her neck and head back and forth in apparent ecstasy, as Beatrix nibbles down to a spot in her right inguinal region. After 1 min Betsy rolls right over and starts to walk away from the camera, but Beatrix clings on, grasping Betsy's belly with one front paw and the small of her back with the other and grooming up over her neck, being somewhat dragged away by Beatrix. Allowing herself to be dragged over backwards, Beatrix now falls into her dorsally reclined position, and we interpret a request for reciprocal attention to her exposed belly. Betsy's initial response is to stand on Beatrix's belly with her left hind foot, and as Betsy literally walks over Beatrix, Beatrix grooms Betsy's belly. Now 1 min 30 s have passed. Briefly separated, Beatrix does another sit-up, spreads her rear legs, and energetically bites at her own groin, then rolls to her feet, and pivots to intercept Betsy, grooming at Betsy's left flank. Again, Beatrix becomes engrossed, pausing only occasionally to scratch her own left armpit with the kicking of her rear leg, dog-like. She works her way down and with her head semi-inverted again nibbles at Betsy's belly, as Betsy stands four-square. Two minutes have passed. Continuing her head inversion into a sinuous roll over her left shoulder, Beatrix again flops onto her back, pawing at her belly, as if to direct Betsy to her point of irritation. Betsy obliges and comes in over Beatrix's right flank and assiduously grooms up and down between her front and back legs the length of her abdomen. After 2 min 40 s, Beatrix has had enough, and sits up onto her feet. Betsy walks away a few paces; Beatrix quickly grooms under Betsy's chin in what we interpret as a 'thank you', and the two trot off in opposite directions.

Observations of mutual grooming piqued our interest for two reasons. First, flea bites not only itch, but worse, while pumping blood in and out through their proboscides, fleas can also inject blood parasites—a consideration not diminished by the fact that the parasites we have studied, trypanosomes, are non-pathogenic (Lizundia et al. 2011) (Chapter 15). After all, flea-borne diseases have shaped human history (e.g. the Black Death, Belich 2016; Montgomery and Macdonald 2020) and might have shaped badgers' too. Second, more immediately, fleas itch. With that in mind, and with the ultimate goal of appreciating the impact of fleas on badgers, their evolution and behaviour, with the badgers sedated we counted fleas as they scuttled ahead of our fingers rifling through their hosts' fur during standard 10 s counts. Flea counts ranged from zero, for about 30% of examinations, to a maximum of >25 (we lost accuracy beyond that point). The average flea burden was 4.96, or 7.19 amongst those badgers that had some fleas to count, with respective modes of 3 and 5. Within individuals, however, there was tremendous variation between captures, although not linked in any obvious way to their body weights.

Flea counts were lower in bad weather when the badgers were soaked and muddy. Perhaps the badgers were skulking less conspicuously, but we doubt it and suspect that, during the rainfall, the fleas had jumped ship in the trap beside the sett, and were lying in wait for a comfortably dried host. Certainly for F1011 we got a zero count only on a wet day, and while she averaged 5.5 fleas on dry days this dropped to 2.5 on wet days. Similarly for M1012, zero counts occurred on 27% of dry days, with an average of 5.12 fleas, but 54% of wet days, averaging just 1.82 fleas.

Counting fleas provided clues to their behaviour, and hence a first step to discovering what badgers do about them. The fleas appeared to respond to researchers' fingers (and perhaps breath) much as they would to a badger's approaching incisors. They switched sides on the badger's body to evade us. If we paused in our 'grooming' for about 40 s, the fleas redistributed themselves so a fresh crop was available for counting when our 'grooming' was resumed (Stewart and Macdonald 2003). By the way, although fleas are adapted to jump when necessary, and to scuttle when discretion pays, only 5% of one sample of 88

fleas jumped when disturbed on a badger, whereas, when removed and placed, exposed and conspicuous, on a sheet of paper, 83% of another sample of 48 fleas jumped before they had scuttled 6 cm, reminiscent of gazelles fleeing a cheetah.

Our second approach was to watch nocturnal videos of badgers grooming in the wild, launching the career of our assiduous flea-counting assistant at the time, Ruth Cox, who went on to become a successful epidemiologist at Prince Edward Island University, Canada. Badgers, like pet dogs, lick themselves not only to groom (they get muddy) but also, with incisors clacketing, to comb for fleas—Ruth's focus was on the latter. She documented that a single grooming badger, despite contortion and writhing, and with much snorting and snuffling while balanced on its bottom, has plenty of itchy nooks and crannies it simply cannot reach, although it can exhaustively pursue fleas on its own chest and belly (Cox et al. 1999).

At this point a solitary carnivore, say a flea-bitten weasel, will have done all it can to alleviate its irritation, in that it has little scope to seek assistance from a peer. The question of what badgers do about their flea burden now splits into: do badgers cooperate in flea hunting, does doing so deal with the 'blind spot' of self-grooming, and how much do they engage in any such cooperation? The give and take of society, *quid pro quo*, is often, and in this case aptly, rendered as 'you scratch my back, and I'll scratch yours'. We saw badgers groom each other more often than we saw them groom alone. Indeed, they went to lengths to recruit partners. For example, a badger might signal its eagerness for grooming by grappling a prospective partner to pull it into a position where it could groom the inviter. Others occasionally placed their head directly below the chin of the partner just prior to grooming, a fairly explicit hint. Sometimes badgers emerging from the sett appeared to be greeted with an initiation of mutual grooming. A badger, seemingly mindful of the inaccessible zone that it wants to have groomed, will sometimes initiate grooming of that same area on a companion. Often the two groom simultaneously; sometimes they alternate. Either grooming simultaneously or alternately with the pendulum of reciprocity, the remarkable outcome is that the combination of self- and mutual grooming results in every part of the body receiving almost equal coverage (Stewart and Macdonald 2003; Figure 8.1).

Insofar as cooperation is defined as the process of working together to the same end, badgers cooperate in mutual grooming, but only in the sense that each is motivated to recruit assistance towards its own objective, rather than being moved specifically to assist its

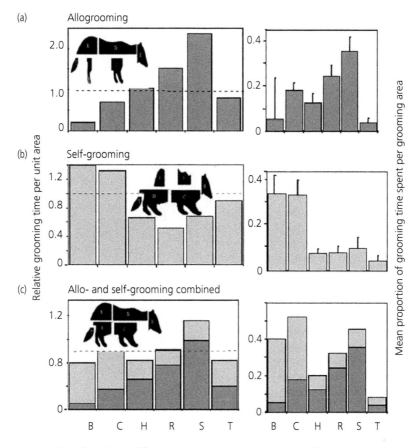

Figure 8.1 The contrasting pattern of attention to different body regions during (a) allogrooming, (b) self-grooming, and (c) both types of grooming combined. The first graph of each row gives the relative proportion of grooming over time per square centimetre of body surface, together with a representation—silhouette scales from the data. The dotted line at 1 represents parity between the proportion of time spent grooming a region and its relative surface area. The second graph of each row gives the proportion of time spent grooming each body region (errors show standard deviations), uncorrected for relative surface area. (c) shows that a combination of self-grooming (light bar) and allogrooming (dark bar) provides a broadly even coverage of the badger's skin surface, with some 'overattention' to the shoulder. Data pooled from 16 individuals (from Stewart and Macdonald 2003, with permission).

companion. So how much time do they devote to this cooperation? From watching all the videos from 1995 Paul saw a total of 1862 allogrooming events amongst 38 badgers. A randomly chosen bout for each of 28 individuals' bout durations ranged from 1 to 19 s, with a median of 9.6 s and a mean of 11 s. The average bout of self-grooming lasted 7.1 s (N = 280 events) similar to allogrooming averaging 6.3 s (N = 305 events). The aggregated total number of self-grooming events and self-grooming durations was broadly similar to the aggregated total number of allogrooming events and allogrooming durations (Stewart and Macdonald 2003, see also Nadin et al. submitted for analysis of a larger sample).

Dividing the total duration of allogrooming events by the number of hours the average badger was on camera, we could determine that each engaged in approximately only 20 s per night. This approximation says nothing about underground interactions, but above ground it could be an order of magnitude wrong without threatening the conclusion that, however important it might be, or whatever its functional significance, badgers are not engaging in mutual grooming to the extent of, for example, the 8 h per day that Vucetich (2021) estimates members of a wolf pack devote to socializing, or, for example, the baboons (*Papio* spp.) that spend from 6 to 23% of their time in mutual grooming (Dunbar 1992).

Does it work? In other words, does cooperation pay? Success might be measured as a reduction in the number of fleas or of flea bites, but realizing that, like us, each badger will suffer differing degrees of itchiness per bite. That is, we presume that badgers are most motivated proximately not by flea bites but by the idiopathic extent to which the individual experiences itchiness (see Chapter 15); a badger soliciting much allogrooming might be driven to distraction by each bite, whereas one with minimal reaction to the bite allergens might eschew allogrooming. This rather obvious distinction may have ramifications insofar as each badger will have its own personal level of immune sensitivity. The consequences of this struck us when we realized that the two border collies, Fly and his sister Qara, owned by David and his wife Dawn, reacted radically differently to flea bites, despite being littermates—whereas Qara itched a bit, Fly was driven to frenzy:[6] deep in their sibling genotypes they had, we guessed, very different CD4+ T-cell and IgE and IgG responses causing inflammation and 'pruritus'—the technical term for itching (Halliwell and Longino 1985). This type of asymmetry may greatly influence the personal stakes each badger places in mutual grooming (and whether or not to engage in grooming at all; see Box 5.3).[7] It might also complicate our interpretations of flea counts, insofar as dogs suffering flea allergy dermatitis (FAD), like Fly, typically have very few visible fleas because they groom obsessively.

One counterfactual is: how many more fleas would an average badger have if there was no mutual grooming (or no grooming at all)? (a second counterfactual, to which we return later, considers that grooming badgers may be on autopilot,[8] so how much grooming would a badger do if it had no fleas?). To judge by our riffling fingers, many fleas are likely to escape a badger's own grooming, seeking refuge in the regions only accessible to a mutual groomer. Do the concerted efforts of self- and mutual groomer crush sufficient fleas to reduce their numbers (remember reinfection can occur from companions and from bedchambers)?

The ideal flea distribution

A cost to sociality is disease transmission. Amongst carnivores, rabies has been modelled to spread like wildfire within groups of red foxes (Macdonald 1980b; Macdonald and Bacon 1982; Ball 1991), and has been seen to do so devastatingly amongst African wild dogs and Ethiopian wolves (Gascoyne et al. 1993; Vial et al. 2006; Randall et al. 2006). Indeed, perhaps yet again farm cats offer a good model: David recorded how all of the kittens of the amalgamated litters of two mothers, while nestled together in a single bedchamber, succumbed to cat flu (Macdonald et al. 2000b. The fitness of those kittens was a brutally obvious trade-off between the communal suckling and shared care they gained from the cooperating mothers, and the fatal infection that resulted: farm cats rear kittens communally, so, *de facto*, on balance, it pays (but it has to be paid for, Wilson et al. 2003; Ezenwa et al. 2016). Writing this at the moment of Covid-19 self-isolation, it is obvious that badgers could cut the risk of flea reinfestation by keeping a flea's jump (10 cm) away from each other, and not sharing dens. Whatever the costs of flea infestations, it appears that they are, to the averagely sensitive individual, less than the benefits derived from communal denning and the extreme proximity necessitated by grooming, allomarking, playing, and sleeping in a heap.

What are the consequences of communal denning in terms of ectoparasites? It seems likely that all members of a cohabiting badger group will be similarly laden with fleas, although different groups might have different flea loads (due, perhaps, to the internal conditions of each sett, such as temperature or humidity, and underlying geology of size and architecture; Butler and Roper 1996). Adapting the language of Fretwell's (1972) ideal free distribution[9] (IFD) we expect fleas to be distributed roughly equally amongst badgers within a group, *ceteris paribus*, according to an *ideal flea distribution* (see Kelly and Thompson 2000). We delved into our database using flea load data from 1993 to 1998, to discover that in 4 of 6 years, variance in flea loads was significantly greater between social groups than within them. Furthermore, both flea load and (log) flea variance were negatively correlated with social group size: members of larger groups each had fewer fleas, more evenly spread amongst them. The epidemiological unit is seemingly the whole

---

[6] Raveh et al. (2011) suggest this is a fatal distraction to foraging gerbils, distracted from their foraging and keeping a lookout for predators.

[7] Amongst people, when experimentally exposed to skin irritants women tend to experience more pain-related sensations and to perceive a sense of burning more than men, leading Hartmann et al. (2015) to conclude that mixed itching and burning sensations are differentially processed by both genders.

[8] Eads et al. (2017) suggest that 25% of the grooming of black-tailed prairie dogs continues even when their fleas are removed.

[9] An IFD is a theoretical way in which a population's individuals distribute themselves among several patches of resources within their environment, in order to minimize resource competition and maximize fitness.

social group: there was no correlation between flea load and the numbers of a temporary subgroup of badgers occupying a particular sett (Johnson et al. 2004). In consequence, every badger in the group had some vested interest in reducing the flea infestation of the collective (remember also the selfish herd): the ideal *'flea'* distribution will make it pay for each badger to contribute to the common good by mutual grooming (Johnson et al. 2004). It must be a Canute-like task, insofar as bedchambers are laden with hatching eggs and younger instars eager to replace those fatally groomed.

A wider perspective

Fleas are an unusual quarry insofar as the rewards of cooperating in their pursuit are not obvious. For two wild dogs joining forces in pursuit of impala the reward can be measured in per capita marginal net gain in calories. The reward is clear (although it took the meticulous work of Scott Creel, some of it analysed while a post-doc at the WildCRU, to measure it), as is the significance of failure: without the calories the wild dogs come closer to death. Do badgers reduce flea numbers (they spend rather little time on the task above ground, and tackle it in rather short bursts)?

Pause to consider some generalizations about mammalian grooming. It is controlled by one of two different mechanisms: first, stimulus-driven response to an itch; and second, preprogrammed hygienic grooming (cats spend about 10% of their waking time grooming), preventatively, irrespective of itching, and in the order face, hind legs, flanks, neck and chest, rump, and tail (Bradshaw et al. 2012); their saliva also discomforts fleas and other epidermal parasites (Benjamini et al. 1963). Long after we began counting fleas, it was discovered in 2013 that mammals perceive itches through a special kind of neuron exclusively devoted to sensing an itch (Han et al. 2013; Johns Hopkins Medicine 2013). Amongst our growing list of known unknowns is: what proportion of badger grooming is stimulus–response or programmed; and how richly supplied are they with itchy neurons? Does grooming, and particularly mutual grooming, just feel nice, enhancing the quality of life (i.e. is it a luxury rather than a matter of fitness)? Luxuries can be worth paying for: Barry Keverne, a Cambridge biologist who pioneered exploration of the hinterland between physiology and ethology, discovered that being groomed releases beta-endorphin (associated with increased serotonin production) (Keverne et al. 1989). Similarly, Oxford's Robin Dunbar (2010) explains how 'social

touch' causes a pleasing release of oxytocin in mammals. However, as even Hercules discovered, there are trade-offs between virtue and pleasure; for instance, blue monkeys are known to risk their lives when their euphoric distraction with allogrooming causes them to take their eyes off predators (Cords 1995) (the same applies to impala; Mooring and Hart 1995). In evolutionary biology, there is really no such thing as a luxury—today's craved indulgence is the answer to yesterday's selective pressure (and remember, mutual grooming is highly, non-randomly, organized between zones of anatomical complementarity in badgers). In the case of grooming badgers might the cementing of social bonds be more valuable than the removal of fleas? Amongst primates, equipped with parasite-picking fingers, allogrooming not only functions reciprocally in the cooperative alleviation of itching, but is offered also in bidirectional exchange for other benefits, such as access to food or coalitionary support (Machanda et al. 2014). For example, (captive) chimpanzees (*Pan troglodytes*) are more tolerant of individuals with whom they engage in bidirectional grooming than of those that simply donate unidirectional grooming to them, requiring nothing in return, with dominants quicker to terminate the simple receipt of grooming than the mutual exchange of grooming (Fedurek et al. 2009). Amongst badgers, as farm cats, mutual grooming flows symmetrically between dyads, but at different bandwidths (Chapter 5): a form of social grease, perhaps carried forward from early maternal care, whose signal of affiliation may outweigh its contribution to hygiene (on the other hand, readers of Box 5.2 will recall that unreciprocated allogrooming diagnoses some rank relationships although these scarcely coalesce into a hierarchy). A step further down that sociological road, Chris had the idea that soliciting grooming has evolved into an appeasement signal, mimicking how cubs approach their mother.

Humble the badger flea may be, but it may also be far from inconsequential. A conceptual breakthrough in the 1970s by Bob May and Roy Anderson established parasites[10] (Anderson and May 1978; May and Anderson 1978) amongst the fundamental drivers of population dynamics; fieldwork confirmed this impact, often with a lag as reproductive debilitation tracked infestation, amongst vertebrates such as the red grouse afflicted by strongyl louping ill (Hudson et al. 2002),

---

[10] Parasite, from the Greek *parasitos*, meaning 'person eating at another's table'.

the Svalbard reindeer emaciated by *Ostertagia gruehneri* and *Marshallagia marshalli* (Albon et al. 2002), or the moose that famously cohabit Isle Royale, in Lake Superior, alongside wolves—these moose are themselves home to the tick, *Dermacentor albipictus*—a heavily infested moose houses 100,000 ticks with cascading impacts on predator–prey cycles and ecosystem function (Vucetich 2021). In evolution, and ecology, seemingly small things (on Isle Royale something as small as a winter tick) can have huge, and ramifying, consequences. What of Wytham Woods and something as small as a badger flea?

The landscape of disgust

This brings us to the 'landscape of disgust', the evocatively named expression of the potential consequences of risks that infectious disease and, particularly, infestation with parasites, can produce (Buck et al. 2018; Weinstein et al. 2018). Remember the tension between the benefits of mutual grooming while cuddling up to a companion to remove fleas, and getting so close that you catch more of them. Bear in mind too the aphorism that prevention is better than cure. This tension is reflected in two types of observation: on the one hand, badgers mutual groom to remove parasites (Stewart and Macdonald 2003), as do social mammals from meerkats to mandrills (Madden and Clutton-Brock 2009; Poirotte et al. 2017); on the other hand, the so-called 'social immune response' (Cremer et al. 2007; Meunier 2015) involves adaptations to maintain collective group health through a behavioural quarantining, even removal, of sickly individuals (Poirotte et al. 2017; Stockmaier et al. 2018). Steering clear of parasites and pathogens can influence how animals arrange themselves in space (Methion and Díaz López 2019; Bachorec et al. 2020; Webber et al. 2020) by the avoidance of sites and sources of contamination (Curtis 2014), something even detected in the hygienic observances of the raccoons, badgers' second cousins, around latrines (Weinstein et al. 2017). So, for the badgers of Wytham Woods, how are the positive benefits of mutual grooming, which tend to increase with local population density (i.e. group size), weighed against the negatives of increased rates of social contact and reduced ability to 'socially distance'?

Enter Greg Albery, whom we met when he was one of those Oxford undergraduates that was clearly going places (the places, in Greg's case, being Edinburgh University for a doctorate and Georgetown University for a post-doc). For Greg's undergraduate project

we investigated patterns of parasitism across the badgers of Wytham Woods; during his post-doc, 5 years later, he returned to tackle the question: do higher badger densities result in higher parasite burdens? We expected the answer to be yes, because of greater exposure, and Greg tested for this using spatio-temporal autocorrelation models (Albery et al. 2020). Then, by linking this to broad mortality rates established from our long-term monitoring (see Chapter 11), we asked whether sickliness was, for these badgers, an avoidable problem.

We assigned each badger to a group, using the residency rules presented in Chapter 3 (Annavi et al. 2014b), and then applied the nattily named 'gambit of the group', which assumes that individuals trapped in the same sett in the same year will have interacted (Franks et al. 2010). And then we were back to counting fleas, using our tried-and-tested 20 s inspection of the badger's full body, with an additional search for ticks (*Ixodes* sp.) and lice (*Trichodectes melis*) counted within a 4×4 cm square of heavily parasitized skin near the groin (per Cox et al. 1999). Recalling Chapter 2, we also looked at our records (1993–1997 and 2009–2017; $N = 1287$ counts) of those more deadly gut parasites, *Eimeria melis* and *Isospora melis*.

Male badgers had more lice than did females, and there was substantial monthly variation in all parasites (Figure 8.2). Cubs had fewer fleas, more lice, and greater *Eimeria* prevalence than did yearlings and adults, and lower *Isospora* prevalence than adults (Figure 8.2). Additionally, body condition was negatively associated with fleas, lice, and *Eimeria* infection in all age/sex classes (Figure 8.2), although it's hard to be sure of causality: possibly parasitism causes a loss of condition (particularly plausible for *Eimeria*, see Chapter 2), whereas badgers in poor condition probably can't shift their burden of fleas and lice. Only louse infection impacted survival probability, and the effect was small. Lice tend to accumulate as the host weakens due to starvation, injury, or other disease, especially amongst cubs.

Parasite burdens measured on successive examinations of the same badger varied much less than they varied between individuals. We deduced that individuals' home ranges (or shared group ranges) differ in terms of the local parasite challenge, and in terms of the local density of badgers residing there. Further exploration of density effects, using spatial and social network analyses, pointed to an overall negative association between local population density and various parasites (Figures 8.3 and 8.4). In short, fleas,

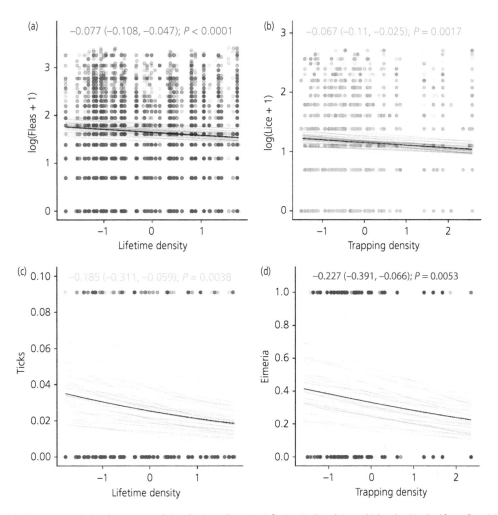

Figure 8.2 Negative associations between population density and parasite infection. Badgers living at higher densities had fewer fleas (A), fewer lice (B), lower tick prevalence (C), and lower *Eimeria melis* prevalence (D). B represents juveniles only. Points represent individual samples, with colours randomized along a colour palette for plotting clarity. Opaque black lines are derived from linear model fits, taking the mean of the posterior distribution. Transparent grey lines represent 100 fits drawn randomly from the posterior estimate distributions of each model, to demonstrate error in the slope and intercepts. Text at the top of the figures communicates slope estimates, with lower and upper 95% credibility intervals in brackets, and *P* values. NB: in panel C, the *y* axis scale and the location of the positive points have been altered for plotting clarity and to visualize the slope better, due to low tick prevalence (from Albery et al. 2020, with permission).

lice, ticks, and *Eimeria* were most prevalent or abundant in areas of lowest badger density, from which you might suspect that more badgers offer greater opportunities for cooperation in grooming (the explanation favoured in some other studies; Almberg et al. 2015; Ezenwa et al. 2016). On the contrary, we clung onto Greg as he led us through an assault course of analyses of co-trapping networks and grouping metrics to arrive at the conclusion that 'direct' social behaviours, such as mutual allogrooming, were unlikely to explain these negative density effects. Furthermore, an exploration

of badger density effects led to the rejection of several explanations for the spatial structuring of the badgers of Wytham Woods,[11] and also to ditching the idea that heavily parasitized individuals were ostracized (as

---

[11] Becker et al. (2019) review candidate hypotheses. In Albery et al (2021) we conclude that the badgers' spatial arrangements did not originate from: (i) differential local mortality; (ii) reduced susceptibility arising from co-habitation and implied cooperation benefits; or (iii) greater local resource availability influencing susceptibility.

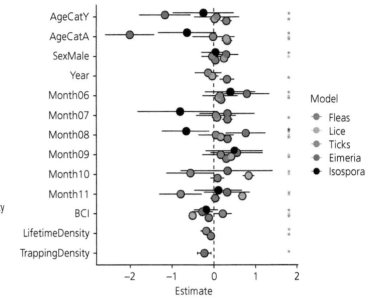

Figure 8.3 Model effect outputs for the full dataset INLA density models. Points represent modes of the posterior distribution of the effect estimates; error bars represent the 95% credibility intervals. Asterisks denote significant results, whose 95% credibility intervals did not overlap with zero. Only the behavioural traits that improved the model fit (ΔDIC < −2) and were kept in the final model are shown (from Albery et al. 2020, with permission).

occurs in honey bees; Baracchi et al. 2012). However, linking the badgers' spatial structuring to the propensity for setts to harbour parasites (Butler and Roper 1996; Cox et al. 1999), the most parsimonious interpretation is that badgers avoid infection behaviourally, preferring to inhabit areas where parasite transmission is reduced (see Table 8.1).

In recognition of the behaviour of individual badgers to avoid areas risky for parasites, and amplifying this to the population level, the pattern emerges in which badger density is inversely related to parasite distribution (Figure 8.3). In short, the 'landscape of disgust' is a driver of how badgers are dispersed throughout Wytham Woods (Weinstein et al. 2018). We had begun by wondering whether groups of badgers were, in part, caused by parasites, imagining that the benefits of mutual grooming drew the badgers together. We had been right, but for the wrong reasons. Parasites are indeed a force for grouping amongst the badgers of Wytham Woods, but not primarily because of the rewards of mutual grouping (although these may exist, as a secondary matter, see the text that follows). Rather, it seems that fleas, along with lice, ticks, and *Eimeria*, 'push' badgers into some areas, while causing them to avoid others. Parasites are, indeed, one reason why some badger groups are large—because the badgers are fleeing from fleas and similar blights (another reason is likely to do with food dispersion, Chapter 10). This selection for sett traits and sites resistant to parasite infestation and transmission

results in fewer setts, with fewer occupants, in more highly parasitized areas. What sort of environmental factors might make a sett good? Probably some sett microclimates will favour fleas, and the oocysts of *Eimeria* (with their faecal–oral transmission) probably do better in moist sett chambers. This may explain the slow decline moving away from the Thames River towards drier parts of the woods. Ultimately, these patterns in parasite risk may lead to parasite avoidance being traded off against foraging success, reproductive success, and survival (Figure 8.5) (Hutchings et al. 2006; Buck et al. 2018). Indeed, when it comes to the positioning of setts, perhaps a further factor (alongside soil type, aspect, etc.—Chapter 4) is the avoidance of parasites, and this may manifest as the preference to cohabit in a 'good sett', rather than to occupy a poorer and less popular one.

Detail matters

The landscape of disgust may be a force driving badgers together, but once they are together, how do they cooperate in pursuit of parasites? We selected two periods: October and November (1995), representative of the non-breeding season; and March–May, the breeding season. These yielded 280 instances of badgers self-grooming (totalling 2146 s) and 305 of mutual grooming (totalling 1569 s) involving 16 individuals (Stewart and Macdonald 2003). On average, mutual sessions lasted a bit over 2 min (137.2 s). The most interesting

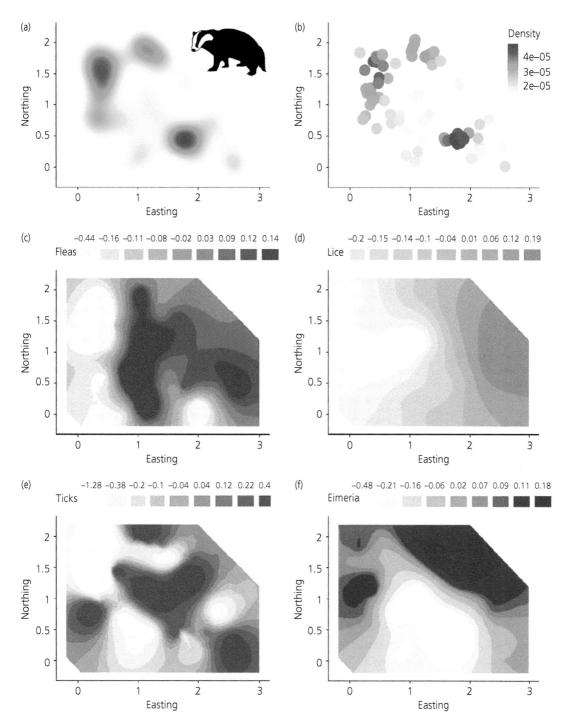

Figure 8.4 Spatial distributions of badger population density and parasites in Wytham Woods, Oxfordshire, between 1989 and 2018. (A) Badger population density distributed across Wytham Woods, calculated based on a space use kernel for individuals' annual centroids. (B) Individual badgers trapped at setts (represented by points) were assigned a local density value based on their location on the rasterized space-use kernel. Darker blue colours in (A) and (B) correspond to greater population density. (C–F) The spatial distribution of the four spatially distributed parasites, estimated using the INLA SPDE effect. Darker colours correspond to increased parasitism. The density values in (B) were fitted as covariates in linear models to explain individual parasite burdens, revealing a negative correlation between density and parasitism. All axes are in kilometres, with the 0,0 point at the bottom left of the study area (from Albery et al. 2020, with permission).

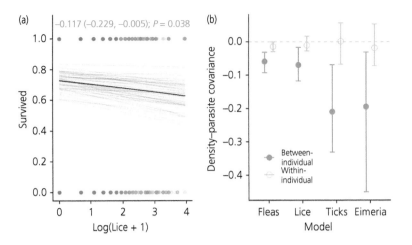

Figure 8.5 Associations of parasitism with survival and density–parasitism covariance partitioning. (A) Louse burden was negatively associated with individual survival probability in juveniles. Opaque black line = linear model fit. Transparent grey lines = 100 fits drawn randomly from the posterior estimate distribution, to demonstrate error in the slope and intercept. Text at the top of panel A communicates slope estimate, with lower and upper credibility interval in brackets and P value. (B) Estimates for within- and between-individual covariance between parasitism and density, taken from MCMCglmm multivariate models. Points = posterior mean effect size estimates; error bars = 95% credibility intervals (from Albery et al. 2020, with permission).

Table 8.1 Reasoning surrounding our hypothesis testing linked to social living in badgers. We rejected all hypotheses for our observed negative density effects other than parasite avoidance.

Potential mechanism	Conclusion	Reason for conclusion
Cohabitation/cooperation benefits	No	No direct social effects or body condition effects
Allogrooming	No	No direct social effects; grooming not possible for endoparasites
Nutrition-associated immune benefits	No	No competition with body condition effects
Local host die-offs	No	Survival associations only discovered for lice
Social ostracism/self-isolation	No	Low within-individual covariance
Encounter dilution	~Yes	Possible for generalist ticks, but not for the other (badger-specific) parasites
Avoidance	Yes	All other possibilities eliminated; consistent with individual-level behavioural responses

feature of these mutual sessions is their start and finish: somebody had to make the first move, and somebody had to be the first to quit—inevitabilities that provide an unusual opportunity to quantify a precious resource: faith.

In Chapter 5 we described the flux of allogrooming amongst the badgers—it was broadly symmetrical, and is now revealed to be, at least in part, a cooperative flea hunt (and, when unreciprocated, a mirage-like vision of rank relationships). Males and females took the roles of both initiator and terminator in mutual grooming bouts with both sexes. Actually, there was often a huddle of groomers—a circumstances known as

'Esther grooming'[12] where A initiates grooming on B, who responds by grooming 'the wrong badger' C (Thelen and Farish 1977). These cases were revealing insofar as A would generally very rapidly terminate, unless,

[12] Acknowledging Esther Theren's work on Piaget's circular reaction, concerning how infants learn circular reciprocity. Interestingly, Esther, who started out studying the supposedly fixed action patterns in the behaviour of a small parasitic wasp, became impressed instead by the individual variation in behaviour of these wasps, and considered individuality important to the evolution of complicated behaviour—a thought worth bearing in mind until we reach the individual energetic strategies of badgers in Chapter 14.

and only unless, C then groomed A in what Paul colourfully called a triad of contentment. In short, these badgers appeared reluctant to donate grooming effort gratis (A's hasty termination in this example illustrates this tendency). In addition, reciprocity in grooming appeared to be a rather grown-up social more: a sample of mothers and cubs of 8–12 weeks old revealed that cubs seldom reciprocated grooming until at least 10 weeks old.

So, what happens when a badger initiates grooming of a recipient? In 148 of 181 instances (the entire sample) 82%, grooming initiation attempts by badger A were met with reciprocation by badger B, but ultimately terminated by A. How quickly did B respond to A's initiation? The median answer, timed for one response each from 19 recipients, was under half a second (or as fast as we could measure it). In short, if grooming was an invitation, the invitee replied fast.

And when one of a pair of groomers quit, what was the reaction of the remaining groomer? A total of 128 of 181 (86%) defections were met by retaliatory termination at similar speed.[13] Sometimes, and this happened at lightning speed, a groomer would shift position—an action that might be mistaken for quitting if the groomer dawdled—the median time for one partner to shift the target zone of its grooming was a fraction of a second. Paul and David measured a sample of 200 such shifts and on 38% (76) of them the other groomer made a correspondingly swift readjustment. Their interpretation was that each cooperating member of a mutually grooming duo had a hair-trigger on quitting if its companion's commitment to the partnership faltered—each badger gave the other the benefit of the doubt for scarcely a second. By analogy, similar impatience would destroy any human dialogue, and nobody would ever hear the punchline to a joke if they were to walk off during the preceding dramatic pause.

Although the timings read from the videos were almost incredibly fine, it seemed that the benefit of the doubt given to a grooming partner was half that allowed for an invitee to RSVP: an initiator allowed an average of 1.2 s for a response in kind before withdrawing the invitation by stopping grooming (although sometimes grooming continued unrequited for up to 30 s; Stewart 1997).[14] How frequent were cases of

more than fleeting totally unreciprocated allogrooming being donated (other than to cubs)? These happened only rarely. During the combined two-season sample in 1995 we logged 1862 bouts of allogrooming involving 38 badgers, of which only 42, that is, 2%, were unreciprocated for >5 s. What were the circumstances of this unreciprocated grooming? Fourteen (of the potential 38) donors experienced this sort of rebuff, and often it was because the receiver was involved in allogrooming a third individual—the Esther case mentioned earlier. In short, invitations to groom seemed close to irresistible, suggesting they were highly valued, but on the other hand credit between badgers was very short-lived (later, with a much larger sample, Nadin et al. (submitted) made the special analysis of unreciprocated grooming presented in Box 5.3). This led us more deeply into game theory (Box 8.2) but rather than become mired in the detail, suffice it to say that we interpret the poor credit rating badgers accord each other as a general reflection that their solitary ancestry has burdened them with a rather distrustful disposition and that even amongst the badgers of Wytham Woods, cooperation does not come naturally.

Fleas, angels, pinheads, and sociality

When we began our research on mutual grooming we might have wondered, with a touch of self-mockery, whether the topic of counting fleas on badgers was perilously close to the famous academic pastime of counting angels on pinheads (as St. Thomas Aquinas put it in the thirteenth century). Now, on the contrary, we see this line of inquiry as a source of questions at the interface of social evolution, immunology, and epidemiology.

A crucial question remains unanswered. How efficiently does grooming (self- or mutual) diminish flea numbers (and associated idiopathies and infection risks), and for how long? We failed to answer this when, with the help of intern Sharon Quinn, we attempted to clean badgers of fleas using a pet 'spot on' (ectopic ivermectin) flea treatment. With rapt attention we watched video footage during the following days—the badgers continued to mutually groom, no less enthusiastically than they had before (but remember that programmed cats groom on autopilot). A

---

[13] The median figures for our samples of initiations and terminations were 0.44 s and 0.48 s, but this gives a spurious sense of accuracy: the practical reality is that as soon as one grooming badger acted, the other responded.

[14] The foregoing account provides medians. In his thesis (pp. 233–234), Paul Stewart compared these figures between

the non-reproductive and reproductive season samples—these were not appreciably different. Mean time to continue an unreciprocated initiation: mating season: 1.17 s (standard deviation (SD) 0.61, $n = 12$) (p. 233); non-mating, 1.25 s (SD 0.57, $n = 12$). Interval to termination following the quitting of a partner: mating, 0.46 s (SD 0.38); non-mating, 0.67 s (0.57).

week later we recaptured them, counted their fleas, and found just as many as pre-treatment. Perhaps the ivermectin hadn't worked, or perhaps it had but the badgers were swiftly reinfested—perhaps it worked in the first few days, prior to reinfestation, but the badgers continued to (allo)groom just as assiduously (and perhaps continued to itch) even in the absence of fleas? Maybe the continual pursuit of ghosts of fleas past is adaptive, or perhaps on a precautionary, prophylactic, basis, prevention is better than cure? We don't know, yet.

And what of the physiological battlefront between badger and flea? A badger's susceptibility to fleas is a cell-mediated response (in other species it is often a sexually dimorphic one, Tschirren et al. 2003). And what of the likely individual differences between badgers in the itchiness they experience and thus their inclination, perhaps their desperation, to seek help—differences illustrated earlier by the sibling collies Fly and Qara (García et al. 2004; Cuellar et al. 2010)? Those in our footsteps might add a skin irritant test to measure sensitivity to flea bites (a.k.a. induced papular urticaria) to the list of 30 measures we routinely recorded from each captured badger.

An uncooperative group

We began this chapter wondering whether, in line with other group-living carnivores, but contrary to first impressions, badgers might after all gain some cooperative benefit from group living that weighed heavily in each individual's assessment of whether to join the family firm or to set up shop alone. No such heavyweight cooperative pay-off has emerged. Indeed on the measures we could make of the quality of their lives in Wytham Woods, badgers did worse in larger

---

Box 8.2 Games badgers don't play

The colloquial tit-for-tatness of these grooming interactions is one thing, but tit-for-tat has a technical and precise meaning in game theory—and at the time Paul Stewart was poring over these videos this topic was much discussed by theoreticians but little explored in the wild. Although the stakes were seemingly low, might cooperative grooming amongst badgers serve as a model exploration of game theory in the wild? To appreciate the context, a brief sortie is required into Rousseau's stag hunt (Talwalkar 2008).

Intrigued then, as we are now, to understand the roots of cooperation, Enlightenment philosopher Jean-Jacques Rousseau (1712–1778) imagined two companions setting out hunting, with the choice of pursuing a rabbit or a stag. While the cooperation of your companion is necessary to secure the rich reward of the stag, each of you, hunting alone, can be sure of catching a rabbit. Imagine that catching a rabbit gives a pay-off of 3, capturing the stag gives a pay-off of 5 to each person, and the reward is 0 for capturing neither. If you pursue a rabbit you safely bank 3 points; however, if you pursue a stag, the pay-off depends on whether the other person chooses stag (each of you gets 5) or rabbit (you get 0). So, tactically, rabbit is a best response to rabbit; stag is a best response to stag. Since the best outcome is that you and your companion both hunt stag, cooperation is favoured. However, if you pick stag, and the other person does not match you, you end up with nothing: you are putting a lot of faith in him, and if the stakes are high you may be trusting him with your life.

In 1974, Sussex University biologist John Maynard-Smith brilliantly co-opted the stag/rabbit idea to explain the evolution and resolution of animal conflicts, and in 1980 Rousseau's stag hunt was re-engineered by Robert Axelrod as the 'Prisoner's Dilemma'[15] (Axelrod 1980). This now famous dilemma involves two prisoners being interviewed separately, each stricken with the decision of whether to support or betray the other.[16] Could the idea be transposed onto a freeloading badger, a recipient of grooming who shirked reciprocation: was it because of this risk that badgers showed so little faith in each other? Why would mutual grooming matter so much to each badger to be so precisely intolerant of freeloading? By happy coincidence, at the time these thoughts took shape, Martin Novak, now Professor of Mathematical Biology at Princeton, was in Oxford launching a meteoric career in game theory.[17]

---

[15] The stag hunt differs from the Prisoner's Dilemma in that the hunt involves two pure strategy Nash equilibria: when both players cooperate and both players defect. In the Prisoner's Dilemma, in contrast, despite the fact that the situation in which both players cooperate is Pareto efficient, the only pure Nash equilibrium is when both players choose to defect.

[16] There is a very clear account of this dilemma, albeit in a different context, by former WildCRU badger student, Dominic Johnson in his 2016 book, *God Is Watching You*.

[17] Nowadays he is steeped in fascinating and unresolved debates about group selection and inclusive fitness (Novak and Highfield 2011). See also Why Evolution is True (2011).

Box 8.2 *Continued*

The resulting collaboration over being short-changed in the mutual grooming game led Martin to contribute pages of elegant algebra to Paul's thesis. The purity of the maths was dazzling, and for a while we thought we had a cut-and-dried case of tit-for-tat cooperation in the wild. Of course, it turned out that nature fits into theory about as readily as a genie, having feasted on assumptions, can be squeezed back into its bottle! Regarding assumptions, to fit those of tit-for-tat, bouts should be of unpredictable length, such that the cooperating pair cannot judge with any degree of certainty whether the other individual is likely to defect—that seemed to apply to mutually grooming badgers. Further, if the risks and benefits of cooperation could be measured in a commensurable currency, such as the numbers of fleas cooperatively crushed, the calculations might be complicated but achievable. However, troublesome realities accumulated. If the pay-offs might be idiopathic measures of relieved itchiness (where one less flea does not equate to an equal measure of relief, depending on individual immunology), the currencies become hopelessly incommensurable. Worse, these badgers know each other well, and build up a track record of creditworthiness (by contrast, prisoners in the dilemma meet only once). The badger's dilemma was not, after all, as simple as the prisoner's, and probably a lot less serious (Johnson et al. 2002c). Furthermore, the early Prisoner's Dilemma was soon augmented by more complex coalition games (Noë et al. 1991), encompassing bargaining and power asymmetries between partners, and now also by a genre of cooperative games.[18] Nonetheless, our sortie into game theory had not been wasted, because these thoughts preyed for a decade on the minds of two of our badger team, Pavel Stopka and Dom Johnson (by then they were, respectively, senior academics at Charles University in Prague and

Harvard University in Massachusetts). Their reflections, published in 2012, led them to two conclusions about mutually grooming badgers—assuming success is measured in the cooperative killing of fleas rather than the prevention of itching—first, that pairwise cooperation was based on little trust, and second that irrespective of whether the individual initiating grooming is rewarded by its companion with a reciprocal act of grooming, it nevertheless reaps a generic benefit because all efforts ultimately contribute to reducing the parasite burden in the group (of course, in reality a new crop of fleas will soon emerge from the eggs and larvae in the soil of the shared sett). Perhaps the ultimate, and important, lesson of our excursion into fleas is that while everything is fascinating, nothing is simple: it's complicated enough that one thing leads to another, that the chaotic consequences of a butterfly beating its wings in Beijing causes a storm in the Caribbean, but the complexity is unimaginable in reverse (try working out whether the storm was caused by the butterfly in Beijing, the stridulating cicada in Myanmar, or the inauspiciously timed sting of a hornet in Spain, or how all three interacted). Remembering Niko Tinbergen's lesson that in biology every question can be answered from four perspectives, to understand the provocations of the humble badger flea it becomes important to know to what extents the badger's reaction is caused by the irritating meanderings through its fur, the moment of impalement, the subsequent itch, the tendency to inflammation, the risk of disease, the benefit of socializing—several more careers-worth of intrigue await.

[18] Cooperative games not only offer insights into how mammalian societies function but have also led to a new line of thought on how conservation might best be advanced (Curry et al. 2019).

---

groups (certainly even fleas seem a discouragement to grouping, their prevalence exhibiting negative density dependence with group size; Alberry et al. [2021]). So, to resolve the cost–benefit analysis of group living we will turn, in Chapter 10, from behaviour to ecology. In passing, recall Sandell's (1989a, p. 164) definition, from Chapter 7, of solitary behaviour in carnivores as 'no collaboration between members in feeding, defence of territory, offspring-rearing, and mating even in cases where ranges overlapped'. By these criteria the rollicking rough and tumble amongst a dozen or more badgers we see not infrequently around setts in Wytham would be solitary behaviour—a paradox. Nonetheless, there is clear-cut reciprocation in mutual grooming,

and it is very finely choreographed, although it is hard to argue that the pay-off weighs heavily amongst the drivers favouring group living. What selective advantage lies behind this fine choreography? Mindful, in the manner of the *Blind Watchmaker*, that natural selection advances through the ratchet of tiny advantages, even a few fleas less may add up to an advantage, as may a few seconds of freeloading prevented,[19] but might there be other functions in the account?

[19] Thoughts of free-loading are formalized in such catchily named models as 'producer–scrounger' (Caraco and Giraldeau 1991) or 'leaders' and 'laggards' (Heinsohn and Packer 1995).

Misunderstandings can have terrible consequences, and correspondingly the communication of goodwill, even appeasement, matters. 'Please don't let me be misunderstood'[20] was the sociologically profound lyric of the Animals song in 1965, and it may have played out in badger evolution. Badgers are dangerous (remember those bite wounds in Chapter 3), and their solitary ancestors were probably aggressive—a short fuse may be included in that phylogenetic baggage. For badgers thrust together by, in evolutionary terms, modern circumstances, a quick signal to quell distrust might avoid expensive misunderstandings (and a ready source for such a signal might build on a mother's comforting grooming of her infants). A loose, but familiar, parallel might be found in the human habit of offering our dominant weapon-wielding hand in handshakes, salutes, pats on the back: evolved signals that 'we come in peace' (Lundmark 2009) (by the way, in terms of honest signals, a sweaty palm is hard to disguise). A quick mutual groom may be an indicator of no intent to bite.

A dead flea is a dead flea whether you killed it in the hope that its host would do the same for you, or because you want to ingratiate yourself. But how did it start? All mammal mothers groom their infants, even tree shrews and hartebeest (both species where mothers steelily eschew contact with their babies, to reduce predation risk), and it seems likely all have done so since the first mother morganucodontid tried it out while getting used to her fused dentary bones. A hundred-million-year-old comforting feeling that brings health benefits is hard to ignore, but the question for badgers is whether health or appeasement led the way as this behaviour graduated from an experience of infancy into an adaptation of adulthood: they probably leap-frogged in primacy, and perhaps eventually pulled in harness towards the present. And why do we use the word appeasement? A nod to the badger's phylogenetic baggage—that a fiercely dangerous, ancestrally aggressive carnivore trapped by circumstance into modern sociality would do well to make its opening remark in any conversation an apology.

Finally, back to the question of how big are the pay-offs. We cannot parse out all the contributors (alloparenthood, communal digging, mutual grooming, and any others) and although none seems plausible as a principal driver of group living, do they in aggregate add up to making it all worthwhile? The ultimate yardstick of the pay-off would surface as breeding success. Do individuals in larger groups benefit from their membership by accruing greater individual breeding success (the question might be phrased with more nuance by asking whether there is any group size, not necessarily the very biggest, that delivers the greatest reproductive success)? If the answer is yes, the quest for something cooperative rewarding group living by Wytham's badgers could not be lightly abandoned. But the answer is no. On the contrary, individual breeding success was 72% when 1–3 females were present in a group, 46% with 4–5 females, 31% with 6–10 females, and 21% with 11–17 females present (Annavi et al. 2014b) (see Dugdale et al. 2008; Chapter 17). In these measures, there is no hint that larger groups pay dividends to individual members—rather, whatever the pay-off to the seemingly rudimentary cooperation amongst badgers, it appears not to redress some general worsening of conditions for those badgers living in the crowded circumstance of big groups (and see Chapter 11, where swapping to more, but smaller groups turned out to be advantageous to the badgers of Wytham Woods). We have searched for evidence for a pay-off attributable to cooperation, and in its absence we are reminded of Lewis Carroll's (1871) judgement: 'Contrariwise, if it was so, it might be; and if it were so, it would be; but as it isn't, it ain't'. Rather than natural selection rushing them towards the rewards of sociality, they seem more to be making the best of a bad job.

A final thought. In comparison to the sociological sophistication of many social carnivores, badger society seems rudimentary. There are reasons, and we will explore them in Chapter 10, for thinking that life in large groups is a rather new evolutionary experience for them. The idea struck us that badgers offer a model, a paradigm, for how carnivore societies got going; a look back in history to how ecological circumstances might once have launched the societies of wolves and meerkats whose intricacy dazzles us today (Macdonald et al. 2015e). There is a philosophical thought here: in the same way that the evolutionary tree of life is not hierarchical, and humans are no better adapted than shrews (each is equally good at being what it is, as judged by the yardstick of its existence), we might ask is badger society less well adapted than meerkat society? Except for two possible circumstances, the answer must be no. Badgers are top notch at being badgers, their adaptations to badgerdom are definitely better than a meerkat's adaptations to being a badger.

---

[20] 'I'm just a soul whose intentions are good, Oh Lord, please don't let me be misunderstood.'

Thus, when we see only minimal evidence of allo-parenthood or cooperative vigilance amongst badgers, two attributes that are strikingly developed amongst meerkats, this does not mean that badgers are shoddy impressions of meerkats. It means they are what they are, and that minimal adaptations to alloparenthood and vigilance are exactly what badgers need. This is not a Pollyanna-ish platitude, and it isn't teleology, either; rather, it is an observation that natural selection adapts species to be the best compromise they can be in their circumstances. It may well be that the circumstances of badgers—at least the badgers of Wytham Woods, and similar high-density populations in the north-west of their European range—and their current adaptations to those circumstances, do mirror aspects of a stage through which other social carnivores once passed en route to their current intricacy (and there is much to learn about the evolution of sociality from that), but that does not mean that contemporary badgers are somehow inadequate.

The two exceptions to this evolutionary reality arise if: (i) badgers, having arrived at their current set of sociological adaptations, thereby create new circumstances, which cause selection to favour further steps in adaptation; or (ii) if badgers, say in Wytham, find themselves in circumstances without evolutionary precedent to which their current adaptations are not a good fit. This latter, deeply unstable circumstance would leave the badgers, in evolutionary terms, running to catch up, and raises a possibility to which we will return in Chapter 10, and beyond. Indeed Chapters 11–14 illustrate how the Red Queen[21] (*sensu* Van Valen 1973) of endless contest with conspecifics, or the 'court jester' (Benton 2009) of a fluctuating environment, constantly threatens survival and fitness. The perennially unfinished business of these evolutionary races leads to various pace-of-life syndromes (Chapter 14), reproductive schedules, and energy expenditure regimens all being played out to test who has the best hand in this evolutionary game of chance.

---

[21] 'Now, *here*, you see, it takes all the running you can do, to keep in the same place'. From *Through the Looking Glass and What Alice Found There*, Lewis Carroll (1871).

CHAPTER 9

Who Goes There: Friend or Foe?

The behavioristic trait manifested by a display of property ownership—a defense of certain positions or things—reaches its highest development in the human species. Man considers it his inherent right to own property either as an individual or as a member of a society or both. Further, he is ever ready to protect that property against aggressors, even to the extent at times of sacrificing his own life if necessary. That this behavioristic pattern is not peculiar to man, but is a fundamental characteristic of animals in general, has been shown for diverse animal groups.

William H. Burt 1943

Individuals are a good starting point. It is individuals that rise or fall, prosper or deteriorate, that interact, and that arrange themselves into societies. And so we have begun with individuals, and for eight chapters we have detailed their behaviour as actors on the stage of badger evolution. Later, we will turn to finer actors on this stage, genes, and to larger aggregations of players, populations. Before that, and crucially, in Chapter 10 we will explore the stage itself, the ecological landscape that determines both the behaviour of individuals and the nature of their societies. But first, one last building block in the edifice of badger behaviour remains to be put in place: the relationship between groups.

There is much to be said in this chapter about neighbourly relationships and the surprisingly difficult question of whether neighbours are friends or foe, but much of it converges on two bald questions that will be the bulwarks of our narrative: first, are the twenty-first-century badgers of Wytham Woods territorial; and second, do these latrines that we have mentioned as bisecting the land between their main setts—often termed border latrines (Photo 9.1)—truly function to mark territorial divides? If the answer to the first question is yes, the answer to the second question cannot be assumed, but if the answer is that these badgers are not territorial (or not in a straightforward way), then almost everything on the wider canvas of accepted badger sociality must be redrawn, including the function of 'border' latrines and the meaning of neighbours.

The average distance between main setts in Wytham is 530 m (standard deviation (SD) ± 393 m, standard error (SE) ± 55 m). They are spaced out—separated—mostly fitting along an exposed strata of suitable geology (Chapter 4). This non-random distribution, spacing out, within suitable habitat, is not only indicative of territoriality; it is, according to some well-argued cases, the very definition of territoriality (Davies 1991; Macdonald and Johnson 2015a). Further, Wytham's badgers defecate at latrines, more than half of which are distributed, like beads on a necklace, along well-trodden paths that bisect the distance between neighbouring setts, and thus create what appears to be a ring-fence at the intersection of neighbouring groups. Not surprisingly, Kruuk (1978a), and almost everybody else who has thought about it (e.g. Neal and Cheeseman 1996; Roper 2010), concluded that these latrines were markers of territorial exclusivity: 'keep out' signs. Consequently they are generally referred to as border or boundary latrines. However, a note on vocabulary. Although the latrines that encircle badger group ranges are generally referred to as border or boundary latrines, the evidence in Chapters 1–8 is that members of different groups often cross at least some of these putative olfactory fence lines, generally with impunity, and are met not with aggression but with various mixes of indifference or sexual favour. This raises doubts about the classical interpretation of the concept of what constitutes a 'border'; that is, in this example, that 'border' latrines proclaim aggressively defended exclusivity: 'Trespassers will be prosecuted'. If this *is* the function of those alleged borders, they would seem to fail badly. Therefore, we will avoid the adjective 'border' with its implication of aggressive

The Badgers of Wytham Woods. David W. Macdonald and Chris Newman, Oxford University Press.
© David W. Macdonald and Chris Newman (2022). DOI: 10.1093/oso/9780192845368.003.0009

Photo 9.1 A latrine pit within a badger latrine in Wytham Wood.

defence, and applied connotation of territorial function, and for the time-being refer to 'perimeter latrines' with the intention of acknowledging their geometric positioning between sett groups, but expressly without any inference as to their function.[1] This juxtaposition of geometries that strongly, even definitially, indicate territoriality, with behaviour that indicates the opposite, demands elucidation.

So, with our focus now on interaction between groups, this chapter begins with the quest to understand the functions of messages that waft from the perimeter latrines between the neighbours whose intersections they define—beyond the fact that badgers have to defecate somewhere, and doing so as far away from your burrow might have a selective advantage in minimizing contamination (Hart 2012). Is a current function of badgers' perimeter latrines in Wytham territorial defence? Happily, olfaction expert Christina Buesching and Neil Jordan have identified six attributes of scent marks (here we focus on latrines) diagnostic of territorial function (Buesching and Jordan 2019). Very briefly (there are elaborations later) these are, in no particular order: (i) signals should carry information that can be matched to the individual that produced it, and thus to the individual, or group, laying claim to the territory; (ii) the spatial distribution of latrines should optimize locations to intercept intruders and (iii) should be situated at sites with substrate and microclimatic conditions to maximize detection; (iv) temporal patterns of use should

peak when the threat of intrusion is most intense; (v) scent profiles should allow association between territory owner(s) and defended areas; and (vi) intruders should avoid scent-marked areas/retreat on encountering scents (scent-fence) or modify their interaction with owner(s) when they match scents in the environment to scents of encountered individual(s). Each of these qualities could bring functional advantage in non-territorial communication, and a territorial function does not necessitate them all, but if badger latrines do not meet at least some of these criteria a defensive function becomes less plausible. Buesching and Jordan (2019) also list alternative, non-exclusive hypotheses for the functions of latrines, which are: (i) information centre/advertisement; (ii) landmarks/orientation; (iii) parasite control; and (iv) predator–prey interactions. Shortly, we will present our evidence regarding these six diagnostic features, but first we will review briefly some general context for the concept of territoriality and the function of latrine marking, including earlier evidence of how both apply to Wytham's badgers.

Territoriality: fundamentals and early insights

There isn't enough to go around, and this truth, evolutionary and economic, as punishingly enforced by natural selection as it is by Adam Smith's market forces,[2] is why territoriality is amongst the most fundamental concepts in ecology. Amongst paradigmatic badgers—our models of ecology and evolution—and all other organisms, survival, and ultimately reproductive fitness, is a question of security, where acquiring a sufficiency of food in a timely manner is a prerequisite to avoiding morbidity and starvation (Bright Ross et al. 2021; Chapter 14). Many animals move about in pursuit of this security and the area each normally uses in day-to-day movements was termed its home range in a remarkably enduring article by Burt (1943). In passing, although the epigraph to this chapter, from Burt's article, illustrates how style has changed since this topic began to preoccupy ecologists, contemporary readers of the original will discover that there was much substantial thinking 80 years ago that is nowadays either ignored, disdained, or reinvented. Anyway, some home ranges are not obviously defended (e.g. Pettett et al. 2018), and may be

---

[1] Humpty Dumpty, a wise egg, remarked, through the pen of Lewis Carroll, 'When I use a word it means exactly what I wish it to mean, neither more nor less'.

[2] Smith, A. (1776). *The Wealth of Nations*.

dispersed without reference to each other (see Grant et al. 1992) (which is not to say that the movements of the occupants do not show dynamic interaction, for example, avoiding proximity via repulsion in the same way that the poles of a compass repel one another; Macdonald et al. 1980a). Exactly how the occupants of undefended home ranges, not spaced out and positioned at random with respect to each other and their resources, limit, or regulate, their numbers to match their resources is an oddly neglected topic in vertebrate spatial ecology. However, amongst vertebrates it is more common for home ranges to be at least partly (in space and/or time) defended, and this exclusive space is termed a territory (Linn 1984). The term territory, with its root in the Latin word *terra* (land) has long been understood colloquially, and geopolitically since Machiavelli (e.g. Elden 2013), but surged into the public understanding of science in 1968 with the publication of Robert Ardrey's *Territorial Imperative*, the year after Desmond Morris' *Naked Ape* (both books were societally important because they interpreted human behaviour in Darwinian terms, i.e. as if humans were 'animals'[3]), and was already much discussed by early twentieth-century biologists in papers now widely ignored. Amongst the earliest, Heape (1931) concluded:

Thus, although the matter is often an intricate one, and the rights of territory somewhat involved, there can, I think, be no question that territorial rights are established rights amongst the majority of species of animals. There can be no doubt that the desire for acquisition of definite territorial area, the determination to hold it by fighting if necessary, and the recognition of individual as well as tribal territorial rights by others, are dominant characteristics in all animals. (Heape 1931, p. 71)

Territoriality is often imagined in its simplest form as being the situation in which the occupant—to take two Oxonian examples, say the robin in your garden, studied by David Lack, or the male stickleback studied by Niko Tinbergen—patrols, fighting as necessary, to defend exclusive rights to their patch. Even in the homogeneous aquarium world the most efficient packing results in beautifully hexagonal fin-swept patterns

in the sand as in the classic photograph in E.O. Wilson's *Sociobiology* (2000) account of territoriality (or indeed as replicated with similar attention to efficient packing in the dimples impressed on the coating of a golf ball).

Territoriality is the arena in which nature is typically reddest in tooth and claw. For example, studies both of wolves (Mech 1977; Cubaynes et al. 2014) and African wild dogs (Woodroffe et al. 2007) reveal that the greatest cause of natural adult mortality, *c.*37–58% and 11%, respectively, was intra-specific conflict. Vucetich (2021) calculates that each alpha wolf typically kills two to four wolves during his or her reign. Although examples in nature that approximate this traditional, absolutist definition of hard-bordered exclusivity are widespread, a more nuanced appreciation focuses less on exclusivity and more on spacing out such that occupants are distributed non-randomly with respect to the resources on which they depend (Davies 1991; see review by Adams 2001). The spacing out evidences behaviour that secures resources, distinguishes haves from have nots, and provides the grist for the mill of population regulation (Lack 1954; McLaren 2017). This behaviour is the essence of territoriality: it may be flexible in time or space, seasonally, inter-annually, or spatially (e.g. von Schantz 1984; Doncaster and Macdonald 1991; Boonstra et al. 2008), but in its nucleus it involves a non-random use of spaces by neighbouring occupants with respect to the resources over which they compete, and access to which will determine their fitness.

In passing, the competition for real estate that leads to territoriality within a species applies also amongst similar species, and amongst Carnivora this intra-guild competition, and associated, often fatal, hostility, is rampant (e.g. Fedriani et al. 2000; Kyaw et al. 2021) and drives the rules of community assembly that are fundamental to niche theory. For example, amongst 'badgers', as a subfamily, we have explored intra-guild competition between sympatric hog and ferret-badger species in China (Zhang et al. 2009) (Chapter 19). Indeed, Peter Apps has identified that carnivores can recognize and respond to key scent signals between species (Apps et al. 2019).

Many territories are occupied by a basic social unit; for example, a breeding pair (back to David Lack's robins). But in configuring their territory to provide the necessary food security, they must plan ahead for bottleneck periods, and one such might arise because of the need to cohabit with a brood of youngsters, ultimately consuming as much as, or more than, their parents. This anticipation can lead to obstinacy in the

---

[3] An anecdote on how quickly times change—as a schoolboy David read both these books in the year of their publication—his school banned Morris' book, as vulgar sedition, which thus had to be smuggled into the dormitory, whereas Ardrey's was merely frowned upon. In 1972 Ashley Montague edited a vehement collection of essays by academics who rejected Ardrey's evolutionary interpretation of human behaviour.

placement of territorial borders, planning for the worst. This turns out to be the case for the Japanese badgers we will discuss in Chapter 19 and is commonplace across solitary rodents (Wolff 1993), ungulates (Owen-Smith 1977), and some carnivores (Bekoff et al. 1984). Further, a female configuring her territory must not only anticipate the appetites of her young, but also those of the male with whom she cohabits, at least during the mating season (although this burden may be spread between several females if the male opts for a polygynous lifestyle socially semi-detached from the females whose ranges he overlaps (Sandell 1989b; Macdonald 1992; Yamaguchi and Macdonald 2003). Some of these pressures may be partly alleviated by dietary differences arising from sexual dimorphism (Dayan and Simberloff 1994) although when we come to badgers, this is minimal (Noonan et al. 2016). These costs of cohabitation would be, all else being equal, greatly multiplied for individuals living as a group, as per Wytham badgers (although, as will become clear in Chapter 10, all else may very well not be equal). Nonetheless, as a starting point, we might predict that badgers, indeed groups of badgers, requiring resources of which some will surely be limiting, will compete in ways that partition the resources, and perhaps also thereby partition space with the consequent creation of 'haves' and 'have nots'.[4] Classically, the 'have nots' disperse (Chapter 3; Greenwood 1980), lurk in the interstices, perhaps waiting their turn to usurp or inherit their own range (Lindström 1986; Macdonald and Courtenay 1996). The options are somewhat different for group-living species, at the confluence of sociological and ecological factors affecting decisions as to whether it is advantageous to stay or to disperse (Macdonald 1983; Kamler et al. 2019). Stamps and Buechner (1985) partition this balance between the availability and dispersion of food and mates: the *resource hypothesis*, which determines the size and geometry to territories, versus the counterweight of the *defence costs hypothesis*, which contends that the number of intruders into territories, or contenders for vacant territories, influences territory sizes, territory overlap, and the acceptance of subordinates, even if resource densities remain constant. Each individual faces the balance sheet of evolutionary fitness, the costs and benefits of staying or going—the latter being determined by the availability of dispersal sinks and the saturation

of suitable habitat (an accountancy detailed in Macdonald and Carr 1989, and explored for badgers in Chapter 10).

Clearly, Wytham's badgers live in groups, so we would expect the groups to space out, but according to what pattern? In Chapter 19 we will discuss the fact that many carnivores adhere to an ancestral system, and, therefore, the badgers of Wytham Woods are amongst a minority that does things differently. The first insight into that difference was Kruuk's (1978a) mapping of spaced-out main setts, and his observations of what seemed to be territorial clashes (Chapter 4). He concluded that the badgers defended classical group territories, each occupied by several females and males. This was exactly the interpretation adopted by others who followed in studies of high-density badger populations in lowland agricultural Britain (Cheeseman et al. 1981; Anderson and Trewhella 1985), and it is the one upheld by most of the many and skilled naturalists who devote themselves to badger watching. So, from our current vantage point, does this reflect the reality in Wytham today? To understand the biological context, read first Box 9.1 on latrine basics.

Readers of Box 9.1 will appreciate that a predominant interpretation of the deployment of scent marks is as a stand-off between neighbours—an updated expression of Hediger's (1949) metaphor of a 'scent-fence' functioning as a 'keep out, trespassers will be prosecuted' sign. So, if the fence may be built cheaply with urine and faeces—perhaps also strung with the more expensive embellishments of customized glandular secretions—where might it most efficiently be erected? We arrive at this question knowing that a possible answer lies in the fact that, where two neighbouring residents interface, a perimeter is thereby defined, and knowing also that Wytham's badgers use latrines, some of which appear to bisect the land between neighbouring setts. Although we already know, from the comings and goings across this perimeter described in Chapter 4, that this interpretation creaks at the seams, we will take it as a starting point.

Wytham and early latrine insights

Hans Kruuk spent the late 1960s in the Serengeti preoccupied by spotted hyaena latrines (Chapter 4). They comprised piles of sun-bleached droppings, each pile an olfactory and visual beacon of tenure, arranged like frontier posts to encircle each of the patchwork of territories that carved up the Ngorongoro Crater

---

[4] 'There are only two families in the world, as a grandmother of mine used to say: the haves and the have-nots'. Miguel de Cervantes, *Don Quixote de la Mancha* (1605).

Box 9.1 Latrine basics

Burdened with the problem of defending a territory, the resident defender faces a harrowing and energy-consuming task of detecting and repelling intruders—the latter risking injury (Gosling and McKay 1990). How then to signal that this property is occupied? Howling wolves provide one answer, but the sound fades immediately—better an enduring sign, a towel left on the sun lounger. As detailed in Chapter 6, the attributes of durability combined with potential complexity of signal make olfactory signals ideal (Brown and Macdonald 1985). That the signal stands for the signaller, *pars pro toto*, the part for the whole, has been a fundamental presumption at the origins of research on social odours (Bilz 1940), so much so that the terms scent marking and territory marking (implying a defensive function) have often been used interchangeably. An early and influential expression of this interpretation was Hediger's (1949) metaphor of a 'scent-fence' functioning as a 'keep out, trespassers will be prosecuted' sign. A latter-day articulation is the active territorial defence hypothesis, which interprets the deployment of scent marks as a stand-off between neighbours; each signpost updated in response to the likelihood of an encounter at the perimeter (Maher and Lott 1995). Human trespassers, say those climbing into an orchard to scrump for apples, do not always obey a keep out sign; such a sign, however, may nonetheless put them on edge (and all the more so if it is freshly painted). In an experiment with red foxes, David found that a hand-reared vixen over-marked a stranger's urine (positioned experimentally) when she encountered it within the area she routinely marked (and seemingly considered to be her own patch), but not outside that area; further when invited to play with a rubber mouse on a string, she normally did so on home ground, but distractedly disdained the toy having encountered the stranger's urine when venturing beyond her own range—seeming to have 'lost her nerve' (Macdonald 1979a; 1989).

If an exclusion fence is to be erected, of what is it to be made? Chapter 6 introduced one candidate source of secretions that is unique to badgers, the subcaudal gland. There are others; many and varied combinations of apocrine, eccrine, and sebaceous glands producing a symphony of odours (Brown and Macdonald 1985) for which evermore nuanced interpretations are regularly discovered (e.g. Buesching and Jordan 2019). Further, faeces and urine are free products that inevitably and unfakeably reveal their depositor's presence at a site (amongst carnivores, faeces are coated in another glandular candidate, AGS). It is but a small behavioural step for these excreta to be deployed, deliberately, at key points of contact with would-be intruders.

The exclusivity of a territory concerns two sorts of outsider, and involves keeping both types out. These are intruding itinerants, intent at least on theft if not usurpation, and also trespassing, possibly expansionist, neighbours, perhaps most likely seeking infidelitous opportunity. There are similarities and differences between these categories, but both jeopardize the ownership of a territory and the benefits it brings—ownership that can be a prerequisite to reproduction if not survival—but at a cost. Fighting itinerant intruders can be dangerous, and so too can be squabbling with neighbours—if escalated, both can cost lives, and so it pays all concerned to know when not to pick a fight, and to know it fast. Remember too that in the case of the badgers of Wytham Woods, while each is likely to be encountered alone (part of the paradox is that while these badgers live in groups they largely operate alone), an itinerant might be a dispersing singleton but a neighbour will be a member of a group, although amongst badgers group members never engage in coordinated attack.

Consider, first, neighbours. They aren't going to go away. The costs can outweigh the benefits when it comes to pursuing feuds, so even if the injunction to love them (Matthew 22: 37–39) might seem a stretch for the average badger (but surprises lie ahead; Chapter 17), it can pay neighbours to observe a truce. This reality is encapsulated in R.A. Fisher's (1954) 'dear enemy' hypothesis, explaining why territory holders may be more tolerant of a familiar neighbour than of an unknown stranger.[5] Morris Gosling, one of the most thoughtful scholars of a generation in the field of olfactory communication, crystallized this appreciation into a late-twentieth-century constellation of ideas. Morris' talent as an artist enabled him to capture on paper much of the postural ethology of scent-marking mammals—most notably the hartebeest, an alcelaphine African antelope on which his doctoral thesis ran to three thick volumes, deflecting much teasing for verbosity from David, who produced a mere two-volume doctorate!

The first step in Gosling's logic is relevant to encounters with both neighbours and itinerants, both at the territory border and in its hinterland. By scent marking their territories, owners dispel ambiguity by providing intruders with a means of assessing their identity (and—see later—by

---

[5] Amongst canids, neighbours are sometimes offspring or other relatives of a breeding pair, which can encourage a good-neighbour strategy (Macdonald and Courtenay 1996; Kamler et al. 2019).

Box 9.1 *Continued*

implication, their quality insofar as territory holders have by definition proven their worth). When an intruder encounters an individual whose odour *matches* that of scent marks in the vicinity, then it has probably just met the territory owner—a deduction based on scent matching.

Badgers live in groups, so it is relevant that scent matching applies equally to identifying at a glance (or sniff) members of a group, if they share a scent, much as a Scotsman's clan is quickly ascertained from his tartan, or a soldier's regiment from his badge or tattoo (Chapter 6). The saving in terms of efficiency, and avoidance of risk, is huge: rather than having to get to know each individual, the odour flags identity, or in the case of a typical carnivore group, allegiance (Gosling 1982, 1985; Gosling and McKay 1990). Of course, this idea transfers readily from insiders and outsiders to the aforementioned dear enemy phenomenon, where neighbours too will have their familiar scent distinguishing them from less familiar strangers. Do badgers respond differently to the scents of their own group, those of their immediate neighbours, and more distant strangers? We undertook field experiments to provide an answer, see the text that follows.

As said, occupying a territory says something about the holder: first, it's probably tough, having vanquished rivals; and second, it has a lot to lose and is likely to fight ferociously to avoid doing so. Gosling and Roberts (2001a) developed these thoughts in the broader context of Resource Holding Potential, the ability of an animal to win an all-out fight if one were to take place. The term was coined by Geoff Parker (1974) to disambiguate physical fighting ability from the motivation to persevere in a fight. This is differently relevant to encounters with footloose itinerants and immovable neighbours. First, itinerants: it's a fair assumption that the territory holder will be higher quality than the itinerant, and it generally pays weaker (low-quality) animals to avoid combat with a stronger (high-quality) individual, which, in this case, also has more to lose. Second, neighbours: the quality factor may cancel out (both are territory holders), but the asymmetry in stakes should alert the trespasser not to push its luck. But amongst badgers, neighbours are members of groups, and groups may differ in the strength of numbers and thus, although they don't act as a pack, may exert different cumulative pressure. This matters if a group has expansionist aspirations: intriguingly, groups of some species of carnivore are expansionist, and others are not (Kruuk and Macdonald 1985); we will investigate this dichotomy for badgers in Chapter 10. For now, the general point is that a scent badge that matches a group may be linked in the recipient's nose to that group's size. These factors could be crucial in determining the merits of escalation:

back to the Scotsman's clan (there are reasons why their neighbours took note of the Macdonalds of Clanranald's ancient motto 'gainsay who dares').

Scent marks have been proposed to help establish a power asymmetry between territory holder and intruder (Smith and Parker 1976; Hammerstein 1981; Gosling 1982) according to the *payoff asymmetry hypothesis* (Dawkins and Krebs 1978; Krebs 1982). Gosling's and Robert's articulation of this fundamental principle was that scent marks provide a cheat-proof signal—even in the absence of the resource holder—to competitors and potential mates, that a resource has been monopolized. Furthermore, not only does ownership signify prowess, the signal itself adds to the intimidation if, like the red deer stag's bellow or the lion's roar, it is expensive to produce. Gosling and Roberts had in mind male mammals, showing off their capacity to synthesize an inexhaustible supply of scent marks to signal their prowess (although, in terms of securing resources for cub rearing, similar dynamics should apply to females).

There are two general summaries of this account of scent matching and Resource Holding Potential to keep in mind. First, decisions on whether to escalate a challenge may be informed via three criteria (Gosling and Roberts 2001a): receivers may (i) detect intrinsic properties of scent marks (e.g. concentrations of volatile components honestly signalling quality); (ii) associate past encounter outcomes with the individual producing that scent; and (iii) remember the smell of marks encountered recently and match this smell with potential opponents that they meet subsequently. In a territorial context, scent marking is cheat-proof as a status signal of ownership because it truly reflects residency during the period of the mark's duration. Second, the foregoing account offers contrasting predictions from two hypotheses regarding scent marking: the scent-fence hypothesis (Hediger 1949) predicts that territory owners should respond with increasing severity on encountering fresh scent marks of recidivist intruders, insofar as the repeat offence is proof that the deterrence of previous responses was insufficient (Sun and Müller-Schwarze 1998). The scent matching hypothesis predicts that the owner's response will stay the same (the *status quo* has not changed) or reduce (the dear enemy adjustment has been established).

Latrine precedents and functions

Latrines are widely used across mammalian orders, including carnivores, primates, rodents, and marsupials, and particularly amongst large herbivorous mammals (Brown and Macdonald 1985). Indeed, it appears that even earlier megaherbivores, dicynodonts, relatives of the reptilian ancestors

Box 9.1 *Continued*

of mammals, used them communally 240 million years ago (Fiorelli et al. 2013). But no mammalian order exceeds the Carnivora in the flamboyance and, to the human nose, smelliness, of its latrines. Selecting examples from the solitary (i.e. semi-detached, intra-sexual territorial, ancestral system), and group-living ends of the social spectrum, both the ocelot (*Leopardus pardalis*) and the meerkat arrange their lives around latrines. Ocelots establish latrines that are used communally by their extended social network: one was visited by 11 males and 6 females over a 6-year study (Rodgers et al. 2015). Whatever messages they were posting, it was to a wide readership: over an average 10-day window for a single latrine, males communicated with a mean of 5.9 other individuals (range 2–14), and females with 4.5 (range 3–12) (see also King et al. 2017). A more cohesively group-living carnivore that, like European badgers, feeds mainly on arthropods and other small prey is the herpestid slender-tailed meerkat, a species that also establishes latrines (Doolan and Macdonald 1996). To appreciate the ferocity of encounters between groups of meerkats, read David's account of near-fatal attacks on a straggler separated from his group, and the theft of females from a defeated coalition; the reactions of a band of meerkats at a border latrine during a territorial battle are frenzied (Macdonald 1992), a musk-laden ferocity matched when banded mongoose bands clash, as Beth Preston craftily instigated by smuggling in alien faeces to simulate a border raid (Preston et al. 2021). Territorial meerkats usually share one latrine with each neighbouring group (Jordan et al. 2007), but also use hinterland latrines concentrated towards the group's core territorial area. Meerkats in the Kalahari Desert occupy large territories, so the chance of intruders missing widely spaced boundary scent marks is high; the concentration, however, of latrines towards the territory's core heightens the likelihood of intercepting them the deeper the intrusion. And meerkats are not alone amongst the smaller, largely insectivorous, group-living carnivores (thinking of badger parallels), and they all use latrines (for banded mongoose see Jordan et al. (2010); dwarf mongooses, see Rasa (1987) and Christensen et al. (2016); and for more on meerkats, Jordan et al. (2007) and Drewe et al. (2009)). However, all of these latrine-using group-living carnivores occupy exclusive territories and defend their borders aggressively (which is not to say they don't push their expansionist luck at every opportunity), whereas the paradox of this chapter is that it appears that badgers do not.

With regard to that paradox, Jordan et al. (2010) raise the interesting point that while the aforementioned territorial species that use perimeter latrines are indeed aggressively defensive of their borders (in which respect they seem to contrast with Wytham's badgers), in each case, and others, there is also evidence of the failure of scent signals to repel territorial intruders. They list the examples of dwarf mongoose (*Helogale parvula*, Rood 1983), African lion (*Panthera leo*, McComb et al. 1994), meerkats (Doolan and Macdonald 1996), and North American beaver (*Castor canadensis*, Sun and Müller-Schwarze 1998). Therefore they suggest that scent (not just at latrines) may not function primarily in territory defence, but in direct competition for mates (Jordan et al. 2007). Insofar as territory and reproduction are almost inseparably entangled, it may be close to impossible to say that one matters more than the other; the key points, however, are that reading, and understanding, a keep out sign is a very different thing to obeying it, and one signpost may relay many different messages.

Moving taxonomically, but not societally, closer to Wytham's badgers, all those species listed in Chapter 1 as sharing the colloquial surname 'Badger' use latrines in one way or another (the particular habits of hog badgers and ferret-badgers are detailed in Chapter 19). What determines which species use perimeter latrines? Addressing this question, Macdonald (1980a) answered that, although faeces may be free they are not unlimited; only group-living species have the capacity to produce sufficient to ring, and replenish, their boundaries with latrines.

But what of latrine functions (Photo 9.2)? Candidate answers have included almost every imaginable variant of communication. There is also reference to the avoidance of endoparasite transmission or reinfestation (Ezenwa 2004); raccoons, considering their relatedness to badgers, are a relevant example in this respect (Hirsch et al. 2014). Latrines closer to range centres tend to be interpreted in terms of within-group messaging (e.g. Dröscher and Kappeler 2014), those closer to borders generally interpeted as territorial—in both cases there has been a move towards seeing them as multi-purpose, what Eppley et al. (2016) call a 'multimodal communicatory signal station'. Indeed, positioning a notice board between your group and your neighbours may optimize internal and external communication. Although the traditional interpretation of border latrines is as a proclamation of private property (see Maher and Lott 2000), one size does not fit all, and increasingly we suspect they function as a message board, displaying a plethora of messages, some of which, to some beholders, may be perceived as beautiful, others offensive or defensive. Think of rival sports teams checking up on the league notice board.

Photo 9.2 Fascinating faeces preoccupy Chris (left), David and Christina.

in Tanzania. Each territory, bounded by its ring of latrines was occupied by a group—he called them clans (a word that was later to infiltrate the lexicon of badger sociology)—6–80+ strong, of spotted hyaenas (Kruuk 1966, 1972). Indeed, in addition to faeces, each hyaena latrine was daubed in the secretions of anal pouches (creamy coloured in the case of spotted hyaenas, and black and white in the case of brown hyaenas; see Mills 1990), and sprinkled with urine and scratched with forepaws, and consequently hootched with odorous information. Whatever else these siren messages might convey, Hans was sure that strident amongst them was: 'Keep out!'. In the open panorama of the Crater, he could see that trespassers were indeed persecuted: neighbouring hyaena clans clashed over these borders and, indeed, even if in hot pursuit of a fleeing wildebeest, might stop dead in their tracks rather than cross the line (see Kruuk and Macdonald 1985).

When, in 1972, Hans Kruuk arrived in Oxford from the Serengeti and began work on Wytham's badgers, he had a strong sense of *déjà vu*. His first observations were that groups of badgers used shared latrines that carved Wytham wood into territories—the parallels with hyaenas were striking, and it was only natural to conclude that the badger latrines were sending the same forceful keep out signals as did spotted hyaena latrines (Kruuk 1978a). As we described in Chapter 4, this Tidy Territorial Model was so appealing, and so closely aligned with the intuitions of many naturalists, that it was widely adopted (Roper et al. 1986, 1993) and entered into the canon of natural history (Drabble 1979; Neal 1986; Neal and Cheeseman 1996).

As it happened, at about the same time that Hans introduced him to the arrangements of badger latrines in Wytham, David found a similar, but very unexpected, arrangement on the shores of the Dead Sea, in that case deployed by a group of over 30 golden jackals (Macdonald 1979b). These jackals defended a tiny

territory of *c.*10 ha around a feeding site maintained by the Israeli government's Nature Reserves Authority, at which carcasses of domestic stock were provisioned to facilitate educational observation by schoolchildren of carnivores of the Judean desert. Remarkably, considering that golden jackals[6] elsewhere had never been observed to use latrines, this group ringed its territory with piles of droppings. Within their territory the jackals followed a more typical pattern for canids, depositing single droppings on visually conspicuous objects— on one memorable occasion as David stooped to collect a scat he noticed, just in time, that it was precisely positioned aloft an anti-personnel mine washed up at the edge of the Dead Sea. Later, working with Claudio Sillero, on endangered Ethiopian wolves, *Canis simensis*, it became clear that they too defecate at border latrines, and that they defend those borders aggressively.

A pattern starts to emerge, a hyaenid and two canids living in large groups in small home ranges, using border latrines—something that they did not do in smaller groups in larger home ranges. And then there were badgers, also living in small territories in what, at the time, seemed like big groups. The die seemed cast. We may have been right at the time, but looking back perhaps we succumbed to a bit of confirmation bias. Consider the history. Hans Kruuk plotted the first bait-marking map of Wytham in 1973, as described in Chapter 4 (David remembers his awe, as a 22 year old, at the large sheet of paper, stuck to pinboard in Hans' office in Niko Tinbergen's Animal Behaviour Research Group, with the colour-coded spokes radiating from each sett to its associated border) (Photo 9.3). That first map has been redrawn a dozen times, our most recent version in 2013 Comparing these first and last maps, bookends to this study over 47 years, led to the understanding of inter-group trespass, amicable visits, and group fission detailed in Chapter 4.

As time passed, the putative scent-fence began to seem more permeable and by the turn of the millennium Paul Stewart and the team concluded of badger latrines that 'The wide range of marking behaviours, compounded by the lack of any clearly sex-limited behaviour at latrines, suggests a multiplicity of roles in the social lives of all age and sex classes of badgers' (Stewart et al. 2002, p. 999). Twenty years' more evidence, and here we return to the question of what these functions might be, and of how they inform our quest to understand relationships between twenty-first century badger groups in Wytham.

---

[6] The point is unaffected by the fascinating discoveries separating golden jackals into two species.

(a)

(b)

(c)

Photo 9.3 Bait-marking preparations (a) preparing to mix colour-coded markers with peanuts, (b) Chris's dog Lycos maintains hopeful guard over the syrup used to bind the peanuts and markers, and (c) Yayoi Kaneko leads the field team in distributing the marked bait to mapped locations.

The evidence

Against this background, let us turn to what we actually saw in the field, in the context of the six tests of territory marking. First, a not atypical observation sets the scene. It was a clear night in July 1997 when Christina Buesching set out on foot to track one of her radio-collared female badgers (F300) resident at the Pasticks sett, roughly central to Wytham Woods (see map, p. xiv). First, badger and tracker traversed cow pasture, then arable fields, before they ambled around Wytham Village, paused for a while in St Mary's cemetery, crossed the old Abbey Orchard, came up past the sawmill, skirted the Chalet, and then plodded wearily back to the Pasticks sett after a journey of around 5 km, taking over 5 h. In so doing, this badger traversed the land occupied by 12 other social groups, and as she did so she seemed unworried, unhurried, and unchallenged. Was she blind (anosmic) to the 'keep

out' signals? Did she have some sort of sociological permit or free pass to wander at will? Was this a special, even aberrant, excursion, defying the odds by dodging the aggressive reprisals of enraged territory holders? What, more generally, is the tenor of encounters documented in Chapter 4 between neighbours at the intersections of their ranges (as defined by those encircling latrines)? A metaphor commonly evoked in the context of scent marking is that of the recipient 'reading the newspaper' written by the signaller. Stretching the metaphor, is the news of the day a simple 'keep out' message or, less unilaterally, a comprehensive message board of everything the signaller wishes to publish or which you wish to read?

Either way, the Buesching and Jordan criteria include some headlines, with prominent subtitles, and some fine print. That is the approximate order in which we'll index the newspaper, starting with

Map 9.1 Group ranges
The home ranges of badger groups hand drawn within the borders of Wytham Woods (the woodland border with adjoining farmland and grasslands is depicted by a heavy black outline). Some group ranges are shown with arrows and dotted lines to spill out of the Woods onto adjoining farmland. This map represents the situation in the latter years of the study. Important setts (named in Map Cluster, pages xiii-xiv) are marked as small white circles. Range name abbreviations given in Map 9.2. The surrounding countryside is depicted by a basemap by OpenStreetMap.

the attributes that are easiest for us to see (certainly not the same as what is most strident in the badgers' olfactory receptors) (attributes 3, 2, and 4), and coming last to the fine print; that is, the detail that the badgers report which they can discern (attributes 1 and 5).

Where are latrines situated?

Amongst their diagnostic symptoms of scent marks that serve in territorial defence, Buesching and Jordan (2019, point 3, p. 95) argue that to deliver this function latrines 'should be situated at sites with substrate and microclimatic conditions to maximise detection (i.e., signal range increase through air flow/elevation) and signal longevity (e.g., by preventing scent marks from drying out)'. Were they? Badger latrines comprise a cluster of shallow pits, each *c.*15 cm in diameter and 10 cm deep, but sometimes twice that size following repeated use. In each pit faeces, from one to several, accumulate. Large latrines encompass a cluster of dozens of these pits (Kruuk 1978a; Buesching et al.

2016a). Within the total extent of a latrine, faeces are placed in a subset of pits, which is used for several consecutive nights. Then, the badgers swap to a different subset of pits—perhaps when the clarity of their olfactory messages gets lost in the *mêlée* of over-marking. But what sources of information are deployed at latrines? Clearly faeces, and some of these bear a blob of dark orange-coloured jelly (the same material is sometimes splattered in a latrine, even without faeces). This is anal sac secretion. We also know from our videos that badgers urinate at latrines (easily discerned as the badger crouches with rump suspended off the ground, and, very occasionally, cocks a leg). The information conveyed by the anal gland secretion and urine (and indeed the subcaudal gland secretions also left at latrines along with everywhere else the badgers go; see Chapter 6) is part of the fine print of the message, to which we return below as parts of criteria 1 and 5.

Latrines are far from randomly positioned. First, the places: within stands, badger latrines are often

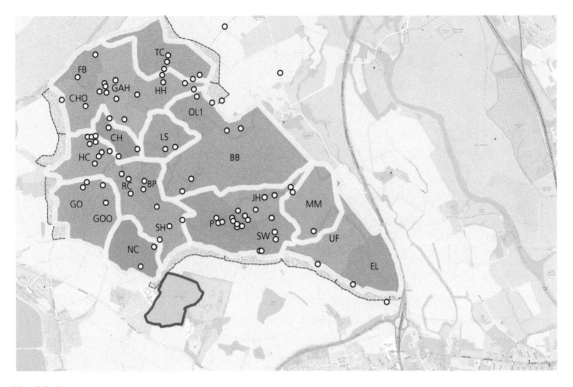

Map 9.2 Super-group ranges
The borders, shown with heavy pale yellow lines, of super-groups formed of the group ranges depicted in Map 9.1. The names of each social group are given by two- or three-letter abbreviations, from north west to south east: FB = Firebreak, GAH = Great Ash Hill, TC = Thorny Croft, HH = Holly Hill, CHO = Chalet Outlier, HC = Hill Copse, CH = Chalet, LS = Lower Seeds, OL1 = Outlier 1, GO = Great Oak, GOO = Great Oak Outlier, RC = Radbrook Common, BP = Burkett's Plantation, BB = Brogden's Belt, NC = Nealing's Copse, SH = Sunday's Hill, PA = Pasticks (in the text we explain the incomplete fission within P involving PO), JH = Jews Harp SW = Singing Way, MM = Marley Main (most recently M2 formed an independent range, shaving off the western third of the MM territory), UF = Upper Follies, BL = Botley Lodge. For further context see pages xiii–xv.

associated with big trees, often an aberrant pine amongst beech, or vice versa. In one detailed survey of 143 latrines in Wytham, they were significantly closer to tree trunks than were random samples, and were more likely to be associated with conifers than broadleafs (Stewart et al. 2002). By chance, 35–37% of latrines would be expected to be situated beside a deciduous tree, and just 4–7% by a conifer, whereas in reality *c.*28% were associated with a conifer and just 16% with a broadleaf. This might be less about the attributes of angiosperms and gymnosperms, and more to do with the placement and salience of the trees: in Wytham stately avenues of red cedars, planted prior to broader reforestation through the twentieth century, grace the edges of woodland rides. Badgers break stride when their paths cross woodland trails, and tend to position latrines at such habitat transitions along their route, so placing latrines beneath the cedars may be more coincidental than causative (other carnivores

such as wolves and foxes also tend to scent mark at trail junctions; Macdonald 1992). That said, certain trees may have special odorous qualities: David and his student Latika Nath (2000) found that tigers select particular species of tree for scratch marking (bears do the same); certainly the soil stays drier under the evergreen umbrella of closely planted cedar branches, in addition to which they may have a special olfactory attribute. Latrines are also associated with other conspicuous features along routes that channel badger movements, on verges where paths intersect woodland trails, along the edge of ditches, or under fence line crossings.

To determine features affecting the choice of boundary latrine site, we first located all 43 latrines (15 hinterland and 28 boundary) in a randomly chosen 1 km² section of Wytham's mixed coniferous/broadleaf woodland and pasture. Then, we generated a random set of map coordinates for 43 'random control

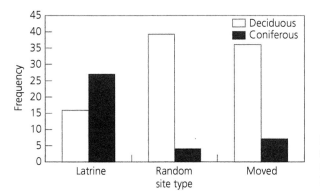

Figure 9.1 Nearest association with a deciduous tree (open squares) and a coniferous tree (filled squares) for latrines, random control sites, and moved control sites (from Stewart et al. 2002, with permission).

Table 9.1 Mean ± SE distances (m) of latrines, random and moved control sites from the nearest linear feature and nearest tree (from Stewart et al. 2002, with permission).

	Linear feature	Tree
Latrine sites	21.6 ± 4.7	1.3 ± 0.3
Random control sites	27.6 ± 4.4	2.3 ± 0.2
Moved control sites	26.0 ± 3.9	2.3 ± 0.3

sites' within this area. Finally, we turned the map of the latrine coordinates through 180 degrees to produce a third set of 43 'moved control sites'.[7] The aim was to test whether latrines were positioned to promote information transmission (e.g. siting near linear features such as paths, fences, or ditches that may channel movement) and information protection (e.g. siting beneath the shelter of tree canopies). Although broadleaf trees dominated the mixed-woodland study area, the nearest tree trunk to a latrine was more likely to be a conifer (Figure 9.1). Latrines were significantly closer to linear features and to the nearest tree trunk (Table 9.1) than were either type of random point. There was no significant difference between the situations of perimeter and hinterland latrines.

What is the spatial pattern of latrine deployment?

The second diagnostic symptom of defensive marking on Buesching and Jordan's list is that the spatial distribution of latrines should optimize the interception

[7] These provided a control for the possibility that the spacing of latrines created an artefactual relationship with other non-random aspects of this woodland such as the distance between paths or trees.

of random intruders and thus 'dependent on costs of patrolling borders and faecal restraint (Buesching et al. 2016a), latrines should be located predominantly along (or near) the perimeter where adjacent groups interface, and/or placed throughout the hinterland' (p. 96). Clearly, latrines, at least perimeter ones, meet this criterion, which is the reason they are prominent in this chapter with its focus on inter-group sociology (from bait marking, Chapter 4, we know this perimeter is the furthest location from their sett in which group members generally defecate). Kruuk's original bait marking revealed the necklace beaded with latrines strung on shared perimeter paths and, more quantitatively, Doncaster and Woodroffe (1993) showed that these shared inter-group latrines generally (and rationally) lie halfway between the main setts of neighbouring groups, coinciding, more than would be expected by chance, with the polygonal edges of Dirichlet tessellations: the shape that most efficiently partitions the land between the constellation of main setts (a partitioning that strongly hints at territorial spacing, reminiscent of the straightforward, and conventionally aggressive, spacing of classic stickleback territories).

Buesching and Jordan's diagnosis mentions also, and in contrast, that hinterland latrines are predicted to be scattered—the point being that scattering (exactly as with the perimeter ring) is patterned according to whatever deployment maximizes the chance of the intended reader encountering the message (a criterion that applies whatever the message might be). Kruuk (1978a) established this pattern, but how is the investment between these different categories of billboards deployed? Remembering the sample of 43 latrines from within just 1 km² of Wytham: 28 were perimeter and 15 hinterland (Stewart et al. 2002). Later, and similarly, our maps for 2000 plotted 244 perimeter and 84 hinterland (Chris's dog, Sam, pioneered the nowadays familiar role of scent dogs in research; more or less trebling our previous historic tally of latrines). In short, about two-thirds of latrines were at what has traditionally been conceived as the perimeter between sett groups. Out of respect for natural selection, rather than teleology, we conclude that this deployment optimizes the probability of the message reaching the right readers. The question is, who are the correct readers; then again, are the intended readers, and the messages, the same at both perimeter and hinterland latrines? Aside from communication, remember a badger must defecate somewhere, so in the context of hygiene and the landscape of disgust (Chapter 8), and the cub-killing coccidial parasites transmitted in faeces (Chapter 2), there may be a health advantage to locating all excrement in a few distinct toilets.

One straw-man hypothesis is that the priority readership consists always and only of neighbours, who are best intercepted by a ring of latrines at the border; still better intercepted, however, by a scattering once they have penetrated the hinterland. An alternative extreme is that all latrines are targetted at all readers, although what they make of those messages may be in the nose of the beholder.

Why then, the concentration, and special configuration, of latrines at the perimeter? Why, too, as revealed by our active radio frequency identification (aRFID) work in Chapter 4, do group members and neighbours alike read so assiduously at both border and hinterland editions? The interpretation that all latrines are targetted at all readers, and that each reader can make what it will of the content—a sort of badger Facebook style of social media aimed at maximum interception—starts to nudge ahead as the most apt metaphor, and is not at odds with the functional interpretation that for some readers the message is hostile.

What is the temporal pattern of latrine deployment?

All the considerations that apply to the spatial deployment of latrines apply similarly to the pattern of their temporal deployment, Buesching and Jordan's fourth diagnostic: 'The patterns of latrine temporal use should peak when intrusion threat is most intense, and should correlate with intruder encounters (especially interlopers threatening territory integrity)' (p. 96). Kruuk (1978a) described a seasonal peak in latrine use in spring, and a secondary one in autumn (a pattern subsequently confirmed by Roper et al. 1986, 1993; Kilshaw et al. 2009). These peaks coincided with the prevailing understanding of the timing of two annual oestrous periods, one post-partum in February–March, and a subsidiary one in August–September—an understanding that samples from 1987 onwards revealed as incorrect in Wytham (Chapter 7; Sugianto et al. 2021b). Nonetheless, if the threat of intrusion—or the advertisement of a desire to encounter suitors or even exotic strangers—is most intense during periods of heightened sexual activity, receptivity, or fertility, then the observed temporal deployment is not at odds with the defensive function hypothesis (but it wouldn't be for any other message that needed to be conveyed at that time, either).

Who goes there? Visit and contribution patterns of individual animals

First on Buesching and Jordan's diagnostic list was that to function defensively a scent mark, in this case at latrines, should carry information that can be matched to the individual that produced it (see Chapter 6), and thus to the individual, or group, laying claim to the territory. This test is explored, later, in our discussion of the fine print, but this diagnostic also encompassed predictions that different individuals would be differently invested in signalling. Specifically, for a defensive function, 'Visits to latrines and scent mark contributions should primarily be made by territory owners, especially representing those individuals with more to gain by territorial defence; for instance, breeding/dominant individuals – and thus contributions may be sex-biased to resist any sex-biased intrusion pressure' (Buesching and Jordan 2019, p. 97).

How do individual badgers deploy their scents at latrines? Are there consistent differences between individuals in their behaviour with respect to latrines? As we shift from the generalities of latrine deployment to the underlying detail of individual behaviour, it is a moment to take stock. All strands of our evidence—the unexpectedly frequent visits by members of one group to the sett of another (Chapter 3), the aRFID-revealed co-proximities of members of neighbouring groups at the latrines between their setts (Chapter 4), the meanderings of radio-tracked animals seemingly undaunted as they passed between home ranges, and the field observations, all too infrequent and fleeting, of encounters between neighbours in what might have been thought of as each other's territories—all these strands converged on the question: how do badgers from neighbouring setts get on?

The problem was that field observations were too infrequent, too fleeting, and often from too far away.

If only these neighbourly relations could be observed in slow motion—and they could, by means of the Paving Slab Experiment. This experiment was part of our investigation of social hierarchy amongst badgers, the topic of Chapter 4 (Macdonald et al. 2002b). However, we present it here because, although it involved encounters at a dinner table, the tenor of the conversation between dining neighbours is revealing. To explain, the work ran between August and September in 1995 and involved five concrete paving slabs (the aforementioned 'dinner tables'), each weighing 11 kg and measuring 45 × 45 × 4 cm. They were positioned as a quincunx (think of the five dots on the face of a die), at c.1 m spacing, at latrines intersecting three setts occupied by the social groups at Sunday's Hill, Burkett's Plantation, and Radbrook Common (of which the first two were members of the same super-group, whereas Radbrook Common was part of the neighbouring super-group to the west).[8] Our intention was that these slabs, beneath each of which was positioned a small pile of peanuts, would slow down feeding, giving us longer than the duration of a hasty gulp to observe the badgers' interactions (of course, in interpreting the results we must be mindful that the food might change the tenor of interactions, like a guard dog distracted when thrown a bone by a thief).

The badgers soon caught on. Actually, they caught on embarrassingly well: another of the great Wytham studies—of winter moths—had been begun in 1950s by the two entomological Georges, Varley and Gradwell (both gave David tutorials—it is only now that he realizes that at the time of the tutorials he was the same age as the study about which they were teaching him!) and was continued at the time of our experiment by Lionel Cole (Varley and Gradwell 1960). Lionel's retreat at Wytham was a small portacabin with a patio of paving

stones at its entrance—during our experiment Lionel was not amused that his entry was hampered nightly by badgers upending these slabs, presumably wondering (rather in the manner of tits opening milk bottles) where their peanuts were. Anyway, in our experiment, a badger encountering a slab for the first time would typically sniff and probe, lifting a corner of the slab and edging it sideways. Subsequently, familiar with the slabs, on a given night the first badger to arrive would begin the time-consuming task of digging under, or upturning, it. As more badgers arrived, each would almost always select a slab of its own. Of 40 occasions when a badger arrived while others were already working on slabs, the newcomer selected an unoccupied slab on 35 (on only 5 did it join another individual already feeding at an upturned slab). The 35 cases in which discretion appeared to prevail over valour gave no hint that some individuals were more or less inclined to avoid particular individuals, whether or not they were from the same or a neighbouring group. Indeed, during these 35 close encounters we could detect no sign that the badgers behaved differently towards group members or neighbours, be they from the same or different super-groups.[9] On the five occasions where a new arrival joined a single badger already feeding at a slab, three involved cubs amicably joining adults from their own social group.

For context, watching badger neighbours at the paving slabs (albeit distracted by peanuts) was strikingly different to watching meerkats, wolves or, indeed, most carnivores. Members of a band of meerkats spotting neighbours from several hundred metres invariably launch an attack, bouncing like stereotypic rocking horses, (or start fleeing, fast, if they belong to the smaller band) (Macdonald 1992; Mares et al. 2012), as do members of a wolf pack catching the whiff of an intruder on the breeze even kilometres away (Vucetich 2021). Both of the foregoing encounters between neighbours can end fatally—indeed, it seems that is exactly how the attackers intend them to end. Whether the meerkats had, moments before, been excavating a gecko, or the wolves pursuing a moose, would have no bearing on the ferocity of their response to neighbours. Nonetheless, mindful of the possible pacifying effect on badger neighbours of the peanuts, there are circumstances where super-abundant food causes a truce amongst neighbours—'martelism', named after the situation in which usually

---

[8] While we were busily trying to understand whether badgers used scent to repel other badgers, another member of the badger team, David's student Sandra Baker—nowadays an established expert on animal welfare—was herself busily trying to repel badgers using scent. Badgers are widely reported as crop pests in the UK. Sandra's experiments investigated whether learned food aversions might be used to protect crops from free-ranging badgers. They demonstrated that badgers could be conditioned, using maize cobs treated with ziram (a fungicide and irritant registered in the UK as a repellent against other taxa), in the presence of a benign odour cue (clove oil), subsequently to avoid untreated baits in the presence of the odour. The most likely mechanism was conditioned taste aversion to the ziram with second-order conditioning of the odour cue. Field-scale trials would be needed to determine whether conditioning could protect growing crops in the presence of an odour cue (Baker and Macdonald 2015).

[9] These were fur-clipped badgers, so we could identify them individually.

territorial pine martens congregate on a carcass; the same phenomenon may be observed when all of the jackals in a neighbourhood feast on a walrus carcass, or grizzly bears gorge together on a run of salmon. With hindsight, should we interpret the indifference of badger neighbours meeting at the paving slabs, beside a perimeter latrine, as just such a temporary truce induced by a rich food patch, or as an insight into prevailing tolerance between badger neighbours? The dispersion of peanuts under the slabs was not a bad approximation of the dispersion of earthworms or beetles for which any of those badgers could otherwise have been foraging in those places, and a few handfuls of peanuts seem unlikely to provoke such an outbreak of peace amongst habitually warring badgers (after all, they didn't fall to battle once the nuts were gone)—more likely, neighbouring badgers tolerate each other, for at least a majority of the time, irrespective of the presence of peanuts. At the time we did not appreciate the badger geopolitics of super-groups, and so did not check whether interactions differed in tenor between individuals that with hindsight we now consider as members of neighbouring groups or neighbouring super-groups. However, none of the interactions were fierce, and only five involved any sign of aggression (two being the occasions when two males from neighbouring groups sought access to the same slab). There were 68 occasions when a single individual affiliated to one group fed in the presence of two, or more, individuals from a neighbouring group. On 65 of these occasions being in the company of a neighbour at the 'border' appeared to evoke only indifference from representatives of both groups. On just three occasions two members of one group chased off their neighbour.

Although our focus here is on neighbourly relations, the Paving Slab Experiment also fortified the impression of Chapter 4 regarding relationships amongst group members. Under these particular circumstances, where five handfuls of awkwardly accessible peanuts were at stake, there were no displacements, usurpations, or even signs of greeting, coalition, or affiliation amongst group members; rather, they resembled ships that pass in the night.

Faeces matching

Are all perimeter latrines used equally? No: the members of each group deploy faeces along different segments of the perimeter in proportion to the number of adjoining neighbours on the other side of that segment—we call this faeces matching—and that proportion is modulated by the distance between neighbouring setts such that the closer the neighbours are, the greater is the deployment of faeces (Stewart et al. 2001). Group size affects faecal capacity, so the matching is not scat for scat, but proportional: a matched ratio of overall group faecal capacity (see also Buesching et al. 2016a). One interpretation of this pattern of latrine deployment and use could be a combative escalation through counter-marking. The absence of overt (feeding) aggression between members of neighbouring groups coinciding at the paving slabs weakens this interpretation, as does the fact that even when the meagre meal was over they did not fall to brawling, but just trotted off. An alternative interpretation is that the messages posted at the latrines are not, or not exclusively, about territorial defence, but more about general communication—the badger Facebook construal mentioned earlier. Amongst the 'posts' might be information on resource depletion, which brings us to the Passive Range Exclusion Hypothesis.

Resource depletion and the Passive Range Exclusion Hypothesis

Food depletion, and the rate of its renewal, is an important factor in foraging theory (Chapter 10) and prompts a way of thinking about badger spatial arrangements in terms of optimally efficient feeding blocks. A useful counter-factual, to which we will return in Chapters 10 and 19, is to imagine the outcomes if the entire population of 300+ badgers lived in one mega-sett atop Wytham Hill. Some individuals might need to walk several kilometres to find unoccupied feeding sites. The energetics of foraging is the topic of Chapter 14, but in terms of resource depletion one can imagine it being more efficient for the 300-strong throng to split into subunits. Paul Stewart conceptualized this by imagining badgers setting out to forage from their sett, consuming such food as is available as they go (Stewart et al. 1997). A gradient of food depletion develops in their wake (Figure 9.2). Ultimately, these foraging group members will reach the advancing front of their neighbours' explorations. Should these neighbours pass each other (amicably) and forage beyond this contact interface, they will enter into a zone of diminishing returns, already depleted by the respective neighbours. Iterative exploration along this front will resolve upon an isopleth (equal contour) of resource depletion between groups. Paul's

(a)　　　　　　　　　　(b)　　　　　　　　　　(c)

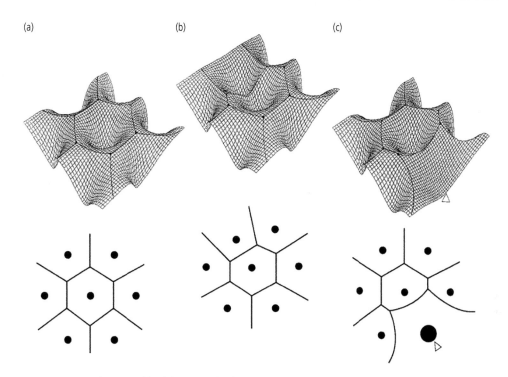

Figure 9.2 Diagrammatic surface plots of food abundance after food density gradient establishment, with a two-dimensional plot of the resulting ridge peaks, for three situations. (a) Equally sized groups that are regularly spaced; (b) equally sized groups but reduced inter-residence distance between two of the groups, note how the ridge is both lower and 'sharper'; and (c) regularly spaced residences with one larger group demonstrating how the larger group skews the now asymmetric ridge towards the smaller neighbouring residences. Filled circles are residence positions. Open triangle points to larger group residence in (c) (from Stewart et al. 1997, with permission).

brainwave was that badgers position their faeces, at latrines, along this isopleth. This highly parsimonious explanation of badger group range geometries is formalized as the 'Passive Range Exclusion Hypothesis', or PREH (Stewart et al. 1997; see also Kilshaw et al. 2009 and Buesching et al. 2016a). Kaufmann (1983) and Davies (1978a) both proposed that this type of mutual avoidance can, in an attempt to alleviate exploitation competition, create range exclusion without active defence.

Applying this line of thought to Wytham's badgers, when assets are shared with other group members there comes a point at which each resident individual meets its minimal level of access, or security, for that resource (worm feeding patch, or fruit bush). To have this proportion eroded further by contest with members of other groups either requires exclusion, or mutual egalitarian acceptance of 'share-and-share-alike' as the fruits of transgressions. This hypothesis rests on the assumption that groups exist as entities experiencing and signalling depletion within and between themselves.

Individuals at borders

Do individuals differ in their pattern of using perimeter latrines? Given that faeces is a limited commodity, and given that the members of a group might gain advantage from sending a strong signal of their Resource Holding Potential (see Box 9.1) to neighbours (and dispersers) on all sides, the question arises: how do they deploy their faecal ammunition? How many latrines to construct, where to position them, and then how to populate them is, by analogy, the sort of question that preoccupied the Romans when, having positioned the 16 forts they dotted along Hadrian's Wall, they considered how to deploy their forces amongst them. The answer may have depended on the shifting pressure of raiding Scots. Edward I faced similar decisions when this thirteenth-century Norman king considered how to man the line of forts beaded along Offa's Dyke to keep the Welsh at bay. Faced with similar decisions, do the members of a group of badgers systematically divvy up the responsibility of marking every latrine bastion, implying considerable coordination and indeed cooperation, or do they instead mark independently or (a different thing) randomly? Or, do they mark at whichever tactically important site is nearest to where they (or their neighbours) happen to be; for example, where they are feeding most that night (badgers forage independently, but all following the same basic feeding algorithm; Chapter 10)? The answer—that at least some individuals within a group are more coordinated in their use of latrines than might have been expected—was revealed by an experiment led by Kerry Kilshaw. Kerry, nowadays best known as WildCRU's expert on Scottish wildcats, began her career by asking what we could learn about the function of badger latrines from the way in which individual badgers deploy their scats at them. Using our individual pot-feeder apparatus (Stewart et al. 2001; Chapter 4), Kerry targetted 8 badgers (5 males, 3 females; recognizable by their fur-clips), each with a unique bait-marking signature (of a particular colour of bead) and all belonging to the Great Oak constellation of setts (which, at that time, numbered 17 residents) (Kilshaw et al. 2009). Simultaneously, we undertook general bait marking of this group and its three adjoining neighbours (Great Oak, Great Oak Outlier, and Chalet) over two 10-day sessions (August and October 2003). These 4 groups shared 76 latrines, of which the 8 target badgers used 36 over the period September 2003 to January 2004, into which they deposited 168 marked faeces of which 85 were incontrovertibly assigned to an individual and usable in the analysis. Often, indeed on 69.4% of nights, more than one of the eight bait-fed badgers deposited their colourful faeces in the same latrine during the same night: considering they had so many latrines to choose from amongst these inter-group sites (there were 105 latrines of which they used 76 during the experiment), and a limited nightly supply of faeces to deploy, this seems to be more than a coincidence. The pairings by sex of these double deposits were common between females/male dyads and commonest between female/male/male triads (female/female 16.9%, 23.7% female/male, 18.6% male/male, 10.2% female/female/male, and 30.5% female/male/male). This suggests not only that the eight group members were not deploying their faeces in the latrines at random, but also that they were not spacing them out (e.g. by seeking collectively to cover the greatest numbers of latrines through a division of labour). On the contrary, they appeared to be motivated to mark in the same places as one another. On particular nights they preferred particular latrines *en masse*. And overall the clusters of latrines they tended to favour tended to be those where the beaded necklace of latrines ran closest to the neighbouring setts (see scent matching in Box 9.1).

Meanwhile, our general bait marking of these social groups returned 491 marked faeces, of a total of 1121 faeces deposited at the76 latrines shared, in various combinations, amongst the 4 groups. Between 21 and 38% of those latrines were used each night. Again, this indicated selection for focused latrine use rather than aiming for maximal coverage—thus it took a 5–10-day cycle for all 76 latrines to receive some refreshed marking, which tended to even up the number of faeces each group deposited per latrine, over time.

How did these badgers deploy their faeces between hinterland and perimeter latrines? The average number of faeces deposited at a latrine per day by a group was significantly greater at latrines closer (<400 m) to that group's main sett than in latrines at the perimeter (Figure 9.3). Similarly, latrines closer to a group's main sett were used more often by that group than those further away. There was a difference between the sexes in the effort (or, more exactly, the faeces) they invested in the hinterland or the perimeter. Male badgers deposited more faeces at latrines closer to the sett than in those further away. In contrast females did not.

Crucially, during this late autumn survey (Photo 9.4), badgers were predominantly eating wheat (% faecal content: wheat 41.2%, blackberries 22.96%, and worms 14.69%, $N = 932$; our bait, in other words, provided only a small supplement to their diet); however, faeces deposited in latrines further away from wheat fields (up to 1000 m) were just as likely to contain wheat as were faeces deposited in latrines close to these fields (<200 m). In short, where the badgers happened to be feeding was not a good predictor of the section of perimeter at which they would defecate (and this was equally true for the group as a whole, with their generically marked faeces, as for the individually marked eight): they were not just defecating at the most convenient latrines, but were making special trips to the latrines they selected. How were they selecting them? Well, apparently not in order to mark the depletion of food (as per PREH), because the food being consumed was nowhere near the latrine being marked. Potentially this marking pattern had a sociological rather than a nutritional basis. There are two possibilities to explain their shared choices: either the badgers were responding, individually, to a stimulus that led them independently but non-randomly to the same latrines, or there was some coordination, perhaps cooperation, in their collective activities.

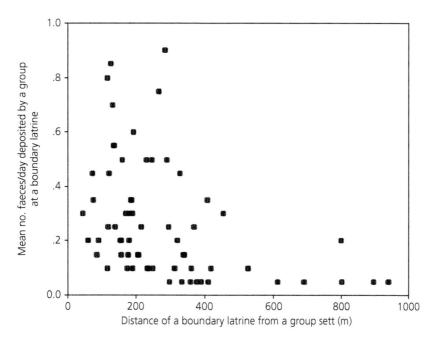

Figure 9.3 Relationship between the number of faeces deposited by a group in a boundary latrine and the distance of that latrine from a group sett (m).

Photo 9.4 David investigating a badger latrine.

What information is signalled? The fine print

Bringing a magnifying glass to bear on the fine print of the message board brings into focus two more of Buesching and Jordan's symptoms of a defensive territorial function. First, as mentioned in the context of individual roles, signals should carry information that can be matched to the individual that produced it, and which lays claim to the territory. Second, and fifth on their list, insofar as that claim may be laid not by an individual but by a group, in facilitating association between the owners and the real estate, scent profiles in gregarious species should be group-specific (but could allow assessment of group size through inclusion of individual-specific characteristics).

So, what does the fine print say? It may be different for different sections of the newspaper, so we'll answer separately for faeces, anal gland secretion, and urine (and later we'll refer back to the answers, in Chapter 6, for subcaudal glands). For faeces, they can reveal that an animal—a territory holder—is present, what it has eaten recently; in addition, hormone metabolites may expose the depositor's sex, reproductive condition (Wielebnowski and Watters 2007), or levels of stress (Palme et al. 2005.

Anal glands

Voicing a pervasive concept in political and military theory, Sun Tzu (sixth century BCE) wrote: 'If you know your enemies and know yourself, you will not be imperilled in a hundred battles ... if you do not know your enemies nor yourself, you will be imperilled'. He may not have appreciated his prescience with regard to the paired anal glands that open into the rectum just internally of the badger's anal sphincter (Stubbe 1971). In Chapter 6 we considered the functions, and information content, of badger subcaudal glands. Now, continuing our journey around the badger's bottom, all the questions we posed about the subcaudal gland can be posed about its cousin (close by juxtaposition but distant by ontogeny), the anal gland. In contrast to the anatomical uniqueness of the subcaudal pouch, anal glands are ubiquitous in carnivores (Brown and Macdonald 1985), the ubiquity being completed by their recent discovery in bears and aardwolves (Apps et al. 1989; Rosell et al. 2011), and familiar to dog owners because of the expensive occasions on which vets have to express their blocked ducts. The first chemical exploration revealed this secretion in badgers to contain mainly low-volatility long-chained fatty acids and water, varying little between individuals, or the sexes, but encoding group membership (Davies et al. 1988), which Roper et al. (1986) interpreted as a long-term territorial defence signal.

Unlike the visually conspicuous movements of bum pressing and kissing, associated uniquely with deploying subcaudal gland secretion, anal gland secretion (AGS) is discharged generally when the badger defecates (or sometimes as an unaccompanied spatter, and has the advantage, to the researcher, of being visible, unlike smears of subcaudal gland secretion). Rarely, it can be expressed as a volatile, such as when badger cubs are alarmed upon their first capture and emit a skunk-like stench resembling burnt rubber. Previously, Davies et al. (1988) had reported that badger AGS profiles exhibited limited variation in chemical composition. However, new techniques are showing distinct differences in semio-chemistry between males and females, as well as variation with age, reproductive status, and individuality (Buesching and Jordan 2018), congruent with studies of anal glands of other mustelids (Zhang et al. 2003; Apps 2013).

This frankly unappealing secretion preoccupies Christina Buesching and our Norwegian collaborator Veronica Tinnesand. To discover how badgers respond to it, they led our study that presented 351 samples of AGS collected, under sedation, from 50 known individuals (22 adult males, 6 yearling males, 18 adult

Photo 9.5 The delicate procedure of everting the anal gland papilla of a sedated badger to allow anal gland secretion collection.

females, and 4 yearling females) to identifiable recipients (n = 187 responses) over 117 field trials at 11 different social groups, staged between 30 May and 8 June 2012 and 3—15 June 2013 (Tinnesand et al. 2015). Acquiring these samples involved Christina gently everting the left anal gland papilla of each sedated badger (Photo 9.5) and palpating it to cause secretion to ooze into 2.0 ml sterile glass vials with Teflon-lined caps, which were then stored at −20°C.

These trials mimicked the methodology we described for exploring the functions of subcaudal gland secretion (Chapter 6). Most immediately, they were designed to answer the question: what information is signalled by AGS? Each trial involved pushing a sample vial containing approximately 0.1 g of AGS into the ground until the rim was level with the surrounding soil. An empty vial was then placed 50 cm (i.e. approximately one badger head–body length) from the test sample, as a control (in fact, badgers never responded to control vials).

Trials were conducted at each of 11 different social groups for 3 days, with trials run at the sett and at the border with an adjacent group, to investigate whether context changed outcomes, with one AGS/blank pair presented between 18:00 and 19:00; AGS from each scent donor was used only once at each trial to avoid habituation or prior context, and treatment order was randomized (Table 9.2). The team wore new latex gloves whenever samples were handled to avoid scent contamination.

AGSs (n = 351) from known individuals were presented to identifiable recipients (n = 187) to assess response variation according to familiarity (own group, neighbours, strangers) and spatial context (in context, at a shared border; out of context, at an

Table 9.2 Scent samples used in the behavioural trials varying in levels of familiarity (own group, neighbour, stranger), spatial context (sett, shared border, unshared border), and sex and age class of the scent donor ($n = 117$).

Treatment	Location	N (total)	N (female adult)	N (female yearling)	N (male adult)	N (male yearling)
Own group	Sett	13	4	1	7	1
	Border	17	6	1	8	2
Neighbour	Sett	17	5	1	8	3
	Shared border	15	5	1	8	1
	Unshared border	13	5	1	6	1
Stranger	Sett	22	6	2	12	2
	Border	20	6	1	11	2

unshared border/the main sett).[10] A *potential response* was defined as a badger ('responder') approaching the sample to *c*.50 cm—we were confident from pilot studies that badgers could detect the odours at that range. Within the sample of these *potential responses* we defined a subset of *actual responses* where the badger progressed to positioning its nose within 10 cm of the sample, 'sniffing' the sample, and during which their noses generally made direct contact with the vial ($n = 188/196$). Sometimes, we could clearly see the badger licking the sample vial, thereby involving the vomero-nasal organ (Lüps et al. 1993).

For each *actual response*, which by definition involved sniffing the sample, we recorded the sniff duration in seconds and the number of times an individual responded with subcaudal marking (i.e. marking under voluntary control, as opposed to an improbable AGS reply) within a 50 cm radius of the sample (they do not respond by anal gland marking, which is typically linked to defecation); when the respondent subcaudal marked, we noted whether this involved over-marking on top of the sample, or proximity marking within 50 cm of it (Buesching and Macdonald 2004).

[10] AGS from three different donor categories were presented: (i) own group OG (i.e. donor resident in the same social group where its AGS was presented; $n = 30$); (ii) neighbour (i.e. donor resident in a neighbouring group to the group where its AGS was presented; $n = 45$); and (iii) stranger (i.e. donor resident in a social group at least 2 territories apart; $n = 42$). AGS was presented either at the main sett or at a border latrine: (i) OG presented at main sett ($n_{Own} = 13$) or at border latrine ($n_{Own} = 17$); (ii) neighbour presented at main sett ($n_{Neighbour} = 17$), at a shared border latrine ($n_{Neighbour-shared} = 15$), or at a latrine on a different sector of the responder's group's border (i.e. not shared with that particular neighbour: $n_{Neighbour-unshared} = 13$); and (iii) stranger presented at main sett ($n_{Stranger} = 22$) or at border latrine (i.e. always unshared: $n_{Stranger} = 20$).

AGS from each scent donor was used only once in each category to avoid habituation. Trials were conducted at each social group for three days, with one AGS sample presented per night at each location. Treatment order was randomized.

Eager to understand the, possibly several, means by which badgers discriminate familiar from unfamiliar individuals, we defined a stranger, in the context of AGS, as a badger living at least two social group territories away (this generally means a member of a different super-group). The *familiarity* hypothesis (Fisher's 1954 'dear enemy'; Temeles 1994) posits that badgers would show a lesser response to AGS from a member of its own group than to secretion from a neighbour or stranger. At the *group level* we found that the durations of sniffing responses to AGS were longer when the sample was collected from less familiar individuals, being briefest to own group AGS ($x = 1.44$ s ± 2.58 SD), intermediate to neighbouring group samples (2.71 s $\pm$ 2.56), and most prolonged to stranger AGS (4.29 s $\pm$ 2.60)[11] (Figure 9.4a). Similarly, the likelihood of over-marking decreased in the order strangers, neighbours, own group (Figure 9.4b). These results are congruent with the dear enemy phenomenon (DEP, Müller and Manser 2007), insofar as briefer sniffing indicates a lower level of concern. The interpretation of sniff duration is much discussed in many chapters of Brown and Macdonald (1985) but in the context of badger odours, Palphramand and White (2007) argue that longer sniff duration is likely related to time required to identify the individual donor, decode individual-specific information (Buesching and Macdonald 2001), and contextualize their implied threat level (Sliwa and Richardson 1998).

Another, non-exclusive, relevant possibility, termed the *threat-level hypothesis* (Temeles 1994), posits that the response is calibrated to the threat indicated by the

[11] From a total of 117 trials, 85 ($n_{Own} = 20$, $n_{Neighbour} = 35$, $n_{Stranger} = 30$) produced 351 potential responses ($n_{Own} = 112$, $n_{Neighbour} = 125$, $n_{Stranger} = 126$), including 196 actual responses involving sniffing ($n_{Own} = 27$, $n_{Neighbour} = 81$, $n_{Stranger} = 88$) and 141 involving subcaudal scent-marking responses ($n_{Own} = 9$, $n_{Neighbour} = 49$, $n_{Stranger} = 83$). When badgers sniffed AGS, their noses generally made direct contact with the sample ($n = 188/196$).

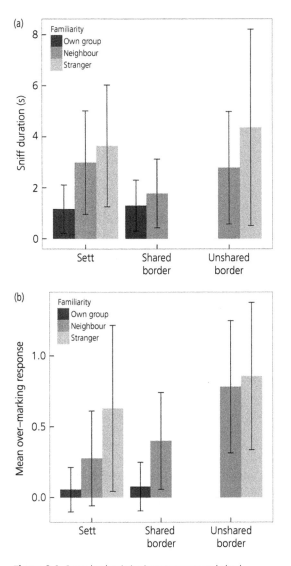

Figure 9.4 Group-level variation in responses to anal gland secretion (AGS) from donors of different levels of familiarity. Mean (±SD) group level of (a) duration of sniffing responses in seconds; and (b) number of subcaudal-marking responses per potential responses to AGS from own group ($n = 20$ trials), neighbours ($n = 35$), and strangers ($n = 30$) presented at either the responding group's sett, or a border (shared or unshared) between the responding group and the scent donor (from Tinnesand et al. 2015, with permission).

odour. For example, the odours of less familiar donors from further away, perhaps perceived as intruders, might demand greater scrutiny (alternatively, and no less plausibly, a really dangerous donor might prompt a very rapid reaction!). Mayeaux and Johnston (2002) explored this sort of idea with hamsters, concluding that small changes in spatial location influence the

salience of conspecific odours. To tease out these possibilities we presented AGS at one of two sites: either at a shared border latrine, where encountering extra-group AGS might not be surprising (although it might be alarming or annoying); or at an unshared border (i.e. from more than one social group distant/a different super-group) or at the main sett, where we anticipated that foreign scent would elicit a greater response (in the context of territoriality, signalling an incursion). The badgers' responses under these two circumstances are revealingly different.

In short, sniff duration differed significantly between donors of nearer or further provenance (i.e. different propinquity). However, the AGS of a neighbour encountered at a shared border evoked no more protracted interest than that presented at the home sett or indeed at a section of border that did not abut that donor's home range (findings paralleling those for subcaudal gland secretion; Chapter 6). Similarly, neighbour AGS presented at, or away from, the border with that neighbour evoked much the same response, in terms of sniffing duration, and the different circumstances of these presentations did not link to differences in the likelihood of over-marking with subcaudal gland secretions either. This might prompt the conclusion that the message, or its interpretation, does not concern defence or territoriality. At the least, considering the purpose of this chapter, it prompts reconsideration of our understanding of neighbours, and indeed borders, in badger society. Despite this seemingly undiscriminating response to samples from same versus neighbouring group, both at the border or close to their sett, the response to the AGS of strangers (which, by definition, are still more unfamiliar if they hail from a different super-group) was very different, sometimes exuberantly so: the sniffing badger might whirl around the odour, sniffing again, then resniffing in apparent disbelief, then dig up the sunken test tube, and repeatedly smear subcaudal secretion over the stranger's sample (this exuberant response is much more reminiscent of the reactions of meerkats to the scent of immediate neighbours, which they loathe, at their border latrines). Overall, the tubes containing AGS samples from strangers were dug out more often (26.7%, $n = 8/30$), compared with neighbours at unshared borders (15.4%, $n = 2/13$) and at setts (7.6%, $n = 1/13$), whereas own group and neighbour AGS at shared borders was never interfered with.

Repeating for *individuals* (rather than groups) all the foregoing analyses, individual-level responses in general followed the same patterns: sniff duration decreased with greater 'familiarity' (Figure 9.5) with

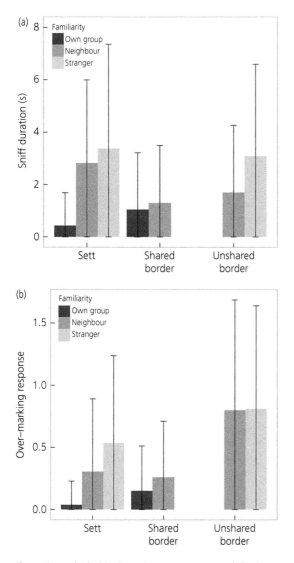

Figure 9.5 Individual-level variation in responses to anal gland secretion (AGS) from donors of different familiarity levels. Individual level (±SD) (a) duration of sniffing responses; and (b) number of subcaudal-marking responses per potential response to AGS from own group ($n = 20$ trials), neighbours ($n = 35$), and strangers ($n = 30$) presented at either the responder's sett, or a border (shared or unshared) between the responder and the scent donor (from Tinnesand et al. 2015, with permission).

At the sett a stranger's experimental AGS sample was more likely to be smeared with subcaudal gland secretion than was a neighbour's and much more so than a sett cohabitant's, but at a shared border we saw not a single instance of over-marking a stranger's AGS.

In all the foregoing, was there a difference in the reactions of males and females to the AGS samples? Slightly; overall, males tended to sniff for nearly a second longer than did females ($x = 2.38$ s ± 3.08 SD, $x = 1.73$ s ± 3.12 SD, respectively) but basically each sniffed for 2 s, give or take, and there was no sex-related difference in the frequency of subcaudal over-marking responses (Figure 9.6). There was also no effect of relative age class in the duration of sniffing or the frequency of subcaudal over-marking responses.

Did the reproductive condition of the sniffer affect their behaviour? This was difficult to test insofar as hormone profiles correlate poorly with external signs that might have been thought of as indicative of reproductive condition (males with very descended testes; females with vulval swelling; Chapter 2). Nonetheless, individuals with these visible signs did tend to sniff AGS for a little bit longer, and over-marked it with subcaudal smears more often. Did the reproductive condition (by the same visible measures) of the sample donor affect the sniffer's response? Yes, males tended to over-mark the secretions of females more if, at recent captures, they had displayed swollen vulvas; however, females exhibited no discernible difference in inter-individual over-marking responses in relation to donor male's reproductive status.

So, as we try to understand the relationships between groups through appreciating the messages left at latrines, whether as warning signs or newspapers, what clues lie in the information conveyed by AGS? Males can clearly identify when the message is left by a female (they sniff the latter for longer), and from females with swollen vulvas (they sniff them for even longer and also over-mark them more than they do messages from females with normal vulvas).[12] The AGS messages also signal, to both sexes, the signaller's group provenance (i.e. they spend longer sniffing the message, and are more likely to over-mark it, when it

no significant effect of the experimental location (at sett or latrine border). Conversely, and strikingly, these individual analyses revealed significant differences in the level of subcaudal over-marking not only between the three categories of familiarity, but also between contextual circumstances of the trial (Figure 9.5).

---

[12] Intriguingly, Palphramand and White (2007) found no evidence of differences in male or female responses in relation to the age or sex of the faeces donor. Was the critical difference somehow methodological, or to do with the association of the secretions with faeces (indeed they mentioned avoidance of samples conspicuously coated in AGS), or because of some difference between the badger populations?

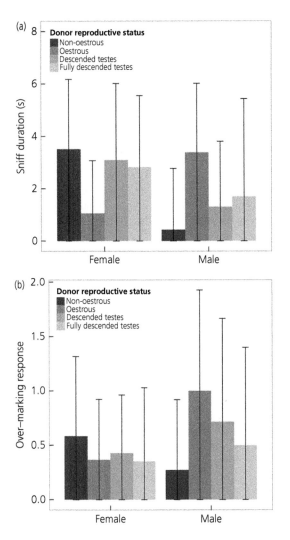

Figure 9.6 Variation in responses to anal gland secretion (AGS) according to the reproductive status of the donor. Individual male and female (±SD) (a) duration of sniffing responses; and (b) number of subcaudal-marking responses per potential response to AGS in relation to the reproductive status of female (levels: oestrous, non-oestrous) and male (levels: descended testes, fully descended tested) scent (from Tinnesand et al. 2015, with permission).

comes from outwith their own group, and even longer if it originates from two ranges distant). The sniffer's interest in this distinction of provenance, as indicated by sniff duration and (at group level) over-marking, does not differ whether the message is posted at their perimeter or close to their main sett. However, individuals, at least some and presumably all, do take account of the location of the message (in that, when analysed through individual responses, they are more likely to

over-mark a stranger's AGS when it is posted close to their sett rather than close to their perimeter).

Urine marks

Badgers urinate in three contexts. First, infrequently, unceremoniously, and alone; second, and more commonly, they combined subcaudal marking with urinating on their setts; and third, at latrine pits, especially perimeter ones. What information is signalled by urine? All mammals produce urine, and the metabolites it carries can encode information such as sex, reproductive state, diet, and health status (for reviews, see Brown and Macdonald 1985; Buesching and Stankowich 2017). We knew that badgers could discriminate sex and age—at least stage of maturity—from urine because shortly before joining our team, Katrina Service had discovered this during her doctorate (Service 1997; Service and Harris 2001). Furthermore, when we asked what other signals our badgers could discern from urine, we had the advantage of knowing the hormonal status of the donors, thanks to Nadine Sugianto. Nadine undertook enzyme immunoassays of 51 urine samples that we gently expressed from the full bladders of sedated badgers during our spring trapping session (24 May to 12 June) in 2017 (Sugianto et al. 2021a). The lab work was done at the Leibniz Institute for Zoo and Wildlife Research (IZW) in Berlin where, to illustrate the tendrils of the badger network, David's first badger student, Heribert Hofer, was by then the Director. Nadine measured oestrodiol metabolites in the urine of females, and testosterone metabolites in that of males—work described in Chapter 17 (Sugianto et al. 2018, 2019c). What matters here is that when we repeated the AGS trials for urine, led by Tanesha Allen and building on Charlotte Ryder's undergrad project, we did so knowing the urinary hormone profiles of the scent donors.[13] Working at the same three setts at which we tested AGS and subcaudal gland (Chapter 6) (Jew's Harp, Pasticks, and Pasticks Outlier), Tanesha explored whether samples with lower or higher concentrations of these sex steroid metabolites triggered different responses from the badgers by comparing the reaction of 7 male and 7

---

[13] The blood plasma and urinary levels of testosterone correlated for individual males across seasons. However, no simple correlation was evident for oestradiol levels in the blood and urine, due to a longer time lag between circulating oestrogen and urinary metabolites, although oestrogen metabolite excretion patterns corresponded with seasonal reproductive patterns.

female responders, for whom urine hormone metabolites were known, to urine donated by 9 male and 14 female donors, including familiar individuals from a responder's own group, from its neighbours, and from strangers (2 or more groups distant) (Allen et al. 2019).

A total of 143 nocturnal videos contained appearances by 26 individually identifiable adults. Possible responses (from 50 cm radius) were recorded in 74 instances (51.7%; 15 with urine only and 8 with water only), while 63 instances (44.0%) included actual responses (within 10 cm) where badgers sniffed only the urine in 24 cases, only the water sample in 23, and both urine and water in the remaining 16 cases. Statistically, therefore, there was no noteworthy difference in the likelihood of their paying attention to urine or water. However, they sniffed urine ($n = 40$, 5.02 s $\pm$ 1.58 s) for a couple of seconds longer than water ($n = 23$, 3.05 s $\pm$ 0.49), a statistically significant difference.

Adult badgers sniffed urine from opposite-sex donors more often than they did same-sex donors regardless of donor familiarity. The effects of sex and provenance interacted: urine from non-resident males triggered more responses than did urine from resident females, non-resident females, and resident males. Adults also over-marked cub urine more often than adult urine, and sniffed cub urine for longer durations.[14] Adult females seemed more interested in urine than were adult males: they sniffed the urine of both adults and cubs ($n = 9$) for longer than did males.

Did the concentrations of excreted sex steroid metabolite (oestradiol and testosterone) levels in urine trigger different responses from the badgers? In general, the answer was no, except that male (testosterone) samples received more attention than female (oestrodiol) samples (conforming with Service and Harris 2001), but otherwise the only effect we could discern linked to the hormone assays was that male responders sniffed urine provisioned from females with low oestrodiol for significantly longer durations. There are several possible explanations. First, perhaps sniffing the urine of females with low oestradiol indicates that the signal was hard to read, and so the reader had to concentrate for longer. Another possibility arises from Sugianto et al. (2021a), where we explored the concentrations of sex hormones between blood plasma and urine hormone samples collected at the same time, for both sexes, and assessed the correlations at both a general population level and for individuals. We found that urinary testosterone metabolite (UTM) and urinary total oestrogen metabolite (UEM) excretion patterns in the urine both corresponded with seasonal badger reproductive patterns, as revealed in the blood, on a *population* level. That was the generalization, but at an individual level male plasma testosterone and UTM levels correlated significantly across seasons, but in contrast there was no short-term correlation for total oestrogen in the blood plasma and urine (UEM) in females. This means that while the overall picture, or more correctly, atmosphere, has the scent of sex in the air for both sexes during the breeding season, and for males the testosterone coursing through their veins is accurately reflected in their urine, for females the link between urinary sex steroid metabolites is a much less reliable means to assess her reproductive mood of the moment. Perhaps that makes male urine more informative and worth a longer sniff.

A final thought about the general absence of an effect of sex steroid metabolites on the recipient badger's behaviour might be explained by laboratory findings that amongst mice urinary proteins reveal individual identity (a role previously attributed instead to the major histocompatibility complex (MHC); see Chapter 15) (Hurst et al. 2001, 2005). Another possibility is thus that the signal quality of urinary proteins (or MHC) is so strident as to render the hormone metabolites the status of supportive footnotes to the headline messages.

Latrines: what are they for?

After these insights, we return to two questions: what is the function of latrines, particularly perimeter ones, and what does their deployment tell us about the relationships between groups of badgers and thus about their social system and its adaptive significance? Interestingly, it is necessary to phrase both questions with respect to the badgers of Wytham Woods in the early twenty-first century. This specificity is necessary, and interesting, because of the growing suspicion that the answers may be different in different places, and perhaps also within Wytham, at different times. Badgers in some other populations do seem to be much more aggressively territorial than those in Wytham today, and even Wytham's badgers seemed more resolutely territorial when Kruuk first documented their behaviour in the early 1970s. More on this topic of variability later, but first, let's take stock.

---

[14] Cubs did not feature in the anal gland and subcaudal gland experiments because they produced none of these secretions, or at least not in sufficient quantities to enable us to sample.

Regarding territoriality, we return to the paradox: while setts are spaced out with respect to suitable habitat, implying territoriality, badgers very frequently visit setts other than, although commonly neighbouring, their main residence (Chapter 3). It is common to find individuals in what would traditionally have been considered neighbouring territory (Chapter 4), and occasional observations of such interlopers, even at perimeter latrines and in the company of their neighbours is generally characterized by 'unconcern' (Chapter 5). Even in the context of sex, neighbours can be seen together without apparent animosity and there is evidence of fatherhood across putative borders (Chapter 7) (and much more evidence of extra-group paternity lies ahead in Chapter 17). So, the paradox is that the fundamental spatial arrangement of contemporary Wytham badgers appears to be territorial, but the observations of their behaviour offer little evidence of exclusive partitioning of defended space. A Martian, even an ecology graduate, presented with anonymized data on badger movements and encounter networks (much as in Chapter 4's aRFID diagrams), and asked to characterize the social system of this hypothetical creature, would not conclude: territorial.

Consider now latrines, and the paradox deepens, or perhaps the confirmation bias strains harder. Perimeter latrines bisect the land between neighbouring main setts, sometimes with the geometric precision of Dirichlet tessellations. These latrines appear magnetically drawn to adjectives like border, boundary, and, yes, territoriality—inevitably, perhaps, given the interpretation of the bait-marking technique used to diagnose where the range of one group stops and another starts. Other species with broadly similar perimeter latrines fight at and across them, bloodily and sometimes fatally. Badgers can inflict on each other gruesome wounds, and in Wytham did so during earlier but not later decades of history (Chapter 4); and although we have almost no evidence of whereabouts, by whom, (or even under what provocation) that wounding occurs, in recent years, at least, abundant evidence has become available of it *not* happening at encounters between neighbouring groups at perimeter latrines. On a point of logic, if latrines are not marking borders, then we have little evidence of where borders are, or if they exist: movement, trespass, visits, and paternity all imply that they don't—a Schengen Zone. If the noun latrine has to be wrenched from its umbilical link to the adjective territorial (a metaphor that correctly captures the birth of the functional interpretation of latrine marking), what better describes their

function? Sticking with metaphors, if the latrine is not a fence, but a newspaper, what is the headline? What devil lies in the detail of the small print? Better yet, if the latrine is a message board, a *Facebook* page, what do readers glean and authors betray?

Of course, badgers have to defecate (and urinate) somewhere. With a sanitary function in mind, doing so tidily in a manner closely analogous to public lavatories could be adaptive, although the thought that these should be as far away from your burrow as your group range permits needs to be reconciled with the density of hinterland latrines around, even at, setts. When it comes to Hediger's fence (or even Maher and Lott's active territorial defence hypothesis) and a defensive, repulsive, territorial function, the evidence passes at least five of Buesching and Jordan's six diagnostic tests. However, none of these same five, each indeed useful, even essential, characteristics of territorial marking, is at odds with signalling for other purposes. The sixth contains one clincher, 'Intruders should avoid scent-marked areas/retreat on encountering scents (scent-fence)' (Buesching and Jordan 2019, p. 95) a box not currently ticked by Wytham's badgers, and then an escape clause: 'or modify their interaction with owner(s) when they match scents in the environment to scents of encountered individual(s)'. Remember David's hand-reared fox, which looked like a fox before, and no less after, sniffing the stranger's urine, but somehow responded differently afterwards, being disinclined to play with the rubber mouse? We can conclude readily that, in the early twenty-first century, the fence does not repel Wytham badgers, and that they seem just as brave after they have crossed it, although we cannot be entirely sure how they feel. If badgers feel on edge having crossed the latrine line, it is not generally obvious, any more than it is obvious that those intruded upon feel outrage. Not only is there a lot of evidence that these badgers are often not aggressive to their neighbours at perimeter latrines, but also the absence of aggression at group borders does not seem indicative that the status quo had stabilized as a dear enemy truce. No. The accumulated evidence is not simply that members of badger groups glower at their neighbours with unexploded hostility, but rather that at least some neighbours wander to and fro through each other's ranges, at the least with indifference and sometimes amiably (to judge by the genetic revelations awaiting in Chapter 17).

Alternatives to defensive functions listed by Buesching and Jordan (2019) include two options: (i) the creation of landmarks to assist with orientation (maybe, given that all latrines lie at path intersections, conceptually, latrines might not only be junctions, but also service stations, where travellers stop to chat, but why bisect neighbouring setts if not to express 'difference'?); and (ii) parasite control (certainly there may be hygiene benefits, but why the deployment pattern?). Another hypothesis concerns (iii) predator–prey interactions (of which PREH is an example), but for now, most resonant with the Facebook message board notions would be the functions of (iv): an information centre/advertisement.

Although by badger standards we have surely just had a superficial sniff at latrines, two conclusions emerge. First, badgers seem able to glean information from latrines about individual identity and group affiliation, group size, timing of occupancy, gender, age, reproductive status, and diet. The Facebook page analogy grows compelling. It seems that each author posts such a wide diversity of information (some 'notes to self', some 'notes to others') that each reader can make of it whatever it chooses. An individual who is a neighbour on one day may be a visitor on another (Chapter 3). What it means to be a neighbour in Wytham's badger society is a topic of Chapter 10, but bear in mind the readership of perimeter latrines: a cluster (super-group) of 2–3 juxtaposed groups, each typically of about 10 individuals, surrounded by an annulus of further similar-sized groups (with another layer of individuals, likely less intimate acquaintances, in the next annulus and so on). The tenor of relationships appears to be determined more by reduced propinquity than by aggressive exclusion.

Second, there appears to be redundancy in the signals (subcaudal and AGSs and urine): identity, and all that is associated with it, would seem to be communicated several times over, and all at the same place. Of course, this is not necessarily surprising to those of us who can, in much less than a second, interpret any and all of the following: taut lips, a furrowed brow, scowling eyebrows, clenched fists, and hunched shoulders as communicating much the same thing. Indeed, in the context of identity, the sound of a voice, the look of a face, and even the smell of a person all give the same answer.

Although it seems that the information is not immediately used in combat, we know from our findings on faeces matching (Stewart et al. 2001) and deployment (Kilshaw et al. 2009) that badgers go to considerable trouble to relay and respond to information about group sizes on either side of the perimeter latrine line, and may even cooperate in deploying that information. As said, faeces are free, but not unlimited. Macdonald (1980a) deduced that while badger groups have the

colonic strength of numbers to produce copious droppings, the availability of faeces may be a limiting factor on groups of different sizes. Buesching et al. (2016a) compared latrine use patterns between badger populations of densities ranging from 5 to 32 adult badgers per km^2. While the number of latrine sites increased with group range size (but not group membership) and thus the area utilized, the number of fresh faeces per latrine was fewer if the area encircled was large, resulting in faeces being meted out more sparingly (Figure 9.7).

In the context of faeces matching and in terms of 'density of message', members of a small group in a large territory may experience faecal constraint, facing the options of maintaining fewer latrines with numerous droppings, or numerous latrines with fewer droppings—the latter seems closer to reality. In contrast, numerous occupants of small territories may have a superfluity of faeces, depositing many at each large latrine. Signalling strength of numbers— for example, howling coyotes (*Canis latrans*) (Gese 2001) and roaring lions (*Panthera leo*) (McComb et al. 1994; Wijers et al. 2020)—and even bluffing about this strength is familiar in the field of vocal signals (territorial male red-winged black birds (*Agelaius phoeniceus*) each have a wide repertoire of songs, which they sing from a profusion of different perches to give intruders the impressions that the area is saturated; Yasukawa 1981a, 1981b). This phenomenon gave John Krebs, while working on great tit territoriality in Wytham in 1977, the idea he called the Beau Geste hypothesis[15] where exaggerated bird song can give the impression of substantial site occupancy. Amidst the ragout of faeces piled in latrine pits, we have no evidence that individual badgers are apportioning their contributions stingily to make a night's supply of faeces stretch as far as possible around the border, or to convey, somewhat Beau Geste-like, that they are in many places at once (Buesching et al. 2016a argue that this is unlikely). The coordinated faeces matching revealed by Kilshaw et al. (2009) indicates that something much more subtle than meets the eye is going on (this is certainly possible when it comes to urine, which can be doled out sparingly by marking carnivores, to increase coverage; Macdonald 1980a).

---

[15] *Beau Geste*, a book by Wren (1934), tells the story of three English brothers enlisted in the French Foreign Legion greatly outnumbered in a battle against a Tuareg army. To create the illusion that their fort was much more heavily defended than it actually was, they propped dead soldiers along its walls and ran around noisily creating the impression of impregnability.

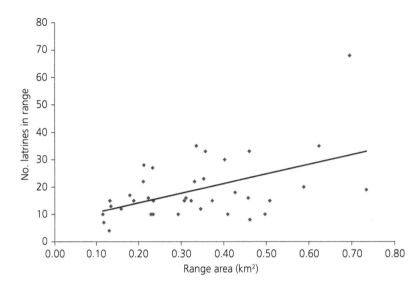

Figure 9.7 The relationship between number of latrines and area of social group range as estimated by minimum convex polygon for the E1 study site in 2002. Number of latrines/group range area: $r = 0.60$, $P < 0.001$. Number of dung pits/group range area: $r = 0.46$, $P < 0.01$ (from Buesching et al. 2016a, with permission).

It seems the answer to the question of what badgers glean from latrines, and it is not trite, is: everything they need to know. Individuals post, and read, self-advertisement signs and *aides memoires* across a wide variety of noticeboards—a badgerine version of Facebook (Tinnesand et al. 2015). Badgers are not silly, so of course the positioning of the signpost signals information, along with the script therein: it would be fatuous to imagine that they are not aware of the difference between, indeed the precise locations of, their setts, their neighbours' setts, and the land between (probably the very antithesis of no man's land). Badgers, as we have seen, scent mark very heavily around their setts—any outsider visiting that sett can be under no misapprehension as to who resides there. Indeed, the sniffer may detect intrinsic properties of scent marks that honestly signal quality, or may be reminded of a previous encounter with the individual producing that scent, which may usefully be borne in mind if the signaller is subsequently encountered in person. Gosling and Roberts (2001b) made a further, more radical proposal: that resource-defence polygyny in mammals may not be economically viable without stakeholders using scent marking to reduce the costs of area defence.

Have things changed, or was Kruuk, followed by ourselves and others, lured by confirmation bias into thinking that badgers behaved in a fashion comparable with spotted hyaenas? That preconception may have made us slow to appreciate the change, but it seems unlikely to have completely misguided Kruuk's interpretation, for several reasons in addition to his prowess, placing him amongst the best naturalist scientists of a generation: (i) in addition to the changes in apparent territoriality and associated latrine function (the focus of this chapter), many other things have changed profoundly in the meantime, most dramatically a three-fold increase in population density (Chapter 11); (ii) the defensive territorial exclusivity that fitted the weight of evidence in the early 1970s typifies the interpretations of other researchers of group-living badgers, all of which live at lower densities than contemporary badgers in Wytham; (iii) those same interpretations are very widely held by large numbers of skilled amateur specialists in badger watching; and (iv) at least some form of territoriality is associated with group living in other mustelids (Chapter 19). But if the odds are that something fundamental has changed in the relationships between neighbouring badgers, we must ask what are the sociological ramifications of that change, and what is their adaptive significance? Those are topics of Chapter 10.

The Ecological Foundations to Badger Group Living

Society exists only as a mental concept; in the real world there are only individuals.

Oscar Wilde

There is no such thing as society... My meaning, clear at the time but subsequently distorted beyond recognition, was that society was not an abstraction, separate from the men and women that composed it, but a living structure of individuals, families, neighbours and voluntary associations.

Margaret Thatcher 1983

When one tugs at a single thing in nature, he finds it attached to the rest of the world.

John Muir

There are, crudely, two categories of selective pressure favouring group living: sociological and ecological. Amongst carnivores the sociological factors vary in importance between, and within, species, but demonstrably often include the benefits of collaborative hunting, border patrolling, vigilance, allogrooming, and a subset of reproductive behaviours ultimately linked to alloparenthood. But intriguingly, as we reported in Chapters 7 and 8, evidence that any of these drives group living in Wytham's badgers is no more convincing now, after half a century of research, than it was when Hans Kruuk (1976) first noticed the badger paradox. To recap briefly, that paradox is that badgers live socially—well, gregariously—but without evidence of cooperative advantages to sociality. Contradictions are rife: social structure is nebulous (Chapter 5); evidence of cooperation is elusive (Chapter 8); territoriality is equivocal (or at least manifests in a non-exclusive way) (Chapter 9); mating appears unexpectedly egalitarian (Chapter 7); and groups are genetically non-exclusive too (Chapter 17). We see badgers together, sometimes writhing with rumbustious sociability, but their paradoxical behaviour compels a new understanding of the meaning, the functional significance, of terms like social group, and probably also of territory, and those refined meanings will involve neither a discrete breeding unit nor an exclusive clique. That paradox deepens with the realization that not all badger populations organize into groups, particularly not the large groups typical in Wytham, which occasionally top 25 members (da Silva et al. 1984; Buesching et al. 2003). At lower densities, across other parts of their continental European range, badgers are often solitary, live in pairs, or form small groups where mothers live with mature, but non-breeding, offspring (Johnson et al. 2002b)—that intra-specific variation will be the opening topic of Chapter 19. This brings us to a view of societies as emergent properties of their ecological circumstances (Macdonald 1983).

The Resource Dispersion Hypothesis (RDH) characterizes a particular, we think pervasive, set of such ecological circumstances. We will elaborate later, but briefly the RDH explains how delayed dispersal of young is facilitated when the dispersion of resources (often food) permits individuals to share a range (cohabit as a group) while still meeting their minimal requirements for (food) security, and without any sociological benefits from cooperation.

This appreciation that sociology has ecological drivers has its roots in the late 1960s in John Crook's seminal work on how habitat, and associated food supplies (e.g. seed- versus insect-eating) and predation

The Badgers of Wytham Woods. David W. Macdonald and Chris Newman, Oxford University Press.
© David W. Macdonald and Chris Newman (2022). DOI: 10.1093/oso/9780192845368.003.0010

risk, determine the social organization and patterns of communication of weaver birds—which vary amongst 105 species from stand-offish pairs to huge colonies (Crook 1964). He went on to describe similar, broad associations between their ecology and five different grades of primate societies (insectivorous, forest-dwelling, pair-living, nocturnal lemurs in Grade I to vegetarian-cum-omnivore savannah-living, single-male harem-forming diurnal patas monkeys in Grade V; Crook and Gartlan 1966). Importantly, this primate model was schematized according to clusters of characteristics, those of the social system, and those of the environment, of which one box in the schematic was, in our opinion, crucial. That box was labelled 'food availability and dispersion'. John Crook's early studies led to a generalized model across mammalian social systems (Crook et al. 1976) linking strategies of resource exploitation to adaptations for predation avoidance, mating, and rearing of young. Crook's broad-brush insights were, importantly, refined for antelope by Peter Jarman (1974), who advanced the paradigm with the realization that feeding styles (e.g. selective pickers, browsers, grazers) amongst African antelope species relate to their maximum group size through the influence of food item dispersion on group cohesion, from monogamous pairs of bud-nibbling dik-dik to lawn-mowing herds of wildebeest. These feeding styles are also related to body size and to habitat choice, both in turn influencing the antelope species' anti-predator behaviour, which, in many species, circles back to influencing minimum group size. Socio-ecological syndromes of this sort will turn out, later, to be important for fat badgers and thin martens but, first, continuing with antelope species: the likelihood of finding a female in a given place at a given time determines the mating strategy a male antelope employs, which is, in turn, a further product of group size and patterns of movement over the home range. The consequences for male mating strategies (spanning inseparably monogamous duikers to exhaustingly polygynous impala) reverberate through sexual dimorphism, adult sex ratio, and differential distribution of the sexes amongst antelope species. Everything is related to everything else; it all starts, however, from ecology (incidentally, almost 50 years on, Jarman's insights emerge unscathed from a recent technical reanalysis; Szemán et al. 2021). The year after Peter Jarman's antelope insights, Tim Clutton-Brock (1975), inspirationally at the time David's housemate on a farm at the edge of Wytham Woods, used an even finer grain of ecological detail to demonstrate that the differences in feeding behaviour between red colobus monkeys (*Colobus badiustephrosceles*), which typically eat a varied daily diet of leaves, shoots, flowers, and fruits from a wide variety of tree species, and black and white colobus (*C. guerezauellensis*), which feed almost exclusively on mature leaves of one or two tree species, could explain why red colobus live in large troops in large ranges while black and white colobus live in small troops in small ranges. Then, 1 more year on, and pivotally, a study of five neotropical species of emballonurid bats in Costa Rica by Jack Bradbury and Sandy Vehrencamp (1976) was to inspire a new understanding of badger society: the bats fed on prey that was variously dispersed and they opted for different social organizations; in particular, some bats fed in clouds on swarms of riparian insects—the locations and sizes of those insect swarms changed from night to night, causing the bats to feed in variously sized groups whose home ranges embraced sufficient meanders of the river always to offer fruitful foraging grounds. This was the first study to consider the effects on social organization of mammalian species of spatio-temporal availability of food patches. There was an interesting parallel with red-winged blackbirds, which fed on patches of lakeside insects that the wind blew hither and thither, so the birds had to configure their territories to ensure they would encompass an insect patch irrespective of wind direction (Horn 1968).

This was the intellectual foment into which David graduated in 1972 (giving up a doctoral place with John Crook in order to throw in his lot with Hans Kruuk). It led to the ideas that still guide our unfolding answer as to why badgers live in groups (the same answer fundamentally affects the reasons for which *all* social species are so organized) (see the text that follows; Macdonald and Johnson 2015a). So, our narrative now turns to resources and, specifically, how the badgers' diet and foraging behaviour affects their propensity to tolerate conspecifics sharing their range, leading to the formation of groups.

Diet and foraging

Picric acid and infrared binoculars were the two tools that revealed the overwhelming annual importance of earthworms to the badgers of Wytham Woods. Although inconveniently explosive, picric acid (2,4,6-trinitrophenol) has the remarkably useful quality of staining bright yellow the chitinous, beautifully sigmoid, undigested chaetae that throng badger dung. Chaetae are the earthworm's crampons. Earthworms have 8 chaetae per segment, arranged in pairs (and 1 oesophageal ring, or clitellum, without chaetae), and around 135–150 segments and thus around 1120 chaetae, which, rendered colourful and conspicuous amidst a matrix of earthy intestinal contents, can readily be counted under a binocular microscope in samples of faeces. Thereby, through a process of back-calculation, offering an estimate of how many worms (each worth c.2.5 calories) the badger had consumed.

While Hans counted the clitella in badger droppings (Kruuk et al. 1979), David counted chaetae in fox droppings, and they converged on the discovery that these two carnivores ate a lot of worms (Macdonald 1980d). Indeed, earthworms are important as prey to multitudinous predators (Macdonald 1984b), even tawny owls (*Strix aluco*; Macdonald 1976b), and recently turn out to be crucial to the ecological energetics of Wytham Woods.

Badgers are omnivores (Neal 1948; Balestrieri et al. 2019). Their diet varies across their vast geographical range from Ireland east to the Volga river (Chapter 1), but in the lowland agricultural landscape of the UK they eat a lot of earthworms whenever these are available. Between 1982 and 1983 David's first badger student, Heribert Hofer, now Director of the Leibniz Institut für Zoo- und Wildtierforschung in Berlin, estimated they comprised 63% of annual diet in Wytham (Hofer 1988). When earthworms aren't available, typically during a summer drought (Chapter 11), the badgers compensate by consuming a range of secondary foods

opportunistically (Bright Ross et al. 2021. A very similar picture had emerged from the Master's thesis of Martin Hancox (1973): earthworms occurred in 91% of over 2000 badger faeces from Wytham Woods, making up 61% of the total volume. The next most important foods were wheat and acorns. Kruuk (1978b) collected another 39 faeces in September/October 1974: 85% contained earthworm remains; 26% wheat; 26% insects; 7% bird remains; and 5% acorns. Only worms and wheat occurred in substantial quantities in these faeces; the mean number of worms per scat was 65.4 (range 0–193) and since badgers defecate two to three times per night (Ryszkowski et al. 1971), Kruuk estimated that at that time they probably ate 130–200 worms per night (amounting to 560–850 g live weight).

To put these findings in wider context, between July 1975 and June 1978 Kruuk and Parish (1981) collected 2159 badger faeces, accumulated at 2-monthly intervals, from latrines close to setts at 6 field sites across Scotland. Mindful of the different information given by different methodologies of scat analysis (Klare et al. 2011), they estimated both relative volume of food ingested (on a 5-point scale) and relative frequency of occurrence, and presented both on the insightful plot reproduced on Figure 10.1.

Across these six sites they found that earthworms accounted for 54% of diet volume, followed by rabbits,

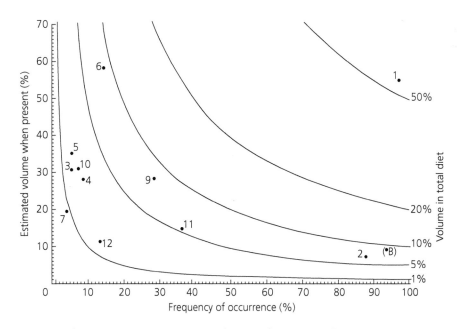

Figure 10.1 Estimated bulk food category whenever it was eaten vs frequency of occurrence, in all study areas. Isopleths connect points of equal relative bulk in the overall diet of the badger. Prey categories: 1 = earthworms; 2 = insects; 3 = amphibia; 4 = birds; 5 = small mammals; 6 = lagomorphs; 7 = carrion; 8 = leaves; 9 = cereals; 10 = fruits; 11 = pignuts; and 12 = fungi (from Kruuk and Parish 1981, with permission).

cereals, insects, and pignuts, in that order, accounting for 28% of food volume. Some earthworms occurred in almost all faeces, and whenever they did they comprised *c*.55% of diet bulk. The predominance of earthworms in the diet varied more between those six Scottish sites than it did within them, and insects were taken mostly in summer, as were amphibia, and there was an autumnal peak in feeding on fruits and cereals.

Against this background, and remembering that we have already mentioned several times that things have changed in Wytham, we offer a recent snapshot of diet. This involved our graduate student Julius Bright Ross, from Easter Island via Harvard, and undergraduate Erin Connelly methodically gathering scats from April to November 2018, and computing monthly relative frequency of occurrence (RFO) for each diet item (Bright Ross et al. 2021). That is, the number of occurrences of a particular dietary item in a month divided by the number of occurrences of all dietary items that month. In addition, updating David's chaetae counting of the 1970s, without the picric acid, we counted

earthworm chaetae present in effluent washed through sieved scats, producing a measure of average chaetae found in 10 1 cm² squares on a Petri dish (after Kaneko et al. 2006; Cleary et al. 2009). Julius and Erin confirmed that earthworms remain the primary prey item for the badgers of Wytham Woods, with 77.9% average monthly frequency of occurrence, FO (and 96% FO in March)—alongside yearlong consumption of arthropods and snails (Figure 10.2b)—but earthworm consumption was substantially lower during the summer months (June–August, Figure 10.2a), with cultivated wheat (*Triticum aestivum var.*, FO = 76% in July) and the annual wild blackberry crop (*Rubus fruticosus*, FO = 84% in August) filling some of this gap. By September, blackberries were supplemented with other fruits (FO = 32% in September) and hazelnuts (*Corylus avellana*), horse chestnuts (*Aesculus hippocastaneum*), and sweet chestnuts (*Castanea sativa*), which continued to be substantial diet components through the resumption of high earthworm consumption in autumn (Figure 10.2b).

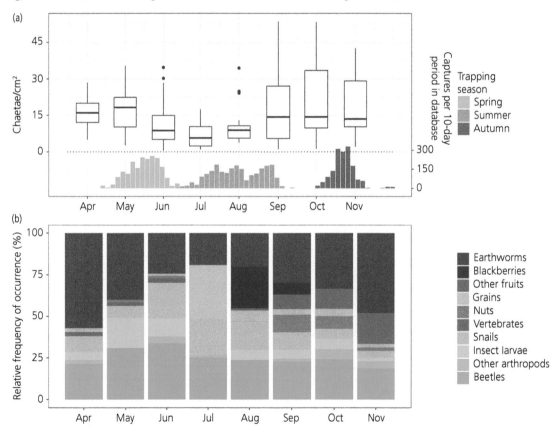

Figure 10.2 Monthly diet for 2018. (a) Average chaetae count in ten 1 cm² squares for each scat sample collected and analysed. (b) Relative frequency of occurrence of diet categories in scats from a month (from Bright Ross et al. 2020, with permission).

To repeat the methodological caveats of Klare et al. (2011; see also Putman 1984), we are hesitant to put too much weight on the differences between this 2018 sample and its predecessors; note, nonetheless, that while worms are still the major component of the diet, they are less overwhelmingly so than in previous years. They account for barely >50% of the RFO in 2 of 8 months. Furthermore, beetles were very popular in 2018, as were summer snails, July grains, and autumnal fruits.

What factors explain this variation in diet from month to month? The answer, illustrated by earthworms, concerns abundance and, more particularly, availability.

Earthworm abundance

With their parallel preoccupations with earthworms, both Hans and David laboriously counted the numbers stimulated to rise to the surface by formalin sampling, working respectively in Wytham and Boars Hill (scarcely 8 km away as the crow flies). The total biomass of worms in Wytham was high: a mean of at least 715 kg of *Lumbricus terrestris* per ha and 255 kg of other worms (Kruuk 1978b). In the vicinity of the Marley and Jew's Harp setts, Hans counted numbers of *L. terrestris*: $1.54\,m^{-2} \pm 0.53$ in young plantations (<15 years); $4.18\ m^{-2} \pm 1.08$ in arable land; $7.0\ m^{-2} \pm 0.71$ in old woodland; and $8.41\ m^{-2} \pm 0.71$ in permanent pasture. This amounts to a colossal biomass of *L. terrestris*; for example, under pasture, the figure is close to 1000 kg per hectare: a massive, delectable vermiforous iceberg of food beneath the surface. Meanwhile, on the agricultural habitats of nearby Boar's Hill David counted *Lumbricus* abundance under four types of field: $14.5\ m^{-2}$ under 20+-year-old pasture grazed by horses; $8.5\ m^{-2}$ under 5+year old grass ley grazed by cattle; and about $1.3\ m^{-2}$ under each of two annually ploughed arable fields—foxes were seen worm hunting in these fields 21 times, 9 times, never, and once, respectively.

Two facts emerge: *L. terrestis* was very abundant, and very variable in its abundance between broad types of habitat (and at a finer scale, e.g. in woodland there were fewer under beech trees than elsewhere). From these two facts, two consequences follow. First, Hans calculated that badgers needed only a small fraction of the *L. terrestris* biomass: a typical 10 kg badger needs about 510 calories day^{-1} (Iversen, 1972; Bright Ross et al. 2021; Chapter 13); if the average worm weighs 4.27 g and contains 17.7% dry matter (Lawrence and Millar 1945) providing 3.983 cal g^{-1} dry weight (Bolton 1969), or 3.01

cal per average worm, it works out that the average badger would need to eat 169 *L. terrestris* per day for its basic energy metabolism.[1] The 25 badgers living, at that time, around Marley and Jew's Harp could live exclusively on earthworms, consuming, collectively 1.9×10^6 or comfortably less than 5% of the *L. terrestris* biomass. Second, while *L. terrestris* abound under these fields of plenty, the badgers face the tantalizing inconvenience that for much of the time they are inaccessibly 2 m deep (see also Box 10.3). Prey abundance is a misleadingly crude measure in the context of predator–prey relationships: the question is neither where the prey is located, nor how many of them there are, but rather where, when, why, and in what numbers do they become available.

Earthworm availability

Earthworms surface only at night—their skin being too sensitive for daytime UV levels. Again, Hans and David worked in parallel, this time armed with torches shielded in red polythene, as they stalked for surfaced worms (in David's case, getting apprehended by the police so often that they ended up bringing him a flask of tea in the patrol car). Summarizing the insights of Kruuk (1978b) and Macdonald (1980d), on Hans' garden lawn, in July–August 1974, the number of *L. terrestris* at midnight varied between 0 and $33\ m^{-2}$. The tally depended on the weather; for example, correlating strongly with rainfall (Figure 10.4) during the previous 24 h. Nearby, on Boar's Hill, David calculated an equation that incorporated hours since rain (h), departure of the temperature from the supposed optimum of 10.5°C (delta T), hours of more than or equal to 95% relative humidity prior to the count (RH) (Figure 10.5), and millimetres of rain that fell during the last shower:[2] the number of worms predicted by this equation correlated closely with the number actually counted on the surface, and the most influential variables were hours since last rainfall and hours of more than or equal to 95% RH. In short, worms were more available as the temperature rose, up to about 17°C: they were scarcely active below −1°C, and most active at 10.5°C (see also Kollmannsperger 1955; Satchell 1967) and they dive underground when even a slight increase in wind

---

[1] Subsequently we settled on an estimate of 2.5cal per *L. terrestris*, but this fine tuning makes no difference to the general principle illustrated here.

[2] Number of worms surfacing, and available to badgers or foxes = 10.31 − 0.3979 (h) − 1.045 (delta T) + 11.43 (RH) − 0.7124 (mm) (the correlation coefficient is 0.73; for the rainfall and RH it is 0.455).

speed causes an exponential decrease of relative air humidity (Monteith 1963). Worms' vigilance and reactivity also affects their availability; the weather, too, impacts on the speed of their reactions: often, a worm venturing onto the surface cautiously keeps its tail anchored firmly in its burrow, while its head prospects for leaves and hermaphroditic liaisons—at the slightest danger it snaps back into its burrow, but the colder the weather the slower its reactions.

The early conclusion, from abundance data, that *L. terrestris* were variably dispersed was writ more clearly in their availability, as microclimate interacts with topography and habitat. In the Marley/Jew's Harp ranges Hans found that in valleys mean minimum temperatures at midnight in November were 2.1°C lower than elsewhere, whereas in July this reversed and valleys were 3°C warmer than elsewhere. Temperature varied by as much as 3.5°C between stands of trees less than 30 m apart, and wind direction threw the entire microclimatic kaleidoscope into turmoil. What were the consequences for predator–prey interactions of this ephemeral landscape of availability? Even a small patch of 20 × 20 m of pasture, at the right place and the right time, could, on bountiful spring or autumn nights, have enough worms on the surface to feed more than 30 badgers for one night; alternatively during a dry summer, it could have none.

So much for nocturnal biologists stalking *L. terrestris*—how do carnivores do it? For David, this story began one cold night in Oslo in 1973 where, above the central railway station (he was visiting Scandinavian carnivore biologists), the illuminated, revolving *Mercedes* news bulletin caught his eye with the fragment '. . .bergen wins Nobe. . .'—several more rotations of the news reel revealed that Niko Tinbergen, Director of Oxford's Animal Behaviour Research Group, had won the Nobel Prize (along with Karl von Frisch and Konrad Lorenz, for their discoveries concerning organization and elicitation of individual and social behaviour patterns). With characteristic generosity, Niko gave part of his prize money to Hans to buy the first pair of Old Delft infrared binoculars from the Dutch army,[3] which Hans, with no less characteristic generosity, in turn shared with David. And so it all began.

The hot-eye, as the infrared binoculars were lovingly known, enabled us to see, in shades of monotonic green, revelations that explained badger (and fox) society in terms of ecological, rather than sociological, selective pressures. The hot-eye revealed that foraging on worms is an intricate business, as David discovered while observing foxes, and investigating their diet, in 1975 (Macdonald et al. 1980b).

Foxes hunt earthworms by walking very slowly, with normal head carriage, ears pricked forward, suggesting they hunt mostly by sound. They pause frequently, often followed by a change in direction. Once the grating of a worm's chaetae on the vegetation is detected, the fox plunges its snout into the grass and grasps the worm between its incisors. If the worm is still partly anchored in its burrow, rather than break the worm the fox pulls its head in a deft accelerating arc to dislodge its quarry. Likely linked to weather and worm availability, sometimes foxes follow a convoluted hunting path: on 28 of 59 observed routes, the fox spent 10–20 min within a 25 m × 25 m plot, catching an average of 2.04 (±1.14) worms per minute. On another 15 occasions the fox took an ongoing foraging path, interrupted by occasional convolutions in particular areas, yielding only 0.63 (±0.39) worms per minute. Thus these tactics were associated with significantly different success rates.[4] This tactical flexibility seems likely to be an adaptation to hunting earthworm prey whose distribution, dispersion, and availability vary from night to night in the same place (reminiscent of those Costa Rican bats).

Badgers do it similarly (Figure 10.3), but principally by scent rather than sound. Hans observed them spending up to 2 h in short-grass pasture foraging at rates between 0.5 and 10 worms per minute. Being lower-slung than foxes, their foraging was more akin to hoovering, picking worms from the surface from immediately below their snouts. The distance between successive captures within a foraging patch could be as little as 20 cm, but usually ranged from 1 to 10 m. Like foxes, badgers caught worms between their incisors, and either pulled them from their burrows with a quick sideways flick of the head and then snapped them up in one movement; or they were pulled out slowly, straight up, over several seconds, then tossed back like a connoisseur with an oyster, with a slight backward and forward movement of the head. Although both badgers and foxes were skilful at extracting worms from their burrow anchorages, 26% of 148 undigested worms examined post-mortem in badger stomachs were missing their tails.

---

[3] Infrared binoculars became relevant to biology when they became irrelevant to the military, due to the invention of image intensifiers.

[4] Sixteen paths did not fall easily into the convoluted/ongoing categories.

(a)　　　　　　　　　　　　(b)

Figure 10.3 Two capture techniques used by badgers to catch worms (a) for lumbricids above the surface and (b) for surface-dwelling species (from Macdonald 1984b, with permission, drawn by the late Priscilla Barrett).

En route from worm hunting to badger society, remember some fundamentals of foraging theory. In the early 1970s, Oxford was the right place to be thinking about foraging—all the citations in this paragraph (not to mention, ultimately, the vast literature on optimal foraging theory; Krebs et al. 1983) spring from the cheery gang of ethologists that populated Niko Tinbergen's Animal Behaviour Research Group, at the time located in a four-storey terraced house in Bevington Road in Oxford, in which it was common to be passed on the stairs by a seagull or a Nobel laureate. There are two fundamental stages in a foraging search: the first being the *search strategy*, which involves a combination of scanning and locomotion effectively to encounter prey with a specific distribution pattern (Smith 1974). The second stage involves *search tactics*—adaptive changes in relation to prey availability (Thomas 1974). One such tactic is *area-restricted searching*[5] (Tinbergen et al. 1967), of which an early demonstration was for carrion crows, *Corvus corone*, by Croze (1970).

We will delve now into the tactics of badgers (and for comparison, foxes) hunting for *L. terrestris*, not only because these prey rank highest in their diet, but also because they illustrate how the predators respond to the spatio-temporal heterogeneity that characterizes, to greater or lesser extents, the availability of almost all their foods. So, while reading about vermivory, bear in mind that it is a metaphor for badger foraging behaviour in general, and that the dispersion of other prey, especially when considered in aggregate as part of an eclectic menu, will have similar implications for badger group living that we illustrate using the vivid case of earthworms (which explains why badgers still live in groups, albeit smaller ones, in places where they eat virtually no worms, e.g. Virgós et al. 2004 and Chapter 19). Within the context of foraging theory, the combined observations on badgers and

foxes (Kruuk 1978b; Macdonald 1980d, 1984b) synthesize to the answers to three questions about the search strategies and tactics of vermivorous carnivores:

1) *How does the abundance and availability of earthworms affect the search strategies and tactics of badgers and foxes?*

In Wytham, the radio-tracked badgers of Marley and Jew's Harp spent most time moving (and presumably foraging) in the habitats in which worm biomass was highest. They made little use of arable fields for catching worms, probably because worm abundance there was low and availability variable. Similarly, 121 sightings of the foxes of Boar's Hill correlated broadly with contemporary worm density (see earlier).

Superimposed onto these coarse search strategies, which took both predators to the habitats richest in this prey, foxes and badgers alike adapted their tactics to microclimate, remembering that relative humidity of >90% and 10.5°C was optimal to encourage worms to surface at night. Thus radio-tracked badgers spent 35% of their night-time activity on pasture (beneath which lived 46% of the *L. terrestris* biomass) during wet weather, but only 2% during dry weather. Overall, badgers foraged especially in the more humid places, where worms were to be found, usually taking to alternatives only when it was dry.

Seasonal rates of capture success paralleled the weather conditions demonstrated to affect *L. terrestris* availability. For badgers, capture rate declined steadily from about three worms per minute in March to under half that in August, creeping back to three per minute in October, and up to a mean of four per minute in November. For foxes, a very similar equation to that which predicted the number of worms available on the surface (see earlier text), also predicted their capture success rate.[6]

---

[5] Area restricted searching involves the forager, having found a food item, slowing down its movements or otherwise behaving so as to remain longer in the seemingly fruitful vicinity.

[6] The number of worms caught per minute (by foxes) closely mirrored the variables predicting worm availability: capture rate per minute = $1.48 - 0.001$ (h) $+ 0.009$ (RH) $- 0.003$ (delta T).

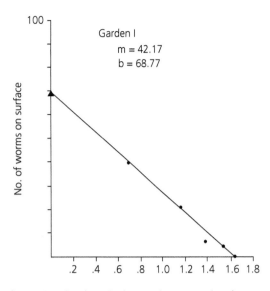

Figure 10.4 The relationship between the mean number of worms (in fifteen 1 × 1 m plots) on the surface and the time since the last rainfall, illustrating the relationship between worm availability and weather (from Macdonald 1980, with permission).

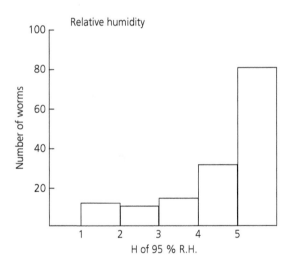

Figure 10.5 Relationship between the number of worms (in fifteen 1 × 1 m plots) on the surface on one lawn with the hours of >95% relative humidity, previous to the count (at about midnight) (from Macdonald 1980, with permission).

Worms are important, but other foods matter too, so both species had alternatives to consider. For example, while radio-tracked badgers visited pasture especially during wet conditions, and while deciduous woodland was always popular, they spent 27% of their time in young conifer plantations, which comprised only 8% of the area and were poorly supplied with worms (although under certain conditions their availability in those locations may have been high). In dry periods they fed on wheat in cereal fields (in recent years agronomic changes led to greater availability of maize). Meanwhile, foxes, in their role here as barometers of earthworm availability alongside badgers, were eating more worms per day as the seasons passed from spring, to summer, autumn, and winter. Figure 10.6 reveals the estimated mean daily intake in grams per fox from several Oxfordshire study sites in and around Wytham where David collected and macerated some 10,452 fox droppings (Macdonald 1980d).

The alternative prey available to both species varied with season. Badgers react to low earthworm availability differently in different seasons: in winter they eat little, and rely on fat reserves and torpor (Chapter 14). In summer, they turn to other foods, principally cereals, but also snails, bees' nests, and blackberries. Although they eat wheat, acorns, and hazelnuts, the undigested remains in scats suggest they digest these poorly. In those days, there were almost no rabbits in Wytham, due to myxomatosis. They abound today but there is no evidence that the badgers consume them (although foxes feast on them), although nowadays they do eat the gralloch of culled deer or moribund squirrels poisoned with warfarin as a woodland pest control measure.

2) Does area-restricted searching occur?

Both carnivores exhibited two search strategies when foraging for earthworms: convoluted and ongoing. The convoluted strategy involved foxes and badgers pacing about within a small hunting area (20 × 30 m) within a large field, exploiting a worm-rich patch. The ongoing strategy involved long winding meanderings.

Badgers following a convoluted strategy typically criss-cross foraging areas, without sharp changes of direction, often searching over the same ground within a few minutes. When badgers foraged for at least half an hour within an area no larger than 1 ha Hans defined it as 'patch feeding', and termed foraging over a larger distance 'long-distance feeding'. David delved into the geometry of foraging, whispering into a Dictaphone as he watched both fox and badger movements through the hot-eye. He recorded every time the predator changed direction—into one of eight 45° sectors relative to its previous direction, noting pauses in movement, vigilance, urine marking, and earthworm captures (this was easier for foxes, whose movements were more angular and punctuated than those of

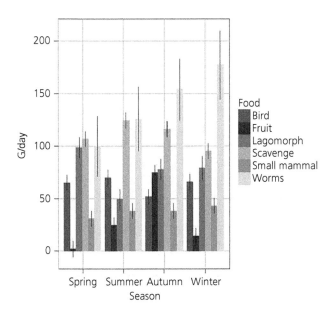

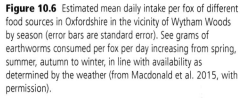

Figure 10.6 Estimated mean daily intake per fox of different food sources in Oxfordshire in the vicinity of Wytham Woods by season (error bars are standard error). See grams of earthworms consumed per fox per day increasing from spring, summer, autumn to winter, in line with availability as determined by the weather (from Macdonald et al. 2015, with permission).

badgers). For foxes this revealed a mean foraging speed of 0.38 m s^{-1}. After capturing a worm, foxes moved a longer beeline distance than they had prior to capture, but did not turn more, pause for longer or move in shorter bursts, irrespective of whether they were, at the time, following ongoing or convoluted tracks. This consistency of foraging geometry was true for the first worm captured in a patch and also for subsequent worms. How did the two predators compare? Observing through the hot-eye and using identical recording techniques, David comparing fox and badger foraging success when using the convoluted tactic in the same field on the same night, under relatively poor microclimatic conditions: a fox caught 2 worms min^{-1}, a badger 1.8. When both used the ongoing tactic, a fox averaged 0.6 worms min^{-1}, a badger 0.4. Foxes sometimes appear to use badgers as foraging guides, feeding on worms successfully in their wake. Badgers will charge foxes feeding in the same area, and foxes will back off, but mischievously resume feeding close by. This might suggest that badgers are better at detecting worm patches (by smell) than are foxes (the latter, by hearing), and that foxes 'cheat' by tracking badgers.

3) Is this foraging behaviour optimal?

When following convoluted tracks, patch foraging, both badgers and foxes depleted the worms in a patch before leaving it. Worm capture success rate usually decreased significantly prior to a badger leaving that patch, after which the badger would generally move on at least 100 m. During a given season, Hans' estimates of the feeding success rates of patch-feeding badgers did not vary significantly between wet (2.28 worms min^{-1}) and dry (2 worms min^{-1}) weather. He concluded that when worms are fewer, badgers try harder.

When a worming fox had depleted a patch, as evidenced from its falling capture rate (worms are both eaten and frightened away by, we guess, the fox's footfall), it would cease its convoluted meanderings and trot to another patch, sometime only 20 m or so away. The fox's 'giving-up time', the interval between swallowing its last worm in a patch and its departure from the patch, varied from night to night: the capture rate at which they gave up was higher when average worm capture rate was higher—that is, on nights with a low capture rate, foxes will persist with foraging in a patch (or even start foraging in a patch) at a rate at which they would have already given up on a night when the average capture rate was higher. This tactic is an adaptive response to prey availability.

When should a worm predator decide to leave a patch? Leaving a patch too soon risks moving to one that is worse, whereas staying too long, for example by following a rule of thumb to persist at each patch for a fixed time, would lead to suboptimal yields from poorer patches. According to Charnov's optimal foraging theorem (1976), the badger or fox should quit if it is either satiated, or when its average success

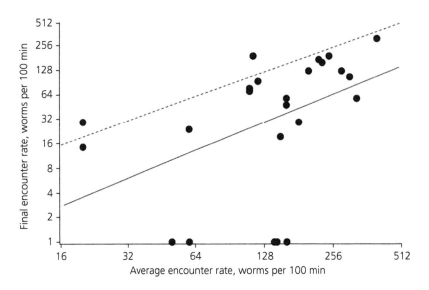

Figure 10.7 Encounter rate with worms before changing foraging location is significantly lower than the average encounter rate for the night. Dotted line shows expectation of the Charnov (1976) marginal value model (from Macdonald et al. 2015, redrawn from Macdonald 1980, with permission).

rate falls below the average success rate offered by patches across the field, while also accounting for travel time between them. Thus when travel time is greater (greater inter-patch distance, or over tougher terrain), the predator should remain longer in its present patch. When considering the time taken to capture the last worm, that was not consistently longer than the average inter-catch time for that patch: in short, the foxes' or badgers' choices on when to quit were not made impetuously on the basis of one prolonged hunt. Rather, it seemed that foxes assessed declining worm captures against the running average success rate. This was revealed by comparing the average time taken to capture five worms with the time taken to catch the last four to six worms: these final capture rates were consistently less than the average, congruent with optimal foraging theory (Figure 10.7). However, according to theory, foxes were staying 'too long' in a patch (regression of success against duration of 0.54 vs the optimal of 1). Possibly foxes know that the number of remaining available patches is few, and so tend to overexploit each one (remembering that the choice ultimately involves also the availability of foods other than worms).

In exactly the same way, when Kruuk and Parish (1981) analysed badger foraging behaviour across the six Scottish study sites, it was at first unexpected that they found no correlation between any indices

of earthworm availability and occurrence in badger diet; and there was no correlation, either, between the number of good 'worm nights' and the volume of worms in the badger diet in the corresponding 2-month period. However, this finding is explicable if badgers do indeed work harder to catch earthworms when fewer are available. Indeed, Kruuk (1978b) noticed that badgers emerge from their setts earlier during dry periods when worms are scarce, under which conditions the badgers swap patch feeding for ongoing tracks, covering longer distances. Further, as David observed during the drought of 1976, badgers have a third tactic: actively dislodging patches of turf with their powerful forepaws to a depth of 5–10 cm to catch surface dwelling *Allobophora* worms (a behaviour that causes conflict with badgers on bowling greens and golf courses; Symes 1989).

In summary, badgers are omnivorous; earthworms, however, when they are available—especially *L. terrestris*—predominate in their diet (see Kowalczyk et al. 2003b). Their search strategies and tactics adapt to the abundance and availability of these prey, which, as likely applies to all their prey, are spatially and temporally variable (heterogeneous) in availability. Part of their search tactics is to opt for area-restricted search when earthworm availability is patchy, and when they exploit patches of these prey they do so in broad accordance with the expectations of optimal foraging theory.

These findings provide a lens through which to appreciate how, with their dietary eclecticism, badger society is an outcome of their foraging behaviour, which itself is an adaptation to the ecological framework created by the resources (particularly food) on which they depend.

Ecology drives behaviour

This brings us to an intellectual turning point in the book—although we are only halfway to the full perspective to be drawn by Chapter 19, we arrive now at the moment for a quantum leap: the weaving together of the strands untangled in the previous nine chapters, to create the understanding of badger behaviour on which the next nine are built.

As a road map, these latter chapters will reveal how the architecture of badger adaptations has been built from the raw material made available to natural selection: the badger's mustelid and mammalian ancestries that set the trajectory of, and the limits to, natural selection's solutions to the challenges badgers face—challenges that are very largely framed by ecology, the topic of this chapter. Consider, first, the analyses of individual movements and how, in Chapter 4, these assemble into subgroups, groups, and supergroups, and the encircling perimeter latrines that once seemed symbolic of aggressive, exclusive territoriality (Chapter 9). Thus far we have phrased many questions, and their unfolding answers, in the context of developmental and behavioural factors, but now comes the moment to link these with ecological factors, and to elaborate on the RDH (Figure 10.8).

RDH in general

Ideas are supposed to begin their journey jotted on the back of an old envelope. This one really did, and the envelope was azure blue. David and Hans Kruuk converged, after a cold night of, respectively, fox- and badger watching, at Hans' home in the Oxfordshire village of Bayworth, for a restorative midnight coffee. Amidst explanation, Hans grabbed that envelope from his desk to draw a series of contorted contours illustrating the indefensibly convoluted shape the territory of a single badger might take if it were to accommodate the patches of earthworms (and other seasonally important food resources), erratically dispersed (i.e. in terms of space and in nightly variability) as these are, which the badger would need to achieve sufficient food security to survive for a year. In contrast, a further flourish of his biro encircled a larger, approximately circular territory with an efficiently short, and easily remembered, perimeter, so much easier to defend, but this territorial efficiency brought with it extra worm patches, such that on any given night it might support a clan of badgers willing to share, although not necessarily to compromise, their individual food security to gain the advantage of reduced food defence costs.

That germ of an idea, much polished in the intervening decades, was formally applied to badgers in 1978 (Kruuk 1978b), foxes in 1981 (Macdonald 1981b) (see also Kruuk and Macdonald 1985), formalized as the Resource Dispersion Hypothesis (Macdonald 1983), and has been variously reviewed (Johnson et al. 2002b); it was still going strong in 2015 (Macdonald and Johnson 2015a), its tendrils reaching from carnivore sociality through human societies, species diversity, and the Black–Scholes model of financial strategies. By that time it had been the subject of several mathematical explorations (Carr and Macdonald 1986; Macdonald and Carr 1989; Bacon et al. 1991a, 1991b; Blackwell 1990, 2007, and been invoked in dozens of papers, including promising applications to such carnivores as brown hyaenas (Mills 1982), giant otters (Groenendijk et al. 2014), lions (Valeix et al. 2012), Blanford's foxes (Geffen et al. 1992), dingos (Newsome et al. 2013), coyotes (Wilmers et al. 2003), and pumas (Elbroch et al. 2016), not to mention a non-carnivore, prairie dogs (Verdolin 2009). Good whisky takes a while to mature, and even in the months of working on this chapter the rate of RDH publications has accelerated: lions (Mbizah et al. 2019), cheetahs (Broekhuis et al. 2021), red foxes (Tolhurst et al. 2021), and even a sideways look at wolves and ravens (Vucetich 2021). Most satisfying, if poignant, Pall Hersteinsson, the first of David's students to become an early adopter of RDH (Hersteinsson and Macdonald 1982), has left a remarkable academic legacy from his distinguished professorial career in Iceland, following his untimely death in 2013—and as this book goes to press his scholarly descendants have published remarkable new findings on Arctic foxes and the 'Hersteinsson emphasis' of RDH moderated by the impact of predation (Erlandsson et al. 2022). Not only has the list lengthened of species to which RDH has been applied, but the hypothesis has been developed in new ways; for example, incorporating the relevance of body fat (Newman et al. 2011), leg length (Macdonald and Newman 2017), fossoriality (Noonan et al. 2015a, fission–fusion (Mbizah et al. 2019) and parasite management (Alberry et al. 2021)—all topics that will surface later in the chapter.

Figure 10.8 How the Resource Dispersion Hypothesis works. (a) If resource patches have a certain probability of availability, then several must be simultaneously defended to guarantee some probability of finding enough food for a primary pair of residents ($2R\alpha$) in a given period. A frequency distribution of availability across all patches (here, arbitrarily, $n = 1$–14) indicates the proportion of nights on which the total amount of resources available will exceed $2R\alpha$. A secondary can join the territory when their own resource needs ($R\beta$) are met on top of those of the primaries (i.e. $2R\alpha + R\beta$). The integral of the distribution illustrates the 'critical probabilities' (Cp), the proportion of times that such conditions occur for primaries (Cpα = 0.95 (upward hatching)) and secondaries (Cpβ = 0.90 (downward hatching)). Wherever these two distributions overlap (i.e. the cross-hatched area), both primaries and secondaries attain their food requirements. Changing the shape of the distribution will not alter Rα and Rβ, but it will alter the critical probabilities associated with them, leading to a different prediction for group size. (b) Two superimposed graphs, similar to that in (a). The taller curve represents a territory in a relatively invariable environment with a low mean resource availability (R1). The flatter curve, by contrast, corresponds to a territory in a more variable environment with a higher mean resource availability (R2). The area under each curve is the same (1.0), and represents the total probability of all the possible levels of availability. In both cases, each curve represents the distribution of resources from the minimum territory required to satisfy a given Cpα. The crucial difference is that Cpβ (the probability of achieving $2R\alpha + R\beta$) is much higher with the flatter curve (90%) than with the taller curve (72%), so secondary animals are more easily supported in the more heterogeneous environment. More variable environments will, therefore, be able to support larger group sizes (from Macdonald and Johnson 2015a, reproduced from Carr and Macdonald 1986, with permission).

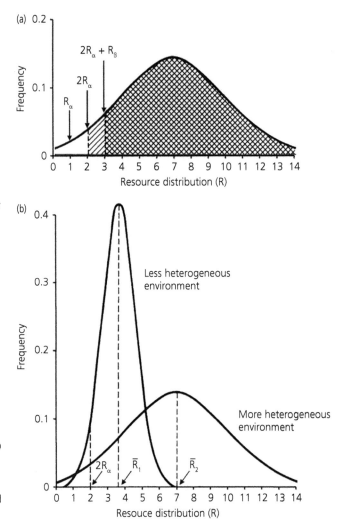

For context, when is a gathering a grouping? As a generalization, greater prey abundance per unit area (or, more correctly, that proportion of it which is available) equates to a higher environmental carrying capacity (see Chapter 11), leading to higher consumer population densities and, pro rata, smaller individual ranging areas (Macdonald 1981a; Gittleman and Harvey 1982; Reiss 1988). Solitary tendencies can, temporarily, be compromised by particular environmental circumstances, leading to a gathering but not a grouping (Eisenberg 1966). Think of the vultures congregating at a kill or, around Wytham, the two dozen or more pair-living red kites assembled from upwards of 10 km away to follow a plough to reap the harvest of earthworms that would moments before have been beyond their, or a badger's, reach. Most beautifully, the second

2015 issue of the *Journal of Zoology* depicts Eric Murray's extraordinary photo of 20 black-backed jackals feeding in a floral array from a Cape fur seal carcass near Walrus Bay in Namibia (Photo 10.1)—jackals that put aside their generally monogamous and territorial animosities to aggregate at one rich food patch, for just as long as it pays each of them to do so (Macdonald and Johnson 2015a; see also Jenner et al. 2011), much like grizzlies at a salmon run or polar bears hauled up together in Churchill (Alaska). Such congregations lack the social structure (of the sort we sought amongst badgers in Chapter 4) and relationships that define a society (Alexander 1974; reviewed for carnivores in Creel and Macdonald 1995). But it is not always easy to diagnose society and distinguish a group from a gathering (Creel and Macdonald 1995), and although

Photo 10.1 black-backed jackals feeding from a rich patch, a cape fur seal carcass (photo by E.J. Murray, with permission).

scavenging around a carcass (Lamprecht 1978) need involve neither coordination nor integration (Leyhausen 1973), beware that seemingly unconnected comings and goings can belie the subtly associated elements of a fission–fusion society (Aureli et al. 2008).[7]

Beware also that spatial groups (those whose members move separately through congruent ranges), may be highly structured but sufficiently discreet in their association to go undetected: red foxes had been held up as the paradigm of solitary (implicitly anti-social) carnivores in both naturalistic hunting lore and the scientific literature, until radio-tracking revealed their spatial groups (Macdonald 1981b, 1987; Macdonald et al. 2015b); moving to musteloids, and thus closer to badgers, similar spatial grouping characterizes some groups of raccoons (Prange et al. 2011). Spatial groups

bring us to the fundamental social unit of an RDH society.

The essence of RDH is that groups can develop where those resources the species utilizes are dispersed such that the smallest economically defensible territory for the primary occupant(s) can also sustain additional animals at a tolerable cost, even in the absence of any direct benefits of group living per se (Macdonald 1983; Carr and Macdonald 1986; Macdonald and Carr 1989; Bacon et al. 1991a, 1991b). As embodied in the title of David's 1989 paper, *The Rewards of Tolerance*, co-authored with Geoff Carr (who went on to be Science Editor of the *Economist*), for each member, group living need not be 'cost-free' but just less costly than the alternative, where benefits (e.g. greater resource security) must simply outweigh costs (e.g. sharing resources) (cf. Roper 2010). Of course, this may require that the species in question has some tolerance for deprivation, where the ability of badgers to store and then burn fat reserves to fuel them through lean times facilitates their RDH society. Indeed, fat turns out to be crucial to explaining not only why musclimorph badgers can adapt to spatio-temporally patchy prey by forming groups whereas their skinny-morph marten cousins, to whom litheness is survival, cannot, despite their similar omnivorous diet (Newman et al. 2011). It is also crucial to the pace at which Wytham's badgers tackle their lives and, remarkably, the different phenotypic strategies they adopt (Chapter 14).

It is axiomatic that, to make membership of a group worthwhile, the fitness benefits obtained by individuals within that group must outweigh the costs of sharing (food) resources, without jeopardizing each other's minimal (food) security (Koening et al. 1992;

---

[7] Mindful of the badger cliques, groups, and super-groups described in Chapter 9, the terms fission–fusion and multi-level are only partly synonymous descriptions of society, because in the latter the lines of fission cleave subgroups of broadly consistent membership (whereas the former can involve fluid regroupings, e.g. lion hunting parties; Mbizah et al. 2019). Papageorgiou et al. (2019) caused surprise by showing that flocks of vulturine guinea fowl emerge from their roost to split up routinely into subgroups of consistent membership—illustrating a complexity of social awareness hitherto not associated with brains as small as those of birds. Similarly, large herds of capybara routinely coalesce from, and then fragment into, the same component groups around dry season water (Macdonald 1981a). The inter-annually stable spatial geometry of badger groups and super-groups, and associated sticky but flexible social networks of their members (including their cliques) put the badgers of Wytham Woods into the realm of a multi-level society.

Cockburn 1998). The accountancy of these costs, and benefits, involves those of a given individual either staying or going, calculated from the viewpoints of each stakeholder (e.g. itself, its parents, its siblings, cohabitants, its mating partner(s) etc.) (Macdonald and Carr 1989; Silk 2007). Membership can be worth it because the costs of going it alone are unsustainable, or involve risk best not taken (thus the RDH can explain why offspring might not disperse (cf. Roper 2010, p. 253). For instance, and as exemplified by giant otters, *Pteronura brasiliensis* (Groenendijk et al. 2014), when ecological constraints (the availability of meanders in which to fish) on independent reproduction promote group living in a landscape of group-held territories (Kruuk and Macdonald 1985; Hatchwell and Komdeur 2000).

At its simplest, resource (e.g. food) dispersion not only in space (the latter existing anywhere on a continuum from homogeneity to heterogeneity, but easily thought of as patches of availability), but also in terms of temporal variability (renewal rate, Waser 1981; even inter-annual cyclicity, Lindstrom 1986) is hypothesized to determine the size of territories, and patch richness, in turn, to determine the number of individuals that can viably cohabit the territory. Insofar as patch richness and patch dispersion are independent, larger groups need not necessarily maintain larger territories (e.g. Macdonald 1983; Carr and Macdonald 1986) if resource patches are tightly packed, whereas small groups may require a large territory if resources are scattered.

In Chapter 8 we asked how social groups might develop in the absence of any, or minimal, cooperative benefits (Kruuk 1976). In answer, the RDH provides a parsimonious functional explanation for how spatial congruity of shared ranges can arise from selective pressures in the absence of any sociological benefit of cooperation (Carr and Macdonald 1986). In its minimalist form, cohabiting adults who do nothing cooperative but merely share a home range are termed a spatial group (Macdonald and Johnson 2015a).

In principle, spatial groups might form and reform with fluid membership from one feeding period to the next, and this may be exactly the case with Bradbury and Vehrencamp's (1976) embellanurid bats. More commonly, at least amongst mammals, origins from maternal care, sibships, social relationships, and genetic ties (offset by inbreeding avoidance; Chapter 17) militate against such fluidity, and the shared purpose in their inclusive fitness will add stability to group membership. The companionship may (but does not have to) then evolve so that the

opportunity for cooperation may facilitate enhanced fitness amongst its members; Chapters 1–9 confirm that while Wytham's high-density badger population has arranged itself into spatial groups, their journey down the road from cohabitation to sociological advantage has not required them to progress beyond the rudimentary (the question of whether that journey is complete, and the implications of the answer for understanding the phenotypic and genotypic elasticity of badger adaptation, is tackled at the end of this chapter). Other regions, with lower environmental carrying capacity, and especially a lower availability of earthworms, support lower population densities of badgers exhibiting less gregariousness (Johnson et al. 2002a) (Chapter 19).

There is a variety of circumstances under which the conditions described by RDH could prevail, but a binomial version of this idea is easily envisioned as a game of dice (see Box 10.1 for detail, but, simply, imagine you can eat in a given patch only if you throw a 6).

Box 10.1 RDH and dice

Imagine a game of dice in which one die is a food patch and one throw determines the food available for one night. Let's say we can only eat if we throw a six, and that throwing a six secures enough food for one animal for that night. Obviously, having six sides, one die gives a one in six or 16.7% chance of eating. It would be imprudent to gamble your fortunes, far less fitness, on such low odds (although your ability to store fat will affect your judgement). Given the choice, you might refrain from joining the game unless you had, say, at least an 80% security of eating each night; that is, an 80% chance of throwing a six. How many food patches, or dice, do you need to secure this level of food security? Because each six-sided die carries a five out of six chance of failure you need nine dice for 80% food security $(1 - (5/6)^9) = 0.806$). Of course, the principle is obvious: that having secured the nine dice needed in order to eat at all on 80% of nights, on many nights extra sixes will be thrown, on which it is possible for primary occupants to share the territory with secondary occupants at little or no cost to themselves. This is the binomial (i.e. all or nothing) version of RDH proposed by Carr and Macdonald (1986). It illustrates how groups may develop where resources are dispersed such that the smallest economically defensible territory for a pair, which we refer to as a 'minimum territory', can also sustain additional animals—in the dice game, feeding periods during which more than one six is thrown (Macdonald and Johnson 2015a).

In terms of how an RDH-based socio-spatial geometry may arise, even a single animal (or a breeding pair, or a mother with offspring) using patchy resources must defend an area large enough to achieve a critical probability that at least one 'ripe' patch will be available to satisfy its resource requirements over any given time (feeding) period.[8] While supporting these *primary occupant(s)*, this area will likely yield an excess of resources at least some of the time, sufficient to sustain additional animals (*secondary occupants*). If access to food is controlled hierarchically, secondary occupants are tolerated on, and are tolerant of being situated on, lower rungs of the ladder. However, in Chapter 5 we show that badger groups in Wytham are scarcely hierarchical, and so the burden of patch depletion may be shared more equitably (of course the naturalistic reality is that a multitude of factors beyond foraging ability govern how fat each badger is; Bright Ross et al. 2021; Chapter 14). Subordinate secondaries, or the throng of compromising badgers, might experience a lower food security than the original residents (or, in a more egalitarian or anarchic model, attain a lower average food security across all residents), but provided that this does not compromise an individual's fitness below the threshold equivalent to the effort of trying to defend individual territories, membership of the group is a better option than dispersal. Minimal food security is likely to be different for different group members (lactating females, pre-dispersal adolescents—which have different body fat tolerances and energetic expenditures; Chapter 14). Members, generally subordinate ones, must be able to sustain periods of food deprivation arising from unlucky throws of the dice (see Box 10.1); a requirement that highlights the aforementioned importance of badger fat (remember a marten must eat at least every 72 h, whereas a badger can, *in extremis*, survive 6 months without food;[9] Newman et al. 2011). This requirement is probably particularly critical for young and old animals that are less able to tolerate food deprivation (see Chapter 14; e.g. Kays and Gittleman 2001)—ultimately setting the ceiling on optimal group size and membership composition.

Furthermore, in philopatric societies (delayed dispersal; Chapter 4), the costs of cohabitation will be offset by benefits arising through vested interests that each member may have in another through their coefficient of relatedness. Inclusive fitness can often drive inflated benefits (e.g. Waser and Jones 1983), potentially involving kin selection (von Schantz 1984), cooperative breeding, and delayed dispersal (Kokko and Ekman 2002), along with any other rewards of cooperation (as reviewed in Chapter 8). Sometimes these rewards, inflated by kinship, outweigh the costs of subordination, which can include suppressed reproduction. Sometimes, too, dominants gain fitness from the efforts of subordinate helpers, an equation that may be solved to the advantage of overworked mothers in need of a little helper (which may include males; Woodroffe and Vincent 1994)—at least, this is true amongst lean, agile species whose lifestyles do not allow them to store fat—but not in badgers (Woodroffe and Macdonald 2000; see also Briga et al. 2019), which are able to accrue capital through body fat storage and can rely on their own reserves to bridge the care gap (see Heldstab et al. 2017). Finally, the decision of whether to stay or go has decisive fitness implications for both the individual concerned and its parents and relatives, because if cohabiting with primary occupants, that is, sharing their territory, is not an option, itinerant individuals that fail to secure their own territories will have zero lifetime reproductive success (for instance, as seen in our work on beavers, *Castor fiber*; Campbell et al. 2017).

The RDH has been criticized for lacking falsifiable predictions (see Macdonald and Johnson 2015a), although this is manifestly incorrect. The RDH provides clear, testable predictions (collated in Johnson et al. 2002b). Amongst them: (i) territory size (TS) can be independent of group size (GS); (ii) TS is determined by the dispersion of resources (increasing with inter-patch distance), whereas; (iii) GS is determined by resource heterogeneity (spatio-temporal heterogeneity means the whereabouts of available food varies from place to place and time to time); and (iv) by the total richness of available resources (i.e. patch richness). In summary, these have been tested (e.g. Mbizah et al. 2019), but they are certainly difficult to investigate under natural conditions. Such difficulties led Revilla (2003a) to argue that the RDH remained a nebulous and unproven explanation for group living (Revilla 2003b). Emilio Revilla is a valued colleague (e.g. Periquet et al. 2016) but in this case Macdonald and Johnson (2015a) took a different view, in part because the RDH has garnered substantial formal mathematical support

---

[8] The metaphor 'ripe' for a patch's availability is exactly apt for the fruiting trees that, according to RDH, facilitate kinkajou group living (Kays and Gittleman 2001). Exactly the same model can apply to frugivorous primates and birds.

[9] In Holland, to close down a badger sett in preparation for building development, an exclusion gates system was used so that animals that left could not return. After 6 months a mechanical digger excavated the sett, and found two badgers alive, albeit thin.

(Carr and Macdonald 1986; Bacon et al. 1991a, 1991b), but more fundamentally because of the strength of its basis in logic. Contrary to Revilla's (2003a, 2003b) supposition, the RDH does not automatically predict that animals will live in groups wherever resources are heterogeneous (it has to pay them to do so, and whether or not it pays is dictated by the availability of alternatives and by the sort of animals they are). The RDH, instead, is a facilitating, rather than a causal, factor leading to group formation. The RDH typically provides the most succinct and economic explanation for group living amongst competing hypotheses. While other, more complicated, solutions may ultimately prove correct, in the absence of certainty, the fewer assumptions that are made, the greater the veracity of the explanation (Macdonald and Johnson 2015a), or, as William of Ockham (*c.*1285–1349) put it '*Pluralitas non est ponenda sine necessitate*' or 'Plurality is not to be posited without necessity' (Sober 1981).

RDH: the badger case

Although often living in quite large groups (in Wytham the largest was 29, recorded in 1997), badger groups are less cohesive, less conspicuously a superorganism than the amoeboid mega-mongoose created by a writhing mob of meerkats or sleep heap of bush dogs (see Chapter 4); for instance, they typically forage alone, unless at particularly rich feeding patches. Their regional specialism on eating earthworms has made them the epitome of an RDH-based society. For context, larger carnivores eat larger prey, and species over a 14–20 kg threshold need to eat much larger prey (Carbone et al. 1999, 2007; Macdonald and Newman 2017). In contrast, small carnivores can balance their energy budgets by eating very small prey. Consequently, operating as a pack is not sustainable amongst small carnivores (Cohen et al. 1993), unless prey richness and renewal rate is very high—a condition usually only fulfilled by insectivory, and exemplified by social mongooses (Macdonald and Newman 2017), and piscivory, amongst social otter species (Lélias et al. 2021) (Chapter 19). Consequently, Johnson et al. (2000) and Noonan et al. (2016) generalize that the greater the contribution small mammals make to the diet, the more solitary and less gregarious small carnivore species tend to be.

Considering badger species further afield, with our Chinese colleagues, we find that diet and food dispersion are precursors to group living amongst various other badger species throughout Asia (Zhou et al.

2008, 2015a, 2015b; Zhang et al. 2009, 2010), with more carnivorous badger species being more solitary (reviewed in Zhou et al. 2017). What then, of vermivorous badgers?

Remember, earthworm availability can fluctuate rapidly in response to prevailing weather and therefore habitat patches variously rich in earthworms, or other seasonally vital foods, become available in different habitats at different times, making patch productivity variable. Of course, important as they are, worms are not the only relevant, patchy prey for the eclectic badger diet. The remaining third or so (seasonally near 100%) comes from including patches of brambles, dead wood replete with beetles, and hazelnut trees in a resource mosaic. The original sketch on the back of that azure envelope depicted a minimum territory for one individual (Kruuk later argued that five food patches would be required); a territory that could be so extensive and highly contorted as to be uneconomic to defend. Kruuk's (1978b) original principle was therefore that group formation leads to less convoluted, more circular territories that encompass food patches that are hard to partition but cost-effective to share (under these circumstances, the serpentine irregularity that weaves to link a fraction of each of five patches becomes more like a tidy quincunx).

A hill top, towards which a chill wind might blow from any direction, perhaps provides an even more intuitive visualization. Thus, on any night, or even any hour, the north, south, east, or west slope of the hill might be windward, and the opposite side be leeward. Earthworms surface and thereby become available only when sheltered from the wind, and so a prudent badger would, in configuring its territory, encircle the hill, thus ensuring it always had access to a leeward, bountiful patch whatever direction the wind took. When, comforted by the microclimate, earthworms surface in that sheltered patch, they do so in such numbers (reaching >20 worms m^{-2}, at 22 kJ g^{-1} fresh biomass—Macdonald 1984b) that several badgers can forage there without getting in each other's way. QED. Under these circumstances of spatio-temporal heterogeneity in the dispersion of available prey, the dispersion of the patches would determine badgers' territory, while their richness determined group size. Further, in drought summer, no worms surface, so a seasonally viable range must contort to include the maize field or the blackberry bushes on the scrub ground—the principle of spatio-temporal patchiness is the same.

Does it hold water?

At a broad scale, mindful that 37% of their diet does not consist of worms, but is patchy, various links have been demonstrated between earthworm biomass (although what really matters is availability), patch dispersion, range size, and group size.[10] Comparing badger populations across Britain, Kruuk and Parish (1982) demonstrated that in areas where worm-rich habitat patches had a more scattered distribution, badger ranges were larger. Kruuk and Parish (1987) also reported that, in a Scottish badger population, as worm biomass decreased over a 5-year period, there was a concomitant (although not statistically significant) decrease in the mean number of badgers per group, but no change either in patch dispersion or in group range size. Similarly, working in predominantly coniferous habitat in upland north-east England, Palphramand et al. (2007) found, consistent with RDH, significant positive correlations between badger range size and the number of grassland patches, but negative correlations with the proportion of grassland.

Analysing long-term data from Woodchester Park (Gloucestershire, UK), Robertson et al. (2015) found that territory size and group size were positively related, and contended that this contradicted a core prediction of the RDH. But this prediction applies only where patch dispersion and richness are themselves uncorrelated. In this case, across their 7.3 km^2 study area, rich food patches were distributed fairly evenly—circumstances forcing group size and territory size to correlate in that instance.

In a meta-analysis, Roper, examining a number of European studies (2010, pp. 217–218), himself unconvinced by RDH (Roper 2010, p. 249), found that higher-density badger populations tend to have larger group sizes, but smaller territory sizes, noting that the relationship between group size and territory size was not linear. The RDH offers a mechanism explaining this interaction: the sizes of groups and territories are determined by different factors (patch richness and dispersion, respectively).

Why was Tim Roper unconvinced by RDH? RDH in its contractionist form, builds a case from the circumstance of primary occupants of a minimum defendable territory size, but Roper (2010) (if we understand him correctly) contended that this starting point was faulty, in that the primary occupant(s), a founder badger, would try to defend an excess of territory as insurance against future lean times. This is the same argument as proposed thoughtfully by Torbjörn von Schantz (1984) with respect to red foxes configuring territories in anticipation of inter-annual troughs in Fennoscandian vole availability, and is therefore an incorrect criticism of RDH for the same reasons (Macdonald 1984a). First, at an extreme where variability in resource availability is entirely temporal, this simply illustrates a temporal, rather than spatio-temporal, variant of the RDH principle—and flushing out this extreme is why von Schantz's intervention was thoughtful (Carr and Macdonald 1986). Second, while the minimum defensible territory is the spatial unit that illustrates circumstances under which secondary occupants can be accommodated alongside primaries at no cost, even if without cooperative advantage, (i) that minimum is defined in terms incorporating the survival of lean, 'bottleneck' periods, and (ii) RDH is untroubled by animals expanding their holding beyond that minimum, as clear in arguments for expansionism (Kruuk and Macdonald 1985; Macdonald and Johnson 2015a) (although that is distinct from the concept of super-territoriality, which is problematic under any model; Verner 1977). As it happens, Roper's construct here assumed (in common with almost everyone, including ourselves in earlier years) active territorial defence and exclusivity, which turns out not to be the case in Wytham (Chapters 4 and 9). Amongst his reasons for rejecting RDH are that latrines are concentrated in the area directly between neighbouring group main setts, traditionally taken to define defended boundaries, whereas, Roper argued, if the asset being defended were rich feeding patches, it would be those patches, rather than the borders, that would be ringed by latrines. Therefore, he concludes that resources are not the basis for group segregation. Difficulties with this argument are: (i) the original thinking behind RDH that it is more efficient to use an approximately circular (actually hexagonal) area than a convoluted one (even if each patch were individually demarcated); (ii) Chapter 9 reveals that it is, anyway, hard to argue that latrines function in the traditional deterrent way

---

[10] Although these results were originally couched in terms of territory size, the revelations about Wytham's badgers in Chapter 4 incline us to use now, instead, the term range size when referring to results from Wytham: we therefore use that term to describe the shared area bounded by border latrines (that was hitherto interpreted as an aggressively defended, and thus classically territorial, border). Discussing results published by authors elsewhere, we retain the use of the word territoriality.

(more likely they are communication outposts or way-markers;[11]) and (iii) defence is not an integral tenet of RDH.[12]

Towards a Wytham synthesis

At Wytham itself, we've attempted to untangle RDH. Changes to the European Union's Common Agricultural Policy (CAP) in the 1980s[13] prompted the first step. Remember that these years saw the Wytham badger population size grow, accompanied by partial fissioning, and in some cases complete fissioning of such social groups as Gates from MT, Great Ash Hill from Great Wood, Firebreak and Chalet Outlier from Great Ash Hill, Chalet from Hill Copse, Great Oak Outlier from Great Oak, Sunday's Hill from Brogden's Belt, Marley 2, Upper Follies, and Botley Lodge from Marley Main, later McBracken from Marley Main—all mapped in Chapter 4. The broader context was macro-economic forces driving land use change with the consequence that earthworm-poor arable land was converted to potentially earthworm-rich pasture. To the earthworm aficionado, amongst whom the badger trumps the ecologist, whole fields are only a crude identifier of food patches (they are often only 10 m or so in diameter, and shaped by topography more than field boundaries), but insofar as fields loosely indicate patch dispersion, the mean inter-patch distance of 270 m in 1974 (from Kruuk and Parish 1982) shrank by 25% to 203 m in 1987 (da Silva et al. 1993). In accord with RDH, mean group range size shrank from 0.9 km^2 in 1974 to 0.3 km^2 in 1987, with associated changes in latrine boundary configurations (remember Chapter 9). Concurrently, the number of social group territories increased from 12 to 16 between 1974 and 1987.

The conversion of worm-poor arable fields to worm-richer pastures was a blunt instrument with which to assess the impact of resource patch dispersion on the badgers of Wytham Woods, but nonetheless the observed changes matched our intuitions in the field about how RDH was working. Further, we expected that these changes in the landscape of resource availability would have reverberating consequences for individual badgers. Was this period of changed land use,[14] resource dispersion, and social group configuration associated with other changes in the badgers' socio-spatial arrangements? Yes. It also revealed that while pasture was an excellent habitat for hunting earthworms when conditions were right (and was good for hunting carabid beetles too), it was a much more specialized, and homogeneous land use than deciduous woodland. Deciduous woodland (in which badgers are much harder to watch than in the open arena of pasture) was also well supplied with earthworms, was topographically, and thus microclimatically, more varied, and was also the source of a much wider menu of alternative foods.[15]

Remember also in terms of earthworm abundance and availability that deciduous woodland is threaded with grassy rides, passing through warm, humid dells and sheltered from the wind—the best of conditions in some pasture patches (indeed at different years, see the text that follows, deciduous woodland housed, by a wide margin, or close to, the greatest abundance of earthworms). It was in deciduous woodland that radio-tracked badgers spent disproportionately most time, and it was the extent of deciduous woodland in a group's range that had the greatest ramifications for its occupants. First, the number of cubs born per territory and the number of breeding females also correlated positively with the proportion of a territory composed of deciduous woodland. This increased cub production appeared to be caused by an increase in the number of females breeding in territories incorporating more deciduous woodland (da Silva et al. 1993). At first it looked as if there were no significant relationship

---

[11] For an interesting instance of waymarking amongst wood mice, see Stopka and Macdonald (2003).

[12] If the function of border latrines is defence through deterrence, then they are sufficiently ineffective in this respect to pass the test of natural selection (the dysteleological argument) with 19.8% of the Wytham badger population changing group association between our seasonal trapping sessions (1987–2003; $n = 5255$ capture events; Macdonald et al. 2008, Chapter 3), i.e. nearly 50% of paternity coming from across the latrine divide (Chapter 17).

[13] Some relevant detail is that the UK joined CAP in 1973, which in the 1980s caused Wytham Estate farms to switch from crops to pasture (background in Macdonald and Feber 2015a, 2015b).

[14] Wytham is a natural laboratory (see YouTube: the Laboratory with Leaves, episode 14, https://www.youtube.com/watch?v=NrQutaWARrw), but the reality of nature is always untidy in comparison to a genuine laboratory. Amongst the myriad things that doubtless also changed during these years of CAP-induced land use change was that the population of fallow deer in Wytham grew to upward of 400, obliterating the woodland understorey until the introduction of an annual cull in 19nn<AU: correct year> reduced the deer to closer to 40. We cannot disentangle exactly how these ecosystem changes impacted the badgers (but see Buesching et al. 2011), but doubtless they did.

[15] By analogy, consider each food type as a 'shop': butcher, baker, grocer, sweet shop, fruiterer—together they provide the necessities for one badger, but each is sufficiently well stocked to sustain its companions too, in a spatial group.

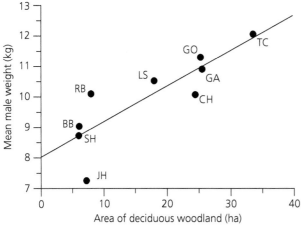

Figure 10.9 The relationship between the mean weight of adult males and the area of deciduous woodland in a territory. Linear regression: $y = 0.12x = 8.03$, $r2 = 0.70$, $F1.7 = 16.6$, $P < 0.005$ (from da Silva et al. 1993, with permission).

between the area of deciduous woodland in a group range and the number of lactating females inhabiting it, but this relationship became apparent when we excluded the Great Ash Hill group from the analysis (as an outlier, this time in the statistical sense).

Were there other sociological corollaries of access to deciduous woodland? Yes, in 1988 a very strong relationship with the average body weight of adult male badgers (9.7 ± 1.6 kg, across 9 social groups), which were clearly heavier in territories with more deciduous woodland (Figure 10.9).

Because we could not find all the earthworm patches, far less measure the availability of earthworms within each (it was still less feasible that we could measure the availability of all the other foods), we could not straightforwardly measure the relationship between group size and patch richness. This thwarts direct testing of the central RDH prediction regarding group size, but there are ways of shining a sideways light on the richness of available prey. First, as a rough proxy, we estimated the abundance of earthworms in each group range (thinking that abundance would be coarsely linked to availability, although indicating little about patch dynamics). These detailed delvings into earthworm counts are summarized in Box 10.2.

Two caveats thread through the results of the Hofer, da Silva, and Johnson theses (and the details in Boxes 10.2 and 10.3): (i) it is difficult to measure group sizes (Chapter 3); difficult, also, to measure earthworm availability, and even harder to measure that of most other foods. These difficulties are multiplicative. (ii) It is not obvious when the right moment arises to measure either group size or food availability, in order to test RDH—the moment that matters, at which resource richness bites into group size, is the bottleneck: the

mean territory productivity or duration of food scarcity at the *leanest* times of the year will determine sustainable group size.

Earthworms are scarce with drought and frost, whereas fruits, nuts, and cereals are highly seasonal. Consequently seasonal and inter-annual weather patterns result in food resource bottlenecks—to which badgers are adapted through fat storage (Chapter 14) and the use of torpor to lower their metabolic rate (Newman et al. 2011). Further, it takes badgers time to adjust to a change in resource availability: they may gain or lose weight faster than they breed or starve; ultimately, group sizes evident at any single moment are not the product of food that is available at that moment, but more likely the product of not only mean annual food productivity but also mean productivity over the inter-generational period for the population (5.6 years; Macdonald and Newman 2002; see Chapter 11).

The question reverberating through Chapter 4 was whether things had been changing. In terms of the spatial relationships of the badgers, the answer in that chapter was clear: yes. Now we have the first evidence that things were also changing extrinsically, in their ecological circumstances. Land use was changing, arguably in favour of worm abundance, although the abundance and size of worms per unit of overexploited land may have declined, but remained so vast that, availability allowing, they could sustain hordes of badgers with only a trifling impact on their numbers.

To date, no study has attempted to measure the richness and dispersion of *all* the food patches relevant to the complete diet of badgers. That prospect is technologically huge, but as a start, given the seasonally prominent role of earthworms, whose availability can reach >20 worms m^{-2} at the soil surface (Macdonald

Box 10.2 Earthworm counts, badger territories, and groups

Earthworm abundance was not related to group size in 1988 (our first full year of trapping, yielding accurate demographic data; da Silva et al. 1993). That year the tally of badgers per group typically ranged from 6 to 26 (mean 11.3; adult range 4 to 19, mean 8.4) and estimated earthworm biomass per group range varied from 14,050 to 33,500 kg (mean 24,445 kg). To estimate the earthworm biomass we used coefficients calculated by Heribert Hofer (1988). He had used formalin flushing to establish crude earthworm abundance for key habitats; for example, 971 kg of earthworms ha^{-1} in pasture, 837 kg ha^{-1} in deciduous woodland, but just 175 kg ha^{-1} in conifer plantations. Because of all the same unknowns regarding food availability that prevented Jack da Silva from testing directly whether group size correlated with patch richness, Heribert was also compelled to explore a proxy—this time the number of badgers per group whose diet was comprised exclusively of earthworms. This was strongly correlated with group size (Figure 10.10). Conversely, but no less importantly, the groups that were most reliant on foods other than earthworms (i.e. foods that were almost certainly less spatio-temporally variable between feeding periods) had smaller memberships.

So, neither Hofer (1988) nor da Silva et al. (1993) tested directly the central RDH prediction on group size, but both uncovered highly significant correlations between group size, associated sociological features, and measures of worm consumption or the use of habitats associated with worm abundance. Wytham badgers eat various foods all of which are at least somewhat patchy (spatio-temporally heterogeneous in availability), but variability in earthworms is the most volatile (other foods, say ripe maize, are unwaveringly available through a short season, but thereafter gone), and it is the consequence of this heterogeneity (i.e. variance in the probability of a given patch being fruitful during a given feeding period) that lies at the heart of RDH (Carr and Macdonald 1986, remember the dice). Indeed, it would stretch scepticism into incredulity to conclude that there is no link between earthworm consumption (dietary eclecticism and the dispersion of seasonally important food types), group size, and other measures of fitness, but that the mechanism remains unproven.

With Dominic's worm samples in hand, we could explore from a different angle the question of whether the abundance of earthworms encompassed by a group's range affects its size. This time we estimated range configurations in two ways: first we mapped the latrines bisecting the ground between main setts and paths, often linked to natural features, to draw the ranges as naturalistically as possible (Chapter 9); and second, with an emphasis on methodological consistency rather than realism, we drew minimum convex polygons (MCP) by drawing straight lines between latrines (Chapter 4). For both methods, we repeated the analyses twice: pooling all the data for 1993–1997, and also analysing each year separately (Johnson et al. 2001a) (Figure 10.11). Remember, these analyses are optimistic in entertaining the hope that a worm-only effect will shine a clear light on an outcome that is actually driven by various food times. Nonetheless, perhaps the most clear-cut result, which supported the RDH prediction that less dispersed patches will be associated with smaller ranges, was

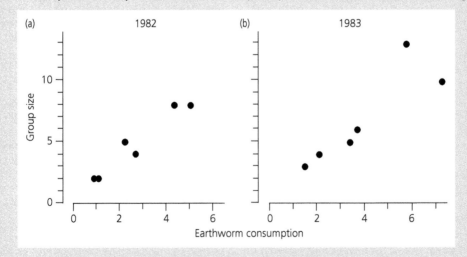

Figure 10.10 The relationship between total group size and total group *L. terrestris* consumption, using the index of *L. terrestris* consumption as listed in table 3, in (a) 1982 and (b) 1983 (from Hofer 1988, with permission).

Box 10.2 *Continued*

that the number of patches (using a proxy of proportions of key earthworm habitat) correlated significantly with group range sizes in six of eight tests (it is not surprising that shifts in group range sizes from year to year did not correlate with between-year contrasts in the richness of those group ranges: things like range sizes are surely sticky, lagging behind ecological changes and buffered against impetuous reorganization).

Beyond that, a miscellany of results tantalized us from the noise: aligned with the RDH prediction that richer patches will be associated with larger groups, there was an association between group range richness and group size for the naturalistic territories, but only in 1993, and for the MCP territories for the pooled 1993–1997 data.

Encouraged by these results, Dom Johnson undertook a more in-depth excavation of earlier sources of data on earthworm abundance between 1974 and 2000 (Johnson et al. 2001b). As predicted by RDH, this revealed no association between group size and the log of group range size but, at odds with the prediction (remembering that the prediction actually concerns availability not abundance), across years group range size neither related consistently to worm dispersion, nor did group size relate consistently to worm richness. The problem, of course, and notwithstanding the imperfect match between earthworm abundance and availability, is that while worms are important, they are only one of several foods.

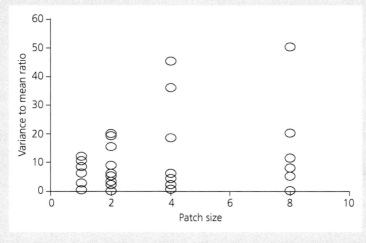

Figure 10.11 Variance/mean ratio plotted against patch size categories (from Johnson et al. 2001a, with permission).

1984b), the foregoing (Box 10.2) three labour-intensive but crude attempts to quantify earthworm availability to badgers on a landscape scale (we had been successful at this, with red torch in hand through sleepless nights, at the scale of a handful of patches; the landscape scale, however, was a different proposition). Despite the crudity of our measures of earthworm availability (and absence of linking these in any detail to the availability of other important foods), at every turn there was evidence of links between earthworms, and the habitats they favour, and badger group size. While the pleasing precision of prediction and falsification was out of reach, the diverse, well-found, coherently inter-digitating building blocks to the story have assembled apace. Of course, there will be critics

who dismiss the aforementioned story as one of the derided *Just So* ilk, an ad hoc fallacy. Others may find the blend of multidimensional, scientifically assembled, and naturalistically supported building blocks compelling. Either way, the building blocks, and the synthesis that now follows, provide a framework for the chapters to come; through population processes and on to deeper insights from badgers as models for behaviour and evolution.

The synthesis and neo-RDH

The brainwave that led Hans Kruuk to that first sketch of RDH on the back of the azure envelope was the

Box 10.3 Have earthworms declined?

Adding a further twist to the worm-counting story, Hofer (1988) found substantially lower mean earthworm biomass per badger territory (29,464 kg) than did Kruuk and Parish during the mid-1970s, even after correcting for a reduction in the mean weight of earthworms measured (from 5 g down to 3.8 g) and for smaller range size. The immediate suspicion that this was due to methodological differences has, with the accumulation of evidence, become less likely. Like Hans and David before him, Heribert had estimated worm abundance using formalin. A decade later, attempting to circumvent shortcomings of this method, Dominic Johnson estimated earthworm biomass by digging them up (Johnson et al. 2001a). Most striking was that $0.3 \times 0.3 \times 0.1$ m (depth) soil samples (10 samples per 10 transects) extracted by spade before dawn (c. 04:00) consistently yielded an order of mag-

nitude fewer earthworms (total dry weight: kg ha^{-1}) across habitat types than had formalin. To what extent was this because formalin revealed worms from deep in the soil, or because worms fled the vibrations of the impending spade, or, most interestingly, because worm numbers had actually declined? We do not know which method provides the better index of worm availability, which is the only measure that matters. However, Figure 10.12, redrawn from Johnson et al. (2001), strongly suggests there was half the biomass of earthworms under Wytham's pastures in 1999 than in 1974, but, importantly, no *relative* decline under deciduous woodland (of course do not know what ups and downs might have occurred between surveys). It is noteworthy that the two land uses evidencing a relative decline in worm abundance over 15 years were both open and agricultural.

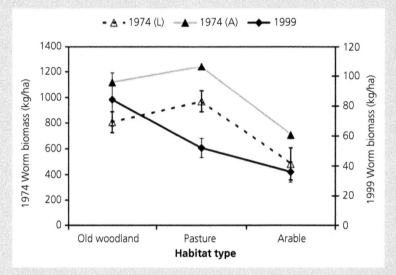

Figure 10.12 Comparison of indices of earthworm abundance in three major habitat types in 1978 and 1999. Data from the 2 years are on different axes because of differences in methodology. 1974 (L) is for *Lumbricus terrestris* only, 1974 (A) denotes all earthworm species. In 1999 figures are for all species. Error bars show ± 1 standard error (from Johnson et al. 2001a, with permission).

prediction that none of the 13 group ranges segmenting Wytham, each occupied by an average of 5.75 badgers, could be partitioned further. Consider a group of 8 occupying a range of 87 ha (the mean in Kruuk's study; 1978b. Why did they not split into 2 groups of 4 in a range of 44, or even a pair occupying 22 ha? They did not cleave because they could not cleave (without jeopardizing food security). In a smaller range they would run low on potentially vital food patches

(worms, hazelnut trees, brambles, maize, etc.) at critical times in poor years. By the way, mindful of the ancestral mustelid system of intra-sexual territoriality (see the text that follows), the range partitioning question can be no less aptly rephrased to apply to groups comprising only adult female badgers (a thought relevant to our discoveries about Japanes badgers in Chapter 19). In RDH terms, the answer was that the spatio-temporal heterogeneity in the dispersion of available resources would not provide the hypothetical members of that smaller group in its smaller range with sufficient food security (think, simply, of the caricature model of earthworms available on the lee of the hill). This indivisibility was the killer prediction, and if it were not upheld the hypothesis would fail.

Problem 1—the apparent divisibility of groups: the 13 progenitor group ranges in Wytham have fissioned into 23 new group ranges. Additional considerations are that: (i) even within the 23 new ranges there are nowadays generally not one, but up to 7 setts (that record was in the range of Outlier 1 group, at its peak in 2003), each partially cleaving the group, within its boundary-marked range, into subgroups; and (ii) nowadays clusters of group ranges form supergroups, also encircled by a necklace of latrines—sections of latrines bounding each component group's range (Chapter 4).

Problem 2—the lack of group exclusivity or defence: the assumption that the badgers of Wytham Woods occupied exclusive, impermeable territories, defended aggressively across latrine-marked borders, had at some point become incorrect. For at least 20 years, and perhaps for 30 or more, the conclusion that the fissioned group ranges at Wytham are aggressively defended territories (as assumed for badgers almost everywhere) could not be sustained. At least in Wytham, as evidenced over the preceding chapters, they are not defended aggressively, at least not consistently so. Indeed trespassing (a term that may be slipping into obsolescence in Wytham along with territoriality) is frequent and, as is already clear (and will become clearer when we quantify extra-group paternity in Chapter 17), often extends beyond platonic indifference. Equally, though, these ranges are not distributed at random with respect to either resources or the congruency in movements that define the spatial groups of their occupants—a non-randomness at the heart of a much more nuanced, if widely ignored, definition of territoriality suggested by Nick Davies in 1981.

On the topic of unsustainable conclusions, and confirmation bias as the enemy of evidence, it is timely to remember how inconveniently facts resist the tidiness of theory. Notwithstanding the generality of approximately circular, even amoeboid, economically shaped ranges in Wytham, consider how awkwardly the messy reality if nature confronts theory, as illustrated by the denizens of the Holly Hill group and Outlier 1 in Box 10.4.

Box 10.4 The messy reality of Holly Hill

The spatial unorthodoxies of some badger groups are illustrated by the relationship from 2004 onwards between the Holly Hill group range in the north-east of Wytham and that of the adjoining Outlier 1 (the name is an anachronism from the 1980s—Outlier 1 has been part of an independent group that encompasses the setts known as Outlier 64, Old Common Piece, New Common Piece, and High Seat). Holly Hill has a typical intersection, with diagnostic latrine-beaded path, with three of its four neighbours: Thorny Croft, Great Ash Hill, and Clearing Outlier groups. However, its spatial arrangements with the fourth neighbour, Outlier 1, are eccentric, in a way reminiscent of a motorway flyover: Holly Hill badgers use a very well-trodden, also latrine-beaded, path that runs straight through Outlier 1 land. This path leads to the fields behind Keeper's Cottage, and bisects the Outlier 1 range. Whereas the arrangements between many other groups more closely approximate a Right to Roam,[16] those between the badgers of Outlier 1 and Holly Hill (both of which feed in the fields beyond Keeper's Cottage) seem to have negotiated a special right of way—a short cut that avoids an otherwise tortuous journey to reach their foraging fields. Neither fits tidily with classical territoriality, but then neither annuls the reality of badger groups that space out their centrally placed, permanent, setts.

[16] Countryside Act, Scotland 2003.

So, do these two problems render the original RDH-based explanation of badger sociology wrong? Actually, a lot has changed and, furthermore, other things may not have been as they appeared in the first instance. The first clue was presented earlier in the form of the European Union CAP. This policy led to the change in farming practice that caused an increase in the number of patches of a habitat—lightly grazed pasture, rather than arable fields—associated with abundant food (earthworms and carabid beetles, under pasture), and decreased their dispersion distance. Of course, this did not manifest in the same way for all groups: consider Great Ash Hill, whose latrine-beaded perimeter lies entirely within the woods; however, the Great Ash Hill badgers visit the Thames-side fields to the north and north-north-east of Wytham and a string of latrines mark their route out of the woods to those worm-rich pastures.

RDH would predict this CAP-induced change in patch dispersion and richness could cause a contraction of the minimum defensible range size necessary to deliver the requisite food security, leading to smaller group ranges able to support equivalent numbers of residents, thereby increasing total population density through group fission. But what if other things were changing simultaneously—things with effects that continue in force long after those of the CAP had run their course? What could cause a change in patch richness—raising the ceiling of potential environmental carrying capacity in Wytham—and alter the geometry of patch dispersion?

There are plenty of candidates, considering the diversity of secondary food types documented in Figure 10.1. For example, suppose the arable crop rotation diminished the dispersion of patches of appetizing maize; or suppose the population of deer in Wytham changed in ways that affected the understorey (which it did, changing impenetrable bramble to more productive and more accessible foraging ground; Buesching et al. 2014), and thus the microclimate, in ways that affected the dispersion of patches of earthworm availability or indeed affected the distribution of brambles and thus blackberries; or suppose some other mysterious factor affected either or both the dispersion of food patches and their richness? These are not castles built on sand—each of them happened and, most importantly, there *was* such a mystery factor. Chapter 11 demonstrates its enormous impact while Chapters 12 and 13 are devoted to untangling its workings, but for the time being the argument is unaffected by what it was. Carr and Macdonald's (1986) binomial version of RDH explores the impact, in principle, on food security, of changing patch richness, renewal rate, and dispersion; for now, however, we'll outline them in natural history.

Hold in mind that RDH posits that as food patch dispersion decreases, that is, the same food security can be provided by fewer patches that are closer together, group ranges can shrink accordingly. Of course, both because the measures of food availability are probabilistic, and will take time to stabilize at a new level, and because during that time, the social relationships of the incumbent badgers will affect the options for social reconfiguration, it would not be surprising if the change takes a while to consolidate. Imagine, for example, an individual, Bertha Badger, who transfers, through progressively prolonged visits, from Group A to Group B; later Bertha B's siblings follow her; later still, their mother and aunts die, and with them the familial allegiance to Group A goes too. Thus social networks shift their centres of gravity. This is, by the way, exactly what happened at the Pasticks sett—for Bertha B was F217, and her sister F221, arriving in 1991 at Pasticks to pair up with Casanova B (M10), who had himself shifted to Pasticks from Brogden's Belt in 1990. History then repeated itself when F380 began shifting her centre of gravity from Jew's Harp to Pasticks. The full detail of this soap opera is related in Chapter 5 (Box 5.1), but this synopsis illustrates the process.

Simultaneously, if patch richness increases so that the available food provides security for a larger number of badgers, group membership can grow accordingly. At its simplest then, if the changing environmental factors cause food patches to be less dispersed, less heterogeneous (i.e. less variable in their schedule of availability, which can also involve faster renewal rates), and richer (i.e. more bountiful)—or to extend productivity longer throughout the year (see Chapter 12)—we would expect to see group ranges shrink while group membership increased. And shrunken ranges, still offering their occupants sufficient food security, permit a greater number of ranges. By what mechanism would the additional ranges be created? Fission of the progenitor groups.

Two old questions resurface, concerning, respectively, the limits to range and group size, to one of which, however, we have an answer that we had not needed to consider before. First, the original Kruukian question: why do group ranges not get smaller? The answer was, and still is, because the dispersion of spatio-temporally variable food patches cannot provide sufficient food security for even one badger if the minimum defendable range contracts and thereby fails to provide sufficient 'patches' to enable the animal to see the year through (e.g. Carr and Macdonald 1986)—according to RDH, for group ranges to shrink, inter-patch dispersion distance must shrink, and/or the frequency with which patches become bountiful must increase. The 2020 reality, with group range size between a half and a fifth of what it was in 1974, can be explained by a decrease in patch dispersion and, probably also, in heterogeneity of patch availability.

The second question is: what limits group membership? Part of the answer was, and still is, ecological: the richness of the patches that facilitates the cohabitation of badgers without intolerable penalty to individual food security (offsetting the duration of bottleneck periods against the time for which a badger can tolerate being famished). This ecological ceiling is solid, tangible, and above it lies hunger, possibly even leading to starvation. But another part of the answer, the part we had not previously had to consider, is sociological, and concerns a lower, less tangible, glass ceiling that caps group size below the ecological limit imposed by patch richness and local environmental carrying capacity; namely, social stress, whose impact was not revealed at Wytham until the factors to be discussed in Chapter 12 radically increased the food security offered by food patches, with the consequence that the minimum economically viable range became able to support groups so large that they headed towards social turmoil. What happens when the two processes, with their two consequences of shrinking ranges and expanding groups, converge?

The question of why Wytham's badger groups did not fission sooner, or now do not fission even more than they did—a question we argue has an ecological answer—can be turned on its head to ask why they do not amalgamate; this time, however, the answer lies at the intersection of energetics (i.e. the costs of travelling further than necessary, as a result of occupying a range larger than the minimum economically viable) and sociology (the chaos of crowding).

If all 300 badgers in Wytham were, as one hypothetically happy family, to function as a single social unit (this is not an absurd proposition: think of a troop of gelada baboons, *Theropithecus gelada*), then the question of how to cohabit would arise. Geladas share cliff ledges requiring no preparation, but badger setts in Wytham occupied by a dozen badgers involve digging at least three or four active holes (Macdonald et al. 2004d). Extrapolation suggests that a warren for the whole herd might involve over 1000 holes and, more particularly, excavation on a scale that would daunt fossorial engineers of the calibre of rabbits, marmots, and naked mole rats—all of whom are much smaller and thus require narrower tunnels. Further, insofar as it is advantageous to know your companions, knowing 300 of them would be a challenge (although geladas manage this adequately) (further considerations are cognitive ability—whether badgers have the social intelligence to cope with a huge group—and reproductive conflict, female banded mongooses, for example, simply can't put up with more than half a dozen reproductive companions of their own sex). A more pressing consideration might be time and distance—the furthest flung foragers would face round trips inflated by a 6 km commute. The temptation to partition and monopolize some nearby, private Elysian fields would, resource dispersion allowing, surely be energetically irresistible. While an astronaut contemplating Wytham might perceive a pair of hills littered at mid-elevation with a vast sinuous sett stretching along its exposed sandy strata, punctuated by occasional gaps between earthworks, we know that the herd segregates into a readily transgressed spatial hierarchy of social contours: cliques, subgroups, groups, and super-groups.

What then if local patch richness can support a large group of badgers, but with closely spaced and relatively homogeneous patch dispersion? Well, this is exactly what happened at the Pasticks sett in 1996 when group size reached 29 individuals. The response was, as we would predict, an attempt at group fission, with the swelling occupancy at their principal 'Pasticks Outlier', and the reoccupation *en masse* of the former main sett in that area, Jew's Harp, that had for some time lain dormant. Our bait-marking survey of 1997 revealed a somewhat incomplete and 'soft' latrine-marked border between Pasticks Main and Jew's Harp, and a strong, but only partial border halfway between the 200 m spaced Pasticks Main and Pasticks Outlier setts, but with that robust 'Siegfried line' persisting only for 100 m or so north and south, fizzling out once it spilled into the open fields. This is the ontogeny of the Pasticks 'super-group'; fundamentally entwined still with Pasticks Outlier, affiliated with Jew's Harp, and occasionally adjunct to the Singing Way sett, along with a host of small outliers scattered through that range, each with one to three holes (Pasticks upper outliers 1 and 2, Pasticks outlier outlier, Pasticks field outlier, Wormscat views 1 and 2, Jew's Harp Annex, Jew's Harp Outlier, etc.).

In the ragged, untidy way of ecological and societal reality, two parallel tensions entwine to heave apart the progenitor 1974 group ranges and their occupants. As it becomes possible to deliver tolerable food security in a smaller range, so too it becomes impossible to manage the social relationships of the larger groups that occupy them.

Nowadays 230 adults and 50 cubs are distributed between 23 groups, giving a mean of about 12 badgers per group (including cubs). This is almost twice the group size estimate in 1974 for the 91 badgers Kruuk estimated to populate the 12–13 ranges of that time (in 1969 Barker estimated 59 badgers, which, all else being equal, would give an average group size of 4 or 5 badgers). But what would be today's mean group size if the current population of badgers were allocated amongst the original progenitor groups? About 23 badgers, combining adults and cubs. Members of groups of a couple of dozen badgers, 3–6 times the membership of the Barker–Kruuk eras, would surely experience very different social dynamics to those of their forebears of 50 years ago.

A thought experiment asks how the benefits and costs of living in a group of 23 badgers (if circumstances allow), compare with those of living in 2 groups of 12 (the modern reality), or 6 groups of 4 (perhaps the 1960s reality)? A related thought experiment weighs the pros and cons of all 23 (or, for the sake of the thought experiment, make it 50—the approximate membership of a contemporary super-group) living in one communal sett, or splitting up (and, most interesting and different amongst thought experiments, one might ask why share a sett at all—although that will be a different topic for later). Badgers have unanimously provided the answer to this sociological accountancy: first, as group memberships grew, cliques developed, perhaps around pregnant females who sought annexes and outliers in which to give birth (away, we guess, from prying, perhaps even infanticidal, 'sisters')—these maternal–offspring groups may attract a few other adults, kin, and males awaiting post-partum oestrus (thus a clique is formed, its structure revealed by the flow of subcaudal allomarking in Chapter 6).

Some of these cliques partially divided the progenitor group ranges: these are the subgroups documented in Chapter 4. Sometimes they spent years striving to fission, erecting perimeter latrines (readers of Chapter 9 may muse on whether their motivation was to exclude or to stay in touch) only for them to dissolve a year later (ecological and behavioural processes in nature are ragged). In the end some fission into separate, latrine-bounded, new social groups occupying new group ranges while others do not. What determines these different outcomes?

In RDH terms, the answer seems obvious: food security in a world of spatio-temporally heterogeneous food patches is probabilistic. When the dice fall badly, even occasionally, all the occupants of a group range may need access to the whole range. As bad rolls of the ecological dice become less probable, so each subgroup is more secure in its own subrange, occasionally criss-crossing the movements of other subgroups; increasingly, however, minding their own business. The dispersion and heterogeneity of the rolling dice of food patch availability become even less challenging to food security and the previously indivisible socio-ecological centre of gravity cleaves into two groups, and does so along the contours of patch availability (each new group needs its supply of worms, blackberries, and hazelnuts) and the fissure lines of societal fragmentation. In fact, another, entirely unexpected, and radical, change in badger behaviour inserted an additional step into the process: very suddenly, over just a year or two (around 1996), they dug about three times more outlier setts.

As we continue, do not miss the drama of this moment. This is no ordinary progression from one paragraph to the next. In our story of the badgers of Wytham Woods this is a Black Swan moment—something, at least from the perspective of Juvenal the Roman poet, who had only ever seen white swans (as would all northern eyes until more than 1000 years passed), unexpected to the point of unimaginable yet of great consequence.[17] Everything changed in the social arrangements of the badgers of Wytham Woods, and not only had we not seen it coming, we're unaware of any students of non-human mammals seeing anything like it before.

This remarkable reconfiguration of the kaleidoscope of badger spatial organization was clearly tied to the consequences for population carrying capacity of the change in environmental circumstances that awaits in Chapter 11. For now, however, the interesting question is what affected the time scale—which was not instantaneous but nonetheless abrupt by ecological standards—of the fissioning of 1970s progenitor groups into millennial contemporary groups? The answer is that, on the one hand, the ecological changes to resource dispersion did not happen overnight, and on the other hand, the bonds of society must be stretched before they snap, or perhaps just drift apart before they wither. As the process unfolds, the occupants of outliers become ever-more hefted to their sett and peel away gradually. From month to month, the

---

[17] In the second century CE Juvenal used the metaphor of something as rare as a black swan ('rara avis in terris nigroque simillima cygno') to indicate the impossible, and no human that had ever lived (outside Australia) before him or after would have disagreed until Willem de Vlamingh's Dutch expedition 'discovered' both black swans and Australia in 1697. In his bitingly beautiful style Nassim Nicholas Taleb—author of Black Swan theory and the book that bears its name—excoriates the 'nurdified fields' (of which we hope we are not an example) that impose Plato-like categories and blinkered induction to explain away the unforeseeable, and is particularly withering about the 'cultivated philistines' (a phrase drawn from Nietzsche) who 'concoct explanations for its occurrence after the fact, making it explainable and predictable', which is precisely what we are about to do. Taleb's criticism of the constrained inductive reasoning that pays 'undue homage to the constructs that inhabit our minds' is illustrated when he rails against 'our tendency to mistake the map for the territory'—this would seem apt as we stare at the shifting map of badger arrangements in Wytham in our effort to understand their 'territory'. As a footnote to this footnote, the expression that the map is not the territory was penned in 1931 by philosopher Alfred Korzybzki in an essay with the encouraging title 'Science and Sanity', and means that an abstraction based on a thing is not the thing—the menu is not the meal (think how various projections of the world distort the juxtapositions of places).

change is scarcely noticeable. Consider the case of F626, a breeder at Great Ash Hill who moved to Firebreak in 1998. Once at Firebreak, F626 mated with M502, himself also an émigré from Great Ash Hill; their son M728 moved back to Great Ash Hill. F626's son of the following year, M787, born in 2000, spent his life vacillating between Firebreak and a little-used outlier of the Great Ash Hill range deep in the north woods, called Thames Outlier, whereas his littermate sister F789 stayed on at Firebreak to become matriarch to subsequent generations in that sett. As this shift in centre of social gravity unfolded, other badgers from Great Ash Hill joined F626; these newcomers spending more of their time at Firebreak until we found that latrine-marked path bisecting Great Ash Hill and Firebreak—a line still then crossed by visitors, as it is today albeit less frequently. This is a ragged mitosis, the stickiness at the metaphorical kinetochore tied to patch indivisibility across the full year and social ties. It is likely that not only food patches and their quality, but also badgers and their quality—that is, young/old, fat/thin, lactating/barren—affect the rate at which chromatids separate. Some types of badger need more security than others, leading us to speculate that two neighbouring groups may be divisible in a year when there are no cubs at one sett, only to become conjoined in a year when there are (i.e. because of their lactating mother's greater need for food security).

What would likely happen if social tensions increasingly distanced the subgroups (perhaps F626 was escaping the suppressive influence of rival females), and the food security just, but only just, allowed them to fission? A plausible answer is that every so often the demands of food security and the unpredictability of spatio-temporal heterogeneity (interacting with the individual's needs) would make it tempting, perhaps essential, to visit a particular food patch in the erstwhile, but now annexed, group range. It might even be advantageous to monitor food availability in that fissioned section as a hedge against future uncertainty. Furthermore, those neighbours comprise kin, lifelong associates, potential mates, and, in the years around fissioning, former cohabitants. The outcome: a system of spatial groups occupying separate group ranges with distinct centres of gravity but whose movements overlap and whose home ranges are highly permeable—perhaps communicating using olfactory messages left at latrines positioned on the nexuses of paths running between the associated groups—a kind of telephone exchange of connectivity. QED, progenitor groups fission into new groups,

and the new groups partly fission into an uneasy instability of subgroups, and progenitor groups become super-groups, all ultimately constrained by habitat saturation.

One important driver of this particular trajectory, mentioned early in this chapter, is now ripe for revisitation: phylogeny, and the baggage it imposes on the badgers' social psyche and physiology. There is remarkable consistency in the ways that taxonomic families of carnivores arrange their social lives (Macdonald 1992). Canids follow a canid model (Macdonald et al. 2004a), felids follow a felid model (Macdonald et al. 2010b), and so too mustelids a mustelid model (Macdonald and Newman 2017). Ecology drives society, and societies are fine-tuned in their adaptations to ecological circumstances, but species are neither equally nor infinitely pliable; a society's starting point influences the route, as an expression of its ecological circumstances, along which it adapts: badgers start as mustelids. Conclusions to be drawn from both inter- and intra-specific variations on the mustelid theme are the topic of Chapter 19, but it suffices here to emphasize the inescapable essence of mustelid sociology: intra-sexual territoriality. That is, as summarized by Powell (1979), amongst mustelids single females maintain exclusive territories, defended aggressively from neighbouring females; single males similarly defend territories, albeit less neatly tessellated, from neighbouring males. However, the males' territories are larger than those of females, so each male therefore overlaps several females in a semi-detached relationship. Take this template and change it in three ways: first, make the food supply more spatio-temporally heterogeneous in its availability (and with relatively low dispersion distances); second, make it progressively richer in its availability; and third, make both the first two changes probabilistically ragged (i.e. as they are in nature). What is likely to happen? Thin species that survive by light-footed agility (martens, weasels, etc.) will simply starve if the probability of securing a meal during each feeding period is too variable, even if meals are rich on bountiful days. But fatter species, with ability to use energy reserves to buffer against food insecurity (Bright Ross et al. 2021; Chapter 14) may well form groups (Newman et al. 2011). So how does this affect their spatial arrangements?

First, remember that the strict mustelid template prevails over species that generally eat very homogeneously dispersed prey (e.g. small mammals) (Macdonald et al. 2017). Imagine that template, but make the prey spatio-temporally heterogeneous (think earthworms or blackberries). Quick reference to the foregoing general account of RDH reveals that each territory can now be occupied by several individuals, to the extent that the cost in food security is bearable to the original occupant (probably their mother, as kin selection comes into play). Low food security is likely most intolerable to maturing adolescents, emerging from maternal indulgence into social subordination (Chapter 3): a combination of factors commonly leading to dispersal in mustelids such as weasels, martens, and mink.

Now imagine that something (Chapters 12 and 13 lie ahead) increases the richness of the food patches. As a result, the group membership of each range increases. As the patches get richer, introduce a change to their heterogeneity—lessen their variability (probability of providing food during each feeding period—weight the die towards throwing a bountiful six on the dice model). Now the larger group can maintain, or almost maintain, its food security while shrinking its range size, and may sustain philopatric offspring breeding in the primary occupant's (probably maternal) range; richer still, and inter-generational groups may persist, even retaining also additional males (a bigger step from the mustelid starting point of larger male ranges, especially if they can find a way to breed beyond female kin). Ultimately, this range may partially partition (doubtless amidst some social tension) between subgroups, each with its own centre of gravity. Finally make everything ragged; that is, like nature. None of the borders is precise, but the original design shimmers through. Groups of females occupy largely congruent spatial groups, but as needs must they forage amidst their previous cohabitants (and still blood kin) next door. Groups of males, similarly in largely congruent spatial groups overlay the females, and make much more frequent, but generally similar, trips amongst their ancestral wider clan. And so you have the badgers of Wytham Woods: moulded by phylogeny, framed by ecology, and assembled by sociology.

From these changes in outline spatial organization, modulated by the intrinsic qualities of the individual (chubbily resilient or lactationally exhausted; Chapter 14) and its species (corpulent musclimorph), how do groups function? Do the members of these groups operate alone within their shared spatial congruence (e.g. red foxes) or cohesively, as a mob whose members forage, feed, or fight together (meerkats, wolves, spotted hyaenas, and lions)? And, whether as a spatial or cohesive group, do the members den together at a stable central refuge (badgers)? These intra-specific comparisons are the meat of Chapter 19, but as a brief taster, the answers key out depending on, first, whether the acquisition of food and avoidance of predation allow, or even favour, group cohesion (cooperative hunting, shared vigilance, and collective defence; Chapter 8), and, second, how resilience to food insecurity (fat) interacts with patch dispersion: if the variably available food patches are widely spaced, there will come a point at which travel costs prohibit a short-legged creature from commuting to and from a central place to cohabit, whether its daily activities are alone or in company. Meerkats, for example, do not occupy a central den, but cycle, always as a cohesive band, between several outposts in different sectors of their vast territories (Doolan and Macdonald 1996; 1997). Raccoons, in contrast, second cousins to badgers, have many short-term acquaintances and a few long-term associations (amongst groups of males) in a fission–fusion society (with each female associating with a particular group of males)—unusually for carnivores, genetic relatedness did not map onto raccoon social networks (Prange et al. 2011; Hirsch et al. 2013). What sets the limits to the social flexibility of each species? This is the topic of Chapter 19.

History repeats itself, or chasing modernity?

The power of natural selection to sculpt adaptations near perfect in their intricacy, delights, indeed often dumbfounds, evolutionary biologists—did you know that there is a tiny Amazonian microhylid frog that has struck an evolutionary deal with an arachnid species, enabling it to shelter between the hairy legs of tarantulas? Adaptations that are miracles of minute compromise in design—the bills of Darwin's finches—are no more striking than those of behaviour: consider the dwarf mongoose throwing itself prostrate—playing dead—at the glance of a hornbill. And consider also the remarkable cooperative behaviours of those dwarf mongooses that determine individual success in their societies: cooperative vigilance and territorial defence, reproductive suppression, and babysitting (Rasa 1977; Rood 1990; Creel and Waser 1994)—societies and their dynamics are adaptations honed by natural selection to deliver survival and fitness with no less precision than Tennyson's eagle's talons[18] clasping the crag in those lonely lands, or the badger's striped face. As we ponder the functional significance of badger society in Wytham, we should not underestimate the power

---

[18] The marvel of the eagle's adaptations inspired Alfred Lord Tennyson's poem in 1815.

of natural selection to match solutions to problems. Furthermore, although most species have had millions of years to get them right, a fact that scarcely diminishes our awe at the intricacy of adaptations, selection can work remarkably fast in refining genotypes: Balaev's foxes (see Chapter 19) were, albeit in a fur farm, selectively honed from timorous to tame in less than a dozen generations (with the pleiotropic shift to piebald pelts, curly tails, and dioestrus packaged along with their new, dog-like, temperament). So, natural selection can engineer remarkable intricacy of adaptation, and it can do it fast.

The foregoing realities have been uppermost in our mind as we have sought to understand the adaptive significance of badger society. We have also borne in mind the fact that students of carnivore sociology have revealed variously extensive flexibility within each species to adapt itself, and its social life, to different circumstances—consider the red foxes living as monogamous pairs or polygynous groups, in territories ranging from <10 ha to >1000 ha, in habitats from sandy or snowy deserts to urban jungles: their evolutionary past has equipped them with a repertoire from which they can perform with success on a remarkable variety of stages. This intra-specific variation in social behaviour concerns the variable phenotypic expression of each species' distinct genotype: such phenotypic variation exists amongst the badgers of Wytham Woods (Chapter 14), and more broadly is the opening topic of Chapter 19. Here we ask, however, whether the badgers' evolutionary past has prepared them (carrying their phylogenetic rucksacks full of mustelid chips on their shoulders) for twenty-first century Wytham Woods? This is a question about the limits to genotypic and phenotypic adaptation, and the process of distinguishing the difference between them.

First, as environments change, genotypes have to catch up. Sometimes, circumstances change fast. Your first schoolroom lesson on natural selection was presciently relevant to the paradox of badger sociality. Remember, in 1848, a naturalist named R.S. Edleston caught a peppered moth, *Biston betularia*, in central Manchester. Peppered moths are not uncommon, but this was the first one ever recorded to be black (most are speckled grey[19]). By 1900, 98% of all peppered moths in English cities were black. The reason was

experimentally demonstrated, rather controversially, by Oxford biologist Bernard Kettlewell (1956), working under the supervision of E.B. (Henry) Ford, a father of ecological genetics. Indeed, Kettlewell and Ford both taught David, who still has a neatly labelled dog's skull that Henry Ford gave him (they occupied adjoining offices, through the walls of which Ford, a man of legion eccentricities, could very often be heard sneezing thunderously and then bellowing in apparent fury at himself). What Kettlewell demonstrated was that soot-blackened urban tree trunks offered protective camouflage from foraging birds to black, but not traditionally peppered, moths, creating a selective pressure that drove the heritable change in moth adaptations to urban melanism. The moths had not been ready for the Industrial Revolution; they had experienced nothing like it before. Insofar as their phenotypes could not cope, their genotypes couldn't either. They had to catch up.

Another revolution, the Agricultural one, including its post-war incarnation as the Agricultural Miracle, has created circumstances that make it suitable for badgers to live in groups on farmland. Of course, some badgers lived in groups in some places beforehand, but studies from the European mainland suggest they were, and generally still are, much smaller in membership (Johnson et al. 2002a; Chapter 19). The patchwork agricultural landscape that favours badgers, and their groups, is probably a phenomenon of only a few hundred years—an evolutionary blink of an eye. Furthermore, the climate that characterizes that landscape has reached its current state in only recent decades (as detailed for Wytham in Chapter 12). Do the ecological circumstances of Wytham in 2021 fit within the envelope of adaptive intra-specific variation (the sociological phenotype) that defines a badger, or might this be the first time badgers (and their genotype) have encountered this particular adaptive problem? Consider, Wytham's badgers now live at a density of >40 per km^2 in groups that sometimes top 30 members. In short, are Wytham's badgers the *Biston betularia* of carnivore social behaviour? Is this sweet spot of agricultural patchiness, climate change, and (only since the 1970s) legal protection, a circumstance that has privileged us to observe badger society grappling with a population density (you might call it crowding) never experienced before by any badger or its genome anywhere in the world? Perhaps, but we should be cautious of falling back too comfortably on exchanging our quest for adaptations for a similar search for pathologies. After all, if biologists do not search assiduously for adaptations, they will surely not find them.

---

[19] Prompted by this line of thought, readers with a penchant for lateral thinking will surely be wondering what it is that favours black leopards prospering amongst spotted ones, and vice versa (e.g. Tan et al. 2015).

On the other hand, the concoction of *Just So* stories, to imagine adaptation where none exists, is no less sloppy.

A metaphor might elucidate the phenotype–genotype distinction. Much of intra-specific variation in social behaviour is like the analogue tuning to frequency of an old wireless radio: as things have changed at Wytham, have the badgers been able constantly to modulate their group dynamics to stay on the optimal wavelength? If so, then the contemporary signal of their societies is coming over loud and clear, the best phenotypic adaptation that badgers can ever make to current circumstances. Or, are they struggling to tune into a radio station that has recently moved to a different bandwidth, where tuning can make the best of a bad job but, without re-engineering the radio, their society will remain out of tune with their circumstances. Such re-engineering takes time and mutations, but market pressure will hasten it, just as it did for the peppered moth.

As you ponder our results thus far, and as you read on, know that badgers are surely adapting as well as their ancient and recent genotypes allow, but bear in mind a new question: are their adaptations as good as they can be, drawing on the experience of a history that has made them and is repeating itself as histories do? Or are they making the best of a bad job, tackling unprecedented circumstances for which they are imperfectly equipped by natural selection? In that case, all else being equal, will they be doing it differently, once their genotypes have caught up in 10, 50, or 100 generations? Time alone will tell, but another species of which that question might usefully be asked is, of course, *Homo sapiens*.

CHAPTER 11

The Economics of Survival

Population Size, and Crashing Through the Ceiling

Death is nature's way of making things continually interesting.

Peter Steinhart

How many? As an opening question this one seems irresistible in almost every context: how many members of the audience, guests at the dinner party, players in the competition, voters in the election, or badgers in the wood? In itself, viewed from an ecological standpoint, 'how many' is a trivial, stamp-collector's question, but the answer is an entry ticket to the most fascinating show on earth: the patterns, more than the sums of their parts, made by animal numbers and the factors governing their population processes. How many might even be the most fundamental question in ecology. Here we ask not simply how many badgers occupied Wytham throughout our studies, but also what were the drivers and, ultimately, the consequences of abundance, such as changes in age profile and sex composition. Fundamentally, this chapter is about patterns drawn by actors on the stage—patterns as wondrous as the swirling movements of a murmuration of starlings, born of the dazzling adaptations of each bird within the flock. Ultimately this becomes a story of environmental carrying capacity and social constraints, and the manifest ways in which badgers can adapt to make the most of what is available.

For much of the UK's native fauna the post-war agricultural miracle has, over the last half-dozen decades, had disastrous consequences for many of the working parts of nature. Birds, butterflies, bumblebees, and bats are amongst the blighted Bs that are down by 50% or so. But for badgers, in 2021, the answer to 'how many?' is: a lot more than we started with. Badger numbers in Wytham had proliferated since Hans estimated there were 91 in 1974 (Kruuk 1978a)—and that was already an increase over the 59 estimated by Barker in 1969. Moreover, the population had apparently grown further by the time our routine catch-ups began in 1987,

and burgeoned thereafter. But by how much, and what constraints set the ceiling for badger abundance?

Of course the primary candidate is the availability of food set against maintenance costs (Chapter 14), but there are also additional limitations, such as disease (Chapter 15) and traffic accidents (a modern incarnation of predation), or, to step back, competition between resident individuals (Chapter 5). All of these may cause badgers to bump their heads on the ceiling of resource limitation as their numbers swell. Crucially, weather conditions play a big role in determining the seasonal availability of earthworms (Chapter 10), and so short-term variability and long-term trends in rainfall, temperature, and humidity are important, governing not only foraging success but also the costs of leaving the warm, dry refuge of the sett. Further, as will become clear, not all badgers (or members of any population) are created equal, with some incurring higher net operating versus productivity costs than others (Chapter 14). Indeed, the ratio of these inputs to outputs changes continually and circumstantially through each individual's life. Ultimately, the actuarial consequences of all these phenomena impact the relative rates of births versus deaths.

Keep in mind too that phylogenetic baggage fundamentally constrains badger population dynamics. European badgers have their roots as a low-density, fairly solitary species, as indeed they remain throughout much of their less productive biogeographical range (far fewer than 1 badger km^{-2} throughout much of continental Europe, Johnson et al. 2002; UK mean 1.39 badger km^{-2}, Wilson et al. 1997; see Chapter 19). Indeed, this is probably how they lived for most of the 2 million years of the ancestral evolutionary history of

The Badgers of Wytham Woods. David W. Macdonald and Chris Newman, Oxford University Press.
© David W. Macdonald and Chris Newman (2022). DOI: 10.1093/oso/9780192845368.003.0011

Box 11.1 The fundamentals of demography

Humans have always needed to understand those factors that determine the availability and success of other species, to hunt, to fish, to raise domestic stock, to avoid competitors and pests, and to cultivate the land. The notion of a 'balance of nature' has been a topic for philosophical contemplation since the time of the ancient Greeks, implicit in the writings of Aristotle, Herodotus, and Plato (Egerton 1968). Notwithstanding the modern marketeer's flagrant abuse of the prefix 'eco-' (Sherman 2012), that balance was clearly in Ernst Haeckel's mind when, in 1869, he first defined ecology as 'The scientific study of the interactions between organisms and their environment'. Considerable advances in the application of scientific rigour to the study of nature were made by Elton (1927), undertaking much of his research in Wytham Woods (Savill et al. 2010), although he too adopted a similarly vague (if evocative) definition of ecology, 'Scientific natural history'. Thirty years later, Andrewartha (1961) refined the doctrine of ecology as 'the scientific study of the *distribution* and *abundance* of organisms', making explicit two useful metrics of interactions between organisms and their environment. Such organisms obviously include people, and it is interesting to contrast how John Graunt (1662) (a haberdasher by trade) had much earlier sought to address the mechanisms underlying these same metrics of distribution and abundance for the human population of London in the seventeenth century in his *Natural and Political Observations Made upon the Bills of Mortality*.[1] It is he whom we must credit with the idea of quantifying birth rate, death rate, sex ratio, age structure (age at first and last reproduction), and migration; vital factors then, as now, underscoring any understanding of population variability (Figure 11.1).

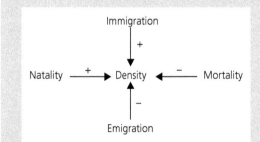

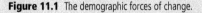

Figure 11.1 The demographic forces of change.

These *rates* are themselves the products of *dynamic interactions*, some intrinsic to the population (what is the maximum litter size of this species?), some extrinsic and the outcomes of environmental processes (what proportion of the litter survives nature's slings and arrows?). It

was another demographic giant, Thomas Malthus, who, in 1798, explained that the ability of a population to grow is inextricably linked to the resources available to it, leading Charles Darwin (1859) to propose that when these resources become limiting a 'struggle for existence' ensues—one in which only the fittest, or 'best fitted', individuals survive (this process, natural selection, provides the greatest, and most elegant, insight into the natural world in the history of scholarship).

Combining the quantitative and dynamic elements, Charlie Krebs, inspired by a lifetime watching the vicissitudes of brutal winters and the hungry predators limiting snowshoe hares in the Canadian Arctic, was ultimately prompted to define ecology as: 'The scientific study of the biotic and abiotic interactions that determine the distribution and abundance of organisms' (Krebs 1985)—a definition that suits the purpose of this chapter nicely, with its emphasis delving beneath simply how many badgers are there, to consider: How is the population composed? How are its members distributed in space? And what processes determine these answers and govern dynamic change?[2]

There are good reasons why such beguilingly basic questions have only infrequently been addressed comprehensively for populations of large-ish mammals; for example, lions (Packer et al. 1991), African buffalo (Sinclair 1977), red deer and Soay sheep (Clutton-Brock 1996), grizzly bears (Mace and Waller 1998), and wolves (Post et al. 1999). Aside from it being difficult, and, because big mammals have long lives, requiring more prolonged data than generally allowed by funding and career structures, the processes (births, deaths, immigration, and emigration) and structures (age and sex ratio) refract through the myriad prisms of the lives of each individual in a population (Dunbar 1988)—lives that our persistent voyeurism has revealed in unusual detail and duration for the badgers of Wytham Woods.

---

[1] Graunt's analysis of the bills of mortality (weekly statistics of deaths) was actually intended to assist Charles II and other officials as they attempted to create an early warning system to alert government to outbreaks of bubonic plague in London. This system was never really put in place; however, Graunt's work did lead to an unprecedented statistical estimation of number and age/sex ratio structure of London's residents. His work ran to five editions by 1676.

[2] It is satisfying to apply these questions to badgers in Wytham, which is where they have their scholarly roots, and where longevity has privileged one of us to have been instructed by both Charles Elton and David Lack: the two Oxford scholars who might be said to have invented, respectively, ecology and density dependence; both, furthermore, having done so in Wytham (see Savill et al. 2010).

Meles species[3] (see Chapter 1), before the Agricultural Revolution created conditions far more favourable to higher badger abundance.

How many badgers?

Jack da Silva was second in the now long lineage of 20 badger doctoral students, amongst whom his distinguishing characteristic was trademark efficiency—an aspiration at odds with the prospect of spending several hundred nights, sleepless and often cold, plodding round the woods listening to the metronomically numbing peep of signals from radio-collared badgers. No, instead Jack reasoned that the badgers should come to him. Further he was a 'how many' man, so he proposed that we should initiate large-scale trap-ups, emulating those conducted routinely by Chris Cheeseman's (by our standards then, huge and well-funded) MAFF (Ministry of Agriculture, Fisheries and Food, now Defra) team at Woodchester Park. Unpersuaded by David's supervisorial arguments to the contrary, which included the details that we neither owned sufficient of the necessary and expensive cage-traps, nor had the funds with which to buy more, Jack was mercifully immovable. We say 'mercifully' because, had he crumpled under the pressure of supervisorly wisdom, we might not have instigated, in 1987, the three-to four-times yearly 'catch-ups' attempting to monitor the entire population on which this chapter, and much of our knowledge about Wytham's badgers, is based.

As of writing, we have amassed 11,488 captures of 1823 individuals (Photos 11.1 to 11.4), monitored 'womb till tomb'. Looking back at 1987[4] we now estimate that 110 adults and 27 cubs comprised the starting population, according to our new enhanced minimum numbers alive (eMNA) calculus (see the text that follows). Do we, however, catch 'all' of the badgers? Short answer 'No', but we catch enough of them often enough[5] to make accurate annual population estimates.

First, how did we catch the badgers? The task was made easier because, thanks to them living in setts, we knew where to find them. Our seasonal capture regime split the woods into quarters (halves prior to 1993), with about 15 setts per sector. Each sector received 80–100 cage-traps, each provisioned with a handful of peanuts,[6] set over 3 nights;[7] thus a trapping session took 12 days. When the badgers are obliging, we often handled 20+ badgers on the first morning of each quadrant, 38 being the daily record on 15 October 1990. Once trapped, most badgers were remarkably calm and, by the time we checked the traps shortly after dawn the next morning, soundly asleep; indeed, it was often difficult to rouse them (for a detailed assessment of trapping welfare, see Sun et al. 2015). Our traps were strong and simple: half-inch steel mesh (any larger and they might try to bite or claw through, risking they injure themselves), with a simple string-trigger door (string, unlike a metal treadle, is easily replaced and can cause no harm to the incumbent).

Traps are heavy (8 kg when empty); in order to ensure that they remained in one place, we bedded them into the soil. It was arduous work, so we didn't want to up-earth traps at every capture. Instead, we transferred each badger to a smaller, 4 kg, holding cage (giving us a mere 16 kg to carry). The transfer typically entailed waking the badger, which, if familiar with the process would generally trot obligingly from one cage to the other, the holding cage door having slid shut behind it. Very occasionally we would catch a non-compliant, snarling adult, and almost always these turned out to be immigrants to the wood, unfamiliar with the procedure. A short ride, shrouded by a black-out tarpaulin, in the back of a pick-up truck, or seated in a rubber mat-lined ATV trailer, delivered the badgers to a barn, to the rear of which a closed room was grandiloquently named our 'Badger Processing Facility' (Photos 11.5). And here each transport

---

[3] Two of the earliest known fossils species of genus *Meles*, *M. polaki* and *M. maraghaus* (Osborn 1910), emerge from the Lower Pliocene of Iraq, but it is not until the Upper Pleistocene (mid-Villafranchian, 2 million years ago (Mya)) that fossils of Thoral's badger, *M. Thorali* Viret (Kurtén 1968), appear in Europe at St-Vallier, near Lyon in France. Deeper into the phylogenetic past, *Melodeon* and *Promeles* were both Upper Miocene—so 11–5 Mya.

[4] These figures are drawn from our most recent (2020) analyses. Small differences between them and figures elsewhere in the text arise where we cite earlier analyses using older methodologies or earlier renditions of recaptures, subsequently augmented (in this case from 2002).

[5] Using the standard MNA calculations we calculated that we caught *c*.85% of the individuals we knew to be alive each year, because eMNA takes into account badgers subsequently caught at a later age, the estimated trapping efficiency using that method falls to 73.1% of the badgers alive that year.

[6] Badgers demonstrably like peanuts; they do not dissolve in the rain, like dog chow, they do not moulder quickly in the sun, like meat, and they are relatively inexpensive. In contrast, our colleague, Yayoi Kaneko, has found that her Japanese badger population (Chapter 19) can be tempted by nothing short of fried chicken and toffee-flavoured popcorn.

[7] Two prior to 1993.

Photo 11.1 A marked badger returns to its sett.

Photo 11.2 Behind the computerfuls of data lies the earthy, physical, and tough task of keeping the equipment going (photo courtesy A.H. Harrington).

cage was positioned up on racks[8] and covered, because darkness fostered their relaxation.

Badgers were then sedated sequentially. Most were easy to inject with anaesthetic (generally while they were, once again, sleeping). Having trialled many alternatives, we found that simple ketamine hydrochloride performed best (see McLaaren et al. 2005; Sun et al. 2015; Sugianto et al. 2019a). It took *c.*3 min for the badger to be fully sedated. Each was then identified from a tattoo on the inner thigh (cubs, and occasional newcomers were freshly tattooed). Thereafter, we logged details including weight, length and fat scores, to blood and subcaudal gland sampling, flea counts, and tooth wear; in addition, we administered enemas to collect

faeces. Some were fitted with tracking collars. Ten minutes later, the processed badger, still unconscious, was returned to its holding cage. They could stand after *c.*30 min, and after 3 h were ready for release back at their exact site of capture (Photos 11.6), where we reset the traps. A badger would often enter cage-traps on 2 or 3 consecutive nights, repeat abductees in this alien process; we recognized recaptures from a squirt of blue sheep dye administered the previous day, and released them immediately.[9]

In passing we should lay to rest the straw man that the overall increase in Wytham's badger population that we documented had, with delusional circularity, been fuelled by the peanuts we used to lure badgers into our cage-traps (Roper 2010). We did not routinely pre-bait our cage-traps, merely provisioning each with *c.*250 g (about a handful) of peanuts on each of the 3 capture nights in each season. This is a trivial snack in terms of badger energetics (see Macdonald et al. 2009; Chapter 14), especially as badgers do not digest peanuts effectively. Furthermore said nuts were variously pilfered by birds and small rodents that squeeze through the trap mesh freely. Ultimately, the low standard error of our long-term mean trapping efficiency (e.g. 6.8% from 1988 to 2016) rules out any notion that the trap bait could drive changes in badger numbers (but see risks to short-term studies in Noonan et al. 2015d; Chapter 12). Each typical trapping session (excepting winter) added over 100 badger capture data

---

[8] Putting the badgers at a better height to sedate, and keeping them off the ground in case of soilage.

[9] All captures were made under Natural England licence (most recently 2019–38,863, Badger Act 1992) and all animal handling procedures were made by qualified Personal Individual Licence (PIL) holders under Home Office licence (most recently PPL 30/3379, Animals (Scientific Procedures) Act 1986).

(a)
(b)
(c)

Photos 11.3 Steps in routine catch-ups, (a) a large cage-trap baited with alluring peanuts, (b) transfer from cage-trap to handling cage, (c) transport to field centre.

Photo 11.4 A marked badger enters its sett and our database.

points to our records (summer 1995 set a record 264 over 12 days).

Nevertheless, our trapping efficiency was imperfect, as a proportion of badgers evade efforts; a fact which, in itself, means that our trapping policy alone does not reveal accurately the number of badgers extant in the population. A better answer is deduced by taking the actual numbers of individuals caught per year and then retrospectively also adding back in badgers that we subsequently captured and which must therefore have

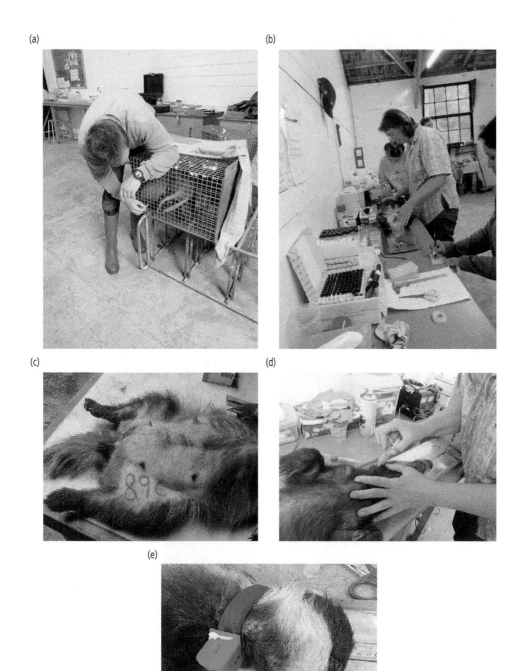

Photos 11.5 Badgers at the field station are kept calmy under a dark drape until (a) anaesthesia, (b) weighted and measured, (c) their tattoo recorded, (d) sampled and (e) instruemented

(a)

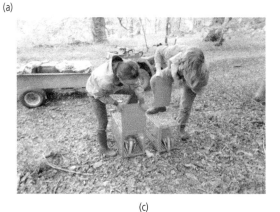

(b)

(c)

Photos 11.6 The moment of release (photos: A.H. Harrington).

been alive during that preceding year—assuming they were present in this relatively closed population (Macdonald et al. 2008; following Macdonald and Newman 2002). This is termed the minimum number alive (MNA) estimator and, while it might lead to an underestimate, we have favoured it—not least because every badger in the population structure has an associated identity number, sex, and age. Other approaches have their drawbacks—for example, earlier approaches to capture–mark–recapture (CMR) statistical models (e.g. Jolly–Seber models; Jolly 1965) consistently overestimated badger numbers in populations where trapping efficiency is high (Rogers et al. 1997) (although new implementations of Cormack–Jolly–Seber are better able to reconstruct a population). Furthermore, CMR techniques infer the presence of individuals that were never caught, and thus for whom we clearly couldn't have the aforementioned additional biometrics, which would be analytically awkward when exploring the corollaries of survival, such as needing to know the badger's weight (Chapter 14), or what parasites lurked in its bloodstream (Chapter 15),

or questions of genetic pedigree (Chapter 17). Even given the tangibility of MNA estimates, sticking with animals we have known personally, some finessing was able to increase our accuracy. Led by Julius Bright Ross, who came to WildCRU from Easter island via Harvard, we adopted a modification based on trapping efficiency. In the population, there are likely to be badgers captured in a previous year that remained alive for some duration but were never trapped again—if, as can occur, one year has particularly low capture efficiency (e.g. 2006 has the lowest capture efficiency of any non-terminal year in our dataset, at 50.0%, due to the disruption of Chris and Christina moving their home to Canada), this can result in artificial drops in MNA estimators. Instead, we can retroactively apply each year's trapping efficiency (the proportion of known-alive individuals captured that year) to estimate the diminishing likelihood that a given 'vanished' badger was still alive but undetected in subsequent years. By summing the likelihoods of these 'shadow' badgers being alive to the individuals demonstrably alive, we derive an 'enhanced' MNA (eMNA) estimate more robust to the vagaries of inter-annual variation in trapping efficiency[10] (Bright Ross et al. 2022[11]). When badgers were not caught initially as cubs, we could further inform our estimates by basing an educated assessment of their age on their tooth wear (see da Silva and Macdonald, 1989) (Box 11.2; Figure 11.2).

How thorough was our counting?

The difference between the number of badgers caught per year (the amalgamation of those caught each season) and the enhanced MNA estimate indicating who was actually around gives us a measure of our annual trapping efficiency, TE. Because badgers are variously preoccupied with foraging and reproduction, and variously lethargic, at different times of the year, we would expect TE to vary with the seasons. Figure 11.3 depicts that per seasonal trap-up variation between 1988 and 2016 peaks at a mean of 50.8% (standard deviation (SD) 10.0%) in summer, enhanced because naïve cubs and dry conditions increased the incentive for badgers to sample the crudités of peanuts in traps, falls off a bit in spring and, more so, autumn, and then being much lower during winter (mean 19.1% SD 9.1%), in major part because badgers lie low during a period of winter-induced lethargy (see Chapter 12). Interestingly too, while these ad hoc winter trappings were primarily directed at screening for pregnancy using ultrasound (see Chapter 2), we caught an inconvenient preponderance of males (mean F:M capture ratio: 0.77:1 ± 0.31 SD), causing us to infer that pregnant females are less inclined to emerge above ground to forage and to enter traps (see Chapter 13).

Taking an annual view of TE, imagine a year where we catch 160 unique individuals (some on several occasions), but later know that the population comprised at least 200 extant badgers, trapping efficiency would be 160/200 = 80%. As mentioned earlier, this simple MNA technique is the one we used to use: applying it from 1988, annual TE remained relatively constant throughout; with means ranging from 83.69% (SE = 1.32%) (1988–2001; Macdonald et al. 2009)

revised down to 81.3% ± 6.1 SD with the addition of later years (1994–2012; Noonan et al. 2015d). This gave us 95% confidence of recapturing any surviving individual badger within 525 days (based on 6193 capture events; Dugdale et al. 2008). The proportion of individuals of known age, that is, tattooed in year of birth, as a cub, was also consistently high: the most recent figures are 1319 of 1823 individuals (72.3%) between 1987 and 2018. Subsequently, as earlier, we've used enhanced MNA, but either way the point is that we caught a large majority of the badgers each year.

Of course, sceptics may argue that MNA (and eMNA) can only account for badgers caught and thereby given an identity on our database, so perhaps a proportion of trap-averse animals may perpetually skulk around undocumented? Not so: to facilitate our infrared video observations (Stewart et al. 1997), there have been periods when we fur-clipped (for visual identification) every badger we trapped, and watching them under video surveillance added to our confidence that no significant numbers of unaccounted ghost badgers were eluding us (Baker et al. 2005, 2007; Buesching et al. 2003; Hewitt et al. 2009) (see Chapter 8). Similarly, our analyses of family trees (detailed in Chapter 17) revealed that the parents and parenthood of each individual could generally be found in our sampled sample, where assignment failures were technical, rather than due to absenteeism (Dugdale et al. 2007, 2008; Annavi et al. 2014a, 2014b). In short, if we missed some individuals, it was very few.

What did our history of annual MNAs reveal in terms of population change? Applying Julius' innovative method we had retrospectively estimated that at the project's inception in 1987 the population comprised 137 adults and 27 cubs (Figure 11.4). Earlier, before eMNA was invented, we'd estimated 93 adults and 30 cubs (Macdonald and Newman 2002) and even that had seemed an extraordinarily high density (c.23 badgers km^{-2} over the 6 km^2 range used by the population), so we were thunderstruck when eMNA reported an increase to an astonishing peak of 328 individuals in 1997 (63 cubs), with a secondary peak of 319 badgers in 2005 (with a whopping 97 cubs that year). Following this secondary peak, the population remained stable at lower numbers, but still very high density compared with other populations; with 242 adults ± 15.14 SD, range = 222–263) and 66 cubs (±8.1 SD, range = 47–97) from 2005 to 2009. Thereafter, following high, and unexplained, mortality across the board in 2010, it settled down to a slightly lower but stable phase through

---

[10] Any discrepancies between figures of this sort, in this book, and those in our earlier cited papers arise where we have updated calculations using refined methodologies, e.g. here we give enhanced MNA equivalents of simple MNA values.

[11] Note that the precision of eMNA is lower for the first year and last 3 years of a study at intermediate trapping efficiency. Technically, newer Cormack–Jolly–Seber probabilistic models are more variable than eMNA is; probabilistic models are more accurate, but less precise.

Box 11.2 Age estimation and worn teeth

Over our complete dataset 1987–2019, the majority of individuals (1319/1823 individuals, 72.3%) were first caught as cubs, and thus their birth years were known absolutely. A further 128 individuals (7.0%) were captured shortly after cubhood and their age was estimated based on their adolescent appearance. For the remaining 376 individuals (20.6%, 100 of which were from the first 3 years of the study) whose age was unknown at first capture, age was inferred using molar tooth wear on a 5-point scale (Hancox 1988) (Photo 11.7).

Although tooth wear in badgers can vary substantially within an age class and between populations (da Silva and Macdonald 1989), our large number of captures for known-age individuals ($n = 9074$) all in the same environment (Wytham) enabled us to use a statistical procedure to approximate inferred age for individuals for whom age was unknown. For each capture, tooth wear is marked independently of previous entries for a given individual, and therefore some error in estimation can occur. These individuals' life records were then updated to the more sophisticated enhanced minimum number alive (eMNA) estimates we describe.

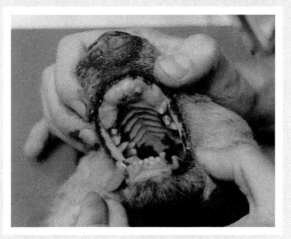

Photo 11.7 An elderly badger with teeth worn to stubs from a lifetime of grinding the earthy gut contents of earthworms.

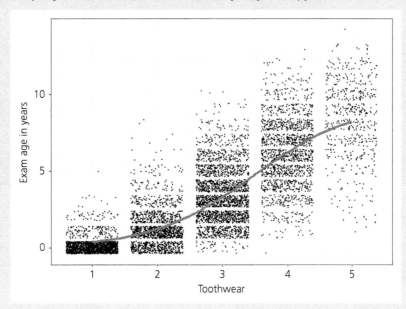

Figure 11.2 Exam age in years as a function of molar tooth wear at time of capture on a scale of 1–5. Points are jittered on both axes for visual clarity. Red line indicates sigmoid fit to the data (from Bright Ross et al. 2020, with permission).

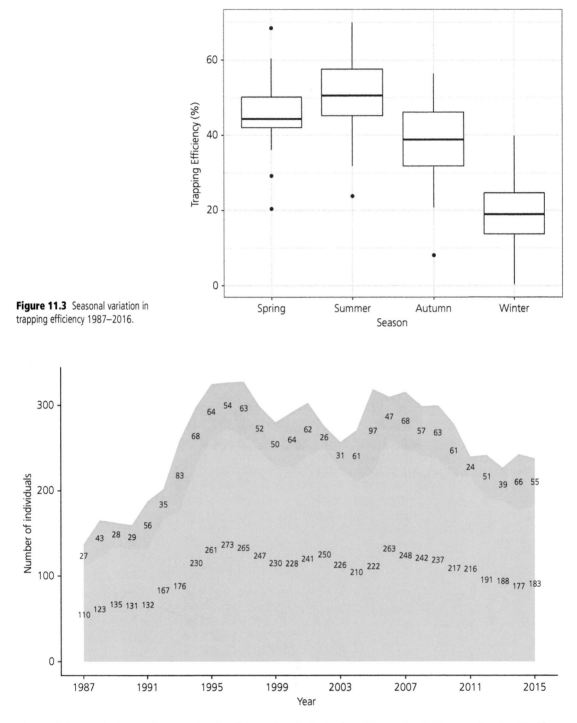

Figure 11.3 Seasonal variation in trapping efficiency 1987–2016.

Figure 11.4 Population MNA graph: applying trapping efficiency-adjusted estimate of population numbers (minimum number alive: MNA) for adults (below/pale green) and cubs (above/dark green) from the first year of the study to 3 years prior to the last capture in our database (adapted from Bright Ross et al. 2020, with permission).

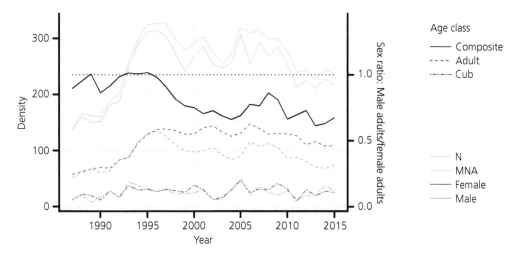

Figure 11.5 Density estimates for the population from 1987 to 2015. N_i (eMNA) and adult male and female density estimates are corrected for trapping efficiency. Sex ratio of male:female adults (black line) is also provided, with a null equal sex ratio indicated by the dotted black line (from Bright Ross et al. 2020, with permission).

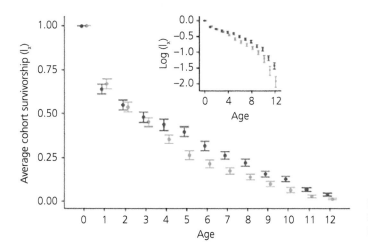

Figure 11.6 Average cohort survivorship on a linear scale and (inset) on a logarithmic scale (from Bright Ross et al. 2020, with permission).

2015, comprising 195 adults (±17.06 SD, range = 177–217) and 49 cubs (±15.47 SD, range = 24–66; Bright Ross et al. 2020).

Concluding our study with over 40 badgers km⁻² in 2019, Wytham has consistently had the highest density of European badgers ever recorded (Macdonald et al. 2009, 2015e; Bright Ross et al. 2020) (Figure 11.5).

This overall variation, however, conceals other important intricacies, including a shift in sex ratio, as will emerge.

Survival dynamics

How do badger annual losses in the population ledger, potentially arising from mortality or emigration, vary with age, and what potential does this have to impact the population? Wytham has a relatively closed badger population, owing to the fact that there is little alternative real estate available in the near vicinity that is not prone to winter flooding (Chapter 4).

Figure 11.6 presents the typical Wytham badger's life assurance policy.[12] Although little more than half (54%; l_x column, Figure 11.6) of the badger cubs emerging

---

[12] Insurance provides financial coverage for unforeseen circumstances surrounding an event, such as fire, theft, or flooding. Assurance, by contrast, provides coverage for events that are certain to occur, such as death. A life assurance policy will always result in a payment being made because the investment is combined with the sum insured.

in time to join our May catch-up survive to breeding age (2 years), which usually commences seriously in their third year (Chapter 2), those that make it become a much better (albeit still unpromising) insurance proposition: 33% of the 10-year sample (1987–2018) (Figure 11.6) were still going strong after their fifth birthday, and 18% survived beyond 8 years. Furthermore, in our early analysis we detected just 1% of badgers surviving to 10+, whereas our later overall analysis found 6% of males achieving this senior citizen status and 12% of females. Females had gained some crucial advantage—a theme to which we'll also return in Chapter 14 (POLS-seg). Only about 2.8% of individuals reached teenage years of 13 or more, and, over 31 years of data, our record-setting longest-lived badger (whose age was known from birth) was F699, succumbing at 15 years of age (born at Chalet in 1997, moving to Hill Copse then to Keeper's Hut (a Hill Copse outlier) where she was last caught 2011: during which time she bore 9 cubs to 4 or more fathers) (see Box 11.3).

Box 11.3 Venerable badgers

Let us digress for a moment to delve into the life history of venerable badger: F699. She had her first cubs (M727 and F232) as a 2-year old in 1999, born at Chalet to an unknown father. Her next litter arrived in 2000 (a singleton F758) also born at Chalet, fathered by M517 from Great Ash Hill. The next year, 2001, she bore another singleton (M862) at Chalet, fathered by M696 from Chalet Outlier. Following a gap of 2 years, in 2004 and aged 7, she bore triplets (M962, F971, and M989), this time born at Hill Copse, and father by M893 from Chalet. A year later, 2005, she had another singleton (M1044), again born at Hill Copse, and again sired by a Chalet Outlier male, but a different one, this time M857. Finally, in 2009, by then aged 12, her last cub (F1266) was born at Keeper's Hut to cohabiting male M1042. This life history is fascinating as a peephole into a badger's private life, but it also provides an extreme illustration of why we shifted our methodology to the enhanced MNA technique: none of three fathers, M517, M857, and M1042, was caught in the conception year (the year prior to birth) or thereafter; their progeny, however, reveal that they were not dead. With eMNA, not only were they scored as alive in those years, but trapping efficiency in subsequent years was used to extend estimates of their continued persistence, uncaught.

But F699 may not have been absolutely our oldest badger: when we first met Badger M32 in 1987 he was already an adult; his worn teeth convinced us he was at least 5 years old and, if so, he too may have been around 15 when he went missing, presumed dead, in 1996. He was steadfastly resident at Chalet sett throughout his life (getting caught 28 times) and assigned just one cub (F85, born to mother unknown, in neighbouring group Great Oak in 1988, and not caught again after 1989, who left him no grandchildren), although likely some of his youthful reproductive activity predated our surveillance.

Based on actuarial data, a life assurance policy would initially cost female cubs slightly more than it would male cubs (Macdonald and Newman 2002). That is, their survival rate (the l_x column in Table 11.1) was marginally lower for youthful females (Figure 11.7), with 66.8% of male cubs surviving their first year, compared with 63.8% of females (despite similar cub sex ratios per annual cohort, see the text that follows). Nevertheless, in an overall analysis (1987–2019; Bright Ross et al. 2020) of total lifespan, males lived on average to 3.48 years and females to 4.08 years (median age at last capture = 2 for males, 3 for females), and survived each year-to-year transition less well than females (lower lx on (Figure 11.7) making males the costlier assurance prospect throughout their adult lives.

This sex differential in mortality probably drives the divergence in sex ratio over time, graphed (Figure 11.5).

The oldest age class (age 8+) showed the steepest decline in survival rates, from 77% (±0.02 SE) in 1988 to just 61% (±0.03 SE) in 2015. Interestingly, too, as the population grew from 1988 to 2015, the proportion of females assigned offspring also decreased in all age classes—but bigger litter sizes must have compensated for this amongst most cohorts, as the per capita rate of cub production did not decrease; in parallel (to a similar extent), the proportion of males assigned offspring decreased amongst those aged 5–7, but for males aged 2, 3–4, and 8+ there was a smaller decline over those same years (indicated by a significantly flatter slope than for females) (Figure 11.7). So, to return to our opening quotation, if death is nature's way of making life interesting, what did, and does, kill badgers?

Palaeolithic cave middens show that people, as well as wolves, were predatory threats to badgers (Charles 1997), and who knows whether their nightmares still recall the cave hyaenas and cave wolves described in the Zoolithen Cave by German palaeontologist Goldfuss in 1823. We have shown that the memory of wolves howling no longer haunts badgers as a ghost

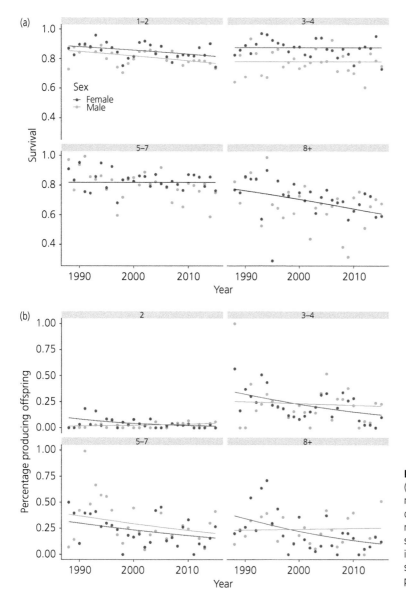

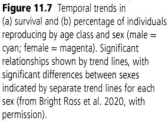

Figure 11.7 Temporal trends in (a) survival and (b) percentage of individuals reproducing by age class and sex (male = cyan; female = magenta). Significant relationships shown by trend lines, with significant differences between sexes indicated by separate trend lines for each sex (from Bright Ross et al. 2020, with permission).

of predators past (Clinchy et al. 2016), but, as detailed in Chapter 5, the sounds of people chatting on Radio 4 causes a timid reaction (and it's not long since these voices might have been those of badger baiters—indeed, illegally, this remains the case in some regions, and some shaving brushes are still made from protected badgers; Domingo-Roura et al. 2006). People still take their toll on badgers: of the adult badgers that disappeared from one year to the next at Wytham, we came across almost half (43.75%) of them on roadsides,

the victims of road traffic accidents, and many victims of vehicles probably limp home to die out of sight, as they do of other causes, making it hard to attribute causes of death; elsewhere, however, we describe those insidious constant companions of debilitating diseases (Chapters 15 and 16), and in Chapter 18 we drill further into the attrition of age. As we shall see, badgers also succumb to the ill-effects of hunger when weather, affecting food supply, is inclement—the topic of Chapters 12–14.

A change of state: through the ceiling

An ecological regime shift, from one state to another, is an unusual thing to see in nature, and frankly we didn't expect to see one. Badger numbers had been nudging up, from estimates of 137 in 1987 to 188 by 1991; as a result, we could almost hear their heads bumping against the ceiling of resource availability. Then everything changed ... by 1996 numbers were to crest 327 (273 adults, 54 cubs)!

An obvious possibility was that perhaps something had held back the population's potential prior to the 1990s. For example, had Wytham's badgers previously been culled? No. Wytham Woods has been a nature reserve managed and maintained by the University of Oxford since 1943 (see Savill et al. 2010), and neither badgers nor their setts had been persecuted on this estate for decades previous—certainly not since Hans Kruuk commenced his studies in 1973—coincidentally this was also the year in which the UK government introduced the original Badger Act (now superseded by the 1992 Act of further protection), prohibiting 'taking, injuring or killing' badgers (see Chapter 16).

No: what we documented, although unexpected, was a change in environmental carrying capacity that precipitated a genuine step change in population abundance and various intrinsic processes from one state of apparently substantial badger occupation (during Hans Kruuk's studies), to a state of even higher density that peaked in 1997.[13] Stenseth and Imms (1993) termed this type of transition a phase-dependent change in demographic responses, or, as Reznick et al. (2002) termed it, state-shifting. It was not that before and after this change, the badgers were lackadaisical in striving to exploit carrying capacity maximally; they strove with equivalent vigour, but simply the ceiling of opportunity was lower.

By analogy, driving a ball out of a steep-sided cup takes effort to overcome 'resistance', and thus a change of state is unlikely given that the ball is inclined to roll back—an ecological tendency to a steady state called 'resilience' (Folke et al. 2004).[14] In Wytham, in the mid-1990s the edges of that metaphorical cup melted away,

allowing the ball to come to rest in another, different, cup; that is, the population achieved an alternative stable state—it broke through a resistance level and then shifted to a different resilient position. But, as we shall see, this is not simply an issue of changes to the stage on which the theatre of population regulation is played out. In addition, the changes we witnessed required the actors to play different roles. In most animal populations, stringent regulatory mechanisms generally mean that individuals continue to behave at high density much as they did at low density—that is, the actors are typecast whatever their numbers—so the change of state we documented in Wytham was as theoretically significant as it was naturalistically remarkable.

Before we describe exactly what changed, we first elucidate the internal mechanisms that affect population change: how do ecologists measure a population's capacity to increase (or decrease)? The answer requires us first to explain some underlying algebra, presented in Box 11.4.

Mindful of the technicalities in Box 11.4, what changed in the circumstances of the badgers of Wytham Woods during this key transitory phase, allowing numbers to reach a previously unrealized peak by 1996? This question was already preoccupying us when we had data only up to 2001 (Macdonald et al. 2009). Then with the change still a novelty, we looked first at the actors—had they escaped a history of typecast roles? Contrasting the period of increase, 1987 and 1996 ($\lambda > 1$), against stuttering decline and restabilization of the population from 1997 to 2001 ($\lambda \leq 1$), we looked amongst a series of key vital parameters (juvenile survival, adult survival, fertility, age at first reproduction, and age at last reproduction) for which might best explain this step change. We did so using yet more sophisticated algebra, some of which we explain in more detail in the text that follows (matrix vs life table response experiments (LTRE)) (Macdonald et al. 2009).

Over our complete early study interval (1987–2001) the actual value that best fitted the population growth pattern was $\lambda = 1.065$ (Table 11.2). Dividing this into 1987–1996, $\lambda = 1.13$ (strong growth), whereas latterly, for 1997–2001, $\lambda = 0.96$ (close to stability/slight decline). Similarly, during this initial 10-year period (1987–1996), we observed a relatively greater survival and fertility rate, compared with the overall picture (Figure 11.8a and b), whereas during the transitionary 5-year period (1997–2001), both survival and fertility rate were lower (Table 11.3).

---

[13] Strictly 1997 had 328 badgers, 1 more than 1996, but fewer adults—63 cubs, 265 adults.

[14] Resistance is the ability for an ecosystem to remain unchanged when being subjected to a disturbance or disturbances. Resilience, by contrast, is the ability and rate of an ecosystem to recover from a disturbance and return to its predisturbed state.

Box 11.4 Quantifying population growth rate

Almost a century ago, Raymond Pearl[15] (Pearl and Parker 1921) introduced ecologists to the 'life table'. This clever device provides an age-specific summary of mortality rates (summarizing those 'rates' first conceptualized by Graunt). Your life assurance company uses something very similar, looking at the age-specific mortality of whatever demographic you fall into, from chain-smoker to sky-diver. Similarly, the sexes are treated separately, for instance, relating to different susceptibility to diseases, or childbirth. We provide a life table illustrating the survival dynamics of Wytham's badgers 1987–2016 in Table 11.1. Each column in the table is depicted by a symbol; symbols used consistently throughout population ecology.

- x = the age interval (in the case of badgers, living from 0 to 10+ years).

- N_x = the number of survivors (members of the starting cohort, often modelled on 1000) at the start of age interval x.

- l_x = proportion surviving to the next age class (so $l_x = n_x/n_0$).

- d_x = number dying over the interval x to $x + 1$ (so $n_2 = n_1 - d_1$).

- q_x = rate of mortality over interval x to $x + 1$ ($q_1 = d_1/n_1$).

- e_x = further expectation of life for individuals alive at the start of interval x.

This final metric involves a more complicated calculation: first we obtain the average number of individuals alive per interval, which is termed L_x (capital L this time), or age structure, so: $L_x = (n_x + n_{x+1})/2$. We then sum all these L_x values together from the bottom of the table, giving us the column T_x. Finally $e_x = T_x/n_x$.

From Table 11.1, we see that the median age for a badger, where proportion of the cohort surviving (l_x) reaches 0.5 (50%), occurs prior to age 3; thereafter some surviving individuals can achieve old age, and, as mentioned earlier, they shift the mean male average lifespan to 3.48 years, and that of females to 4.08 years with a long survivorship tail. This pattern is, for no obvious reason, called a Type I survivorship. Type I or convex curves are characterized by high age-specific survival probability in early and middle life, followed by a rapid decline in survival in later life—the long tail indicating that a relatively few individuals soldier on way after their contemporaries. This sort of life history is typical of species that produce few offspring but care for them well, including humans and many other large mammals.

Knowing these values positioned us to take a second step, to build a survival table. This time the credit goes to Alfred Lotka,[16] who in 1925 derived a function he called the *natural rate of population increase*. Sparing too much complexity, this approach adds age-specific fecundity to the survival schedule (incidentally, facilitating the evaluation of comparative reproductive value by age). It is typically modelled for females only, because only females actually produce offspring.

In our 2002 analysis we ran a very basic population growth projection based on a starting point of 40 badgers—imagining a scenario in which Hans Kruuk may have stumbled on a medium-density badger population in the early 1970s, to see where this would lead. Realistically, we took half of these badgers to be female. Based on the life table, 14 of these 20 females would thus be of breeding age, and if, at this low density starting point, they produced a near maximal litter size of 3 (see Chapter 17), of which 1.5 cubs would be female, they would give rise to 21 female cubs. This gives us a column symbolized as b_x: the number of female offspring per female aged x. Mortality rates were projected at 30% for cubs and 20% for adults.

[15] An American biologist from Johns Hopkins University, Pearl was a prolific author of journal articles and a committed science communicator. He went on to found the subject of biogerontology, to which we will return in Chapter 18.
[16] Lotka was a physical chemist by training, and one of the first transdisciplinary population ecologists, as he sought to bring the quantitative rigor of physics and chemistry to biology.

Box 11.4 *Continued*

Table 11.1 Vital population parameters calculated from the full mortality schedules for cohorts from 1987 to 2006 and from the partial schedules for 2007–2016, extrapolated onto a starting population of 1000.[17]

Overall

Age	n_x	l_x	d_x	q_x	L_x	T_x	e_x
0	1000	1	342.44	0.34	828.78	3807.82	3.81
1	657.56	0.66	104.78	0.16	605.16	2979.04	4.53
2	552.77	0.55	85.04	0.15	510.25	2373.87	4.29
3	467.73	0.47	75.07	0.16	430.19	1863.62	3.98
4	392.66	0.39	65.45	0.17	359.93	1433.43	3.65
5	327.21	0.33	63.13	0.19	295.65	1073.5	3.28
6	264.08	0.26	48.63	0.18	239.77	777.85	2.95
7	215.45	0.22	34.41	0.16	198.25	538.08	2.5
8	181.04	0.18	50.36	0.28	155.86	339.83	1.88
9	130.69	0.13	37.26	0.29	112.06	183.96	1.41
10	93.43	0.09	43.05	0.46	71.91	71.91	0.77

Males only

Age	n_x	l_x	d_x	q_x	L_x	T_x	e_x
0	1000	1	330.75	0.33	834.63	3467.86	3.47
1	669.25	0.67	128.88	0.19	604.81	2633.24	3.93
2	540.37	0.54	90.06	0.17	495.34	2028.42	3.75
3	450.31	0.45	101.28	0.22	399.67	1533.08	3.4
4	349.03	0.35	90.08	0.26	303.98	1133.41	3.25
5	258.94	0.26	45.18	0.17	236.36	829.43	3.2
6	213.77	0.21	41.48	0.19	193.03	593.07	2.77
7	172.28	0.17	29.15	0.17	157.71	400.05	2.32
8	143.13	0.14	46.1	0.32	120.08	242.34	1.69
9	97.03	0.1	36.36	0.37	78.85	122.26	1.26
10	60.67	0.06	34.53	0.57	43.41	43.41	0.72

Females only

Age	n_x	l_x	d_x	q_x	L_x	T_x	e_x
0	1000	1	353.64	0.35	823.18	4133.67	4.13
1	646.36	0.65	81.72	0.13	605.5	3310.49	5.12
2	564.64	0.56	80.24	0.14	524.52	2704.99	4.79
3	484.4	0.48	49.55	0.1	459.62	2180.47	4.5
4	434.85	0.43	42.05	0.1	413.82	1720.85	3.96
5	392.8	0.39	81.15	0.21	352.22	1307.02	3.33
6	311.64	0.31	55.46	0.18	283.91	954.8	3.06
7	256.18	0.26	38.71	0.15	236.83	670.89	2.62
8	217.47	0.22	54.59	0.25	190.18	434.06	2
9	162.88	0.16	37.88	0.23	143.94	243.88	1.5
10	125	0.12	50.11	0.4	99.94	99.94	0.8

[17] Note, this life table is calculated using only known-age individuals, rather than using eMNA estimates. This is due to a peculiarity of the eMNA estimation process: as there are only five tooth wear scores, when we catch an adult badger with no tattoo, there are only five ages we can estimate for it—if, for example, two of those ages are 6 and 8 years (as is the case, see Figure 11.2), then there will actually be some of those individuals that were age 7. Over the long run, new badgers are caught that plug those holes and smooth out the population estimate (hence the use of these eMNA metrics in Figure 11.3), but because cohorts are of different sizes when they are born, it is better to study the specific drop-off of the known-age individuals rather than risk skewing it with individuals that were assigned erroneously to that cohort.

Box 11.4 *Continued*

From this we need to get to the net reproductive rate, R_0, which is the number of daughters born in generation $t + 1$ divided by the number of daughters in generation t; that is, the multiplication rate per generation (by the way, if readers feel that R_0 seems familiar, that is because it is exactly the same R_0 that became familiar in national news about Covid, but here applied to an animal population rather than a virus). Mathematically:

$$R_0 = \sum l_x b_x$$

But how long would a generation (G) take under such a scenario? Crudely, that time elapsing between the birth of parents (mother) and their offspring (daughter)? This is calculated as:

$$G = \frac{\sum l_x b_x x}{\sum l_x b_x}$$

Finally, innate capacity for population change (r_m), that is, its biotic potential for maximal increase under ideal conditions (unlimited resources, no hindrance), is calculated as:

$$r_m = \frac{log_e(R_0)}{G}$$

From this, we can model how much larger the population will be in the future (time N_t) than at the start (N_0), rather like projecting the future value of an investment portfolio based on an interest rate (r_m).

Population size at time $N_t = N_0 e^{rmt}$.

Where e = a natural log constant of 2.71828.

From this simple projection of how a relatively unconstrained and expanding population would perform, we estimated a generation length at around 3.09 years,[18] net reproductive rate around 2.34 and r_m approximating 0.27; converting r_m to a finite rate of increase λ, where $\lambda = e^{rm}$ suggested that that population *could* expand at a rate[19] of 1.31 per individual (131%), per year (Macdonald and Newman 2002). Based on these maths our starting population of 20 female badgers (14 breeders) could grow to 104 (females), the number the woods typically support, in only 6 years.

A vital realization from all this is that the badgers of Wytham Woods have consistently had the capacity, if unchecked by intrinsic and extrinsic mortality factors, to increase well beyond the numbers we observed—much as, if unchecked by social distancing and other containment measures, the homologous R-value metric now familiar from Covid-19 would have exceeded one and had the capacity to run even more rife through society. Without these constraints, a simplistic Malthusian nightmare imagines that badger numbers would rocket on, if unchecked, to crest 1000 females (1229) in just 15 years (Macdonald and Newman 2002), vastly exceeding its consistent range of 220–230 individuals that persisted through the last decade of our study (Figure 11.5). So why isn't the earth covered knee-deep in badgers? Much as Darwin (1859) demonstrated for the reproductive potential of elephants (see Podani et al. 2018), the answer is that ultimately each species becomes bounded by the niche space available to it. Remember, not all available earthworms are eaten by badgers, a proportion also goes to foxes, owls, moles, shrews, and a host of others. Even within a species, eventually there are too many mouths for the food supply to feed—at that point the population encounters density-dependent regulation (Murdoch 1994), which ultimately curbs exponential growth. These circumstances create competition between individuals (or species), in which some will have selective advantages over others—all considerations to which we will return (Chapter 14).

[18] This is a crude prediction of how a fecund population could potentially expand, below the carrying capacity ceiling. Later we report that the true value turns out to be closer to 6, because not all individuals actually perform to this colonizing level.

[19] For readers who became familiar, during Covid-19, with the so-called contact rate of infectious disease (R, for reproductive number), you will recognize this as the same parameter (which needs to be above one for the disease (or in this case, the badger population) to spread), so if R is two, two infected people will, on average, infect four others, who will, in turn, infect eight others, and so on.

Box 11.4 *Continued*

Table 11.2 Projected Wytham badger population growth, modelled from a starting population of 40 badgers of which 20 are female. Of these 14 are of reproductive age, bearing typically 1.5 female cubs each, thus giving rise to 21 offspring (for full details see Macdonald and Newman 2002).

X	n_x	n_b	l_x	b_x	$(l_x{:}b_x)V_x$	$l_x{:}b_x{:}X$	Year	Population (N_t)
0.00	20.00	0.00	1.00	0.00	0.00	0.00	1974	20.00
1.00	6.00	1.50	0.69	0.25	0.17	0.17	1975	26.32
2.00	5.00	6.00	0.55	1.20	0.66	1.33	1976	34.64
3.00	4.00	6.00	0.42	1.50	0.63	1.89	1977	45.58
4.00	3.00	4.50	0.35	1.50	0.52	2.08	1978	59.98
5.00	2.00	3.00	0.23	1.50	0.35	1.76	1979	78.94
6.00							1980	103.88
7.00							1981	136.71
8.00							1982	179.90
							1983	236.75
							1984	311.56
							1985	410.01
							1986	539.56
							1987	710.06
							1988	934.42
							1989	1229.69
							1990	1618.25

See text for meaning of symbols.
$R_0 = 2.34$; $l_x b_x X = 7.22$; $G = 3.09$; $\log e(R_0) = 0.85$; $r_m = 0.27$; $N_t = N_0 e^{rmt}$; $N_0 = 20$.

The rate of growth, or shrinkage, of a population can be plotted against the population density, the slope indicating the extent to which the population is below the carrying capacity of its environment, or has overshot it, to which it is then nudged back. Of course, identifying that equilibrium point in itself identifies this carrying capacity: population expansion decreases as resources become scarce (or at which some other constraint, such as the effects of predation, parasitism, and disease, starts to bite).

Density dependence is evidenced by a logistic or 'S'-shaped growth curve (geometric growth, a 'J'-shaped curve, would suggest unconstrained expansion, as per Malthusian extrapolation to the 1000 badgers hypothetically thronging Wytham) (Figure 11.9).

Our attempt to identify carrying capacity, K, for our badger population (Figure 11.10) was spearheaded by Parisian maths whiz Pierre Nouvellet (now Reader in Evolution, Behaviour and Ecology at the University of Sussex).

The overall 15-year mean carrying capacity for females (the offspring-bearing sex); that is, the point at

which annual per capita population growth equalled zero, was K = 147. Breaking this down revealed that between 1987 and 1996 carrying capacity had been about 132 adult females, whereas from 1996–2001 it rose to 174.[20] Did the population completely manifest this potential? Almost, where, assuming an equal sex ratio, we saw a peak of 273 adults in 1996.

In the late 1970s Richard Sibly had been a dauntingly quantitative student of seagulls in Oxford's Animal Behaviour Research Group, and became a friend to whom David had realized it was wise to listen. So, listen, or at least read carefully, we did when in 2005 Richard published on the subject of analysing population growth (Sibly et al. 2005). His paper revealed that there was much to learn from the shape of the curves

[20] Simulating the dynamics of the population using best-fit parameters (from least square regression using data from the first decade) (Figure 11.5), and using a change in mean carrying capacity from K = 173.6 to 132.1, for the two periods respectively, we observed a very good fit of the expected measures of carrying capacity.

THE ECONOMICS OF SURVIVAL

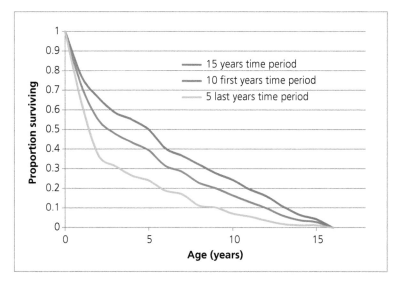

Figure 11.8 Age-specific survival of cohorts over the first 15 years of our study, highlighting differences between 1987–1996 (population growth) and 1997–2001 (population restabilization).

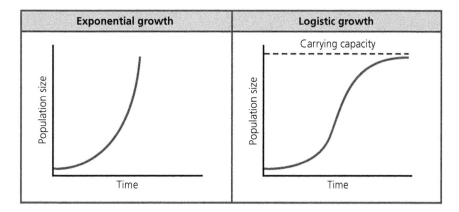

Figure 11.9 (a) Unconstrained population growth; (b) density-dependent constrained population growth.

(Figure 11.12), where this time we were plotting numbers of badgers against their population growth rate (Figure 11.11). The shape of the curve diagnoses the type of density dependence at work.

The pivotal moment

A statistic called 'theta' characterizes these typical density dependence relationships.[21] The Sibly generalization was that territorial animals, such as badgers are classically considered, would respond quickly to the

---

[21] Linear (logistic) when $\theta = 1$, logarithmic when $\theta = 0$, convex when $\theta > 1$, and concave when $\theta < 0$ (Sibly et al. 2005).

effects of density dependence, the case when theta (θ) < 1. Did the badgers follow this expectation? Yes: sure enough, the best fit to our badger data had $\theta = 6 \times 10^{-3}$, a tiny value where theta was so much less than one that it was only slightly greater than zero. This indicates a strong (in this case logarithmic) regulatory effect of density long before badgers approach the ceiling of carrying capacity (so between the blue and orange plots on Figure 11.12; closer to orange). This was a pivotal moment of explosive significance: juxtapose this discovery with the revelation in Chapter 4 that in 1996 we observed a sudden and radical spatial rearrangement of the badgers. Coincidence? No. The stage of our theatre had been reset: over just a few

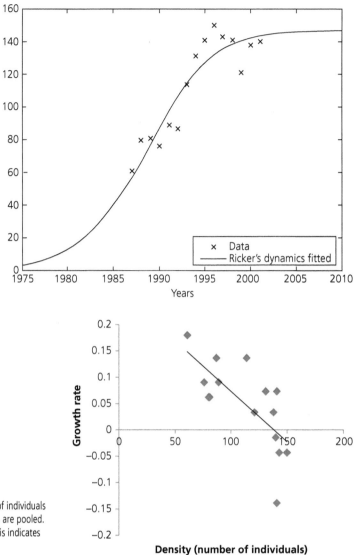

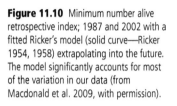

Figure 11.10 Minimum number alive retrospective index; 1987 and 2002 with a fitted Ricker's model (solid curve—Ricker 1954, 1958) extrapolating into the future. The model significantly accounts for most of the variation in our data (from Macdonald et al. 2009, with permission).

Figure 11.11 Linear regression depicting density of individuals plotted against the population growth rate. All years are pooled. The intersection between the trend line and the x-axis indicates estimated carrying capacity.

months, excavations doubled the number of available setts while extending others, allowing more actors to fit upon this expanded stage.

To push our analogy, no matter how much improved the size of the stage, the cast would grow no larger if the actors (the badgers) refused to tread its boards beyond some fixed limit on their numbers. In order for a larger cast to squeeze into the performance the actors had to find a new performance presentation that enabled more of them to share the enlarged stage simultaneously. That new way was to partition the stage into more, but smaller, roles, or setts. It was in this way that

the badgers smashed through a ceiling and then their numbers soared. The Black Swan had landed.

With the restrictions of socio-spatial regulation relaxed in parallel with improved environmental conditions (see Chapter 12), exactly as our earlier projection had suggested, badgers could, in principle, manage a much higher per capita growth rate (a situation analogous to the higher growth rate experienced by founder populations in a new range or after a population perturbation (see Chapter 16). *Quod erat demonstrandum*, or as close to it as we are likely to get with an observational study: the changes we had witnessed in

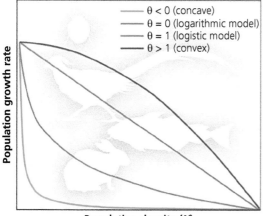

Figure 11.12 Models of how different types of the growth rate of populations (species) typically react to increasing density. (with permission).

badger numbers were at the interface of a sociological response to a physical opportunity. Had the badgers (actors) stuck to their former roles, a ceiling would have restrained their potential for population growth, but witnessing animals change their social geometries and escape typecasting so quickly and adeptly on the cusp of new opportunity is more than unusual, although illustrative of the social plasticity that is emerging as a hallmark of European badgers (Chapter 19).

The late Jerry Wolff, a gifted American field biologist also equipped with witheringly direct theoretical logic, was a frequent visitor to our project at Wytham, to which he brought his compendious expertise on the interaction of intrinsic population reactions to environmental (extrinsic) circumstances (Wolff 1997). Under his influence, our next question was to understand not just what had happened in our population, but also why. The discovery of stages of recalibration over three decades implied three synergistic things: (i) Wytham's carrying capacity for badgers had increased; (ii) population vital rates had responded in kind; and (iii) the badgers' society—their social group geometries—had reconfigured as they adapted to exploit this opportunity. *Alea iacta est*.[22] To explore and understand the intricate way these processes unfolded, we first here consider which intrinsic demographic variables had adapted—as cause or effect—to enable the change in population density. Then, in Chapter 12 we will

address the extrinsic component on which this demographic shift capitalized.

Through the looking glass

'Now, here, you see, it takes all the running you can do, to keep in the same place!' observed the Red Queen to a doubtless athletic, if breathless, Alice, through the pen of Lewis Carroll in 1871. A century later, Van Valen (1977) gave her name to the notion that animals are in a perpetual arms race to adapt and reproduce in ever-changing environments; that is the 'Red Queen hypothesis'. All that running is exhausting, and it turns out that badgers, metaphorically, have been doing a lot of it in Wytham. Back to the socio-spatial capitalization on new opportunity: towards the end of the first decade of our study, the badger population was enabled to achieve a manifest intrinsic change in its age and sex structure, its socio-spatial distribution, and its survival dynamics. Bang—a population explosion. But exactly what did these changes in demographic rates look like? Earlier in this chapter we eliminated the hypothesis that something had held back the population's potential prior to the 1990s. No, this was not an escape from suppression, it was a genuine ecological regime shift.

In 1996, our then 10-year run of data was already long by the standards of mammal studies, and revealed that the annual tally of adults aged 2 or more (i.e. excluding non-breeding yearlings) was a significant predictor of the number of cubs that would be born the following year (Macdonald and Newman 2002), but cub recruitment into the adult population was not in direct proportion to population growth. At that time, our dataset gave us a mean of 95 adults over the first 5 years (1987–1991) producing c.30 cubs, whereas by the second 5-year interval (1992–1996) adult numbers increased by a factor of 2, while cub production increased only by a factor of 1.5. This hints at overproductivity, where the constraints that affect cub survival (such as coccidiosis, Chapter 2, and vulnerability to harsh conditions, Chapter 12) are different, or differently savage, to those affecting adults. Remembering too that there is an inherent lag factor in cub production due to delayed implantation (Sugianto et al. 2021b; see Chapter 7, and, so circumstances at conception may be very different to those that obtain by the time cubs are born (something termed a 'predictive adaptive response', which we elucidate in Chapters 15 and 18).

So, remembering that male survivorship worsened quite substantially after the observations of the 1990s, whereas female survival to age classes 3 and 4 improved slightly, our quest to quantify which

---

[22] 'The die is cast', Julius Caesar.

(a)

				A					n
0	F(M_1)	F(M_2)	F(M_3)	F(M_4)	F(F_1)	F(F_2)	F(F_3)	F(F_4)	*Cubs*
ρ	P(M_1)	0	0	0	0	0	0	0	M_1
0	G(M_1)	P(M_2)	0	0	0	0	0	0	M_2
0	0	G(M_2)	P(M_3)	0	0	0	0	0	M_3
0	0	0	G(M_3)	P(M_4)	0	0	0	0	M_4
1−ρ	0	0	0	0	P(F_1)	0	0	0	F_1
0	0	0	0	0	G(F_1)	P(F_2)	0	0	F_2
0	0	0	0	0	0	G(F_2)	P(F_3)	0	F_3
0	0	0	0	0	0	0	G(F_3)	P(F_4)	F_4

(b)

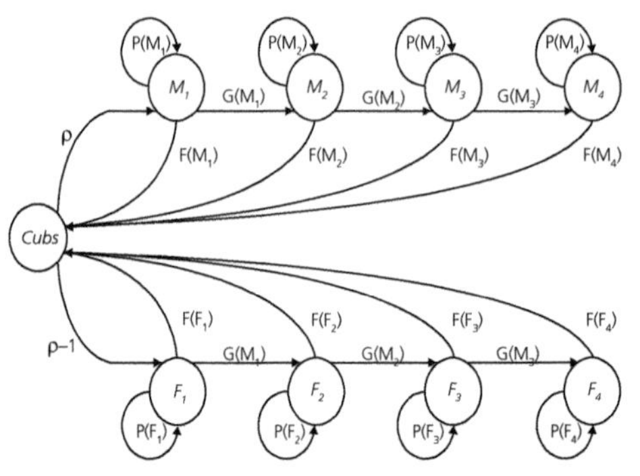

Figure 11.13 A visual representation of a two-sex, five-stage Lefkovitch matrix. On the left (a), the matrix of transition probabilities (*A*) is multiplied with a population state vector (*n*) for time *i*; on the right (b), these probabilities are illustrated alongside the transitions they represent. Note that cubs, as non-reproducing individuals, are considered asexual. Males are depicted on the top row of panel (b), females on the bottom row (from Bright Ross 2021, with permission).

intrinsic regulatory mechanisms best accounted for this step change led us to Leslie matrices (Leslie 1945, 1948). These matrices formalize life cycle graphs, summarizing transitions between ages, or stages (in which stage-specific cases they are termed 'Lefkovitch' matrices) loaded with survival values of the transition between ages (t to $t + 1$) taken from life tables. From our tables these age classes were: prereproductive (cub, then juvenile), young adult, prime adult, and elderly (Chapter 18). Simultaneously, arrows loop back per age (class) to the production of cubs, loaded with age-specific fecundity (Figure 11.13).

An often painful distinction lies in the difference between what might have been and what actually transpired, and parsing apart these two is the purpose of so-called LTRE. Two techniques, described in Box 11.5, advance from the foundation of the Lefkovitch matrix, using the actuarial data from Wytham, to discover first how much each transition rate in the life stages *could* have influenced population dynamics (this is the task of sensitivity/elasticity analyses), and second to identify within that envelope of possibilities which rates actually *did* cause the observed changes.

By doing both, we intended to identify which of five vital demographic parameters had the greatest potential to influence population growth rate (λ) and produce changes in population size, and then to distinguish those processes that actually had influenced the population prior to and after the marked Wytham badger population density change (Macdonald et al. 2009). By way of analogy,[23] the child of very tall parents might have had the potential to grow very tall, but having gone hungry throughout his/her adolescence, might have achieved only medium height.[24]

Less mathematical readers are not likely to find much amusement in further discourse on the algebra behind the matrix and LTRE analyses; for enthusiasts, however, Oli and Dobson (2003) explain these methods fully: indeed, Professor Dobson kindly helped us with these analyses as used in our paper (Macdonald et al. 2009), an introduction to which is set out in Box 11.5.

What could change—what did change?

So, using the matrix approach for the period 1987–2001, we calculated a finite per capita growth rate estimate (λ) of 1.065, with a mean generation time of 5.8 years.[25] This was considerably lower than in our unconstrained projection models (Macdonald and Newman 2002),

[23] The analogy, while apt, is loose, because height is a polygenic character.

[24] Indeed, it turns out that 60–80% of the difference in height between people is determined by genetic factors, whereas 20–40% can be attributed to environmental effects, mainly nutrition (Chao-Qiang Lai 2006).

[25] This is the seeming incongruity mentioned on p. 238 Table 11.2, where we estimated 3 years as the prediction of how an unconstrained population could maximize its reproductive potential when below carrying capacity; so, the shortest generation time that badgers can experience is 3 years, whereas here we show that the reality is nearly twice that.

Box 11.5 Matrix models and LTRE with badger demography

Distinguishing what might have been from what was

Building on the foundation of the Lefkovitch matrix, things get more technical, involving matrix population modelling (see Caswell 2001; Oli and Zinner 2001). The useful facility afforded by matrix models is that they can demonstrate how strongly changes in each transition rate between stages (or age classes) affect the total rate of population growth (λ)—an attribute termed 'sensitivity', along with how proportional changes in each vital rate change λ—termed 'elasticity'. That is, sensitivity measures the absolute contribution, while elasticity is a relative measure of that contribution.

Sensitivity of λ to perturbations in different life history variables may not be comparable with each other, because they are measured in different units; for example, survival rates P_a and P_j are probabilities and can have values only between 0 and 1, whereas fecundity F is not restricted in

this way. The concept of 'elasticity' addresses this problem, where elasticities are proportional sensitivities, and quantify potential changes in λ with respect to proportional changes in life history variables. Elasticities are scaled dimensionless quantities, and thus directly comparable over time within a population or between populations and species.

Sensitivity/elasticity analyses provide a *prospective*, that is, forward-looking, projection of how changes in demographic vital rates (such as death rates at given ages) would influence population size. This approach enabled us to discern which badger population vital rates had undergone a state shift to catapult Wytham's badger population from one stable state to the other that sustained a greater population density. Matrix model parameters, however, can also be used to inform LTRE that test *retrospectively* the extent to which growth rate λ was actually affected by each demographic vital rate. So we did this too (Macdonald et al. 2009).

Table 11.3 Life history trait values, parameter values, and results of sensitivity and elasticity analyses. The population data are pooled either over the 15-year study interval, the first 10-year period, or last 5-year period (from Macdonald et al. 2009, with permission).

	λ_1 From age-classified model			λ_2 From stage-classified model	
Life history trait					
15 years	1.065			1.066	
First 10 years	1.113			1.11	
Last 5 years	0.964			0.981	
Estimated parameters in the partial matrix model	α	P_j	P_a	F	ω
15 years	2	0.717	0.837	0.267	15
First 10 years	2	0.767	0.876	0.269	15
Last 5 years	2	0.627	0.762	0.262	15
Sensitivity of λ to change in parameter					
15 years	−0.110	0.275	0.724	0.987	0.002
First 10 years	−0.100	0.266	0.726	1.002	0.002
Last 5 years	−0.110	0.293	0.717	0.954	0.002
Elasticity of λ to change in parameter					
15 years	−0.100	0.185	0.568	0.247	0.028
First 10 years	−0.090	0.184	0.573	0.243	0.029
Last 5 years	−0.120	0.187	0.557	0.255	0.026

Table 11.4 Life table response experiment: analysis illustrating the contribution to change in λ of key population parameters (excluding for age at maturity and reproductive life span where change = 0) between the lower- and higher-density population phases.

$\Delta\lambda = 1.110 − 0.981 = 0.128$

		Change	Mean
Period 1–Period 2		In trait	Sensitivity contribution
$\Delta P_j = 0.767 − 0.627$	=	0.140	(*0.275 = 0.0385)
$\Delta P_a = 0.876 − 0.762$	=	0.114	(*0.724 = 0.0825)
$\Delta F = 0.269 − 0.262$	=	0.007	(*0.987 = 0.0069)
		Summed contributions = 0.1279	

and calculated a true generation length twice the theoretical minimum (Macdonald et al. 2009).

Over the first 10 years of this study λ was around 0.13 greater than during the latter 5 years, showing that population growth had ceased, indeed was hinting towards a negative value (and keep this 0.13 in mind; it becomes pertinent again, shortly). Our forward-looking 'prospective' matrix model (i.e. what could change) revealed that the influence of the five vital demographic parameters potentially influencing population growth rate (λ) ranked in descending order of importance as: adult survival > fertility > juvenile survival and age at first reproduction > age at last reproduction.

However, *retrospectively* we could see that what could have changed is *not* what actually had changed. Rather the LTRE analysis agreed that change in adult survival (0.83, SE 0.01) accounted for the greatest component, in fact 64% (i.e. contributions from Table 11.4 in Box 11.5: 0.0825/0.1279 = 0.645) of the total change in λ, but with juvenile survival (0.67, SE 0.03) ranked second in importance at 30% (0.0385/0.1279 = 0.301); fertility was demoted to third place, with a rather minor (5%, 0.0069/0.1279 = 0.054) influence. Neither age[26] at maturity nor reproductive life span had any influence at all (see Macdonald et al. 2009). Note that the sum of all contributions (bottom line of table 11.4 = 12.8) once more closely approximates the change in λ between the two periods.

Thus while our matrix modelling had suggested that changes in fertility (ΔF) might have had a big impact on population growth, the retrospective LTRE approach, observant of the anchor that evolution has destined badgers to small litter sizes, revealed that this was not such an important factor for badger population growth rate (Δω, age of last reproduction, mattered even less at the population level, being similarly invariable). What really mattered was variation in adult (P_a) and juvenile survival (P_j). And so we had our culprits (readers wondering why fertility did not have a big impact should detour to Box 11.6).

---

[26] As with generation time, p. 237, this is another potential confusion between the minimum possible—in this case, age at sexual maturity—and the actual age at first rep/primiparity—where this influenced longevity etc., and is documented in Chapter 14.

Box 11.6 Why fertility did not transform Wytham's badger population

The profit side of the ledger for any animal population gets its income from births and immigration—that is, recruitment. A crucial component of recruitment is fecundity, which in badgers is constrained by a litter size that is relatively small and invariant compared with that of most other carnivores (Creel and Creel 1991; Creel and Macdonald 1995).

The carnivoran record holder is the Arctic fox, which, under the right circumstances, can churn out litters of 18 (with the record held at 25; Tannerfeldt and Angerbjörn 1998); vixens can have up to 14 teats (Hildebrand 1952). This potential for productivity is adapted so that Arctic foxes can produce litters according to the boom and bust, peaks and troughs, in the lemming cycle (Hersteinsson and Macdonald 1996; Angerbjörn et al. 1999). Even the remarkably badger-like raccoon dog, *Nyctereutus procyonoides*, a canid sympatric (co-resident) with the European badger in Norway and with the congeneric Japanese badger, *Meles anakuma*, can produce up to 16 pups, with a mean of 6.7.

How do badgers compare with other mustelids? Litter sizes vary from as few as 2–3 (often, indeed, just 1) in martens, wolverines, and sea otters (98% of sea otter litters are singletons), to 6 in Asian short-clawed otters and up to 14 pups (average 8.7) in the mountain weasel (*Mustela altaica*). Amongst the 9 species of the subfamily *Melinae* (Zhou et al. 2017), litter size never exceeds 5. From our own data (see Chapter 2) litter sizes for Wytham's badgers average 1.62 ± 0.62 SD (Macdonald and Newman 2002) to more recently 1.46 ± 0.65 (Macdonald et al. 2009; Annavi et al. 2014a). To look at it another way, amongst 378 litters we recorded (1988–2010; Annavi et al. 2014a), 36.8% (139; n = 378) were polytocous (117 twins, 20 triplets, 2 quadruplets) and 93% included fewer than 3 cubs (range 1–5). The age of lactating females averaged 5.81 years (SE 0.17), with 37.67% (SE = 2.02) of sexually mature females (i.e. >2 years old) lactating per year. We also found patterns in litter sizes, where in the driest, most food scarce years we recorded on average one cub per lactating female (early in our dataset; Macdonald and Newman 2002), whereas, in the wettest and most food abundant years, litter size rose to an average of 2.5 cubs; more of that, however, in Chapter 12. Nevertheless, differential fecundity does not have the capacity to act as a lightning rod for badger population regulation and is therefore acquitted as a suspect prompting the phase shift.

We might, however be glossing over an important related detail here (we do not know this for sure, however): we get to know new cubs of the year individually only when they are old enough to enter our cage-traps—generally in May when they are 12–14 weeks old. Therefore, any succumbing before that date never enter our data. How many of these anonymous dead slip through our actuarial net? Pooling the limited literature from southern England, Neal and Cheeseman (1996) give a mean litter size *at birth* of 2.9. Similarly, from post-mortem study of the uterine scars of a huge sample of 1522 female badgers, Anderson and Trewhella (1985) reported a mean birth litter size of 2.7 cubs. Insofar as quite of few of those badgers came from as far afield as Switzerland and Austria, it may be unwarranted to assume that this average applies in Wytham, but if it does then we deduce from this that in Wytham pre-emergence cub mortality averaged 36.4% (SE = 10.2; ranging from maximum of 60.36% in 1996 and a minimum of 6.15% in 1993) (Macdonald and Newman 2002).

Clearly cub mortality had a significant impact on the population, and accounted for up to 40% of the losses in any one cohort. During the population growth characterizing the first 10 years of the study, cub mortality was lower than the overall 15-year average. From 1997 to 2001, however, as the population shrank back slightly and stabilized at its new higher level, cub mortality exceeded the 15-year average. Therefore, despite the lower capacity for changes in juvenile mortality to affect population dynamics (as shown by our elasticity analysis), the changes in this vital rate were large enough and consistent enough to overwhelm the term's elasticity and heavily influence population dynamics between periods.

Before we leave demographic parameters that had some, albeit not overriding, impact on the rate of population change, consider the proportion of females that suckled young. As population size increased through the 1990s, we recorded an increase in the proportion of lactating females: between 1987 and 1996 a yearly average of 29% lactated, a proportion that crept up to 37.67% (SE = 2.02) when we subsequently examined a longer data interval 1987–2012 (Annavi et al. 2014a). Since the second, longer period ending in 2012, incorporates the earlier, shorter run of data ending in 1996, it follows that the percentage of lactating females during those latter 16 years was much higher—again, another social metric had changed after 1996.

In overview, the first decade of data suggested a net reproductive rate of 0.22 female cubs/adult female (SE = 0.03), or gross reproductive rate of 0.24 cubs/adult (SE

= 0.05). This, in turn, yields a productivity rate of 1.62 (SE = 0.28) cubs/reproducing female. These are average figures, but the interest, along with the devil, resides in the detail: there was a huge peak in cubs born in 1993 where a whopping 90.3% of these cubs survived the year with no compensatory increase in adult deaths; a state of affairs that could have only one outcome—the population peak we observed by the time this cohort matured in 1996.

Before leaving this dissection of the five vital demographic parameters impacting the population growth rate (λ) of the badgers of Wytham Woods, we want to situate the badgers' reality in the wider context of mammalian life history theory. Species with age at first reproduction $\alpha = 1$ and fertility $F > 1$ have a greater elasticity for α and F, while species with $\alpha > 1$ and $F < 1$ have a greater elasticity for P_a and P_j. Some theorists had argued that, generally, age at first reproduction, α, should have the largest relative influence on λ (Cole 1954; Leowontin 1965); this was certainly not so, however, for Wytham's badgers, which were, instead, proportionately more susceptible to the influence of juvenile survival. In Wytham, survival and recruitment of their relatively 'expensive' badger cubs, as well as the continuing survival of adults has proven to be the principal driver of change in badger numbers (Macdonald et al. 2009). A better fit with theory came from an alternative generalization, proposed for mammals by Oli and Dobson (2003), who had drawn attention to the ratio of fertility to age at first reproduction; termed F/α. Amongst the badgers of Wytham Woods, age at first successful mating was close to 3 years for both males and females, whereas fertility was 0.395 (estimated using pedigree, which necessarily is an underestimate but nevertheless shows it is categorically different to, say, that of an Arctic fox, at 15; Bright Ross et al. 2020), giving an F/α of 0.134, a value which is intermediate along the continuum of values known amongst mammal species. The relatively late onset of reproduction ($\alpha > 1$) in badgers, and their relatively low fertility ($F < 1$), likely arise from a trade-off between the need to invest urgently in juvenile growth (i.e. they're too small to sustain torpor at 10 months of age, while a hard winter will end a short life; Sugianto et al. 2019b; Chapter 2) and reproductive effort, which can be postponed with less immediate peril (Oli and Dobson 2003; see also Dugdale et al. 2011; Campbell et al. 2017)—a topic to which we will return in Chapter 14, once we have explored those extrinsic factors driving the responses we discuss here.

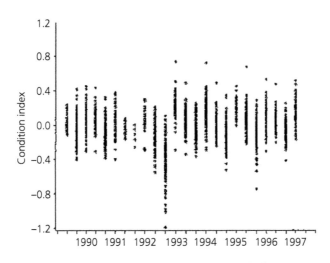

Figure 11.14 Body condition trend of Wytham Woods badgers, 1990–1997. The overall trend is for condition to improve over the sample period. Summer is the season with poorest condition scores each year, with lowest scores occurring in summer 1993 (from Macdonald et al. 2002, with permission).

Thomas Malthus (1798) had said that: 'The superior power of population cannot be checked without producing misery or vice', or can it?[27] Can this grim prediction be circumvented through some adaptive twist on how effectively resources are exploited, or competition mitigated, much as human society avoided, or at least postponed, this Malthusian catastrophe over the past two centuries though improved food production technology and healthcare? In Chapter 10 we described how groups became crowded. So too, inevitably, did the whole population. As Aristotle (350 BCE) forebodingly observed: 'Yet it is clear that if a number of sons are born and the land is correspondingly shared there will inevitably come to be many poor men'.[28] Poverty, in badgers, is measured in the currency of fat. We had devised an index of mean body condition (BCI) (Figure 11.14). Initially, between spring 1990 and autumn 1992, as the population increased there was a steady and substantial decline in this body condition index.

Unexpectedly, however, the trend reversed from 1993 to 1997. Bizarre? Certainly incongruent with theory on density dependence (Fowler 1987; Wolff 1997). A yet more bizarre fact, however, was that

this reversal was so strong that, over the entire 1990–1997 period, mean badger residual condition indices actually increased (Figure 11.14). If individual quality declines with group size, but increases with total population abundance, surely this could only mean that the larger population comprised more, smaller groups? Exactly—this was precisely what we found in Chapters 4 and 10, linking the sociological with the demographic.

So, back to the shift in circumstances, and demography, that occurred in the early 1990s. Now we know that while the number of territories reached an early plateau, the number of setts continued to climb. As setts proliferated there was a marked change in cub survival. Based on our (Macdonald and Newman) 2002 analysis (and the original simple MNA cohort size estimates we used at that time), immediately prior to sett proliferation, just 10 of the 23 cubs born in 1990 survived until 1991 (43.5%), whereas 37 of 46 cubs born in 1991 survived until 1992 (80.4%). In 1993, after sett proliferation, 56 of the 62 cubs did so that year.

It thus seems that the precursor to greater population abundance was a spike in juvenile survival. Similarly, it was a spike in juvenile mortality in 1996 that brought to an end this golden age of population success, with 39.1% of 54 cubs surviving just prior to a plunge in density through 1998 (Figure 11.5).

What is the solution to our population mystery? Populations constantly tend to push against the ceiling of environmental carrying capacity (Hui 2006). For the Wytham badger population, carrying capacity

---

[27] As we work on this chapter on Tuesday 27 July 2021 it was declared 'tipping day'—when humankind officially started to consume more than the gross primary production of the earth, tipping us into deficit.

[28] Aristotle's *Politics* (350 BCE), book 2, section 1270b (Aristotle n.d.).

apparently increased in 1987–1996, but greater badger density brought the Malthusian curse of stress and competition. The badgers responded to this stress by adapting their group configurations (Chapter 10)—further evidencing the group-living flexibility of this facultatively social species—generally forming a more peaceable society, better optimized to an improved food resourcescape. Insofar as the success of any strategy may be tested against its outcome, aside from a blip in 2006 (see Figure 11.5), population size has settled to a steady 220 adults and c.45 cubs per year, spread between a consistent number of social groups. Modelling population growth potential from observed natality and survival we can project that with 147 female individuals present (the offspring-bearing sex) a population growth rate of nil would occur—hence 294 badgers (based on an even sex ratio (95% confidence interval, 124–170) would be sustainable, which is almost exactly the peak population (i.e. c.270–300) (Figure 11.5).

With hindsight the population biology of Wytham's badgers since 1987 has been a story of two halves.

Using eMNA, between 1987 and 1991 we estimate a yearly mean of 126.3 (SE = 4.56) adults producing 36.6 (SE = 5.66) cubs. During the next 5 years (1992–1996) mean adult numbers increased by a factor of 1.75; cub production, however, increased only by a factor of 1.66.[29] The growth of the population was therefore not just a simple matter of cub recruitment—something else was at play. It lies in the distinction between the demographic and the sociological. We can consider density as a measure of individuals per unit area, and as such it increased in Wytham through the early 1990s and now rests at a factor of 1.37 above our starting point. But if we consider density as a measure of individuals per unit available resources, the improvement in body condition we noted from 1993 reveals suggests that badgers found a more efficient way of making use of every morsel available to them, or found ways to expend less energy cost acquiring food. What changed? The answer lies in Chapter 12, which will reveal why the population carrying capacity changed, and the socio-demographics of the badgers jumped into a different category of resilience.

---

[29] Not surprisingly, the number of cubs alive in one year strongly predicts the number of adults alive during the next year (no. adults = 128.8 + 1.6 × no. cubs), for the obvious reason that when population density is high more cubs are born, and most of the adults that produced them are still alive a year later.

CHAPTER 12

Weather

Actuarial Insights

As environments alter, so fauna and flora either adapt or die out; nature is very unsentimental.

Martin Keeley

What really matters to badgers is food security, and that's the crux of it. In Chapter 10 we explained that in the context of the Resource Dispersion Hypothesis food security has a technical, probabilistic meaning that, in Chapter 14, will boil down to having sufficient energy intake to fuel expenditures for enough of the time. Happily, the notion is colloquially intuitive: dictionary.com defines food security as 'an economic and social condition of ready access by all members of a household to nutritionally adequate and safe food'—helpfully mentioning both economic (our currency, in this context, is energy) and social elements. Further-more, Chapter 11 revealed that overall the richness of the badgers' food supply in Wytham and reduced costs of obtaining it, entwined with the changes in social group dynamics it precipitated, had, remark-ably, increased environmental carrying capacity by 32% by 1996. But what were the roots of this change? Had some insidious constraint previously stifled the badgers of Wytham Woods, preventing their popula-tion from capitalizing on an environmental carrying capacity that had always been there? And/or might that carrying capacity have increased, and, if so, why?

Regarding the first possibility, two plausible cat-egories of mortality are direct persecution, which we dismissed in Chapter 11, and pollution. Between 1962 and 1969 there was concern amongst badger watchers in certain parts of Britain that cub pro-duction had declined (Neal and Cheeseman 1996). Coincidentally, Don Jefferies (1969) reported that at least 6, and perhaps 12, of 17 badgers examined post-mortem between 1964 and 1968 had died of 'dieldrin' poisoning. Dieldrin is a persistent organo-chlorine insecticide, alongside its more infamous sibling DDT (dichlorodiphenyltrichloroethane), which killed a horrific range of wildlife around this time (Holmes et al. 1967; Porter and Wiemeyer 1970); also causing severe human illness (Wade et al. 1997; Høyer et al. 1998). Residues of these pesticides bio-accumulated in watercourses (killing otters; Jefferies and Hanson 2002; see also Yamaguchi et al. 2003), and, ominously for badgers, permeated earthworm tissue (Haimi et al. 1992). Had the consumption of toxic earthworms prevented Wytham's badgers from fulfilling the environmental carrying capacity? Probably not. First, the symptoms of dieldrin poisoning are dramatic (Shankland 1979), and neither we, nor Hans Kruuk before us, saw any badgers thus afflicted. Furthermore, dieldrin was banned in the UK and European Union (EU) from 1975, with DDT banned in 1986. Even with a half-life of 5 years, and DDT contamination remaining around 145% above pre-war levels in 1991 (Crick and Ratcliffe 1995), this danger was largely gone long before Wytham badger numbers eventually burgeoned.

Shifting the searchlight from impinging factors to enhancing factors, what of the change in land use we described in Chapter 10 (da Silva et al. 1993)? This occurred between 1974 and 1988, but during the 1989–1996 window in which we saw the Wytham badger population surge there were no substantial changes in areas of deciduous woodland, plantation, and rough grassland within the woods. But what about the area beyond the woods on the adjacent Northfield farm? During this 1989–1996 lead up to population growth, between 26 and 32 of 64 fields were under pasture

The Badgers of Wytham Woods. David W. Macdonald and Chris Newman, Oxford University Press.
© David W. Macdonald and Chris Newman (2022). DOI: 10.1093/oso/9780192845368.003.0012

(mean 28.9, standard error (SE) ± 0.77), and 87% of them were permanent (agricultural rotation affecting the remaining 13%). We tested whether the number of adult badgers each year (with various lags) related to the number of pasture fields but detected not a glimmer of a link (Macdonald and Newman 2002). More widely, Wilson et al. (1997) had found that no single land class change could have accounted for the substantial coincident growth of the English badger population over the 1990s (see also Chapter 19).

Land use may not have changed but what of the bountifulness (to foraging badgers, at any rate) of that land? Specifically, were earthworms, and perhaps other foods, more available or available for a greater part of the year, perhaps at a lower energetic cost to foraging? And could such an ultimate cause of the enhanced juvenile and adult survival that drove the population increase (Chapter 11) link what happened to why it happened?

Environments comprise two components, the habitat itself and the climatic conditions that prevail over it, which help to shape the ecosystem. Having largely ruled out any landscape change explaining increased environmental carrying capacity, we are left with a change in the weather (Laidre et al. 2008). We British can be mocked for our endless fascination with the weather, but such a preoccupation might be well founded for British badgers, for which the barometer of their food availability rises and falls depending on the productivity and predictability of summers and the duration and severity of winters: this is what transposes the landscape into the badgers' energy-scape.

Box 12.1 Climate context: How does weather impact animals?

By way of context: while climate change caused by people is shamefully unique, and adds woefully to the dreadful litany of anthropogenic threats to biodiversity, in the theatre of evolution climate change has always been a major driver of adaptation (Hoffmann and Sgro 2011). There is thus nothing new in the way that Wytham's badgers must face the adversity of inclement weather, day on day (of course, what is new is the cause of the extremes they face). The badgers know nothing of the IPCC's forecasts or that they face climate change; they take the weather one night at a time.[1] In this context, Chris' critical scalpel, honed by his undergraduate degree in geology is ever-sharp to excise naïve ideas on climate stability, and has preserved our thinking about badgers and the climates of their evolutionary past, from the snares of three myths: on a palaeontological time scale the earth today is not especially hot (indeed it's quite close to ice-age conditions), in a geological context CO_2 levels are not high, and rapid rates of climate change are not unprecedented. 'Jurassic Park' would have been around 6°C warmer than current global temperatures. Indeed, the mega-mammal explosion of the Eocene Thermal Optimum saw temperatures 12°C hotter and sea levels almost 100 m higher than today, with CO_2 levels of exceeding 1000 ppm—well in excess of today's much feared 400 ppm (Newman et al. 2017).

It is of course neither excuse nor consolation for the current causes that even archaeological and historical records show that across the generations of ancestral badgers dramatic and rapid swings in temperature have occurred before; for instance, Dansgaard–Oeschger (D-O) events (Dansgaard,

1993), which occur quasi-periodically on a 1470-year cycle in the northern hemisphere. D-O events are characterized by episodes of rapid warming, followed by gradual cooling over centuries. For example, the Greenland ice sheet warmed by 8°C within just 40 years, around 11,500 BP (the Bølling-Allerød interstadial; Bégeot et al. 2000)—just one of several extreme palaeo-climatic events that exceeded the amplitude of temperature change observed in Greenland over the past century by an order of magnitude (Alley 2000a, 2000b). Unusually cold episodes, called 'Heinrich events' (Bond et al. 1999), which lead to an expansion of the polar front and increased ice cover in the North Atlantic (Bond et al. 1999), usually herald D-O events. It seems this cyclicity is driven by 'binge–purge' dynamics: as polar ice sheets accumulate mass they affect the Atlantic Ocean circulation causing deep ocean currents to overturn, altering the flow of the gulf stream, until ultimately this ice mass becomes unstable (Zhang et al. 2014). Given that the average mammalian species survives for 1.5 million years (Barnosky 2001), the ancestors of Wytham's badgers, and of many species alive today, have coped with at least 65 Milankovitch events,[2]

---

[1] We use weather to describe short-term atmospheric conditions that may face the badgers with changes over hours, days, or weeks, in temperature, humidity, cloudiness, rainfall, and wind. Climate is the long-term average of the weather—how long is long-term? The answer is arbitrary, but it is common for people to evaluate climate over a period of, say, 30 years.

[2] Named for Serbian geophysicist and astronomer Milutin Milanković, Milancovitch events relate to three cycles (eccentricity, obliquity, and precession) that relate to how much solar radiation the earth is exposed to, influencing global weather.

Box 12.1 *Continued*

and around 1000 D-O events (Dansgaard et al. 1993) during this 1.5 million-year portion of the Pleistocene epoch (see Chapter 19).

Such rapid, natural swings in climate were felt in the UK by the rapid cooling defining the Younger Dryas stadial period (12,800–11,500 BP), which caused mean annual temperature to drop to 5°C in under a decade (Bakke et al. 2009). More recently, between 950 and 1250, during the medieval warming period, the North Atlantic region was at least as warm as conditions today (McIntyre and McKitrick 2003), followed by a swing during the so-called Little Ice Age (Mann et al. 2009), between 1650 and 1850 (with three peak cold events around 1650, 1770, and 1850), when the River Thames, skirting Wytham, regularly froze over (and there were 'Frost Fairs' in London and Paris).

This long-term history is relevant because it shows that how badgers adapt to climate change is a relevant evolutionary topic, despite lamentable factors currently exacerbating it: climate change is a constant driver of adaptation in evolutionary ecology. What is it about climate change today that raises such deep concern over the health of ecosystems? As we work on this chapter while the IPCC convenes despairingly in August 2021, catastrophic fires the size of London rage in California, Greek islands are evacuated, and temperatures in Sicily reached a historical record of 47.8°C. A major part of the alarm is because humans—and all other animals—find unpredictable and uncontrollable change or extremes stressful, even unmanageable. Irregular weather jeopardizes control and disrupts optimal strategies (e.g. Reed et al. 2010), such as when trying to farm the land or, as a badger, live off it. Of course, there is a risk that climate change becomes a scapegoat for other pressures humanity places on ecosystems, but less readily acknowledges (see human induced rapid environmental change (HIREC) factors, *sensu* Sih 2013). As Jean Jacques Rousseau put it 'Nature never deceives us; it is always we who deceive ourselves', making it doubly important to avoid verisimilitude—the flawed appearance of being true or real (Niiniluoto 2016; see Noonan et al. 2015d; Chapter 13).

Given the current rapid rate of global warming, correlation and causation become hard to disentangle; as we turn to weather effects (which may be on their way to climate effects) on the badgers of Wytham Woods, relentless rigour will be needed to untangle them (Lorenzoni et al. 2007).

For more immediate context, NERC[3] commissioned us to assess the potential impacts of climate change on the ecology and viability of the UK's native mammals (Newman and Macdonald 2015). For most species data were inadequate; badgers, however, provided a sensitive model (see Newman et al. 2017). With over 30 years of detailed population records, we could develop and test hypotheses not only relevant to how badgers react to climate change, but extrapolating more widely (e.g. Campbell et al. 2012; 2013; Zhou et al. 2013a). When it comes to climate change, badgers serve as a proverbial 'canary in a coal mine', informing on how more fragile species might be affected.

The influence of weather as an evolutionary force behind population change for badgers, and more generally, depends on two components: (i) changeability in physical properties of weather regimes; and (ii) the behavioural and physiological adaptability of the species (Parmesan et al. 2000). Mean weather conditions exhibit variability and extremes, over a range of time scales (daily, seasonal, annual, millennial), ultimately comprising the constant flux of climate. To paraphrase Heraclitus,[4] when it comes to climate the only constant is change, but the rate of change can vary. What matters is how adaptable or vulnerable an individual (or species) is to both the direct (e.g. hot/cold) and indirect (e.g. food supply etc.) stresses (or benefits) caused by weather, as well as effects manifesting at the ecosystem level—and all these topics, with the badgers of Wytham Woods as a model for evolution and behaviour, are focal to this, and the succeeding chapters.

[3] The Natural Environmental Research Council, UK.
[4] Heraclitus, sixth century bce: 'Nothing endures but change is the greatest virtue in keeping with its nature'. That is, the fear of change is as constant as change itself.

Weather as a driver of badger population dynamics

British badgers have experienced a substantial increase in numbers since the late 1980s. Looking at the situation beyond the microscope lens of Wytham, Wilson et al.'s (1997) extrapolation, using setts as a proxy for

badger social groups, deduced a 77% increase in badger numbers between 1988 and 1997. A more recent national sett survey, commissioned by Defra, estimates 71,600 (66,400–76,900) social groups in England and Wales, giving a mean density of 0.485 km^{-2} (95% confidence interval 0.449–0.521) (Judge et al. 2014). This equates to an 88% (70–105%) increase in the number of

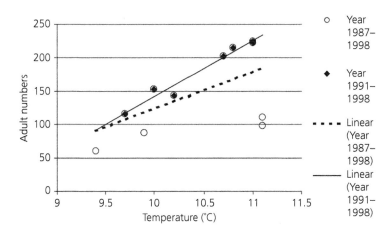

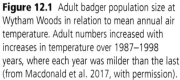

Figure 12.1 Adult badger population size at Wytham Woods in relation to mean annual air temperature. Adult numbers increased with increases in temperature over 1987–1998 years, where each year was milder than the last (from Macdonald et al. 2017, with permission).

social groups inferred across England and Wales since the first national survey (Wilson et al. 1997), a rate of increase at 2.6% (2.2–2.9%) per year.

In comparison, our initial insight into how the dynamics of the Wytham badger population were being driven by weather patterns was spurred by the rise in numbers of badgers alive (minimum number alive (MNA)) we described in Chapter 11: a huge increase between 1987 and 1996 to between 2.39- and 3.80-fold of original numbers (the larger increase based on our 2002 publication using MNA, the lesser on our Bright Ross et al. 2022 publication using enhanced minimum number alive (eMNA), which suggested, respectively, numbers increasing from 60 to 228, the latter from 137 to 327)—the exact numbers are really a detail; what matters is that there was a huge increase over a decade that warrants explanation (Macdonald and Newman 2002; Bright Ross et al. 2021). What caught our attention in particular was that the increase in population density from *c*.10 adult badgers km^{-2} to 38 adults km^{-2} (or, recalculated using eMNA as 18.3 to 45.5 adults km^{-2}) that we observed through the 1990s coincided with a trend towards warmer conditions (Figure 12.1).

Could the warming climate be the uniting factor across the national and local scales of change? Yes and no: this beguiling linear relationship between warming weather and the upward trajectory of population growth up to 1998 did not persist into the 2000s. This is because the benefit of warming was a step change (likely less winter frost), beyond which further warming had no further effect (there is no benefit to less than no frost), so this is a tale of tipping points, not linear relationships. Instead, we became increasingly aware of the effects of balance between *good* and *bad* weather conditions in relation to the likelihood of

earthworm availability; that is, worms surface, and become available to foraging badgers, only when the micro-climate is right (Macdonald 1980d; Newman et al. 2017)—optimally at high surface humidity and a temperature of 10°C (not lower than 2°C) (Jiménez and Decaëns 2000; Curry, 2004). Might some net tendency for greater overall clemency of the weather (the ameliorated drought of summer and the reduced duration and extent of winter chill) explain the burgeoning badger population? Ultimately the balance between these odds hinged on the exploitability of the 'productive' summer season, and how to tolerate the less (or un-) productive winter season. As we will unfold through the remainder of this chapter, and through Chapters 13 and 14, this was to become a major investigation, unfolding in stages, as the answer to one question spawned a host more.

Spring and summer

To start at the beginning, between 1987 and 1991 we had noticed an alluringly convincing relationship between August rainfall and cub mortality (Woodroffe and Macdonald 1995a). This tidy relationship seemed to make perfect sense because dry summer weather decimates earthworm availability on the baked surface soil (and some earthworms aestivate—the summer equivalent of hibernating[5]). Conversely, wetter summers slithering with available worms would give especially cubs an advantage, better enabling them to put on the weight necessary to survive the lean months of winter lethargy. QED: less summer rain would mean

---

[5] Some species, e.g. *Lumbricus terrestris* do not *aestivate* but remain active in soils that have water contents as low as 10% or less (Zicsi 1958; Gerard 1967)

fewer cubs surviving until the next spring. As with the neat initial linear relationship between temperature and population growth (Figure 12.1), perhaps we should have quit while we were ahead, because while it might have been that simple, it wasn't!

True, in 1990, with just 26 mm of August rainfall cub survival was merely c.30%, whereas in the August of 1998 with 43 mm of rain, cub survival soared to c.88%. Perhaps we were lucky that during these contrasting years the summer droughts fell tidily into August, affirming our naturalistic suspicion that weather was important, but risking an illusion of simplicity that was soon dispelled when we delved deeper. By the time our dataset extended to 2001 we could see that life-threateningly dry conditions often began to bite long before August, with the predominant effect of spring and summer (April–September) rainfall drifting towards a greater overall impact of May rainfall on cub survival (Figure 12.2) (Macdonald and Newman 2002).

A consequence of this effect was that annual rainfall was an overarching predictor of cub (but not adult) survivorship. It's not surprising that cubs are a weak link in the population's survival—after all, when recently weaned cubs first enter our traps (and database) in May healthy ones weigh 4 kg, but withering ones weigh as little as 1.5 kg. This diminutive stature brings thermoregulatory challenges (Chapter 14) at exactly the juncture when they have least fat reserves, and least experience in foraging (Newman et al. 2011; Macdonald and Johnson 2015a). Faced with baked summer ground and a shortage of worms, they quickly starve. Worse, this is when juvenile coccidiosis strikes (causing dehydration and electrolyte loss; Chapter 2) and if coccidial infection coincides with shortages of food and water during dry spring weather, annual cub cohort survival rates can plummet to 10% (Figure 12.3; Newman et al. 2001; Nouvellet et al. 2013).

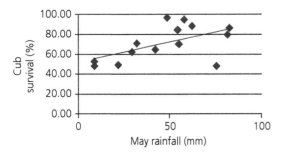

Figure 12.2 May rainfall correlates with cub survival, 1998–2001. In wetter years cub survival rate is higher (from Macdonald et al. 2017, with permission).

Autumn and winter

There is something druidically portentous about equinoxes; this is so for good reason when it comes to reading the runes of badger fortunes. The autumnal equinox in Wytham Woods could scarcely be more achingly beautiful in its auburn hues, but just as the copper leaves turn so too do the lives of the badgers. In an annual pivot point, temperature in September correlates significantly and *negatively* with badger weight gain; a month later, weight gain correlates *positively* with rainfall (Macdonald et al. 2010c). It's a case of, out of the frying pan; into the ... freezer. The toasted ground of summer may soften in autumn, but by midwinter the soil is cold and sometimes frozen. In general, due to weather systems, the colder the winter the lesser the rainfall: circumstances that combine to lessen earthworm availability at a time when alternative foods are scarce (Roper 1994). Worse, those nuts, fruits, and cereal, the muesli that supplement badger diet through the autumn, are also in short supply through winter, leaving slim pickings even for an opportunistic omnivore (Bright Ross et al. 2021); the longer the winter, the worse the deprivation. Even attempting to forage risks a wasteful loss of condition, fat squandered unproductively on keeping warm while fruitlessly out and about (see Chapter 14). In response, badgers conserve energy during freezing, unproductive conditions by spending more time underground, relying upon fat reserves and torpor (Newman et al. 2011; Noonan et al. 2014; see also Box 12.1). Comparing winters of varying severity, early on we found that badgers in Wytham were about 20% heavier during the warmest Januaries than the coldest: the difference for males being between averages of 9 kg in the coldest winters to 11.5 kg in the warmest, and for females between 8.1 and 10.1 kg (Macdonald and Newman 2002).

Winter conditions also affect reproductive success. In Chapter 7 we introduced delayed implantation; here it is sufficient to remember that blastocysts fertilized during the post-partum oestrus in February remain suspended *in utero* until they implant as the days shorten into winter (Thom et al. 2004a; Sugianto et al. 2021b). To discover when precisely they implant (in turn determining ultimate birthdays) we used portable ultrasound machines to monitor pregnancies. Over the Christmas–New Year periods between 1987 and 1996 (and occasionally in years thereafter) our Yuletide entertainment was to trap and scan—hopefully pregnant—females; transfixed as we watched the greyish grainy images of little embryonic hearts beating. The scanner enabled us to count the unborn litter sizes and estimate exactly the implantation date from plotting embryo lengths against established growth curves. Thus we established that for corpulent females implantation commences around 21 December (17 December is the earliest we ever recorded), whereas their scrawnier contemporaries implanted as late as 31 December (11 January for the latest individual) (Woodroffe 1995; see Macdonald et al. 2015e). This spread likely has implications for the timing of parturition at the end of roughly 56 days of gestation, with implications for cub survival.

Prevailing winter weather also influences the numbers of youngsters born: annual cub cohorts have tended to be larger following milder winters (Macdonald et al. 2015e), and to include a slight preponderance of male cubs (Dugdale et al. 2003; see Chapter 3). What mechanism links milder winters and larger cub cohorts? Obvious possibilities include a greater proportion of females taking pregnancies to full term, relative effects on the duration of gestation and premature birth,[6] larger litters per breeding female, and greater survival rate amongst cubs during the 2 months between their birth and their first capture and recruitment into our dataset. Candidate mechanisms could also involve the intermediary of an absolute effect of maternal body condition, such that emaciated females abort their litters (or could potentially die themselves)—another topic for Chapter 14.

Annual overview

So, does net annual temperature, as the siren signal for 'global warming', affect cohort size? Taking the year as a whole and thus combining both positive and negative effects incurred during summer and winter, the average number of cubs produced by each adult female peaked at intermediate temperatures (10.7°C) decreasing following the warmest and the coolest. In years with a more dramatic amplitude between summer heat and winter cold the likelihood of females' reproducing was reduced (Bright Ross et al. 2020). There were also more subtle weather effects on reproduction, linked to age of primiparity, where greater rainfall variability was linked to higher likelihood of latecomer, previously nulliparous (or, at least, with no assigned offspring) females beginning reproduction (41.8% ± 16.7 SE vs. 6.1% ± 3.3 SE). More females produced their

first litter at age 4 (a prime reproductive age) following warmer years (38.3% ± 16.0 SE in the warmest years vs 3.3% ± 2.2 SE in the coolest), modelling a constant sex ratio at parity. For males, a greater proportion bred successfully (i.e. chose to mate females that bred successfully) in years with greater temperature variability, siring more offspring following years with greater rainfall variability (a remarkable genetic corollary of this finding emerges with respect to paternal heterozygosity in Chapter 17).

Annual trends, of course, conceal detail, where the complexity of ecological systems makes straight lines unusual (Green and Sadedin 2005). Of course, binary on/off responses, such as reproduce/or not, live/die, pivot around thresholds to which individuals may differ in their sensitivity. Consequently even factors that vary on a continuum, such as temperature, will not have linear effects (Jackson et al. 2009). Frost, as mentioned above, is one such on/off factor: what matters is whether a winter is frost-free or not, giving the advantage of the change, not so much a graded relationship where winter warmth above the threshold will not so readily translate into heavier winter weights (Macdonald and Newman 2002; Macdonald et al. 2010c; Bilham et al. 2018). Commonly, population processes are most affected by brutal switching between there being not enough of a good thing on the one hand, and an excess of it on the other, especially if this switch oscillates quicker than tactics can usefully respond (Chapter 14). Thus, while summer showers bring earthworms to the surface and produce chubby badger cubs, torrential downpours saturate the soil and result in hungry and diseased cubs, soaked and chilled when compelled to forage (Webb and King 1984). And so, to Goldilocks.

Variability—shifting constraints

In nature, much as in the stock market, predictability might provide little scope to make your fortune—what's needed is volatility and the opportunity for a big win. Nevertheless, stability, unlike variability, won't kill you, or wipe out your investments, either (Timmermann 1993). More particularly, how individuals can prospect risk (McDermott et al. 2008) within populations and play out varied life history strategies at different ratios may be the key to optimal population success and resilience (Chapter 14).

With this in mind, we became preoccupied with exactly how badgers modify their annual routines (Feró et al. 2008) to fit with variability and seasonality in conditions, subject to the frequency and amplitude of abnormality (Newman et al. 2017).

---

[6] This topic remains under-investigated, partly because welfare concerns precluded us from catching badgers during this period, fearful of the risk of separating mothers from their dependent cubs.

Think of an encircling contour, an envelope of conditions that encompasses the 'normal' or 'optimal' range of weather that characterizes a species' bioclimatic niche—some spokes of the vividly named *n-dimensional hyperspace* that defines the Hutchinsonian niche (Parker and Maynard-Smith 1990). This is the 'Goldilocks zone', in deference to Joseph Cundall's retelling of the fable of Goldilocks and the three bears (1850[7]) in which irresistible porridge was 'neither too hot, nor too cold, but just right'. Variability (how extreme and/or rapid these fluctuations or oscillations are) and unpredictability (how unusual they are) then conspire to spike into this bioclimatic envelope, stealing niche space.

Beware the tyranny of averages: for instance, in the UK, the month of June might have an average daytime temperature of 20°C not only if every day is exactly 20°C, but also if alternate days are 10°C and 30°C; or 0°C and 40°C, and so on, *ad absurdum*. Furthermore, Jensen's Inequality, with which readers will become intimate in Chapter 14, shows that travelling the same route in two directions need not lead you back to your starting point: if a population of 100 declines by 50% there will be 50 badgers, but if these 50 bounce back by 50% the following year, their numbers will recover to just 75. The idea that *variability* in the weather might be at least as important as the *average* weather was intuitive enough, but the mathematics needed to probe it were daunting. And so back to Pierre Nouvellet with his expertise in Bayesian statistics, a talent that enabled him to combine analysis of nuanced influences of the weather on badgers with that of the navigational aptitudes of ants. Pierre joined us on loan from Sussex University when our dataset spanned only 21 years, from 1987 to 2008. This amounted to 1125 life histories, which enabled us to examine how cub and adult survival rates, as well as recruitment (i.e. reaching maturity, defined here as 2 years of age) were associated with the extent of daily variability in the weather concealed within annual and seasonal means (Nouvellet et al. 2013).

Mean annual temperature, although rather a blunt measure of the weather, varied quite a lot from year to year, but from that point in our study we found it had no overriding influence on the annual survival rates of either adults or cubs (mean adult survival, 81%; mean juvenile, 79%;[8] 1988–2016; see also Macdonald et al. 2009, 2010c). It likewise did not influence cub recruitment into adulthood. However, the highest survival rates for both cubs and adults occurred in years with the least daily temperature variation from the long-term daily mean; that is, for those years in which, on a higher number of days, temperature was 'typical'. Recruitment was also significantly, if to a lesser extent, influenced by temperature deviation from the long-term mean, with the trend, however, in the opposite direction: recruitment was lower in the least eccentric years (Figure 12.3), linked to confounding density-dependent effects resulting from larger surviving cub cohort sizes (Figure 12.2).

Other Nouvelletian discoveries included higher cub survival in years when the jolt from summer to winter was less pronounced; that is, with lower amplitude of temperature change between winter and summer seasons (Figure 12.3). In short, while warmer weather (one might hazard, climate warming) exerts a potent selection pressure on Wytham's badgers, it is not net residual annual 'hotness' that matters, but rather the extent to which temperature oscillates from normative values; that is, the evolutionarily familiar envelope of variation testing the badgers' capacity to cope.

What about rainfall? The broad-brush answer is that variation was even more influential; of substantial importance to understanding how climate change might affect populations generally, we discovered that cub survival and recruitment rates were lower in years that were either wetter *or* drier than normal; that is, they exhibited a quadratic response—a groundbreaking discovery for mammalian climate ecology studies (Nouvellet et al. 2013). In contrast, adult survival rates were simply higher in wetter years; that is, those with higher total cumulative rainfall, likely reflecting the earthworm–rainfall link. The coefficient of rainfall variability had no influence on either adult or cub survival.[9] Presumably, then, badgers, like people, are well adapted to the chance of rain on any day in the UK. As with temperature variability, density dependence bit into recruitment more when less

---

[7] In 1837 Robert Southey published the tale of three civilized bears terrorized by an old-lady protagonist who stole their porridge. In 1850 Joseph Cundall rewrote the tale, transforming the protagonist to a sweet golden-haired girl. There was a 1903 adaptation by William Wallace. But, strictly it was 1904 before Flora Annie Steel's adaptation literally named 'Goldilocks' (formerly known as 'the girl with golden hair' or 'the old woman with silver hair' in *The Story of the Three Bears*).

[8] The eMNA technique estimates adult survival more accurately than juvenile survival.

[9] For aficionados, we used this coefficient rather than standard deviation because the latter was highly correlated with mean rainfall; see Nippert et al. (2006).

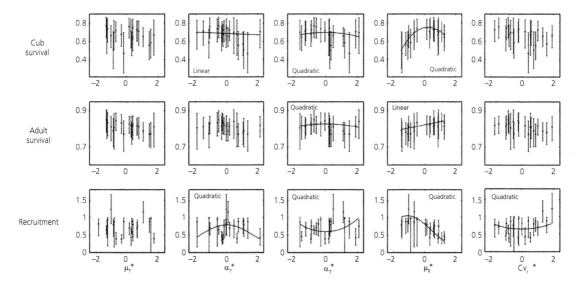

Figure 12.3 Survival rate estimates, for juveniles and adults, and recruitment rate as a function of climate metrics, with 95% confidence intervals (error bars, based on model averaging). The solid curve represents the statistically significant link between life history parameters and climate metrics, for which we indicate whether the linear or quadratic (or both) component(s) was (were) significant. Importantly, a significant linear component does not imply a straight line in the representation shown, as the relationship is defined as linear within a logistic transformation (for survival rates), or within a log transformation (for recruitment) (from Nouvellet et al. 2013, with permission).

variable rainfall led to a greater abundance of surviving, competing cubs (Figure 12.3).

Of course, these are blanket responses for broad age classes. Inevitably, all immature cubs, scarcely off the nipple for long enough to develop their own life strategies, are vulnerable to the vagaries of weather acting through food supply, although recall from Chapter 2 that this hits male cubs harder than it does females,[10] presumably because it takes them a few months longer to reach adult size (Sugianto et al. 2019c). Thereafter, no two individuals will follow exactly the same trajectory through adult life, in terms of when they breed, how often, how they divvy up investments through their lives, and thus how they age—all details to which we return in Chapters 14 and 18. Ultimately then, in evolution as in our daily lives, the devil is in the detail.

Biogeographic responses to weather

Although our hindsight nowadays is extended by more than a decade, these data we had gathered by 2008 already provided a sophisticated understanding of the effects of weather variability on the badgers of Wytham Woods (Nouvellet et al. 2013). Were similar patterns

[10] Females respond more to group size (Chapter 2).

evident at larger spatial scales? Peeking through the keyhole of Wytham into the wider world of badgers, we turned first to Ireland, where badgers typically live at densities of 1–3 badgers km^{-2} (Byrne et al. 2012a, 2012b). There, we collaborated with Andrew Byrne, a government research scientist and specialist in the epidemiology of bovine tuberculosis, to establish how rainfall and temperature metrics influence badger body weights on a national scale.

Ireland

Anybody with even a glancing acquaintance with the raging 40-year-long debate about bovine tuberculosis will know that government interventions in both the UK and the Republic of Ireland have involved killing tens of thousands of badgers. The efficacy of these culls is the topic of Chapter 16; their relevance here, however, is the light which the Irish sample of post-mortem examinations shines on the drivers of badger population dynamics (and diseases in Chapter 15). Specifically, between 2009 and 2012, the Department of Agriculture, Food and the Marine (DAFM), killed 15,878 badgers (captured at 7224 setts across 15,000 km^2). Galvanized by our Goldilocks findings, Andrew applied a similar analytical framework to examine the effects of weather variability on these badgers. Some hailed from areas where the habitat seemed ideal (improved grasslands/pasture, interspersed with hedgerows and woodlands; described in Byrne et al. 2014) whereas others came from less lush countryside, the latter being, as a result, on average almost half a kilo lighter (0.45 kg; 95% CI 0.29–0.61 kg). Overall, the average Irish badger weighed 9.7 kg (SD 1.86; Females 9.4 kg (SD 1.86); Males 9.9 kg (SD 1.84)) with the heaviest badger weighing 17.5 kg. By the way, these Irish badgers were around 2 kg heavier than those from Wytham. Moreover, the weather varies between Irish counties, and the badgers' body weights correlated positively and significantly with inter-county variation in temperature (and with all lag metrics of temperature) over the last 3 months of their lives. Delving further into the summer–autumn sample, Byrne et al. (2015a) found that the influence of temperature during just the 1 month prior to capture nudged close to a significant positive effect on body weight. In winter, however, the badgers had been significantly heavier when the weather had been warmer in the month prior to their death. A general pattern starts to emerge when we take together this result from Irish badgers in winter and our discovery that in Wytham, badgers were nearly 2 kg heavier when winters were warmer (standardized 2°C above long-term mean) (and were correspondingly lighter during those years when standardized temperature was 2°C below the long-term mean) (Macdonald and Newman, 2002; see also Noonan et al. 2014).

In Wytham, we had found *mean* rainfall to be the principal driver of survival rate (Macdonald and Newman 2002); the *variability* of that rainfall, however, had no effect (Nouvellet et al. 2013). In contrast, in the Irish sample, when the effects of rainfall or temperature in the 1, 3, or 12 months prior to the badger's capture were tested, although badgers were heavier when a greater *mean* rainfall occurred during the 1 or 2 months prior to capture, the effect was subtle and unrelated to rainfall variability indices; thus less influential than the effects of temperature overall. Irish badgers, it seems, are at home with Irish rain (Byrne et al. 2015a).

The United Nations' Intergovernmental Panel on Climate Change's (IPCC) fifth Assessment (2014) proposes 1.4–1.8°C mean temperature increase in the Republic of Ireland by the 2050s, relative to the 1961–1990 trend, with a 10% increase in winter rainfall (Sweeney et al. 2008). If so, these findings would predict that Irish badgers are likely to become even more portly. Furthermore, extrapolating from our calculations in Wytham, the projected weather in Ireland

may enhance fecundity, recruitment, and survival rates due to improved food availability and energy budgets (Nouvellet et al. 2013). However, these gains could be wiped out if future warming is linked to greater temperature variability (*sensu* Nouvellet et al. 2013).

Scotland

Wings, or even long legs, make it easier to follow a receding (or expanding) climate envelope (Parmesan 2006), but badgers have neither. How species of limited mobility, tied, as a result, to central-place foraging, respond to a changing world is poorly understood (Jump and Penuelas 2005; see also our work on lodge-bound Eurasian beavers, *Castor fiber*, Campbell et al. 2012). Badgers face this problem at the extremes of their geographical range. In Chapter 19 we focus on intraspecific variation, examining at an extreme badgers' adaptations in Portugal at the southern edge of their distribution (Rosalino et al. 2005a). Here, we attend to the northern edge, which, due to climate warming, is a moving target, badgers struggling more than many species that move with it (Van Der Wal et al. 2013), but extending northwards albeit gradually, through Finland (Kauhala 1995). Although not so close to the frigid isotherms as Finland, badgers become sparse as one moves northwards through Scotland, where the severity and duration of winter continuous to test the limits of their adaptive winter lethargy strategy in ways now rarely seen in Wytham. Indeed, in 1997 Wilson et al. had estimated the density of badgers in the Scottish Highlands to be <0.1 badgers km^{-2}. Furthermore, although they described a 77% estimated increase in badger numbers across the entire UK for the period 1988–1997, they reported no change in Scotland.

As a collateral benefit of a detailed camera-trapping study of Scottish wildcats (Kilshaw et al. 2016; Senn et al. 2019) we had the opportunity to study the impact of weather conditions on two other carnivores, pine martens (Moll et al. 2016) and badgers. Our intern André Silva investigated whether elevation, temperature, and/or human activity affected whether badgers were photographed by the camera traps during the brutal weather of autumn and winter (Silva et al. 2017). This involved what is called an occupancy analysis (Figure 12.4) (in simple terms, whether the badgers were observed in that location, or not) using photos hailing from 168 camera-trap stations, deployed across 11 study areas (averaging *c*.21.3 km^2) spread across 50,000 km^2 of Northern Scotland.

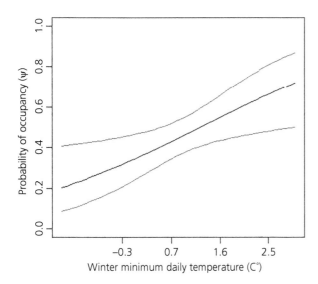

Figure 12.4 Response plot of badger occupancy (or site use) probability as a function of winter mean minimum daily temperature in northern Scotland (from Silva et al. 2017, with permission).

Over 8610 effective camera-trap days (i.e. 24-h periods, including nights) through 2010–2013, 527 photos depicted active badgers on autumn or winter nights. That works out at 6.12 captures/100 trap days. Badgers were recorded in all 11 study areas and at 68 (40.5%) of the 168 camera-trap locations. Once again, the chance of a badger wandering around to get itself photographed depended on mean minimum daily winter temperature. The less bitterly cold was the minimum daily winter temperature, the more likely we were to photograph a badger. Moreover, the photo capture rate was 0.518 ± 0.035 in warmer areas, but only 0.261 ± 0.041 where it was cooler. Delving further, the detection probability also depended on elevation[11] and human disturbance, an effect which interacted with that of climate. In areas with higher than average mean minimum winter temperature (>1.2°C) at lower (less exposed) elevation (<133 m), badgers were associated with sites further away from settlements and main roads, whereas in areas with lower mean minimum winter temperatures (<0.3°C) at higher elevation (>246 m) badgers were associated with agricultural sites. This exemplifies how species can exhibit different preferences (or avoidances) of other environmental cofactors between 'good' and 'bad' climate areas (Wong and Condolin 2015). Of course, what it doesn't tell us reliably is how many badgers could have been there (with surely some wise fat badgers ensconced in their setts), but it does corroborate our postulate

that warmer conditions are associated with higher rates of winter activity, apparently tempting badgers from winter lethargy. Indeed, in Wytham and Ireland the phenomenon of heavier winter weights linked to milder winters suggests that some temperature tipping point marks the difference between activity, successful foraging, and weight maintenance, versus conserving energy with lethargic inactivity, depleting fat reserves in the process (see Chapter 14).

The IPCC's Special Report on Emissions Scenario (SRES) A1B (i.e. medium emissions scenario, MES) projects a 4°C temperature increase for northern Scotland by 2050. If so, badgers seem likely to benefit by staying fatter (Silva et al. 2017). Of parallel relevance is that Weir et al. (2017) discuss how the northern distribution margin of the North American badger, *Taxidea taxus*, in Canada, may also be limited by frigid climatic conditions. There, worms are not an issue, but rather the availability of unfrozen, diggable soil across the grassland habitat from which colonial rodents can be excavated.

How will badgers, amongst others, in Wytham and elsewhere, adapt to climate change?

As explained in Box 12.1, climate change is not new to the badger lineage, in either historical or palaeontological time. The travails of their lineage are, however, little consolation to individuals experiencing contemporary changeable weather, to which they are ill-adapted in their own lifetimes. In Europe, badgers

---

[11] Where temperature and elevation are linked through the adiabatic lapse rate; Brunt (1933).

survived the *Eemian*[12] interglacial period (127,000–106,000 years before present)—with average temperatures being 4°C warmer, at that time, than they have been during our current Holocene, even with recent anthropogenic temperature enhancements—and also the *Weichselian* glaciation (called the *Wisconsin* glaciation in North America) that followed it, which abated only 12,000 years ago, which pushed them south to avoid ice sheets over a kilometre thick blanketing northern Europe. Clearly, they can weather the

storm of see-sawing climate (although not without costs), surely by drawing adaptation from their toolbox of dietary flexibility, body fat storage, winter torpor, and delayed implantation (with emigration a last resort where feasible). In periods of shorter, milder winters the ceiling of their environmental carrying capacity rises, notwithstanding any risks wrought by greater weather variability and summer drought—the mechanism through which badgers achieve this adaptation being the topic of Chapter 13.

[12] The Eemian interglacial period (name taken from the Eem River in the Netherlands) was the penultimate warm period to which the land was subjected before the Holocene (current period). This period began 127,000 years ago and extended to 106,000 years ago. That same period being referred to as the *Sangamon* in North America.

Weather

Badgers Adapt or Die

Whether the weather be fine, Or whether the weather be not, Whether the weather be cold, Or whether the weather be hot; We'll weather the weather, Whatever the weather, Whether we like it or not.

Anonymous

Correlation is not causation—true enough, and a sage antidote to the seduction of coincidence,[1] but it can be a clue. That is why, in Chapter 12, we took a correlative perspective on the tandem changes of weather patterns and badger population dynamics. But causality is king, its coronation confirmed only by proof of how animals actually adapt to the challenge of change—knowledge that dismisses coincidence and titrates cause from effect, between evolution's cat and mouse, between the Red Queen and the Court Jester (Chapter 11; Benton 2009). So, to reveal those correlations that are causal, in this chapter we ask how individuals amongst the badgers of Wytham Woods deploy adaptive tactics, both behavioural and physiological, to outsmart the weather, and with what consequences for the population's resilience (Chapter 11; Capdevila et al. 2020).

There are, ubiquitously, four things organisms can change when faced with trends in climatic conditions: their distribution, phenology (i.e. the timing of life history events), behaviour, and physiology (Gardner et al. 2011). Which of the four predominates depends strongly on whether climatic conditions constitute a trend, the emergence of a new normal, or whether unseasonable weather is merely a blip within the established weather pattern. Judging correctly whether climate is genuinely changing presents an adaptive gamble that could risk an individual prematurely making permanent adaptations to conditions that subsequently turn out to be extraordinary, but temporary. So, what is 'normal'? What is 'extraordinary'?

Over one, or several, generations of badgers variation around the norm can include drought, flood, heatwave, and/or polar vortex, causing violent disruption to optimal routines. Climate is what we expect; weather is what we experience. To begin then, let's delve deeper into those categories of adaptations that species, such as badgers, can use to cope with what nature throws at them.

The adaptive cascade

When things change erratically and unpredictably in the short term the issue is how to cope quickly and immediately. One urgent response would be to leave: to disperse, or migrate. But perhaps these changed circumstances prevail so far and wide (in relation to realistic dispersal distances) that relocation is not feasible. The only option, then, is *in situ* adaptation, of which the first tool is behavioural plasticity, a tactic that has the advantage of being reversible (Bradshaw and Hardwick 1989) (although complicated by epigenetic tweaks, Chapter 19). Oddly, although adaptable behaviour has always intrigued naturalists, responses at this fine scale have generally been overlooked in predictive models of climate change (Noonan et al. 2018a). For instance, and illustrating the risks of generalization, Riddell et al. (2021) report how the strategic capacity of small rodents in the Mojave desert enables them to dodge the impacts of climate change through deft use of shade, whereas under these same conditions bird populations, less able to behave their way out of overheating, show marked decline. A similar life-saving nuance, more granular than accommodated in general models, is that amongst orcas there

[1] An insight dated to 1880 and the pen of British statistician Karl Pearson.

The Badgers of Wytham Woods. David W. Macdonald and Chris Newman, Oxford University Press.
© David W. Macdonald and Chris Newman (2022). DOI: 10.1093/oso/9780192845368.003.0013

are three ecotypes each with a different temperature-related feeding specialization (Moore and Huntingdon 2008).

If altered conditions persist, or trend further, flexible behaviour may not be enough. The next tier of options involves revisions hinging not only on choices, but on physiology, and those necessitate micro-evolutionary genetic changes, either mutations conserved by selection, or a reshuffling of genes already present within the population through advantageous mate choice, and letting natural selection sieve the progeny. In due course we'll illustrate for badgers this selection for 'fitter' genotypes in Chapter 17 in terms of subtle immunological advantages of heterozygosity, and in Chapter 14 in the context of different metabolic approaches to the accumulation of body fat. Such changes evolve over generations, leaving the Red Queen breathless (Chapter 11; Barnosky and Kraatz 2007). For species with longer inter-generational times it inevitably takes longer to catch up with moving environmental goalposts (Rosenheim and Tabashnik 1991). In Chapter 10 we raised the possibility that in this regard the neuroendocrine physiology of badgers, linked to their ancestral solitary habits, may explain why their society has lagged behind the new opportunities offered by group living. These opportunities have arisen due to a different (but linked) aspect of man-made environmental change—the Agricultural Revolution.

And so back to worms and the costs of finding them, the benefits of eating them, and the scramble for alternatives when they become scarce.

The conservation of energy

According to Sir Isaac Newton's laws on the conservation of energy (*Principia* 1687) energy can be neither lost nor gained: what goes in must come out. It is this very principle that governs the flux of energy through the lives of badgers, as of all things living. How the energy that goes into a badger re-emerges depends on manifold factors, and is ultimately limited by the intake each badger has to work with; the trick of it is that the sum of expenditures must equate with the initial income, and/or any banked capital, in one way or another. You can't spend what you don't have, and just as armies march on their stomachs, for Wytham's badgers food, along with the cost of acquiring it, is at the foundation of their life history strategies.

Grabbing a 2.5-calorie strand of animated spaghetti might seem a lesser predatory challenge than that represented by, for instance, the teamwork of African

wild dogs panting at the brink of exhaustion in pursuit of a sharp-horned impala; the same bottom line, however, measures the differential success with which individuals secure the energy to ensure their survival and reproductive fitness—investments each individual may make in a different way (Chapter 14). To reveal this inter-individual adaptability required much more personal detail than our flicker book quarterly encounters with trapped badgers provided (Chapter 11); more even than could be achieved by trying to radio-track hordes of badgers simultaneously (Chapter 4). Observing the number of worms going into a badger—one badger—is hard enough (Chapter 10); quantifying the complete foraging harvest of a useful sample of individuals is uncomfortably close to impossible or, if attempted through faecal analysis, wearisome. Therefore, we turned the puzzle on its head: rather than measuring 'food in', we developed a technique for measuring 'energy out'.

Our approach involved re-engineering technology that had not been designed for ecological research. Assembling a team of computing and engineering wizards, and galvanized by a grant from the Engineering and Physical Science Research Council, we set about combining the active radio frequency identification (aRFID) transponder technology described in Chapter 4 with pioneering developments in tri-axial accelerometry (Dyo et al. 2012; Noonan et al. 2014, 2018a; Ellwood et al. 2017). Recall from Chapter 4 that aRFID tags are small devices that use electromagnetic fields to locate themselves against a receiver beacon: the clothing tag is the aRFID, the pillars either side of the shop door are the detectors. But what are accelerometers? Basically they are devices to record how something (here, a badger) is moving around in terms of the G-forces that the subject experiences—your Fitbit watch uses accelerometry to monitor your jogging or those 3–4 G of exhilaration you might experience on a rollercoaster.

This new system was to be at the core of Rhodes Scholar Mike Noonan's doctoral thesis. Mike, now an associate professor at the University of British Columbia, was not only a talented biologist and an accomplished international wrestler, but also an expert baker of sourdough bread that sustained us through wintry fieldwork—perhaps the resulting appreciation of the importance of energetic balances preadapted him to battling the millions of data points that poured from our badger tracking system.

By measuring the energy each badger put into the tri-axial trajectories of pitch, yaw, and roll (think acceleration, braking, and cornering forces on the

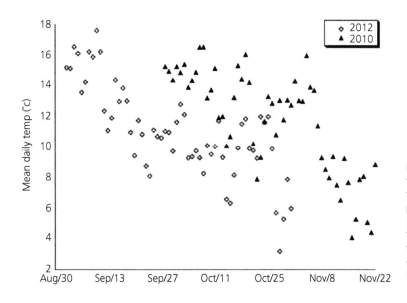

Figure 13.1 Mean daily temperature over the study periods. Mean daily air temperature over the course of the study periods in 2010 and 2012. Between the 2 years we found no difference in daily temperature, or between temperature variability, but 2012 proved significantly colder over matched calendar dates (from Noonan et al. 2014, with permission).

aforementioned rollercoaster), Mike was able to use these rock 'n' roll tags to summarize the total of mechanical energies exerted as ODBA: *overall dynamic body acceleration* (Noonan et al. 2014; see also Wilson et al. 2006; McClune et al. 2015). In due course (in Chapter 14) we'll explore variation in how individuals spend their budgets, but, for now keep in mind that a badger's propensity to spend on movement relates to the amount of capital it has at its disposal, in terms of its income (foraging success) and savings (body fat reserves), and the importance of the purchase (activity). The balance in this account reflects each individual's propensity to take risk—a so-called 'state-dependent response' (Lima 1998).

Remember from Chapter 12 that autumn can be a make-or-break season for badgers; ideally they finish this season up to 4 kg heavier than they were when it started, as they attempt to fatten for winter. We therefore next asked how the interaction between weather through the autumn equinox and body fat reserves might influence their activity regimens. Serendipitously, the weather gods provided Mike with two contrasting autumnal equinoxes, the relatively dry, mild autumn of 2010 (Mean T°C 13.34 ± 2.11; rain 2.21 ± 3.81 mm) versus the relatively wet autumn of 2012 (T°C 9.41 ± 2.24; rain 2.98 ± 5.68 mm) that averaged 4°C cooler (Figure 13.1). Temperature declined steadily in 2012, but this trajectory was more variable than it had been throughout 2010, skewed strongly by a cold 16-day interlude through 7–22 November that saw daily (24-h mean) temperatures suddenly tumble from 16°C to 8.5°C; after this interval temperatures

recovered to a more seasonable 13.4°C daily average. In terms of the factors likely to make earthworms available (Chapter 10) soil temperature was similarly 3.5°C cooler at 10 cm depth in 2012 than 2010. Temperature varied less at 30 cm soil depth, but was still 2.6°C cooler in 2012 than 2010. There was, however, no significant difference in rainfall between years, although 2012 was more humid, leading to the soil being damper.

How did the average badger's activity vary between these years with different weather? As these prototype energy-measuring devices rolled off the workbench, Mike was able to instrument four badgers in 2010 (three males and one female) and four different individuals in 2012 (three males and one female); these eight first movers were to yield him 3.5 hundred million (3.5 × 10^8) accelerometer data points.[2]

Our efforts to partition their activities and ascribe motivations were complicated by the badgers' aptitudes for multi-tasking. This versatility means it is commonplace to see an individual scoff a beetle, deposit a scent mark, and groom a companion within seconds, which makes it very hard to allocate time, and thus energy budgets, to each activity. Nonetheless, using the rock 'n' roll tags we could determine a threshold of ODBA above which we were confident that the

---

[2] In 2010, data were collected from 28 September until batteries ran out on 24 November; that is, for a total of 58 days. All batteries lasted for the same duration except for those in one collar (N30), which ran out after only 45 days. In 2012, data were collected from 1 September until all batteries ran out on 30 October; that is, for a total of 60 days.

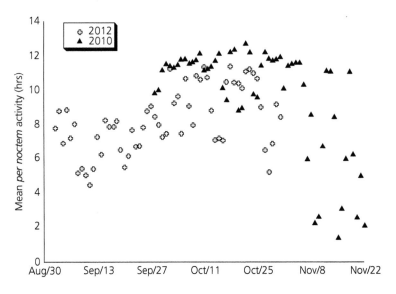

Figure 13.2 Mean *per noctem* activity over the study periods. Mean *per noctem* activity over the course of the study periods in 2010 and 2012. The duration of *per noctem* activity was significantly more variable in 2010 than it was in 2012; this was also true when compared on the same calendar dates, to control for night length, significantly longer in 2010 (from Noonan et al. 2014, with permission).

badger had been moving—deploying energy—and it is to this which we shall refer henceforth as 'Activity'.

Using Bayesian inference,[3] in both years, Activity increased initially, as expected, with night length through October, but then tipped into decrease until late November when autumn nuts and fruits ceased to be as available and the badgers were recaught, reweighed, and collars removed (Figure 13.2). Badgers were significantly more Active, and more variable in their Activity, in 2010 (11.25 ± 1.02 h) than in 2012 (9.22 ± 1.17 h), when compared, that is, on the same calendar dates (28 September–30 October) between years, in order to control for night length (Figure 13.2). Two interpretations come to mind: either that badgers accomplish their nightly agenda (i.e. eating enough worms while also scent marking, grooming, socializing, etc.) in a shorter time under cooler, drier autumnal conditions, or, conversely, with few earthworms and high thermoregulatory stress, Activity was sufficiently unrewarding that it paid them to stay snug underground. In 2012, badgers were significantly more Active with greater rainfall and humidity, perhaps securing a greater return for effort due to greater earthworm availability. Air temperature during 2012 (the cool year) did not affect Activity (Figure 13.3b), although there was less Activity when soil conditions at 10 and 30 cm (with less soil temperature variation) were warmer—possibly enabling badgers to meet earthworm-consumption requirements more quickly.

In 2010, the hotter, drier year with greater Activity, when worms were probably harder to come by, we detected no correlation between mean badger Activity and relative humidity or rainfall, although there was a positive effect of warmer air temperature (Figure 13.3a), soil temperature at 10 and 30 cm, and soil water content. It seems likely that a paucity of earthworms under these conditions forced the badgers to turn their attention to seasonal fruits and nuts, whose availability was unaffected by rain. Clearly then, our pilot study was pointing to complex effects of either temperature and/or rainfall on activity—effects we will disentangle in Chapter 14.

While these generalizations were a satisfying match to our naturalistic expectations, what most caught our attention was that Activity differed a lot between individuals, particularly so in the autumn of the more variable year, 2010. We began to think about what different options were open to individuals. Of course, if poor weather and low food availability persist, badgers can opt out of Activity, retreat to their setts, and, up to a point, fall back on their fat reserves, thereby conserving energy through torpor (Newman et al. 2011; Zhou et al. 2017). Keeping an eye on the weather before deciding to set out on a foray makes sense, but it would be prudent to make the decision while mindful of fat in the bank[4] particularly during the autumn when packing on weight may determine survival over winter.

---

[3] Bayes' theorem is used to update the probability for a hypothesis as more evidence or information is added.

[4] Thinner, however, does not necessarily equate with hungrier—the topic of appetite regulation emerges in Chapter 14.

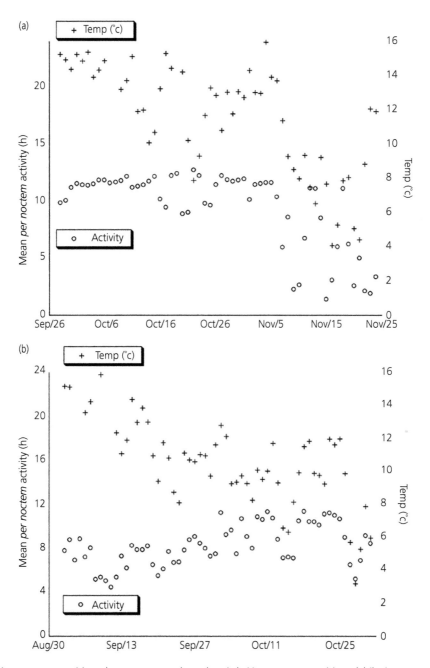

Figure 13.3 Mean *per noctem* activity and temperature over the study periods. Mean *per noctem* activity and daily air temperature over the course of the study periods in 2010 (a) and 2012 (b). In 2010 activity correlated significantly and positively with temperature, where in 2012, there was no correlation between activity and temperature (from Noonan et al. 2014, with permission).

To investigate, we analysed body condition index (BCI: log weight divided by log length), a proxy for nutritional state, asking whether this influenced individual activity regimens. In 2010, when drier conditions likely diminished earthworm availability, in our small sample those chubbier badgers with good BCI engaged in less nightly activity, while scrawnier badgers with poor BCI were the most active (Figure 13.4a). Only one (N30) of our four tagged badgers had a BCI below the contemporary population mean at the outset of the study, whereas two others, a male (N31) and a female (N34), had BCIs significantly above that population mean. Of these four badgers, it was thin N30 who engaged in the longest bouts of night-time activity, being active for an average of 30 min more per night than the two fat badgers (N31 and N34). We were puzzled that the fourth individual, N33, having been fairly chubby when we fitted the tag, was also generally active for 30 min longer than N34—that mystery was solved, however, when we recaptured these two badgers in November, and found that their BCIs had reversed: N33 had lost weight while N34 had gained it (Noonan et al. 2014).

In contrast, in 2012, with wetter colder conditions, favouring earthworm availability but exerting greater thermoregulatory costs, the plumpest badgers engaged in the longest nightly Activity (Figure 13.4b). When instrumented in September, both badgers N48 and N49 had BCIs above the then population mean, but by November only N48's BCI remained above the then mean. This fat badger was, on average, active for 20 min more each night than thinner N49, whose BCI had fallen. A third badger, N44, had a lower BCI than either N48 or N49 in both September and November and was, on average, active for 40 min less per night than N48 and 20 min less than N49.

Meanwhile, for a wider perspective, in Poland badger body weight in autumn was almost twice that in spring, and there Goszczyński et al. (2005) found a level of idleness amongst the rotund badgers that would have made even the most sedentary fat Wytham badger in 2010 seem hyperactive. These Polish badgers were inactive for an average of 96 days each winter, their activity being strongly linked to the temperature (Kowalczyk et al. 2003a). Predictably, in warmer parts of the European badgers' range, such as in the Mediterranean region, this seasonal variation in body fat reserves and conservation of winter activity is less extreme (Rosalino et al. 2005a), showing the biogeographical adaptability of *Meles* (Chapter 19).

Two important points emerge from these eight rock 'n' rolling badgers over the two contrasting years. First,

each badger did its own thing, with respect to its energy bills. Second, the consequences of plumpness were essentially opposite under the two contrasting weather regimens: when the weather favoured earthworm availability in 2012, the fatter badgers made the most of it and were most active; by contrast, however, when the weather hindered earthworm availability in 2010, the fatter badgers cut their losses and were least active. Mike led us in summarizing these ODBA findings in a paper we subtitled 'one size does not fit all' (Noonan et al. 2014, 2018a): a subtitle that acknowledged our growing fascination, beyond generalizations, with the individual tactical adaptations. Overall, we conceived of badger foraging behaviour as following an internal algorithm of potential energy gain (e.g. earthworm availability) against perceived risk of energy loss (foraging effort), weighed against body fat reserves and propensity to take risk.

This breakthrough in understanding how badgers respond to weather interested us, but did it help foretell their future? Insofar as badgers might be representative of some generalist mammals, and predictions abound for how climate change will threaten mammal populations (Pacifici et al. 2015), what prognosis do these data portend for badgers?

Predicting the future

Crystal balls can be opaque, but by drawing upon the clarifying framework of UK Climate Projections 2009 (Murphy et al. 2009) we were able to simulate how badger Activity for the 25 km^2 area around Wytham Woods might change under two warming scenarios (Noonan et al. 2018a) (Figure 13.5).

The complex and non-linear responses of our pilot sample made modelling challenging; however, from our data, we would predict that, on average, badger responses to climate change would be small. Modestly warmer conditions under the low emissions scenario could cause a net increase in the duration of nightly activity (median = 21.6% increase, 95% confidence intervals (CIs) 21.5–22.0), with activity being maintained into the winter months (Figure 13.5a). Although the net amount of energy expended while active would be only marginally greater than current levels (median = 0.46% increase, 95% CIs 0.43–0.49; Figure 13.5b), we projected that ODBA through November could be 15.6% greater (95% CIs 15.5–15.7). In contrast, under a high-emissions scenario, warmer and drier conditions might bring about a far greater (median = 98.9%, 95% CIs 98.6–99.2) increase in the duration of nightly activity (Figure 13.5d), but with negligible net changes in the

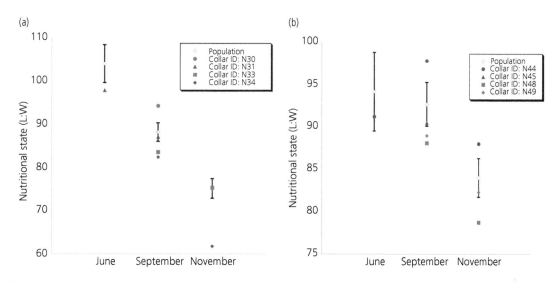

Figure 13.4 Variation in individual nutritional states from the population. Individual measures of nutritional state (L:W) in comparison with the population mean (PM) 6 the standard deviation (SD) in June, September, and November in 2010 (A) and 2012 (B). Red points indicate badgers with comparatively long activity, where blue points indicate badgers with comparatively short activity (from Noonan et al. 2014, with permission).

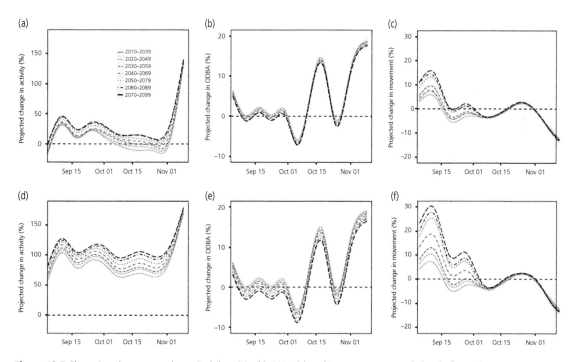

Figure 13.5 The projected percentage change in daily activity (h); ODBA (g); and inter-sett movements (ha) under future climate projections over seven overlapping time periods, in relation to present day trends. Panels a–c depict projected change in behaviour under a low-emissions scenario (IPCC SRES B1); d–f depict the projected change behavior under a high-emissions scenario (IPCC SRES A1F1). The trends in projected responses were generally consistent between scenarios, but the magnitudes of the changes were greater under the high-emissions scenario (from Noonan et al. 2018, with permission).

amount of energy spent while active (median = −0.16% increase, 95% CIs −0.19 to −0.13; Figure 13.5e); however, again, we would predict that late autumn/winter ODBA could be 15.0% (95% CIs 14.88–15.14) greater by November.

Were these modest net changes an indication that badgers exist comfortably within their adaptive envelope, with plenty of adaptive space left available to them, or were they the result of them living on the edge of their behavioural and physiological plasticity, with little capacity for further flexibility? Certainly any strategy that resulted in badgers faced with weather/food stress opting for torpor would backfire if winters became so warm that they were physiologically unable to hold down their metabolic rates, forcing arousal (as speculated for dormice (*Muscardinus avellanarius*; Pretzlaff and Dausmann 2012) and microchiropteran bats (Foley et al. 2011)). This scenario is further complicated insofar as the once reliable endocrinological signal of day length as a regulator of metabolism, mediated by the pineal hormone melatonin, may no longer be dependable (Bechtold et al. 2012). This mixture of factors and mechanisms makes it hard to predict the future, but the badgers' nonlinear responses exemplify why it can be misleading to assume that the effects of climate change on species or ecosystems will be straightforward (Doak and Morris 2010; Oliver and Morecroft 2014).

From these pilot studies our focus shifted to population resilience and the extent to which individuals cut their coats according to their cloth—an analysis of energetics that leads, in Chapter 14, to remarkable insights into badger individuality. Before turning to that, a note on trappability.

Trappability

The link between the weather, and the emergence and risk aversion in badgers, especially fat ones, has implications beyond badgerology (Noonan et al. 2015d; Martin et al. 2017). Capturing (or trapping, observing, or photographing) and recapturing individuals consistently and reliably is central to the accuracy of any and all population monitoring (Noonan et al. 2015d, see also Gopalaswamy et al. 2015), because detection is ultimately a function of an animal's activity and denning habitats (e.g. Rowe and Moll 1991). Coyotes (*Canis latrans*), for example, are less likely to be spotted on camera traps during harsh weather (Madsen et al. 2020), a finding from western Nebraska that parallels our results for badgers in the Scottish Highlands (Chapter 12). By way of analogy, if we were to deduce

the human population of Oxford from a survey of cyclists pedalling to work in the city, we should surely be mindful of differing enthusiasm for bicycling when the sun is beaming or the rain pouring;[5] if a badger does not emerge, it can neither be seen nor captured. Factors like this affect what is known as *detection likelihood*, and have been too rarely considered—where Kellner and Swihart (2014) lamented that only 23% of 537 relevant articles they reviewed accounted for imperfect detection, while Gopalaswamy et al. (2015) have emphasized that this can lead to big trouble—hence, for example, the smouldering debate about the reliability of counting tigers from detecting their footprints or spoor (applicable to lions too; Dröge et al. 2020).

We opened this chapter with reference to the rookie error of confusing correlation with causation, and rarely is this temptation more obvious than in the context of climate change: the climate has changed, mobile phones have proliferated, social media has burgeoned, and the badgers of Wytham Woods have increased in number: beware 'false positives' (Araújo et al. 2005; Parry et al. 2007) and the logical fallacy dubbed classically as '*cum hoc ergo propter hoc*' ('with this, therefore because of this') (Noonan et al. 2015d).[6] To dig down to the underlying mechanisms that truly link climate to badger numbers, it seemed plausible that it might also affect their propensity to be captured (Photo 13.1), undermining the simple assumption that trap success straightforwardly reflects badger numbers. Mindful of Charles Darwin's (1879) sage observation: 'To kill an error is as good a service as, and sometimes even better than, the establishment of a new truth or fact', we sensed an error to be killed.

During the 19 years from 1994 to 2012 we had encountered 1179 individual badgers through 3288 capture/recapture events (Photo 13.2). We calculated trapping efficiency (TE; calculated as the percentage of badgers caught that we later knew to be alive in the population from our long-term monitoring—based here on simple MNA procedures see Chapter 11) (Noonan et al. 2015d). Then we asked what factors predicted the likelihood of catching particular individuals? This game of analytical 'catch me if you can' (which involved principal component analysis and multi-model inference) did indeed reveal the most

---

[5] Newman et al. (2016).
[6] This has been neatly satirized by Bobby Henderson's internet spoof (n.d.), which shows a statistically significant relationship between increasing temperatures and the shrinking numbers of pirates since the 1800s.

Photo 13.1 Catching badgers in the rain.

Photo 13.2 Fieldwork in torrential rain—an all-weather pursuit in which the weather impacts success (photo: A.L. Harrington).

potent predictors of TE as fat (i.e. mean BCI), season, and time-lagged measures of temperature and precipitation. Amongst these, BCI trumped the other factors: the more corpulent the badger, the less reliably we could catch it (Figure 13.7). Moreover, a seasonal interaction with this effect of body condition meant that the greatest proportion of badgers was caught in summer (Figure 13.6). In short, mild, wet conditions that promote earthworm availability result in well-fed badgers not much tempted by peanuts to enter traps. Therefore, when the population is generally in good body condition, fewer badgers get caught, and vice versa. Taken uncritically, the resulting skewed subsample of the catchable could give the impression that badgers are thinner when the prevailing weather is wet, whereas the opposite is true. The very conditions that militate

against capture success are broadly associated with the best survival and thus greatest abundance of badgers.

But, so what? Well, projecting best-/worst-case weather conditions and BCI for each trapping session, over time, made a big difference to calculated trapping efficiency; that is, 8.6% ± 4.9 SD difference in the seasonal TE between the two scenarios.

So were our analyses undermined? Thankfully not, due to the benefits of doggedness and thus the longevity of our records, because this didn't affect our population estimates built on the confidence of long-term data collection (Figure 13.8); nevertheless, it could have badly misled a snapshot project. A short-term project, even if repeated on a schedule, might well have lacked sufficient inter-annual weather variation to resolve population dynamic responses for a mammal as long-lived as badgers (e.g. Pocock et al. 2004). Indeed, abundance estimates derived from our best/worst seasonal projections (Figure 13.9) established that weather-induced variability had the potential to generate a four-fold difference in estimates, with a population size underestimation as high as 55.0%, in comparison with long-term MNA estimates, under the worst-case scenario, and an over-estimation, still high at 38.6%, under the best case. Indeed, it would take 2 years (i.e. six trapping sessions) following our regime under the most optimal weather (itself an unlikely scenario) to arrive at the reality revealed by our long-term MNA estimates. These sources of error are why accounting for imperfect TE on a year-to-year basis in estimating density and other population parameters has become an important part of our methodological toolbox in recent studies (Chapter 11; Bright Ross et al. 2020, 2022).

Accuracy almost always matters when it comes to counting animals—and, furthermore, counting animals almost always matters when it comes to planning their conservation. Indeed, there are particularly practical, even political, reasons why it matters when counting badgers. Chapter 16 reviews the festering topic of bovine tuberculosis (bTB) in the UK, and policy attempts to manage its prevalence in dairy cattle by killing badgers. All these attempts involve at least some reference to the numbers of badgers present in the target area, and since 2014 these hinged on an aggregating suite of killing zones that were established against declared yardsticks of success, foremost amongst which was the proportion of badgers killed (Defra 2014; IEP 2014). In an article published in 2015, we described these interventions as deeply unpromising from the outset (Macdonald et al.

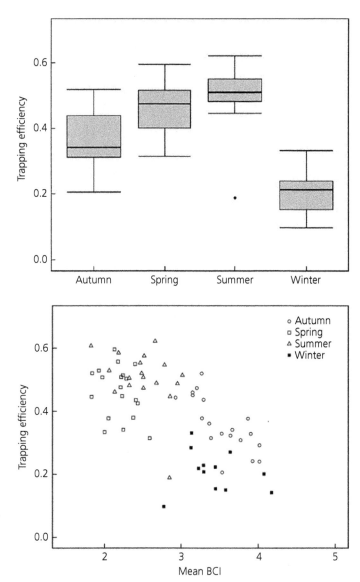

Figure 13.6 Box plots showing the proportion of adult badgers trapped in each season (trapping efficiency) in relation to contemporaneous minimum number alive (MNA) estimates (from Noonan et al. 2015, with permission).

Figure 13.7 Scatter plot representing the relationship between the proportion of adult badgers trapped within each trapping session (trapping efficiency) and the mean BCI of captured animals—a subcutaneous fat score scaled 1 (thin)—5 (fat) (from Noonan et al. 2015, with permission).

2015e)—a view previously attributed to David in the *Guardian* newspaper.[7] Whatever the efficacy, whether anticipated or judged with hindsight, there is no dodging the fact that evaluation of the badger removal (culling) progress hinges on counting reliably the numbers of badgers before and after the killing measured against the declared criterion of success. In this instance, the British government's criterion for success was that by the end of the culling operation 70% of the badgers should be dead. Against this criterion, the

early results were mixed. In 2014, while 'enough' were killed in Somerset (341 of the targeted 316–435 badgers (108–78%)), only 174 of the targeted 615–1091 (28–16%) were killed in Gloucestershire (Defra 2014). These culls occurred during a rather cold, wet period, and from our 'catch me if you can' study these are conditions under which we'd expect the badgers to be rather inactive and reluctant either to enter traps or to present themselves as targets to marksmen. According to our analysis, had luck provided, instead, a run of warmer and drier weather the Somerset cull would likely have achieved closer to its upper quota target, although

---

[7] Carrington (2013).

Figure 13.8 Estimated population numbers and possible errors, based on uncorrected raw data.

even under these conditions the Gloucestershire cull would still have fallen short of minimum quota targets. This merits further attention because Martin et al. (2017) subsequently noted that weather conditions in Ireland also influenced badger captures, albeit in different ways, being highest in drizzle, rain, and heavy rain, and also when minimum temperatures ranged from 3 to 8°C.

In summary, amongst the many lessons to take from Wytham's badgers, one is to beware of too uncritically succumbing to inductive bias (Hume 1748);[8] the risks posed by climate change to species and ecosystems, as well as to human enterprises, are too serious to get things wrong. Verisimilitude, a pleasingly polysyllabic name for the semblance of truth, is increasingly problematic in the light of ever-more variable weather patterns (Hulme 2005; Harrison et al. 2006; IPCC 2007

and 2012). For example, let us briefly describe a cryptic, and damaging, interaction between weather and badgers that is contingent upon these emergence and activity patterns.

Tertiary influences of weather: road traffic accidents

We have described how mild winter conditions should be associated with greater badger winter activity—a trigger switched on or off by the occurrence, or not, of frost.[9] Congruent with this, mild winter weather was *beneficial* to survival rates in our early data (Macdonald and Newman 2002), but our longer run of data turned this result on its head, and, counterintuitively, in subsequent analyses mild winters appeared to have a *detrimental effect* on over-winter survival (Macdonald et al. 2010c); in short, something seemed to be causing these active badgers to die. This *volte-face* caused

---

[8] Hume's criticism of inductive bias is that it is inappropriate to predict that instances of which we have no experience resemble those of which we have had experience—an objection that aligns closely with the Black Swan theory mentioned in Chapter 12.

[9] This genre of switch is called a *pejus tipping point*, where performance begins to decline but less fatally than at a critical limit (Doak and Morris 2010).

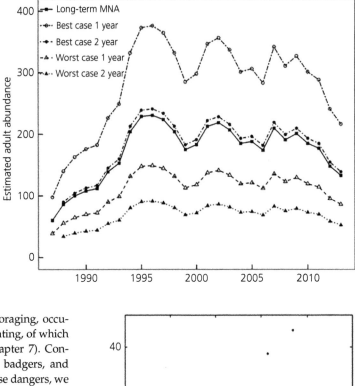

Figure 13.9 The annual abundance of adult badgers present in Wytham Woods, estimated from both long-term capture records (MNA), and the predicted number of (re)captures based on the trapping efficiency of each trapping session under theoretical best- and worst-case weather scenarios. One-year estimates (1 year) were generated from the weather scenarios for each season projected over a full year; 2-year estimates (2 years) from the same scenarios projected 2 years (from Noonan et al. 2015, with permission).

us to ponder on what, in addition to foraging, occupies badgers in winter: the answer is mating, of which the main spate is in February (see Chapter 7). Considering which risks imperil amorous badgers, and how mild weather might exacerbate these dangers, we first dismissed fighting—insofar as badgers mate every year, we saw no reason why they should be more cantankerous in mild years—and indeed the frequency of bite wounds gouged into protagonists' rumps was not correlated with more benign weather (Macdonald et al. 2004c). Nonetheless, we knew that mild weather encouraged badgers to spend more time above ground (Noonan et al. 2014), and there was even a hint that it was associated with them visiting neighbouring social groups more often (Noonan et al. 2018a). This led us to hypothesize that mild conditions combined with lascivious distractions might increase badger exposure to the deadliest of their contemporary foes—the motor car.

Through a study we conducted in the 1990s we documented that around 44% of annual mortality of the badgers of Wytham Woods was attributable to road traffic accidents (RTAs) (Macdonald and Newman 2002). Similarly, the last time anybody checked, national figures showed that around 50,000 badgers were killed on British roads each year in the 1980s, equivalent to 30–66% of annual adult and post-emergence cub mortality (Skinner et al. 1991; Clarke et al. 1998). To investigate this more fully, Defra helpfully gave us the information it had logged on the deaths of over 700

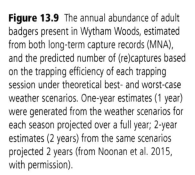

Figure 13.10 RTAs occurring in February in relation to temperature, based on over 700 RTSa gathered in south-west England between 1972 and 1999 (from Macdonald et al. 2010, with permission).

badger RTAs from 1972 to 1999, mainly in south-west England. These data showed that accidents peaked when crepuscular emergence times coincided with evening and morning rush hours, in early and late winter. More interestingly, RTA rate doubled for every 1°C increase in the mean daily temperature during the February mating peak (Figure 13.10; Macdonald et al. 2010c). We infer that warmer weather leads to wider-ranging courtships and libidinous distractions from road safety, resulting in a heavier RTA toll.

With warming trends projected to continue (Murphy et al. 2009; IPCC 2014) these effects, compounded by population fragmentation caused by traffic and expanding road networks, risk elevating badger mortality rates (Newman and Macdonald 2015). More widely, we knew road traffic was a major risk to other carnivores ranging from otters in England (Philcox et al. 1999) to raccoon dogs in Japan (Saeki and Macdonald 2004). Indeed, our own work has documented the disruption brought by various kinds of roads and trails to animal activity (Zhou et al. 2013) and scaling up, the Chinese Belt and Road Initiative (Kaszta et al. 2019), bodes ill for the future of clouded leopards across Asia; scaling down, we have seen how roads impact seed dispersal by rodents (Cui et al 2018), all demonstrating the wider perils of roads for conservation. Glista and colleagues' (2008) conclusion that temperature is the primary predictor of RTA-related mortality for more than 60 vertebrate species reinforces this bleak futurology.

Conclusion

Mindful of the tyranny of words, we have been diligent in attributing the badgers' physiological, behavioural, and population responses to the weather or to climate 'changeability', with scarcely a mention of 'climate change'. That circumspection is neither because we are climate change deniers—far from it—nor indeed because we do not suspect a major role of anthropogenic factors tipping the weather into a rapidly changing climate, but rather because we caution that climate is constantly changing and has never been stable throughout the evolution of badgers, or any species (Chapter 11), and so we expect the badgers' responses to changeable weather to herald the impact of whatever climate hindsight will in coming years reveal to be upon them.

What are the prospects for Wytham's badgers? Of nature's swings and roundabouts, they seem likely to do slightly 'better' under the anticipated warmer climate with milder winters, and microclimatic conditions favouring earthworm availability, and lower costs of thermoregulation. The flip side of that coin is that springs and summers may become problematically dry and hot (stressful for the portly), and weather extremes uncomfortably variable (Nouvellet et al. 2013). Reflecting on how this might play out in life history vulnerabilities, remember that population density and viability are not the products of the strongest individuals, but rather of the most fragile. Cubs, followed by the elderly, are the weakest link, with lowest tolerance for deprivation, and greater susceptibility to inclement conditions (Newman et al. 2011). Might such frailty, under an altered climate, tip Wytham's badgers onto the slippery slope to population decline (Newman et al. 2017), while in the north of Scotland a parallel shift in the weather might see them flourish?

In 1940, a US Airforce scientist, Gilbert Daniels, was tasked to find out why aircraft controls, and cockpits, seemed so badly, often fatally, fitted to pilots. He measured the hands, and ultimately 10 dimensions, of 4000 pilots to construct the average pilot. Almost none of the pilots matched the average, less than 3.5% matched even three of the dimensions, so a cockpit designed for the average pilot fitted none of them— hence the expression that there's no such thing as an average person—and you're not likely to meet an average badger, either. Each badger operates as an individual, and each is locked into a Darwinian struggle for existence in competition with its peers. Natural selection can apportion advantage to fortitude or guile or even genetic immunity to diseases (more of that in Chapter 15), but each throw of the dice brings with its tangle of luck, good or bad, the right place and the right time. In part it's about the hand you're dealt, but beyond that, it matters how you play the game. Technically, this is the grist of life history theory, and the topic of Chapter 14.

The Game of Life

Ecological patterns, about which we construct theories, are only interesting if they are repeated. They may be repeated in space or in time, and they may be repeated from species to species. A pattern which has all of these kinds of repetition is of special interest because of its generality, and yet these very general events are only seen by ecologists with rather blurred vision. The very sharp-sighted always find discrepancies and are able to say that there is no generality, only a spectrum of special cases. This diversity of outlook has proved useful in every science, but it is nowhere more marked than in ecology.

Robert MacArthur 1968

All animals have objectives. Delivering them, even just to stay alive, has costs, that ultimately are measured in energy. So it was that in Chapter 13 we learnt how individual badgers must balance their energy books, behaving in such a way as to ensure that they can secure enough energy intake—food—to deliver on their life history objectives (survival and reproduction). From this individual perspective you will notice a motif that recurs societally: that food (energy again) security is, as we argued in Chapter 10, the fundamental force shaping badger society according to the Resource Dispersion Hypothesis (RDH). As we plumb the flow of energy through population, society to individual, we return to Gilbert Daniels' rarity of the average (Chapter 14) and the realization that while those features that pilot badgers through their lives may appear as an average fit, that appearance is an artefact—a potential verisimilitude where, actually, how each individual lives its life is unique, and certainly subject to changing criteria as that individual navigates its lot. In this chapter we therefore ask whether all badgers march to the same drum, or, more likely, do we see MacArthur's plethora of special cases?

The historical emphasis of ecology has focused on species as the minimal unit of description (e.g. Pearson and Dawson 2003), examining those typical ways in which species characteristically respond to environmental conditions, often considering variation as noise that annoyingly obscures generality. This is despite the long-held appreciation that, fundamentally, natural selection operates on the individual (Darwin 1859; but see Wade and Kalisz 1990; Mayr 1997). Ruel and Ayres

(1999), who reviewed ecological papers published from 1887 to 1971 and found that mentions of 'variance' (individuality) featured in only around 40 (appearing just 10 times in abstracts from the Ecological Society of America through the 1960s), lamented this tendency to ignore individual variation. A new view increasingly recognizes the pervasive role of individual variation, or 'heterogeneity', in animal population dynamics (Bolnick et al. 2011): the nub of it is that individuals may exhibit different optimal responses to the same conditions, departures from the average arising due to personal sensitivity to stressors (Harding et al. 2004). Different populations of one species can be characterized by different social behaviour—the spotted hyaenas of the Ngorongoro Crater configure well-defined territories to partition their sedentary prey base, while those in the Serengeti cover vast less-defined ranges in the wake of their migratory prey—intra-specific variation that reflects ecological circumstances (Macdonald 1983; Chapter 19); within each population, moreover, individuals may be variously eccentric, each with its particular life history and personality traits, emergent from circumstances and genetics. And so, what might previously have been perceived as noise is now revealed as a basis for selective advantage (Réale et al. 2010),[1] as we

---

[1] Evolutionary biologists realize that not only is standing genetic variation the substrate of evolution, but also the degree and patterning of genetic and phenotypic variation can determine the direction and outcome of natural selection. In parallel, ecologists increasingly recognize individual variation as an important factor affecting intra- and interspecific competition and the structure and dynamics of an ecological network.

The Badgers of Wytham Woods. David W. Macdonald and Chris Newman, Oxford University Press.
© David W. Macdonald and Chris Newman (2022). DOI: 10.1093/oso/9780192845368.003.0014

turn now to a quest for individual variation between the life histories of badgers.

Nature is a fastidious book keeper, a cruel accountant that punishes profligacy with its common currency: energy (Chimienti et al. 2020), or as Thomas, Duke of Norfolk, expressed it in a letter to Thomas Cromwell, 1538, 'You can't have your cake and eat it'. The penalty for overexpenditure can be death; the executioner, natural selection (Anholt and Werner 1995; Monteith et al. 2013). How to adapt to this exacting regime?

As the old adage goes 'there's more than one way to skin a cat',[2] and equally there can be many pathways to success in nature, both in response to immediate circumstances, and according to lifetime opportunity. The question might be asked of any organism, but it turns out that the answers given by the badgers of Wytham Woods are unexpected and have far-reaching implications for evolutionary principles (Pontzer 2015), including our own evolution (Cordain et al. 1998). From the ethological perspective, these inter-individual strategic differences have been termed 'animal personalities' (Gosling 2001). A badger's personality, just like yours and ours, would boil down to a repeated and predictable way of responding to situations, and could involve traits ranging from 'reactive' (shy, docile, sedentary) to 'proactive' (bold, aggressive, exploratory) (Careau et al. 2008; Réale et al. 2010). In the case of humans, Gerlach et al. (2018) categorize personality as the blend of five traits: neuroticism, extraversion, openness, agreeableness, and conscientiousness, which distil to four predominant personality types labelled: reserved, self-centred, role model, and, heaven forbid, average. It's interesting to muse on how these traits might translate into an assessment of badger character (anybody who has reared puppies, or in David's case, fox cubs (to mention a wild carnivore), will need no convincing that their personalities differ radically on loosely similar traits). Anyway, these idiosyncratic ways of reacting would influence each individual's navigation between the siren appeals of risk and prudence, obviously affecting the odds regarding mortality and reproduction. Proactive individuals rev faster, with higher resting metabolic rates, than do reactive ones (Metcalfe et al. 2016; Pettersen et al. 2016).

This individual-based perspective will not come as a surprise to anybody who has reflected on their own way of navigating life in the context of their society. Each person does what they can, based on their experience, skill set, personality, and circumstances, and each of us strives, in Frank Sinatra's words, 'to do it my way'. Individual strengths are neither the same nor equal; outcomes aren't, either. And of badgers in the mating game: perhaps a strapping male might impress his way to paternity, whereas the stripling might rely more on charm or chance; one might throw everything into grasping the moment, *carpe diem*, another might bide his time, accruing capital for the right moment (and remember, in the case of capybaras, these two strategies were associated with different ratios of interstitial to spermatogenic tissues in their testes; Chapter 7).

Enigma variations

You may puzzle at an emergent paradox: if the evolutionary actuary is so remorseless in meting out penalties, why has it not canalized all variation to the 'perfect' fit (much as produced by asexual reproduction)? The answer is that even under stable conditions, the target is moving, and, to extend the metaphor, the rifling of each gun barrel is different too. Combine these variances in both environment and organism, and the result is populations comprising various individuals pursuing varied tactics in their quests for the same strategic goal: maximizing lifetime reproductive success (ultimately, their fitness). Individuals, each equipped with different raw material in their unique genome, dance as best they can to the dynamic tune of their surroundings, ending up with a spectrum of life history profiles rather than a single, optimal one (Tuljapurkar et al. 2009b). In fact, this flexibility is key for the population's resilience, where, in many cases, change occurs too quickly for genetic adaptation[3] (Berteaux et al. 2004; Merilä and Hendry 2014). Consequently, the capacity to respond more quickly to environmental cues ('plasticity') is often achieved tactically (Parmesan 2006; Chevin et al. 2010; Merilä and Hendry 2014).

This soup of diversity is vital to the ability of populations to endure disruption from whatever unexpected direction it may come (e.g. Beever et al. 2017; Ogura et al. 2017)—an amalgam termed the 'portfolio effect'. We discussed in Chapter 13 how a population must

---

[2] 'There's more than one way to skin a cat' is an aphorism which conveys the fact that there are many ways to do something; there are many ways to achieve a goal. The oldest known use of the phrase dates back to 1854, in the work *Way down East; or, Portraitures of Yankee Life*, by Seba Smith.

[3] We consider epigenetic change in Chapter 19.

either exhibit resistance and/or resilience in order to persist through repeated disruptions (Harrison and Taylor 1997). Recall from Chapter 10 that resistance is the ability of a population to respond to stressors (disturbance) without experiencing demographic losses, and these safety nets typically include such pre-adaptation as conservative reproductive investment (Svensson et al. 2005) or behavioural plasticity (Snell-Rood 2013). Resilience is different; it is a population's marble-like ability to roll around the cup in different demographic states and recover its former stability, without cresting the lip and tumbling into extinction. The measure of how far an ecosystem can be shifted from its previous state and still return to normal is called its amplitude, and depends on the intrinsic capacity of individuals to adapt, for example by producing more offspring (Capdevila et al. 2016); sometimes circumstances can give advantages to a phenotype that is generally suboptimal—a phenomenon termed 'conservative bet-hedging' (Ogura et al. 2017)[4] within a population. Just like a city trader playing the stock market, individual organisms can play for either high-reward (but high-risk) strategies, or opt for seemingly inferior 'safe bet' strategies that play the long game, which for the patient player can offer lower variance in fitness in the long term than strategies that match more closely the current environmental optimum in the short term (Starrfelt and Kokko 2012; Rajon et al. 2014).[5]

---

[4] 'Diversified' bet hedging equates, in biological terms, to producing a variety of offspring phenotypes, banking that at least one of them will fit the environment; 'conservative' bet hedging is adopting a suboptimal phenotype because it may do well in the longer run.

[5] Mathematically, the kind of natural variability within a population that arises from bet hedging is summarized by 'Jensen's inequality' (Ruel et al. 1999). Jensen's inequality is a mathematical law that can pithily be distilled to: the average output for a variety of inputs to a non-linear function is not the same as the output of the average input. In practical terms, this is easily grasped by the example that a 50% population decline in one year is not fully compensated by a 50% increase in the next (although the arithmetic mean of the population multiplication rate would be $(0.5 + 1.5)/2 = 1$, the geometric mean is only $(0.5 \times 1.5)^{1/2} = 0.866$, and so the average population growth rate is negative because the mean of the non-linear function of population growth $(r = \ln\lambda)$ is below one). This means that, if there is some relationship between fitness and an environmental variable with an optimum value (and there almost always is—consider, for example, the twin unpleasantnesses of a sweltering hot summer day and a below-zero winter storm), the fact that the environment invariably fluctuates means that a 'slow and steady' phenotype will (when these fluctuations are sufficiently large) have an average long-run

What sort of factors might rock the environmental boat, and destabilize a badger, or any other, population? Such shocks could include any random perturbations to the abiotic, physical environment, ranging from climate change, fire, or flood, to tectonic events, or even meteorite impacts. These 'Court Jesters' (Barnosky 2001) stand in opposition to the Red Queen: remember, she embodies the constant biotic pressure driving the adaptation and evolution of species (or individuals) pitted against ever-evolving opposing species (or individuals) (Van Valen 1977; Strotz et al. 2018).[6] The horses-for-courses aphorism that renders familiar, indeed obvious, this newfound academic acknowledgement of behavioural differences has deep roots. Similar ideas thread through the fables of Aesop, the fifth-century BCE Greek slave, who thought any or all of the skills of a carpenter, bricklayer, or tanner might be useful when the need arose to defend a besieged city.

And so we ask how do the aptitudes, and life history choices, of individuals coalesce within the functioning of badger society?

Energy dynamics: bearish, bullish, and banking badgers

On the same day, in the same place, while all ultimately motivated to maximize their survival and reproductive fitness, each badger in the population will likely make different decisions on how to budget its activity, influenced by its fixed costs—unavoidable energy expenditure dictated by environmental conditions that impact all members of a population (e.g. weather, Nilsson et al. 2016; parasitism, Hicks et al. 2018, *inter alia*), but also by its personal life history traits (Stearns 1977; Nylin and Gotthard 1998) and bodily (somatic) condition (Monteith et al. 2013) (see Box 14.1).

The wider context of these energy budgeting decisions starts with the fact that few wild animals die comfortably in their beds, having fulfilled their potential lifespan (Collins and Kays 2011). This may be due to predators (Arendt 2009; including humans, Collins and Kays 2011), accidents (Macdonald et al. 2010c), or

---

fitness that is higher than a high-risk, high-reward phenotype that might be more competitive in the short term.

[6] Specialists may wish to dissect this quick summary into relevant hypotheses such as the stability hypothesis of Stenseth and Maynard Smith (1984), Vrba's habitat theory (1992), Vrba's turn-over pulse hypothesis (1985), Vrba's traffic light hypothesis and relay model (1995), Gould's tiers of time (1985), Brett and Baird's coordinated stasis (1995), and Graham and Lundelius' coevolutionary disequilibrium (1984) theories.

pathogens (see Chapter 15). But even without these pressures, wild animals are continually preoccupied with avoiding simple starvation; balancing nutritional homeostasis against residual body condition (Simpson et al. 1995), and thus ultimately trading off the investments they make in survival and reproduction (Stearns 1992). Different species make these trade-offs in different ways (Doughty and Shine 1997), according to their lifestyles. Some, such as bats (Culina et al. 2019), rely solely on daily 'income', living, proverbially 'hand to mouth'. In contrast, others, such as badgers, bears, seals, etc., can either carry a reserve of body fat or cache food for hard times (Macdonald 1976a; Pond 2001)—using this 'capital' (Smith and Reichman 1984) to buffer mortality risk (Humphries et al. 2003) and meet future reproduction costs (Pond 1978; Fowler and Williams 2017).

To find out exactly how each individual badger deployed the energy it ingested—that is, whether it was, in effect, a victim to circumstance or master of its own destiny—we turned again to that tri-axial accelerometry pioneered with Mike Noonan and Andrew Markham (Chapter 13). This time, however, Julius Bright Ross led us as we instrumented a bigger sample over more seasons, with the specific goal of discovering how widely individuals differed in their tactics for energy deployment, on what basis, and with what consequences. Elective activity costs comprise not only those of foraging behaviour (Chapter 10), but also the costs of scent marking (Chapters 6 and 9; see Rosell et al. 2008; Mares et al. 2012) and competition or collaboration with conspecifics (Chapter 8; see Portier et al. 1998; Sheppard et al. 2018), including, of course, mating (Chapter 7). Further, there is a broader range of demands on energy budgets than those arising from movement (mechanical expenditure), namely the largely involuntary yet obligatory costs of simply ticking over, homeostasis (Rey-Rassat et al. 2002; Toïgo et al. 2006). These costs include cell repair, thermoregulation (Oelkrug et al. 2015), and immune responses (Lochmiller and Deerenberg 2000; Cutrera et al. 2010), etc. Even sleep, as a passive alternative to activity, plays an obligate role in energy budgeting (Lesku et al. 2006) and then there's ovulation/spermatogenesis, gestation (Atkinson and Ramsay 1995), and lactation (Zhu et al. 2015)—all of which have to be paid for.

How were we to achieve the forensic accountancy necessary to unravel these cryptic running costs inherent in being a badger? Measuring the calorific *intake* of each badger per day—every worm, every hazelnut—was logistically beyond us. Likewise measuring their

oxygen consumption, as one would that of athletes on a treadmill, was unpromising—although Chris was tempted to attempt this, nostalgically recalling his design, for his undergrad dissertation, of just such a treadmill–respirometer system for measuring the serpentiform energy expenditure of captive garter snakes, *Thamnophis sirtalis*. Once his feet were back on the ground, however, we opted for 'doubly labelled water' (DLW), a metabolic isotope marker technique (Butler et al. 2004). Once again, it fell to Julius to deliver the solution. DLW is crafty stuff—it looks, feels, even tastes (harmlessly) like normal water, but its molecules have been rejigged to feature higher representation of heavy H^2 and O^{18} (hence the name doubly labelled (with 650,765 ppm O^{18} and 342,395 ppm H^2)) rather than conventional H^1 and O^{16} isotopes.[7]

In a nutshell, the procedure works by injecting a harmless dose of DLW into the subject whose energetic bank balance is under audit; then, after about 3 h to give the isotope time to equilibrate in the blood, drawing a first blood sample. A week or so of everyday life in the wood then passes before a second blood sample provides a comparison of the ratios of natural to weighty isotopes. That ratio will have changed according to how the individual has spent its energy in the meantime, as the two injected isotopes are eliminated as a result of different activities: H^2 is depleted through water loss (urination, panting) and O^{18} eliminated through both water loss and CO_2 respiration. By comparing the loss of the two, one can therefore deduce the total CO_2 respired and, consequently, the calories burned. Enter John Speakman, DLW aficionado from Aberdeen University, with whom David had previously collaborated in a comparison of the energy expenditure of red and grey squirrels (Bryce et al. 2001) and later on the energy problems of hedgehogs (Pettett et al. 2017a). Led particularly by Catherine Hambly, the Aberdonian team estimated the total CO_2 each badger had respired over the intervening days (Speakman 1997). For us fieldworkers, this is nerve-jangling stuff, because if we failed to recapture the badger at just the right moment within the 2-week deadlines, the isotopes will have been purged from its bloodstream, taking our data (and money) with them. Catching particular badgers to order, and on time, is not easy.

[7] All oxygen atoms have 8 protons, but the nucleus might contain 8, 9, or 10 neutrons. 'Light' oxygen-16, with 8 protons and 8 neutrons, is the isotope most commonly found in nature, followed by much lesser amounts of 'heavy' oxygen-18, with 8 protons and 10 neutrons.

The clinically crisp scientific publication of our results belies the frenetic choreography of the fieldwork. The easy bit was calculating the DLW dose (1.46 + (0.31 × weight of badger in kilograms) = grams administered per badger), accomplished by subtracting the known weights of Julius and the holding cage from the weight registered on bathroom scales on which he balanced while holding the target caged badger. Injecting exactly that dose, to a required accuracy of 0.0001 of a gram into that badger[8] while unanaesthetized was, however, an entirely different proposition. The routine injection of lucid badgers with anaesthetic, while easy enough with experience, involved acceptable imprecision as the patient moved its rump unhelpfully—imprecision that would have been ruinous to the delicate DLW calculations. Furthermore, once provoked one irritated badger can awaken all of his companions, also waiting in line for their turn, making for a wearing day, and so stealth and subtlety count; the subject of our continuous drive to improve each badgers' handling experience (Sun et al. 2015).

This DLW protocol required each badger to be injected with DLW 3 h before sedation, and logistics necessitated treating all participating badgers in quick succession. Worse, the volume of DLW required, in the c.4 ml range, was more than could be injected into one site, so each badger needed two syringes, each containing half of the dose (and the second injected into a now-wary badger). An added excitement arose when departmental maintenance disconnected the power, and with it the lighting, to our facility, thus on the cold November mornings when we faced a queue of up to 30 badgers and with no room for error we had to work in near total darkness, noting that badgers do not like to be dazzled by head-torches. And here's the kicker: DLW makes Perrier look cheap, costing £33.18 per gram and adding a budget-draining £110–151 for each 6–10 kg badger. Practice makes perfect, and Chris has had plenty, so very nearly every badger got exactly its full dose; only two managing to bite the syringe and send its costly contents splashing to the processing room floor. Despite these complications, during 2018–2019, it felt like a triumphant crescendo to the project that we secured recovery of 63 of 65 overall dynamic body acceleration (ODBA) collars simultaneously fitted to these isotoped badgers, and despite a catastrophic failure of SD cards in the autumn of 2018 and some early hiccups in the soldered wire connections, we produced 41 datasets for 19 different

individuals, some of which participated in several seasonal trials (n spring = 15, summer = 15, autumn = 7; see Bright Ross et al. in prep). Of the DLW data, 57 of the desired 65 recaptures were achieved within the 2-week window of DLW isotope elimination, although only 41 samples spanning 30 individuals were reliable enough to analyse back at the lab (n spring = 14, summer = 14, autumn = 13).[9] Just as a 'smart meter' reveals a household's electricity consumption, these data enabled us to see each badger's personal energetic story: just how much do individuals in the wild differ in their energy expenditure? And just as the smart meter revs when the toaster is switched on against the background cost of the central heating, so we could document each badger's relative allocation to the active component of energy expenditure (ODBA) as opposed to other expenses (all of which, plus activity, make up daily energy expenditure (DEE)). Might some individual badgers even have characteristics that constrain or liberate them to adopt a particular ODBA/DEE tactic? And if so, running ahead of ourselves while remembering our account of badger sociology in Chapter 10, what do the answers tell us about the way in which badger society is structured?

First, we asked whether each badger's individual traits, such as sex, age, and body weight, affected either its active component of energy expenditure (ODBA) or whole DEE. Then we asked how each individual's expenditure on its energy account in one season carried over to its bank balance of body condition in the next, and ultimately to its risk of mortality. To characterize the variability in individual energetic tactics, and thus the variability between them, we used the fraction of energy spent on activity based on ODBA/DEE, and called this measure 'OD'. If all individuals partitioned voluntary and involuntary spending similarly, irrespective of their different life history traits, both ODBA and OD should respond to the same drivers. Conversely, if individuals under the same conditions budgeted energy differently, that would indicate that badgers opt for a diversity of energetic tactics.

We found that ODBA differed substantially between individuals in all seasons, with the most active badger expending 1.6 times as much as the least active in spring, 2.2-fold in summer, and 2.1-fold in autumn. Similarly, DEE varied 1.3-fold between individuals in

---

[8] Once sedated the badger could be weighed exactly, and the precise dosing could be calculated retrospectively.

[9] Because cage restraint on trapping nights and on the night of post-handling release compromises badger behaviour and activity, we omitted first and last days of deployment from each average ODBA calculation (mean days used for each ODBA deployment = 5.3).

spring (averaging = 3013 kJ day^{-1}, SD = 240 kJ day^{-1}[10]), and 2-fold in both summer (averaging 3049 kJ day^{-1} ± 630 kJ day^{-1}) and autumn (averaging 3483 kJ day^{-1} ± 767 kJ day^{-1}). There was, however, no overall difference between patterns in 2018 versus 2019 in either ODBA or DEE.

Box 14.1 Size matters

Size matters, a lot, when it comes to energy. Double the weight of an animal and it won't need precisely twice the calories, but it will suffer some substantial consequences. This non-linearity is termed 'allometry' or allometric scaling. For example, consider the difficulty Jack (of Beanstalk fame) had with the giant: if a 6 foot (1.8 m) 175 lb (80 kg) person was scaled to be twice as big, being twice the height (12 feet/3.6 m), breadth, and depth, they would end up (2 × 2 × 2) eight times heavier, or 1400 lb or 640 kg—this would have consequences! For an introduction to the consequences of allometry for mammals, see Macdonald 1995. While weight gain cubes with a doubling in size, the strength of a muscle relates not to its length but only to its cross-sectional area, a square function. Consequently, while at eight times our weight, the giant might huff and puff 'Fee-fi-fo-fum',[11] he would actually be only four times as strong or, embarrassingly, proportionately only half as strong as a standard person (represented by Jack). So it is that squirrels can make extraordinary leaps, but elephants can't jump at all. Another, crucial, allometry is that surface area (skin) is a square function, so bigger animals have relatively less outside to inside and thus dissipate heat less quickly than small ones. Thus bigger species, or individuals, have lower central heating costs (and conversely, more difficulty losing overproduced heat—elephants need those ears for cooling) and thus energy requirements per unit mass—formalized in Kleiber's square-cube scale law (Hulbert 2014), or a two-thirds allometry for heat loss.[12]

Empirically, it turns out that for mammals metabolic rate actually scales nearer to an exponent of 0.64 (Hudson et al. 2013; see also Macdonald and Newman 2017), some further cost savings to largeness arising from the way allometries of contributory organ systems inter-relate (Hulbert et al. 2014). As an aside, these individual running costs reverberate fascinatingly through to a series of 'biogeographical rules', which, working with colleagues in China (see Chapter 19), we explored—specifically,

how they impacted the physical proportions of Chinese pygmy dormice (*Typhlomys cinereus*) with elevation (Cui et al. 2020): Bergmann's rule states that organisms in colder climates (so at higher latitudes/elevations, or during ice ages) should be larger and stockier than those closer to the equator/lower down better to conserve heat, while Allen's rule states that they will have shorter and thicker limbs at higher latitudes/elevation, and Hesse's— also known as the heart–weight rule—states that species inhabiting colder climates have a larger heart in relation to body weight than do closely related species inhabiting warmer climates. For the dormice, it turned out that none of these rules could trump overarching selection pressure for functional adaptations in this blind, arboreal, echolocating, ancient species.

[11] The famous words of the giant in Joseph Jacob's 1890 retelling of *Jack and the Beanstalk*, although the earliest printed version of *Jack and the Beanstalk* was published in England in the 1730s as The *Story of Jack Spriggins and the Enchanted Bean*, in a satirical collection of folktales.

> Fee-fi-fo-fum,
> I smell the blood of an Englishman,
> Be he alive, or be he dead
> I'll grind his bones to make my bread.

Pleasingly, this seemingly nonsense expression, can, at a stretch, be interpreted as Gaelic for 'Behold food, good to eat, sufficient for my hunger!' (Mackay 1877).

[12] Metabolic rates for larger animals (over 10 kg) typically fit to three-quarters much better than two-thirds; for smaller animals (including invertebrates), the reverse holds. three-quarter-scaling arises because of efficiency in nutrient distribution and transport throughout an organism (West et al. 1997). Thus, as a consequence of scale law, over the same time span, a cat weighing 100 times more than a mouse will consume only about 32 times the energy the mouse uses.

Mindful that a hair's breadth can tip the balance between success and failure, can even mean the difference between life and death, how did these Wytham badgers differ in their outgoings on energy? In both summer and autumn, a substantial amount of energy expenditure variation stemmed from simply 'how much badger' each individual comprised: heavier badgers expended up to *c.*1500 kJ day^{-1} more in their DEE than did lighter badgers. Pooling all seasons, DEE scaled to the 0.62 power (dashed line in Figure 14.1a), very close to the mammalian mean (although this varied between the −0.14-power in spring, 0.90-power in summer, and 0.65-power in autumn). The springtime allometric deviation, to the power of −0.14, was driven substantially by lactating females, which had higher than average DEE combined with lower than average bodyweight, costing them around an

---

[10] Equivalent, in real terms, to 720 calories, where, in Chapter x, we estimated that each earthworm provides around 2.5 calories.

extra 350 kJ day^{-1}. Turning to ODBA, heavier badgers expended significantly less mechanical energy in spring and autumn (but not summer) (Figure 14.1b) than lighter badgers. Conversely, it seemed that lighter badgers (with their fundamentally lower DEE) spent a higher fraction of their total energy expenditure on ODBA (with high OD ratios, Figure 14.1a and c), and stayed light—likely as a consequence (Figure 14.3); this included all the females that had lactated during the spring, which had ODBA values on average 1.3 times higher than other badgers (Figure 14.1).

Aspects of life history affected energy budgets far beyond the obvious contrast of whether a badger was breeding or not. Older readers may also empathize with the fact that propensity to expend energy, and to do so efficiently, changes with age, relative to declines in basal metabolic rate (Pontzer et al. 2001). Age-related metabolic changes affect an ageing mammal's ability to maintain 'normothermia' (i.e. to thermoregulate autonomically; Terrien et al. 2011; Nybo et al. 2014), due to diminished capacity both to generate heat (McDonald and Horwitz 1999) and to dissipate it (Larose et al. 2014). This can lead to vulnerability to overheating—hyperthermia (Kenney and Munce 2003), thus elderly people suffer high mortality during heat waves (Benmarhnia et al. 2017). Accordingly, older badgers exhibited lower DEE than did younger ones (e.g. on average, 9 year olds expended 627 kJ day^{-1} less than 2 year olds; Figure 14.1b).

A lot is known about heat stress amongst the elderly from research into risks to human health (e.g. Kim et al. 2012), especially in light of global warming (e.g. Hanna and Tait 2015). Hyperthermia, modulated by changes in dopamine and serotonin levels (Nybo et al. 2011), triggers central fatigue. In turn this reduction in activity translates into less heat generation—thermogenesis (McKinley et al. 2018). Appetite is also reduced, mediated by leptin in mammals (Meier et al. 2004), amounting to a physiological response rather than a choice, ultimately leading to reduced body fat levels.

Strikingly, in summer, older badgers (9 year olds) engaged in only around 70% of the ODBA activity recorded for younger (2 year olds) badgers; we deduce that they thereby sought to minimize heat generation during this warmer season (we found no effect of sex on this pattern). Relevant too is that sarcopenia (the age-related deterioration of muscle tissue; Rolland et al. 2008) reduces the efficiency of activity in older individuals; that is, an identical task consumes more overall energy in the elderly. On top of that, the impact of any infirmity should not be underestimated (Grémillet et al. 2018). For instance, badger M1417 developed a limp and subsequently expended only 73% of the average population ODBA but 98.5% of the average

population DEE for the season (he did less moving about but presumably did his moving less efficiently, and was probably expending calories on healing too; we therefore excluded him from further analyses). Of course, in contrast, winter risk of hypothermia can be obviated through staying warm underground (Noonan et al. 2015a), but only for badgers plump enough to afford this indulgence (Terrien et al. 2011; Tsunoda et al. 2018). These age-related effects on energy expenditure mirror what Shakespeare might have deemed the sixth, pantaloon, age of badgers (see Chapter 18), for which we found a lower average body condition ('He is a shell of his former self') (Bright Ross et al. 2020, Bright Ross et al. 2021) (see the text that follows). In short, we demonstrate that older badgers expend more energy when active, and cannot dissipate heat as effectively—under these circumstances, not carrying a heavy, insulating layer of body fat unburdens them in the summer. Without this adaptive slimming tactic, we project they would have suffered a 44.9% reduction in survival probability (Figure 14.4). The counterpoint risk then becomes whether they can replenish their fat reserves sufficiently to prepare for winter, to which we will return later.

What of declining reproductive performance with age? Venerable female badgers (age 9 and older) exhibit heterogeneous endocrinological profiles (Sugianto et al. 2020), with some seemingly more resistant than others to the reproductive constraints of age. Furthermore, as we will explain later in this chapter with regard to pace of life (POL) syndromes, older female badgers that remain reproductive typically enjoy higher survival probability than do their non-reproductive contemporaries—not, of course, because rearing a cub is so wonderfully rejuvenating, but rather we presume because something about these females makes them more likely both to survive and to reproduce at these ages (Van Noordwijk and de Jong 1986; Bright Ross et al. 2021). In line with this, we detected a broader spread of variability in individual ODBA scores in spring (this being the season most closely associated with mating) in badgers aged 8 and above (Figure 14.1d), suggesting some old females—particularly those assessed as reproductive—were more active than others.

So what chunk of DEE were badgers spending electively on ODBA, as opposed to spending on cryptic drains on energy or banking as body fat? Importantly, there was no simple answer: we could find no unifying pattern, either overall or within seasons, even after accounting for the allometric effects of body size differences between individuals. That is, the ratio of ODBA to DEE (OD) suggests that each badger 'did its own thing'—each following some unknown (to us)

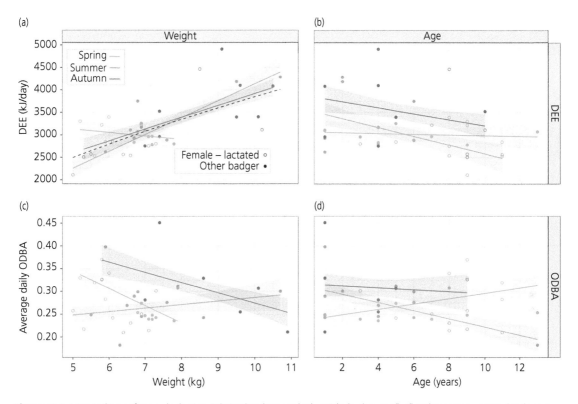

Figure 14.1 Intrinsic drivers of energy budgeting. Relationships between body weight (a, c) or age (b, d) and two energy metrics (total energy expenditure (a, b) and mechanical energy expenditure (c, d)) are shown (± SE). Dashed line in panel (a) shows overall (all seasons together) allometric scaling relationship, DEE $\propto W^{0.62}$. Open circles show females that had lactated during the year studied, while closed circles represent all other badgers. Notice how, even when following strong relationships, the scatter around the line remains substantial—in other words, representing variation beyond simply how much badger is being moved around (from Bright Ross et al. in prep.).

individual roadmap of energy expenditure. Between individuals, the spread in allocation on energy to activity spanned a 1.6-fold range in spring, 2.2-fold in summer, and 2.4-fold in autumn (Figure 14.2). Once more, lactating females greatly influenced the spring result, explaining 71% of OD variance in this season; females that had lactated expended 30% more (350 kJ day^{-1} above allometric expectations; Figure 14.1a) of their total energy budget on ODBA than others. Either or both of two explanations might be at play: first, reproductive individuals were probably more social badgers, visiting more setts during the spring; and, second, reproductive females probably have to move around more to secure more food due to the somatic costs of reproduction. Either way, these same females appeared to be worn out by the effort, insofar as they were more likely to exhibit lower than average body condition indices (BCIs) in subsequent seasons.

Were the badgers banking residual energy as body fat or exploiting previous reserves to make ends meet?

Again, there was no simple answer when we looked for consistent trends for ODBA to predict body condition in the following season (a residual—in the statistical sense meaning approximately what's left over unexplained—from a model controlling for effects of age, sex, and day of year on BCI: BCI$_{res}$; Figure 14.3a). The big spenders, that is, badgers with higher DEE (particularly in summer), were actually in better condition the following season (Figure 14.3b)—to a certain degree, a larger badger will require more energy and have to seek out more food to supply its bulk. More markedly, however, there was a strong negative correlation between OD and next-season BCI$_{res}$ (Figure 14.3c); that is, while ODBA spending seemed relatively unrelated to body condition, proportional overspending on activity by those that could least afford it caused their bank balances to run low.

Furthermore, the range of next-season body condition values predicted by the OD had ramifications for the predicted probabilities of the survival of certain

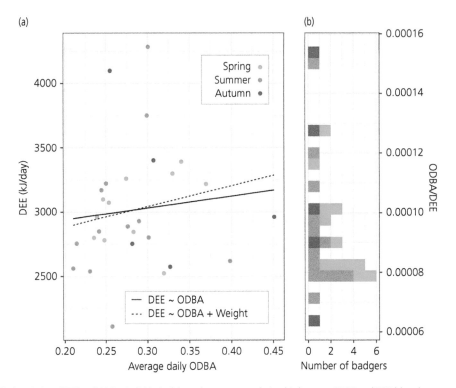

Figure 14.2 Covariation of DEE and ODBA. Individuals did not show a strong relationship between ODBA and DEE (a) and consequently exhibited widely varying ratios of ODBA to DEE (b) (from Bright Ross et al. in prep.).

individuals, especially for older badgers (Figure 14.4). In particular, and perhaps counterintuitively, those badgers engaging in riskier energetic strategies, that is, those with high relative ODBA, were predominantly the individuals whose life history (age or reproductive status) put them at least risk of mortality as a result—a sort of blessed elite of the 'to those that hath shall be given' genre. For instance, we calculated that in summer, the survival probability (to the next year) of the fattest 2 year olds was 9.6% greater than it was for the thinnest, increasing to a 23.1% advantage in autumn. In comparison, for 9 year olds, the fattest would reap a 29.8% advantage in survival odds over the thinnest in summer, increasing to a massive 44.9% advantage in autumn. Is this starting to evidence that not all badgers suffer the same consequences from the same action? Is life as a badger fundamentally unfair, where under a certain set of conditions some individuals (in this case, those ODBA overspenders) are simply intrinsically better—fitter—than others?

Furthermore, we also discovered a rather curious effect of reproduction (as diagnosed from lactation): for any given body condition, breeding females had a higher probability of survival than their nulliparous (or male) counterparts (Figure 14.4; see also Chapter 11). Note, this does not mean that a female emaciated by reproduction is immune to the risk incurred by her loss of condition; indeed, as mentioned, females reproducing in one year (and presumably worn down by the effort) were more likely to exhibit lower than average BCIs through the following summer. What it does mean is that she would still not be at as much mortality risk as a similarly emaciated female that had not attempted reproduction, suggesting that there's something special (an intrinsically superior quality) about the breeders that enables them to recuperate (Hamel et al. 2018). Expressing this in terms of badger tactics: a female will only continue to invest her bodily (somatic) condition in reproduction if there is a good chance that she can earn it back. This discovery aligns with our finding, in the context of POL theory (see the text that follows), that females who breed successfully are more likely subsequently to breed again (Bright Ross et al. 2020). Again, this hints at inherent inequalities between individuals: some females are less debilitated by breeding than others.

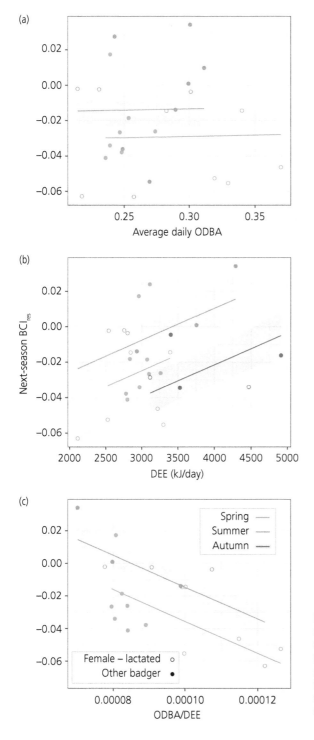

Figure 14.3 Energetic drivers of next-season BCI_res. Lines show seasonal relationships (± SE) between (a) average daily ODBA, (b) daily energy expenditure, and (c) the ratio of ODBA to DEE and next-season BCI_res, while points show actual values. Open circles represent female badgers that had lactated during the year in which they were observed. BCI_res values are residuals from a GAM model of BCI as a function of sex, age, and calendar date (from Bright Ross et al. in prep.).

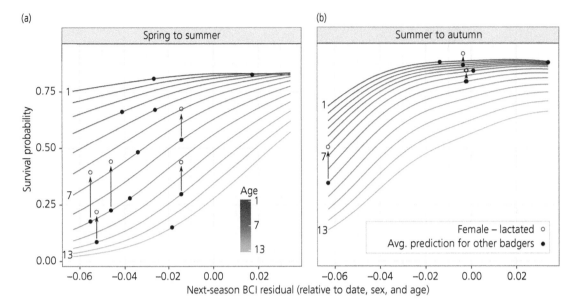

Figure 14.4 Survival costs from OD-driven range of BCI$_{res}$. Lines represent how a badger of a given age (NB we're talking here about a median animal, i.e. one representing the average of males with seasonally predominant scrotal testes descent condition; likewise, the average of non-reproductive females) relates survival probability to (a) summer BCI$_{res}$ and (b) autumn BCI$_{res}$. Points show individual position on these curves from this study; arrows to open circles show how much better survival odds are for the reproductive females in the study with these ages and BCI$_{res}$ values (from Bright Ross et al. in prep.).

How might this reproductive skew—some females producing more young than others—arise? Well, it is clearly similar, but less complete, than the total despotism observed amongst other social carnivores (Wasser and Barash 1983), from meerkats (Clutton-Brock et al. 2008; Hodge et al. 2008) to giant otters (Groenendijk et al. 2014); indeed, it is reminiscent of the lower-density breeding system seen amongst Japanese badgers (Kaneko et al. 2014; Chapter 19), which we might conceive of as following more the ancestral breeding format. The puzzle is that while females of these other elitist carnivores, like meerkats, wolves, or wild dogs, exhibit rampantly explicit dominance, readers of Chapter 5 will recall that rank amongst badgers was, to use our oft-repeated metaphor, more of a shimmering mirage, albeit coming somewhat and centripetally into focus with subcaudal gland marking (Chapter 6). Nonetheless, the observations of the last few paragraphs corroborate our strengthening interpretation that female badgers suppress the reproductive efforts of other competitor females (as suspected by Woodroffe and Macdonald 1995b), with the outcome that while all mature females mate (Chapter 7) only a minority produce surviving cubs (Dugdale et al. 2008; Chapter 17). In Chapters 4, 10, and 11 we saw how socio-spatial redistribution—an unprecedented

Black Swan event—appeared to alleviate this intra-female reproductive suppression: more setts resulted in females becoming more spread out and, we deduce, less opportunity for bullying. In turn, this intrinsic adaptation allowed the population to respond to the extrinsic opportunity of higher environmental carrying capacity (Chapter 11). So, while deduction leads strongly to the conclusion of reproductive suppression amongst the female badgers of Wytham Woods, it's puzzling that we have seen next to no behavioural evidence of the mechanism behind it.

Why would these inferior (with respect to breeding) females exist in a population, and not be excised by the scalpel of selection? Sure, selection can be messy, and lower-quality individuals can slip through the net, but for more than half of the breeding-age females within a population to consistently exhibit low reproductive quality, suggests either that natural selection was careless—a contradiction in terms—or that something more complex may be occurring. We propose that the more complex goings-on is bet-hedging across the population, where phenotypes with lower fitness (females that aren't consistently reproduction) in the shorter term can become an asset, enabling population growth (and in the process increasing their genes' representation in the population), if good times (opportunities) arrive (Starrfelt and Kokko 2012). This is exactly as we saw in Chapter 11. Remember the constraint on fecundity amongst badgers whereby high-quality females could not themselves take up this slack, because they were limited by the phylogenetic inheritance of small litter sizes and the restriction of having only one litter each year. By this same token, if the going gets bad, these 'über-females' are the last to stop trying, making them vulnerable to demise versus females that simply call off their efforts early (Liedtke et al. 2015; Krause et al. 2017).

Herein, the fluctuating selection caused by unexpected (stochastic) throws of the environmental dice jolting wild populations is important (Tuljapurkar et al. 2009a, 2009b). The result is that populations continue to comprise individuals with more than one approach to energy-budgeting tactics, some of which are better adapted to one set of circumstances, while others cope better with different circumstances (Nilsson and Nilsson 2016). In what might be termed the 'horses for courses' appreciation of animal populations, the constant rearrangement of the playing field, and thus of the rules, means that as the game of life unfolds players with different aptitudes find that their day has come, or gone. These aptitudes go beyond individual versatility: it's true that some pentathletes can excel under a spellbinding diversity of circumstances, and the same applies to many carnivorans—think especially of master opportunists such as red foxes, coyotes, palm civets, and raccoons. Here we are considering individual differences beyond the scope of individual prowess: the heavyweight wrestler and the pole-vaulter can both excel when in the right competition but both can resemble ducks out of water if they find themselves on the wrong playing field. And so it is amongst the badgers of Wytham Woods, where tactical diversity arises because members of wild populations are not monoclonal identikits with equal opportunities in life. Not least, some individuals are inevitably older than others, but beyond that each will have its individual-specific sensitivity to stressors resulting from its particular life history (Bateson et al. 2014) and contemporary energetic circumstance (Monteith et al. 2013). This plethora of factors, and more, combines to cause individuals to experience different energy ceilings (summit metabolic throughput; Elliott et al. 2014) and floors (basal requirements; Ricklefs et al. 1996). The result is a multitude of concurrent paths to success: it takes all sorts, and therefore all sorts stick around. After all, metabolism is heritable (particularly active metabolism; Pettersen et al. 2018), and so a temporarily successful strategy may persist through lineages, provided that environmental

conditions continue to favour particular traits (Anderson and Jetz 2005).

What if environmental conditions change? Changes in food availability, foraging costs, or basic running costs due to environmental conditions can all force animals not only to adjust their reproductive tactics but even more fundamentally to attempt to rebalance their energy budgets accordingly—analogous to people adapting their household accounts to circumstances (Elliott et al. 2014). How badgers should best rebalance their energy budgets will vary depending on whether such changes are predictable, stochastic, or catastrophic. We have identified that weather—through food supply and activity costs—is the principal agent of change afflicting Wytham's badgers (Chapters 12 and 13). Naturally this led us to ask whether the badgers adjusted their seasonal expenditure on energy (measured as ODBA) in response to weather conditions. Of course we expected the answer to be yes, but what was much less obvious was whether they would do so in unison, a herd of one-trick ponies, or to what extent each individual, whether pole-vaulter or heavyweight wrestler or even chess player, would attempt to outplay the others according to its unique strengths and circumstances.

Building on the approach of Chapter 13, we once more computed the simple 'Activity' metric, denoting when ODBA (averaged over each minute) exceeded a threshold (0.28; see Bright Ross et al. subm.) for more than 30 min in any hour. Crudely, this gives an idea of the 'activation energy' under different conditions— did the badger get out of his or her metaphorical bed and go out? We also paired this with an analysis of the absolute amount of ODBA expended by each badger exerted (as a measure of the actual amount of mechanical energy expended in an hour, not just whether or not it got out of bed but whether it ventured beyond the fridge) under the same conditions.

In spring

Temperature affected both Activity and ODBA, but in different ways. Activity increased linearly with temperature, but for ODBA the effect was quadratic, meaning that it increased with temperatures below an optimum value. Individual BCI determined the specific value of that optimum (a matter to which we will return), but as temperatures rose higher than that optimum the badgers cut back on their ODBA expenditure (Figure 14.5b). In spring, rainfall also had an effect, which paralleled that of temperature, with ODBA (but not Activity) increasing up to 0.8 mm h[-1], then falling back beyond that optimum. Any naturalist knows that

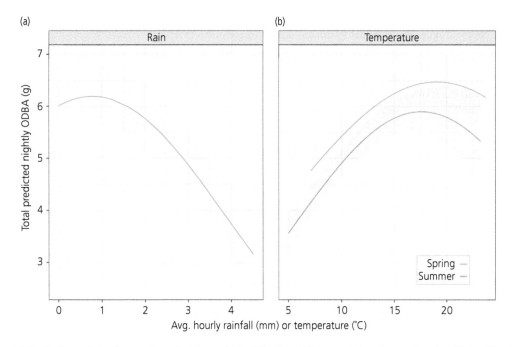

Figure 14.5 Non-linear effects of temperature and rainfall on nightly ODBA. Total ODBA expended per diem as a function of Rain or Temp (from Bright Ross et al. in prep.).

wind is important, but oddly it is often overlooked in weather analyses. We found it had strong effects. In spring, wind had a negative effect on both Activity and ODBA, with the highest speeds (3.8 m s^{-1}) coinciding with a 24.2% reduction in ODBA (12.1–34.7% SE) compared with still conditions (0.3 m s^{-1}). A blustery wind can not only be chill, it can make it hard to hear the earthworms' chaetae grating on the grass, make them harder to smell, and, worst of all, drive them underground. Also probably best interpreted through the lens of worm hunting, shallow soil temperature (10 cm depth: Soil T10) didn't influence either ODBA or Activity, but at greater depths (30 cm down), warmer soil was associated with an increase of up to 23.1% in ODBA. Although wind and soil temperature were linear effects, the non-linear weather effects, each with its inflection at the point of too much of a good thing, bring to mind the influences of weather on earthworm availability (Chapter 10), and mirror closely the weather effects we reported on badger condition and survival in Chapter 12 (Nouvellet et al. 2013; Bright Ross et al. 2021).

In summer

In summer the relationship with temperature was similar to that in spring, and the peak depends on BCI.

Further, in summer, but unlike spring, there was a tipping point in Activity too—it decreased above 13.1°C and depending very little on BCI. There was no summer effect of rainfall, but, interestingly badgers seemed to appreciate a breeze on summer nights, with ODBA increasing by up to 19.4% with winds up to 6.8 m s^{-1} (SE 9.2–28.7%). Warm, dry Soil T10 was associated with a fall of up to 28.1% in ODBA relative to the nights with the moistest Soil T10.

In autumn

Due to the heart-breaking failure of the inappropriately named SD[13] data storage cards, we had data from just 1 year's worth of autumnal deployment. While this revealed no strong effect of air temperature, there was a very marked effect for Soil T10, with 2.3-fold greater ODBA and 44% greater Activity under the warmest conditions. Once again, badger activity seemed to reflect the conditions under which earthworms are likely to surface. Again, too, badgers avoided heavy autumn showers, which reduced

[13] SD stands for secure digital, an irony in which we could take no pleasure when they all failed.

Activity by up to 25.6% (± SE, 2.2–47.4%)[14] (see also Chapter 13; Noonan et al. 2014). As autumn draws on, earthworms become scarce on frosty nights, and badger foraging success plummets (Chapter 10)—to which the best solution is torpor.

Weaving together these observations, remembering the badgers' stocky build and, in some cases extreme, fat layer, we began to appreciate that they might feel the heat of summer at least as keenly as they do the cold of winter. Further, in summer, earthworms tend to aestivate far out of reach, and the badgers of Wytham Woods turn to a more diverse diet (Figure 10.2) including cereals and berries, the night to night availabilities of which the weather does not greatly affect. We began to suspect that in summer badger activity may be substantially guided by a concern to stay cool. For mammals in general, overheating can be just as stressful as chilling (Speakman and Król 2010), especially for the elderly, and can leave them worn out (a condition portentously known as a hyperthermic lethargy response; Marino 2004; Nybo et al. 2011). More broadly, badgers evolved originally to cope with cold climates (Mardurell-Malaperia 2011; see Chapters 1 and 19), and, furthermore, for them the option of avoiding heat by becoming more nocturnal— a common strategy amongst the diurnal (Levy et al. 2019)—is a card they have already played (probably to avoid diurnal dangers and access nocturnal prey): they have nowhere to go when it is troublingly hot, except underground.

How badgers in general adapt their mean energy budgets according to seasonal weather makes for interesting natural history, but what really caught our attention were the differences—perhaps reflecting phenotypic tactics—that we discovered amongst individuals. The breakthrough grew from the realization that individually different energy expenditures (ODBA) were linked to individual body condition. In spring, thinner badgers consistently had a high ODBA (range 5.9–7.2 predicted), whereas fatter badgers lay low under cooler conditions and their ODBA increased

with temperature (range 3.9–6.4; Figure 14.6a). We deduce that in spring fatter badgers enjoy an energetic buffer that allows them to avoid expending energy on cool spring nights (such nights are poor for foraging; a deduction that bookends our parallel behavioural observations for autumn, reported in Chapter 13; Noonan et al. 2014), whereas thinner individuals have no choice but to prioritize foraging activity irrespective of suboptimal temperature conditions (Figure 14.6a; Monteith et al. 2013). Things are different in summer, when fatter badgers clocked up a fairly consistent ODBA (around a mean of 5.9), whatever the temperature, whereas amongst thinner badgers ODBA ranged widely from 2.6 on colder nights to 7.1 with warmer conditions (Figure 14.6). The evidence was mounting that well-insulated, chubby individuals seek to avoid overheating (and its pathophysiological effects; Horowitz and Hales 1998), maintaining instead a steady expenditure that probably corresponds with the time needed to feed their bulk.

For carnivores one way to cool down is to pant (Robertshaw 2006). To state the obvious (ask any sweaty sportsman), reducing activity reduces the need (in the badger's case). This is important because in summer badgers eat relatively few moist worms, but rather dry cereals, and any rehydration trip in search of water yields neither energetic return nor social benefit (see also Fuller et al. 2021); this thirsty constraint greatly impacts the distribution of badgers in arid Mediterranean regions (Rosalino et al. 2005a; Chapter 19).

You might have thought that in the uncertain world of wild animals 'fat is [always] good' would apply as universal insurance. Our findings challenge that orthodoxy. Whether fat is good, or burdensome, actually depends on season, age, and life history details that determine each individual's sensitivity to food insecurity (*sensu* RDH, Chapter 10). There is simply no need to carry fat, a burden which carries with it the associated risk of overheating, when food is readily at hand (Atkinson and Ramsay 1995; Chang and Wiebe 2016). Also working in Wytham, Andy Gosler and Louise Gentle discovered that great tits (*Parus major*) shed weight in favour of manoeuvrability in flight when at risk from sparrow hawks (*Accipiter nisus*) (Gentle et al. 2001). On the other hand, thinness is a gamble, and for the badgers of Wytham Woods it is essential to rebuild fat reserves in preparation for the unproductive winter months, especially amongst the elderly (Bright Ross et al. 2021): a finding that supports the hypothesis that frailty becomes an ever-greater constraint as individuals age (Varadhan et al. 2018), with the consequence

---

[14] Readers who think hard about this will be puzzled and ask how it is possible that ODBA didn't fall when Activity did—after all, how would ODBA be expended if not in Activity? Here we are at the mercy of the statistical model, which tells us the result (ODBA was not identified as reducing during heavy autumn showers) but gives us no hint of why. At such counterintuitive moments one remembers that a model is just a model—a mathematical quest for a pattern—it's not a naturalist (although it might inspire one). Perhaps badgers that are active when it's rainy are very active, their fervor cancelling out (in the model) the inactivity of those sleeping their way through the shower.

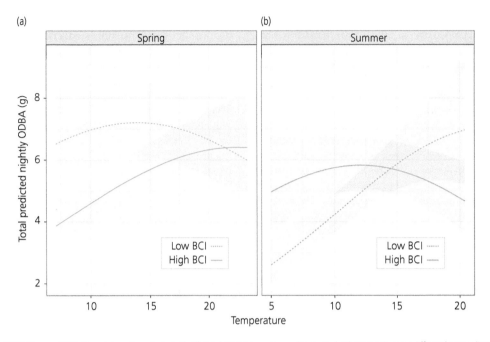

Figure 14.6 Seasonal BCI–Temp interaction. An individual's total ODBA in spring and summer (with 95% CI) varies differently according to temperature for low-BCI (dashed) and high-BCI (solid) individuals (from Bright Ross et al. in prep.).

that populations biased towards an older age distribution are more fragile than those with a bottom-heavy, youthful distribution.

The weather, the seasons, fat, and age all contributed to a smörgåsbord of explanatory factors for badger energy budgets, but in concert they did not account fully for all the variation in ODBA we found between individuals even on the same night, which varied 1.1–5.9-fold. Of the variance that remained unexplained by the best mechanistic seasonal predictors (both weather and individual), 37%, 27%, and 38% in spring, summer, and autumn, respectively, was captured by steady individual differences. By definition, this variation between individuals was also associated with strands of consistency; within a seasonal deployment window, some individuals were consistently active, some were consistently lethargic. Was this, perhaps, a further hint of badgers in the population having 'different jobs in the same market place'? Strikingly, these characterizations did not persist across seasonal deployments of DLW: a low-activity badger in one deployment might be unrecognizably active in another, and vice versa. It seemed that each badger's energetic disposition had something to do with its circumstances. Minimally, then, badgers show flexibility in their energy profiles (within age and sex constraints) when it comes to activity.

In this chapter we have so far conceptualized activity and energy expenditure largely in terms of foraging costs invested into survival and reproduction in relation to life history traits. Reality, of course, is more complicated—badgers may differ, for example, in how much they choose to invest in self- and allogrooming (Chapter 5), cooperation (Chapter 8), territorial defence (Chapter 9), communication (Chapter 6), sexual advertisement (Chapter 6) *inter alia*. For instance, low-BCI badgers engage in fewer inter-sett visits than high-BCI badgers (Noonan et al. 2018a), signifying that non-essential energy expenditure is curtailed when it can be least afforded. Sleeping (Lesku et al. 2006) and torpor (Walker and Berger 1980; Boyles et al. 2020) may also be deployed tactically to manage energy expenditure. Group dynamics (Chapter 4) may also be influential, driving greater foraging specialization (Sheppard et al. 2018). The notion of feeding specializations amongst individuals of a carnivore species might seem fanciful, but not at all: another archetypally 'solitary', but highly endangered, mustelid that Hans and David worked on with powerhouse Belarussian scientist Vadim Sidorovich was the European mink—in the same population some individuals specialized on frogs, others on crayfish (Sidorovich et al. 2001). Ultimately such specializations drive greater cumulative population niche breadth (the niche variation hypothesis (NVH); Van Valen 1965; Svanbäck and Persson 2004; Bolnick et al. 2007). Atop this, there are drains on metabolism about which each badger has no choice—fixed costs such as running the immune responses to infection, or healing from injuries.

Staring into the crystal ball of the species' future, what do these energy management tactics mean for the resilience of the Wytham badger population? Badgers are generalist omnivores with a variable social system in a seasonally temperate mosaic habitat: in short, they occupy a multi-dimensional niche (Machovsky-Capuska et al. 2016), no one size fits all their circumstances; no silver bullet solves all the problems of being a badger in any one place at any one time, far less of being all badgers in all places. Nevertheless, if environmental anomalies coalesce into a trend, and erratic weather matures into climate change (Shao 2016), fluctuating selection pressures may turn into directional selection, nudging the population bias towards the most optimal phenotype.

Alternatively, if conditions deteriorate into ever-greater unpredictability, this chaos may bring about disruptive selection, under which alternative extreme phenotypes may give an advantage, representing an 'each-way bet' (King 1985), with a higher assured pay-off than a single bet on a 'winning' strategy (Gaillard and Yoccoz 2003; Wilbur and Rudolf 2006).[15]

So an interim stock-take is that beyond the generalizations of life history traits, individual badgers differ substantially in their energy tactics. Their body fat reserves characterize their ability to expend energy on activity, insofar as the demands of thermoregulation permit. Our insights into the role of BCI, however, were glimpsed through small data windows of just a couple of weeks. Our interest piqued, we asked whether there were grander patterns woven through the many thousands of BCI records in our broader database. Did these patterns reveal how fat management might assuage stressors over a badger's lifetime?

A fat chance of survival

Body fat levels are personal. You might be the, nowadays enviable, person with 'hollow legs' who, stuffed with pies and puddings, gains not an ounce of weight; unfettered indulgence may not dent your health as you maintain low cholesterol and modest blood sugar. Or

you might be the one for whom that surreptitious slice of gateau on the lips spends a lifetime on the hips. But view these tendencies through a Palaeolithic lens, when your ancestors were 'wild', and living with the daily risk of hunger. Then, a racing metabolism may have been less enviable, litheness (or lack of reserves) being a step on the road to rapid starvation (Bellisari 2008). A few extra pounds could have been a life-saver. Now, perhaps as then, that neutral range of equivalently sized men (2000–3000 kcal day^{-1}) or women (1600–2400 kcal day^{-1}) engaging in the same activity regimen can easily vary over 50%. Add to this variation additional differences apparent between individuals, such as sex, age, and skeletal size allometries, and it becomes clear that we all start out as being far from equal (Wejis et al. 2008). The revised Harris–Benedict equation for basal metabolic rate (BMR) is, for men: BMR = 88.36 + (13.4 × weight in kilograms) + (4.8 × height in centimetres) – (5.68 × age in years); and for women BMR = 447.59 + (9.25 × weight in kilograms) + (3.1 × height in centimetres)—(4.33 × age in years). Take Brian Shaw, who, at the time of writing, is the World's Strongest Man, standing 6 foot 9 inches (205 cm) and weighing in at a fairly solid 445 lb (202 kg, or just shy of 32 stone)—every day he must consume a minimum of 4000 calories just to exist, actually eating more like 12,000 to support his intense training regimen (the difference made by, yes, a high-ODBA activity regime). Despite his incredible strength, he'd probably fare badly if food ran short (although you might decide not to argue with him over the last few scraps). Health too affects calorie requirements: a raging fever can add 50% to what's typical (for every degree Fahrenheit of rise in body temperature, BMR increases by c.7%; see Baracos et al. 1987). And then there's the cost of activity, on which expense we have just shown the badgers to vary greatly: the average person will only need about 50 calories per hour to sleep, whereas peak athletic performance, for example, skipping rope at 120 skips min^{-1}, tops out at 1000 calories per hour; mechanically lifting consumes about 0.24 calories to move 100 kg through 1 m (if you wanted to burn 1000 (k)*calories*, you'd realistically have to bench press 100 kg about 830 reps (i.e. given all the muscles and elevated breathing involved)).

So, back to the badgers—what are the costs and benefits of body fat reserves? Distinguishing cause and effect in our correlative evidence (Figure 14.3) is difficult: where does the chicken–egg cycle begin within the equally plausible possibilities that fatter badgers can reduce activity because they don't need to forage or that their lassitude causes them to grow podgy.

---

[15] Species with a more specialized diet, or benefiting from group hunting, are likely to have more similar activity and energy expenditure patterns (Prugh et al. 2008) (see Chapter 19).

Despite excess fat hindering agility (Gentle and Gosler 2001; Newman et al. 2011) and presenting a health risk in humans (and, though less studied, more broadly; see Kopelman 2007; Garaulet et al. 2010), this topic is somewhat neglected in wild animal ecology.

When running out of currency, in this case energy, changing behaviour to economize has its limits, beyond which the next option is to rely on physiological mechanisms to endure fasting—how this is achieved is elaborated in Box 14.2, and involves links

between BMR (Johnstone et al. 2005), endocrine profiles, and immunological state (Norris and Evans 2000).

So, given the tactical variation amongst badgers in how they spend energy, how do they bank their savings over the longer term (Macdonald and Johnson 2015a; Gangloff et al. 2018)? Savings are needed as a safety net for periods—predictable and stochastic alike—of food scarcity.

There are variously sophisticated ways to judge a badger's condition. We've already introduced our ratio

---

Box 14.2 Energy

Even under the same conditions, levels of circulating hormones can vary by 5- to 15-fold in free-living animals. This is the basis for variation in energetic expenditure, both between individuals and over time in the same individual. In the brain, thyrotropin-releasing hormone (TRH) is released from the hypothalamus, which stimulates the anterior pituitary gland to produce thyroid-stimulating hormone (TSH, Harris et al. 1978). TSH, in turn, stimulates thyroxine (T_4) production in the thyroid gland (Mullur et al. 2014). In tissues, 5'-deiodinase type 2 (D2) enzymes convert T_4 to triiodothyronine (T_3), while another variant (D3) performs the opposite role (Gereben et al. 2008). T_3 is the 'active' form of thyroid hormone (TH), promoting expression of gene products that quicken the metabolism, as by generating heat (i.e. thermogenesis, Silva 1995), and breaking down lipids or carbohydrates (i.e. catabolism)—the direct underpinning of how much energy badgers will have expended. There are many points along this metabolic axis that give scope for the individual variation at which our data begin to hint, with concentrations of readily activated circulating plasma T_4 shaping each individual's metabolism (Johnstone et al. 2005). Differences in energy expenditure between the sexes, for instance, are partially explained by different relationships between TSH and serum lipids between males and females (implying differing feedbacks to lipid metabolism, Meng et al. 2015); age effects on metabolic rate, certainly amongst older humans, also can relate to higher TSH levels (Wang et al. 2018), which probably compensate for reduced capacity to oxidize fat in the respiring tissues of the aged (Toth and Tchernof 2000).

Developmentally too, the TH axis is very closely tied to the hormonal regulatory systems for various forms of energetic expenditure. This time the biological truism that everything is related to everything else is manifest in the role of TH in up-regulating the transcription of the growth

hormone (GH) axis, playing a role in development and muscle deposition (Evans et al. 1982). Remember from Chapter 2 that male badger cubs grow quicker and larger than do females, but so we can also imagine different badgers of the same sex growing quicker than others if their TH profiles differ: achieving sexual maturity sooner may sometimes have advantages, but comes with costs (Chapter 2); in any case, TH variation could drive some of the underlying differences that eventually affect population dynamics. TH triggers cardiac vasodilation and increased blood flow to skeletal muscle (McAllister et al. 1995), providing one of the primary links between organ size, basal metabolic rate (BMR: formally, the energy expended by an adult, inactive, post-absorptive, non-reproductive, and thermoneutral animal), and maximum metabolic rate (MMR: the maximum energy an animal can expend while exercising) (Meerlo et al. 1997)—the structural differences underpinning differential individual athletic ability and energy needed for activity. And yet another link: the immune system (Chapter 15), an oft-neglected energetic expenditure (Cutrera et al. 2010), is also dependent on TH action—diminished circulating T_4 inhibits T-cell responses (Silberman et al. 2002). And perhaps the most diverse category of energetic expenditures—those associated with reproduction (Chapter 7)—also interacts with TH: production of testosterone and luteinizing hormone (LH) (both drivers of energetic responses, Day et al. 2005; Kim et al. 2021) are responsive to T_4 levels (Chiao et al. 1999).

A curiosity we frequently encounter in Wytham in summer is slim badgers amidst a landscape bountiful with delicacies—fields of maize to thickets of fruit-laden blackberry bushes. Clearly, it's more than the availability of food that whets a badger's appetite. On short time scales, blood glucose concentrations (available for metabolism) are maintained in a narrow homeostatic band by the opposing effects of insulin (anabolic) and glucagon (catabolic), which

Box 14.2 *Continued*

regulate glycogen storage in the liver (Gerich et al. 1976). Appetite, or hunger, arises from the hormone ghrelin, produced in the stomach. Over longer time scales, fat (adipose) tissue produces the opposing hormone leptin, which evokes a satiation effect, and adiponectin, which increases insulin sensitivity. Consequently, fatter, even slightly fatter, badgers have the option of reducing their food consumption by switching to adipose catabolism (Meier and Gressner 2004). This suggests that when food security is satisfactory, it may not be advantageous to accumulate more fat than

necessary: those slim badgers may be optimally adapted. In anticipation of predicable seasonal cycles of food shortage, however, some species at certain times of the year have evolved reduced sensitivity to leptin and thus may never feel full (Rousseau et al. 2003; Florant et al. 2004), which is how badgers can double their weight as winter approaches. That pays for badgers reaping the fat-fuelled rewards of winter lethargy, but the same leptin insensitivity can cause failure in the satiation response of people, leading to obesity (Klok et al. 2007).

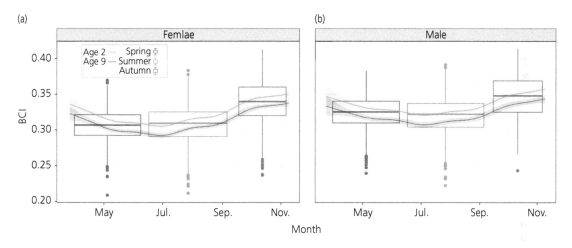

Figure 14.7 Variation in body condition. Boxplots show variation in BCI by season (1990–2019); lines show population averages according to a best-fit GAM regression at age 2 (blue) or 9 (red). Boxplot width corresponds to the period considered in each season. Model controls for individual via random effect and permits different splines by sex for age and day of year (from Bright Ross et al. in prep.).

approach (ln(body mass)/ln(body length)), developed from an idea pioneered for otters by Hans Kruuk (Kruuk and Conroy 1991) and refined by Noonan et al. (2014, 2018a) to deliver our BCI. But there's a more agricultural alternative, fit for some purposes and developed from David's days as a sheep farmer: when we handle badgers, we feel for a waistline, with the badger lying on its side, as we judge fat score criteria used for grading stock for market (Speedy 1980). A sharp dip between ribs and pelvis indicates emaciation, which we score as 1, whereas a plump hump scores a corpulent 5. While the BCI gives us a measure of 'density' to include muscle mass, along with fat (the massive Mr Shaw, mentioned earlier, would score exceptionally), these fat scores are indicators of

current surplus. Nevertheless, for the 8615 waistline measurements we made of badgers between 1990 and 2019, this stockman's measure of subcutaneous fat correlated closely with BCI. To judge who is fatter than whom, we centre BCI on the mean, for each sex and for each season: a badger below that mean was thinner than its contemporaries; above the mean, fatter; either way, we secured a residual deviation from the mean, BCI_r.

Seasonality in food availability and diet

Considering the seasonal lifestyles of badgers, it is not surprising that we found substantial cyclical BCI variation through the year (Figure 14.7) (Bright Ross et al.

2021). BCI was lowest in both sexes through the spring and early summer, with badgers at their trimmest in July, mirroring the low availability of earthworms in summer (Figure 10.2) (earthworm chaetae remain at low frequency in scats until September). Blackberry (*Rubus fruticosus*) consumption fuelled BCI gains throughout late summer; the latter continuing into autumn as earthworm availability peaks, along with that of hazelnuts (*Corylus avellana*), horse chestnuts (*Aesculus hippocastaneum*), and sweet chestnuts (*Castanea sativa*). BCI peaked by the latest date trapped, 28 November (Figure 14.7).

Effects of body condition on survival and reproduction

Do all badgers tackle the problem of fat storage in the same way? No, as with energy expenditure, even at a given moment at Wytham individual badgers exhibit widely differing body fat storage tactics (Figure 14.7). Was body condition a life and death matter? Generally speaking yes, high body condition (either BCI or fat score) was associated with both a higher probability that an individual would survive, and greater reproductive success. However, this effect mostly manifested through a minimum threshold below which a badger's survival was no longer 'safe' (within the bounds of luck, of course)—we term all body condition values above this threshold as within the 'safe zone' (Figure 14.8; the stockman's fat score, being a coarser metric, was generally associated with a more robustly defined safety zone in relation to survival and reproduction than was BCI; Tuomi et al. 1983; Simpson et al. 1995). In other words, no further mortality-avoidance benefit was derived from additional fat storage above a badger's specific safe zone.

As intricacy piles upon intricacy, this relationship was inconsistent across seasons, ages, or BCI ranges. Yet again, as any experienced animal breeder would expect, individual badgers differ and, when it comes to body fat, 'one (tactical) size was not fitting all'. That variation was clearly apparent between the sexes and age groups. Prime-age badgers were relatively safe from BCI-associated mortality (except in autumn, when being about two standard deviations below normal came with elevated mortality risk), but older badgers had clear, strong relationships between BCI and mortality (Figure 14.8). However, unlike younger badgers, older badgers did not suffer such a detrimental effect of low BCI during warmer, drier summers

(Figure 14.8). This narrower safe zone for the elderly in harsh summers is linked to their down-regulation of metabolism to reduce energy expenditure, minimize overheating, and avoid the danger of starvation (see Box 14.2), all of which combine in explaining their lower autumnal BCIs relative to younger badgers (Figure 14.8). Furthermore, during years of high population density thin old badgers suffer disproportionate additive scrawniness compared with youngsters in summer (Figure 14.8), although by autumn, under these highly competitive conditions, these older badgers regained corpulence at faster rates than were typical for their age class in lower-density years. This suggests that these older badgers expertly regained reserves in preparation for winter food scarcity and torpor, and that part of their summer slim-down was tactical, rather than unavoidable (see ODBA section in 'Energy dynamics: bearish, bullish, and banking badgers' and notes on appetite in Box 14.2). Geriatric emaciation is familiar in other group-living animals: it can arise through reduced competitiveness (Sharp and Clutton-Brock 2011), or exclusion from group feasts (e.g. wolves; Jordan et al. 1967). However, neither of these factors is likely to be influential amongst badgers because: (i) they forage solitarily (da Silva et al. 1994); (ii) feeding hierarchies are not strongly expressed (Macdonald et al. 2002b; Chapter 5); and (iii) foraging patches are typically super-abundant when available (e.g. worm patches, cereal fields, blackberry bushes; Macdonald 1984b; da Silva et al. 1993; Chapter 10). Cell division slows in old age, risking lipotoxicity from the degeneration of adipose tissue (Slawik and Vidal-Puig 2006), which may make it even more beneficial for older badgers to hold back on calories (although clearly not to the point of malnutrition), also reducing age-related cancer risk (Weindruch 1992; Anderson and Weindruch 2010). On the up-side, regaining fat reserves when they're needed as winter approaches may be aided by some senior savviness in food acquisition (Atkinson and Ramsay 1995; Chang and Wiebe 2016). Some social species gain dominance, and its perks, simply from growing older (Spong et al. 2008), but we could find no evidence of this amongst badgers, but their familiarity with a wide set of within- and between-group members, along with a potential set of related descendants may offset or exceed any decrease in feeding competitiveness (perhaps due to loss of teeth and chewing ability) and senescent declines in anabolism (Valenti and Schwartz 2008).

So much for the interaction of age and corpulence—what of reproduction? To set the scene, a first and

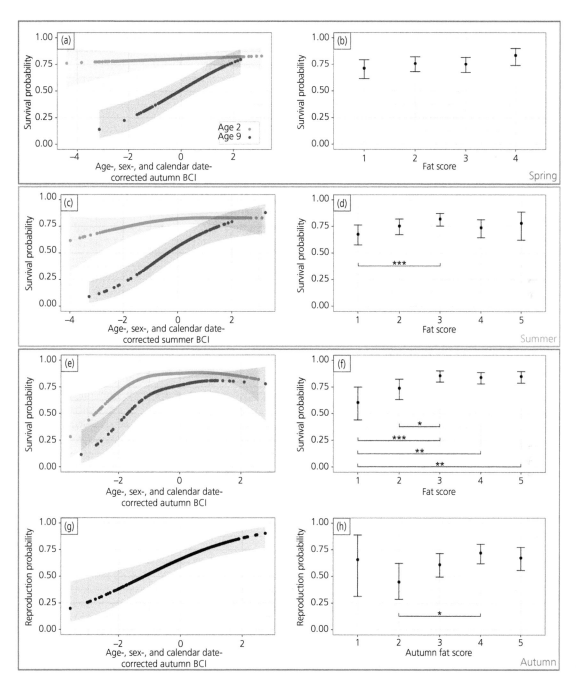

Figure 14.8 Effect of body condition on individual fitness. Relationship between seasonal BCI_r (a, c, e, g) or fat scores (b, d, f, h) and the proportion of individuals surviving (a–f) or females reproducing (g–h) under average conditions (0 = 0%, 1 = 100%), with 95% confidence intervals. BCI_r–survival relationship is shown for young (age 2, blue) and old (age 9, red) badgers to demonstrate the changing effects of body condition according to intrinsic factors; to illustrate BCI_r range coverage of the broader dataset, points show the (scaled) BCI_r values of individuals aged 1–8 (on the Age 2 line) or 9+ (on the Age 9 line). Standard asterisk notation shown for Tukey *post-hoc* comparison of fat score groups (P 0.001***; 0.001 ≤ P < 0.01**; 0.01 ≤ P < 0.05*) (from Bright Ross et al. in prep.).

surprising paired fact is that in spring and summer reproductive females had lower BCI than ones that had not reproduced that year, but that females that bore cubs were neither more nor less likely to survive than non-reproductive ones. At first it's surprising that these results suggest that reproduction carries no (detectable) extra risk-related cost, but on reflection it is consistent with our finding that for any given body condition value reproducing females had a higher probability of survival than their non-reproductive counterparts (Figure 14.4). Put another way, the survival probability of non-reproductive females (specifically those that invested nothing in trying, vs those that tried, spent, but failed) was linearly enhanced by BCI, with the highest BCI providing a 39% (spring, SE = 23–54%) or 55% (summer, SE = 38–68%) advantage over the lowest BCI; reproductive females exhibited no such relationship.

However, there was a straight-line relationship between high BCI in the year of conception and the greater likelihood of successful reproduction in the following year (Figure 14.8g); there was a similar, but non-linear trend for fat score, with females with intermediate plumpness (fat score 2–4) achieving a significant reproductive benefit. While survival probability was determined by a shifting BCI threshold (Figure 14.8a, c, and e, with shifts in the size of the safe zone according to life history factors and weather), higher autumnal BCI was associated *pro rata* with greater female reproduction (with only fat scores showing a threshold effect on reproduction, Figure 14.8g–h). Ultimately then, and in contrast to the situation regarding survival, there was no BCI safe zone that guaranteed female reproduction. This, too, explains the lack of a relationship between BCI and survival for reproductive females: above some level of bodily (somatic) security, each additional unit of energy simply measures another small chance of reproducing, and a female could, if necessary, simply trade off that energy to survive instead. There was some evidence that the same elite females are getting to make these reproductive investments over and over, while others simply fought to hold on: a female was substantially (24%, SE = 15–33%) more likely to reproduce in a year if she had also done so in the previous year.

In comparison, whether or not his testes were descended had no detectable effect on the probability of a male surviving, although it did cost condition (i.e. males with descended testes had seasonal BCIs 5.0% (spring), 3.6% (summer), and 4.0% (autumn) lower than those with ascended testes).

Seasonal weather-related drivers of individual BCI

How is body condition impacted by weather between the seasons? Remember, from Chapter 11,[16] the metrics of annual weather and weather variability which impacted survival. Now we used those same metrics to reveal that these correlations with weather operated, mechanistically, through impairing body condition. That is, when it gets hot badgers do not, of course, simply drop dead. Rather, they first grow desperately thin. Temperature variability (σ_T) had a negative effect on BCI in most seasons, while warmer winter and spring temperatures (correlates of higher earthworm availability) were associated with higher spring BCI. However, BCI gains accrued only up to 2 mm rain day^{-1}, and daily spring temperatures only improved BCI above 9.5°C (once more, effects were quadratic—too much, as well as too little, of a good thing). In years with higher temperature seasonality (α_T; i.e. extreme seasonal temperature differences) plumpness conferred a survival advantage; in particular, fatter autumnal body condition was beneficial for survival in rainier years (μ_R)—thus those badgers destined to remain skinny in these wetter years survived poorly (Figure 14.9d). Plumpness (higher BCI) also benefited survival in years with low rainfall variability (CV_R), but not in years with more variable rainfall. Summer drought was, of course, problematic. When the driest 30-day summer period (associated with minimal earthworm availability) was warmer than average, or consistently dry (low CV_R), younger badgers were bony. Interestingly, this did not apply to older badgers, possibly because elderly individuals were destined to be relatively thin in summer regardless—or because they compensated for it in some way, such as by reducing expenditure through activity. Hot summers did, however, lead to older badgers having particularly low autumnal BCI (Figure 14.9b, perhaps as a trade-off from whatever compensatory mechanism enabled their fortitude during hot summers); these same older individuals also put on less fat during wet autumns than did youngsters (Figure 14.9c), and emerged from

[16] (i) Mean temperature (μ_T), (ii) mean rainfall (μ_R), (iii) annual temperature seasonality (α_T), the amplitude of a sinusoidal curve constructed for a year's temperature, (iv) temperature variability (σ_T), calculated as the squared sum of deviations from the year's sinusoidal temperature curve, and (v) rainfall variability (CV_R), calculated as the coefficient of variation (SD/mean). See Nouvellet et al. (2013) for a full accounting of these metrics.

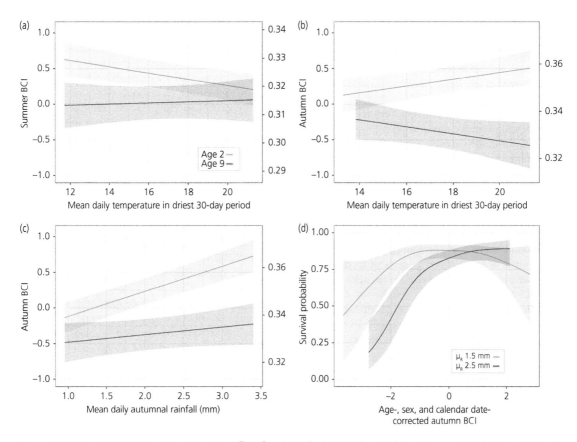

Figure 14.9 Extrinsic-intrinsic interaction relationships. Effect of μ_T during the driest 30-day period in a given summer on (a) summer BCI and (b) autumn BCI for badgers of age 2 (young) and 9 (geriatric). Age-specific effects of autumn μ_R on BCI is shown in (c), while (d) illustrates the different effects of BCI_r on survival in relatively dry or relatively wet autumns (predicted for average-age badger, 4.5 years old). BCI for (a–c) is provided in both SDs (left axis) and real values (right axis) (from Bright Ross et al. in prep.).

wet winters with particularly low BCI. Comparing these weather effects across seasons, it emerged that for the badgers of Wytham Woods autumn weather was particularly influential, with higher mean temperature (μ_T) and mean rainfall (μ_R) strongly associated with storing more body fat. Badgers also benefited from autumns during which rainfall was consistent (low CV_R) with higher BCI when rainfall was either particularly variable or particularly invariable (in this case, both a little or a lot of variability were beneficial relative to a middling amount).

Did these associations between weather and BCI have consequences for individual survival probability? The breadth of the BCI safe zone was narrower in rainy years (Figure 14.9d), and in years with higher temperature seasonality (α_T). Higher winter rainfall and low winter temperature variability were positively correlated with female reproductive success. Reproductive females, however, had even lower BCI

than other sectors of the population after warm summers.[17]

And what of individual heterogeneity? Once again, demographic and weather-related variables accounted for less than half of the variance we observed in BCI across all seasons. The rest of the difference between individuals was down to those individual-specific traits in energy budgeting that loomed large in our earlier discussion.

Body condition and climate change

In Chapter 13 we extrapolated to predict the future fortunes for Wytham's badgers based mostly on correlative analyses. Now we can revisit that crystal ball,

[17] Although winter temperature variability (σ_T) did not negatively affect reproductive females' BCI, unlike the rest of the population.

using body condition as the pathway to starvation—where any mismatch between the timing of energetic demands and extrinsic energetic supply risks a slippery slope into extinction (Stenseth and Mysterud 2002, Durant et al. 2007; Newman and Macdonald 2015).

Alarm bells ring for badgers because hot summers and high seasonality not only drove low BCI amongst most individuals (Figure 14.4a), but also pushed marginal individuals beyond their safe zone of BCI variation. The IPCC's RCP 8.5 pathway (Riahi et al. 2011) predicts that within 20 years at least 50% of summers in the UK will resemble the hottest summers currently on record (Li et al. 2017). This will narrow the bottleneck of summer food supply and/or alter trophic choices and foraging dynamics. In contrast, shorter and less severe periods of food scarcity in winter may reduce the need for fat accumulation and/or influence the fine-tuning of torpor. Crucially, as an adaptation, torpor is linked to an expectation that food will soon be in short supply (Kato et al. 2018). On the whole, warmer winter temperatures will likely result in badgers extending their activity deeper into the winter (Noonan et al. 2018a); if this increased activity is not met with elevated food availability, the costs of rousing from torpor may outweigh the benefits of foraging (Humphries et al. 2003; Caro et al. 2013). Two interesting questions arise: how much of a badger's decision to deposit fat is a plastic, behavioural choice, easily adapted to prevailing conditions; and how much is mandated by day-length change through endocrine regulation (Caro et al. 2013)? Time will tell, but being blindly tied to an obsolete adaptation could quickly become disadvantageous for badgers, and other hibernating species (Newman and Macdonald 2015).

So, another stock-take becomes necessary: individual badgers exhibit a diversity of energy tactics, beyond that accountable through life history traits, and use body fat reserves to buffer periods of scarcity, and also to aid in thermoregulation. But these are short-term measures, which we revealed through intensive analysis of small windows of a couple of weeks of data collection per season. What of the longer term? Surely each badger's objective is, proverbially, to look after the economic pennies so that the pounds can take care of themselves?[18]

Pace of life syndromes: 'know when to hold 'em and when to fold 'em'

We have seen that badgers differ in how their lives are scheduled: some produce more offspring (Chapter 17), some mature earlier than others (Chapter 3), and, as will emerge in Chapter 18, some continue reproducing for longer into old age, and, at least in the case of reproducing females, some seem able to bounce back from post-reproduction fat loss. These are different phenotypes (remember, phenotypes are the observable characteristics of an animal), so what maintains them side by side amongst the badgers of Wytham Woods? Is it simple chance, or is it more than happenstance that some badgers live their lives differently to others? Might they have different strategies when it comes, in the immortal words of the Kenny Rogers song 'The Gambler', might they know 'when to hold 'em and when to fold 'em'?[19]

Wild animals that invest disproportionately in reproduction rather than looking after themselves (technically termed somatic conservation; see also Chapter 18) tend to 'live fast' at the cost of 'dying young' (Promislow and Harvey 1990; Hall et al. 2015). The lemming and the wood mouse are both small, r-selected rodent species,[20] yet the female lemming's racing metabolism allows her to have her first dozen offspring by the time she's 42 days old. In contrast, with a more conventional metabolism for its size, the wood mouse reproduces once or twice (to a maximum of four times) per year, with a litter size of four to seven young. Although both are insectivorous, shrews live much more frenetic lives than their chiropteran cousins (bats; Barclay et al. 2003; Culina et al. 2019). Similarly, intra-specifically, populations differ (Chapter 19): contrast red foxes scrambling to capitalize on a vole cycle, and those comfortably fattening up in suburban gardens. What, though, of differences in life history strategies amongst individual badgers in the same population?

Might individuals opt to reproduce earlier and/or to produce more (maybe poorer quality/less competitive) offspring sooner? Typically, such 'faster' strategies occur when either a population is recovering from depression below environmental carrying capacity (e.g. after badger culling (Chapter 16), or Ethiopian wolves bouncing back from a rabies outbreak (Marino

---

[18] Wisdom attributed to Old Mr Lowndes, the Secretary of the Treasury, during the reigns of English monarchs King William, Queen Anne, and King George I.

[19] 'The Gambler', sung by Kenny Rogers (1978), written by Don Schlitz (1976).

[20] The terms r-selected and K-selected contrast two broad evolutionary approaches to life histories, emphasizing quantity (r) versus quality (K) of offspring. Insofar as investment in quality limits quality, the two are traded off.

et al. 2013)), or if, say, improved food availability enhances that carrying capacity. In contrast, in populations experiencing a contraction in available resources, say due to habitat loss, or a worsening of environmental conditions, individuals tend to delay age at first reproduction and, in longer-lived species, wait out these substandard conditions, and/or invest more heavily in offspring quality. The latter approach results in 'slower' life history dynamics (Dammhahn et al. 2018; Montiglio et al. 2018). As a consequence, adaptive fitness optimization occurs along a continuum in which metabolic needs (Metcalfe et al. 2016), behavioural traits (Careau et al. 2009), and lifetime reproductive success (Réale et al. 2010; Dammhahn et al. 2018) must be balanced and matched carefully to circumstances. But does this theoretically appealing perspective actually operate in nature, and specifically amongst the badgers of Wytham Woods?

No, at least not to perfection, any more than any person can plan their life perfectly. Circumstances change, making the adaptive optimum an ever-shifting target (Wright et al. 2019). Furthermore, the more complex the lifestyle, the more combinations of options present themselves; a complexity manifesting itself, perhaps, in eclectic omnivorous food choices, social choices about dispersal, group affiliation, or with whom or when to breed over a potentially long iteroparous reproductive lifespan. However, this elusive compromise can be achieved at the population level, wherein different individuals following a diversity of life history trajectories (trades[21]) leading to the outcome that 'on average'

the population covers all its bases, returning us once again to another facet of the concept of bet-hedging (see Starrfelt and Kokko 2012; Rajon et al. 2014).

Furthermore, this spectrum of who chooses to chance their hand at what can reflect immutable serendipity stemming from early-life conditions. Whether an individual gets a good start in life or a ropey one can lead to irreversible differences in their quality (van Noordwijk and de Jong 1986). This matter of luck leads to the so-called 'silver spoon' hypothesis (Stamps 2006), and the topic of predictive adaptive responses (PAR, Bateson et al. 2014), both of which pitch that particular individual's genetics against the roulette wheel of fortune in different ways (Réale et al. 2003). We will return to these procedural generators of diversity in Chapters 17 and 18, but for now a brief introduction is: (i) the silver spoon hypothesis has it that individuals which grow up privileged by good circumstances will gain fitness benefits throughout their lives; whereas (ii) PAR refers to a form of developmental plasticity in response to environmental cues that does not confer an immediate advantage to the developing organism; however, if the PAR correctly anticipates the postnatal environment it will prove advantageous in later life. As Patrick Bateson and his colleagues (2014) ominously put it, 'When the predicted and actual environments differ, the mismatch between the individual's phenotype and the conditions in which it finds itself can have adverse consequences for Darwinian fitness and, later, for health'. What might this generality mean in terms of individual badgers? Consider a high energy-type badger (i.e. one that lives in the fast lane, eats a lot, is very active, reproduces much), which finds itself pitched into a productive year: this promises to be a propitious match; however, if that same badger's destiny lands it in a meagre year, the mismatch grates against natural selection. In contrast, in this poor year a low energy-type badger (i.e. one that eats little, sleeps a lot, reproduces little) might leave descendants when its high-flying counterpart would not; however, in a good year, it is the unambitious strategy that will be outclassed.

So far in this chapter, we have seen that there is considerable variation in how individual badgers choose (or are compelled) to accrue and spend energy capital in relation to those two basic evolutionary metrics: survival and reproduction. Yet they all ultimately share

---

[21] In the Introduction to *The Velvet Claw: A Natural History of the Carnivores*, David crafted the *Parable of the Two Potters*: 'A father bequeathed a pot mould to his son, Potter A, and a potter's wheel to his other son, Potter B. They both made pots, but A could make them faster and cheaper with his mould than B with his laborious wheel. So, B, at a competitive disadvantage, was soon on the verge of bankruptcy. Then Potter B discovered that his wheel could also be used to make plates, cups, bowls, and a whole range of marketable items that A could not make with his rigid mould. By diversifying, Potter B was able to stay in business alongside Potter A. Some time later Tupperware pots came into vogue and the market in clay pots crashed. This was a disaster for Potter A, who could make nothing else with his mould. Potter B suffered from the loss of the pot trade but could still market his plates, cups and so on. Potter B survived while Potter A went out of business. Then fashions changed again, and clay pots were once again in demand. Potter B noticed the gap in the market, and started making clay pots on his wheel, as he had done in the past.'

Similarly, lineage or phenotype A of animals highly adapted to a particular niche will be able to exploit that niche more efficiently than lineage or phenotype B containing less specialized members. But this does not mean that lineage/phenotype A will ultimately survive or that B will be ousted completely. The

question we face in the context of badger society is one of collective action: how do the aptitudes, and life history choices, of individuals coalesce within the functioning of badger society?

the goal of leaving as many descendants as possible—and certainly more than their rivals. At the life history scale, these intricate trade-offs play out in what are termed 'pace of life' (POL) profiles (Dammhahn et al. 2018; Bright Ross et al. 2020); basically, should 'Belinda badger' live fast, reproduce early, and die young from burning her candle twice as bright, for half as long, or should she smoulder, bide her time, and offset reproductive costs until later in life—remembering that a consequence of this latter strategy is that any multiplication of her riches in the form of grandchildren and beyond will be offset into a more distant, and thus more uncertain future? Of course, the amount and predictability of resources an individual can glean from its environment (while dodging diseases and predators), weighted against the costs of foraging, provide not only a constraint to how much it can invest in reproduction beyond simply surviving, but also a clue as to how its investment may pan out. This trade-off between current and future reproduction often results in polymorphic populations in which some individuals put more emphasis on future fitness returns than others—the outcome being a community in which there would certainly be a diversity of individuals, such that it is partly as a result of chance, partly of design as to who turns out to be superior and who inferior. Life history theory predicts that such differences in fitness expectations should result in systematic differences in risk-taking behaviour. Individuals with high future expectations (who still have much to lose) are expected be more risk averse than individuals with low expectations (Wolf et al. 2007).

That's the theory, but was there any signal for variation in pace of life syndromes in our Wytham dataset? This question has rarely been asked in relation to a wild mammal, because few studies achieve the necessary continuity of data on lifetime reproductive success over successive generations. But, by this point, we had 31 years of data and 28 years of genetic pedigree (Chapter 17)—a mammoth task, again masterminded by ever-ingenious Marshall Scholar Julius Bright Ross (Bright Ross et al. 2020). To investigate lifetime reproductive success requires that individuals have completed their reproductive lives: for us, a long wait. This limited us to 269 individuals (128 females, 141 males), born in 2007 or earlier (8 years before the end of the pedigree), which produced at least 1 detected cub.

To measure POL variation we used: (i) age at first reproduction, α; (ii) annual reproductive success; and (iii) two composite metrics that embody both early- and later-life life history traits—lifetime reproductive success (LRS), defined as the total number of offspring

assigned to an individual, and the ratio of fertility to age at first reproduction (F/α), with fertility (F) defined as the number of offspring produced per potential reproductive year of an individual's lifespan (as per our preliminary use of this metric in Chapter 11). Aside from characterizing individual strategies, this latter metric is also used to position species along the fast–slow life history continuum (Oli 2004; Dobson and Oli 2007).

Strategic, and ultimately energetic, investment into reproduction is obviously fundamentally dependent on whether an individual is male or female (Royauté et al. 2018)—starting points likely to affect tactical life history variation in different ways (Hämäläinen et al. 2018). Female mammals must commit their energy to gestation, lactation, and a degree of post-partum care (Aloise King et al. 2013), whereas males can potentially mate with multiple females, hedging the risk that any female partner could fail to produce offspring against a typically higher investment in mating plurally (Kleiman and Malcolm 1981; Plard et al. 2018). Amongst Wytham's badgers the age at which males first mate (i.e. mating in the year prior to becoming a father) to produce a cub, assigned from the genetic pedigree, was taken to indicate their first reproductive success, while, because of the 11 months of delayed implantation, for females we took age at primiparity as that at which their first genotyped cub was born. This recognizes that males (but not females) may die after mating and still sire a viable litter the following spring. Building on this, we also examined how within-sex density (as a metric of competition) and sex ratio of the population (expressed as male:female) might affect POL. To ascertain whether badgers were prioritizing looking after themselves (somatic self-preservation) over reproductive investment, we again turned to our BCI. To characterize weather conditions, we used our now familiar array of sophisticated metrics set out in Chapter 12,[22] and revisited earlier in the context of fat deposition.

Remarkably, we discovered that the earlier an individual bred, the shorter its life, compared with those that delayed the start of their reproductive endeavours. Each year of delay to starting reproduction benefited males by 0.93 ($\pm$0.09) years of longevity and females by 0.38 ($\pm$0.14) years (Figure 14.10). How was this effect meted out? Each cub produced had a discernible and significant impact on female survival to the next year—a truth also evident, but less significant, for males.

---

[22] Of the weather covariates used, only μ_T exhibited a linear increase over time (1987–2016: 0.03°C year^{-1}).

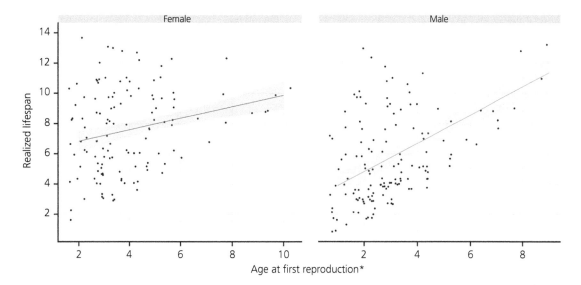

Figure 14.10 Trade-off between age at first reproduction and realized lifespan by sex. Regression line shown (for males, with average intercept for cohorts), with error bar of 1.96 standard error (from Bright Ross et al. 2020, with permission).

However, in both sexes, the impacts of reproduction incurred against survival became smaller in proportion to the parent's increasing age, with, for instance, 2-year-old females losing 8.2% survival probability per cub, while 5 year olds lost just 5.8% survival probability, on average, per cub.

Living fast or slow: the social context

Did the social milieu in which a badger existed affect POL? Yes, markedly. Zooming out to the population scale, social factors that prevailed in any given year through the individual's lifespan (population density, within-sex density, and sex ratio) explained around 52% of changes in the population's ratio[23] of POL for males and 91% for females.

In short, the social climate had a greater impact on the badgers' POLs than did the meteorological climate. That is, the social milieu influenced POL more than even the direct effects of prevailing weather—although, indirectly, the weather obviously determines the food supply and foraging costs that underlie these influential population attributes. In the way that

everything interacts with everything else, the weather can interact with population density and foraging efficiency in ways that influence reproduction, often in different fashions, for the sexes (Weladji et al. 2005).

It is well known that the cohort into which a social mammal is born can have lifelong repercussions: famously, the number of contemporary brothers with which a male lion cub grows up affects almost everything about its fortunes. Although lion society is very different to that of Wytham's badgers, we found a powerful underlying similarity insofar as the cohort size into which a badger was born was particularly important amongst the sociological parameters affecting its POL. Males born in larger, and by deduction more competitive, cohort years tended to attain higher LRS—remember, of course, that this is only measuring those badgers that succeeded in producing at least one cub; we can infer, then, that more competition leads to more competitive males, with a smaller portion of them monopolizing more of the pie. Measures of POL (high LRS in males; high F/α and LRS in females) were highest when individuals were born at a sex ratio of c.0.8 (i.e. 10 females for every 8 males), slowing either side of this optimum. For instance, males born when the sex ratio in Wytham was at its lowest (0.66) averaged a LRS of 2.47 (±0.32 SE) cubs, whereas when the sex ratio was 0.8, males averaged almost twice as many, 4.21 (±0.46 SE), cubs over their lifetimes. The proportion of sexually mature individuals (of both sexes) reproducing successfully also declined with increasing

[23] Where we calculated a simple ratio of the proportion of the population in each year that was above the long-term average F/α value—in other words, a measure of how 'fast' the population was in any given year. We did this by sex, so as not to mix up the different processes leading to pace of life variation in the two sexes.

ratio of males to females (in the case of this study, this meant a higher per capita reproduction rate at sex parity, but a lower rate when the population was female-skewed). While higher female densities were typically associated with smaller litter sizes amongst breeding females (another reason to deduce competition between females), higher male densities were, counterintuitively, associated with a greater likelihood that a male would breed (under these conditions, when the population was skyrocketing, it seems all sorts of males could secure mating opportunities as the carrying capacity was lifted).

Once upon a time it was probably the case for ancestral badgers that a shortage of males, or of encounters with such males as were available, may have been a problem for females (see also Chapter 19). However, as was vividly clear in Chapter 7, nowadays this is not a limiting factor in the reproduction of Wytham's badgers. However, with more females present, the proportion of females living in the shadow of others of their sex inevitably rose. Was that shadow protective or oppressive? For many it seems the outcome was suppression, as we deduced, albeit puzzledly, earlier (Mallinson et al. 1992; Woodroffe and Macdonald 1995b; see Chapter 8). Consequently, by the latter decades of our study, when a female-skewed sex ratio prevailed in Wytham in the aftermath of the sett proliferation and socio-spatial reorganization detailed in Chapter 10, the unforeseen but highly consequential event we consider to be a socio-ecological Black Swan, there may have been more reproductive opportunities for the population but fewer per capita. In those later years, 15% fewer females began reproduction early, and we saw overall declines in per capita female reproductive rates. Females were, it seemed, pushed to reproduce ever later, with as many as 40% of them only beginning reproduction at 4 or 5 years old. As we fretted over earlier, while evidence accumulates in support of the deduction that there is competition, indeed suppression, amongst female badgers with respect to breeding, the mechanism is elusive. It need not involve active bullying, but perhaps a swirling of stress hormones amongst individuals that differ in the susceptibilities of their adrenal glands—endocrinological outcomes that, while still mysterious in badgers, have been documented amongst canids (van Kesteren et al. 2013).

Digging deeper, we also noted POL effects linked to the *density* of same-sex individuals present: a greater proportion of males mated successfully for the first time at age 1 at higher male densities (10.8% ± 0.46 SE at the highest density, but only 1.0% ± 0.5 SE at

the lowest—again, perhaps chaos or anarchy was rife in the relationships amongst males in those times). Opposing this pattern, but in line with the deduction of female competition, fewer females commenced reproduction at age 1 at higher female densities (2.4% ± 1.3 SE at the highest densities; 17.1% ± 8.7 SE at the lowest). Around sex ratio parity, females tended to produce their first litter around age 4–5 (36.6% ± 10.3 SE at age 4 and 37.8% ± 12.9 SE at age 5).

Returning again to the population's shifting ratio of strategists, a balanced sex ratio was associated with a greater proportion of individuals (of both sexes) exhibiting fast life history strategies; that is, exceeding the long-term median F/α. Slow life histories were associated with higher same-sex density. A high F/α ratio signals what we might think of as a 'faster' population, in which individual badgers produce more cubs per year of life, and start breeding earlier on average. High female density (low sex ratio), which was characteristic of the latter part of our study, from 1997 onward, was associated with substantially slower population-wide female pace of life (Figure 14.11a). In males, low same-sex density from 1997 to 2007 was associated with a relatively consistent F/α distribution (Figure 14.11b).

When it comes to the speed at which they live their lives, in the wider spectrum of mammals badgers are a rather plodding species. As a rule, for species with disposition for life in the slow lane, population growth generally depends strongly on adult survival (Macdonald et al. 2009; Chapter 11). Where POL relates to adult survival (which it almost invariably does), the details can shed light on wider population dynamics: amongst badgers, the greater proportion of 'fast-living' and, therefore, fast-dying, males (see Figures 14.3 and 14.4) in the earlier years of our study (prior to 1997) likely contributed to the sex ratio switch to up to 60.2% female observed following the population peak in 1997 (Chapter 11). If we characterize these fast-living males as following 'quicker', riskier strategies (which were advantageous while the population was rapidly ballooning), then this strategy was selected against when these racy males comprised a greater proportion of the supportable population (and as this population bumped its head against its new, albeit higher, carrying capacity). Under these latter circumstances 'slow and steady' strategists prospered. This seesaw of advantage supports the view that the selection of life history strategy is risk sensitive (a theory proposed by Bårdsen et al. 2008).

To what degree did these population changes produce 'equity' in reproductive opportunity in badger

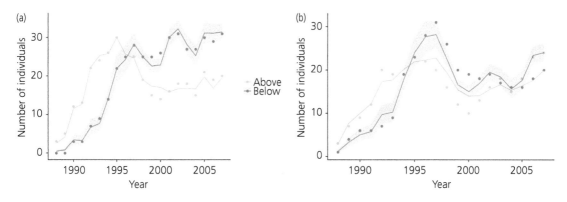

Figure 14.11 Number of reproductive (a) females and (b) males in the population above (blue) and below (red) the long-term (1986–2007) individual median F/α value. Points represent real data, lines are model predictions based on population sex ratio (males/females) and sex-specific population density, with 95% confidence intervals (from Bright Ross et al. 2020, with permission).

society? Not much: some badgers were winners; others found themselves facing a new inequality. These 'haves' and 'have nots' mirror inequalities across various animal, including human (Cushing et al. 2015; Davies et al. 2017), societies. People are painfully familiar with battling with their lot, making the best of things even when the omens are unpromising. In other social mammals, where a nuanced flick of an ear, or odorous molecule astray, can instantly discriminate a 'have' from a 'have not', it seems likely that each individual has some awareness of its own standing, including the probability of being a 'never going to have'. Yet badgers seem to strive, and fate is fickle. Variance in reproductive success within animal populations is widespread in the wild (Cabana and Kramer 1991; Tatarenkov et al. 2008). Recently, contributions to this distribution skew have been dissected more strictly to distinguish variation in individual quality ('individual heterogeneity', Hamel et al. 2018) and luck (rather weightily termed 'dynamic heterogeneity' by Snyder and Ellner 2018). Moreover, the precise shape of variation in reproductive success (which we first tackled in Dugdale et al. 2008) is becoming appreciated as an underlying evolutionary force in wild populations (Tuljapurkar et al. 2020). As yet, however, the extent of differences between individuals in a population has not been comprehensively studied in the context of population persistence in the face of the vagaries of environmental change.

The Gini index (GI), often used in the social sciences to measure inequality in societal resource distribution, particularly wealth and access to health care (e.g. Wan 2001), has also been employed in demography to characterize inequality of mortality rate in both human (Shkolnikov et al. 2003) and non-human populations (Archer et al. 2018), and even to compare inequality of reproductive success across populations (Dobson 1986). The number of offspring supported by a population's (non-static) carrying capacity is limited (Skogland 1985); therefore, offspring can be thought of as a limited resource for fitness. With regard to a given resource, the GI for a population is proportional to the concavity of that resource's Lorenz curve, which graphically represents the distribution of the resource amongst individuals in the population (Figure 14.12). In the case of perfect egalitarianism, GI = 0, whereas if one despot sequesters 100% of the resource, GI = 1.

For the badgers of Wytham Woods, Julius led us in computing annual (1988–2015) cub production GI using the population pedigree (Annavi et al. 2011 updated pers. comm.). This revealed very high reproductive inequality: mean annual GI for males was 0.87, and 0.84 for females. Over the years 2000–2010, this corresponds to the top 10% of fathers producing 74.5% of cubs, and the top 10% of mothers producing 67.3% of cubs.[24]

How did circumstances impact these inequalities? One important circumstance in Wytham was the changing population density of the badgers (Chapter 11), and it turned out that for females the GI inequality was statistically higher at higher female densities, but this was not so for males (Figure 14.13). Substantial reproductive marginalization existed for both sexes in all years, but the trend towards egalitarianism

---

[24] Because the genealogy is incomplete, these percentages are approximations, but the message is clear.

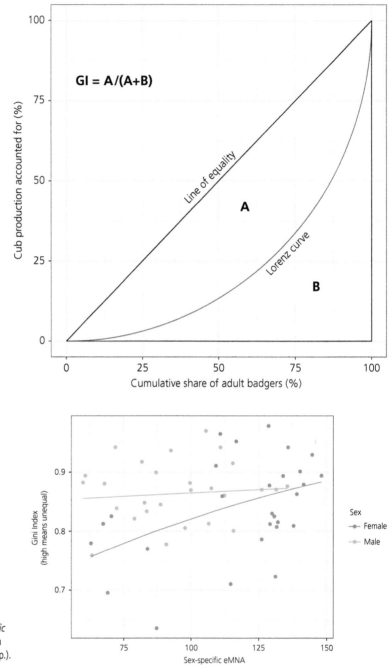

Figure 14.12 Illustration of Lorenz curve and relationship to GI. The Lorenz curve (red line) for a resource's distribution can be derived by sorting individuals from lowest to highest resource controlled, and plotting the cumulative resources (in this non-empirical example, cubs produced) controlled by each percentile of controllers. The Gini index is equal to the proportion of the area below the line of equality (along which, no individual controls more resource than any other) that lies above the Lorenz curve (from Bright Ross 2021, with permission).

Figure 14.13 *GI over different sex-specific densities.* Trend lines show predictions from β-regression (from Bright Ross et al. in prep.).

at lower female densities suggests that at least some marginal females were able to breed more easily at lower population densities, which we interpret as the result of those subordinate females experiencing lower competition when populations were more sparsely distributed.

It is commonplace in mammalian societies for male dominance to be clear-cut: from wolves to gorillas the dominant male is so obvious to even an untrained eye that the terms top dog, alpha male, and silverback have entered everyday language. The reproductive consequences of competitive exclusion are also clear; for example, in the harem defence in red deer *Cervus elaphus* (Carranza et al. 1990). However, the early chapters of this book documented our quest for similarly clear-cut sociological structures in badgers, with the result that they were so elusive that we even discussed elements of anarchy. The greater degrees of male reproductive inequality revealed by the GI over during all years in Figure 14.13 therefore warrant discussion. The generality, beyond badgers, is that the mechanisms underscoring the higher reproductive success of some males over others can be cryptic. It may be largely in a male's control to knock the stuffing out of his competitors, and even to guard his trophy after inseminating her, but he has little control over whether a female carries their offspring to term; ultimately, amongst Wytham's badgers, it is female body condition and the prevailing conditions that determine cub production. Moreover, while in other species mate choice might permit males to choose females more likely to carry offspring to term (males can be choosy too, Edward and Chapman 2011), the highly promiscuous polygynandrous sex lives of the badgers of Wytham Woods (Chapter 7), with something of a scrum when it comes to who mounts whom, seem to put choosiness in the background. Similarly, female badgers' capacity for superfoetation results in widespread multiple paternity (Chapter 17) so who knows (even amongst the badgers themselves) which male in the Rabellaisian sequences scores the fertilization—certainly, being first in the queue is not an obvious prerequisite of fatherhood. Indeed, as we emphasize in Chapter 7, because female badgers (like all mustelids) are induced ovulators, conceptive matings cannot occur until 48–72 h after first intromission, so females must solicit nonconceptive matings to ensure fertilization. As we pondered Julius' GI result, two non-exclusive mechanisms came to mind that may explain the success of some male badgers over others. The first is post-mating selection through cryptic female choice (Firman et al. 2017): in badgers, high rates of extra-group paternity may indicate that cryptic female choice provides a way to ensure high heterozygosity and low inbreeding depression (linked to cub survival, Annavi et al. 2014b, 2014a). The second is high investment in reproduction by males (although this is variable between individuals), either through sperm competition (Preston et al. 2003) or the enthusiasm with which they dedicate themselves to maximizing total mating opportunities (Canal et al. 2012). In this wider context, whereas the cost of reproduction to females is obvious and widely appreciated, the cost to male mammals is often an afterthought in ecology (Bleu et al. 2016).

Dodging the weather

Weather may have mattered less than society in setting the POL amongst Wytham's badgers, but did it have any influence at all? Yes—and even subtle effects (the butterfly famously flapping its wings in Beijing[25]) can have far-reaching consequences—ultimately, the environment is the backdrop against which the theatre of animal society plays out. Proportionally fewer females bred in years with seasonally extreme temperature. This result is both intuitive (seasonal extremes are challenging as readers who have experienced heatstroke or chilblains will affirm), and important (not least because extremes are likely to become more normal), so the obvious question is: how big is the effect? The nature of our analytical models diagnoses the effect with confidence, but makes the magnitude difficult to estimate, as it is always combined with other confounding factors in real life. Nonetheless, if everything else (age, a variety of weather conditions, and body condition) were averaged, we can obtain an estimate from our models: for that quartile of most extreme seasonality (α_T) 9.5% of females might breed, whereas in that quartile with least radical seasonality more like 41.1% would do so. Interestingly, the likelihood of a female breeding was also lower in warmer years, conditions also associated with smaller litter sizes per reproducing female. This supports the idea that females adapt reproductive investment to their likelihood of being fit enough to carry a litter to term (Mallinson et al. 1992). In contrast, a greater proportion of males bred in years with more variable temperature than when temperature was less variable.

Temperature also affected the badgers' entry points for reproduction, with more females deciding to throw caution to the wind and produce their first litter at age 4 (which we considered as latecomer first breeders) following warmer years (our models predict 38.3% ± 16.0 SE in the warmest years vs 3.3% ± 2.2 SE in the coolest, holding everything else constant at average values[26]). Exactly the opposite applied to abstinent middle-aged males: fewer such males were assigned their first cubs at 4 years of age in warmer years (39.4% ± 15.5 SE in the coldest years, 2.1% ± 1.9 SE in warm years). As a male can only have a cub assigned to him if he inseminates a successful female, there is clearly something

---

[25] Edward Lorenz created chaos theory and illustrated it with the 'butterfly effect', famously rendered in the film *Jurassic Park* as 'A butterfly can flap its wings in Peking, and in Central Park, you get rain instead of sunshine'.

[26] This holding constant of all other factors applies to all other similar estimates reported in this section.

going on here. Plausibly, spinster virgin females were being snapped up by the cadre of previously successful males. Latecomer (age 4) females were more likely to commence reproduction when they experienced more variable annual rainfall (41.8% ± 16.7 SE vs 6.1% ± 3.3 SE), while, as with temperature, the opposite was true of latecomer males.

Environmental carrying capacity, heavily determined by the weather, and the number of mouths competing for those food resources, becomes tangible in terms of how relatively fat, or slim, individuals are: did corpulence relate to POL? Fatter badgers did not just have the higher chance of reproduction we documented earlier, but also began reproducing earlier in life (females at age 3: 42.0% in highest condition, 6.3% in lowest; males at age 2: 32.3% ± 9.5 SE in highest condition, 2.2% ± 1.7 SE in lowest; males at age 3: 39.4% ± 13.6 SE in highest condition, 1.9% ± 1.9 SE in lowest). In particular, the likelihood of 3-year-old females achieving primiparity in the following year was also correlated with BCI. In males, fatter individuals tended to sire more cubs, and 30–40% of younger males reproduced as young as age 2 or 3 if they had a chubbier BCI (vs 2.2% ± 1.7 SE at age 2 for the lowest BCI; 1.9% ± 1.9 SE at age 3). These youngsters were probably making use of the surplus energy they had acquired: male badgers must achieve a minimum body size for the onset of puberty (Sugianto et al. 2019c; Chapter 3). Consequently, when weather and foraging conditions affect juvenile male growth, their 'race to the starting line' is postponed (Bright Ross et al. 2020). In contrast, impacts on females manifest more as perturbations in their otherwise steadier POL once they've commenced active reproduction.

The different POL tendencies between the sexes stem from the energy costs of reproduction. Earlier breeding (where conditions allow) can lead to a higher relative fitness advantage in males (*sensu* Oli and Dobson 2003). But male age at first breeding is, predictably, more sensitive to environmental conditions than that of females, because polygynous males can achieve a faster overall POL if they invest in multiple litters simultaneously (Hämäläinen et al. 2018)—something females can't do. Conversely, year-to-year weather conditions affected female reproduction more significantly. Of course, as is always the case with badgers, the dichotomy between males and females is not quite so clear-cut: body condition also affects females' likelihood of primiparity at age 3, while the variability of temperatures in any given year affects the likelihood of male reproduction from year to year. In summary, the preponderance of the evidence points to the notion that, insofar as weather is concerned, males are preoccupied with the race to the starting line, while females are more concerned with their pace once in the race.

A generalization from Julius' measures of the speedometers of badger lives is that even if a population comprises the same number of individuals at two points in time or two populations occur at similar density, neither of these situations implies equality. Ideally a population (to imbue a population with a false sentience—more precisely we mean the force of natural selection acting on individuals in the population) will maintain a proportion of both 'fast' and 'slow' strategists (Starrfelt and Kokko 2012), and this diversity brings the consequence of an insurance policy. Over the long term, maximizing proximate fitness will ultimately not provide an evolutionarily stable strategy in light of inexorable environmental stochasticity (Ogura et al. 2017), resulting in an ecological 'trap' for the population (Le Galliard et al. 2005).The ratio of 'fast' to 'slow' individuals, and the ratio of males to females, is maintained by shifting selection and provides the population with substantial fodder with which to adopt exit routes from such a seemingly inevitable fate.

Predicted adaptive responses and the weather—an epilogue

As an epilogue, just as we go to press, the project's last student, freshly minted as Dr Ming-shan Tsai, uncovered a new link between our findings on early life adverse weather conditions and offspring herpesvirus reactivation later in life (Tsai et al. 2022, and further discussion in Chapter 15). Remember, adversity in early life can cause permanent changes in individual phenotypes (and one mechanism is through epigenetic modification, such as DNA methylation, which causes the modification of gene expression rather than alteration of the genetic code itself. Chapter 19), and this prenatal adversity can take various forms (Lu et al. 2019), amongst which prevailing weather can exert substantially different effects between cohorts of, for example, red deer (Albon et al. 1992). Recall also that during harsh years Wytham's badgers have no choice but to invest more energy when they can afford it the least (Bright Ross et al. 2021). A meta-analysis of 111 animal studies examining early-life stress found generally negative effects of developmental stress on phenotypes (Eyck et al. 2019), supporting Monaghan's (2008) silver spoon hypothesis discussed earlier. However, Gluckman et al.'s (2005b) predicted adaptive response (PAR)

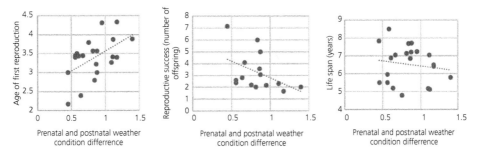

Figure 14.14 The relationship of prenatal–postnatal mismatch in weather conditions and offspring performances, revealing a highly significant effect on age at first reproduction, but no statistically significant effect on lifetime reproductive success or lifespan (there was no indication of differences between the sexes), for 21 cohorts from 1987 to 2007 (from Tsai et al. in prep).

hypothesis envisages that individuals will prosper most (e.g. live longer or produce more young) if the prenatal conditions they experience match the post-natal ones they subsequently encounter (whereas unmatched environmental conditions lead to poorer performance and such disadvantages as metabolic syndromes and, in people, psychiatric illness (Bateson et al. 2014; Vaiserman 2015)).

So, led by Ming-shan and Christina, we divided the badgers into 21 completed cohorts by birth year (from 1987 to 2007) and measured the mismatch for each cohort between the weather conditions (considering deviation of temperature and rainfall to long-term averages) during their prenatal *in utero* development (October to November when, critically, mother badgers accumulate fat for winter) and their postnatal/neonatal development (April to May when cubs emerge from the sett). We then tested, using linear regression, whether the extent of mismatch affected any or all of age at first reproduction, lifetime reproductive success or offspring lifespan. As background, we found no significant effect of the weather, whether good or bad, prevailing at either the prenatal or cubhood stage. However, badgers that had experienced a greater mismatch between the weather when they were in the womb[27] and that prevailing during the first months of their foraging lives went on to breed for the first time later, and this effect was highly significant (with an adjusted $R^2 = 0.25$). In contrast, we detected no significant effect of mismatch on cohort

lifespan or lifetime offspring. Were these findings, which fit with PAR, associated with epigenetic programming? As we apply the full stop to this sentence, Ming-shan is busily at work on the epigenetic effects on gene expression that may provide the answer (Figure 14.14).

On reflection: exceptions that prove rules

The plethora of routes through life, and the varied paces at which badgers follow them, make one thing clear: individual badgers in Wytham Woods differ greatly in their tactics for coping with life there. Against the punchline that individuality is rampant, two interesting questions percolate through the swarm of variation that we have shown shapes badgers' destinies: to what extent does personality or, perhaps better, aptitude, talent, even character or quirkiness, defeat generalizations in setting each badger's course through life, and are some badgers simply better than others? Like archaeologists we have dug through finer and finer layers of analysis, sifting the seemingly chaotic variation of badger lives in Wytham into generalizations dictated by their sex, age, the seasons, weather, and food supply. At each level in this dig, another layer of seeming randomness falls into line as a new rule of thumb. But at the end, even the forest of causes and effects that populate this chapter leave unexplained something like 40% of the variation between individuals. Does that mean the unexplained is inexplicable, other than by the chance quirks of individuality? Of course not. It means, at least in part, that our archaeology has not dug deep enough, our sieves not fine enough, to reveal the next layers of generalization. Perhaps, for example, some of that unexplained variation would be explained if we unearthed relationships between the sex ratio of an individual's siblings, perhaps in

---

[27] This effect probably occurred during delayed implantation rather than post implantation, in line with Ming-shan's results using sliding window analysis to explore the effect of prenatal weather conditions at different time windows on the incidence of later-life herpesvirus reactivation (see Chapter 15 and Tsai et al., 2022).

their birth chamber or even *in utero*, or the balance of acorns to blackberries laden with antioxidants it ate in its first autumn, or the grumpiness of its mother. Does this mean that there's always another rule awaiting discovery, that the exception to every rule is merely the fodder for the next one, *ad infinitum*? Not really. There is indeed always an explanation, and while we might look forward to further excavations that reduce the 40% of unexplained variation to, say, 20%, there is eventually going to be a level of granularity that, while explicable by genetics, development, and circumstance, is most usefully thought of as individuality, whether we are considering a clan of badgers, a litter of puppies, or a schoolroom of children. Does that mean, then, turning to the second question, that some badgers are simply better than others—better at being badgers? There's no denying that some badgers may indeed be stupid or lazy or smart or energetic, but what is more interesting is that amongst the residue of individuality, while it may be so that at any one moment in history some individuals are better at being badgers than others, when history blinks its eye the tables can be turned, the race to the tortoise not the hare, to the plodding not the agile, to the conservative not the exploratory, indeed, even to the meek. That is the lesson of POL theory: that the playing field will change and, when it does, different sorts of players will win the day. And the lesson of this chapter is that while we have been excited to reveal ever-more nuanced generalities, we are no less excited to see that even amongst badgers individuality reigns. A cohort of badgers in Wytham Woods is not a monoculture any more than is a classroom of children of the same age in the same village, and to understand either it is hard to know which matters most, the rules or their exceptions, so it's best to appreciate both. In ecology, preoccupation with the beauty of patterns should not blind us to the marvel of the individual.

CHAPTER 15

In Sickness and in Health

And I looked, and beheld a pale horse: and his name that sat on him was Death, and Hell followed with him. And power was given unto them over the fourth part of the earth, to kill with sword, and with hunger, and with death, and with the beasts of the earth.

Revelation 6:8

Nature is cruel, and while such eighteenth-century poets as Keats, Shelley, and Wordsworth may have romanticized the concept of wildness, the reality for any wild-living creature is closer to Tennyson's characterization[1] of 'Nature, red in tooth and claw', and not so pleasantly cosseted as the post-penicillin, vaccinated world of modern humans. Badgers, like all wild species, must constantly dodge the arrows of competition, starvation, and environmental calamity, while also facing the insidious apocalyptic horseman of infectious disease. We turn now to the impact of illness and its doppelgänger, stress, and the attempts badgers make to resist morbid demise.

Back in Chapter 2, when the behavioural ecology of the badgers of Wytham Woods still seemed fairly straightforward, we documented how coccidiosis plays a callous role in determining which cubs survive to maturity by exacerbating their struggle to find enough food to fuel their own development while bearing their parasitic burden. But surviving coccidiosis is just the start of the ever-present handicap of pathologies that will shape their lives, and involve risky trade-offs between investment in immune responses and antioxidant systems versus body condition and, ultimately, reproduction (see Chapter 14).

Diseases afflicting badgers are, potentially, many and varied, and our work with them has been eclectic. The infectious ones range from viral, bacterial, or protozoan to multicellular pathogens. The non-communicable ones (NCDs)[2] include cancers, heart disease, and genetic disorders.

And in turn, both suites of disease cause, and are exacerbated by, physiological (and psychological) stress. Infectious diseases not only cost lives, but also energy, through the labour of the host's immune system (to appreciate the impacts of disease it's helpful to understand the immune system, so we provide a primer in Box 15.1). Infectious diseases may smoulder largely unseen in a population (termed enzootic disease), or blaze through it (epizootic), sometimes causing catastrophic population crashes that flash bright on the conservationist's radar; for example, fungal white-nose disease in bats (Blehert et al. 2009), bacterial *Pasteurella multocida* in saiga antelope (causing haemorrhagic septicaemia; Feridouni et al. 2019), or viral rabies in Ethiopian wolves (Haydon et al. 2006). Usually it is only then that the significance of disease becomes integrated into an understanding of the species' population dynamics (Newman and Byrne 2017). In practical terms, outbreaks of disease in wildlife tend to be noticed too late to stop them (Bacon and Macdonald 1981).

Much as predators single out the weak or lame, or freezing weather weeds out the malnourished, the axe of infectious disease doesn't fall at random, either. Individuals differ in their immunological resilience, sometimes because of unavoidable factors such as their age or genetics, sometimes because

---

[1] Tennyson's Canto 56: 'Who trusted God was love indeed, And love Creation's final law, Tho' Nature, red in tooth and claw, With ravine, shriek'd against his creed'.

[2] According to the World Health Organization, the principal NCDs afflicting people include cardiovascular diseases (such as heart attacks and stroke), cancers, chronic respiratory diseases (such as chronic obstructive pulmonary disease and asthma), and diabetes.

The Badgers of Wytham Woods. David W. Macdonald and Chris Newman, Oxford University Press.
© David W. Macdonald and Chris Newman (2022). DOI: 10.1093/oso/9780192845368.003.0015

Box 15.1 Fundamentals of the immune system

To understand the importance of disease in animal ecology, and specifically to badgers, one must understand the basics of an immune response. Mammalian immunity fundamentally comprises three systems (Hoebe et al. 2004). The innate response provides the first line of defence against pathogens (any invader that is non-self) as well as offering an immediate response to injury (i.e. stopping pathogens from gaining entry). This is where reactive oxygen species (ROS) play a key role in the progression of the inflammatory response. This involves phagocytic cells such as neutrophils, macrophages, and dendritic cells. These can detect ('swallow up and digest') microbes and use an oxidative burst of ROS to incinerate them (Akira et al. 2006; Weiskopf et al. 2009). Innate cells are activated when pathogens elicit the acute inflammatory response, which is accompanied by systemic vasodilation and vascular leakage, allowing leukocyte emigration (thus phagocytic cells can permeate tissues under attack). While this is broadly advantageous to the organism, if unchecked it can also lead to the four cardinal signs of localized acute inflammation that were first described almost 2000 years ago by the Roman physician Celsus: *Calor* heat, *Rubor* redness, *Tumour* swelling, and *Dolor* pain, leading to *Functiolaesa*, that is, the loss (or impairment) of function (Mittal et al. 2014).

Within a short period of innate immune activation, the acute inflammatory response triggers the secretion of various signalling molecules (cytokines and chemokines) that in turn trigger other parts of the immune system and summon additional leukocytes and lymphocytes (Nathan 2006; Mantovani et al. 2011; Vivier et al. 2011). Neutrophils then secrete further vaso-active and pro-inflammatory mediators including histamine (which is why antihistamine drugs reduce swelling when you get hay fever or a mosquito bite), which are followed by fluid accumulation (oedema).

These phagocytic leukocytes are also known as antigen-presenting cells because after engulfing the invading microbe, they display pieces of protein from the microbe, called antigens, like trophies on their surface—it seems reminiscent of a gamekeeper hanging his victims on his gibbet,

on what is called their major histocompatibility complex (MHC) molecule (this, topically at the time of writing, is one part of the immune response that SARS-CoV-2, the causative agent of Covid-19, successfully evades). All this can end in the 'cytokine storm' that became widely discussed during Covid-19, if the immune system is overstimulated, and/or if antibodies are not raised to move immunity on to its second step. That second step involves adaptive immunity, which includes a two-pronged attack: cell-mediated immunity and humoral immunity. Adaptive immunity is antigen-specific and elicited by the stimulation of both T and B lymphocytes;[3] the humoral response is controlled by activated B lymphocytes that can differentiate to produce antibodies (a.k.a. immunoglobulins) against specific antigens; the cell-mediated response involves T-cells (Iwasaki and Medzhitov 2010; Mantovani et al. 2011).

A crucial fact is that cortisol, a steroid hormone with many regulatory duties in the metabolism and immune system, plans ahead by suppressing the release of cytokines in anticipation (it does this even if the assault is psychological), where swelling could be debilitating. As readers have probably already discovered to their cost, stress thus compromises the efficacy of the immune system, thereby leaving the body more vulnerable to infection; chronic stress is thus often associated with disease (Cohen and Hamrick 2003).

Finally, the oft forgotten Cinderella sister of immunity is the complement system—the 'sweeper-upper' behind its better-recognized siblings—which clears out disabled pathogens, using proteases (mainly produced in the liver) to cleave and dissolve spent cells.

The relevance of all this to the badgers of Wytham Woods is that between the challenges of their social behaviour and wider environment, they are likely to encounter a wide array of pathogens, against which the only thing that will keep them alive for long enough to deliver on their inclusive fitness is the fortitude of their immune system.

[3] T and B lymphocytes stand for thymus-derived cells and bone marrow-derived cells, respectively.

---

they took a gamble and weakened their constitution (e.g. by (over-)investing in reproduction), sometimes because of bad luck such as poor nutrition or co-infections, and sometimes because of contagion contingent upon where and with whom they associate (e.g. Albery et al. 2020; see Chapter 8). Furthermore, these intrinsic differences between individuals

in their susceptibility to disease play out in the contexts of habitat, diet, phylogeny, lifestyle, and body type (Newman and Byrne 2017). Ultimately, however, there may be no defence against highly pathological diseases, felling indiscriminately the toughest or feeblest, fleetest or slowest, most assertive or most meek.

Importantly, disease is not only a problem when it is roiling or festering. Even when kept subpatent by the host's immune system, the body may be working hard to keep illness in check, a competency that incurs energetic costs. For example, for the average person consuming 2000 calories per day, around 400 cal are spent on maintaining immune system defences, increasing beyond 500 cal with even mildly febrile infections (Straub 2017; see Chapter 14). Therefore, the potential for infection to drive population decline not only due to sickness, but also due to the demands this places on limited local food supply, is a broad phenomenon in animal population dynamics (Hudson et al. 2002). Sterling and Eyer (1988) termed this 'allostatic load'—'allostasis' being the process by which the body responds to stressors in order to regain homeostasis;[4] specifically 'the wear and tear on the body' that accumulates as an individual is exposed to repeated or chronic stress (McEwan 1998). Historically, zoonotic disease has shaped human history, and continues to do so (Montgomery and Macdonald 2020). Diseases confined to wild hosts unsurprisingly tend to attract less interest than zoonotic conditions able to spread to humans (Zhou et al. 2020; Xiao et al. 2021), or even spill over into livestock (Macdonald and Laurenson 2006). Badgers have been at the eye of just such a storm, in what is, globally, arguably the most vexed wild mammal disease debate, certainly in British history: bovine tuberculosis (bTB)—a disease brought to badgers by the dairy farming industry, and the topic of Chapter 16 (e.g. Macdonald et al. 2015; Abdou et al. 2016). But what of the badgers' other natural pathogens—so often overlooked—and the implications of mounting an immune or stress response?

What makes badgers sick?

To understand the socio-ecological, genetic, and evolutionary implications of badger disease, first we must know what we're dealing with, and so, briefly, let us explore the burden of diseases with which badgers must cope most commonly.

Viruses

We start with their viruses, motivated by the preeminence of viruses in other WildCRU studies, ranging from rabies in Ethiopian wolves (Sillero-Zubiri et al. 1996b) to feline leukaemia, calicivirus, and coronavirus in Scottish wildcats (Daniels et al. 1999). We began screening the Wytham badgers for Aleutian mink disease virus (AMDV), given that we had found antibodies to it nearby amongst American mink in the Thames Valley (Yamaguchi and Macdonald 2001). Mink, like badgers, are mustelids, all broadly prone to this parvovirus variant. Our suspicions were further roused when we came across some seemingly feverish, lethargic badger cubs with visibly weak hind legs—all candidate symptoms of AMDV. So, we tested for AMDV antibodies in serum in the late 1990s; the results were negative for the badgers of Wytham Woods (but *canine* parvovirus enteritis has been detected in badger cubs elsewhere in the UK; Barlow et al. 2012).

Equally conspicuous by its absence was canine distemper virus, CDV, insofar as a very broad range of carnivores is susceptible to this RNA paramyxovirus (think measels, mumps, and bronchitis in humans). Despite being present in other UK wildlife, 468 badgers tested by Delahay and Frölich (2000) were seronegative, suggesting that this is not a major badger disease. Similarly, in Germany, no CDV antibodies were detected in badger sera in a study by Frölich et al. (2000), although CDV was detected by PCR in lung tissues tested (suggesting exposure at some point in life, but not recent infection).

Having started out being concerned by the diversity of viruses to which mustelids are susceptible, our perspective was changing to one of surprise at just how few were present in Wytham's badgers. Nonetheless, our health screening turned up one new discovery: a polyomavirus, tentatively named *Meles meles polyomavirus 1* (MmelPyV1), first reported in Cornish and French badgers (Hill et al. 2015), although without evidence of any pathophysiology linked to infection.

Mustelid herpesvirus

Our duck was broken,[5] however, when our student, Yungwa 'Simon' Sin, now a distinguished Associate Professor in Hong Kong University (via a post-doc at Harvard), detected a variant of Mustelid herpesvirus (MHV; or specifically here variant 'MusGHV-1': hereafter MHV) in 354 of 361 blood samples we collected from 218 Wytham badgers in 2014, using a genetic

---

[4] The tendency towards a relatively stable equilibrium between interdependent elements, especially as maintained by physiological processes.

[5] By the way, the English idiom of scoring a 'duck', meaning getting zero, or nowhere in any of life's endeavours, derives from the similarity between the appearance of a duck's egg and a score of 0 on a cricket score board.

amplification and nucleic acid sequence detection technique called qPCR (quantitative polymerase chain reaction) (Sin et al. 2014; see also King et al. 2004).[6] This extremely high prevalence indicates that most of the badgers have acquired MHV, which lurks, latent, in their lymphocytes.[7]

Once contracted, MHV, like all other herpes viruses, can remain latent for a lifetime in the host cells. Positive individuals exhibited no clinical symptoms associated with infection, although strictly speaking, a PCR diagnosis only really reveals the presence of a pathogen, not the fact that it may or may not be currently causing sickness. Nevertheless, we were intrigued by such high prevalence. As is the case with all herpes viruses, while MHV remains subpatent most of the time, rendering badgers asymptomatic, it can flare up (undergo 'reactivation') during periods of immune-compromise (as per resurgent cold sores around the mouth of stressed people infected with herpes simplex virus-1) causing the individual not only to fall ill, but also to become an infectious spreader.

The reproductive tract is one important route of spread for herpes viruses,[8] with the gruesome consequence that reactivation localized there can cause serious venereal disease to its hosts (Spano et al. 2004; Knowles et al. 2012). Herpes viral infection can occur during mating, via the semen of an infectious father (Neofytou et al. 2009), if it is not already present in the mother, or vice versa (François et al. 2013). Vertical transmission from mother to offspring is possible but rare (occurring through contact with virus shed into the vaginal tract during birth, or extremely rarely via the placenta). Equally insidious, if the mother acquires primary infection while she's pregnant, the foetus can catch the disease via the placenta, causing abortion. More commonly, if primary infection occurs during birth, the consequence to the neonates can be severe, including mortality, or damage to the developing central nervous system and brain (Avgil and Ornoy 2006). Postnatal infections probably arise

from the mother shedding virus, or through nursing on virus-contaminated milk. By whichever route MHV arrives in a new-born, even if disease is not sufficiently severe to cause the infant's immediate death, survivors may carry neurological deficits likely to hasten their demise. Vertical and venereal transmission can play a significant role in the epidemiology of herpes viruses generally, such as herpes simplex virus 2 (HSV-2) in humans (Looker et al. 2015), murine herpes virus 4 in mice (François et al. 2013), and bovine herpes virus in cattle (Graham 2013).

These generalities about herpes virus acting as a sexually transmitted infection (STI) and potentially causing sterility (i.e. reduction in host fecundity due to infection; McLeod et al. 2019) prompted us to think about the striking difference between the proportion of female badgers we knew to mate each year (almost all of them; Chapter 7), and the much smaller proportion (a third) to which, from May captures onwards, living cubs are assigned (Chapter 17). Back to the speculation on reproductive suppression, but might some of that discrepancy be caused by venereal MHV?

Led by WildCRU intern Alice Kent, we conducted pan-herpes PCR testing using genital swabs to screen the vaginas of 71 females (51 adults and 20 cubs) and the penile prepuce of 27 males (26 adults and 1 cub), specifically testing for MusGHV-1 (Kent et al. 2018). Sure enough, we found herpes infections in 54 of 98 genital swabs (39/77 adults and 15/21 cubs) tested. Once more DNA sequencing confirmed this to be MusGHV-1. This shows that in these individuals, MusGHV-1 was not only present (as per Simon's PCR survey) but had replicated and shed in the genital tract, rendering adults infectious and thus indeed capable of spreading infection horizontally during coitus and vertically to cubs during pregnancy.

Two pennies dropped. First, cubs, even though sexually immature, had a higher prevalence of virus shedding than adults, likely due to primary infection.[9] Second, MHV infection intensity was associated with thinness (low weight/length ratio). Of course, in the manner of horses and carts, it was equally possible that this was because skinny badgers more frequently experienced herpesvirus reactivation, or because a greater

---

[6] Aside from genital herpes caused by herpes simplex 2, herpes viruses infecting humans include Epstein–Barr Virus (EBV), causing infectious mononucleosis, known affectionately by many adolescents as 'kissing disease'; and *Varicella zoster*, causing chicken pox and shingles.

[7] Unlike alpha-HV, gamma-HV hide in lymphocytes, so testing blood can reveal the true prevalence.

[8] As background, although herpes viruses are typically spread amongst animals by bite wounding (and the inhalation and ingestion of infected sputum, herpes virus infections are sometimes spread sexually and/or passed vertically from mother to offspring.

[9] Venereal MusGHV-1 shedding was widespread in this population. All 98 swabs tested negative for alpha- and beta-HVs, both in the pan-HV generic PCR as well as the MusAHV-1-specific PCR. However, 55% (54/98) of those samples tested with pan-HV generic PCR were positive for gamma-HV identified as MusGHV-1 by sequencing. The positive samples were from 43% (22/51) of adult females, 65% (17/26) of adult males, and 75% (15/20) of female cubs.

proportion of virus-shedding badgers were skinny. Taken together, these two early insights hinted that frailty—vulnerability to immune-compromise resulting from stressors—was a factor. So we asked whether MHV was a factor in the population processes of the badgers of Wytham Wood, and particularly whether it might shed light on that high proportion of females not producing cubs.

Sociologically, of the 18 social groups tested, 17 had amongst their members at least some badgers with genital virus shedding. At this point, induced ovulation becomes a factor, linked as it is to the adaptive significance of multiple matings and promiscuity (Chapter 7). The more numerous the partners, and more repeated the matings, the more likely individuals are to contract MusGHV-1 at some point in their life; it is easy to see how promiscuity, characterizing the badgers of Wytham Woods, through frequency-dependent transmission, could push infection towards bedded-in endemicity (Rudolf and Antonovics 2005). We knew that MHV was widespread within Wytham's badger population in a venereal form. Also that none of these infectious individuals, of either sex, showed any fever or genital discharge—in a way this makes matters worse insofar as the lack of symptoms could mean that partners cannot detect the risk of their intimacy (although who knows what they might smell) (Newman and Buesching 2019).

The possibility that MusGHV-1 contributed to the explanation of high rates of reproductive failure felt like a breakthrough-in-waiting: enter doctoral student Ming-shan Tsai, joining us from Taiwan, where she had accumulated expertise in animal health, via Bristol, where she had taken her Master's degree in global wildlife health, and gained experience with infectious diseases in small carnivores in captivity and the wild.

Ming-shan made the full exploration of MusGHV-1 the topic of her thesis; one facet of her studies led her to revitalize our collaboration with colleagues in Ireland (Chapter 12), and we will detour there first. Working with post-mortem data from badgers culled as part of the Republic's bovine TB control programme, Ming-shan used PCR to test genital swabs collected from 144 wild badgers (71 males, 73 females) for MusGHV-1 DNA (Tsai et al. 2020). Overall, these genital swabs revealed that most, indeed almost two-thirds, of the badgers had MusGHV-1.[10]

Infection didn't seem to compromise male fertility (in terms of testes volume or sperm quality, noting that one of five sperm samples collected tested positive for MusGHV-1). The rate of infection amongst this Irish population was considerably higher than that revealed, using the same swab technique, amongst the badgers of Wytham Woods. Prevalence amongst those 40 males culled and collected during the post-partum mating season was 82.5%. Moreover, at the onset of the mating season, MusGHV-1 reactivation in the male genital tract (82.5%, 33/40) was 6.7 times[11] more likely than in females (47.5%, 19/40),[12] and 4.7 times more likely in young adults (72.4%, 21/29) than in middle-age adults (50%, 10/20). This may have serious consequences: during pregnancy primary herpes virus infection, or reactivation from latency, can cause various sorts of reproductive failure, including infertility, embryonic reabsorption, foetal abortion, pre-term birth, stillbirth, poor neonatal condition, or neonatal death.

Dissecting deeper, Ming-shan also found inflammatory lymphoid hyperplasia—a condition that sounds, and is, ominous. It involves proliferation of lymphocytes (white blood cells; see Box 15.1) contained in lymph nodes. This probably means that the afflicted badgers had been battling infection, likely due to virus, with an active immune response. This inflammatory lymphoid hyperplasia was diagnosed in mucosa of the genital tract in 48.4% (30/62) females and 52.5% (21/40) males.

The breakthrough seemed to draw closer with the finding that females experiencing genital herpes virus reactivation during the breeding season are 3.7 times more likely to have had unsuccessful pregnancies than those without genital infection. Given that pregnancy is energetically costly, Ming-shan further separated females into pregnant and non-pregnant, and found that it was only amongst pregnant females that lighter bodyweight was marginally associated with propensity for genital MusGHV-1 reactivation.

In the winter dataset 71.8% (28/39) of females implanted successfully while 61.5% (24/39) reproduced 'successfully' (success, in this instance, rather mournfully defined as still being pregnant or having recently been so as evidenced by fresh placental scars at the time when they were killed) with an average litter size of 2.46 ± 0.72 (min. = 1, max. = 4). In the spring dataset, 51.6% of females (16/31) showed

---

[10] The prevalence of MusGHV-1 in these genital swabs was 61.8% (95% confidence intervals (CI): 53.7–69.3%, 89/144 swabs).

[11] Calculated by odds ratio, a measure of risk of disease.

[12] This much greater rate of reactivation in males raises the question of whether males have more sexual partners than females—that had not been apparent in Wytham (Chapter 7).

signs of recent pregnancy with an average of 2.41 ± 0.9 (min. = 1, max. = 4) placental scars (representing most recent litter sizes); blastocysts (the unimplanted embryos much discussed in Chapter 7) were present in the uteri of 52.2% (12/23) of adult females with an average of 2.08 ± 0.63 blastocysts (min. = 1, max. = 3, n = 12) per pregnant female.

The major result is that, in this Irish population, females with genital MusGHV-1 reactivation had lower rates of successful pregnancy (41.2%, 14/34) than MusGHV-1 negative females (72.2%, 26/36). Breaking this down by season, the effect is only significant when the samples were gathered in winter, when females have given birth or are about to do so, rather than in spring, when mothers were nursing cubs and about half have conceived blastocyst(s). We deduced that the sterility caused by MusGHV-1 impacts during gestation rather than afflicting the blastocysts during the pre-implantation stage when they are in suspended animation. Furthermore, during spring there was no difference in MusGHV-1 prevalence in the genital tracts of females with (45.4%, 5/12) or without (41.7%, 5/11) blastocysts during spring. A practical consequence is that finding the impact of the virus depends partly on looking for it at the right time. However, in relation to vertical transmission of infection from mother to cubs, all 43 foetuses Ming-shan dissected from 18 well-preserved pregnant females appeared normally developed and healthy, irrespective of their mothers' herpes status. Three out of the 10 placentas we tested were positive for MusGHV-1 DNA (one from a genital MusGHV-1 negative mother; two from mothers with partial foetal loss—one tested positive for genital MusGHV-1, one negative); however, no MusGHV-1 DNA was detected in any foetal tissues, further supporting our deduction that vertical transmission via the placenta is unlikely.

What is reactivation?

Herpes viruses have evolved to outsmart the immune system, with gamma herpes virus hiding away as a latent infection in lymphocytes (we don't know if that's where MusGHV-1 hides, but its close relatives do), from which they reactivate repeatedly throughout life.[13] It's only when they are reactivated, and thus in a non-latent state, that herpes can be spread between hosts. What might cause such a scenario?

The answer involves various triggers (for more on immuno-suppression, see the text that follows) that prompt unwelcome reactivation, including localized tissue trauma, physical or emotional stress, fever and microbial infection, hormone imbalance, allergic reactions, cross-infections, and even UV exposure (Stoegar and Adler 2019). Reactivation of gamma herpes viruses occurs in plasma cells and epithelial cells of the mucosa that function as doorways to external contact (e.g. mouth, nose, eyes, and genital tract), and thus facilitate transmission.

Does this scupper our hypothesis that MusGHV-1 was contributing to reproductive failure? No, not exactly, but we do have to phrase it more precisely. Gamma herpes virus rarely leads directly to spontaneous abortion (unlike alpha herpes virus). However, the added pressure that infection places on the maternal immune system does lead to a substantially increased susceptibility to ascending cervical and uterine secondary bacterial infections (Racicot et al. 2013), which in turn cause abortion. Also, while transplacental vertical transmission seems unlikely, it is highly plausible that new-borns contract MusGHV-1 from their mothers through direct contact during delivery through the infectious virons present in her vagina.

Amazing ourselves at the intricacy of our journey, we charted a route through stress to ageing (which will eventually bring us to senescence in Chapter 18). A first turn on the road came when Ming-shan and Christina noticed that rates of genital herpes virus reactivation were lowest amongst middle-aged adults (5–6 years), but start to climb amongst badgers older than 6 years in our Wytham data. On the point of stress, we alluded, earlier, to an effect where thinness was associated with venereal virus shedding, and so our story next coalesces around body condition, as featured in Chapter 14, and ultimately how badger foraging success and capacity to store excess energy as body fat are dependent on the weather (Chapters 12 and 13).

Returning to Wytham's badgers, Ming-shan and Christina collected 251 genital swabs from 150 individuals, caught in May, September, and November 2018, which they screened for the presence of MusGHV-1 DNA using PCR targeting the DNA polymerase gene (Tsai et al. 2021). The overall detection rate of genital MusGHV-1 DNA across these swabs was 35.5% (89/251, 95% CI: 30–41.6%), peaking in summer (43.8%, 35/80) versus in spring (34.4%, 33/96) and autumn (28%, 21/75). Age was also influential (Figure 15.1) with genital tract infection detected in 45.2% (33/73

---

[13] In comparison, human herpes simplex hides in peripheral neurons, waiting for any abatement in immune control to burst free or 'reactivate', recausing disease.

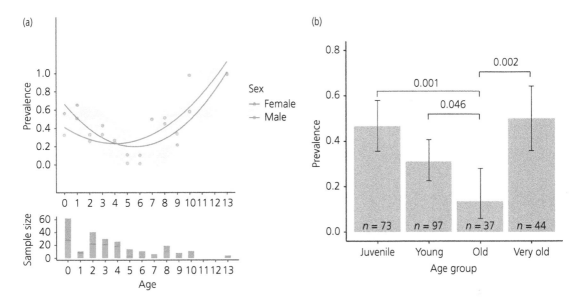

Figure 15.1 Prevalence of genital MusGHV-1 across different ages (a) and age groups (b) revealing statistically significant difference (from Tsai et al. 2021, with permission).

of swabs from juveniles (<2 years)) and 47.7% (21/44) of older badgers (>8 years), whereas rates were only 30.9% (30/97) in 2–5 year olds, and just 13.5% (5/37) amongst prime aged badgers aged 5–8 years.

Drilling deeper with multivariate models, consistent with emergent themes across badgerdom, we found that juveniles and adults exhibited key vulnerabilities to different sets of risk factors, likely reflecting whether virus shedding was due to primary infection or reactivation from latency (Figure 15.2). Other than the effect of ageing, in adults, reactivation rate can also be affected by season—virus shedding is more likely in summer than in spring and autumn. In addition, poor body condition, indicative of nutritional or physical stress, was a strong risk factor of MusGHV-1 reactivation. Unexpectedly, amongst juveniles, virus shedding was higher amongst females than males, and amongst those in the best (i.e. above average) body condition. This mirrors experimental results from mice afflicted with the MHV-4 variety of herpes for which genital reactivation occurs after primary infection only in females, whereas males acquire infection during sex with virus-shedding females (François et al. 2013). In the way that one thing leads to another, with sometimes unwelcome consequences, young male badgers in better body condition enter puberty earlier, and these young bucks are thereby exposed to a higher risk of venereal infection (Sugianto et al. 2019c). Interestingly, and consistent with our findings on fleas in

Chapter 8 (see Albery et al. 2020), living in social groups with more cubs (>30% of the membership are cubs) was a risk factor of MusGHV-1 reactivation for all badgers. Simon Sin had found a higher MusGHV-1 blood viral load in cubs, so perhaps members of groups swarming with cubs face a higher bombardment of their immune systems, leading to greater reactivation of MusGHV-1 amongst them (Sin et al. 2014).

Having uncovered the link between MusGHV-1 reactivation and physical stress, we wondered whether weather conditions (Figure 15.3), established in Chapters 12–14 as the most influential environmental factor for badger survival, might also impact herpesvirus reactivation in individual badgers, perhaps seeded as early as from the stage of prenatal immune system programming. Ming-shan and Christina then led us in the application of a technique called the 'sliding windows approach' (van de Pol and Cockburn 2011; Hindle et al. 2019).[14] Its purpose was to test which, if any, weather conditions experienced by mothers during delayed implantation and pregnancy

---

[14] The sliding window bioinformatic method involves a 'window' of specified length, {Len}, that is slid over the data, sample by sample; the statistic is then computed over the data in that window. The output for each input sample is the statistic over the window of the current sample and the {Len − 1} of previous samples.

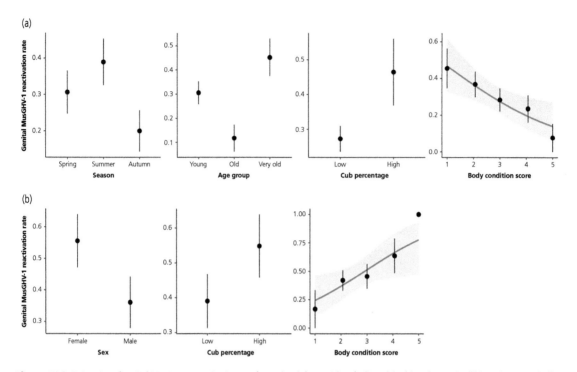

Figure 15.2 Estimation of genital MusGHV-1 reactivation rate for each risk factors identified in adults (a) and juveniles (b) based on mixed-effect models constructed using 251 genital swab samples taken from 151 badgers in different seasons (spring, summer, and autumn) in 2018 (from Tsai et al. 2021, with permission).

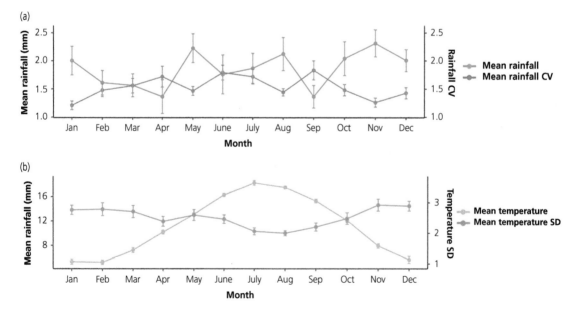

Figure 15.3 Monthly rainfall (a) and temperature (b) mean and variation from 2004 to 2018, with error bars indicating standard errors across years (from Tsai et al. 2022, with permission).

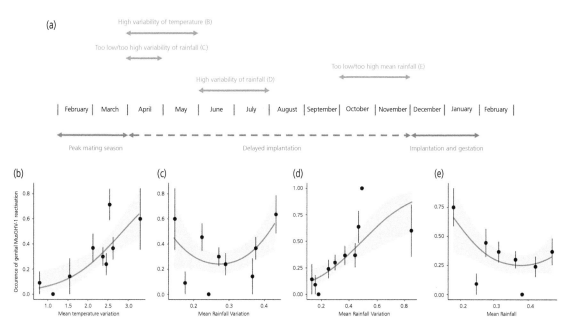

Figure 15.4 (a) Climate windows in the year before affected offspring subsequent lifetime susceptibility to genital MusGHV-1 infection. The blue arrows illustrate typical badger reproductive events, and orange arrows indicate the influential windows, with effect directions shown in (b), (c), (d), and (e) (from Tsai et al. 2022, with permission).

influence the incidence of MusGHV-1 in offspring later in life (Tsai et al. 2022).

A reward to doggedly long-term study is that an archaeological dig in our freezers delivered into our frost-bitten fingers 164 genital swabs collected from 95 identifiable individuals (50 males and 45 females) born between 2005 and 2016. With these we could observe the inter-annual effects of weather up to and including 2018. The result: again a rather similar 31.1% (51/164) of swabs tested positive for MusGHV-1 DNA (Figure 15.4).

Mothers' circumstances may affect the destinies of their unborn babies; for the badgers in our sample, of all the weather conditions they experienced during their lifespans, those most influential on their subsequent genital MusGHV-1 status all occurred when they were still ensconced in their mother's womb. Pre-birth neonatal effects were at play (Figure 15.4a), and once again they were both linear and quadratic (remember the 'Goldilocks zone' and porridge that is neither too hot nor too cold; Chapter 12). Most influential was temperature variability (not temperature mean) (see Chapter 12) in the late spring; this was so in the year prior to birth, however, when these individuals existed as *in utero* blastocysts, awaiting implantation (see Nouvellet et al. 2013). Similarly we once again

saw an effect of rainfall variability in the spring (April, quadratic, Figure 15.4c) and summer (June to July, linear, Figure 15.4c), also during the year preceding birth. Low, and, to a slightly lesser extent, high mean rainfall during late autumn (October to November) prior to birth date contributed a positive quadratic effect (Figure 15.4e).

Considering that badgers suffer lower body condition in harsh and frosty winters (Macdonald and Newman 2002; Chapter 12), we were surprised not to find any post-implantation effect of winter weather during gestation proper. This prompted us to speculate that foetal programming—the name given to the phenomenon where an event occurring during a critical stage of pregnancy may affect foetal phenotypes and persists in later life (Godfrey and Barker 2001)—might mainly occur in badgers during the pre-implantation period (i.e. before implantation in December). How might maternal stress during pregnancy influence embryonic development? Well, it is generally the case that placental mammals undergo some degree of phenotypic adaptation in response to environmental conditions experienced by their mothers during pregnancy (Bateson et al. 2014). This plasticity generally involves epigenetic changes (see Chapter 19) that increase the probability that the offspring's phenotype will be well matched to the environmental conditions that await it (Godfrey et al. 2007; Bateson and Gluckman 2012). For example, Lee and Zucker (1988) discovered that autumn-born vole pups (*Microtus pennsylvanius*) have much thicker coats than those born in spring—cued hormonally *in utero* as a result of the mother experiencing changes in day length—this is an example of a predictive adaptive response (PAR) of the sort introduced in Chapter 14 (Gluckman and Hanson 2004). Provided the environment doesn't change between the development of these adaptations and the time at which the individual deploys them to help it get through life, PARs offer selective fitness advantages. Another hypothesis is that phenotypes developing under good early-life conditions (with abundant resources) are typically larger, more fecund, and longer-lived compared with those developing in poor early-life conditions (with scarce resources), irrespective of the adult environmental conditions (Pigeon et al. 2019). Such blessed youngers enjoy the benefits of a metaphorical 'silver spoon'[15] (Grafen 1988; Lindström 1999; Monaghan 2008). Returning to the 95 badgers we used to test early-life weather conditions on herpesvirus reactivation, we found whose mothers had experienced more benign weather when 'pregnant' with them (during delayed implantation) turned out to be at lower risk of MusGHV-1 reactivation in later life—sipping from a silver spoon while they were blastocysts appeared to have benefited them. Although more frequent herpesvirus reactivation is considered disadvantageous given our previous findings on its negative link to female reproductive success (Tsai et al. 2020), higher reactivation rate may simply indicate that a badger is more sensitive to stress that can cause immunosuppression and herpesvirus replication (whether this mechanism induces disease or decreases survival is a topic for future research). In Chapter 14 we reported the link between early-life weather adversity on offspring outcomes: we could detect no effect on offspring performance of adverse weather conditions (deviation from long-term average of monthly rainfall and temperature) during either the 2 months (October and November) before implantation or the 2 months after emergence (April and May). However,

greater mismatch of weather conditions between the two periods is significantly associated with later reproduction in both males and females. Therefore, although more adverse early-life weather conditions can lead to higher occurrence of herpesvirus reactivation later in life, the outcome for the offspring may still be affected by the match, or mismatch, between the embryonic and subsequent environmental conditions.

It can be difficult to distinguish the burden of straightforward disadvantage from an adaptation prompted by a disadvantaged gestation that equips an individual to poor conditions; indeed, the generality is that PAR has not often been demonstrated in empirical studies, probably because of the irreversible constraints on the health and development in later life of those blighted by poor nutrition in early life (Rickard and Lummaa 2007; Hayward et al. 2013). As for other mammals, amongst humans too, poor maternal nutrition can impair neonatal birth weight and/or cause non-adaptive disruptions to development, potentially compromising the future success of the individual (Bateson et al. 2004; Gluckman et al. 2005a; Ghalambor et al. 2007). The outcome can be children with inferior immune systems, more prone to diseases, especially when there is a mismatch between conditions during development and those experienced later in life (Godfrey et al. 2007). In human medicine this is termed the Developmental Origins of Health and Disease (DOHaD) paradigm. Barker et al. (1993, 1997) were the first to formalize that undernutrition during gestation reprogrammes the relationship between glucose and insulin, and between growth hormone and IGF (insulin-like growth factor).

The adaptive capacity of nature leaves no stone of opportunity unturned: specific weather conditions during foetal development later come to influence the adult life history of that individual (consider pace of life syndromes in Chapter 14 and the fraying of telomeres in Chapter 18). Are these findings widespread amongst the 130 mammal species experiencing some degree of delayed implantation? Nobody knows, but Nakanishi et al. (2012) found *de novo* methylation[16] during up to 4 days of experimental diapause in mice. In mink, another mustelid species

---

[15] An English idiom, meaning to be born into affluence or under lucky auspices, and arising from wealthy aristocrats feeding babies off silver (not wooden) spoons; there also being a tradition for wealthy godparents to give a silver spoon to their godchildren at christening ceremonies. The expression 'born with a *silver spoon* in his mouth' first appears in print in Cervantes' *Don Quixote* (published in 1605). See also Chapter 14.

[16] DNA methylation is a common epigenetic signal that functions to 'lock' expression of a gene. In humans, methylation signals on the genome are lost immediately after fertilization. However, remethylation occurs from the blastocyst stage until birth and during this time environmental cues can affect the methylation process with the result that the gene is expressed differently. The importance of epigenetic selection is discussed in Chapter 19.

exhibiting delayed implantation, females that eat insufficient protein during pregnancy produce offspring with lower birth weight and lower insulin- and leptin-related mRNA expression (Matthiesen et al. 2010); and, in the case of their sons, lower protein oxidation. However, these inadequacies caused by the mother's diet can be remedied if the youngsters have an adequate post-weaning diet (Vesterdorf et al. 2014).

Everything we have so far discovered about badger herpesvirus indicates long-term co-evolution between host and pathogen. MusGHV-1 spreads widely in badgers but seems not to make them very unwell. This seeming evolutionary harmony requires continuing balance in an arms race between host and pathogens as both accumulate mutations, for better or worse, in relevant genes (e.g. the MHC genes for the host, or virulence-related genes in the pathogens such as the spike protein). Once again, the badgers of Wytham Woods find themselves in the court of the Red Queen (Clay and Kover 1996).

One way to reveal the process of co-evolution is to look for fixed mutations in virus genomes sampled from the host population. Ming-shan therefore led us in sequencing part of the DNA polymerase gene of MusGHV-1 (694 base pairs long; Tsai et al. 2021), which she excavated from genital, oral and rectal swabs, and blood samples) scooped from 66 individuals in 2018 (Tsai et al. submitted). Intriguingly, she found not one but two variants of herpes virus, differentiated by mutations found at nine nucleotide positions. One group of sequences occurred in 89.4% (n = 59) of the samples and the other in 16.7% (n = 11), so we gave them the imaginative names of common and novel variants, respectively. Four hapless badgers had both. Of course the pressing question is what distinguishes these two viral genotypes in terms of the badgers' biology. First, we found that badgers infected with the common variant are widespread across the woods, whereas infection with the novel variant is clustered in only three social groups: Radbrook Common, Chalet, and Chalet Outlier (Figure 15.5). Those three groups, each bordering the next, had the highest prevalence of genital MusGHV-1 throughout the woods. Wondering whether the two viral variants differed in their virulence, Ming-shan tested the viral load in each badger's blood stream.[17] We found juveniles generally had higher MusGHV-1 blood viral loads than

adults, and that samples collected in summer generally had higher blood viral load than at other times of year. Most excitingly, there were two levels of virulence: badgers infected with the novel variant had higher blood viral loads than those infected with the common variant (Tsai et al. 2022).

Nowadays there are libraries of viral sequences, one of which is the National Center for Biotechnology Information (NCBI) GenBank. Using a program called BLAST it is possible to reference a given genotype against the library, and in this way we confirmed that the common genotype is identical to the previously banked MusGHV-1 sequence,[18] originally isolated from a badger in Cornwall but also recorded in Irish badgers (Tsai et al. 2020). What selective pressures have favoured the emergence of the more virulent viral strain in the Wytham badger population? We don't know, yet. That said, later in this chapter we will report that Simon Sin led us to discover that badgers with particular MHC genotypes have different blood viral loads of MusGHV-1 (Sin et al. 2014), which indicates that the hosts differed in their resistance to this pathogen: this situation could provide grist to the mill of natural selection, favouring the emergence of a new strain of the virus. It could be revealing to link the host MHC genotype to the MusGHV-1 genotype with which they are infected. In particular, do members of the social groups with novel MusGHV-1 genotype infection have a different MHC allele frequency from social groups infected with the MusGHV-1 common genotype? A question for the future.

MusGHV-1, much like other herpes strains, appears to have achieved co-evolutionary success, insofar as asymptomatic infection is the general condition of immune-competent individuals. However, remembering that physical stress can facilitate MusGHV-1 shedding and onward transmission, we wonder whether chronic stressors, such as from climate change, could disrupt the current host–pathogen balance. To reveal consequences such as the herpes-related cancer that has been reported in European mink and other wild mammals (Nicolas de Francisco et al. 2020; Gulland et al. 2020), or sterility, will require monitoring of infection dynamics, disease development, and emergence of more virulent strains in badgers as the unfolding impact of MusGHV-1 on the badgers of Wytham Woods emerges.

---

[17] We assessed viral load by measuring relative quantity of MusGHV-1 genome copies by real-time PCR from 40 samples collected from 24 individuals at different locations known to be infected with a given genotype.

[18] GenBank accession number AF275657, Irish badgers MT332102.

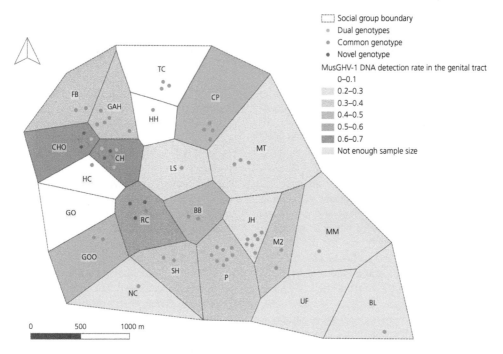

Figure 15.5 Map of badger social group distribution, genital tract MusGHV-1 occurrence, and MusGHV-1 genotype distribution in Wytham Wood in 2018. The name of each social group is highlighted with yellow (see Map 9.2). The grey areas signify sample sizes of more than three. The points of MusGHV-1 genotype distribution are approximate, located around the setts at which the host badgers were sampled (from Tsai et al. 2022, with permission).

Rabies

Before leaving viruses, a brief note on rabies. Through the 1980s, David was heavily engaged in researching the epizootiology of rabies amongst wildlife hosts in Europe (Macdonald 1980b; Macdonald and Bacon 1982; Macdonald and Voigt 1985; Macdonald 1988)—a trail that also led to jackal rabies in Africa (Rhodes et al. 1998; Macdonald 1993; Loveridge and Macdonald 2001;). Badgers were linked to this work in two ways: first, because they were collateral damage in attempts to eradicate rabies; and second, because it was the failure of the policy of killing foxes in order to control rabies, on the one hand, contrasted with the success of a campaign of vaccinating the vulpine population against the virus, on the other, that planted the seed that became the perturbation hypothesis as an explanation for the failure of killing badgers to control bTB in cattle—the topic of Chapter 16. Although rabies afflicted badgers prior to the effective elimination of this disease in western Europe, other species, such as foxes, raccoon dogs, and stone martens, were always more susceptible and important to regional disease dynamics (Wandeler et al. 1974). This, however, did not save badgers from sett gassing, which decimated

Continental populations (Wandeler et al. 1974). Today, rabies continues to be a problem in the badger's Asian cousin, *Meles leucurus*, where, in collaboration with colleagues in China, we have reported (Zhou et al. 2017; Chapter 19) on the death of seven people bitten by sluggish rabid badgers they attempted to capture (two of these badgers were then eaten) during an epizootic in Zhejiang Province from 2002 to 2004.

Protozoa

Extending our search beyond viruses, and mindful of the profound role of *Eimeria melis* in cub mortality and subsequent retardation of development amongst survivors (Chapter 2), it seemed important to screen for protozoan pathogens more extensively; indeed, what of other coccidians that might be present?

Most coccidians are species-specific, but a close relative of *Eimeria*, *Toxoplasma gondii*, is an exception. *Toxoplasma gondii* can only complete the sexual reproduction phase of its life cycle in cats (Robert-Gangneux 2014), but can nevertheless cause toxoplasmosis across mammal species, including people (22–33% of adults in the UK show sero-prevalence, which increases with age (and thus exposure) and, amongst other things, risks congenital foetal malformation; Joynson 1992). *Toxoplasma gondii* was a familiar topic for us; others in the WildCRU team spent years researching its prevalence and consequences in rats (Macdonald et al. 2015a). Remarkably, infection led to parasite-altered behaviour, where kamikaze-like rats wilfully exposed themselves to cats, which obliged the parasite by eating the infected rodents; thereby enabling the parasite to complete its life cycle (Webster et al. 1994; Berdoy et al. 1995b, 2000). We expected to find *T. gondii* in badgers and, with the help of our Persian parasitologist colleague, Professor Ali Anwar, we found antibodies in 63 of 90 badgers tested; we could detect, however, neither symptoms nor behavioural corollaries of infection (Anwar et al. 2006).

This finding prompted us to investigate whether badger blood teemed with other protozoan parasites, and uncovered another insidious passenger in the bloodstream, *Trypanosoma pestanai*—a relative of *Trypanosoma bruciei*, which, with the complicity of tsetse flies, has influenced the course of history by causing sleeping sickness in Africa, and Chagas disease in South America (*T. cruzi*—vectored by blood-sucking triatomine bugs). Of 718 blood smears collected from 263 badgers (1989–1991) then stained and examined under a microscope (Figures 15.6 and 15.7), 33 (4.6%) tested positive for this host-specific protozoan, representing 20 (7.7%) individuals (Macdonald et al. 1999. Low prevalence weakened our statistical analyses, but active blood infection presented almost exclusively in badgers under 3 years old, peaking at 13% amongst 23 cubs caught in 1989. In this sample the prevalence was unaffected by whether the badger group was large or small. Following up on this, working with colleagues at the Royal Veterinary College, but this time using the more rigorous technique of genetic (18S rRNA) PCR primers to detect *T. pestanai*, we found that 60 of 207 (29%) individuals tested positive, with significantly higher prevalence in males (42%) than females (27%). As with coccidiosis, there was a tendency for juveniles to be more susceptible to infection (40% prevalence) than either young adults (1–5 years old, 35% prevalence) or mature adults (>5 years old, 16% prevalence) (Lizundia et al. 2011).

Although we did not detect any signs of illness amongst these badgers, we were curious to know what vector spread *T. pestanai* amongst them. Naturally, our attention returned to that itchy infestation of host-specific fleas that badgers harbour (Chapter 8). Sure enough fleas collected from these same badgers also tested PCR-positive for *T. pestanai*. Furthermore, we could actually see the protozoans swimming about in the fleas' Giemsa-stained hindguts. There was, however, no evidence for an association between a badger's flea burden and presence of *T. pestanai* in its blood—reinforcing our suspicion from Chapter 8 that the immediate motivation for badger grooming is itchiness, with little impact on disease risk.

This same initial blood screening also led us to be the first to discover another blood parasite in badgers, an apicomplexan called *Babesia missiroli* (a more familiar member of the family might be *Plasmodium* spp., which causes malaria; Macdonald et al. 1999). This parasitaemia characteristically involved *Babesia* embedded within red blood cells (i.e. intra-erythrocytic infection). A total of 430 of 718 (59.9%) blood samples screened positive, representing 203 of 263 (77.2%) individuals. Badgers aged 2 years old or younger had the highest rates of infection (70–90% between years amongst cubs vs 35% in 4 year olds) and 71.7% of females were infected (1991) compared with 53.3% of males (a departure from pattern of higher incidence in males that had typified the other pathogens we've mentioned). Again, the size of the badger social group appeared to have no effect; however, it was nonetheless the case that badgers testing positive weighed less and were in poorer condition than those uninfected.

Yet again we could detect no pathophysiology; this was surprising insofar as babesia—usually tick-borne—has nasty symptoms in puppies and foals (of horses, donkeys, and zebras; Chhabra et al. 2012), including 'piroplasmosis' (haemolysis—that is, bursting—of red blood cells, which leads to acute anaemia) with often fatal levels of leukopaenia (lack of white blood cells) (Fukumoto et al. 2005).[19,20] Intriguingly though, there was evidence that infection with *B. missirolii* affected the probability of recapturing a badger in the year following a positive test. Per year, a higher proportion of badgers that had tested positive in the previous year were trapped compared with those that had tested negative. This was statistically significant in 1991, when 56 of the 80 adult badgers (70.0%) known to be infected in 1990 were trapped; only 7 of the 16 badgers uninfected in 1990 (43.8%) were trapped in 1991. Putting aside the frivolous thought that babesia is good for badgers, this greater likelihood of retrapping infected badgers the following year could not

---

[19,] It's hard to distinguish babesia from thieleria, so both used to be referred to as 'piroplasma'. Equine pyroplasmosis is a killer of racehorses in the sub-tropical United States. It also kills feral dogs in Italy.

[20] It is possible that cubs die neonatally of piroplasmosis before we can sample them.

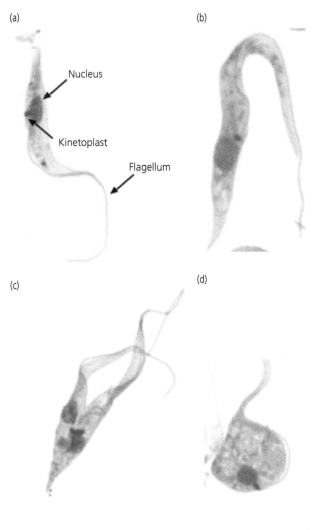

Figure 15.6 Trypanosomes in the blood of Wytham's badgers. (A) Epimastigote-like long slender form. (B) Epimastigote-like swollen form. (C) Dividing epimastigote. (D) Large pear-shaped form ('degenerative' form) (from Lizundia et al. 2011, with permission).

10 µm

be attributed to higher survival amongst the infecteds because infection in 1989, 1990, and 1991 had no effect on the likelihood of a badger being alive during 1995. This trappability effect was curiously reminiscent of *T. gondii* making rats less cautious.

Babesia parasites are invariably transmitted via ticks (analogous to Lyme disease; Spielman et al. 1985; Scharlemann et al. 2008), as well as through transplacental routes (Joseph et al. 2012). As we ran our fingers through the fur of Wytham's badgers, it was clear that ticks are surprisingly rare, detected in just 8% of examinations. Those we did encounter were predominantly of the hedgehog variety (*Ixodes hexagonus*), usually nestled in the groin region, although, very occasionally (i.e. in only 1–2 badgers per 100 examined),

we found sheep (*I. ricinus*) and dog (*I. canisuga*) ticks in the animals' ears (see Sin et al. 2014). The badgers of Wytham Woods have orders of magnitude fewer ticks than we have observed when handling badgers elsewhere; for instance, compared with the lower-density North Nibley (Gloucestershire) study site where we studied bTB (after considering viruses and protozoa (and their agents) here, we will postpone our consideration of bacteria, strikingly embodied by bTB, to Chapter 16). Similarly, while the badger-specific mite, *Trichodectes melis*, is common in Wytham, we found only mild cases of acariasis, or 'mite rash' (Cox et al. 1999; Chapter 8). A different sort of mite (predominantly *Sarcoptes scabiei canis*) that notoriously causes mange, or scabies in foxes (Balestrieri et al. 2006), can

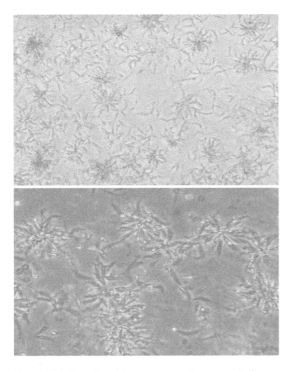

Figure 15.7 Formation of Trypanosome rosettes as a result of incomplete separation of daughter cells observed by inverted phase contrast microscopy at different magnifications (upper panel, ×20; lower panel, ×40) (from Lizundia et al. 2011, with permission).

be far more grave than a bad itch—in the case of the rare Mednyi Island Arctic foxes it threatened their extinction (Goltsman et al. 1996). These mites were rare in Wytham, as were *Otodectes cynotis* mites (the cause of otodectic mange), although we did find rare instances (<1/500 examinations) around the ears of badgers that were already in debilitated condition.

With this somewhat eclectic *dramatis personae* of badger pathogens assembled on the stage (with bTB ahead in Chapter 16), we now consider how these, and the various other stressors we have mentioned in the lives of badgers, from inclement weather to contesting resources, affect the immune system; thereby determining a badgers' capacity to cope.

Coping with disease and stresses on the body

Adult badgers appear to cope well with microparasites in their blood, although cubs cope less well with those in their bowels. How do badgers manage 'dis-ease', in the broadest sense of the loss of 'ease' to infection? Furthermore, does psychological stress, for example perhaps provoked by living in a particularly

dense population, compound the risks to good health amongst Wytham's badgers? Notably in this regard, badgers, like all living things, are not only vulnerable to pathogenic attack, but also to damage to tissues, organs, and DNA arising through 'oxidative stress'. What is oxidative stress? This is a complicated topic, highly relevant to badgers, that reaches surprisingly far back into the evolution of life. Box 15.2 summarizes it.

So, Box 15.2 describes how leakage of reactive oxygen species (ROS, a.k.a. free radicals) along the electron transport chain causes cellular imbalance, which in excess is a bad thing known as oxidative stress. For the mechanically minded, think of your car engine, with its carefully balanced ratio of air to fuel for optimum combustion; if this mixture gets out of balance, the engine splutters. Similarly if mitochondria are stressed (essentially overtaxed) they too produce an imbalance of all kinds of nasty emissions,[21] comprising not only ROS, but also reactive nitrogen species (RNS), and carbon-centred and sulfur-centred radicals (Salim 2014). This is as unwelcome to badgers as it is to you.

Challenged by the risks that these emissions cause to vertebrate cells, natural selection has equipped organisms with an antioxidant response to mitigate any potentially abnormal increase in ROS (think of the antioxidant response as a catalytic converter if you'll indulge the car engine analogy a little further). Some level of ROS in the cell is beneficial (as well as attacking bacteria, they serve as secondary cell messengers, especially regulating cardiac and vascular function). But if too many ROS persist without being 'quenched' by antioxidants (which should maintain redox (reduction–oxidation) balance), they can damage the individual's DNA[22] (and its proteins, lipids, and sugars). Subsequent modifications to amino acids ultimately render proteins non-functional, diminishing cellular detoxification and repair/regeneration capacity (this is especially bad for brain function, and associated with depression and anxiety). Indeed, a surplus of ROS is not only dangerous from the point of view of one's day-to-day functionality, but also in terms of the broader tendency for the wear and tear caused by ROS[23] to accumulate with age underlies the process of senescence, prompting the 'free radical theory of aging'

---

[21] Free radicals are atoms or groups of atoms with an unpaired number of electrons, which are highly reactive substances that can result in chain reactions, with each step forming a free radical.

[22] Disturbing the maintenance of normal adenine and pyridine nucleotide status.

[23] It's not that ROS worsen with age (although they do—linked to the 'inflammation–ageing' hypothesis), it's that—to extend the motoring metaphor—they choke up your inlet valves, stopping fuel from getting into your cylinders.

Box 15.2 What is oxidative stress?

Life runs on electricity, and oxygen is toxic! To elaborate—all key biological processes essential to life, such as photosynthesis, (cellular-) respiration, and detoxification, depend on energy conversion (the cellular mechanics of paying the body's household energy bills that we explored in Chapter 14). This is achieved by the transfer of electrons between charged ions[24] (i.e. electricity). To jog the memories of those whose last encounter with biochemistry was in school: in nature the molecular currency of intracellular energy transfer is a molecule called adenosine triphosphate (ATP). ATP is able to store and transport chemical energy within cells. Evolution has come up with various routes to synthesize ATP, but basically all involve stripping away electrons to yield positively charged hydrogen ions, which, in a reaction catalysed by ATP synthase, stack an extra inorganic phosphorus atom (Pi) onto an adenosine diphosphate (ADP) precursor to create ATP. Recalling that classroom equation: $ADP + P_i + H^+_{out} \rightleftharpoons ATP + H_2O + H$. Subsequently, when the energy stored in this terminal phosphate linkage in ATP is broken, using water, 30.5 kJ mol^{-1} of energy is released. It may not sound much, but it is worth having; without it, the lights go out.

To appreciate how this is relevant to stress in badgers, go back 3.5 billion years into deep palaeontological time. Then, our earliest unicellular ancestors, termed 'prokaryotes' (i.e. lacking a cell nucleus) lived in an anoxic (oxygen-deficient) environment and experimented with shunting electrons down through a reaction to create two ATP energy molecules.[25] They achieved this using charged molecules such as sulfate (SO_4^{2-}), nitrate (NO_3^-), or sulfur (S) as the terminal electron acceptor (this process is nonetheless termed 'respiration'—not to be confused with breathing). These 'chemoautotrophs' got life started (and still exist today as nitrogen-fixing cyanobacteria, but their methods of energy conversion were, so to speak, steam driven). Subsequently, around 2.3 billion years ago, some innovative 'photoautotrophic' prokaryotic cells evolved a new system called 'photosynthesis', which uses solar energy to split water molecules (photolysis), and then combine the resultant liberated hydrogen atoms with carbon dioxide (CO_2) to produce a hydrocarbon 'glucose' molecule and storing, for future use when metabolized, 12 ATP (of which 6 ATP are used to regenerate catalysts essential to this chemical process). This process uses a nattily named molecule, nicotinamide adenine dinucleotide phosphate—NADP$^+$ for short—as the final electron acceptor, yielding free oxygen as a waste product.

So what were the consequences of all of this oxygen waste product gradually building up in the Earth's atmosphere, where formerly there had been none (touching $c.\,5\%$ by 2.5 billion years before present (Bybp), vs 21% today)?

Well, as any wine maker knows when using anaerobic yeast to ferment alcohol from sugar, oxygen can be a bad thing. In the absence of suitable enzymes to stop it happening, oxygen can be reduced to the cheerily named but highly destructive (as we shall see later) 'free radical' superoxide ion:[26] $O_2^{e-} \longrightarrow O_2^-$.

The point is, oxygen started out as a pollutant,[27] but evolution, always the opportunist, came up with organisms that could put it to good use: the aerobic heterotrophs able to use it in a further respiratory process. This new exploitation of oxygen produces 38 ATP over three steps: glycolysis, the Krebs[28] cycle, and the electron transport chain, with oxygen sitting at the end to pick up positively charged hydrogen atoms to form water (the schoolroom equation was: $C_6H_{12}O_6 + 6O_2 \rightarrow 6CO_2 + 6H_2O$, i.e. Glucose + Oxygen -> Carbon dioxide + Water). Now, to provide a very brief synthesis of the seminal work of biologist Lynn Margulis in the 1960s, 'eukaryotic' cells (i.e. those with a nucleus and other internal organelles) evolved through a process of 'endosymbiosis' starting around 2 billion years ago. Some species of prokaryotes that could respire oxygen got 'swallowed up' and evolved into cellular engines, termed mitochondria, while other (plant) cells also swallowed up photosynthetic organelles, termed chloroplasts. These are the ancestors from which all current eukaryotic cells—and thus all multicellular readers of this book—ultimately evolved.

This brief history of life (and purists, please forgive our simplifications) illustrates that the oxygen respiration we take for granted is actually a highly evolved state. Half of the entire history of life on Earth had passed before oxygen respiration was invented. This history provides an anchor to understanding oxidative stress—which turns out to be

---

[24] A theory of electron transfer tracing back to Rudolph A. Marcus in 1956.

[25] $C_6H_{12}O_6 = 2C_2H_5OH + 2CO_2 + 2ATP$.

[26] Superoxide anion reacts with hydrogen peroxide generating free hydroxyl radical (most potent biological oxidant) that can attack virtually any organic substance in the cell. $O_2^- + H_2O_2 \rightarrow OH^- + OH^. + O_2$ (Haber–Weiss reaction).

[27] In passing, oxygen was also bad for photosynthesis (inhibiting certain enzyme reactions), termed the 'Warburg effect', while chloroplasts can reduce oxygen to form dangerous peroxide (H_2O_2) ions, termed the Mehler reaction (Bjokman 1966).The Warburg effect is also a diagnostic tool in oncology, where tumours produce energy predominantly through glycolysis, due to oxygen supply limitation, leading to lactic acid fermentation in cytosol.

[28] Hans Krebs, in this case the father and discoverer of what is also known as the citric acid cycle (CAC) or the tricarboxylic acid cycle (winning the 1953 Nobel Prize for Medicine); he was also father of John Krebs, originator of the Krebs Trial to test the role of badgers in the transmission of bTB to badgers (Chapter 16), providing a genealogical link between stress and badger disease.

Box 15.2 *Continued*

highly relevant to badgers, as it is to humans. The key to anchoring this history to an understanding of oxidative stress is the fact that the process of oxidative phosphorylation involves five enzymatic exchanges taking place along the respiratory (electron transport) chain to yield 34 of the 38 ATP (the other 4 coming from glycolysis and the Krebs cycle), where each exchange involves dumping electrons that are ultimately accepted by terminal oxygen. This process is not perfect, and by-products 'leak' out of this electron transport chain, liberating into the cell charged ROS (a species only in the chemical derivative sense). These ROS are a double-edged sword, necessary for cell signalling and immune defence, but also dangerous if present in too large quantities. These ROS are counterbalanced by antioxidant systems, which essentially 'mop up' ROS to avoid them causing damage to tissues. If this process becomes chemically unbalanced—and it does if the individual, including a badger, is stressed—there will be an excessive amount of ROS and not enough mopping up capacity, which will eventually cause damage to tissues. Do not be beguiled by their liberal-sounding alternative name, 'free radicals'—these ROS are powerful cellular players whose overproduction has relevance to impaired health for badgers, and people alike, and is something that we will be discussing shortly (with the help of Box 15.3 on antioxidant capacity).

(FRTA; Hartman 1956). This adds further importance to understanding ROS and stress (see also Chapter 18).

So what does cellular stress resulting in a redox imbalance and the build-up of ROS mean for badgers? Part of the answer lies in white blood cells, more properly called 'leukocytes', particular variants of phagocytic leukocytes called neutrophils, monocytes, and macrophages, which put the incinerating capacity of the ROS they produce to work positively by cauterizing infections to limit their spread through the body (Paiva and Bozza 2014).[29] This process is vital to the body's first line of 'innate' immune defence.[30] By analogy, phagocytic leukocytes use their ROS like lasers to zap pathogens, and thereby provide a first line of immune protection. But if they overdo it, without sufficient antioxidant fire extinguisher on hand to prevent oxidative damage from friendly fire, the result is tissue damage.

If danger signals, such as those generated by bacterial wall compounds,[31] stimulate these phagocytic white blood cells, the latter instigate and regulate the body's so-called inflammatory response. Specifically, these leukocytes engulf 'phagocytose' invading pathogens and use nicotinamide adenine dinucleotide phosphate (NADPH) oxidase to reduce O_2^- to an oxygen free radical and then H_2O_2, to destroy invading culprits, and thus protect against infection.[32] This functional response, termed an oxidative burst, contributes to host defence but it can also result in collateral damage of host tissues[33] (Chen and Yunger 2012). The critical key link here is that in

[32] In more detail, how do the different types of leukocytes destroy invading bacteria? The neutrophils and monocytes use myeloperoxidase to further combine H_2O_2 with Cl^- to produce hypochlorite, which plays a role in destroying bacteria. Similarly, macrophages engulf (phagocytose) bacteria and use NADH-dependent phagocytic oxidase to produce superoxide, which dismutes to form H_2O_2 to zap the ensnared pathogen. Macrophages also use reactive free radical (unpaired electron) nitrogen species (RNS), mainly nitric oxide (NO) and nitrogen dioxide to produce bacteria-busting hydroxyl radicals and nitrogen dioxide radicals—a defence system to which we will return in Chapter 16, in the context of badgers' bTB defences. In addition to oxidation, macrophages use vesicle-mediated delivery of various antimicrobial effectors, which include proteases, antimicrobial peptides, and lysozyme, further to ensure pathogen destruction.

[33] To explain that collateral damage: Covid-19 is an unhappily familiar example (Ragab et al. 2020). In response to Covid-19 virus, pattern recognition receptors (PRRs) on different leukocyte variants recognize pathogen-associated molecular patterns (PAMPs) on the virus surface (the corona 'spikes'; Leth-Larsen 2007). As PRRs bind to PAMPs this triggers the production of pro-inflammatory cytokines, intended to combat further viral invasion, but causing the now infamous 'cytokine storm' in sensitive patients. Cytokines are produced by several immune cells including innate macrophages, dendritic cells, natural killer cells, and the adaptive T and B lymphocytes. This increase in cytokines causes the influx of various immune cells such as macrophages, neutrophils, and T cells into the site of infection with destructive effects on tissue resulting from destabilization of endothelial cell-to-cell interactions, damage of vascular barrier, capillary damage, diffuse alveolar damage, multi-organ failure, and in some cases death (see Tay et al. 2020).

[29] These cells contain lysosome organelles, packed with enzymes, or 'lysozymes' formed through a particular redox chain variant tailored to result in acidification along with other ROS-based enzymes throughout the cytoplasm (in all animal cells), such as superoxide dismutase (SOD) and catalase (CAT), and malondialdehyde (MDA) (Nohl and Gille 2005).

[30] Antibodies come later and are proteins synthesized by B-lymphocytes that bind to pathogens via the immune cascade providing 'acquired immunity' (see Chapter 16).

[31] lipopolysaccharide (LPS) and lipopeptides (or, in the case of bTB exposure, Toll-like receptor 4 or 2; see Chapter 16).

addition to invading pathogens triggering the immune systems, so too psychological stress can trick it into activity, with damaging consequences. So, we were eager to understand the collateral damage that could be caused to badgers by ROS released in response either to pathogens or to other—for example, sociological—causes of cellular stress. We were aware, also, that the assault could be met by a defensive oxidative burst measured as antioxidant capacity. Therefore, we set about measuring both the problem (namely ROS) and the solution (the capacity to neutralize the problem—AntiOXidant capacity, termed, AOX) and the resulting damage (oxidative damage) in badgers of different ages, stages, and under various circumstances. Our findings follow, but, to understand more fully how antioxidant capacity works, see Box 15.3.

Box 15.3 Antioxidant capacity

'Too much of a good thing', as Shakespeare noted in *As You Like It* (1600), can cause harm. While a burst of ROS can incinerate invading microbes, and ROS are essential for cell signalling, the persistent or repeated firing of this defence can damage bodily tissues and DNA (Kehrer 1993). It is not that oxidative stress kills the badger (or you), but rather that the resulting oxidative damage leads to a variety of diseases, from neuro-degenerative disorders to cancers and telomere shortening (Uttara et al. 2009; Chapter 18). So how does the body deal with this problem? Excess ROS are counteracted by the production of antioxidant molecules and enzymes that 'quench' or, metaphorically, 'mop up' excessive ROS (McGraw et al. 2010). Mammals utilize two sources of antioxidants, which can either be synthesized endogenously in the body in the form of molecules or enzymes[34] or acquired exogenously via direct consumption (as is the case for antioxidant vitamins C and E, carotenoids, and polyphenols).

Species, and indeed individuals, vary in their capacities to produce antioxidants (Yu 1994), bearing in mind that antioxidant systems are costly to implement and maintain (Selman et al. 2012). Therefore, in evolutionary terms, investment in these defences is often traded off against other developmental traits such as growth (McKinney and MacNamara 1991; Khan and Black 2003; Fontagné et al. 2008; Chapter 2).

---

[34] The five main examples being superoxide dismutase (aka SOD), alpha lipoic acid (ALA), coenzyme Q10 (CoQ10), catalase, and glutathione peroxidase (Gpx).

This led us to ask if badger antioxidant capacity might be inhibited under the periods of food shortage that inevitably occur in nature. Our first breakthrough was the discovery that badgers faced with food shortage did indeed exhibit impaired plasma antioxidant capacity, which we measured initially rather crudely in units of vitamin E analogue (VEA) equivalents (Montes et al. 2011). Specifically, from a sample of badgers studied in 2002, we found (Figure 15.8) that emaciated badgers of all ages, and equally in both sexes, mounted significantly lower antioxidant responses than their plump contemporaries. *Whosoever hath not, from him shall be taken away* (Matthew 13:12, the Matthew effect), and so, piled on the disadvantage of emaciation, these starving badgers are at further risk as they suffer oxidative damage caused by unquenched ROS when their cellular system reacts to stress.

Next we turned to more subtle, but more general, impacts on the immune system, as seasonal lows and highs in food availability interacted with variation in body condition (Chapter 14). Season did affect the badgers' antioxidant capacity. Seasonality, linked to the annual cycle in body fat, had the effect that corpulent badgers in autumn generally exhibited greater plasma antioxidant capacity than they did when slimmer in summer.

These early eco-physiological findings prompted a flurry of more intricate questions, starting with how does the antioxidant capacity of cubs develop as they mature? Insofar as producing antioxidants is energetically expensive to the body (remember the trade-offs posited in Box 15.2), we might expect that it is a capacity that takes a while to develop at a stage of life when a priority investment is surely in somatic growth (Chapter 2). This expectation was, initially, turned on its head when, in 2011, new graduate student Kirstin Bilham (now a laboratory manager at Oxford's Wellcome Centre for Human Genetics working on nation-wide Covid antibody levels) surprised us with the discovery that, in that year, the mean antioxidant capacity of 16-week-old cubs was already equivalent to that of prime-age adults (1–5 years old) (Bilham et al. 2013). Furthermore, these cubs exhibited significantly higher mean total plasma antioxidant capacity than did elderly individuals aged ≥6 years.

Was this evidence of a developmental trade-off, in which cubs investing in antioxidant defences could cope better with ROS arising from catabolic stress resulting, in turn, from a meagre food supply? Intrigued, we gathered another 568 blood samples (422 adult and 146 cub samples) from the 280 individuals (of which 102 were cubs) during spring, summer,

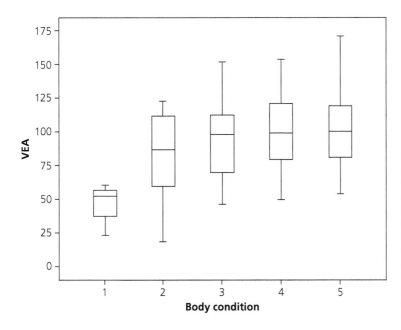

Figure 15.8 Vitamin E analogue equivalent units (VEA units) in badgers (*Meles meles*) of different body condition (body condition 1 *n* = 3; body condition 2 *n* = 17; body condition 3 *n* = 19; body condition 4 *n* = 13; body condition 5 *n* = 18). The bar shows the average, the top and bottom of the box show the 75th and 25th percentiles, respectively, and the top and bottom whiskers show the largest and smallest value that is not an outlier, or an extreme score (from Montes et al. 2011, with permission).

and autumn 2012–2014; analysing these was a logistical challenge for Kirstin because the tests have to be performed immediately on freshly collected blood, requiring her to spend long nights in the company of centrifuges after long days capturing badgers in the field. Further, in addition to VEA, we tested these samples for a wider (and subtler) array of oxidative stress biomarkers.[35]

Curiously, the results of this second study at first seemed to contradict those from 2011. In 2012–2014 cubs in early spring had *lower* antioxidant capacity than adults, and during these years cubs suffered significantly *greater* oxidative damage (versus AOX capacity) than adults (this was true for both prime (1–5 years), and elderly badgers (6+ years)). Furthermore, cubs with poorer body condition, and those that died prior to adulthood, tended to exhibit worse cellular damage, as revealed by higher levels of lipid peroxidation, a biomarker of cellular damage. In short, as we had expected, but not found in 2011, in these 2012–2014

measures of oxidative stress, and its consequences, we could see the burdensome physiological costs incurred by cubs in battling with the environment.

Torn by the conflicting evidence from the two study periods, we reconsidered the question of whether individual badgers actually do trade-off between investing in the immune system or in growth. Did cubs that had invested more heavily in antioxidant defences and survived (in this second larger sample) pay a cost? Yes: they grew to have shorter adult body lengths[36] than their contemporaries that chanced their luck with oxidative stress and got away with it.

Badger cubs are, naturally, substantially smaller than adults, and this has consequences. Metabolic rate is scaled to mass, fundamentally because of the surface area to volume law (explained in Chapter 14), and so due to their size alone cubs, at 1.7–3.0 kg in early spring, will have faster metabolisms per unit mass compared with adults; thereby generating more ROS than adults per unit weight (McClune et al. 2015). Moreover, growth rate and the production of growth hormones, which call for even greater metabolic activity, are generally linked to higher ROS production (Holzenberger et al. 2003), potentially worsening oxidative damage (e.g. Alonso-Alvarez et al. 2007).

[35] Amongst these was a generic measure, total antioxidant capacity (TAC, total non-enzymatic antioxidant capacity. These are usually the exogenous ingested ones: *non-enzymatic antioxidants* are vitamin C, vitamin E, plant polyphenol, carotenoids, and glutathione. Enzymatic ones = superoxide dismutase, peroxidase and catalase.); peroxidase enzymatic reactivity (PER) to provide a measure of enzymatic antioxidant capacity; red blood cell (RBC) half-life, which measures resistance to oxidative stress as the capacity of cell walls to respond to an oxidative challenge; and lipid peroxidation (LP), which indicates oxidative damage (OD) to tissues.

[36] Four centimetres shorter than those investing less in antioxidants (this analysis took account of sexual dimorphism; see Noonan et al. 2016; Sugianto et al. 2019a; Chapter 2).

Corroborating theory, cubs, especially when very young, were more prone to oxidative damage, indicated by their greater levels of lipid peroxidase than adults.[37] Why then would badger cubs not invest unreservedly in antioxidant capacity? Because, as always, there's a countervailing selective pressure: youngsters that grow faster can gain, thereby, a survival advantage in badgers (Sugianto et al. 2019c; Newman et al. 2001) and generally across species (Taborsky 2006; Dmitriew 2011; Chapter 14). We have shown, and this was not previously obvious, that some cubs invest more heavily in protection from oxidative damage rather than growth, whereas others 'take the risk' of prioritizing growth over defences. Are these differences just a matter of natural variation, or do they represent different strategies? Certainly, the consequences will vary with circumstances: prioritizing growth while being more cavalier towards oxidative stress defences might pay off during years characterized by lower risk, perhaps in terms of food availability or social stability. How long did it take cubs to acquire an adult capacity to produce antioxidants? Although they had a lesser antioxidant capacity than adults in the early spring, cubs caught up with, and surpassed, adults by late summer (the adults' capacity having remained constant).

Against this broader background, what explains our seemingly opposing results from 2011 versus 2012–2014? An obvious possibility was that the cubs we sampled in 2011 had been older than those tested in 2012–2014: we can dismiss this explanation because both sets of samples were collected around May/June. What else could explain the contradiction? Well, cubs born in 2011 had been unusual in two respects, and surely these were linked: first, they had faced an unusually dry spring; and second, the surviving portion of the cohort we caught in May/June was very few in number, just 12 cubs; well below the average cohort for the preceding decade of 44.55 ± 5.37 (range 16–86, years 2000–2010). We deduce that the 2011 cohort, with its high antioxidant capacity, was the 'cream of the crop'; select survivors of a drought-induced shortage of earthworms (Chapter 10) during the critical period of early growth and development (Chapter 2). Consequently, the subset that comprised our 2011 sample were those individuals strong enough to have survived (a filtering termed 'selective mortality'; Gaillard and Yocoz 2003; see Chapter 18). Selective mortality sorts the wheat from the chaff. Cohorts initially comprise

individuals of 'good' and 'bad' quality for various traits, the latter dragging down the average. The poorly fitted strategists are, however, quickly weeded out by natural selection (a process termed 'canalization'; Van Buskirk and Steiner 2009), thus those individuals contributing to the calculation of average attributes later in a cohort's history reflect a greater proportion of better-quality individuals (we have argued the same proposition for beaver survival patterns; Campbell et al. 2017).

We suspect that the much larger cohorts that Kirstin examined during the less punishing springs of 2012–2014 (respectively, n = 39, 25, and 38) had not yet undergone intense juvenile mortality at the time we gathered and tested the samples, in contrast to those in the exceptionally dry spring of 2011. Nevertheless, by later in the summers of 2012–2014 a similar weeding out had taken place, selecting for cubs as robust to oxidative stress as were the adults.

Having established that harsh environmental conditions outside could be reflected in the cellular stress inside the badgers, we considered what else might be stressful. An obvious answer is reproduction, and sure enough we also found that oxidative damage was significantly higher in females that had reproduced in a given year, compared with females that had not. However, we could detect no change through time in antioxidant defences or oxidative damage amongst adults sampled repeatedly during their lifetimes. Sex also had no distinct effect on oxidative stress biomarkers, implying that life was no tougher for males or females, both seemingly experiencing similar levels of oxidative stress and damage, despite the differences in their life history stressors.

A further puzzle lingered in our measures of oxidative damage and antioxidant capacity in the 2012–2014 samples; that is, we could detect no direct link between body condition index (BCI) and levels of oxidative damage or antioxidant defences (although plenty of interactions between body condition and age or season). This counterintuitive result was the opposite to our more naturalistically explicable finding in 2011 (Montes et al. 2011). Perhaps the environmental stresses of 2012–2014 were insufficient to push any individuals (whatever their BCI) into the danger zone of oxidative damage. A possible explanation may lie in the findings of Strohacker et al. (2009) that oxidative stress can generally arise from repeated cycles of weight loss and gain, as might be caused by a sporadic food supply. Perhaps it was this repeated challenge of binge–starve in 2011 that had led to the link between body condition and oxidative damage,

---

[37] Although both cubs and adults had a similar RBC half-life, which measures resistance to oxidative stress as the capacity of cell walls to respond to an oxidative challenge.

which, however, did not recur under the more benign weather of 2012–2014; that is, when the oscillation in condition is greater, the effect is worse. Furthermore, cubs have a lower tolerance for enduring the ups and downs of food availability and fluctuating body condition (Chapter 10) than do adults and also are typically slimmer, placing them at greatest risk (Newman et al. 2011; Macdonald and Johnson 2015a).

Our field observations and demographic data had demonstrated that the weather was the major driver of differences in cub survival from year to year (Chapters 12 and 13). Here measures of oxidative stress reaffirmed the danger cubs can face from spring drought. Using the internal barometer of oxidative stress to explore mechanisms underscoring associations, we asked next: which aspects of the weather are most stressful to the cubs?

Remembering the Goldilocks zone of Chapter 12, an inverted U-shaped curve of oxidative damage emerged: higher cub lipid peroxidation (our measure of cellular damage) was associated with intermediate rainfall and warmer temperatures. These conditions of greatest oxidative stress were not the same conditions associated with highest mortality. It is tempting to wonder whether at least some of these results resonate with Friedrich Nietzsche's enduringly pertinent remark that what doesn't kill you makes you stronger—perhaps badgers that survive some of the greatest stressors go on to be survivors. In any event, higher rainfall seemed to be stressful, but not necessarily fatal, and was associated with increased inter-individual variation in lipid peroxidation levels (Bilham et al. 2018).

In conclusion, the balance between oxidative stress and the weather (as an indicator of food availability) was different for adults and cubs. For adults, the biomarkers indicating a greater ability to resist oxidative stress varied in accord with their survival dynamics (outside appearances mirrored inside conditions, and the biomarkers simply repeated what we knew from our actuarial tables): wetter (and slightly cooler) conditions were associated with higher antioxidant and PER levels, indicative of a greater ability to resist oxidative stress. In contrast, amongst cubs the 'Goldilocks zone' of benign weather (technically, the negative quadratic weather effect on cub mortality; Chapter 12) was not associated with propitious measures of oxidative stress and damage—something else was overridingly important for their cellular stress (for the future, our thoughts would turn to sociological factors). What does stress and ensuing ROS build-up mean for badgers? While there may

be a straightforward mechanistic relationship between drivers of oxidative stress and damage and the mortality of adults, amongst cubs the situation is more complicated, as other co-factors are likely at work.

Our demographic analyses (Chapter 12) had revealed that an important measure of the weather was the extent of inter-annual variation (Nouvellet et al. 2013). Another means for assessing weather effects was to contrast if inter-annual variation affected measures of oxidative stress. Kirstin found that antioxidant capacity was least variable between individuals during the harshest year (i.e. 2013, of the 2012–2014 dataset; Bilham et al. 2018). We interpreted this as further support for canalization reinforcing the phenomenon that in harsh springs 'poor-quality' cubs die before we get the chance to sample them—the weather fatally undermining their capacity to withstand the parasite *Eimeria* (Chapter 2). Conversely, in milder years, when more cubs survived until the spring trapping (2012 and especially 2014), there was considerable inter-individual variation in oxidative stress measurements.

Returning to our trade-offs question regarding how each cub strategically deploys its energy between growth versus immunity and antioxidant defences in order to maximize its chances of survival: the answer depends on how that cub's own constitution interacts with the severity of the annual weather, as an extrinsic influence. Investing in defence against oxidative stress and associated damage might pay survival dividends in stressful years, whereas ploughing that energy instead into development may pay greater returns when the weather is benign (some remarkable differences between the fates of cubs in good and bad years will become apparent in the context of paternal heterozygosity in Chapter 17).

Ecologists know that the effects of weather in one year can carry over to the next. Indeed, the effects of weather in one generation can reverberate across generations: there is a grandmother effect in white-tailed deer (*Odocoileus virginianus*) such that the effects of winter snow on nutrition in the face of wolf predation can be detected on survival over three generations (Mech et al. 1991). We saw in Chapters 12 and 13 how winter weather can be critical for badgers, with moment-to-moment changes in frost making earthworms unavailable (Newman et al. 2017). But what of longer-term impacts of the weather in one season on the badgers' oxidative stress during the next?

We couldn't have planned it, but by good fortune, winter weather had been highly variable during the years of Kirstin's study: 2012–2014. The number of winter frost days ranged from 50 in 2013 to just 5 in 2014 (2012 equalled the long-term average of 26), while winter rainfall in 2014 was twice the long-term average (other years had normal winter rainfall). Carry-over effects (Harrison et al. 2011) were apparent amongst the badgers: drier and milder winter conditions were associated with 'higher' oxidative damage (measured by lipid peroxidation) in the following spring amongst surviving adults and cubs born in the preceding year and transitioning to yearling status through that winter. Not only did signs of winter damage carry over to the following spring, so too did signs of greater investment in defences: badgers that had survived drier and milder winters had, in the following spring, higher antioxidant capacity (RBC half-life and higher AOX) (Bilham et al. 2018).

Why should milder, drier winter weather prove stressful? Well, as we discussed in Chapter 12, badgers use periods of torpor to mitigate food scarcity in winter (Newman et al. 2011). But this off-switch to badger activity is likely controlled by temperature—if it's warmer, badgers awake and become active (Noonan et al. 2014). Exercise is one way to induce oxidative stress (Radak et al. 2008); conversely, reduced activity and metabolic rate during torpor seems likely to lessen the risk of oxidative damage (Heldmaier and Ruf 1992). Perhaps then, counter-intuitively, warmer conditions promoted more badger activity (less torpor) but at a cost. And since it was dry, these higher metabolic costs were expended at a time when food supply, especially that consisting of earthworms, might have been limited.

Having established that weather can have both an immediate, and a carry-over, effect on the badgers' oxidative stress responses, the question arises as to what lies in store with climate change. Evolutionarily novel stressors arising from human-induced rapid environmental change (HIREC; Sih 2013) exacerbate the physiological burdens faced by wild-living species (Sies 1997). How might the increasingly unseasonable, variable and extreme, conditions predicted under UK climate change scenarios (IPCC 2014) interact with oxidative stress and investment in antioxidants? Worryingly for badgers (and probably some other species too) the generally warmer conditions predicted under both high- and low-emission scenarios seem likely to promote substantial increases in oxidative damage (lipid peroxidation) (Figure 15.9).

Conversely, we simultaneously project that badger antioxidant coping capacity would likely increase, as evidenced by the trends for greater AOX (Figure 15.9a–c) and longer RBC half-life (Figure 15.9j–l), although with multi-directional responses in peroxidase concentrations (Figure 15.9a–c)—a good thing from the badgers' viewpoint, showing badgers still have some residual response space left within their bioclimatic niche adaptations. Ultimately, therefore this conclusion (based on looking from the inside-out at the badgers' adaptability) agrees with our conclusion drawn from looking from the outside-in at broader demographic interpretations of population change (Chapter 13). The probable demands of medium-term climate change are unlikely, on the basis of insights from oxidative stress coping capacity or population dynamics, to push badgers beyond their adaptive envelope (Smit et al. 2000; Newman and Macdonald 2015).

An assay for measuring acute stress

Whatever the cause, the mammalian (human or badger) body's 'stress response' involves a complex combination of cellular, metabolic, neuro-endocrine, and behavioural changes, which collectively reduce the defence provided by the immune system. Moreover, we mentioned earlier that even psychological stress can trigger this cascade, leading to an increased risk of infection or disease (people get sick when stressed; Segerstrom 2007). This debility adds allostatic straws to a metaphorical camel's back that is already buckling under the challenges of survival in the wild. To discover whether this impacted badgers we needed to measure acute stress, beyond retrospective measures of the oxidative damage it can wreak on cells. Could the explosive force of the 'oxidative burst' evolved to blast invading pathogens reveal if badgers (or even humans) were registering an acute assault?

By way of background: various methods exist for testing for an acute stress response, such as examining cortisol (stress hormone) levels (e.g. Beerda et al. 1996; Palme 1997; Harper and Austad 2000), haematological values (e.g. Millspaugh et al. 2000), and physiological parameters; also, observing aberrant behaviour (for comparison, see WildCRU's work on vicuñas; Bonacic and Macdonald 2003; see also McLaren et al. 2007). Each method portrays only a partial story, however; the ideal indicator would be comprehensive. In response, David, and his remarkable colleague Dr Rubina Mian, invented a test which they called leukocyte coping capacity, or LCC for short, which harnesses the body's oxidative burst arising through its leukocyte response.

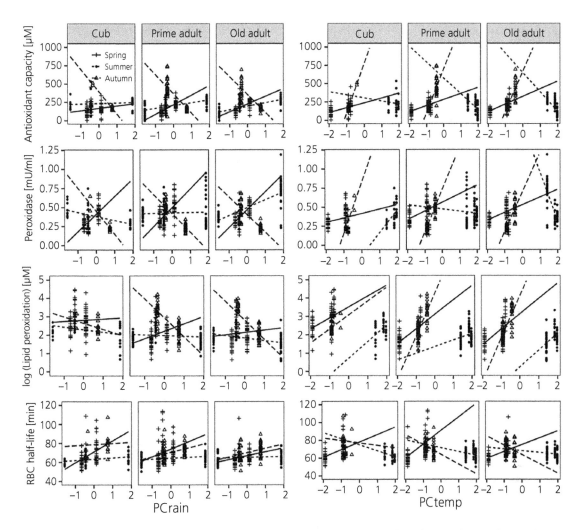

Figure 15.9 Scatterplots depicting the relationships between biomarkers measured in badgers (*Meles meles*) in Wytham Woods, UK, and weather variables across all age categories (cubs, <1 year; prime adults, 1–5 years; old adults, ≥6 years). The left-hand panels depict the biomarkers as a function of seasonal mean weather metrics in the reduced dimension space of PC rain; in the right-hand panels, weather metrics are reduced according to the dimension space of PC temp. RBC = red blood cell (from Bilham et al. 2018, with permission).

Leukocyte coping capacity

The idea of the LCC test arose from a chance meeting, actually on an animal welfare (Home Office) course, between David and Rubina, which led to this now widely applied stress test; first described in McLaren et al. (2003). The rationale stems from a fundamental fact about stress: whatever its cause (it might be infection, cigarette smoke, or bullying), an immune response is precipitated. That is, the body reacts as though to pathogenic attack, even if the actual stress is social or psychological (e.g. post-traumatic stress

disorder, PTSD), inducing increased respiratory oxygen intake and metabolic turnover (anxiety), which takes energy and generates ROS (even if, in the context of psychological stress, they have no real pathogens to fight; Gutteridge and Halliwell 1993). Furthermore, stress induces the release of pro-inflammatory molecules. The bodily wear and tear (the allostatic load) caused by psychologically (as well as physically) induced oxidative stress and inflammation has knock-on consequences that play a major role not only in the development of many diseases, from coronary heart disease, through inflammatory bowel disease to

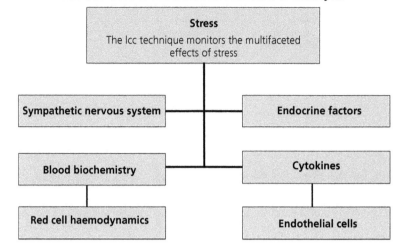

Factors believed to affect the activation state of leukocytes

Figure 15.10 Flow diagram showing factors believed to affect the activation state of leukocytes.

neurodegenerative disorders, but also, in ways highly relevant to badgers, in ageing (the topic of Chapter 18). Even prolonged working hours, workload, fatigue, and writing books about badgers cause oxidative injury in human subjects, as evidenced by significantly increased formation of 8-hydroxydeoxyguanosine (8-OH-dG), another marker of oxidative DNA damage.

So, the relevance of the LCC here is that it provides us with a measure of the stress a badger (or any other mammal) is experiencing, whatever the equanimity of its outward appearance. And so to the mechanics of the test: the LCC test is based on the fact that once leukocytes (mainly neutrophils) have been exposed to real, *in vivo* stressors (Figure 15.10) their subsequent *in vitro* capacity to produce ROS in response to further artificial stimulation is reduced or exhausted. Such a stress can be induced artificially by exposure to PMA (phorbol 12-myristate 13-acetate): a bacterial wall component used in immunology to induce an innate immune reaction[38] (Atanackovic et al. 2002).

First, we needed to test whether the LCC technique worked under field conditions. The transportation phase of our handling protocol provided an ideal

opportunity. Working with former WildCRU team members, Graeme McLaren and Inigo Montes, we asked: can the LCC test distinguish the stress response of badgers caught in our peanut-baited cage-traps and anaesthetized on the spot from the response of those additionally transported, unsedated, on an ATV trailer for 10 min to our central processing room prior to anaesthesia? To be clear, we were not trying to discover whether the ATV ride was stressful—of course it was, although we strive to minimize that stress by using rubber mats and cage coverings (Sun et al. 2015). Rather, the question was: is the LCC technique sufficiently sensitive to detect that stress? Yes, we could distinguish the weaker LCC response of eight transported badgers from that of the eight control badgers processed at their setts, without transport (see Figure 15.11).

This important proof of concept of the LCC test was further supported when, on examining simple blood smears, we found that the transported individuals not only had significantly fewer circulating leukocytes, but also exhibited different varieties of them (Montes et al. 2003). Leukocytes come in various forms (some readers will have had a white cell (leukocyte) count to investigate their infection status) and transported animals had a higher percentage of neutrophils (innate immune, ROS-producing cells) and a correspondingly lower percentage of lymphocytes (antibody-secreting acquired immune cells).[39]

---

[38] In terms of the technicalities of how the LCC test works, the aforementioned bacterial wall isolate PMA is mixed *in vitro* with blood samples; the extent of the ensuing respiratory burst (or lack thereof) as leukocytes try to attack this faked pathogen can be measured as emitted light, using a chemiluminescent marker chemical (luminol:5-amino-2,3-dihydrophthalzine; Sigma A8511). Simply, if the sample glows brightly, the coping capacity of the leukocytes when challenged by the PMA is high; the dimmer the glow of the luminol, however, the more the individual's blood has already been exhausted due to recent stress of one form or another.

[39] For further explorations of the LCC technique see Montes et al. (2004).

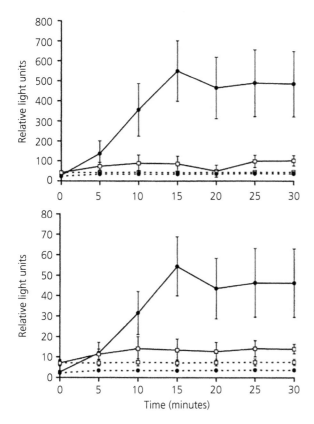

Figure 15.11 The *in vivo* leukocyte coping capacity (mean ± standard error (SE)) of badgers after transport stress. Badgers that were not transported had the highest coping capacity (•), whereas badgers transported and then aneasthetised (□) had the lowest LCC. The dashed line represents the mean basal value for all individuals. The lower graph presents LCC per neutrophil, where LCC has been calculated as relative light units divided by the number of neutrophis (1091–1). Lines and symbols are as in upper graph (from McLaren et al. 2003, with permission).

Encouraged that the LCC test was usefully diagnostic, and further intrigued by the increased activation of neutrophils amongst the transported badgers, Inigo led us in a further analysis of 28 badgers split amongst three transport regimes: transported (n = 9), transported and rested for at least 30 min before anaesthesia (n = 11), and not transported (n = 8). This time, to ensure we covered all our bases by using a more established test, we used nitro blue tetrazolium (an effective immune response turns a colourless dye blue; Montes et al. 2004). The NBT colour change confirmed that transported badgers had significantly higher percentages of activated circulating neutrophils, with mean activation highest immediately after transport and lowest in non-transported animals (Figure 15.12)

Studying a further 18 badgers we found that a 60-min pause prior to being sedated and sampled allowed for a near complete recovery in their LCC (depicted in Figure 15.13). This rest resulted in these 18 transported badgers displaying oxidative bursts of similar vigour to those produced by a sample of 19 badgers that were sampled without transport (there was no difference between males and females in these immunological responses). An immediate practical consequence was that we adapted our field practices to rest badgers after delivery to our field lab prior to sedation, balancing the welfare benefits of having them safe and sound in a processing room against minimizing the effects of transport.[40]

---

[40] A brief cautionary note re interpreting endocrinology. Transport stress activates the hypothalamic–pituitary–adrenal axis (HPA), leading to short-term increased concentrations of plasma cortisol in transported livestock (e.g. Kent and Ewbank 1986). David and his student Cristian Bonacic measured this in vicuña (Bonacic et al. 2003). So we asked, how did the cortisol response of badgers compare to our LCC-derived measures of transport stress? Graduate student Inigo Montes exposed a further 30 badgers (15 males, 15 females) to a typical 10-minute ride in our ATV trailer, again comparing those anaesthetized at the sett or after being transported to our handling facility (Montes 2007). Contrary to expectation, we observed a statistically significant effect *in the opposite direction*: non-transported badgers had higher mean plasma cortisol concentration (cortisol concentration pg ml^{-1} = 1689; SE = 359) than transported

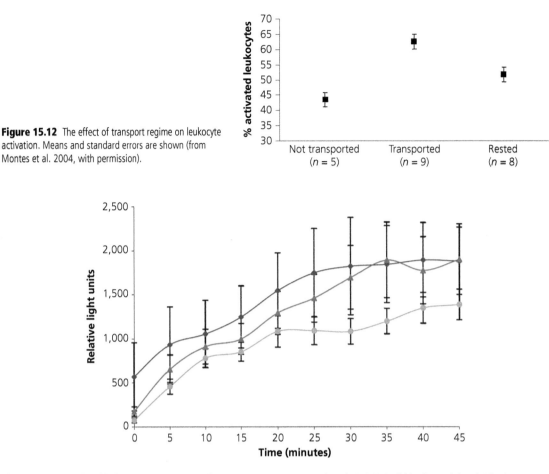

Figure 15.12 The effect of transport regime on leukocyte activation. Means and standard errors are shown (from Montes et al. 2004, with permission).

Figure 15.13 Combined leukocyte coping capacity values using PMA at a concentration of 10–3 M. Dark blue line and rhomboid points represent non transported badgers. Lavender line and circular points represent transported badgers. Red line and triangular points represent transported and rested badgers. Error bars represent the standard errors (from Montes et al unpublished).

Subsequently, the LCC has become part of the field ecologist's toolbox for measuring acute stress

badgers (m = 1059; SE = 116), but with no effect of sex. The lesson is that studies of endocrinology must be alert to the elaborate feedback system that controls mammalian cortisol secretion and gluconeogenesis (Hiller-Sturmhöfel and Bartke 1998). In this case the probable explanation is, given the rapidity of cortisol responses, badgers had likely undergone their initial 'fight or flight' response the moment the trap door dropped behind them, possibly as much as 12 hours before we sampled their blood, and irrespective of transport regime. A further complication would arise if, by virtue of their nocturnality, badgers have an inverted circadian rhythm of cortisol production—this would lead to low levels around dawn sampling, when badgers would generally settle down for their daily rest period. These complexities cause us strongly to favour LCC over cortisol as a measure of immediate stress in badgers.

(McLaren et al. 2003). For example, Merryl Gelling led a WildCRU team using this stress test to improve the handling and husbandry of water voles (*Arvicola terrestris*, now renamed *A. amphibius*) prior to reintroduction to the wild (Gelling et al. 2010; see also Gelling et al. 2012). Merryl also used LCC to assess the impact of live-trapping and handling on bank voles (*Clethrionomys glareolus*, now *Myodes glareolus*) and wood mice (still *Apodemus sylvaticus*) (Gelling et al. 2009). Further afield, Cristian Bonacic (a remarkable WildCRU alumnus now Dean of the distinguished Pontificia Universidad Catolica de Chile) and David also used it to evaluate the most welfare-friendly method of handling guanacos, a south American camelid, for wool shearing (Bonacic and Macdonald, 2003). Other wildlife teams have also adopted it; for example, to assess capture-

and handling-induced stress in Scandinavian brown bears (*Ursus arctos*) (Esteruelas et al. 2016). Not to mention its many uses beyond ecology; for instance, assessing human stresses from professional athletes to car drivers and even cancer patients (e.g. Shelton-Raynor et al. 2010, 2012)—a long journey from badger beginnings.

How does the immune system operate?

Throughout our tour of the diseases of Wytham's badgers we repeatedly mention the immune system, and how the badger's body deals with assaults against it, be these from chronic and acute pathogens, malnutrition (Chapter 11), bites, wounds and injuries (Chapter 3), or the general backcloth of social stress arising from competition (Chapter 7). Next we ask how the badgers of Wytham Woods fare in the battle with infection. To appreciate the answer, some readers may need to brush up their understanding of how the immune system works: see Box 15.1.

Mate selection for optimal offspring immunity: the major histocompatibility complex and acquired immunity

It wasn't really clear in Chapter 7 whether badgers selecting a mate do so with any overt criterion of quality in mind. Soon, in Chapter 17, it will become clear that their choices actually reflect some more deeply seated genetic considerations. En route, we now add an immunological perspective on mate selection. How do badgers (or any vertebrate species) keep pace with an ever-evolving variety of pathogens poised to attack them in a hostile world? This returns us to the MHC, this time focused on its role in pathogen recognition (the immunology involved is summarized in Box 15.4).

How then does MHC link badger pathogen resistance and mate choice? Natural selection has made this detection system comprehensive. MHC diversity (of which each individual's MHC set is termed a 'repertoire') is encoded genetically, so there is an opportunity to reconfigure this repertoire for the next generation each time an offspring is conceived. The effectiveness of the offspring's repertoire depends on propitious parental pairings (Sin et al. 2015; see theory in Penn and Potts 1999; Penn et al. 2002; Milinski et al. 2005). Nevertheless, there is always compromise, due

to the competing selective pressures of seeking to compose offspring capable of responding to both common and occasional epidemic diseases. The blending of the MHC pot is achieved through sexual reproduction. Sexual reproduction functions to deliver the right diversity of MHC genes, and thus the most impenetrable immune defences.

How does a badger (or any species) recognize and 'read' the repertoire of a potential mate, and judge its compatibility with their own MHC genes? It is as extraordinary as it is fascinating that, when choosing a mate, individuals can generally detect the MHC profile of a candidate partner, either through its appearance (e.g. in the plumage of the common yellowthroat (*Geothlypis trichas*); Dunn et al. 2013) or scent (from mice to mandrills; (Setchell et al. 2010; Eklund et al. 1991) (Chapters 6 and 9). There is strong evidence that even people make subliminal choices about reproductive partners based on their MHC: Wedekind et al. (1995) famously showed that women prefer the body odours of men whose MHC differs from their own, suggesting that MHC or linked genes influence human mate choice (see also Hamilton and Zuk 1982; Thornhill et al. 2003; Sin et al. 2016).

So, are badgers ultimately making competent mate choices with respect to defence against pathogens, optimizing offspring MHC diversity, in ways beyond our visual scrutiny? To investigate this, we first had to characterize the class I and class II MHC genes involved. This task, a long way from our muddy-booted background in behavioural ecology, fell once more to Yungwa (Simon) Sin (Sin et al. 2012b, 2012a; Sin et al. 2014). What defines the most adaptive MHC characteristics is a moving target, depending on circumstances. In environments subject to a wide variety of pathogens, MHC heterozygosity (gene variety) should be advantageous, binding a wider range of antigens than would be possible for homozygotic individuals (see Box 15.4). However, where there are occasional outbreaks of epidemic diseases, a rare allele may be life-saving (rare-allele advantage hypothesis; Takahata and Nei 1990). Simon therefore asked three questions: First, whether particular MHC haplotypes or alleles are associated with lower or higher pathogen burden (prevalence and infection intensity)? Second, whether, in comparison to MHC homozygotes, MHC heterozygotes exhibit a lower prevalence and/or intensity of individual pathogen infection and/or lower number of co-infecting pathogens? Third, whether rare alleles conferred any detectable advantage?

Box 15.4 MHC

Although nowadays familiar, organ transplants were amongst the greatest achievements of the twentieth century. They involve not only masterful surgical skill but also overcoming the body's rejection of a foreign organ due to an extreme immune antigen response. This requires aligning MHC (in humans also called human leukocyte antigen, HLA) molecules on the surface of the donor cells with those of the host. Antibody (immunoglobulin) responses are very destructive (which is normally a good thing for keeping you safe), so it is crucial that they be directed at only those molecules that are foreign to the host and not, self-destructively, to the molecules of the host itself: beware friendly fire. The MHC allows the body to recognize self from non-self cells (antigens): the process is called 'self-tolerance'. In this way pathogens, tumours and also transplanted organs and grafts can be targeted.

Marking pathogens with these MHC molecular tags (glycoproteins) enables them to be recognized and targeted by cytotoxic T-cells (of which there are two types, CD4 and CD8) and helper T-cells. Regarding the antigen tags, consider shepherds (actually two shepherds applying different criteria) putting MHC eartags in those sheep from their flock destined for the slaughterhouse, so the butcher T-cells can readily pick them out, while meticulously leaving the untagged sheep unharmed. T-cells belong to a group of white blood cells known as lymphocytes (another variant of the leukocytes that are the operating system of the LCC test described earlier). MHC molecules come in two varieties: class I MHC molecules have subunits (eartags) called β2 and these can be recognized only by the receptors of CD8[41] butcher; and class II MHC molecules have β1 and β2 subunits (eartags) and can be recognized by the co-receptors of both CD8 and CD4 T-cells. The existence of two types of MHC molecules helps T-cells distinguish self-cells from invading pathogens.[42]

Next, in a further cascade, activated B cells produce long-acting humoral antibodies (immunoglobulins, Ig). In a mechanism analogous to a molecular lock and key, these humoral antibodies attack the antigen's epitope (key) (Figure 15.14) with their paratope (lock), whereupon the resulting immune complex can be eliminated from the body by the complement system (Khan and Salunke 2014). In a feedback loop, antibody binding also marks invading pathogens for destruction, mainly by making it easier for phagocytic cells of the innate immune system to ingest them.

If this invader-detection system is too specific, then unidentified pathogens will slip through, leading the individual (here, a badger) to disease; even to death. If, on the other hand, the system is too trigger-happy, self-harm will follow.

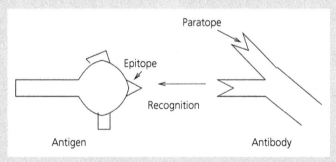

Figure 15.14 The trigger-happy extreme is all too familiar: if the invader-detector system is undiscerning, the lock and key of the IgE antibody is hypersensitive, causing (as some readers may personally regret) an extreme immune response to allergens such as pollen, cat dander, wheat germ, etc. Even more extreme hypersensitivity causes anaphylaxis; for instance, in response to shellfish, nuts, penicillin, or bee stings. In these instances the overproduction of IgE complexes trigger massive, and potentially life-threatening releases of histamine and trypase into the blood stream, causing hives (urticaria) and soft-tissue swelling (angioedema) that can cause the throat to close (from Allen and Koplin 2012, with permission).

[41] CD8 (cluster of differentiation 8) is a transmembrane glycoprotein that serves as a co-receptor for the T cell receptor (TCR).

[42] Although the body's targeting system can sometimes fail to achieve self-tolerance, causing a broad class of autoimmune diseases (Garcia 2012).

We tested, *in vitro*, whether particular MHC haplotypes or alleles[43] were associated with a lower or higher burden (prevalence and intensity) of infection from a wide panel of 13 diverse potential pathogens:[44]

> Enteric parasites—*Eimeria* (Chapter 2), *Isospora*, *Capillaria*, *Strongyloides* (number of badgers n_b = 185; numbers of samples n_s = 217).

> Blood infections—*Trypanosoma* (above: n_b = 217; n_s =360), MusGHV-1 (above: n_b = 218; n_s = 361).

> Infectious gut (enteric) bacteria (*Salmonella*, *Yersinia*, *Campylobacter*; n_b = 99; n_s = 150).

> Along with badger fleas, lice, and ticks (Chapter 8; n_b = 226; n_s = 418).

> Amongst these we also had overlapping individuals and samples, allowing us to look at co-infections in n_b = 90; n_s = 120.

In answer to Question 1: we found that the associations between badgers' MHC class II–class I haplotypes and particular pathogens were typically due to the presence of a resistant or susceptible allele: indeed, two particular MHC alleles conferred qualitative and/or quantitative resistance to certain pathogens, while a further three alleles were associated with pathogen susceptibility (Sin et al. 2014). For instance, all MHC haplotypes with the allele Meme-DRB*01 were associated with higher MusGHV-1 infection intensity, except for the haplotype that carried Meme-MHCI*03, which was associated with lower MHV infection intensity (see Figures 15.15 and 15.16). Similarly, MHCI*04, and the only haplotype that carried it (DRB*01–MHCI*04[45]) was associated with higher *Yersinia* prevalence. Both haplotype DRB*04–MHCI*02 and the DRB*04 allele were associated with lower *Yersinia* infection intensity.

In answer to Question 2: we found that there was evidence of heterozygote advantage against individual pathogen infection, although MHC heterozygosity was of no benefit to co-infection status.

However, with regard to Question 3, in accord with the rare-allele advantage hypothesis, we also noted cases where haplotypes, but not allele presence, were associated with infection. For example, the haplotype DRB*01–MHCI*03 was associated with lower *T. pestanai* prevalence and higher *Eimeria* intensity.

Overall we revealed a dauntingly complex immunogenetic facet to badger mate choice. We demonstrate that rare-allele advantage and/or fluctuating selection, and heterozygote advantage are major selective forces shaping MHC diversity in badgers. The mate choices made by parent badgers appear, *post hoc*, from our MHC probings, to be adapted to optimize the MHC diversity of their offspring with respect to the prevailing and ever-evolving disease-scape in which they are destined to live (or die). This indicates that badgers are waging an immune defence against a limited, if potentially broad set of enemies, rather than taking a scatter-gun approach to a world full of untold pathogenic dangers. Simultaneously, it seems that no single pathogen was driving any homozygotic, focused immune advantage.

Given the severe singular pathophysiology of eimerian coccidiosis (Chapter 2), this conclusion might seem unexpected; however, the window during which coccidiosis mainly causes disease is during early cubhood, prior to cubs developing, or 'acquiring' antibody resistance—the process in which MHC is so vital—leaving coccidiosis primarily under innate immune regulation. Indeed, acquired antibody resistance, combined with the selective mortality of coccidiosis weaklings, is likely why *Eimeria* is almost undetectable amongst adults of normal immune health. Why has evolution not adapted badgers to err on the side of caution and have a diverse MHC, just in case? Well, high intra-individual MHC diversity can result in a depletion of the mature T-cell repertoire because, during T-cell maturation in the thymus, a negative selection process eliminates T-cells with T-cell receptors (TCRs) that would otherwise react strongly with self-peptide–MHC complexes and cause auto-immune diseases—back to the dangers of a hair-trigger. For example, the oversensitivity of the *DRB-01* gene is strongly associated with rheumatoid arthritis in people (Scally et al. 2013). The depletion of the TCR repertoire, due to high MHC diversity, also degrades immunocompetence.

Do badgers trade off reliance on innate (as measured by LCC) and acquired (MHC) immunity to deal with pathogens? Earlier (p. 350), we explained how the

---

[43] An *allele* is a variant form of a gene. Some genes have a variety of different forms, which are located at the same position, or genetic locus, on a chromosome.

[44] Heterozygote advantage should be more readily detectable when multiple pathogens are considered simultaneously (Westerdahl et al. 2012).

[45] The protein produced from the *HLA-DRB1* gene, called the beta chain, attaches (binds) to another protein called the alpha chain, which is produced from the *HLA-DRA* gene. Together, they form a functional protein complex called the HLA-DR antigen-binding heterodimer. This complex displays foreign peptides to the immune system to trigger the body's immune response.

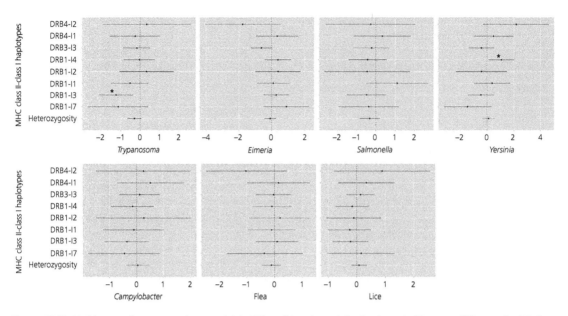

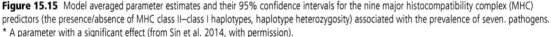

Figure 15.15 Model averaged parameter estimates and their 95% confidence intervals for the nine major histocompatibility complex (MHC) predictors (the presence/absence of MHC class II–class I haplotypes, haplotype heterozygosity) associated with the prevalence of seven. pathogens. * A parameter with a significant effect (from Sin et al. 2014, with permission).

innate immune system's respiratory burst of ROS provides an effective primary defence to sizzle invading pathogens; at the potential cost, however, of considerable energy and also at the risk of damaging host tissues through oxidative damage. That respiratory burst is short lived. This prompts the question of whether there could be selective advantage to downgrading the LCC response, swapping it for an upgraded cell-mediated MHC response? Three-spined sticklebacks (*Gasterosteus aculeatus*) illustrate the point[46]—in their case, intermediate MHC diversity, rather than fewer or many MHC alleles, correlates with a lower reliance on LCC. Badgers (like all mammals), however, have far fewer loci than do highly duplicated fish MHC genes; and so with badgers as a wild mammal paradigm, our next question was whether we could detect in practice, for the first time in the order Mammalia, any compensatory relationship between LCC and MHC systems (Sin et al. 2016).

Simon led us to the discovery that there was no interaction between the two systems in the contexts of

these same five most promising pathogens and parasites:[47] *E. melis, T. pestanai*, MHV, as well as badger fleas (*Paraceras melis*) and lice (*Trichodectes melis*) (Sin et al. 2014). To add a little detail to that negative answer, we found no association between LCC and infection intensities with the five pathogens, although we did identify (Figure 15.17) seasonal and annual variation of LCC,[48] likely due to physiological trade-offs, or temporal variation in pathogen infections.

Crucially, we found that not only was LCC not inversely related to MHC diversity, it was actually not related in any way to specific MHC haplotypes or MHC alleles, or MHC diversity. We tentatively conclude that, unlike sticklebacks, badgers evidence no compensation between these two prongs of immunological attack. Further, even when pathogen-susceptible MHC alleles were present, or when the MHC diversity was low, LCC did not step in to alleviate the

[46] Sticklebacks might seem an odd species to compare with badgers, but it is one of the only other species that has been studied in the wild.

[47] This analysis took account of the effects of age class, sex, body condition, season, year, neutrophil and lymphocyte counts, and intensity of infection.

[48] The variation in LCC was there, but is difficult to characterize simply. LCC was higher in summer than in spring in 2009 but lower in summer than in spring in 2010. The difference between LCC in autumn compared with spring was higher in 2010 than 2009, where LCC was much lower in autumn than in spring in 2010 and than in 2009.

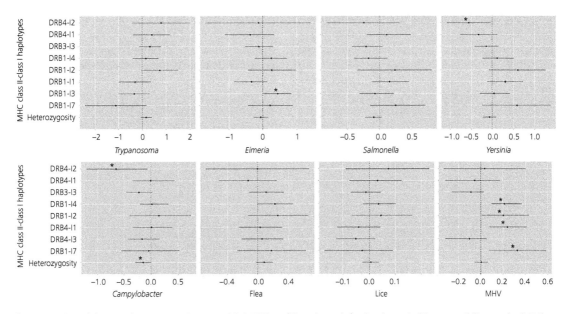

Figure 15.16 Model averaged parameter estimates and their 95% confidence intervals for the nine major histocompatibility complex (MHC) predictors (the presence/absence of MHC class II–class I haplotypes, haplotype heterozygosity) associated with the infection intensity of eight pathogens. * A parameter with a significant effect (from Sin et al. 2014, with permission).

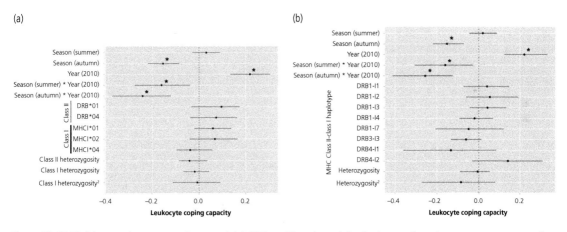

Figure 15.17 Model averaged parameter estimates and their 95% confidence intervals for the three predictors (season, year, season × year), and (a) presence/absence of three MHC class I and two MHC class II genes, linear effect of class II heterozygosity, and linear and quadratic effects of class I heterozygosity; or (b) presence/absence of MHC class II–class I haplotypes, and linear and quadratic effects of haplotype heterozygosity associated with the leukocyte coping capacity. * indicates a parameter with a significant effect. Spring and year 2009 were the reference categories (from Sin et al. 2014, with permission).

burden on its inadequate partner system. Why not? This could again be due to the high energetic trade-off costs of mounting an innate immune response and/or due to the immuno-pathological effects of oxidative stress.

On reflection

Early in this chapter, having started with the expectation that badgers would be battling diseases left, right, and centre, we soon began to wonder whether the most interesting question was why so *few* diseases afflict them, especially as adults. Perhaps, like the American house finches, *Haemorhous mexicanus*, studied experimentally by Bonneaud et al. (2019), they have evolved both resistance to, and tolerance of, pathogens—thinking back to those pustulent bite wounds by which they seemed so little affected during the social turbulence of the 1980s (Chapter 6), to the wimpish human onlooker badgers seem supremely tough. Perhaps also they have benefited from the relatively isolated situation of Wytham Woods, although, as we shall see in the next chapter, a new and unwelcome pathogen has arrived on the neighbouring farm.

CHAPTER 16

The Story of Badgers and TB

Perturbation and Beyond

Science is organized common sense where many a beautiful theory was killed by an ugly fact.

Thomas H. Huxley

Bovine tuberculosis (bTB) is, as the name suggests, primarily a disease of cattle. Question: So you want to protect cattle from bovine tuberculosis? Answer: Then vaccinate the cattle. That answer, offered by Rosie Woodroffe, distinguished alumna of the Wytham badger project, might seem glib, but is one possible part of the solution that has been sought since the debate about the badgers' role as a wildlife reservoir of *Mycobacterium bovis* began in 1971. It seems like a sensible answer. Might it work? Another part of the solution that has seemingly been rolled out unnoticed by those blinded by badgers is more effectively to detect, and remove, infected cattle. Might that work? Now, as we write in 2022, both these cattle-oriented solutions are increasingly recognized as the targets that matter. England is now concentrating more on testing cattle and tougher cattle measures, and advances in molecular veterinary medicine may bring an effective cattle vaccine closer, but how close? With the realization that the solution to bTB lies first and foremost in measures to control the disease in cattle, perhaps our earlier reviews of the role badgers play (Macdonald 1984c; Macdonald et al. 2006, 2015e), may soon be of interest only to historians of mismatch between evidence and policy. Nonetheless, a third approach might also be part of the solution: vaccinating badgers. Could that work? These are amongst the myriad questions, large and small, to which some answers lie ahead.

Our purpose in this chapter is to arrive at a scientifically informed judgement on what should, and should not, be done henceforth about bTB in badgers. To arrive at that judgement, and answering the three foregoing questions, while tackling a fourth—has killing badgers worked?—we will provide readers with enough, but no more, of the history of the disease, and of attempts to control it in humans, cattle, and badgers, to enable you to appreciate the evidence and arguments. En route, we will summarize WildCRU's work in areas where officials have killed badgers in attempts to control the disease in cattle, and our discoveries in Wytham, before offering a critical assessment of recent policy and what should not have been done, along with some personal reflections on David's involvement with policy-making.

First, some deeper history. The symptoms of TB are all too familiar to humans as 'consumption' and its close relative, milkmaid's disease. The human pathogen *Mycobacterium tuberculosis* was discovered in 1882 by Robert Koch, who developed the first tuberculin[1] 'skin' test, used diagnostically in 1890 (Gradmann 2001). According to the Registrar General's statistical reviews of England and Wales, in 1920 rates of death from pulmonary TB were as high as 735 per million in rural communities and up to 974 per million in towns[2] (a topical comparison is that, as of February 2021, the death rate from Covid-19 in the UK was 1650 per million).

Antibiotic treatment is relatively new—Chris was raised by his Great Aunt Mabel, a nurse in York during the pre-antibiotic era,[3] who would often recount the

---

[1] A small amount of the tuberculin (*M. bovis* or the same family of bacteria) is injected into skin and if the animal is infected, the skin shows an allergic reaction.

[2] Registrar General's statistical review of England and Wales (Annual), HMSO, London.

[3] Selman Waksman first discovered that streptomycin acted against *M. tuberculosis* in 1943, with the first patient cured by antibiotic treatment in 1949. Subsequently more refined antibiotics came into use, first isoniazid in 1952, which was cheap and well tolerated; then ethambutol in the early 1960s. From

The Badgers of Wytham Woods. David W. Macdonald and Chris Newman, Oxford University Press.
© David W. Macdonald and Chris Newman (2022). DOI: 10.1093/oso/9780192845368.003.0016

treatment of consumptive patients sent to the Fairfield Sanatorium and the Grassington Hospital in the belief that fresh air and better nutrition might alleviate symptoms. David's Glaswegian Great Aunt Louise travelled with her consumptive brother to Cape Town in the belief that the sun and sea would cure him—it didn't (but she stayed and her son, by odd coincidence, became the South African veterinary expert on rabies and other wildlife diseases; S2A3 Biographical Database of Southern African Science n.d.). Indeed, physician, demographer, and Oxford Rhodes Scholar Thomas McKeown famously argued (the 'Mckeown Thesis' 1976) that decline in mortality in England (and the USA) from 1901 owed far more to improvements in living conditions and especially nutrition, delivered by agricultural productivity, than to improved medical interventions.

In 1953 the BCG, or 'Bacillus Calmette–Guérin', was introduced in the UK, initially offered to school-leavers at age 14, because young adults were most susceptible to infection. Conditions less conducive to the spread of *M. tuberculosis*, vaccination, and antibiotics gradually reduced incidence, so by 2005 the programme of vaccinating 10–14 year olds was largely terminated in the UK[4]—although TB is now worsening once more in London, and continuing to infect *c.*10 million people annually worldwide; it killed 1.4 million in 2019. For perspective, a BCG vaccine dose currently costs only around £1.12–1.70 (US$2–3).

When we come, later, to consider badgers, remember that like many humans before them, they live in moist labyrinths in which respiratory bacteria can proliferate and spread amongst them, and beyond. Indeed, there is an irony that part of the solution for people arose from good nutrition, including dairy produce, which itself grew from a huge increase of dairy cattle, on the altar of whose well-being badgers have been sacrificed as collateral damage.

Milkmaid's disease is a close relative of consumption, this time caused by *Mycobacterium bovis*, and spread to humans in infected milk. Dairy herds were often kept within crowded Victorian and Georgian cities to aid milk supply, and in 1907, the interim report of the Royal Commission on Tuberculosis demonstrated that *M. bovis* was a zoonotic disease (i.e. transmitted from cattle to people). Consequently, in addition to *M. tuberculosis*, milk contaminated with *M. bovis* posed a significant threat of human infection, causing *c.*65,000 deaths between 1912 and 1937 in England and Wales, especially amongst children (Wilson 1943; Good and Duignan 2011). Following the Milk and Dairies Act in 1922, pasteurization was introduced in Britain, leading to a dramatic decline in human cases of zoonotic *M. bovis* infection, but both mycobacteria continued to cause disease in their respective primary hosts. In 1934, the Committee on Cattle Disease, the Gowland Hopkins Committee, estimated that 40% of cattle in dairy herds were infected (Economic Advisory Council 1934). Nowadays, almost nobody in Britain contracts bTB from cattle (perhaps no more than die from misadventure while routinely testing cattle for bTB), so hold in mind, while considering the huge investment in attempted control of this disease that it is not currently a human health problem (Torgerson and Torgerson 2008).

The Mantoux tuberculin skin test (TST; a.k.a. the reactor test or single intradermal comparative cervical tuberculin (SICCT) test in cattle[5]) came next in 1947, which, since 1973, cattle must pass before their milk can be sold legally and to comply with European Union (EU) rules regarding certification of TB testing exported animals. The basis of these skin tests is the induction of a delayed-type hypersensitivity reaction in response to the intradermal injection of antigenic substances (purified protein derivative, PPD) derived from an attenuated laboratory strain of *Mycobacterium* spp. (*M. bovis* or *M. tuberculosis*, respectively). Remembering the immunology from Box 15.3, individuals that have been infected by TB bacteria[6] will display

---

the 1970s onward rifampin became the keystone tuberculosis therapy of choice.

[4] In 2018, 4655 people were diagnosed with TB in England, an 8.2% decline compared with 5070 in 2017. The rate of TB reached an all-time low of 8.3 per 100,000 population in 2018, and has therefore been, since 2017, below the 10 per 100,000 World Health Organization definition of a low-incidence country (Tuberculosis in England: Executive Summary, HMSO, 2019).

[5] Be alert to the distinction between the SICCT and SICT: the additional C, for comparative, refers to comparing two lumps, the comparator being the avian one. The SICCT is used in the UK, Republic of Ireland and Portugal, whereas most other parts of the world use the SICT, which is loosely equivalent to the 'severe interpretation' of the SICCT, which exerts a tough filter on infected animals.

[6] To develop an immune response an animal must be infected with *M. bovis*; i.e. it will not develop an immune response (IR) merely by passive exposure without infection. This is because the IR needs two signals to develop: presentation of antigens by competent antigen-presenting cells to T-/B-cells in the context of the major histocompatibility complex (MHC), and a 'danger' signal through engagement of antigens with Toll-like receptors. Without this two-stage process no IR develops. For an animal to have antibodies means it is infected, it has been infected, or it's been vaccinated (or, perhaps benefited from maternal transfer).

a cell-mediated response[7] to the SICCT test, causing fluid and cells to accumulate in the dermis due to this hypersensitivity, creating swelling—usually simply a lump—at the injection site, sometimes characterized by erythema[8] (and occasionally induration[9] and ulceration). In TB-positive responses, this intradermal skin reaction develops within 24 h of injection, peaking around 48–72 h.[10] How good is this test? That is one amongst the many questions for which we will, in answer, cite a compendious critical review paper as a source: Godfray et al. (2013). It provides such a convenient shortcut to consensus, and its lead author will later have such influence, that it merits a brief digression.

In 2012 the Royal Society, vigilant to the depth of controversy and scientific dissent surrounding bTB policy, commissioned a critical review of the evidence to be led by Oxford's Sir Charles Godfray—a man with incomparable knowledge of tiny parasitic wasps, who, in that capacity, has lured David into a variety of escapades, of which by far the least plausible was the collecting of agromyzid flies. Anyway, Charles' brainwave was to assemble a small team—it comprised names that will become familiar during this chapter, including the Oxford badger aficionados, David, Rosie Woodroffe, and statistician Christl Donnelly—and require them to rank, numerically, the quality of the evidence (i.e. from watertight experiment to anecdote) available to answer every imaginable question about bTB. That rigour is why, in the summary that follows, we often default to the compendium of Godfray et al. (2013) rather than the thousands of other bTB publications. The approach was so successful that Charles went on to apply it under the series title *Restatements* to topics from neonicotinoids afflicting bees to endocrine disruptors in wild mammals. Importantly, the premise behind the bTB review and its successors was that, not all evidence being equal, policy-makers needed help to sort the wheat from the chaff.

So, what was Godfray and colleagues' (2013) summary regarding the tuberculin test? Its sensitivity and specificity can vary considerably at both the individual and herd levels, but typically the SICCT has >99.9% *specificity*[11] at an individual level (Strain et al. 2011)—this means a positive test can be trusted. However, in Godfray et al. (2013) herd-level *sensitivity* is cited at just 49% (95% credible interval (CrI): 27–74%). Nonetheless, Defra continues to cite (TBhub n.d.a) a median individual sensitivity of 80% (range 52–100%) (De la Rua-Domenech et al. 2006), whereas evidence has subsequently accumulated that even at severe interpretation, SICCT can have a low to moderate sensitivity (e.g. Lahuerta-Marin et al. 2018 document the herd sensitivity of the 'standard' SICCT as 40.5–57.7% and the severe SICCT as 49.0–60.6%).[12] This crucial fact may partly explain why 38% of herds cleared from movement restrictions experience a recurrent incident within 24 months (Karolemeas et al. 2011).[13] Conlan et al. (2012) estimate that 50% of such recurrent breakdowns are due to infections missed during prior tuberculin testing. Clegg et al. (2018) retested SICTT negative animals using a more modern test, interferon gamma (IFN-γ) and of those IFN-γ positive, 18.9% were positive at post-mortem—no wonder they suggest the 'prompt removal' of SICTT negative/IFN-γ positive animals, to reduce future transmission. Indeed, before the use of IFN-γ was increased in the last few years an almost theological faith in the SICCT meant that after infected herds have been tested and judged officially TB free (OTF) in high risk areas (HRAs), between 20 and 50% of them could actually still harbour the time bomb of an infected animal missed by the testing. Even now the combination of SICCT and IFN-γ still has major limitations and leaving many infected cattle in herds deemed to be OTF. This is important, but technical, so see footnote[14] to appreciate that the sensitivity of

---

[7] This DTH (type IV) response involves an interaction of T-cells, monocytes, and macrophages.

[8] Erythema—redness of the skin or mucous membranes, caused by hyperaemia (increased blood flow) in superficial capillaries.

[9] Induration—an increase in the fibrous elements in tissue commonly associated with inflammation and marked by loss of elasticity and pliability: sclerosis.

[10] Some animals infected with *M. bovis* fail to respond to the test (false negatives); this being a constraint of test sensitivity. Some animals not infected with *M. bovis* react to the test (false positives); this being a feature of test specificity.

[11] The specificity of a test refers to the proportion of uninfected animals correctly identified. In this case, the test has 99.9% specificity at individual level, so the test identifies correctly 99.9% of individuals without the disease (true negatives) and 0.01% are incorrectly identified as testing positive (false positives).

[12] Nunex-Garcia et al. (2018, their table 4) using meta analysis estimated SICCT at standard interpretation according to two models, one of which yielded a median sensitivity of 0.50 (95% CrI 0.26, 0.78); the other of 0.64 (95% CrI 0.48, 0.78).

[13] Other reasons include the survival of the bacterium in soil; the practice of culling only reactors while—unlike the policy in New Zealand—sparing the rest of the herd, and the fact that farmers have been known to conceal infected cattle.

[14] A yet more modern test, called Enferplex, and awaiting large-scale test, appears to undermine faith in the SICCT, and even in the SICCT/IFN-γ combination (Watt et al., 2021). It reveals (World Organisation for Animal Health (OIE) validation) with a diagnostic sensitivity of 94.2% for *M. bovis* culture positives, and a specificity of 98.4–99.7%. Enferplex appears to reveal the sensitivity to SICCT to be even poorer than feared.

even the SICCT/IFN-γ combination might be as low as 25%. Considering the size of the cattle population, this has startling implications: not just an inconvenient handful, but thousands of infected cattle may have been missed; many of them going on to be amongst the 2.8 million cattle movements occurring annually. Not so much a leak of infection, but more a running tap.

Using a fully World Organisation for Animal Health (OIE)-validated[15] serology test, Enferplex, Neil Watt (pers. comm., 18 May 2021) first compared the routine SICCT (used almost ubiquitously until 2014) with more severe interpretations of SICCT, which revealed, across three English herds, the former detected 112 infected animals but missed a further 800 subsequently diagnosed by the latter, more severe, interpretation. Horrifyingly, when 1500 of the same cattle were then tested using the seemingly much more sensitive Enferplex antibody test, a further 250 bTB infected beasts that had been missed even by the severe interpretation of the SICCT test were detected (Watt et al. 2021[16])— perhaps the leakage is less of a running tap and more of a geyser.

As the percentage of cattle SICCT/IFN-γ negative but Enferplex positive animals goes up the RSe of SICCT then IFN-γ on a population goes down, and that of SICCT then Enferplex goes up. These early results suggest it is common to find herds that have been tested SICCT/IFN-γ negative but include 25% Enferplex positive (N. Watt, pers. comm., 18 May 2021). At this level RSe of SICCT followed by Enferplex catches about 93% of infected cattle, whereas SICCT followed by IFN-γ detects about 25%.

[15] OIE approval number: 20190113; date of registration: May 2019. On 25th May 2022 OIE became World Organisation for Animal Health.

[16] At the time of writing, the work cited in N. Watt (pers. comm.) and Watt et al. (2021) on Enferplex, an OIE-validated test, has been presented at conferences, but not yet published following peer review. The Enferplex test appears to detect bTB infection throughout the disease process and to identify infected animals missed by skin and IFN-γ tests. The test involves testing for reaction on 11 *M. tuberculosis* var *bovis* specific antigen spots of which a variable number can react giving a traffic light score (0–1 = green, 2–4 yellow, 5–7 amber, 8+ red). Progression from one category to the next can take 12–18 months, so while a 'red' reactor should be killed immediately, a farmer with an 'amber' reactor might have time, for example, to wean a calf before killing the cow. The evidence is that sooner or later almost all Enferplex positives progress to become SICCT or IFN-γ positives (if they are not culled for commercial reasons first), so turning a blind eye to them delays the inevitable (in the early stages that delay does buy time and save money, but latterly it becomes progressively more foolhardy, risking further transmission). To confirm that all Enferplex positives are genuinely infected will ultimately depend on definitive PCR tests for *M. bovis* DNA, but is often strongly indicated by a very severe field test.

Enferplex is not the only source of evidence that SICCT, even with IFN-γ and severe interpretation, misses reservoirs of cattle infection. Another test, Actiphage,[17] detects *M. bovis* in blood samples within 6 h and without involvement of the host's immune system, but instead directly detects the *M. bovis* bacterium and according to early trials is almost 100% sensitive and specific. It uses a bacteriophage[18] to enter the *M. bovis* bacterium and lyse it, leaving material that can then be identified.[19] Because this phage-based test has not been OIE-validated at the time of writing (but the process is underway), farmers have to pay for it, but like David's neighbour David Christensen, judge it's worth doing so (PBD Biotech 2021). In one, albeit non-peer-reviewed, analysis, 161 high-risk TB cows were tested as least once with Actiphage, and although none had been classified as reactors under the standard interpretation of the SICCT, about 80% were positive to Actiphage. The conclusion was that many cattle within this endemically infected herd were latently infected (as detected by Actiphage) and a significant proportion of them went on to shed significant numbers of bacteria in their faeces[20] 'and so became infectious and a significant risk to further new infections' (Sibley 2018). Running tap? Geyser? It's hard to see where the metaphor of dysfunctional plumbing goes next.

There are several reasons for test failures. Jen Claridge and her colleagues (2012) estimated an underdetection rate of *M. bovis* of one-third with the SICCT test when cattle are co-infected with the common liver fluke *Fasciola hepatica*,[21] which is likely to decrease the

[17] Developed commercially by PBD Biotech. And on 25.5.22 yet another test was announced by iOmics.

[18] A bacteriophage is a type of virus that invades and replicates within a bacterium, the name coming from the Greek for bacteria eater.

[19] This involves isolating white blood cells (some of which are macrophages, some of which may be infected by *M. bovis*), lysing them to release any *M. bovis*, and stirring in the bacteriophage, which binds to and enters any living *M. bovis*. Having washed off any unattached phage, the remainder is dropped onto a lawn of *Mycobacterium smegmatis*. The phage will then lyse (effectively explode) any *M. bovis* it has infected, and infect the smegmatis, leaving a hole in the lawn. The material in the hole is then subject to PCR to determine the brand of tuberculosis bacterium: avian, vole, or cattle TB. This technique identifies active infection as rather than an immune response. It has revealed that low-level bacteraemia is much more common, and occurs earlier in infection/pathogenesis, than previously thought.

[20] Most infection between cattle is aerosol.

[21] *Fasciola hepatica*, also known as the common liver fluke or sheep liver fluke, is a parasitic trematode (fluke or flatworm, a type of helminth) that can infect the liver of various mammals, including humans.

immune response to the test, causing a false diagnostic result. Other pathogens may have similar effects. Second, certain genetic lines of cattle (and indeed any pregnant cows) may react less strongly to the test, and hence avoid slaughter even when infected (Amos et al. 2013)[22]. In addition, the interpretation in the farmyard of the skin response to the SICCT is subjective, determined by evaluating the sizes of two lumps generated as immunological responses to the *M. bovis* and *M. avium* components of the tuberculin test (Godfray et al. 2013) (that this evaluation involves some worrisome subjectivity was indicated by Enticott's (2012) observation that the judgement is affected by the gender of the vet making it). The flip side of this disheartening demolition of faith in SICCT, even when used in combination with the IFN-γ test, is that the Enferplex bovine TB antibody test used as a follow-up 4–30 days after the SICCT (which boosts sensitivity of antibody tests) may achieve a sensitivity of 90–94%, and thus offer a hugely important answer to the second of our opening questions: might bTB be controlled, or even eradicated, by better detection and removal of infected cattle? To judge by the results of Watt et al. (2021), the answer is yes. So, back to the history. On the basis of applying this skin test, the Attested Herd Scheme programme for milk production was introduced in 1935,[23] aimed at eradicating the bovine form of TB. Farmers were encouraged to join the scheme with bonus payments of one penny per gallon of milk from an attested herd (replaced by £1 per animal in 1938; Defra 2011). Figure 16.1 depicts the 1930s distribution of bTB in cattle—a picture worth remembering as it is so radically different to the contemporary whereabouts of infected badgers and cattle (see Figure 16.2).

The year 1950 saw the introduction of a national compulsory TB Area Eradication Plan applied to herds that were then much smaller than today's, using the tuberculin skin test and, for cases of severe disease, immediate slaughter of the whole herd to which the reactor[24] cattle belonged, for which farmers were compensated. By October 1960, the whole country had been attested. This strategy, applied at a time when herds were smaller and movements between them far fewer,

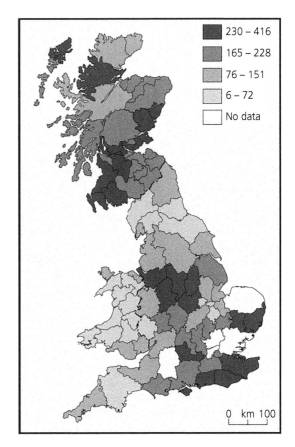

Figure 16.1 Reactors per thousand cattle tested in 1938 (quartiles) (from Francis 1947, with permission).

was a tremendous success and by the late 1960s bTB had all but disappeared from cattle herds. But, problematically, in the mid-1970s some herds cleared of bTB in Cornwall, Gloucestershire, and west Wales again started to disclose reactors. From 1986/87 the numbers of reactor herds started to rise and spread once more, with an annual increase of around 18% in the 1990's, rising to a 24% increase and 40,000 reactors per year in 2008 across the UK. This was really the first sign of a leak in the bucket. Where was this bTB coming from?

Broadly, there are two families of hypotheses, two buckets from which, despite all these control efforts, infections were leaking. One, call it the cattle paradigm, observes how difficult it is (see earlier, regarding the poor performance of even severe interpretation SICCT tests, and limited introduction of IFN-γ tests only from 2008) to ensure all infected cattle have been identified and removed from herds, to ensure that there is no further transmission within the herd or onward transmission to other herds. A key point here is that when

---

[22] An evolutionary possibility is that slaughtering immunopositive cattle puts selection pressure on cattle for immune-tolerance and on *M. bovis* for strains that do not stimulate an immune response.

[23] Administered by the Ministry of Agriculture and Fisheries in England and Wales, and by the Department for Agriculture for Scotland.

[24] Reactor cattle are cattle that had a positive test result—an allergic reaction to the skin test injection—and are therefore considered to be infected with bTB.

herds were smaller it was feasible to kill an infected herd, but nowadays many herds are vast making it economically and ethically difficult to kill every animal: the result *de facto* is that individual cattle, not herds, are treated as the epidemiological unit. We'll return to this hypothesis later.

The second, although not exclusive, hypothesis, brings us to badgers. In 1971 the post-mortem of one badger ignited the suspicion that this species was a reservoir of infection, plunging the then Ministry of Agriculture, Fisheries and Food (MAFF, now Defra) from a position of optimism to a state of panic, launching a complement, even a competitor to the cattle paradigm, the badger paradigm, and a succession of campaigns of badger culling that has continued ever since (see Cassidy 2019).[25]

Thus far, the story feels like history, but by 1972 badgers had become a current affair for David: in the early 1980s he was on the Consultative Panel on bTB with such notables as MAFF's badger man, Chris Cheeseman, from what was then known as the Central Science Lab (CSL, now replaced by the Animal and Plant Health Agency); veterinary epidemiologist, John Wilesmith; and Harry Thompson, myxomatosis expert and choreographer of the influential Thornbury Trial, and known at the time, variously affectionately, as Silver-tongued Harry (Thompson 1978). The externals included Ernest Neal, David, and, latterly as Chairman, David's friend from the different world of the Farming and Wildlife Group, Mark Tomasin-Foster, and stakeholder vets generally; the latter contingent exhibiting the traditional veterinary opinion that if a creature is dead, it won't cause trouble. The year 1984 saw David's first publication on the topic (Macdonald 1984c) and, growing from his earlier experiences with fox-borne rabies, the birth of the perturbation hypothesis (see the text that follows). This no longer feels like history, but a stressfully clear and present danger—indeed, for those interested in the passage of the present into the past, perhaps thinking that the story must at last be over, in 2015 the Wellcome Trust published an oral history of reflections by the survivors of the then 50-year saga (i.e. since 1960; Overy and Tansey 2015).

During the mid-twentieth century the number of test reactors fell to less than 0.49% of herds

(0.02% of all cattle) tested in 1979: the number of individual reactor cattle fell from 15,000 in 1961 to 569 in 1982. Problem solved? Not quite: pockets of disease persisted in the south-west of England, Gloucestershire, and west Wales, and by the mid-1980s herd breakdowns started to rise in those regions. The future had begun, and the proportion of total herds with reactors in the south-west increased from 0.75% in 1986 to 2.6% in 1996 (vs 0.10% to 0.51% for the same period in the rest of England; Krebs 1997), and by 2019, in an era of markedly increased cattle movements between farms, reaching an estimate (with wide variation) of 6.2% of British herds in 2019 (e.g. Figure 16.2). The torrential extent of cattle movements between farms will be hard for readers outside the cattle business to comprehend, but is made startlingly vivid for Ireland in a University College Dublin infographic in which the colourful tangle of movement arrows completely blankets the country month after month (UCD 2016; see also Christley et al. 2005). At this point it is worth recalling some facts material to the cattle paradigm. The high intensity of cattle testing during the Area Eradication Scheme of the 1950s and 1960s was scaled back, and was only really regained during the last 5-7 years—during the intervening time, testing went to 2-, 3-, or 4-year intervals, potentially providing less of a leaky bucket and more of a watering can for undetected infection spreading amongst cattle from hotspots in Cornwall, Gloucestershire, and west Wales to ever increasing areas across England and Wales.

For perspective on the timeline, the news broke when David was an undergraduate, and Chris was just (re-)learning to walk (having broken his femur on his first attempt): it was a Cotswold farmer from Wortley Valley who, in April 1971, had brought that dead badger to local MAFF veterinarian Roger Muirhead. He, mindful that farmers had speculated on badger involvement since 1962, performed a postmortem examination at his surgery in Wotton-under-Edge (Figure 16.3). This necropsy revealed pathological lesions caused by tuberculosis, confirmed in fluids taken from the badger's lymph glands.

Inevitably there was a raging thirst for a non-agricultural scapegoat, so the temptation to slake it with the blood of badgers was unsurprising—as this history unfolds some readers may wonder if that temptation became an addiction. Anyway, how did that Wortley Valley badger come to have tuberculous lesions? There had to be a first badger to catch bTB, although this assuredly wasn't it, and while it might at a stretch have done so from a deer, a farm cat, or even a vole, that first badger—like many after

---

[25] The first record of tuberculosis in a European badger (*Meles meles*) came from the Basle region of Switzerland in 1957 and there was no indication that this was connected with bTB in cattle; roe deer and chamois had previously been found suffering from bTB in the same area and were thought to have been the source of the badger's infection.

THE MAPS

Geographical distribution of new breakdowns in 1992 *(left)*, 2004 *(centre)* and 2015 *(right)*.

ON THE 2015 MAP:
● = Low risk area breakdown
● = Edge area breakdown (intermediate area in Wales)
● = High risk area breakdown

Sources: *LAMBERT, LEONARD AND MAY;*
TB ADVISORY SERVICE

Figure 16.2 Geographical distribution of new breakdowns 1992, 2004, and 2015 (from Defra, with permission).

it—probably caught the disease from cattle.[26] How? *Mycobacterium bovis* can potentially spread, whether within or between species, through direct contact, as well as indirectly via urine, faeces, and sputum contaminating the shared environment. Bacteria in the

environment potentially can remain viable for months (Young et al. 2005; Courtenay et al. 2006). Amongst badgers, infection can also spread through bite wounds such as in the gory photos of Chapter 3 (Jenkins et al. 2012). Anyway, MAFF's urgent response was to attempt to kill infected badgers by gassing their setts with sodium cyanide, a technique already in use for rabbit control. Perversely, this book owes much to this tragic convergence of disease and wildlife

[26] The first badger diagnosed with bTB was in Switzerland in the 1960s, but the first to catch it might have been around when cattle were first domesticated.

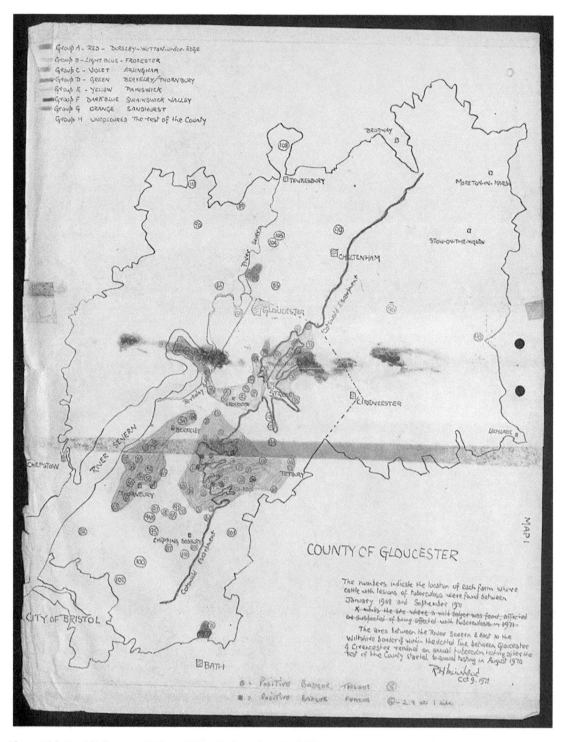

Figure 16.3 Map initially prepared by Roger Muirhead in September 1971, held in MAFF FT 41/88. Note strikethrough of original 'X marks the spot' notation and replacement with multiple outbreaks (from Cassidy 2019, with permission).

management, in that it prompted David to launch our project, believing that a study of the presumably TB free badgers of Wytham Woods would inform thinking about infected populations.

The Nature Conservancy (shortly to be renamed Nature Conservancy Council, or NCC, and then English Nature and latterly Natural England, of which body David was founding Chairman of the Science Advisory Committee (NE-SAC), and hence at the eye of the storm) was, however, not so convinced that culling was the way forward, and in August 1971 expressed concern that 'a widespread purge of Badgers will occur'. It was this concern, alongside ongoing persecution of badgers from digging and baiting that recast badgers from verminous to victim in need of protection—a Jekyll and Hyde schizophrenia that continues to this day. The ensuing socio-political history is worthy of a book in itself (such a book has, in fact, been written by Cassidy 2019), and holds jarring lessons regarding the deeply pot-holed road that lies between evidence and policy (see Policy postscript). Here, we can afford space for only a tantalizing digest to set the scene for some personal reflections on how our own work fits in.

Although the Cruelty to Animals Act 1835,[27], had prohibited badger (and bear) baiting, the image of the badger as a pest, or game animal, had continued until the 1970s.[28] It was around this time that David found himself as the new boy amongst a remarkable group that comprised the original, and with hindsight, revolutionary, RSPCA Wild Animals Committee: people like Bill Jordan, later to found both Care for the Wild and the Peoples' Trust for Endangered Species (which, years later would fund the Wytham badger study for more than a decade), Mike Stoddart (whose work presaged WildCRU's work on water voles (Moorhouse et al. 2015) and who was latterly Professor of Zoology in Hobart, Tasmania), Lord Gathorne Medway (naturalist extraordinaire, who had carved his name on the same guano-besmirched bunk in Borneo's Niah Caves in which, decades later, as recounted in *Expedition to Borneo*, David slept amidst a hoard of cockroaches), and Richard Ryder (clinical psychologist, David's neighbour, and convener of memorable parties, who coined

the powerful term 'speciesism' and became one of the most influential twentieth-century thinkers on animal rights; see Paterson and Ryder 1979). Looking back, these were characters at the well-spring of a step change in attitudes towards human–wild animal relationships in British society; attitudes that continue to whirl, like a cyclone, around badgers (Cassidy 2019). Meanwhile, the allegorical image of 'wise old badger' presented by Kenneth Grahame's *Wind in the Willows* (1908) was given naturalistic detail by Ernest Neal's first popular monograph on the species, *The Badger* (1948); David still treasures the creased copy that he acquired as a schoolboy. In 1976 David's BAFTA finalist film, *The Night of the Fox (World About Us)*, made television history by including the first ever broadcast-quality infrared nighttime video footage, using spare parts scavenged from old cameras by BBC technology wizards Paul Townsend and John Noakes; a year later that same technology was adapted in Ernest Neal's *Badger Watch* to cement the badger's position in the nation's heart.

That cyclone of societal change, framed by the shifting of wider environmental tectonic plates such as the Blueprint for Survival, Stockholm Convention, Club of Rome, and publication of Paul Ehrlich's *Population Bomb* (see Macdonald et al. 2007a for some history), included the recruitment of the then 500,000-strong Women's Institute to the badger's cause by a Cheshire housewife, Jane Radcliffe, whose enthusiasm for badger photography progressed to rescuing badgers under threat, catalysing a 1971 Mammal Society Symposium on the need for badger protection. This gained prominence in the national press, along with broader debate on 'blood sports' (see Macdonald and Johnson 2015b). In 1972, as David first encountered the badgers of Wytham Woods, Lord Arran introduced a wide-ranging protective bill in the House of Lords, supported by a Private Member's bill proposed by Peter Hardy in the Commons, which morphed into the Bill that passed into law in July 1973, protecting badgers while introducing a legal framework to license sanctioned badger control and culling. As regards killing badgers in the hope of limiting the spread of bTB to cattle, in 1975 MAFF implemented the use of gassing with 'Cymag' (sodium cyanide—releasing cyanide gas when in contact with moisture) as a less inhumane alternative to snaring. Soon, however, Wildlife Trust groups reported disoriented and sickly badgers; that is, those that Cymag had failed to kill outright.

Entering the scene was Eunice Overend, a badger advocate in whose spartanly cold caravan David

---

[27] Other progressive acts through the nineteenth century included the establishment of the Royal Society for Prevention of Cruelty to Animals (RSPCA) (1824), the National Trust (1884), and the Royal Society for the Protection of Birds (RSPB) (1889), alongside movements supporting temperance, the abolition of slavery, and universal suffrage.

[28] Badger baiting continues (*Belfast Telegraph* 2009) see Patrick Barkham's (2014) *Badgerlands*.

sipped tea and with whom Chris enjoyed many fascinating conversations at National Federation of Badger Groups AGMs through the 1990s. A trained biologist, Eunice argued that, like 'consumption' in people, bTB infection could likely linger in an infected badger in a latent form for many years, only to become actively 'infectious' when stress suppressed its immune system (see Chapter 15).[29] She argued that culling was disrupting badger groups, causing distressed infectious survivors to move around more, exacerbating the incidence of badger bTB—a line of thought matching that reached by David from his experiences of attempts to control rabies by killing foxes (e.g. Bacon and Macdonald 1980), which was also exactly his argument to the Consultative Panel, later formalized as the 'perturbation hypothesis'. The ensuing debate led MAFF to fund David's first research into perturbation, conducted at North Nibley (see p. 356) in collaboration with close friend and colleague, Chris Cheeseman of MAFF's Woodchester Park study. Meanwhile, Eunice (along with Ian Beales, Editor of the *Western Daily Press*) was to pioneer a major anti-gassing campaign, notoriously bringing a dead badger along as evidence to the enquiry led by Lord Zuckerman—then Chief Scientific Advisor to the British Government and President of the Zoological Society of London and *inter alia* designer of the Zuckerman civilian defence helmet of World War II (HMSO 1980). Zuckerman's initial 1980 conclusion, that gassing was humane, was reversed when it turned out that badgers were 'unusually resistant to cyanide poisoning'. In 1982 MAFF therefore turned to a trap-shoot policy, drawing upon an earlier trap design developed by Ruth Murray and the Universities Federation for Animal Welfare (UFAW), —but ingeniously upgraded through understanding of both engineering (from flying gliders) and badgers (from working in Wytham with Hans and David) by Peter Mallinson, who invented a nifty door-spring and lock while working with Chris Cheeseman's Woodchester team in the 1970s. Badgers had gained further general protection when, in 1981, the NCC led the passing of the Wildlife and Countryside Act, outlawing the use of snares, poisons, and guns to kill a list of native British fauna and allowing, for the first time, the prosecution of those killing badgers without a licence. Trapping badgers proved far more time consuming and

thus expensive than gassing, causing badger culling to be scaled back considerably following a review led by Aberdonian fulmar expert, George Dunnet,[30] published in 1986: 'We believe that it is not necessary, and would be a waste of resources, to seek further confirmation for the transmission of tuberculosis from badgers to cattle'.

And thus the 'badger debate' was born, pitting naturalists, and very often scientists, against government in the ever-failing campaign to stamp out bTB in cattle by undertaking some version of badger culling at varying scales. The objective in each case was to remove badgers in the vicinity of TB-affected farms in the hope of avoiding reinfection of cattle. Between 1982 and 1986 the 'clean ring' strategy was deployed, in which samples of badger carcasses were taken from all setts used by social groups within the breakdown area. If an infected individual was detected, then the entire group was (in theory but not practice) cage-trapped and shot (in contrast, amongst cattle herds, only animals testing positive or with dangerous contacts are killed). This extended out on the surrounding area until a 'clean ring' of uninfected groups was found; the mean removal area was 9 km². In the so-called 'interim' strategy (1986–1998) badgers were trapped and killed, whether infected or not, on the land that had been used by reactor cattle over a mean area of 1 km². Then came the live test trial (1994–1996), aiming to remove all individuals from setts in which one or more infected badgers had been detected by a blood test, within all land potentially used by badgers and extending up to cattle herds that had been at risk from these badgers, with a mean area of 9 km² (Krebs 1997; Smith et al. 2001).

Although, in principle, culling-induced reductions in badger density might have been expected to reduce contact between infected and uninfected badgers and thus between infected badgers and susceptible cattle,[31] the unhappy fact is that bTB rates in cattle rose steadily from the mid-1980s (don't forget the cattle paradigm and poor sensitivity of the standard SICCT test). In 1986 the National Federation of Badger Groups (NFBG; now the Badger Trust) was officially born, unifying opposition to culling from a national

---

[29] Subsequently it became clear that bTB does not necessarily kill infected badgers, although mortality rates of badgers excreting *M. bovis* are higher than those of test-negative individuals (0.667 excreting vs 0.315 test-negative for males; 0.480 vs 0.243 for females; Wilkinson et al. 2000).

[30] George M. Dunnet, Regius Professor of Natural History at the University of Aberdeen, expert on fulmars, whose career began at Oxford's Bureau of Animal Populations.

[31] Under Kendell's Threshold theorem the reproduction rate of an infectious disease (R_0) might be expected to decline beneath the threshold of persistence ($R_0 = 1$) if numbers of infectious individuals are reduced.

non-governmental organization (NGO) platform, and prompting, under John Major's government in 1992, the Protection of Badgers Act, making it an offence to disturb badgers or their setts, putting further obstacles in the path of culling. In the face of successive failed policies, an independent review in 1997, chaired by Oxford's Professor (now Lord) John Krebs, made

a recommendation historical in the context of wildlife management: a randomized controlled field trial to quantify the extent to which badger culling might reduce TB in cattle (Krebs 1997). The intention to inform policy with evidence was at the cutting edge, but it necessitated killing badgers and so was vociferously opposed by NFBG.

Box 16.1 Significant dates in the history of bTB management

1920s: The tuberculin skin test is introduced, allowing routine testing of cattle for bTB.

1935: The pasteurization of milk largely protects people from bTB.

1935–1937: The Tuberculosis–Attested Herd Scheme is rolled out by the Ministry of Agriculture.

1950s: Compulsory TB testing of cattle introduced, which progressively reduces the number of reactors.

1960: All cattle in the UK have been tested at least once and all reactors removed.

1970s: The south-west is identified as having a higher rate of bTB recurrence.

1971: TB first discovered in badgers.

1973: The Badgers Act is introduced to protect badgers against baiting.

1975–1981: Strategic culling using gassing is employed. The 104 km^2 Thornbury Trial took place and a reduction in cattle TB was observed until 1992. However, this was not a controlled experiment and other factors, such as the use of a new tuberculin, more efficient cattle testing, increased cattle culling removing silent carriers, could have influenced the outcome; at a time when there was a similar drop in bTB reactor incidence elsewhere too.

1980: Zuckerman Report results in temporary halt to culling.

1981: Gassing stops as a culling method.

1982–1986: Clear ring policy: social groups of badgers living on and around every breakdown farm were identified and trapped. Sample carcasses from these groups were then examined. Where infection was found, an attempt was made to remove all members of that social group. The ring was then extended outwards until all badgers caught were free of infection. Trapping took place in cleared areas for a further 6 months, in order to keep them 'clean'.

1986: Dunnet Report: Partial trapping policy replaces clean ring policy as interim strategy.

1986–1993: Interim strategy: this involved removing and culling badgers only on farms where bTB had been confirmed. During its operation, the incidence of bTB increased

in south-west England, and spread to areas with no recent history of infection, including the West Midlands and South Wales.

1987: bTB outbreaks resume a year-on-year increase, with a recent hint of decline since 2017 (namely the rate of OTF herds having their status withdrawn (OFTW) in England was >7.0 for 2015–2017, then 6.0, 5.8, and 5.6 for 2018, 2019, 2020 – any improvement associated with enhanced surveillance in high risk areas (HRAs), but little change in low risk areas (LRAs)).[32]

1992: The Protection of Badgers Act introduced.

1994–1996: A trial study of a diagnostic test on live badgers was begun but then suspended due to the poor sensitivity of the test, and problems with the trial itself.

1996: The Krebs report: given that the interim strategy was failing Professor John Krebs was commissioned to review policy. He recommended a randomized block experiment to determine the effectiveness of different culling strategies in reducing the incidence of TB breakdown in cattle herds.

1997: Ongoing culls continued but no new culls started.

1998–2007: The Randomised Badger Culling Trial (RBCT; see main text): proactive vs reactive culling; benefits to cattle, impacts on badgers.

2001: Tuberculin testing suspended due to foot and mouth disease (FMD), and movement controls abandoned during restocking. No badgers were culled during May 2001–January 2002 due to the suspension of fieldwork caused by FMD. Restocking following mass slaughter due to FMD correlates closely with the increase in both range and number of breakdowns post-2001.

---

[32] Apparent variations from year to year are desperately difficult to interpret because of confounding factors that have changed in the meantime, and because of the unreliability of the SICCT test. For example, the metric herd years at risk is based on the assumption that an officially TB free (OTF) herd is indeed free of disease and at risk of getting infected when in reality many of them are already infected, especially in HRAs and west Wales.

Box 16.1 *Continued*

2003: The Independent Scientific Group on Cattle TB (ISG)/RBCT—reactive culling component terminated prematurely based on evidence that this was counter-productively increasing bTB incidence in these treatment areas.

2005 (December): Pre-movement testing in England and Wales implemented to reduce cattle-to-cattle bTB spread.

2005 (December): The Independent Scientific Group on Cattle TB (ISG)/RBCT—publication of results from proactive culls.

2007 (July): ISG reviewing the RBCT concluded that 'badger culling can make no meaningful contribution to cattle TB control in Britain'.

2008: Stricter cattle movement controls introduced to try and reduce transmission to other herds.

2010: Badger vaccination first deployed; spreading to 18 counties with programmes; vaccinated 890 badgers in 2019.

2011: The Welsh Assembly's plans for a mass badger cull in west Wales were abandoned, and a vaccination programme introduced. Strain et al. suggest 'an increasing probability that the true figure of SICCT sensitivity is near 51%', and subsequent work suggests it may be half that (Watt et al. 2021).

2013 (October): 'Pilots'[33] of the culling/licensing regime, within which 'controlled shooting' was trialled in areas of Gloucestershire and Somerset. Cattle testing using SICCT in England organized according to three separate risk areas established: (i) LRA—4 yearly; (ii) HRA—annual; and iii) edge—increased from 2 or 4 yearly to annual. First Godfray review cites 49% mean herd-level sensitivity of the SICCT test.

2014: All breakdowns in the edge area and some in LRA to be subject to SICCT at severe interpretation from January 2014. All confirmed breakdowns in the edge and LRAs to be tested supplementarily with IFN-γ.

2014–2015: Licensed culls repeated and expanded to include three additional areas in Dorset in 2015, with seven more in 2016 (two in Cornwall, two in Devon, one each in Dorset, Gloucestershire, and Hereford), and annually thereafter (see Box 16.4 for the chronology of new sites).

2014: IEP report on pilot badger culls in Somerset and Gloucestershire and the humaneness and effectiveness of the shooting.

2015: Declaration of the 25-year bTB eradication strategy; publication by Wellcome Trust Overy and Tanseya History of Bovine TB *c*.1965–*c*.2000.

2016–2017: Culling policy expanded further. Severe interpretation of SICCT extended across the HRA for all breakdowns from April.

2017: Compulsory increase in IFN-γ testing across the HRA for confirmed breakdowns which meet any of three criteria.

2017: SBC introduced. Intended to preserve the presumed benefit of 4-year intensive culls, allowing a 5-year extension, but arguably risking local eradication and without precedent in RBC trials. 2021 update: almost all intensive culls completing 4 years to proceed to SBC of 2 years' duration. Intensive culls introduced in 2021, 2022 will not proceed to SBC.

2018: Godfray Report: emphasizes that culling badgers can have only a 'modest impact', recommending improvements to the current cattle testing regime, on-farm biosecurity, and trading restrictions. Laheurta-Marin et al. (2018) estimate median sensitivity of the SICCT test sensitivity at standard interpretation to be 40.5–57.7%, and at severe interpretation 49.0–60.6%.

2020: Plans to extend culling zones to 11 new areas include Avon, Oxfordshire, Derbyshire, Somerset, Wiltshire, Cornwall, Herefordshire/Worcestershire, Leicestershire, Warwickshire, Shropshire, Lincolnshire, and Derbyshire. Six-monthly surveillance testing introduced in 2020 to the HRA counties of Staffordshire and Shropshire.

2021: Defra launches Call for Views on measures primarily designed to reduce the risk of undisclosed infection being transmitted to other herds and other areas.

2021: In late January, Secretary of State George Eustice judged badger culling to be unacceptable.

2026: Although this chronology is written in 2022, we can anticipate that the licences issued last year will run until 2026 and thus estimate that between the start of intensive culling in 2013 and its supposedly anticipated end in 2026, some 225,000 badgers will have been killed across 25% of England. Time will tell.

---

[33] Although these culls were termed pilots, following the IEG 2013 report, rejected by Defra, there was no published analysis of their effect (Independent Expert Panel on Badger Culling Pilots 2014).

The smoking gun

What was the evidence that badgers with bTB can pass the disease to cattle? At first, it rested largely on guilt by association. Broadly, there tended to be more badgers in places where the incidence of bTB was highest in cattle (Figure 16.2), but where badgers were thin on the ground in the eastern and northern regions of England, and throughout Scotland, the incidence of bTB was low.[34] Nationally, between 1972 and 1978, Wilesmith (1983) reported a higher incidence of bTB in cattle in 10 km × 10 km grid squares that were thought to support higher densities of badger setts (see also Griffin et al. 1993). However, subsequent studies have consistently failed to detect associations between badger density and cattle bTB risk, whether through case control studies (Johnston et al. 2005, 2011), or correlations across Randomised Badger Culling Trial (RBCT) areas (Donnelly and Hone 2010), or transects through bTB 'hotspot' areas (Mathews et al. 2006a). Indeed, there is evidence that prevalence of infection is lower at higher badger population densities (Woodroffe et al. 2005). The national-level correlation may occur because badgers thrive in mixed pasture and woodland landscapes where most cattle farming occurs; in short, however, the links between badger population density, prevalence of bTB amongst them, and the incidence of bTB in cattle have not been proven and do not seem to be straightforward. At a finer scale, infections in both species were spatially associated at a scale of 1–2 km (Woodroffe et al. 2005), and where there were clusters of high incidence the same strains of *M. bovis* were usually diagnosed in both cattle and badgers (Smith et al. 2003). These strains are characterized through genotyping using spacer oligonucleotide typing ('spoligotyping'; Kamerbeek et al. 1997), which allows bacteria to be classified to one of the small number of readily identifiable *M. bovis* clones that occur in Britain (Smith et al. 2003). Highly pertinent insight comes from the Badger Found Dead Study undertaken by Nottingham and Surrey Universities, led by Malcolm Bennett. Briefly, their post-mortem analyses of badgers killed by cars revealed that it would be wise to be wary of generalizations: in Cheshire, there was a high frequency of bTB in both badgers and cattle, and it was the same strain (spoligotype) in both species (suggesting they

were part of the same epidemic, although allowing no conclusion as to the direction of transmission). However, in other counties, within the so-called northern edge region: in Derbyshire there was a high prevalence of bTB in cattle but not in badgers; low prevalence in both species in Nottinghamshire and Northamptonshire, but moderate prevalence in both in Leicestershire and Warwickshire, but in the latter counties the two species had different spoligotypes. Their powerful conclusion, therefore, was that 'With the exception of in Cheshire, therefore, this study found little evidence to directly link the expanding epidemic in cattle in England to widespread badger infection' (Swift et al. in press).

The obvious question, can bTB persist amongst badgers without reinfection from cattle?—in other words, are badgers a maintenance host for the disease?—has remained, remarkably, uncertain.[35] The most avant garde technique applied to this question is phylodynamic modelling, which combines information on the genetics of the bacterium and the epidemiology of the disease. Two recent papers, sharing several co-authors, have applied this exquisitely sophisticated technique to bTB. Akhmetova et al. (2021) used data from a study area in Ireland, whereas O'Hare et al. (2021) used data from an area in England, and they arrived at different emphases regarding the likely importance of badgers—the different outcomes suggest that the transmission dynamics of the disease are situation dependent, varying with circumstances. Akhmetova et al. use the genomic techniques to reveal that in their Irish study area cattle-to-badger transmission is much more common than vice versa. In their English area O'Hare et al. used only data from cattle to conclude that there must be another mystery reservoir to generate the observed genetic diversity of the pathogen, and they deduced that the most parsimonious candidate is badgers (another might be environmental contamination, e.g. from slurry, which, like badgers,

[34] Scotland gained officially tuberculosis free (OTF) status in 2009 by the European Standing Committee for Food Chain and Animal Health. This status is only awarded after meeting certain criteria; for example, for 6 consecutive years, the percentage of bovine herds proven to be infected has been <0.01% (British Veterinary Association 2014).

[35] Atkins and Robinson (2013) puzzle over the disjunction between the early- to mid-twentieth-century distribution of bTB in cattle and its current distribution in badgers. In 1938 in Cheshire there were >400 positive cattle per 1000 tested, some of which were at an advanced stage of disease never seen today; shedding, as a result, copious bacteria on pasture. In 1960 the disease was eradicated from cattle in Cheshire. Between 1972 and 1990 389 road traffic accident badgers were tested; however, only 1 was positive. Clearly the cattle had not caught bTB from the badgers, but why had the badgers not caught it from the cattle, of which, in some herds in the 1930s, 40% were tuberculous?

persists in the absence of the cows) (other papers combining genomics and epidemiology from this remarkably innovative team, all led by Rowland Kao, include Crispell et al. (2019) in yet a different study area where they concluded that infections from badgers to cattle might be 10-fold more frequent than vice versa). Perhaps it is unsurprising that badgers play some role, but that it varies with circumstances. Further, and with slurry in mind, as a sidebar respectful of Lucretius' exhortation with which we opened the Prologue, to think outside the box, it might be worth listening to a drum beaten with solitary stoicism by Malcolm Bennett of Nottingham University: that in looking for the 'environmental' source of bTB it might be wise to look beyond badgers. Many mycobacteria are 'environmental', not just living in soil but in a range of protozoa. *Mycobacterium bovis* itself is adept at this lifestyle, surviving and growing in a range of environmental protozoa, some of which, for example, *Acanthamoeba*, can form cysts that survive for months or more in the sort of mud that sticks to the tread of tyres driven between farms (e.g. Taylor et al. 2003; Sanchez-Hidalgo et al. 2017). Indeed, infection of protozoa has been suggested as an 'evolutionary training ground' for many pathogens that end up with intracellular careers, particularly those that infect macrophages.

Returning to the behaviour of bTB in badgers, one of the best insights, which also revealed that cattle and infected badgers can co-exist for long periods without the disease necessarily being transmitted to cattle, came from the long-term study of infection amongst the Woodchester Park badgers. There, Cheeseman et al. (1988b) found that the temporal spread of infection between social groups in the study area was slow and restricted; also that pseudo-vertical (mother to suckling cub) transmission may be important in the maintenance of bTB where it infects badger populations,[36] as may be horizontal transmission by bite wounding and aerosol infection. However, there was neither an apparent relationship between badger population density within the Woodchester study area (they were pretty numerous everywhere) and the prevalence of *M. bovis* infection, nor was there evidence that *M. bovis* infection depresses badger population density significantly below disease-free levels.

---

[36] The fact that Malcolm Bennett (pers. comm., 5 September 2021) found similar prevalence amongst young and adult badgers might also suggest young badgers get infected in the sett.

Box 16.2 Legalities prevail: why wasn't cattle vaccination always the focus?

Member states of the European Commission are required (via a number of complex directives) to control bTB through routine testing, towards the objective of achieving officially TB free (OTF) status. Vaccination of cattle against bTB is prohibited[37] because vaccinated cattle test positive to the reactor skin test (above): BCG and the tuberculin test share the same antigens, and hence this test cannot distinguish a vaccinated individual from an infected one (DIVA[38] in Waters et al. 2012); for example, 6 months after administering BCG, 80% of experimentally vaccinated cattle tested positive to the tuberculin test[39] (Whelan et al. 2011). The point here is that cattle/meat exports from the UK into EU member states have been governed by this requirement for a negative test and previously this was cited by the NFU and Defra as the block to a cattle vaccination policy. Two things have changed: an effective DIVA test may be coming closer,[40] and the UK is no longer a member of the EU and hence can

decide its own policy (although in the absence of a free-trade deal the UK may well have to choose between being able to export cattle and cattle products to the EU and aligning aspects of TB control policies[41]).

[37] Vaccination against bTB is explicitly forbidden in the EU legislation on disease control (Council Directive 78/52/EEC) and implicitly also in intra-Union trade legislation, as vaccination is not compatible with the provisions for testing and herd qualification (Council Directive 64/432/EEC). EU legislation is fully in line with OIE standards on international trade and can be changed only by the European Parliament and the Council.

[38] Differentiate Infected and Vaccinated Animals (DIVA).

[39] Ninety per cent revert to test negative within a year.

[40] Although, insofar as this might rely on adaptations of SICCT and IFN-γ. it is not obvious how the same sensitivity shortcomings that limit their use in diagnosis would be circumvented.

[41] This alignment should not affect meat exports, and so presumably focuses on sperm and breeding stock, which would have pre-movement SICCT and be from TB-free herds.

Box 16.2 *Continued*

The Tuberculosis (England) Order 2007 (or TB Order) requires cattle farmers to have their herds routinely bTB tested at frequencies depending on the level of disease in the area: since 1 July 2021 6-monthly surveillance testing of cattle herds in the HRA of England; once a year or even once every 6 months in 'edge areas', to arrest spread; and once every 4 years in 'LRAs' (which would seem a long interval if the objective is to stop the epidemic expanding).

A herd breakdown occurs if: at least one animal fails the TB skin test (a 'reactor' animal); at least one animal returns two consecutive inconclusive skin tests; the slaughterhouse reports typical bTB lesions in the carcass; an animal shows possible signs of bTB (clinical case), is slaughtered before testing, and TB lesions are identified on the carcass; or if the bTB test becomes overdue.

A herd suffering a breakdown is put under 'TB2' movement restrictions, to prevent onward transmission to another herd. Farmers then receive compensation for cattle lost to TB, if they have the correct ear tags and passports, under the 2006 Order, in which values are assigned according to a sort of Hammurabi code that assigns different values to male or female, beef or dairy, pedigree or non-pedigree, age group and whether females have calved, etc. Nevertheless, lesion-free parts of cattle slaughtered for a positive TB test can still enter the human food chain, provided carcasses don't show tuberculosis lesions in more than one organ or body part.[42]

After a breakdown, Defra inspectors retest cattle every 60 days (except calves younger than 42 days). A herd is certified as officially TB free (OTF[43]) (a technical, and, even with recent mandatory supplementary IFN-γ testing, eccentric if not downright misleading use of the word 'free', insofar as officially TB free is no guarantee that it is actually TB free) if it clears two skin tests at least 60 days apart.

The IFN-γ test (see Box 16.3) has been mandatory for all confirmed breakdowns in the edge and LRAs as a supplementary test since 2014 (this was expanded under certain criteria to the HRA in 2017—and further ramped up in 2018 and 2019). While it can detect infected cattle not identified by the skin test (with 49% herd sensitivity (p4) due to its higher individual-level sensitivity (90%[44]) its lower specificity (estimated median of 96.5%, vs 99.9% for SICCT (p4; in Godfray et al. 2013) means that the high number of false positives make it politically unacceptable as a primary diagnostic text (Strain et al. 2011).

---

[42] Before consumption, the viscera (the digestive system including liver, stomach, bladder, and intestines and the reproductive organs), pluck (thoracic contents including heart and lungs), and kidneys are removed.

[43] OTF means officially tuberculosis free and is used in EU legislation (Directive 64/432) to describe those cattle herds that may undertake intra-community trade. A herd is OTF if: all animals undergo a TB test with negative results annually, and there are no clinical signs or suspicion of TB infection in the herd.

[44] TBhub (n.d.b)

Wytham: a bTB-free population?

For the first 40 years or so of studies, Wytham's badgers were apparently bTB free. There had been no breakdowns amongst the large dairy herd on the adjoining farms, and during the 1980s cultured blood samples gave no evidence of infection. Then, in July 2014, came the dreaded email from Wytham's Conservator, Nigel Fisher: a beef steer on the university's farm had tested positive. Second-round testing revealed 8 cattle amongst 100 breeders testing positive that summer. There were two suspects: a breeding bull imported from an attested herd owned by the same farm business in the Lake District or our badgers. Aided by financial support from FAI (Food Animal Initiative), we tested the badgers (Box 16.3).

Which were infected first in Wytham, the badgers or the cattle? The evidence points to the cattle. The sequence of events was that using the Statpak blood test, and in collaboration with the Animal and Plant Health Agency (APHA; then AHVLA, part of Defra), in August 2014 we found that 13/112 badgers were Statpak positive. However, AVHLA was wary of the likelihood of false positives with Statpak and tested a subsample of the 13 Statpak positive badgers with IFN-γ: all were negative, convincing APHA that probably none of the badgers had genuinely been positive. However, in November we found that 5/66 badgers tested positive by both Statpack and IFN-γ. Two of these individuals had been IFN-γ negative in August, suggesting (notwithstanding the sensitivities of the tests) that infection broke out between then and November, and therefore after bTB was first diagnosed amongst FAI cattle.

Our next opportunity to test the badgers was in the following spring, May/June 2015—had the situation

Box 16.3 Testing badgers

The two gold-standard diagnoses for bTB require either 6 weeks, for culture testing or for the badger to be necropsied (revealing lesions).[45] Mention of gold standards makes diagnosis sound straightforward—it's not. Challenges change in complicated ways with the progression of infection as the serological and cell-mediated responses of the immune system swing into operation (Chapter 15)—as simply schematized in Figure 16.4.

Other diagnostic tools partly circumvent these difficulties. One is the *Brock TB ELISA test*, available in a *Statpak* formulation usable at the capture site. This humoral antibody test cannot distinguish animals actively excreting bacilli

(Newell et al. 1997), has 93% specificity; sensitivity is poor, however, at 49%, especially for animals with subclinical infections (those with evident lesions at post-mortem were correctly identified in 66–78% of cases; Chambers et al. 2008). This low sensitivity is problematic, and in the live test trial led to the decision to group badgers by sett, rather than social group; bearing in mind that each social group is likely to occupy multiple setts—this had the unwelcome consequence that badgers which tested negative and were released at one sett were regularly culled at another (Woodroffe et al. 1999; Tuyttens et al. 2000b).

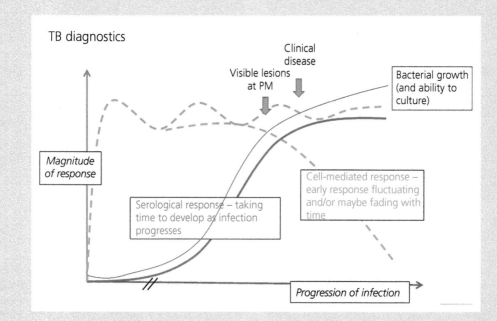

Figure 16.4 Challenges to diagnosing bovine TB as infection progresses (based on a schematic by M. Bennett, with permission).

[45] Samples of blood or faeces, urine, bite wounds or tracheal aspirates, and other tissues are treated with 10% oxalic acid (this gets rid of the many fast-growing bacteria that might prevent the slow-growing mycobacteria from growing), centrifuged and pelleted; these pellets are then used to inoculate petri dishes or tubes of special culture media, such as modified Middlebrook 7H11 agar. Cultures are then incubated at $37°C \pm 2°C$ for at least 8 weeks. Any growth of organisms characteristic of mycobacteria are identified as *M. bovis* by PCR and spoligotyping (Kamerbeek *et al.* 1997). Interestingly, according to Malcolm Bennett's research, most lesions (macro and micro), and the best places for culturing, are not the lungs (unlike cattle and humans) but head and neck and other non-respiratory tract lymph nodes; TB in badgers may be, as in deer and camelids, enteric as well as being transmitted by biting, rather than a respiratory infection.

Box 16.3 *Continued*

The interferon gamma, or IFN-γ test is more sensitive but requires immediate, lengthy, laboratory analysis (Drewe et al. 2010). Along with IDEXX ELISA, IFN-γ is the only blood test approved in the UK and the EU to supplement the skin test for TB in cattle, and measures the levels of the IFN-γ (cell-mediated immunity, Box 15.3), a cytokine, which is released by white blood cells (memory T-cells) when blood from cattle or badgers infected with bTB is stimulated with bovine and avian tuberculins (the same as used in the

tuberculin skin test) (Sawyer et al. 2007). The amount of IFN-γ is measured using sandwich ELISA or by PCR. Relative changes in the levels of IFN-γ mRNA in response to bovine tuberculin and specific antigens are greater amongst bTB culture positive badgers and cattle. In badgers the IFN-γ EIA (enzyme immunoassay) based on a monoclonal pair (mEIA) had a specificity of 93.6%, and 80.9% sensitivity (compared with 48.9% for ELISA) (Dalley et al. 2008).

---

changed? By then, we had abandoned the Statpak test in favour of the IFN-γ, which indicated that 3/66 badgers tested positive. Once an individual has had bTB (or the BCG vaccine), it would subsequently be expected to test positive, so it is perplexing that three of the individuals IFN-γ positive the previous November now tested IFN-γ negative. DiNardo (2020) however explains that 'anergy' (i.e. a lack of reaction by the body's immune defence system to foreign substances) can occur in some clinical cases if patients are underweight or suffering hypoproteinaemia, which badgers often are; thus a heterogeneous response to stimulation with IFN-γ even amongst infected individuals isn't too surprising (especially considering that the same test can apparently be differently reliable in humans from Europe, Africa, and Asia; Centers for Disease Control and Prevention (CDC) 2016).

Meanwhile, down on university's farm, selective slaughter of cattle had seemingly solved the immediate problem, with the beef herd passing its second bTB clear test, and declared as OTF in November 2014 (see testing, Box 16.3). Of the 400 dairy cows on a neighbouring, there was a moment of apprehension on 21 August 2015 when a single cow gave an inconclusive (positive) reaction; it passed on retest 60 days later, however.

The infected badgers all belonged to only three, conterminous, badger groups (Thames, Great Ash Hill (part of the same super-group), and Lower Seeds), located towards the centre of the woods, slicing down from Wytham's northern edge, adjacent to the River Thames (see map, p. xiii). None of these group ranges actually bordered FAI pasture, and, curiously, badgers from those setts (notably Thorny Croft, Common Piece, and Brogden's Belt) that did adjoin the pastures where cattle had tested positive all tested negative for bTB

(but remember (Chapter 9) how permeable the territorial fences that latrines were once thought to enforce really are).

Importantly, the outbreak involved a genetic variant, or 'spoligotype', of *M. bovis*: a mutant strain different to, but plausibly evolved from, the 'A10' type associated with both endemic badgers and cattle in the Oxfordshire region.[46] This mismatch—and, to an even greater extent, evidence—that bTB was apparently not infecting Wytham's badgers until some months after the cattle outbreak, raises the possibility that infection did not originate from the badgers, but rather from cattle, perhaps transmitted to a badger snuffling through cowpats in search of a dung beetle dinner.

How could we respond to this situation? Our best option was to vaccinate the badgers with BCG (see badger vaccination, Box 16.2). However, as bad luck would have it, a fire at the production lab in Holland put the vaccine in short supply. Nonetheless, under this first run of Defra's 'Badger Edge Vaccination Scheme' (BEVS-I), we were able to secure 50 doses.

As it happens, previous experience with vaccinating Ethiopian wolves against rabies had taught us that it was possible to stop the spread of an epizootic by vaccinating fewer individuals than might conventionally

---

[46] N. Smith (APHA, pers. comm., 20 April 2015) considered the strain to differ from A10 by a single mutation at the ETR-B locus (from a 5 allele to a 3 allele). Knowing the rate of loci change, three separate occurrences of this genotype could occur in a predominantly A10 area. We were unable to isolate a genotype from a badger carcass, but had we been able to do so, and had it turned out to be A10, this would not have confirmed the direction of infection. That could have been achieved only if the badgers had been shown to carry the barcode variant from the Lake District.

have been calculated by taking their spatial organization into account (Haydon et al. 2006; see also Randall et al. 2006). Therefore, the localized distribution of bTB positive badgers, confined to three groups, enabled us to deploy this 'gold-dust' vaccine to maximum effect amongst members of those groups and their immediate neighbours, in the hope that they would thereby form a 'cordon sanitaire'. Vaccine was completely unavailable in 2016–2017, but supply resumed in 2018, although the University of Oxford could not qualify under Defra's BEVS-II requirements, so we secured our own import licence for the Sofia BCG vaccine produced in Bulgaria. Until finally thwarted by Covid-19, in 2018 we vaccinated 157 badgers, of 187 individuals trapped[47] (and 223 estimated as alive from minimum number alive (MNA), justified in Chapter 11);[48] then in 2019 we (re)vaccinated 144 individuals, including 41 new ones not vaccinated in 2018, of 175 individuals trapped. MNA could not be calculated in 2019 because we were unable to trap in 2020; however, observations suggest the population remained in the 220 range. This gives an approximate population vaccination coverage (accounting for typical mortality rates of vaccinated individuals 2018–2019) of c.75%. We have not continued testing. Why not? Because badgers, like cattle, once vaccinated, would all now test positive, although in future testing cultured blood would be an option.

Other mammals

What of other mammals? This was the topic of a huge study led by WildCRU alumna Fiona Mathews, nowadays President of the UK's Mammal Society and Professor of Environmental Biology at the University of Sussex, who had the benefit of training in both medical epidemiology and ecology, and was assisted by Merryl Gelling, one of our then experts on small mammals. The WildCRU team focused on small mammals and other carnivores, while in tandem Dez Delahay, a CSL scientist, led a parallel project focusing particularly on deer. The answer? Suffice it to say that infection was very rare in other British mammal species (e.g. we found 1 positive bank vole out of 1307), although some pockets of prevalence have been detected in farmed

and park deer (although not in Wytham),[49] which are thought puzzlingly not to be good at passing the infection on[50] (Mathews et al. 2006a; Delahay et al. 2007; Crispell et al. 2020). This situation in 2004–2006 didn't mean it was impossible that cattle were catching bTB from a small mammal, but it was very unlikely.

What is it about badgers that can cause this cattle disease to be particularly prevalent amongst them? Led by Kirstin Bilham (introduced in Chapter 15), this question took us back to white blood cells, this time 'macrophages' (Bilham et al. 2017). Macrophages have proteins on their cellular membranes, which act like 'Velcro lock and key' entangling specific microbes. These proteins are referred to as Toll-like receptors (or TLRs), and once triggered they activate one component of the body's innate (inflammatory) immune responses—to remind readers of Box 15.3, this response produces a range of signalling chemicals (called cytokines) that recruit other facets of the immune response, meanwhile macrophages ingest or 'phagocytose' the offending microbe (along with the action of neutrophils) and lyse it with a battery of enzymes and toxic peroxidases. In particular, macrophages provide a primary line of defence against mycobacteria, which they tackle using an immunological weapon called 'inducible nitric oxide (NO). NO is another of those free radicals discussed in Chapter 15 (actually it is a gas), which oxidizes ('burns') target cells. When key macrophage TLRs recognize the enemy in the form of mycobacterial PAMPs (pathogen-associated molecular patterns[51]), they should produce the cytokine interferon gamma (the same IFN-γ that plays a starring role in the diagnostic test), activating the macrophage's inducible nitric oxide synthase (*iNOS*) gene to produce NO (and recruiting other immune cells). Interestingly, however, lab mice deficient in iNOS are highly susceptible to TB. And so we asked, do badgers have some kind of fundamental impairment of iNOS synthesis?

[47] We did not boost the same individual within years, only between years, where some individuals would otherwise receive four or five expensive and unnecessary vaccines per year.
[48] Missing out badgers receiving doubly labelled water in case response to vaccine enhanced their metabolic rate in ways hard to incorporate into our experiment.

[49] *Mycobacterium bovis* can also infect goats, pigs, sheep, camelids (e.g. farmed alpacas and llamas), as well as domestic mammals. These species could cause infections in wildlife species and companion animals that could affect humans, and although they are not a major source of infection for cattle, there could be rare and occasional localized transmission (Godfray et al. 2013). It is unknown whether increasing incidence in cattle has translated into an increased risk of infection in small mammals and other wildlife.
[50] A factor that has coincided with the northward spread of bTB has been the spread of muntjac, but no causal link is known.
[51] Such as, lipopolysaccharide (LPS), lipoteichoic acid (LTA), lipoproteins (LP), and glycophosphatidylinositol (GPI).

Forming an exciting collaboration with Oxford immunologist Adrian Smith and, testing blood samples in his lab, we discovered that badger macrophages differ from those of most mammals. They fail to produce NO, or to up-regulate[52] iNOS following treatments with chemical stimulants (agonists; such as IFN-γ) aimed at key TLRs. Furthermore, badger macrophages also failed to make NO even when stimulated with a PAMP component of bacterial cell walls (called lipopolysaccharide, LPS) that should naturally initiate this immune cascade (this stimulation did, however, cause macrophages to produce other cytokines (interleukins, tumour necrotic factor alpha/TNFα), and so other parts of the badger's immune pathway worked as expected. The most notable, unexpected deficiency was that badger TLR 9 was very resistant to stimulation, and this is the very receptor that is especially important in recognizing TB mycobacteria across mammals. Without this innate immune macrophage activation, other white blood cells are subsequently not activated in the cascade, and so, downstream, antibodies are not produced so efficiently.

Put simply, badger macrophages fail to do what they should and deliver a primary line of defence against mycobacteria infection. Ultimately, these cellular immunological deficits begin to explain why badger macrophages are unable to resist bTB. But why should this immunological vulnerability have evolved in badgers? Or, how have they survived with this unusual immunological deficiency? To speculate, presumably this was a much less powerful selective pressure until badgers met up with high-density tuberculous cattle (insofar as a half-million-year-old individual of *Homo erectus* had TB, early cattle may, in turn, have got the disease from people; Kappelman et al. 2008). Symptoms of 'disease' generally arise from the severity of the body's own immune reactions (termed 'immunopathology')—and so if the pathogen can be tolerated, there can be advantages in not becoming sick.[53] This may explain why badgers can carry TB for years, without becoming morbid or infectious. It seems

then, that bTB does not exert particularly strong selective pressure immunologically on badgers (although indirectly, culling is now a major mortality factor). An alternative speculation, suggested to us by Malcolm Bennett, is that because badgers, living in and eating soil, spend their lives buffeted by high densities of environmental mycobacteria, it may be advantageous to them not to have an immune response that responds to such challenge by producing lesions. Either way, our fruitful collaboration with immunologist Adrian Thomas, and the resultant discovery of the badger's feeble TLR9 response, opens up the possibility of developing more targeted vaccines to stimulate alternative badger immune pathways to enable the species to mount an immune response to bTB.

Beyond Wytham

Returning to ecology, and the fundamental question of how it shapes social organization, it is obvious that social behaviour is central to the spread of infectious disease, where David and the aforementioned Eunice Overend (1980) shared the insights that converged on what David named the perturbation hypothesis. This postulates that 'killing individuals may affect the survivors in ways (behavioural, physiological, immunological) that cause a disproportionate, and perhaps counter-productive, effect, such as increasing the ranging behaviour and susceptibility[54] of the survivors, so that in this instance transmission of disease to cattle was unintentionally, worsened' (Macdonald et al. 2006[55]). In an early model, we (Swinton et al. 1997) showed that perturbation had the potential to increase bTB transmission, rather than reduce it, if removal of badgers was insufficiently complete (the conservationist's dilemma is that this problem is obviated if the badgers are simply annihilated (More 2019), e.g. as achieved in the Republic of Ireland, using 'stopped' snares (Byrne et al. 2015c)). Of course, untimely deaths of companions, for example, through road traffic accidents—the latter-day analogue of predation—perturb survivors all the time. So, in the context of attempted control of bTB, perturbation refers to the additional consequences of killing larger numbers of badgers, over and above the reverberating

---

[52] Signalling pathway *up-regulation* is a process by which the availability of molecules involved in the signalling pathway—such as proteins, mRNA, or even energy—is increased in the cell. Depending on the environmental cues, the cell can increase the availability of certain molecules to carry out specific functions.

[53] For example, in the now tragically well-known example of Covid-19, it is a person's own innate immune cytokine storm, leading to excessive and pervasive inflammation, which kills them.

[54] A related unintended consequence is that stress (cortisol) lowers immune response, causing latent TB to recrudesce into active bacilli production.

[55] The perturbation hypothesis is relevant to conservation in contexts distant to epizootiology, see Tuyttens and Macdonald (2000).

consequences of background mortality. Further, per-turbation affects both those badgers that escape culling within that zone, and those impacted by the disruption of their neighbours, and beyond, outside the culled zone.

David, having been a midwife at the birth of the perturbation hypothesis, and a strident voice for its relevance to policy, positioned the WildCRU team to be deeply involved in researching evidence to test it in nature. Insofar as the badgers of Wytham Woods were seemingly free of bTB until recently, their contribution was as a baseline comparator, and our field research in infected populations was done elsewhere. Because that work is summarized in print, and given that we want, here, to devote our space to newer insights, we direct readers to Macdonald et al. (2015e and the references therein), and provide here only a skeletal summary.

The perturbation effect

This work was undertaken at two sites, North Nibley in Gloucestershire (Swinton et al. 1997; Tuyttens et al. 1999, 2000a, 2000b, 2000c; Macdonald et al. 2006, 2015e) and so-called Triplet E in Wiltshire (Macdonald et al. 2006; Riordan et al. 2011). The embattled histories of the badgers prior to each study are almost as revealing as the research findings.

The village of North Nibley sits amongst a Gloucestershire farming community blighted by bTB. David directed a radio-tracking study of the badgers there, led by doctoral student Frank Tuyttens (Frank, always engagingly philosophical and then characteristically tousle-haired, is now a Senior Researcher with the Belgian Institute for Agricultural and Fisheries Research and a visiting Professor of Ethology at the University of Ghent). The research at North Nibley compared the movements of badgers that were subjected to a MAFF (subsequently Defra) Badger Removal Operation (i.e. a euphemism for shooting animals caught in cage-traps; the sanitized acronym is a BRO) with those in two relatively undisturbed populations: at Woodchester Park, Gloucestershire, where badgers were infected but culling had ceased 18 years earlier in 1978; and, of course, at our undisturbed and, at that point, uninfected site, Wytham Woods (Delahay et al. 2000; Macdonald et al. 2006). Population trends for all three are shown in Figure 16.5 (NB the scale is different for North Nibley).

David and Frank started work in 1995 immediately after a tuberculin test identified a local bTB outbreak in cattle. The area's history makes vivid the patchwork impact of bTB and BROs in the countryside. In the

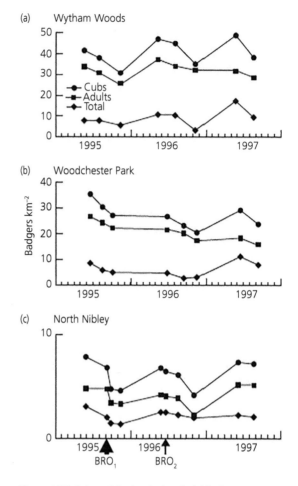

Figure 16.5 Estimated density of cub and adult badgers at (a) Wytham Woods, (b) Woodchester Park, and (c) North Nibley between 1995 and 1997. Note that the density of adult badgers in North Nibley in autumn 1996 is believed to be an underestimate (see text) (from Tuyttens et al. 2000, with permission).

17 km² study area, 12 badger-culling interventions (at the time under the live test and interim strategies—see Box 16.1) had taken place over the previous 20 years (most recently just a year earlier in 1994). Clearly, the badgers had already been serially disturbed, but were they perturbed? Between January and September 1995, before culling started, the combined WildCRU/CSL team identified 42 setts, trapped and tattooed 62 individuals, radio-tracked 14, and matched 558 latrines to their social groups by bait marking (see Chapter 9). In September 1995, officials started live-trapping badgers in a 6.5 km² area around the farm and tested their bTB status by ELISA (enzyme-linked immunosorbent assay) live test (see Box 16.3) for antibodies to

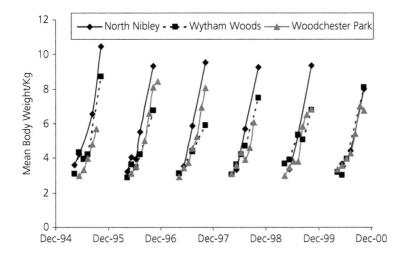

Figure 16.6 Growth curves for badger cubs at North Nibley, Wytham Woods, and Woodchester Park over the course of each year in which they were born (from Macdonald et al. 2006, with permission).

M. bovis, while we monitored infection status by culture of samples of faeces, urine, tracheal aspirate, and pus from wounds (868 sampling events[56]). Through ELISA (remember the diagnostic shortcomings) and culture tests, six groups, frequenting eight setts, were identified as infected; furthermore, 27 badgers were killed by government officials in 1995 (live test) and two in 1996 (Interim strategy). In close agreement with this, the population estimate for the 17 km² area fell from 145 animals in 1995 to 118 in 1996.

Badgers at North Nibley were generally younger, heavier, and in better condition than those in Wytham Woods and Woodchester Park, presumably because the culling relieved survivors of density-dependent constraints relative to environmental carrying capacity (Figure 16.6) (see Chapter 11).

WildCRU's engagement with Triplet E was set within a much grander plan: the RBCT commissioned by the Blair government and widely known as the 'Krebs Trial' in deference to its conception by John Krebs. The trial involved 30 replicate areas of 100 km², each with a high incidence of herd breakdowns over the 3 previous years and comparable densities of badgers (Donnelly et al. 2006). WildCRU's role against the backcloth of the coarse grain of this vast enterprise was to shed light on the fine-grain ecological outcomes in just one site to help interpret the less granular monitoring at the other 29. The 30 areas were grouped into 10 sets of 3 (Triplets A–J, see the text that follows),

and areas within each triplet were subject to a randomized intervention strategy: proactive, reactive, or no culling. In proactive culling areas an attempt was made to kill all the badgers on all accessible land,[57] and to keep the population suppressed throughout the trial period by roughly annual follow-up culling operations (Bourne et al. 2007). In reactive culling areas the badger population received one round of culling on and around farms with a confirmed recent bTB outbreak in cattle (much as had happened at North Nibley). Bedevilled by the tongue-tying ambiguity of the word 'control'—meaning on the one hand the killing of badgers, and on the other, the sparing of badgers as a basis for comparison—areas without culling (scientific 'controls') are here termed 'survey-only'. The trial was probably the biggest such controlled intervention in the history of wildlife management, and cost *c.*£50 million between 1998 and 2007 (the last badger was killed on 31 October 2005). It reveals how killing a large proportion of badgers, proactively or reactively, affects the incidence of bTB in cattle herds, although the exact relationship between the proportion of badgers killed and the outcome was difficult to assess because badgers are hard to count and no one really knew how many were present at the start.[58]

Between 2000 and 2004, and in a research partnership with MAFF/Defra (once again embodied by Chris

---

[56] Trapping success was low at Nibley, and particularly so for adult badgers (Tuyttens et al. 1999), rendering methods such as MNA less suitable in that location for estimating true population size.

[57] The proportion of accessible land was an important factor, because some farmers refused access: a criterion for siting the triplet was that >70% of the land be accessible to officials.

[58] Both treatment and control areas were surveyed in autumn 2000. A total of 71 and 38 active setts were identified in the treatment and control areas, respectively, giving a density of approximately two setts km⁻² in both areas.

Cheeseman of CSL), David directed a detailed study of the 37.3 km^2 reactive (51° 27'N 2° 25'W) and 18.9 km^2 survey-only (51° 27'N 2° 25'W) elements of one Triplet—known as 'E', near the village of Cold Ashton and 30 km from North Nibley. The field team was led by Philip Riordan, now Head of Conservation Science at Marwell Wildlife. The aim was to compare badger behaviour, and patterns of bTB infection, before and after culling in the reactive area, in comparison with the survey-only area.

As per our approach at North Nibley, Triplet E badgers were bait marked to establish social group affiliations, live-trapped, of which 50 and 20, respectively, were radio-collared in the reactive and survey-only areas (gathering 8311 fixes). The field teams made 663 badger captures: 481 in the reactive culling area (71 active setts) and 182 in the non-culling (38 active setts) area, involving a total of 423 individuals (each identified by a tattoo), from which 466 samples were collected for bTB culture, but yielding only 40 bTB positive individuals—a prevalence by lab culture of 9.5%.

The work started in 2000, but as poet Robbie Burns famously observed, 'the best laid plans o' mice and men [including, and perhaps particularly, those working with nature] gang oft agley' [go often astray]; and so it was that fieldwork had to be suspended due to the 2001 epidemic of foot and mouth disease, but recommenced in October 2001, continuing until December 2003. Defra officials undertook the culling according to their routine procedures, and in the reactive area they enacted three culls over 2002–2003, with 77 badgers trap-shot,[59] representing 34–43% of the individuals estimated to be resident in the target setts.

What did we learn—especially regarding perturbation?

The crucial discovery from both North Nibley and Triplet E was that perturbation occurred. There were several strands of evidence—trap records, bait marking, radio-tracking, and bite wounding—and they all pointed to the same conclusion: that the killing caused social perturbation amongst the survivors (later, we turn to the question of whether or not the killing affected the incidence of bTB in the survivors).

First, from trapping, we learnt that badgers moved between social groups more often after a culling operation than before it. Although 12 individual inter-group

movements[60] were observed amongst the 663 trapping events at Triplet E (and thus movement was consistently very rare), 9 of these (75%) occurred following culling in the reactive area, and all involved badgers moving into removed groups either from other culled groups or non-culled neighbouring groups. Although there are legion technical biases to capture[61] and movement data (Tuyttens et al. 2000b) it seems that a vacuum effect was drawing neighbouring individuals into culled areas (see also Carter et al. 2007). At North Nibley young females were the first to recolonize vacated areas, but there was no sex bias in post-culling rearrangements at 'E'.

The next strand of evidence was bait marking: at both North Nibley and in Triplet E culling was followed by a wider dispersion of bait-marking returns, with the result that the juxtaposition of bait-marking ranges changed from loosely tessellating towards chaotically overlapping. Consequently the estimated social group ranges at Nibley enlarged by 68% after the culling (Figure 16.7a, b), using latrines 35% further away (Figure 16.7c).

The team worked at North Nibley for long enough to be able to evaluate how quickly survivors in this perturbed population returned to 'normal'. Of course, following two decades of repeated perturbation, normality is a slippery notion, but while the population took only 3 years to recover numerically to its previous abundance (Fig. 16.5), the distribution of badgers across the landscape had not returned to its initial state after 4 years (Figure 16.8).

Further evidence of social perturbation came from radio-tracking. The 50 badgers radio-tracked in Triplet E (reactive) revealed that inter-range overlap in summer increased a year after culling, from an average of 7.5 to 13.0 ha in neighbouring groups, and from 8.2 to 10.8 ha in removed groups; overlap did not change significantly amongst the 20 individuals tracked in the survey-only area. Radio-tracking also revealed that in the year following culling the number of shared

---

[59] Culling November 2002, 38 badgers from 14 social groups; January 2003, 6 badgers from 4 social groups; August 2003, 33 badgers from 13 social groups (Riordan et al. 2011).

[60] Inter-group movement was defined for the purposes of the study as occurring when a badger was trapped at a sett belonging to a different group than the one in which it had previously been captured, or when the badger was located at a sett belonging to a different group by night-time radio-tracking or day-time positioning.

[61] The exact interpretation of movements, as evidenced by captures at setts, before and after culling, is horribly complex. For example, if culling reduced the numbers of badgers, yet trapping effort remained the same before and after culling, then there would be a greater number of traps available in which to catch each surviving badger.

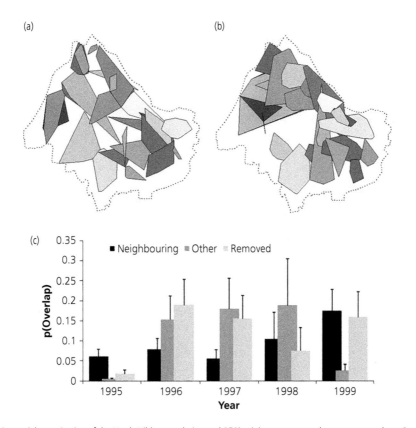

Figure 16.7 Socio-spatial organization of the North Nibley population and 95% minimum convex polygon range overlaps. Outer limits of each group range as determined by bait marking: (a) in 1995, before the main cull, showing a relatively well separated population with most groups forming largely exclusive ranges (although less so than at Woodchester or Wytham). Where significant overlaps do occur, these mostly show social groups utilizing more than one sett; and (b) following the cull in 1996, the situation shows an appreciable breakdown of the social group delineations; (c) Proportion of overlaps across the study area within the three categories of Nibley badger social groups: 'removed' at which culling took place in 1995, neighbouring groups adjacent to removed groups, and 'other' groups, which are at least one social range away from any removed group (adapted from Macdonald et al. 2006, with permission).

group ranges with which each individual badger overlapped in summer increased for neighbouring and other groups (from an average of 0.8 to 2.7 groups) but declined (from 0.8 to 0.2) in removed groups (Riordan et al. 2011).

The final strand of evidence of perturbation following culling came from Triplet E following culling in the winter 2002/3, when the proportion of badgers with fresh bite wounds increased significantly in the neighbouring and removed social groups (the overall rate at which male individuals received fresh bite wounds in the reactive area was 22.6%). In contrast, culling-induced social perturbation was not associated with higher levels of bite wounding at North Nibley or Woodchester; however, there was evidence that female badgers endured higher rates of biting during recolonization (Delahay et al. 2006b).

What of disease?

While the ultimate purpose of these badger culls at North Nibley and Triplet E was to reduce bTB in cattle, the immediate intention was to eliminate infected badgers. Did it work? Prior to the culling (in September 1995) at North Nibley, 14.5% of badgers tested positive for bTB using ELISA tests (but only two to bacterial culture, so culture prevalence was 3.2%; Macdonald et al. 2006). Between 1996 and 1999, after culling, none of the 258 badgers tested positive by culture,[62] so perhaps culling had eliminated bTB. How long did the badgers remain TB free? By 2000, 3 badgers (of 64

---

[62] Culture test has a sensitivity of only c.8%. One of the culture negative badgers tested positive using ELISA, which is a more sensitive test.

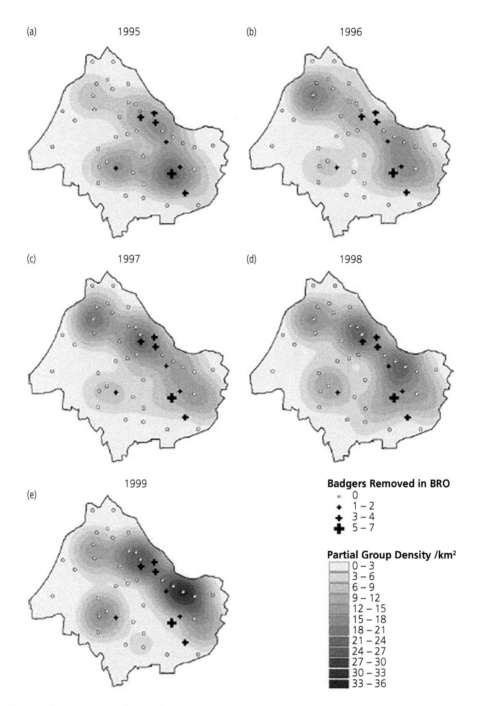

Figure 16.8 Interpolated visualization of badger density across the North Nibley study area between 1995 and 1999. Social groups removed in 1995 are shown as black crosses, with the symbol size indicating the number of animals culled. Social groups not culled in 1995 are shown as white dots (from Macdonald et al. 2006, with permission).

tested) were culture positive (culture prevalence 4.7%). What of the ultimate purpose? Sadly, despite indications that the badger population was bTB-free, two farms in 1998, and one in 1999, suffered herd breakdowns. If the badger population was indeed bTB-free, then cattle must have become infected in the absence of infected badgers, while badgers, for their part, subsequently became infected when the only source of the disease was within cattle.

Outcomes were even less promising for the culling strategy at Triplet E. Before the cull, the detected prevalence of badgers excreting *M. bovis* was not statistically different between the treatment and control sites, with estimates of 8.1% and 7.1%, respectively. It didn't differ, either, between the sites after the cull, although the proportion of badger groups that included some surviving culture positive individual increased significantly in both sites: at the reactive site infected groups increased from 5% (3 groups) before the intervention to 15% (18 groups) afterwards, whereas at the control site group positivity increased from 2.7% to 15.9%. In short, things got worse at both sites, which suggests not only that culling didn't work, but also raises a puzzle over perturbation insofar as there was culling in the reactive area but supposedly not in the other area. The widely whispered hearsay on the ground was that badgers in the control (no cull) zone were heavily culled illicitly, rendering both sites *de facto* 'reactive'. We do not know. However, what was dramatically clear was that in the reactive area culling was associated with a significant increase in detected *M. bovis* in cubs, from one cub excreting bacilli amongst 101 pre-culling to 8[63] of 76 cubs post-culling (Figure 16.9) (Riordan et al. 2011).

It requires unshakeable Pollyannaishness[64] to interpret these results as evidence that culling improved the situation of badger infection. Furthermore, turning from individuals to groups, the increased prevalence of *M. bovis* detected in surviving badgers following culling within the Triplet E reactive area was greatest within neighbouring social groups, from 1% (standard deviation (SD) ± 4%) to 14% (SD ± 12%), and was greatest of all amongst groups whose members had been involved in inter-group movements.

From our local-level analyses at Triplet E and North Nibley our conclusion was that reactive culling at the spatial and time scales undertaken in the treatment areas was of short-lived benefit at Nibley, and none at Triplet E for controlling bTB in badgers, and at Triplet E may have made it worse (Macdonald et al. 2006). At the larger level of the RBCT, reactive culling was associated with an increase both in the prevalence of *M. bovis* in badgers across all triplets and the frequency and duration of bTB incidents in cattle herds. Therefore, in November 2003 the government stopped the reactive treatment. This outcome, entirely in accord with the perturbation hypothesis, is elaborated in Woodroffe et al. (2006a), Bourne et al. (2007), Vial and Donnelly (2012), and Karolemeas et al. (2011, 2012), and was first flagged by Donnelly and colleagues'(2003) report of evidence of a 22% increase in the incidence of new confirmed bTB cattle herd breakdowns (95% confidence intervals (CI): 2.5% to 45% increase) across all 10 reactive elements of RBCT triplets. It may therefore be surprising that (see the text that follows) a version of reactive culling has been the government's default strategy in the low-risk areas of England (Cumbria 2018, Lincolnshire 2020).

With our microscope ratcheted down on the detail, by 2004 the findings at North Nibley and Triplet E had convinced David that culling badgers induced social perturbation amongst the surviving badgers, that this seemed to be associated with either a worsening of bTB prevalence amongst the survivors or, at best, only short-lived improvements, and that the corresponding evidence for the rate of breakdowns amongst cattle herds was similarly unpromising. This was exactly the conclusion of other ecologists on the ground: CSL's badger ecologist, Chris Cheeseman, reminds us of his experience, in the mid-1970s, that small-scale badger removal operations around breakdown farms were followed by a rash of new breakdowns in the neighbouring farms.

But these conclusions stemmed from only fragments of the landscape—as time passed it was possible to see the bigger picture revealed by the whole RBCT. Many of these analyses were led by ecologist Rosie Woodroffe and also by Christl Donnelly, the latter now Professor of Applied Statistics in Oxford; whose commendable fascination with the reality behind the numbers brought her and her sons to visit our Wytham catch-ups. Their analyses later revealed that, across the whole RBCT, infection prevalence in badger populations rose after *both* reactive *and* proactive culling (Woodroffe et al. 2006a, 2009b).

---

[63] Four cubs tested positive for *M. bovis* from urine samples, three from sputum, and one from a faecal sample.

[64] The word comes from a 1913 children's book by Eleanor H. Porter, *Pollyanna*, about a young girl who tries to find something positive in every situation—a trick she calls 'the Glad Game'.

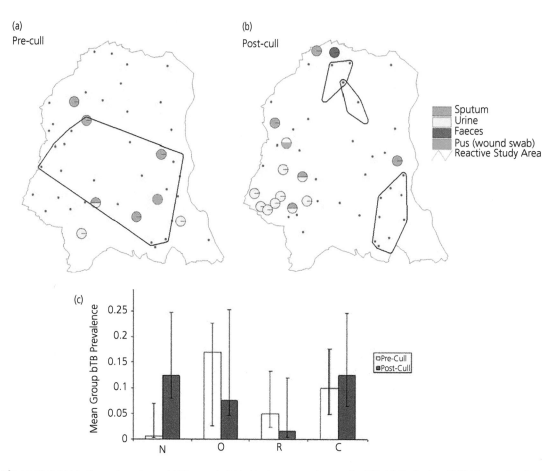

(a) Pre-cull

(b) Post-cull

Sputum
Urine
Faeces
Pus (wound swab)
Reactive Study Area

(c)

Figure 16.9 Distribution and prevalence of bTB among individual badgers pre- and post-cull at TEE. (a) Distribution of bTB (a) pre-cull and (b) post-cull. Infection status of badgers was determined by bacterial culture of samples of urine, faeces, sputum, and pus (n = 466 individuals). (c) Mean bTB prevalence in Removed (R; culling had taken place), Neighbouring (N; groups adjacent to removed groups), Other (O; at least one social group range away from any removed group), and Control (C) badger social groups before and after culling. Error bars show 95% binomial confidence intervals (from Riordan et al. 2011, with permission).

With the failure of the reactive strategy—a failure that had been predicted by the perturbation hypothesis, although of course it had not necessarily been for that reason alone—focus turned to proactive culling. As proactive culls occurred, an increasing proportion of captures was in the periphery of the culling areas (Woodroffe et al. 2006a) and an increasing proportion had genotypes indicating they were likely immigrants (Pope et al. 2007), both facts compatible with the perturbation hypothesis. The key results from the proactive treatment of the RBCT were: (i) while culling was occurring, the number of confirmed herd breakdowns within the culling area was 23% lower than that inside non-culling areas; but (ii) herd breakdowns increased by 25% in the 2 km-wide perimeter surrounding the core culling area (where culling had perturbed badger populations; Donnelly et al. 2007;[65] Woodroffe et al.

[65] The two proxy measures of herd bTB incidence (itself unknown) are the number of confirmed breakdowns and 'total' breakdowns. Confirmed breakdowns are those where a positive SICCT test is confirmed by the presence of lesions post mortem; if no lesions are found, the case is unconfirmed. The most quoted results of the RBCT used confirmed incidence (it was not until 2018 that the specificity of SICCT was known to approximate 100% (Nunez-Garcia et al. 2018) so this measure may have cautiously avoided the then perception of possible inaccuracy of misallocation owing to SICCT positive animals whose absence of lesions was because they were false positives). This measure reveals the 23.2% decrease

2006b; (Table 16.1). In the longer term, beneficial effects of culling within culled areas increased after culling ended and persisted for several years, albeit diminishing over time (Jenkins et al. 2010 and associated updates). And what of perturbation, the early lessons of failed fox rabies control, Eunice Overend's freezing caravan, the tendency of some vets to dismiss ecologists, and of politicians to dismiss both? With the hindsight of the RBCT, it was widely considered that the increased prevalence in cattle in the perimeter zone, and the general ineffectiveness of previous BRO strategies, were at least partly attributable to perturbation.

What were the policy implications of this work? The formal conclusions of the Independent Scientific Group on Cattle TB (ISG), headed by veterinarian John Bourne included: 'badger culling cannot meaningfully contribute to the future control of cattle TB in Britain'; further, that 'some policies under consideration are likely to make matters worse rather than better'; and that 'weaknesses in cattle testing regimes mean that cattle themselves contribute significantly to the persistence and spread of disease in all areas where TB occurs' (Bourne et al. 2007). What next? Considering the conclusion of the ISG, any readers who have subsequently been living on Mars might imagine that that was that—debate resolved, with the result that which action should be taken, and which not, was abundantly clear. Indeed, that is how it appeared to David and his colleagues on the Board of Natural England as they strove to guide policy with evidence. Not only a founding member of Natural England's Board (and, previously, on English Nature's Board), David was also founding Chairman of Natural England's Science Advisory Committee (NE-SAC), and so in the thick of guiding policy. A critical conduit for that guidance was Dr Tom Tew, then Natural England's Chief Scientist, responsible for Science Advice (for the connoisseur, Andrew Wood was responsible for Science Policy). Tom, tall, laconic, prematurely grey,

and unusually clever, had trained with David first as a doctoral student and latterly as a post-doctoral researcher in WildCRU, as an expert on small mammals on farmland; he was gifted, also, with exceptional clarity of thought, making him the ideal vessel by which to relay the Board's considerations to Hilary Benn, then Secretary of State for Environment, Food and Rural Affairs. On 7 July 2008, Mr Benn made a statement to parliament that the government's policy would be neither to cull badgers nor to issue any licences to farmers to do so. However, culling was not dead, and shortly we will return to what happened next, and whether it was a good idea. First, some reflections on perturbation.

Reflections on perturbation

How might the insights from North Nibley and the RBCT alongside those on the, in some ways, chaotic sociology of the badgers of Wytham Woods (notably in Chapter 4 and 9) inform the perturbation hypothesis? Three topics merit reflection: What characterizes undisturbed badger populations on the one hand, and disturbed populations on the other? What distinguishes the tenor of interactions between unperturbed and perturbed badger communities? Which elements of perturbation contribute most to epidemiological outcomes?

What characterizes undisturbed and disturbed populations? It was already clear to Tuyttens et al. (2000a, 2000b) that the badgers at Nibley had been serially culled (12 BROs in previous 20 years) (Figure 16.10), potentially explaining why, when the killing began there in 1995, the badger group home ranges already tessellated less neatly than those at Wytham or Woodchester. The changes documented in badger behaviour pre- and post-BRO were, therefore, not between undisturbed and disturbed, but, rather, between those recovering from disturbance on the one hand, and those recovering from a greater degree of disturbance, on the other; the same applied to Triplet E, with its similar history of culling. Indeed, some culling in Triplet E was so inefficient that it constituted mortality at a level (34–43%) only about twice that caused by road traffic accidents at Wytham and some otherwise undisturbed populations (Rogers et al. 1997; Macdonald and Newman 2002; see Chapter 11).

So, while there seems no doubt that the counterproductive effect of BROs predicted by the perturbation hypothesis does occur (even when the culling is undertaken at a very small scale; Bielby et al. 2014), the manner in which percentage mortality, and the

---

($P < 0.001$) in herd breakdowns in the Proactive culled area, and the 24.5% increase ($P = 0.057$) in the 2 km perimeter. However, More and McGrath (2015) repeated these analyses using 'total' unconfirmed incidence, which led, respectively, to a 11.7% decrease ($P = 0.063$) and a 13.5% increase ($P = 0.17$), neither effect being statistically significant. While the confirmed incidences therefore show an effect of badger culling on the number of herd breakdowns, and thus evidence of badger to cow transmission, including a perturbation effect, the unconfirmed incidences suggest neither. Insofar as the specificity of the SICCT approximates 100% one might expect many of the unconfirmed cases in the 'total' measure to have been real, and thus the analysis using unconfirmed incidence, while seemingly less conservative, to be more accurate.

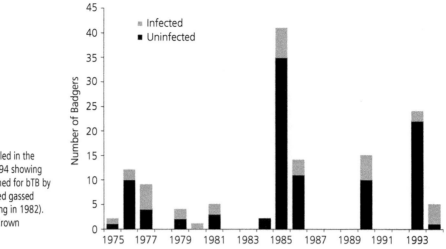

Figure 16.10 Badgers culled in the North Nibley area 1975–1994 showing numbers caught and examined for bTB by MAFF (excluding unrecovered gassed animals before end of gassing in 1982). Data provided by Defra. ©Crown Copyright.

expanded network and increased rate of contacts, translate to epidemiological consequence is less clear.

Clues may lie in answers to the second question: what distinguishes the tenor of interactions between unperturbed and perturbed badger communities? Although the perimeter latrines, at least until about 2007 (when we last bait marked), deployed by the higher-density badger populations of Wytham (Chapter 9) and Woodchester (Delahay et al. 2000) tessellated much more tidily than those of the serially perturbed lower-density badger population of North Nibley or Triple E, we have seen that these crowded badgers from Wytham and Woodchester (Carpenter et al. 2005; see also Marjamäki et al. 2019, 2021) very often stray widely, meet neighbours one, two, or three setts away from their own, and often conceive offspring while doing so. If, as the results of the RCBT suggest, perturbation is associated with a post-BRO expansive, highly interactive social network, what is the critical dimension, epidemiologically, that distinguishes their interactions?

Two non-exclusive possibilities come to mind: the much lower population density of badgers in most cull areas may mean that prior to a BRO they experienced fewer social contacts than those crowded into Wytham (the sociology of populations of different density may be profoundly different (see Chapter 19), e.g. in terms of the number of oestrous periods per year (Chapter 7)). Equally, the tenor of the interactions, more importantly than their number, may change radically after a BRO, as may the susceptibility to infection of the badgers. Remember that even as the abundance of badgers in Wytham almost trebled, so the frequency of

bite wounding declined to nearly zero (Chapter 11), whereas following the reactive cull at Triple E (but not at Nibley) the number of bite wounds soared (see also Jenkins et al. 2012). Perhaps while strangers fight, crowded familiars merely bicker. This leads to the third question: which elements of perturbation contribute most to epidemiological outcomes? There may be a very high frequency of interactions both amongst the relatively few survivors of a BRO, and amongst the swarming crowd of an undisturbed high-density population; they may, however, be of an epidemiologically crucially different tenor, involving, for example, more bite wounding. They may also, in the bereaved population, involve animals immuno-compromised by pervasive stress. The depleted immuno-competence of survivors is a largely unquantified[66] but potentially significant aspect of perturbation, as too might be the calibre of survivors if culling removes primarily the uncautious, meaning that survivors are harder to catch (and may also have been the more sickly, less adventurous individuals that skulked below ground). These are speculations, which are, however, compatible with evidence in Tuyttens et al. (1999 and 2001). Survivors may also behave differently with regard to where and when they urinate and defecate (see Chapter 9), with implications for transmission to cattle (Hutchings and Harris 1999).

So, when all is said and done with caveats and complexities, is perturbation relevant to bTB policy? Yes, it confounds culling efficacy, except in cull zones with

---

[66] Stress associated with infection status, but not perturbation, is tackled by George et al. (2014).

hard boundaries, provoking counter-productive consequences that can make a marginal gain diminishingly worthwhile.

For those interested in the interplay between politics and science, politicians and scientific advisers—a subject currently topical, courtesy of Covid-19—David recalls the political shenanigans, not all of them laudable, of the badger TB debate at the end of this chapter (see Policy postscript).

Anyway, the question 'what next?' becomes more pressing, as we consider first vaccination, before returning to the implications of the sensitivity of diagnostic tests.

Vaccination

Covid-19 may have reminded a wide public of the great power of vaccination to control infectious disease. Further, from the moment the perturbation hypothesis first formed around the sociobiological benefits of vaccinating foxes against rabies[67], the idea was that the very same sociology that undermines the efficacy of killing badgers as a way to stop them infecting cattle with bTB might work in favour of vaccinating them for the same purpose. But to end where we began: might it work? What to vaccinate (badgers and/or cattle), and with what? The key review of this topic is Buddle et al. (2018).

Vaccinating badgers

Defra has invested over £16 million on research into badger vaccines since 1994. Although the paperchase of licences[68] and the practical expense[69],[70] still remain, BadgerBCG has been authorized and is available (Defra 2013). Further, it works, at least well enough to be a sensible alternative to the generally unsuccessful and—to speak in broader

terms—ethically problematic, culls. Sandrine Lesellier and colleagues at Defra (2006) vaccinated captive badgers with commercial BCG using two consecutive doses (impractical in the wild)—a first dose of strong vaccine preparation (ranging from 16 to 22×10^7 colony-forming units (CFU)), followed 15 weeks later by a weaker dose (4–7×10^5 CFU). Did the vaccine elicit an immunological response? Yes, just 13 days after injection, badgers exhibited strong cellular responses (see Box 15.3), which persisted for 76 days—in practical terms, sadly not long (the lower the booster dose, the more reduced and shorter lived the cellular response).

Although BadgerBCG, like most vaccines, does not prevent infection, its intra-muscular injection reduces disease progression both in captive individuals by 74% (decreasing its severity and the excretion of *M. bovis* pathogens, which might sensitize or infect cattle) and in social groups of wild badgers (Chambers et al. 2011; Lesellier et al. 2011). In addition, Carter et al. (2012) found that when more than a third of a social group had been vaccinated, the risk of infection to unvaccinated cubs was reduced by 79% through an apparent herd immunity effect. Models suggest annual BCG vaccination of 70% of badgers could eradicate bTB from their population in 20–30 years (depending presumably on spill-over risk from cattle)—this is obviously a long time; however, it is at least a shorter duration than that during which culling badgers has been associated with increasing bTB prevalence amongst them (Wilkinson et al. 2004).

What of the costs of vaccination? BCG is not expensive, would not require the expensive policing associated with culling, and much of the fieldwork can be shouldered by volunteers, although the latter do need to be trained.[71] All this sounds promising, but it isn't easy. The badgers must be cage-trapped and then competently injected while lucid (remember, when we do it we are anaesthetizing the badgers for other reasons anyway, which makes vaccinating them much easier, but much more expensive, and is a luxury that lay vaccination groups would not enjoy). Worse, previous culling will have weeded out the gullible, easily trapped individuals, leaving a greater proportion of evasive individuals (at Woodchester infected

---

[67] An interesting possible ecological cascade is that the Swiss fox population increased following the elimination of rabies, with an associated trebling of the human disease alveolar echinococcus caused by the parasite *Echinococcus multilocularis*, which infects Continental (but not British) foxes.

[68] This vaccine must be (i) prescribed by a veterinarian under the Veterinary Surgery (Vaccination of Badgers against Tuberculosis) Order 2010, (ii) requires a trapping licence from Natural England along with the use of authorized traps, and (iii) vaccinators must have a Certificate of Competence (with an annual fee, currently £700), accredited through an APHA training course (currently £750).

[69] Each live-trap retails at *c*.£140. Government estimates a cost of around £2250 per square kilometre per annum to vaccinate badgers.

[70] Defra has grant funding schemes, up to £250,000 per year, under the BEVS-I, prioritized to high-risk areas.

[71] Government policy enables farmers and landowners to cull and/or vaccinate badgers under licences, i.e. under the Protection of Badgers Act 1992 and Wildlife and Countryside Act 1981. Defra's 'Guidance to Natural England on the implementation and enforcement of a badger control policy' (issued under section 15(2) of the Natural Environment and Rural Communities Act 2006 ('the NERC Act')) sets out what is required.

badgers were four times less likely to be re-trapped than were uninfected ones - so cage trapping would preferentially remove uninfected ones). Furthermore, it needs to be repeated yearly to 'boost' the immunity of previously inoculated adults, and to include new cubs (although a single BCG vaccination gives a person some, although perhaps not much, protection for up to 15 years against *M. tuberculosis*).[72] In short, while killing badgers is more expensive, and more costly in currencies in addition to money, vaccination is far from cheap. In 2015 *Farmers Weekly* estimated the full economic cost of vaccinating a Welsh badger was £825. Mark Chambers and colleagues (2014) reported that 6788 *badger* BCG doses were delivered in England and Wales between 2010 and 2013. The single largest vaccination project to date was initiated in 2012 by the Welsh government. In the first year they vaccinated 1424 badgers over 241 km^2 in west Wales at a cost of approximately £945,000 (Welsh Government 2013). In the second year (2013), 1352 badgers were vaccinated over 258 km^2 at a cost of c.£927,000 (which works out at just shy of £700 per badger) (Welsh Government 2014). Costs per badger of this order of magnitude push hope in the direction of fantasy, but the task would be less daunting, and the costs could tumble, if vaccine could be delivered orally in baits (Chambers et al. 2014). That raises questions about badgers eating too little or too much, or cattle eating any,[73] and technical questions about getting the oral vaccine into a lipid matrix (used successfully for possums in New Zealand) (Godfray et al. 2013; see also Robinson et al. 2012). A field trial in Ireland delivered lipid-encapsulated BCG to the backs of the throats of one sample of anaesthetized badgers, and a lipid-coated placebo to another, and these treatments were applied to three experimental zones where 0, 50%, and 100% of the badgers were vaccinated (Gormley et al. 2017). Badgers re-examined between 18 and 36 months suggested a vaccine efficacy of 84%. What is more, only 9% of badgers in the 100% zone presented *M. bovis* lesions, compared with 26% in the non-vaccinated zone. The target dose of BCG to vaccinate badgers orally is still unknown, and is likely to be high, but that is not in itself a problem.

Badgers having originally been infected with bTB by cattle, there is likely to be a continuing need to reduce the disease in badgers to prevent them passing it back

to cattle. Might badger vaccination work? If pursued assiduously, probably, yes.[74] If accomplished orally, then a useful contribution in principle would become an excellent one in practice. That said, it would, to recall the opening of this chapter, be simpler to vaccinate cattle—they are, after all, routinely vaccinated against blackleg, tetanus, 'husk' (lungworm disease), rotavirus, and a swarm of other infections including, topically, coronavirus.[75] But would that work?

Vaccinating cattle

Does BCG work for cattle? Yes, although it's only 70% effective at best, and in individual cattle or herds with other disease issues (e.g. Johnes, BVD, liver fluke burden) it's much less effective than that. In an important paper Srinivasan et al. (2021) suggest an overall vaccine efficacy of 25% (95% CI: 18, 32) as measured by the presence of visible lesions and/or culture, and taking account of direct benefits and herd immunity. BCG vaccination reduces the pathophysiology of infection in cattle and can reduce the numbers of bacteria, the progression, severity, and excretion of bTB under experimental conditions (e.g. Wedlock et al. 2007; Hope et al. 2011; Tompkins et al. 2013), and in calves it works equally orally or parenterally (Buddle et al. 2005). A key point is that while experimental challenge trials had shown that vaccination only reduced the severity of disease, not its transmission, field trials show that it can markedly reduce the number of infected cattle and therefore is reducing transmission (Lopez-Valencia et al. 2009; Ameni et al. 2010). This reduction in transmission really matters, because it, in turn, reduces *pro rata* the crippling costs to the farmer associated with slaughtering reactor cattle. Srinivasan et al. (2021) entitle their meta-analysis 'is perfect the enemy of good?', and conclude BCG is good enough to accelerate control of bTB. The answer may differ between wealthy and poor parts of the world and the 'good enough' conclusion may, importantly, be especially strident in middle- and lower-income countries. Here, our focus is on the UK.

In the UK, and if culling is to continue alongside vaccination, none of this will matter unless there is a

---

[72] This protection is far from complete. Despite years of intensive research, BCG remains the only licensed vaccine and has variable efficacy (Fatima et al. 2020). One recent paper estimates vaccine effectiveness at 30% (Katelaris et al. 2020).

[73] Rendering them positive to the tuberculin skin test.

[74] The ideal would be to reduce prevalence in badgers to a level that minimizes risk to cattle, while achieving a prevalence in cattle that minimizes risk to badgers. Now that more sensitive diagnostic tests are emerging, modelling the route to this biosecurity sweet spot would be a worthwhile priority for research.

[75] Cattle carry their own bovine coronavirus (BCV)—a major cause of diarrhoea in young calves in winter. They can be protected against this via an intranasal vaccine.

test to differentiate infected and vaccinated animals (DIVA). It was the absence of such a test, and the confounding effect of BCG on the tuberculin SICCT test, that prompted the World Health Organization/Food and Agriculture Organization (WHO/FAO) to conclude in 1959 that 'generally speaking, vaccination has no place in the eradication of tuberculosis in cattle' (WHO/FAO 1959). Times may change, and immunologically, the most promising approach to distinguishing vaccinated and infected cattle involves the same reagent that has improved diagnosis: IFN-γ (the DIVA interferon gamma test, or DIT, to compound acronyms). Currently DIT tests are under field trials by a company, perhaps unexpectedly, previously specialized in slaughterhouse inspection (Eville & Jones n.d.). If the DIT tests perform to OIE standards the test could be available to use in 5 years: game over. . .? Well, only if the BCG vaccine works operationally in tandem with the surveillance tests: is that likely on an imminent time scale?

Sadly, while the time to firing the starting pistol is uncertain, the time thereafter to reach the finishing line is over the horizon. While the scenario explorations of Srinivasan et al. (2021) suggest that rolling out BCG cattle vaccination within the next 10 years would be broadly helpful, their assessment of averting cumulatively 50–95% of cases (where prevalence is <15%) is computed over 50 years. Meanwhile, the difficulty with perfecting the DIT is that BCG vaccination sensitizes cattle to bovine tuberculin, compromising the current surveillance tests (SICCT and IFN-γ release assay) that use this test reagent. Jones et al. (2017) circumvented that by replacing the tuberculin reagents in the IFN-γ release assay with two peptide cocktails (ESAT-6–CFP-10 and Rv3615c) that induce responses in blood from *M. bovis*-infected cattle but not from BCG vaccinates. Work at APHA on DIT tests is being led by Professors Glyn Hewinson and Martin Vordermeier (Aberystwyth University; see Vordermeier et al. 2016). Currently its sensitivity is 56% (CI: 37–78%); better, in other words, than the SICCT.[76] Therefore, some vaccinated animals will be negative in the DIT but positive on the SICCT, so it is not obvious when, how, and where the vaccine/DIT test would be used (e.g. ubiquitously or selectively, perhaps to suppress infection in persistent chronic herds or in the low risk areas (LRAs) to provide immunity); then again, not every

infected animal responds to the DIT antigens themselves, so if an animal is infected it may still be negative to the DIT antigens (i.e. it would appear that it had been vaccinated rather than infected). In addition there is still the issue of how the use of BCG and these tests would affect trade and export of live animals, meat, milk, and their products, and how it would affect OTF status for England, Scotland, and Wales. However deeply farmers and badger conservationists alike long for cattle vaccination against bTB to be the panacea, both may have to continue longing for quite a while if they rely on the immunological DIT-type DIVA test. However, they may not have to: although not yet OIE-validated as a legal test, it seems a DIVA test exists in the form of Actiphage (the non-immunological, phage-based test mentioned earlier): if validated, this would be a game-changer as it clearly overcomes the problems of identifying infected cattle, as opposed to those that have only responded to vaccination.

There is an interesting issue of redundancy when comparing the outcomes of BCG vaccination of cattle versus potentially more sensitive diagnosis (e.g. Enferplex or Actiphage): both approaches could reduce the bankrupting and heartbreaking costs of slaughter, although in the end killing infected cattle will be necessary to eradicate the disease, and will cost a fortune. The comparison is thus between the marginal benefits of the number of cattle saved from premature slaughter by the reduced transmissibility due to vaccination versus those saved by much earlier diagnosis and the associated slower staging of slaughter facilitated by the traffic-light system (footnote 16). It would be useful to model whether vaccination adds much to the benefit gained by much more sensitive diagnosis, and to calculate the annual costs (surely north of £100 million) and years to eradication of both.[77] So, vaccinating cattle seemed like the obvious answer, but will it work? Probably not soon enough.

Meanwhile, what of the collateral damage?[78] If, as seems likely, badgers caught bTB from cattle, and with

---

[76] With 50% sensitivity and 100% specificity, of 1000 infected cattle tested with SICCT, the result will be negative in about 500.

[77] A newcomer to the field might ask, why not treat the tuberculous cattle, and try to cure them, considering that there are several CDC-approved treatments for people. The answer, aside from the practical logistics of doing so (the treatment might last 6 months), is that there is already a huge problem with resistant human TB, and the risks of wider environmental effects of the drugs being excreted into the environment would require extreme scrutiny.

[78] A form of collateral damage from the badger cull, which we have not detailed here, involves cascading ecosystem effects. For example, culling badgers is associated with increasing abundance of foxes and hedgehogs, leading to speculation on added threats to ground-nesting birds.

the disease's control in cattle it is left festering in badgers, is there a moral responsibility to right a wrong wrought by agriculture on nature? A widespread view in conservation is that it is inappropriate for people to intervene in natural processes such as disease or predation, but nowadays natural processes often have unnatural consequences due to derailment by the human enterprise; for example, exposing wild animals to the perils of extreme endangerment. Where the threat to wildlife originated, even indirectly, from people, people may be less sanguine about being passive onlookers. Two obvious questions come to mind about bTB in badgers in a hypothetical new world of vaccinated cattle: first, will infection in badgers be sustained in the absence of reinfection from cattle; and second, how great would be the continuing mortality, morbidity, and suffering? Answers to those questions might be entered in the balance sheet of the costs of a badger vaccination programme, which, if it really did eradicate badger TB in 20 or 30 years, might then facilitate reducing the costs of cattle vaccination.

What now?

In the meantime, how's it going? In England in 2020, 27,852 cattle (reactors and direct contacts) were slaughtered (Defra 2017); in the 12 months to April 2020, 35,034 badgers were culled (in a process designed such that it is unlikely we will ever know its consequences). Meanwhile the compulsory slaughter in Britain of reactor cattle or their direct contacts increased from fewer than 7000 in the late twentieth century to 32,791 in 2013 (incidence rate of 4.5% of new herds infected[79]) (Defra and AHVLA 2014), to 44,656 in 2018 (House of Commons Briefing Paper, 2019).

It's time to return to the two candidate leaking buckets for spreading infection to cattle: badgers and undiagnosed cattle. Insofar as the badger culling has been implemented in a way that renders its impact uninterpretable,[80] what of diagnosing cattle? Remember the

poor sensitivity of the SICCT test and reducing frequency of testing. Obviously, relying on a test that is scarcely 50% sensitive is like staking your future in a game of Russian roulette where half the bullets really are blanks but the other half may not be. There are the makings of an indicative scientific comparison between Wales and England, albeit poorly controlled. Remember that exactly the same evidence that persuaded England to push for the badger cull, in 2011, following a judicial review, convinced Wales not to. Instead, from 2010, the Welsh opted for annual cattle testing across Wales, wider use of severe interpretation of the SICCT test, supported, from 2014, by IFN-γ testing, and IDEXX[81] ELISA since 2018. Perhaps preoccupied with killing badgers, the English were slower to adopt these improvements to diagnosis, and also to a suite of other cattle-directed restrictions. England only applied severe interpretation of SICCT to all breakdowns in the HRA in April 2016, where at that time 90% of confirmed breakdowns occurred, and adding IFN-γ tests to 73,268 cattle in 2016, increasing to 98,529 in 2017, 193,638 in 2018, 263,774 in 2019 and 221,766 during covid in 2020. These circumstances might permit two cautious interpretations of trends in bTB breakdowns between Wales and England (mindful of the host of confounding differences between the two countries, their farms, and badgers): unless a reduction in the incidence of herd breakdowns is much greater in England than Wales, it cannot be attributed to badger culling rather than improved diagnostic testing; further, if any improvement in England is found to be not much greater than that in Wales, it will seem likely that the badger culling has been fruitless.

The longest running West End show, aptly titled in this context, *Les Miserables*, began the year after David began writing about bTB; sadly, however, there's all too great a chance that the badger and TB debate will outlast it. How are the recent acts, played in parallel in England and Wales, unfolding? According to Defra (2020):[82]

- In England the herd incidence rate for the 12 months to end December 2020 was 9.4, the same as the

---

[79] New herd incidents with officially TB free status withdrawn (OTFW) divided by the number of tests on officially TB free herds (OTF).

[80] A conclusion echoed by Defra's Chief Scientific advisor (Boyd 2019) in an advice note: 'The method adopted by Defra involves a set of different interventions applied in concert with each other and this includes badger culling. It is not possible to examine any single measure, such as supplementary badger culling, alone as having either a positive or negative effect. Only the whole set can be considered together. Therefore I agree with the letter which states that it is not possible to attribute causation of any effect directly to supplementary badger culling.'.

[81] OIE data suggest test sensitivity of 65% and a specificity of 98% for the IDEXX antibody test in cattle. To maximize the sensitivity of the test, a prior tuberculin skin test is required to boost *M. bovis*-specific antibody levels in TB-infected cattle. Antibody tests are less sensitive overall compared with the tuberculin skin test and IFN-γ blood test, but can be useful for identifying small numbers of infected cattle that are skin and IFN-γ test negative.

[82] See also Defra and APHA (n.d.) and 15/6/22 Latest National Statistic.

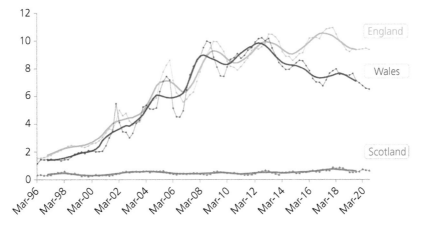

Figure 16.11 New herd incidents per 100 herd years at risk of infection during the year—GB, per quarter (from Defra 2020, with permission).

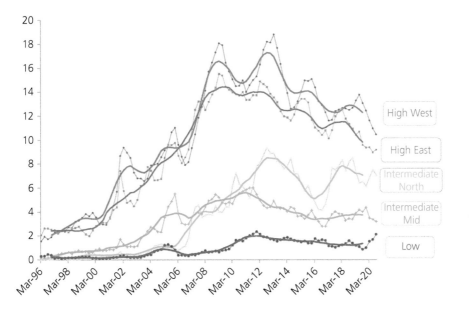

Figure 16.12 New herd incidents per 100 herd years at risk of infection during the year—Wales, per quarter (from Defra 2020, with permission).

rate for the previous 12 months, while in Wales it decreased. (by June 2022 incidence was declining in England, and was still lower but increasing in Wales).

- Total animals slaughtered due to a TB incident in England in the 12 months to December 2020 decreased 11% on the previous 12 months, while in Wales the number decreased 20%.

Data from the same Defra source show a similar contrast in recent years (Figure 16.11), with a particularly impressive decline in new herd incidents

achieved in Wales in both the High West and the High East areas (Figure 16.12) compared with England (Figure 16.13). The west Wales area has also been designated an intensive action area—with 6-monthly testing, concentration of IDEXX tests, herd management plans, and a badger vaccination programme that finished in 2017. Another metric of success is the incidence of OTFW—year on year, 2019 to 2020, this declined by −8% for England and −19% for Wales (and by −28% and −18% in the High West and High East areas, respectively): the greater the reduction the better, and there is little evidence here that England, with

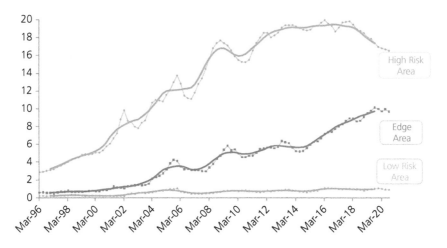

Figure 16.13 New herd incidents per 100 herd years at risk of infection during the year—England, per quarter (from Defra 2020, with permission).

its badger cull, is doing better than Wales, in which there is no badger cull. It is not clear what evidence National Farmers' Union (NFU) Deputy President Stuart Roberts had in mind when concluding: Badger culling is currently the best tool available to tackle bovine TB (NFU February 2021).

What do the numbers say? Defra report 38,799 cattle prematurely slaughtered in GB during 2021 (down from 43,742 in 2019), against a general picture of increase since the millennium, associated with more and better testing. This sounds, and is, awful. Lest our sympathies are misconstrued, Chris is currently a registered farmer in Canada and David, who has spent most of his career amongst farmers and was, for more than 20 years, himself a registered farmer, albeit of sheep, and a member of the NFU throughout his years on the Natural England Board, recalls prefacing countless interjections to the Board with expressions of sympathy for the farmer and also of distress that they were being cruelly given false hope that the cull would solve their problems. However keenly concern for badgers is felt, it does not diminish that for farmers on the ruinous diagnosis of a reactor in a herd that their forebears may have nurtured for generations. Nonetheless the context is different to that of a century ago. Pasteurization of milk and regular testing and slaughter of infected cattle mean that during our lifetimes bTB has scarcely been a threat to the health of humans in the UK: from 5200 culture-confirmed cases of human TB in 2012, only 0.7% were *M. bovis*, compared with 97% by *M. tuberculosis* (Public Health England 2013).

The financial cost has soared: in 2013 the average cost was about £30,000 per cattle incident[83] (Defra 2013), two-thirds paid as government compensation for compulsory slaughter and testing, the remainder falling on the farmer (Defra 2011a). The surveillance, testing, and compensation costs the English taxpayer £100 million year[−1]. Add the cost of culling: Defra (2019a) estimates £670,000 per area in its middling scenario, set against an estimated benefit of between £420,000 and £2.85 million (an incomplete estimate is £48m 2013–2020).

And the badger vaccinators? In 2019, licence holders vaccinated 890 badgers across 20 counties of England (Natural England 2019).[84] The volunteers did best in Cornwall, where they tallied 159 vaccinations, but in 10 of the 20 counties in which vaccination was licensed that tally was below 50. In Oxfordshire it was 23—for perspective, in Wytham Woods we vaccinated 198.[85]

Now the emphasis turns to cattle vaccination; the aim being, within 5 years, to incorporate cattle vaccination and DIVA testing and also a raft of cattle-related measures that emerged from the 2021 Call for Views, with the intention of eradicating bTB from cattle in the UK by 2038. Few could dissent from Charles Godfray's conclusion in 2018 (few, for that matter, would differ

---

[83] About a third of this cost was borne by the farmer; the rest by the tax-payer (Defra 2013). The average cost of a herd having TB free status suspended but not withdrawn is estimated at £9400 (Defra 2011a).
[84] See also Natural England (2021a).
[85] Under our Animals (Scientific Procedures), 1986, rather than Defra, licence.

with it at any other time in the past 30 years) that 'TB is a complex and difficult disease to control, both in humans and in animals', and not many would disagree, either, with his realism that 'this is a disease control campaign with a clear objective and, unfortunately, requiring sacrifices to be successful'. Godfray concludes that the current incidences of bovine TB in cattle and in badgers 'cannot be allowed to continue' and insofar as 'There are no easy answers to reducing disease levels', 'what is required is new drive and a concerted and concentrated effort by all sectors involved' (Godfray et al. 2018).

Part of the 'concentrated effort' Charles Godfray's team might have had in mind was some experimental test of the efficacy of intensive badger culls. Indeed, they proposed that after a 4-year licensed run, cull zones might be allocated to one of two 4-year treatments: (i) badger vaccination; and (ii) a 2-year moratorium followed by 2-year resumption of culling. This would allow a comparison, in terms of herd breakdowns, of vaccination and intermittent culling, and while a modest experiment in comparison to the RBCT, it would have been infinitely more informative than no experiment at all. It is not clear that this suggestion will be taken up, although in April 2021 a tender was let for a 5-year trial of a farmer-delivered vaccination programme in a 250 km^2 area of east Sussex (Bidstats 2021).

In late January 2021, reflecting on the 38,642 badgers killed during intensive culls[86],[87] in 2020 (by which time the culling area totalled more than 20-fold that subjected to proactive culling under the RBCT), George Eustice, the sixth Secretary of State to drink from this poisoned chalice, announced a move from culling after 2022. Mr Eustice judged badger culling to be 'unacceptable', but suppurating beneath the headline was confirmation that 10 or more cull zones could be added to the existing 54 (Defra 2011c) that are currently undergoing Supplementary Badger Control (SBC), the remaining 31[88] 4-year licences will continue to run their course then proceed to 2 years SBC finally concluding in 2025/26.[89] Figure 16.14 depicts the forecast numbers of badgers to be killed as culling is phased out—graphs

that may not be what many listeners presumed Mr Eustace had in mind when he spoke of badger culling being unacceptable.

One wonders how many of the new cull zones will have the perplexing attribute of Derbyshire in 2020, where a cull was licensed in an area in which badgers had been vaccinated. Mr Eustice, honouring a longstanding tradition, launched a public consultation. The judgement that badger culling is 'unacceptable', whether referencing effectiveness or ethics, would seem discomfitingly similar to the position described by John Bourne's Expert Group after the RBCT, albeit 140,000 dead badgers later. The Secretary of State's parliamentary statement offers the vindicating innuendo, 'We are now seeing sustained improvements in the high-risk area', from which the uncritical reader might conclude some causality between the improvements and the badger killing. But what evidence points to that causality? Battling the blurring of uncontrolled comparisons, two studies have sought an answer: at first, Downs et al. (2019) scrutinized the first 4 years of culling, 2013–2017, and emphasized a 66% decrease in incidence in comparison with non-culled areas in Gloucestershire (and a 37% decrease in Somerset), but by 2018, with another year's data, McGill and Jones (2019) find trends in prevalence and incidence going in every imaginable direction—striving for a generalization they predict that the cohort of herds that had been culled since the outset of the culls should, if the badger culling had delivered benefits, at least have better outcomes than those herds recruited more recently; in all three cull zones, however, the opposite outcome prevailed. Further, referencing the RBCT's prediction that if badger culling was going to work, its impact should be apparent by the time perturbation subsided in the third consecutive year of culling: in fact in 2018 the sixth year in Gloucestershire delivered a 130% increase in herd incidents (10 OTFW cases in 2017 increased to 23 in 2018)—the improvement in the HRAs mentioned by the Secretary of State seems more likely to be due to the greater use of severe interpretation since 2016 and much wider use of IFN-γ. McGill and Jones conclude 'that the main factor contributing to the ongoing bTB epidemic is the high number of infected cattle not detected by the SICCT skin test'. The fact that England with culling has done no better than Wales without culling, and in terms of new herd incidence is not doing as well, as discussed earlier, is perhaps the most revealing comparison. With each passing year the possibility of systematically comparing bTB herd incidence rates between culled and non-culled areas grows. As we go to press, this is exactly what Langton et al. (2022) have

---

[86] Under 4-year licences. Figures have yet to be released for 10 supplementary culls during 2020 (Natural England 2021b).

[87] The average cull rate is 3.18 badgers killed per km^2 (range 1.81–7.21 badgers km^{-2}).

[88] $54 - 2 \times LRAs - 21 \times SBC = 31$.

[89] LRAs are the subject of a different policy to HRAs: the period of time over which culling takes place is open ended although licences are issued for 4 years initially. Cumbria combined culling with vaccination in year 3.

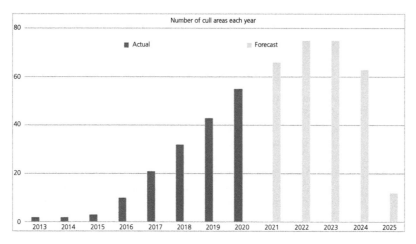

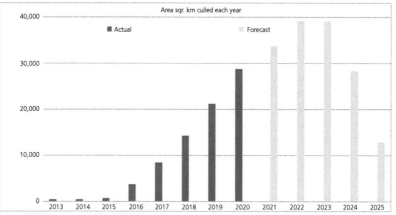

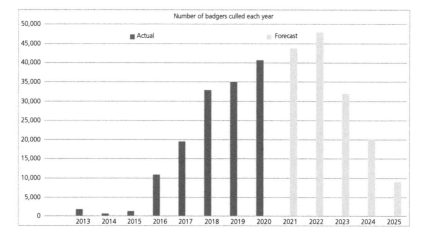

Figure 16.14 Extrapolations based on the assumption that all areas conduct supplementary culls on the conclusion of the first 4 years culling until 2021 but not thereafter. Predictions based on average values for culls between 2013 and 2019 (from Langton et al. (2022), with permission).

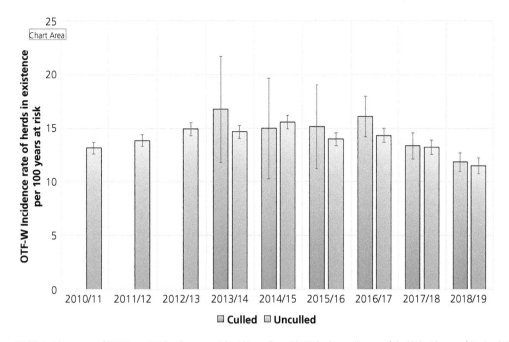

Figure 16.15 Incidence rate of OTFW per 100 herd years at risk within and outside 30 badger cull areas of the high risk area of England, during badger cull years (September to August) 2013/14–2018/19 (from Langton et al. 2022, with permission).

been able to do for the decade up to and including 2020. They compared the incidence rate (OTFW per 100 herd years at risk (HYAR)) for areas with and without badger culling in the HRA. The culled areas had completed 1–6 years of culling and observations[90] They found no statistically significant differences in OTFW herd incidence between the culled and unculled areas per 100 years at risk, and if anything, in most years, the incidence levels were slightly higher where the badgers had been culled (Figure 16.15). And as for cause and effect, in 9 of the 10 areas, the peak in OTFW incidence occurred before badger culling began in that area, so rates were falling before culling started and, in the exception (Somerset), culling was followed by an increase in incidence. Indeed, imagining that proponents of badger culling might argue that a longer run of culling would produce a better outcome, Langton et al. note that the proportion of the HRA that had undergone a full 4-year cull by the end of 2020 was still less than a quarter of the geographic area, making it hard to see how badger culling could claim responsibility for the decreases in OTFW incidence over the wider HRA during this period.[91]

The Langton et al. (2022) paper was published as we checked the proofs to this book, and was greeted by an intemperate blog from Defra (2022), which began 'This paper has been produced to fit a clear campaign agenda' and continued 'it is disappointing to see it published in a scientific journal'—invective that doubtless left the journal's editor looking askance. Anticipating the revelation of some statistical malfeasance commensurate with the vitriol of their reaction, we scrutinized the Defra critique (Middlemiss and Henderson 2022) and the subsequent Langton et al. (2022b) rebuttal. It appears that Middlemiss and Henderson's finding of no change in OTFW/100 HYAR over the period 2015/16 to 2019/20 in unculled areas (seemingly undermining Langton et al.'s result that the decline in incidence was as great in unculled as in culled areas) arose from changing the definition of 'unculled' to include only areas *never* culled, thereby excluding the two-thirds of the data that were from areas that at the time had not yet been culled. The data left out by Middlemiss and Henderson show a clear decline in the OTFW/100 HYAR temporal trend before culling began (Langton et al. 2022b). On the other hand, there is no dodging the fact that the Langton et al. analysis uses data from culled areas

[90] Comprising up to 4 years of intensive and in some cases further years of supplementary badger culling.

[91] See also footnote 54. Further, the proportion of IFN-γ testing undertaken in areas of badger culls has increased from

46.5% in 2018 to 84.5% in 2020, presumably affecting the rate of bTB decline.

that aggregates places where culling has been ongoing for mostly 1–3 years, in some of which culling badgers may not have had sufficient time to deliver an effect that Middelmiss and Henderson say 'takes some time to appear'; nonetheless Middlemiss and Henderson's graph (based on disaggregated, finer-scale spatial information not yet in the public domain) depicts a large decline in OTFW incidence by the second year of culling, so either the effect of culling had had sufficient time to be manifest or that early decline was due to disease control from cattle measures as Langton et al. had concluded. In Langton et al.'s analysis no effect of culling was detectable, but even accepting the rather unintuitive definition of 'unculled' that Middlemiss and Henderson propose as the right comparison for culled areas, it is hard to see that Middlemiss and Henderson land a knock-out punch on Langton et al.'s analysis in the context of an arena awash with conflation from the huge complexity and variation of the agricultural/ecological systems and considering the impacts on financial, human, and natural capital. Contemplating the balance of benefits and harms, one can only wonder whether an effect that is so hard to demonstrate has been worth the cost of bringing it about.

So, to return to the fourth question of our opening: has killing badgers worked to reduce the incidence of bTB in cattle? Probably not. Of course, the question can be phrased in importantly different ways, and costs can be measured in different currencies. For example, to the extent that killing badgers has yielded any benefit to cattle farmers, has it been cost effective? Almost certainly not.

When the Secretary of State announces the end of badger culling, does that mean that badger culling will end? No. it means that current intensive and supplementary culls will be phased out, but Defra still plans to identify sources of infection and remove them. This will lead to continuing reactive culls for an unspecified time into the future. Indeed, in work that Defra states 'was commissioned to identify areas with a likely reservoir of TB in badgers', which one might imagine could be targeted for further culls, in an effort to identify which edge areas, between HRAs and LRAs, Downs et al. (2021) sought to identify 'badger associated *M. bovis* reservoirs'. In the absence of data on infection in badgers—and it seems from the Swift et al. (in press) road kill work cited earlier, failure to find evidence of infected badgers—the authors defined the potential reservoir areas in terms of the recent history of TB in the cattle herds (e.g. at least one OTFW incident in a cattle herd not attributed to cattle movement in the previous 2 years). Considering all that is documented of the poor sensitivity of SICCT and IFN-γ cattle testing, defining the risk to cattle from badgers on the basis of unexplained infection in cattle might be thought a self-fulfilling prophecy of giddying circularity[92]. The result was that 20% of the edge area was thereby classified as having a local bTB reservoir in badgers, although it would be surprising if this was considered a substantive basis for further badger culling. Indeed, although APHA modelling had predicted that bTB in cattle in the edge area was fuelled by badgers, Swift et al. (in press) found very little bTB in badgers in Derbyshire and Nottinghamshire—it would seem extraordinary not to have checked that before starting to kill them (nor even, usually, having secured their corpses).

Box 16.4 The roll-out of badger cull zones

2013 1 Gloucester 2 Somerset
2015 3 Dorset
2016 4 Cornwall 5 Cornwall 6 Devon 7 Devon 8 Dorset 9 Gloucester 10 Hereford
2017 11 Cheshire 12 Devon 13 Devon 14 Devon 15 Devon 16 Dorset 17 Somerset
    18 Somerset 19 Wiltshire 20 Wiltshire 21 Wiltshire
2017 Gloucester 1 and Somerset 2 commence 5-year SBC
    Note SBC is licensed from June 1 to January 31 the following year
2018 22 Cornwall 23 Devon 24 Devon 25 Devon 26 Devon 27 Devon 28 Devon
    29 Gloucester 30 Somerset 31 Staffordshire 32 Cumbria
2019 33 Avon 34 Cheshire 35 Cornwall 36 Staffordshire 37 Devon 38 Devon
    39 Dorset 40 Hereford 41 Staffordshire 42 Wiltshire 43 Wiltshire
2019 Dorset 3 commences 5-year SBC
2020 44 Avon 45 Derbyshire 46 Gloucester 47 Hereford 48 Leicester 49 Oxfordshire
    50 Shropshire 51 Somerset 52 Warwickshire 53 Wiltshire 54 Lincolnshire
2021 Areas 11–21 commence 2-year SBC licensed from 1 June1 to 31 January 2022 (year 1)

The counties of Devon, Dorset, and Wiltshire have reached cull saturation.

---

[92] Donnelly and Nouvelet (2013) estimate that only 5.7% of the transmission to cattle herds is badger-to-cattle.

THE STORY OF BADGERS AND TB

Shropshire culled 50% of its area 2020, the other 50% is expected in September, along with probably another 10/11 large zones. Sixty-eight per cent HRA is currently undergoing either intensive or SBC. The aim is to reach 90% over the next 2 years. HRA = 37,000km^2.

So when all is said and done, what do we think? With the unattractive wisdom of hindsight, but the redeeming fact that it differs little from the foresight of 10 or even 20 years ago, the gains from culling badgers with the purpose of curtailing the spread of bTB in cattle seem to have been marginal—close to the point of futility. The impossibility of decisively parsing out the impacts of different interventions on the incidence of herd breakdowns follows from abrogating scientific design to the point of irresponsibility, and, insofar as it led to killing over a hundred thousand badgers on a suck-it-and-see basis, the morality of this policy is, to err on the side of understatement, debatable.

On the topic of moral discomfort, a personal reflection born of old intimacies: in the 1970s David radio-tracked foxes in Westmorland—morphed to Cumbria in 1974—perilously stumbling by night across brutally sharp limestone pavements as he radio-tracked Orton and Asby, two vixens hunting rabbits and weaving amongst the sheep to pick up the amputated tales of lambs (the bright orange lambing rings later a tell-tale sign in their dropping; Macdonald 1987). Foxes, being foxes, led him to many a badger sett, and while writing this chapter David learned of badger setts culled between 2018 and 2020 in the Cumbrian cull area, colourlessly named HS21:[93] several were sadly recognizable, 46 years later, as 'his setts'. Then, reading Rossi et al. (2020a, 2020b), it emerges that this bTB outbreak in an area of former low risk to cattle, and in which bTB was unknown in badgers, could be traced back, through genomic analysis of the bacterium, to an infection from a single imported cow from northern Ireland, first detected in 2014. The infection passed from cow to cow, becoming entrenched amongst them, before passing to the badgers. Three of 52 road-killed badgers tested positive between 2016 and 2018. A cull

---

[93] Hot spot 21. What is a hot spot? A 'potential bovine TB hotspot' area is identified when one or more lesion and/or culture positive TB breakdowns of obscure origin occurs in the LRA. Obscure origin means that, following investigation by APHA, the breakdown cannot be attributed to purchased/brought-in cattle or to spread from other cattle herds or non-bovine livestock in the locality. If *M. bovis* infection is confirmed by the wildlife survey, the potential hotspot becomes a confirmed hotspot.

licence was issued that year, and of the first 602 badgers killed that autumn 11% were infected. This study is remarkable in its sophistication and, using machine learning to estimate retrospective patterns of transmission, the authors show that most of the transmissions in this system likely happened within species; that is 20 cattle-to-cattle and 29 badger-to-badger, with 12 cattle-to-badger incidents, and 3 badger-to-cattle. There is much to be learnt from these analyses, but it does little to quell the feeling of moral squeamishness that at the time of writing hundreds of cattle and 1051 badgers have been killed because some people were insufficiently careful about the movement of one cow from Ireland—a feeling made very personal by those all-night vigils around those same setts watching the progenitors of those badgers 46 years ago.

Of the four questions posed in the opening paragraphs, we have assembled answers as this chapter progressed, but as a quick summary of what might work: (i) vaccinating cattle—the most attractive option, solving the problem at a stroke, but sadly blocked until the prerequisite DIVA test exists (as it may already do with Actiphage) and is validated, and thereafter, to gain significant traction could take decades; (ii) better diagnosis and regulation to weed out infected cattle—feasible and apparently immediately within reach by combining SICCT and Enferplex antibody test or Actiphage; (iii) vaccinating badgers—feasible if undertaken assiduously, although labour intensive, logistically challenging, and prohibitively expensive unless by oral delivery using baits (and if cattle-to-cattle transmission is solved by better cattle diagnosis and regulation or vaccination, necessary only to limit reinfection of cattle); and (iv) killing badgers—marginal benefit unconvincing, undermining the financial (and ethical) justification. Anything less than total and absolute commitment to eradicating cattle-to-cattle transmission brings any marginal benefit of killing badgers closer to fruitlessness and correspondingly greater indefensibility.

It may be obvious why the debate has been so irreconcilable between those with a Kantian conviction in the rightness of one view over another, or between those who function as uncritical lobbyists, crudely for farmers or badgers. A much more interesting question, in that it sheds light on the reality of all of modern conservation science, that is, on the journey from evidence to science and on the demons that writhe beneath the WildCRU's alluringly benign mission *to achieve practical solutions . . . through scientific research*, is why well-informed, decently motivated, pragmatic people have scrutinized evidence for 50 years and failed to fall into comfortable alignment. Putting aside implacable deontology, however open-minded the Benthamite balance sheet of pros and cons, however logical the delving into virtue ethics (in this case we lean heavily to a non-anthropocentric position that attributes intrinsic value to the badger alongside the quest for social justice for the farmers; Vucetich et al. 2018), the transdisciplinary reality of modern conservation denies any simple conclusion a clear-cut victory: there's always another side to the story; another hilltop from which to view the landscape (Macdonald 2019). This reality transcends the badger debate to suffuse every cranny of conservation: the appropriateness of hunting foxes for sport (Macdonald and Johnson 2015b), of lions for trophies (Macdonald et al. 2017; Vucetich et al. 2018), of killing mink to save water voles (Moorhouse et al. 2015e), the neutering of domestic cats to preserve the genetics of wild ones (Macdonald et al. 2010d), the ranching of ball pythons to satisfy thousands of American pet-keepers (Harrington et al. 2020), or the disallowing, or not, of farmers harvesting lion bones to satisfy millions of Chinese users of traditional medicine (Coals et al. 2020)—to cite from our own stable just a smattering of examples. Conservationists would not be working on these topics if they were tidily soluble, so we should not be surprised at the dissent, even acrimony, that emerges from the almost inevitable realization that they are not. A realization that draws conservation from the comfortable sanctuary of evidence perilously close to the uncomfortable exposure of politics—a process where evidence demands judgement and hopes for wisdom. Why were the debates of Natural England's Board so fraught, just like the debates large and small on all these conservation dilemmas? It is because, whatever balance of probabilities emerges from the slithering spaghetti strands of evidence, there is always the gnawing doubt that the other fella might be right.

Policy postscript: the political story; some personal reflections

The purpose of the RBCT was to produce evidence to answer the question of what to do. Those charged with interpreting the evidence decided that what should not be done was more culling. What happened? More culling. How did this surprising outcome arise? After the trial ended in 2005, the elevated bTB risks to cattle on adjoining land soon disappeared but, and strikingly, even 5 years later, the overall rate of herd breakdowns for farmers in the 2 km perimeter was, on average over the entire 9 years, worse than it would have been if no badger culling had occurred (Godfray et al. 2013; Table 16.1). That is, the relative benefits accrued by core farmers (their absolute situation was little improved[94]) came at a cost to the (arbitrarily spared) farmers surrounding them. Doubtless affected by the personal distress of farming families impacted by the disease, and also surely influenced by the NFU, the government opted to try again. Some may have thought charitably of Robert the Bruce's commendable spider;[95] others, less charitably, were reminded of Einstein's barbed observation that, 'Insanity is doing the same thing over and over again and expecting different results'.

When, on 7 July 2008 the Secretary of State decided against culling badgers, he understandably left ajar the window that this position could be revisited under exceptional circumstances, or if new scientific evidence were to become available. It didn't, but by 2010, there was a new government, the Conservative/Liberal Democrat coalition: the Tory manifesto had promised a badger cull, and many Lib Dem MPs had constituencies in the badly affected south-west. They faced the bleak reality that c.25,000 cattle had been slaughtered in England that year due to bTB (costing £90 million) and 23% of farms in the west and south-west had been unable to move stock off their premises at some point (Defra 2011a). David recalls a year of hair-pullingly intense discussions between Defra officials, the Natural England Board, and the NFU, culminating in impasse. Something had to be done, so they launched a consultation.[96] Extrapolating from RBCT, and making some hopeful assumptions, Defra estimated that 5 years of proactive culling of badgers within a 150 km² circular area would cause a 20–34% relative reduction in confirmed cattle herd breakdowns within the culled areas over 9 years. In the small print,

---

[94] To be clear, the very approximately 20% reduction (one in five farms) is in comparison with what might have been and is relative to a background increasing epidemic such that overall TB increased. The result was thus a decline in the rate of increase of herd breakdowns, not a reduction in the number of breakdowns.

[95] Dejected after his disastrous first year as King of Scots, while waiting out the winter of 1306 Robert Bruce watched a spider on the cave wall try time and again to spin its web. Every time the spider fell, it rose to begin again.

[96] We are reminded of pioneering ecologist, Charles Elton's, observation: 'In 1891 and 1892 voles multiplied on grazing areas in the Border countries of Scotland, with the result that lambs starved and Parliamentary Committee was set up to see what should be done about it. After the Committee had sat, the voles disappeared.' Charles Elton (1946) *The Ecology of Animals*, 2nd edn.

Table 16.1 Percentage change in herd breakdowns within the RBCT culling area and in a 2 km-wide perimeter surrounding it (from Godfray et al. 2013, with permission).

Time period	Average % change			
	Proactive culling area		Areas surrounding cull	
	Central estimate (%)	95% CI	Central estimate (%)	95% CI
During trial	−23	−12 to −33	+25	−1 to +56
After trial*	−28	−15 to −39	−4	−26 to +24
Entire period	−26	−19 to −32	+8	−14 to +35

* After trial period includes averages over 5 years of data.

however, is the awkward fact that taking account of the surrounding, perturbation-prone, 2 km-wide perimeter, the estimated overall benefit falls to 3–21%, and so Defra (2011a) settled on an anticipated average net benefit of a 16% reduction[97] (Table 16.2). There is a matter of statistical propriety at the point where estimates meet policy: the confidence interval (CI) around the estimated outcome spanned 3–21%—any figure between these limits would not therefore be considered implausible.[98]

It is a brutal societal conundrum whether this relative benefit is worthwhile, but politically, something had to be done, so a new generation of culls was proposed, which was intended to involve killing about 1500 badgers over 4 years in each area of at least 150 km² (i.e. culling over a larger area reduces the relative size of the perturbed perimeter). This would rely on a new methodology, which combined both cage-trapping and shooting and free (subsequently renamed, controlled) shooting.[99] The killing would be done by licensed groups of farmers or their agents. The

Table 16.2 Percentage change in herd breakdowns of core and perimeter over 9 years (five annual culls and then 4 more years) extrapolated from estimates in Table 16.1 (estimates by C. Donnelly in Macdonald and Burnham 2011). (i) 5 years from the initial proactive cull (assuming annual culling, with five such culls); (ii) The following 4 years: from 12 to 60 months after the final proactive cull; (iii) After 9 years: from the completion of the initial proactive cull until 60 months after the fifth annual cull (if the removal area had 'hard edges' completely impermeable to badgers and thus no perimeter perturbation, the reduction in bTB incidence might be 20–34%).

Extrapolated consequences of badger culling in minimum licenceable cull area for predicted herd breakdowns		
Time period	Average % change 150 km² circular culling area plus 2 km-wide perimeter surrounding it	
	Central estimate (%)	95% CI
(i) Cull years	−4	+9 to −17
(ii) Post-cull years	−21	−8 to −33
(iii) Entire period	−12	−3 to −21

Natural England Board's launching into the fray with its most Whitehall-savvy Executive, Andrew Wood, who had in 2010 taken over the role from Tom Tew after the latter had been reorganized out, expressed its broiling disquiet at almost every detail of this proposal. Andrew, honed by a lifetime in diverse crannies of government, was, by nature, less of an Exocet and more of a tactical snap-trap; he inserted his wedge into the gap between clear-cut RBCT 'best practice' and the proposed controlled shooting, and pushed, quietly but firmly.[100] The result was a modicum of restraint: two

---

[97] An average of 16% reduction equates to preventing 47 out of 292 breakdowns over 9 years, over the 150 km² culled area and 2 km surrounding ring.

[98] To be precise, 95% of estimates calculated in the same way would be expected to contain the true population effect (to stretch the statistical niceties surrounding conventional frequentist statistics in the interests of clarity, we are 95% confident the true effect lay within this range). Even this rather wide interval is conservative in the sense that its reliability depends on the unlikely state of affairs that a series of assumptions were upheld. In everyday life, where policy decisions matter, it is arguable that prudence should lead us to focus on the worst case, not a middling one: invest your life savings in a bond that offers a 3–21% return, or agree to your child taking a dangerous medicine that has a 3–21% chance of working—which end of the spectrum of possibilities would you think it responsible to focus on?

[99] The original adjective 'free' had conjured too vivid an image of men bearing high-powered rifles with a range of over a mile, roaming the countryside, and shooting at night at small moving targets. 'Controlled' sounded, well, more controlled.

[100] Andrew and David had more than one shared enthusiasm in their manoeuvres to advance nature conservation through Natural England's influence—you can read more about Andrew's behind-the-scenes impact in Isobel Tree's remarkable book *Wilding*.

1-year pilots of the proposal (and the establishment of an independent group to evaluate them).

As the national debate became febrile, the intolerable pressures faced by the Natural England Board due to its schizophrenic role (as licensor of a campaign it could not endorse) required dextrous chairmanship—a task that fell to Poul Christensen. Poul's credentials for impartiality were impeccable: he was an uncommonly successful dairy farmer who had pioneered Britain's farming and wildlife movement from his prize-winning demonstration farm in Oxfordshire (across which, a lifetime before, David had tracked foxes), and he had the distinction of being one of the most decent men alive, a quality that did not lessen his adroitness in debate. The Board gave Andrew Wood a mandate to engage in conversations with Defra, as their team formulated advice to Ministers. The Secretary of State was painstaking in evaluating the arguments and seemed open-minded, but the political sticking point was that it was impossible to establish 'the innocence' of the badger—a very different plea to the Board's, which was that the planned intervention was unlikely to work. Politically, something would have to be done; eventually the coalition rulers of the day opted for culling.

The government estimated that the rewards of this policy if applied to a 350 km^2 zone for 5 years' culling with 4 years' post-culling, was a saving of £3.68 million in reduced cattle TB, at a cost of £4.56 million to deliver the cull (Defra 2011b). As an investment proposition this was not immediately convincing, but the currency was politics, not money. Worse, it turned out to be a serious underestimate of the cost (the benefit is unknown and probably unknowable): public protest more than doubled the cost of policing, and the free-shooting farmers had to be replaced by labour-intensive cage-trapping professionals.

Things became very uncomfortable for the Board, and probably most acutely so for David—conspicuous as the Board's badger expert—who recalls interminable besuited London meetings, circling ever-more senior Defra officials, like cats around hot porridge, and, mentally drained, increasingly despondent train journeys home with Poul (they are neighbours in Oxfordshire). It is easy to pontificate about translating science into policy—indeed the WildCRU's proud mission since 1986 has been to 'achieve practical solutions to conservation problems through original scientific research'— but the anguished reality of delivering that aspiration merits a moment's reflection. Against the readily caricatured polarity of the actors on the stage, for the most part the reality was of well-intentioned people plunged into a wicked problem, most of them mindful, although in different proportions, of the horrors of farming families losing more than their livelihoods, badgers losing their lives, the vicissitudes of political fall-out, and the fear of getting it wrong when there was no right answer, only a balance of probabilities. For David, that balance fell against the cull, because at best the marginal gain to farmers would likely be small whereas the cost to badgers and society was certain to be huge; if, however, for government (with its view of the same landscape from a different hilltop) the balance were to fall in favour of a cull, David's sticking point was that it should be done in a way that offered scientifically interpretatable data: then, even if a cull failed, it would at least yield learning—you might consider that knowledge a memorial to the badgers and a return to the taxpayer. In short, David's wishes were that there would be no cull, but if there was to be one, it should be undertaken, as the RBCT before it, in a way interpretable to science; however, he got neither wish. What of Natural England? The Board of this statutory body is charged with evaluating and, ultimately, is under legal obligation to carry out, government's instructions. NE was required, by statute, to give advice to Defra; advice that, in this context, those involved confirm was 'largely shaped' by David's interventions in Board meetings. Weeks, months, years of meetings proliferated. The temperature rose. David recalls private meetings with Defra's guileful Director General Peter Unwin; Defra's Chief Scientist, Ian Boyd; and the government's Chief Scientist, John Beddington—friends from earlier lives but now thrown into various levels of ill-ease as positions hardened. At the time David was also Chairman of the UK's Darwin Initiative for the Conservation of Biodiversity and between the two roles his visits to Defra's headquarters at Nobel House near Whitehall were so frequent that he was issued a staff pass! He recalls also the public haranguing from endless conference floors by strident pro- and anti-cull voices, the interrogation by furious farmers in the backroom of a Gloucestershire pub, the endless phrasing and rephrasing of the small print, the bombarding calls from journalists, call after late night call with wise counsels Andrew Wood, Poul Christensen, John Krebs, Rosie Woodroffe, Chris Cheeseman, Christl Donnelly, and the visits from the anti-terrorism unit to inspect his house and car, and the hotline safety numbers, following death threats. Such, it turns out, is the reality of translating science into policy. And where did it lead? Considering the facts, unknowns, and assumptions, the Board concluded that the prospects of Defra's proposed cull succeeding, far less being judged worthwhile by society, were, 'at best unpromising'—words that had been polished until they glistened in preparation for an earlier meeting with the Secretary of State,[101] and a judgement given by David in an article authored by *Guardian* journalist Damian Carrington (2013).

Natural England thus found itself in an anguished position when instructed to license two pilot culls, one in Somerset, another in Gloucestershire—a decision in 2012 said by Defra's Chief Scientific Adviser Ian Boyd, and Chief Veterinary Officer Nigel Gibbens, to be based on 'the best available scientific evidence after more than 15 years of intensive research'. Although, as the government's then Chief Scientist, Sir John Bedding-ton, wryly remarked, it was 'interesting' that exactly the same scientific evidence was used to decide, on the one hand, for a cull in England but, on the other, against a cull in Wales. In 2011, the Natural England Board, appalled at the prospect of being compelled to exercise its licensing role over a policy in which it had little faith, and had advised against, came to the brink of resignation.[102] Andrew Wood turned to the Board for a decision: would it implement its obligations to Defra as a licensor (along with its other wildlife licensing responsibilities) or not, and take the consequences? Some Board members were against any involvement with the cull policy, and at first David was amongst them; he was persuaded, however, by Andrew's government-wise argument that if Natural England wasn't involved a less sympathetic scrutineer would be put in its place (and the entire machinery of conservation in England would risk being dismantled in revenge). The decision was taken that Natural England could not shirk from involvement.

As 2011–2012 passed, the shape of the licensing regime and, therefore, of the cull operation resolved.

[101] Well over 100 additional person-hours of anguish subsequently burnished 'at best unpromising' into the Natural England judgement that the proposed cull offered 'a low level of confidence'.

[102] The complexity of this dilemma illustrates the choke-chain of bureaucracy around the throat of conservation. Wildlife licensing in England is a function of the Secretary of State, delegated to Natural England by means of an 'agreement' under the 2006 NERC Act that created Natural England. Licensing the cull was considered by Defra (which, one can only imagine, was exercising a twisted sense of humour) to be analogous to Natural England's other licensing activities (translocation of misplaced water voles, inconvenient great crested newts, or errant edible dormice, or even the killing of gulls threatening airstrikes) and therefore within the agreement. Natural England's only recourse would have been to abrogate the entire agreement and hand the wildlife licensing function back.

But frustration grew in Natural England's Board that Defra was straying from the hard-won lessons of the RBCT experience, which was the only approach to culling for which there were evidence-based outcomes, and which consequently shaped the methodology most likely to deliver the reductions in incidence in cattle that had followed the RBCT interventions. David's recollection is of the Board wearily striking its collective forehead as Defra constantly pushed for compromises that departed from the original and, therefore, rendered the success of the cull even less likely. To remind you of these boundaries and how they were pushed (listed in descending order of adherence):

- 70% of the licence area must be capable of being shot over—this was forcefully imposed on the consortia.
- Cull zones should be at least 100 km². The licence set, and generally exceeded, these minimum area requirements.
- 70% of the badgers must be shot—this was translated into minimum (to ensure 70% met) and maximum (to avoid local extinction) acceptable numbers to be culled ('targets').
- Cull zones should have 'hard boundaries' to minimize perturbation. This was encouraged, not required, and progressively relaxed: the original Gloucestershire cull had two motorways and the Severn, whereas Somerset had much smaller roads or nothing.
- Culls should be completed quickly, to minimize perturbation. Natural England's opening bid was for 3 weeks. Defra, pressed hard by the NFU, started at 4 and ended at 6 (by which time, in the first Gloucestershire area, about 35% of the estimated population had been killed).
- Badgers should be cage-trapped and shot. The first proposal was for 'free-shooting', which involved shooting at night by free-roaming marksmen, soon morphed on the licence to 'controlled' shooting (at bait points over night-sights), which failed dismally until cage-trapping and shooting became the norm.

As a measure of its exasperation, the Board resolved to submit statutory advice to the Secretary of State. To rank this bland-sounding statement on the Richter Scale of insubordination, Natural England had never used this weapon before, and has never done so since (save, latterly, on the Marine Conservation Zones series); English Nature (its previous name) had never used it, either. The result of this grand gesture, and the hours of consideration that went into its every syllable, was that the Secretary of State thanked the Board politely for its advice, which was then tucked away until being exhumed by a Freedom of Information request in 2015.

Disarray began to swirl. The scientific disagreement focused on the numbers of badgers to be culled (Defra officials, clutching calculations based on counting badger setts and multiplying by the average number of badgers, appeared before hastily convened Board members whose long faces signalled scepticism—it turned out farmers were bad at identifying badger setts and bad at diagnosing whether they were active; additionally, Tuyttens et al. 2000b discuss how the number of badgers per sett varies widely), but more prosaically the cull companies couldn't source enough of the equipment they needed. The planned cull of 2012 was aborted, and finally begun in 2013. After the 6 weeks of culling permitted by the licence, both cull zones were well short of their target numbers (numbers designed to ensure that 70% of badgers were shot in year 1). The Secretary of State challenged over this fiasco famously observed that the fault lay not with Defra, nor even with the whole idea, but arose because 'the badgers had moved the goalposts'. Defra asked for an extension of 3 weeks to both licences. David argued strongly, and implacably, against this—cognisant, as he was, of what is known amongst evolutionary biologists as the Concorde fallacy,[103] which crudely translates into the aphorism of not throwing good money after bad. An alternative aphorism might be not spoiling a job for a h'penny worth of tar, and this fear weighted heavily with the Board. The decision, perhaps ultimately skilful but more immediately undistinguished, was to leave the decision to Andrew Wood.

Again, there's more to this than meets the eye in terms of the *realpolitik* of where the rubber (evidence) meets the road (policy). By the end of week 4 it was clear that both pilots would miss their targets by miles, so Defra asked Andrew (Science Policy) to extend the licences. He responded that the licence conditions did not allow extensions. They, most notably the Chief Vet, asked again. Andrew responded that he would consider an application for a supplementary licence that did allow for an extension. Both culls duly applied for an additional 3 weeks, and Andrew indicated that he was minded to sign. Poul Christensen put the matter to the Board, with the question of whether it wished to retrieve the delegated right of signature from the Executive, and then refuse to sign. Another bureaucratic

choke-chain tightened: the formal scheme of delegated duties gave the licensing role (all those bats, newts, and edible dormice) to officers (in this case, Andrew)—if the Board chose to abrogate this delegation for extending the cull licence, then Defra's obvious response would be to generalize that same abrogation to all 1500 or so contentious licence decisions each year: the Board would be suffocated within days. The delegation was left with Andrew, and he signed. Meanwhile, David was quoted in the *Guardian* as saying that extending the cull was 'not easily reconciled with the evidence' and in the *Times* that it offered 'Little confidence' (Carrington 2013; Webster 2013). Why did Andrew sign? While, in language familiar to aficionados of the TV series *Yes Minister*, Andrew could not possibly comment, David, for his part, arrived at the interpretation that by granting the extension he was offering the culls rope with which terminally to hang themselves by even more stark demonstration of a shambles: neither cull had come close to its target numbers, it was nearly December (when badgers grow lethargic) and worse weather was predicted (so in fact only dozens, not hundreds, more badgers were shot). If this was Andrew Wood's thinking, history makes plain that it was predicated on a reasonable but false assumption: it seems that Defra was not much troubled if the failure was even more obvious—the culls have carried on regardless. Which brings us back to that article in the *Guardian* and what became known as the 'Carrington leak', and led to demands for David's sacking.

Damian Carrington's article, published on 21 October 2013, reminding readers of David's knowledge of badgers and role as Chairman of Natural England's Science Advisory Committee, was remarkably well informed and tightly argued, and ended by quoting David as opining that the cull was 'unpromising'. That morning an outcry of civil servants descended on Andrew Wood and Dave Webster (the acting CEO). A day or so later, the Secretary of State phoned Poul and demanded that he impose discipline, and in a pincer movement Andrew Wood was summoned to see Defra's Director General and told 'in no uncertain terms' that David should be reminded that he was a Secretary of State appointee and could be dis-appointed in the same way. Both Poul and Andrew found themselves in increasingly stern meetings with their opposite numbers, but steadfastly argued that David had expressed views that were in keeping with Natural England's statutory advice to Ministers. Defra may have hoped that David would do the decent thing and resign, but adopting the Andrew Wood's philosophy that it is more useful to be in than

---

[103] Known to economists as the sunk cost fallacy.

out, and calculating that if he were sacked it would be much more damaging to the cull, he sat tight. Poul issued an oblique reminder about discipline to Board members (mentioning neither David nor the Carrington leak) and life returned to normal. Normal, that is to say, in the sense that the cull continued and in 2014–2015 a third cull zone, in Dorset, was added and the independent group was subtracted.

After the storm of the Carrington leak, another environmental journalist, Jonathan Leake, quoting David's judgement that the cull was 'an epic failure' reported in the *Sunday Times* (6 July 2014) that the tensions between the government and its NDPB,[104] Natural England, meant that David would have to resign—the mood seemed to be that either he or the Secretary of State had to go. David nervously awaited an imminent Board meeting, knowing that he was top of the agenda under the ominous heading 'burning issues'; days before the meeting, however, the coalition sprang a cabinet reshuffle and the Secretary of State was replaced (culled, as the press put it; Carrington 2013). As the Board meeting opened in the awful concrete drabness of Hercules House, David, quickly grabbing the opportunity to interject while the new Chair, Andrew Sells, cleared his throat in preparation, stated: 'I believe there was some question about who would go, myself or the Secretary of State, but it appears he has gone, so nothing remains for discussion'. Indeed, nothing was discussed.

Another complication rumbled on: the Convention of European Wildlife and Natural Habitats (1979), known to its familiars as the Bern Convention, under which badgers are a protected species. Basically, it would be a breach of the Convention's Article 7 to jeopardize an animal population. Article 7 refers to appendix 3 species (my italics):

This article obliges the Contracting Parties to ensure the protection of the fauna listed in Appendix III. Nevertheless, considering that these species may all, in varying degrees, be legitimately subject to exploitation in a particular State, the Convention does not exclude the possibility for each Contracting Party to authorise such exploitation on condition that this affects only those species not threatened on its territory and that *such exploitation does not jeopardise the animal population concerned.* In so doing, the Contracting Party must supervise the exploitation and, if necessary, impose stricter measures. The article has been drafted in this way in order to provide States with flexibility with regard to species that may from time to time not be directly threatened.

Article 9, however, allows parties to make exceptions but only providing that they are not detrimental to the survival of the local or national population concerned. Natural England insisted that Defra refer itself to the Bern Secretariat, and its case was accepted on the grounds that it would only cull in discrete areas and only 70% of the badgers would be shot. This was, perhaps flimsily arguable in 2012/13; by 2020/21, however, with many contiguous cull zones and over 100,000 badgers shot, 'flimsy' sounds like an exaggeration—it becomes important to define what constitutes a badger population.

To bookend the personal reflection: Poul Christensen stepped down as Chair of Natural England at the beginning of 2014, and later that year David ended his 11-year run successively on the Boards of English Nature (2003–2005) and Natural England (2005–2014), and 9-year Chairmanship of the NE-SAC, throughout all of which time the badger debate had relentlessly preoccupied these quangos.[105] In 2015 Andrew Wood resigned, and became an anti-cull campaigner.

Was NE-SAC right in its pessimism? First, operationally, the culls got off to an unpromising start with the postponement from 2012 to November 2013 because the NFU was unable to survey baseline badger numbers from which the requisite 70% cull efficiency could be calculated. In the end this was done using DNA identification of individuals based on hair-traps. However, the uncertainty in the estimates of population size meant that the minimum number to be killed exceeded the lower confidence limit on estimated population size in both areas. At the time, we thought the Bern Convention was a serious consideration, and that this lack of confidence therefore put marksmen in the awkward position that they might contravene it by eliminating badgers from the area while simultaneously contravening their licence conditions by failing to kill enough badgers and thereby causing more perturbation (Donnelly and Woodroffe 2012). Further, badgers had to be shot from 70% of the culling area,[106] raising questions about the willingness of sufficient landowners to participate for 4 years (their invitation to do so came with the price tag of paying for the night-shooting, hiring of the trained

---

[104] Non-departmental public body.

[105] A body that has a role in the processes of national government, but is not a government department or part of one, and which accordingly operates to a greater or lesser extent at arm's length from Ministers.
[106] The two 70% figures caused confusion throughout. Seventy per cent of the total population across the entire area must be shot and at least 70% of the area must be shot over.

marksman, and for the equipment to dispose of the carcasses) (Moody 2013), work to be completed within 6 consecutive weeks each year (essentially to quell a perturbation effect on the rate of infection; Woodroffe et al. 2006b). So, costs to both farmers and badgers looked onerous, when weighted against those unpromising odds of worthwhile reduction of bTB in cattle on the one hand and of burgeoning public opposition on the other.

Hindsight is the arbiter of whether history judges doubters as presciently wise or naysaying curmudgeons, so what happened? The Independent Expert Panel established to oversee the pilot culls concluded that the 2013 culls had failed to meet their targets for both effectiveness and humaneness (they were not asked to opine in 2014 and were disbanded in 2015). Informed by this judgement of operational failure, and remembering that there was no monitoring of the status of diseases allegedly borne by badgers, and that any evidence of an impact on cattle infections would be years off, the incredulous reader might no longer be surprised at the next decision: that culls be continued. In 2014 and 2015 the free, now called controlled, shooting conducted in Gloucestershire and Somerset was repeated, and expanded to ever larger areas from 2016 onwards. According to Defra (2011b) this was because 'we still need to tackle TB in order to support high standards of animal health and welfare, to promote sustainable beef and dairy sectors, to meet EU legal and trade requirements and to reduce the cost and burden on farmers and taxpayers'. These taxpayers will have watched agog as over the period 2012–2014, badger culling cost the government £16.8 million, during which time 2476 badgers were culled, equating to approximately £6700 for every badger killed, with no published evaluation,[107] indeed no possibility of evaluating unambiguously the payback in terms of reduced bTB in cattle. The government estimates[108] the cost of vaccinating badgers to be $c.$£2500 per km^2.

Perhaps still mindful of Robert the Bruce's encouragement to the aforementioned spider 'if at first you don't succeed, try, try, try again', Defra officials pressed on, killing 1467 badgers in 2015. Their redoubled energy would surely have left the spider

dumbstruck: in August 2016, officials confirmed that they had issued licences for 7 new cull zones, taking the total to 10 cull zones across 6 counties: Cornwall, Devon, Dorset, Gloucestershire, Herefordshire, and Somerset (Natural England estimates 2300–3300 badgers would be killed in each zone over the 4-year licence). The government estimated that between 9800 and 14,200 badgers would be targeted during the 2016 cull. And so this trend continued: through 2019 35,034 badgers were killed in 43 culling zones stretching from Cornwall to Cumbria—for perspective, more in 2019 than the 25,000 or so cattle slaughtered due to bTB. This brings the total number of badgers culled during this latest phase of bTB intervention to 102,349, at a cost to the taxpayer of over £60m; a number representing $c.$20% of the total former UK badger population of 562,000, as estimated by Mathews et al. (2018). The impact on bTB in cattle remains to be seen, and in the absence of scientific controls, may never be agreed. It may have been a triumph of hope over expectation that led Stuart Roberts, then Deputy President of the NFU, to conclude in March 2019: 'In areas where TB in badgers is endemic, we must retain culling as a vital tool enabling industry to get on top of the disease quickly and reduce further transmission.'

Any scientist looking at the current programme, whatever their perspective (utilitarian, deontological, anthropocentric or not) will be unsettled that it is neither designed nor implemented to provide unambiguous answers to primary questions such as what are the impacts on prevalence in cattle, or badgers; or, indeed, to second-tier questions such as the impacts on the numbers of badgers and their behaviour; or, yet again, to provide a full accounting of wider ecological, agricultural, economic, and societal effects, all considered in terms of a full life cycle analysis. Of course, doing it properly would be expensive, but so too would be doing it shoddily. Moreover, a balance must be struck, but our evaluation is that it has not, in this post-RBCT phase, been struck wisely.

Nevertheless, some questions can, at least partly, be answered by the accumulating evidence. Amongst these, what proportion of the badgers killed have tested positive (assuming the killing is random; the answer should tell us what proportion of the badger population was infected in the cull areas). Unfortunately, this question cannot be answered adequately because not only were badgers culled in the 'pilot culls' not tested (about 60 were autopsied to evaluate humaneness), those killed subsequently are collected for disposal but the data from their carcasses are largely

---

[107] Aside from the 60 or so autopsied in 2013 to provide evidence of humaneness to the IEP. Two reports mention collecting and testing some badgers (2016 badgers removed from 9 cull zones for post-mortem examination (Defra 2018a), and 2018 badgers removed from Gloucestershire area 29 for PME (Defra 2019b).

[108] Costs and benefits table 1 (Defra 2011d).

squandered.[109] Going back to 2006–2009 the answer for about 300 badgers triple tested each year in Gloucestershire, the worst county in England for cattle TB, was that about half the badgers were infected (having adjusted for the sensitivity—61–86%—of the triple test). A reliable benchmark comes from culture testing of road-killed badgers led by Malcolm Bennett. A historic sample from Cheshire between 1972 and 1990 (Atkins and Robinson 2013): the incidence was 0.25%, but by the time southern Cheshire was designated a HRA in 2014 it was 22% (Sandoval Barron et al. 2018). Recently, analysis of 611 samples from edge counties revealed that the average was 8% (11% amongst males, 5.4% of females) (M. Bennett, pers. comm., 5 September 2021).

So, where does this leave us? Charles Godfray provided the most recent answer. Having already led an independent review of bTB in 2013, he was off to a running start when asked to lead a second in 2018 (Defra 2018b). The government's digest of this review was (Defra 2020b):

Professor Godfray suggests that culling badgers can have a 'modest impact', his review emphasises limitations in the current cattle testing regime, the poor take-up of on-farm biosecurity measures to reduce the spread of bovine TB on farms, and the lack of adequate trading restrictions to prevent potentially infected cattle spreading the infection within or between cattle herds, as major factors limiting the effectiveness of the current policy.

In short, a barrowload of loose ends that thwart the hope of accomplishing the aforementioned full life cycle analysis of the pros and cons on this, or alternative, interventions. One can imagine the hours of draftsmanship that went into coining the phrase 'modest impact' (David is reminded of the effort Natural England's Board devoted to selecting the word 'unpromising' to describe the cull plan at the outset). Whether modest is too modest as to be worthwhile is, at least to the utilitarian, a question of whether society judges these gains to merit the costs (financially and in badgers), and over what time scale. That evaluation will now consider more than 30% of the English countryside, the killing of about 100,000 badgers, and whatever judgement society makes (ours is that an unpromising idea was misguidedly extended in a way

that guaranteed its interpretation would be controversial); it seems, however, that such judgement will be made with hindsight because in March of 2020 the government announced that culling would be phased out in favour of vaccination.[110] This was a rather perverse interpretation of the expression 'phased out' as it coincided with the release of plans to expand the culling zones in the counties of Avon, Derbyshire, Gloucestershire, Herefordshire, Leicestershire, Oxfordshire, Shropshire, Somerset, Warwickshire, Wiltshire, and Lincolnshire, with authorization dates set from 1 September 2020 (Natural England 2020). According to the Badger Trust (reporting a leaked government document; as corroborated by the BBC) this phasing out could result in the killing of a further 64,500 badgers per year, with costs of up to £100 million per year.

A forensic accountant evaluating whether the past 15 years of badger deaths (to say nothing of the 20 years' worth of deaths that preceded them) have repaid the farmers' hopes or the taxpayers' investment would surely struggle to approve a scientific audit of policy on what remains fundamentally a cattle problem. Bovine TB still circulates in the cattle population due to the deficiencies of the TB testing system and the frequent transport of animals between farms. Badgers have been collateral damage.

Where does this Policy postscript leave us? We have already concluded that subsequent to the RBCT successive culls have, with hindsight, been as much of a mistake as they appeared likely to be with foresight. We have also, at the end of this chapter's main narrative, concluded why it has proven so difficult for people of good will—a cap that has not fitted comfortably all actors on this stage—to find alignment. What remains to be said from this saga, at a higher level? We answer with a comment on government, illustrated by nightingales, not because of their beauteous song, but as a memorial to the bravest use of science in policy. In July 2012, while buffeted by the exhausting corrosion of the badger debate, the entire Board of Natural England decamped to Kent to bear the scrutiny of a judicial review. The issue was that 87 pairs of nightingales—that is, slightly over 1% of the entire, but dwindling, British population of the species—was resident on some scrubby land called Lodge Hill—exactly the same piece of scrubby land on which the Medway Council wished to build 5000 houses. The decision, and

---

[109] There are exceptions, e.g. HS21 in Cumbria. In 2020 in one control area in Cumbria all of 134 culled badgers were examined post-mortem; all being found to be negative (in 2019 and 2018, respectively, the prevalences had been 0.6% and 11.1%). In one Lincolnshire area 24.5% of 139 culled badgers in 2020 were positive (Defra 2021).

[110] In the devolved administrations vaccination is the preference policy in Wales; Scotland was officially declared bTB-free under EU rules in autumn 2009, with no current interventions.

the tens of millions of pounds that rested on it, drew a national spotlight to that meeting. David shudders as he recalls the ferocity and despair of the interlocutors, as the Natural England Board stood firm to its statutory duty under section 28 of the Wildlife and Country-side Act 1981 'to notify as an SSSI any land which is in its opinion of special interest by reason of its flora, fauna, etc.'. With a Herculean effort of draftsmanship equivalent to the 'unpromising' or 'low level of confidence' expressed in the badger cull, David recalls the sweat dripped into the words 'place a very substantial question over the soundness of this development'. The development was blocked, and with sympathy, not triumphalism, towards those whose hopes were dashed; in other words, an independent arm's length advisory body served government and nature well. But neither nightingales nor badgers won, nor ultimately did Natural England. As the years have passed the NDPB's voice grew more falsetto and independence has been confused with insolence, even treachery. The higher-level lesson, then, is not that wildlife and the environment provide problems easily solved by evidence—on the contrary, these problems grow more wicked[111]—but rather that government will govern better if it nurtures the independence of those charged with evaluating the evidence.

[111] In planning and policy, a wicked problem is a problem that is difficult or impossible to solve because of incomplete, contradictory, and changing requirements that are often difficult to recognize.

CHAPTER 17

Genetic Mate Choice—Quality Matters

The wise man must remember that while he is a descendant of the past, he is a parent of the future.
Herbert Spencer

In the survival game, luck—serendipity—matters, a lot, but then so does aptitude. And so a mixture of quantitative opportunism entwined with qualitative traits, honed by millennia of evolution, add up to the inequalities that substantially write the runes of individual success, as it does the success of the populations those individuals comprise. Collectively, fewer, smaller subpopulations, faced with a run of bad luck, are more likely to blink out (individually and collectively) than are larger more numerous subpopulations. As the goshawk smashes into a flock of pigeons, it may largely be a numbers game that determines whether a given individual dies—but the larger the flock, the lower the likelihood that it's your number that will be up. But, as Ernest Hemingway observed, 'you make your own luck' and thus the intrinsic *quality* of each individual matters too—the pigeon with the sharper eyes, quicker reflexes, and more powerful flight muscles may just have the edge over its companion when it comes to dodging that goshawk's talons. How then is the relative ability of individual badgers, within those population fluctuations described in Chapter 11, to cope with environmental stressors (Chapter 12) and disease (Chapter 15) determined? Darwin's (1859) phrase 'natural selection' emphasizes the selective advantage of being better adapted—a better fit—for the immediate, local environment. But complementing this thought was Darwin's other great insight: 'the advantage which certain individuals have over other individuals of the same sex and species solely in respect of reproduction', resulting in an intraspecific reproductive competition. Not only did Darwin recognize that sexual selection (through mate choice or, ultimately as it would transpire, through genetic compatibility) allow parents the opportunity to give offspring a leg-up in the evolutionary race, but also that prospective mating partners might advertise their superior quality in ways leading to the existence and preservation of traits running counter to natural selection. Thus he (Darwin 1859, p. 88) proposed that sexual selection 'depends not on the struggle for existence, but on the struggle between males for possession of females' (see Hosken and House 2011).

It was not until Darwin's *Variation of Animals and Plants under Domestication* (1868, p. 6) that he combined these concepts into that singular but renowned phrase, 'the Survival of the Fittest', acknowledging that he borrowed this term from Herbert Spencer's *Principles of Biology* (1864), in which Spencer, too, had had in mind not just the importance of survival, but also that of phenotypic characteristics enhancing reproduction to leave more or better descendants.[1] It is this relative rate of reproductive output amongst a class of genetic variants that has become the criterion of 'fitness' central to modern evolutionary biology (Ariew and Lewontin 2004). The final neo-Darwinian component of this trifecta is the genetic heritability of traits evident in the subsequent fitness of trans-generational offspring, to form what Mousseau and Roff (1987) termed a parent-favouring lineage; that is, parents begetting children, grandchildren, and so on. In the vernacular, it's all about sex; the failure to produce offspring equals zero fitness,[2] irrespective of any other metric of lifetime success. This leads to Fisher's (1930) fundamental axiom of natural selection 'the rate of increase in fitness of any organism at any time is equal

---

[1] 'This preservation, during the battle for life, of varieties which possess any advantage in structure, constitution, or instinct, I have called Natural Selection; and Mr. Herbert Spencer has well expressed the same idea by the Survival of the Fittest.'

[2] We return later to the contribution of alloparental behaviour to inclusive fitness (Hamilton 1964).

The Badgers of Wytham Woods. David W. Macdonald and Chris Newman, Oxford University Press.
© David W. Macdonald and Chris Newman (2022). DOI: 10.1093/oso/9780192845368.003.0017

to its genetic variance [from other members of its population] in fitness at that time'.

But as we saw in Chapter 7, behavioural 'mate choice' is an arena fraught with complexities: available mates are a limited commodity; the optimal partner might simply not exist in the population, or within the social clique of the selectee. Even being in the right place at the right time can lead to competing for prime opportunities, whereas not being in the right place at the right time can necessitate the compromise of 'making do' (Gibson and Langen 1996). The nexus here lies between behavioural choice (preference, social bonds), opportunity (overlapping spheres of activity; unopposed access), and genetic compatibility (aligning one partner's genes with the best available complement of genes from the other). Remembering too that, for females especially, reproduction is such a huge investment that it can sometimes be better to abort gestation than to produce offspring unlikely to be able to compete and survive, such as can arise through inbreeding (Lihoreau et al. 2007).

The dilemma at the foregoing nexus takes us back to the Vignette of the Five Bs in Chapter 7. Colloquially, Bill badger might fancy Betty, after months of flirting and ingratiation; he is, perhaps, however, at a disadvantage, insofar as she prefers apples to the blackberries he favours, for which he forages in the orchard. As a result, he sees and smells her only infrequently compared to Bob; the latter, Bill's arch nemesis, is also courting Betty and happens to share her predilection for the forbidden fruit. Perhaps, worse too, Betty is a small badger, as is Bill, raising the probability of their offspring being diminutive, and therefore uncompetitive; might big Bertha make a much better mate proposition for Bill, and big Bob for Betty? Complexities abound in the mating game: if Bill consorts often with Betty it might suit her that a first mating with him induces her ovulation, but she might seek a more exotic father when her eggs ripen a couple of days later. So as each soap opera romance unfolds and intersects, the complexity grows.

There is an additional counterforce too: can an individual promote a proportion of its own genes without ever mating? Yes: as former Wytham village resident and Oxford's Royal Society Research Professor Bill Hamilton (Chapter 8) famously proposed (1963; developing a concept theorized by J.B.S. Haldane in *The Causes of Evolution*, 1932), if an individual contributes selflessly to the survival of other (typically juvenile) group members, it is serving the interests of its community—a concept Hamilton termed *inclusive fitness* (remember Chapter 8). A year later, John Maynard-Smith took this idea one step further, formulating the related idea of *kin selection*, specifying the added benefit that if the donor of these good graces is related to the recipient (aunt, uncle, grand-parent, older sibling), it is *de facto* favouring the frequency of those genes, shared by the pair, within the population (we return to grand-parenting in the context of senescence in Chapter 18). In a sense these complementary theories reconcile the conflict between 'selfish' genes being perpetuated through the survival of the fittest, with group—and colony—cooperative behaviours for the greater good. For instance, in eusocial insects (bees, ants), the non-mating drones serve their queen in the interests of shared genetics[3] (Nowak et al. 2010).

Similarly, in many cooperatively breeding carnivores (see Macdonald and Moehlman 1982; Jennions and Macdonald 1994), from wolves (Ruprecht et al. 2012) to meerkats (Macdonald and Doolan 1997), the contribution of non-breeding 'helpers' is altruistic only in the sense that, despite their subordinate status, those genes that they share with the progeny of the group's dominant individuals (or pair) will flourish (e.g. siblings are, on average, as closely related to each other as parents are to their offspring[4]).

But why might an individual sacrifice its own outright breeding potential? Often because local resource and territory availability will discourage all individuals from breeding, or from dispersing effectively; philopatry, as a result, offers greater ultimate fitness than going it alone. Further, there are other advantages to staying at home, as many parents of adult children have noticed, including, along with benefits to inclusive fitness, gaining experience in raising young, and possibly even inheriting the home. As for badgers, do fellow group members contribute positively to the survival of kith and kin? No, they do not; recalling from Chapter 8 that competition between rival mothers puts

---

[3] In hymenopteran insects their haplodiploid sex-determination system causes females to be more closely related to their sisters than to their own (potential) offspring.

[4] John Maynard Smith mentioned in *New Scientist* magazine in August 1975 that while chatting with J.B.S. Haldane in the now-demolished Orange Tree off the Euston Road, Haldane, after much scribbling (inevitably on the back of an envelope) announced that *he was prepared to lay down his life for eight of his cousins or two of his brothers*. As Maynard Smith noted 'This remark contained the essence of an idea which W. D. Hamilton, a lecturer in zoology at Imperial College, London, was later to generalise'.

their respective cubs at risk, and that little measurable benefits can be attributed to 'helpers' (Woodroffe and Macdonald 2000; Dugdale et al. 2010).

Sexual selection thus sits alongside natural selection (and kin selection) as a filter through which genes must pass if they are to be perpetuated, risking being extinguished not only by the failure of their current host to survive, but also as a result of the risk that their traits will fail to attract a breeding partner, resulting in those pace of life trade-offs we documented in Chapter 14. And when we say all this is a complicated business, consider, for example, which of the two following scenarios would be preferable? Proven genes to maintain the types of offspring traits that benefited their demonstrably successful parents, or different (i.e. assortative) genes to give the next generation more flexibility? The answer depends on whether the environment seems likely to stay the same and exert similar selection pressures between generations, or to change, so that selection favours novel adaptations. The odds will differ for different selective pressures—that is, the weather is changeable (Chapter 12); the threat of endemic (but not epidemic) diseases, however, may be constant (Chapter 15).

How then do badgers go about this intricate process of optimizing offspring quantity and quality, while selecting from a limited pool of breeding partners, and betting on the future demands that changing circumstances may place on their descendants (Charlesworth 1988)? The answer ties together the remarkable observations of earlier chapters, and spawns the next generation of questions. Key ingredients include:

1. Almost all adult badgers mate, and appear to do so indiscriminately, with blatant promiscuity (Chapter 7).
2. Although there is a peak in mating in February, Wytham's badgers mate in every month of the year (Chapter 7); nonetheless, given the single oestrous cycle we found in Wytham's badger population (Sugianto et al. 2019b), a lot of mating appears to be non-conceptive.
3. There is remarkable tolerance of 'territorial' intruders, and an intriguing pattern of frequent short-term visits to neighbouring groups (phenomena discussed in Chapters 4 and 9).
4. Accumulating evidence of expanded, seemingly super-territorial, social networks (Chapter 10).
5. Insofar as a male's reproductive fitness is heavily contingent on the female with whom he mates (i.e. if she fails, he fails), much sexual selection

may be shifted into the post-copulatory arena wherein the badgers' speciality of delayed implantation (Chapter 7) offers a substantial period during which the cryptic choice of which blastocysts to implant, and which to abort, can be exercised; superfecundation offers yet more scope for ongoing re-evaluation of what genes are available— processes termed 'cryptic mate choice' and which include phenomena such as sperm competition (which we argue in Chapter 7 may be less compelling when applied to badgers than one might first thing) and immuno-compatibility at conception and implantation (Andersson and Simmons 2006).

Doubting that badgers are cognisant of the fact that mating leads to offspring, and having examined the social interaction that brings together parents in Chapter 7, here we focus more on the consequences of that liaison for the next generation, how parental choices and compatibilities lead to offspring fitness benefits—or not. Evolution leading badgers by the nose to deliver on its grand outcome ... continuing a particular detail of life, a gene, through an ever-changing environment.

Investigating the implications of sexual selection for fitness across the generations heralded a period in our project's history led by doctoral students Simon (Yung Wa) Sin and Geetha Annavi, who built on the earlier insights of Hannah Dugdale; let us, however, begin with the basic theory.

Pedigree

Understanding how active, as well as cryptic, mate choice might lead to better adapted offspring depends on ability to match cubs to their parents—the stuff of paternity suits and ancestry websites. In the remarkably complicated social and, it turns out, sexual, networks that exist amongst the community of Wytham's badgers, we have discovered not so much a family tree, but more of 'a family bush'; entangled with such complexities as inter-generational mating, multiple paternity within litters, and a degree of inbreeding amongst kin.

The prospect of working out the parenthood of every cub in Wytham's population seemed enticingly real in the early days of our study. It was Alec Jeffreys' experiment, at Leicester University, on 10 September 1984, that first exposed on X-ray film similarities in the DNA of his technician's family members (Jeffreys et al. 1985). Responding to the potential for genetic finger-printing to revolutionize forensic science, not to mention paternity and immigration disputes (Gill et al. 1985), an academic industry speedily, and enthrallingly, sprang up matching minisatellites, microsatellites, and short tandem repeats (STRs) to family relationships. As paternity studies proliferated, especially of birds, so did widespread evidence of extra-pair matings and cuckoldry.[5] David eagerly bounded towards collaborations, only to find, time after time, that the DNA probes of the day failed to reveal sufficient variation to disentangle the genetics of not only Wytham's badgers, but any badgers on a meaningful geographical scale. In those days David occupied an office deep in the bowels of the (now demolished) Department of Zoology's Tinbergen Building. In the adjoining office Peter Evans, subsequently a noteworthy figure in cetacean conservation, was unravelling the genetics of starlings. David despairingly recounted his succession of failed sorties into badger DNA, whereupon Peter, notoriously optimistic, offered to process some badger bloods through his allozyme system. The findings were to lead, in 1989, to a publication (Evans et al. 1989) that, as it was to turn out more than 30 years later, correctly foresaw an intriguing revelation of our studies: super-groups. That foresight went largely unrealized for the next three decades until, with a combination of modern genetic techniques and cutting-edge technologies, the pieces of the puzzle fell into place. What, then, are allozymes, and what did Peter and David learn from them?

This first successful dalliance with badger DNA involved the examination of blood samples from 170 individual badgers, collected during 1984–1985 across 13 social groups at Woodchester Park (the then MAFF, Ministry of Agriculture Fisheries and Foods, now Defra) study site in Gloucestershire. Allozymes (or alloenzymes) are allelic variants (i.e. versions of the same gene at the same place (locus) on a chromosome that control the same characteristic, such as blood type or colour blindness) of genes encoding enzymes that vary structurally, but not functionally. Using 23 of these alloenzyme encoding gene loci, they found that only two (EST-1, PT-3) were sufficiently polymorphic (varied) to be scored consistently and accurately, while only EST-1 exhibited enough variation to be useful in comparing gene frequencies within this population.

The frequencies of three EST-1 alleles deviated significantly from a Hardy–Weinberg equilibrium[6] with a marked deficit of heterozygotes—where a heterozygote is an individual having two different alleles at a genetic locus (a homozygote is an individual having two copies of the same allele at a locus). This 'uniformity' or 'sameness' implied much less genetic variation across the population than was expected from similar studies of other mammals at that time, thereby implying assortative mating, or a Wahlund effect (i.e. the combination of individuals from more than one genetically distinct population). In short, the Woodchester badger population segregated genetically, evidenced at one locus, into four 'super-groups' each incorporating several badger groups. The partitioning of Woodchester Park into these four alliances, each composed of several groups, might usefully be thought analogous to the partitioning of the Scottish Highlands between clans,[7] each comprising several septs.

This vision of super-groups, based on allozymes, shimmered like a mirage until the accumulation of sociological evidence documented in Chapters 4, 9, and 10, brought it into inescapably sharp focus, along with the priority to understand genetic mate choice amongst badgers. The atmosphere of interpretation had also changed, with examples of multi-level societies emerging from vulturine guineafowl (*Acryllium vulturinum*; Papageorgiou et al. 2020) to hamadryas baboons (*Papio hamadryas*; Schreier and Swedell 2009) and Grévy's zebra (*Equus grevyi*), where Dan Rubenstein and Mace Hack (2004) highlight that higher levels of social organization provide individuals with social options for solving ecological and social problems that cannot be solved by adjusting relationships within core social groups.

A factor that may be key for the badgers of both Woodchester and Wytham is that both were living at high density, which got higher—alleviating inbreeding even without other mechanisms (Woodchester increased from 7.8 adult badgers per km² in 1978 to 25.3 in 1993 (Rogers et al. 1997), Wytham, from 18.3 in 1987

---

[5] Literally from the cuckoo, which sometimes lay their eggs in other birds' nests, cuckoldry implies that a male's female partner surreptitiously has offspring with another male; see Møller and Birkhead (1993).

[6] A principle that states that allele and genotype frequencies in a population will remain constant from generation to generation in the absence of other evolutionary influences.

[7] Note, this is a slightly different use of the clan analogy to that originally invoked by Hans Kruuk, who originally used the word clan to describe each individual badger group in Wytham Woods.

to 45.5 in 1996; i.e. as calculated using the eMNA calculation of Bright Ross et al. 2020 (Chapter 11)). Members of these populations lacked places to disperse to, and so their local numbers swelled.

Despite the rarity of juvenile dispersal at Woodchester (as at Wytham), with 93% of cubs remaining philopatric (over 1984–1985 only five males dispersed permanently at Woodchester, Evans et al. 1989) badgers in this Gloucestershire population were less inbred than four of the five other species of social mammals that had been studied at the time (i.e. free-tailed bats, *Tadarida brasiliensis*, McCracken and Bradbury 1977; yellow-bellied marmots, *Marmota flaviventris*, Schwartz and Armitage 1980; McCracken and Bradbury 1981; pocket gophers, *Geomyidae*, Patton and Feder 1981; black-tailed prairie dogs, *Cynomys ludovicianus*, Chesser 1983). Amongst these and other social mammals (Greenwood 1980), the dispersal of juveniles, especially males, reduces the heterogeneity of gene frequencies within groups. In contrast, in Woodchester's badgers, at the time it was believed that stable group membership and preferential recruitment of young born into the group promoted gene frequency heterogeneity, which was then reduced either by occasional dispersal between groups or by adult males mating outside their own group (but predominantly within their wider clan). With hindsight, and given that our conclusions were based on just one allozyme locus, the deduction of segregation into super-groups was prescient. And regarding this early evidence of philandering beyond the group, 16 years later researchers at Woodchester park confirmed 50% extra-group paternity from subsequent genetic pedigree refinements (Carpenter et al. 2005).

That group memberships swollen to the point of fission, but constrained by a lack of opportunity for dispersal, might become only semi-detached—each sept retaining an affiliation to its roots—makes intuitive sense. Nevertheless, it is satisfying to have foreseen that this resulted in '. . . members of one social unit may frequently be mating with those of another' (Evans et al., 1989, p. 593), which in turn promoted gene flow, at least within the clannish super-groups.[8]

The allozyme study was a breakthrough. It was the first evidence of multiple male badgers siring young in a single litter; it indicated that a minority of breeders monopolized cub production; it revealed temporary movements (Chapter 3), with 27 males and 15 females being caught at some point away from their principal group affiliation; and it provides the first hint that there were distinct conception patterns within the apparent randomness of promiscuity, which caused the persistence of distinct genetic, as well as spatial (Chapter 4) super-groups, although it would be another 20 years before we were able fully to explore the basis of badger mate selection.

We needed more genetic markers—more distinguishing features that could aid in precise recognition and parentage assignment. In a slightly bruising illustration of the uncertainties of research, to build on the discoveries made with Peter Evans we recruited three successive doctoral students, Jack da Silva (1986), Rosie Woodroffe (1988) and Paul Stewart (1992) with the firm conviction that further breakthroughs in DNA technology would materialize in time for them to benefit from our longed-for genealogy—all of whom completed excellent doctoral theses without a scintilla of progress being made on these genetic family trees. It was 1995 when doctoral student Xavier Domingo-Roura joined us to work exclusively on DNA, which saw us nudge forward: Xavier, a diversely gifted Catalonian polymath who was to die tragically young, tested primers (short single strands of RNA or DNA, generally about 18–22 bases, that serve as a starting point for DNA synthesis) that enabled the amplification of 12 microsatellites[9] by PCR. However, although these showed variability across badgers as a species,

---

[8] For historical interest it is noteworthy how close to the truth Evans et al got when, in 1989, they wrote, 'Studies of territorial, highly stable groups of wild Eurasian badger, Meles meles, revealed that *more than one adult of each sex may breed within a group*, and that *extra-territorial movements may occur within clusters of territories*. Although there is some genetic structuring within a local population and a deficiency of heterozygotes, due probably to minimal juvenile dispersal, heterogeneity of gene frequencies is reduced by: (a) adults transferring between adjacent groups, and (b) matings between males of one group and females of another. Marked changes in gene frequencies between generations indicate that a minority of males have a strong influence on the genotypes of the offspring, being either polygynous or promiscuous. Within one generation, the *young of a given group may be sired by two or more males*, and these males may not necessarily be members of that group.'

[9] There are predictable inheritance patterns at certain locations (called loci) in the genome. Microsatellites are di-, tri-, or tetra nucleotide short tandem repeats (STRs) in DNA sequences, prone to a higher rate of mutation than other DNA regions. These STRs differ in length and sometimes sequence and this allelic diversity that can be used to map out locations within the genome and infer patterns of parentage, or 'pedigree' within a population. If variation within microsatellite loci

only five of them were variable within Wytham. So, frustratingly, while this was a huge breakthrough,[10] it wasn't sufficient (Domingo-Roura et al. 2003): we needed yet more polymorphic loci.

We continued doggedly, convinced by Xavier's breakthrough that one day we'd solve the problem of insufficient marker variation. That day came, but it took until the new millennium. In those days David ran a fortnight-long conservation module for Oxford's Master's course on integrative biology, during which Chris organized a 'Badger Day'—in 2000 one particularly talented student caught the badger bug and threw herself into a project on badger sex ratio (Chapter 2), fledged into a research assistant before starting a doctorate with us in 2002. Thus it was that Hannah Dugdale became the next doctoral student in the genetic sequence, and applied a set of 22 polymorphic microsatellite loci. Hannah could thus delve into our archived goldmine of 1080 individuals sampled between 1987 and 2005: she managed to genotype 915 of them, deriving mother and father for 331 of 630 cubs (53%) at $\geq$95% confidence (both parents were assigned to 595 cubs (94%) at 80% confidence), along with inferred sibship.[11] It's important to bear in mind that this applies only to cubs surviving until May/June, when we are first able to catch and sample them.

From this dataset and after years of suspicion and innuendo about badger promiscuity, Hannah's revelations were to confirm just how promiscuous they were. First, and somewhat unusually amongst carnivores (although similarly, in this respect, to banded mongooses and lions) Hannah quantified how several adults of each sex within the same group could sire cubs in a given year—indeed, in each social group the mean number of mothers bearing cubs was 5.6 ± 0.4 (median = 5), and 5.8 ± 0.4 (median = 5) for fathers. Nonetheless, remembering that we had watched as almost all badgers in a group mated (Chapter 7), we had been puzzled that our demographic analyses found that only 29% of females ultimately lactated. This was pleasingly close to Hannah's genetic

confirmation that 31 ± 6% (range = 8–51%, median = 28%; 95% confidence) of females, and only 27% ± 6% (range = 8–60%, median = 24%; 95%) of adult males were assigned offspring each year, which confirmed that mating was no guarantee of breeding (or at least weaning) success. It also confirmed the inequality, skew, in female badger breeding success, which was to take us further down the road of deducing competition, and perhaps suppression, amongst females.

The building of a pedigree from these parentage assignments was technical, and involved inputting genotypic data on variations at these 22 loci into a pair of computer programs (for those familiar with them, these programs were 'Cervus' (Marshall et al. 1998; Kalinowski et al. 2007), and 'Colony' (Wang 2013)).[12] Analyses were run systematically according to strict rules (Figure 17.1), enhanced by a sprinkling of demographic logic, following Carpenter et al. (2005): females were only electable as mothers if over the age of 2 in the year when the cub was born (due to delayed implantation; Chapter 7) and present in the cub's natal group, whereas electable fathers had to be at least 1 year old, but could reside anywhere in the woods. Other, surely uncontroversial qualifying credentials were that there had to be a high probability that both potential parents were still alive at the time of conception (i.e. at that time we presumed a badger was dead, and thus disallowed, if not caught within a 525-day window—a time frame that typifies the maximum duration (95% confidence) within which an extant badger might evade capture (Dugdale et al. 2007), and echoing the eMNA logic in Chapter 11).

This pedigree revealed that the mean post-emergence litter size amongst females assigned at least one cub was 1.3 ± 0.06 (range = 1–3, mode = 1,

---

is low, more markers (sites) are needed to produce a sensitive, reliable family tree.

[10] Three loci, which were variable in the Netherlands and Denmark, were monomorphic in the Wytham population.

[11] Separately, maternity was assigned to an additional 7 and 5 cubs, resulting in a total of 602 (96% at $\geq$80%) and 336 (53% at $\geq$95%) cubs that were assigned a mother. Paternity was assigned to a further 16 and 7 cubs, resulting in 611 (97% at $\geq$80%) and 338 (54% at $\geq$95%) cubs that were assigned a father.

[12] Genotypes were first analysed in Cervus, which uses a multi-allelic likelihood based approach to assign a parent pair, or if a pair could not be assigned then we assigned either maternity or paternity alone. Cervus can incorporate the presence of relatives, genotyping error, and the proportion of unsampled individuals to improve accuracy, although it can only assign parents from among sampled individuals. Cubs that were not assigned a mother and/or father using Cervus were therefore included in a sibship inference using Colony (Wang 2013), which clusters (half-)siblings sharing the same non-genotyped parent to substantially increase the number of within-cohort pedigree links (Walling et al. 2010). The likelihood of a particular pedigree configuration is thus the probability of observing genotypes, conditional on the genotypes of the assigned parents, multiplied over all individuals and, when loci are assumed to be independent, multiplied over all loci (Huisman 2017).

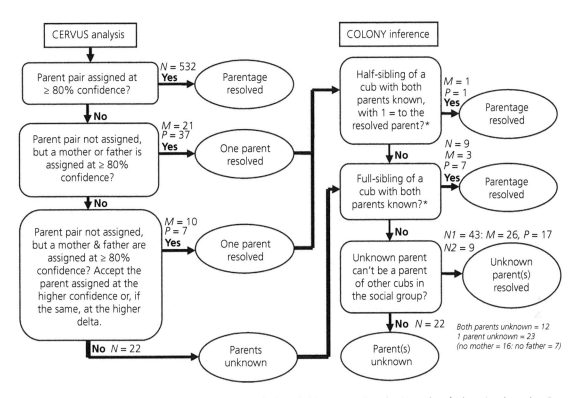

Figure 17.1 Flow chart of the cervus parentage assignment and colony sibship reconstruction rules. *M*, number of cubs assigned a mother; *P*, number of cubs assigned a father. *N*, number of cubs that followed that path: *N1*, with one parent inferred, and *N2*, with both parents inferred (from Dugdale et al. 2007, with permission).

$n = 262$), while males, to which the paternity of at least one cub in a given year had been assigned, sired a mean of 1.5 ± 0.12 cubs (range = 1–6, mode = 1, $n = 221$) (Dugdale et al. 2007)—again pleasingly close to what we had inferred from teat measures in the absence of pedigree in our demographic explorations (Chapter 11).

Thus far, our description might have been of a plural breeding group with occasional infidelities, but from where did paternity hail? The generality is that lions breed, for the most part, within their prides (Packer et al. 1991), banded mongooses within their bands (Cant et al. 2013), and so in these rather closed carnivore societies kleptogamy is a minority achievement. Even amongst the interestingly unfaithful Ethiopian wolves, whose females seek unselective liaisons with any available male at the territory boundary, extra-pack progeny are a minority (Sillero et al. 1996), so if you are unsettled by departures from convention, sit down and draw a deep breath. Fifty per cent of 569 cubs (assigned parentage with 80% confidence) had

extra-group fathers, or 42% of 331 cubs assigned at 95%. Of these foreign males, it was 'the boy next door' that gained 74% of extra-group paternities at 80% confidence (37% of total paternity assignments) or 86 (36% of all assignments) at 95%. Furthermore, of 143 multiparous (i.e. with more than 1 cub; prerequisite for this metric) litters, 64 (45%) involved more than 1 father, i.e. multiple paternity: this is a very high occurrence of what Whitehead (1994) delicately termed 'alternative reproductive tactics' (ARTs) (Figures 17.2 and 17.3).

Interesting too, was the fact that, despite the orgy of mating activity we observed each spring, over their lifetimes 67.7% of males and 70.0% of females never produced cubs (from cohorts 1986–2007; we were restricted to this date range because only once individuals are certainly dead can one deduce lifetime reproductive success, LRS). Furthermore, 18.3% of males and 14.1% of females produced exactly one litter, while just 14.0% of males and 16.0% of females produced more than one litter in their lifetimes (Figure 17.4).

(a)

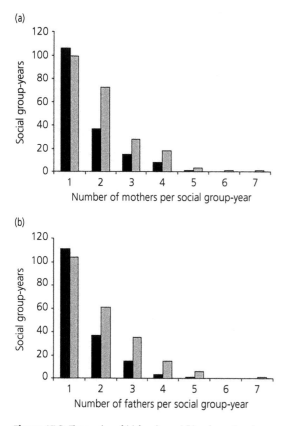

(b)

Figure 17.2 The number of (a) females and (b) males assigned as parents within each social-group year with an assigned parent(s). Grey bars represent parentage assignment at ≥80% confidence ($n = 222$), and black bars at ≥95% confidence ($n = 167$). Fewer parents were assigned at ≥95% confidence, resulting in more social groups with just one mother or father than with the ≥80% confidence assignments (from Dugdale et al. 2007, with permission).

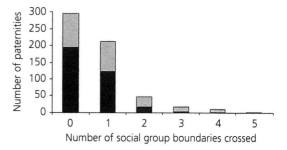

Figure 17.3 Number of paternities assigned according to the number of social-group boundaries that must have been crossed in order to gain paternity at ≥80% (grey bars, $n = 585$) and ≥95% (black bars, $n = 338$) confidence (from Dugdale et al. 2007, with permission).

of which Geetha was able to genotype 1020. Using an additional likelihood-based parentage assignment and pedigree construction program, MasterBayes,[14] Geetha's additional polymorphic loci allowed inferences from second-order relatives; that is, grandparents, aunts, and uncles, which increased the discrimination power between full-siblings and half-siblings from 71 to 88% (at ≥80% confidence). In Geetha's era we were able to assign both parents to 378 litters, of which 36.8% (139; $n = 378$) were multiparous (117 twins, 20 triplets, 2 quadruplets); and of these, 27% (38; $n = 139$) had multiple paternity(methodologies matter, and in this case Geetha's, applied to the expanded genealogy, never detected more than two fathers per litter, whereas in Hannah's day we had thought the maximum was four). Again, Geetha confirmed that far fewer badgers bore cubs than mated (Chapter 7), with just 45.42% of 502 candidate mothers (228) and 32.37% of 612 candidate fathers (201) actually breeding successfully during their lifetimes—so much for equal opportunity in badger society.

More years passed, yielding more individuals to genotype and, as we explained over Chapters 11–13, times changed. To dig deeper we needed even more polymorphic microsatellites at our disposal, in order to make the parentage and sibship assignments more powerful. As the years passed, we accumulated this larger set, with the development—by the next genetics student in the sequence, Geetha Annavi (2008)—of 15 more primers, boosting the total from 22 to 35 (Annavi et al. 2011).[13] Between 1987 and 2010, we had accumulated DNA from 1247 badgers,

were polymorphic (Table 17.1), 6 were monomorphic, and 8 failed to amplify or amplified non-specific products.

[14] Bayesian approach that simultaneously estimates the parentage of a sample of individuals and a wide range of population-level parameters in which we are interested (Hadfield et al. 2006). We show that joint estimation of parentage and population-level parameters increases the power of parentage assignment, reduces bias in parameter estimation, and accurately evaluates uncertainty in both.

The primary aim of MasterBayes is to use MCMC techniques to integrate over uncertainty in pedigree configurations estimated from molecular markers and phenotypic data. Emphasis is put on the marginal distribution of parameters that relate the phenotypic data to the pedigree.

[13] We isolated 432 new badger microsatellite sequences: Mel-80–Mel-611(FR745442–FR745873). Of the 35 loci tested, 21

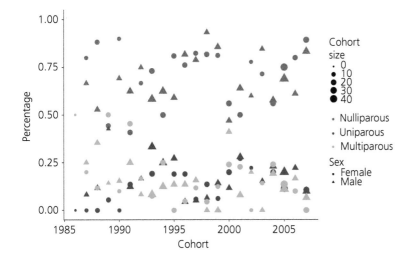

Figure 17.4 Depicting the percentage of adults engaging in one of the following reproductive options per year: red = nulliparous; purple = uniparous; and blue = multiparous. The figure shows what proportion of a given cohort's cubs went on to produce 0, 1, or 2+ litters in their lifetime (the size of the point/triangle shows how many individuals were in the cohort to compete with) (from Bright Ross 2021, with permission).

Lifetime reproductive success

Strikingly, reproductive success is distributed far from evenly between breeders—further supporting our observations of fundamental inequalities between individuals (Chapter 14) (Figures 17.5 and 17.6: remember, LRS can be calculated only at the end of the lifetime). Six males each fathered in excess of 10 cubs (all between 1988 and 2001), an additional 12 fathered more than 5 cubs (more generally across the dataset), and the rest fewer. The paternity record goes to Casanova B (M90) (see Chapter 7) with 19 offspring, over his 6-year lifespan (Figure 17.5); see Box 17.1.

To the sociobiologist, not to mention the geneticist, immunologist, or other voyeur, the obvious question is what made Casanova B special? Was it simply that he was the only eligible mate in the vicinity? No. Over the years he cohabited with up to nine, and never fewer than three, other adult males, all of which, to human eyes at least, looked passably eligible (during the same period between four and six adult females also inhabited these two setts).

While we logged the paternities of quite a few high LRS males in the early years, these Lotharios became uncommon as the years passed. For instance, amongst post-millennial badgers, M901 (from Thorny Croft sett) apparently died at only 3 years of age, having, however, enjoyed a lifetime reproductive success of just one daughter, F944, born in 2004 to F843 (also from Thorny Croft). Was this also her only child? Yes. Or M1097 (Lower Seeds), who mated F767 (primarily resident at Lower Seeds) to produce his only child M1276 in 2009 (but born at Chalet sett, suggesting F767 moved temporarily because she was back at Lower Seeds in

2010, although born at Radbrook Common in 2000 and taking 7 years to form any real fidelity to a sett, being caught at various setts through the north-east quarter of Wytham until then); this was to be her last of six cubs.

Our working hypothesis was that it became increasingly difficult for males, for whom the lifetime reproductive success window is wider because they can sire multiple litters in a good year, to monopolize productivity, as population density rose and badger socio-spatial distribution became more diffuse.

Amongst females too, some were super-breeders, although the limitation of producing only a single litter annually constrained them. Two females, F699 and F790, shared the record for lifetime reproductive success, with nine cubs each. They were amongst only five females (also Delta, F227 and F229) to manage seven or more litters between 1987 and 2000 (Figure 17.5). As with males, these super-breeders were a feature only of our earlier years of study. An example is Dowager B (F227) (see Box 17.2).

Of course, potential to produce offspring is also a feature of longevity, especially for females limited to but one litter per year (whereas exceptional males, like Casanova B, can father several per annum). This prompted us to investigate how LRS was distributed as a function of age (Figure 17.6).

We see that it is not simply that the longer the badger lives the greater its number of accumulated offspring (although the case of female Dowager B (F227) who, by the age of 14, had had 8 cubs, did take full advantage of the opportunities of longevity (Box 17.2)). Actually, amongst badgers, being average seems the best recipe for peak LRS; that is, living to the modest age of

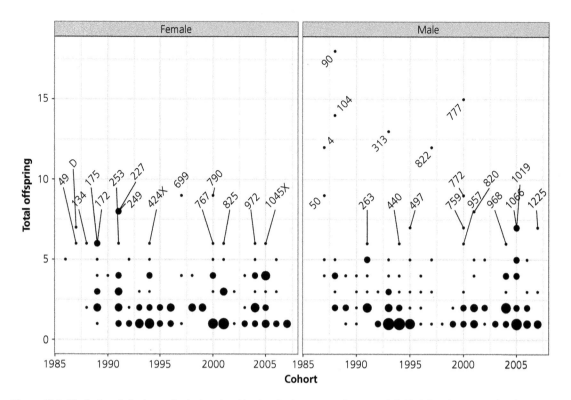

Figure 17.5 Distribution of offspring production by cohort. The size of a dot represents how many individuals born in year *x* produced *y* offspring. Labels provide the tattoo numbers of all badgers that produced at least six cubs in their offspring. Only individuals assigned at least one cub by the pedigree are shown (from Bright Ross 2021, with permission).

6–9, which 13–26% of the population achieve. Dying before their prime is disadvantageous; then again, living to advanced years offers no substantial LRS benefit, and in Chapter 18 we will explore the reproductive demise of aged badgers, traversing menopause and andropause. Indeed, if LRS is going to remain relatively constant even by living another 3 years, then this approach is inferior reproductively to producing the same number of cubs over a shorter lifespan—the old solution to Cole's paradox. This is reminiscent of pace of life syndromes in Chapter 14, where notwithstanding the occasional merits of the tortoise or the hare, a middle pace, perhaps intuitively, appears optimal for LRS.

But how does this distribution of reproductive success translate across annual cohorts over the course of the study? That is, viewing things from the population rather than the individual perspective, how did the average badger's lifetime reproductive success change based on when he or she was born? Well, a model comprising the cohort in which an individual was born, an individual's sex, and their maximum realized lifespan

captures 23.2% of the variation in individual LRS for those badgers that produced at least one cub in their life (keeping in mind as always that it is difficult to truly estimate nulliparity for badgers assigned zero offspring by the pedigree). Setting aside the variation in maximum age realized (in fact, assuming each cohort reached the average maximum age in this data subset of 6.6 years old), we see that LRS was highest for those individuals that reproduced at the very beginning of the study—particularly for males. Thereafter, as density rose, individual badgers lost reproductive potential, but males born right around the population peak in the late 1990s exhibited a slight resurgence of their reproductive fates (Figure 17.7).

Parents matter: direct and indirect benefits

So what might discriminating badgers—indeed any mammal capable of discernment in the heat of courtship—look for in a mating partner? And just how discerning are badgers, especially males? By definition,

---

Box 17.1 The dynasty of Casanova B (M90)

Casanova B, or M90 as he was more prosaically coded, was special. He sired more cubs than any other male in our study: 19 that we genotyped.

Born in 1988 at the Upper Follies sett, Casanova B, upon achieving maturity, quickly moved 250 m south-east downhill to the neighbouring Botley Lodge sett (see map p. xx), but thereafter vacillated between the two until his last capture in 1993 (we now know that these two groups are affiliated to the same super-group). His first fatherhoods were in 1990, when he conceived one cub (F226) with F180 and two cubs (F231 and F261) with an unknown mother (and, as these setts lie on the periphery of Wytham, plausibly she could have been a farm girl). All of these cubs were born a year later (1991) at Botley Lodge sett; they could therefore be considered as within-group offspring insofar as Casanova B was caught only at Upper Follies in 1990. In the 1991 breeding year/1992 birthing year, during which Casanova B was caught only at Botley Lodge, he returned again to favoured mistress F180, who bore him a daughter (F283) at Botley Lodge. But in 1993 (having been conceived in 1992) he was very prolific, breeding once again with F180 to produce another daughter (F368) at Botley Lodge, also fathering a cub (F353) to F86, a resident of MacBracken sett, some 300 m due east across a pasture field (which, with hindsight again, formed part of the same super-group). Spouse number three for that year was F214, who bore him two cubs (M329 and F352), born at Botley Lodge, where Casanova B was living in the conception year. Spouse number four was F226, also producing two of his cubs (F321 and M328)

at Botley Lodge. Spouse five was F279, mother to his additional three cubs (M355, M370, and M378), also born at Botley Lodge. So, in 1993 alone, Casanova B fathered nine offspring. Although not caught after 1993, his legacy continued (perhaps he evaded capture, or moved to a less-often trapped adjacent farmland sett): in 1994 (conceived 1993) he consorted once more with F180 with whom he fathered a daughter (F431) born at Botley Lodge, and also hooked up with F231 (an incestuous mating with his own daughter from 1991), who went on to bear him two sons (M412 and M433), also at Botley Lodge; he also returned to former mate F226, who added to his tally of daughters (F428) at Botley Lodge. The year 1995 saw his breeding success continue with a daughter (F467) born to an unknown mother at Upper Follies. As an aging *roués*, his final offspring assignment came in 1996 (possibly posthumously given the <5% probability he could have evaded capture for 3 years)—his only extra-super-group progeny, a son (M531), born at the neighbouring Singing Way sett (Casanova B was last resident at Upper Follies sett) to mistress F227 (she was never caught beyond her Singing Way super-group). In total, his 11 daughters and 8 sons, over 13 litters, were born to 8 mothers (including the incestuous episode with his own daughter), noting that Casanova B mated successfully with F180 four times (each yielding a singleton litter; she had no cubs with any other male), and with F226 twice (she did have one additional cub, M610 in 1997 at Botley Lodge, after Casanova B's presumed death; the latter having been born to M366, also of Botley Lodge, while M610 died in infancy).

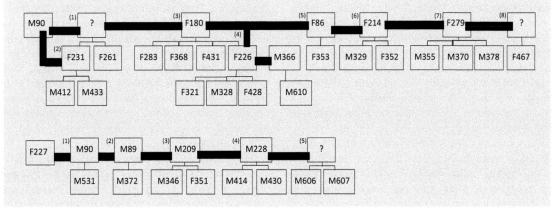

---

the qualities that make for a propitious choice are those most likely to enhance unfolding contributions through generations to come. One problem, amply illustrated through Chapters 11–13, is the breathless

presence of the 'Red Queen'—the adaptations that matter most are continually changing. Will the next generation of offspring grow up in a world most burdened by drought, disease, or social competition?

Box 17.2 The dynasty of Dowager B (F227)

Of the female super-breeders, consider the case of Dowager B (née F227) (she lived to be 14; see pane in Figure 17.5), with her 8 lifetime offspring. What is noteworthy is not where her cubs were born (all in setts within the Singing Way super-group: 1993 litter = M346, F351, and M372, born at Singing Way; 1994 litter = M414 and M430, born at Marley 2; 1996 litter = M531, born at Singing Way; and 1997 litter = M607 and M607, born at Singing Way), rather, the striking fact is the variety of locations from which their fathers hailed. Dowager B obviously earned her soubriquet only later in life; she had been born as F227 in 1991 at Marley 2 sett, within the Singing Way super-group. She had her first litter in 1993 (having, unusually, conceived as a yearling, in 1992). Her three cubs had two different fathers, one extra-group, from Jew's Harp, M89 (fathering M372) and

one within group, from Marley 2 M209 (fathering M346 and F351). Of these three cubs, only F351 lived beyond her first year, surviving to age 4. Conceived with M228 of the Marley 2 Outlier in 1993, Dowager B had two cubs (M414 and M430), born in 1994 at a time when she resided between M2 and the M3 outliers—both within 200 m of the main Singing Way sett. In 1995, Dowager B skipped a breeding year, bearing no surviving cubs; in 1995, however, she mated with Casanova B (see Box 17.1), producing cub M531 (Casanova B's only extra-group offspring, as mentioned earlier), born at Singing Way in 1996. Her final litter was born in 1997, comprising two sons (M606 and M607), whose father we could not identify. In total, Dowager B produced four litters comprising seven sons and a single daughter.

Circumstances are brutally mercurial (and very occasionally a Black Swan lands, as in Chapter 10), so there's no one overarching answer, or at least not one that lasts for long. Natural selection would favour badgers whose crystal ball-gazing led them to make the right predictive adaptive response (a notion introduced in Chapters 14 and 15).

Given that females are limited to a maximum of one litter per breeding cycle (per year, in badgers), they tend to be the choosy sex, whereas males, which are potentially able to fertilize several females in polygynandrous systems (Chapter 7), and have no commitments to gestation and lactation, tend to compete with one another for the opportunity to mate with receptive females (Hunt et al. 2009). This creates an important dichotomy: in Chapter 7, we highlighted polygynandry in badgers—many females mating with many males—but no matter how many males a female mates, she is inescapably bound to the size of her own, solitary litter per breeding cycle. In contrast, plural mating allows males to father litters with multiple females in good years, 'making hay while the sun shines'. Consequently, unlike mating, 'breeding' can only ever be polygynous, aside from perhaps a female splitting paternity within her single litter (through cryptic female choice of blastocycts). Typically then, this leads either to female choosiness in carefully assessing the relative attributes a potential partner could contribute towards producing and/or maintaining viable, ideally good quality, offspring, or that females have a tight sieve of cryptic post-copulatory filters that weed

out poor sperm or poor zygotes, favouring those with immuno-genetic advantages (Schwensow et al. 2008. Remembering too the pause for 'thought' that delayed implantation allows amongst badgers: Orr and Zuk (2014) argue that any such hiatus provides opportunity for discrimination or, as Macdonald (1994) put it, an opportunity for second thoughts under new circumstances.

In some mammal species, of course, males offer emphatic 'direct' benefits, such as protection from predators or paternal care (Kokko et al. 2003); for instance, in Patagonian maras (*Dolichotis patagonum*) the male stands vigilant while the mother of his pups grazes (Taber and Macdonald 1992a) (this paternal protectiveness is one of several features also seen commonly in birds; for example, Hadfield et al. 2006). Importantly though, some evident phenotypic trait must signal how likely a male is to make a substantive contribution, else how would the female anticipate the fatherly capability of her mate at the point of coitus (Candolin 2003)?

This ability to discriminate paternal qualities before committing to a particular father is particularly intriguing in species in which material advantages attached to particular mate options are less apparent (Zelano and Edwards 2002). Even more so for superfecund badgers (Chapter 7) where potential mothers get to sample many males before sorting through the blastocysts to which each contributed. A few pages earlier, in Box 17.1, we detailed the success of Casanova B, amongst other males that achieved particularly high

Separated by maximum age attained

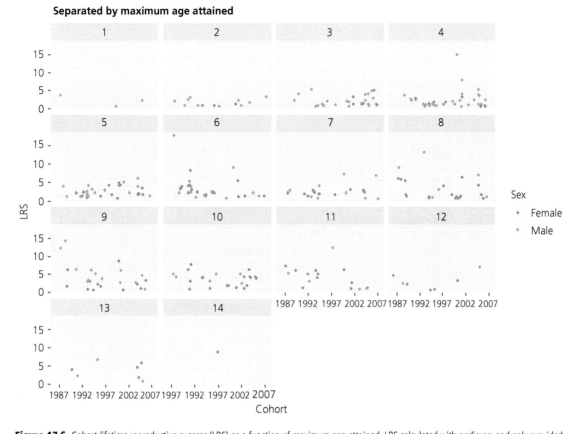

Figure 17.6 Cohort lifetime reproductive success (LRS) as a function of maximum age attained. LRS calculated with pedigree, and only provided for individuals with at least one cub produced in their lifetime. The first plot has a separate panel for individuals whose maximum realized lifespan match (labelled in the grey at the top of each panel), and shows LRS as a function of the badger's cohort (limited to 2007 because of our 2015–2018 rule). Readers can see that, generally, females don't change very much, but there's a dip amongst the males. This is despite the fact that there aren't necessarily more reproductive males dying earlier in later cohorts—this plot is solely of individuals that reproduced at least once. Considering the patterns of males dying earlier overall (including LRS = 0 males), we know that increases in the late 1990s. The second and third plots show the modelled effect of cohort on LRS, on the exponential scale, after controlling for the effect of maximum age on LRS—the effect is clear for males but not females (from Bright Ross 2021, with permission).

reproductive success (see also Dugdale et al. 2008). But by looking at, and measuring, them we could not detect any overt attribute that gave these males their edge. Perhaps we, noseless, humans are simply blind, visually or osmically, to what makes a male badger irresistible (see Chapter 6), or perhaps the *je ne sais quoi* of these males really is cryptic, an 'indirect benefit' accruing to certain mate pairings, enhancing offspring genetic fitness and thereby increasing the parents' inclusive fitness (Kokko et al. 2003; Consuegra and Garcia-de-Leaniz 2008)? For instance, it has long been recognized (Beherman et al. 1960) that a major factor in infertility amongst anatomically and physiologically healthy human couples lies in the (in)compatibility of their ABO(H) blood types, wherein incompatible antigens carried by spermatozoa are blocked or immobilized by the antibodies present in the cervical secretion.

Biologists have identified two categories of such potential indirect (and, in addition—cryptic) benefits, leading to either (i) high(er) quality genes amongst offspring (adaptive indirect benefits, which can arise through such physiological mechanisms as post-copulatory conception filters); or to (ii) offspring carrying genes that make them more attractive (arbitrary indirect benefits that can arise through behavioural mate choice) (Dawkins and Guilford 1996)—a phenomenon that Ronald Fisher (1915) coined the 'sexy son hypothesis'. This posits that if females choose

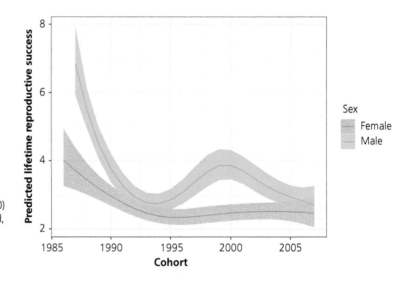

Figure 17.7 Predicted lifetime reproductive success as a function of cohort. Curves show the sex-specific predictions (+/− standard error) from a model of reproductive individuals (LRS > 0) born throughout the Wytham study period, assuming individuals reached the average maximum realized lifespan (from Bright Ross 2021, with permission).

physically attractive males, they will tend to beget physically attractive sons, and thus more grandchildren, because subsequent choosy females will prefer their attractive, sexy sons, and so on (Fisher 1930). So long as the trait in question is heritable and variable between individuals, it will function regardless of, even in opposition to, any more tangible direct benefits, because it is the possession of the trait per se that makes the male attractive, and not the qualities of the trait in itself. Moreover, the ability of the male to thrive despite handicap may advertise this attractiveness. We were alert to this possibility because, in the early 1970s when David was a doctoral student in Niko Tinbergen's Animal Behaviour Research Group, he was privileged to make friends with Amotz Zahavi, a brilliantly insightful and mischievously provocative evolutionary biologist, who was then a regular visitor to Oxford. Indeed, that inspiring friendship was part of the reason why, in 1976, as a Churchill Memorial Fellow, David worked on foxes and jackals in Israel (Chapter 9), and was introduced by Amotz to the Arabian babblers (*Argya squamiceps*) residing near the settlement of Hatzeva in the Arava Valley south of the Dead Sea (Zahavi 1981). These small communally nesting passerine birds live in groups with strict rank order, and their show-off behaviour had led Zahavi to the idea of the handicap principle (Zahavi 1975). Somewhat analogous to the peacock's tail plumage, or the deer's antlers— ornamentation that would be maladaptive if it were not so alluring to females—male babblers gave food away as evidence of their superiority. These handicaps, akin to conspicuous consumption, signal the ability of the male to squander a resource while inferior males, for

their part, cannot afford to produce such extravagant appendages. Another Oxford academic at the time, first introduced on p. 145, was Bill Hamilton, a gifted naturalist and fan of Wytham's badgers, described memorably by Richard Dawkins as the greatest biologist since Darwin (and also notable for a shortage of fingers resulting from schoolboy curiosity about explosives in the chemistry class). In 1982, Bill, and Marlene Zuk, advanced the handicap idea further, with the eponymous Hamilton-Zuk hypothesis (1982). This proposes that sexual ornaments can co-indicate resistance to parasites and diseases, which can be important drivers of sexual selection through mate choice. Furthermore, sexual selection can accelerate preference for evolutionary 'bling' in a positive feedback cycle, in which desirable traits are embellished in offspring, leading to a form of self-reinforcing co-evolution (Andresson and Simmons 2006), which Fisher (1915, 1930) had formalized as his 'runaway' hypothesis. Indeed, given the immunosuppressive effects of testosterone (Chapter 15), the 'immunocompetence handicap hypothesis' extends this concept further; where secondary sexual characteristics and traits indicative of masculinity indicate that the male in question can contend with health hazards successfully (Rantala et al. 2013). Insofar as male attributes lead to greater vulnerability to pathogen or parasite attack (see Chapter 15), one might expect only high-quality males to be able to afford to display sexual characteristics fully without suffering more severe parasite loads (Roberts et al. 2004). Badges that concentrated testosterone courses in their veins include the wattle size of cock pheasants (*Phasianus colchicus*) (Briganti et al. 1999), the intensity of nuptial skin colouration in male Algerian sand racer lizards (*Psammodromus algirus*) (Salvador et al. 1996), and the antler size of white-tailed deer (*Odocoileus virginianus*) (Ditchkoff et al. 2001). Moreover, these signs of machismo cannot be bluffed—cheaters could not invade this trait because in the attempt to punch above their weight their immunocompetence would be so compromised by the effects of testosterone that, far from being rewarded, their lifetime reproductive success would be reduced (Folstad and Karter 1992). Even amongst humans, men show greater susceptibility to infections than do women (Klein 2000) (although the weaker male immune response makes them less susceptible to auto-immune disorders than are women; Grossman 1990). If females might be attracted to males that can bear the cost of an evolutionary folly, the handicap, might they not also be discerning about traits that straightforwardly indicate prowess in survival stakes? That is exactly the proposition codified in the so-called good genes hypothesis, crystallized by Møller and Alatalo (1999), which makes the naturalistically intuitive proposition that the choosy sex will mate with individuals which possess traits that signify overall genetic quality (Neff and Pitcher 2005).

The genetic compatibility hypothesis (Tregenza and Wedell 2000; Kempenaers 2007), under which it is not sufficient simply for genes to be 'good' (in terms of fitness through LRS), completes this suite of ideas about possible indirect benefits underlying mate choice. It posits that, in addition, the genes contributed from each parent should be compatible and function together to provide optimal offspring fitness by self-reference to their own genotype (Puurtinen et al. 2005). These genetic differences in compatibility can be no small matter—recall that the incompatibility between blood types mentioned above doesn't just take the competitive edge off the offspring, it kills them. The genetic compatibility hypothesis requires the female not only to be a good judge of male genes, but also to have, in the unconscious way of evolution, a sense of self-knowledge. In the quest for compatibility, the choosy female must somehow assess her own genotype relative to the genotypes of available partners (Penn and Potts 1999; Newman and Buesching 2019). A promising means of making these assessments, both of potential mates and of self, lies in olfactory cues present in body odours (Yamazaki et al. 1976; Milinski et al. 2005)—something with which badgers are richly endowed (see Chapters 6 and 9). This prompts yet a further hypothesis: the microflora hypothesis (Yamazaki et al. 1990; Lanyon et al. 2007), which involves the major histocompatibility complex (MHC)

genes we introduced in Chapters 9 and 15.[15] In the context of revealing body odours, MHC genes may control odour indirectly by influencing the growth of microbiota commensal on sweat and scent gland secretion, which contribute to variation in individual odour (a topic we introduced, in Chapter 6, in the context of group recognition and the subcaudal gland).

Indeed, congruent with the sensory bias hypothesis (Fuller et al. 2005), it seems that badger subcaudal gland secretion, a trait evolved in a different sociobiological context over millions of years (Chapter 6; Buesching et al. 2002a, 2000b), can be exploited as a criterion for advertisement and selection to solicit optimal opportunities for conception. Insofar as sensory biases can trend quickly, considerable trait signalling differences can emerge between populations and closely related species, leading to reproductive isolation (Smadja and Butlin 2009).

This evolutionary logic and associated plethora of hypotheses lay the foundation upon which we pose our next question: having failed to detect any compelling evidence in Chapter 7 that badgers are overtly selective about with whom they mate, based on criteria such as body size, do they instead select for covert genetic and immunological traits in their mates, as evidenced by the genotype of their offspring? The answer was yes, in Chapter 15, for MHC genes, but we ask the question now with regard to heterozygosity; that is, the possession by a single individual of two different alleles of a particular gene or genes.

Selecting mates for heterozygosity

Gamblers hedge their bets, hoping for a strong hand from the reshuffled pack (Chapter 14). Life is a gamble in the casino of natural selection and in 1997 Jerram Brown proposed that choosing mates for genome-wide heterozygosity, combined with outbreeding, should lead to offspring with greater allelic diversity, leading to greater vigour (*heterosis*), involving traits such as higher offspring survival rates (Cohas et al. 2009; Mainguy et al. 2009a), reproductive success (Slate et al. 2000; Harrison et al. 2011), developmental stability, and disease resistance (Coltman et al. 1999; Whiteman et al. 2006).

---

[15] In a controversial but well-known experiment college women were then asked to rate odours from T-shirts slept in by various men, some with similar MHC genes to their own and others with dissimilar genes. Women tended to rate the odours higher if the men's genes were more dissimilar to their own, suggesting consequences for human mate choice (see Wedekind et al. 1995).

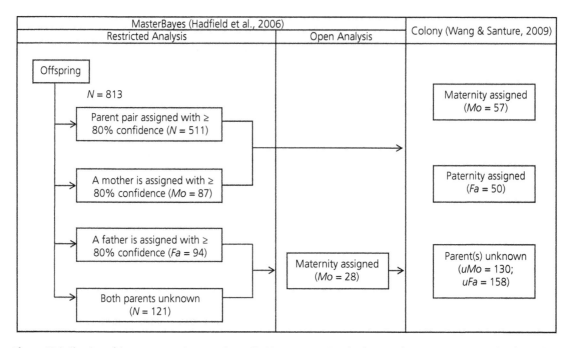

Figure 17.8 Flowchart of the parentage assignment rules used in MasterBayes 2.47 and Colony 2.0. The MasterBayes restricted analysis only included females aged ≥2 years and present in the cubs' natal group as candidate mothers, whereas the open analysis included all females in the population aged ≥2 years. N = total number of cubs; Mo = number of cubs with an assigned mother; Fa = number of cubs with an assigned father; uMo = number of cubs with an unassigned mother; uFa = number of cubs with an unassigned father. Parentage was assigned with ≥80% confidence (from Annavi et al. 2014a, with permission).

If females follow this strategy, known as 'good genes as heterozygosity', they would be expected to seek mates contributing alleles to conceive maximal heterozygosity. So, is there a direct relationship between each badger's genetic diversity and its fitness? Geetha Annavi led the investigation of this question, undertaking a parentage analysis of 35 microsatellites of 813 individuals (1988–2010[16])[17] to reveal genomic heterozygosity–fitness correlations (HFCs) (Annavi et al. 2014a) (Figure 17.8). To identify each cub's mother we assessed as candidates all females, aged ≥2 years, in any given cub's natal group, along with all males of breeding age (>1 year old) across the entire population as candidate fathers.

Although HFCs can arise through single or multi-locus effects, for simplicity, here we present results for standardized heterozygosity (SH), which is estimated by dividing the proportion of heterozygous loci

per individual by the population mean. These general HFC effects (Hansson and Westerberg 2002) arise due to inbreeding depression (i.e. breeding between related parents), which increases the probability that deleterious mutations are expressed.[18] Here we use the inbreeding coefficient, f (i.e. the degree to which offspring and parents were related), to measure any reduction in heterozygosity on a genome-wide scale.[19]

Geetha's explorations of the impact of heterozygosity and inbreeding on offspring fitness led to several exciting discoveries. Very interestingly, inbred badgers (defined as $f \geq 1.25$) experienced a disadvantage in that they averaged lower lifetime breeding success than those that were outbred;

[16] Excluding 25 cubs caught in 1987 due to low confidence in assignments because parents were not on record before trapping began in 1986.

[17] These analyses used pedigree programs MasterBayes (v2.47; Hadfield et al. 2006) and Colony v2.

[18] Pedigree Viewer 6.3 (Kinghorn and Kinghorn 1994) was used to calculate f for 561 of the 813 genotyped cubs to which we reassigned both parents with ≥0.8 probability. We followed the approach of Szulkin et al. (2007) by restricting our dataset to 420 (52%) cubs that had at least one grandparent assigned, and to 88 (11%) with all four grandparents assigned.

[19] The alternative of local effects, driven by single-locus HFCs would have implied assortative overdominance, due to certain loci being in linkage disequilibrium with functional loci (David 1998).

although first-year survival probabilities were not correlated with the inbreeding coefficient f, which was principally governed by the weather (Annavi et al. 2014b).

These inbred badgers also had lower individual SH (0.77 vs 1.01), consistent with broader evidence of the costs of inbreeding depression (Pusey and Wolf 1996). Despite the routine forays and philandering beyond the social group (19.8% of the population at any given time was found in a social group other than its principal group), because of high natal group philopatry (recapping from Chapter 3: 35.8% of badgers only ever being caught in their natal group, 39.9% being caught in no more than two groups; $n = 267$ individuals, 5255 trapping events), low permanent dispersal rates (19.1%; Macdonald et al. 2008), and the tendency of siblings to reside together, co-resident badgers nonetheless tended to be more related to each other than they were to members of neighbouring groups.

In terms of extra-group matings, over their reproductive lifespans around 30% of badgers of both sexes exhibited consistent within-group mating fidelity, whereas slightly more (40%) always played the away field, the remaining third engaging in a mixed tactic, home and away. This is a pivotal, and curious, insight, insofar as it reveals some meline tacticians devoted to group segregation, others to societal integration, and yet others that played a mixed hand. Could it be that opposing evolutionary forces balance out these strategies according to annual circumstances and the extent to which the population enjoyed stability?

Does heterozygosity matter?

A striking result, dove-tailing with the important contribution to badger ecology made by weather patterns (Chapters 12 and 13) was that there was a pattern whereby heterozygosity amongst cubs, their mothers, and their fathers was all associated with cubs tending to survive their first year better (Figure 17.9); however this effect was statistically significant only in relation to the heterozygosity of a cub's father and only in in wetter summers. What caused this curious weather effect? In drier years, rates of indiscriminate (i.e. unrelated to heterozygosity) mortality (such as coccidiosis) appear to mask differential heterozygosity-related survival effects. In those drier years, having a heterozygous father does not help a cub survive to its first birthday. We deduce that the risks to badger cubs of either an extreme soaking, chilling them to the bone, or of drought (with, in the latter scenario, consequent reduced availability of worms), are beyond the niceties

of adaptation: these are aspects of environmental Russian roulette that, when the bullet is in the chamber, can have only one outcome. All the good genes in the world can't deflect the bullets of hypothermia or starvation.

But why was this effect only significant in relation to paternal heterozygosity? At first we found this result both stunning and baffling. After all, in a species with no paternal care, how can these heterozygosity advantages of the father be visited upon offspring if said cub does not itself inherit that same high extent of heterozygosity? Plausibly, this may relate to probability. Each litter of cubs has but one mother, but may have at least two fathers (multiple paternity, earlier). Consequently more males get input into the genetics of cubs born per year than do females, potentially tipping the paternal effect into statistical significance. This is one speculative hypothesis for future testing. Another would be that mothers invest in their offspring differentially (Burley 1986) according to the heterozygosity of the offspring's father, translating into survival differences in good years. If so, not only would we have to ask why: what would be the functional advantage, but also this would require that promiscuous females do actually know who the father of their offspring is.

Returning to more solid ground and our fundamental question about the possible benefits of extra-group paternity (EGP), it seems reasonable to speculate that the near 50% per annum EGP rates we discovered amongst Wytham's badgers might impart a selective advantage to offspring fitness. However, the profusion of mate selection hypotheses and mechanisms evidence that vivid imaginations are generally more available than data (especially data describing the activities of a wild-living medium-sized mammal). Testing this plethora of ideas required that we also test and subsequently reject a common denominator: the null hypothesis that mate pairings reflect random chance through the probability of casual encounter. We therefore asked next if the occurrence of EGP was related to opportunity alone?

Relatedness, contact rates, and extra group paternity

Conceptually, the advantage to males of mating beyond their social group is clear-cut, insofar as it extends their mating opportunities, potentially allowing even inferior males to 'steal' some kleptogamous mating opportunities (Young et al. 2007). As we explained earlier, insofar as the number of litters

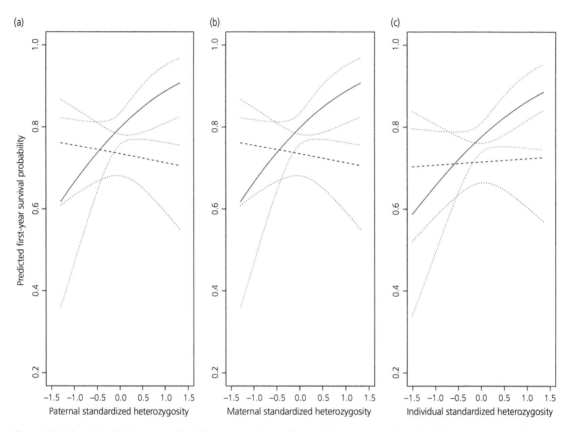

Figure 17.9 The relationship between predicted first-year survival probability and: (a) paternal standardized multilocus heterozygosity (SHPat); (b) maternal heterozygosity (SHMat); and (c) an individual's own heterozygosity (SHInd). Standardized total summer rainfall (SR) was categorized for ease of visualization; SR > 0 and SR ≤ 0 were years with above and equal to or below mean rainfall, respectively. Probabilities are plotted under mean conditions of high (solid line; SR = −0.4) and low (dashed line; SR = 0.6) total summer rainfall (May–October). The dotted lines represent the 95% confidence intervals (from Annavi et al. 2014a, with permission).

(multiparous) and the litter sizes (polytocous) that females can produce per breeding season limits (and also invests) them to a greater extent than males, they are predicted to strive the hardest for offspring of the best quality (not just quantity). It may therefore be advantageous to females to seek out desirable genes from whomever they can conceive with, whether within or beyond their social unit. However, there could be a sociological problem with such unfettered free love: while promiscuous mating within social groups tends to confuse paternity and alleviate the risk of infanticide (Agrell et al. 1998), EGP could result in the rejection of extra-group offspring if resident adults could detect them (Westneat and Stewart 2003; Borries et al. 2011). Consequently, females might strive to trade off the co-option of good genes (if they can recognize them) against possible disadvantages of mating beyond their group—a candidate disadvantage

being the provocation of infanticidal attacks by cuckolded resident males (and maybe even by competitive females). This possibility conforms with what Rawls (1971) called the 'veil of ignorance', under which decision-makers are blind to the provenance of the actors involved, promoting impartiality and cooperation. Certainly communally breeding banded mongooses, ignorant of the individual parentage of their communal young, allocate post-natal care in a way that reduces inequality amongst offspring, just as predicted by a Rawlsian model of cooperation (Marshall et al. 2021). Furthermore, while uncertainty over relatedness in insect societies promotes cooperative behaviour insofar as workers cooperate to raise the offspring of other workers when they are ignorant of their relatedness to them, but when they can discriminate worker-laid eggs from queen-laid eggs they kill them (Queller and Strassmann 2013).

Don't underestimate the unconscious. While the insights of the ethologist's eye are remarkable, not even the keenest eye can discern those aspects of mate choice beyond active behavioural selection processes. Out of sight, the encounter between sperm and egg, and the genes they carry, is where the reproductive rubber meets the road, and can skid. At the molecular level, the genes expressed on sperm and ova can be fundamentally incompatible, notably the 'Izumo' protein on the sperm, and 'Juno' on the egg (Bianchi et al. 2014). These boy-meets-girl incompatabilities can involve three restrictive mechanisms, summarized by Springate and Friaser (2017):

1. Sperm competition, where intense competition between sperm to win the fertilization contest can lead to the rapid evolution of advantageous genes.[20]
2. Sexual selection, where particular sperm–egg combinations have higher success rates than others, resulting in the continual co-evolution of genes expressed on the gametes of both sexes.
3. Sexual conflict, where selection on ova to ensure only one sperm wins the race, and intense competition amongst sperm to be that winner, create conflicting selection pressures on ova to make multiple fertilization difficult, and on sperm to more rapidly fertilize the ova.

So what of badgers? In this context, remember too, from Chapter 7, the badgers' reproductive specialisms of superfecundation, superfoetation, and delayed implantation (Sugianto et al. 2019b). These potentially extend the opportunity for females to (re-)select advantageous mates, either within- or extra-group, through post-copulatory mate choice (Chapter 7; Andersson and Simmons 2006; Fisher et al. 2016). Indeed, promiscuous females may be mating with multiple males as a hedge against genetic incompatibility, potentially arising at the molecular and cellular level at the point of conception (Zeh and Zeh 1997).[21] This would be possible if the female's physiology were capable of dictating which blastocysts implant and gestate, and

which are aborted from amongst a suite of options to which the female holds on, *in utero* (Yamaguchi et al. 2006); indeed, as Orr and Zuk (2014) write 'By delaying implantation, females may allow for the accumulation and comparison of zygotes'. They argue that this has been a major selection pressure for the evolution of reproductive delays in many sorts of mammals, and could be especially advantageous to those whose life histories constrain them to small litters—consider badgers and bats (it has been known since 1945 that in the bat *Perimyotis subflavus*, up to seven eggs are fertilized but only one implants (Wimsatt 1945).

This physiological possibility that unpromising embryos will not implant, or be aborted is further tempered by social conditions: given a *free* choice, females are generally predicted to exercise mating discretion to select for perceived quality and compatibility (Clutton-Brock and McAuliffe 2009), but, of course, as the 'Rolling Stones' sagaciously counsel: 'You can't always get what you want, but if you try sometimes you'll find you get what you need'. Thus the reality is that the world is rarely ideal and choice is rarely *free*; further, the best range (Yanagimachi 1978) of most suitable mates is unlikely to be present (Millstein 2002; Walsh et al. 2002). Consequently females are limited not only by the genes available in the population, but also by access to the males carrying them; be these co-resident, or extra-group.

This logic demands that females be able to reject unsuitable mates, or, at least, to exercise post-copulatory abortion of ill-conceived offspring (Kempenaers 2007). Are these conditions met? Physiologically, it would seem badgers probably have access to a variety of post-copulatory filters (Orr and Zuk 2014), and behaviourally, our impression from the observations in Chapter 7 is that if a female badger mates with a male, it is because she chooses to do so and not because of coercion (Smuts and Smuts 1993). Exploring further the balance of promiscuity and coercion amongst females in a classically semi-detached mustelid species (mindful of the aforementioned phylogenetic baggage influencing modern badger behaviour), with characteristically raised eyebrow, laconic Australian graduate student Mike Thom led on the development of the mink invertabrothel. Mike who ascended to be a professor at Plymouth University, and died prematurely while we were writing this book, carried out two experiments to explore what happened when female mink were able to choose which males they mated with: he designed things so that sylph-like female American mink had the opportunity to enter, or not, different chambers, each

---

[20] When the sperm cell attaches to the zona pellucida (a glycoprotein coating protecting the ovum), it is triggered to undergo the acrosome reaction allowing the hyper-activated motile sperm cell to drill through the zona pellucida (secondary zona pellucida binding coinciding with sequential local zona pellucida digestion and rebinding). Multimeric zona recognition complexes (MZRC) may provide a block to polyspermy (Reid et al. 2011).

[21] It is these profound sperm-ova incompatibilities that limit or preclude hybridization between otherwise similar species (Yanagimachi 1978).

of which housed variously hefty male mink: the doors to these chambers were of a diameter that allowed the slender females to come and go at will, but were too narrow for the hulking males to exit their chambers (Thom et al. 2004b).

In short, each female mink voluntary sought the company of, and mated with, multiple males. In the first experiment, where eight females each had access to three (sampled from a pool of eight) males over 18 days of the mating season, seven females chose to mate with all three males, and the remaining female mated with two. The second experiment involved 16 females, of which one declined to mate at all but the other 15 opted to mate with between 1 and 7 partners, with an average of about 3. Clearly, when these mink, with their traditional mustelid social system, were given a choice, females actively sought to mate with several partners. Does promiscuous mean undiscerning? No, the female mink were selective: they favoured males with larger testes. They also tended to mate for longer with larger males (those with longer bodies and tails), and they mated last in the sequence with males that had more siblings. In similar vein, we have several times mentioned that not all matings are equal, and timing can be critical with regard to weighing the odds in terms of particular sperm winning the race at ovulation: although taxonomically more distant to badgers, Cecile Roland used a similar experimental design to show that female house mice behaved much like female mink, but saved the best male until last. These female mice always switched male after the first mating, they accepted more intromissions and ejaculations from the dominant than from the subordinate male, and in a sequence they always gave the final mating to the dominant (Rolland et al. 2003; Brandt and Macdonald 2011 found that female harvest mice favoured the most familiar male—propinquity again). Anyway, all this convinces us that female badgers can make up their own minds about their partners, and that badgers are at least capable of inbreeding avoidance and reducing inbreeding depression (as seen in other mammals, e.g. Sillero-Zubiri et al. 1996a; Kamler et al. 2013a), notwithstanding any further cryptic processes, and that any tendency for Wytham's badgers to pursue EGP mating strategies is by design.

Of course, an alternative hypothesis is that EGP may be neither more nor less advantageous than within-group paternity, WGP (especially within a society founded on inter-group gene exchange, where extra-group genes may offer little functional variety from intra-group). This possibility would align with the conclusion of Forstmeier et al (2011) that extra-group

(or extra-pair) paternity occurs in many species without any apparent benefit. Apparently female badgers mate frequently and perhaps at random to induce ovulation (Chapter 7). Are they then any more selective about with whom they mate when they ovulate, and it really matters? Or do they continue to mate with whichever males they happen to encounter and that happen to appeal to them (a probability that propinquity, and also the distance and the relative density of home versus away suitors, would affect) when opportunity presents itself (Kokko and Rankin 2006), and then let post-copulatory filters sort out the wheat from the chaff?

We investigated these alternatives by again using the metric of SH, deriving SH values per social-group year,[22] as well as pairwise[23] relatedness between assigned mothers and candidate within-group mates. But first, just how common is EGP amongst the badgers of Wytham Woods?

Of course, each badger starts as somebody's offspring. Those individuals that had been fathered by males residing within their natal group in the year of their conception were termed 'within-group offspring' (WGO). WGOs numbered 340 (52%), and were assigned to 125 WG fathers; badgers fathered by non-resident males. More remarkably, despite group living, 'extra-group offspring' (EGO) numbered 315 (48%), assigned to 140 EG fathers.[24] Once again, whatever badger groups may be, they are not exclusive breeding cliques. Moreover, of these EGO, 85% were attributable to fathers residing in the neighbouring group (268 of 315), where this greater tendency to choose a bordering extra-group mate, rather than a likely rarer more exotic one, supports those landscape genetic patterns linking groups into super-groups. However, all neighbours are not equal: members of one group may breed with neighbours on one flank extensively, but rarely with neighbours on the opposite flank—the breaks in continuity across the landscape that fracture the population into super-groups. Mean litter size, amongst this sample fathered by the 'boy next door', was

---

[22] Models with mean and maximum within-group candidate fathers' heterozygosity (SH) produced comparable results overall; therefore, we present results from mean SH models in the main text.

[23] The commonly used Queller and Goodnight's pairwise estimator (QG; Queller and Goodnight, 1989), and the Lynch and Ritland's pairwise estimator (LR; Lynch and Ritland, 1999),

[24] We excluded 158 cubs from this analysis because we were unable to assign their paternity.

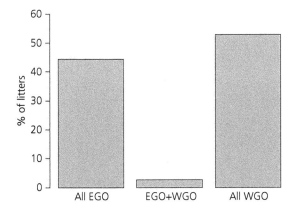

Figure 17.10 Percentage of litters with only within-group offspring (WGO), only extra-group offspring (EGO) and having both WGO and EGO (from Annavi et al. 2014b, with permission).

1.46 (ranging from 1 to 5). Expressed in other ways, we detected EGP in 64% of social-group years, and 47% (178 of 378; Figure 17.10) of litters (considering only those litters in which all offspring were assigned fathers). Drilling deeper, amongst these 178 litters, 64% included one EGO (94% of them with litter size of one), 32% included two EGOs and 4% three (Annavi et al. 2014b); thus there was a mean of 0.65 EGO per litter (range 0–5). While 152 litters (including at least one EGO-bearing litter) were fathered exclusively by a single extra-group male, 26 involved 2, and never more than 2, extra-group fathers (16 litters included two extra-group fathers and 10 involved 1 extra-group and 1 within-group father). In terms of infidelity, this meant that 64% of females mated with an extra-group male at some stage in their lives (max. 5 different partners).

What did these infidelitous matings mean in terms of avoiding inbreeding? The more related the potential parents within a group, the more likely was EGP (measured as EGO or extra-group mate pairs, EGMP, per litter) Annavi et al. 2014b). Our next step was to assess whether the relatedness of prospective pairs had any bearing on whether successful liaisons tended to be within or without group. It did. The absolute number of EGO and the relative proportion of EGO and EGMP per litter increased with mean pairwise relatedness between within-group assigned mothers and candidate fathers (although it tailed off slightly at very high relatedness Figure 17.11), but there was no significant effect on EGMP.

However, whatever caused this tendency to mate beyond the group when candidate pairings were more closely related, it was not a quest for heterozygosity: mean within-group candidate father SH was not related to either the absolute or relative chance of gaining an EG paternity. So despite subtle advantages to heterozygosity, it was not apparent, at least amongst individuals' fathers, in wetter summers (earlier), that this was being sought out as the motive for extra-group matings.

Given this lack of any SH effect, and the near 50:50 split between within- and without-group paternal provenance, was there evidence in support of the alternative random encounter hypothesis? Well, opportunity proved pivotal: the presence of higher numbers of within-group candidate fathers was associated with a lower proportion of EGO and EGMP per litter, irrespective of pairwise relatedness. Simultaneously, driving this effect from the other direction, the number of neighbouring-group candidate fathers had a linear positive effect on both the absolute number and relative proportion of EGO and EGMP, per litter (Figure 17.12). In short, and given the flagrant promiscuity we described in Chapter 7, this was a ratios game: simply, the more likely a female was to meet each category of male, the more likely that category was to be represented in the paternity of her litters.

What of mothers? Did the number of candidate mothers in a group affect the proportion of EGO they bore? Yes. The more females assigned cubs per natal group, the higher the relative proportion of EGO born to those females—that is, the greater the extent to which this cadre of reproductively ripe females (in condition to mate; Chapter 14) had relied on extra-group males to deliver fertilization. From a meta-analysis of 26 mammal species, Isvaran and Clutton-Brock (2007) drew the generalization that when more within-group females are present (and when the breeding season is extended) within-group males are less able to mate guard and prevent kleptogamy. From our naturalistic observations of mating, it was plain that although male badgers do not guard the female they just mated (Dugdale et al. 2007) (rather the next male will follow the former without any conflict; Chapter 7) no mechanism appears to exist in Wytham badger society to inhibit females from seeking extra-group contacts, especially as they certainly trespass into adjoining territories and feeding ranges (Chapter 9). One might speculate that an exhausting attempt at defending females just wasn't worth the effort under the circumstances of Wytham's badgers (and defending females is difficult: Adams and Wilkinson (2020) found that males of the harem-forming bat, *Phyllostomus hastatus* threw themselves

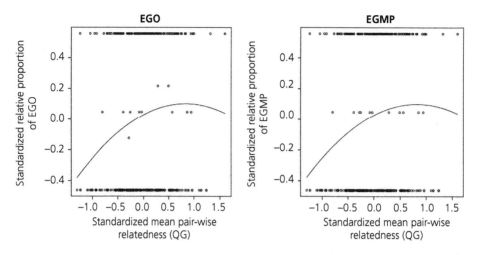

Figure 17.11 The relationship between the mean pairwise relatedness (Queller and Goodnight's estimator; QG) of assigned mothers and candidate fathers within each social-group year, and the relative proportion of extra-group paternity measured as extra-group offspring (EGO) and extra-group mate pairs (EGMP) per litter. Data points represent the standardized raw data from which the regression lines are derived (from Annavi et al. 2014b, with permission).

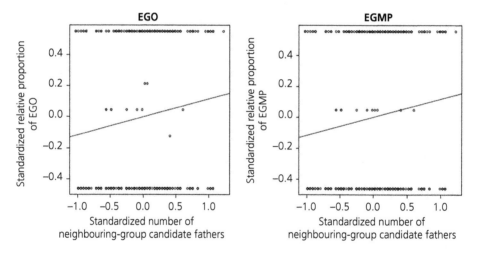

Figure 17.12 The relationship between the number of neighbouring-group candidate fathers and the relative proportion of extra-group paternity measured as extra-group offspring (EGO) and extra-group mate pairs (EGMP) per litter. Data points represent the standardized raw data from which the regression lines are derived (from Annavi et al. 2014b, with permission).

into aggressive harem defence, but nonetheless over 70% of harems still included extra-group offspring.

In review then, three strands of complementary evidence converge on an explanation of EGP: first, relatedness within groups promotes EGP (although post-copulatory mechanisms and neonatal mortality could also filter out any overly unpropitious inbred offspring genotypes); second, within-group males make no attempt to prevent cuckoldry (where they invest nothing, so have nothing at stake); and third, while

encountering foreign males less often than home males, female badgers certainly do have extensive opportunity to strike up liaisons while out foraging, a mating tactic widespread amongst other social carnivores (Kokko and Rankin 2006). And keep in mind too, from Chapter 7, that female badgers require intromission in order to ovulate at all, leading them to promiscuity. We speculate that if behavioural mate choice occurs, then perhaps female badgers select 'Mr Right' when the moment is right for conception, but if she is unaware of her own fertility window, she may not be discriminating at all.

In accounting the pros and cons of female extra-group mating from a male's perspective, remember this is a two-way street: males resident in one group are not, on balance, disadvantaged if 'their' females are mated by neighbouring males, provided they have the equal opportunity to mate with their neighbours' females. Of course, all else is unlikely to be equal, and there are reasons, such as sexually transmitted diseases (STDs) (Chapter 15), why some individuals may be prudent to avoid certain others, or that STDs might extinguish those blastocysts conceived—noting too that any mating with a female that fails to wean cubs (e.g. poor body condition, Chapter 14) will extinguish any evidence that a male 'chose' her in the first place.

Ultimately, why might Wytham's badgers not behave to capitalize more fully on the hypothesized benefits offered by discriminating mate choice? One possibility is that this sort of plurality of promiscuity might be expected to effect, specifically to reduce, the antagonism and risks typically associated with aggressive encounters with jealous same-sex competitors: why fight if there is nothing to fight over? Remember the recurring theme of phylogeny (developed further in Chapter 19). The libidinous polygynandry we observed might have its roots in adaptive advantage originally accrued in low-density badger populations, as a fertility assurance mechanism (Sheldon 1994; Vedder et al. 2011); choosiness might not have been an option under the circumstances of ancestral female badgers, exposed to far fewer potential suitors than Wytham offers today, but with intromission as a prerequisite to induce ovulation (Chapter 7). A further possible advantage of extra-group mating is that it obviates the need for permanent dispersal, with its attendant risks and the draining of genes from the natal community (Van Vuren and Armitage 1994).

How then might natural selection sift the next generation born to broadly undiscriminating parents? Remember those cryptic post-copulatory mechanisms of sperm competition, fertilization incompatibility, selective implantation, and even neonatal mortality, which act as filters to ensure that progeny from less auspicious mating partners are weeded out, and Zuk and Orr's (2014) proposal that delayed implantation extends this opportunity for second thoughts (it is worth reflecting on the advantages offered if, in the case of badgers, a metaphorical 'morning after' pill remains an option for the better part of 10 months).

Laden with phylogenetic baggage, contemporary Oxonian badgers may fit uncomfortably into

socio-ecological circumstances that facilitate hundreds of them cohabiting in the same 1000 acres. In Chapter 10 we reflected on the extent to which Wytham's badgers are perfectly adapted to their current circumstances; the theoretical implications of the answer are far-reaching (Connell 1980; Chapter 19).

Do extra-group offspring (EGO) perform better than within-group offspring (WGO)?

According to the fitness benefits hypothesis (Sardell et al. 2011, 2012), in a non-random setting, females should behave so as to produce EGO only if these outperform WGO. Performance here refers to the survival and reproduction of these offspring over their subsequent lifespan, contributing to the mother's lifetime reproductive success. In short then, did EGO perform better than WGO, in terms of repaying their mother's investment? Intriguingly, and with far-reaching theoretical consequences, the answer is, only partly. Irrespective of sex, EGO survived their first year less well: they had $c.34\%$ lower first-year survival probability (probability of first-year survival 0.55) than did WGO (0.74) (Figure 17.13b). Worse, they had shorter lives: EGO lived on average 1.3 years less than WGO (average 6.14 vs 4.84 years; Annavi 2012); from this dataset, females typically survived almost 1 year longer than did males irrespective of parentage (5.56 vs 4.61 years) (Fig. 17.13a). Even more remarkably, WGO females delivered greater reproductive success to their mothers: WGO females were assigned significantly more litters over their lifetime than EGO females (2.4 versus 1.6) and produced significantly more offspring over their lifetimes (3.4 versus 2.1). Furthermore, WGO females exhibited marginally higher lifetime reproductive success (40.5% of EGO, 42.7% of WGO), as did males (35.2% of EGO, 39.3% of WGO; cohorts 1988–2008).

Two questions arise stridently. First, why do WGO outperform EGO; and, second, if WGO are a better bet, why do females so frequently bear EGO cubs? To answer the first question we might turn to the superior natal fit hypothesis (Forstmeier et al. 2011): namely, perhaps WGO are genetically better suited to their birth environment (e.g. better equipped to deal with local endemic diseases; remember the predictive adaptive response, PAR, from Chapters 14 and 15). However, this seems scarcely plausible at the scale of badger social organization in Wytham (although there could have been some local variation in exposure to pathogens; Albery et al. 2020). Instead, we might

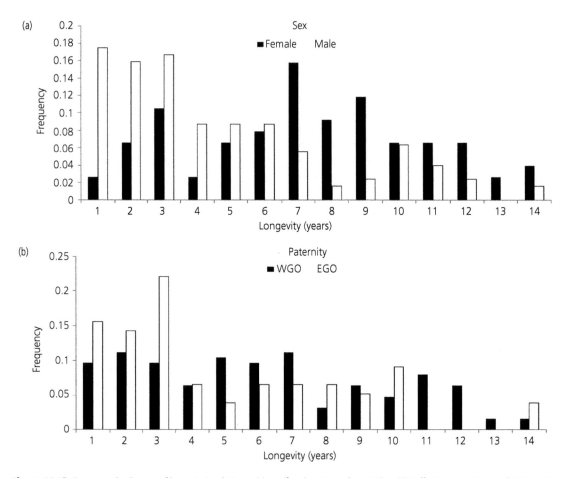

Figure 17.13 Frequency distributions of longevity in relation to (a) sex (female = 76, male = 126) and (b) offspring-paternity type (WGO = 125, EGO = 77) (from Annavi 2012, with permission).

turn to the EGO sociological disadvantage hypothesis: perhaps extra-group paternity attracts active discrimination amongst birth group residents. This would be consistent with the concept of 'opposing selection' (*sensu* Foerster et al. 2007), such that drawbacks are selected against, rather than advantages selected for. That said, neither explanation is particularly compelling.

Returning to the second question, is there more to be said about why females behave in a way that results in about half of their offspring being EGO, when it appears that this category on average offers reduced fitness benefits? An additional fact, of crucial significance, may expand our earlier answer: males with extra-group fathers went on to father more offspring in their lifetime than did males who had within-group fathers (3.4 versus 2.3). This lends support to the fitness benefits hypothesis; however, it only applies to one sex: males. Could it be that, like father, like

son, if promiscuous tendencies are inherited (reminding us of Fisher's sexy son hypothesis, back on p. xx); that is, in the 'candy store' of available females at high density, this male trait outpaces that of males that don't philander away from home (see Forstmeier et al. 2011): remember too, 35.8% of individuals were never caught away from their natal setts throughout their lives. Furthermore, females may increase their indirect reproductive success by choosing males that will ultimately produce a greater number of descendants, provided that females can detect these 'sexy son' phenotypic traits through cues (i.e. sexy son hypothesis; Fisher 1930), or alternatively, but more cryptically, these males may produce more competitive sperm (Annavi 2012).

Interestingly, considering females overall, irrespective of whether the fathers of their cubs were within- or extra-group, as we learnt in Chapter 14, those that bred at least once had a significantly higher lifetime number

(a)

(b)

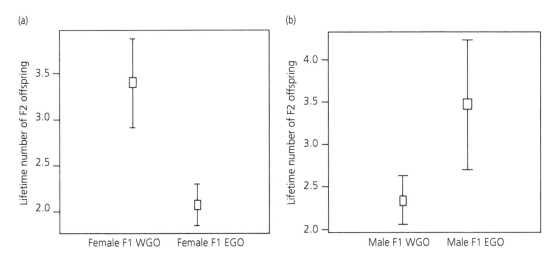

Figure 17.14 The lifetime number of F2 offspring (a) produced by F1 WGO females ($N = 20$) and F1 EGO females ($N = 14$) and (b) sired by F1 WGO males ($N = 32$) and F1 EGO males ($N = 17$) (from Annavi 2012, with permission).

of litters ($Dataset_{1988-1996}$ mean = 2.06; $Dataset_{1988-2008}$: mean = 2.05) than did males ($Dataset_{1988-1996}$: mean = 1.71; $Dataset_{1988-2008}$: mean = 1.69). However, if mothers produced EGO they fared worse in terms of producing both offspring and grand-offspring, whereas males benefited in terms of grand-descendants if they produced EGO (Annavi et al. 2014b) (Figure 17.14).

From a sociological perspective, playing a mixed strategy might be a work of evolutionary genius by female badgers. Through plural WG mating they maintain social bonds and status and avoid harassment from amorous suitors. Simultaneously, by engaging in EG matings, females extend the gene pool to which they have access, despite the fact that this most appropriate mate choice, in terms of genetic offspring fitness, might not be the best sociological option. All the while, they confuse paternity, cauterizing the risk that genetically incompatible/'inferior' males might—especially if not mated with—commit infanticide (Wolff and Macdonald 2004).

Selecting mates for pathogen resistance through major histocompatibility complex type

In Chapter 15 we introduced both the MHC and Simon (Yung Wa) Sin. Remember, in that immunological context, MHC encodes cell surface glycoproteins that bind, and present antigens to, T-cells and trigger an immune cascade (Swain 1983) conferring pathogen resistance. We showed that it was important for a badger to have

an MHC adapted to those pathogens it might meet (an exemplar was MusGHV-1). By extension, when we ask why each parent in our genealogy selected the partner it did—or, at least, went on to produce viable offspring from that pairing, maybe through a post-copulatory filter—a possible answer might lie in the animals' concern for their offspring's' capacity to fight disease—or, at least, that these post-copulatory immunocompatibility filters might weed out those least able to fight disease—and to have the right MHC profile.

Optimizing the MHC profile of offspring comes under powerful sexual selection (Milinski 2006); that is, the need is to pick the right partner to perfect the best possible profile, or for cryptic post-copulatory filters to hone that choice. Does this remarkably intricate possibility actually happen in nature? And can some element of active mate choice be demonstrated? Seemingly, yes, at least in the handful of species for which there are data. In a few instances females apparently prefer MHC-similar males (MHC-assortative).[25] This would be adaptive for individuals in populations facing a consistent pressure from endemic pathogens (e.g. Yamazaki et al. 1978; Bollmer et al. 2012). More typical, however is MHC-disassortative

---

[25] Assortative mating (also referred to as positive assortative mating or homogamy) is a mating pattern and a form of sexual selection in which individuals with similar phenotypes mate with one another more frequently than would be expected under a random mating pattern.

mating[26] (Kamiya et al. 2014), adaptive to changing pathogen pressures and epidemics in a landscape of constantly changing disease challenges.[27] This causes changes in genotype/MHC-type frequencies between generations (in the absence of complicating evolutionary pressures), resulting in deviation from the status quo, or Hardy–Weinberg equilibrium (Chen 2010).

This level of intricacy in behavioural mate selection seems almost unbelievable in the earthy blunderings of badger society, but recall how scent profiles may signal MHC suitability and the sweaty T-shirt experiment we described in Chapter 6. Yet more subtle are the cryptic post-copulatory mechanisms. With Simon Sin's skills, we set out to discover whether badgers had their MHC in mind while selecting a mate. As described in Chapter 15, we began by characterizing the evolution of MHC class I and class II genes in badgers (Sin et al. 2012b, 2012c). We used the alleles we discovered to test for sexual selection in the reproductive patterns of 366 individuals for class I[28] and 356 for class II.[29,30]

While an amorous badger sniffing gustily at potential mates, and doing so in the mating throngs we have witnessed, seems very unlikely consciously to be evaluating the nuances of a handful of genes buried deep within the nuclei of its interlocutor's cells, was it, in reality, eyeing up whether that mating would be MHC-assortative or MHC-disassortative? As difficult as this may be to believe, prepare, nonetheless, to be astonished again. Parent pairs exhibited smaller MHC class II amino acid distances (excepting those of class I) and smaller functional distances than would have been expected if matings had occurred at random. This suggests that MHC functional similarity, that is, MHC-assortative mating, was a target of mate choice or, at least, of which embryos develop. In addition, and no less remarkable, neighbouring group parent pairs (although not within-group parent pairs) had MHC class II DRB alleles more similar than would have been expected if random matings with candidates from

neighbouring groups had occurred. Crucially, however, and in contrast, we found that matings were disassortative for general microsatellite loci (based on 35 microsatellite loci; Annavi et al. 2011), interpretable as avoidance of inbreeding depression (Pusey and Wolf 1996).

Why did we not find evidence of strongly disassortative parenthood, selecting mates with partners with highly divergent MHC? Parenthood with dissimilar mates will typically increase the heterozygosity of offspring, and individuals more heterozygotic for MHC might have the advantage of being susceptible to fewer pathogens than those more homozygotic (McClelland et al. 2003) (Chapter 15). Assortative matings (birds of a feather) might provide resulting cubs with welcome protection from those local pathogens and diseases to which they more than likely will be exposed, under stable environmental conditions. In this case, disrupting co-adapted gene assemblages by disassortative parenthood might reduce fitness (Neff 2004). This would provide an explanation for mate choice being targeted at producing offspring with 'optimal' rather than 'maximal' diversity at MHC loci (Wegner et al. 2003; Milinski 2006).[31] Flipping back to the cryptic mechanisms side of the coin, MHC class II incompatibility at the *HLA-DQA1* allele[32] is a major cause of foetal abortion in humans (Beer et al. 1981; Ober et al. 1993). Stepping back, sperm rejection at the zona palucida tends to be stronger in HLA-dissimilar male–female pairs than in HLA-similar combinations (Jokiniemi et al. 2020). Consequently, if 'choice' is not causing these selective effects then potentially wastage caused by immune incompatibility during foetal development is.

Our behavioural observations in Chapter 7 revealed that badgers mated extensively within and between their groups, especially within super-groups. Indeed, the paternity tests reported in the first part of this chapter, also suggested a free-love model of paternity governed mostly by opportunity and propinquity, although we saw many repeating pairings between the same male and female, as described earlier for Casanova B with F180. The resulting conceptions, how-

---

[26] A recent meta-analysis of 116 effect sizes from 48 studies showed MHC-dissimilar mate choice only in species with multiple MHC loci examined (Kamiya et al. 2014).

[27] Disassortative mating reduces the genetic similarities within the family. Positive assortative mating occurs more frequently than negative assortative mating. In both cases, the non-random mating patterns result in a typical deviation from the Hardy–Weinberg principle (which states that genotype frequencies in a population will remain constant from generation to generation in the absence of other evolutionary influences, such as 'mate choice' in this case).

[28] A total of 201 assigned parent pairs and their cubs.

[29,] A total of 186 assigned parent pairs and their cubs.

[30] Years 1993, 2004, 2005, 2008, 2009, and 2010.

[31] This survival disadvantage potentially operates through highly divergent MHC genes associated with a smaller T-cell repertoire (Lawlor et al. 1990; Nowak et al. 1992), increased risk of autoimmune diseases, and disruption of local adaptations of co-adapted gene complexes (Kaufman 1999; Hendry et al. 2000; Neff 2004).

[32] Human leukocyte antigen (HLA) is a term used in medicine synonymous with MHC.

Table 17.1 MHC-based and relatedness-based mate-choice results across 6 years (1993, 2004, 2005, 2008, 2009, and 2010), calculated using Fisher's method of combining probabilities (from Sin et al. 2015, with permission).

Parent pairs	MHC gene	Allele sharing, χ^2_{12}	Smaller a.a. distance (all), χ^2_{12}	Smaller a.a. distance (ABS), χ^2_{12}	Smaller functional distance, χ^2_{12}	Lower R, χ^2_{12}
Combined	Class I	18.62 (higher)	25.42	24.55	16.43	**29.78**[*]
	Class II	12.87 (higher)	**40.90**[***]	**40.43**[***]	**40.64**[***]	**28.36**[*]
Within-group	Class I	20.54 (lower)	21.12	20.97	26.86	**44.81**[***]
	Class II	13.04 (lower)	21.90	21.35	21.38	**42.12**[***]
Neighbouring group	Class I	16.27 (higher)	20.7	20.79	11.49	10.72
	Class II	12.23 (higher)	**32.32**[**]	**39.00**[***]	**36.48**[***]	9.86

[*]$P < 0.005$; [**]$P < 0.002$; [***]$P < 0.001$. From Sin et al. 2015 with permission.

ever, were revealed by our MHC studies, not to be indiscriminate: either only certain MHC Class II similar zygotes were conceived or implanted, or they outcompeted MHC heterozygotes at some developmental stage.

Having discovered that the MHC profiles for offspring per parent pairs were not random for MHC class II loci, we asked next: does this lack of randomness in the MHC profiles of offspring apply equally to within- and between -group pairings? No, it does not: only neighbouring-group pairs showed MHC-assortative mating, due to similarity at MHC class II loci.

How to interpret this difference in the criteria for parenthood selection/compatibility between and within groups? One possibility is that the badgers are adopting two bet-hedging strategies, likely levied not simply by mate choice, but also by the relative success of progeny equipped with different immune profiles: the cubs born to within-group matings should have a better MHC repertoire for coping with familiar, local pathogens, thus maintaining co-adapted gene complexes (Kaufman 1999; Hendry et al. 2000; Neff 2004), enhancing the chances of their survival over extra-group parented cubs that remain philopatric; whereas cubs conceived by extra-group pairings should have a more diverse immune repertoire, making them more competitive when faced with novel epidemics, remembering that the ecological reality in Wytham was that in Chapter 8 we reported that four out of five parasite taxa exhibited consistent spatial hotspots of infection, which peaked amongst badgers living in areas of low local population density (Albery et al. 2020).

Thus, in the greater scheme of badger population resilience (Chapter 11) and evolution, both bases are covered; resistance against local endemic diseases amongst the philopatric, and a genotype better able to cope with exotic rare pathogens which they may face upon more extensive dispersal; or if circumstances change for the population, *in situ*. Linked to this, genomic inbreeding avoidance will be greater within groups, when candidate males are more related to co-resident females, potentially over-riding within-group MHC-assortative mating; thereby, limiting or forcing choices to include extra-group partners. Furthermore, there was inter-annual variation in the strength of these patterns, with smaller amino acid distances of the MHC class II gene than would be expected solely as a result of random chance; parenthood only being apparent in 4 of the 6 years of data examined (Figure 17.15; Table 17.1).

On reflection

With whom should one mate? This is a preoccupation that vexes mammals constantly, based on a range of genetic criteria, but also on who's available, and to what extent they are compatible, detrimentally related, or desirable. For most mustelids, the choice is simple: the female gets to choose from—at best—one or two males that hold breeding territories encompassing her own. But for badgers, thrust into high-density living in the British farmscape, suddenly (i.e. and increasingly, over the past few hundred years) a wealth of opportunities present themselves. While making the optimal choice may be adaptive and yield

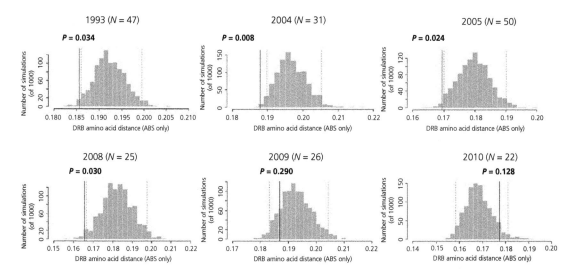

Figure 17.15 Mean MHC class II DRB amino acid distances (antigen-binding sites only) of assigned within-group and neighbouring-group parent pairs (solid line), compared to random within-group and neighbouring-group parent pairs. The frequency distributions (grey bars) are mean values generated from 1000 simulations of random pairings between all assigned mothers and all candidate fathers from both within the group and neighbouring groups. The two-tailed 95% confidence intervals (dashed lines) indicate cut-offs for significant departures from random mating. Randomizations were performed separately for each year: 1993, 2004, 2005, 2008, 2009, and and 2010. N indicates the number of assigned parent pairs. Significant P-values are shown in bold (from Sin et al. 2015, with permission).

more, or fitter, offspring, we also see evidence that sociology interacts with heredity, where females mix the currency of mating with the dividend of lifetime breeding success. Ultimately, sexual selection can be a cruel mistress, especially when accompanied by the mysteries of cryptic post-copulatory filtering, and so perhaps the most curious aspect is the fact that we could not reveal mate choices that would seem optimal to us; although not, perhaps, to the badgers. Yet, we did reveal parenthood outcomes that indicate astonishing selectivity, and/or effects of compatibility, down to a few genes influencing the immune system, and only in the context of a subset of sociological circumstances. And that nuanced selectivity of parenthood occurred within a society where cuckolding, of sorts, seems rife, with around half of all offspring fathered not by co-resident males, but rather

by neighbouring males—all extending the range of genes from which the female has to pick. Ultimately, it seems an intricate equilibrium exists amongst various competing selective pressures, gambling on what traits will enable offspring to fit best and perpetuate their lineages.

The fine-tuning of mate choice in badgers will be flexible according to circumstances (Stacey and Ligon 1991). Again, this raises the issue of whether what we see in high-density UK badger populations is the polished outcome of natural selection, or rather is a work in progress struggling to catch up; still ironing out the wrinkles to resolve upon the maximum advantages, where climate and agricultural practices conspire to give the UK and Ireland much higher carrying capacity than has typified the evolutionary history of these mustelids.

CHAPTER 18

Senescence, Telomeres, and Life history Trade-offs

Old age is not a disease—it is strength and survivorship, triumph over all kinds of vicissitudes and disappointments, trials and illnesses.

Maggie Kuhn

A parody of Jacques' melancholic monologue from Act II Scene VII of Shakespeare's *As You Like it* would seem neatly to encapsulate badgers' fate: 'Wytham is a stage, and all the badgers merely players; They have their exits and their entrances . . . an act of seven ages'. How well do Shakespeare's seven ages[1] match the badger's reality? Quite well. Each begins the mewling infant (Chapter 1), passes through the whining 'schoolchild' vying for their group's attention (Chapter 2). By Chapter 7 we find them in the furnace of love; by Chapters 9 and 10 perhaps the 'devoted soldier', staking a claim through 'strange oaths' considering the many questions raised by the badger's system of land tenure. In Chapter 14 perhaps we find the 'wise judge', adapting to life's risks and vicissitudes, striving

to pace its life optimally for maximum fitness. Here, in this chapter, we consider Shakespeare's last two ages, the old badger in slippered pantaloon still alert but heading towards senility and finally, and fascinatingly because it is so rare amongst mammals, a phase of obsolescence to the evolutionary imperative of reproductive productivity, but, unusually for a wild animal, prolonged dotage.

Few badgers survive to old age, even fewer than amongst Shakespeare's contemporaries in Elizabethan England, whose lifespans averaged 42. No, for badgers as a species the generality is to *live fast and die young* (although for many individual badgers in Wytham a more forlorn reality might be to live rather slowly and nonetheless die young). The tyranny of averages was, in Chapter 14, defeated by the quirks of individuality, where we described how the costs of living can come home to roost in different ways not only between species, but between individuals within each species. An individual's youthful stresses can bite in older age—even jeopardizing the chance of surviving to old age at all (Bright Ross et al. 2020). Remember that the younger badgers are when they first breed, the shorter their lives are likely to be (Chapter 14). Conversely, taking care of oneself can pay a dividend, although the physical benefits of excessive temperance must ultimately be weighed against the lack of genetic or social investment in descendants that might one day have been beholden to their grandparents. Failure to leave a lineage of offspring equates to zero reproductive fitness, but timing how to minimize the physical stress of when to breed, while maximizing output of descendants, is a gamble beset with risk and subtlety (Chapter 14).

---

[1] 'All the world's a stage, And all the men and women merely players; They have their exits and their entrances, And one man in his time plays many parts, His acts being seven ages. At first, the infant, Mewling and puking in the nurse's arms. Then the whining schoolboy, with his satchel And shining morning face, creeping like a snail Unwillingly to school. And then the lover Sighing like furnace, with a woeful ballad Made to his mistress' eyebrow. Then a soldier Full of strange oaths and bearded like the pard, Jealous in honour, sudden and quick in quarrel, Seeking the bubble reputation Even in the cannon's mouth. And then the justice, In fair round belly with good capon lined, With eyes severe and beard of formal cut, Full of wise saws and modern instances; And so he plays his part. The sixth age shifts Into the lean and slippered pantaloon, With spectacles on nose and pouch on side; His youthful hose, well saved, a world too wide For his shrunk shank, and his big manly voice, Turning again toward childish treble, pipes And whistles in his sound. Last scene of all, That ends this strange eventful history, Is second childishness and mere oblivion, Sans teeth, sans eyes, sans taste, sans everything."

The Badgers of Wytham Woods. David W. Macdonald and Chris Newman, Oxford University Press.
© David W. Macdonald and Chris Newman (2022). DOI: 10.1093/oso/9780192845368.003.0018

This general realization that early life expenditure can incur grievous debts to be repaid in later life owes much to another Oxford personality, Paul Harvey, and his colleague Dan Promislow. They undertook a comparative analysis of variation in life histories across all mammals, while controlling for the phylogenetic effects of shared ancestry (Promislow and Harvey 1990). Some species, it turned out, live fundamentally faster lives than others. For instance, for tiny insectivorous shrews, life is typically over within 13 months, whereas similarly tiny and insectivorous bats, such as the pipistrelles, can live 13 years (a male Brandt's myotis bat holds the known record for a small mammal in the wild, at over 41 years old; Culina et al. 2019).

Musing on the deterioration associated with old age, French singer Maurice Chevalier quipped, in 1952, 'Old age isn't so bad when you consider the alternative', but it nonetheless heralds the kind of system failure one might have expected natural selection to postpone, ideally indefinitely. Coincidentally, 1952 was also the year in which Peter Medawar seeded the ideas that underpin much modern thinking on old age. First, by way of clarification, distinguish two non-synonymous terms: 'senescence' refers to declining performance with age, whereas 'ageing' marks the passage of time: two phenomena that are not necessarily shackled together. Indeed amongst sharks, and some reptiles, ageing is not accompanied by senescence, simply more years exposed to mortality factors that inevitably bite in the end (Stenvinkel and Shiels 2019).[2] Mammals, birds, and lizards do deteriorate physically, and reproductively, with age (Brunet-Rossinni and Austad 2006). Their rate of senescence is a product of evolutionary trade-offs—the insight of Peter Medawar (1952), crystallized, as so often in twentieth-century evolutionary theory, by Bill Hamilton (1966) (see Charlesworth 2000) and subsequently developed by Tom Kirkwood and Michael Rose (1991). From this scholarly genealogy emerges three established theories of senescence, all resting on purifying selection[3] acting more strongly on genes in earlier, rather than later, life (Figure 18.1).

The mutation accumulation hypothesis (Medawar 1952), which proposes that weakly deleterious mutations build up in the gene pool to act later in life because negative selection against mutations that exhibit harmful effects only during old age will not be sieved efficiently out of the gene pool. The antagonistic pleiotropy hypothesis, which proposes that traits selected early in life, aimed at enhancing reproduction, can have deleterious effects later on when selection is weaker (Williams 1957). Finally, the disposable soma hypothesis, in which Kirkwood (1995) integrated these prior hypotheses, positing that, given a limited amount of available resources, a trade-off occurs between investing in reproduction and investment in the 'soma' or body (i.e. growth and the maintenance of tissues and organs).

We build here on our exploration of pace of life theory in Chapter 14, which revealed that there is no single best way for a badger to live its life in the face of the almost infinite variability of the intrinsic machinations of population density change, social dynamics, and stresses, as well as the extrinsic vicissitudes of environmental conditions and food supply (Bright Ross et al. 2020). But exactly how does this inevitable wear and tear of life afflict badgers, and with what consequences?

Our story once more returns to tactical heterogeneity—strategists, phenotypes, shifting ratios in game play, and the way individuals attempt to stay one step ahead of the relentless Red Queen (van Valen 1977); for, as Alice said through Carroll's Looking-Glass, 'It takes all the running you can do, to keep in the same place'. However much running a badger might do though, they all eventually grow old, and with age comes all too familiar deterioration, but in degrees, dependent on the tear resulting from the wear. So we offer, as an exemplar, a vignette of the

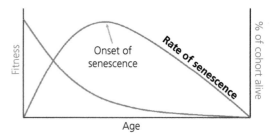

Figure 18.1 Senescence and the strength of natural selection. The age trajectory of an individual with fitness against age. After an increase in fitness in early life, fitness starts to decline, defining the onset of senescence. The slope with which it declines is the rate of senescence. The forces of natural selection decline with age due to pressure of age-independent extrinsic mortality, with mutations in early life under stronger selection since the number of individuals in a cohort declines with age (from van Lieshout 2021, with permission).

---

[2] The oldest Greenland shark (*Somniosus microcephalus*) investigated is estimated to be 392 years old ± 120 years, making it the longest lived vertebrate known (Burke 2016).

[3] Purifying selection is the selective removal of alleles that are deleterious, and can result in stabilizing selection through the purging of deleterious genetic polymorphisms that arise through random mutations.

(a)

(b)

(c)

Photos 18.1 The toothless 13-year-old M478 ('Little B').

elderly badger affectionately known to us as Little B (Photos 18.1), more formally called Male 478. Little B was amongst the 42% of males whose entire lifespan was bereft of recorded reproductive success. Born in 1995 at the Lower Seeds sett (central to the woods; see map), and subsequently caught 28 times, following his 10th birthday he began almost nightly raids on the Chalet in Wytham Woods, where Chris then lived. Thus began the habit of standing on his back legs to tap impatiently on Chris's kitchen window. By then, Little B was already almost toothless, but Chris was nonetheless cautious and so rewarded the elderly badger with long thin items that kept human fingers and badger jaw apart; for example, sausages or, occasionally favoured, bananas. By his 13th year Little B would sit with Chris by night in the garden. One night, WildCRU's water vole aficionado, Tom Moorhouse, visited the Chalet so that he and Chris could adjust the brakes on Tom's car. Up on ramps, Tom had slithered beneath the car, his arms pinioned in the confined space beneath the chassis, when he found himself joined by Little B, near blind with cataracts, snuffling his face (a worrying intimacy brought to a happy close by some distracting peanuts).

Little B's decline defines the process of badger senescence: loss of body condition (the sarcopenia we described in Chapter 14), tooth decay (Chapter 11), cataracts—coarse symptoms of underlying compromises to hormones (Chapter 7), immune-competence, antioxidant capacity (Chapter 15), and, as we shall see, even the unravelling of the DNA encoding cellular generation.

Reproductive senescence

Running out of reproductive steam is a key feature of senescence. As elder generations acquire injuries, ailments, infections, gynaecological wear and tear, etc., there comes an age beyond which these individuals have little left to offer, as selection becomes too weak to prevent the onset of reproductive senescence (Croft et al. 2015). Remember that, notwithstanding the promiscuity that characterizes Wytham's badgers (Chapter 7), 42% of individuals that reached breeding age were never assigned any offspring, and even amongst those achieving some lifetime breeding success, in any 1 year only 36% (±4%) of males and 29% (±4%) of females produced offspring. With the notable exception of some highly social mammals, such as elephants, toothed whales, humans, and bonobos (and there are reasons, given in the text that follows, not too readily to dismiss the possibility of badgers joining this list), natural selection usually acts as if post-reproductive individuals are just baggage, consuming resources that might be put to more effective use by younger individuals (vom Saal et al. 1994; Austad 1997). Ultimately, the evolutionary dealership wants you to buy a new car, not endlessly repair the old one.

Even within age classes, in a study led by Hannah Dugdale et al. (2011) we found that offspring were not distributed evenly, but were skewed. Recall from Chapter 14 that everything for badgers gets harder as they get old, and they become much more susceptible to falling outside the 'safe zone' of body condition. How then is reproductive success levied between age classes?

Hannah worked with 233 Wytham badgers (126 males and 107 females) that we first encountered as cubs between 1987 and 2005, and were therefore of known age and for which we had complete lifetime reproductive success,[4] we used pedigree assignments from a total of 735 cubs to ask if breeding success was (i) age-specific, with an initial increase with age; followed by a subsequent, again age-related, decrease, in breeding success. We then tested whether our results are consistent with any of four hypotheses, asking whether breeding success is correlated: (ii) positively with age of last breeding (selective disappearance)— the selection hypothesis (Curio 1983; Nol and Smith 1987); (iii) positively with the number of years during which individuals had previously raised offspring successfully to independence—the constraint hypothesis (Curio 1983); (iv) positively with residual reproductive lifespan—the restraint hypothesis (Williams 1966; Pianka 1976); and (v) negatively with old age—the senescence hypothesis (Nussey et al. 2008).[5]

Our analysis revealed (Figure 18.2a) that annual cub assignments were highest at intermediate ages, with peak reproductive performance coming at age 3 for females and 5 for males, but thereafter decreased with age. Unsurprisingly, the younger badgers were when they started breeding (α), the greater the lifetime

reproductive success they tended to accumulate— provided they survived their first attempt; however, the later they started, the more prone they were to deteriorating breeding success with age. To recap briefly from Chapter 14, it seems that, rather than the rigours of reproduction invariably depleting females, a class of elite individuals sustained it well. Moreover, success begets success, and once badgers, especially females, embarked on a sequence of weaning surviving cubs year-on-year, they were thereafter more likely to reproduce than other females coming late to breeding (Bright Ross et al. 2020). Importantly, within this trend there was still a lot of inter-individual variation, so age does not explain everything (indeed we know from Chapter 14 that body condition has far-reaching effects on reproduction).

At the other end of reproductive life, an intermediate age at last reproduction (ω) of around 7 years in females and 8 in males (remembering that delayed implantation offsets mating and birth by over a full year in badgers) was associated with maximal lifetime breeding success. This demonstrates that because purifying selection (above) allows natural selection to sort continuously the wheat from the chaff, and refine the quality of the surviving proportion of each cohort, badgers performing less well tend to get eliminated. Certainly, at the population level, selective disappearance and reappearance in the record of which individual contributes to reproduction each year was a factor; so, as a cohort ages, it includes a progressively greater percentage of more productive individuals. This fits with *the selection hypothesis* (ii; Curio 1983; Nol and Smith 1987), which predicts differential survival of individuals based on their phenotype, such that a cohort is expected to consist of proportionally more 'good-quality' (i.e. fitter) individuals over time (Forslund and Pärt 1995).

The constraint hypothesis (iii; Curio 1983) posits that individuals may initially be constrained from breeding due to limitations of their physiological condition or inadequate skills. For example, individuals may gain experience from breeding or foraging, such that over time they improve their breeding performance (Nol and Smith 1987).[6] So was there any impact of previous reproductive success on badger breeding histories? Yes, the more previous breeding experience (annual success) both males and females had, the later the age at which they last bred. However, the number of

---

[4] This was a subsample of the 630 genotyped cubs, from a total of 735 cubs born between 1988 and 2005 (Dugdale et al. 2007).

[5] See also Bright Ross et al. (2020) Chapter 17 analysis of age-specific breeding success based on a subsequent sample of 269 cubs.

[6] As has been observed in grey seals *Halichoerus grypus* (Bowen et al. 2006) and captive chimpanzees *Pan troglodytes* (Fessler et al. 2005).

cubs per year sired by males decreased as the number of years since an individual was last assigned paternity increased; that is, males 'wore out'.

Intriguingly, though, we detected neither an abrupt terminal decline, nor a final fling of investment; instead, breeding success underwent gradual decline with residual reproductive lifespan (Figure 18.2)—that is, there was no clear menopause or andropause (see the text that follows). Remember that in Chapter 11, the take-home message is that given average conditions breeding females have an edge in survival (Bright Ross 2020. There was, however, some interaction with age at first breeding in females, suggesting that refraining from breeding for those first crucial years may lead to a more successful long life of reproduction (again, there is a parallel with beavers, amongst which females 'accrue capital'; Campbell et al. 2018). This is congruent with *the restraint hypothesis* (iv; Williams 1966; Pianka 1976) that proposes that reproductive effort varies according to what remains of an individual's residual reproductive lifespan, such that it may be advantageous for younger individuals to refrain from breeding or reduce their effort, to improve their chances of survival and later reproduction (where this is possible). In contrast, older individuals, with nothing left to lose, should invest more in reproduction due to their diminished residual reproductive lifespan. However, considering only badgers that lived to be at least 5 years old, these clearly demonstrated a decrease of reproductive output with advancing age for both sexes; in accord with hypothesis iv, that is, reproductive senescence.[7]

So our population-level probing revealed patterns of reproductive performance with age, but what mechanism determined those patterns? In Chapter 14 we learnt that body condition interacts with reproductive performance in complex ways, both as a predictor and a response. But crucially, when it comes to the underlying physiology that might explain reproductive decline, hormones are clearly pivotal, with the well-known cessation of reproductive potential

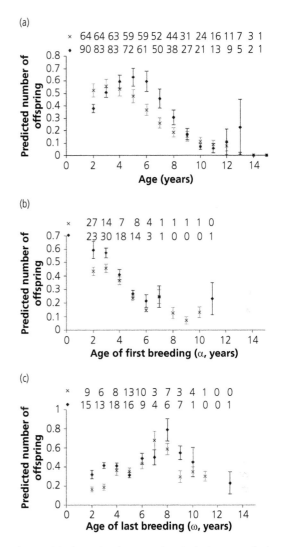

Figure 18.2 The predicted mean number of cubs, from generalized linear mixed models (GLMMs), assigned to females and males according to: (a) age; (b) age of first breeding (α); and (c) age of last breeding (ω). Parentage was assigned with 80% confidence, and was based on 251 and 185 cubs whose fathers and mothers, respectively, had complete lifetime breeding success data. Error bars display the standard errors of the means. Data labels display the number of badgers present in each age category; the GLMMs control for repeated measures on the same individual. crosses = female; filled diamonds = male (from Dugdale et al. 2011, with permission).

associated with abrupt menopause and more gradual andropause.

It is worth highlighting here that human experience is atypical: that is to say, more usually, ageing female mammals die shortly after they experience the reproductive *constraint* of menopause; similarly, male

---

[7] Of course, to return yet again to our oft-repeated caveat, we never truly know how many cubs are born each year. We know only how many survive long enough to be counted in our database records (see, particularly, discussion of F/α in Chapter 14). Consequently, we under-observe reproductive inputs and their potential trade-off consequences for subsequent reproductive performance, physical condition, and life expectancy. Perhaps a female exhibiting premature reproductive decline had put extreme effort into trying to keep her previous litter alive, but ultimately failed to wean them, and so we were unable to count them?

mammals undergo the *restraint* of reduced competitive ability, beyond which their days are numbered (Bribiescas et al. 2012). Indeed, humans are one of a tiny minority (established robustly for around five species) of mammals that live on in reproductive obsolescence. Due to the specific social evolution of modern humans, where grandparenting and the wisdom of elders can pass down traditional knowledge to benefit the clan (Lahdenerä et al. 2004), this post-reproductive lifespan (PRLS; Cohen 2004; see Medawar 1952) can amount to almost half a lifetime in post-menopause, or functional andropause. Those other few exceptional mammal species that live on after they cease breeding include some other primates (Alvarez 2000), elephants (Lee et al. 2016), whales (McAuliffe and Whitehead 2005), and, from our own group's studies, perhaps giant otters, *Pteronura brasiliensis*—which, like badgers, are members of the *Mustelidae* (Groenendijk et al. 2014).

As with humans, in these other instances, geriatric, post-menopausal matriarchs may bring the selective advantage that their wisdom helps their descendants to survive (Croft et al. 2015; Gavrilov and Gavrilova 2001); often through grand-mothering benefits (Hawkes et al. 1998; Ward et al. 2009). Typically this PRLS manifests as a slowing of somatic senescence, but without extension of the fertile span (Hawkes et al. 1998). Another factor is that, as the Grim Reaper looms, individuals that do anticipate an ongoing life sometimes make a last ditch disproportionate reproductive effort, sacrificing subsequent survival (Clutton-Brock 1984; see Campbell et al. 2017). But the success of this 'terminal investment' (TI) is far from guaranteed (McNamara et al. 2009). Reduced body condition, or lowered social status, can impede, or 'restrain' the return on TI (Chapter 14; McNamara et al. 2009; Croft et al. 2015). We were alerted to a further twist in this tale, by our collaborative work on Eurasian beavers, which revealed a type of heterochrony (Campbell et al. 2017; see McNamara 2012). Heterochrony is the phenomenon in which reproductive senescence (Finch, 1990) and somatic senescence (Reed et al. 2008) in both sexes follow different asynchronous trajectories, influenced by resource availability (Bondurianski et al. 2008). Rather counter-intuitively, considering their semi-aquatic habits, beavers do badly in the rain: for them, rainfall turns out to be a negative proxy for annual resource availability. This was a surprising discovery insofar as one might have expected rainfall to be good for plant growth (and thus for the beavers that eat them), and in general it is. However, just above the waterline, in the area where beavers are safest from wolves, the opposite applies: heavy rain leads to waterlogging of the aspen trees, which therefore grow

less. This explained our discovery that higher rainfall was consistently associated with lower reproductive output for female beavers of all ages. In contrast, the quality of beaver breeding territories was associated with differences in individual patterns of reproductive senescence: beavers from lower-quality territories grew old at a younger age; and in poorer (i.e. wetter), but not in better, years, older mothers produced larger offspring than did younger mothers (Campbell et al. 2017).

How does this finding about beavers, fascinating in itself, link to badgers? Recall, from Chapter 2, that male badgers too show heterochrony in fast or slow trajectories towards puberty, linked to the body size they attain during their first year of life, and the body size these young males attain is, in turn, linked to the social milieu in which they mature (Sugianto et al. 2019b). Amongst Wytham's badgers, the effects on badger cubs of extrinsic factors on somatic development (growth rate and final body size) are different for each sex: males are more sensitive to social factors whereas females are more susceptible to adverse weather (Sugianto et al. 2019c). Further, at the population level (Chapter 14), a badger's social circumstances, and the weather during its first year, impacted its pace of life; explaining *c*.10% of variance in the ratio of fertility to age at first reproduction (F/α) and lifetime reproductive success, with sex ratio (SR) and sex-specific density explaining 52.8% of male and 91.0% of female F/α ratio variance.

So there was plenty of theory in the air, and precedents in our data, to prompt our next question: did the sex steroid profile of badgers change throughout their lives? Once again, Nadine Sugianto, our Indonesian former veterinary student whose endocrinological insights informed Chapters 1, 2, and 7, had the expertise. In the context of senescence, the key was if, or when, sex steroid concentrations drop below the hormonal 'constraint' threshold, indicating a full stop to oestrous cycling or a semi-colon to spermatogenesis—termed 'reproductive cessation' (RC; Cohen 2004). We diagnosed this point as dropping back to pre-pubescent hormone levels.

The answer came from a sample of 67 males and 49 females analysed by Nadine. From about their sixth year[8] onward, we found a gradual (quadratic) decline in the sex steroid levels in the badgers' blood during the spring mating season (Figures 18.3, 18.4, and 18.5) (Sugianto et al. 2020). Although Wytham's badgers can live to be 14 years old (Chapter 11), only

---

[8] Females reached their reproductive peak/ prime-age at 5.5 years, slightly earlier than males, which peaked at 6 years.

Female fitted line Spring

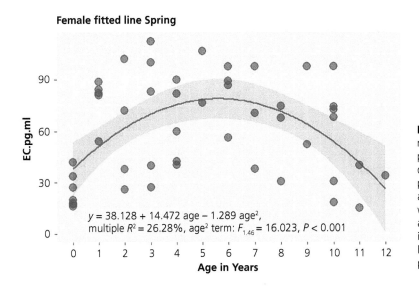

$$y = 38.128 + 14.472 \text{ age} - 1.289 \text{ age}^2,$$
multiple $R^2 = 26.28\%$, age^2 term: $F_{1,46} = 16.023$, $P < 0.001$

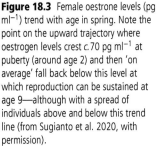

Figure 18.3 Female oestrone levels (pg ml^{-1}) trend with age in spring. Note the point on the upward trajectory where oestrogen levels crest *c.* 70 pg ml^{-1} at puberty (around age 2) and then 'on average' fall back below this level at which reproduction can be sustained at age 9—although with a spread of individuals above and below this trend line (from Sugianto et al. 2020, with permission).

Male fitted line Spring

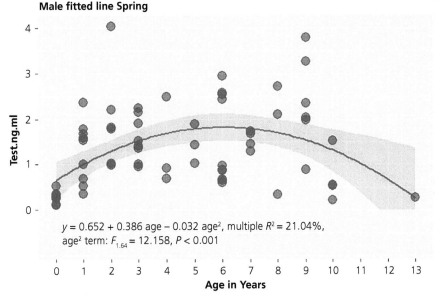

$$y = 0.652 + 0.386 \text{ age} - 0.032 \text{ age}^2, \text{ multiple } R^2 = 21.04\%,$$
age^2 term: $F_{1,64} = 12.158$, $P < 0.001$

Figure 18.4 Male testosterone levels (ng ml^{-1}) trend with age in spring (from Sugianto et al. 2020, with permission).

31.8% of females and 37.8% of males born, or 55.8% and 20.9% of those attaining sexual maturity reached the age of reproductive cessation (Bright Ross et al. 2020). Nonetheless, this waning in sex hormones turns out to be relevant to a noteworthy proportion of badgers.[9]

For both sexes, the onset of decline in somatic condition began early—a phenomenon familiar to human athletes—detectable in badgers from around age 3 years (Figure 18.6), substantially preceding the decline in reproductive hormones that began at about 5–6 years of age: there was, in the jargon, heterochrony between rates of somatic and hormonal senescence in

---

[9] Similar declines in sexual advertisement with age can be seen in the curvilinear decline of antler growth linked to dominance status and reproductive success in male red deer (*Cervus elaphus*; Kruuk et al. 2002) and white-tailed deer (*Odocoileus*

virginianus; Scribner et al. 1989). Similarly, older male house mice (*Mus musculus domesticus*) produce lower concentrations of involatile signalling proteins in their urine (MUPs), which serve in mate attraction (Garratt et al. 2011).

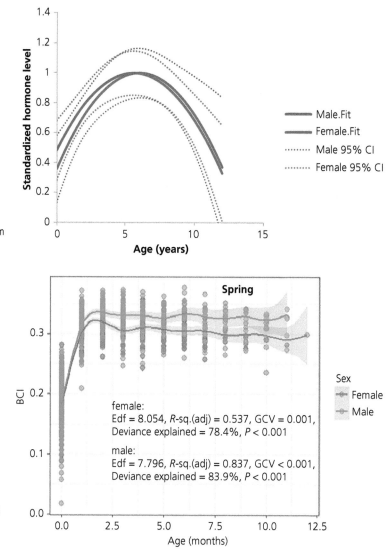

Figure 18.5 Male and female quadratic hormone curves with age during spring (from Sugianto et al. 2020, with permission).

Figure 18.6 Body condition index (BCI) against age in males and females during spring (Edf = empirical distribution function; GCV = generalized cross-validation) (from Sugianto et al. 2020, with permission).

both sexes. These heterochronic declines were highly synchronous between males and females. This is in contrast to what is known of most polygynous mammals, for which males typically senesce earlier than females; leading to more rapid rates of male mortality (Nussey et al. 2009). This is the generality for mammals as diverse as black-tailed prairie dogs (*Cynomys ludovicianus*), red deer (*Cervus elaphus*), and African lions (*Panthera leo*) (Clutton-Brock and Isvaran 2007).

Paralleling this ebb, flow, and ebbing again of sex steroids, subcaudal gland secretion volume—a plausible proxy for investment in reproductive advertisement (see Chapter 6)—also gradually declined, especially after age 5; more abruptly so in males than females (Figure 18.7). This mirrors how antler growth,

a sexual advertisement linked to dominance status and reproductive success in male red deer (*C. elaphus*; Kruuk et al. 2002) and white-tailed deer (*O. virginianus*; Scribner et al. 1989) undergoes a curvilinear decline with age.

Insofar as it is one thing to be able to breed, but another actually to do so, our next question was which of these endocrinologically competent but ageing individuals actually produced offspring? Based on the blood samples from those 67 males and 49 females, we learned that cub assignments on average decreased amongst older badgers: in practice this means that some badgers manage just one litter and fail thereafter. A general conclusion is that, amongst badgers, physical deterioration in body condition index (BCI) does not

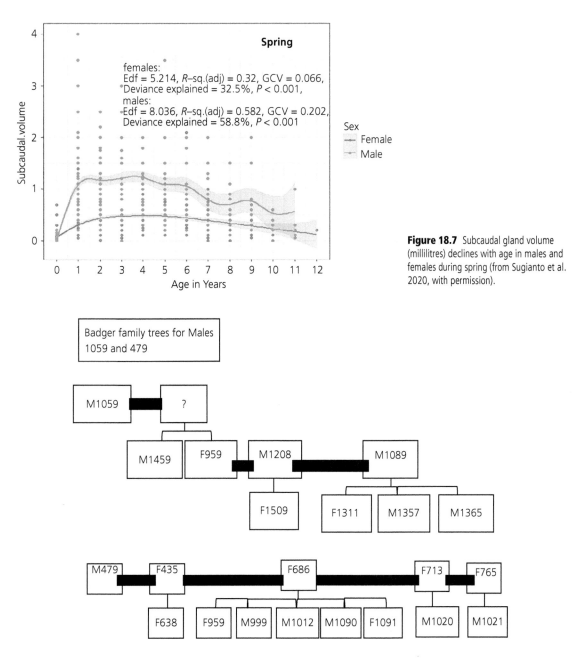

Figure 18.7 Subcaudal gland volume (millilitres) declines with age in males and females during spring (from Sugianto et al. 2020, with permission).

bite before hormonal constraint. But what about the effects of the general dip below sub-puberty hormone levels around age 9?

Elderly males

First, elderly males: some males were still fathering cubs up to the age of 13 years (Dugdale et al. 2011; Sugianto et al. 2020). Earlier, Christina had led a study that revealed two endocrinological phenotypes amongst especially older males (Buesching et al. 2009). Building on this, amongst Nadine's subsample of hormone-profiled badgers, we found that of the 11 males living to be 9 or older, five maintained testosterone levels within 80% of their youthful peak (i.e. <1.76 ng ml^{-1}); of which two were assigned cubs: M1059 (born 2005, resident at Common Piece), was aged 9 when his son M1459 was born in 2014[10] at neighbouring group Thornicroft; the second reproductively successful, prolonged-testosterone male was M497. Until he was 9 years old M479 had been assigned only a single cub (F638), born when he was 3 years old (with spouse F435 of Sunday's Hill, then a 4 year old). However, in his ninth year M479 got into his mating stride. Aged 9, in 2004, he fathered a notable daughter F959 (Pastick's Outlier), then in 2005 he fathered M999 and M1012—remarkably, all of these offspring were with F686, consistently of Pastick's Outlier, where she had been born in 1999 (and was thus aged 5 when M479 was 9); in addition in 2005 M479 fathered M1020 (with F713, at Pasticks) and M1021 (with F765, at Pastick's Outlier), finally fathering M1090 and F1091, back with F686 as mother, when he was aged 11 in 2006.

What of the other six elderly males for which testosterone levels dropped below 80% to pre-pubescent levels? Well, on average they survived a further 2.02 ± 1.50 more years (range 0.42–4.67 years). But did they also continue to reproduce? While low testosterone impairs, it does not extinguish, spermatogenesis (Adkins-Regan 2005), and despite his low testosterone, 9-year-old M528 (then resident at Jack Bracken, but ambling between that location and the adjacent groups at Jew's Harp and Marley 2) was nevertheless assigned two cubs (F1009 born at Jack Bracken and M1021 born at Pastick's Outlier) in 2005; thereby, throughout his life, helping to create the components of a super-group.

In combination these observations suggest that a complex swirl of factors is involved in achieving breeding success for males, such as how testosterone influences mating effort, reproductive advertisement, roaming, and sperm quantity and quality, not simply absolute fertility (Cohen 2004; Clutton-Brock and Isvaran 2007). Our interpretation of this overall picture is that there is a mixed strategy whereby around half of males continue to make a TI in maintaining later-life testosterone potential, of which, in our subsample, about one-third of these hormonally capable badgers was successful. That is, they continued at roughly the same rate of annual reproductive success typical of younger males (see Chapter 17; Annavi et al. 2014b)—remembering that the success of the females inseminated by males will always constrain the latter's reproductive achievement. In contrast, the other half of the elderly male contingent had largely

thrown in the hormonal towel, with the exception of lucky M528.

Elderly females

So what of reproductive success amongst older females? While low testosterone doesn't entirely stifle sperm production in males, for mammalian females ovulation fails below a certain oestrone threshold (Packer et al. 1998; Cohen 2004; Adkins-Regan 2005). Female badgers start to ovulate when their oestrone levels reach 70 pg ml^{-1} at c.11 months old, and, in a very few cases, genetic analysis reveals youngsters conceiving at that age (remember, however, delayed implantation) (see Figure 18.3, earlier). This 70 pg ml^{-1} cut-off indicates that, on average, those females in our endocrinology subsample dipped into functional menopause at around 9 years of age. Not for the first time in these pages, the tyranny of the mean risked obscuring the all-important inter-individual variation (Chapter 14): 4 of the 11 sampled females aged 9–12 actually maintained oestrone levels above the 70 pg ml^{-1} threshold during the spring conception season (for a reminder of the function of oestrone, check back to p. 134, Chapter 7). It seems that females, similarly to males, also have two ways of doing things; that is, two reproductive phenotypes. Returning to the pedigree for corroboration, the oldest assigned mother in this subsample was F959 (yes, daughter of the late-life reproducing M1059). F959 was, as earlier, born in 2004 at Pasticks Outlier (where we noted that as a cub she was producing an unusual volume of sub-caudal gland secretion). She had her last litter (of 4 in her lifetime) aged 10 in 2014. That last litter comprised a daughter, F1509, born at Pasticks Outlier and fathered by M1208, who vacillated between Jew's Harp and Pasticks through the prior conception year; her previously favoured younger mate, M1089, born at Pasticks in 2006, who was father to her litters in 2009 (cub F1311) and 2010 (cubs M1357 and M1365) was not caught beyond the 2009 conception year. Crucially, however, none amongst the seven elderly females with low oestrone levels produced cubs, evidencing female reproductive cessation, where PRLSs averaged c.2.59 ± 1.29 years (range 1.25–4.75 years). Why was only one of these older, but endocrinologically competent females assigned offspring? Torrents of hormones amount to nothing if body condition is too poor to sustain pregnancy (most likely manifesting through failure to implant conceived embryos after delayed implantation; Woodroffe 1995; Sugianto et al. 2020), given the narrower BCI safe zone for reproduction in

---

[10] Considered to be an example of delayed implantation, from a mating in 2013.

seniors (Bright Ross et al. 2021). This outcome was paralleled, yet again, in our study of beavers (Campbell et al. 2017).

Given the rarity of a PRLS amongst non-primate or cetacean mammals, were these post-reproductive badgers in our population the standard bearers for an adaptive strategy (like people or, perhaps, elephants), or merely the lucky tail of a distribution (Tully and Lambert 2011)? A first, if cautious, thought is that finding PRLS may partly be a function of looking for it with sufficient care over sufficient time (Peccei 2001), where PRLS insures against the risk of dying by chance before reproductive cessation closes the door (Tully and Lambert 2011). For further reflections, drawing parallels with Cecil the Lion, see Box 18.1.

So how do our observations sit with the profusion of theories proposed to explain the phenomenon of extended reproductive lifespans in some individuals and PRLS in others? *The patriarch hypothesis* (Marlowe 2000) posits that older males secure more successful mating opportunities (as seen in elephants; Hollister-Smith et al. 2007), but falls in the face of the few badger cubs assigned to geriatric fathers (Dugdale et al. 2011; Chapter 17). *The adaptive kinship hypothesis*, based on the inter-generational cessation of breeding by older females as their daughters supersede them (i.e. Cant and Johnstone 2008) is also a nonstarter for badgers: if anything, prime-age female badgers suppress the reproductive capacity of younger females (Woodroffe and Macdonald 1995), and while

---

Box 18.1 No country for old men

Readers may recall Cecil the lion, whose shooting in an area adjoining Hwange National Park, Zimbabwe, in 2015 by Minnesotan bow-hunting dentist, Walter Palmer, created what may have been the most viral wildlife news story in history (Macdonald et al. 2016b). Following a TV show in the USA during which the normally acerbic Jimmy Kimmel wept on screen while describing WildCRU's work on Cecil, that night 4.4 million of his fans visited WildCRU's website: it melted! Oxford University began a period that it soon recognized as its greatest engagement with the global public in its almost 1000-year history. A search in 125 languages revealed that in the traditional editorial media there were a total of 94,500 mentions of Cecil over a 3-month period. However, our point here is not about the media explosion and the deep questions of ethics it prompts (Vucetich et al. 2019); rather it is about Cecil's journey to old age. Both species are Carnivora, both group living, both in multi-male, multi-female societies with polygynandrous mating habits, but is the reproductive trajectory of their exits from Shakespeare's stage different?

A WildCRU team, led by Andrew Loveridge and David, had been studying Hwange's lions since 1999, so we knew a lot about Cecil (Loveridge 2018). He was really unusual, not because of how he died (that was all too commonplace; Loveridge et al. 2016) but because of how he lived or, more particularly, survived into old age. The WildCRU team first tagged Cecil in 2008, shortly after he and his putative brother had wandered into the study area. This duo's reproductive career began when they displaced the surviving resident male of the Ngweshla pride, which numbered between 6 and 10 lionesses, where they reigned

supreme until 2009 when Cecil was ousted, and his brother killed, during an attack by a coalition of four males from the north—one of these lethal aggressors was nicknamed Jericho—a name to remember for what follows. Had Cecil read the right text books, that would have been that, but unusually, aged 7 he launched a new and single-handed career, taking over the Back Pans pride when the fickle, and in this case ironical, hand of fate led the incumbent male to be shot by a trophy hunter. Cecil remained sole monarch of Back Pans until 2013 when an incoming coalition ejected him. Deposed and facing disaster again, his meanderings led him into the company of Jericho, himself recently deposed from the Ngweshla pride where his father and two of his brothers had killed Cecil's brother 4 years earlier. But Jericho's father had died of wounds sustained while killing Cecil's brother, and then, one after the other, Jericho's brothers were picked off by trophy hunters. His supporters all dead, Jericho himself became itinerant, scarcely maintaining a foothold with the pride until, unusually, in what was probably Cecil's 11th year, the two veteran males, previously mortal enemies, palled up to create a new dynasty at Ngweshla, where the six lionesses soon produced a new crop of cubs. Looking back on his history, Cecil's reproductive career has left 13 surviving sons and daughters and 15 surviving grand-offspring. In what was his third period of primacy, he and Jericho had seven young cubs at Ngweshla when, as he often had before, on 1 July 2015, aged 13, Cecil strayed onto the neighbouring hunting concession and was shot, seemingly illegally, at bait by the bowhunter.

One can hardly call his death, albeit squalid, premature: although lions can live to 30 years of age in zoos, for a male

Box 18.1 *Continued*

lion to survive long after being on the losing side of a take-over is unusual. The point of this excursion into lion biology is to draw a contrast with male badgers. Male lions reach the Fourth Age—the soldier—and their last battle does not long precede their last breath. Their post-reproductive life span approximates zero (death, not meno- or andro-pause, prunes their family tree). In contrast, ageing male badgers can hold their own during the Fifth Age and even then slide gradually into the prolonged reproductive retirement of the Sixth, and even, remember Little B, Seventh Age. Lions do not die quietly in their beds (lionesses don't either), but quite a few badgers do.

In short it seems that lions fall off a cliff, whereas badgers slither down it. Because lions live soldierly lives, once deposed their days are numbered. For badgers we see little aggressive competition, and their rather meandering lifestyle means that unlike the otter which can no longer swim or the pine marten which can no longer leap, they can simply amble into decrepitude. However, badgers are big enough

and tough enough (and there are no wolves left in Wytham, or elsewhere in the UK) not to be killed quickly by predators as they decline, so they may inhabit an unusual sweet spot, in which they can wither on the vine rather than being picked off it.

So much for the sociology of the tail of the distribution, but is that tail unnaturally long compared with other species? Have badgers been selected for prolonged PRLS, and if so, to what advantage? Remember, there's little evidence of cooperation (although the notion of information exchange remains tantalizingly unexplored). Also, ostracism or exclusion are unlikely to kill an old badger, which, anyway, may still remain dangerous, toothlessness aside. Indeed elderly badgers may even be venerated by inter-generational companions. While the idea of functional significance to a prolonged PRLS is attractive, and not implausible, it is probably wrong. Given that the gerontology of badgers appears genuinely unusual the tail of their age distribution wags their emergent society for far longer than is typical in most carnivore societies.

all females mate, only about a quarter[11] are assigned offspring each year (Macdonald et al. 2015e). Furthermore the tell-tale social flux of subcaudal scent marking reveals young females anointing older prime-aged females, not the reverse (Buesching et al. 2003; Chapter 6). A further, considerable, complication to interpreting intra-sex breeding competition arises because our observations from Chapters 7 and 17 suggest it is hard for older females to control younger females' access to mating opportunities because they will mate, and breed, so readily with whichever male they meet, including extra-group liaisons. Similarly, the minimal allomaternal care in badgers (Fell et al. 2006; Chapter 8) with no food provisioning or discernible benefits from helpers (Woodroffe and Macdonald 2000, but see Dugdale et al. 2010) further precludes the *mother hypothesis* (Williams 1957), which proposes that females terminate reproduction in mid-life to gain fitness advantages by

investing in previous matured litters of offspring; or the *grandmother hypothesis* (Hawkes et al. 1998), earlier, which proposes that post-reproductive females can increase their inclusive fitness by supporting weaned grand-offspring—something Grandma badger does not do. Finally, because badgers exhibit high inter-generational relatedness (inbreeding coefficient f = 0.010; Annavi et al. 2014b) and inter-generational overlap (generation length for the Wytham population is at least 5.8 years (longer according to the enhanced MNA methodology of Bright Ross et al. (2020)), Cant and Johnstone's (2008) *inter*-generational reproductive conflict hypothesis does not apply.

In short, we found no firm evidence for PRLS in badgers being adaptive, although at a population level it is advantageous that a proportion of elders continues reproducing 'just in case' (Rogers 1993). In other social carnivores such as lions (see Box 18.1), obsolete elders are cast aside, killed, or starved, whereas in badgers the eclectic, patchy natures of food supply makes it harder to exclude individuals from food than it would be to exclude them from a kill (Chapter 10), with the result that PRLS may be impossible to select against, rather than its being selected for. In the absence of a better idea, perhaps the most parsimonious explanation for why elderly badgers run out of reproductive

---

[11] Our earlier analyses suggested 21–29% were assigned offspring each year—the figure varies depending on the period of years considered and the ages of the females, but our most recent analysis using Julius Bright Ross's eMNA calculation across the whole dataset yielded a figure of 17.3% of females age 2 or above known to be alive assigned pedigree cubs in a given year (weighted average by the number of females in the population that year).

steam—in line with a general thesis that sociality in badgers is facultative and linked to resource dispersion, not cooperation—is that by ceasing to reproduce, females escape the increased risk of mortality associated with late-life pregnancy, both sexes escaping the demands that maintaining reproductive condition places on their bodily (somatic) condition (Penn and Smith 2007).

Now, of course, there's much more to getting old than failing reproductive performance, even if that metric is natural selection's prime arbiter. No, as we increasingly can vouch, in people ageing brings with it a distressing array of gradual failures: from teeth to heart valves and from grey hair to arthritis. What is it that goes wrong, or impedes repair at the cellular level? The answer lies, once again, in DNA, and the implications for badger gerontology of the selective disadvantages that accrue with age occupy the remainder of this chapter.

Cellular senescence and telomere attrition

According to the old saw, attributed to Charles H. Spurgeon,[12] one should always *start as one means to go on*, and when it comes to longevity there is a theoretical basis for the aphorism that permeates to an individual's pre-programming at the cellular level. DNA is arranged into chromosomes (we have 23 pairs, badgers have 22 pairs). The functional genes, transcribing protein building blocks, arranged along chromosomes, are vulnerable to corruption as cells replicate, particularly due to entanglement (i.e. end-to-end fusion) of the chromosomal thread. 'Telomeres'—protective caps that bookend the library of information archived along the chromosomal library shelf[13]—prevent this mitotic calamity (Figure 18.8). The bookend role of the telomeres is to ensure that as wear erodes the ends of chromosomes, it is not the functional genes that disappear but the protective caps that are sacrificed.

Due to incomplete DNA-replication at the 3'-end of the DNA-strand, telomeres shorten with age until the cell line can no longer undergo further mitosis

and becomes progressively quiescent.[14] Consequently, telomere length can function as a biomarker of senescence (Blackburn 2000; de Lange 2004) shortening with age until finally tissues and organs are so packed with defunct cells that either chronic or catastrophic system failures occur (Armanios and Blackburn 2012; Campisi 2005). However, as will become clear, in badgers telomeres turn out to be less of a biomarker of senescence and more a reflection, even a logbook, of the physiological consequences of stressors.

In the lab, principally amongst mice, the more demanding the lifestyle (e.g. growth, reproduction, coping with stress/disease), the faster the rate of cellular division and the greater the attrition of telomeres.[15] Furthermore, the amount of telomeric DNA lost in each cell division depends on cellular conditions (Monaghan and Ozanne 2018), especially levels of oxidative stress—remember the dangerous free radicals of Chapter 15 (Wickens 2001; Reichert and Stier 2017). Those by-product reactive oxygen species (ROS) produced during oxygen metabolism in the mitochondria that can cause cellular damage (Beckman and Ames 1998; Selman et al. 2012) are especially damaging to guanine, the most prevalent nucleobase within telomeres (Oikawa and Kawanishi 1999). So, while there is no conclusive evidence that telomeres are impacted by oxidative stress—the villain of Chapter 15—it seems likely.

So much for laboratory mice—what of telomeres in nature? Study of these phenomena in the wild, and especially amongst carnivores, is scant, so we turn again to the badgers of Wytham Woods as models for evolution and behaviour (pioneering research on the telomeres of wild mammals focused on Soay sheep (Fairlie et al. 2016), red deer (Wilbourn et al. 2017), various primates (e.g. Herbig et al. 2006), and follows the lead of a wealth of work on birds that has largely informed theories on this subject (e.g. Hall et al. 2004)).

First, a brief primer on telomeres. Even when the starting pistol of life fires, individuals can, from birth, have inherently different telomere starting lengths (Fairlie et al. 2016) arising from a poor environment during early—even prenatal—life (Boonekamp et al. 2014; Nettle et al. 2015; Cram et al. 2017),

---

[12] 'Begin as you mean to go on, and go on as you began, and let the Lord be all in all to you.' In: 'All of Grace' 1915.

[13] Telomeres are essentially non-coding hexameric repeats (5'-TTAGGG-3') that, along with associated shelterin proteins, act as protective caps at the chromosome's end, maintaining genomic stability.

[14] This is a state functionally different from apoptosis (cell death) that characterizes a stage in normal cell line replication and regeneration in healthy individuals (Campisi and di Fagagna 2007).

[15] Typically, in humans, we max out at 50× Hayflick cycles, each of 2 years, defining ultimate life expectancy.

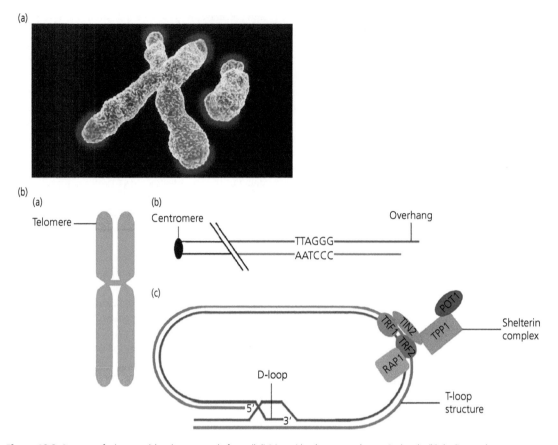

Figure 18.8 Structure of telomeres: (a) a chromosome before cell division with telomeres at the terminal ends; (b) the linear telomere structure and sequence in eukaryotic chromosomes, where the DNA strand has an overhang at the 3'-end due to the end-replication problem; and (c) the telomere T-loop that prevents fusion of linear chromosomes through the single strand by invading double-stranded telomeric DNA (D-loop). The shelterin complex aids in T-loop formation and telomere functioning and thus maintaining genomic integrity (from van Lieshout 2021, with permission).

compounded by additive genetic effects,[16] potentially even inheriting slightly abbreviated telomeres from stressed parents (the so-called 'Lansing effect'; see Noguera et al. 2018) (Dugdale and Richardson 2018; Box 18.3). Further, when it comes to interpreting losses to telomere length, a complication that is increasingly topical is that they can also regain length through the actions of a telomere-elongating enzyme called 'telomerase' (Blackburn et al. 1989)—to the limited extent it can keep pace. The obvious solution to ageing would seem to be to evolve a continuing capacity to make telomerase (see Box 18.2), but it has, until recently,

been believed that, outside of gametes and stem cells, telomerase production is transcriptionally repressed later in development.[17] The consensus has been that, in most species, telomerase production is stopped at birth, because its overproduction can lead to continuous cell replication and grave risks from malignant cells and tumours (see Box 18.2). Certainly in human somatic cells proliferation potential is strictly limited and senescence follows approximately 50–70 cell divisions, whereas in contrast replication potential in most

---

[16] Additive genetic effects occur when two or more genes source a single contribution to the final phenotype, or when alleles of a single gene (in heterozygotes) combine so that their combined effects equal the sum of their individual effects.

[17] Transcriptional repressors are proteins that bind to promoters (a sequence of DNA to which proteins bind that initiate transcription of a single RNA from the DNA downstream of it) in a way that impedes subsequent binding of RNA polymerase, the enzyme that follows a strand of DNA and copies that DNA sequence into an RNA sequence, during the process of transcription.

Box 18.2 Telomerase

After passing the embryonic life stage, why don't badgers, and indeed most other organisms (with notable exceptions including lobsters, quahog clams, and even bristlecone pines), produce levels of telomerase in non-germ or stem cells sufficient to continually rebuild telomeres and avoid cellular senescence? Well, while the road to immortality may be paved with good intentions, they may backfire. Taking the long view, immortality can be a curse rather than a blessing—as Tithonus learned to his cost. This mythical Trojan prince was so handsome that he bewitched Eos, the goddess of dawn. She successfully petitioned Zeus to grant Tithonus immortality so she could be with him forever. But Zeus took her too literally. Tithonus didn't die, but he did age. He lost his good looks and his faculties, and Eos lost her interest. She eventually shut him away in a room where he babbled endlessly. A different, but no less unwelcome, consequence of prolonged dousing of the chromosomes in telomerase is that it engenders the risk of developing cancers. So, at least in humans, there is evidence that this constitutes a selective advantage for repressing telomerase production: consider Henrietta Lacks who died in 1951 from cervical cancer, her name immortalized in the 'HeLa tumour' caused by an excess of telomerase (Robin et al. 2014; Kim et al. 2016). A further fact is that germline cells (gametes: sperm and ova) are fairly impervious to telomere attrition, where gonad-specific telomerase activity maintains telomere length, especially through the extensive production of sperm (Ozturk 2015). Without this, offspring would be born carrying all the losses already suffered by their parents—'Dolly the sheep', for example, was cloned from mammary gland tissue subject to telomere attrition,

and hence was born 'biologically old'. Furthermore, amongst mammalian species, there is a size effect where telomerase activity co-evolves with body mass, but not lifespan (Risques and Promislow 2018). This is relevant to badgers (adults in Wytham average 8.7 kg) in two respects. First, species below 2 kg as adults typically express some telomerase, while larger ones do not (Gorbunova and Seluanov 2009), and, second, species heavier than 5 kg generally have shorter telomeres than smaller species. To put faces to these generalities, the cells of small rodents, such as mice (Myomorpha), are immortal when cultured at physiological oxygen concentration, whereas those of large rodents such as beavers (also Myomorpha) or capybaras (Hystricomorpha) repress telomerase activity. Why should this be so? Back to oncology: long and replenished telomeres, multiplied by the number of cells vulnerable to mutation, would increase cancer risk. The so-called 'Peto's paradox',[18] in which the likelihood of cancer does not, counterintuitively, actually increase with the number of cells comprising more massive species (Caulin and Maley 2011),[19]

characterizes the link between telomerase repression and the evolution of tumour suppressor mechanisms in larger mammals.

[18] Peto's paradox is the observation, named after Richard Peto, that at the species level, the incidence of cancer does not appear to correlate with the number of cells in an organism. For example, the incidence of cancer in humans is much higher than the incidence of cancer in whales.

[19] Remember too, from Chapter 14 that as cell division slows in old age, adipose tissue leads to lipotoxicity (Slawik and Vidal-Puig 2006), where the risk of age-related cancers can be alleviated through caloric restriction (Weindruch 1992; Anderson and Weindruch 2010).

---

tumour cells is unlimited (Zvereva et al. 2010). This creates an awkward trade-off between telomere maintenance and tumour suppression; Tian et al. 2018), upon which the badgers of Wytham Woods add some important new insights.

To what extent might variability in the length of telomeres link life experiences to cellular deterioration, defining individual badger performance and longevity? This question became the *raison d'être* of Henricus Johannes van Lieshout, notable not only for his portentous moniker (thankfully, if mysteriously, abbreviated to Sil), but also because he, as a talented example of continuity if not longevity, was a sort of supervisorial grandson, being the graduate student of former badger project student, Hannah Dugdale,

now a Professor in Groningen University. Diving into our deep-freeze, Sil selected 1248 blood samples from 612 individual badgers (308 from males and 304 from females). They were selected to represent individuals of varying lifespans, ranging from 14 to 233 months (mean ± standard error (SE) = 97.2 ± 1.88 months).[20] How did the lengths of their telomeres (specifically those capping chromosomes in their leukocytes[21]— the white cells introduced in Chapter 15) vary with age? Early in life, and as expected with exposure to

[20] Individuals were either sampled once ($n = 163$) or more ($n = 449$ badgers; two to nine times per individual).

[21] In mammals leukocytes have nuclei and were thus a convenient source of chromosomes for our study of telomeres.

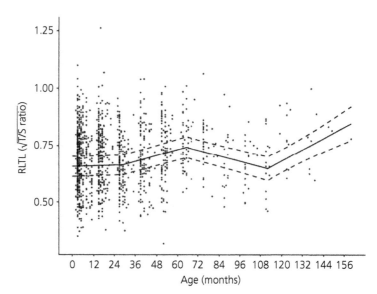

Figure 18.9 Age-related variation in relative leukocyte telomere length (RLTL), with inflection points at 38, 74, and 112 months old. Raw data points (*n* = 1248) are shown with fitted lines representing the model prediction for RLTL (T/S ratio) with 95% confidence intervals within the four age groups (from van Lieshout et al. 2019, with permission).

developmental stress, telomere lengths decreased (van Lieshout et al. 2019). However, soon after their third birthday (specifically, on average at 38 months old, and thus neatly matching the point at which somatic ageing begins), Sil was surprised to find that across this sample cross-section, badger telomeres increased in relative length (we recorded RLTL, i.e. the relative leukocyte telomere length[22]). This *increase* in RLTL continued up to 112 months of age, essentially rejuvenating to Pan-like chromosomal youthfulness, whereupon it *decreased* again into old age (Figure 18.9).

Why the inflection points in Figure 18.9? Well, this pilot investigation was cross-sectional, i.e. not predicated on tracking an individual's telomeres through its life. It therefore included any effect of individuals with short telomeres dying out as the sampling continued (i.e. selective mortality, see p. xxx), leaving a longer average amongst survivors. So while this winnowing of the sample offers a partial explanation, it was not the full story. Sil's next approach was longitudinal—to monitor changes in telomere length over the lifetimes of a sample of individuals; again, these longitudinal data clearly revealed that RLTL can increase as the badger ages. This was very surprising, but there was no doubting it: the rate of increase was in the range of 0.004–5.829% per month, and it was evident in 56.4% of the 449 individuals for

which we had two or more measurements of telomere length (Figure 18.10). Against the background of an average shortening of telomeres over an individual's lifetime, we found spurts of increases interspersed with decreases in telomere length (Figure 18.11).

Not only was this reversal atypical for the textbook story for vertebrate ageing, especially when contrasted with a recent study of another carnivore also notable for intense early-life competition meerkats (*Suricata suricatta*), whose telomeres follow the convention and progressively shorten in relative length with age (Cram et al. 2017). And to add to the puzzle, according to Peto's paradox because larger animals have more cells and thus would be at more risk of cancer from enhanced telomerase activity they suffer greater relative telomere attrition with age than do smaller animals that can synthesize telomerase with less oncological risk, and meerkats (and their ancestors) are less than a tenth the size of badgers.

We discovered that the destiny of those surviving infant badgers that were able to avoid or resist the vicissitudes of misfortune was written indelibly in their telomeres. The RLTL of youngsters <1 year old predicted not only the likelihood of their survival to adulthood (Figure 18.12), but also their lifespan (Figure 18.13). Measuring youthful telomere length relative to that of a control gene (called IRPB—giving a measure known as the T/S ratio unit), the bell tolled after a predictably 8.6% longer life for those badgers blessed with one additional relative unit of telomere length (two T/S units gives 17.2% greater longevity, etc.). This remarkable effect, however, only applied

[22] Technically this involves quantitative polymerase chain reaction, and measures the length of telomeres relative to a reference gene (IRBP) in leukocytes.

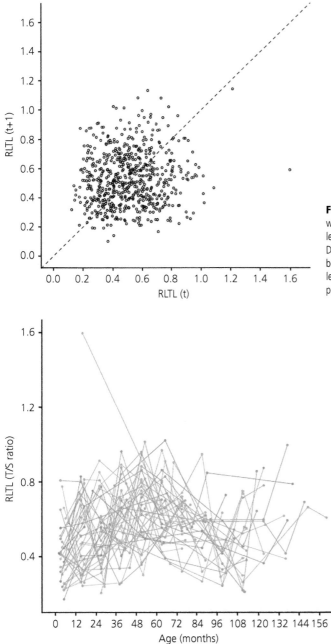

Figure 18.10 Telomere dynamics in European badgers: within-individual variation in relative leukocyte telomere length (RLTL) over consecutive time points (t and t + 1). Dashed line represents parity, thus data points above and below this line represent increases and decreases in telomere length, respectively (from van Lieshout et al. 2019, with permission).

Figure 18.11 Telomere dynamics in European badgers: longitudinal telomere dynamics for 41 individuals that were measured at least four times (from van Lieshout et al. 2019, with permission).

when extrapolating from RLTL measured amongst cubs; measuring age-specific RLTL at a more mature age point thereafter revealed no such relationship with residual lifespan. This seems to be partly because, while early life provides a consistent measure of telomere length, the later measures are beset by the rollercoaster lengthening and shortening revealed in Figure 18.11 such that at any given time in its extension or contraction the telomere's length reflects less well the relationship with lifespan. In summary, statistically, a cub with lengthy telomeres (i.e. when first measured in May) should be destined to live a long life; the effect, however, was not significant if measured at an older age—another factor being the scale effect,

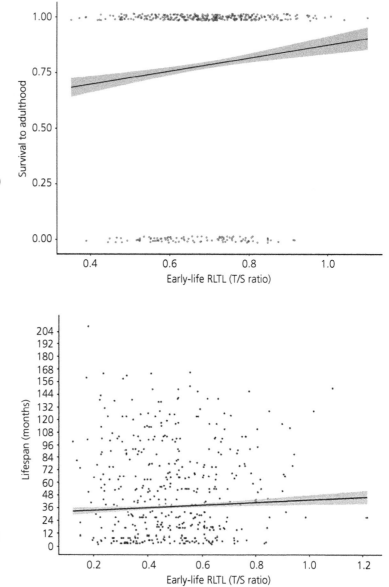

Figure 18.12 Survival to adulthood (>1 year old) predicted by early-life (<1 year old) relative leukocyte telomere length (RLTL). The regression line from a binomial generalized linear mixed model (GLMM) is shown, with associated 95% confidence interval as a shaded area, and raw jittered data as open circles (*n* = 435) (from van Lieshout et al. 2019, with permission).

Figure 18.13 Early-life (<1 year old) relative leukocyte telomere length (RLTL) predicts lifespan. Raw data (*n* = 435) are shown as open circles, the regression from the generalized linear mixed model (GLMM) as a black line, and the 95% confidence interval as the shaded area (from van Lieshout et al. 2019, with permission).

whereby the less life a badger had, in all probability, remaining to it, the smaller became the relative advantage of longer telomeres. Just to be clear on that scale effect—if, as you read this, you work out assiduously in the gym, and are 30 years old, probably your fitness will correlate with your remaining lifespan; if, however, you are 95 there's not enough scope for variability remaining in your lifespan for your exemplary fitness to pay much dividend. The long and the short of telomere length in badgers is that it can encode longevity, but not immortality.

Although we don't meet the cubs until May, and so they have already run one leg of their life's race by the time we get our RLTL measure, we deduce that being born with superior telomeres counts. Crucially too, because within-individual patterns were similar, irrespective of age of last reproduction, this demonstrates that although selective disappearance of individuals with shorter telomeres did contribute to our initial cross-sectional findings it could be ruled out as the reason behind lengthening as seen in our longitudinal analyses (Figure 18.11). Figures 18.10 and 18.11 offer different lessons: 18.11 shows how chaotic telomere length can be across lifetimes, whereas 18.10 shows that clearly for about half the time during which lengths increase; likewise for about half the time during which they decrease. By the way, Sil ruled out the possibility that our results could have arisen through error due to faulty measurements because variance amongst duplicates (of the same sample) was smaller than variance amongst different samples from the same individual.

But why, and indeed how, might telomeres get longer in meline middle age? Remember, the ability to produce telomerase outside of embryonic development is generally switched off (transcriptionally repressed[23]), especially in somewhat larger mammals, and so before we could unequivocally call this 'regeneration' we wanted to explore the alternative. One possible competing explanation for this phenomenon arose out of our previous work on white blood cells, leukocytes. Remember (from Chapter 15) that there are different types of leukocytes.[24] Could different cell types simultaneously have different telomere lengths? If so, could changes in blood composition explain shifting mean RLTL? As to the first question, yes, each type of leukocyte cell line does wear through its telomeres at a characteristically different rate due to their different capacities to proliferate and express the telomerase enzyme (Aubert and Lansdorp 2008; Weng 2001). Amongst these white blood cells, neutrophils (along with lower-volume eosinophils and basophils) contain granules packed with antimicrobial agents, lysozyme and acid hydrolase enzymes (also the reactive oxygen and nitrogen 'species' we discussed in Chapter 15), and so are termed granulocytes. These cells function in innate immunity (Box 15.1) and phagocytose (engulf) antigens/pathogens in the circulation. In contrast, lymphocytes (the T-cells, B-cells (producing antibodies) and monocytes mentioned in Chapter 15) are agranulocytes and function in adaptive immunity. The jokers in this pack are macrophages (see Chapter 16 for their role in bTB immunity), which are agranular (although with spherical nuclei vs the lobed nuclei of leukocytes), yet innate immune cells

(although they phagocytose antigens mainly in the tissue). In mammals, granulocytes make up around 70% of circulating white blood cells (90% neutrophils), with 30% agranulocytes (of which macrophages can comprise 30%). In humans and baboons, at least, granulocytes have longer telomeres than agranulocytes, and in longitudinal human studies telomeres shorten faster in agranulocytes (Rufer et al. 1999).

As to whether blood composition changes with age, it does. We already knew a lot about badger blood composition (Domingo-Roura et al. 2001) and had demonstrated previously that badger leukocyte cell composition varies not only amongst similar aged cubs, but also throughout their lives, as a result of changes in immune system activation due to repeated pathogen challenge (McLaren et al. 2003). Those involved in the innate response (Box 15.1) are principally phagocytic cells such as neutrophils (granular), macrophages and dendritic cells, able to detect novel antigens (Akira et al. 2006; Weiskopf et al. 2009). Cell-mediated adaptive immune responses involve those agranular leukocytes, called T- and B-lymphocytes (Chapter 15).

Sure enough, Sil found that, specifically, the agranulocyte-to-granulocyte (adaptive immunity-to-innate immunity) ratio in Wytham's badgers did indeed decrease with age (van Lieshout et al. 2020). This suggests a decline in the antibody and T-cell responses, consistent with a general phenomenon known dishearteningly as 'immunosenescence', which results in a greater proportion of longer telomered granulocytes in a blood sample from an elderly individual.[25]

The flip side is that, in humans and in mammals generally, as granular innate cells start to predominate this can lead to chronic inflammation (Franceschi et al. 2018). This suggests that early-life immune challenges (such as the coccidiosis rampant in Chapter 2) prime adaptive immunity (there being a phase where juveniles develop antibodies rapidly due to exposure to pathogens), whereas the T-cell (agranular lymphocyte) repertoire reduces with age (becoming biased towards CD8+ effector memory cells[26]).

---

[23] Transcriptional repressors are proteins that bind to promoters (a sequence of DNA to which proteins bind that initiate transcription of a single RNA from the DNA downstream of it) in a way that impedes subsequent binding of RNA polymerase, the enzyme that follows a strand of DNA and copies that DNA sequence into an RNA sequence, during the process of transcription.

[24] We refer here to mammalian leukocytes; in birds, erythrocytes are nucleated and used as the source of telomeric DNA: erythrocytes are all of one kind and all short-lived cells.

[25] We were unable to undertake flow cytometry during this study, and consequently do not know the absolute white blood cell count. There is thus a possibility that lymphocyte numbers were unchanged whereas granulocytes increased. In this case the badgers would nonetheless have experienced 'inflamaging', but without a decline in their antibody response.

[26] Memory CD8+ T-cells are a long-lived population that are antigen-specific and provide an enhanced protective response when the same antigen is encountered again.

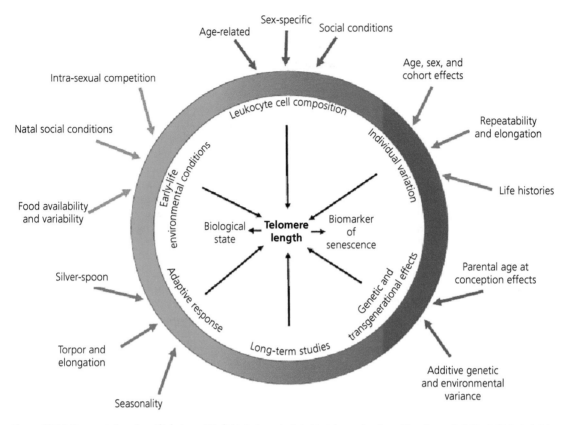

Figure 18.14 Representation of specific factors within fields that may be linked to telomere length and therefore an individual's biological state or senescence patterns. It is basically all things that may impact telomere length (Adapted from van Lieshout 2021, with permission).

Intriguingly, this agranulocyte-to-granulocyte effect was greatest amongst males living in smaller groups. This may relate to the 'hygiene hypothesis', which posits that exposure to a lesser diversity or intensity of pathogens during early life—as might be expected to occur in smaller groups—can cause the adaptive immune system to be undertrained, leading to a propensity for auto-immune and extreme allergic responses (Garn and Renz 2007; Albery et al. 2020b; Chapter 8). Possibly this arises because the composition of smaller badger groups is more male-biased in Wytham, due to male philopatry (Chapter 3).

Which is it, then, that explains telomere-lengthening within individual badgers: telomerase activity or a change in immune cell composition? Well, we can confirm a change in immune cell composition, and a process of deduction indicates strongly that there is a resurgence of telomerase activity when the telomere RLTLs of individual undergo a spurt. We deduce this because we found no sex effect in variation in telomere length, whereas there was a clear-cut sex effect for variation in immune cells. In short, something other

than shifts in the community composition of immune cells was involved in driving telomere length variation, and we demonstrated that that 'something' was not measurement error. The best bet is that a resurgence of telomerase[27] (or other telomere-elongating pathway) was at work. Whatever the mechanism, the overarching points remains that, while the future cannot be read in tealeaves, it can in part be deduced from telomeres (Table 18.14): the longer they are during a badger's youth, so too the longer that individual's natural lifespan is likely to be—*ceteris paribus*.

An early suggestion regarding senescence—that is, why natural selection allows it to happen at all, when, after all, death is bad for you—was that, due to energy being allocated differently between early and later life, being brought up in an adverse environment might result in faster senescence (Williams 1957; Medawar 1952; Kirkwood and Rose 1991). The strength

[27] Technical insights into alternative telomere elongating pathways are considered by Cesare and Reddel (2010) and Mendes-Bermudez (2012).

of selection decreases with age, a situation which predicts slower adaptation of traits expressed late in the life cycle compared with those expressed in early life. Why does the strength of selection decrease with age? Because in early youth more individuals are alive, providing more variation for selection, which is therefore stronger. Correspondingly, as the variation that has survived this early filter inevitably becomes narrower, the strength of selection decreases with age and thus traits expressed in later life are less strongly selected for. This imbalance in the strength of selection on youthful or ageing contingents creates circumstances favouring the existence of enticingly named antagonistic pleiotropic genes. What are they? Well, antagonism is all too familiar, whereas pleiotropy means the production by a single gene of two or more apparently unrelated effects, a sort of split personality: putting the two together, the two sides of the personality don't like each other. In this context the idea is that genes that have beneficial effects early in life can be selected for strongly (when selection is fierce), but can carry with them unwelcome detrimental (pleiotropic) effects (the ones causing senescent deterioration), which the weaker selection that prevails in these sunset days fails to weed out (Hamilton 1966). Therefore, exposure to stress during a particularly sensitive developmental period might trigger a faster pace of life (Kirkwood and Rose 1991; Lemaitre et al. 2015; Chapter 14). Indeed, in Chapter 2 we showed how social and environmental stressors influenced the rate of cub maturation to puberty (Sugianto et al. 2019b, 2019c). So, mindful that for Wytham's badgers an adverse environment generally means bad weather (Chapters 12, 13, and 14), we asked: is there a link between the weather, and particularly its variability outside the Goldilocks comfort zone (Chapter 12), and telomeres?

Based on genomic DNA extracted from whole blood samples (n = 841 samples; 562 badgers), and our now familiar weather metrics[28] (Chapter 12), we discovered that cubs experiencing higher mean daily temperatures and higher mean daily rainfall with low variability, did indeed have longer early-life telomere lengths (Figures 18.16, 18.17, and 18.18).[29] Why so? These are the conditions associated with high earthworm availability in spring and low foraging costs. It seems that this benign weather allows cubs to achieve a positive energy balance, subsequently invested in either growth or somatic maintenance, causing slower senescence

schedules (e.g. Hammers et al. 2013). What of cubs experiencing relatively colder conditions? They had shorter early-life RLTLs (Figures 18.16) (van Lieshout et al. 2021b).

So, what of the Goldilocks zone? Did variability in rainfall impact telomere length? Yes, more rainfall was beneficial to cub RLTL, and so too was lower variability in rainfall (Figures 18.17 and 18.18). In short, we found that living outside the envelope of historically normal weather variation was uncomfortable for badgers foraging for earthworms, even impacting their population dynamics (Chapter 12) and pace of life (Chapter 14), and now we see, also, that these weather stresses manifest in the fraying of their telomeres.

Curiously though, and contrary to the current consensus that contest during early-life social conditions is predicted to be bad for RLTL (Boonekamp et al. 2014; Nettle et al. 2015; Cram et al. 2017), cubs born in groups with a *greater number* of siblings had longer early-life telomere lengths (van Lieshout et al. 2021b). The general supposition would be that membership of a larger cohort would be indicative of greater competition, but in badger society it is not immediately clear that this would be the case—rather, larger cohorts may be indicative of generally better ecological circumstances for the group; that is, cub cohort size is less restricted in the most lenient years. This is an intricate thought, so it may help to express it another way: in a hard year, cohort size is constrained; in a good year, it may be unconstrained by the weather, but is nonetheless bound by limits set by fecundity and litter size, dictated by phylogeny rather than weather—thus there is still a ceiling, although it is set differently.

Why and how does this intricate interplay between environment and the replicative capacity of cells come to be?

Eroded telomeres—does this arise through weathering?

Given that individual badgers arrive at life's starting line bearing fundamental telomeric inequalities, and given the way in which weather (Chapters 12, 13, and 14), and social (Chapter 8) conditions intricately affect badger—especially cub—success, we asked next whether early-life environmental condition could explain the differences in RLTL we observed in Wytham's badgers?

Using the same set of blood samples as earlier, we found substantial intra-annual changes in telomere length during the first 3 years of life, during which telomeres were shorter in the first winter each

[28] Mean temperature and rainfall, plus temperature variability and the coefficient of variation in daily rainfall over the period of early cub development February–May (1987–2010).
[29] <1 year old.

individual experienced but then elongated again into their yearling spring (Figure 18.15) (van Lieshout et al. 2021b).

What might explain this? At first we thought a change in leukocyte composition was an unpromising explanation for these relatively young individuals, and besides, any immunity linked to fighting juvenile coccidiosis (Chapter 2) would primarily recruit innate (granular, shorter-telomered) cells. However, we subsequently discovered that the same pattern occurred as the badgers entered their second winter—telomeres shortened—and then passed into their second spring—telomeres lengthened. The pattern emerged as being less about age and more about the seasons (Figure 18.15). This drew our attention to the way badgers can use winter lethargy, along a trajectory towards hibernation, to alleviate their energetic struggles due to reduced winter food availability (Chapter 12). Generally, facultative winter torpor conserves energy and reduces oxidative stress (Chapter 13), permitting investment into somatic maintenance; also, telomere restoration/elongation (as earlier, it seems likely that a redistribution of immune cell types and

a reignition of telomerase activity are both at work). Once again, however, weather interactions are rarely simple: to judge from other species, what matters may be the frequency of arousal. The more often, and the bigger, the switch between torpid and active body temperature, the greater the damaging oxidative stress incurred. Such repeated cycles shorten telomeres in arctic ground squirrels (*Urocitellus parryii*) (Wilbur et al. 2019) and in edible dormice (*Glis glis*) (Turbill et al. 2013), although the latter, along with its garden dormouse (*Eliomys quercinus*) cousin, suffers less than do ground squirrels because their core temperature transition is smaller (Nowack et al. 2019). In contrast though, non-hibernating juvenile garden dormice that more frequently undergo fasting-induced torpor experience greater telomere shortening than do individuals undergoing torpor less frequently (Giroud et al. 2014).

Plausibly blinking in and out of torpidity in some years might explain the badgers' winter telomere shortening (see also Bilham et al. 2018). In contrast, old-fashioned frosty winters that keep plump badgers dozy and their cores up to 8.9°C below euthermic levels (Fowler and Racey 1988; Geiser and Ruf 1995) would

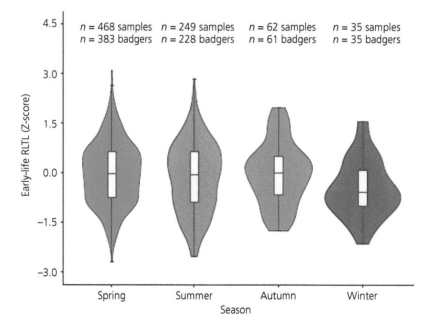

Figure 18.15 Variation in early-life relative leukocyte telomere length (RLTL) among seasons in European badgers. The data distributions and probability densities are shown (*n* = 814 samples; 533 badgers—the sum of badgers in the plot is >533 due to repeated measures). Data were collected in 19 years, across 59 trapping periods. The line in the boxplot represents the median, with first and third quartiles, and whiskers represent 1.57 times the interquartile range. When controlling for age, weight, and body length, we found a significant effect of season on RLTL with badgers having shorter RLTL in winter compared with spring. After partitioning the within- and between-individual effects we found that there was a within-individual effect of shorter RLTL in winter than in spring and a significant between-individual effect. From spring to winter there was a within-individual decline in RLTL, whereas from winter to the following spring there was a marginally non-significant within-individual increase in RLTL (from van Lieshout et al. 2021, with permission).

also function to reduce mitosis (rate of cell division; Kruman et al. 1988) and therefore potentially reduce telomere shortening. Just to give a further crank to the ratchet of speculation, winter weather predictions by the Intergovernmental Panel on Climate Change (IPCC 2014) anticipate more of the erratic, milder winters, so might the penalty of climate change for badgers be an increased allostatic load and accelerated telomere shortening?

The big question, however, is what explains the spurts of telomere lengthening that recurred through an individual's life, at least sometimes associated with

springtime. We know that something more than a shift in leukocyte types is at work, and since it probably involves telomerase or other telomere-lengthening pathways, the intersection between emergence from torpor and the spring weather seems like a fruitful place to start. In terms of weather conditions during the productive portion of the year, and in line with our general findings on what makes for benign early-life conditions (see Chapter 11), cubs born in warmer (Figure 18.16), wetter (Figure 18.17) springs with low rainfall variability (Figure 18.18) typically had longer early-life (3–12 months old) telomeres; ergo,

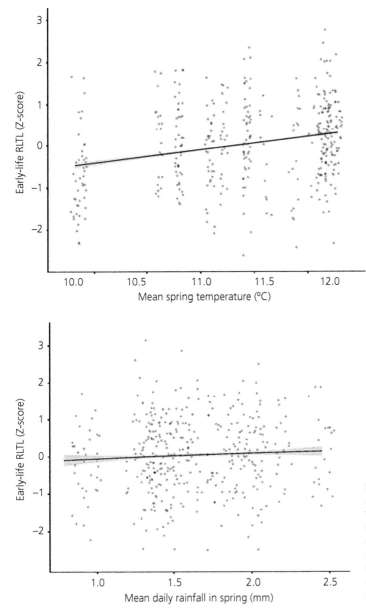

Figure 18.16 The association between mean spring temperature and early-life relative leukocyte telomere length (RLTL). Raw data points ($n = 406$ samples; 406 badgers) are shown, and jittered for clarity on the amount of data. The fitted line represents the regression from the mixed model, and the 95% confidence intervals as shaded areas (from van Lieshout et al. 2020a, with permission).

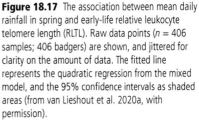

Figure 18.17 The association between mean daily rainfall in spring and early-life relative leukocyte telomere length (RLTL). Raw data points ($n = 406$ samples; 406 badgers) are shown, and jittered for clarity on the amount of data. The fitted line represents the quadratic regression from the mixed model, and the 95% confidence intervals as shaded areas (from van Lieshout et al. 2020a, with permission).

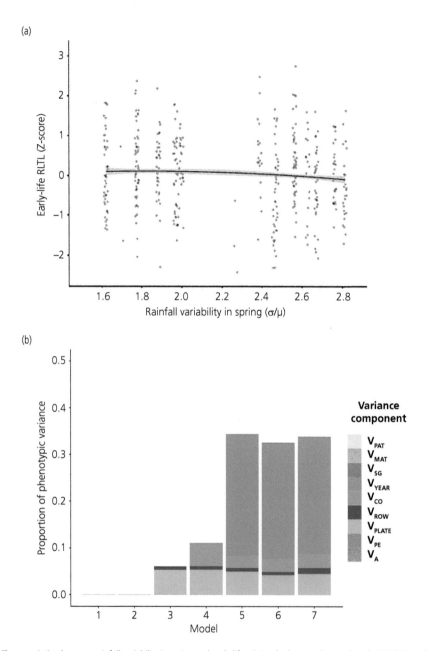

Figure 18.18 The association between rainfall variability in spring and early-life relative leukocyte telomere length (RLTL). Raw data points (*n* = 406 samples; 406 badgers) are shown, and jittered for clarity on the amount of data. The fitted line represents the quadratic regression from the mixed model, and the 95% confidence intervals as shaded areas. Proportion of variance explained in relative leucocyte telomere length (RLTL; models 1–7) in European badgers of all ages. Variance components: VA, additive genetic; VPE, permanent environment; VPLATE, plate; VROW, row; VCO, cohort; VYEAR, year; VSG, social group; VMAT, maternal; and VPAT, paternal (from Lieshout et al. 2021, with permission).

cubs born in cooler, drier springs had shorter early-life RLTL (paralleling the association between lower spring productivity and worsened coccidial infection amongst cubs (Chapter 2) and of course likely availability of earthworms).

At this point we wondered about the social circumstances of the cubs, and whether, weather aside, this had any bearing on their telomere profiles. At first it seemed that the answer was no. After controlling for weather effects, there was no evidence for any association between the number of adults, same-sex group size, or total number of individuals in the natal group with early-life telomere length. However, further statistical excavations then revealed a marginal effect, strong enough to shine through even when all the weather variables were controlled, such that the more cubs there were in a natal group, the *longer* their early-life RLTL. One might have thought that more cubs means more competition—after all, amongst

birds the feeding rate of chicks tends to decrease with clutch size in the nest (e.g. Vedder et al. 2017). But we already knew that amongst Wytham's badgers, suckling rate per offspring increases with litter size (Dugdale et al. 2007), so it seems that a large cohort of cubs may be the hallmark of bountiful conditions and a luxurious youth. And what lies behind these prosperous conditions? Well, it seems like it's the weather again. Summarizing all these findings about early-life telomere dynamics, it seems as if the weather plays a major role: pump-priming that component of life expectancy attributable to telomeres and thus bequeathing individuals different lifetime prospects straight out of the gate.

In sum, we found considerable variation in telomere length amongst the badgers of Wytham Woods, but little or none of it was explained by either genetic variance or parental (or maternal) age at conception (Figure 18.19). However, for the meline fortune-teller,

---

Box 18.3 Does he have his father's . . . 'telomeres'?

In 1947 Albert Lansing published his observation (*The Lansing Effect*) that (simply stated[30]) the offspring of old parents tend to have shorter lifespans compared with the offspring of young parents, and in both cases these tendencies are transmitted to successive generations (Lansing 1947). As is so often the case, it turns out to be not quite that simple; although Noguera et al. (2018) recently reported that for the zebra finch (*Taeniopygia guttata*), at least, embryos fathered by old males have shorter telomeres than those produced by the same mothers but with younger fathers. This takes us to the issue of heritability: how much of the variation in offspring telomere length is due to genetic effects (and what can be attributed to other factors), and in this case examples point in opposite directions: Asghar et al. (2015) found high RLTL heritability in great reed warblers (*Acrocephalus arundinaceus*; 0.480 ± 0.120 SE) whereas there was no such effect for white-throated dippers (*Cinclus cinclus*; 0.007 ± 0.013 SE; Becker et al. 2015). These avian examples have caused theorists to consider the possibility of increased likelihood of inheriting adverse germ-line mutations with advancing parental age, or of changes in the genome stability of germ cells or of changes in the epigenome with age (see Marasco et al. 2019). So much for avian inspiration, what about badgers? Sil led us on tackling both these questions, which are conceptually different: the parental age phenomenon is a correlative effect, whereas heritability is all about variances.

Sil's explorations returned to those 471 RLTL measurements from 240 offspring (121 females and 119 males; with 108 unique fathers and 120 unique mothers) for whom parentage was known (van Lieshout et al. 2021b). Was telomere length heritable, as in the reed warblers, or not, as in the dippers? It turns out that it was not (Figures 18.18 and 18.19).

We could detect neither heritability[31] of telomere length from either parent, nor any longitudinal parental age (Lansing) effects on offspring telomere length at any age. But while parentage explained only 2% of variation in telomere length, a much more strident result emerged from 'year' of birth. Of course, the Gregorian calendar year itself is of no consequence to badgers; it serves, however, as a proxy for things that we would by now expect to matter a lot: prevailing environmental and social conditions, which explained a considerable amount, 25%, of the phenotypic variation of telomere length (these weather and social effects are disentangled in the main body of this chapter, Figures 18.16, 18.17, and 18.18; van Lieshout et al. 2021a).

---

[30] A more precise statement of the Lansing effect is that isogenic (similar genotype) lines derived from young parents tend to persist for more generations than lines derived from old parents (King 1983).

[31] Heritability is the proportion of phenotypic variance explained by additive genetic variance.

Box 18.3 *Continued*

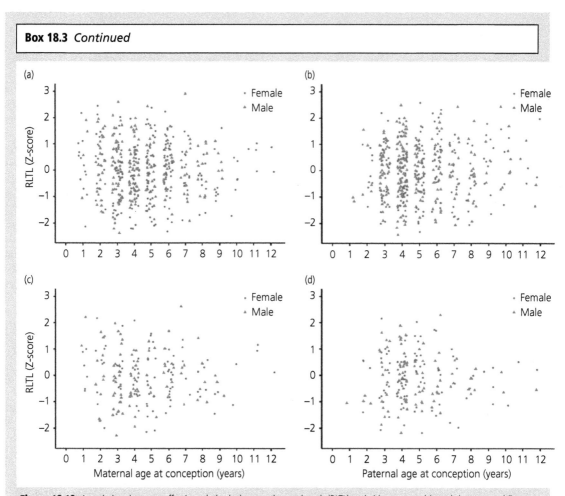

Figure 18.19 Associations between offspring relative leukocyte telomere length (RLTL) and either maternal (a and c) or paternal (b and d) age at conception (years) in European badgers. Scatterplots show raw data (blue for females and brown for males) for all ages (a and b; *n* = 417 measurements; 240 badgers) or only offspring measured as cubs (<1 year; c and d; 194 measurements; 194 badgers), and jittered for clarity (from van Lieshout et al. 2020a, with permission).

Box 18.4 Badger burial

There are various flamboyant folkloric accounts of badgers 'burying their dead'. For example Vesey-Fitzgerald (1942; also Hampton 1947) recounts badgers being dragged long distances and taken down burrows deliberately. Tim Roper (1994) interprets this evocative account prosaically as an extension of carrion caching behaviour (although if badgers cache food, it is rarely recorded; Macdonald 1976a). Undeniably, sett excavations can expose what might fancifully be construed as the revered ossuaries. More likely, sick or injured individuals limp off to the security of their underground chamber and die therein. We imagine that other badgers avoid such tombs—which, being rife with decay, unpleasant odours, and pathogens, may ultimately cave in to form a mausoleum, until some future descendant unwittingly extends the sett and exhumes its ancestor's bones. Badger skulls are tough and preserve well, and we find them with surprising regularity adorning excavated spoil heaps.

telomere length was strongly related to survival (the striking result of our heritability analyses). Further statistical untangling partitioned this into a strong effect of weather and small effect of sociology. Further variation in telomere length, which includes not only the anticipated shortening but also the rejuvenating elongation is explained, likewise, by seasonal patterns. What is the source of this elixir of life? These changes cannot be explained by measurement error or leukocyte cell composition. So as badgers see the tantalizing prospect of eternal youth emerge in propitious springs, only to recede in the broken sleep of winter, they might be eager to know the provenance of these seasonal increases. Of course, we all have a personal interest in the answer, and John Muir might have been on the right lines when he wrote 'Take a course in good water and air; and in the eternal youth of Nature you may renew your own'.

The badger's epitaph

And so, our play is almost done and our badger leaves its seventh age. Its journey from 'womb to tomb' (Chapter 1) may have been long and productive until finally its telomeres unravelled and its cells ceased to replicate effectively, or cut short abruptly and prematurely by starvation or accident. While half of all individuals born die before their third birthday (Chapter 11), a few do make it to be old, with c.3%

achieving teenage seniority (Chapter 11; applying the 7:1 'dog years' rule,[32] think of this, in crude terms, as corresponding to a person aged 90). Indeed, our very oldest badgers, F699 and M32, surviving to be at least 15, might be considered the equivalent of human centenarians. There is a lot of luck involved in meline longevity. Accidents of birth—where, with whom and under what conditions—take a toll on early-life telomeres, linked to potential lifespan; moreover unfortunate exposure to juvenile coccidiosis during a harsh, dry birth year can beckon the Grim Reaper prematurely for any individual—even those with otherwise beneficial heterozygosity (Chapter 17). Bad luck remains a threat throughout life, but an element of quality shines through. This is not some Faustian deal with the devil. Whether blessed by genes conferring resistance to diseases or oxidative stress (Chapter 15), or hedging the right bets to optimize when to breed and when to hold back and preserve body condition (Chapter 14), by the time some 13% of badgers surmount age 9 (notionally approximating human retirement age, according to our dubious conversion), we do see those that 'have', and those that 'have not'[33] succumbed to Shakespeare's seventh age: 'Sans teeth, sans eyes, sans taste, sans everything'.

Alas, poor Little B! I knew him well; a fellow of infinite jest, of most excellent fancy.

—*Hamlet*, Act 5, Scene 1

---

[32] This mythology of the '7-year rule' popped up in the 1950s, proposing that a dog aged 10 mirrors the sort of geriatric decline typical of a person aged 70; however, it is in truth not that simple, where a 1 year old dog has the maturity of a 15-year-old human. The origins of the '7-year myth' are actually somewhat mysterious. A Judgement Day calculation inscribed in 1268 at Westminster Abbey declared a ratio of 9 to 1—that is, humans live to age 81, and dogs to age 9. Georges Buffon, the eighteenth-century French naturalist, had more or less the same theory: humans can live to 90 or 100 years, and dogs 10 or 12.

[33] As the down-to-earth Sancho Panza, in a conversation with the idealistic Don Quixote, said, 'As a grandmother of mine used to say, there are only two families in the world, the haves and the have-nots'.

CHAPTER 19

Of the Same Stripe, or Not—Exceptions That Prove Rules

I have called this principle, by which each slight variation, if useful, is preserved, by the term of Natural Selection.

Charles Darwin

As our interpretative hot-air balloon soars higher, beyond the Petri dish that is Wytham Woods, the detailed machinations of the prolific F217, F221, M10, M4, and even ancient No Ears and embattled Little B become but one piece of the jigsaw puzzle of badger populations throughout the UK, and eastwards across the Bosphorus and below the Caucasus where their spatially nearest relative, *Meles canescens*, hoves into view; then beyond the Volga River to intersect with their phylogenetically closest cousin the Asian or (from the Greek) white-tailed badger, *M. leucurus*. That is the perspective of this chapter, asking first what intra-specific variation exists within this mosaic of badgerdom, and thereafter why? Second, soaring higher, what can our observations of European badgers reveal about broader ecological paradigms playing out amongst their meline and wider musteloid kin, and yet more broadly amongst the Carnivora, and beyond. This latter question, about patterns permeating the behavioural ecology of carnivores, could sensibly be tackled from the vantage point of any of the order's nine terrestrial families or 256 or so species, but ours is a book about badgers, so badgers are the hilltop from which we view the panorama of carnivore evolution—specifically, we scan that panorama from the hilltop of Wytham Woods.

We prepare for this interpretative journey, using badgers as models of behaviour and evolution, torn by mixed emotions: energized by the unfolding understanding of their kith and kin, and the natural processes that bind them, to which the badgers of Wytham Woods have led us—optimistic that we have made a useful start—but desolate at the peril, made more vivid by that understanding, that these marvellous species face due to humanity's influence on many of them, their habitats, and their ecosystems. Wytham may remain comfortable, but the view over the horizon threatens a cataclysm of anthropogenic extinction. The earth is shrouded in too much carbon—Warren et al. (2018) reckon it could lead to the extinction of up to half the animals and plants in some of the world's most biodiverse areas by 2100—a date so soon that in the sanctuary of some library shelves, this book might survive much of the biodiversity that intrigued and delighted its authors. According to Wiens (2016) 47% of 976 species assessed, across habitats and climatic zones, are already experiencing insidious local demise attributable to climate change. Other destructive forces already rampage ahead of climate so that many carnivores, and many mammals, will be gone from the ring before climate change lands its knock-out punch (Macdonald and Kays 2005; Ripple et al. 2017; Macdonald 2019).

This sounds bleak—and it is—but Nature will not go gentle into that far-from-good night: remember also van Valen's (1977) Red Queen whose never-ending marathon has traversed many chapters of this book. Her endless cat-and-mouse game[1] will continue between a hostile world and those organisms struggling to survive in it. This Darwinian process of evolution relies upon adaptation: those individuals

---

[1] A metaphor that aptly reminds us of the analyses of Sandom et al. (2017a, 2017b) analyses of how felid conservation depends on conservation of their prey—see also Ripple et al. 2015).

The Badgers of Wytham Woods. David W. Macdonald and Chris Newman, Oxford University Press.
© David W. Macdonald and Chris Newman (2022). DOI: 10.1093/oso/9780192845368.003.0019

and populations able to cope with transformed conditions, or lucky enough to miss the worst of them. Those less fortunate will succumb (Scheffers et al. 2016). Within this adaptive continuum, populations under one extreme of environmental circumstances may adapt one way, while their fellows elsewhere are selected for different traits. Some individuals, and ultimately thus the populations and species they comprise, will make it, others won't: it has always been so, and these workings of nature have always fascinated naturalists and scholars, before the weight of responsibility for these processes came to rest so shoddily on humanity's shoulders. Scholars can remain fascinated by the process, while practitioners grow evermore appalled at the outcomes, and we may hope that both the fascination and the understanding will both help to forestall tragedy. Therefore, as a roadmap to the journey ahead, navigating to our concluding synthesis of intra- and inter-specific variation, and the solutions this provides to some of the puzzles accumulated in previous chapters, we will, in preparation, make a brief pit stop for a reminder of how adaptation works.

How does adaptation work?

Adaptation comes in two fundamental types, behavioural and physical, and several different flavours. It arises from the never-ending cat-and-mouse game between a hostile world and those organisms struggling to survive in it—jogging along with the Red Queen who ran tirelessly through Chapters NN, 14, and 18 (van Valen 1977). Recent rounds of this game have been played under a changing climate—a circumstance threatening biodiversity globally but which has been to the recent advantage of the *Lumbricus*-eating badgers of Wytham Woods (Chapter 12) (Scheffers et al. 2016).[2]

Badgers, and carnivores in general, are a bit like chameleons insofar as each one can express itself in a variety (Chapter 14), but not an infinity, of hues: there are limits to variation—between individuals, populations, and species—and in this chapter we will explore the different extents to these limits, and what sets them. One constraint, introduced in our opening pages, is the legacy we have called phylogenetic baggage—that is, the influence of ancestral inheritance. This restriction of options, or predispositions towards solutions, is technically termed phylogenetic inertia, and refers to a limit on future evolutionary pathways as a consequence of constraints imposed by previous adaptations in the ancestral lineage (Blomberg and Garland 2002). This reality was already apparent to Charles Darwin in the *Laws of Conditions of Existence* (1859), in which he noted that organisms do not start with a clean slate, but rather they are equipped with the characteristics of their precursors (Shanahan 2011). It was, however, Huber who, in 1939, formalized the modern appreciation of this phenomenon, which Richard Dawkins, in his 1982 book, *The Extended Phenotype*, explained by likening natural selection to the flowing of a river. Reptiles lay eggs; birds are descended from reptiles, and thus are inextricably launched on this reproductive pathway, just as the primate's arm, horse's forelimb, bat's wing, and seal's flipper are homologous expressions of the five-fingered, pentadactyl, ancestral limb. Similarly, all species of badger carry in their evolutionary rucksacks their ancestors' powerful digging limbs, propensity for fossorial living, and a carnivorous dental plan. Importantly, much of that baggage may be stowed away unseen (we might guess, for example, that all *Meles* species share the immunological deficiency of up-regulating inducible nitric oxide synthase (iNOS) (Chapter 16), and thus are all prone to respiratory infections). Here, as we consider badgers, we keep in mind the heirlooms of their inheritance, what they can change, and to what extent; to which end results, also, their predispositions direct them.

Box 19.1 'Mechanisms of adaptation' is a brief primer on the various mechanisms delivering adaptation, relevant to interpreting the intra- and inter-specific variations presented in this chapter.

Reminded of how adaptation works, and bearing in mind the relationships amongst badger species that were explained in Chapter 1, we will now explore variation as we move, first, progressively further from Wytham (while remaining amongst European badgers); then, through inter-specific variation within the *Melinae*, and beyond.

[2] Carbon emissions risk rapidly elevating global temperatures by 4.5°C, resulting in up to half the animals and plants in some of the world's most biodiverse areas going extinct by 2100 (Warren et al. 2018). Even limits set by the Paris Climate Accord of 2°C would see the Amazon and the Galapagos lose one-quarter of their species diversity. As mentioned earlier, *c*.47% of 976 species studied, across habitats and climatic zones, are already experiencing attrition of local populations (Wiens 2016).

Box 19.1 Mechanisms of adaptation

Behavioural adaptation provides immediate response to the trials of life, and is daily seen in the fruits of human reliance on our wits (Chagnon 2017). You do it constantly, as you put on your winter woollies or summer sunglasses. Behavioural plasticity[3] enables some badgers to eat predominantly earthworms, and to live in a group several dozen strong, whereas others rarely eat an earthworm, and live principally alone. Behavioural adaptations, like ideas, can spread and persist for generations, as illustrated by the great tits who open milk-bottle tops or the matriarchal elephants who lead the way to a far-off water hole (in 1997 the badgers of Lower Seeds group adopted the habit of tipping over our cage-traps to access the peanuts without entering the trap—this local culture was, happily for us, lost from their collective memory in subsequent years). However, behavioural adaptation has limits, and these limits are different for different species: it can enable both badgers and wolves to live solitarily or in large groups; unlike the behaviour emblematic of wolves, however, the adaptability of badgers has never been known to stretch to allowing them to hunt cooperatively.

In contrast, *physical adaptation* involves permanent structures, encoded in DNA (genotypic adaptation) like the fish's gill or the bird's wing, while also including phenotypic adaptive responses such as the tanned skin or callused hands that stem, respectively, from life outdoors or hard manual labour. There are no studies of badger phenotypic adaptation, but considering their huge geographical and topographical extent, there may be lessons to learn from altitude sickness in humans, as a generalized example of the spectrum from temporary to permanent phenotypic adaptation.

At around 2500 m (8200 ft) the thinning air causes most of us to experience some degree of hypoxia, resulting in headache, nausea, and even cerebral oedema. WildCRU's Geraldine Werhahn recently discovered that Himalayan wolves gain a competitive edge through a genetic adaptation to cope with the hypoxic stresses in high-altitude habitats, associated with their divergence from the Holarctic grey wolf complex 691,000–740,000 years before present (Werhahn et al. 2018, 2020). This involves three genes, *EPAS1*, *ANGPT1*, and *RYR2*, linked to hypoxia adaptation— these genes are also differentiated in high-altitude human populations.

For people living at 4000 m (13,000 ft) every lungful of air has only 60% of the oxygen molecules present at sea level, a circumstance particularly dangerous for pregnancy due to elevating maternal blood pressure (one wonders how pregnant Himalayan wolves are impacted). How do mountain peoples adapt to thin air? Although they all do

so through a mixture of adaptive physiological processes that include elevated resting ventilation, hypoxic ventilatory response, oxygen saturation, and haemoglobin concentration, the interesting pattern is that populations in different regions have different suites of adaptations, some more permanently encoded than others. The physiological details are riveting (Beall et al. 2002; Beall 2007), but the key point here is that Andean people, Tibetan highlanders, and Ethiopian highlanders all do it differently. So human evolution reveals three divergent adaptive strategies to address the same environmental stressor: altitude hypoxia (i.e. phenotypes involving a range of distinct genetic mutations; Bigham 2016). We should expect similar complexity when it comes to intra-specific variation in how European badgers have adapted to their varied circumstances across the 10 million or so square kilometres of their range, and in the inter-specific comparisons we make later in this chapter.

The straightforward distinction between phenotypic plasticity and genotypic adaptations became much less straightforward with the discovery of *epigenetics*, the phenomenon whereby *environmental* factors can cause genes to be switched on or off, to express adaptations. Although the term epigenetics is credited to Waddington in 1942, modern appreciation of the subject is only a few decades old— it wasn't even mentioned when we were undergraduates. It arises through organic compounds (methyl or phosphate groups) attaching to DNA, or modifying the proteins around which DNA is wound (as described in Chapter 15). Although these changes in whether genes are expressed or not may be passed on to the next generation, they are, importantly, also reversible, should the causal environmental stressor be alleviated—a type of phenotypic plasticity akin to but less pliable in the short term than behavioural adaptive traits. Through this mechanism individuals can rapidly adapt the expression of their current genotype in response to environmental change. This has the effects of buying time for natural selection to work on advantageous mutations to evolve the genotype, or allowing phenotypic flexibility in changeable environments where a more permanent genotypic change might be premature. Neither effect implies that natural selection has any sort of foresight—as Richard Dawkins pithily put it in *The Ancestor's Tale*, 'Evolution, or its driving engine natural selection, has no foresight.

---

[3] Behavioural plasticity allows animals to adjust their behaviour in response to complex environmental conditions and, as expressed by Mery and Burns (2010), is the result of 'a complex interaction between evolutionary pre-programmed cue-response behaviour (innate behavioural response) and cumulated lifetime experience (learning)'.

> **Box 19.1** *Continued*

In every generation within every species, the individuals best equipped to survive and reproduce contribute more than their fair share of genes to the next generation. The consequence, blind as it is, is the nearest approach to foresight that nature permits.' The epigenetic option can be especially advantageous for individuals of longer-lived species (where inter-generational genetic selection is slow relative to the pace of environmental change) and generalist species (where individuals may experience a wide variety of conditions), which, as a result, usually exhibit more potential for heritable plasticity (Murren et al. 2015).

A familiar example of epigenetics is the effect of temperature on the sex ratio of hatchling crocodiles and turtles (i.e. those reptiles that lack sex chromosomes): on the Great Barrier Reef this once adaptive epigenetic effect of 'thermal sex determination' has been driven askew by climate change with the disastrous result that on the warmest beaches female hatchling turtles now outnumber males 116:1 (see Jensen et al. 2018).

In 1959 Russian geneticist Dmitri Belyaev began an experiment, which continues to this day at the Institute of Cytology and Genetics at Novosibirsk, selecting silver foxes (melanistic red foxes, *Vulpes vulpes*, farmed for their coats) for those tractable behavioural traits inherent in domestic dogs (Driscoll and Macdonald 2010). Belyaev, and his colleague Lyudmila Trut, selected from each generation of captive-bred foxes the 10% that exhibited the least aggression and fearfulness towards people, and within three generations they had created highly tractable 'pet-able' individuals. Notably, this occurred with only minimal contact with people, with foxes still living in fur farm-type cages, and so could not be attributed to habituation. By the fourth generation pups wagged their tails, whined eagerly to solicit human contact, and licked researchers. Furthermore, these domesticated foxes exhibited floppier drooping ears, and curlier tails and two oestrous periods per year—traits seen in dogs and other domesticated species. Their legs, tail, and snout all shortened, their skull widened, and their coat became fluffier and more mottled. Interestingly, they achieved sexual maturity around a month earlier than wild-type foxes.[4]

Why the change in behaviour? Well, domesticated foxes had far higher levels of serotonin than did their ancestors—the 'affection hormone', mediating aggressive behaviour, and especially important in shaping an animal's early development (Matsunaga et al. 2013). But what genetics might

underscore a process of domestication reminiscent of how wild wolves became our domestic companions 25,000 years ago in the Palaeolithic? Pörtl and Jung (2017) advance the active social domestication (ASD) hypothesis, in which conventional genetic selection alone is not a sufficient explanation. They propose an epigenetic 'switch' causing immediate changes in the hypothalamic–pituitary–adrenal axis and the 5-hydroxytryptamine (5HT) system, a precursor in serotonin synthesis. Prosocial behaviour causes an epigenetic enhancement of hippocampal glucocorticoid receptor (hGCR) genes, increasing serotonin and simultaneously inducing demethylation of the GRexon1;7promotorbloc, leading to decreased cortisol production. Lowered cortisol promotes social learning capability, and greater prefrontal cortex activity, contributing to better executive functioning, making domesticated wolves more amenable to human-directed behaviour.

Of course, writing as domesticated apes, humans have much in common with Belyaev's foxes. Indeed, Heranado-Herraez et al. (2013) found, amongst all great apes, 170 genes that have a methylation pattern unique to humans that might help to explain our reduced aggression and enhanced sociability. Indeed, the vulnerable ape hypothesis has it that our ancestors became more sociable before they became smart, rather than the other way around (Macdonald et al. 2019 advance a similar argument as one benefit arising from monogamy for members of the dog family). But why is this information useful in preparing us to interpret the badgers of Wytham Woods? Given that chronic stress responses are conserved across vertebrates (Cunliffe 2016), it seems highly plausible that ultra-high population density in Wytham (Chapter 12) caused allostatic overload (Chapter 18), so we wonder whether epigenetic switching in response to psycho-social crowding stress has contributed to atypical tolerance amongst Wytham's badgers.

---

[4] Whether all these traits were linked pleiotropically as part of a domestication syndrome has been called into doubt by Lord et al. (2020) because the Russian foxes originated from Canadian fur farms established in the early 1800s, where the stock had already been selected for friendliness and several others of the altered traits before some were shipped to Russia. There is therefore likely a strong founder effect, and genetic drift associated with a small original population. For our purposes Belyaev's foxes nonetheless illustrate the power of behavioural genetics and selection operating at a speed that may be relevant to Wytham's badgers. For perspective, selection on the original Canadian foxes began at roughly the time people began selecting for behavioural traits in different breeds of dogs.

Box 19.1 *Continued*

Epigenetics is such a new branch of science that readers may raise an eyebrow at the thought of such rapid adaptation. Sceptics should reflect on the remarkable case of guinea pigs, which, in the wild, usually mate at temperatures below 5°C. In an experiment, some males were allowed to breed at this temperature before being transferred to a tropical 30°C for 2 months, and then allowed to breed again: in the meantime, these males had developed significant methylation of 10 genes linked to regulating their body temperature. What of their offspring? Well, the offspring born to males that mated at 5°C were normal with respect to the genetic control of their temperature, whereas the offspring conceived when their fathers were hot inherited methylation, but it was expressed in a different pattern to the 10 genes epigenetically altered in their fathers. Amazingly, the epigenetic 'preparedness' of the father to a warmed climate was passed on via modified sperm, and instigated a further batch of epigenetic adaptation in their offspring (see Weyrich et al. 2016). Anyway, as we explore intra- and inter-specific variation amongst badgers and their carnivore kin, bear in mind that evolutionary changes can happen quickly on ecological rather than just palaeontological time scales (Schoener 2011). Anticipating what lies ahead, have in mind that different species in the badger's wider family have adopted different solutions to the same genre of ecological problem. Thus, to meet the needs of defence against predators, skunks took a morphological route, developing the ability to spray predators with repellent and colouration to warn them of this ability, whereas sea otters took the purely behavioural route, forming groups where more eyes can look out for orcas and sharks. Similarly, faced with seasonal food shortages some mustelid species opt for torpor, and others use dietary switching to alternative foods.

Intra-specific variation in badger behaviour

What socio-ecological variation exists amongst European badgers in the UK and beyond? As we turn to the broader canvas, an estimated 1.5 million European badgers live in Europe, with over half a million in the UK representing the largest national share[5] (Mathews et al. 2018). Across this stage their population densities vary substantially from the Wytham peak of 50 badgers km^{-2} across three orders of magnitude to records of a thousandfold less at below 0.05 badgers km^{-2} in Lithuania, Estonia, and Poland (see Balestrieri et al. 2016). But let us begin our exploration of the intra-specific warp and weft of badger society in the UK (Box 19.2), with a summary from Hans Kruuk's 1989 book (see also Kruuk and Parish 1982). In brief, Hans studied European badgers in three areas in Scotland and one in England (Wytham), to explore variation in their population densities, territory sizes, group sizes, and the biomass and distribution of their main prey species, earthworms. Territory size was not correlated with earthworm biomass (per unit area) but increased with the distance required to reach areas in which

earthworms were available to badgers (i.e. with patch dispersion, in RDH terms). The number of animals per group increased with the biomass of earthworms in the territory, and with the biomass of worms per feeding area (i.e. possible measures of patch richness). There was no correlation between territory size and group size, but the overall badger density increased strongly with earthworm biomass. Hans Kruuk concluded that it is likely that badger population density is adjusted to food availability through a mechanism regulating group size, in the absence of other limiting factors such as lack of suitable sett sites or predation or persecution. Of the latter, the overlapping spatial organization of the badgers of North Nibley where persecution was high (see Chapter 16) clearly documents the impact of persecution. To this variation we can add urban badgers. Seventeen badgers from six social groups in a 1 km^2 area tracked by Davison et al. (2009) in an urban study in Brighton between 2005 and 2007 foraged mainly in gardens; travelling through scrub and allotment in the tiniest ranges so far recorded for badgers: 4.91 ha for individuals and 9.26 ha for groups (100% minimum convex polygon)—males had significantly larger core areas than did females. Most scent marking was in the vicinity of setts, and there was no evidence of territorial scent marking.

---

[5] Estimates for Sweden are *c.*350,000, Germany *c.*142,000, and European Russia 30,000.

Box 19.2 Badgers in Britain

In the years following Kruuk's research, badger numbers increased not just in Wytham, but across the British Isles—a generality partly explained by our findings regarding climate change in Chapter 12, alongside better protection and less persecution. Between 1988 and 1997 Wilson et al. (1997) deduced a 77% increase in UK badger numbers (using setts as a proxy for badger social groups). Subsequently, Judge et al. (2014, 2017) estimated 71,600 (66,400–76,900) social groups in England and Wales (2011–2013), yielding a population estimate between 391,000 and 581,000 badgers across England and Wales (density = 0.485 badgers km^{-2}, 95% confidence interval (CI) = 0.449–0.521); an 88% (70–105%) increase from Wilson et al. (1997) and in the same ball park as the Mathews et al. (2018) national estimate of 760,000 (95% CI = 528,000–1,370,000).

Rainey et al. (2009) reported between 7300 and 11,200 badger main setts across the Scottish mainland between 2006 and 2009, most in the Borders and Lothian regions, few in the Highlands and none on the isles (they are also absent from the Isle of Man, the Isles of Scilly and the Channel Islands, although they thrive on the Isle of Wight, where they were probably introduced). In Northern Ireland, Reid et al. (2012) estimated 7600 (range 6200–9000) groups, equating to 34,100 (range 26,200–42,000) individuals. Beyond the UK, in Eire, Byrne et al. (2015b) estimated 19,200 (range 12,200–27,900) badger social groups, and Sleeman et al. (2009) estimate 84,000 badgers.

Animal numbers are the outcome of the patterns in which resources and death impact their populations, and the outcomes for European badgers stretch over more than three orders of magnitude in population density, with Wytham's record of 50 badgers km^{-2} (Chapter 11) double the score of its *proxime accessit* at Woodchester Park (Rogers et al. 1997), and with no other continental report even making double figures. Considering food, shelter, and water as candidate resources, Johnson et al. (2002b) attributed much intra-specific variation in badger numbers to variations in the food available to them.

How, then, does badger diet vary across landscapes? Viewed through the lens of Wytham Woods the pivot is the earthworm, but it is not everywhere thus. Indeed at high (37–40°N) and low (Mediterranean) latitudes, badgers consume almost none: a latitudinal band of earthworm consumption increases to 40–70% of biomass consumed between 55 and 63°N (Goszczyński et al. 2000; Revilla and Palomares 2002; Fischer et al. 2005). In a review of 69 studies of diets of badgers from the former Soviet Union, Roper and Mickevicius (1995) found that earthworms never exceeded 5% of the diet by volume (insects were 30% by volume, small mammals 20%). Rosalino et al. (2005b) reviewed the entwining of habitats and foods across Europe, and found the pattern is patchy: comparing study sites in two adjoining eastern European countries, earthworms were not significant in badger diet in Belarus (Sidorovich et al. 2011), while in Poland they comprised 82–89% of biomass consumed in spring (56% in summer and autumn in pristine forest, 24% in a mosaic habitat of forests, fields, and orchards).

Whether or not earthworms, and specifically *Lumbricus*, are on the local menu, listing the diversity of badger foods leaves you feeling rather like a breathless rapper: preferred insects include chafers, dung beetles, ground beetles, caterpillars, tipulid larvae, leatherjackets, ants, bees, and stinging wasps (Cleary et al. 2009). Of fruits: windfall apples, pears, plums, blackberries, bilberries, raspberries, strawberries, and elderberries (badgers can be a pest to fruit cultivation). And nuts include sweet chestnuts, horse chestnuts, acorns, hazelnuts, beechmast, pignuts, and wild arum corms. Add a muesli of maize, wheat, oats, and more rarely barley (badgers can damage cereal production). Not forgetting carnivory and the delights of small vertebrates such as frogs and small reptiles, ground-nesting birds and their eggs (Hounsome and Delahay 2005), scavenged afterbirths (Symes 1989), and rodents and rabbits, *Oryctolagus cuniculus*—generally dug from their burrows in true Oligocene style (see Chapter 1). The latter are a mainstay for badgers in xeric Mediterranean regions (Martín et al. 1995), where they are garnished with olives, *Olea europaea* (Kruuk 1981), and cultivated fig, *Ficus carica* (Barea-Azcón et al. 2010). Hedgehogs, *Erinaceus europaeus*, despite their spines, are easily unrolled and consumed (remember that forelimb dexterity). Patrick Doncaster, a founder member of WildCRU's ancestor, the Oxford Foxlot, and now a Professor in Southampton University, was the first to demonstrate that where badgers abound hedgehogs are absent (Doncaster 1992; Pettett et al. 2017b; Williams et al. 2018). David and Patrick recruited doctoral student Jane Ward to explore the mechanism. Patrick, in those days implausibly tall and thin, added the invention of stink bombs made of liquefied badger faeces to his existing notorieties (which included adapting a London taxi to radio-track urban foxes, and breaking his back falling off a cliff in Chile while watching marine otters). David remembers the stealth and aim required as Jane and he stalked radio-tagged hedgehogs and, from a discreet distance, lobbed the faecal stink bombs ahead of them: the hedgehogs beat a hasty retreat from badger odour (while remaining entirely sanguine about the explosion of water-laden bombs, as controls, and pressing on undaunted). In enclosure trials, the hedgehogs almost always avoided feeding at sites tainted with badger faeces and continued this avoidance for 2 days, but not after 4. The wild hedgehogs, scuttling away from lobbed badger perfume, their thoughts doubtless preoccupied by a landscape of fear, reduced their foraging effort over the ensuing 5–30 min (Ward et al. 1996). However, they got over these anxieties sufficiently to revisit the sites after 24 h: this, in contrast to the behaviour of the hedgehogs in the enclosures, which had been given alternative food; the difference being that the wild hedgehogs probably had no option but to return to their favoured foraging sites (Ward et al. 2000).

The abundance of each badger food has its particular biogeographical distribution within the constraints of bioclimatic zones. In the north, unproductive winters (Macdonald and Newman 2002) limit badger distribution (Bevanger and Lindström 1995; Silva et al. 2017) but this limit is being loosened by climate change (as it has been for the northward and uphill spread of red foxes, to the detriment of Arctic foxes; Hersteinnson and Macdonald 1992). Badgers crept northwards in Finland as the late twentieth century warmed (Kauhala 1995). Likely, they are making their way up the Italian Alps (Balestrieri et al. 2009), much as their occupancy in the Scottish Highlands was higher in warmer, lower-elevation sites than in higher, cooler ones (Silva et al. 2017; Chapters 12 and 13). Conversely, to the south, warmer, drier conditions restrict the European badgers' southern distribution (Virgós and Casanovas 1999).

A further filter between the badger and access to abundant food is competition and fear of predation (linked, in the text that follows, to the risk of mortality). Superimposed on the distributions of their foods are those of species that would eat those foods first, and/or eat the badgers. For example, in Poland's Białowieża Forest, badgers used setts in areas most frequented by wolves c.60% less often than those in areas where perceived lupine risk was lowest (and it was only in those safer areas that cubs were found) (Diserens et al. 2021).

The topic of badger fearfulness preoccupied our sabbatical visitors, renowned Canadian field experimentalists, Mike Clinchy and Lianne Zanette. The badgers of Wytham Woods no longer need fear native

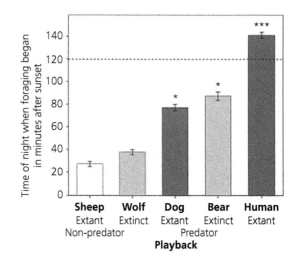

Figure 19.1 Effects of playbacks on the time of night when badgers began foraging at the first food patch, measured in minutes after sunset, comparing extant predators (dog, human) with extinct predators (wolf, bear) and the non-predator control (sheep). The horizontal dashed line indicates 2 h after sunset, when the playbacks turned off. Asterisks signify significant differences (*$P < 0.05$; ***$P < 0.001$) in Dunnett's tests comparing treatments and the control (sheep). Values are means +/− standard error (from Clinchy et al. 2016, with permission).

large carnivores (bears, wolves, or lynx),[6] but might have cause to fear humans more. Led by Mike and Lianne, we tested (Figure 19.1) the fearfulness badgers demonstrated to audio playbacks of extant (dog) and extinct (bear and wolf) large carnivores, and humans, by assaying the suppression of foraging behaviour. Hearing human voices (of the BBC Radio 4 ilk) affected latency to feed, vigilance, foraging time, number of feeding visits, and number of badgers feeding. Hearing dogs (a barking Labrador) and bears, by contrast, had a far lesser effect on latency to feed, and hearing wolves had no effect—humans, it turns out, are far more frightening.

Badgers in the UK interact with a guild of mesopredators, and a factor that varies across their geo-

[6] Vadim Sidorovich's studies in low earthworm density, and hence low badger density, habitats in Belarus, suggest that lynx and, especially, wolf, are important predators of badgers (e.g. Rotenko and Sidorovich 2017)—this state of affairs, gone with the passing of larger predators from the UK and western Europe, should be at the forefront of our mind when thinking about the circumstances under which contemporary European badgers evolved. As an aside in the context of predation, both Indochinese leopards in Thailand and clouded leopards in Laos, preyed heavily on hog badgers, (Grassman 1999; Rasphone et al. 2022).

graphical range is the combination of fellow guild members and predators. For example, in Białowieża Primeval Forest (eastern Poland) in winter, 88% of badger setts were occupied by both badgers and raccoon dogs, 4% by badgers and red foxes, and 4% by all three species. In summer, only 10% of badger setts were cohabited by raccoon dogs and 10% by foxes (Kowalczyk et al. 2008). When wintering in the same sett, badgers and raccoon dogs used different parts of the sett (although their combined body heat might nonetheless add to the underground central heating). In a fascinating investigation of carnivore community ecology led by our colleague Vadim Siorovich in Belarus the spread of invasive raccoon dogs appears to have disadvantaged badgers due to competition over carrion during a winter period of scarcity, alongside sett cohabitation in which raccoon dogs may fatally block torpid badgers into their sleeping chambers (the implication being that they suffocate); as their numbers have fallen, male badgers have travelled further from the safety of setts in the quest for females, and thereby been exposed to heavy predation by wolves and lynx, with the outcome that in northern Belarus only 12.9% of adult badgers are male (Rotenko and Sidorovich 2017). Indeed, Sidorovich et al. (2011) conclude that in the low-density circumstances typical of Belarus, with few earthworms, badger numbers are limited top-down by wolves. These low-density populations, affected by natural predation, may give important insight into the significance of impregnable setts in badger evolution.

As a further variation on community structure, from camera-trapping in Italy, Mori et al. (2015) observed badgers sharing their setts with a total of eight mammal species including red fox, pine marten, and stone marten (along with crested porcupine, coypu, eastern cottontail, brown rat, and wood mouse). In the Stara Planina Mountains of Bulgaria, camera-trapping revealed interactions between European badgers, golden jackal, European wildcat, stone marten, and red fox (Tsunoda et al. 2020). The recurrence of red foxes on all these lists makes it interesting to ask how they interact with badgers. In answer, badgers are clearly dominant and displace feeding foxes (Macdonald et al. 2004a). David led an analysis of 135 encounters between at least 35 different badgers and at least 5 red foxes observed in Wytham through video surveillance at a feeding site, and in the vicinity of six badger setts. Badgers fed in longer bouts than foxes and were less visibly vigilant. The only instances of aggression involved a badger charging towards a fox and swiftly displacing it. At the feeding site the duration of fox vigilance increased significantly,

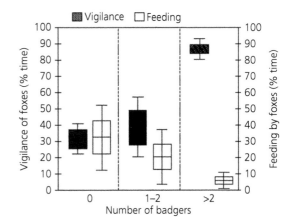

Figure 19.2 Relationship between number of European badgers present at the feeding site and percentage of time foxes spent feeding and being vigilant (from Macdonald et al. 2004a, with permission).

and the percentage of time that they spent feeding decreased significantly, with the number of badgers present (Figure 19.2). Similarly, recalling the exploits of diminutive but indomitable 'Little L' from Chapter 5, Chris would observe foxes skulking at the margins of his peanut-strewn driveway, sneaking to peripheral piles of nuts only when Little L was seemingly distracted. Unexpectedly, having met at the sett, the two species seemed strangely attracted to each other's company, and regardless of whether or not food was available at the time.

Mortality

When it comes to humans persecuting them, the European badger has been hunted by people for millennia in sufficient numbers to impact their populations and societies (see Chapter 16). They are hunted not only for meat (with the downside being the potential risk of infection with the nematode *Trichinella spiralis*), leather, and pelts, but also for bristles used for shaving brushes (Domingo-Roura et al. 2006); ditto its recently split-off congener the Asian badger, whose fat is also used for cooking, water-proofing clothes, and even to manufacture soap, along with Asian traditional medicine in the treatments of burns, scalds, gastric ulcers, and even haemorrhoids (Zhou et al. 2017). A taxon further east, the Japanese badger is heavily hunted during the peculiar sport of digging torpid individuals from their winter dens, although waning enthusiasm for this dismal pastime probably explains why annual tallies of 7000 in the 1970s fell to less than 2000 in the late 1980s, and to 1050 in 2012. More recently, many thousands of Japanese badgers have been trapped and skewered with traditional spears on the island of Kyushu, in retribution for them taking melons (Kaneko et al. 2017). Many different ethnic groups in Southeast Asia eat the similarly sized but distantly related greater hog badger—some even esteem it (Zhou et al. 2017); and it is even farmed for food in China (Duckworth et al. 2016)—which became relevant to discussion of the origins of the Covid pandemic (Xiao et al. 2021).

Variation in badger social organization

So, against this précis of man-made factors that affect European (and Asian congeneric) badger numbers, and the impacts of food resources, additional mortality (and probably fear) upon those numbers, our principal fascination is with the extent (and source) of variation in how they are organized. Generally, the answer varies from groups of as many as, or even more than, 30 individuals in Wytham to variations on a solitary lifestyle at low density (Rodríguez et al. 1996; Brøseth et al. 1997; Remonti et al. 2006; see Revilla and Palomares 2002; Johnson et al. 2002b). What drives this? It seems mostly tied to diet: while there is a loose, but non-linear association between population density and group size (Delahay et al. 2006a; see Roper 2010), larger groups are generally associated with a diet rich in earthworms (Chapter 10). In less gregarious badger societies they feed mainly on fruits, cereals, invertebrates, supplemented with small mammals, and amphibians. Low-density European badgers are variously described as quite solitary in central Italy (Pigozzi 1987), 'pair-living with dependent cubs sharing the parental territory' in Poland and something similar across much of their continental range (Rodriguez et al. 1996; Revilla and Palomares 1999, 2002; Kowalczyk et al. 2000, 2003b), and forming maternal–offspring groups until reaching sexual maturity in Switzerland (DoLinh San et al. 2007). Although its exact meaning may have had little consequence to some of the authors that have used it, the expression 'pair living' has great interest in the context of the argument we are developing here. Does it mean a male and a female overlapping somewhat, perhaps within the context of the semi-detached relationship between the intrasexually territorial sexes that approximates the ancestral mustoloid spatial arrangements? Or, does it mean something closer to the tight congruency of overlap that is characteristic of some behaviourally monogamous species (some populations of red foxes would

be good examples). Why this distinction is important to our argument will become clear in the text that follows, and it prompts us now to delve more carefully into the accounts that led to mention of 'pair-living' badgers.

The seven badgers radio-tracked by Pigozzi (1992a) in the Maremma Natural Park in the Apennines of central Italy lived solitarily, the territories of males were 4–5 times the expanse of the 30–40 ha non-overlapping territories of each female—that is, each male's territory, entirely separate from those of its male neighbours, overlapped at least a little those of several females. This would seem to be a somewhat spaced-out version of the ancestral mustoloid intra-sexual territorial system and semi-detached relationship between males and females (and, no doubt, during the breeding season each male would make sure its range overlapped those of the females sufficiently to secure matings).

In the Cota Doñana, a Mediterranean region of south-west Spain, Rodríguez et al. (1996) found badgers living at low population density, feeding on rabbit kittens, and forming small groups (gamekeepers dug up many setts and very rarely found more than two adults in occupation—although this mortality is in itself a confounding factor—see Chapter 16). Furthermore, it was rare for the researchers to catch more than one badger of the same sex in the same territory. The tentative conclusion of two adult badgers per group may be what has led some subsequent authors to write of badgers at low densities living in 'pairs', but the original paper does not allow the spatial arrangements of the male and female to be positioned decisively on the spectrum we define as varying from semi-detached (as per most mustelids) to congruent (as per modern red foxes). Our colleagues Eloy Revilla and Francisco Palomares (1999, 2002) added fascinating extra detail on the diet of these rabbit-hunting badgers of the Doñana National Park in Andalusia, Spain, but comment no further on the fine detail of their spatial arrangements.

In a different, but also low productivity, biome, the Białowieża Primeval Forest of Poland, Kowalczyk et al. (2000) found an average of 3.8 adult and young per sett and, in a subsequent study a mean of 2.4 adults per social group (Kowalczyk et al. 2003b). These badgers occupied ranges that were huge by badger standards, averaging 12.8 km^2. We sense some opacity in the exact details that have been interpreted as 'pair living', but there is no doubting that in many places badgers live at low densities and in small 'groups', which may be tantamount to a male and female overlapping, together with some youngsters—her offspring.

These maternal–offspring associations would seem to be the launchpad for larger badger groupings. Population densities, while still low, were a little higher in the Swiss Jura, where small social units of one to five adults, subadults, and cubs partly overlapped: the dispersion of arable soils and fields and the perturbation caused by heavy human mortality (Do Linh San et al. 2007) influenced this arrangement somewhat chaotically.

In this continental context, motivated to explore badgers at their ecological extremes, David established a collaboration with Margarida Santos Reis. The idea, having studied the highest-density population of European badgers, in Wytham, was to compare them with those at the lowest density, which is why Margarida led us to the montado cork oak woodlands of Portugal's Serra de Grândola at the south-western extreme of the species' distribution. Margarida and David recruited two industrious students, Miguel Rosalino and Filipa Loureiro—nowadays Miguel is Assistant Professor at the University of Lisbon and continues to work with us on our Scottish badgers (Silva et al. 2017). Between 1999 and 2000 Miguel and Filipa harvested 450 badger droppings, which revealed that fruits, mainly olives, pears, and figs, and adult and larval arthropods comprise 89% of the biomass they ate (Rosalino et al. 2005a). Obviously the availability of these foods varied seasonally—variation that led the team to deduce that food selection was driven by the pattern of olive availability. Filipa further investigated the responses of badgers to the temporal pattern of resource availability (Loureiro et al. 2009); comparison of diet with the phenology of the availability of food resources showed strong parallels for practically all of the principal items that were consumed. This tracking by badgers of primary resources with temporally differing peaks of availability and abundance appeared to involve trade-offs with energy and water requirements: acorns peaked simultaneously with olives, which, although not as abundant, had a higher fat content, whilst other minor fruits (e.g. loquats and figs) also seemed to be tracked and were important resources for badgers during summer when air temperatures were high. This broader perspective reveals that despite their earthworm specialism in Wytham, European badgers more broadly are generalist foragers with seasonal specialities.

Miguel found 0.36–0.48 badgers km^{-2}, one of the lower population densities recorded where badgers occur in western Europe, with individuals using seasonally stable home ranges that averaged 4.46 km^2 and were occupied by three or four adults plus three or four cubs of the year (Rosalino et al. 2004). More broadly, Miguel established that badgers lived in 35% of a nearly 7000 km^2 study area—with badger occurrence linked to middling rainfall (800–1000 mmyear^{-1}) for the region and mild mean temperatures (15–16°C), and associated with herbaceous fields and shrublands (5–10%), <15% *Eucalyptus* plantations, >50% of podzols and exposed rocks, >4 individuals km^{-2} of sheep/goat and no cattle are also more likely to host badgers (Rosalino et al. 2019).

Using 20 months of radio-tracking data from six badgers, Miguel revealed lower sett fidelity than that reported for the badgers of Wytham Woods—that is, only on 36% of occasions did they come back to sleep in the same sett, with females returning more often than males to the same sett (Rosalino et al. 2005b). Assuming that badger population density was related to home range sizes and that this in turn was influenced by food and water availability and the existence of substrate suitable for sett construction, Miguel led the team in exploring the relationship between these parameters (Rosalino et al. 2005c); revealing that in these Mediterranean montado woodlands of Serra de Grândola, the main factor limiting badger's density where there was sufficient water for them to occur at all is the availability of suitable sites for setts.

Intra-specific variation and RDH

So, in Chapter 10 we developed an ecological explanation for the group-living sociology of the badgers of Wytham Woods, the resource dispersion hypothesis (RDH). Does that hypothesis fit with what we have now presented for European badger sociology across the ecological diversity of their geographical range? We will take a first glance at that question now, before going on to ask whether the RDH provides a framework for understanding inter-specific variation in the behavioural ecology of other species of badger, and their evolutionary cousins that all descend from that common ancestor in the Oligocene grasslands.

As a brief reminder regarding the evolution of sociality, it is almost tautologously obvious that group living can evolve only when the fitness benefits (of membership) outweigh the costs (of sharing key resources, and contracting infections; Chapter 15) (e.g. Alexander 1974), and the common route in, for at least one sex and for at least some time, is to cohabit long term with parents (Emlen 1982; Waser 1988; Koenig et al. 1992). The RDH provides a mechanism to facilitate that cohabitation (Macdonald 1984a; Macdonald and

Johnson 2015a), an option for which individuals will opt if it pays (as described in a paper entitled 'The rewards of tolerance' by Macdonald and Carr 1989), and without invoking any benefits of cooperation (see also Rasmussen et al. 2008). The cardinal principle is that groups may develop where resources are dispersed such that the smallest economically defensible territory for a pair (or the basic breeding unit, termed the primary occupants) can also sustain additional animals (and where individuals of the species have the capacity to tolerate whatever lowered food security that entails). In Chapter 10 we detailed how the spatio-temporal heterogeneity, patchiness in space and time, of badger food, especially earthworms, creates in Wytham Woods exactly those conditions that facilitate cohabitation. Whether the option is advantageous for a given individual will depend on the options elsewhere (dispersal), and any added benefits of staying (which might be delivered by kin selection, reciprocity, or mutualism, through diverse opportunities for cooperation; Chapter 8).

Does the RDH plausibly explain the variation in European badger socio-ecology across their range? In short, yes, and much in the manner anticipated by Kruuk (1989). The phylogenetic starting point is the basic musteloid spatial blueprint of intra-sexual territoriality, with a tessellating mosaic of smaller female territories overlain by a similar mosaic of larger territories occupied by larger males, the combined pattern creating a polygynous geometry (Powell 1979) (this partial spatial dislocation between a polygynous male's territory somewhat overlapping those of several separated females we call, for short-hand, a semi-detached relationship). This makes clear why we were so keen to ascertain, a few pages earlier, where on the spectrum from semi-detached to congruent were the relationships of males and females in those low-density populations of European badgers. That foundation may be most closely approximated by Georgio Pigozzi's Apennine badgers. This foundational starting point, the norm in most mustelids (indeed most Carnivora), such as typified by American mink (Yamaguchi and Macdonald 2003), is probably rather uncommon amongst European badgers because their omnivorous diet tends towards spatio-temporal heterogeneity, and hence facilitates some infilling of the single-occupancy territories of the ancestrally solitary females (for a more typical contrasting mustelid diet consider martens; Zhou et al. 2011). As our comparisons extend from intra- to inter-specific, the pattern of that infilling, and resultant juxtaposition of male and female ranges will emerge as significant.

The notion of 'infilling' will be helpful as this synthesis unfolds: imagine each female territory as a beaker or measuring cup of diameter sufficient to supply food for a female and, seasonally, her young (this might be the world of a female mink or weasel, populated somewhat homogeneously by rodent prey). That beaker is always filled to the depth of one adult female (cohabiting partially with her share of a semi-detached, and often hefty—sexually dimorphic—male; Yamaguchi and Macdonald 2003), and seasonally it is filled to a greater depth with the hungry mouths of a crop of youngsters, before they disperse.[7] But if resource dispersion is such that foods are sufficiently patchy in their availability that the smallest economically defensible territory for that solo female to accommodate also one or two biding, non-dispersing daughters (secondary occupants in the vocabulary of RDH) in the long term, and if it pays all concerned for them to stay (which depends partly on their tolerance of the ensuing level of food [in]security), then the metaphorical beaker of territory occupation will always be filled the depth of two, three, or even more adults. It is in this sense of increasing the number of occupants resident in the minimum territory that we use the word infilling, and the RDH argument is that it is the pattern in which resources are available within that minimum territory (contractionist *sensu* Kruuk and Macdonald 1985) that facilitates the depth of that infilling.

For many populations of European badgers, that infilling extends such that each territory is never occupied by just one female, but always contains at least her non-dispersed young and sometimes also, one or more companions and their non-dispersed young too. In these cases patch richness is sufficient to sustain a few secondary occupants, but not the many for whom the enormous, but patchy, richness of available earthworms in the pastures of Wytham, Woodchester Park, North Nibley, and the like facilitates cohabitation. This low-level infilling seems scarcely detectable amongst the low-density badgers studied by Pigozzi (1992a) and Gradzinki et al. (2002), but becomes apparent in the alpine ones studied by DoLinh San et al. (2007) and, from our own portfolio, the cork–oak woodlands of Portugal. The latter case is

---

[7] As an aside, for large species the young of one litter may take more than the season in which they were born to mature, in which case a single mother's territory must be configured to accommodate them for more than 1 year—this would certainly be true for, say, a tiger or polar bear, and thus even if a such mothers are defined as 'solitary' their territories must *de facto* always sustain a 'group'.

doubly interesting because the critical resource dispersion includes not only food, but also diggable den sites (because the ground was so stony that opportunities for sett-digging were probably limiting; Rosalino et al. 2005c).[8]

Along a continuum of patch richness, and thus group sizes, as the patchiness, that is, variance in spatio-temporal heterogeneity, increases, and facilitates larger memberships of minimum defendable, contractionist territories, the Portuguese system culminates in the Wytham one. But notice that this continuum in increasing group sizes spans what may be a step change from semi-detached males to fully resident ones whose home ranges are geometrically congruent, fully overlapping, with those of just one group of females with their non-dispersed offspring of both sexes. We say, what *may* be a step change because the inevitably small sample sizes of tracked badgers (this is demanding work), the inaccuracy of the plotted movements, and the almost inevitable seasonal and inter-annual variation make it difficult to discern from the aforementioned studies of low-density European badgers whether they most closely approximate the ancestral semi-detached relationship between males and females, or, rather, a somewhat fuzzily defined, disarticulated version of the congruency.[9] This will come to matter, insofar as it determines whether the fundamental spatial system of European badgers is still the ancestral mustelid one, or whether all European badgers are nowadays programmed to approximate the shared, congruent spatial groups typical of high-density populations. Why this distinction is intriguing will become apparent in the text that follows.

Is there a step change of male European badgers mapping onto one cluster of females rather than several, or do they all start out that way? The answer, and its consequences, will lie in wider, inter-specific comparisons, to which we now turn.

Inter-specific variation: close cousins

What then of intra-generic variation? Until 2005 there could have been none: *Meles* was a monospecific genus monopolized from Honshu to Holyhead by the Eurasian badger.[10] This offered a marvellous paradigm for the study of intra-specific carnivoran variation (trumped of course by the Holarctic red fox, which had, in turn, usurped the persecuted wolf as the most widely distributed terrestrial carnivore, encompassing, as it did, badgers in habitats as diverse as forest, desert, taiga, urban Oxford, and, of course, the astonishing population densities achieved by the badgers of Wytham Woods. Then, in 2005, the area occupied by this one species was cut by *c*.60%, when the Asian moniker of the Eurasian badger was lopped off with the recognition that the one species was actually three (Wilson and Reeder 2005), and then four (Abramov and Puzachenko 2005; Kinoshita et al. 2017) (Chapter 1, Box 1.2 'Badger phylogeography'): the Asian (*M. leucurus*) and Japanese (*M. anakura*) stood independently alongside the European (*M. meles*) (Linnaeus 1758) and latterly the hoary badger (*M. canescens*; Blandford 1875; see Abramov and Puzachenko 2013) that resides south-east of the arid Karakum and Kyzylkum dark deserts of central Asia (largely Turkmenistan), as well as allegedly on the Mediterranean islands of Crete and Rhodes). At a stroke, the studies we had begun of badgers in Japan, with a view to intra-specific comparisons, were morphed into an inter-specific (intra-generic) comparison.

Are all *Meles* badgers of the same stripe? Just as, intra-specifically, we had opted to compare the highest-density European badgers of Wytham with the lowest and most marginal ones of Serra de Grândola, so now we opted to focus our inter-specific comparison on the geographically and phylogenetically most distant cousin, *M. anakuma*, the Japanese badger. While unmistakably badger-like in physique, facially the Japanese badger has no stripe at all, but rather a bandit's eye-patches set on a blandly blond face—with an orangey tint reminiscent of some pale ferrets (Photo 19.1).

The door to 'anakuma' (literally 'the bear in the cave') was opened to us some 20 years ago by the ever-cheerful Yayoi Kaneko, who spent a postdoctoral year in Wytham, and is now a professor at the Tokyo University of Agriculture and Technology. A little smaller than its European cousin (4.2–9 kg; Hinode population,

---

[8] It is important to be vigilant for seasonal bottlenecks: for example, despite hugely productive summers, Scandinavian and central continental badgers endure long winters, just as Mediterranean ones may suffer summer drought. Seasonality can create bottlenecks in annual productivity and strain the badgers' body fat reserves (as exemplified even for Wytham in Chapter 14).

[9] Similarly, more sophisticated tracking technology is revealing that wolverines (Copeland et al. 2017) and American badgers (Weir et al. 2017) are much more socially attached than previously thought.

[10] *Meles* colonized Japan from the north via the Korean peninsula, and to the west was reintroduced to Anglesey in the 1970s.

Photo 19.1. Japanese badger.

compared with 7.5–10 kg in the Wytham Woods population; see Chapter 1), the Japanese badger's diet also contains a significant proportion of worms (although different species to those in Wytham), as Yayoi and colleagues documented between 1992 and 1998 in the outskirts of Hinode town (near Tokyo)—an area comprising a mosaic of woodland, grassland, paddy fields, farmland, irrigation ponds and canals, and including small settlements—a landscape termed a *Satoyama* in Japanese (a term applied to the border zone or area between mountain foothills and arable flat land; literally, *sato* means village, and *yama* means hill or mountain; Kaneko et al. 2006). During spring and summer, earthworms (*Megaseolocidae* spp.) occurred in 80% of 82 faecal samples analysed, supplemented with berries (30%) (*Rubus* spp.), beetles (35%), and persimmon (*Dymopyrus kaki*) (29%). Interestingly, when persimmon became abundant in autumn, badgers swapped to these berries and away from worms (remember the Portuguese badgers of Serra de Grândola). This strategy, termed dietary switching, involves animals consuming the most profitable food even if it is not the most abundant, thereby optimizing energy gain against foraging and food acquisition expenditure (this is a general aspect of animal behaviour (Charnov 1976), and one that we documented for another mustelid, the yellow-throated marten (*Martes flavigula*), in China (Zhou et al. 2011); for an example among the *Canidae*, see Atkinson et al. 2002). Torpor and living off fat reserves circumvents winter food scarcity.

Turning to their social lives, and radio-tracking 21 *anakuma*, Yayoi found that females with cubs had stable home ranges of 15.2 ± 6.3 ha and occupied small setts with an average of 3.8 (±3.1 standard deviation (SD); *n* = 12) holes (Kaneko et al. 2014). The home ranges of adult females did not overlap and were configured around areas rich in food resources. This

intra-sexual female territoriality is the convention for mustelids (Macdonald and Newman 2017). Males had flexible home ranges that expanded from 33.0 ± 18.1 ha in the non-mating season, to 62.6 ± 48.2 ha in the mating season, to encompass several females—again, approximating the norm for solitary mustelids. Japanese badgers used latrines that looked just like those of European ones and, during the mating season, males made frequent visits to latrines used by females, presumably to sniff out their oestrous status, and post their own messages (see Chapters 6 and 9). Aspects of *M. anakuma*'s reproductive biology mirrored that of European badgers: superfoetation and superfecundation promoted polygynandrous mating opportunities through the summer, followed by delayed implantation (Kaneko 2001) with post-partum mating occurring in April.

Female offspring remained with their mother for around 14 months (Yayoi found one instance of matriarchal territory inheritance), whereas male offspring delayed dispersal for longer, cohabiting with their mother until age 26 months, albeit with decreasing contact from age 15 to 19 months (this pattern of delayed dispersal of sons is mirrored in at least one other 'badger' species, *Mellivora*—see the text that follows and also Zhou et al. 2017). Why? Because it is daughters that risk imposing on their mother's territory the burden of resident grandchildren, whereas sons leave their progeny in some other grandmother's territory.

It has been 2.6 million years since the common ancestor of the European and oriental strands of *Meles* diverged, so what part of the differences in behaviour between contemporary European and Japanese badgers arises from their evolutionary separation and what part from their different circumstances? Do ecological conditions alone offer sufficient explanation for the quasi-sociality in Japan in contrast to the intergenerational groups in Wytham? It is noteworthy that a second study of Japanese badgers, in western Honshu, by Tanaka et al. (2002), mirrored exactly our findings of a semi-detached, slightly infilled approximation to the ancestral musteloid spatial arrangements. Are these two species, appearances aside, essentially the same thing—interchangeable behavioural ecological analogues? This brings us back to the question of whether the spatial arrangements of the low-density populations of European badgers approximate the ancestral semi-detached state (as those of Japanese badgers manifestly do), or some fuzzy oscillation around congruency? Would a Wytham badger transported to Hinode behave exactly as its oriental congener, and vice versa if the stripeless Japanese species woke up one day in Wytham? This is the inter-specific question that would flush out the role of phylogeny—not of that baggage both species have carried with them since *M. thorali* and perhaps back to the Oligocene, but rather whether the European badger has picked up some additional baggage on its more recent journey (Box 19.3). We cannot exclude the possibility that in unexplored corners of Europe or Japan there are flourishing badgers behaving in a way beyond the scope of their congener, but a lot depends on whether there is a real difference in spatial arrangements between Hinode's badgers from those of Mediterranean Portugal. Either way, neither the *anakuma* of Hinode, nor the *meles* of Serra de Grândola, has encountered anything like the conditions of Wytham Woods, but if they did, would they be equally equipped to cope?

The Wytham to Hinode comparison, albeit conflated by intense badger-hunting in Japan (as throughout Asia), of *meles* and *ankuma* spans the east–west extremes of badgerdom, but what of the sister taxa that lie between: *M. leucurus* and *M. canescens*? Both Asian and hoary badgers look much like European badgers and although they inhabit harshly arid and cold continental interior habitats the feeding ecology of *M. leucurus* seems similar to that of *M. meles* (Goszczyński et al. 2000). *Meles leucurus* also hibernates, for example, in Mongolia, where WildCRU's Jed Murdoch and Suuri Buyandelger (2010) reported inactivity from November through to April.

If we had another lifetime available, we'd prioritize a systematic comparison of the sociology of all four species. Insofar as the European badger is well known, a first step would be to select two habitats at extremes of productivity for each of the three remaining species, and compare intra- and inter-specific variations in their behaviour. Meanwhile, the question regarding the other three species is: do they form social groups? Well, in some regions, Asian badgers use large setts, extending up to 300 m², with up to 65 entrances, whereas elsewhere they are said to live in groups of 1–3 (although this may be affected by hunting pressure and habitat loss (Hao et al. 2009; Li and Jiang 2014), and intra-guild competition (Murdoch et al. 2006)). In northern China, Asian badgers mate from the end of March to the beginning of May and give birth a year later, around late March to early April, suggesting that delayed implantation is conserved across the genus (Zhou et al. 2017).

Hoary badgers *M. canescens*, a species recognized only since 2013, range over the Caucasus as well as the foothills of the western Tien Shan Mountains in China and the Pamir-Alai Mountains in Russia (Del Cerro et al. 2010; Abramov and Puzachenko 2013). Aridity likely restricts their diet, and nothing is known

of their society beyond their general similarity to their Asian cousins. In south-east Uzbekistan, where the occidental and oriental strands of badgerdom abut, and *M. canescens* and *M. leucurus* are sympatric, *M. canescens* occupies mountain biotopes, whereas *M. leucurus* inhabits plains and semi-deserts. A study of this sympatry would surely be rewarding.

First cousins and RDH

Turning to inter-specific comparisons, and other members of the genus *Meles*: so far, only the Japanese badger affords sufficient detail to describe reliably one snapshot of the social system, and it was clearly at the slightly infilled (mother and cohabiting offspring) semi-detached stage. To be clear, the larger ranges of male Japanese badgers overlapped several small territorial groups of females and their non-dispersed young—exactly the geometry of the original mustelid model, but partly infilled by non-dispersers. Whether some populations of Japanese badgers are arranged with large infilled groups, and the males coalescing to spatial congruity with the females, as is the case with the badgers of Wytham Woods, we do not know. As a step towards shedding light on whether *M. anakuma* (and indeed *M. leucurus* and *M. canescens*) are behaviourally interchangeable when confronted with identical conditions of resource dispersion, or whether each has a different genetic limit to its societal phenotypes, such an exploration of intra-specific variation in Japanese (Asian and hoary) badgers strikes us an interesting focus for future study. The same could fruitfully be asked of the only other members of the subfamily *Melinae*, the hog badgers; thereby pushing the most recent common ancestor one branch further back up the family tree. So with our colleagues Liang Zhang and Youbing Zhou, then with the Chinese Academy of Sciences, we set out to study hog badgers in the subtropical forests of Houhe Nature Reserve.

When we turned our attention to hog badgers (Photo 19.2) they were still considered to belong to only one species, *Arctonyx collaris* (Cuvier 1825). Now, they are three (Helgen et al. 2008). And so, retrospectively, it turned out we had actually been studying the northern hog badger *A. albogularis*, which occurs in China, Bangladesh, and north-eastern India (see Zhang et al. 2009; Chen et al. 2015; Zhou et al. 2015a, 2015b). Nowadays there is also a Sumatran hog badger, *A. hoevenii*, which is restricted to that island (Zhou et al. 2017), and the original *A. collaris*, now called the greater hog badger, which occurs throughout mainland Southeast Asia.

Photo 19.2. Hog badger.

Arctonyx spp. are similar to *Meles* spp. in body sizes, cycles of seasonal weight gain/loss, and in delaying implantation (possibly along with superfecundation and superfoetation)—they even possess the same sub-caudal gland (Chapter 6; as presumably did their last common ancestor, *M. thorali* and *Arctomeles*). However, their society was unknown until, over 3 years, Youbing led us on the trail of the northern hog badger, in a subtropical forest in central China (Houhe National Nature Reserve, Hubei Province; 30° 5′N, 111° 42′E, henceforth 'HNNR') (Zhou et al. 2015b), using camera-trapping (Zhou et al. 2017). Despite a generalist diet, of which earthworms (families *Moniligastridae*, *Megascolecidae*, and *Lumbricidae* (*L. rubellus*); principal amongst these, *Amynthas hupeiensis* and *Metaphire guillelmi*; Zhang et al. 2010) comprised *c.*40% of annual biomass consumed, peaking at *c.*70% in spring, giving a trophic niche breadth index[11] of 3.34 in summer (indicative of extensive omnivory, falling to 1.14 as they gorged on autumnal fruit; Zhou et al. 2015b). They nonetheless appeared to be solitary. For comparison, in the south of England, badgers typically have a trophic niche breadth varying seasonally from 3.05 to 2.5. That is, hog badgers were less specialized as earthworm consumers and more inclined to fruit, with earthworms underutilized in their diet relative to environmental abundance (remembering that abundance may not be proportional to availability). The opposite was true of insect consumption; that is, a large number were eaten, even

[11] Trophic niche breadth is defined as the degree of similarity between the frequency distribution of resources used by members of a population and the frequency distribution of resources available to them.

when their abundance was low. In autumn, hog badger diet was predominantly (73%) fruit (Figure 19.3), although in proportion to its high environmental abundance, with no evidence of selection preference.

They used latrines, but seemingly not in territorial defence, but rather as a type of book-keeping of foraging success (Henry 1977; Zhou et al. 2015a). The number of faeces per latrine peaked in early summer (4.2 faeces per latrine), was lowest in autumn (0.7), and was significantly higher in logged and selectively logged forest, and lowest in farmland; that is, latrines were used most by hog badgers in the habitats with the lowest food abundance.[12] That pattern of seasonal latrine use was reminiscent of an anthropogenic Allee effect where value rises steeply with rarity:[13] a notion as relevant to philately as it is to the high prices fetched by rare species in wildlife trade[14] (Valavanis-Vail 1954; Patterson 1998). Interesting, and doubtless general, as this aspect of foraging theory (and human psychology) might be, what preoccupies us here is a link between hog badger feeding ecology and any tendency, or seeming lack of thereof, to group formation: our field study revealed no evidence of them living in groups.

The northern hog badgers' single-occupancy setts with only 1.57 ± 0.76 (1–3) entrances (Zhang et al. 2010 in which they hibernate, apparently alone, from

[12] Something similar has been reported for European badgers in central Italy, where latrines were more active in years when the autumnal availability of fruit (their most important food) was low (Pigozzi 1992b).
[13] Further evidence that hog badgers were marking most when resources were most precious was apparent in the trade-off between energy expended to acquire food and its reciprocal energetic value (Weiner 1992). In spring and early summer, when they fed mostly on 'expensive' food items (e.g. energetically digging up a particular species of earthworm), latrines were used more actively than in the autumn, when badgers consumed foods that were 'cheap' to acquire (such as fallen fruit) (Zhou et al. 2015a). The argument is that that at times of food superabundance (e.g. autumn), it would probably prove unnecessary and uneconomical for an individual to signal a right of access to these resources.
[14] A sinister variant of this phenomenon where the rate of increase in price with rarity (e.g. 24 cent stamps known as Red Jennies now sell for more than US$900,000) was illustrated by Macdonald (2013) as leading to the demise of the prophetically named giant yellow croaker (*Bahaba taipingensis*). This fish, discovered only in 1937, had swim-bladders that were so valued in Traditional Chinese Medicine that soon 50 tonnes were being landed annually. But by 2000, over 100 boats managed to catch just 10 specimens, while a single fish caught early in 2010 fetched US$500,000. The croaker may have croaked, but the enduring point is that its value was inflated so spectacularly *by* rarity that its exploitation remained economically worthwhile even as it approached extinction.

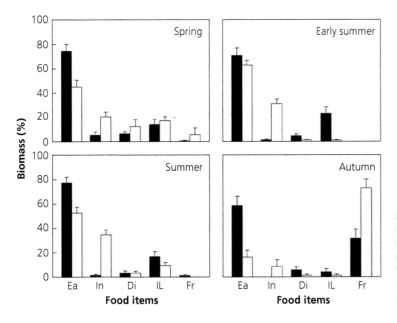

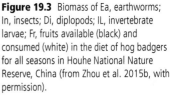

Figure 19.3 Biomass of Ea, earthworms; In, insects; Di, diplopods; IL, invertebrate larvae; Fr, fruits available (black) and consumed (white) in the diet of hog badgers for all seasons in Houhe National Nature Reserve, China (from Zhou et al. 2015b, with permission).

December through to March; at least in subtropical and temperate areas (Zheng et al. 1988; Zhou et al. 2015a, 2015b), reinforced the conclusion that they were largely solitary. If these hog badgers do not form groups, why not? Is it because in Houhe National Park the resources upon which they depend are dispersed in a pattern that does not facilitate group formation (Chapter 10), or is it because they are hog badgers, and groups are beyond the limits of intra-specific flexibility in their societies? Developing a little the first of these two speculations, Zhou et al. (2015b) and Ross et al. (2017) noted that hog badger habitat throngs with a substantial guild of trophic competitors (including *M. leucurus*) whose competitive presence may tip the hog badgers' food security below a threshold that might, in the absence of these competitors, have facilitated group living.

While mindful that competitors may stand between hog badgers and their resources, it is also relevant to remember that bushmeat hunting for badgers is likewise prevalent. Lee et al. (2004) reported that up to 620 hog badgers (and 520 ferret badgers, see the text that follows) were sold each day in markets in Guangzhou and Shenzhen (China). Lau et al. (2010) consider this a cause of rapid decline in hog badgers, risking regional extirpation. This prompted us to investigate the effect of hunting on the genetic diversity of a hog badger population in Hubei Province, using samples from animals confiscated from illegal traffickers and local hunters by forest authorities in the market towns of Wufeng and Yuguan (Chen et al. 2015). Using novel hog badger

genetic microsatellite markers which we developed, we found that, despite only 40 km separation between towns, hog badgers seized by the regulatory authorities exhibited a high level of genetic diversity.[15] The exploitation of large catchment areas to supply relatively small rural markets led us to conclude that the exploitation of this species was likely unsustainable.

Second cousins

Having considered the *Melinae*, what is the next step to interpreting inter-specific variation through the lens of the badgers of Wytham Woods? Broadly, our approach will be the climb backwards through the musteloid and then carnivore family tree,[16] pausing at branches to scrutinize successive 'most recent common ancestors' in turn while considering how their genealogy maps onto their ecology and, specifically the aforementioned continuum of spatial arrangements from semi-detached to congruent. But as we begin the climb

[15] Mean alleles per locus [A] were 8.33 and mean expected heterozygosity [H_E] was 0.77, compared with just H_E = 0.54 in a British population of *M. meles* (Domingo-Roura et al. 2003) and H_E = 0.62 in a Swiss population (Frantz et al. 2010).

[16] Readers as pedantic as ourselves will feel the need to observe that we are actually climbing backwards, down the tree, towards its roots, with European badgers at its apical growing tip, and ancestral carnivores located close to those roots. Yet more pedantic readers will appreciate that we are comparing branch tips—the growing points on which extant species linger.

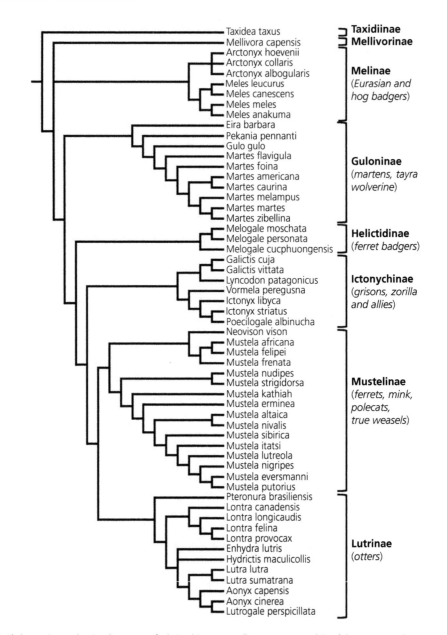

Figure 19.4 Phylogenetic tree showing the pattern of relationships among all extant genera and 54 of the 63 proposed, or recognized, species within the family *Mustelidae*. Bracketed clades or lineages define the eight recognized subfamilies. The following six species are not depicted in the tree due to absence of formal phylogenetic studies (but see Nyakatura and Bininda-Emonds 2012): *Melogale everetti*, *Melogale orientalis* (*Melinae*); *Martes gwatkinsii* (*Guloninae*); *Mustela lutreolina*, *Mustela subpalmata* (*Mustelinae*); and *Aonyx congicus* (*Lutrinae*) (from Koepfli et al. 2017, with permission).

(down), out of deference to their shared surname (whether due to physique or stripiness), we turn first to other species of so-called badgers. To save readers repeatedly flicking back to Chapter 1, we reproduce here, as Figure 19.4, the relevant family tree.

Because they were once considered a genus of 'true badgers'—a term that nowadays makes sense only if applied to the *Melinae*—we turned first to the ferret badgers. Now classed as the subfamily *Helictidinae*, they have recently been split into three species

Photo 19.3 Chinese ferret badger.

(Sato et al. 2004; Koepfli et al. 2008; Wolsan and Sato 2010; Sato et al. 2012) of which, once again in Houhe NNR with Youbing and colleague Liang Zhang, we studied the Chinese ferret badger, *Melogale moschata* (Photo 19.3).

Given their small, marten-like small body size (at 1–3 kg) and shape, we initially predicted a solitary lifestyle for Chinese ferret badgers, suspecting they were too small to tolerate the lowered food security likely to limit opportunity for secondary members of RDH-based groups (Newman et al. 2011). Instead, radio-tracking 9 (6 males, 3 females) of 22 ferret badgers caught, revealed that they were arranged into three groups (Zhang et al. 2010). The largest group comprised at least seven members, including at least four adults (including two radio-tracked males and two females) whose ranges overlapped congruently and which cohabited in the same sett. Setts had 3.36 ± 3.15 (1–12) entrances (Zhang et al. 2009). Camera traps recorded juveniles accompanying an adult female (plausibly their mother) when foraging (implying extended natal philopatry; Noonan et al. 2015a). Neither home range sizes (128.3 ± 131.9 ha 100% MCP

± standard deviation (SD)), nor movement distances or daily activity patterns differed between sexes. They deposited their faeces in shared group latrines (Zhou et al. 2008). Chinese ferret badgers mate from February to April and, for a carnivore of their size, should only require a gestation period of *c.*50–60 days. Instead, around 60–80 days elapse before cubs are born (Pei and Wang 1995) suggesting a short delay to implantation (Chapter 7; see Sugianto et al. 2021b). The ferret badgers were least active in December (33.1%) and progressively more so through the spring and summer to peak in June (52.6%), and despite snowy conditions remained active all year round (Zhang et al. 2010; Zhou et al. 2013b), although they do appear to pack on fat for winter. From 163 droppings collected through autumn and winter 2004–2006, we found that in Houhe NNR *M. moschata* ate mostly insects and fruits (Zhou et al. 2008), notably fleshy- and seed-pulp rich fruits. In contrast, studies of ferret badger diet from southern China (60 stomach contents; Qian et al. 1976) and Taiwan (16 stomach and 60 faecal sample; Chuang and Lee 1997; Wu 1999) report mostly the consumption of earthworms, arthropods (insects), and amphibians (frogs). When it comes to deciding whether it is purely ecological selective pressures, that is, resource dispersion, that facilitate group living and have led to congruent spatial groups, or whether sociological selective pressures have added compound interest to the inclusive fitness accrued by group-living ferret badgers, it will become very important to know whether they cooperate, but we don't yet know—so that's another topic for future research.

So now, we have to consider why ferret badgers live in groups while hog badgers living alongside them in Houhe NNR do not, even though their diet also consists principally of insects and fruits (Zhou et al. 2008), while in southern China (Qian et al. 1976) and in Taiwan (Chuang and Lee 1997; Wu 1999) ferret badgers ate a slightly different blend of earthworms, insects, and frogs. The puzzle deepens because, despite their patchy, shareable diet, their small body size (at 1–3 kg; see the consequences of fat), rather lithe marten-like physique (Newman et al. 2011), and seeming absence of hibernation (Zhang et al. 2010) might, as we will argue in the text that follows, militate against group living. Be that as it may, while both hog badgers and ferret badgers are both omnivorous with broadly similar diets, we hypothesize that the dispersion of the particular menu favoured by ferret badgers facilitates multimale/multi-female groups, whereas that favoured by hog badgers does not.

Box 19.3 What of the baggage, whether inherited or picked up?

In interpreting the behaviour of the badgers of Wytham Woods within the context of intra- and inter-specific variation, aside from solving for the algebra of ecology (ultimately the dispersion of resources and risk), what constraints, or baggage, may frame adaptation? That baggage may be picked up from the deep past, phylogeny, or acquired along the way from convergence to similar ecological trades. The obvious starting point is that the badger lineage has diagnostic dentition adapted to the move from carnivory towards omnivory. Three less obvious constraints are worth bearing in mind too.

Allometric size constraints on fossoriality:

1. Bears are rather like badgers in their omnivory, but imagine the task for a bear-sized animal to dig a burrow for itself; then extrapolate from that the task involved in digging a burrow large enough to accommodate a whole (badger-sized) group of them. Burrow size also scales allometrically, which puts an upper limit on their architecture: even the most consolidated soil types cannot support the large and extensive excavations that would be needed to house, for example, a group of bear-sized individuals (White, 2005). In an analysis of the evolution and function of fossoriality we found that very few extant fossorial eutherian mammals that dig their own burrows have an adult mass greater than 15 kg (the aardvark and giant armadillo are exceptions)—the energetic cost of tunnelling just becomes too great (Noonan et al. 2015a). Giant carnivores of the past, such as cave bears (*Ursus spelaeus*) and cave hyaenas (*Crocuta crocuta spelaea*), found another solution for sheltering, and the clue is in their names: they used caves. Musteloids are also relatively small among the Carnivora, where the cost associated with burrow construction scale with body mass (M) according to $\alpha M^{0.54}$ (Vleck 1981) and the force an animal can exert scales according to $\alpha M^{0.25}$. Actual burrow size also scales allometrically ($\alpha M^{0.33}$, see Vleck 1981) setting an upper limit on burrow size imposed by structural stability (White 2005). As a consequence, medium and large canids tend quickly to outgrow natal dens (Noonan et al. 2015b), although many small canids use dens throughout adulthood (Kamler et al. 2012, 2013b), whereas while the felids, procyonids, and viverrids evolved along trajectories incompatible with digging (Macdonald 1992; Andersson 2004) although some species in these groups, like the miniature black-footed cat, use burrows made by other species (Kamler et al. 2015).

2. Forelimb anatomy. Digging requires robust humeri and the ability to supinate (turn in) large (semi-)plantigrade (flat-footed) front paws to grip and pull at soil and rocks—ankle and shoulder adaptations that substantially preclude running. Long claws also hamper running. Compared with cats, which retract their weaponry, and the generally shorter nails of dogs, the adaptive complex of digging comes at the cost to badgers of less flexible backs than, for example, martens, fishers, and wolverines, making them less adept at climbing (Kitchener et al. 2017). So badgers simply cannot emulate the loping stride, with digitigrade (tip-toe) feet extending limb length, that makes canids fast and enduring, and cheetah fast. Indeed, although forelimb supination can support large body sizes, as seen in bears weighing in north of 300 kg (Andersson 2004), as weight increases, cursoriality becomes more economical (Kram and Taylor 1990); however, this is at the expense of fossorial capacity (McNab 1979; Vleck 1981): if they are to dig, again it pays digger-like badgers to be below the 15 kg threshold (Noonan et al. 2015a).

3. Facial masks. All badger species, and their cousins, have at least shadows of a pair of black stripes running over their white face from behind the ears, over the eyes, intersecting at the nose. In *M. meles*, these facial bands are c. 15 mm (0.6 in.) wide, and widen to 45–55 mm (1.8–2.2 in.) in the ear region (Newman and Buesching in press). This striking 'aposematic' facial mask (Newman et al. 2005) likely serves to warn larger predators that badgers are unusually fierce, with a formidable bite. That said, the Japanese badger has somewhat paler and fainter stripes, but with more intense 'spots' over the eyes; *M. leucurus* tends to have bold, straight stripes inboard of its ears, while *M. canescens* often appears to wear a light mascara around its eyes. Hog badgers, *Arctonyx* spp. sometimes, but not always, have much darker faces overall, with broader stripes. Straying further from the meline lineage, ferret badgers, *Melogale* spp., are more ornamented in appearance, with a dark crown, bisected by a white line and a dark nose, dark eye mascara, and a faint line linking these features.

Box 19.3 *Continued*

Self-confidence that their advertisement is not a bluff may explain why musclimorph badgers, unlike other small group-living carnivores, invest little effort in vigilance, and post no sentinels. In a similar vein, oxytocin, the affection or 'anti-psychotic' hormone (Magon and Kalra 2011; Rubin et al. 2014), must be inhibited for delayed implantation (Douglas et al. 1998), which is another meline hallmark (see Chapter 7). This may well destine badgers, through reproductive demands, to a lower propensity for sociality than other group-living carnivores (Caldwell 2017. Indeed, this is almost the opposite predisposition to that encouraged by canid pseudopregnancy, which hormonally primes sociality and alloparental care (Macdonald et al. 2019).

Another consideration is size, which intersects with nocturnality, and thus with the dangers of being seen and the benefits of seeing. Of all the musteloid species that form groups, ferret badgers are the smallest. Consequently, they are likely to face the largest number of bigger competitors and predators, and this may increase the relevance of various rewards accruing to strength of numbers and the benefits of many eyes making safe work. The general point is that in carnivore communities, or guilds, it is, in the words of renowned palaeontologist Blaire van Valkenburgh (2001), a 'dog eat dog world'—a colloquial expression of what is technically termed intra-guild competition (Palomares and Caro 1999), which can blend into predation. Cooperation, which presupposes group living, can be an antidote in various ways; for example, defence of prey (or of territory or young), or defence against becoming prey. Exemplifying the former, Chris Carbone—a former WildCRU post-doc— and colleagues (1997) identify defence of prey against thieving spotted hyaenas as one amongst a suite of rewards (Courchamp and Macdonald 2001) to larger pack sizes amongst African wild dogs (others include keeping their pups off the menu of bigger guild members; Courchamp et al. 2002). Similarly, Jan Kamler, a long-term WildCRU stalwart, found that bat-eared foxes in South Africa formed larger groups primarily to defend against predation by jackals (Kamler et al. 2013) and that larger groups suffered less predation from jackals than did smaller groups (Kamler et al. 2012). Defining the predatory guild more broadly, John Vucetich, also a distinguished research associate of WildCRU, reveals pack size as a determinant of the success of wolves in protecting their kills from ravens (Vucetich et al. 2004). Exemplifying the latter, the avoidance of becoming prey, amongst the musteloids, coatis in larger groups benefit from both collective defence and collective vigilance against predators (larger carnivores or raptors), as do all the sociable mongooses, which are about the same size as ferret badgers. A crucial difference is that coatis and social mongooses are diurnal, can spot an eagle, literally a mile off, and gain obvious benefit from operating as tight-knit mobs—benefits that will not apply to ferret badgers operating under the cover of darkness. A general point affecting intra- and inter-specific variation in social organization is the blend of competitors and predators faced by a particular population. Thus, while the badgers of Wytham Woods have essentially no competitors or predators, badger-like species in tropical regions may face many (see Zhou et al. 2015b); for example, in Japan *M. anakuma* competes not only with native raccoon dogs but also with introduced North American raccoons and masked palm civets, for both food and den sites (Zhou et al. 2017). These are amongst the points to which we will return in a synthesis later, but they are useful in the meantime to bear in mind as we extend our review to other badgers.

So, what of other musteloids that are badgers by name, if only variously by genealogy? The literature offers insights. The American badger, *Taxidea taxus*, neatly conforms with fundamental mustelid geometry occupying the semi-detached, intra-sexual, and sexually dimorphic territories (Goodrich and Buskirk 1998), configured to partition one-animal helpings of prairie dog colonies (Weir et al. 2017). Male American badgers are about a quarter again bigger than females and so too, on the other side of the world, amongst honey badgers, *Mellivora capensis*, males are a third larger than females. They too consume an eclectic range of generally small prey (Kruuk and Mills 1983), which, particularly in the bottleneck of the cool dry season, are thin on the ground (Begg 2001; Begg et al. 2003) and not dispersed in shareable ways. Despite their shared surname, and mustelid inheritance, American and honey badgers are not close relatives; in keeping, however, with their similar trades their social organizations are comparable today (Begg et al. 2005 and it's tempting to speculate that they have been for the last 30 million years since the days of their last common ancestor. Receptive female honey badgers have no fixed breeding season and are a scarce resource, scattered unpredictably within a tessellating mosaic of 138 km² territories, overlain by a layer of extensively overlapping ranges (548 km²) of polygynous males (Begg 2005. Cubs, of which 1–2 are born after little or no delayed implantation following a gestation of 6–8 weeks (Begg et al. 2005), bide with their mother for at least 12–16 months (Prinsloo 2016), loosely paralleling the behaviour of Japanese badgers; during this apprenticeship they seemingly acquiring hunting proficiency amidst a large guild of competitors (Carter et al. 2017).

Although also blessed with the surname badger, and thus attracting Hans Kruuk's attention (Kruuk 2000), stink badgers, *Mydaus marchei* or *M. javanensis*, are actually skunks (Hass and Dragoo 2017), separated from the meline line since the days of *Miomephitis* 22 million years ago (Mya). Would we expect them to live in groups? On the basis of their omnivorous diets (Hwang and Lariviere 2003, 2004), perhaps; all the evidence, however, is that stink badgers are strictly solitary, females raising two or three cubs single-handedly (Long and Killingley 1983) in the polygynous, semi-detached sociality typical of skunks from the Americas. Omnivory spans various extents of the spatio-temporal, rapidly renewing food dispersion that towards an extreme facilitates group living, and it may be that the invertebrates, eggs, and carrion eaten by the Sunda stink badger (supplemented by freshwater crabs for the Palawan stink badger) are too far from the shareable end of that spectrum. There may also be yet another factor at play, and a strong candidate is fat.

The consequences of fat

We will argue later that it is sometimes helpful to ask not just why things are as they are, but also why they are not otherwise. To introduce that thought experiment, consider why stink badgers aren't, in sociological terms, ferret badgers—the answer being that instead they are more marten-like (Newman et al. 2011).

Martens are fairly close taxonomic relatives of badgers, in the loose category of second cousins; they shared a last common ancestor 11 Mya according to molecular clock estimates (Koepfli et al. 2008—see Figure 19.4). Furthermore, they live across similar temperate forest habitats and they both eat an omnivorous diet (although martens eat more tree squirrels and other rodents, whereas badgers eat more earthworms): compare, for example, the European badger and the European pine marten, *Martes martes*. Yet martens are solitary. Why? Because they must remain lean in order to hunt effectively their rodent prey (e.g. squirrels). Pine martens weigh 1–2 kg, males outweighing females by 12–30%; were either sex to store fat, as badgers do, their impaired agility would be fatal. The flip side of that coin being that, without fat, they cannot tolerate variable food security; an unfed marten can starve within 72 h, and consequently cannot risk sharing resources with a secondary occupant (in RDH terms) of their territory. This same necessity for litheness also rules out torpor/hibernation, into which the corpulent *M. meles* can retreat. Sitting behind these generalizations is, again, the pattern in which available food is dispersed: squirrels and other rodents are available to martens year round, so they don't need the off-season that nudges badgers, through the winter dearth of earthworms and other foods, towards torpor. American badgers pose a conundrum here: they build up body fat and go into winter torpor, yet feed almost entirely on rodents. Their prey, unlike the arboreal squirrels on which martens largely depend, are fossorial, and although active in winter are often under a snowy blanket.

Appreciating, now, as we do, the consequences of fat, let us return to why ferret badgers, also weighing 1–2 kg (but with no sexual dimorphism) form groups. The fundamental answer is that fruit, insects, and worms occur in the subtropics in spatio-temporal patches that are more shareable, and less divisible, than squirrels (or indeed than the omnivorous prey of the portly but solitary sympatric hog badger). But a secondary answer is that, insofar as they are neither predatory nor arboreal, ferret badgers may be somewhat exempted from the agility demanded of martens and thus be able to combine small body size with sufficient portliness for fat reserves to bolster their food security, and perhaps even make the species more competitive among the guild of carnivores sympatric with both ferret and hog badgers. The same might be asked of stink badgers, which also weigh 1–2 kg, and thus have limited capacity to endure sustained food insecurity. Something caused stink badgers (remember that they are actually skunks) to fall on the marten side of the divide; to adhere to sexual dimorphism and semi-detached intra-sexual territoriality, whereas ferret badgers fall on the badger side (that 'something' is not climate, since both occur in the tropics or subtropics and thus neither is troubled with the need to hibernate while fuelled with fat reserves).

Zhou et al. (2017) distilled from this account of badgers, *sensu stricto* and beyond, some generalizations about intra- and inter-specific variations in social organization. A tendency toward group living is promoted by: (i) eating a diverse diet of heterogeneously dispersed resources (especially earthworms); (ii) being large enough to cope with the lessened food security associated with being a secondary occupant in the face of variable daily and seasonal food availability (i.e. being able to store body fat), and being thereby able to reduce the effect of seasonal food constraints through torpor/hibernation; and (iii) using subterranean dens/setts, a characteristic that (see the text that follows) turns out to be associated with extended natal philopatry. Factors seemingly in opposition to group formation include being: (i) predatory/carnivorous—eating prey that cannot easily be shared; (ii) in less productive/more seasonal habitats; and (iii) a member of a larger guild (likely associated with highly productive/tropical areas of high biodiversity), where competitors interfere with food security of the badger species in focus. Keep in mind these factors as we turn to the European badger's metaphorical third cousins—other musteloid families and beyond.

Third cousins

For perspective, the time travel that separates the badgers of Wytham Woods and those of Japan from their most recent common ancestor is not far off that separating us from the first known remains, in Ethiopia, of the genus *Homo*. Mindful that the last common ancestor of the European badger and the martens lived at roughly the time as the last common ancestor of humans and chimpanzees, our exploration of inter-specific variation is expanding to involve more distant relatives—metaphorically, third cousins. We will now work backwards to shared common ancestors of badgers and other branches within the *Mustelidae* and *Musteloidea*. That shared progenitor with the ferret badgers, polecats, weasels, and mink was at least 2 million years earlier than the 10–11 Mya old split with the martens, that with the honey badger and American badger some 15 and 25 Mya, respectively, whereas to find the common ancestor of *Meles* and the skunks and stink badgers requires peering into evolution's rear-view mirror for some 30 My. To prepare for that journey we need first to delve somewhat into a trait that we have several times mentioned as varying greatly between the species under discussion, as for example between the European badger, in which it is minimal, and the honey badger, in which it is considerable—that is, sexual size dimorphism.

Sexual size dimorphism

One conspicuous aspect of musteloid biology is that, in common with many mammal species, males tend to be larger than females (Moors 1980). The variation in sexual size dimorphism (SSD) of the musteloids ranges from parity up to males being more than twice the size of females. Extreme sexual dimorphism runs through musteloid history from its roots—fossils of, for example, *Aleurocyon*, *Megalictis*, and *Paraoligobunis* suggest that similar sexual politics have characterized the super-family for at least 20 million years (Hunt and Skolnick 1996).

Early work speculated that SSD was a special case of character displacement (Hendrick and Temeles 1989), with dimorphism serving to reduce competition for suitably sized prey types between sexes. In her meta-analysis, however, Fairbairn (1997) concluded that it is unlikely that inter-sex niche divergence plays more than a subsidiary role in the evolution of SSD (see also Gittleman and Van Valkenburgh 1997). Evidence has swung towards the importance of sexual selection, arising in polygynous mating systems because larger males are more competitive in terms of securing access to mates. In instances where females' mate choice is also influential, the prospect of larger, more competitive sons would be preferred, leading to positive sexual selection (Soulsbury et al. 2014). In short, sexual selection is now the most widely accepted explanation of SSD in mammals (e.g. Isaac 2005).

The tendency for SSD to increase with body mass in taxa in which males are larger, and to decrease when females are larger, is known as Rensch's rule. In mammals, where the trend occurs, it is believed to be the result of a competitive advantage for larger males, while female mass is constrained by the energetics of reproduction. When, later, it becomes critical to our argument to compare badgers, and other musteloids, with felids and canids we will return to an exploration of Rensch's rule for those families (Johnson et al. 2017), but here we first explore the allometry of SSD specifically within the Musteloidea, in an analysis led by Paul Johnson and Mike Noonan in which we demonstrated a hypo-allometry contrary to Rensch's rule, with lower SSD associated with larger body size (Figure 19.5) (Noonan et al. 2016). We argue, from a position that will soon lead us back to the resource dispersion hypothesis (Chapter 10), that feeding ecology explains this reversal of Rensch—where diet promotes

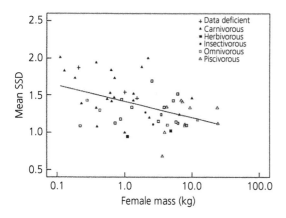

Figure 19.5 Scatter plot depicting the relationship between sexual size dimorphism (SSD) and female mass (kilograms) in musteloids, showing dietary classes. Data-deficient species were those for which no accurate dietary data were available (from Noonan et al. 2016, with permission).

group living, and this tends to be the case for the larger, least carnivorous, musteloids feasting on earthworms, fruit or fish; males seeking to access several females need not configure large ranges to overlap several smaller single-occupancy female ranges from which they benefit from maximized brawn to fight off rival males. Rather, the pattern of resource availability that allows several females to cohabit presents these as a ready-made harem, within one territory, to a male (or even males) that cohabit with them.[17] We'll build on this idea below, but the general point is that the effect of feeding ecology on mating systems may be a hitherto neglected factor explaining variation in SSD.[18]

[17] As will become clear, most of these larger patch-feeding musteloids, like badgers, live in groups with several members of each sex. An important variable is how closely the patchiness of resource dispersion maps onto the clumping of the females. Where food dispersion makes females a clumped, defensible resource, this is likely to lower females' availability and increases mating skew towards the few males that can monopolize a harem. However, in the case of badger-like societies, although patchiness of available food facilitates group living, the patches are spatio-temporally dispersed, making it hard for one male to monopolize all the females.

[18] A topic for future research is whether this patchy food supply favours females growing bigger because they are less energetically constrained, and/or because males do not need to grow larger to compete for females—one consideration is that insofar as being larger reduces surface area for heat loss and accommodates more fat for insulation, there might be a winter-time energy conservation incentive for both sexes to grow larger, and indeed, on the hot-water-bottle analogy, to favour huddling with larger companions (see the text that follows).

Continuing to climb backwards through the family tree (Figure 19.4), amongst other members of the *Mustelidae*, what evidence is there of infilling of the ancestral semi-detached, sexually dimorphic, solitarily occupied intra-sexual territoriality? The mustelids exist as six subfamilies in addition to the *Melinae*, of which we've just met the *Helictidinae* or ferret badgers, the remainder being the *Mustelinae* (weasels), *Guloninae* (wolverines), *Ictonychinae* (grisons), *Lutrinae* (otters), and two outlying but convergent monospecific 'badgers', *Taxidiinae* (American badgers) and *Mellivorinae* (honey badgers[19])—previously an eighth subfamily comprised the *Mephitinae* (skunks), but nowadays they are elevated to their own family (itself partitioned into the stink badgers that we also met a few pages earlier, the hog-nosed, spotted, and striped skunks).

Of the 19 *Mustelinae*, the true weasels, all adhere to the ancestral, solitary semi-detached society. Similarly, none of the following: the 10 species of *Guloninae*; the wolverine and martens; the 7 species of *Ictonychinae*; the zorillas; and the American badger (this last-named being the single representative of the *Taxidiinae*) evidence any infilling of the ancestral system (the 11 species of skunks, the *Mephididae*—more distant though they may be—don't, either). From the smallest to the largest, the least weasel to the wolverine, males have almost exclusive intra-sexual territories that encompass those of one to four females (e.g. Copeland et al. 2017). Against this 38 (49 counting the skunks) majority, come the group livers, the formerly one, now three, species of Helictidinae ferret badgers, the mixed bag of seven *Melinae* of which four *Meles* are sometimes group living, which brings us last to the otters, and a qualitative sociological difference. It is amongst the 13 species of otters that we find the greatest evidence of infilling and, at extremes, a partial shift, paralleling the European badger at Wytham and the Chinese ferret badgers of Houhe, towards group occupancy of congruent ranges. And what is the qualitative sociological difference documented only amongst the otters?

Group living occurs amongst five diurnal species (the African and Congo clawless otters, the African spot-necked otter, Asian smooth-coated otter, and neotropical giant otter) in which cooperative care is demonstrated and cooperative hunting (and vigilance) suspected amongst the giant otter, while cooperative

[19] Although generally considered solitary, recent film of 10 honey badgers together has been recorded in Botswana (http://movspec.mus-nh.city.osaka.jp/ethol/showdetail-e.php?movieid=momo170622mc01b).

vigilance may benefit not only the giant otter but also the four smaller group-living species (a fuller review is given in Barocas and Ben-David 2021). Group living occurs in some coastal populations of European, and North American, river otters, and cooperative hunting is suspected amongst all five habitual group livers. In summary, 60 species (81%) of the 74 species of *Mustelidae* and *Mephitidae* steadfastly approximate the ancestral semi-detached sociology; one, possibly two, more (European and northern river otters) sometimes form infilled spatial groups, as more routinely do four (formerly one) species of true badger (which, put another way, means that almost 90% of the species in the lineage live, at least often, solitarily). It looks like three (formerly one) species of ferret badgers habitually form slightly cohesive groups, as more dedicatedly do five species of otter, of which one is demonstrably cooperative. To summarize for emphasis, of 74 or so species of the mustelid/mephitid lineage, 60 appear entirely conventional, while 8 usually form groups (and 6 do sometimes); of the 14 that at least sometimes form groups, 9 are spatial and 5 cohesive, and only one is demonstrably cooperative (although 7 others might be, at least to a certain extent).

We add a further 15 species to these comparisons by including the 2 other musteloid families, 1 species of *Ailuridae*, the red panda, and 14 species of *Procyonidae* (Macdonald et al. 2017a). Interestingly, and in line with the shareability of their foods, they group into a cluster of the ringtail and the cacomistle, which are small, nocturnal, solitary hunters of small vertebrate prey, living in classic intra-sexual territories (although occasionally up to eight cacomistles have been seen to congregate in a frugivorous extravaganza at a fruiting tree (Poglayen-Neuwell 1989), a cluster of three species of sometimes solitary individuals in intra-sexual, semi-detached territories, sometimes spatial grouping (Zeveloff 2002) (thereby exhibiting variation reminiscent of European badgers), highly omnivorous nocturnal raccoons, a cluster of omnivorous, but more heavily frugivorous, coatis of which all three species form cohesive diurnal female bands, several of which are overlapped by the larger ranges of solitary males, and a single species of heavily frugivorous nocturnal kinkajou, weighing about 3 kg, that forms multi-female, multi-male spatial groups that convene at rich fruiting trees (Brooks and Kays 2017). To these are added the four mysterious species of nectivorous/frugivorous nocturnal olingos, each reminiscent of a mini-kinkajou but at *c.*1 kg only a third the size, which are usually seen alone—it is tempting to suspect that their frugivorous diet

might facilitate spatial groups, but their sociology is unknown.

So, the argument coalesces. This survey of socio-ecological variation amongst European badgers, and between badgers of genus *Meles* and subfamily *Melinae*, and the miscellany of badger-like *Mustelidae*, offers a cross-section of sociologies of which some adhere to the classic semi-detached model of polygynous intra-sexual territoriality and others involve some infilling of single-occupancy territories to create groups. As first became clear in Chapter 10, we frame the answer in terms of resource dispersion, but is this principle valid? Obviously, the answer would emerge from experimental manipulation, but sadly, as lamented by Macdonald and Johnson (2015), the conflating realities of carnivore ecology make these extremely difficult to conduct in nature, and so in the discussion that follows we will rely on the accumulation of indicative comparisons. First, however, and in passing, we record that the only experimental manipulation thus far attempted on musteloid society did provide evidence that the dispersion of available food altered social organization: this comforting, if unsurprising, result came from the heroic attempt at manipulating resource dispersion carried out by Morgan Wehtje and Matt Gompper (2011). They compared the sizes, and measures of overlap, between home ranges of two adjoining communities of raccoons, each provisioned with identical quantities of the same food but dispersed according to two different patterns. The food was presented in clumps to 22 raccoons, whereas it was presented to 19 raccoons nearby in a non-clumped and spatially–temporally unpredictable pattern. These experimental differences in food dispersion did not affect home range sizes, but the movements of females experiencing the clumped treatment overlapped twice as much.

And so to solving the puzzle by deductive comparison of observations in the field. Small mustelids rely on even smaller vertebrate prey (often rodents) that tend to be homogeneously dispersed and thus divisible into discrete single-occupancy territories. Consequently, the greater the contribution small mammals make to their diet, the more solitary and less sociable musteloid species tend to be (Johnson et al. 2000). Furthermore, for these hunters of homogeneously dispersed prey, the generality is that greater prey abundance per unit area (of which only a proportion may be available) equates to a higher environmental carrying capacity, leading to higher population densities and, *pro rata*, smaller individual ranging areas (Macdonald 1981b; Gittleman and Harvey 1982; Reiss

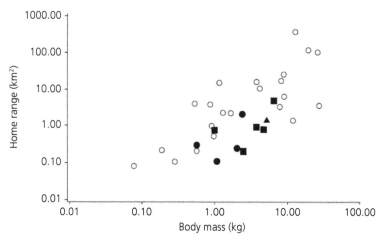

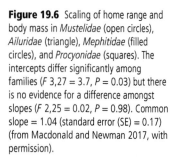

Figure 19.6 Scaling of home range and body mass in *Mustelidae* (open circles), *Ailuridae* (triangle), *Mephitidae* (filled circles), and *Procyonidae* (squares). The intercepts differ significantly among families ($F_{3,27} = 3.7$, $P = 0.03$) but there is no evidence for a difference amongst slopes ($F_{2,25} = 0.02$, $P = 0.98$). Common slope = 1.04 (standard error (SE) = 0.17) (from Macdonald and Newman 2017, with permission).

1988). However, as we reviewed in Macdonald and Newman (2017):

Many musteloids eat few, if any, rodents, favouring instead frugivorous and insectivorous diets, often linked to opportunistic omnivorous generalism (e.g. Zhou et al. 2011). For example, *Martes* spp. (martens), *Meles* spp. (old world badgers), *Procyon* spp. (raccoons), *Bassariscus* spp. (ringtails), *Potos flavus* (kinkajou), *Bassaricyon* spp. (olingos), and *Ailurus fulgens* (red panda) consume significant seasonal proportions of everything from insects, worms, molluscs, eggs, fruits, seeds, and nuts, through to cereal crops, and even honey. These food types, along with the fish eaten by piscivorous otters (Lutrinae), tend to be locally abundant and rapidly renewing, constituting clumped, widely dispersed resources.

Notice that while not all medium to large musteloids form groups, all group-forming musteloids are medium to large (ferret badgers are the smallest). Why? The answer falls out of an analysis of how musteloid home ranges scale across the striking range of body size across the super-family (Macdonald and Newman 2017). All else being equal, the greater metabolic demands of being larger lead to the expectation that home ranges should scale with body mass to the power of 0.75 (McNab 1963). The expected scaling is not 1.0 because larger bodies have lower surface area to volume ratios—a fact that reduces the rate of heat loss, and thus overall energy requirements (LaBarbera 1989). In reality, across a wide range of taxa and types of feeding ecology, home ranges actually scale with body mass with an exponent consistently higher than 0.75 (Glazier 2005). For the musteloids as a whole, the scaling exponent is close to 1.0 (Figure 19.6). The reason why is not well understood, but there is an inevitable tendency for larger ranges to overlap more (Jetz et al. 2004).

Plotting body size against home range size, we examined the intercepts for best fit lines and found they differed between the families comprising the *Musteloidea* (Figure 19.6) with the mustelids tending, for a given body mass, to have the largest home ranges—perhaps because of their reliance on dispersed small rodents (Macdonald and Newman 2017). Delving deeper, we classified social systems into four types: solitary, pairs, variable groups, and groups to reveal the eureka moment that group-living species have smaller home ranges, relative to their body size, than do solitary ones (Figure 19.7).

Thus group-living populations of the European badger, and of both the giant otter and sea otter, lie well below the best fit line for solitary species. The diet of all three species comprises small prey items occurring in rich patches: invertebrates, fish, and bivalves, respectively. In summary, both the size of home ranges and the extent to which ranges can be shared (and collectively defended) by conspecifics depend on the quality and dispersal of resources—once again an RDH-based explanation of societies. Variation in the availability and distribution of food resources (as well as other resources, such as dens, water, and mates) can facilitate infilling of single-occupant territories (those metaphorical beakers described earlier), which necessitates social tolerance of conspecifics and, in circumstances where the net marginal benefits of mutual associations exceed the costs of competition (Alexander 1974), favours group living (Macdonald and Carr 1989; Silk 2007; Huchard and Cowlishaw 2011). But what are the rewards amongst musteloids of this new-found social tolerance? In a few cases, the answer is obvious, notably amongst the lutrines (Groenendijk et al. 2017) and certain procyonids (Hirsch and Gompper

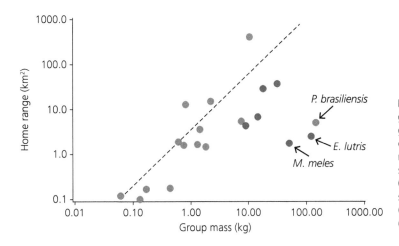

Figure 19.7 Scaling of home range with group mass for solitary (red circles), pair or group living (blue circles) and one conspicuously cooperative (green circle) mustelids. Best fit line for solitary species shown (slope = 1.40, standard error (SE) = 0.24). The scaling exponent was steeper in solitary species compared with the others (adapted from Macdonald and Newman (2017), with permission).

2017; Brooks and Kays 2017)—amongst which, species display a repertoire of social behaviour associated with group living (which we will review briefly later). That some sociable species should behave sociably, even cooperatively, is not surprising; that other group-living species do not, is surprising, and reminds us why, in Chapter 1, we had followed Hans Kruuk through this question to the badgers of Wytham Woods, whose groups, in the paradox of Chapter 8, can grow larger than those of any other mustelid recorded anywhere in the world, with scarcely a hint of cooperation.

Cooperative musteloids

So, we argue that every instance where musteloids form groups can be interpreted as an adaptation to resource dispersion (although the mirror image statement is not universally true: sometimes, as illustrated by martens, resources may be dispersed in ways that could facilitate group formation, but lithe body form and phylogenetic inertia preclude it). Macdonald and Carr (1989, see also Macdonald and Johnson 2015) recognized this ecological force as one of two categories of selective pressure shaping the societies of carnivores. Thus far in this chapter we have dwelt on the ecological factors: the dispersion of food may facilitate group formation by reducing the costs of multiple occupancy, whereas the occupation of nearby territories may restrict it by thwarting dispersal. Of sociological factors, various forms of cooperation, as reviewed by Creel and Macdonald (1995) may add per capita benefit to group membership. The audit of the costs and benefits of these ecological and sociological pressures will determine the nature of society. On the profit side, cooperation may involve any and all of:

hunting and defending prey; creating and defending dens; defending territory; collective vigilance for, and defence against, predators and rival conspecifics; communal care, including allonursing, of young; and, as WildCRU's Liz Campbell has argued is too often overlooked, the benefits of huddling—social thermoregulation (Campbell et al. 2018).

What evidence is there of sociological selective pressures adding benefits to group living amongst the *Mustelidae*, or more broadly the *Musteloidea*?

Cooperative hunting

In Chapter 8 we concluded that amongst the many activities on which the badgers of Wytham Woods do not cooperate, hunting is conspicuous. A phylogenetic history launched by hunting as a digger, burrow pursuit, and stealth offers few opportunities for cooperating in the chase (Macdonald and Newman 2017), which explains the complete absence of cooperative hunting permeating seven of the eight mustelid subfamilies (although Newman et al. (2011) conclude that scavenging martens can tolerate close contact, and in South America the grison, *Galictis cuja*, and tayras, *Eira barbara*, may forage in groups (Diuk-Wasser and Cassini, 1998; Ferrari. 2009)). It is only amongst the *Lutrinae* that evidence of cooperative hunting becomes convincing: groups of smooth-coated otters herd fish into the shallows—indeed, fishermen train them to herd fish into nets (Feeroz et al. 2011). The strongest candidate is the giant otter, 'porpoising' together in a focused onslaught on a shoal of fish, although no study has quantified their per capita energetic gain when hunting together rather than separately. For quantitative evidence we can cite just one species, the white lipped coati, of one genus, *Nasua*, amongst the three remaining musteloid families (*Procyonidae*, *Ailuridae*, and *Mephitidae*) to find the only quantitative evidence of cooperative hunting amongst the 90 species of musteloids: coatis appear to forage more efficiently for lizards when in groups (Gompper 1996).[20]

Vigilance and defensive behaviour

Whereas cooperative hunting is generally a habit of larger carnivores, group-living small carnivores can gain selective advantage from collective vigilance, and even cooperative defence—but to do so, of course, they must not only live as a spatial group (facilitated by resource dispersion) but also travel together as a band (unlike the solitary foraging badgers of Wytham Woods). Mustelids are notorious for their ferocity (and associated aposematic facial masks, a flamboyant dichromatism taken to extremes by skunks (Buesching and Stankowich 2017). Evidence of cooperative vigilance is nonetheless sparse for musteloids, but documented for smooth-coated otters (*Lutrogale perspicillata*) (van Helvoort et al. 1996) and colonial sea otters when resting on land (Maldini et al. 2012). Cooperative defence is also rarely documented, although giant otters combine forces to see off their rivals and marauding caiman (Ribas et al. 2012), and Cape clawless otter groups defend against conspecifics. A plausible, if unproven, suggestion is that large rafts of sea otters may confuse cruising sharks (Garshelis et al. 1986) and orca (Hatfield et al. 1998).

For more dramatic evidence of cooperative vigilance and defence, one must move beyond the *Mustelidae* to procyonid coatis, which position juveniles in the centre of a foraging group, sharing vigilance, alarm calling, and mobbing and attacking predators (Di Blanco and Hirsch 2006). Similarly, coatis at waterholes risked the distraction of more drinking bouts when in larger groups, and members of larger groups lingered longer at the waterhole and were individually less vigilant than members of smaller groups (Burger and Gochfeld 1992). Large groups of coati may better repel attacks from predators (see Blundell et al. 2002a), and male raccoons may combine forces in territorial defence (Pitt et al. 2008). The reality for most musteloid species, with their digging proclivities, is that the best form of defence is retreat, to an underground den.

Alloparental care

Father wolverines have been seen interacting with their cubs (Copeland et al. 2017), but the only mustelid for which there is clear-cut quantitative evidence of the functional significance of alloparental behaviour is, again, the giant otter: non-breeding adults of both sexes provision the pups born to their group's breeding pair with fish (Groenendijk et al. 2017).[21] The greater the number of these non-breeding helpers, the greater the number of pups that survive to become juveniles. Furthermore, larger groups occupy larger core areas, and pups produced on these larger core areas are more likely to disperse and breed successfully. Interestingly too, this society not only cares for the young, but also for the old, with Davenport (2010) reporting assistance given to the declining group matriarch. The phylogenetically next closest instance of alloparental care lies beyond the mustelids with the kinkajous, in which extended families offer protection to vulnerable offspring (Brooks and Kays 2017).

In conclusion, our argument is that all musteloid groups (actually, probably all vertebrate groups) arise because resource dispersion reduces the costs to individuals of cohabiting as a spatial group, allowing low-cost infilling of the ancestral single-occupancy territories through extended natal philopatry, further promoted by the use of natal dens driving cohabitation. As Kruuk and and Macdonald (1985) first argued, it can happen that there is no need to invoke sociological selective pressures to explain group-living in contractionist territories. However, launched towards sociality by resource dispersion, spatial groups, especially of matrilineal kin, are well placed to benefit sociologically from cooperation and would be disadvantaged by mutual antagonism.

The foregoing brief review suggests that for most group-living musteloids such sociological benefits are few—or at least they have not yet been established. But where they do arise, Kruuk and Macdonald's (1985) original argument was that the benefits of cooperation could launch groups on the trajectory to expansionism: recruiting more members to commandeer more resources, to recruit more members, up to the point at which larger groups no longer paid their members greater per capita fitness dividends. This almost biblical scenario of, 'To them that hath shall be given' (Matthew 25:29), has, amongst the musteloids been demonstrated only for the giant otter: larger groups,

---

[20] Both American badgers and honey badgers form cooperative hunting partnerships with other species, respectively, coyotes and black-backed jackals (Minta et al. 1992; Gorta 2020).

[21] Some groups contain otters that aren't related to the other members, so are non-kin helpers (Ribas et al. 2016).

including more non-breeding helpers, produce proportionately more cubs annually, which results in more recruits to create a yet larger group, and of those that disperse their chances of success are greater when originating from a larger group.

A carnivore perspective

How does the balance of ecological and sociological selective pressures apparent in badger, and more broadly musteloid, group living compare with that in other carnivore families? Detailed explorations of the socio-ecology of canids and felids are given in Macdonald et al. (2004b; 2010b) and Macdonald and Moser (2010), respectively, so here we offer only a brief overview sufficient to inform the synthesis of carnivore society viewed from the vantage point of Wytham Woods and considering the badgers therein as models of evolution and behaviour. There is little to say here about the *Viverridae*, and linked *Prionodontidae* and *Nandiniidae*, or the *Eupleuridae* (all on the cat branch of the Carnivora) or the *Ursidae*, because, in common with the musteloid families, all adhere to the solitary ancestral polygynous intra-sexual pattern of semi-detached, sexually dimorphic, territoriality, as too do the great majority of *Herpestidae*, and the majority of *Felidae*, and one of the four *Hyaenidae*. But it is in these last four families, albeit amongst only a minority of their members, that we find the cooperators. When felid species form multi-adult groups they, like mongooses and hyaenas, typically exhibit collaborative behaviour, notably cooperative hunting of large prey and alloparental care (Macdonald et al. 2010b; Macdonald and Moser 2010; the same is true of canids but from a pairwise starting point) and collective vigilance and defence. We argue that all carnivore societies are fundamentally RDH-based (Macdonald and Johnson 2015), but in those closer to their ecological roots, the group living facilitated by ecological factors may not yet enjoy the compound interest to sociality delivered by cooperation; rather these spatial groups are 'groups of convenience', ecologically the most economic socio-spatial geometries. This latter situation characterizes not only the badgers of Wytham Woods but also, as we saw earlier, most cases of that small minority of musteloid examples of group living. In 2017 we published a review of cooperative behaviour amongst musteloids (Macdonald and Newman 2017), and it was populated by fewer examples than in equivalent reviews for canids and felids (Macdonald et al. 2004b, 2010b). The groups formed by members of all three taxa can be powerfully interpreted as being facilitated by an RDH framework, but it seems that the mustelids in particular, and musteloids in general, have (so far) capitalized the least on sociological selective advantage built from that ecological foundation.

At this point we need to remember another sociologically relevant parameter that we already know varies in an upside-down way amongst the musteloids: SSD. When we come to compare inter-specific variation in musteloid sociology with that in other carnivores, specifically felids and canids, it will be insightful to have in mind variation in SSD amongst them too. Paul Johnson led us in an examination of the allometry of SSD (Johnson et al. 2017): in felids, there is positive allometric scaling, while there is no such trend in canids. Once again, we attribute these differences ultimately to variation in feeding ecology, where the dispersion of resources for omnivorous and insectivorous species precludes mating systems in which defending access to multiple females in single-occupancy territories is a viable male strategy. The observation that felids are predominately solitary, and all are obligate carnivores, supports this. Carnivory, however, does not contribute to a Rensch effect as dietary variation occurs across the mass spectrum. Observed inter-familial differences are also consistent with reduced constraints on female mass in the canids, in whom litter size increases with body mass (Geffen et al. 1996), versus the situation amongst felids, in which no similar allometry exists between litter size and body mass. We concluded that diet and resource dispersion can promote social and mating systems that undermine the advantage of large male size, by reducing the extent to which contest competition contributes to male reproductive success.

Why is the roster of cooperative behaviour so short amongst musteloids, and particularly mustelids? Adapting the famous quip of boxer Joe Louis, musteloids can't run, but they can hide[22] (and indeed can sneak into all manner of crevices in order to secure competitive space alongside representatives of rival carnivore families). That is, at Oligocene root, the strategic divergence that separates the musteloids from other branches of the order Carnivora. While felids and canids evolved longer legs and a more fluid gait to run down, stalk, and pounce upon large prey in the proliferating Oligocene grasslands, the musteloids (the terriers and dachshunds of the wild) evolved to

---

[22] Commenting on his 1946 heavyweight title bout, Joe Lewis said of his opponent, Billy Conn, 'He can run but he can't hide'.

hunt small rodent prey by either digging them out of (musclimorph) or by following them into (skinny-morph) burrows; subsequently they radiated into myriad strategic niches not monopolized by their felid and canid cousins (the trades of musteloids, which accomplished much of their diversification in Eurasia following the opening up of forest glades and savannahs 30 Mya, parallel those of herpestid mongooses (amongst whom group living may have coincided with the copious herbivore dung produced during another period of opening plains 6–7 Mya), and those of the viverrids of which the oldest known are 24 Mya *Semigenetta* from Europe and the roughly contemporary *Kichechia* from Africa). The diversification of some musteloids into omnivory, and the extended dietary breadth this entails, creates a reliance upon trophic resources that have dispersion and renewal characteristics that can facilitate group living in these species. Moreover, as we will elaborate shortly, the making and utilization of burrows by many musteloid species facilitates the perpetuation of philopatric natal groups, sometimes into adulthood, congregating together at the group den (Noonan et al. 2015a). But much of this happens with little need of cooperation. Further, musteloids have a proclivity for aposematic colouration, suggesting they are particularly able to fend for themselves (Newman et al. 2005).

Amongst musteloids, It is only really the coatis and some otters, that interact with the level of communal enthusiasm seen amongst African wild dogs (Creel and Creel 2002), bush dogs, *Speothos venaticus* (Macdonald 1996) and meerkats, *Suricata suricatta* (Doolan and Macdonald 1999) and others among the exuberantly sociable Carnivora (Macdonald 1992; Creel and Macdonald 1995). Spatial groups arising purely from resource dispersion need not acquire any direct group-living benefits, beyond achieving adequate food security. In Macdonald and Newman (2017) we described the societies of group-living musteloids, and showed how each could be interpreted within the framework of the RDH. Here, a brief précis highlights the points relevant to the development of the current argument. Specifically, in addition to the earthworms that as rich, shareable, but not readily divisible patches are so influential on badger society, two entirely different food sources share these characteristics and are influential in facilitating group living in other musteloids: fruit and fish. All three not only facilitate infilling of single-occupant territories with additional group members; at an extreme they become associated with a transfer from semi-detached to congruent spatial arrangements.

Frugivorous groups

First, fruit. Consider first a society retaining the ghost of ancestral polygynous semi-detached geometry; that is, that of the white-nosed coati. Infilling of female territories is made inexpensive by the rich, shareable, and not readily divisible food patches created by individual fruiting trees (Hirsch and Gomper 2017). Bands comprise 6 to over 30 highly social, related and unrelated females and their immature offspring. Adult males (34% larger than females) are solitary except during a 2-week synchronous breeding season during which each courts several bands of females. One population of white-nosed coatis also provides a different illustration of RDH, where the bottleneck resource is water not food: Valenzuela and Macdonald (2002) linked the home range sizes of these coatis in Mexico to the dispersion of dry season drinking water. A step further down the road, travelling from semi-detached geometry towards congruency between female and male ranges, we arrive at the kinkajou, whose social units comprise one adult female, her pre-dispersal offspring, and two adult males (Brooks and Kays 2017). Members of these groups spend around 20% of their time feeding together in large fruiting trees (although they travel and feed singly, as a classic spatial group, when feeding on the fruit of small trees). The larger the feeding tree (the richer the ephemeral food patch in RDH terms[23]), the larger the group. Yellow-throated martens, which also feed extensively in the crowns of subtropical fruit trees but generally neither feed nor den in groups and abide by the ancient semi-detached spatial arrangements (Zhou et al. 2008, 2011), provide a revealing contrast to the coati–kinkajou trajectory.[24] The difference is that kinkajous, and even coatis, are a bit portly whereas martens are lithe arboreal acrobats. The reason is that, in addition to gorging on rich patches of fruit, yellow-throated martens also actively hunt vertebrate prey: as mentioned, they have to be lean, so they cannot tolerate reduced food security during the intervals when rich patches of fruit are unavailable (Newman et al. 2011).

---

[23] Of course, fruiting trees, like at least some other ephemeral food patches, are likely to reappear in the landscape on later, sometimes cyclical, occasions, such as in the next fruiting season.

[24] There are cases when yellow-throated martens forage in groups (Chutipong et al. 2016), and even hunt cooperatively.

Piscivorous groups

Second, fish (and shellfish). A textbook example of infilling is provided by the coastal Eurasian otters in Scotland studied by Kruuk and Hewson (1978). Solitary males fished in large, 2.7–4.5 km linear offshore territories, each of which overlapped several smaller territories, each occupied by two to five females. Female groups were configured around fresh water and rich fishing patches, which in RDH terms were both indivisible and shareable. Without any sign of cooperation, all members of these female spatial groups bred each summer, each female of the clique spending more than half her time in her own core area. Similar infilled groups of females and their philopatric offspring, with little evidence of cooperation, overlain by semi-detached males, occur in some populations of the closely related North American river otters (*Lontra canadensis*), amongst which young males disperse from such family groups earlier than females (an intriguing variant is that groups of over 10 gregarious adult males sometimes form; Hansen et al. 2009).

It was this type of system that was anticipated by Rowe-Rowe (1977), when he proposed that sociality in otters may have arisen almost by default; there being no selective pressures against group foraging, particularly on crabs or in muddy rivers in the tropics. Thus, although perhaps as much influenced by molluscs and crustacea as by fish, several other species of otter also form groups: African clawless otters (*Aonyx capensis*) (Somers and Nel 2004), hairy-nosed otters (*Lutra sumatrana*) (Long 2000), Asian short-clawed otters (*Amblonyx cinerea*), and smooth-coated otters (*Lutrogale perspicillata*). At least the latter two, and perhaps also spotted-necked otters (*Hydrictis maculicollis*), have moved a critical stage further in their group living, in that they have abandoned the ancestral, albeit somewhat infilled, semi-detached spatial arrangement in favour of congruent ranges occupied by monogamous pairs and, sometimes, their undispersed young. This spatial congruency reaches its zenith with piscivorous giant otters (*Pteronura brasiliensis*) whose ranges are configured around patches of superabundant fish such that, in line with patch richness, territories with larger cores supported groups of up to 13 individuals (but never more than a single breeding male), while territories with smaller cores supported only breeding pairs (Groenendijk et al. 2017).[25] Why this step

change to congruent ranges, and a switch from a polygynous, spatially semi-detached, to seemingly monogamous, spatially congruent, fundamental structure? The answer will become clear shortly in the wider context of carnivore societies.

As we place badgers, and the musteloid lineage in the wider context of carnivore societies, mindful of the phylogenetic baggage that we expect to influence their sociological responses to ecological circumstances, it is timely to ask: what is the genealogy of the order Carnivora? We answer here only in outline, but for a general introduction to carnivore evolution see Macdonald (1992), from which a crucial warning emerges: how things are now is not necessarily how they always were—families have come and gone, niches have changed unrecognizably, trades been swapped, continents have drifted, and biogeographies transmogrified. This caveat undermines interpretations based solely on the blinkered snapshot glimpse of current realities.[26]

Family trees are best considered tracking back to the last common ancestor, which binds together a lineage. For all Carnivora this great great great[n] grandparent was a marten-like miacid that lived in the Palaeocene about 60 Mya, not long after the demise of the dinosaurs and when North America, Greenland, and Eurasia were drifting apart as Laurasia fissioned. A bit earlier, more than 65 Mya, another rather marten-like predator, *Cimolestes*, produced two lineages with similarly innovative but differently positioned shearing teeth, one of which led to the creodonts, who monopolized the carnivorous trade for the next 20 million years; the other, through miacids, becoming the

hunting within them. However, the reality will be more complicated because fish are found also in the rivers that link the lakes, and shoals of fish move throughout the whole system (perhaps analogous to the way patches of earthworm availability move through badger ranges). From period to period, the availability of prey throughout the giant otters' territory is probably best conceived as contours of richness. In this system while there are relationships between group size and the size of oxbow lake core areas, there is no relationship between group size and total territory size. Furthermore Barocas et al. (2021) show there is no simple relationship between oxbow lake size and fish abundance (remembering that what matters to the otters is fish availability). Further observations on giant otter sociality are given by Leuchtenberger and Mourão (2008) and Leuchtenberger et al. (2013).

[26] This advice applies to conservation too: using examples of big cats disappearing in the wake of their extinct prey, Chris Sandom led an analysis that illustrates how an appreciation of extinctions in the past informs predictions of their consequences in the future and thus suggests where priorities for intervention lie (Sandom et al. 2017).

[25] Groenendijk et al. (2017) indicate that part of the story of giant otter group sizes, and reproductive success, is linked to the size of oxbow lakes, and probably the patches of good fish

Carnivora and usurping the creodonts at around the time of two major phenomena—global climate change starting about 39 Mya, and the Bering Land Bridge reconnecting Eurasia and North America 30 Mya. Since that miacid progenitor, the intervening generations of the Carnivora have been so buffeted by chance and opportunity, much like balls finding routes through a pinball machine, that pretty well the only thing that nowadays unites that miacid's great great great[n] grandoffspring is possession of those revolutionary carnassial (scissor) teeth. However, intermediate notables through the generations have bequeathed their particular evolutionary innovation to their branch of the genealogy. So it was that about 60 Mya, as the denizens of the northern continents watched their companions inch over the skyline, the vine that is the order Carnivora split into two branches: those stranded in the New World became the *Canoidea* or vulpavines, dog branch, and those adrift on the Old World became the *Feloidea* or viverrines, or cat branch. From each, further splits produced, like bunches of grapes, clusters of closely related species—many of which prospered on the vine for a while only to wither, but some of which, moulded and remoulded by changing circumstances, survive today. By defining each clade with respect to its common ancestor these clusters form progressively closer relatives (infra-orders and super-families)—on the dog branch five extant families comprise the *Musteloidea* (the *Mustelidae*, *Procyonidae*, *Ailuridae*, *Procyonidae*, and *Mephitidae*), linked most closely to the surviving bear family, *Ursidae*, and slightly less so the surviving dog family, *Canidae*. On the cat branch the super-family Herpestoidea links the hyaena family, *Hyaenidae*, and the mongoose families, *Herpestidae* and *Eupleridae* (the latter from Madagascar), both then united with the civet family, *Viverridae* under the infra-order Viverroidea, on the parallel vine leading to the super-family *Feloidea* whose only survivors are the cat family, *Felidae*, and linsang family, *Prionidae*, with that entire cluster of cat-branch taxa sitting alongside their most distant surviving relatives, the African palm civets of the family *Nandiniidae*. Importantly, it was only 30 Mya that representatives of the dog and cat branches began trotting, respectively, east and west across the Bering land bridge and encountered each other for the first time, and only in the following 10 million years that they managed to get the creodonts off their backs and to rise to be top predators. Despite the passage of unimaginable numbers of generations, astoundingly the members of each family share characteristics much more obvious than the architecture of a tooth—so much so that their blood lines are identifiable

at a glance—a fact that speaks to the enduring influence of ancestry and the innovative power of their evolutionary characters; characters that suffuse their social behaviour, which is why they are crucially relevant to our argument. This enduring influence of ancestry deserves a moment's reflection. David is a member of a Scottish clan, the Macdonalds of Clanranald, who descend from a twelfth-century Highland hero, Somerled, credited with regaining Gaelic control of the western seaboard of Scotland from Norse kings. Oxford geneticists Bryan Sykes and Jayne Nicholson (2000) analysed the non-recombining segment of the Y-chromosome, passed from father to son, and found four chromosomes shared by all five contemporary clan chiefs of this lineage, one traceable to a chromosomal mutation 900 years ago, and they calculated, aghast, that not only do all five chiefs still carry Somerled's genes, but so too do 19% of Clanranald men with the surname Macdonald (by extrapolation, a quarter of a million men with the surname Macdonald may still carry Somerled's Y-chromosome). It may be staggering to see the influence of Somerled's genes after 85 generations, but pause to consider that the pivotal mutations of the founders of carnivoran families continue their conspicuous impact after 20 million generations. So it was, in characterizing the *Musteloidea*, that we wrote, only a little flippantly, 'the procyonids are the brains of the Musteloidea, the otters the beauty, while the brawny badgers exemplify the mentality that all problems inevitably yield to force. They are an ensemble cast of contrarians that do things that bit differently, unified by the major distinction that they are not cats, or dogs; and by the epithet of being "long, thin, and stinky"' (Macdonald et al. 2017a, p. 188).

Against this canvas, our interest here in these familial characters is in how a taxonomy of social organizations maps onto this phylogeny. Again, we will be brief, but an introductory overview exists in Macdonald (1984a), with detail reviewed in Creel and Macdonald (1995), and for three taxa specially relevant here, canids, felids, and musteloids, very detailed reviews and case studies in Macdonald and Sillero (2004), Macdonald and Loveridge (2010), and Macdonald et al. (2017b). In summary, what are the ancestral social organizations of these 13 extant families of terrestrial carnivores? Remarkably, they are almost all the same: most members of most carnivore taxonomic families live solitarily, and for most families (all except two) the ubiquitous spatial arrangements approximate intrasexual territoriality, characterized by what we have termed semi-detached relationships between males and females. The ubiquity of these arrangements, from least weasel (the smallest carnivore ever) to polar bear in the dog branch, and slender mongoose to tiger in the cat branch, is as impressive, and more conspicuous, than their shared carnassial teeth, and one may presume that similar semi-detached relationships characterized great great great[n] grandparent miacids (and probably most other early mammals too; Lukas and Clutton-Brock 2012a). And the exceptions? All 37 members of the extant *Canidae* and all four of the *Hyaenidae* unanimously opt for congruent spatial arrangements, in this case exceptions that prove (both demonstrate and test) two rules.

And what are those rules? The first is that for carnivores the starting point for group living involves the infilling of additional group members into a spatial mosaic dictated by phylogeny. That infilling occurs when conditions align at the intersection of body size, the distribution of food sizes, resource dispersion, and phylogeny. The members of all modern carnivore families, except two, inherit from their respective last common ancestors socio-spatial arrangements involving intra-sexual, sexually dimorphic territories between which males maintain semi-detached relationships with the females whose ranges they overlap. These ancestrally single-occupant territories are variously filled in with larger memberships as resource dispersion facilitates. This process works first for females, with males being more partitioned by sexual rivalry. Thus, whether amongst dog-branch musteloids or cat-branch feloids, the much larger ranges of solitary males overlay several small groups of territorial female Japanese badgers or coastal European otters, or white-lipped coatis, or matrilines of farm cats.

This infilling appears to be on a continuum: sometimes males join forces too—lions and cheetahs are differently revealing cases, for both of which females represent patchily dispersed resources sought by coalitions of males (Loveridge et al. 2010 Broekhuis et al. 2020). Under conditions where the population sex ratio is skewed such that males are few (conditions brought about 'experimentally' by trophy hunting), the territories of single males, or small coalitions, overlap the territories of two, three, or even four small prides of females. However, as male numbers increase (i.e. under a trophy hunting moratorium), the size of their coalitions grows, as does competition between rival coalitions until the kaleidoscope rearranges such that each larger male coalition maps congruently onto the range of a single female pride. The difference between the European badgers of the Mediterranean and those of Wytham Woods mirrors this intra-specific journey from semi-detachment to congruency. The

comparison between the semi-detached relationships between male coastal otters overlapping somewhat infilled groups of females and the entirely congruent ranges of male and female giant otters also mirrors it, inter-specifically.

So what prompts the congruency that creates mixed-sex spatial groups of lions, giant otters, meerkats, and, of course, the badgers of Wytham Woods? The generic answer is a switch in the return, in inclusive fitness, to males from cohabiting with several social units of females to cohabiting with just one. That switch occurs where the costs of fighting for several widely dispersed female units becomes punitive relative to the rewards of devoting most energy to just a single female unit. For most species of carnivore, in all save two families, this never happens when females are in single-occupancy territories, and it only happens where resource dispersion facilitates rather large spatial groups of females, which, in addition to the economics of defence, may open the door to the rewards of social selective pressures. These cooperative rewards are apparent for the giant otters and lions, but not (according to Chapter 8) for the badgers of Wytham Woods, suggesting that the economics of defence alone have been sufficient to drive them to congruency (but remembering the reproductive sacrifice is minimal, with the *c.*50% frequency of extra-group paternity; Chapter 17).

This brings us to the second rule to emerge from this synthesis. It is that there are, phylogenetically, only two foundations to the relationships between male and female carnivores: the huge majority builds on the semi-detached foundation; a minority on congruency; and the sole difference between them is the evolution of paternal care, itself made possible by prey characteristics at the aforementioned interface between body size, distribution of food sizes, and resource dispersion.[27] Of the 37 species of *Canidae*, most generally live as behaviourally monogamous pairs in congruent territories; several facultatively develop generally rather small infilled spatial groups as resource dispersion allows, and although it seems that all canids can survive as pairs, three to five species are variously close to obligate in group living. Insofar as every single one of these 37 species opts for the exception to the carnivoran rule by occupying congruent territories, we deduce this is a trick handed down from their last common ancestor—perhaps *Leptocyon*, whose name means thin

---

[27] Under both scenarios, the general route, that is the ontogenetic pathway, to group formation, is delayed dispersal, and even the reproduction, of stay-at-home offspring, sometimes centred on a shared den.

dog, 5–7 Mya ago, which lived in Eurasia; or perhaps even back towards the evocatively named evening dog, *Hesperocyon*, 35 Mya when canids lived only in North America.

Why did *Leptocyon* or its ilk eschew the ancestral semi-detached mating system? Of course, because it paid, and insofar as pair-living canids do it, and probably did it before group-living ones existed, the benefits must accrue even to males with only one mate. By deduction, the answer must be paternal care—a trait that is today the defining hallmark, along with behavioural monogamy, of all canid species (Macdonald et al. 2019). The rewards of paternal care pay the male canid better than those of semi-detached philandering, and have done so for at least six million years (which is when the vulpine and lupine lineages split: both practise paternal care now and so probably their last common ancestor did so then). That pay-off is easy enough to understand, being after all the argument for the evolution of monogamy (Lukas and Clutton-Brock 2012b; Macdonald et al. 2019), but why did this accountancy not apply to all those other carnivore families? One, less interesting, answer is that it may have done so, but only recently: the rewards of paternal care, feeding young otters, or protecting young lions from infanticide may contribute to the congruent groups of these species, as exceptions amongst their semi-detached kin. But the fact is, to judge by the general ubiquity of semi-detached intra-specific, sexually dimorphic territoriality in each family, in each case their great great great[n] grandfather, the last common ancestor, was not inclined to paternal care.

The more interesting answer to why most other carnivores have not opted for canid congruency lies again at the intersection of their body sizes, the distributions of the sizes of their food, and resource dispersion. Canids have a unique brand of omnivory and associated cursorial anatomy (remember the ancestral thin dog: they did not start out as bow-legged muscle-bound diggers like badgers). Canid omnivory emphasizes the 'omni'—they can be almost entirely vegetarian, even vegan; insectivorous, vermivorous, or frugivorous (it's hard to resist the trite—although ultimately erroneous—thought that anything a badger can do a fox can do better), and on top of that they can be serious cursorial predators of medium-sized vertebrates. It is the medium size of their vertebrate prey that made paternal care so attractive to the carnivores that became canids: had they eaten only earthworms or persimmon fruits, the task of ferrying these back to nursing females would have been profitless (Macdonald et al. 2019). Had they somehow started their

evolutionary journey from a point at which they were already killers of massive prey (as runners not stalkers this was anyway not their proclivity, and at odds with the risk of overheating) the booty would have been too big to haul home (contrast this with the fact that a father jackal can conveniently ferry home to his mate and youngsters a gazelle fawn, a springhare, even a mouthful of rats).[28] To speculate a step further, why did the ancestral canid opt for these medium-sized prey in the first place—partly because ancestral felids had already monopolized the trade of pulling down huge prey (and remember, that while a giant felid can sit in wait interminably,[29] a giant canid could not run far without its blood boiling[30]), which is why when they hunt big prey canids do it as a pack and, of course, a pack of wolves can decide to split up when only medium-sized prey is available—an option of fissioning not open to a hungry tiger (but see Mbizah et al. 2019 regarding fissioning of lion prides). Subsequently, amongst contemporary canids, some, such as wolves, wild dogs, dholes, even bush dogs, do hunt large prey, or at least prey much larger then themselves, but they have evolved this trait within a canid congruent spatial system, within which the large (and cooperative) memberships of their groups have been facilitated by resource dispersion.

And so to a final nuance of the canid congruency story: the distinction between the dozen species of vulpine canid and 21 species of lupine canid, which split some 6 Mya (the raccoon dogs, bat-eared foxes, and grey foxes are sisters to them both). Both adhere to the tradition of canid congruency in their phylogenetic infilling as resource dispersion allows, but they do it differently. The lupines (*Canini*), including the South American species that split off from them across the Panamanian land bridge 3 Mya, infill with multi-female, multi-male societies (Geffen et al. 1996) converging in this sense with badgers and giant otters, although from a different starting point,

---

[28] While, without the trick of regurgitation, a father badger would be frustrated in bringing home earthworms or blackberries to his litter, in a 'what if' sort of way one might wonder why natural selection hasn't nudged them towards carrying home the occasional pheasant or rabbit. One impediment to starting down that road might be delayed implantation and the associated multiple paternity, which raises doubt as to which are his progeny.

[29] Fossils of the rhino-hunting giant Smilodon sabre tooth (*Machairodus lahayishupup*) that lived 9–6 Mya may have approached 400 kg.

[30] Indeed, under the canid, or even hunter-gathering human, tactic of relentless chase, it is the blood of the giant prey that boils first.

whereas vulpine foxes almost always infill with just a single male cohabiting with several females (an interesting exception involves more than one male in some groups of Arctic foxes enjoying super-rich patches; Kruchenkova et al. 2009); these females are generally recruited as non-dispersing daughters. Why this vulpine/lupine divergence? Both opt for spatial congruency, and so on our argument both are wedded to paternal care; they differ, however, in body size. *Canis* species (and their close relatives *Cuon* and *Lycaon*) are invariably bigger than *Vulpes* species, and thus in the sizes of their prey. Vulpine foxes are smaller, and always have been, and their prey are smaller, although nonetheless large enough to make it worthwhile carrying back some chunky vertebrate delicacies (e.g. a leveret) to a nursing mate, but not large enough for it to be worthwhile to hunt them cooperatively. Therefore, we posit that the modern vulpine model was adopted by the earliest canids, carrying with them the single male polygynous semi-detached heritage of almost all other carnivores. For them, the progressive eschewing of semi-detached, intra-sexual territoriality in favour of congruency was a journey made worthwhile almost entirely by paternal care. We guess it was made worthwhile enough, but not critically so—many modern red fox vixens, after all, rear their cubs as widows. But some other canids, slightly bigger ones, hunting slightly bigger prey, perhaps found that the benefits of cooperative hunting offered a greater return in inclusive fitness to delaying dispersal not only of daughters, but of sons too, and the *Canis* way of doing things took hold.[31] In these scenarios, two other traits become important: reproductive suppression and regurgitation.

Reproductive suppression is seemingly universal amongst female canids with their penchant for strict hierarchies, but this socially mediated but ultimately endocrinological birth control is strikingly absent in most other carnivore families (although variously apparent in social herpestid mongooses: meerkats do it, banded mongooses don't). However, it is fiercely evident in giant otters, which, thanks to rich, renewing spatio-temporally patchy shoals of fish, have converged on canid-like sociology.[32] Second, regurgitation occurs only amongst the *Canidae*, and specifically within *Canis*, *Cuon*, and *Lycaon*, not vulpine species (observations of foxes regurgitating are so rare as to be dubious—they carry food for their cubs, a behaviour linked to the avidness with which they also cache food (Macdonald 1976a), although it reportedly occurs in swift foxes (Poessel and Gese 2013)). These traits—reproductive suppression and the associated fitness advantage of alloparental care and regurgitation—both enhance care of the young, and of their mother, whereas food caching evens out irregularities in food availability without storing it as fat at the cost of agility. Together, these are part of the syndrome of paternal care that defines canids.

We mentioned that one other carnivore family, and it is so small nowadays as to make generalizations difficult, seems to opt directly for congruent ranges and the paternal care that make them profitable: the *Hyaenidae*. Although stemming from the cat branch of the Carnivora, in the Miocene, following the appearance of the first of their ilk in Eurasia, the African civet-like *Plioviverrops* 20 to 15 Mya dog-like hyaenas became specious. *Ictitherium viverrinum* was representative of many dog-like hyaenas that dominated the dog trade until 5–7 Mya when the true dogs, which had been perfecting their trade in North America, crossed the Bering land bridge. Shortly before the arrival in the Old World of these North American émigrés, climate change reduced the dog-like hyaenas to a low ebb, and by the time the climate stabilized and they might have bounced back, the incoming New World dogs were diversifying to put their Old World hyaenid analogues out of business. In those times when hyaenas were dogs, at least 30 species of jackal-like hyaenas outnumbered the ancient canid species of the day—indeed their fossils outnumber those of all other carnivores put together—whereas nowadays, 22 species of Old World canid replace them. Dog-like Miocene hyaenas, like real monogamous dogs, were not much sexually dimorphic in size. Perhaps that explains why congruent spatial arrangements and paternal care shimmer through the behaviour of the four surviving species of modern hyaenas (although neither reproductive

---

[31] For canid aficionados, the rather annoying aphorism that exceptions prove the rule may come to mind regarding the maned wolf and bush dog. The maned wolf is larger than *Cuon* and several *Canis* species, yet its social system is still 'vulpine'—likely because its diet is fox-like (e.g. fruits, insects, and small rodents). Its large body may have simply been an adaptation to allow it to see above the grass, either for food acquisition or to avoid predation. In contrast, the bush dog is the same size as red foxes, yet it follows the *Canis* way of forming large groups with both adult males and adult females, likely because its diet is lupine-like (i.e. hypercarnivorous).

[32] The complexities of carnivore societies increase with the effort devoted to studying them, and thus it starts to emerge that giant otter society is also nuanced. Ribas et al. (2012) found that although groups are largely composed of relatives, there are transfers between groups, extra-group copulations, and subordinate reproduction.

suppression nor regurgitation form part of the hyaena package, and their caching is an underwater variant). Nowadays, of the four, only brown and striped hyaenas approximate dog-like trades, operating in pairs or small spatial groups (Mills 1990), while the termitivorous aardwolf had to sacrifice carrying food home when it gave up the canid-type hallmark of medium-sized prey, and spotted hyaenas are now the wolves of the feliform lineage, where prey dispersion allows (famously in the Ngorongoro Crater—remember those cross-suborder comparisons between spotted hyaenas and the badgers of Wytham Woods in Chapter 1?), occupying congruent multi-female, multi-male clan ranges, with assiduous biparental care (somewhat complicated by reversed sexual dimorphism and penile mimicry, but that's a different story).

To parse out the influences of body size, the distribution of prey sizes, resource dispersion, and phylogeny on the social systems of the Carnivora when viewed through a badger lens, consider three questions that are not as frivolous as they might at first seem: why is a badger not (i) a lion, (ii) a meerkat, or (iii) a jackal? The imaginary transmogrifications reveal much.

A badger, specifically the badgers of Wytham Woods, are not lions, specifically those of the Ngorongoro Crater, solely because of the size of their prey. Both descend from ancestries committed to semi-detached relationships between the single-occupant residents of sexually dimorphic territories that have encountered resource dispersion, which has allowed female matrilineal groups to develop by infilling of female ranges;[33] then facing, however, such rich, shareable but not divisible food patches that they progressed to groups so big that it became profitable for adult males to be spaced congruently with a single female range (Valeix et al. 2012). The fact that badgers operate solitarily and lions do so collectively is merely a detail arising from the differential packaging of the rich patches on which both prosper: for the badger these are usually numerous small clustered items—earthworms; for the lion they are huge items, each a rich patch in itself, but often aggregated as a herd of rich patches.

The answer to why badgers are not meerkats is similar, but this time pivots around the size of the carnivore rather than the size of its prey. Like lions, and also launched from the feliform side of the order, meerkats decend from the herpestid semi-detached intra-sexual polygynous sexually dimorphic territoriality, again infilled as facilitated by resource dispersion, and, as in the case of badgers, these rapidly renewing patches comprise very small prey to suit their diminutive predators (see Carbone et al. 1999, to appreciate further energetic consequences of feeding on very small prey). The situation closely parallels that of the badgers of Wytham Woods, except that meerkats, each perhaps a fifteenth the weight of a badger, forage by day in constant trepidation of jackals and eagles, whereas the musclimorph badger forages by night, defiantly ready to flash its warning stripes at the ghosts of wolves. Finding itself alone, a meerkat becomes almost hysterically anxious. Together, many eyes make light work of the existential task of vigilance and, when the chips are down, the whole seething mega-mongoose mob can fling itself at a marauding jackal.[34]

This brings us back to the question of why badgers are not jackals. As argued earlier, the ancestral spatial

---

[33] The tendencies of sons or daughters to postpone dispersal vary between species: e.g. amongst lions, daughters tend to stay indefinitely, whereas sons disperse before they fall into fatal conflict with their fathers (or incoming males); amongst Japanese badgers it is seemingly daughters that leave first, before their reproduction throws them into conflict with their mothers (remembering that badgers, unlike canids, do not exhibit the clear-cut, despotically determined, reproductive suppression of subordinates typical of alpha-dominated carnivore societies).

[34] In passing, the answer to the question of why badgers are not banded mongooses, while almost the same, is that they almost are! Everything that can be said about diminutive body size and dread of diurnal predators applies equally to both banded mongooses and meerkats, but whereas reproduction amongst meerkats is generally the prerogative of a despotic alpha pair, amongst both bandeds and badgers there is multi-male, multi-female, communal breeding. Why are bandeds more similar to Wytham's badgers than to meerkats in this reproductive regard? Not for the first time, the answer may lie at two levels, an ecological one framing a sociological one. Ecologically, meerkats live in the much harsher, and variable, environment of the Kalahari sands, subsist on lower primary productivity, and hence have smaller body size (much as Arctic foxes are smaller than red foxes), and exist closer to the edge of a perilous energetic cliff. Probably they need all hands on deck to raise a litter, and hence opt for reproductive suppression and assiduous helpers. In contrast, bandeds operate under less demanding energetic strictures, and can get away with a more relaxed, egalitarian approach to reproduction. Nonetheless, dominant bandeds too attempt reproductive suppression whenever possible, engaging in meerkat-like brutalities such as infanticide and eviction of pregnant subordinates, but suppression is nonetheless thwarted by their remarkable synchrony, often to the day, in births—the result is that the adults, even, the mothers, don't seem to know whose pups are whose (and when more females give birth per capita survivorship of young in the den increases) (Marshall et al. 2021). Suppressed female meerkats might well try to get away with similar synchronicity and secrecy if their ecology (and associated food security) would let them, but our guess is that it generally doesn't.

organization of canids differs from the semi-detached sociality at the roots of almost all other carnivore families by being congruent—the minimum social unit is behaviourally monogamous not polygynous (Macdonald et al. 2019). Therefore, exactly the same intersection of body size, distribution of prey sizes, and resource dispersion would lead a mustelid and a canid to infilling according to a different pattern. En route to congruency and a spatial group, the mustelid has an intermediate step, embodied by the Japanese badger (along with the wolverine and American badger) where one male overlaps, in a semi-detached fashion, several female territories each with a small number of female occupants. In contrast, female and male canids, exemplified by jackals (Kamler et al. 2019, 2021), but equally illustrated by Ethiopian wolves (Marino et al. 2012), infill their territories congruently as the dispersion of available prey facilitates the addition of extra members of both sexes. So there are two respects in which badgers are, sociologically, not jackals. First, with respect to a continuum of intra-specific variation, badgers arrive at spatial groups occupying congruent ranges via infilling of semi-detached society and a spatial rearrangement, whereas jackals get there with no step change, but only incremental infilling with additional members. Second, the journey for jackals starts from profound cooperation even between the monogamous members of the minimum social unit, pivoting decisively on paternal care (Macdonald et al. 2019), cooperation that can be built on by communal sharing of alloparental care (e.g. Ethiopian wolves; Sillero-Zubiri et al. 2004), and even close-to-obligate cooperative hunting (e.g. African wild dogs; Courchamp and Macdonald 2001). The journey for badgers starts devoid of cooperation and, to judge by the badgers of Wytham Woods, pretty much stays that way, fuelled only by the ecological selective pressure of resource dispersion. Of course, some badger cousins, notably otters, dabble in paternal care that propels them to congruent ranges, just as it did ancestral canids, and thus culminates fulsomely in the tight-knit groups of giant otter (Groenendijk et al. 2017); they began, however, from a semi-detached starting point. Some canids in some places eat similar diets to badgers—in Boars Hill, just outside Oxford, red foxes eat masses of earthworms, just as badgers do, and they form spatial groups too (Macdonald et al. 2015b: the difference? Paternal and allomaternal care.

It becomes important to consider burrow dwelling—the fossoriality that is a hallmark of badgers—because this is linked to why badgers cannot be either lions or jackals. Lions and jackals are both big enough to hunt large prey and, crucially in the case of jackals, to provision their young. Badgers' current diet is too small to provision young, and their fossorial morphology constrains them from being able to exploit larger prey.[35] As one evolutionary door opens, another closes: by shoe-horning themselves into being fossorial with their more specialized morphology, badgers have limited their ability to expand their niche by, for example, hunting larger prey. And so we turn to fossoriality.

Fossoriality

Perhaps the most obvious thing about badgers, widely known by all naturalists, is that they shelter, communally, in underground setts, and we have argued that the essence of musteloids for the past 30 million years has its roots in burrowing in the Oligocene grasslands. An obvious question, then, is how the use of subterranean dens, fossoriality, intersects, as cause or effect, with their group living, and whether the answer is part of a pattern within the Carnivora. Our former graduate student, Mike Noonan, led our analysis of these questions to discover that fossoriality is an additional factor promoting group living within the Carnivora (Noonan et al. 2015a). Group living had been associated with fossoriality in other taxa (e.g. rodents; Ebensperger 2001), but this link had been insufficiently evaluated in the Carnivora. Hypothesizing (i) that fossoriality will be associated with carnivoran sociality; and (ii) that this association will be most evident in those species making extended use of subterranean dens, we demonstrated from a meta-analysis of key behavioural, ecological, ontological, and trophic traits that three-quarters of carnivore species exhibit some reliance on underground dens, and, furthermore, that congruence between life-history traits and metrics of fossoriality shows that: (i) there are phylogenetic, and morphological constraints on wholly fossorial life histories (you need suitable forelimb anatomy to dig, and you can be too big to squeeze into a burrow); (ii) fossoriality correlated positively with the extent of offspring altriciality, linked to the use of natal dens (helpless

---

[35] Lest we gloss over the untidiness of nature's patterns, remember the conundrum of the American badger: fat, fossorial, hypercarnivorous on burrow-dwelling rodents, hibernating, yet solitary, neither regurgitating nor provisioning food to its young—features that straddle conventional trajectories to both solitary and group living. We interpret this mismatch of characteristics as revealing the constraining power of evolutionary baggage, overlain on the presumed relatively homogeneous dispersion of their rodent prey, which, in an RDH framework, does not facilitate group formation (Chapter 10).

babies need shelter—remember our periscopic views of badger cubs in Chapter 1); (iii) burrow use increased with latitude (it gets colder); and (iv) insectivorous carnivores are more fossorial than predatory carnivores. This last point is the clincher—it arises for two reasons, and it results in species utilizing subterranean natal dens being 2.5 times more likely to form groups than those that do not. The two reasons are that the hunters of very small prey tend themselves to be small, and therefore to be both weedy and to have many larger enemies from which they must seek shelter (e.g. social mongooses); and, second, those very small prey, often invertebrates, tend to occur in spatio-temporally heterogeneous, often rapidly renewing patches that, according to the RDH, facilitate group formation. As explained earlier regarding the distinction between meerkats and badgers, small diurnal ones band together to face their perils, whereas large robust ones can operate alone, but nonetheless choose to cohabit, as badgers do. Our comparative analyses thus showed support for an evolutionary relationship between diet, fossoriality, and sociality (Noonan et al. 2015a). Fossorial dens obviously act as a safe haven but they may also promote territorial inheritance and cooperative breeding, and among smaller (<15 kg) den-using carnivores, and especially for omnivorous/insectivorous species for which food resource dispersion is favourable (such as Wytham's badgers), continued cohabitation at natal dens is associated with cohabitation among adults in philopatric (not necessarily cooperative) spatial groups. Functionally, group philopatry therefore requires that species are big (fat) enough to tolerate periods of food deprivation, but not too big to share a den. But which comes first? Does den occupancy lead to group living, or group living lead to den occupancy? Insofar as group formation is launched by offspring prolonging cohabitation with their mother, their continued sharing of a den is a first step. But when it comes to groups larger than the nuclear mother–offspring hub, the question of what comes first is akin to that posed regarding chicken versus egg, to which the answer is that they are inseparable—part of a syndrome.

At this point it is worth remembering that cuddling is cosy. Indeed, this was in their minds when David and his student Andrew Taber, now a senior scientist in the United Nations Food and Agriculture Organization, studied the communal denning of otherwise strongly standoffish monogamous maras, *Dolichotis patagonum*, in Patagonia—a study that caused David to prompt a less cuddly reaction from the Buenas Aires immigration authorities as he became one of the first Brits to enter Argentina after the Falklands war—a gambit made possible by an amiable Argentinian consul in Paris, but that's another story. Anyway, as elaborated in Box 19.4, one explanation for why these generally hostile pairs of caviomorph rodents declared a truce to deposit their babies in an underground communal crèche was to share warmth by huddling in the chill Patagonian night (Taber and Macdonald 1992a, 1992b; Macdonald et al. 2007b) (an idea previously invoked for burrow-nestling warthog piglets). The benefits of huddling had already been preoccupying David's graduate student, Liz Campbell, who argued that it was a potent force for sociality in Barbary macaques (Campbell et al. 2018). While Liz and David discussed the aforementioned chicken-and-egg question, Liz linked points 3 and 4 of the attributes of burrowers—they predominate in northern latitudes and are insectivorous—into one 'energy conservation' theory: insectivores are those that may face the greatest seasonal energy limitations, more so at higher latitudes, and so their adoption of den use could lend itself to incidental discovery of the benefits of group living for cosy huddling (social thermoregulation)—a benefit that could itself become a selective force for group living and den sharing where ecological circumstances allow, and leading, as Liz suggested, to increased sociality facilitating greater inactivity over winter by reducing heat loss and homeostatic energy costs, possibly leading to coevolution between gregariousness, hibernation, and body size. However, we have no evidence that badgers do huddle, or even that their bodies function collectively, like animated but torpid (Fowler and Racey 1988) hot-water bottles, to raise burrow temperatures, but what we do know thanks to our work with Marie Tsunoda is that badger setts offer their occupants thermal benefits (Tsunoda et al. 2018) (as, we deduce, do those of the solitary American badgers in much harsher climates). Nonetheless, a fruitful future study would use the magnetic resonance technology described in Chapter 4 to detailed underground associations insofar as they effect and reflect the energetic consequences of the thermal microclimate.

A comparison of corpulent communal den diggers (see Box 19.4 'A wisdom of wombats and large burrowers') sheds light on what unites them. All eat patchily available food that results in the smallest economically defendable territory required by a minimum breeding unit being able to accommodate additional secondary occupants at little or no cost. While large, all nonetheless have very good reason to fear, as a threat to themselves and their young, larger predators, from which

---

Box 19.4 A wisdom of wombats and large burrowers

What can we learn from the comparative socio-ecology of communal burrowers? Consider hefty ones comparable, from a tunnelling perspective, to badgers: they all face a substantial allometric biomechanical challenge because burrowing does not scale linearly—the bigger you are the disproportionately tougher it is[36] (Vleck 1981; Noonan et al. 2015a). The engineering problem is unaffected by what you eat, so the largest fossorial insectivores at about 60 kg are aardvarks; next at 30 kg are giant armadillos (*Priodontes maximus*) and finally giant pangolins (*Smutsia gigantea*), weighing in at much the same weight as the biggest burrowing herbivores, marsupial wombats. Of these three, however, it is only the last-named that live socially, and so are of interest here, and encouragingly their collective is known as a wisdom of wombats. They make interesting contrasts in convergence with the similarly herbivorous 8 kg maras, *Dolichotis patagonum*, of which pairs form social petal constellations around a communal den (50 kg capybaras can't burrow; Macdonald et al. 2007b) and, in comparison with 10 kg badgers, the biggest digging carnivores, spotted hyaenas and northern wolves, push the limits at about 50 kg.

First, to a wisdom of wombats, and the intriguingly Wytham badger-like arrangements of the southern hairy-nosed wombat, *Lasiorhinus krefftii*. Communities of these wombats share large, complex warrens. Like badger setts, these excavations become a big investment, with their labyrinthine complexity increasing as they are handed down across generations (Shimmin et al. 2002. Chambers budded off from tunnels are apparently used in part to park juvenile offspring. A colony (or wisdom) of southern hairy-nosed wombats may collectively use 10–20 warrens, of which a single individual may use several, perhaps 10, so each wombat might utilize a different portion of its home range over a number of nights depending on where, within the collective range, it is currently camped (Matthews and Green 2012). Several hairy noses share a warren at a given time, and around each warren their grazing mows a *c.* 1.3–7.8 ha lawn—fantasize on a perfect but biogeographically improper symbiosis, if only badgers were there to access the earthworms surfacing on the lawn and to share the warren excavation (Finlayson et al. 2005; Walker 2006). Underground, each wombat is thought to keep to its own burrow (Taggart 2008), with the interesting occasional exception of sharing by closely related males (Walker 2008). There are gaps between wisdoms, because each occupies a grassy patch separated by an otherwise arid landscape—this system is intriguingly reminiscent of the badger super-groups in Wytham, and as we asked of the badgers in Chapter 10, so too we now ask of the southern hairy noses why, within

the extent of their grassy patch, do they not each occupy a single burrow, rather than cohabiting in these labyrinthine warrens (think setts), or, if they favour a collective, why do they not all occupy one warren?

These questions may be helpfully rephrased with respect to the spatial arrangements of another species of wombat, less like those of Wytham's badgers but more like those of Serra de Grândola. Wisdoms of the northern hairy-nosed wombat, *Lasiorhinus krefftii*, comprise a loose cluster of small warrens, each much like a small badger sett with between one and seven entrances connected by surface trails (Johnson 1991). Up to 10 small warrens may be loosely clustered in a few hectares. The third variation on the wombat theme, colourlessly named the common wombat, *Vombatus ursinus*, with one of which David formed an amiable understanding when it fell snoringly asleep on his lap in Perth Zoo, involves a colony, up to 160 strong, living in ground pockmarked like a Dutch cheese with 20 or so burrows per hectare. Individual common wombats operate alone, each using, and thus sharing, several burrows but in a sort of game of musical chairs; each burrow being occupied by one individual at a time. The particular constellation of burrows used by a particular common wombat perforates a small (*c.* 10 ha) range, traversed by well-worn paths along which a network of wombats lumbers but seldom approaches each other closer than 3 m (Jackson 2003).

So why are badgers not wombats?

There are several more things it is useful to know about wombats, before comparing them with other sizeable communal burrowers. Although they are strict herbivores (eating grasses, sedges, forbs, roots, and bulbs), the name *ursinus* betrays convergence with the bear-like badger (remember Linnaeus's *Ursus meles*, p. xxx). All dig extensive burrow systems, using short, sturdy limbs and powerful claws (cleverly their marsupium pouches open *backwards*, so as not to fill with dirt). While Wytham's badgers may inherit nightmares about wolves, wombats may shudder beneath ground at the ancestral memory of now extinct predators such as *Thylacoleo* and possibly the thylacine. Temperature is an issue for wombats (as it is for badgers; Chapter 14), and their burrows help them avoid getting either too hot or too cold (Jackson 2003); also—again like badgers—they can drop their core body temperature to save energy (and, like badgers, but even more so (grass is less nutritious than worms), are also frugal in producing

---

[36] The generalized relationship between burrow cross-sectional area (A_b) and body mass (M, g) across a variety of burrowing mammals scales as: $A_b = 1.34M^{0.65}$ (White 2005).

Box 19.4 *Continued*

offspring—for wombats, one every 2 years). They rely extensively on olfactory communication (Triggs 1996; Steele and Temple-Smith 1998), using paired cloacal glands that produce a brown secretion (they can distinguish the odours of familiar co-colony members from strangers, to which they react aggressively and, nightly, deposit up to 100 oddly cubic faecal pellets (to allow stacking and avoid rolling)).

On the other side of the world, but also in an arid (this time, cold and stony) desert punctuated by a patchwork of grassy lawns, lives the Patagonian mara (Taber and Macdonald 1992a, 1992b). They too are grazers and use a communal burrow or *cueva*. Reminiscent of Wytham's badgers, this burrow is, like many badger main setts, centrally placed in the community's collective range. When David, and his student Andy Taber (later senior Wildlife Conservation

Society biologist for South America), studied maras in the 1980s, he likened this central den to the calyx of a flower, surrounded by petals—the analogy is apt because each *cueva* is used by several adult pairs (occasionally polygynous trios), who store their infants in the collective den while foraging standoffishly from other pairs in their own 'petal' home range. Each pair visits the *cueva* to nurse its young and, as they grow, to lead them off into the family petal to graze; pairs coinciding at the *cueva*, however, are hostile to each other, and to each other's pups. Nonetheless—and this time in contradistinction to the badgers of Wytham Woods—it pays to be a member of the *cueva* collective: the larger the crèche of young, the better their per capita survival—probably because of both collective diurnal vigilance for, and sanctuary from, predators, and the thermal benefits of huddling.

---

underground sanctuary offers protection. But exactly because they are large, digging those sanctuaries to a safe depth is a mammoth labour. It is a labour most efficiently done collaboratively, and once the investment has been made it becomes an heirloom. QED: from phylogenetically, and dietetically, vastly different starting points, these hefty group livers can form spatial groups because the dispersion of their foods make it cheap (and/or advantageous) to do so, and they den communally because, despite their size they are fearful—fearful especially for their young, which need to be safely (and cosily) stashed while their parents are at work (excepting marsupials who take them in their shopping baskets)—owing to their size, however, the architectural challenge benefits from cooperation, and then becomes an enduring monument to cooperation across generations. A monument so expensive[37] that once built, it is better, if possible, to inherit than to start again.

In a very different context, actually in an essay penned in 2009 as a festschrift for an inspirational correspondent of his student days, fox ethologist Günter Tembrock,[38] David argued that an inter-generational pact was essential to deliver global biodiversity conservation alongside the equitable well-being of future generations of people (Macdonald 2013).[39] How ironic that this, perhaps most abstract and challenging, form of cooperation—perhaps beyond the capacity of humanity as a whole—where the reward spans generations (remembering that the badgers of Wytham Woods still benefit from setts dug by their eleventh-century forebears and recorded in the Doomsday book; Chapter 1), should turn out to be at least partly the basis of the evolutionary success of badgers, which are otherwise so steadfastly uncooperative. This, at least, may be a lesson the current generation of humans could learn from the badgers of Wytham Woods.

---

[37] This argument is calibrated by, but not at odds with, the remarkable digging ability, bestowed by the musclimorph morphology that we saw, in Chapter 10, enables badgers very quickly to excavate new setts (namely the sett dispersion hypothesis). First, we surmise that amongst setts bigger is better, and bigger obviously takes more work. Second, the best sites may be geologically limited (constrainingly so in habitats like Serra de Grândola), and in some places may still be limited by a landscape of fear (e.g. neighbouring wolves). Finally, while badgers may be good at it, there's no denying that even in good soil digging a sett costs work that must be paid for in energy. Nor does the fact that frustrated planners will tell you that badgers displaced by housing developments quickly, and repeatedly, dig new setts ahead of the urban sprawl diminish

the fact that it takes them work to do so and may anyway result in a new-build that is inferior to a centuries old manor.

[38] See Tembrock, G. (1957). Zur Ethologie des Rotfuches (*Vulpes vulpes* L.) unter besonderer Berücksichtigung der Fortpflan zung. *Der Zoologica Garten, 23*, 289–532.

[39] In that essay, David wrote, 'In undertaking a journey it is well to keep in mind the destination... Taking the long view, where might humanity hope to reach with the biodiversity that supports it? One answer might be a human population enjoying a healthy, equitably high and sustainable standard of living, alongside functioning ecosystems populated with "natural" levels of biodiversity.' It was this goal that required an inter-generational pact.

References

Abdou, M., Frankena, K., O'Keeffe, J., and Byrne, A.W. (2016). Effect of culling and vaccination on bovine tuberculosis infection in a European badger (*Meles meles*) population by spatial simulation modelling. *Preventive Veterinary Medicine, 125*, 19–30.

Abramov, A.V. and Puzachenko, A.Y. (2005). Sexual dimorphism of craniological characters in Eurasian badgers, *Meles* spp. (Carnivora, Mustelidae). *Zoologischer Anzeiger—A Journal of Comparative Zoology, 244*(1), 11–29.

Abramov, A.V. and Puzachenko, A.Y. (2013). The taxonomic status of badgers (Mammalia, Mustelidae) from Southwest Asia based on cranial morphometrics, with the redescription of *Meles canescens. Zootaxa, 3681*(1), 44–58.

Adams, D.M. and Wilkinson, G.S. (2020). Male condition and group heterogeneity predict extra-group paternity in a Neotropical bat. *Behavioral Ecology and Sociobiology, 74*(11), 1–12.

Adams, E.S. (2001). Approaches to the study of territory size and shape. *Annual Review of Ecology and Systematics, 32*(1), 277–303.

Adams, G.P. and Ratto, M.H. (2012). Ovulation-inducing factor in seminal plasma: A review. *Animal Reproduction Science, 136*(3), 148–156.

Adams, G.P., Ratto, M.H., Huanca, W., and Singh, J. (2005). Ovulation-inducing factor in the seminal plasma of alpacas and llamas. *Biology of Reproduction, 73*(3), 452–457.

Adkins-Regan, E. (2013). *Hormones and Animal Social Behavior*. Princeton University Press, Princeton, NJ.

Agrell, J., Wolff, J.O., and Ylönen, H. (1998). Counter-strategies to infanticide in mammals: costs and consequences. *Oikos, 83*(3), 507–517.

Agutter, P.S. and Wheatley, D.N. (2004). Metabolic scaling: consensus or controversy? *Theoretical Biology and Medical Modelling, 1*(1), 1–11.

Ahnlund, H. (1980). Sexual maturity and breeding season of the badger *Meles meles* in Sweden. *Journal of Zoology, 190*(1), 77–95.

Akhmetova, A., Guerrero, J., McAdam, P., Salvador, L.C., Crispell, J., Lavery, J., Presho, E., Kao, R.R., Biek, R., Menzies, F., and Trimble, N. (2021). Genomic epidemiology of *Mycobacterium bovis* infection in sympatric badger and cattle populations in Northern Ireland. *bioRxiv*. https://doi.org/10.1101/2021.03.12.435101

Akira, S., Uematsu, S., and Takeuchi, O. (2006). Pathogen recognition and innate immunity. *Cell, 124*(4), 783–801.

Alberts, S.C. (2012). Magnitude and sources of variation in male reproductive performance. In *The Evolution of Primate Societies* (J.C. Mitani, J. Call, P.M. Kappeler, R.A. Palombit, and J.B. Silk, eds, pp. 412–431). University of Chicago Press, Chicago, IL.

Alberts, S.C., Watts, H.E., and Altmann, J. (2003). Queuing and queue-jumping: long-term patterns of reproductive skew in male savannah baboons, *Papio cynocephalus. Animal Behaviour, 65*, 821–840.

Albery, G.F., Newman, C., Ross, J.B., MacDonald, D.W., Bansal, S., and Buesching, C. (2020a). Negative density-dependent parasitism in a group-living carnivore. *Proceedings of the Royal Society of London. Series B: Biological Sciences, 287*(1941), 20202655.

Albery, G.F., Watt, K.A., Keith, R., Morris, S., Morris, A., Kenyon, F., Nussey, D.H., and Pemberton, J.M. (2020b). Reproduction has different costs for immunity and parasitism in a wild mammal. *Functional Ecology, 34*(1), 229–239.

Albery, G.F., Morris, A., Morris, S., Kenyon, F., Nussey, D.H., and Pemberton, J.M. (2021). Fitness costs of parasites explain multiple life-history trade-offs in a wild mammal. *The American Naturalist, 197*(3), 324–335.

Albon, S.D., Clutton-Brock, T.H., and Langvatn, R. (1992). Cohort variation in reproduction and survival: implications for population demography. *The Biology of Deer*, 15–21. doi:10.1007/978-1-4612-2782-3_2

Albon, S.D., Stien, A., Irvine, R.J., Langvatn, R., Ropstad, E., and Halvorsen, O. (2002). The role of parasites in the dynamics of a reindeer population. *Proceedings of the Royal Society of London. Series B: Biological Sciences, 269*(1500), 1625–1632.

Albone, E. (1984) *Mammalian Semiochemistry: the Investigation of Chemical Signals between Vertebrates*. Wiley, Chichester.

Albone, E.S., Gosden, P.E., and Ware, G.C. (1977). Bacteria as a source of chemical signals in mammals. In *Chemical Signals in Vertebrates*, Vol. 1 (D.M. Müller-Schwarze and M.M. Mozell, eds, pp. 35–43). Springer, Boston, MA.

Albone, E.S., Gosden, P.E., Ware, G.C., Macdonald, D.W., and Hough, N.G. (1978). Bacterial action and chemical signalling in the red fox (*Vulpes vulpes*) and other mammals. In *Flavour Chemistry of Animal Foods*, Vol 67 (R.W. Bullard, ed., pp. 78–91). ACS Symposium Series. American Chemical Society, Washington, DC.

Alexander, J. and Stimson, W.H. (1988). Sex hormones and the course of parasitic infection. *Parasitology Today*, 4(7), 189–193.

Alexander, R.D. (1974). The evolution of social behavior. *Annual Review of Ecology and Systematics*, 5(1), 325–383.

Allen, A., Guerrero, J., Byrne, A., Lavery, J., Presho, E., Courcier, E., O'Keeffe, J., Fogarty, U., Delahay, R., Wilson, G., and Newman, C. (2020). Genetic evidence further elucidates the history and extent of badger introductions from Great Britain into Ireland. *Royal Society Open Science*, 7(4), 200288.

Allen, K.J. and Koplin, J.J. (2012). The epidemiology of IgE-mediated food allergy and anaphylaxis. *Immunology and Allergy Clinics*, 32(1), 35–50.

Allen, T.M. (2020). Honest fitness advertisement in European badgers (*Meles meles*). DPhil thesis, University of Oxford.

Allen, T.M., Sugianto, N.A., Ryder, C., Newman, C., Macdonald, D.W., and Buesching, C.D. (2019). Encoded information within urine influences behavioural responses among European badgers (*Meles meles*). In *Chemical Signals in Vertebrates 14* (C.D. Buesching, ed., pp. 38–59). Springer, Cham.

Allen, T., Newman, C., Macdonald, D.W. and C.D. Buesching (in press). The rules of effective advertisment for European badgers (Meles meles). *Chemical Signals in Vertebrates*, 15.

Alley, R.B. (2000a). Ice-core evidence of abrupt climate changes. *Proceedings of the National Academy of Science*, 97(4), 1331–1334.

Alley, R.B. (2000b). The Younger Dryas cold interval as viewed from central Greenland. *Quaternary Science Reviews*, 19(1), 213–226.

Almberg, E.S., Cross, P.C., Dobson, A.P., Smith, D.W., Metz, M.C., Stahler, D.R., and Hudson, P.J. (2015). Social living mitigates the costs of a chronic illness in a cooperative carnivore. *Ecology Letters*, 18, 660–667. doi:10.1111/ele.12444

Aloise King, E.D., Banks, P.B., and Brooks, R.C. (2013). Sexual conflict in mammals: consequences for mating systems and life history. *Mammal Review*, 43(1), 47–58.

Alonso-Alvarez, C., Bertrand, S., Faivre, B., and Sorci, G. (2007). Increased susceptibility to oxidative damage as a cost of accelerated somatic growth in zebra finches. *Functional Ecology*, 21(5), 873–879.

Alt, G.L., Matula Jr, G.J., Alt, F.W., and Lindzey, J.S. (1980). Dynamics of home range and movements of adult black bears in northeastern Pennsylvania. *Bears: Their Biology and Management*, 4, 131–136.

Alvarez, H.P. (2000). Grandmother hypothesis and primate life histories. *American Journal of Physical Anthropology: The Official Publication of the American Association of Physical Anthropologists*, 113(3), 435–450.

Ameni, G., Vordermeier, M., Aseffa, A., Young, D.B., and Hewinson, R.G. (2010). Field evaluation of the efficacy of *Mycobacterium bovis* bacillus Calmette-Guerin against bovine tuberculosis in neonatal calves in Ethiopia. *Clinical and Vaccine Immunology*, 17(10), 1533–1538.

Amlaner, C.J. and Macdonald, D.W. (eds) (1980). *A Handbook on Biotelemetry and Radio Tracking: Proceedings of an International Conference on Telemetry and Radio Tracking in Biology and Medicine, Oxford, 20–22 March 1979*. Elsevier, Abingdon.

Amos, W., Brooks-Pollock, E., Blackwell, R., Driscoll, E., Nelson-Flower, M., and Conlan, A.J. (2013). Genetic predisposition to pass the standard SICCT test for bovine tuberculosis in British cattle. *PLoS ONE*, 8(3), e58245.

Amstislavsky, S. and Ternovskaya, Y. (2000). Reproduction in mustelids. *Animal Reproduction Science*, 60, 571–581.

Anderson, K.J. and Jetz, W. (2005). The broad-scale ecology of energy expenditure of endotherms. *Ecology Letters*, 8(3), 310–318.

Anderson, R.M. and May, R.M. (1978). Regulation and stability of host-parasite population interactions: I. Regulatory processes. *The Journal of Animal Ecology*, 47(1), 219–247.

Anderson, R.M. and Trewhella, W. (1985). Population dynamics of the badger (*Meles meles*) and the epidemiology of bovine tuberculosis (*Mycobacterium bovis*). *Proceedings of the Royal Society of London. Series B: Biological Sciences*, 310(1145), 327–381.

Anderson, R.M. and Weindruch, R. (2010). Metabolic reprogramming, caloric restriction and aging. *Trends in Endocrinology & Metabolism*, 21(3), 134–141.

Andersson, K.I. (2004). Elbow-joint morphology as a guide to forearm function and foraging behaviour in mammalian carnivores. *Zoological Journal of the Linnean Society*, 142, 91–104. doi: 10.1111/j.1096-3642.2004.00129.x

Andersson, M. and Simmons, L.W. (2006). Sexual selection and mate choice. Trends in Ecology & Evolution, 21(6), 296–302.

Andrén, H., Persson, J., Mattisson, J. and Danell, A.C. (2011). Modelling the combined effect of an obligate predator and a facultative predator on a common prey: lynx *Lynx lynx* and wolverine *Gulo gulo* predation on reindeer *Rangifer tarandus*. *Wildlife Biology*, 17(1), 33–43.

Andrewartha, H.G. (1961). *Introduction to the Study of Animal Populations* (No. 591.5 A53). Springer, Berlin, Heidelberg.

Andrewartha, H.G. (2012). *Introduction to the Study of Animal Populations*, reprint edition. Springer Science & Business Media, Berlin, Heidelberg.

Andrewartha, H.G. and Birch, L.C. (1986). *The Ecological Web: More on the Distribution and Abundance of Animals*. University of Chicago Press, Chicago, IL.

Angerbjörn, A., Tannerfeldt, M., and Erlinge, S. (1999). Predator–prey relationships: arctic foxes and lemmings. *Journal of Animal Ecology*, 68(1), 34–49.

Angulo, E., Rasmussen, G.S., Macdonald, D.W., and Courchamp, F. (2013). Do social groups prevent Allee effect related extinctions?: the case of wild dogs. *Frontiers in Zoology*, 10(1), 11.

Anholt, B.R. and Werner, E.E. (1995). Interaction between food availability and predation mortality mediated by adaptive behavior. *Ecology*, 76(7), 2230–2234.

Annavi, G. (2012). Genetic, socio-ecological and fitness correlates of extra-group paternity in the European badger (*Meles meles*). PhD thesis, University of Oxford.

Annavi, G., Dawson, D.A., Horsburgh, G.J., Greig, C., Dugdale, H.L., Newman, C., Macdonald, D.W., and Burke, T. (2011). Characterisation of twenty-one European badger (*Meles meles*) microsatellite loci facilitates the discrimination of second-order relatives. *Conservation Genetics Resources*, 3(3), 515–518.

Annavi, G., Newman, C., Buesching, C., Burke, T., Macdonald, D.W., and Dugdale, H.L. (2014a). Heterozygosity–fitness correlations in a wild mammal population: single locus, paternal and environmental effects. *Ecology and Evolution*, 4(12), 2594–2609. doi:10.1002/ece3.1112

Annavi, G., Newman, C., Dugdale, H.L., Buesching, C., Sin, Y.W., Burke, T., and Macdonald, D.W. (2014b). Neighbouring-group composition and relatedness drive extra-group paternity rate in the European badger (*Meles meles*). *Journal of Evolutionary Biology*, 27(10), 2191–2203. doi:10.1111/jeb.12473

Anwar, M.A., Newman, C., MacDonald, D.W., Woolhouse, M.E.J., and Kelly, D.W. (2000). Coccidiosis in the European badger (*Meles meles*) from England, an epidemiological study. *Parasitology*, 120(3), 255–260.

Anwar, A., Knaggs, J., Service, K.M., McLaren, G.W., Riordan, P., Newman, C., Delahay, R.J., Cheesman, C., and Macdonald, D.W. (2006). Antibodies to *Toxoplasma gondii* in Eurasian badgers. *Journal of Wildlife Diseases*, 42(1), 179–181.

Apps, P.J. (2013). Are mammal olfactory signals hiding right under our noses? *Naturwissenschaften*, 100, 487–506. doi:10.1007/s00114-013-1054-1 PMID: 23674106

Apps, P.J., Viljoen, H.W., Richardson, P.R.K., and Pretorius, V. (1989). Volatile components of anal gland secretion of aardwolf (*Proteles cristatus*). *Journal of Chemical Ecology*, 15(5), 1681–1688.

Apps, P., Rafiq, K., and McNutt, J.W. (2019). Do carnivores have a world wide web of interspecific scent signals? In *Chemical Signals in Vertebrates 14* (C.D. Buesching, ed., pp. 182–202). Springer, Cham.

Aramini, J.J., Stephen, C., and Dubey, J.P. (1998). *Toxoplasma gondii* in Vancouver Island cougars (*Felis concolor vancouverensis*): serology and oocyst shedding. *The Journal of Parasitology*, 84(2), 438–440.

Araújo, M.B., Whittaker, R.J., Ladle, R.J., and Erhard, M. (2005). Reducing uncertainty in projections of extinction risk from climate change. *Global Ecology and Biogeography*, 14(6), 529–538.

Archer, C.R., Basellini, U., Hunt, J., Simpson, S.J., Lee, K.P., and Baudisch, A. (2018). Diet has independent effects on the pace and shape of aging in *Drosophila melanogaster*. *Biogerontology*, 19(1), 1–12.

Archie, E.A., Altmann, J., and Alberts, S.C. (2012). Social status predicts wound healing in wild baboons. *Proceedings of the National Academy of Sciences USA*, 109, 9017–9022.

Arendt, J.D. (2009). Influence of sprint speed and body size on predator avoidance in New Mexican spadefoot toads (*Spea multiplicata*). *Oecologia*, 159(2), 455–461.

Ariew, A. and Lewontin, R.C. (2004). The confusions of fitness. *The British Journal for the Philosophy of Science*, 55(2), 347–363.

Aristotle (n.d.). *Politics, Book 2*. http://data.perseus.org/citations/urn:cts:greekLit:tlg0086.tlg035.perseus-eng1:2.1270b (accessed 4 March 2022).

Armanios, M. and Blackburn, E.H. (2012). The telomere syndromes. *Nature Reviews Genetics*, 13(10), 693–704.

Arnqvist, G. and Kirkpatrick, M. (2005). The evolution of infidelity in socially monogamous passerines: the strength of direct and indirect selection on extrapair copulation behavior in females. *The American Naturalist*, 165(S5), S26–S37.

Asa, C.S. and Valdespino, C. (1998). Canid reproductive biology: an integration of proximate mechanisms and ultimate causes. *American Zoologist*, 38(1), 251–259.

Asghar, M., Bensch, S., Tarka, M., Hansson, B., and Hasselquist, D. (2015). Maternal and genetic factors determine early life telomere length. *Proceedings of the Royal Society. Series B: Biological Sciences*, 282(1799), 20142263.

Atanackovic, D., Atanackovic, D., Brunner-Weinzierl, M.C., Kröger, H., Serke, S., and Deter, H.C. (2002). Acute psychological stress simultaneously alters hormone levels, recruitment of lymphocyte subsets, and production of reactive oxygen species. *Immunological Investigations*, 31(2), 73–91.

Atkins, P.J. and Robinson, P.A. (2013). Bovine tuberculosis and badgers in Britain: relevance of the past. *Epidemiology & Infection*, 141(7), 1437–1444. doi:10.1017/S095026881200297X.

Atkinson, R.P.D., Macdonald, D.W., and Kamizola, R. (2002). Dietary opportunism in side-striped jackals *Canis adustus*. *Journal of Zoology*, 257, 129–139.

Atkinson, S.N. and Ramsay, M.A. (1995). The effects of prolonged fasting of the body composition and reproductive success of female polar bears (*Ursus maritimus*). *Functional Ecology*, 9(4), 559–567.

Aubert, G. and Lansdorp, P.M. (2008). Telomeres and aging. *Physiological Reviews*, 88(2), 557–579.

Aureli, F., Schaffner, C.M., Boesch, C., Bearder, S.K., Call, J., Chapman, C.A., and Holekamp, K. (2008). Fission-fusion dynamics. *Current Anthropology*, 49(4), 627–654.

Austad, S.N. (1997). Comparative aging and life histories in mammals. *Experimental Gerontology*, 32(1–2), 23–38.

Avgil, M. and Ornoy, A. (2006). Herpes simplex virus and Epstein-Barr virus infections in pregnancy: consequences of neonatal or intrauterine infection. *Reproductive Toxicology*, 21(4), 436–445.

Axelrod, R. (1980). Effective choice in the prisoner's dilemma. *Journal of Conflict Resolution*, 24(1), 3–25.

Bachorec, E., Horáček, I., Hulva, P., Konečný, A., Lučan, R.K., Jedlička, P. et al. (2020). Spatial networks differ when food supply changes: foraging strategy of Egyptian fruit bats. *PLoS ONE*, 15(2), e0229110. https://doi.org/10.1371/journal.pone.0229110

Bacon, P.J. (1985). *Population Dynamics of Rabies in Wildlife*. Academic Press, Cambridge, MA.

Bacon, P.J. and Macdonald, D.W. (1980). To control rabies: vaccinate foxes. *New Scientist*, 87(1216), 640–645.

Bacon, P.J. and Macdonald, D.W. (1981). Habitat classification, fox populations and rabies spread. Nature, 289(5799), 634–635.

Bacon, P.J., Ball, F., and Blackwell, P. (1991a). Analysis of a model of group territoriality based on the resource dispersion hypothesis. *Journal of Theoretical Biology*, 148(4), 433–444.

Bacon, P.J., Ball, F., and Blackwell, P. (1991b). A model for territory and group formation in a heterogeneous habitat. *Journal of Theoretical Biology*, 148(4), 445–468.

Badyaev, A.V. (2002). Growing apart: an ontogenetic perspective on the evolution of sexual size dimorphism. *Trends in Ecology & Evolution*, 17(8), 369–378.

Baer, D.J., Gebauer, S.K., and Novotny, J.A. (2016). Walnuts consumed by healthy adults provide less available energy than predicted by the Atwater factors. *The Journal of Nutrition*, 146(1), 9–13.

Baker, K.C. and Aureli, F. (2000). Coping with conflict during initial encounters in chimpanzees. *Ethology*, 106(6), 527–541. doi:10.1111/j.1439-0310.2000.00553.x

Baker, R. (2014). *Human Sperm Competition*. eBook Partnership, London.

Baker, S.E. and Macdonald, D.W. (2015). Managing wildlife humanely with learned food aversions. In *Wildlife Conservation on Farmland, Volume 2: Conflict in the Countryside* (D.W. Macdonald and R.E. Feber, eds, pp. 260–275). Oxford University Press, Oxford.

Baker, S.E., Ellwood, S.A., Watkins, R., and MacDonald, D.W. (2005). Non-lethal control of wildlife: using chemical repellents as feeding deterrents for the European badger Meles meles . *Journal of Applied Ecology*, 42(5), 921–931.

Baker, S.E., Johnson, P.J., Slater, D., Watkins, R., and Macdonald, D.W. (2007). Learned food aversion with and without an odour cue for protecting untreated baits from wild mammal foraging. *Applied Animal Behaviour Science*, 102(3–4), 410–428.

Bakke, J., Lie, Ø., Heegaard, E., Dokken, T., Haug, G.H., Birks, H.H., Dulski, P., and Nilsen, T. (2009). Rapid oceanic and atmospheric changes during the Younger Dryas cold period. *Nature Geoscience*, 2(3), 202–205.

Bakker, J. and Baum, M.J. (2000). Neuroendocrine regulation of GnRH release in induced ovulators. *Frontiers in Neuroendocrinology*, 21(3), 220–262.

Balestrieri, A., Remonti, L., Ferrari, N., Ferrari, A., Valvo, T.L., Robetto, S., and Orusa, R. (2006). Sarcoptic mange in wild carnivores and its co-occurrence with parasitic helminths in the Western Italian Alps. *European Journal of Wildlife Research*, 52(3), 196–201.

Balestrieri, A., Remonti, L., and Prigioni, C. (2009). Exploitation of food resources by the Eurasian badger (*Meles meles*) at the altitudinal limit of its Alpine range (NW Italy). *Zoological Science*, 26(12), 821–827.

Balestrieri, A., Cardarelli, E., Pandini, M., Remonti, L., Saino, N., and Prigioni, C. (2016). Spatial organisation of European badger (*Meles meles*) in northern Italy as assessed by camera-trapping. *European Journal of Wildlife Research*, 62(2), 219–226.

Balestrieri, A., Remonti, L., Saino, N., and Raubenheimer, D. (2019). The 'omnivorous badger dilemma': towards an integration of nutrition with the dietary niche in wild mammals. *Mammal Review*, 49(4), 324–339.

Ball, F.G. (1991). Dynamic population epidemic models. *Mathematical Biosciences*, 107(2), 299–324.

Balme, G.A. and Hunter, L.T. (2013). Why leopards commit infanticide. *Animal Behaviour*, 86(4), 791–799.

Baracchi, D., Fadda, A., and Turillazzi, S. (2012). Evidence for antiseptic behaviour towards sick adult bees in honey bee colonies. *Journal of Insect Physiology*, 58, 1589–1596. doi:10.1016/j.jinsphys.2012.09.014

Baracos, V.E., Whitmore W.T., and Gale, R. (1987). The metabolic cost of fever. *Canadian Journal of Physiology and Pharmacology*, 65(6), 1248–1254. doi:10.1139/y87-199

Barclay, R.M., Harder, L.D., Kunz, T.H., and Fenton, M.B. (2003). Life histories of bats: life in the slow lane. *Bat Ecology*, 209, 253.

Bårdsen, B.J., Fauchald, P., Tveraa, T., Langeland, K., Yoccoz, N.G., and Ims, R.A. (2008). Experimental evidence of a risk-sensitive reproductive allocation in a long-lived mammal. *Ecology*, 89(3), 829–837.

Barea-Azcón, J.M., Ballesteros-Duperón, E., Gil-Sánchez, J.M., and Virgós, E. (2010). Badger *Meles meles* feeding ecology in dry Mediterranean environments of the southwest edge of its distribution range. *Acta Theriologica*, 55(1), 45–52.

Barker, D.J. (1997). Maternal nutrition, fetal nutrition, and disease in later life. *Nutrition*, 13(9), 807–813.

Barker, D.J., Godfrey, K.M., Gluckman, P.D., Harding, J.E., Owens, J.A., and Robinson, J.S. (1993). Fetal nutrition and cardiovascular disease in adult life. *The Lancet*, 341(8850), 938–941.

Barkham, P. (2014). *Badgerlands: the Twilight World of Britain's Most Enigmatic Animal*. Granta, London.

Barlow, A.M., Schock, A., Bradshaw, J., Mullineaux, E., Dastjerdi, A., Everest, D.J., McGowan, S., Steinbach, F., and Cowen, S. (2012). Parvovirus enteritis in Eurasian badgers (*Meles meles*). *Veterinary Record*, 170(16), 416.

Barnosky, A.D. (2001). Distinguishing the effects of the Red Queen and Court Jester on Miocene mammal evolution in the northern Rocky Mountains. *Journal of Vertebrate Paleontology*, 21(1), 172–185.

Barnosky, A.D. and Kraatz, B.P. (2007). The role of climatic change in the evolution of mammals. *Bioscience*, 57(6), 523–532.

Barocas, A. and Ben-David, M. (2021). Inter and intraspecific variation in the social structure of marine otters. In *Ethology of Sea and Marine Otters and the Polar Bear* (R. Davis and A. Pagano, eds, pp. 83–105). Springer International, Berlin, Heidelberg.

Barocas A., Golden, H.N., Harrington. M., McDonald, D.B., and Ben-David, M. (2016). Coastal latrine sites as social information hubs and drivers of river otter fission-fusion dynamics. *Animal Behavior*, 120, 103–114.

Barocas, A., Flores, J.A., Pardo, A.A., Macdonald, D.W., and Swaisgood, R.R. (2021). Reduced dry season fish biomass and depleted carnivorous fish assemblages in unprotected tropical oxbow lakes. *Biological Conservation*, 257, 109090.

Barot, S., Rossi, J.P., and Lavelle, P. (2007). Self-organization in a simple consumer–resource system, the example of earthworms. *Soil Biology and Biochemistry*, 39(9), 2230–2240.

Barrett, L. and Henzi, S.P. (2006). Monkeys, markets and minds: biological markets and primate sociality. In *Cooperation in Primates and Humans* (C.P. Van Schaik and P.M. Kappeler, pp. 209–232). Springer, Berlin, Heidelberg.

Bateson, P. (1979). How do sensitive periods arise and what are they for? *Animal Behaviour*, 27, 470–486.

Bateson, P. and Gluckman, P. (2012). Plasticity and robustness in development and evolution. *International Journal of Epidemiology*, 41(1), 219–223.

Bateson, P., Barker, D., Clutton-Brock, T., Deb, D., D'Udine, B., Foley, R.A., Gluckman, P., Godfrey, K., Kirkwood, T., Lahr, M.M., and McNamara, J. (2004). Developmental plasticity and human health. *Nature*, 430(6998), 419–421.

Bateson, P., Gluckman, P., and Hanson, M. (2014). The biology of developmental plasticity and the predictive adaptive response hypothesis. *The Journal of Physiology*, 592(11), 2357–2368.

Beall, C.M. (2007). Two routes to functional adaptation: Tibetan and Andean high-altitude natives. *Proceedings of the National Academy of Sciences*, 104(Suppl. 1), 8655–8660.

Beall, C.M., Decker, M.J., Brittenham, G.M., Kushner, I., Gebremedhin, A., and Strohl, K.P. (2002). An Ethiopian pattern of human adaptation to high-altitude hypoxia. Proceedings of the National Academy of Sciences, 99(26), 17215–17218.

Bechtold, D.A., Sidibe, A., Saer, B.R., Li, J., Hand, L.E., Ivanova, E.A., Darras, V.M., Dam, J., Jockers, R., Luckman, S.M., and Loudon, A.S. (2012). A role for the melatonin-related receptor GPR50 in leptin signaling, adaptive thermogenesis, and torpor. *Current Biology*, 22(1), 70–77.

Becker, D.J., Albery, G.F., Kessler, M.K., Lunn, T.J., Falvo, C.A., Czirják, G., Martin, L.B., and Plowright, R.K. (2019). Macroimmunology: the drivers and consequences of spatial patterns in wildlife immune defense. *Journal of Animal Ecology*, 89, 972–995. doi:10.1111/1365-2656.13166

Becker, P.J., Reichert, S., Zahn, S., Hegelbach, J., Massemin, S., Keller, L.F., Postma, E., and Criscuolo, F. (2015). Mother–offspring and nest-mate resemblance but no heritability in early-life telomere length in white-throated dippers. *Proceedings of the Royal Society. Series B: Biological Sciences*, 282(1807), 20142924.

Beckman, K.B. and Ames, B.N. (1998). The free radical theory of aging matures. *Physiological Reviews*, 78(2), 547–581.

Beehner, J.C. and Lu, A. (2013). Reproductive suppression in female primates: a review. *Evolutionary Anthropology: Issues, News, and Reviews*, 22(5), 226–238.

Beehner, J.C., Gesquiere, L., Seyfarth, R.M., Cheney, D.L., Alberts, S.C., and Altmann, J. (2009). Testosterone related to age and life-history stages in male baboons and geladas. *Hormones and Behavior*, 56(4), 472–480.

Beeler, J.A., Frazier, C.R., and Zhuang, X. (2012). Putting desire on a budget: dopamine and energy expenditure, reconciling reward and resources. *Frontiers in Integrative Neuroscience*, 6, 49.

Beer, A.E., Quebbeman, J.F., Ayers, J.W., and Haines, R.F. (1981). Major histocompatibility complex antigens, maternal and paternal immune responses, and chronic habitual abortions in humans. *American Journal of Obstetrics and Gynecology*, 141(8), 987–999.

Beerda, B., Schilder, M.B., Janssen, N.S., and Mol, J.A. (1996). The use of saliva cortisol, urinary cortisol, and catecholamine measurements for a noninvasive assessment of stress responses in dogs. *Hormones and Behavior*, 30(3), 272–279.

Beever, E.A., Hall, L.E., Varner, J., Loosen, A.E., Dunham, J.B., Gahl, M.K., Smith, F.A., and Lawler, J.J. (2017). Behavioral flexibility as a mechanism for coping with climate change. *Frontiers in Ecology and the Environment*, 15, 299–308.

Bégeot, C., Richard, H., Ruffaldi, P., and Bossuet, G. (2000). Palynological record of Bolling/Allerod Interstadial climatic changes in eastern France. *Bulletin de la Societe Geologique de France*, 171(1), 51–58.

Begg, C.M. (2001). Feeding ecology and social organisation of honey badgers (*Mellivora capensis*) in the southern Kalahari. DPhil thesis, University of Pretoria, Pretoria.

Begg, C.M., Begg, K.S., Du Toit, J.T., and Mills, M.G.L. (2003). Sexual and seasonal variation in the diet and foraging behaviour of a sexually dimorphic carnivore, the honey badger (*Mellivora capensis*). *Journal of Zoology*, 260(3), 301–316.

Begg, C.M., Begg, K.S., Du Toit, J.T., and Mills, M.G.L. (2005). Spatial organization of the honey badger Mellivora capensis in the southern Kalahari: home-range size and movement patterns. *Journal of Zoology*, 265(1), 23–35.

Behrman, S.J., Buettner-Janusch, J., Heglar, R., Gershowitz, H., and Tew, W.L. (1960). ABO (H) blood incompatibility as a cause of infertility: a new concept. *American Journal of Obstetrics & Gynecology*, 79(5), 847–855.

Bekoff, M., Daniels, T.J., and Gittleman, J.L. (1984). Life history patterns and the comparative social ecology of carnivores. *Annual Review of Ecology and Systematics*, 15(1), 191–232.

Belfast Telegraph (2009). Crackdown appeal after swoops on badger-baiting ring. https://www.belfasttelegraph.co.uk/news/crackdown-appeal-after-swoops-on-badger-baiting-ring-28467973.html (accessed 11 March 2022).

Belich, J. (2016). *The Prospect of Global History*. Oxford University Press, Oxford.

Bellemain, E., Swenson, J.E., and Taberlet, P. (2006). Mating strategies in relation to sexually selected infanticide in a non-social carnivore: the brown bear. *Ethology*, 112(3), 238–246.

Bellisari, A. (2008). Evolutionary origins of obesity. *Obesity Reviews*, 9(2), 165–180.

Benjamini, E., Feingold, B.F., Young, J. D., Kartman, L., and Shimizu, M. (1963). Allergy to flea bites. IV. In vitro collection and antigenic properties of the oral secretion of the cat flea, *Ctenocephalides felis felis* (Bouche). *Experimental Parasitology*, 13(2), 143–154.

Benmarhnia, T., Kihal-Talantikite, W., Ragettli, M.S., and Deguen, S. (2017). Small-area spatiotemporal analysis of heatwave impacts on elderly mortality in Paris: a cluster analysis approach. *Science of the Total Environment*, 592, 288–294.

Benton, C.H., Phoenix, J., Smith, F.A., Robertson, A., McDonald, R.A., Wilson, G., and Delahay, R.J. (2020). Badger vaccination in England: progress, operational effectiveness and participant motivations. *People and Nature*, 2, 761–775.

Benton, M.J. (2009). The red queen and the court jester: species diversity and the role of biotic and abiotic factors through time. *Science*, 323(5915), 728–732.

Berdoy, M. and Macdonald, D.W. (2005). Multiple male mating in brown rats: male coercion or female promiscuity. In *The Second WildCRU Review: Another Ten Years of Conservation Research* (D.W. Macdonald, ed., pp. 250–251). WildCRU, Oxford.

Berdoy, M., Smith, P., and Macdonald, D.W. (1995a). Stability of social status in wild rats: age and the role of settled dominance. *Behaviour*, 132, 193–212.

Berdoy, M., Webster, J.P., and Macdonald, D.W. (1995b). Parasite-altered behaviour: is the effect of *Toxoplasma gondii* on *Rattus norvegicus* specific? *Parasitology*, 111(4), 403–409.

Berdoy, M., Webster, J.P., and Macdonald, D.W. (2000). Fatal attraction in rats infected with *Toxoplasma gondii*. *Proceedings of the Royal Society of London. Series B: Biological Sciences*, 267(1452), 1591–1594.

Berteaux, D., Réale, D., McAdam, A.G., and Boutin, S. (2004). Keeping pace with fast climate change: can arctic life count on evolution? *Integrative and Comparative Biology*, 44(2), 140–151.

Bevanger, K. and Lindström, E.R. (1995, January). Distributional history of the European badger *Meles meles* in Scandinavia during the 20th century. *Annales Zoologici Fennici*, 32, 5–9.

Bianchi, E., Doe, B., Goulding, D., and Wright, G.J. (2014). Juno is the egg Izumo receptor and is essential for mammalian fertilization. *Nature*, 508, 483–487.

Bidstats (2021). Project 30072—Farmer-delivered, East Sussex Badger Vaccination Pilot. https://bidstats.uk/tenders/2021/W14/748421618 (accessed 10 March 2022).

Bielby, J., Donnelly, C.A., Pope, L.C., Burke, T., and Woodroffe, R. (2014). Badger responses to small-scale culling may compromise targeted control of bovine tuberculosis. *Proceedings of the National Academy of Sciences*, 111(25), 9193–9198.

Biggins, D.E. (2012). Use of multi-opening burrow systems by black-footed ferrets. *Western North American Naturalist*, 72, 134–139.

Bigham, A.W. (2016). Genetics of human origin and evolution: high-altitude adaptations. *Current Opinion in Genetics and Development*, 41, 8–13.

Bilham, K., Boyd, A.C., Preston, S.G., Buesching, C.D., Newman, C., Macdonald, D.W., and Smith, A.L. (2017). Badger macrophages fail to produce nitric oxide, a key antimycobacterial effector molecule. *Scientific Reports*, 7(1), 1–10.

Bilham, K., Sin, Y.W., Newman, C., Buesching, C.D., and Macdonald, D.W. (2013). An example of life history antecedence in the European badger (*Meles meles*): rapid development of juvenile antioxidant capacity, from plasma vitamin E analogue. *Ethology Ecology & Evolution*, 25(4), 330–350.

Bilham, K., Newman, C., Buesching, C.D., Noonan, M.J., Boyd, A., Smith, A.L., and Macdonald, D.W. (2018). Effects of weather conditions on oxidative stress, oxidative damage, and antioxidant capacity in a wild-living mammal, the European badger (*Meles meles*). *Physiological and Biochemical Zoology*, 91(4), 987–1004.

Bilz, R. (1940). *Pars pro toto Ein Beitrag zur Pathologie menschlicher Affekte und Organfunktionen*. Georg Thieme Verlag, Leipzig.

Björkman, O. (1966). The effect of oxygen concentration on photosynthesis in higher plants. *Physiologia Plantarum*, 19(3), 618–633.

Blackburn, D. and Evans, H. (1986). Why are there no viviparous birds? *The American Naturalist*, 128(2), 165–190.

Blackburn, E.H. (2000). Telomere states and cell fates. *Nature*, 408(6808), 53–56.

Blackburn, E.H., Greider, C.W., Henderson, E., Lee, M.S., Shampay, J., and Shippen-Lentz, D. (1989). Recognition and elongation of telomeres by telomerase. *Genome*, 31(2), 553–560.

Blackwell, P. (1990). Deterministic and stochastic models of social behaviour based on the resource dispersion hypothesis. *Mathematical Medicine and Biology: A Journal of the IMA*, 7(4), 261–279.

Blackwell, P.G. (2007). Heterogeneity, patchiness and correlation of resources. *Ecological Modelling*, 207(2–4), 349–355.

Blackwell, P.G. and Macdonald, D.W. (2000). Shapes and sizes of badger territories. *Oikos*, 89(2), 392–398.

Blanckenhorn, W.U., Preziosi, R.F., and Fairbairn, D.J. (1995). Time and energy constraints and the evolution of sexual size dimorphism—to eat or to mate? *Evolutionary Ecology*, 9(4), 369–381.

Blehert, D.S., Hicks, A.C., Behr, M., Meteyer, C.U., Berlowski-Zier, B.M., Buckles, E.L., and Stone, W.B. (2009). Bat white-nose syndrome: an emerging fungal pathogen? *Science*, 323(5911), 227–227.

Bleu, J., Gamelon, M., and Sæther, B.E. (2016). Reproductive costs in terrestrial male vertebrates: insights from bird studies. *Proceedings of the Royal Society. Series B: Biological Sciences*, 283(1823), 20152600.

Blomberg, S.P. and Garland Jr, T. (2002). Tempo and mode in evolution: phylogenetic inertia, adaptation and comparative methods. *Journal of Evolutionary Biology*, *15*(6), 899–910.

Blundell, G.M., Ben-David, M., and Bowyer, R.T. (2002a) Sociality in river otters: cooperative foraging or 562 reproductive strategies? *Behavioral Ecology*, *13*, 134–141.

Blundell, G.M., Ben-David, M., Groves, P., Bowyer, R.T., and Geffen, E. (2002b). Characteristics of sex-biased dispersal and gene flow in coastal river otters: implications for natural recolonization of extirpated populations. *Molecular Ecology*, *11*(3), 289–303.

Böhm, M., Hutchings, M.R., and White, P.C. (2009). Contact networks in a wildlife-livestock host community: identifying high-risk individuals in the transmission of bovine TB among badgers and cattle. *PLoS ONE*, *4*(4), e5016.

Bollmer, J.L., Dunn, P.O., Freeman-Gallant, C.R., and Whittingham, L.A. (2012). Social and extra-pair mating in relation to major histocompatibility complex variation in common yellowthroats. *Proceedings of the Royal Society. Series B: Biological Sciences*, *279*(1748), 4778–4785.

Bolnick, D.I., Svanbäck, R., Araújo, M.S., and Persson, L. (2007). Comparative support for the niche variation hypothesis that more generalized populations also are more heterogeneous. *Proceedings of the National Academy of Sciences*, *104*(24), 10075–10079.

Bolnick, D.I., Amarasekare, P., Araújo, M. S., Bürger, R., Levine, J.M., Novak, M., and Vasseur, D.A. (2011). Why intraspecific trait variation matters in community ecology. *Trends in Ecology & Evolution*, *26*(4), 183–192.

Bolton, P.J. (1969). Studies in the general ecology, physiology and bioenergetics of woodland Lumbricidae. PhD thesis, Durham University.

Bonacic, C., & Macdonald, D.W. (2003). The physiological impact of wool-harvesting procedures in vicunas (Vicugna vicugna). *ANIMAL WELFARE-POTTERS BAR THEN WHEATHAMPSTEAD-*, *12*(3), 387–402.

Bonacic, C., Macdonald, D.W., and Villouta, G. (2003). Adrenocorticotrophin-induced stress response in captive vicunas (*Vicugna vicugna*) in the Andes of Chile. *Animal Welfare*, *12*(3), 369–386.

Bond, G.C. and Lotti, R. (1995) Iceberg discharges into the North Atlantic on millennial time scales during the last glaciation. *Science*, *267*(5200), 1005–1010.

Bond, G.C., Showers, W., Elliot, M., Evans, M., Lotti, R., Hajdas, I., Bonani, G., and Johnson, S. (1999). The North Atlantic's 1–2 kyr climate rhythm: relation to Heinrich events, Dansgaard/Oeschger cycles and the little ice age. In *Mechanisms of Global Change at Millennial Time Scales*. Geophysical Monograph (112) (P.U. Clark, R.S. Webb, and L.D. Keigwin, eds, pp. 59–76). American Geophysical Union, Washington, DC.

Bonduriansky, R., Maklakov, A., Zajitschek, F., and Brooks, R. (2008). Sexual selection, sexual conflict and the evolution of ageing and life span. *Functional Ecology*, *22*(3), 443–453.

Bonneaud, C., Pérez-Tris, J., Federici, P., Chastel, O., and Sorci, G. (2006). Major histocompatibility alleles associated with local resistance to malaria in a passerine. *Evolution*, *60*(2), 383–389.

Bonneaud, C., Tardy, L., Giraudeau, M., Hill, G.E., McGraw, K.J., and Wilson, A.J. (2019). Evolution of both host resistance and tolerance to an emerging bacterial pathogen. *Evolution Letters*, *3*(5), 544–554.

Bonnin, M., Canivenc, R., and Ribes, C.L. (1978). Plasma progesterone levels during delayed implantation in the European badger (*Meles meles*). *Reproduction*, *52*(1), 55–58.

Boonekamp, J.J., Mulder, G.A., Salomons, H.M., Dijkstra, C., and Verhulst, S. (2014). Nestling telomere shortening, but not telomere length, reflects developmental stress and predicts survival in wild birds. *Proceedings of the Royal Society. Series B: Biological Sciences*, *281*(1785), 20133287.

Boonstra, R., Lane, J.E., Boutin, S., Bradley, A., Desantis, L., Newman, A.E., and Soma, K.K. (2008). Plasma DHEA levels in wild, territorial red squirrels: seasonal variation and effect of ACTH. *General and Comparative Endocrinology*, *158*(1), 61–67.

Borgia, G. (1995). Why do bowerbirds build bowers? *American Scientist*, *83*(6), 542–547.

Borries, C., Savini, T., and Koenig, A. (2011). Social monogamy and the threat of infanticide in larger mammals. *Behavioral Ecology and Sociobiology*, *65*(4), 685–693.

Bos, D.H., Williams, R.N., Gopurenko, D., Bulut, Z., and Dewoody, J.A. (2009). Condition-dependent mate choice and a reproductive disadvantage for MHC-divergent male tiger salamanders. *Molecular Ecology*, *18*(15), 3307–3315.

Bourne, F.J., Donnelly, C., Cox, D., Gettinby, G., McInerney, J., Morrison, I., and Woodroffe, R. (2007). Bovine TB: the scientific evidence, a science base for a sustainable policy to control TB in cattle, an epidemiological investigation into bovine tuberculosis. Final report of the Independent Scientific Group on Cattle TB. Defra, London.

Bowden, S. and Sadler, A. (2015). Getting it right? Lessons from the interwar years on pulmonary tuberculosis control in England and Wales. *Medical History*, *59*(1), 101–135.

Bowen, W.D., Iverson, S.J., McMillan, J.I., and Boness, D.J. (2006). Reproductive performance in grey seals: age-related improvement and senescence in a capital breeder. *Journal of Animal Ecology*, *75*(6), 1340–1351.

Bowman, J., Jaeger, J.A., and Fahrig, L. (2002). Dispersal distance of mammals is proportional to home range size. *Ecology*, *83*(7), 2049–2055.

Boyd, I.L. (2019). Chief Scientific Adviser. Advice about Supplementary Badger Culling, June 2019. Note to Defra in relation to a pre Action Protocol letter issued on behalf of Thomas Langton, in relation to issue of badger culling licences in 2019. Published on Badger Crowd online blog March 2020. https://thebadgercrowd.org/2020/03 (accessed 12 May 2022).

Boydston, E.E., Morelli, T.L., and Holekamp, K.E. (2001). Sex differences in territorial behavior exhibited by the spotted hyena (Hyaenidae, *Crocuta crocuta*). *Ethology*, *107*(5), 369–385.

Boyles, J.G., Johnson, J.S., Blomberg, A., and Lilley, T.M. (2020). Optimal hibernation theory. *Mammal Review*, 50(1), 91–100.

Brabin, L. and Brabin, B.J. (1992). Parasitic infections in women and their consequences. *Advances in Parasitology*, 31, 1–81.

Bradbury, J.W. and Vehrencamp, S.L. (1976). Social organization and foraging in emballonurid bats. *Behavioral Ecology and Sociobiology*, 1(4), 337–381.

Bradshaw, A.D. and Hardwick, K. (1989). Evolution and stress—genotypic and phenotypic components. *Biological Journal of the Linnean Society*, 37(1–2), 137–155.

Bradshaw, J.W.S., Casey, R.A., and Brown, S.L. (2012). *The Behaviour of the Domestic Cat*. CABI, Wallingford.

Bratt, C.L. and Dayan, N. (2011). Corynebacterium species and their role in the generation of human malodor. In *Innate Immune System of Skin and Oral Mucosa: Properties and Impact in Pharmaceutics, Cosmetics, and Personal Care Products* (N. Dayan and P.W. Wertz, p. 333). John Wiley & Sons, Hoboken, NJ.

Brett, C.E., Baird, G.C., Erwin, D.H., and Anstey, R.L. (1995). Coordinated stasis and evolutionary ecology of Silurian to Middle Devonian faunas in the Appalachian Basin. In *New Approaches to Speciation in the Fossil Record* (D. Erwin, ed., pp. 285–315). Columbia University Press, New York.

Bribiescas, R.G., Ellison, P.T., and Gray, P.B. (2012). Male life history, reproductive effort, and the evolution of the genus Homo: new directions and perspectives. *Current Anthropology*, 53(S6), S424–S435.

Briga, M., Jimeno, B., and Verhulst, S. (2019). Coupling lifespan and aging? The age at onset of body mass decline associates positively with sex-specific lifespan but negatively with environment-specific lifespan. *Experimental Gerontology*, 119, 111–119.

Briganti, F., Papeschi, A., Mugnai, T., and Dessi-Fulgheri, F. (1999). Effect of testosterone on male traits and behaviour in juvenile pheasants. *Ethology Ecology & Evolution*, 11(2), 171–178.

Bright Ross, J. (2021). The effects of energetic expenditure tactics and life-history variability on European badger (*Meles meles*) ecology. PhD thesis, University of Oxford.

Bright Ross, G., Markham, A.C. Buesching, C.D., Hambley, C. Speakman, J.R., Macdonald, D.W. and Newman, C. (2022, submitted) Links between somatic condition, life-history, and energy budgets reveal heterogeneous energy management tactics in a social mesocarnivore. *American Naturalist*.

Bright Ross, J.G., Newman, C., Buesching, C.D., and Macdonald, D.W. (2020). What lies beneath? Population dynamics conceal pace-of-life and sex ratio variation, with implications for resilience to environmental change. *Global Change Biology*, 26(6), 3307–3324.

Bright Ross, J.G., Newman, C., Buesching, C.D., Connolly, E., Nakagawa, S., and Macdonald, D.W. (2021). A fat chance of survival: body condition provides life-history dependent buffering of environmental change in a wild mammal population. *Climate Change Ecology*, 2, 100022.

Bright Ross, J.G., Newman, C., Buesching, C.D., and Macdonald, D.W. (2022). Preserving identity in capture–mark–recapture studies: increasing the accuracy of minimum number alive (MNA) estimates by incorporating inter-census trapping efficiency variation. *Mammalian Biology*, 1–14.

Broekhuis, F., Elliot, N.B., Keiwua, K., Koinet, K., Macdonald, D.W., Mogensen, N., Thuo, D., and Gopalaswamy, A.M. (2021). Resource pulses influence the spatio-temporal dynamics of a large carnivore population. *Ecography*, 44(3), 358–369.

Brooks, M. and Kays, R. (2017). Kinkajou: the tree-top specialist. In *Biology and Conservation of Musteloids* (D.W. Macdonald, C. Newman, and L.A. Harrington, eds, pp. 493–501). Oxford University Press, Oxford.

Brøseth, H., Knutsen, B., and Bevanger, K. (1997). Spatial organization and habitat utilization of badgers *Meles meles*: effects of food patch dispersion in the boreal forest of central Norway. *Zeitschrift fur Saugetierkunde*, 62(1), 12–22.

Brown, J.L., Bush, M., Packer, C., Pusey, A.E., Monfort, S.L., O'Brien, S.J., and Wildt, D.E. (1991). Developmental changes in pituitary–gonadal function in free-ranging lions (*Panthera leo leo*) of the Serengeti Plains and Ngorongoro Crater. *Reproduction*, 91(1), 29–40.

Brown, L., Shumaker, R.W., and Downhower, J.F. (1995). Do primates experience sperm competition? *The American Naturalist*, 146(2), 302–306.

Brown, R.E. and Macdonald, D.W. (eds) (1985). *Social Odours in Mammals*. Vols 1 and 2. Oxford University Press, Oxford.

Brownell, R.L. Jr and Rails, K. (1986). Potential for sperm competition in baleen whales. *Report of the International Whaling Commission (Special Issue)*, 8, 97–112.

Brunet-Rossinni, A.K. and Austad, S.N. (2006). Senescence in wild populations of mammals and birds. *Handbook of the Biology of Aging*, 1, 243.

Brunt, D. (1933). The adiabatic lapse-rate for dry and saturated air. *Quarterly Journal of the Royal Meteorological Society*, 59(252), 351–360.

Bryan, H.M., Darimont, C.T., Paquet, P.C., Wynne-Edwards, K.E., and Smits, J.E. (2014). Stress and reproductive hormones reflect inter-specific social and nutritional conditions mediated by resource availability in a bear–salmon system. *Conservation Physiology*, 2(1).

Bryant, H.N., Russell, A.P., and Fitch, W.D. (1993). Phylogenetic relationships within the extant Mustelidae (Carnivora): appraisal of the cladistic status of the Simpsonian subfamilies. *Zoological Journal of the Linnean Society*, 108(4), 301–334.

Bryce, J.M., Speakman, J.R., Johnson, P.J., and Macdonald, D.W. (2001). Competition between Eurasian red and introduced Eastern grey squirrels: the energetic significance of body-mass differences. *Proceedings of the Royal Society of London. Series B: Biological Sciences*, 268(1477), 1731–1736.

Buchanan, G.D. (1966). Reproduction in the ferret (*Mustela furo*) I. Uterine histology and histochemistry during pregnancy and pseudopregnancy. *American Journal of Anatomy*, 118(1), 195–215.

Buck, J.C., Weinstein, S.B., and Young, H.S. (2018). Ecological and evolutionary consequences of parasite avoidance. *Trends in Ecology and Evolution*, 33(8), 619–632. doi:10.1016/j.tree.2018.05.001

Buckley, N.J. and Ruxton, G.D. (2003). The resource dispersion hypothesis and the 'future value' of food. *Trends in Ecology & Evolution*, 18(8), 379.

Buddle, B.M., Aldwell, F.E., Skinner, M.A., De Lisle, G.W., Denis, M., Vordermeier, H.M., Hewinson, R.G., and Wedlock, D.N. (2005). Effect of oral vaccination of cattle with lipid-formulated BCG on immune responses and protection against bovine tuberculosis. *Vaccine*, 23(27), 3581–3589.

Buddle, B.M., Vordermeier, H.M., Chambers, M.A., and de Klerk-Lorist, L.-M. (2018). Efficacy and safety of BCG vaccine for control of tuberculosis in domestic livestock and wildlife. *Frontiers in Veterinary Science*, 5, 259. doi:10.3389/fvets.2018.00259

Buesching, C.D. (ed.). (2019). *Chemical Signals in Vertebrates 14*. Springer International Publishing, Cham, Switzerland.

Buesching, C.D. and Jordan, N. (2019). The social function of latrines: a hypothesis-driven research approach. In *Chemical Signals in Vertebrates 14* (C.D. Buesching, ed., pp. 94–103). Springer, Cham.

Buesching, C.D. and Macdonald, D.W. (2001). Scent-marking behaviour of the European badger (*Meles meles*): resource defence or individual advertisement? In *Chemical Signals in Vertebrates 9* (A. Marchlewska-Koj, J.J. Lepri and D. Müller-Schwarze, eds, pp. 321–327). Springer, New York.

Buesching, C.D., Waterhouse, J.S., and Macdonald, D.W. (2002a). Gas-chromatographic analyses of the subcaudal gland secretion of the European badger (*Meles meles*) part I: chemical differences related to individual parameters. *Journal of Chemical Ecology*, 28(1), 41–56.

Buesching, C.D., Waterhouse, J.S., and Macdonald, D.W. (2002b). Gas-chromatographic analyses of the subcaudal gland secretion of the European badger (*Meles meles*) part II: time-related variation in the individual-specific composition. *Journal of Chemical Ecology*, 28(1), 57–69.

Buesching, C.D., Newman, C., and Macdonald, D.W. (2002c). Variations in colour and volume of the subcaudal gland secretion of badgers (Meles meles) in relation to sex, season and individual-specific parameters. *Mammalian Biology*, 67(3), 147–156.

Buesching, C.D. and Macdonald, D.W. (2004). Variations in scent-marking behaviour of European badgers *Meles meles* in the vicinity of their setts. *Acta Theriologica*, 49(2), 235–246.

Buesching, C.D. and Stankowich, T. (2017). Communication amongst the musteloids: signs, signals, and cues. In *Biology and Conservation of Musteloids* (D.W. Macdonald, C. Newman, and L.A. Harrington, eds, pp. 149–166). Oxford University Press, Oxford.

Buesching, C., Stopka, P., and Macdonald, D.W. (2003). The social function of allo-marking behaviour in the European badger (*Meles meles*). *Behaviour*, 140(8–9), 965–980.

Buesching, C.D., Heistermann, M., and Macdonald, D.W. (2009). Seasonal and inter-individual variation in testosterone levels in badgers *Meles meles*: evidence for the existence of two endocrinological phenotypes. *Journal of Comparative Physiology A*, 195(9), 865–871.

Buesching, C.D., Clarke, J.R., Ellwood, S.A., King, C., Newman, C., and Macdonald, D. W. (2010). The mammals of Wytham Woods. In *Wytham Woods* (pp. 173–196). Oxford University Press, Oxford.

Buesching, C., Newman, C., Jones, J.T., and Macdonald, D.W. (2011). Testing the effects of deer grazing on two woodland rodents, bankvoles and woodmice. *Basic and Applied Ecology*, 12(3), 207–214. doi:10.1016/j.baae.2011.02.007

Buesching, C.D., Newman, C., and Macdonald, D.W. (2014). How dear are deer volunteers: the efficiency of monitoring deer using teams of volunteers to conduct pellet group counts. *Oryx*, 48(4), 593–601.

Buesching, C.D., Slade, E.M., Newman, C., Riordan, P., and Macdonald, D.W. (2015). Do 'many hands make light work'? A critical evaluation of citizen science in Wytham Woods. In *Wildlife Conservation on Farmland Volume 2: Conflict in the Countryside* (D. W. Macdonald, and R. E. Feber, eds, pp. 293–328). Oxford University Press, Oxford.

Buesching, C.D., Newman, C., Service, K., Macdonald, D.W., and Riordan, P. (2016a). Latrine marking patterns of badgers (*Meles meles*) with respect to population density and range size. *Ecosphere*, 7(5), e01328.

Buesching, C.D., Tinnesand, H.V., Sin, Y., Rosell, F., Burke, T., and Macdonald, D.W. (2016b). Coding of group odor in the subcaudal gland secretion of the European badger *Meles meles*: chemical composition and pouch microbiota. In *Chemical Signals in Vertebrates 13* (B.A. Schulte, T.E. Goodwin, and M.H. Ferkin, eds, pp. 45–62). Springer, Cham.

Buesching, C.D., Jordan, N.R., San, E.D.L., Sato, J.J., Belant, J.L., and Somers, M.J. (2018). The function of small carnivore latrines: case studies and a research framework for hypothesis-testing. In *Small Carnivores: Evolution, Ecology, Behaviour & Conservation* (E.D.L. San, J.J. Sato, J.L. Belant, and M.J. Somers, eds). Wiley Publishing, Hoboken, NJ.

Burgener, N., Dehnhard, M., Hofer, H., and East, M.L. (2009). Does anal gland scent signal identity in the spotted hyaena? *Animal Behaviour*, 77(3), 707–715.

Burger, J. and Gochfeld, M. (1992). Effect of group size on vigilance while drinking in the coati, *Nasua narica* in Costa Rica. *Animal Behaviour*, 44(6), 1053–1057.

Burke, K.L. (2016). Longest-lived vertebrate. *American Scientist*, 104(6), 331.

Burley, N. (1986). Sexual selection for aesthetic traits in species with biparental care. *The American Naturalist*, 127(4), 415–445.

Burns, C.E., Goodwin, B.J., and Ostfeld, R.S. (2005). A prescription for longer life? Bot fly parasitism of the white-footed mouse. *Ecology*, 86, 753–761.

Burt, W.H. (1943). Territoriality and home range concepts as applied to mammals. *Journal of Mammalogy*, 24(3), 346–352.

Butler, J.M. and Roper, T.J. (1996). Ectoparasites and sett use in European badgers. *Animal Behaviour*, 52(3), 621–629.

Butler, P.J., Green, J.A., Boyd, I.L., and Speakman, J.R. (2004). Measuring metabolic rate in the field: the pros and cons of the doubly labelled water and heart rate methods. *Functional Ecology*, 18(2), 168–183.

Buyers, J.A. (1985). Olfaction related behaviour in peccaries. *Zeitschrift für Tierpsychologie*, 70, 201.

Byrne, A.W., Sleeman, D.P., O'Keeffe, J., and Davenport, J. (2012a). The ecology of the European badger (*Meles meles*) in Ireland: a review. *Biology and Environment: Proceedings of the Royal Irish Academy*, 112b(1), 105–132.

Byrne, A.W., O'Keeffe, J., Green, S., Sleeman, D.P., Corner, L.A., Gormley, E., Murphy, D., Martin, S.W., and Davenport, J. (2012b). Population estimation and trappability of the European badger (*Meles meles*): implications for tuberculosis management. *PLoS ONE*, 7(12), e50807.

Byrne, A.W., White, P.W., McGrath, G., James, O., and Martin, S.W. (2014). Risk of tuberculosis cattle herd breakdowns in Ireland: effects of badger culling effort, density and historic large-scale interventions. *Veterinary Research*, 45(1), 1–10.

Byrne, A.W., Fogarty, U., O'Keeffe, J., and Newman, C. (2015a). In situ adaptive response to climate and habitat quality variation: spatial and temporal variation in European badger (*Meles meles*) body weight. *Global Change Biology*, 21(9), 3336–3346.

Byrne, A.W., Kenny, K., Fogarty, U., O'Keeffe, J.J., More, S.J., McGrath, G., Teeling, M., Martin, S.W., and Dohoo, I.R. (2015b). Spatial and temporal analyses of metrics of tuberculosis infection in badgers (*Meles meles*) from the Republic of Ireland: trends in apparent prevalence. *Preventive Veterinary Medicine*, 122(3), 345–354.

Byrne, A.W., O'Keeffe, J., Fogarty, U., Rooney, P., and Martin, S.W. (2015c). Monitoring trap-related injury status during large-scale wildlife management programmes: an adaptive management approach. *European Journal of Wildlife Research*, 61(3), 445–455.

Byrne, R. and Whiten, A. (1989). *A Machiavellian Intelligence – Social Expertise and the Evolution of Intellect in Monkeys, Apes, and Humans*. Oxford University Press, Oxford.

Cabana, G. and Kramer, D.L. (1991). Random offspring mortality and variation in parental fitness. *Evolution*, 45(1), 228–234.

Caillol, M., Mondain-Monval, M., and Rossano, B. (1991). Gonadotrophins and sex steroids during pregnancy and natural superfoetation in captive brown hares (*Lepus europaeus*). *Reproduction*, 92(2), 299–306.

Calderone, J.B. and Jacobs, G.H. (2003). Spectral properties and retinal distribution of ferret cones. *Visual Neuroscience*, 20(1), 11–17.

Caldwell, H.K. (2017). Oxytocin and vasopressin: powerful regulators of social behavior. *The Neuroscientist*, 23(5), 517–528.

Campbell, L.A., Tkaczynski, P.J., Lehmann, J., Mouna, M., and Majolo, B. (2018). Social thermoregulation as a potential mechanism linking sociality and fitness: Barbary macaques with more social partners form larger huddles. Scientific Reports, 8(1), 1–8.

Campbell, R.D., Nouvellet, P., Newman, C., Macdonald, D.W., and Rosell, F. (2012). The influence of mean climate trends and climate variance on beaver survival and recruitment dynamics. *Global Change Biology*, 18(9), 2730–2742.

Campbell, R.D., Newman, C., Macdonald, D.W., and Rosell, F. (2013). Proximate weather patterns and spring green-up phenology effect Eurasian beaver (*Castor fiber*) body mass and reproductive success: the implications of climate change and topography. *Global Change Biology*, 19(4), 1311–1324.

Campbell, R.D., Rosell, F., Newman, C., and Macdonald, D.W. (2017). Age-related changes in somatic condition and reproduction in the Eurasian beaver: resource history influences onset of reproductive senescence. *PLoS ONE*, 12(12), e0187484. doi:10.1371/journal.pone.0187484

Campbell-Palmer, R., Puttock, A., Graham, H., Wilson, K., Schwab, G., Gaywood, M.J., and Brazier, R.E. (2018). Survey of the Tayside Area Beaver Population 2017–2018. Scottish National Heritage Commissioned Report No. 1013. Scottish National Heritage, Inverness.

Campisi, J. (2005). Senescent cells, tumor suppression, and organismal aging: good citizens, bad neighbors. *Cell*, 120(4), 513–522.

Campisi, J. and Di Fagagna, F.D.A. (2007). Cellular senescence: when bad things happen to good cells. *Nature Reviews Molecular Cell Biology*, 8(9), 729–740.

Canal, D., Jovani, R., and Potti, J. (2012). Multiple mating opportunities boost protandry in a pied flycatcher population. *Behavioral Ecology and Sociobiology*, 66(1), 67–76.

Candolin, U. (2003). The use of multiple cues in mate choice. *Biological Reviews*, 78(4), 575–595.

Canivenc, R. and Bonnin, M. (1981). Environmental control of delayed implantation in the European badger (*Meles meles*). *Journal of Reproduction and Fertility Supplement*, 29, 25–33.

Cant, M.A. and Johnstone, R.A. (2008). Reproductive conflict and the separation of reproductive generations in humans. *Proceedings of the National Academy of Sciences*, 105(14), 5332–5336.

Cant, M.A., Vitikainen, E.I.K., and Nichols, H.J. (2013) Demography and social evolution of banded mongooses. In *Advances in the Study of Behavior* (H.J. Brockmann et al., eds, pp. 407–445). Academic Press, Cambridge, MA. doi:10.1016/B978-0-12-407186-5.00006-9

Capdevila, P., Hereu, B., Riera, J.L., and Linares, C. (2016). Unravelling the natural dynamics and resilience patterns of underwater Mediterranean forests: insights from the demography of the brown alga *Cystoseira zosteroides*. *Journal of Ecology*, 104(6), 1799–1808.

Capdevila, P., Stott, I., Beger, M., and Salguero-Gómez, R. (2020). Towards a comparative framework of demographic resilience. *Trends in Ecology & Evolution*, 35(9), 776–786.

Caraco, T. and Giraldeau, L.A. (1991). Social foraging: producing and scrounging in a stochastic environment. *Journal of Theoretical Biology*, 153(4), 559–583.

Carbone, C., Du Toit, J.T., and Gordon, I.J. (1997). Feeding success in African wild dogs: does kleptoparasitism by spotted hyenas influence hunting group size? *Journal of Animal Ecology*, 66(3), 318–326.

Carbone, C., Mace, G.M., Roberts, S.C., and Macdonald, D.W. (1999). Energetic constraints on the diet of terrestrial carnivores. *Nature*, 402(6759), 286.

Carbone, C., Teacher, A., and Rowcliffe, J.M. (2007). The costs of carnivory. *PLoS Biology*, 5(2), e22.

Careau, V., Thomas, D., Humphries, M.M., and Réale, D. (2008). Energy metabolism and animal personality. *Oikos*, 117(5), 641–653.

Careau, V., Bininda-Emonds, O.R.P., Thomas, D.W., Réale, D., and Humphries, M.M. (2009). Exploration strategies map along fast–slow metabolic and life-history continua in muroid rodents. *Functional Ecology*, 23(1), 150–156.

Caro, S.P., Schaper, S.V., Hut, R.A., Ball, G.F., and Visser, M.E. (2013). The case of the missing mechanism: how does temperature influence seasonal timing in endotherms? *PLoS Biology*, 11(4), e1001517.

Carpenter, P.J., Pope, L.C., Greig, C., Dawson, D.A., Rogers, L.M., Erven, K., Wilson, G.J., Delahay, R.J., Cheeseman, C.L., and Burke, T. (2005). Mating system of the Eurasian badger, *Meles meles*, in a high density population. *Molecular Ecology*, 14(1), 273–284.

Carr, G.M. and Macdonald, D.W. (1986). The sociality of solitary foragers: a model based on resource dispersion. *Animal Behaviour*, 34(5), 1540–1549.

Carranza, J., Alvarez, F., and Redondo, T. (1990). Territoriality as a mating strategy in red deer. *Animal Behaviour*, 40(1), 79–88.

Carrington, D. (2013). Stop badger cull immediately, says Natural England science expert. https://www.theguardian.com/environment/2013/oct/21/stop-badger-cull-immediately-natural-england-scientist (accessed 6 May 2022).

Carroll, L. (1871). *Through the Looking-Glass, and What Alice Found*. Macmillan & Co., London.

Carter, S., Du Plessis, T., Chwalibog, A., and Sawosz, E. (2017). The honey badger in South Africa: biology and conservation. *International Journal of Avian Wildlife Biology*, 2(2), 00091.

Carter, S.P., Delahay, R.J., Smith, G.C., Macdonald, D.W., Riordan, P., Etherington, T.R., Pimley, E.R., Walker, N.J., and Cheeseman, C.L. (2007). Culling-induced social perturbation in Eurasian badgers *Meles meles* and the management of TB in cattle: an analysis of a critical problem in applied ecology. *Proceedings of the Royal Society. Series B: Biological Sciences*, 274(1626), 2769–2777.

Carter, S.P., Chambers, M.A., Rushton, S.P., Shirley, M.D., Schuchert, P., Pietravalle, S., Murray, A., Rogers, F., Gettinby, G., Smith, G.C., and Delahay, R.J. (2012). BCG

vaccination reduces risk of tuberculosis infection in vaccinated badgers and unvaccinated badger cubs. *PLoS ONE*, 7(12), e49833.

Case, T.J. (1978a). On the evolution and adaptive significance of postnatal growth rates in the terrestrial vertebrates. *The Quarterly Review of Biology*, 53(3), 243–282.

Case, T.J. (1978b). A general explanation for insular body size trends in terrestrial vertebrates. *Ecology*, 59(1), 1–18.

Cassidy, A. (2019). *Vermin, Victims and Disease: British Debates over Bovine Tuberculosis and Badgers*. Springer Nature, Cham.

Caswell, H. (2001). *Matrix Population Models: construction, Analysis, and Interpretation*, second edition. Sinauer Associates Inc., Sunderland, MA, USA.

Caulin, A.F. and Maley, C.C. (2011). Peto's Paradox: evolution's prescription for cancer prevention. *Trends in Ecology & Evolution*, 26(4), 175–182.

Centers for Disease Control and Prevention (2016). Tuberculosis (TB). Fact sheets. https://www.cdc.gov/tb/publications/factsheets/testing/igra.htm (accessed 10 March 2022).

Cesare, A.J. and Reddel, R.R. (2010). Alternative lengthening of telomeres: models, mechanisms and implications. *Nature Reviews Genetics*, 11(5), 319–330. https://doi.org/10.1038/nrg2763

Chae, C., Kwon, D., Kim, O., Min, K., Cheon, D.S., Choi, C., Kim, B., and Suh, J. (1998). Diarrhoea in nursing piglets associated with coccidiosis: prevalence, microscopic lesions and coexisting microorganisms. *Veterinary Record*, 143(15), 417–420.

Chagnon, N. (2017). *Adaptation and Human Behavior: An Anthropological Perspective*. Routledge, Abingdon.

Chambers, M.A., Crawshaw, T., Waterhouse, S. Delahay, R.J., Hewinson, R.G., and Lyashchenko, K.P. (2008). Validation of the BrockTB Stat-Pak assay for detection of tuberculosis in Eurasian badgers (*Meles meles*) and influence of disease severity on diagnostic accuracy. *Journal of Clinical Microbiology*, 46(4).

Chambers, M.A., Rogers, F., Delahay, R.J., Lesellier, S., Ashford, R., Dalley, D., Gowtage, S., Davé, D., Palmer, S., Brewer, J., and Crawshaw, T. (2011). Bacillus Calmette-Guérin vaccination reduces the severity and progression of tuberculosis in badgers. *Proceedings of the Royal Society. Series B: Biological Sciences*, 278(1713), 1913–1920.

Chambers, M.A., Carter, S.P., Wilson, G.J., Jones, G., Brown, E., Hewinson, R.G., and Vordermeier, M. (2014). Vaccination against tuberculosis in badgers and cattle: an overview of the challenges, developments and current research priorities in Great Britain. *Veterinary Record*, 175(4), 90–96.

Chang, A.M. and Wiebe, K.L. (2016). Body condition in snowy owls wintering on the prairies is greater in females and older individuals and may contribute to sex-biased mortality. *The Auk: Ornithological Advances*, 133(4), 738–746.

Chapman, T., Arnqvist, G., Bangham, J., and Rowe, L. (2003). Sexual conflict. *Trends in Ecology & Evolution*, 18(1), 41–47.

Charles, R. (1997). The exploitation of carnivores and other fur-bearing mammals during the north-western European Late and Upper Paleolithic and Mesolithic. *Oxford Journal of Archaeology, 16*(3), 253–277.

Charlesworth, B. (1988). The evolution of mate choice in a fluctuating environment. *Journal of Theoretical Biology, 130*(2), 191–204.

Charlesworth, B. (2000). Fisher, Medawar, Hamilton and the evolution of aging. *Genetics, 156*(3), 927–931.

Charlton, B.D., Newman, C., Macdonald, D.W., and Buesching, C.D. (2020). Male European badger churrs: insights into call function and motivational basis. *Mammalian Biology, 100*, 429–438.

Charnov, E.L. (1976). Optimal foraging, the marginal value theorem. *Theoretical Population Biology, 9*(2), 129–136.

Charnov, E.L., Los-den Hartogh, R.L., Jones, W.T., and Van Den Assem, J. (1981). Sex ratio evolution in a variable environment. *Nature, 289*(5793), 27.

Chase, I.D. (1986). Explanations of hierarchy structure. *Animal Behaviour, 34*(4), 1265–1267.

Chase, I.D. and Seitz, K. (2011). Self-structuring properties of dominance hierarchies: a new perspective. *Advances in Genetics, 75*, 51–81.

Chase, I.D., Bartolomeo, C., and Dugatkin, L.A. (1994). Aggressive interactions and inter-contest interval: how long do winners keep winning? *Animal Behaviour, 48*, 393–400. doi:10.1006/anbe.1994.1253

Cheeseman, C.L., Jones, G.W., Gallagher, J., and Mallinson, P.J. (1981). The population structure, density and prevalence of tuberculosis (*Mycobacterium bovis*) in badgers (*Meles meles*) from four areas in south-west England. *Journal of Applied Ecology, 18*(3), 795–804.

Cheeseman, C.L., Cresswell, W.J., Harris, S., and Mallinson, P.J. (1988a). Comparison of dispersal and other movements in two badger (*Meles meles*) populations. *Mammal Review, 18*(1), 51–59.

Cheeseman, C.L., Wilesmith, J.W., Stuart, F.A., and Mallinson, P.J. (1988b). Dynamics of tuberculosis in a naturally infected badger population. *Mammal Review, 18*(1), 61–72.

Chen, B.X., Yuen, Z.X., and Pan, G.W. (1985). Semen-induced ovulation in the Bactrian camel (*Camelus bactrianus*). *Journal of Reproduction and Fertility, 74*(2), 335–339.

Chen, J.J. (2010). The Hardy-Weinberg principle and its applications in modern population genetics. *Frontiers in Biology, 5*(4), 348–353.

Chen, W., Newman, C., Liu, Z., Kaneko, Y., Omote, K., Masuda, R., Buesching, C.D., Macdonald, D.W., Xie, Z., and Zhou, X. (2015). The illegal exploitation of hog badgers (*Arctonyx collaris*) in China: genetic evidence exposes regional population impacts. *Conservation Genetics Resources, 7*(3), 697–704.

Chen, Y. and Junger, W.G. (2012). Measurement of oxidative burst in neutrophils. In *Leucocytes. Methods in Molecular Biology (Methods and Protocols)*, vol. 844 (R. Ashman, ed.). Humana Press, Totowa, NJ. https://doi.org/10.1007/978-1-61779-527-5_8

Chesser, R.K. (1983). Genetic variability within and among populations of the black-tailed prairie dog. *Evolution, 37*(2), 320–331.

Chevin, L.M., Lande, R., and Mace, G.M. (2010). Adaptation, plasticity, and extinction in a changing environment: towards a predictive theory. *PLoS Biology, 8*(4), e1000357.

Chhabra, S., Ranjan, R., Uppal, S.K., and Singla, L.D. (2012). Transplacental transmission of *Babesia equi* (*Theileria equi*) from carrier mares to foals. *Journal of Parasitic Diseases, 36*(1), 31–33.

Chiao, Y.-C., Lee, H.-Y., Wang, S.-W., Hwang, J.-J., Chien, C.-H., Huang, S.-W., Lu, C.-C., Chen, J.-J., Tsai, S.-C., and Wang, P.S. (1999). Regulation of thyroid hormones on the production of testosterone in rats. *Journal of Cellular Biochemistry, 73*, 554–562.

Chimienti, M., Desforges, J.P., Beumer, L.T., Nabe-Nielsen, J., van Beest, F.M., and Schmidt, N.M. (2020). Energetics as common currency for integrating high resolution activity patterns into dynamic energy budget-individual based models. *Ecological Modelling, 434*, 109250.

Christensen, C., Kern, J.M., Bennitt, E., and Radford, A.N. (2016). Rival group scent induces changes in dwarf mongoose immediate behavior and subsequent movement. *Behavioral Ecology, 27*(6), 1627–1634.

Christian, J.J. (1970). Social subordination, population density, and mammalian evolution. *Science, 168*(3927), 84–90.

Christian, S.F. (1993). The behavioural ecology of the Eurasian badger (Meles meles): space use, territoriality and social behaviour. PhD thesis, University of Sussex.

Christian, S.F. (1993). The behavioural ecology of the Eurasian badger (Meles meles): space use, territoriality and social behaviour. PhD thesis, University of Sussex.

Christiansen, P. and Wroe, S. (2007). Bite forces and evolutionary adaptations to feeding ecology in carnivores. *Ecology, 88*(2), 347–358.

Christley, R., Robinson, S., Lysons, R., and French, N. (2005). Network analysis of cattle movement in Great Britain. Proceedings of the Society for Veterinary Epidemiology and Preventive Medicine, Nairn, Inverness, Scotland, pp. 234–244.

Chuang, S.A. and Lee, L.L. (1997). Food habits of three carnivore species (*Viverricula indica, Herpestes urva,* and *Melogale moschata*) in Fushan Forest, northern Taiwan. *Journal of Zoology, 243*(1), 71–79.

Chutipong, W., Duckworth, J.W., Timmins, R.J., Choudhury, A., Abramov, A.V., Roberton, S., Long, B., Rahman, H., Hearn, A., Dinets, V., and Willcox, D.H.A. (2016). *Martes flavigula*. In *The IUCN Red List of Threatened Species* 2016: e.T41649A45212973. https://dx.doi.org/10.2305/IUCN.UK.2016-1.RLTS.T41649A45212973.en (accessed 30 April 2022).

Claridge, J., Diggle, P., McCann, C.M., Mulcahy, G., Flynn, R., McNair, J., Strain, S., Welsh, M., Baylis, M., and Williams, D.J. (2012). Fasciola hepatica is associated with the failure to detect bovine tuberculosis in dairy cattle. *Nature Communications, 3*(1), 1–8.

Clarke, G.P., White, P.C., and Harris, S. (1998). Effects of roads on badger *Meles meles* populations in south-west England. *Biological Conservation*, *86*(2), 117–124.

Clarke, P. and Fredin, E. (1978). Newspapers, television and political reasoning. *Public Opinion Quarterly*, *42*(2), 143–160.

Clauset, A., Newman, M.E., and Moore, C. (2004). Finding community structure in very large networks. *Physical Review E*, *70*(6), 066111.

Clay, K. and Kover, P.X. (1996). The Red Queen Hypothesis and plant/pathogen interactions. *Annual Review of Phytopathology*, *34*(43), 29–50.

Cleary, G.P., Corner, L.A., O'Keeffe, J., and Marples, N.M. (2009). The diet of the badger *Meles meles* in the Republic of Ireland. *Mammalian Biology*, *74*(6), 438–447.

Clinchy, M., Zanette, L.Y., Roberts, D., Suraci, J.P., Buesching, C.D., Newman, C., and Macdonald, D.W. (2016). Fear of the human 'super predator' far exceeds the fear of large carnivores in a model mesocarnivore. *Behavioral Ecology*, *27*(6), 1826–1832.

Clutton-Brock, T.H. (1975). Feeding behaviour of red colobus and black and white colobus in East Africa. *Folia Primatologica*, *23*(3), 165–207.

Clutton-Brock, T.H. and Isvaran, K. (2007). Sex differences in ageing in natural populations of vertebrates. *Proceedings of the Royal Society. Series B: Biological Sciences*, *274*(1629), 3097–3104.

Clutton-Brock, T. and McAuliffe, B.K. (2009). Female mate choice in mammals. *The Quarterly Review of Biology*, *84*(1), 3–27.

Clutton-Brock, T.H., Albon, S.D., and Guinness, F.E. (1984). Maternal dominance, breeding success and birth sex-ratio in red deer. *Nature*, *308*, 358–360.

Clutton-Brock, T.H., Albon, S.D., and Guinness, F.E. (1985). Parental investment and sex differences in juvenile mortality in birds and mammals. *Nature*, *313*(5998), 131–133.

Clutton-Brock, T.H., Stevenson, I.R., Marrow, P., MacColl, A.D., Houston, A.I., and McNamara, J.M. (1996). Population fluctuations, reproductive costs and life-history tactics in female Soay sheep. *Journal of Animal Ecology*, *66*(6), 675–689.

Clutton-Brock, T.H., Hodge, S.J., and Flower, T.P. (2008). Group size and the suppression of subordinate reproduction in Kalahari meerkats. *Animal Behaviour*, *76*(3), 689–700.

Coals, P., Dickman, A., Hunt, J., Grau, A., Mandisodza-Chikerema, R., Ikanda, D., Macdonald, D.W., and Loveridge, A. (2020). Commercially-driven lion part removal: what is the evidence from mortality records? *Global Ecology and Conservation*, *24*, e01327.

Cockburn, A. (1998). Evolution of helping behavior in cooperatively breeding birds. *Annual Review of Ecology and Systematics*, *29*(1), 141–177.

Cohas, A., Bonenfant, C., Kempenaers, B., and Allaine, D. (2009). Age-specific effect of heterozygosity on survival in alpine marmots, *Marmota marmota*. *Molecular Ecology*, *18*(7), 1491–1503.

Cohen, A.A. (2004). Female post-reproductive lifespan: a general mammalian trait. *Biological Reviews*, *79*(4), 733–750.

Cohen, J.A. and Fox, M.W. (1976). Vocalizations in wild canids and possible effects of domestication. *Behavioural Processes*, *1*(1), 77–92.

Cohen, J.E., Pimm, S.L., Yodzis, P., and Saldaña, J. (1993). Body sizes of animal predators and animal prey in food webs. *Journal of Animal Ecology*, *62*(1), 67–78.

Cohen, S. and Hamrick, N. (2003). Stable individual differences in physiological response to stressors: implications for stress-elicited changes in immune related health. *Brain, Behavior, and Immunity*, *17*(6), 407–414.

Cole, L.C. (1954). The population consequences of life history phenomena. *The Quarterly Review of Biology*, *29*(2), 103–137.

Collins, C. and Kays, R. (2011). Causes of mortality in North American populations of large and medium-sized mammals. *Animal Conservation*, *14*(5), 474–483.

Coltman, D.W., Pilkington, J.G., Smith, J.A., and Pemberton, J.M. (1999). Parasite-mediated selection against Inbred Soay sheep in a free-living island population. *Evolution*, *53*(4), 1259–1267.

Conaway, C.H. (1971). Ecological adaptation and mammalian reproduction. *Biology of Reproduction*, *4*(3), 239–247.

Conlan, A.J., McKinley, T.J., Karolemeas, K., Pollock, E.B., Goodchild, A.V., Mitchell, A.P., Birch, C.P., Clifton-Hadley, R.S., and Wood, J.L. (2012). Estimating the hidden burden of bovine tuberculosis in Great Britain. *PLoS Computational Biology 8*(10), e1002730. https://doi.org/10.1371/journal.pcbi.1002730

Connell, J.H. (1980). Diversity and the coevolution of competitors, or the ghost of competition past. *Oikos*, *35*(2), 131–138.

Consuegra, S. and Garcia de Leaniz, C. (2008). MHC-mediated mate choice increases parasite resistance in salmon. *Proceedings of the Royal Society. Series B: Biological Sciences*, *275*(1641), 1397–1403.

Coombes, H.A., Stockley, P., and Hurst, J.L. (2018). Female chemical signalling underlying reproduction in mammals. *Journal of Chemical Ecology*, *44*(9), 851–873.

Coombes, M.A. and Viles, H.A. (2015). Population-level zoogeomorphology: the case of the Eurasian badger (*Meles meles* L.). *Physical Geography*, *36*(3), 215–238.

Copeland, J.P., Landa, A., Heinemeyer, K., Aubry, K.B., van Dijk, J., May, R., Persson, J., Squires, J., and Yates, R. (2017). Social ethology of the wolverine. In *Biology and Conservation of Musteloids* (D.W. Macdonald, C. Newman, and L.A. Harrington, eds, pp. 493–501). Oxford University Press, Oxford.

Cordain, L., Gotshall, R.W., and Eaton, S.B. (1998). Physical activity, energy expenditure and fitness: an evolutionary perspective. *International Journal of Sports Medicine*, *19*(5), 328–335.

Cords, M. (1995). Predator vigilance costs of allogrooming in wild blue monkeys. *Behaviour*, *132*(7), 559–569.

Corner, L.A., Stuart, L.J., Kelly, D.J., and Marples, N.M. (2015). Reproductive biology including evidence for superfetation in the European badger *Meles meles* (Carnivora: Mustelidae). *PLoS ONE*, *10*(10), e0138093.

Corner, L.A.L., Murphy, D., and Gormley, E. (2011). *Mycobacterium bovis* infection in the Eurasian badger (*Meles meles*): the disease, pathogenesis, epidemiology and control. *Journal of Comparative Pathology*, *144*(1), 1–24.

Courchamp, F. and Macdonald, D.W. (2001). Crucial importance of pack size in the African wild dog *Lycaon pictus*. *Animal Conservation Forum*, *4*(2), 169–174.

Courchamp, F., Rasmussen, G.S., and Macdonald, D.W. (2002). Small pack size imposes a trade-off between hunting and pup-guarding in the painted hunting dog *Lycaon pictus*. *Behavioral Ecology*, *13*(1), 20–27.

Courtenay, O., Reilly, L.A., Sweeney, F.P., Hibberd, V., Bryan, S., Ul-Hassan, A., Newman, C., Macdonald, D.W., Delahay, R.J., Wilson, G.J., and Wellington, E.M.H. (2006). Is *Mycobacterium bovis* in the environment important for the persistence of bovine tuberculosis? *Biology Letters*, *2*(3), 460–462.

Cox, R., Stewart, P.D., and Macdonald, D.W. (1999). The ectoparasites of the European badger, *Meles meles*, and the behavior of the host-specific flea, *Paraceras melis*. *Journal of Insect Behavior*, *12*(2), 245–265.

Craighead, J.J. and Craighead, F.C. (1967). *Management of Bears in Yellowstone National Park*. Montana Cooperative Wildlife Research Unit, Missoula, MT.

Cram, D.L., Monaghan, P., Gillespie, R., and Clutton-Brock, T. (2017). Effects of early-life competition and maternal nutrition on telomere lengths in wild meerkats. *Proceedings of the Royal Society. Series B: Biological Sciences*, *284*(1861), 20171383.

Creel, S. (2005). Dominance, aggression, and glucocorticoid levels in social carnivores. *Journal of Mammalogy*, *86*(2), 255–264.

Creel, S.R. and Creel, N.M. (1991). Energetics, reproductive suppression and obligate communal breeding in carnivores. *Behavioral Ecology and Sociobiology*, *28*(4), 263–270.

Creel, S. and Creel, N.M. (2002). *The African Wild Dog: Behavior, Ecology, and Conservation* (Vol. 65). Princeton University Press, Princeton, NJ.

Creel, S.R. and Macdonald, D.W. (1995). Sociality, group size, and reproductive suppression among carnivores. *Advances in the Study of Behavior*, *24*, 203–257.

Creel, S.R. and Waser, P.M. (1994). Inclusive fitness and reproductive strategies in dwarf mongooses. *Behavioral Ecology*, *5*(3), 339–348.

Cremer, S., Armitage, S.A.O., and Schmid-Hempel, P. (2007). Social immunity. *Current Biology*, *17*, 693–702. doi:10.1016/j.cub.2007.06.008

Cresswell, W.J., Harris, S., and Cheeseman, C.L. (1992). To breed or not to breed: an analysis of the social and density-dependent constraints on the fecundity of female badgers (*Meles meles*). *Philosophical Transactions of the Royal Society of London, Series B*, *338*, 393–407.

Crick, H.Q.P. and Ratcliffe, D.A. (1995). The Peregrine *Falco peregrinus* breeding population of the United Kingdom in 1991. *Bird Study*, *42*(1), 1–19.

Crispell, J., Cassidy, S., Kenny, K., McGrath, G., Warde, S., Cameron, H., and Gordon, S.V. (2020). *Mycobacterium bovis* genomics reveals transmission of infection between cattle and deer in Ireland. *Microbial Genomics*, *6*(8).

Crispell, J., Benton., C.H., Balaz, D., de Maio, N., Ahkmetova, A., Allen, A., Biek, R., Presho, E.L., Dale, J., Hewinson, G., Lycett, S., Nunez-Garcia, J., Skuse, R., Trewby, H., Wilson, D.J., Zadoks, R., Delahay, R.J., and Kao, R.R. (2021). Combining genomics and epidemiology to analyse bi-directional transmission of *Mycobacterium bovis* in a multi-host system. *eLIFE*, *8*, e45833.

Croft, D.P., Brent, L.J., Franks, D.W., and Cant, M.A. (2015). The evolution of prolonged life after reproduction. *Trends in Ecology & Evolution*, *30*(7), 407–416.

Crook, J.H. (1964). The evolution of social organisation and visual communication in the weaver birds (Ploceinae). *Behaviour*, Suppl. 10.

Crook, J.H. and Gartlan, J.S. (1966). Evolution of primate societies. *Nature*, *210*(5042), 1200–1203.

Crook, J.H., Ellis, J.E., and Goss-Custard, J.D. (1976). Mammalian social systems: structure and function. *Animal Behaviour*, *24*(2), 261–274.

Croze, H. (1970). *Searching Image in Carrion Crows: Hunting Strategy in a Predator and Some Anti-predator Devices in Camouflaged Prey* (No. 5). Parey, Verlam Singhofen, Germany.

Cubaynes, S., MacNulty, D.R., Stahler, D.R., Quimby, K.A., Smith, D.W., and Coulson, T. (2014). Density-dependent intraspecific aggression regulates survival in northern Yellowstone wolves (*Canis lupus*). *Journal of Animal Ecology*, *83*(6), 1344–1356.

Cuéllar, A., Rodríguez, A., Halpert, E., Rojas, F., Gómez, A., Rojas, A., and García, E. (2010). Specific pattern of flea antigen recognition by IgG subclass and IgE during the progression of papular urticaria caused by flea bite. *Allergologia et Immunopathologia*, *38*(4), 197–202.

Cui, J., Chen, W., Newman, C., Han, W., Buesching, C.D., Macdonald, D.W., and Zhou, Y. (2018). Roads disrupt rodent scatter-hoarding seed-dispersal services: implication for forest regeneration. *Perspectives in Plant Ecology, Evolution and Systematics*, *34*, 102–108.

Cui, J., Lei, B., Newman, C., Ji, S., Su, H., Buesching, C. D., and Zhou, Y. (2020). Functional adaptation rather than ecogeographical rules determine body-size metrics along a thermal cline with elevation in the Chinese pygmy dormouse (*Typhlomys cinereus*). *Journal of Thermal Biology*, *88*, 102510. https://doi.org/10.1016/j.jtherbio.2020.102510

Culina, A., Linton, D.M., Pradel, R., Bouwhuis, S., and Macdonald, D.W. (2019). Live fast, don't die young: survival–reproduction trade-offs in long-lived income breeders. *Journal of Animal Ecology*, *88*(5), 746–756.

Cunliffe, V.T. (2016). The epigenetic impacts of social stress: how does social adversity become biologically embedded? *Epigenomics*, *8*(12), 1653–1669. https://doi.org/10.2217/epi-2016-0075

Curio, E. (1983). Why de young birds reproduce less well? *Ibis*, *125*(3), 400–404.

Curry, J.P. (2004). Factors affecting the abundance of earthworms in soils. *Earthworm Ecology*, *9*, 113–113.

Curry, O., Whitehouse, H., and Mullins, D. (2019). Is it good to cooperate? Testing the theory of morality-as-cooperation in 60 societies. *Current Anthropology*, *60*(1).

Curtis, V.A. (2014) Infection-avoidance behaviour in humans and other animals. *Trends in Immunology*, *35*, 457–464. doi:10.1016/j.it.2014.08.006

Cushing, L., Morello-Frosch, R., Wander, M., and Pastor, M. (2015). The haves, the have-nots, and the health of everyone: the relationship between social inequality and environmental quality. *Annual Review of Public Health*, *36*, 193–209.

Cutrera, A.P., Zenuto, R.R., Luna, F., and Antenucci, C.D. (2010). Mounting a specific immune response increases energy expenditure of the subterranean rodent *Ctenomys talarum* (tuco-tuco): implications for intraspecific and interspecific variation in immunological traits. *Journal of Experimental Biology*, *213*(5), 715–724.

Cuvier, F. (1825). *Dictionnaire des sciences naturelles*, Vol. 35. FG Levrault, Paris.

da Silva, J. and Macdonald, D.W. (1989). Limitations to the use of tooth wear as a means of ageing Eurasian badgers, *Meles meles*. *Revue d'Ecologie, Terre et Vie*, *44*(3), 275–278.

da Silva, J., Woodroffe, R., and Macdonald, D.W. (1993). Habitat, food availability and group territoriality in the European badger, *Meles meles*. *Oecologia*, *95*(4), 558–564.

da Silva, J., Macdonald, D.W., and Evans, P.G. (1994). Net costs of group living in a solitary forager, the Eurasian badger (*Meles meles*). *Behavioral Ecology*, *5*(2), 151–158.

Dahle, B. and Swenson, J.E. (2003). Family breakup in brown bears: are young forced to leave? *Journal of Mammalogy*, *84*(2), 536–540.

Dalley, D., Davé, D., Lesellier, S., Palmer, S., Crawshaw, T., Hewinson, R.G., and Chambers, M. (2008). Development and evaluation of a gamma-interferon assay for tuberculosis in badgers (*Meles meles*). *Tuberculosis*, *88*(3), 235–243.

Dammhahn, M., Dingemanse, N.J., Niemelä, P.T., and Réale, D. (2018). Pace-of-life syndromes: a framework for the adaptive integration of behaviour, physiology and life history. Behavioral Ecology and Sociobiology, *72*(3), 1–8.

Daniels, M.J., Golder, M.C., Jarrett, O., and MacDonald, D.W. (1999). Feline viruses in wildcats from Scotland. *Journal of Wildlife Diseases*, *35*(1), 121–124.

Dansgaard, W., Johnsen, S.J., Clausen, H.B., Dahl-Jensen, D., Gundestrup, N.S., Hammer, C.U., Hvidberg, C.S., Steffensen, J.P., Sveinbjörnsdottir, A.E., Jouzel, J., and Bond, G. (1993). Evidence for general instability of past climate from a 250-kyr ice-core record. *Nature*, *364*(6434), 218–220.

Darwin, C. (1859). *On the Origin of Species by Means of Natural Selection, or Preservation of Favoured Races in the Struggle for Life*. John Murray, London.

Darwin, C. (1871). *The Descent of Man, and Selection in Relation to Sex*. John Murray, London.

Darwin, F. (ed.) (1887). *The Life and Letters of Charles Darwin, Including an Autobiographical Chapter*, Volume 2. John Murray, London [Letter 752, 5 March 1879, to A. Stephen Wilson].

Davenport, L.C. (2010). Aid to a declining matriarch in the giant otter (*Pteronura brasiliensis*). *PLoS ONE*, *5*, e11385.

David, P. (1998). Heterozygosity–fitness correlations: new perspectives on old problems. *Heredity*, *80*(5), 531–537.

Davies, J.B., Lluberas, R., and Shorrocks, A.F. (2017). Estimating the level and distribution of global wealth, 2000–2014. *Review of Income and Wealth*, *63*(4), 731–759.

Davies, J.M., Lachno, D.R., and Roper, T.J. (1988). The anal gland secretion of the European badger (*Meles meles*) and its role in social communication. *Journal of Zoology*, *216*(3), 455–463.

Davies, N.B. (1978a). Ecological questions about territorial behaviour. In *Behavioural Ecology: An Evolutionary Approach*, 2nd edn (J.R. Krebs and N.B. Davies, eds, pp. 317–350). Blackwell Scientific Publications, Oxford.

Davies, N.B. (1978b). Territorial defence in the speckled wood butterfly (*Pararge aegeria*): the resident always wins. *Animal Behaviour*, *26*, 138–147.

Davies, N.B. (1991). Mating systems. In *Behavioural Ecology: An Evolutionary Approach*, 3rd edn (J.R. Krebs and N.B. Davies, eds, pp. 263–294). Blackwell Scientific Publications, Oxford.

Davison, J., Huck, M., Delahay, R.J., and Roper, T.J. (2009). Restricted ranging behaviour in a high-density population of urban badgers. *Journal of Zoology*, *277*(1), 45–53.

Davison, K.E., Hughes, J.L., Gormley, E., Lesellier, S., Costello, E., and Corner, L.A. (2007). Evaluation of the anaesthetic effects of combinations of ketamine, medetomidine, romifidine and butorphanol in European badgers (*Meles meles*). *Veterinary Anaesthesia and Analgesia*, *34*(6), 394–402.

Dawkins, M.S. and Guilford, T. (1996). Sensory bias and the adaptiveness of female choice. *The American Naturalist*, *148*(5), 937–942.

Dawkins, R. and Krebs, J.R. (1978). Animal signals: information or manipulation. *Behavioural Ecology: An Evolutionary Approach*, *2*, 282–309.

Day, D.S., Gozansky, W.S., Van Pelt, R.E., Schwartz, R.S., and Kohrt, W.M. (2005). Sex hormone suppression reduces resting energy expenditure and β-adrenergic support of resting energy expenditure. *The Journal of Clinical Endocrinology & Metabolism*, *90*(6), 3312–3317.

Dayan, T. and Simberloff, D. (1994). Character displacement, sexual dimorphism, and morphological variation among British and Irish mustelids. *Ecology*, *75*(4), 1063–1073.

de Boer, R.A., Vega-Trejo, R., Kotrschal, A., and Fitzpatrick, J.L. (2021). Meta-analytic evidence that animals rarely avoid inbreeding. *Nature Ecology and Evolution*, *5*, 949–964. https://doi.org/10.1038/s41559-021-01453-9

De la Rua-Domenech, R., Goodchild, A.T., Vorder-meier, H.M., Hewinson, R.G., Christiansen, K.H., and Clifton-Hadley, R.S. (2006). Ante mortem diagnosis of tuberculosis in cattle: a review of the tuberculin tests, γ-interferon assay and other ancillary diagnostic techniques. *Research in Veterinary Science*, 81(2), 190–210.

De Lange, T. (2004). T-loops and the origin of telomeres. *Nature Reviews Molecular Cell Biology*, 5(4), 323–329.

De Vries, H. (1995). An improved test of linearity in dominance hierarchies containing unknown or tied relationships. *Animal Behaviour*, 50(5), 1375–1389.

de Vries, H., Stevens, J.M.G., and Vervaecke, H. (2006). Measuring and testing the steepness of dominance hierarchies. *Animal Behaviour*, 71, 585–592.

de Waal, F.B.M. (1984). Coping with social tension: Sex differences in the effect of food provision to small rhesus monkey groups. *Animal Behaviour*, 32(3), 765–773. doi:10.1016/S0003-3472(84)80152-80159

De Waal, F.B. (1995a). Bonobo sex and society. *Scientific American*, 272(3), 82–88.

De Waal, F.B. (1995b). Sex as an alternative to aggression in the bonobo. In *Sexual Nature, Sexual Culture* (P.R. Abramson and S.D. Pinkerton, eds, pp. 37–54). University of Chicago Press, Chicago, IL.

Defra (2014). Information on badger culls of 2013 and 2014. https://www.gov.uk/government/publications/information-on-badger-culls-of-2013-and-2014 (accessed 7 March 2022).

Defra (Department for Environment, Food and Rural Affairs) (2006).

Defra (Department for Environment, Food and Rural Affairs) (2011a). The government's policy on bovine TB and badger control in England. https://assets.publishing.service.gov.uk/government/uploads/system/uploads/attachment_data/file/69463/pb13691-bovinetb-policy-statement.pdf (accessed 10 March 2022).

Defra (Department for Environment, Food and Rural Affairs) (2011b). Bovine TB eradication programme for England. https://assets.publishing.service.gov.uk/government/uploads/system/uploads/attachment_data/file/69443/pb13601-bovinetb-eradication-programme-110719.pdf (accessed 10 March 2022).

Defra (Department for Environment, Food and Rural Affairs) (2011c). Guidance to Natural England: preventing spread of bovine TB. https://www.gov.uk/government/publications/guidance-to-natural-england-preventing-spread-of-bovine-tb?%20utm_medium=email&utm_campaign=govuk-notifications&utm_source=4596e95e-d925-41fc-9d50-368f743c5322&utm_content=daily (accessed 10 March 2022).

Defra (Department for Environment, Food and Rural Affairs) (2011d). Impact assessment. Measures to address bovine TB in badgers. https://assets.publishing.service.gov.uk/government/uploads/system/uploads/attachment_data/file/182452/bovine-tb-impact-assessment.pdf (accessed 10 March 2022).

Defra (Department for Environment, Food and Rural Affairs) (2013). Bovine TB vaccination. https://publications.parliament.uk/pa/cm201213/cmselect/cmenvfru/writev/bovine/m23.htm (accessed 10 March 2022).

Defra (Department for Environment, Food and Rural Affairs) (2017). Statistical data set. Tuberculosis (TB) in cattle in Great Britain. https://www.gov.uk/government/statistical-data-sets/tuberculosis-tb-in-cattle-in-great-britain (accessed 10 March 2022).

Defra (Department for Environment, Food and Rural Affairs) (2018a). TB surveillance in wildlife in England. https://assets.publishing.service.gov.uk/government/uploads/system/uploads/attachment_data/file/787588/tb-surveillance-wildlife-england-2017.pdf (accessed 10 March 2022).

Defra (Department for Environment, Food and Rural Affairs) (2018b). Bovine TB strategy review. October 2018. https://assets.publishing.service.gov.uk/government/uploads/system/uploads/attachment_data/file/756942/tb-review-final-report-corrected.pdf (accessed 10 March 2022).

Defra (Department for Environment, Food and Rural Affairs) (2019a). Bovine TB epidemiology and surveillance in Great Britain, 2019. https://www.gov.uk/government/publications/bovine-tb-epidemiology-and-surveillance-in-great-britain-2019 (accessed 10 March 2022).

Defra (Department for Environment, Food and Rural Affairs) (2019b). An update on TB surveillance in wildlife. https://assets.publishing.service.gov.uk/government/uploads/system/uploads/attachment_data/file/830810/surveillance-wildlife-2018.pdf (accessed 10 March 2022).

Defra (Department for Environment, Food and Rural Affairs) (2020a). Quarterly publication of national statistics on the incidence and prevalence of tuberculosis (TB) in cattle in Great Britain—to end September 2020. https://assets.publishing.service.gov.uk/government/uploads/system/uploads/attachment_data/file/967695/bovinetb-statsnotice-Q3-quarterly-16dec20.pdf (accessed 11 March 2022).

Defra (Department for Environment, Food and Rural Affairs) (2020b). A strategy for achieving Bovine Tuberculosis Free Status for England: 2018 review—government response. Defra, London.

Defra (Department for Environment, Food and Rural Affairs) (2021). An update on wildlife TB surveillance in low risk area hotspots. https://assets.publishing.service.gov.uk/government/uploads/system/uploads/attachment_data/file/974351/tb-surveillance-in-wildlife-mar2021.pdf (accessed 10 March 2022).

Defra (Department for Environment, Food and Rural Affairs) (2022). Rebuttal of claims on TB cull effectiveness. https://deframedia.blog.gov.uk/2022/03/18/rebuttal-of-claims-on-tb-cull-effectiveness/?s=03 (accessed 12 May 2022).

Defra (Department for Environment, Food and Rural Affairs) and AHVLA (Animal Health and Veterinary Laboratories Agency) (2014). The efficacy of badger population reduction by controlled shooting and cage trapping, and the change in badger activity following culling from 27/08/2013 to 28/11/2013. https://assets.publishing.service.gov.uk/government/uploads/system/uploads/attachment_data/file/300385/ahvla-extension-efficacy.pdf (accessed 10 March 2022).

Defra (Department for Environment, Food and Rural Affairs) and APHA (Animal and Plant Health Agency) (n.d.). Latest national statistics on tuberculosis (TB) in cattle in Great Britain—quarterly. https://www.gov.uk/government/statistics/incidence-of-tuberculosis-tb-in-cattle-in-great-britain (accessed 10 March 2022).

Del Cerro, I., Marmi, J., Ferrando, A., Chashchin, P., Taberlet, P., and Bosch, M. (2010). Nuclear and mitochondrial phylogenies provide evidence for four species of Eurasian badgers (Carnivora). *Zoologica Scripta*, 39(5), 415–425.

Delahay, R. and Frölich, K. (2000). Absence of antibodies against canine distemper virus in free-ranging populations of the Eurasian badger in Great Britain. *Journal of Wildlife Diseases*, 36(3), 576–579.

Delahay, R.J., Brown, J.A., Mallinson, P.J., Spyvee, P.D., Handoll, D., Rogers, L.M., and Cheeseman, C.L. (2000). The use of marked bait in studies of the territorial organization of the European badger (*Meles meles*). *Mammal Review*, 30(2), 73–87.

Delahay, R.J., Carter, S.P., Forrester, G.J., Mitchell, A., and Cheeseman, C.L. (2006a). Habitat correlates of group size, bodyweight and reproductive performance in a high-density Eurasian badger (*Meles meles*) population. *Journal of Zoology*, 270(3), 437–447.

Delahay, R.J., Walker, N.J., Forrester, G.J., Harmsen, B., Riordan, P., Macdonald, D.W., Newman, C., and Cheeseman, C.L. (2006b). Demographic correlates of bite wounding in Eurasian badgers, *Meles meles* L., in stable and perturbed populations. *Animal Behaviour*, 71(5), 1047–1055.

Delahay, R.J., Smith, G.C., Barlow, A.M., Walker, N., Harris, A., Clifton-Hadley, R.S., and Cheeseman, C.L. (2007). Bovine tuberculosis infection in wild mammals in the south-west region of England: a survey of prevalence and a semi-quantitative assessment of the relative risks to cattle. *The Veterinary Journal*, 173(2), 287–301.

Devine, D. (n.d.). How long is a generation? Science provides an answer. Available at: https://isogg.org/wiki/How_long_is_a_generation%3F_Science_provides_an_answer (accessed 1 April 2022).

Dhir, N., Wood, F., Vákár, M., Markham, A., Wijers, M., Trethowan, P., and Macdonald, D. (2017). Interpreting lion behaviour with nonparametric probabilistic programs. 33rd Conference on Uncertainty in Artificial Intelligence, Sydney, Australia, 1–15 August.

Di Blanco, Y. and Hirsch, B.T. (2006). Determinants of vigilance behavior in the ring-tailed coati (*Nasua nasua*): the importance of within-group spatial position. *Behavioral Ecology and Sociobiology*, 61(2), 173–182.

DiNardo, A. (2020). Heterogenous immune response to TB is not a false negative. *The American Journal of Tropical Medicine and Hygiene*, 102(3), 698–698.

Dingemanse, N.J. and Wolf, M. (2013). Between-individual differences in behavioural plasticity within populations: causes and consequences. *Animal Behaviour*, 85(5), 1031–1039.

Diserens, T.A., Bubnicki, J.W., Schutgens, E., Rokx, K., Kowalczyk, R., Kuijper, D.P.J., and Churski, M. (2021). Fossoriality in a risky landscape: badger sett use varies with perceived wolf risk. *Journal of Zoology*, 313(1), 76–85.

Ditchkoff, S.S., Spicer, L.J., Masters, R.E., and Lochmiller, R.L. (2001). Concentrations of insulin-like growth factor-I in adult male white-tailed deer (*Odocoileus virginianus*): associations with serum testosterone, morphometrics and age during and after the breeding season. *Comparative Biochemistry and Physiology Part A: Molecular & Integrative Physiology*, 129(4), 887–895.

Diuk-Wasser, M.A. and Cassini, M.H. (1998). A study on the diet of minor grisons and a preliminary analysis of their role in the control of rabbits in Patagonia. *Studies on Neotropical Fauna and Environment*, 33(1), 3–6.

Dixon, D.R. (2003). A non-invasive technique for identifying individual badgers *Meles meles*. *Mammal Review*, 33(1), 92–94.

Dmitriew, C.M. (2011). The evolution of growth trajectories: what limits growth rate? *Biological Reviews*, 86(1), 97–116.

Do Linh San, E., Ferrari, N., and Weber, J.M. (2007). Socio-spatial organization of Eurasian badgers (*Meles meles*) in a low-density population of central Europe. *Canadian Journal of Zoology*, 85(9), 973–984.

Doak, D.F. and Morris, W.F. (2010). Demographic compensation and tipping points in climate-induced range shifts. *Nature*, 467(7318), 959–962.

Dobson, A.P. (1986). Inequalities in the individual reproductive success of parasites. *Parasitology*, 92(3), 675–682.

Dobson, F.S. and Oli, M.K. (2007). Fast and slow life histories of mammals. *Ecoscience*, 14(3), 292–299.

Domingo-Roura, X., Newman, C., Calafell, F., and Macdonald, D.W. (2001). Blood biochemistry reflects seasonal nutritional and reproductive constraints in the Eurasian badger (*Meles meles*). *Physiological and Biochemical Zoology*, 74(3), 450–460.

Domingo-Roura, X., Macdonald, D.W., Roy, M.S., Marmi, J., Terradas, J., Woodroffe, R., Burke, T., and Wayne, R.K. (2003). Confirmation of low genetic diversity and multiple breeding females in a social group of Eurasian badgers from microsatellite and field data. *Molecular Ecology*, 12(2), 533–539.

Domingo-Roura, X., Marmi, J., Ferrando, A., López-Giráldez, F., Macdonald, D.W., and Jansman, H.A. (2006). Badger hair in shaving brushes comes from protected Eurasian badgers. *Biological Conservation*, 128(3), 425–430.

Doncaster, C.P. (1992). Testing the role of intraguild predation in regulating hedgehog populations. *Proceedings of the Royal Society of London. Series B: Biological Sciences*, 249(1324), 113–117.

Doncaster, C.P. and Macdonald, D.W. (1991). Drifting territoriality in the red fox *Vulpes vulpes*. *Journal of Animal Ecology*, 60(2), 423–439.

Doncaster, C.P. and Woodroffe, R. (1993). Den site can determine shape and size of badger territories: implications for group-living. *Oikos*, 66(1), 88–93.

Donnelly, C.A. and Nouvellet, P. (2013). The contribution of badgers to confirmed tuberculosis in cattle in high-incidence areas in England. *PLoS Currents Outbreaks*, 5.

Donnelly, C.A. and Hone, J. (2010). Is there an association between levels of bovine tuberculosis in cattle herds and badgers? *Statistical Communications in Infectious Diseases*, 2(1). https://doi.org/10.2202/1948-4690.1000

Donnelly, C.A. and Woodroffe, R. (2012). Reduce uncertainty in UK badger culling. *Nature*, 485(7400), 582–582.

Donnelly, C.A., Woodroffe, R., Cox, D.R., Bourne, J., Gettinby, G., Le Fevre, A.M., McInerney, J.P., and Morrison, W.I. (2003). Impact of localized badger culling on tuberculosis incidence in British cattle. *Nature*, 426(6968), 834–837.

Donnelly, C.A., Woodroffe, R., Cox, D.R., Bourne, F.J., Cheeseman, C.L., Clifton-Hadley, R.S., Wei, G., Gettinby, G., Gilks, P., Jenkins, H., and Johnston, W.T. (2006). Positive and negative effects of widespread badger culling on tuberculosis in cattle. *Nature*, 439(7078), 843–846.

Donnelly, C.A., Wei, G., Johnston, W.T., Cox, D.R., Woodroffe, R., Bourne, F.J., Cheeseman, C.L., Clifton-Hadley, R.S., Gettingby, G., Gilks, G., Jenkins, H., Le fever, A.M., McInerney, J.P., and Morrison, W.I. (2007). Impacts of widespread badger culling on cattle tuberculosis: concluding analyses from a large-scale field trial. *Journal of Infectious Diseases*, 11, 300–308.

Doolan, S.P. and MacDonald, D.W. (1996). Diet and foraging behaviour of group-living meerkats, *Suricata suricatta*, in the southern Kalahari. *Journal of Zoology*, 239(4), 697–716.

Doolan, S.P. and Macdonald, D.W. (1997). Band structure and failures of reproductive suppression in a cooperatively breeding carnivore, the slender-tailed meerkat (*Suricata suricatta*). *Behaviour*, 134, 827–848.

Doolan, S.P. and Macdonald, D.W. (1999). Co-operative rearing by slender-tailed meerkats (*Suricata suricatta*) in the Southern Kalahari. *Ethology*, 105(10), 851–866.

Doughty, P. and Shine, R. (1997). Detecting life history trade-offs: measuring energy stores in "capital" breeders reveals costs of reproduction. *Oecologia*, 110(4), 508–513.

Douglas, D.A., Houde, A., Song, J.H., Farookhi, R., Concannon, P.W., and Murphy, B.D. (1998). Luteotropic hormone receptors in the ovary of the mink (*Mustela vison*) during delayed implantation and early-postimplantation gestation. *Biology of Reproduction*, 59(3), 571–578.

Downs, S.H., Prosser, A., Ashton, A., Ashfield, S., Brunton, L.A., Brouwer, A., Upton, P., Robertson, A., Donnelly, C.A., and Parry, J.E. (2019). Assessing effects from four years of industry-led badger culling in England on the incidence of bovine tuberculosis in cattle, 2013–2017. *Scientific Reports*, 9(1), 1–14.

Downs, S.H., Ashfield, S., Arnold, M., Roberts, T., Prosser, A., Robertson, A., Frost, S., Harris, K., Avigad, R., and Smith, G.C. (2021). Detection of a local Mycobacterium bovis reservoir using cattle surveillance data. *Authorea*, 19 May. doi:10.22541/au.162142253.37669744/v1.

Drabble, P. (1971). The function of mutual grooming in badgers (*Meles meles*). *Journal of Zoology*, 164, 260.

Drabble, P. (1979). *No Badgers in My Wood*. Michael Joseph, London.

Dragoo, J.W. and Honeycutt, R.L. (1997). Systematics of mustelid-like carnivores. *Journal of Mammalogy*, 78(2), 426–443.

Drewe, J.A., Madden, J.R., and Pearce G.P. (2009). The social network structure of a wild meerkat population: 1. Intergroup interactions. *Behavioral Ecology and Sociobiology*, 63(9), 1295–1306.

Drewe, J.A., Tomlinson, A.J., Walker, N.J., and Delahay, R.J. (2010). Diagnostic accuracy and optimal use of three tests for tuberculosis in live badgers. *PLoS ONE*, 5(6), e11196.

Drews, C. (1993). The concept and definition of dominance in animal behaviour. *Behaviour*, 125(3–4), 283–313.

Driscoll, C. and Macdonald, D.W. (2010). Top dogs: wolf domestication and wealth. *Journal of Biology*, 9(2), 10.

Driscoll, C.A., Macdonald, D.W., and O'Brien, S.J. (2009). From wild animals to domestic pets, an evolutionary view of domestication. *Proceedings of the National Academy of Sciences*, 106(Suppl. 1), 9971–9978.

Dröge, E., Creel, S., Becker, M.S., Loveridge, A.J., Sousa, L.L., and Macdonald, D.W. (2020). Assessing the performance of index calibration survey methods to monitor populations of wide-ranging low-density carnivores. *Ecology and Evolution*, 10(7), 3276–3292.

Dröscher, I. and Kappeler, P.M. (2014). Maintenance of familiarity and social bonding via communal latrine use in a solitary primate (*Lepilemur leucopus*). *Behavioral Ecology and Sociobiology*, 68(12), 2043–2058.

Duarte, M.D., Henriques, A.M., Barros, S.C., Fagulha, T., Mendonca, P., Carvalho, P., Monteiro, M., Fevereiro, M., Basto, M.P., Rosalino, L.M., and Barros, T. (2013). Snapshot of viral infections in wild carnivores reveals ubiquity of parvovirus and susceptibility of Egyptian mongoose to feline panleukopenia virus. *PLoS ONE*, 8(3), e59399.

Dubey, J.P. (1998). *Toxoplasma gondii* oocyst survival under defined temperatures. *The Journal of Parasitology*, 84(4), 862–865.

Duckworth, J.W., Timmins, R., Chutipong, W., Gray, T.N.E., Long, B., Helgen, K., Rahman, H., Choudhury, A., and Willcox, D.H.A. (2016). *Arctonyx collaris*. In *The IUCN Red List of Threatened Species 2016*: e.T70205537A45209459. http://dx.doi.org/10.2305/IUCN.UK.2016-1.RLTS.T70205537A45209459.en (accessed 20 February 2017).

Dugatkin, L.A. (1997). *Cooperation Among Animals: An Evolutionary Perspective*. Oxford University Press, Oxford.

Dugdale, H.L. and Richardson, D.S. (2018). Heritability of telomere variation: it is all about the environment! *Philosophical Transactions of the Royal Society. Series B: Biological Sciences*, 373(1741), 20160450.

Dugdale, H.L., Macdonald, D.W., and Newman, C. (2003). Offspring sex ratio variation in the European badger, *Meles meles*. *Ecology*, *84*(1), 40–45.

Dugdale, H.L., Macdonald, D.W., Pope, L.C., and Burke, T. (2007). Polygynandry, extra-group paternity and multiple-paternity litters in European badger (*Meles meles*) social groups. *Molecular Ecology*, *16*(24), 5294–5306.

Dugdale, H.L., Macdonald, D.W., Pope, L.C., Johnson, P.J., and Burke, T. (2008). Reproductive skew and relatedness in social groups of European badgers, *Meles meles*. *Molecular Ecology*, *17*(7), 1815–1827.

Dugdale, H.L., Ellwood, S.A., and Macdonald, D.W. (2010). Alloparental behaviour and long-term costs of mothers tolerating other members of the group in a plurally breeding mammal. *Animal Behaviour*, *80*(4), 721–735.

Dugdale, H.L., Davison, D., Baker, S.E., Ellwood, S.A., Newman, C., Buesching, C.D., and Macdonald, D.W. (2011a). Female teat size is a reliable indicator of annual breeding success in European badgers: genetic validation. *Mammalian Biology*, *76*(6), 716–721.

Dugdale, H.L., Griffiths, A., and Macdonald, D.W. (2011b). Polygynandrous and repeated mounting behaviour in European badgers, *Meles meles*. *Animal Behaviour*, *82*(6), 1287–1297.

Dugdale, H.L., Pope, L.C., Newman, C., Macdonald, D.W., and Burke, T. (2011c). Age-specific breeding success in a wild mammalian population: selection, constraint, restraint and senescence. *Molecular Ecology*, 20(15), 3261–3274.

Dunbar, R.I. (1988). *Primate Societies*. Chapman and Hall, London.

Dunbar, R.I. (1992). Time: a hidden constraint on the behavioural ecology of baboons. *Behavioral Ecology and Sociobiology*, *31*(1), 35–49.

Dunbar, R.I. (1998). The social brain hypothesis. *Evolutionary Anthropology: Issues, News, and Reviews: Issues, News, and Reviews*, *6*(5), 178–190.

Dunbar, R.I. (2010). The social role of touch in humans and primates: behavioural function and neurobiological mechanisms. *Neuroscience & Biobehavioral Reviews*, *34*(2), 260–268.

Dunmartin, B., Ribes, C., Souloumiac, J., Charron, G., and Canivenc, R. (1989). Experimental study of foetal growth in the European badger *Meles meles* L. *Mammalia*, *53*, 279–285.

Dunn, P.O., Bollmer, J.L., Freeman-Gallant, C.R., and Whittingham, L.A. (2013). MHC variation is related to a sexually selected ornament, survival, and parasite resistance in common yellowthroats. *Evolution: International Journal of Organic Evolution*, *67*(3), 679–687.

Durant, J.M., Hjermann, D.Ø., Ottersen, G., and Stenseth, N.C. (2007). Climate and the match or mismatch between predator requirements and resource availability. *Climate Research*, *33*(3), 271–283.

Dyo, V., Ellwood, S.A., Macdonald, D.W., Markham, A., Trigoni, N., Wohlers, R., Mascolo, C., Pásztor, B., Scellato, S., and Yousef, K. (2012). WILDSENSING: design and deployment of a sustainable sensor network for wildlife monitoring. *ACM Transactions on Sensor Networks (TOSN)*, *8*(4), 1–33.

Eads, D.A., Biggins, D.E., Jachowski, D.S., Livieri, T.M., Millspaugh, J.J., and Forsberg, M. (2010). Morning ambush attacks by black-footed ferrets on emerging prairie dogs. *Ethology Ecology & Evolution*, *22*(4), 345–352.

Eads, D.A., Biggins, D.E., and Eads, S.L. (2017). Grooming behaviors of black-tailed prairie dogs are influenced by flea parasitism, conspecifics, and proximity to refuge. *Ethology*, *123*(12), 924–932.

Ebensperger, L.A. (1998). Strategies and counterstrategies to infanticide in mammals. *Biological Reviews*, *73*(3), 321–346.

Ebensperger, L.A. (2001). A review of the evolutionary causes of rodent group-living. *Acta Theriologica*, *46*(2), 115–144.

Economic Advisory Council (1934). Committee on cattle diseases: report. Her Majesty's Stationery Office, London.

Edward, D.A. and Chapman, T. (2011). The evolution and significance of male mate choice. *Trends in Ecology & Evolution*, *26*(12), 647–654.

Edwards, J.C. and Barnard, C.J. (1987). The effects of *Trichinella* infection on intersexual interactions between mice. *Animal Behaviour*, *35*(2), 533–540.

Edwards, C., Rasmussen, G., Riordan, P., Courchamp, F., and Macdonald, D.W. (2013) Non-adaptive phenotypic evolution of the endangered carnivore *Lycaon pictus*. *PLoS ONE*, *8*(9), e73856.

Egbert, A.L., Stokes, A.W., and Egbert, A.L. (1976). The social behaviour of brown bears on an Alaskan salmon stream. In *Bears: Their Biology and Management*, Vol. 3. *A Selection of Papers from the Third International Conference on Bear Research and Management, Binghamton, New York, USA, and Moscow, U.S.S.R., June 1974*. IUCN Publications New Series no. 40, pp. 41–56.

Egerton, F.N. (1968). Studies of animal populations from Lamarck to Darwin. *Journal of the History of Biology*, *1*(2), 225–259.

Eisenberg, C. (2014). Wolverine (*Gulo gulo luscus*). In *The Carnivore Way: Coexisting with and Conserving North America's Predators* (C. Eisenberg, ed., pp. 147–169). Island Press, Washington, DC.

Eisenberg, J.F. (1966). The social organisation of mammals. *Handbuch der Zoologie: Eine Naturgeschichte der Stamme des Tierreiches*, *8*, 1–92.

Eklund, A., Egid, K., and Brown, J.L. (1991). The major histocompatibility complex and mating preferences of male mice. *Animal Behaviour*, *42*(4), 693–694.

Elbroch, L.M. and Quigley, H. (2017). Social interactions in a solitary carnivore. *Current Zoology*, *63*(4), 357–362.

Elbroch, L.M., Lendrum, P.E., Quigley, H., and Caragiulo, A. (2016). Spatial overlap in a solitary carnivore: support for the land tenure, kinship or resource dispersion hypotheses?. *Journal of Animal Ecology*, *85*(2), 487–496.

Elden, S. (2013). How should we do the history of territory? *Territory, Politics, Governance*, *1*(1), 5–20.

Ellwood, S.A., Newman, C., Montgomery, R.A., Nicosia, V., Buesching, C.D., Markham, A., Mascolo, C., Trigoni, N., Pasztor, B., Dyo, V., and Latora, V. (2017). An active-radio-frequency-identification system capable of identifying co-locations and social-structure: validation with a wild free-ranging animal. *Methods in Ecology and Evolution*, 8(12), 1822–1831.

Elton, C. (1927). *Animal Ecology*. Sidgwick and Jackson, Ltd, London.

Emlen, S.T. (1982). The evolution of helping. I. An ecological constraints model. *American Naturalist*, 119, 29–39.

Endler, J.A. (1992). Signals, signal conditions, and the direction of evolution. *The American Naturalist*, 139, S125–S153.

Enquist, M. and Leimar, O. (1987). Evolution of fighting behaviour: the effect of variation in resource value. *Journal of Theoretical Biology*, 127, 187–205.

Enquist, M. and Leimar, O. (1990). The evolution of fatal fighting. *Animal Behaviour*, 39, 1–9.

Enticott, G. (2012). Regulating animal health, gender and quality control: a study of veterinary surgeons in Great Britain. *Journal of Rural Studies*, 28, 559–567.

Eppley, T.M., Ganzhorn, J.U., and Donati, G. (2016). Latrine behaviour as a multimodal communicatory signal station in wild lemurs: the case of *Hapalemur meridionalis*. *Animal Behaviour*, 111, 57–67.

Erlandsson, R., Hasselgren, M., Norén, K., Macdonald, D., and Angerbjörn, A. (2022). Resources and predation: drivers of sociality in a cyclic mesopredator. *Oecologia*, 198(2), 381–392.

Esteruelas, N.F., Huber, N., Evans, A.L., Zedrosser, A., Cattet, M., Palomares, F., Angel, M., Swenson, J.E., and Arnemo, J.M. (2016). Leukocyte coping capacity as a tool to assess capture-and handling-induced stress in Scandinavian brown bears (*Ursus arctos*). *Journal of Wildlife Diseases*, 52(2s), S40–S53.

Estes, R.D. (1974). Social organization of the African Bovidae. *The Behaviour of Ungulates and Its Relation to Management*, 1, 166–205.

Evans, A.C.O. and O'Doherty, J.V. (2001). Endocrine changes and management factors affecting puberty in gilts. *Livestock Production Science*, 68(1), 1–12.

Evans, P.G.H. (1980). Population genetics of the European starling. PhD thesis, University of Oxford.

Evans, P.G.H., Macdonald, D.W., and Cheeseman, C.L. (1989). Social structure of the Eurasian badger (*Meles meles*): genetic evidence. *Journal of Zoology*, 218(4), 587–595.

Evans, R.M., Birnberg, N.C., and Rosenfeld, M.G. (1982). Glucocorticoid and thyroid hormones transcriptionally regulate growth hormone gene expression. *Proceedings of the National Academy of Sciences*, 79(24), 7659–7663.

Eville & Jones (n.d.). Diversifying into agricultural research with lead role in bovine TB cattle vaccination trials. https://eandj.co.uk/news/diversifying-into-agricultural-research-with-lead-role-in-bovine-tb-cattle-vaccination-trials/ (accessed 10 March 2022).

Eyck, H.J.F., Buchanan, K.L., Crino, O.L., and Jessop, T.S. (2019). Effects of developmental stress on animal phenotype and performance: a quantitative review. *Biological Reviews*, 94, 1143–1160. doi:10.1111/brv.12496

Ezenwa, V.O. (2004). Selective defecation and selective foraging: antiparasite behavior in wild ungulates? *Ethology*, 110(11), 851–862.

Ezenwa, V.O., Ghai, R.R., McKay, A.F., and Williams, A.E. (2016). Group living and pathogen infection revisited. *Current Opinion in Behavioral Sciences*, 12, 66–72.

Fairbairn, D.J. (1997). Allometry for sexual size dimorphism: pattern and process in the coevolution of body size in males and females. *Annual Review of Ecology and Systematics*, 28(1), 659–687.

Fairlie, J., Holland, R., Pilkington, J.G., Pemberton, J.M., Harrington, L., and Nussey, D.H. (2016). Lifelong leukocyte telomere dynamics and survival in a free-living mammal. *Aging Cell*, 15(1), 140–148.

Farine, D.R. and Whitehead, H. (2015). Constructing, conducting and interpreting animal social network analysis. *Journal of Animal Ecology*, 84(5), 1144–1163.

Fatima, S., Kumari, A., Das. G., and Dwivedi, V.D. (2020). Tuberculosis vaccine: a journey from BCG to present. *Life Sciences*, 252, 117594.

Fedriani, J.M., Fuller, T.K., Sauvajot, R.M., and York, E.C. (2000). Competition and intraguild predation among three sympatric carnivores. *Oecologia*, 125(2), 258–270.

Fedurek, P., Dunbar, R.I., and British Academy Centenary Research Project (2009). What does mutual grooming tell us about why chimpanzees groom? *Ethology*, 115(6), 566–575.

Feeroz, M.M., Begum, S., and Hasan, M.K. (2011). Fishing with otters: a traditional conservation practice in Bangladesh. *IUCN Otter Specialist Group Bulletin*, 28, 21.

Fell, R.J., Buesching, C.D., and Macdonald, D.W. (2006). The social integration of European badger (*Meles meles*) cubs into their natal group. *Behaviour*, 143(6), 683–700.

Fereidouni, S., Freimanis, G.L., Orynbayev, M., Ribeca, P., Flannery, J., King, D.P., and Kock, R. (2019). Mass die-off of saiga antelopes, Kazakhstan, 2015. *Emerging Infectious Diseases*, 25(6), 1169.

Feró, O., Stephens, P.A., Barta, Z., McNamara, J.M., and Houston, A.I. (2008). Optimal annual routines: new tools for conservation biology. *Ecological Applications*, 18(6), 1563–1577.

Ferrari, S.F. (2009). Predation risk and antipredator strategies. In *South American Primates: Comparative Perspectives in the Study of Behavior, Ecology, and Conservation* (P.A. Garber, A. Estrada, J.C. Bicca-Marques, E.W. Heymann, and K.B. Strier, eds, pp. 251–277). Springer, New York.

Ferris, C. (1986). Mating and early maturity of badgers in Kent. *Journal of Zoology. Series A*, 209(2), 282.

Ferris, C. (1986). *The Darkness Is Light Enough*. Michael Joseph, London.

Ferris, C. (1988). *Out of the Darkness: Revelations of a Badger Watcher*. Unwin Hyman, Glasgow.

Fessler, D.M., Navarrete, C.D., Hopkins, W., and Izard, M.K. (2005). Examining the terminal investment hypothesis in humans and chimpanzees: associations among maternal age, parity, and birth weight. *American Journal of Physical Anthropology: The Official Publication of the American Association of Physical Anthropologists*, 127(1), 95–104.

Festa-Bianchet, M., Jorgenson, J.T., and Wuhart, W.D. (1994). Early weaning in bighorn sheep, *Ovis canadensis*, affects growth of males but not of females. *Behavioral Ecology*, 5(1), 21–27.

Finarelli, J.A. and Flynn, J.J. (2009). Brain-size evolution and sociality in Carnivora. *Proceedings of the National Academy of Sciences*, 106(23), 9345–9349.

Finch, C.E. (1990). *Longevity, Senescence and the Genome*. University of Chicago Press, Chicago, IL.

Finlayson, G.R., Shimmin, G.A., Temple-Smith, P.D., Handasyde, K.A., and Taggart, D.A. (2005). Burrow use and ranging behaviour of the southern hairy-nosed wombat (*Lasiorhinus latifrons*) in the Murraylands, South Australia. *Journal of Zoology*, 265(2), 189–200.

Fiorelli, L.E., Ezcurra, M.D., Hechenleitner, E.M., Argañaraz, E., Taborda, J.R., Trotteyn, M.J., and Desojo, J.B. (2013). The oldest known communal latrines provide evidence of gregarism in Triassic megaherbivores. *Scientific Reports*, 3(1), 1–7.

Firman, R.C., Gasparini, C., Manier, M.K., and Pizzari, T. (2017). Postmating female control: 20 years of cryptic female choice. *Trends in Ecology & Evolution*, 32(5), 368–382.

Fischer, C., Ferrari, N., and Weber, J.M. (2005). Exploitation of food resources by badgers (*Meles meles*) in the Swiss Jura Mountains. *Journal of Zoology*, 266(2), 121–131.

Fisher, D.N., Rodríguez-Muñoz, R., and Tregenza, T. (2016). Comparing pre-and post-copulatory mate competition using social network analysis in wild crickets. *Behavioral Ecology*, 27(3), 912–919.

Fisher, J.B. (1954). Evolution and bird sociality. In *Evolution as a Process* (J. Huxley, A.C. Hardy, and E.B. Ford, eds, pp. 71–83). Allen & Unwin, London.

Fisher, R.A. (1915). The evolution of sexual preference. *The Eugenics Review*, 7(3), 184.

Fisher, R.A. (1930). *The Genetical Theory of Natural Selection*. Clarendon Press, Oxford.

Fitzgerald, J. and Butler, W.R. (1982). Seasonal effects and hormonal patterns related to puberty in ewe lambs. *Biology of Reproduction*, 27(4), 853–863.

Florant, G.L., Porst, H., Peiffer, A., Hudachek, S.F., Pittman, C., Summers, S.A., Rajala, M.W., and Scherer, P.E. (2004). Fat-cell mass, serum leptin and adiponectin changes during weight gain and loss in yellow-bellied marmots (*Marmota flaviventris*). *Journal of Comparative Physiology B*, 174(8), 633–639.

Flynn, J.J., Neff, N.A., and Tedford, R.H. (1988). Phylogeny of the Carnivora. In *The Phylogeny and Classification of the Tetrapods* (M.J. Benton, ed., pp. 73–116). Oxford University Press, New York.

Foerster, K., Coulson, T., Sheldon, B.C., Pemberton, J.M., Clutton-Brock, T.H., and Kruuk, L.E. (2007). Sexually antagonistic genetic variation for fitness in red deer. *Nature*, 447(7148), 1107–1110.

Foley, J., Clifford, D., Castle, K., Cryan, P., and Ostfeld, R.S. (2011). Investigating and managing the rapid emergence of white-nose syndrome, a novel, fatal, infectious disease of hibernating bats. *Conservation Biology*, 25(2), 223–231.

Folke, C. (2006). Resilience: the emergence of a perspective for social–ecological systems analyses. *Global Environmental Change*, 16(3), 253–267.

Folke, C., Carpenter, S., Walker, B., Scheffer, M., Elmqvist, T., Gunderson, L., and Holling, C.S. (2004). Regime shifts, resilience, and biodiversity in ecosystem management. *Annual Review of Ecology, Evolution, and Systematics*, 35, 557–581.

Folstad, I. and Karter, A.J. (1992). Parasites, bright males, and the immunocompetence handicap. *The American Naturalist*, 139(3), 603–622.

Fontagné, S., Lataillade, E., Breque, J., and Kaushik, S. (2008). Lipid peroxidative stress and antioxidant defence status during ontogeny of rainbow trout (*Oncorhynchus mykiss*). *British Journal of Nutrition*, 100(1), 102–111.

Foreyt, W.J. and Lagerquist, J.E. (1994). Experimental infections of *Eimeria wapiti* and *E. zuernii*-like oocysts in Rocky Mountain elk (*Cervus elaphus*) calves. *Journal of Wildlife Diseases*, 30(3), 466–469.

Forslund, P. and Pärt, T. (1995). Age and reproduction in birds—hypotheses and tests. *Trends in Ecology & Evolution*, 10(9), 374–378.

Forstmeier, W., Martin, K., Bolund, E., Schielzeth, H., and Kempenaers, B. (2011). Female extrapair mating behavior can evolve via indirect selection on males. *Proceedings of the National Academy of Sciences*, 108(26), 10608–10613.

Fortes, G.G., Grandal-d'Anglade, A., Kolbe, B., Fernandes, D., Meleg, I.N., García-Vázquez, A., and Frischauf, C. (2016). Ancient DNA reveals differences in behaviour and sociality between brown bears and extinct cave bears. *Molecular Ecology*, 25(19), 4907–4918.

Fowler, C.W. (1987). A review of density dependence in populations of large mammals. In *Current Mammalogy* (H.H. Genoways, ed., pp. 401–441). Springer, Boston, MA.

Fowler, M.A. and Williams, T.D. (2017). A physiological signature of the cost of reproduction associated with parental care. *The American Naturalist*, 190(6), 762–773.

Fowler, P.A. and Racey, P.A. (1988). Overwintering strategies of the badger, *Meles meles*, at 57 N. *Journal of Zoology*, 214(4), 635–651.

Franceschi, C., Garagnani, P., Parini, P., Giuliani, C., and Santoro, A. (2018). Inflammaging: a new immune–metabolic viewpoint for age-related diseases. *Nature Reviews Endocrinology*, 14(10), 576–590.

Francis, J. (1947). *Bovine Tuberculosis*. Staples Press, London.

François, S., Vidick, S., Sarlet, M., Desmecht, D., Drion, P., Stevenson, P.G., and Gillet, L. (2013). Illumination of murine gammaherpesvirus-68 cycle reveals a sexual transmission route from females to males in laboratory mice. *PLoS Pathogens*, 9(4), e1003292.

Franks, D.W., Ruxton, G.D., and James, R. (2010). Sampling animal association networks with the gambit of the group. *Behavioral Ecology and Sociobiology*, 64, 493–503. doi:10.1007/s00265-009-0865-8

Frantz, A.C., Pope, L.C., Etherington, T.R., Wilson, G.J., and Burke, T. (2010). Using isolation-by-distance-based approaches to assess the barrier effect of linear landscape elements on badger (*Meles meles*) dispersal. *Molecular Ecology*, 19(8), 1663–1674.

Frantz, A.C., McDevitt, A.D., Pope, L.C., Kochan, J., Davison, J., Clements, C.F., Elmeros, M., Molina-Vacas, G., Ruiz-Gonzalez, A., Balestrieri, A., and Van Den Berge, K. (2014). Revisiting the phylogeography and demography of European badgers (*Meles meles*) based on broad sampling, multiple markers and simulations. *Heredity*, 113(5), 443–453.

Franz, M., McLean, E., Tung, J., Altmann, J., and Alberts, S.C. (2015). Self-organizing dominance hierarchies in a wild primate population. *Proceedings of the Royal Society, Series B: Biological Sciences*, 282(1814), 20151512.

Fredrich, E., Barzantny, H., Brune, I., and Tauch, A. (2013). Daily battle against body odor: towards the activity of the axillary microbiota. *Trends in Microbiology*, 21(6), 305–312.

Freeland, W.J. (1976). Pathogens and the evolution of primate sociality. *Biotropica* 8(1), 12–24. doi:10.2307/2387816.

Fretwell, S.D. (1972). *Populations in a Seasonal Environment*. Princeton University Press, Princeton, NJ.

Frisch, K.V., Wenner, A.M., and Johnson, D.L. (1967). Honeybees: do they use direction and distance information provided by their dancers? *Science*, 158(3804), 1072–1077.

Frölich, K., Czupalla, O., Haas, L., Hentschke, J., Dedek, J., and Fickel, J. (2000). Epizootiological investigations of canine distemper virus in free-ranging carnivores from Germany. *Veterinary Microbiology*, 74(4), 283–292.

Fukumoto, S., Suzuki, H., Igarashi, I., and Xuan, X. (2005). Fatal experimental transplacental *Babesia gibsoni* infections in dogs. *International Journal for Parasitology*, 35(9), 1031–1035.

Fuller, A., Mitchell, D., Maloney, S.K., Hetem, R.S., Fonsêca, V.F., Meyer, L.C., and Snelling, E.P. (2021). How dryland mammals will respond to climate change: the effects of body size, heat load and a lack of food and water. *Journal of Experimental Biology*, 224(Suppl. 1).

Fuller, R.C., Houle, D., and Travis, J. (2005). Sensory bias as an explanation for the evolution of mate preferences. *The American Naturalist*, 166(4), 437–446.

Furuichi, T., Connor, R., and Hashimoto, C. (2014). Nonconceptive sexual interactions in monkeys, apes, and dolphins. In *Primates and Cetaceans* (J. Yamagiwa and L. Karczmarski, eds, pp. 385–408). Springer, Tokyo.

Gadagkar, R. (2003). Is the peacock merely beautiful or also honest? *Current Science*, 85(7), 1012–1020.

Gaillard, J.M. and Yoccoz, N.G. (2003). Temporal variation in survival of mammals: a case of environmental canalization? *Ecology*, 84(12), 3294–3306.

Gallagher, J. and Nelson, J. (1979) Cause of ill health and natural death in badgers in Gloucestershire. *Tuberculosis*, 10(12), 14–16.

Gangloff, E.J., Sparkman, A.M., and Bronikowski, A.M. (2018). Among-individual heterogeneity in maternal behaviour and physiology affects reproductive allocation and offspring life-history traits in the garter snake *Thamnophis elegans*. *Oikos*, 127(5), 705–718.

Garaulet, M.O.J.M.J., Ordovas, J.M., and Madrid, J.A. (2010). The chronobiology, etiology and pathophysiology of obesity. *International Journal of Obesity*, 34(12), 1667–1683.

García, E., Halpert, E., Rodríguez, A., Andrade, R., Fiorentino, S., and García, C. (2004). Immune and histopathologic examination of flea bite-induced papular urticaria. *Annals of Allergy, Asthma and Immunology*, 92(4), 446–452.

Garcia, K.C. (2012). Reconciling views on T cell receptor germline bias for MHC. Trends in Immunology, 33(9), 429–436.

Gardner, J.L., Peters, A., Kearney, M.R., Joseph, L., and Heinsohn, R. (2011). Declining body size: a third universal response to warming? *Trends in Ecology & Evolution*, 26(6), 285–291.

Garn, H. and Renz, H. (2007). Epidemiological and immunological evidence for the hygiene hypothesis. *Immunobiology*, 212(6), 441–452.

Garratt, M., Stockley, P., Armstrong, S.D., Beynon, R.J., and Hurst, J.L. (2011). The scent of senescence: sexual signalling and female preference in house mice. *Journal of Evolutionary Biology*, 24(11), 2398–2409.

Garshelis, D.L. and Hellgren, E.C. (1994). Variation in reproductive biology of male black bears. *Journal of Mammalogy*, 75(1), 175–188.

Garshelis, D.L., Garshelis, J.A., and Kimker, A.T. (1986). Sea otter time budgets and prey relationships in Alaska. *The Journal of Wildlife Management*, 50(4), 637–647.

Gascoyne, S.C., Laurenson, M.K., Lelo, S., and Borner, M. (1993). Rabies in African wild dogs (*Lycaon pictus*) in the Serengeti region, Tanzania. *Journal of Wildlife Diseases*, 29(3), 396–402.

Gavrilov, L.A. and Gavrilova, N.S. (2001). The reliability theory of aging and longevity. *Journal of Theoretical Biology*, 213(4), 527–545.

Geffen, E., Hefner, R., Macdonald, D.W., and Ucko, M. (1992). Habitat selection and home range in the Blanford's fox, *Vulpes cana*: compatibility with the resource dispersion hypothesis. *Oecologia*, 91(1), 75–81.

Geffen, E., Gompper, M.E., Gittleman, J.L., Luh, H.K., Macdonald, D.W., and Wayne, R.K. (1996). Size, life history traits, and social organization in the Canidae: a re-evaluation. *American Naturalist*, 147(1), 140–160.

Geiser, F. and Ruf, T. (1995). Hibernation versus daily torpor in mammals and birds: physiological variables and classification of torpor patterns. *Physiological Zoology*, 68(6), 935–966.

Gelling, M., McLaren, G.W., Mathews, F., Mian, R., and Macdonald, D.W. (2009). Impact of trapping and handling on leukocyte coping capacity in bank voles (*Clethrionomys glareolus*) and wood mice (*Apodemus sylvaticus*). *Animal Welfare*, 18(1), 1–7.

Gelling, M., Montes, I., Moorhouse, T.P., and Macdonald, D.W. (2010). Captive housing during water vole (*Arvicola terrestris*) reintroduction: does short-term social stress impact on animal welfare? *PLoS ONE*, 5(3), e9791.

Gelling, M., Johnson, P.J., Moorhouse, T.P., and Macdonald, D.W. (2012). Measuring animal welfare within a reintroduction: an assessment of different indices of stress in water voles *Arvicola amphibius*. *PLoS ONE*, 7(7), e41081.

Gende, S.M. and Quinn, T.P. (2004). The relative importance of prey density and social dominance in determining energy intake by bears feeding on Pacific salmon. *Canadian Journal of Zoology*, 82(1), 75–85.

Gentle, L.K. and Gosler, A.G. (2001). Fat reserves and perceived predation risk in the great tit, Parus major. *Proceedings of the Royal Society of London. Series B: Biological Sciences*, 268(1466), 487–491.

George, S.C., Smith, T.E., Mac Cana, P.S., Coleman, R., and Montgomery, W.I. (2014). Physiological stress in the Eurasian badger (*Meles meles*): effects of host, disease and environment. *General and Comparative Endocrinology*, 200, 54–60.

Gerard, B.M. (1967). Factors affecting earthworms in pastures. *The Journal of Animal Ecology*, 36(1), 235–252.

Gereben, B., Zavacki, A.M., Ribich, S., Kim, B.W., Huang, S.A., Simonides, W.S., Zeold, A., and Bianco, A.C. (2008). Cellular and molecular basis of deiodinase-regulated thyroid hormone signaling. *Endocrine Reviews*, 29(7), 898–938.

Gerich, J.E., Charles, M.A., and Grodsky, G.M. (1976). Regulation of pancreatic insulin and glucagon secretion. *Annual Review of Physiology*, 38(1), 353–388.

Gerlach, M., Farb, B., Revelle, W., and Nunes Amaral, L.A. (2018). A robust data-driven approach identifies four personality types across four large data sets. *Nature Animal Behaviour*, 2, 735–742.

Gese, E.M. (2001). Territorial defense by coyotes (*Canis latrans*) in Yellowstone National Park, Wyoming: who, how, where, when, and why. *Canadian Journal of Zoology*, 79(6), 980–987.

Ghalambor, C.K., McKay, J.K., Carroll, S.P., and Reznick, D.N. (2007). Adaptive versus non-adaptive phenotypic plasticity and the potential for contemporary adaptation in new environments. *Functional Ecology*, 21(3), 394–407.

Gibson, R.M. and Langen, T.A. (1996). How do animals choose their mates? *Trends in Ecology & Evolution*, 11(11), 468–470.

Gill, P., Jeffreys, A.J., and Werrett, D.J. (1985). Forensic application of DNA 'fingerprints'. *Nature*, 318(6046), 577–579.

Ginsburg, L. and Morales, J. (2000). Origin and evolution of Melinae (Mustelidae, Carnivora, Mammalia). *Proceedings of the Academy of Sciences. Series IIA: Earth and Planetary Science*, 330(3), 221–225.

Giroud, S., Zahn, S., Criscuolo, F., Chery, I., Blanc, S., Turbill, C., and Ruf, T. (2014). Late-born intermittently fasted juvenile garden dormice use torpor to grow and fatten

prior to hibernation: consequences for ageing processes. *Proceedings of the Royal Society. Series B: Biological Sciences*, 281(1797), 20141131.

Gittleman, J.L. and Harvey, P.H. (1982). Carnivore home-range size, metabolic needs and ecology. *Behavioral Ecology and Sociobiology*, 10(1), 57–63.

Gittleman, J.L. and Van Valkenburgh, B. (1997). Sexual dimorphism in the canines and skulls of carnivores: effects of size, phylogency, and behavioural ecology. *Journal of Zoology*, 242(1), 97–117.

Gladwell, M. (2008). *Outliers: The Story of Success*. Little, Brown and Company, London.

Glazier, D.S. (2005). Beyond the '3/4-power law': variation in the intra-and interspecific scaling of metabolic rate in animals. *Biological Reviews*, 80(4), 611–662.

Glista, D.J., DeVault, T.L., and DeWoody, J.A. (2008). Vertebrate road mortality predominantly impacts amphibians. *Herpetological Conservation and Biology*, 3(1), 77–87.

Glocker, M.L., Langleben, D.D., Ruparel, K., Loughead, J.W., Valdez, J.N., Griffin, M.D., Sachser, N., and Gur, R.C. (2009). Baby schema modulates the brain reward system in nulliparous women. *Proceedings of the National Academy of Sciences*, 106(22), 9115–9119.

Gluckman, P.D. and Hanson, M.A. (2004). Developmental origins of disease paradigm: a mechanistic and evolutionary perspective. *Pediatric Research*, 56(3), 311–317.

Gluckman, P.D., Hanson, M.A., and Pinal, C. (2005a). The developmental origins of adult disease. *Maternal & Child Nutrition*, 1(3), 130–141.

Gluckman, P.D., Hanson, M.A., Spencer, H.G., and Bateson, P. (2005b). Environmental influences during development and their later consequences for health and disease: Implications for the interpretation of empirical studies. *Proceedings of the Royal Society. Series B: Biological Sciences*, 272, 671–677. doi:10.1098/rspb.2004.3001

Gluckman, P.D., Hanson, M.A., Cooper, C., and Thornburg, K.L. (2008). Effect of in utero and early-life conditions on adult health and disease. *New England Journal of Medicine*, 359(1), 61–73.

Godfray, C., Donnelly, C., Hewinson, G., Winter, M., and Wood, J. (2018) Bovine TB strategy review. Defra, London.

Godfray, H.C.J., Donnelly, C.A., Kao, R.R., Macdonald, D.W., McDonald, R.A., Petrokofsky, G., Wood, J.L., Woodroffe, R., Young, D.B., and McLean, A.R. (2013). A restatement of the natural science evidence base relevant to the control of bovine tuberculosis in Great Britain. *Proceedings of the Royal Society. Series B: Biological Sciences*, 280(1768), 20131634.

Godfrey, K.M. and Barker, D.J. (2001). Fetal programming and adult health. *Public Health Nutrition*, 4(2b), 611–624. https://doi.org/10.1079/phn2001145

Godfrey, K.M., Lillycrop, K.A., Burdge, G.C., Gluckman, P.D., and Hanson, M.A. (2007). Epigenetic mechanisms and the mismatch concept of the developmental origins of health and disease. *Pediatric Research*, 61(7), 5–10.

Goessmann, C., Hemelrijk, C., and Huber, R. (2000). The formation and maintenance of crayfish hierarchies:

behavioral and self-structuring properties. *Behavioral Ecology and Sociobiology*, 48(6), 418–428.

Goltsman, M., Kruchenkova, E.P., and Macdonald, D.W. (1996). The Mednyi arctic foxes: treating a population imperiled by disease. *Oryx*, 30(4), 251–258.

Good, M. and Duignan, A. (2011). Perspectives on the history of bovine TB and the role of tuberculin in bovine TB eradication. *Veterinary Medicine International*, 2011, 410470.

Goodrich, J.M. and Buskirk, S.W. (1998). Spacing and ecology of North American badgers (*Taxidea taxus*) in a prairie-dog (*Cynomys leucurus*) complex. *Journal of Mammalogy*, 79(1), 171–179.

Gopalaswamy, A.M., Delampady, M., Karanth, K.U., Kumar, N.S., and Macdonald, D.W. (2015). An examination of index-calibration experiments: counting tigers at macroecological scales. *Methods in Ecology and Evolution*, 6(9), 1055–1066.

Gorbunova, V. and Seluanov, A. (2009). Coevolution of telomerase activity and body mass in mammals: from mice to beavers. *Mechanisms of Ageing and Development*, 130(1–2), 3–9.

Gorman, M.L. and Mills, M.G.L. (1984). Scent marking strategies in hyaenas (Mammalia). *Journal of Zoology*, 202(4), 535–547.

Gorman, M.L., Kruuk, H., and Leitch, A. (1984). Social functions of the sub-caudal scent gland secretion of the European badger *Meles meles* (Carnivora: Mustelidae). *Journal of Zoology*, 203(4), 549–559.

Gormley, E., Ní Bhuachalla, D., O'Keeffe, J., Murphy, D., and Aldwell, F.E. (2017). Oral vaccination of free-living badgers (*Meles meles*) with Bacille Calmette Guérin (BCG) vaccine confers protection against tuberculosis. *PLoS ONE*, 12, e0168851. doi:10.1371/journal.pone.0168851.

Gorta (2020). What goes around comes around: complex competitive interactions between two widespread southern African mesopredators. *Canid Biology and Conservation* 22, 8–10.

Gosling, L.M. (1982). A reassessment of the function of scent marking in territories. *Zeitschrift für Tierpsychologie*, 60(2), 89–118.

Gosling, L.M. (1985). The even-toed ungulates: order Artiodactyla. *Social Odours in Mammals*, 2, 550–618.

Gosling, L.M. and McKay, H.V. (1990). Competitor assessment by scent matching: an experimental test. *Behavioral Ecology and Sociobiology*, 26(6), 415–420.

Gosling, L.M. and Roberts, S.C. (2001a). Scent-marking by male mammals: cheat-proof signals to competitors and mates. *Advances in the Study of Behavior*, 30, 169–217.

Gosling, L.M. and Roberts, S.C. (2001b). Testing ideas about the function of scent marks in territories from spatial patterns. *Animal Behaviour*, 62(3), F7–F10.

Gosling, S.D. (2001). From mice to men: what can we learn about personality from animal research? *Psychological Bulletin*, 127(1), 45.

Goszczyński, J., Jedrzejewska, B., and Jedrzejewski, W. (2000). Diet composition of badgers (*Meles meles*) in a pristine forest and rural habitats of Poland compared to other European populations. *Journal of Zoology*, 250(4), 495–505.

Goszczyński, J., Juszko, S., Pacia, A., and Skoczyńska, J. (2005). Activity of badgers (*Meles meles*) in Central Poland. *Mammalian Biology*, 70(1), 1–11.

Gouat, J. (1985). Notes sur la reproduction de *Ctenodactylus gundi* rongeur Ctenodactylidae. *Zeitschrift für Säugetierkunde*, 50(5), 285–293.

Gould, S.J. (1985). The paradox of the first tier: an agenda for paleobiology. *Paleobiology*, 11(1), 2–12.

Gould, S.J. and Lewontin, R.C. (1979). The spandrels of San Marco and the Panglossian paradigm: a critique of the adaptationist programme. *Proceedings of the Royal Society of London. Series B: Biological Sciences*, 205(1161), 581–598.

Gould, S.J. and Vrba, E.S. (1982). Exaptation—a missing term in the science of form. *Paleobiology*, 81(1), 4–15.

Gradmann, C. (2001). Robert Koch and the pressures of scientific research: tuberculosis and tuberculin. *Medical History*, 45(1), 1–32.

Grafen, A. (1988) On the uses of data on lifetime reproductive success. In *Reproductive Success. Studies of Individual Variation in Contrasting Breeding Systems* (T.H. Clutton-Brock, ed., pp. 454–471. University of Chicago Press, Chicago, IL.

Graham, D.A. (2013). Bovine herpes virus-1 (BoHV-1) in cattle—a review with emphasis on reproductive impacts and the emergence of infection in Ireland and the United Kingdom. *Irish Veterinary Journal*, 66(1), 1–12.

Graham, R. W. and Lundelius, Jr, E.L. (1984). Coevolutionary disequilibrium and Pleistocene extinctions. In *Quaternary Extinctions: A Prehistoric Revolution* (P. S. Martin and R. G. Klein, eds, pp. 223–249). University of Arizona Press, Tucson, AZ.

Grahame, K. (1994). *The Wind in the Willows*. 1908. Scribners, New York.

Grandage, J. (1972). The erect dog penis: a paradox of flexible rigidity. *The Veterinary Record*, 91(6), 141–147.

Grant, J.W.A., Chapman, C.A., and Richardson, K.S. (1992). Defended versus undefended home range size of carnivores, ungulates and primates. *Behavioral Ecology and Sociobiology*, 31(3), 149–161.

Grassman, L.I. (1999). Ecology and behavior of the Indochinese leopard in Kaeng Krachan National Park, Thailand. *Natural History Bulletin of the Siam Society*, 47(1), 77–93.

Grayson, A.J. and Jones, E.W. (1955). Notes on the history of Wytham Estate with special reference to the woodlands. Imperial Forestry Institute, Oxford University, Oxford.

Green, D.G. and Sadedin, S. (2005). Interactions matter—complexity in landscapes and ecosystems. *Ecological Complexity*, 2(2), 117–130.

Green, P.A., Van Valkenburgh, B., Pang, B., Bird, D., Rowe, T., and Curtis, A. (2012). Respiratory and olfactory turbinal size in canid and arctoid carnivorans. *Journal of Anatomy*, 221(6), 609–621.

Greenwood, P.J. (1980). Mating systems, philopatry and dispersal in birds and mammals. *Animal Behaviour, 28*(4), 1140–1162.

Grémillet, D., Lescroël, A., Ballard, G., Dugger, K.M., Massaro, M., Porzig, E.L., and Ainley, D.G. (2018). Energetic fitness: field metabolic rates assessed via 3D accelerometry complement conventional fitness metrics. *Functional Ecology, 32*(5), 1203–1213.

Griffin, J.M., Hahesy, T., Lynch, K., Salman, M.D., McCarthy, J., and Hurley, T. (1993). The association of cattle husbandry practices, environmental factors and farmer characteristics with the occurrence of chronic bovine tuberculosis in dairy herds in the Republic of Ireland. *Preventive Veterinary Medicine, 17*(3–4), 145–160.

Griffiths, H.I. and Thomas, D.H. (1993). The status of the badger *Meles meles* (L., 1758) (Carnivora, Mustelidae) in Europe. *Mammal Review, 23*(1), 17–58.

Groenendijk, J., Hajek, F., Johnson, P.J., Macdonald, D.W., Calvimontes, J., Staib, E., and Schenck, C. (2014). Demography of the giant otter (*Pteronura brasiliensis*) in Manu National Park, south-eastern Peru: implications for conservation. *PLoS ONE, 9*(8), e106202.

Groenendijk, J., Hajek, F., Schenck, C., Staib, E., Johnson, P.J., and Macdonald, D.W. (2015). Effects of territory size on the reproductive success and social system of the giant otter, south-eastern Peru. *Journal of Zoology, 296*(3), 153–160.

Groenendijk, J., Hajek, F., Johnson, P.J., and Macdonald, D.W. (2017). Giant otters: using knowledge of life history for conservation. *Biology and Conservation of Musteloids, 1*, 466–486.

Grolemund, G. and Wickham, H. (2011). Dates and times made easy with lubridate. *Journal of Statistical Software, 40*(3), 1–25.

Grossman, C.J., Roselle, G.A. and Mendenhall, C.L. (1991). Sex steroid regulation of autoimmunity. *The Journal of Steroid Biochemistry and Molecular Biology, 40*(4–6), 649–659.

Guilford, T. and Dawkins, M.S. (1991). Receiver psychology and the evolution of animal signals. *Animal Behaviour, 42*(1), 1–14.

Gulland, F.M.D., Hall, A.J., Ylitalo, G.M., Colegrove, K.M., Norris, T., Duignan, P.J., Halaska, B., Acevedo Whitehouse, K., Lowenstine, L.J., Deming, A.C., and Rowles, T.K. (2020). Persistent contaminants and herpesvirus OtHV1 are positively associated with cancer in wild California sea lions (*Zalophus californianus*). *Frontiers in Marine Science, 7*(December), 1–13.

Gutteridge, J.M. and Halliwell, B. (1993). Invited review free radicals in disease processes: a compilation of cause and consequence. *Free Radical Research Communications, 19*(3), 141–158.

Hadfield, J.D., Richardson, D.S., and Burke, T. (2006). Towards unbiased parentage assignment: combining genetic, behavioural and spatial data in a Bayesian framework. *Molecular Ecology, 15*(12), 3715–3730.

Haimi, J., Salminen, J., Huhta, V., Knuutinen, J., and Palm, H. (1992). Bioaccumulation of organochlorine compounds in earthworms. *Soil Biology and Biochemistry, 24*(12), 1699–1703.

Haldane, J.B.S. (1932). *The Causes of Evolution*. Princeton University Press, Princeton, NJ.

Hall, M.E., Nasir, L., Daunt, F., Gault, E.A., Croxall, J.P., Wanless, S., and Monaghan, P. (2004). Telomere loss in relation to age and early environment in long-lived birds. *Proceedings of the Royal Society of London. Series B: Biological Sciences, 271*(1548), 1571–1576.

Hall, M.L., van Asten, T., Katsis, A.C., Dingemanse, N.J., Magrath, M.J., and Mulder, R.A. (2015). Animal personality and pace-of-life syndromes: do fast-exploring fairy-wrens die young? *Frontiers in Ecology and Evolution, 3*, 28.

Halliwell, R.E.W. and Longino, S.J. (1985). IgE and IgG antibodies to flea antigen in differing dog populations. *Veterinary Immunology and Immunopathology, 8*(3), 215–223.

Hämäläinen, A., Dammhahn, M., Aujard, F., Eberle, M., Hardy, I., Kappeler, P.M., Perret, M., Schliehe-Diecks, S., and Kraus, C. (2014). Senescence or selective disappearance? Age trajectories of body mass in wild and captive populations of a small-bodied primate. *Proceedings of the Royal Society of London. Series B: Biological Sciences, 281*(1791), 20140830.

Hämäläinen, A., Immonen, E., Tarka, M., and Schuett, W. (2018). Evolution of sex-specific pace-of-life syndromes: causes and consequences. *Behavioral Ecology and Sociobiology, 72*(3), 1–15.

Hamel, S., Gaillard, J.M., and Yoccoz, N.G. (2018). Introduction to: Individual heterogeneity—the causes and consequences of a fundamental biological process. *Oikos, 127*(5), 643–647.

Hamilton, W.D. (1963). The evolution of altruistic behavior. *The American Naturalist, 97*(896), 354–356.

Hamilton, W.D. (1964). The genetical evolution of social behavior, Part I and II. *Journal of Theoretical Biology, 7*, 1–52.

Hamilton, W.D. (1966). The moulding of senescence by natural selection. *Journal of Theoretical Biology, 12*(1), 12–45.

Hamilton, W.D. (1971). Geometry for the selfish herd. *Journal of Theoretical Biology, 31*(2), 295–311.

Hamilton, W.D. and Zuk, M. (1982). Heritable true fitness and bright birds: a role for parasites? *Science, 218*(4570), 384–387.

Hammers, M., Richardson, D.S., Burke, T., and Komdeur, J. (2013). The impact of reproductive investment and early-life environmental conditions on senescence: support for the disposable soma hypothesis. *Journal of Evolutionary Biology, 26*(9), 1999–2007.

Hammerstein, P. (1981). The role of asymmetries in animal contests. *Animal Behaviour, 29*(1), 193–205.

Hampton, C. (1947). The badger's funeral. *Field Sports, 7*, 24.

Han, L., Ma, C., Liu, Q., Weng, H. J., Cui, Y., Tang, Z., Kim, Y., Nie, H., Qu, L., Patel, K. N., and Li, Z. (2013). A subpopulation of nociceptors specifically linked to itch. *Nature Neuroscience, 16*, 174–182.

Hancox, M.K. (1973). Studies on the ecology of the Eurasian badger (*Meles meles* L.). Master's thesis, Elton Library, Oxford University.

Hancox, M. (1988). Field age determination in the European badger. *Revue d'écologie*, 43(4), 399–404.

Hanna, E.G. and Tait, P.W. (2015). Limitations to thermoregulation and acclimatization challenge human adaptation to global warming. *International Journal of Environmental Research and Public Health*, 12(7), 8034–8074.

Hansen, H., McDonald, D.B., Groves, P., Maier, J.A., and Ben-David, M. (2009). Social networks and the formation and maintenance of river otter groups. *Ethology*, 115(4), 384–396.

Hansson, B. and Westerberg, L. (2002). On the correlation between heterozygosity and fitness in natural populations. *Molecular Ecology*, 11(12), 2467–2474.

Hao, H., Chu, K.L., Pei, E.L., and Xu, H.F. (2009). Activity patterns of badgers in suburbs of Shanghai. *Sichuan Journal of Zoology*, 28(1), 111–114.

Harcourt, A.H. (1997). Sperm competition in primates. *The American Naturalist*, 149(1), 189–194.

Harding, E.J., Paul, E.S., and Mendl, M. (2004). Cognitive bias and affective state. *Nature*, 427(6972), 312–312.

Härdling, R. and Kaitala, A. (2005). The evolution of repeated mating under sexual conflict. *Journal of Evolutionary Biology*, 18(1), 106–115.

Harman D. (1956). Aging: a theory based on free radical and radiation chemistry. *Journal of Gerontology*, 11(3), 298–300. doi:10.1093/geronj/11.3.298

Harper, J.M. and Austad, S.N. (2000). Fecal glucocorticoids: a noninvasive method of measuring adrenal activity in wild and captive rodents. *Physiological and Biochemical Zoology*, 73(1), 12–22.

Harrington, L.A., Green, J., Muinde, P., Macdonald, D.W., Auliya, M., and D'Cruze, N. (2020). Snakes and ladders: a review of ball python production in West Africa for the global pet market. *Nature Conservation*, 41, 1.

Harris, A.R., Christianson, D., Smith, M.S., Fang, S.L., Braverman, L.E., and Vagenakis, A.G. (1978). The physiological role of thyrotropin-releasing hormone in the regulation of thyroid-stimulating hormone and prolactin secretion in the rat. *The Journal of Clinical Investigation*, 61(2), 441–448.

Harrison, P., Berry, P.M., Butt, N., and New, M. (2006). Modelling climate change impacts on species' distributions at the European scale: implications for conservation policy. *Environmental Science & Policy*, 9(2), 116–128.

Harrison, S. and Taylor, A.D. (1997). Empirical evidence for metapopulation dynamics. In Metapopulation Biology: Ecology, Genetics, and Evolution (I. Hanski and M.E. Gilpin, pp. 27–42). Academic Press, Washington, DC.

Harrison, X.A., Blount, J.D., Inger, R., Norris, D.R., and Bearhop, S. (2011). Carry-over effects as drivers of fitness differences in animals. *Journal of Animal Ecology*, 80(1), 4–18.

Harrod, E.G., Coe, C.L., and Niedenthal, P.M. (2020). Social structure predicts eye contact tolerance in nonhuman primates: evidence from a crowd-sourcing approach. *Scientific Reports*, 10(1), 1–9.

Hart, B.L. (2012). Defenses against parasites. In *Parasites and Pathogens: Effects On Host Hormones and Behavior* (N.E. Beckage, ed., p. 210). Springer Science & Business Media, Berlin, Heidelberg.

Hartmann, E.M., Handwerker, H.O., and Forster, C. (2015). Gender differences in itch and pain-related sensations provoked by histamine, cowhage and capsaicin. *Acta Dermato-Venereologica*, 95(1), 25–30.

Hass, C.C. and Dragoo, J.W. Competition and coexistence in sympatric skunks. In *Biology and Conservation of Musteloids* (D.W. Macdonald, C. Newman, and L.A. Harrington, eds, pp. 464–477). Oxford University Press, Oxford.

Hatchwell, B.J. and Komdeur, J. (2000). Ecological constraints, life history traits and the evolution of cooperative breeding. *Animal Behaviour*, 59(6), 1079–1086.

Hatfield, B.B., Marks, D., Tinker, M. T., Nolan, K., and Peirce, J. (1998). Attacks on sea otters by killer whales. *Marine Mammal Science*, 14(4), 888–894.

Hausfater, G., Altmann, J., and Altmann, S. (1982). Long-term consistency of dominance relations among female baboons (*Papio cynocephalus*). *Science*, 217(4561), 752–755.

Havlíček, J., Winternitz, J., and Roberts, S.C. (2020). MHC-associated odour preferences and human mate choice: near and far horizons. *Philosophical Transactions of the Royal Society B*, 375, 20190260

Hawkes, K., O'Connell, J.F., Jones, N.B., Alvarez, H., and Charnov, E.L. (1998). Grandmothering, menopause, and the evolution of human life histories. *Proceedings of the National Academy of Sciences*, 95(3), 1336–1339.

Hawley, P.H. (1999). The ontogenesis of social dominance: a strategy-based evolutionary perspective. *Developmental Review*, 19(1), 97–132.

Haydon, D.T., Randall, D.A., Matthews, L., Knobel, D.L., Tallents, L.A., Gravenor, M.B., Williams, S.D., Pollinger, J.P., Cleaveland, S., Woolhouse, M.E.J., Sillero-Zubiri, C., Marino, J., Macdonald, D.W., and Laurenson, M.K. (2006). Low-coverage vaccination strategies for the conservation of endangered species. *Nature*, 443(7112), 692–695.

Haydon, D.T., Randall, D.A., Matthews, L., Knobel, D.L., Tallents, L.A., Gravenor, M.B., Williams, S.D., Pollinger, J.P., Cleaveland, S., Woolhouse, M.E.J., and Sillero-Zubiri, C. (2006). Low-coverage vaccination strategies for the conservation of endangered species. *Nature*, 443(7112), 692–695.

Hayward, A.D., Rickard, I.J., and Lummaa, V. (2013). Influence of early-life nutrition on mortality and reproductive success during a subsequent famine in a preindustrial population. *Proceedings of the National Academy of Sciences*, 110(34), 13886–13891. https://doi.org/10.1073/pnas.1301817110

Heape, W. (1931) *Emigration, Migration and Nomadism*. W. Heffer & Sons Ltd, Cambridge.

Hediger, H. (1949). Mammalian territories and their marking. *Bijdragen tot de Dierkunde*, 28(1), 172–184.

Hedrick, A.V. and Temeles, E.J. (1989). The evolution of sexual dimorphism in animals: hypotheses and tests. *Trends in Ecology & Evolution*, 4(5), 136–138.

Heffner, R.S. and Heffner, H.E. (1985). Hearing in mammals: the least weasel. *Journal of Mammalogy*, 66(4), 745–755.

Heffner, R.S. and Heffner, H.E. (1987). Localization of noise, use of binaural cues, and a description of the superior olivary complex in the smallest carnivore, the least weasel (*Mustela nivalis*). *Behavioral Neuroscience*, 101(5), 701.

Heinsohn, R. and Packer, C. (1995). Complex cooperative strategies in group-territorial African lions. *Science*, 269(5228), 1260–1262.

Heldmaier, G. and Ruf, T. (1992). Body temperature and metabolic rate during natural hypothermia in endotherms. *Journal of Comparative Physiology B*, 162(8), 696–706.

Heldstab, S.A., van Schaik, C.P., and Isler, K. (2017). Getting fat or getting help? How female mammals cope with energetic constraints on reproduction. *Frontiers in Zoology*, 14(1), 29.

Heldstab, S.A., Müller, D.W., Graber, S.M., Bingaman Lackey, L., Rensch, E., Hatt, J.M., and Clauss, M. (2018). Geographical origin, delayed implantation, and induced ovulation explain reproductive seasonality in the Carnivora. *Journal of Biological Rhythms*, 33(4), 402–419.

Helgen, K.M., Lim, N.T., and Helgen, L.E. (2008). The hog-badger is not an edentate: systematics and evolution of the genus Arctonyx (Mammalia: Mustelidae). *Zoological Journal of the Linnean Society*, 154(2), 353–385.

Henazi, S.P. and Barrett, L. (1999). The value of grooming to female primates. *Primates*, 40(1), 47–59. doi:10.1007/bf02557701

Henderson, B. (n.d.). Open letter to Kansas School Board. https://www.spaghettimonster.org/about/open-letter (accessed 6 May 2022).

Hendry, A.P., Wenburg, J.K., Bentzen, P., Volk, E.C., and Quinn, T.P. (2000). Rapid evolution of reproductive isolation in the wild: evidence from introduced salmon. *Science*, 290(5491), 516–518.

Henry, J.D. (1977). The use of urine marking in the scavenging behavior of the red fox (*Vulpes vulpes*). *Behaviour*, 61(1–2), 82–105.

Herbig, U., Ferreira, M., Condel, L., Carey, D., and Sedivy, J.M. (2006). Cellular senescence in aging primates. *Science*, 311(5765), 1257–1257.

Hernando-Herraez, I., Prado-Martinez, J., Garg, P., Fernandez-Callejo, M., Heyn, H., Hvilsom, C., Navarro, A., Esteller, M., Sharp, A.J., and Marques-Bonet, T. (2013). Dynamics of DNA methylation in recent human and great ape evolution. *PLoS Genetics*, 9(9), e1003763. https://doi.org/10.1371/journal.pgen.1003763

Herrera, E.A. and Macdonald, D.W. (1989). Resource utilization and territoriality in group-living capybaras (*Hydrochoerus hydrochaeris*). *Journal of Animal Ecology*, 58(2), 667–679.

Hersteinsson, P. and Macdonald, D.W. (1982). Some comparisons between red and arctic foxes, *Vulpes vulpes* and *Alopex lagopus*, as revealed by radio tracking. *Symposia of the Zoological Society of London*, 49, 259–289.

Hersteinsson, P. and Macdonald, D.W. (1992). Interspecific competition and the geographical distribution of red and arctic foxes *Vulpes vulpes* and *Alopex lagopus*. *Oikos*, 64(3), 505–515.

Hersteinsson, P. and Macdonald, D.W. (1996). Diet of arctic foxes (*Alopex lagopus*) in Iceland. *Journal of Zoology*, 240(3), 457–474.

Hewitt, S.E., Macdonald, D.W., and Dugdale, H.L. (2009). Context-dependent linear dominance hierarchies in social groups of European badgers, *Meles meles* . *Animal Behaviour*, 77(1), 161–169.

Hewitt, S.E., Macdonald, D.W., and Dugdale, H.L. (2009). Context-dependent linear dominance hierarchies in social groups of European badgers, *Meles meles*. *Animal Behaviour*, 77(1), 161–169.

Hicks, O., Burthe, S.J., Daunt, F., Newell, M., Butler, A., Ito, M., Sato, K., and Green, J.A. (2018). The energetic cost of parasitism in a wild population. *Proceedings of the Royal Society. Series B: Biological Sciences*, 285(1879), 20180489.

Hildebrand, M. (1952). The integument in Canidae. *Mammal*, 33, 419–428.

Hill, S.C., Murphy, A.A., Cotten, M., Palser, A.L., Benson, P., Lesellier, S., Gormley, E., Richomme, C., Grierson, S., Bhuachalla, D.N., and Chambers, M. (2015). Discovery of a polyomavirus in European badgers (*Meles meles*) and the evolution of host range in the family Polyomaviridae. *The Journal of General Virology*, 96(6), 1411.

Hiller-Sturmhöfel, S. and Bartke, A. (1998). The endocrine system: an overview. *Alcohol Health and Research World*, 22(3), 153.

Himelright, B. (2014). Sequential ovulation, fertility of polyestrus, and induced ovulation in American black bears (Ursus americanus). Master's dissertation. California State University, USA.

Hindle, B.J., Pilkington, J.G., Pemberton, J.M., and Childs, D.Z. (2019). Cumulative weather effects can impact across the whole life cycle. *Global Change Biology*, 25(10), 3282–3293.

Hirotani, A. (1994). Dominance rank, copulatory behaviour and estimated reproductive success in male reindeer. *Animal Behaviour*, 48(4), 929–936.

Hirsch, B.T. and Gompper, M.E. (2017). Causes and consequences of coati sociality. In *Biology and Conservation of the Musteloids* (D.W. Macdonald, C. Newman, and L.A. Harrington, eds, pp. 515–526). Oxford University Press, Oxford.

Hirsch, B.T. and Maldonado, J.E. (2011). Familiarity breeds progeny: sociality increases reproductive success in adult male ring-tailed coatis (*Nasua nasua*). *Molecular Ecology*, 20(2), 409–419.

Hirsch, B.T., Prange, S., Hauver, S.A., and Gehrt, S.D. (2013). Raccoon social networks and the potential for disease transmission. *PLoS ONE*, 8(10), e75830.

Hirsch, B.T., Prange, S., Hauver, S.A., and Gehrt, S.D. (2014). Patterns of latrine use by raccoons (*Procyon lotor*) and implication for *Baylisascaris procyonis* transmission. *Journal of Wildlife Diseases*, 50(2), 243–249.

HMSO (1980). Badgers, cattle and tuberculosis by Lord Zuckerman. HMSO, London.

Hodge, S.J., Manica, A., Flower, T.P., and Clutton-Brock, T.H. (2008). Determinants of reproductive success in dominant female meerkats. *Journal of Animal Ecology*, 77(1), 92–102.

Hoebe, K., Janssen, E., and Beutler, B. (2004). The interface between innate and adaptive immunity. *Nature Immunology*, 5(10), 971–974.

Hofer H. (1988). Variation in resource presence, utilization and reproductive success within a population of European badgers (*Meles meles*). *Mammal Review*, 18(1), 25–36.

Hoffmann, A.A. and Sgro, C.M. (2011). Climate change and evolutionary adaptation. *Nature*, 470(7335), 479–485.

Hogeweg, P. and Hesper, B. (1983). The ontogeny of the interaction structure in bumble bee colonies—a MIRROR model. *Behavioral Ecology and Sociobiology*, 12, 271–283.

Holekamp, K.E. (2007). Questioning the social intelligence hypothesis. *Trends in Cognitive Sciences*, 11(2), 65–69.

Holekamp, K.E., Dantzer, B., Stricker, G., Yoshida, K.C.S., and Benson-Amram, S. (2015). Brains, brawn and sociality: a hyaena's tale. *Animal Behaviour*, 103, 237–248.

Holling, C.S. (1973). Resilience and stability of ecological systems. *Annual Review of Ecology and Systematics*, 4(1), 1–23.

Hollister-Smith, J.A., Poole, J.H., Archie, E.A., Vance, E.A., Georgiadis, N.J., Moss, C.J., and Alberts, S.C. (2007). Age, musth and paternity success in wild male African elephants, Loxodonta africana . *Animal Behaviour*, 74(2), 287–296.

Holmes, D., Simmons, J.H., and Tatton, J.G. (1967). Chlorinated hydrocarbons in British wildlife. *Nature*, 216(5112), 227–229.

Holzenberger, M., Dupont, J., Ducos, B., Leneuve, P., Géloën, A., Even, P.C., Cervera, P., and Le Bouc, Y. (2003). IGF-1 receptor regulates lifespan and resistance to oxidative stress in mice. *Nature*, 421(6919), 182–187.

Hope, J.C., Thom, M.L., McAulay, M., Mead, E., Vordermeier, H.M., Clifford, D., Hewinson, R.G., and Villarreal-Ramos, B. (2011). Identification of surrogates and correlates of protection in protective immunity against *Mycobacterium bovis* infection induced in neonatal calves by vaccination with M. bovis BCG Pasteur and M. bovis BCG Danish. *Clinical and Vaccine Immunology*, 18(3), 373–379.

Horn, H.S. (1968). The adaptive significance of colonial nesting in the Brewer's blackbird (*Euphagus cyanocephalus*). *Ecology*, 49(4), 682–694.

Hornocker, M.G. (1962). Population characteristics and social and reproductive behavior of the grizzly bear in Yellowstone National Park. Thesis, University of Montana.

Horowitz, M. and Hales, J.R.S. (1998). Pathophysiology of hyperthermia. Physiology and Pathophysiology of Temperature Regulation (C.M. Blatteis, ed., pp. 229–245). World Scientific Publishing, Ltd, Singapore.

Hosken, D.J. and House, C.M. (2011). Sexual selection. *Current Biology*, 21(2), R62–R65.

Hounsome, T. and Delahay, R. (2005). Birds in the diet of the Eurasian badger *Meles meles*: a review and meta-analysis. *Mammal Review*, 35(2), 199–209.

House of Commons Briefing Paper (2019). Bovine TB statistics: Great Britain. Number 6081, 22 July 2019. https://researchbriefings.files.parliament.uk/documents/SN06081/SN06081.pdf (accessed 11 March 2022).

Howard, R.D., Martens, R.S., Innis, S.A., Drnevich, J.M., and Hale, J. (1998). Mate choice and mate competition influence male body size in Japanese medaka. *Animal Behaviour*, 55(5), 1151–1163.

Høyer, A.P., Grandjean, P., Jørgensen, T., Brock, J.W., and Hartvig, H.B. (1998). Organochlorine exposure and risk of breast cancer. *The Lancet*, 352(9143), 1816–1820.

Hrdy, S.B. (1979). Infanticide among animals: a review, classification, and examination of the implications for the reproductive strategies of females. *Ethology and Sociobiology*, 1(1), 13–40.

Hsu, Y.Y., Earley, R.L., and Wolf, L.L. (2006). Modulation of aggressive behaviour by fighting experience: mechanisms and contest outcomes. *Biological Reviews*, 81, 33–74. doi:10.1017/s146479310500686x

Huber, B. (1939). Siebrohrensystem unserer Baume und seine jahreszeitlichen Veranderungen. *Jahrbücher für Wissenschaftliche Botanik*, 88, 176–242.

Huchard, E. and Cowlishaw, G. (2011). Female–female aggression around mating: an extra cost of sociality in a multimale primate society. *Behavioral Ecology*, 22(5), 1003–1011.

Hudson, L.N., Isaac, N.J., and Reuman, D.C. (2013). The relationship between body mass and field metabolic rate among individual birds and mammals. *Journal of Animal Ecology*, 82(5), 1009–1020.

Hudson, P.J., Dobson, A.P., and Newborn, D. (2002). Parasitic worms and population cycles of red grouse. In Population Cycles: *The Case* for *Trophic Interactions* (A. Berryman , ed., pp. 109–130). Oxford University Press, Oxford.

Hui, C. (2006). Carrying capacity, population equilibrium, and environment's maximal load. *Ecological Modelling*, 192(1), 317–320.

Huisman, J. (2017). Pedigree reconstruction from SNP data: parentage assignment, sibship clustering and beyond. *Molecular Ecology Resources*, 17(5), 1009–1024.

Hulbert, A.J. (2014). A sceptic's view: 'Kleiber's Law' or the '3/4 Rule' is neither a law nor a rule but rather an empirical approximation. *Systems*, 2(2), 186–202.

Hulbert, A.J., Kelly, M.A., and Abbott, S.K. (2014). Polyunsaturated fats, membrane lipids and animal longevity. *Journal of Comparative Physiology B*, 184(2), 149–166.

Hulme, P.E. (2005). Adapting to climate change: is there scope for ecological management in the face of a global threat? *Journal of Applied Ecology*, 42(5), 784–794.

Hume, D. (1748) *Enquiries Concerning the Human Understanding*. A. Millar, London.

Hume, D.J., Yokum, S., and Stice, E. (2016). Low energy intake plus low energy expenditure (low energy flux), not energy surfeit, predicts future body fat gain. *The American Journal of Clinical Nutrition*, 103(6), 1389–1396.

Humphries, M.M., Thomas, D.W., and Kramer, D.L. (2003). The role of energy availability in mammalian hibernation: a cost-benefit approach. *Physiological and Biochemical Zoology*, 76(2), 165–179.

Hunt, J., Breuker, C.J., Sadowski, J.A., and Moore, A.J. (2009). Male–male competition, female mate choice and their interaction: determining total sexual selection. *Journal of Evolutionary Biology*, 22(1), 13–26.

Hunt, R.M. and Skolnick, R. (1996). The giant mustelid Megalictis from the early Miocene carnivore dens at Agate Fossil Beds National Monument, Nebraska; earliest evidence of dimorphism in New World Mustelidae (Carnivora, Mammalia). *Rocky Mountain Geology*, 31(1), 35–48.

Hunter, J.R. and Thomas, G.L. (1974). Effect of prey distribution and density on the searching and feeding behaviour of larval anchovy *Engraulis mordax* Girard. In *The Early Life History of Fish* (E. Kamler, ed., pp. 559–574). Springer, Berlin, Heidelberg.

Hurst, J.L., Fang, J., and Barnard, C.J. (1993). The role of substrate odours in maintaining social tolerance between male house mice, *Mus musculus domesticus*. *Animal Behaviour*, 45(5), 997–1006.

Hurst, J.L., Payne, C.E., Nevison, C.M., Marie, A.D., Humphries, R.E., Robertson, D.H., and Beynon, R.J. (2001). Individual recognition in mice mediated by major urinary proteins. *Nature*, 414(6864), 631.

Hurst, J.L., Thom, M.D., Nevison, C.M., Humphries, R.E., and Beynon, R.J. (2005). MHC odours are not required or sufficient for recognition of individual scent owners. *Proceedings of the Royal Society of London. Series B: Biological Sciences*, 272(1564), 715–724.

Hutchings, M.R. and Harris, S. (1999). Quantifying the risks of TB infection to cattle posed by badger excreta. *Epidemiology & Infection*, 122(1), 167–174.

Hutchings, M.R., Service, K.M., and Harris, S. (2002). Is population density correlated with faecal and urine scent marking in European badgers (Meles meles) in the UK? *Mammalian Biology*, 67(5), 286–293.

Hutchings, M.R., Judge, J., Gordon, I.J., and Athanasiadou Sandkyriazakis, I. (2006). Use of trade-off theory to advance understanding of herbivore-parasite interactions. *Mammal Review*, 36, 1–16. doi:10.1111/j.1365-2907.2006.00080.x

ibis, O., Tez, C., Özcan, S., Yorulmaz, T., Kaya, A., and Mohradi, M. (2015). Insights into the Turkish and Iranian badgers (the genus *Meles*) based on the mitochondrial cytochrome b gene sequences. *Vertebrate Zoology*, 65(3), 399–407.

IJzerman, H., Coan, J.A., Wagemans, F., Missler, M.A., Beest, I.V., Lindenberg, S., and Tops, M. (2015). A theory of social thermoregulation in human primates. *Frontiers in Psychology*, 6, 464.

Independent Expert Panel (IEP) (2014). Pilot badger culls in Somerset and Gloucestershire: Report by the Independent Expert Panel. Independent Expert Panel, London.

Intergovernmental Panel on Climate Change (IPCC) (2007). *Climate Change 2007: Synthesis Report. Contribution of Working Groups I, II and III to the Fourth Assessment Report of the Intergovernmental Panel on Climate Change* (core writing team, R.K. Pachauri, and A. Reisinger, eds). IPCC, Geneva.

Intergovernmental Panel on Climate Change (IPCC) (2012). Summary for Policymakers. In *Managing the Risks of Extreme Events and Disasters to Advance Climate Change Adaptation. A Special Report of Working Groups I and II of the Intergovernmental Panel on Climate Change* (C.B. Field, V. Barros, T.F. Stocker, D. Qin, D.J. Dokken, K.L. Ebi, M.D. Mastrandrea, K.J. Mach, G.-K. Plattner, S.K. Allen, M. Tignor, and P.M. Midgley, eds, pp. 1–19). Cambridge University Press, Cambridge.

Intergovernmental Panel on Climate Change (IPCC) (2014). *Climate Change 2014: Synthesis Report. Contribution of Working Groups I, II and III to the Fifth Assessment Report of the Intergovernmental Panel on Climate Change* (Core writing team, R.K. Pachauri and L.A. Meyer, eds). IPCC, Geneva.

Isaac, J.L. (2005). Potential causes and life-history consequences of sexual size dimorphism in mammals. *Mammal Review*, 35(1), 101–115.

Isvaran, K. and Clutton-Brock, T. (2007). Ecological correlates of extra-group paternity in mammals. *Proceedings of the Royal Society. Series B: Biological Sciences*, 274(1607), 219–224.

Iversen, J.A. (1972). Basal energy metabolism of mustelids. *Journal of Comparative Physiology*, 81(4), 341–344.

Iwasaki, A. and Medzhitov, R. (2010). Regulation of adaptive immunity by the innate immune system. *Science*, 327(5963), 291–295.

Jackson, S. (2003). Wombats. In *Australian Mammals: Biology and Captive Management* (S. Jackson, ed., pp. 145–182). CSIRO Publishing, Collingwood, Victoria.

Jackson, S.T., Betancourt, J.L., Booth, R.K., and Gray, S.T. (2009). Ecology and the ratchet of events: climate variability, niche dimensions, and species distributions. *Proceedings of the National Academy of Sciences*, 106(Suppl. 2), 19685–19692.

Jameson, J.L., de Kretser, D.M., Marshall, J.C., and De Groot, L.J. (2013). *Endocrinology Adult and Pediatric: Reproductive Endocrinology*, 6th edn. Elsevier Health Sciences, Hayana, India.

Jarman, P. (1974). The social organisation of antelope in relation to their ecology. *Behaviour*, 48(1–4), 215–267.

Jefferies, D.J. (1969). Causes of badger mortality in eastern counties of England. *Journal of Zoology*, 157(4), 429–436.

Jefferies, D.J. and Hanson, H.M. (2002, September). The role of dieldrin in the decline of the otter (*Lutra lutra*) in Britain: the analytical data. In *Proceedings of the First Otter Toxicology Conference* (J.W.H. Conroy, P. Yoxon, and A.C. Gutleb), Isle of Skye, September 2000 (pp. 95–144).

Jeffreys, A.J., Wilson, V., and Thein, S.L. (1985). Individual-specific 'fingerprints' of human DNA. *Nature*, *316*(6023), 76–79.

Jenkins, H.E., Woodroffe, R., and Donnelly, C.A. (2010). The duration of the effects of repeated widespread badger culling on cattle tuberculosis following the cessation of culling. *PLoS ONE*, *5*(2), e9090. https://doi.org/10.1371/journal.pone.0009090

Jenkins, H.E., Cox, D.R., and Delahay, R.J. (2012). Direction of association between bite wounds and *Mycobacterium bovis* infection in badgers: implications for transmission. *PLoS ONE*, *7*(9), e45584.

Jenner, N., Groombridge, J., and S.M. Funk (2011) Coimmuting, territoriality and variation in group and territory size in a black-backed jackal population reliant on a clumped, abndant food resource in Namibia. *Journal of Zoology*, *284*, 1–238.

Jennings, D.J., Carlin, C.M., Hayden, T.J., and Gammell, M.P. (2011). Third-party intervention behaviour during fallow deer fights: the role of dominance, age, fighting and body size. *Animal Behaviour*, *81*(6), 1217–1222.

Jennions, M.D. and Macdonald, D.W. (1994). Cooperative breeding in mammals. *Trends in Ecology & Evolution*, *9*(3), 89–93.

Jensen, M.P., Allen, C.D., Eguchi, T., Bell, I.P., LaCasella, E.L., Hilton, W.A., Hof, C.A., and Dutton, P.H. (2018). Environmental warming and feminization of one of the largest sea turtle populations in the world. *Current Biology*, *28*(1), 154–159.

Jessup, R.S. (1970). *Precise Measurement of Heat Combustion with a Bomb Calorimeter*. Monograph No. 7. US Bureau of Standards, Washington, DC.

Jetz, W., Carbone, C., Fulford, J., and Brown, J.H. (2004). The scaling of animal space use. *Science*, *306*(5694), 266–268.

Jiménez, J.J. and Decaëns, T. (2000). Vertical distribution of earthworms in grassland soils of the Colombian Llanos. *Biology and Fertility of Soils*, *32*(6), 463–473.

Johns Hopkins Medicine (2013, 2 January). Itchiness explained: Specific set of nerve cells signal itch but not pain, researchers find. *Science Daily*. www.sciencedaily.com/releases/2013/01/130102104548.htm (accessed 5 April 2022).

Johnson, C.N. (1991). Utilization of habitat by the northern hairy-nosed wombat *Lasiorhinus krefftii*. *Journal of Zoology*, *225*(3), 495–507.

Johnson, D. (2016). *God Is Watching You: How the Fear of God Makes Us Human*. Oxford University Press, New York.

Johnson, D.D. and Macdonald, D.W. (2003). Sentenced without trial: reviling and revamping the resource dispersion hypothesis. *Oikos*, *101*(2), 433–440.

Johnson, D.D., Macdonald, D.W., and Dickman, A.J. (2000). An analysis and review of models of the sociobiology of the Mustelidae. *Mammal Review*, *30*(3–4), 171–196.

Johnson, D.D., Baker, S., Morecroft, M.D., and Macdonald, D.W. (2001a). Long-term resource variation and group size: a large-sample field test of the resource dispersion hypothesis. *BMC Ecology*, *1*(1), 1–18.

Johnson, D.D., Macdonald, D.W., Newman, C., and Morecroft, M.D. (2001b). Group size versus territory size in group-living badgers: a large-sample field test of the Resource Dispersion Hypothesis. *Oikos*, *95*(2), 265–274.

Johnson, D.D., Jetz, W., and Macdonald, D.W. (2002a). Environmental correlates of badger social spacing across Europe. *Journal of Biogeography*, *29*(3), 411–425.

Johnson, D.D., Kays, R., Blackwell, P.G., and Macdonald, D.W. (2002b). Does the resource dispersion hypothesis explain group living? *Trends in Ecology & Evolution*, *17*(12), 563–570.

Johnson, D.D., Stopka, P., and Bell, J. (2002c). Individual variation evades the Prisoner's Dilemma. *BMC Evolutionary Biology*, *2*(1), 15.

Johnson, D.D., Stopka, P., and Macdonald, D.W. (2004). Ideal flea constraints on group living: unwanted public goods and the emergence of cooperation. *Behavioral Ecology*, *15*(1), 181–186.

Johnson, P.J., Noonan, M.J., Kitchener, A.C., Harrington, L.A., Newman, C., and Macdonald, D.W. (2017). Rensching cats and dogs: feeding ecology and fecundity trends explain variation in the allometry of sexual size dimorphism. *Royal Society Open Science*, *4*(6), 170453.

Johnston, W.T., Gettinby, G., Cox, D.R., Donnelly, C.A., Bourne, J., Clifton-Hadley, R., Le Fevre, A.M., McInerney, J.P., Mitchell, A., Morrison, W.I., and Woodroffe, R. (2005). Herd-level risk factors associated with tuberculosis breakdowns among cattle herds in England before the 2001 foot-and-mouth disease epidemic. *Biology Letters*, *1*(1), 53–56.

Johnston, W.T., Vial, F., Gettinby, G., Bourne, F.J., Clifton-Hadley, R.S., Cox, D.R., Crea, P., Donnelly, C.A., McInerney, J.P., Mitchell, A.P., and Morrison, W.I. (2011). Herd-level risk factors of bovine tuberculosis in England and Wales after the 2001 foot-and-mouth disease epidemic. *International Journal of Infectious Diseases*, *15*(12), e833–e840.

Johnstone, A.M., Murison, S.D., Duncan, J.S., Rance, K.A., and Speakman, J.R. (2005). Factors influencing variation in basal metabolic rate include fat-free mass, fat mass, age, and circulating thyroxine but not sex, circulating leptin, or triiodothyronine. *The American Journal of Clinical Nutrition*, *82*(5), 941–948.

Jokiniemi, A., Kuusipalo, L., Ritari, J., Koskela, S., Partanen, J., and Kekäläinen, J. (2020). Gamete-level immunogenetic incompatibility in humans—towards deeper understanding of fertilization and infertility? *Heredity*, *125*(5), 281–289.

Jolly, G.M. (1965). Explicit estimates from capture-recapture data with both death and immigration-stochastic model. *Biometrika*, *52*(1/2), 225–247.

Jones, G.J., Coad, M., Khatri, B., Bezos, J., Parlane, N.A., Buddle, B.M., Villarreal-Ramos, B., Hewinson, R.G., and Vordermeier, H.M. (2017). Tuberculin skin testing boosts interferon gamma responses to DIVA reagents in *Mycobacterium bovis*-infected cattle. *Clinical and Vaccine Immunology*, *24*(5), e00551–16.

Jordan, N.R., Cherry, M.I., and Manser, M.B. (2007). Latrine distribution and patterns of use by wild meerkats: implications for territory and mate defence. *Animal Behaviour*, 73(4), 613–622.

Jordan, N.R., Mwanguhya, F., Kyabulima, S., Rüedi, P., and Cant, M.A. (2010). Scent marking within and between groups of wild banded mongooses. *Journal of Zoology*, 280(1), 72–83.

Jordan, P.A., Shelton, P.C., and Allen, D.L. (1967). Numbers, turnover, and social structure of the Isle Royale wolf population. *American Zoologist*, 7(2), 233–252.

Joseph, J.T., Purtill, K., Wong, S.J., Munoz, J., Teal, A., Madison-Antenucci, S., Horowitz, H.W., Aguero-Rosenfeld, M.E., Moore, J.M., Abramowsky, C., and Wormser, G.P. (2012). Vertical transmission of babesia microti, United States. *Emerging Infectious Diseases*, 18(8), 1318.

Joynson, D.H. (1992). Epidemiology of toxoplasmosis in the UK. *Scandinavian Journal of Infectious Diseases Supplement*, 24, 65–65.

Judge, J., Wilson, G.J., Macarthur, R., Delahay, R.J., and McDonald, R.A. (2014). Density and abundance of badger social groups in England and Wales in 2011–2013. *Scientific Reports*, 4, 3809.

Judge, J., Wilson, G.J., Macarthur, R., McDonald, R.A., and Delahay, R.J. (2017). Abundance of badgers (*Meles meles*) in England and Wales. *Scientific Reports*, 7(1), 1–8.

Jump, A.S. and Penuelas, J. (2005). Running to stand still: adaptation and the response of plants to rapid climate change. *Ecology Letters*, 8(9), 1010–1020.

Kaburu, S.S. and Newton-Fisher, N.E. (2015). Egalitarian despots: hierarchy steepness, reciprocity and the grooming-trade model in wild chimpanzees, Pan troglodytes . *Animal Behaviour*, 99, 61–71.

Kalinowski, S.T., Taper, M.L., and Marshall, T.C. (2007). Revising how the computer program CERVUS accommodates genotyping error increases success in paternity assignment. *Molecular Ecology*, 16(5), 1099–1106.

Kamerbeek, J., Schouls, L.E.O., Kolk, A., Van Agterveld, M., Van Soolingen, D., Kuijper, S., Bunschoten, A., Molhuizen, H., Shaw, R., Goyal, M., and van Embden, J. (1997). Simultaneous detection and strain differentiation of *Mycobacterium tuberculosis* for diagnosis and epidemiology. *Journal of Clinical Microbiology*, 35(4), 907–914.

Kamiya, T., O'Dwyer, K., Westerdahl, H., Senior, A., and Nakagawa, S. (2014). A quantitative review of MHC-based mating preference: the role of diversity and dissimilarity. *Molecular Ecology*, 23(21), 5151–5163.

Kamler, J.F., Stenkewitz, U., Klare, U., Jacobsen, N.F., and Macdonald, D.W. (2012). Resource partitioning among cape foxes, bat-eared foxes, and black-backed jackals in South Africa. *The Journal of Wildlife Management*, 76(6), 1241–1253.

Kamler, J.F., Gray, M.M., Oh, A., and Macdonald, D.W. (2013a). Genetic structure, spatial organization, and dispersal in two populations of bat-eared foxes. *Ecology and Evolution*, 3(9), 2892–2902.

Kamler, J.F., Stenkewitz, U., and Macdonald, D.W. (2013b). Lethal and sublethal effects of black-backed jackals on cape foxes and bat-eared foxes. *Journal of Mammalogy*, 94(2), 295–306.

Kamler, J.F., Stenkewitz, U., Sliwa, A., Wilson, B., Lamberski, N., Herrick, J.R., and Macdonald, D.W. (2015). Ecological relationships of black-footed cats (*Felis nigripes*) and sympatric canids in South Africa. *Mammalian Biology*, 80(2), 122–127.

Kamler, J.F., Stenkewitz, U., Gharajehdaghipour, T., and Macdonald, D.W. (2019). Social organization, home ranges, and extraterritorial forays of black-backed jackals. *Journal of Wildlife Management*, 83, 1800–1808.

Kamler, J.F., Minge, C., Rostro-García, S., Gharaje-hdaghipour, T., Crouthers, R., In, V., Pay, C., Pin C., Sovanna, P., and Macdonald, D.W. (2021). Home range, habitat selection, density, and diet of golden jackals in the Eastern Plains Landscape, Cambodia. *Journal of Mammalogy*, 102(2), 636–650.

Kaneko, Y. (2001). Life cycle of the Japanese badger (Meles meles anakuma) in Hinode Town, Tokyo. 哺乳類科学, 82, 53–64.

Kaneko, Y., Maruyama, N., and Macdonald, D.W. (2006). Food habits and habitat selection of suburban badgers (*Meles meles*) in Japan. *Journal of Zoology*, 270(1), 78–89.

Kaneko, Y., Newman, C., Buesching, C.D., and Macdonald, D.W. (2010). Variations in badger (*Meles meles*) sett microclimate: differential cub survival between main and subsidiary setts, with implications for artificial sett construction. *International Journal of Ecology*, 2010, 859586.

Kaneko, Y., Kanda, E., Tashima, S., Masuda, R., Newman, C., and Macdonald, D.W. (2014). The socio-spatial dynamics of the Japanese badger (*Meles anakuma*). *Journal of Mammalogy*, 95(2), 290–300.

Kaneko, Y., Buesching, C.D., and Newman, C. (2017). Unjustified killing of badgers in Kyushu. *Nature*, 544(7649), 161–161.

Kappelman, J., Alçiçek, M., Kazancı, N., Schultz, M., Özkul, M., and Sen, S. (2008). Brief communication: first Homo erectus from Turkey and implications for migrations into temperate Eurasia. *American Journal of Physical Anthropology*, 135, 110–116.

Karolemeas, K., McKinley, T.J., Clifton-Hadley, R.S., Goodchild, A.V., Mitchell, A., Johnston, W.T., Conlan, A.J.K., Donnelly, C.A., and Wood, J.L.N. (2011). Recurrence of bovine tuberculosis breakdowns in Great Britain: risk factors and prediction. *Preventive Veterinary Medicine*, 102(1), 22–29.

Karolemeas, K., Donnelly, C.A., Conlan, A.J.K., Mitchell, A.P., Clifton-Hadley, R.S., Upton, P., Wood, J.L.N. and McKinley, T.J. (2012). The effect of badger culling on breakdown prolongation and recurrence of bovine tuberculosis in cattle herds in Great Britain. *PLoS ONE*, 7(12), e51342. https://doi.org/10.1371/journal.pone.0051342

Karubian, J. and Swaddle, J.P. (2001). Selection on females can create 'larger males'. *Proceedings of the Royal Society of London. Series B: Biological Sciences*, 268(1468), 725–728.

Kaszta, Ż., Cushman, S.A., Hearn, A.J., Burnham, D., Macdonald, E.A., Goossens, B., Nathan, S.K., and Macdonald, D.W. (2019). Integrating Sunda clouded leopard (*Neofelis diardi*) conservation into development and restoration planning in Sabah (Borneo). *Biological Conservation, 235,* 63–76.

Katelaris, A.L., Jackson, C., Southern, J., Gupta, R.K., Drobniewski, F., Lalvani, A., Lipman, M., Mangtani, P., and Abubakar, I. (2020). Effectiveness of BCG vaccination against *Mycobacterium tuberculosis* infection in adults: a cross-sectional analysis of a UK-based cohort. *The Journal of Infectious Diseases, 221*(1), 146–155.

Kato, G.A., Sakamoto, S.H., Eto, T., Okubo, Y., Shinohara, A., Morita, T., and Koshimoto, C. (2018). Individual differences in torpor expression in adult mice are related to relative birth mass. *Journal of Experimental Biology, 221*(12), jeb171983.

Kaufman, J. (1999). Co-evolving genes in MHC haplotypes: the 'rule' for nonmammalian vertebrates? *Immunogenetics, 50*(3), 228–236.

Kaufmann, J.H. (1983). On the definitions and functions of dominance and territoriality. *Biological Reviews, 58*(1), 1–20.

Kauhala, K. (1995). Changes in distribution of the European badger *Meles meles* in Finland during the rapid colonization of the raccoon dog. *Annales Zoologici Fennici, 32,* 183–191.

Kaushic, C., Ashkar, A.A., Reid, L.A., and Rosenthal, K.L. (2003). Progesterone increases susceptibility and decreases immune responses to genital herpes infection. *Journal of Virology, 77*(8), 4558.

Kavaliers, M., Fudge, M.A., Colwell, D.D., and Choleris, E. (2003). Aversive and avoidance responses of female mice to the odors of males infected with an ectoparasite and the effects of prior familiarity. *Behavioral Ecology and Sociobiology, 54*(5), 423–430.

Kays, R.W. and Gittleman, J.L. (2001). The social organization of the kinkajou *Potos flavus* (Procyonidae). *Journal of Zoology, 253*(4), 491–504.

Keeble, E. (1999). Surgical treatment of a territorial fight wound in a badger (Meles meles): a novel technique. In *Proceedings of the BVZS Autumn Meeting, 20–21 November,* pp. 53–54. British Veterinary Zoological Society, London.

Kehrer, J.P. (1993). Free radicals as mediators of tissue injury and disease. *Critical Reviews in Toxicology, 23*(1), 21–48.

Kelliher, K.R., Baum, M.J., and Meredith, M. (2001). The ferret's vomeronasal organ and accessory olfactory bulb: effect of hormone manipulation in adult males and females. *The Anatomical Record: An Official Publication of the American Association of Anatomists, 263*(3), 280–288.

Kellner, K.F. and Swihart, R.K. (2014). Accounting for imperfect detection in ecology: a quantitative review. *PLoS ONE, 9*(10), e111436.

Kelly, D.W. and Thompson, C.E. (2000). Epidemiology and optimal foraging: modelling the ideal free distribution of insect vectors. *Parasitology, 120*(03), 319–327.

Kelly, J.B., Kavanagh, G.L., and Dalton, J.C. (1986). Hearing in the ferret (*Mustela putorius*): thresholds for pure tone detection. *Hearing Research, 24*(3), 269–275.

Kempenaers, B. (2007). Mate choice and genetic quality: a review of the heterozygosity theory. *Advances in the Study of Behavior, 37,* 189–278.

Kenney, W.L. and Munce, T.A. (2003). Invited review: aging and human temperature regulation. *Journal of Applied Physiology, 95*(6), 2598–2603.

Kenny, G.P., Yardley, J., Brown, C., Sigal, R.J., and Jay, O. (2010). Heat stress in older individuals and patients with common chronic diseases. *Canadian Medical Association Journal, 182*(10), 1053–1060.

Kent, A., Ehlers, B., Mendum, T., Newman, C., Macdonald, D.W., Chambers, M., and Buesching, C.D. (2018). Genital tract screening finds widespread infection with mustelid gammaherpesvirus 1 in the European badger (*Meles meles*). *Journal of Wildlife Diseases, 54*(1), 133–137.

Kent, A., Ehlers, B., Mendum, T., Newman, C., Macdonald, D.W., Chambers, M., and Buesching, C.D. (2018). Genital tract screening finds widespread infection with mustelid gammaherpesvirus 1 in the European badger (*Meles meles*). *Journal of Wildlife Diseases, 54*(1), 133–137.

Kent, J.E. and Ewbank, R. (1986). The effect of road transportation on the blood constituents and behaviour of calves. III. Three months old. *British Veterinary Journal, 142*(4), 326–335.

Kettlewell, H.B.D. (1956). A resume of investigations on the evolution of melanism in the Lepidoptera. *Proceedings of the Royal Society of London. Series B: Biological Sciences, 145*(920), 297–303.

Keverne, E.B., Martensz, N.D., and Tuite, B. (1989). Beta-endorphin concentrations in cerebrospinal fluid of monkeys are influenced by grooming relationships. *Psychoneuroendocrinology, 14*(1–2), 155–161.

Khan, J.Y. and Black, S.M. (2003). Developmental changes in murine brain antioxidant enzymes. *Pediatric Research, 54*(1), 77–82.

Khan, T. and Salunke, D.M. (2014). Adjustable locks and flexible keys: plasticity of epitope–paratope interactions in germline antibodies. *The Journal of Immunology, 192*(11), 5398–5405.

Khokhlova, I.S., Spinu, M., Krasnov, B.R., and Degen, A.A. (2004). Immune responses to fleas in two rodent species differing in natural prevalence of infestation and diversity of flea assemblages. *Parasitology Research, 94*(4), 304–311.

Kilshaw, K. and Macdonald, D.W. (2011). The use of camera trapping as a method to survey for the Scottish wildcat. Scottish Natural Heritage Commissioned Report, *479.*

Kilshaw, K., Newman, C., Buesching, C., Bunyan, J., and Macdonald, D. (2009). Coordinated latrine use by European badgers, *Meles meles*: potential consequences for territory defense. *Journal of Mammalogy, 90*(5), 1188–1198.

Kilshaw, K., Montgomery, R.A., Campbell, R.D., Hetherington, D.A., Johnson, P.J., Kitchener, A.C., Macdonald, D.W., and Millspaugh, J.J. (2016). Mapping the spatial configuration of hybridization risk for an endangered population of the European wildcat (*Felis sylvestris silvestris*) in Scotland. *Mammal Research*, 61(1), 1–11.

Kim, S.H., Park, J.J., Kim, K.H., Yang, H.J., Kim, D.S., Lee, C.H., Jeon, Y.S., Shim, S.R., and Kim, J.H. (2021). Efficacy of testosterone replacement therapy for treating metabolic disturbances in late-onset hypogonadism: a systematic review and meta-analysis. *International Urology and Nephrology*, 53, 1–14.

Kim, W., Ludlow, A.T., Min, J., Robin, J.D., Stadler, G., Mender, I., Lai, T.P., Zhang, N., Wright, W.E., and Shay, J.W. (2016). Regulation of the human telomerase gene TERT by telomere position effect—over long distances (TPE-OLD): implications for aging and cancer. *PLoS Biology*, 14(12), e2000016.

Kim, Y.M., Kim, S., Cheong, H.K., Ahn, B., and Choi, K. (2012). Effects of heat wave on body temperature and blood pressure in the poor and elderly. *Environmental Health and Toxicology*, 27, e2012013.

King, C.E. (1983). A re-examination of the Lansing Effect. In *Biology of Rotifers* (P. Starkweather and T. Nogrady, eds, pp. 135–139). Springer, Dordrecht.

King, D.P., Mutukwa, N., Lesellier, S., Cheeseman, C., Chambers, M.A., and Banks, M. (2004). Detection of mustelid herpesvirus-1 infected European badgers (*Meles meles*) in the British Isles. *Journal of Wildlife Diseases*, 40, 99–102.

King, K.M. (1985). Gambling: three forms and three explanations. *Sociological Focus*, 18(3), 235–248.

King, T.W., Salom-Pérez, R., Shipley, L.A., Quigley, H.B., and Thornton, D.H. (2017). Ocelot latrines: communication centers for Neotropical mammals. *Journal of Mammalogy*, 98(1), 106–113.

Kinghorn, B.P. and Kinghorn, S. (1994, August). Pedigree Viewer—a graphical utility for browsing pedigreed data sets. In *5th World Congress on Genetics Applied to Livestock Production* (Vol. 22, pp. 85–86). University of Guelph, Guelph, Ontario.

Kinoshita, E., Kosintsev, P.A., Raichev, E.G., Haukisalmi, V.K., Kryukov, A.P., Wiig, Ø., Abramov, A.V., Kaneko, Y., and Masuda, R. (2017). Molecular phylogeny of Eurasian badgers (*Meles*) around the distribution boundaries, revealed by analyses of mitochondrial DNA and Y-chromosomal genes. *Biochemical Systematics and Ecology*, 71, 121–130.

Kinoshita, E., Abramov, A.V., Soloviev, V.A., Saveljev, A.P., Nishita, Y., Kaneko, Y., and Masuda, R. (2019). Hybridization between the European and Asian badgers (*Meles*, Carnivora) in the Volga-Kama region, revealed by analyses of maternally, paternally and biparentally inherited genes. *Mammalian Biology*, 94(1), 140–148.

Kirk, E.C. (2006). Eye morphology in cathemeral lemurids and other mammals. *Folia Primatologica*, 77(1–2), 27–49.

Kirkwood, T.B. and Rose, M.R. (1991). Evolution of senescence: late survival sacrificed for reproduction. *Philosophical Transactions of the Royal Society of London. Series B: Biological Sciences*, 332(1262), 15–24.

Kirkwood, T.B.L. (1995). The evolution of aging. *Reviews in Clinical Gerontology*, 5(1), 3–9.

Kitchener, A.C., Meloro, C., and Williams, T.M. (2017). Form and function of the musteloids. In *Biology and Conservation of Musteloids* (D.W. Macdonald, C. Newman, and L.A. Harrington, eds, pp. 92–128). Oxford University Press, Oxford.

Klare, U., Kamler, J.F., and Macdonald, D.W. (2011). A comparison and critique of different scat-analysis methods for determining carnivore diet. *Mammal Review*, 41(4), 294–312.

Kleiman, D.G. and Malcolm, J.R. (1981). The evolution of male parental investment in mammals. In *Parental Care in Mammals* (D.J. Gubernick and P.H. Klopfer, eds, pp. 347–387). Springer, Boston, MA.

Klein, S.L. (2000). The effects of hormones on sex differences in infection: from genes to behavior. *Neuroscience & Biobehavioral Reviews*, 24(6), 627–638.

Klok, M.D., Jakobsdottir, S., and Drent, M.L. (2007). The role of leptin and ghrelin in the regulation of food intake and body weight in humans: a review. *Obesity Reviews*, 8(1), 21–34.

Knowles, S.C.L., Fenton, A., and Pedersen, A.B. (2012). Epidemiology and fitness effects of wood mouse herpesvirus in a natural host population. *Journal of General Virology*, 93(11), 2447–2456.

Koenig, W.D., Pitelka, F.A., Carmen, W.J., Mumme, R.L., and Stanback, M.T. (1992). The evolution of delayed dispersal in cooperative breeders. *The Quarterly Review of Biology*, 67(2), 111–150.

Koepfli, K.P., Deere, K.A., Slater, G.J., Begg, C., Begg, K., Grassman, L., Lucherini, M., Veron, G., and Wayne, R.K. (2008). Multigene phylogeny of the Mustelidae: resolving relationships, tempo and biogeographic history of a mammalian adaptive radiation. *BMC Biology*, 6(1), 1–22.

Koepfli, K.P., Dragoo, J.W., and Wang, X. (2017). The evolutionary history and molecular systematics of the Musteloidea. In *Biology and Conservation of Musteloids* (D.W. Macdonald, C. Newman, and L. Harrington, eds, pp. 75–90). Oxford University Press, Oxford.

Kokko, H. and Ekman, J. (2002). Delayed dispersal as a route to breeding: territorial inheritance, safe havens, and ecological constraints. *The American Naturalist*, 160(4), 468–484.

Kokko, H. and Jennions, M.D. (2008). Sexual conflict: the battle of the sexes reversed. *Current Biology*, 18(3), R121–R123.

Kokko, H. and Johnstone, R.A. (1999). Social queuing in animal societies: a dynamic model of reproductive skew. *Proceedings of the Royal Society of London. Series B: Biological Sciences*, 266(1419), 571–578.

Kokko, H. and Mappes, J. (2005). Sexual selection when fertilization is not guaranteed. *Evolution*, 59(9), 1876–1885.

Kokko, H. and Rankin, D.J. (2006). Lonely hearts or sex in the city? Density-dependent effects in mating systems. *Philosophical Transactions of the Royal Society. Series B: Biological Sciences*, 361(1466), 319–334.

Kokko, H., Brooks, R., Jennions, M.D., and Morley, J. (2003). The evolution of mate choice and mating biases.

Proceedings of the Royal Society of London. Series B: Biological Sciences, 270(1515), 653–664.

Kollmannsperger, F. (1955). About rhythms in lumbricides. *Decheniana, 180,* 81–92.

Komdeur, J. (1996). Facultative sex ratio bias in the offspring of Seychelles warblers. *Proceedings of the Royal Society of London. Series B: Biological Sciences, 263*(1370), 661–666.

Kopelman, P. (2007). Health risks associated with overweight and obesity. *Obesity Reviews, 8,* 13–17.

Kotrschal, K. (2012). Emotions are at the core of individual social performance. In *Emotions of Animals and Humans* (S. Watanabe and S. Kuczaj, eds, pp. 3–21). Springer Science & Business Media, Berlin and Heidelberg.

Kowalczyk, R., Bunevich, A.N., and Jędrzejewska, B. (2000). Badger density and distribution of setts in Białowieża Primeval Forest (Poland and Belarus) compared to other Eurasian populations. *Acta Theriologica, 45*(3), 395–408.

Kowalczyk, R., Jędrzejewska, B., and Zalewski, A. (2003a). Annual and circadian activity patterns of badgers (*Meles meles*) in Białowieża Primeval Forest (eastern Poland) compared with other Palaearctic populations. *Journal of Biogeography, 30*(3), 463–472.

Kowalczyk, R., Zalewski, A., Jędrzejewska, B., and Jędrzejewska, W. (2003b). Spatial organization and demography of badgers (*Meles meles*) in Bialowieza Primeval Forest, Poland, and the influence of earthworms on badger densities in Europe. *Canadian Journal of Zoology, 81*(1), 74–87.

Kowalczyk, R., Zalewski, A., and Jędrzejewska, B. (2004). Seasonal and spatial pattern of shelter use by badgers *Meles meles* in Białowieża Primeval Forest (Poland). *Acta Theriologica, 49,* 75–92.

Kowalczyk, R., Jędrzejewska, B., Zalewski, A., and Jędrzejewski, W. (2008). Facilitative interactions between the Eurasian badger (*Meles meles*), the red fox (*Vulpes vulpes*), and the invasive raccoon dog (*Nyctereutes procyonoides*) in Białowieża Primeval Forest, Poland. *Canadian Journal of Zoology, 86*(12), 1389–1396.

Kram, R. and Taylor, C. R. (1990). Energetics of running: a new perspective. *Nature, 346,* 265–267. doi:10.1038/346265a0

Krause, E.T., Krüger, O., and Schielzeth, H. (2017). Long-term effects of early nutrition and environmental matching on developmental and personality traits in zebra finches. *Animal Behaviour, 128,* 103–115.

Krause, J., Ruxton, G.D., Ruxton, G., and Ruxton, I.G. (2002). *Living in Groups.* Oxford University Press, Oxford.

Krebs, C.J. (1985). *Ecology: the Experimental Analysis of Distribution and Abundance,* third edition. Harper & Row, New York.

Krebs, J. (1997). Bovine tuberculosis in cattle and badgers. MAFF Publications, London.

Krebs, J.R. (1982). Territorial defence in the great tit (*Parus major*): do residents always win?. *Behavioral Ecology and Sociobiology, 11*(3), 185–194.

Krebs, J.R., Stephens, D.W., and Sutherland, W.J. (1983). Perspectives in optimal foraging. In *Perspectives in Ornithology* (A.H. Brush and G.A. Clark, Jr, eds, pp. 165–216). Cambridge University Press, Cambridge.

Kruchenkova, E.P., Goltsman, M., Sergeev, S., and Macdonald, D.W. (2009). Is alloparenting helpful for Mednyi Island arctic foxes, *Alopex lagopus semenovi*? *Naturwissenschaften, 96*(4), 457–466.

Kruman, I.I., Kolaeva, S.G., Rudchenko, S.A., and Khurkhulu, Z.S. (1988). Seasonal variations of DNA-synthesis in intestinal epithelial cells of hibernating animals—2. DNA-synthesis in intestinal epithelial cells of ground squirrel (*Citellus undulatus*) during autumn and late hibernation season. *Comparative Biochemistry and Physiology. B, Comparative Biochemistry, 89*(2), 271–273.

Kruuk, H. (1966). Clan-system and feeding habits of spotted hyaenas (*Crocuta crocuta* Erxleben). *Nature, 209*(5029), 1257–1258.

Kruuk, H. (1972). *The Spotted Hyena: A Study of Predation and Social Behavior.* University of Chicago Press, Chicago, IL.

Kruuk, H. (1976). Feeding and social behaviour of the striped hyaena (*Hyaena vulgaris* Desmarest). *African Journal of Ecology, 14*(2), 91–111.

Kruuk, H. (1978a). Spatial organization and territorial behaviour of the European badger Meles meles. *Journal of Zoology, 184*(1), 1–19.

Kruuk, H. (1978b). Foraging and spatial organisation of the European badger, *Meles meles* L. *Behavioral Ecology and Sociobiology, 4*(1), 75–89.

Kruuk, H. (1981). Food and habitat of badgers (*Meles meles* L.) on Monte Baldo, northern Italy. *Zeitschrift fur Saugetierkunde, 46,* 295–301.

Kruuk, H. (1986). Dispersion of badgers *Meles meles* (L., 1758) and their resources: a summary. *Lutra, 29,* 12–16.

Kruuk, H. (1989). *The Social Badger.* Oxford University Press, Oxford.

Kruuk, H. (1992). Scent marking by otters (*Lutra lutra*): signaling the use of resources. *Behavioral Ecology, 3*(2), 133–140.

Kruuk, H. (2000). Notes on status and foraging of the pantot or palawan stink-badger, *Mydaus marchei. Small Carnivore Conservation Newsletter and Journal of the IUCN/SSC Mustelid, Viverrid, and Procyonid Specialist Group, 22,* 11–12.

Kruuk, H. and Conroy, J.W.H. (1991). Mortality of otters (*Lutra lutra*) in Shetland. *Journal of Applied Ecology, 28*(1), 83–94.

Kruuk, H. and Hewson, R. (1978). Spacing and foraging of otters (*Lutra lutra*) in a marine habitat. *Journal of Zoology, 185*(2), 205–212.

Kruuk, H. and Macdonald, D.W. (1985). Group territories of carnivores: empires and enclaves. In *Behavioural Ecology: Ecological Consequences of Adaptive Behaviour* (R.M. Sibly and R.H. Smith, eds, pp. 521–536). Blackwell, Oxford.

Kruuk, H. and Mills, M.L. (1983). Notes on food and foraging of the honey badger *Mellivora capensis* in the Kalahari Gemsbok National Park. *Koedoe, 26*(1), 153–157.

Kruuk, H. and Parish, T. (1981). Feeding specialization of the European badger *Meles meles* in Scotland. *The Journal of Animal Ecology, 50*(3), 773–788.

Kruuk, H. and Parish, T. (1982). Factors affecting population density, group size and territory size of the European badger, *Meles meles*. *Journal of Zoology*, 196(1), 31–39.

Kruuk, H.H. and Parish, T. (1987). Changes in the size of groups and ranges of the European badger (*Meles meles* L.) in an area in Scotland. *The Journal of Animal Ecology*, 56(1), 351–364.

Kruuk, H., Parish, T., Brown, C.A.J., and Carrera, J. (1979). The use of pasture by the European badger (*Meles meles*). *Journal of Applied Ecology*, 16, 453–459.

Kruuk, H., Gorman, M., and Leitch, A. (1984). Scent-marking with the subcaudal gland by the European badger, *Meles meles* L. *Animal Behaviour*, 32(3), 899–907.

Kruuk, L.E., Slate, J., Pemberton, J.M., Brotherstone, S., Guinness, F., and Clutton-Brock, T. (2002). Antler size in red deer: heritability and selection but no evolution. *Evolution*, 56(8), 1683–1695.

Kummer, H. (1971). *Primate Societies: Group Techniques in Ecological Adaptation* University of Chicago Press, Chicago, IL.

Kurtén, B. (1968). *Pleistocene Mammals of Europe*. Weidenfeld and Nicolson, London.

Kutsukake, N. and Clutton-Brock, T.H. (2010). Grooming and the value of social relationships in cooperatively breeding meerkats. *Animal Behaviour*, 79(2), 271–279.

Kyaw, P.P., Macdonald, D.W., Penjor, U., Htun, S., Naing, H., Burnham, D., Kaszta, Ż., and Cushman, S.A. (2021). Investigating carnivore guild structure: spatial and temporal relationships amongst threatened felids in Myanmar. *ISPRS International Journal of Geo-Information*, 10(12), 808.

LaBarbera, M. (1989). Analyzing body size as a factor in ecology and evolution. *Annual Review of Ecology and Systematics*, 20(1), 97–117.

Lack, D. (1954). *The Natural Regulation of Animal Numbers*. Clarendon Press, Oxford.

Lahdenperä, M., Lummaa, V., Helle, S., Tremblay, M., and Russell, A.F. (2004). Fitness benefits of prolonged post-reproductive lifespan in women. *Nature*, 428(6979), 178–181.

Lahuerta-Marin, A., Milne, M.G., McNair, J., Skuce, R.A., McBride, S.H., Menzies, F.D., McDowell, S.J.W., Byrne, A.W., Handel, I.G., and Bronsvoort, B.D.C. (2018). Bayesian latent class estimation of sensitivity and specificity parameters of diagnostic tests for bovine tuberculosis in chronically infected herds in Northern Ireland. *The Veterinary Journal*, 238, 15–21.

Laidre, K.L., Stirling, I., Lowry, L.F., Wiig, Ø., Heide-Jørgensen, M.P., and Ferguson, S.H. (2008). Quantifying the sensitivity of Arctic marine mammals to climate-induced habitat change. *Ecological Applications*, 18(sp2), S97–S125.

Lammers, A.R., Dziech, H.A., and German, R.Z. (2001). Ontogeny of sexual dimorphism in *Chinchilla lanigera* (Rodentia: Chinchillidae). *Journal of Mammalogy*, 82(1), 179–189.

Lamprecht, J. (1978). The relationship between food competition and foraging group size in some larger carnivores. *Zeitschrift für Tierpsychologie*, 46(4), 337–343.

Langton, T.E., Jones, M.W. and McGill, I. (2022). Analysis of the impact of badger culling on bovine tuberculosis in cattle in the high-risk area of England, 2009–2020. *Veterinary Record*, 190(6), p.e1384.

Langton, T.E., Jones, M.W. and McGill, I. (2022b). Badger culling to control bovine TB. *Veterinary Record*, 190(10), pp. 419–420 .

Lansing, A.I. (1947). A transmissible, cumulative, and reversible factor in aging. *Journal of Gerontology*, 2(3), 228–239.

Lanyon, C.V., Rushton, S.P., O'Donnell, A.G., Goodfellow, M., Ward, A.C., Petrie, M., Jensen, S.P., Morris Gosling, L., and Penn, D.J. (2007). Murine scent mark microbial communities are genetically determined. *FEMS Microbiology Ecology*, 59(3), 576–583.

Larivière, S. and Ferguson, S.H. (2002). On the evolution of the mammalian baculum: vaginal friction, prolonged intromission or induced ovulation? *Mammal Review*, 32(4), 283–294.

Larivière, S. and Ferguson, S.H. (2003). Evolution of induced ovulation in North American carnivores. *Journal of Mammalogy*, 84(3), 937–947.

Larose, J., Boulay, P., Wright-Beatty, H.E., Sigal, R.J., Hardcastle, S., and Kenny, G.P. (2014). Age-related differences in heat loss capacity occur under both dry and humid heat stress conditions. *Journal of Applied Physiology*, 117(1), 69–79.

Lau, M.W.N., Fellowes, J.R., and Chan, B.P.L. (2010). Carnivores (Mammalia: Carnivora) in South China: a status review with notes on the commercial trade. *Mammal Review*, 40(4), 247–292.

Lavelle, P. (1988). Earthworm activities and the soil system. *Biology and Fertility of Soils*, 6(3), 237–251.

Lawlor, D.A., Zemmour, J., Ennis, P.D., and Parham, P. (1990). Evolution of class-I MHC genes and proteins: from natural selection to thymic selection. *Annual Review of Immunology*, 8(1), 23–63.

Lawrence, R.D. and Millar, H.R. (1945). Protein content of earthworms. *Nature*, 155(3939), 517–517.

Le Galliard, J.F., Fitze, P.S., Ferrière, R., and Clobert, J. (2005). Sex ratio bias, male aggression, and population collapse in lizards. *Proceedings of the National Academy of Sciences*, 102(50), 18231–18236.

Leberg, P.L. and Smith, M.H. (1993). Influence of density on growth of white-tailed deer. *Journal of Mammalogy*, 74(3), 723–731.

LeBlanc, M., Festa-Bianchet, M., and Jorgenson, J.T. (2001). Sexual size dimorphism in bighorn sheep (*Ovis canadensis*): effects of population density. *Canadian Journal of Zoology*, 79(9), 1661–1670.

LeBlanc, P.J., Obbard, M., Battersby, B.J., Felske, A.K., Brown, L., Wright, P.A., and Ballantyne, J.S. (2001). Correlations of plasma lipid metabolites with hibernation and lactation in wild black bears *Ursus americanus*. *Journal of Comparative Physiology B*, 171(4), 327–334.

Lee, K., Lau, M., and Chan, B. (2004). Wild animal trade monitoring at selected markets in Guangzhou and Shenzhen, south China, 2002–2003. Kadoorie Farm & Botanic Garden Technical Report No. 2. KFBG, Hong Kong.

Lee, P.C., Fishlock, V., Webber, C.E., and Moss, C.J. (2016). The reproductive advantages of a long life: longevity and senescence in wild female African elephants. *Behavioral Ecology and Sociobiology*, 70(3), 337–345.

Lee, P.L. and Hays, G.C. (2004). Polyandry in a marine turtle: females make the best of a bad job. *Proceedings of the National Academy of Sciences*, 101(17), 6530–6535.

Lee, T.M. and Zucker, I. (1988). Vole infant development is influenced perinatally by maternal photoperiodic history. *American Journal of Physiology-Regulatory, Integrative and Comparative Physiology*, 255(5), R831–R838.

Lélias, M.L., Lemasson, A., and Lodé, T. (2021). Social organization of otters in relation to their ecology. *Biological Journal of the Linnean Society*, 133(1), 1–27.

Lemaître, J.F., Berger, V., Bonenfant, C., Douhard, M., Gamelon, M., Plard, F., and Gaillard, J.M. (2015). Early-late life trade-offs and the evolution of ageing in the wild. *Proceedings of the Royal Society. Series B: Biological Sciences*, 282(1806), 20150209.

Lesellier, S., Palmer, S., Dalley, D.J., Dave, D., Johnson, L., Hewinson, R.G., and Chambers, M.A. (2006). The safety and immunogenicity of Bacillus Calmette-Guerin (BCG) vaccine in European badgers (*Meles meles*). *Veterinary Immunology and Immunopathology*, 112(1–2), 24–37.

Lesellier, S., Palmer, S., Gowtage-Sequiera, S., Ashford, R., Dalley, D., Davé, D., Weyer, U., Salguero, F.J., Nunez, A., Crawshaw, T., and Corner, L.A. (2011). Protection of Eurasian badgers (*Meles meles*) from tuberculosis after intra-muscular vaccination with different doses of BCG. *Vaccine*, 29(21), 3782–3790.

Lesku, J.A., Roth II, T.C., Amlaner, C.J., and Lima, S.L. (2006). A phylogenetic analysis of sleep architecture in mammals: the integration of anatomy, physiology, and ecology. *The American Naturalist*, 168(4), 441–453.

Leslie, P.H. (1945). On the use of matrices in certain population mathematics. *Biometrika*, 33(3), 183–212.

Leslie, P.H. (1948). Some further notes on the use of matrices in population mathematics. *Biometrika*, 35, 213–245.

Leth-Larsen, R., Zhong, F., Chow, V.T., Holmskov, U., and Lu, J. (2007). The SARS coronavirus spike glycoprotein is selectively recognized by lung surfactant protein D and activates macrophages. *Immunobiology*, 212(3), 201–211.

Leuchtenberger, C.M. and Mourão, G. (2008). Social organization and territoriality of giant otters (Carnivora, Mustelidae) in a seasonally flooded savanna in Brazil. *Sociobiology*, 52(2), 257–270.

Leuchtenberger, C., Oliveira-Santos, L.G.R., Magnusson, W., and Mourão G. (2013). Space use by giant otter groups in the Brazilian Pantanal. *Journal of Mammalogy*, 94(2), 320–330.

Levy, O., Dayan, T., Porter, W.P., and Kronfeld-Schor, N. (2019). Time and ecological resilience: can diurnal animals compensate for climate change by shifting to nocturnal activity? *Ecological Monographs*, 89(1), e01334.

Lewontin, R.C. (1965). Selection for colonizing ability. In *The Genetics of Colonizing Species* (H.G. Baker and G.L. Stebbins, eds, 77-91). Academic Press, New York.

Leyhausen, P. (1973). *Verhaltensstudien an Katzen*, 3rd edn. Parey, Berlin.

Li, C., Zhang, X., Zwiers, F., Fang, Y., and Michalak, A.M. (2017). Recent very hot summers in Northern Hemispheric land areas measured by wet bulb globe temperature will be the norm within 20 years. *Earth's Future*, 5(12), 1203–1216.

Li, F. and Jiang, Z. (2014). Is nocturnal rhythm of Asian badger (*Meles leucurus*) caused by human activity? A case study in the eastern area of Qinghai Lake. *Biodiversity Science*, 22(6), 758.

Liedtke, J., Redekop, D., Schneider, J.M., and Schuett, W. (2015). Early environmental conditions shape personality types in a jumping spider. *Frontiers in Ecology and Evolution*, 3, 134.

Lihoreau, M., Zimmer, C., and Rivault, C. (2007). Kin recognition and incest avoidance in a group-living insect. *Behavioral Ecology*, 18(5), 880–887.

Lima, S.L. (1995). Back to the basics of anti-predatory vigilance: the group-size effect. *Animal Behaviour*, 49(1), 11–20.

Lima, S.L. (1998). Stress and decision-making under the risk of predation: recent developments from behavioral, reproductive, and ecological perspectives. *Advances in the Study of Behavior*, 27, 215–290.

Lindenfors, P., Dalèn, L., and Angerbjörn, A. (2003). The monophyletic origin of delayed implantation in carnivores and its implications. *Evolution*, 57(8), 1952–1956.

Lindsay, D.S., Dubey, J.P., and Blagburn, B.L. (1997). Biology of *Isospora* spp. from humans, nonhuman primates, and domestic animals. *Clinical Microbiology Reviews*, 10(1), 19–34.

Lindsay, I.M. and Macdonald, D.W. (1985). The effects of disturbance on the emergence of Eurasian badgers in winter. *Biological Conservation*, 34(4), 289–306.

Lindström, E. (1986). Territory inheritance and the evolution of group-living in carnivores. *Animal Behaviour*, 34(6), 1825–1835.

Lindström, J. (1999). Early development and fitness in birds and mammals. *Trends in Ecology & Evolution*, 14(9), 343–348.

Linn, I. (1984). Home ranges and social systems in solitary mammals. *Acta Zoologica Fennica*, 171(1), 245–249.

Lizundia, R., Newman, C., Buesching, C.D., Ngugi, D., Blake, D., Sin, Y.W., Macdonald, D.W., Wilson, A., and McKeever, D. (2011). Evidence for a role of the host-specific flea (*Paraceras melis*) in the transmission of *Trypanosoma* (*Megatrypanum*) *pestanai* to the European badger. *PLoS ONE*, 6(2), e16977.

Lloyd, S. (1983). Effect of pregnancy and lactation upon infection. *Veterinary Immunology and Immunopathology*, 4(1–2), 153–176.

Lloyd, S. and Smith, J. (1997). Pattern of *Cryptosporidium parvum* oocyst excretion by experimentally infected dogs. *International Journal for Parasitology*, *27*(7), 799–801.

Lochmiller, R.L. and Deerenberg, C. (2000). Trade-offs in evolutionary immunology: just what is the cost of immunity? *Oikos*, *88*(1), 87–98.

Lodé, T. (2001). Mating system and genetic variance in a polygynous mustelid, the European polecat. *Genes & Genetic Systems*, *76*(4), 221–227.

Loehle, C. (1987). Hypothesis testing in ecology: psychological aspects and the importance of theory maturation. *The Quarterly Review of Biology*, *62*(4), 397–409.

Long, B. (2000). The hairy-nosed otter (*Lutra sumatrana*) in Cambodia. *IUCN Otter Specialist Group Bulletin*, *17*(2), 91.

Long, C.A. and Killingley, C.A. (1983). *The Badgers of the World*. Charles C. Thomas Publisher, Springfield, IL.

Looker, K.J., Magaret, A.S., Turner, K.M., Vickerman, P., Gottlieb, S.L., and Newman, L.M. (2015). Global estimates of prevalent and incident herpes simplex virus type 2 infections in 2012. *PLoS ONE*, *10*(1), e114989.

Lopez-Valencia, G., Renteria-Evangelista, T., Del Real, L.M., De La Mora, A., Medina-Basulto, G., Williams, J., and Licea, A. (2009). Field evaluation of the Mycobacterium bovis-BCG vaccine against tuberculosis in Holstein dairy cows. *Journal of Animal and Veterinary Advances*, *8*(11), 2171–2176.

Lord, K., Larson, G., Coppinger, R.P., and Karlsson, E.K. (2020). The history of farm foxes undermines the animal domestication syndrome. *Trends in Ecology and Evolution*, *35*(2), 125–136.

Lorenz K. (1943). Die angeborenen Formen moeglicher Erfahrung. *Zeitschrift für Tierpsychologie*, *5*, 235–409.

Lorenzoni, I., Nicholson-Cole, S., and Whitmarsh, L. (2007). Barriers perceived to engaging with climate change among the UK public and their policy implications. *Global Environmental Change*, *17*(3–4), 445–459.

Lotka, A.J. (1925). *Elements of Physical Biology*. Williams & Wilkins, PA.

Loureiro, F., Rosalino, L.M., Macdonald, D.W., and Santos-Reis, M. (2007). Use of multiple den sites by Eurasian badgers, *Meles meles*, in a Mediterranean habitat. *Zoological Science*, *24*(10), 978–985.

Loureiro, F., Bissonette, J.A., Macdonald, D.W., and Santos-Reis, M. (2009). Temporal variation in the availability of Mediterranean food resources: do badgers *Meles meles* track them? *Wildlife Biology*, *15*(2), 197–206.

Loveridge, A.J. (2018) *Lion Hearted: The Life and Death of Cecil & the Future of Africa's Iconic Cats*. Simon and Schuster. Riverside, NJ.

Loveridge, A.J. and Macdonald, D.W. (2001). Seasonality in spatial organization and dispersal of sympatric jackals (*Canis mesomelas* and *C. adustus*): implications for rabies management. *Journal of Zoology*, *253*(1), 101–111.

Loveridge, A.J., Hemson, G., Davidson, Z., and Macdonald, D.W. (2010). African lions on the edge: reserve boundaries as 'attractive sinks'. In *The Biology and Conservation of Wild Felids* (D.W. Macdonald and A.J. Loveridge, eds, pp. 283–304). Oxford University Press, Oxford.

Loveridge, A.J., Valeix, M., Chapron, G., Davidson, Z., Mtare, G., and Macdonald, D.W. (2016). Conservation of large predator populations: demographic and spatial responses of African lions to the intensity of trophy hunting. *Biological Conservation*, *204*, 247–254.

Løvlie, H., Gillingham, M.A., Worley, K., Pizzari, T., and Richardson, D.S. (2013). Cryptic female choice favours sperm from major histocompatibility complex-dissimilar males. *Proceedings of the Royal Society. Series B: Biological Sciences*, *280*(1769), 20131296.

Lu, A., Petrullo, L., Carrera, S., Feder, J., Schneider-Crease, I., and Snyder-Mackler, N. (2019). Developmental responses to early-life adversity: evolutionary and mechanistic perspectives. *Evolutionary Anthropology*, *28*, 249–266. doi:10.1002/evan.21791.

Lukas, D. and Clutton-Brock, T. (2012a). Life histories and the evolution of cooperative breeding in mammals. *Proceedings of the Royal Society. Series B: Biological Sciences*, *279*(1744), 4065–4070.

Lukas, D. and Clutton-Brock, T. (2012b). Cooperative breeding and monogamy in mammalian societies. *Proceedings of the Royal Society. Series B: Biological Sciences*, *279*(1736), 2151–2156.

Lundmark, T. (2009). *Tales of Hi and Bye: Greeting and Parting Rituals Around the World*. Cambridge University Press, Cambridge.

Lüps, P.J. (1993). Gattung Meles Brisson, 1762. *Handbuch der Saugetiere Europas*, *5*(2), 855.

Lüps, P.J. and Wandeler, A.I. (1993). *Meles meles* (Linnaeus, 1758) Dachs. In *Handbuch der Säugetiere Europas 5* (M. Stuebbe and J. Niethammer, eds, pp. 856–906). Aula Verlag, Wiesbaden.

Lynch, M. and Ritland, K. (1999). Estimation of pairwise relatedness with molecular markers. *Genetics*, *152*(4), 1753–1766.

Lynn, M. (1991). Scarcity effects on value: a quantitative review of the commodity theory literature. *Psychology & Marketing*, *8*, 43–57.

MacArthur, R.H. (1968). The theory of the niche. In *Population Biology and Evolution* (R.C. Lewontin, ed., pp. 159–176). Syracuse University Press, Syracuse, NY.

Macdonald, D.W. (1976a). Food caching by red foxes and some other carnivores. *Zeitschrift für Tierpsychologie*, *42*(2), 170–185.

Macdonald, D.W. (1976b). Nocturnal observations of tawny owls *Strix aluco* preying upon earthworms. *Ibis*, *118*(4), 579–580.

Macdonald, D.W. (1978). Observations on the behaviour and ecology of the striped hyaena, *Hyaena hyaena* in Israel. *Israel Journal of Zoology*, *27*(4), 189–198.

Macdonald, D.W. (1979a). Some observations and field experiments on the urine marking behaviour of the red fox, *Vulpes vulpes* L. *Zeitschrift für Tierpsychologie*, *51*(1), 1–22.

Macdonald, D.W. (1979b). The flexible social system of the golden jackal, *Canis aureus*. *Behavioral Ecology and Sociobiology*, *5*(1), 17–38.

Macdonald, D.W. (1979c). 'Helpers' in fox society. *Nature*, 282(5734), 69–71.

Macdonald, D.W. (1980a). Patterns of scent marking with urine and faeces amongst carnivore communities. *Proceedings of the Symposia of the Zoological Society of London, 45*, 107–139.

Macdonald, D.W. (1980b). Rabies and Wildlife. A Biologist's Perspective. Oxford University Press, Oxford.

Macdonald, D.W. (1980c). Social factors affecting reproduction behaviour among red foxes (*Vulpes vulpes* L., 1758). In *The Red Fox: Behaviour and Ecology* (Vol. 18, E. Zimen, ed., pp. 131–183). Boston, The Hague.

Macdonald, D.W. (1980d). The red fox, *Vulpes vulpes*, as a predator upon earthworms, *Lumbricus terrestris*. *Zeitschrift für Tierpsychologie, 52*(2), 171–200.

Macdonald, D.W. (1981a). Dwindling resources and the social behaviour of capybaras, (*Hydrochoerus hydrochaeris*) (Mammalia). *Journal of Zoology, 194*, 371–391.

Macdonald, D.W. (1981b). Resource dispersion and the social organisation of the red fox (*Vulpes vulpes*). Paper presented at the The First International Worldwide Furbearer Conference, Frostburg, Maryland.

Macdonald, D.W. (1983). The ecology of carnivore social behaviour. *Nature, 301*(5899), 379–384.

Macdonald, D.W. (1984a). Carnivore social behaviour—does it need patches? (reply). *Nature, 307*(5949), 390–390.

Macdonald, D.W. (1984b). Predation on earthworms by terrestrial vertebrates. In *Earthworm Ecology* (J.E. Satchell, ed., pp. 393–414). Chapman and Hall, London.

Macdonald, D.W. (1984c). Badgers and bovine tuberculosis—case not proven. *New Scientist, 104*(1427), 17–20.

Macdonald, D.W. (1987). *Running with the Fox*. Unwin Hyman, London.

Macdonald, D.W. (1988). Rabies and foxes: the social life of a solitary carnivore. In *Vaccination to Control Rabies in Foxes* (symposium coordination P.P. Pastoret, B. Brochier, I. Thomas, and J. Blancou, pp. 5–13). Commission of the European Communities, Brussels.

Macdonald, D.W. (1989). *Running with the Fox*. Unwin Hyman, London.

Macdonald, D.W. (1992). *The Velvet Claw*. BBC Books, London.

Macdonald, D.W. (1993). Rabies and wildlife: a conservation problem? *Onderstepoort Journal of Veterinary Research, 60*, 351–351.

Macdonald, D.W. (1995). Wildlife rabies: the implications for Britain. Unresolved questions for the control of wildlife rabies: social perturbation and interspecific interactions. In *Rabies in a Changing World. Proceedings of the British Small Animal Veterinary Association Held at The Royal Society of Medicine, London, UK, on Wednesday 3rd May 1995* (P.H. Beynon and A.T.B. Edney, eds, pp. 33–48). British Small Animal Veterinary Association, Cheltenham.

Macdonald, D.W. (1996). Social behaviour of captive bush dogs (*Speothos venaticus*). *Journal of Zoology, 239*(3), 525–543.

Macdonald, D.W. (2004). *Biology and Conservation of Wild Canids*. Oxford University Press, Oxford.

Macdonald, D.W. (2013). From ethology to biodiversity: case studies of wildlife conservation. *Nova Acta Leopoldina, 111*(380), 111–156.

Macdonald, D.W. (2019). Mammal conservation: old problems, new perspectives, transdisciplinarity, and the coming of age of conservation geopolitics. *Annual Review of Environment and Resources, 44*, 61–88.

Macdonald, D.W. and Bacon, P.J. (1982). Fox society, contact rate and rabies epizootiology. *Comparative Immunology, Microbiology and Infectious Diseases, 5*(1–3), 247–256.

Macdonald, D.W. and Barrett, P. (2001). *Mammals of Europe*. Princeton University Press, Princeton, NJ.

Macdonald, D.W. and Burnham, D. (2011). The state of Britain's mammals 2011. Peoples' Trust for Endangered Species and WildCRU, London and Oxford.

Macdonald, D.W. and Carr, G.M. (1989). Food security and the rewards of tolerance. In *Comparative Socioecology: the Behavioural Ecology of Humans and Other Mammals* (V. Standen and R.A. Folley, eds, pp. 75–99). Blackwell Scientific Publications, Oxford.

Macdonald, D.W. and Courtenay, O. (1996). Enduring social relationships in a population of crab-eating zorros, *Cerdocyon thous*, in Amazonian Brazil (Carnivora, Canidae). *Journal of Zoology, 239*(2), 329–355.

Macdonald, D.W. and Doolan, S.P. (1997). Band structure and failures of reproductive suppression in a cooperatively breeding carnivore, the slender-tailed meerkat (*Suricata suricatta*). *Behaviour, 134*(11–12), 827–848.

Macdonald, D.W. and Feber, R.E. (eds) (2015a). *Wildlife Conservation on Farmland Volume 1: Managing for Nature on Lowland Farms*. Oxford University Press, Oxford.

Macdonald, D.W. and Feber, R.E. (eds) (2015b). *Wildlife Conservation on Farmland Volume 2: Conflict in the Countryside*. Oxford University Press, Oxford.

Macdonald, D.W. and Johnson, P.J. (2008). Sex ratio variation and mixed pairs in roe deer: evidence for control of sex allocation? *Oecologia, 158*, 361–370. https://doi.org/10.1007/s00442-008-1143-7

Macdonald, D.W. and Johnson, D.D.P. (2015a). Patchwork planet: the resource dispersion hypothesis, society, and the ecology of life. *Journal of Zoology, 295*(2), 75–107.

Macdonald, D.W. and Johnson, P.J. (2015b). Foxes in the landscape: hunting, control, and economics. *Wildlife Conservation on Farmland, 2*, 47–64.

Macdonald, D.W. and Kays, R.W. (2005). Carnivores of the world: an introduction. In *Carnivores of the World* (R.M. Nowak, ed., pp. 1–67). Johns Hopkins University Press, Baltimore, MD.

Macdonald, D.W. and Laurenson, M.K. (2006). Infectious disease: inextricable linkages between human and ecosystem health. *Biological Conservation, 131*(2), 143.

Macdonald, D.W. and Loveridge, A.J. (eds) (2010). *The Biology and Conservation of Wild Felids*. Oxford University Press, Oxford.

Macdonald, D.W. and Moehlman, P.D. (1982). Cooperation, altruism, and restraint in the reproduction of carnivores. In *Ontogeny* (P.P.G. Bateson and P.H. Klopfer, eds, pp. 433–467). Springer, Boston, MA.

Macdonald, D.W. and Mosser, A. (2010). Felid society. In *The Biology and Conservation of Wild Felids* (D.W. Macdonald and A.J. Loveridge, eds, pp. 125–160). Oxford University Press, Oxford.

Macdonald, D.W. and Newman, C. (2002). Population dynamics of badgers (*Meles meles*) in Oxfordshire, UK: numbers, density and cohort life histories, and a possible role of climate change in population growth. *Journal of Zoology*, 256(1), 121–138.

Macdonald, D.W. and Newman, C. (2017). Musteloid sociality: the grass-roots of society. In *Biology and Conservation of Musteloids* (D.W. Macdonald, C. Newman, and L.A. Harrington, eds, pp. 167–188). Oxford University Press, Oxford.

Macdonald, D.W. and Sillero-Zubiri, C. (eds). (2004). *The Biology and Conservation of Wild Canids*. Oxford University Press, Oxford.

Macdonald, D.W. and Tattersall, F. (1996). *The WildCRU Review: the Tenth Anniversary Report of the Wildlife Conservation Research Unit at Oxford University*. WildCRU, Oxford University, Oxford.

Macdonald, D.W. and Voigt, D.R. (1985). Biological basis of rabies models. In Population Dynamics of Rabies in Wildlife (P.J. Bacon, ed., pp. 71–108). Academic Press, London.

Macdonald, D.W., Ball, F.G., and Hough, N.G. (1980a). The evaluation of home range size and configuration using radio tracking data. In *A Handbook on Biotelemetry and Radio Tracking* (D.W. Macdonald and C.J. Amlaner Jr, eds, pp. 405–424). Pergamon Press, Oxford.

Macdonald, D.W., Boitani, L., and Barrasso, P. (1980b). Foxes, wolves and conservation in the Abruzzo mountains. In *Red Fox* (E. Zimen , ed. pp. 223–235). Springer, Dordrecht.

Macdonald, D.W., Apps, P.J., Carr, G.M., and Kerby, G. (1987). Social dynamics, nursing coalitions and infanticide among farm cats, *Felis catus*. Advances in Ethology, 28(Suppl.), 1–66.

Macdonald, D.W., Anwar, M., Newman, C., Woodroffe, R., and Johnson, P.J. (1999). Inter-annual differences in the age-related prevalences of Babesia and Trypanosoma parasites of European badgers (*Meles meles*). *Journal of Zoology*, 247(1), 65–70.

Macdonald, D.W., Stewart, P.D., Stopka, P., and Yamaguchi, N. (2000a). Measuring the dynamics of mammalian societies: an ecologist's guide to ethological methods. In Research Techniques in Animal Ecology: Controversies and Consequences (M.C. Pearl, L. Boitani, and T.K. Fuller, eds, pp. 332–388). Columbia University Press, New York.

Macdonald, D.W., Yamaguchi, N., and Kerby, G. (2000b). Group-living in the domestic cat: its sociobiology and epidemiology. In *The Domestic Cat: The Biology of Its Behaviour*, 2nd edn (D.C. Turner and P. Bateson, eds, pp. 95–118). Cambridge University Press, Cambridge.

Macdonald, D.W., Newman, C., Stewart, P.D., Domingo-Roura, X., and Johnson, P.J. (2002a). Density-dependent regulation of body mass and condition in badgers (*Meles meles*) from Wytham Woods. *Ecology*, 83(7), 2056–2061.

Macdonald, D.W., Stewart, P.D., Johnson, P.J., Porkert, J., and Buesching, C. (2002b). No evidence of social hierarchy amongst feeding badgers, *Meles meles*. *Ethology*, 108(7), 613–628.

Macdonald, D.W., Buesching, C.D., Stopka, P., Henderson, J., Ellwood, S.A., and Baker, S.E. (2004a). Encounters between two sympatric carnivores: red foxes (*Vulpes vulpes*) and European badgers (*Meles meles*). *Journal of Zoology*, 263(4), 385–392.

Macdonald, D.W., Creel, S., and Mills, M.G. (2004b). Canid society. In *The Biology and Conservation of Wild Canids* (D.W. Macdonald and C. Sillero-Zubiri, eds, pp. 85–106). Oxford University Press, Oxford.

Macdonald, D.W., Harmsen, B.J., Johnson, P.J., and Newman, C. (2004c). Increasing frequency of bite wounds with increasing population density in Eurasian badgers, *Meles meles*. *Animal Behaviour*, 67(4), 745–751.

Macdonald, D.W., Newman, C., Dean, J., Buesching, C.D., and Johnson, P.J. (2004d). The distribution of Eurasian badger, *Meles meles*, setts in a high-density area: field observations contradict the sett dispersion hypothesis. *Oikos*, 106(2), 295–307.

Macdonald, D.W., Tew, T.E., and Todd, I.A. (2004e). The ecology of weasels (*Mustela nivalis*) on mixed farmland in southern England. *Biologia*, 59(2), 233–239.

Macdonald, D.W., Riordan, P., and Mathews, F. (2006). Biological hurdles to the control of TB in cattle: a test of two hypotheses concerning wildlife to explain the failure of control. *Biological Conservation*, 131(2), 268–286.

Macdonald, D.W., Collins, N.M., and Wrangham, R. (2007a). Principles, practice and priorities: the quest for 'alignment'. In Key Topics in Conservation Biology (D.W. Macdonald and K. Service, eds, pp. 271–290). Blackwell, Oxford.

Macdonald, D.W., Herrera, E.A., Taber, A.B. and Moreira, J.R. (2007b). Social organization and resource use in capybaras and maras. In *Rodent Societies: An Ecological and Evolutionary Perspective* (J.O. Wolff and P.W. Sherman, eds, pp. 393–402). The University of Chicago Press, Chicago, IL.

Macdonald, D.W., Newman, C., Buesching, C.D., and Johnson, P.J. (2008). Male-biased movement in a high-density population of the Eurasian badger (*Meles meles*). *Journal of Mammalogy*, 89(5), 1077–1086.

Macdonald, D.W., Newman, C., Nouvellet, P.M., and Buesching, C.D. (2009). An analysis of Eurasian badger (*Meles meles*) population dynamics: implications for regulatory mechanisms. *Journal of Mammalogy*, 90(6), 1392–1403.

Macdonald, D.W., Loveridge, A.J., and Nowell, K. (2010a). Dramatis personae: an introduction to the wild felids. *Biology and Conservation of Wild Felids*, 1, 3–58.

Macdonald, D.W., Mosser, A., and Gittleman, J.L. (2010b). Felid society. In *Biology and Conservation of Wild Felids* (D.W. Macdonald and A. Loveridge, eds, pp. 125–160). Oxford University Press, Oxford.

Macdonald, D.W., Newman, C., Buesching, C.D., and Nouvellet, P. (2010c). Are badgers 'under the weather'? Direct and indirect impacts of climate variation on European badger (*Meles meles*) population dynamics. *Global Change Biology*, 16(11), 2913–2922.

Macdonald, D.W., Yamaguchi, N., Kitchener, A.C., Daniels, M., Kilshaw, K., and Driscoll, C. (2010d). Reversing cryptic extinction: the history, present and future of the Scottish wildcat. In Biology and Conservation of Wild Felids (D.W. Macdonald and A.J. Loveridge, eds, pp. 471–492). Oxford University Press, Oxford.

Macdonald, D.W., Berdoy, M., and Webster, J.P. (2015a). Brown rats on farmland: ecological citizens or subsidised carpet-baggers? In *Wildlife Conservation on Farmland. Volume 2: Conflict in the Countryside* (D.W. Macdonald and R.E. Feber, eds, pp. 222–244). Oxford University Press, Oxford.

Macdonald, D.W., Doncaster, P., Newdick, M., Hofer, H., Matthews, F., and Johnson, P.J. (2015b). Foxes in the landscape. Ecology and sociality. In *Wildlife Conservation on Farmland. Conflict in the Countryside* (D.W. Macdonald and R.E. Feber, eds, pp. 20–27). Oxford University Press, Oxford.

Macdonald, D.W., Fenn, M.G.P., and Gelling, M. (2015c). The natural history of rodents: preadaptations to pestilence. In Rodent Pests and Their Control, 2nd edn (A.P. Buckle and R.H. Smith, eds, pp. 1–18). CABI, Wallingford.

Macdonald, D.W., Harrington, L.A., Yamaguchi, N., Thorn, M.D.F., and Bagniewska, J. (2015d). Biology, ecology, and reproduction of American mink *Neovison vison* on lowland farmland. In *Wildlife Conservation on Farmland Volume 2: Conflict in the Countryside* (Vol. 2, D.W. Macdonald and R.E. Feber, eds, pp. 126–147). Oxford: Oxford University Press.

Macdonald, D.W., Newman, C. and Buesching, C.D. (2015e). Badgers in the rural landscape—conservation paragon or farmland pariah? Lessons from the Wytham Badger Project. *Wildlife Conservation on Farmland*, 2, 65–95.

Macdonald, D.W., Burnham, D., Dickman, A., Loveridge, A.J., and Johnson, P.J. (2016a). Conservation or the moral high ground: siding with Bentham or Kant. *Conservation Letters*, 9(4), 307–308.

Macdonald, D.W., Jacobsen, K.S., Burnham, D., Johnson, P.J., and Loveridge, A.J. (2016b). Cecil: a moment or a movement? Analysis of media coverage of the death of a lion, Panthera leo. *Animals*, 6(5), 26.

Macdonald, D.W., Harrington, L.A., and Newman, C. (2017a). Dramatis personae: an introduction to the wild musteloids. In *Biology and Conservation of Musteloids* (D.W. Macdonald, C. Newman, and L.A. Harrington, eds, pp. 3–74). Oxford University Press, Oxford.

Macdonald, D.W., Newman, C., and Harrington, L.A. (eds) (2017b). *Biology and Conservation of Musteloids*. Oxford University Press, Oxford.

Macdonald, D.W., Loveridge, A.J., Dickman, A., Johnson, P.J., Jacobsen, K.S., and Du Preez, B. (2017c). Lions, trophy hunting and beyond: knowledge gaps and why they matter. *Mammal Review*, 47(4), 247–253.

Macdonald, D.W., Campbell, L.A., Kamler, J.F., Marino, J., Werhahn, G., and Sillero-Zubiri, C. (2019). Monogamy: cause, consequence, or corollary of success in wild canids? *Frontiers in Ecology and Evolution*, 7, 341.

Mace, R.D. and Waller, J.S. (1998). Demography and population trend of grizzly bears in the Swan Mountains, Montana. *Conservation Biology*, 12(5), 1005–1016.

Machanda, Z.P., Gilby, I.C., and Wrangham, R.W. (2014). Mutual grooming among adult male chimpanzees: the immediate investment hypothesis. *Animal Behaviour*, 87, 165–174.

Machovsky-Capuska, G.E., Senior, A.M., Simpson, S.J., and Raubenheimer, D. (2016). The multidimensional nutritional niche. *Trends in Ecology & Evolution*, 31(5), 355–365.

Mackay, C. (1877). *The Gaelic Etymology of the Languages of Western Europe: And More Especially of the English and Lowland Scotch, and Their Slang, Cant, and Colloquial Dialects*. Trubner, London.

Madden, J.R. and Clutton-Brock, T.H. (2009). Manipulating grooming by decreasing ectoparasite load causes unpredicted changes in antagonism. *Proceedings of the Royal Society. Series B: Biological Sciences*, 276, 1263–1268. doi:10.1098/rspb.2008.1661

Madsen, A.E., Corral, L., and Fontaine, J.J. (2020). Weather and exposure period affect coyote detection at camera traps. *Wildlife Society Bulletin*, 4(2), 342–350.

Madurell-Malapeira, J., Alba, D.M., Marmi, J., Aurell, J., and Moyà-Solà, S. (2011a). The taxonomic status of European Plio-Pleistocene badgers. *Journal of Vertebrate Paleontology*, 31(4), 885–894.

Madurell-Malapeira, J., Martínez-Navarro, B., Ros-Montoya, S., Espigares, M.P., Toro, I., and Palmqvist, P. (2011b). The earliest European badger (*Meles meles*), from the late villafranchian site of Fuente Nueva 3 (Orce, Granada, se Iberian Peninsula). *Comptes Rendus Palevol*, 10(8), 609–615.

Maher, C.R. and Lott, D.F. (1995). Definitions of territoriality used in the study of variation in vertebrate spacing systems. *Animal Behaviour*, 49(6), 1581–1597.

Maher, C.R. and Lott, D.F. (2000). A review of ecological determinants of territoriality within vertebrate species. *The American Midland Naturalist*, 143(1), 1–29.

Mainguy, J., Cote, S.D., and Coltman, D.W. (2009a). Multilocus heterozygosity, parental relatedness and individual fitness components in a wild mountain goat, *Oreamnos americanus* population. *Molecular Ecology*, 18(10), 2297–2306.

Mainguy, J., Côté, S.D., Festa-Bianchet, M., and Coltman, D.W. (2009b). Father–offspring phenotypic correlations suggest intralocus sexual conflict for a fitness-linked trait in a wild sexually dimorphic mammal. *Proceedings of the Royal Society. Series B: Biological Sciences*, 276(1675), 4067–4075.

Majolo, B., Lehmann, J., de Bortoli Vizioli, A., and Schino, G. (2012). Fitness-related benefits of dominance in primates. *American Journal of Physical Anthropology*, 147(4), 652–660.

Malcolm, J.R. (1985). Paternal care in canids. *American Zoologist*, 25(3), 853–856.

Maldini, D., Scoles, R., Eby, R., Cotter, M., and Rankin, R.W. (2012). Patterns of sea otter haul-out behavior in a California tidal estuary in relation to environmental variables. *Northwestern Naturalist*, 93, 67–78.

Mallinson, P.J., Cresswell, W.J., Harris, S., and Cheeseman, C.L. (1992). To breed or not to breed: an analysis of the social and density-dependent constraints on the fecundity of female badgers (*Meles meles*). *Philosophical Transactions of the Royal Society of London. Series B: Biological Sciences*, 338(1286), 393–407.

Malthus, T.R. (1798). *An Essay on the Principle of Population As It Affects the Future Improvement of Society, with Remarks on the Speculations of Mr. Goodwin, M. Condorcet and Other Writers*. J. Johnson in St Paul's Church-yard, London.

Mann, J., Connor, R.C., Barre, L.M., and Heithaus, M.R. (2000). Female reproductive success in bottlenose dolphins (*Tursiops* sp.): life history, habitat, provisioning, and group-size effects. *Behavioral Ecology*, 11(2), 210–219.

Mann, M.E., Zhang, Z., Rutherford, S., Bradley, R.S., Hughes, M.K., Shindell, D., Ammann, C., Faluvegi, G., and Ni, F. (2009). Global signatures and dynamical origins of the Little Ice Age and Medieval Climate Anomaly. *Science*, 326(5957), 1256–1260.

Mantovani, A., Cassatella, M.A., Costantini, C., and Jaillon, S. (2011). Neutrophils in the activation and regulation of innate and adaptive immunity. *Nature Reviews Immunology*, 11(8), 519–531.

Marable, M.K., Belant, J.L., Godwin, D., and Wang, G. (2012). Effects of resource dispersion and site familiarity on movements of translocated wild turkeys on fragmented landscapes. *Behavioural Processes*, 91(1), 119–124.

Marasco, V., Boner, W., Griffiths, K., Heidinger, B., and Monaghan, P. (2019). Intergenerational effects on offspring telomere length: interactions among maternal age, stress exposure and offspring sex. *Proceedings of the Royal Society. Series B: Biological Sciences*, 286, 28620191845

Mares, R., Young, A.J., and Clutton-Brock, T.H. (2012). Individual contributions to territory defence in a cooperative breeder: weighing up the benefits and costs. *Proceedings of the Royal Society. Series B: Biological Sciences*, 279(1744), 3989–3995.

Marino, F.E. (2004). Anticipatory regulation and avoidance of catastrophe during exercise-induced hyperthermia. *Comparative Biochemistry and Physiology Part B: Biochemistry and Molecular Biology*, 139(4), 561–569.

Marino, J., Sillero-Zubiri, C., Gottelli, D., Johnson, P.J., and Macdonald, D.W. (2013). The fall and rise of Ethiopian wolves: lessons for conservation of long-lived, social predators. *Animal Conservation*, 16(6), 621–632. doi:10.1111/acv.12036

Marino, J., Sillero-Zubiri, C., Johnson, P., and Macdonald, D. (2012). Ecological bases of philopatry and cooperation in Ethiopian wolves. *Behavioral Ecology and Sociobiology*, 66(7), 1005–1015. doi:10.1007/s00265-012-1348-x

Marjamäki, P.H., Dugdale, H.L., Dawson, D.A., McDonald, R.A., Delahay, R., Burke, T., and Wilson, A.J. (2019). Individual variation and the source-sink group dynamics of extra-group paternity in a social mammal. *Behavioral Ecology*, 30(2), 301–312.

Marjamäki, P.H., Dugdale, H.L., Delahay, R., McDonald, R.A., and Wilson, A.J. (2021). Genetic, social and maternal contributions to *Mycobacterium bovis* infection status in European badgers (*Meles meles*). *Journal of Evolutionary Biology*, 34(4), 695–709.

Marlowe, F. (2000). The patriarch hypothesis—an alternative explanation of menopause. *Human Nature and an Interdisciplinary Biosocial Perspective*, 11(1), 27–42.

Marmi, J., López-Giráldez, J.F., and Domingo-Roura, X. (2004). Phylogeny, evolutionary history and taxonomy of the Mustelidae based on sequences of the cytochrome b gene and a complex repetitive flanking region. *Zoologica Scripta*, 33(6), 481–499.

Marmi, J., López-Giráldez, F., Macdonald, D.W., Calafell, F., Zholnerovskaya, E., and Domingo-Roura, X. (2006). Mitochondrial DNA reveals a strong phylogeographic structure in the badger across Eurasia. *Molecular Ecology*, 15(4), 1007–1020.

Marshall, H.H., Johnstone, R.A., Thompson, F.J., Nichols, H.J., Wells, D., Hoffman, J.I., Kalema-Zikusoka, G., Sanderson, J.L., Vitikainen, E.I.K., Blount, J.D., and Cant, M.A. (2021). A veil of ignorance can promote fairness in a mammal society. *Nature Communication*, 12, 3717. https://doi.org/10.1038/s41467-021-23910-6

Marshall, T.C., Slate, J.B.K.E., Kruuk, L.E.B., and Pemberton, J.M. (1998). Statistical confidence for likelihood-based paternity inference in natural populations. *Molecular Ecology*, 7(5), 639–655.

Martin, L.E., Byrne, A.W., O'Keeffe, J., Miller, M.A., and Olea-Popelka, F.J. (2017). Weather influences trapping success for tuberculosis management in European badgers (*Meles meles*). *European Journal of Wildlife Research*, 63(1), 30.

Martín, R., Rodríguez, A., and Delibes, M. (1995). Local feeding specialization by badgers (*Meles meles*) in a Mediterranean environment. *Oecologia*, 101(1), 45–50.

Mathews, F., Kubasiewicz, L.M., Gurnell, J., Harrower, C.A., McDonald, R.A., and Shore, R.F. (2018). A review of the population and conservation status of British mammals. http://publications.naturalengland.org.uk/publication/5636785878597632 (accessed 11 May 2022).

Mathews, F., Lovett, L., Rushton, S., and Macdonald, D.W. (2006a). Bovine tuberculosis in cattle: reduced risk on wildlife-friendly farms. *Biology Letters*, 2(2), 271–274.

Mathews, F., Macdonald, D.W., Taylor, G.M., Gelling, M., Norman, R.A., Honess, P.E., Foster, R., Gower, C.M., Varley, S., Harris, A., and Palmer, S. (2006b). Bovine tuberculosis (*Mycobacterium bovis*) in British farmland wildlife: the importance to agriculture. *Proceedings of the Royal Society. Series B: Biological Sciences*, 273(1584), 357–365.

Matsunaga, M., Isowa, T., Yamakawa, K., and Ohira, H. (2013). Association between the serotonin transporter polymorphism (5HTTLPR) and subjective happiness level in Japanese adults. *Psychology of Well-Being: Theory, Research and Practice*, 3(1), 5.

Matthews, A. and Green, K. (2012). Seasonal and altitudinal influences on the home range and movements of common wombats in the Australian Snowy Mountains. *Journal of Zoology*, 287(1), 24–33.

Matthiesen, C.F., Blache, D., Thomsen, P.D., Hansen, N.E., and Tauson, A.H. (2010). Effect of late gestation low protein supply to mink (*Mustela vison*) dams on reproductive performance and metabolism of dam and offspring. *Archives of Animal Nutrition*, 64(1), 56–76.

May, R.M. and Anderson, R.M. (1978). Regulation and stability of host-parasite population interactions: II. Destabilizing processes. *The Journal of Animal Ecology*, 47(1), 249–267.

Mayeaux, D.J. and Johnston, R.E. (2002). Discrimination of individual odours by hamsters (*Mesocricetus auratus*) varies with the location of those odours. *Animal Behaviour.*, 64, 269–281.

Mayr, E. (1997). The objects of selection. *Proceedings of the National Academy of Sciences*, 94(6), 2091–2094.

Mbizah, M.M., Valeix, M., Macdonald, D.W., and Loveridge, A.J. (2019). Applying the resource dispersion hypothesis to a fission–fusion society: a case study of the African lion (*Panthera leo*). *Ecology and Evolution*, 9(16), 9111–9119.

Mbizah, M.M., Farine, D.R., Valeix, M., Hunt, J.E., Macdonald, D.W., and Loveridge, A.J. (2020). Effect of ecological factors on fine-scale patterns of social structure in African lions. *Journal of Animal Ecology*, 89(11), 2665–2676.

McAllister, R.M., Delp, M.D., and Laughlin, M.H. (1995). Thyroid status and exercise tolerance. *Sports Medicine*, 20(3), 189–198.

McAuliffe, K. and Whitehead, H. (2005). Eusociality, menopause and information in matrilineal whales. *Trends in Ecology & Evolution*, 20(12), 650.

McClelland, E.E., Granger, D.L., and Potts, W.K. (2003). Major histocompatibility complex-dependent susceptibility to *Cryptococcus neoformans* in mice. *Infection and Immunity*, 71(8), 4815–4817.

McClune, D.W., Kostka, B., Delahay, R.J., Montgomery, W.I., Marks, N.J., and Scantlebury, D.M. (2015). Winter is coming: seasonal variation in resting metabolic rate of the European badger (*Meles meles*). *PLoS ONE*, 10(9), e0135920.

McComb, K., Packer, C., and Pusey, A. (1994). Roaring and numerical assessment in contests between groups of female lions, *Panthera leo*. *Animal Behaviour*, 47(2), 379–387.

McCracken, G.F. and Bradbury, J.W. (1977). Paternity and genetic heterogeneity in the polygynous bat, *Phyllostomus hastatus*. *Science*, 198(4314), 303–306.

McCracken, G.F. and Bradbury, J.W. (1981). Social organization and kinship in the polygynous bat *Phyllostomus hastatus*. *Behavioral Ecology and Sociobiology*, 8(1), 11–34.

McCulloch, S.P. and Reiss, M.J. (2017). Bovine tuberculosis and badger control in Britain: science, policy and politics. *Journal of Agricultural and Environmental Ethics*, 30. 10.1007/s10806-017-9686-3

McDermott, R., Fowler, J.H., and Smirnov, O. (2008). On the evolutionary origin of prospect theory preferences. *The Journal of Politics*, 70(2), 335–350.

McDonald, R.B. and Horwitz, B.A. (1999). Brown adipose tissue thermogenesis during aging and senescence. *Journal of Bioenergetics and Biomembranes*, 31(5), 507–516.

McElligott, A.G., Gammell, M.P., Harty, H.C., Paini, D.R., Murphy, D.T., Walsh, J.T., and Hayden, T.J. (2001). Sexual size dimorphism in fallow deer (*Dama dama*): do larger, heavier males gain greater mating success? *Behavioral Ecology and Sociobiology*, 49(4), 266–272.

McEwen, B.S. (1998). Stress, adaptation, and disease: allostasis and allostatic load. *Annals of the New York Academy of Sciences*, 840(1), 33–44.

McGann, J.P. (2017). Poor human olfaction is a 19th-century myth. *Science*, 356. https://www.science.org/doi/10.1126/science.aam7263

McGill, I. and Jones, M. (2019). Cattle infectivity is driving the bTB epidemic. *Veterinary Record*, 185(22), 699–700.

McGraw, K.J., Cohen, A.A., Costantini, D., and Hõrak, P. (2010). The ecological significance of antioxidants and oxidative stress: a marriage between mechanistic and functional perspectives. *Functional Ecology*, 24, 947–949.

McIntyre, S. and McKitrick, R. (2003). Corrections to the Mann et al. (1998) proxy data base and northern hemispheric average temperature series. *Energy & Environment*, 14(6), 751–771.

McKeown, T. (1976). *The Modern Rise of Population*. Edward Arnold, London.

McKinley, M.J., Martelli, D., Pennington, G.L., Trevaks, D., and McAllen, R.M. (2018). Integrating competing demands of osmoregulatory and thermoregulatory homeostasis. *Physiology*, 33(3), 170–181.

McKinney, F. (1985). Primary and secondary male reproductive strategies of dabbling ducks. *Avian Monogamy*, 37, 68–82.

McKinney, M.L. and McNamara, K.J. (1991). Heterochrony. In *Heterochrony: the Evolution of Ontogeny* (M.L. McKinney and K.J. McNamara, eds, pp. 1–12). Springer, Boston, MA.

McKitrick, M. (1993). Phylogenetic constraint in evolutionary theory: has it any explanatory power? *Annual Review of Ecology, Evolution, and Systematics*, 24, 307–330.

McLaren, G., Bonacic, C., and Rowan, A. (2007). *Animal Welfare and Conservation: Measuring Stress in the Wild*. Blackwell Publishing, Melbourne.

McLaren, G.W., Macdonald, D.W., Georgiou, C., Mathews, F., Newman, C., and Mian, R. (2003). Leukocyte coping capacity: a novel technique for measuring the stress response in vertebrates. *Experimental Physiology*, 88(4), 541–546.

McLaren, G.W., Thornton, P.D., Newman, C., Buesching, C.D., Baker, S.E., Mathews, F., and Macdonald, D.W. (2005). The use and assessment of ketamine–medetomidine–butorphanol combinations for field anaesthesia in wild European badgers (*Meles meles*). *Veterinary Anaesthesia and Analgesia*, 32(6), 367–372.

McLaren, I.A. (2017). *Natural Regulation of Animal Populations*. Routledge, Abingdon.

McLeod, D.V. and Day, T. (2019). Why is sterility virulence most common in sexually transmitted infections? Examining the role of epidemiology. *Evolution*, 73(5), 872–882.

Mcloughlin, P.D., Ferguson, S.H., and Messier, F. (2000). Intraspecific variation in home range overlap with habitat quality: a comparison among brown bear populations. *Evolutionary Ecology*, 14(1), 39–60.

McNab, B.K. (1963). Bioenergetics and the determination of home range size. *American Naturalist*, 97, 133–139.

McNab, B.K. (1979). The influence of body size on the energetics and distribution of fossorial and burrowing mammals. *Ecology*, 60, 1010–1021. doi:10.2307/1936869

McNamara, J.M., Houston, A.I., Barta, Z., Scheuerlein, A., and Fromhage, L. (2009). Deterioration, death and the evolution of reproductive restraint in late life. *Proceedings of the Royal Society. Series B: Biological Sciences*, 276(1675), 4061–4066.

McNamara, K.J. (2012). Heterochrony: the evolution of development. *Evolution: Education and Outreach*, 5(2), 203–218.

Mead, R.A. (1981). Delayed implantation in mustelids, with special emphasis on the spotted skunk. *Journal of Reproduction and Fertility. Supplement*, 29, 11–24.

Mead, R.A. (1989). The physiology and evolution of delayed implantation in carnivores. In *Carnivore Behavior, Ecology, and Evolution* (J.L. Gittelman, ed., pp. 437–464). Springer, Boston, MA.

Mead, R.A. (1993). Embryonic diapause in vertebrates. *Journal of Experimental Zoology*, 266(6), 629–641.

Mead, R.A. and Eik-Nes, K.B. (1969). Oestrogen levels in peripheral blood plasma of the spotted skunk. *Reproduction*, 18(2), 351–353.

Mead, R.A., Seal, U.S., Thorne, E.T., and Bogan, M.A. (1989). Reproduction in mustelids. In *Conservation Biology and the Black-footed Ferret* (U.S. Seal, ed., pp. 124–137). Yale University Press, New Haven, CT.

Mech, L.D. (1977). Wolf-pack buffer zones as prey reservoirs. *Science*, 198(4314), 320–321.

Mech, L.D., Nelson, M.E., and McRoberts, R.E. (1991). Effects of maternal and grandmaternal nutrition on deer mass and vulnerability to wolf predation. *Journal of Mammalogy*, 72(1), 146–151.

Medawar, P.B. (1952). On growing old. *Nature*, 170(4320), 260–260.

Meerlo, P., Bolle, L., Visser, G.H., Masman, D., and Daan, S. (1997). Basal metabolic rate in relation to body composition and daily energy expenditure in the field vole, *Microtus agrestis*. *Physiological Zoology*, 70(3), 362–369.

Mehlman, P.T. and Chapais, B. (1988). Differential effects of kinship, dominance, and the mating season on female allogrooming in a captive group of *Macaca fuscata*. *Primates*, 29(2), 195–217.

Meier, U. and Gressner, A.M. (2004). Endocrine regulation of energy metabolism: review of pathobiochemical and clinical chemical aspects of leptin, ghrelin, adiponectin, and resistin. *Clinical Chemistry*, 50(9), 1511–1525.

Mendez-Bermudez, A., Hidalgo-Bravo, A., Cotton, V.E., Gravani, A., Jeyapalan, J.N., and Royle, N.J. (2012). The roles of WRN and BLM RecQ helicases in the alternative lengthening of telomeres. *Nucleic Acids Research*, 40(21), 10809–10820. https://doi.org/10.1093/nar/gks862

Meng, Z., Liu, M., Zhang, Q., Liu, L., Song, K., Tan, J., Jia, Q., Zhang, G., Wang, R., He, Y., and Ren, X. (2015). Gender and age impact on the association between thyroid-stimulating hormone and serum lipids. *Medicine*, 94(49), e2186.

Merilä, J. and Hendry, A.P. (2014). Climate change, adaptation, and phenotypic plasticity: the problem and the evidence. *Evolutionary Applications*, 7(1), 1–14.

Mery, F. and Burns, J.G. (2010). Behavioural plasticity: an interaction between evolution and experience. *Evolutionary Ecology*, 24, 571–583. https://doi.org/10.1007/s10682-009-9336-y

Metcalfe, N.B., Van Leeuwen, T.E., and Killen, S.S. (2016). Does individual variation in metabolic phenotype predict fish behaviour and performance? *Journal of Fish Biology*, 88(1), 298–321.

Methion, S. and Díaz López, B. (2019). Individual foraging variation drives social organization in bottlenose dolphins. *Behavioral Ecology*, 31, 97–106. doi:10.1093/beheco/arz160

Meunier, J. (2015). Social immunity and the evolution of group living in insects. *Philosophical Transactions of the Royal Society. Series B: Biological Sciences*, 370, 19–21. doi:10.1098/rstb.2014.0102

Milinski, M. (2006). The major histocompatibility complex, sexual selection, and mate choice. *Annual Review of Ecology, Evolution, and Systematics*, 37, 159–186.

Milinski, M. and Parker, G.A. (1991) Competition for resources. In *Behavioural Ecology: an Evolutionary Approach*, 3rd edn (J.R. Krebs and N.B. Davies, eds, pp. 137–168). Blackwell Scientific, Oxford.

Milinski, M., Griffiths, S., Wegner, K.M., Reusch, T.B., Haas-Assenbaum, A., and Boehm, T. (2005). Mate choice decisions of stickleback females predictably modified by MHC peptide ligands. *Proceedings of the National Academy of Sciences*, 102(12), 4414–4418.

Mills, L.S., Zimova, M., Oyler, J., Running, S., Abatzoglou, J.T., and Lukacs, P.M. (2013). Camouflage mismatch in seasonal coat color due to decreased snow duration. *Proceedings of the National Academy of Sciences*, 110(18), 7360–7365.

Mills, M.G. (1982). Factors affecting group size and territory size of the brown hyaena, *Hyaena brunnea* in the southern Kalahari. *Journal of Zoology*, 198(1), 39–51.

Mills, M.G.L. (1990). *Kalahari Hyaenas: Comparative Behavioral Ecology of Two Species*. Unwin Hyman, London.

Millspaugh, J.J., Coleman, M.A., Bauman, P.J., Raedeke, K.J., and Brundige, G.C. (2000). Serum profiles of American elk, *Cervus elaphus*, at the time of handling for three capture methods. *Canadian Field-Naturalist*, 114(2), 196–200.

Millstein, R.L. (2002). Are random drift and natural selection conceptually distinct? *Biology and Philosophy*, 17(1), 33–53.

Minta, S.C., Minta, K.A., and Lott, D.F. (1992). Hunting associations between badgers (*Taxidea taxus*) and coyotes (*Canis latrans*). *Journal of Mammalogy*, 73(4), 814–820.

Mittal, M., Siddiqui, M.R., Tran, K., Reddy, S.P., and Malik, A.B. (2014). Reactive oxygen species in inflammation and tissue injury. *Antioxidants & Redox Signaling*, 20(7), 1126–1167.

Miyazawa, E., Seguchi, A., Takahashi, N., Motai, A., and Izawa, E. (2020). Different patterns of allopreening in the same-sex and opposite-sex interactions of juvenile large-billed crows (*Corvus macrorhynchos*) *Ethology*, 126(2), 195–206. https://doi.org/10.1111/eth.12992.

Moehlman, P.D. (1979). Jackal helpers and pup survival. *Nature*, 277(5695), 382–383.

Moll, R.J., Kilshaw, K., Montgomery, R.A., Abade, L., Campbell, R.D., Harrington, L.A., Millspaugh, J.J., Birks, J.D.S., and Macdonald, D.W. (2016). Clarifying habitat niche width using broad-scale, hierarchical occupancy models: a case study with a recovering mesocarnivore. *Journal of Zoology*, 300(3), 177–185.

Møller, A.P. (1988). Ejaculate quality, testes size and sperm competition in primates. *Journal of Human Evolution*, 17(5), 479–488.

Møller, A.P. and Alatalo, R.V. (1999). Good-genes effects in sexual selection. *Proceedings of the Royal Society of London. Series B: Biological Sciences*, 266(1414), 85–91.

Møller, A.P. and Birkhead, T.R. (1989). Copulation behaviour in mammals: evidence that sperm competition is widespread. *Biological Journal of the Linnean Society*, 38(2), 119–131.

Møller, A.P. and Birkhead, T.R. (1993). Cuckoldry and sociality: a comparative study of birds. *The American Naturalist*, 142(1), 118–140.

Møller, A.P. and Höglund, J. (1991). Patterns of fluctuating asymmetry in avian feather ornaments: implications for models of sexual selection. *Proceedings of the Royal Society of London. Series B: Biological Sciences*, 245(1312), 1–5.

Monaghan, P. (2008). Early growth conditions, phenotypic development and environmental change. *Philosophical Transactions of the Royal Society. Series B: Biological Sciences*, 363(1497), 1635–1645.

Monaghan, P. and Ozanne, S.E. (2018). Somatic growth and telomere dynamics in vertebrates: relationships, mechanisms and consequences. *Philosophical Transactions of the Royal Society. Series B: Biological Sciences*, 373(1741), 20160446.

Monteith, J. and Unsworth, M. (2013). *Principles of Environmental Physics: Plants, Animals, and the Atmosphere*. Academic Press, Cambridge, MA.

Monteith, J.L. (1963). Gas exchange in plant communities. *Environmental Control of Plant Growth*, 95, 95–112.

Monteith, K.L., Stephenson, T.R., Bleich, V.C., Conner, M.M., Pierce, B.M., and Bowyer, R.T. (2013). Risk-sensitive allocation in seasonal dynamics of fat and protein reserves in a long-lived mammal. *Journal of Animal Ecology*, 82(2), 377–388.

Montes, I., McLaren, G.W., Macdonald, D.W., and Mian, R. (2003). The effects of acute stress on leukocyte activation. *Journal of Physiology*, 548(170), 3.

Montes, I., McLaren, G.W., Macdonald, D.W., and Mian, R. (2004). The effect of transport stress on neutrophil activation in wild badgers (*Meles meles*). *Animal Welfare*, 13(3), 355–359.

Montes, I., Newman, C., Mian, R., and Macdonald, D.W. (2011). Radical health: ecological corollaries of body condition, transport stress and season on plasma antioxidant capacity in the European badger. *Journal of Zoology*, 284(2), 114–123.

Montgomery, R.A. and Macdonald, D.W. (2020). COVID-19, health, conservation, and shared wellbeing: details matter. *Trends in Ecology & Evolution*, 35(9), 748–750.

Montgomery, R.A., Macdonald, D.W., and Hayward, M.W. (2020). The inducible defences of large mammals to human lethality. *Functional Ecology*, 34(12), 2426–2441.

Montiglio, P.O., Dammhahn, M., Messier, G.D., and Réale, D. (2018). The pace-of-life syndrome revisited: the role of ecological conditions and natural history on the slow-fast continuum. *Behavioral Ecology and Sociobiology*, 72(7), 1–9.

Moody, O. (2013). Controversial badger culling to go ahead this summer. https://www.thetimes.co.uk/article/controversial-badger-culling-to-go-ahead-this-summer-h9hm6m6lkht (accessed 12 May 2022).

Moore, S.E. and Huntington, H.P. (2008). Arctic marine mammals and climate change: impacts and resilience. *Ecological Applications*, 18(sp2), S157–S165.

Moorhouse, T.P., Gelling, M., and Macdonald, D.W. (2015). Water vole restoration in the Upper Thames. In *Wildlife Conservation on Farmland Volume 1: Managing for Nature on Lowland Farms* (D.W. Macdonald and R.E. Feber, eds, pp. 255–268). Oxford University Press, Oxford.

Mooring, M.S. and Hart, B.L. (1992). Animal grouping for protection from parasites: selfish herd and encounter-dilution effects. *Behaviour*, 123(3–4), 173–193.

Mooring, M.S. and Hart, B.L. (1995). Costs of allogrooming in impala: distraction from vigilance. *Animal Behaviour*, 49(5), 1414–1416.

Moors, P.J. (1980). Sexual dimorphism in the body size of mustelids (Carnivora): the roles of food habits and breeding systems. *Oikos*, 34(2), 147–158.

More, S.J. (2019). Can bovine TB be eradicated from the Republic of Ireland? Could this be achieved by 2030? *Irish Veterinary Journal*, 72(1), 1–10.

More, S. and McGrath, G. (2015). Randomised badger culling trial: interpreting the results. *Veterinary Record*, August, 129.

Moreira, J.R.A., Clarke, J.R., and Macdonald, D.W. (1997a). The testis of capybaras (*Hydrochoerus hydrochaeris*). *Journal of Mammalogy*, 78(4), 1096–1100.

Moreira, J.R.A., Macdonald, D.W., and Clarke, J.R. (1997b). Correlates of testis mass in capybaras (*Hydrochaeris hydrochaeris*): dominance assurance or sperm production? *Journal of Zoology*, 241, 457–463.

Mori, E., Menchetti, M., and Balestrieri, A. (2015). Interspecific den sharing: a study on European badger setts using camera traps. *Acta Ethologica*, 18(2), 121–126.

Morton, E.S. (1977). On the occurrence and significance of motivation-structural rules in some bird and mammal sounds. *The American Naturalist*, 111(981), 855–869.

Moshkin, M., Litvinova, N., Litvinova, E.A., Bedareva, A., Lutsyuk, A., and Gerlinskaya, L. (2012). Scent recognition of infected status in humans. *The Journal of Sexual Medicine*, 9(12), 3211–3218.

Mousseau, T.A. and Roff, D.A. (1987). Natural selection and the heritability of fitness components. *Heredity*, 59(2), 181–197.

Mowat, G. and Heard, D.C. (2006). Major components of grizzly bear diet across North America. *Canadian Journal of Zoology*, 84(3), 473–489.

Mueller, H.C. (1990). The evolution of reversed sexual dimorphism in size in monogamous species of birds. *Biological Reviews*, 65(4), 553–585.

Müller, C.A. and Manser, M.B. (2007). 'Nasty neighbours' rather than 'dear enemies' in a social carnivore. *Proceedings of the Royal Society. Series B: Biological Sciences*, 274(1612), 959–965.

Mullur, R., Liu, Y.Y., and Brent, G.A. (2014). Thyroid hormone regulation of metabolism. *Physiological Reviews*, 94(2), 355–382.

Munger, J.C. and Karasov, W.H. (1994). Cost of the bot fly infection in white-footed mice: energy and mass flow. *Canadian Journal of Zoology*, 72, 166–173.

Murdoch, J.D. and Buyandelger, S. (2010). An account of badger diet in an arid steppe region of Mongolia. *Journal of Arid Environments*, 74(10), 1348–1350.

Murdoch, J.D., Munkhzul, T., Amgalanbaatar, S., and Reading, R.P. (2006). Checklist of mammals in Ikh Nart Nature Reserve. *Mongolian Journal of Biological Sciences*, 4(2), 69–74.

Murdoch, W.W. (1994). Population regulation in theory and practice. *Ecology*, 75(2), 271–287.

Murie, A. (1944). *The Wolves of Mount McKinley* (No. 5). US Government Printing Office.

Murphy, J.M., Sexton, D.M.H., Jenkins, G.J., Booth, B.B., Brown, C.C., Clark, R.T., Collins, M., Harris, G.R., Kendon, E.J., Betts, R.A., and Brown, S.J. (2009). UK climate projections science report: climate change projections. Met Office Hadley Centre, Exeter.

Murphy, S., Collet, A., and Rogan, E. (2005). Mating strategy in the male common dolphin (*Delphinus delphis*): what gonadal analysis tells us. *Journal of Mammalogy*, 86(6), 1247–1258.

Murren, C.J., Auld, J.R., Callahan, H., Ghalambor, C.K., Handelsman, C.A., Heskel, M.A., and Pfennig, D.W. (2015). Constraints on the evolution of phenotypic plasticity: limits and costs of phenotype and plasticity. *Heredity*, 115(4), 293–301.

Nadin, C.E., Macdonald, D.W., Baker, S.E., Buesching, C.D., Ellwood, S., Newman, C., and H.L. Dugdale (2021). Unreciprocated allogrooming hierarchies in a population of wild group-living mammals. Submitted.

Nadler, T., Streicher, U., Stefen, C., Schwierz, E., and Roos, C. (2011). A new species of ferret-badger, Genus *Melogale*, from Vietnam. *Der Zoologische Garten*, 80(5), 271–286.

Nakanishi, M.O., Hayakawa, K., Nakabayashi, K., Hata, K., Shiota, K., and Tanaka, S. (2012). Trophoblast-specific DNA methylation occurs after the segregation of the trophectoderm and inner cell mass in the mouse periimplantation embryo. *Epigenetics*, 7(2), 173–182.

Nath, L. (2000). Conservation management of the tiger, *Panthera tigris tigris*, in Bandhavgarh National Park, India. PhD thesis, University of Oxford.

Nathan, C. (2006). Neutrophils and immunity: challenges and opportunities. *Nature Reviews Immunology*, 6(3), 173–182.

National Farmers' Union (NFU) (n.d.). Government report confirms TB strategy retains culling option. https://www.nfuonline.com/news/latest-news/government-report-confirms-tb-strategy-retains-culling-option/ (accessed 10 March 2022).

Natural England (2019). Summary of badger vaccination in 2020. https://www.gov.uk/government/publications/bovine-tb-summary-of-badger-control-monitoring-during-2020/summary-of-badger-vaccination-in-2020 (accessed 10 March 2022).

Natural England (2020). Bovine TB: authorisation for badger control in 2020. https://www.gov.uk/government/publications/bovine-tb-authorisation-for-badger-control-in-2020 (accessed 10 March 2022).

Natural England (2021a). Policy paper. Summary of badger vaccination in 2020. https://www.gov.uk/government/publications/bovine-tb-summary-of-badger-control-monitoring-during-2020/summary-of-badger-vaccination-in-2020 (accessed 10 March 2022).

Natural England (2021b). Policy paper. Summary of 2020 badger control operations. https://www.gov.uk/government/publications/bovine-tb-summary-of-badger-control-monitoring-during-2020/summary-of-2020-badger-control-operations (accessed 10 March 2022).

Neal, E. (1948). *The Badger*. Collins, London.

Neal, E. (1977). *Badgers*. Blandford Press, London.

Neal, E. (1986). *Natural History of Badgers*. Croom Helm, London.

Neal, E.G. and Cheeseman, C.L. (1996). *Badgers*. T & AD Poyser Ltd, London.

Neal, E.G. and Harbison, R.J. (1958). Reproduction in the European badger (*Meles meles* L.). *The Transactions of the Zoological Society of London*, 29(2), 67–130.

Neff, B.D. (2004). Stabilizing selection on genomic divergence in a wild fish population. *Proceedings of the National Academy of Sciences*, 101(8), 2381–2385.

Neff, B.D. and Pitcher, T.E. (2005). Genetic quality and sexual selection: an integrated framework for good genes and compatible genes. *Molecular Ecology*, *14*(1), 19–38.

Neofytou, E., Sourvinos, G., Asmarianaki, M., Spandidos, D.A., and Makrigiannakis, A. (2009). Prevalence of human herpes virus types 1–7 in the semen of men attending an infertility clinic and correlation with semen parameters. *Fertility and Sterility*, *91*(6), 2487–2494.

Neri, I., Bassi, A., and Patrizi, A. (2015). Streptococcal intertrigo. *The Journal of Pediatrics*, *166*(5), 1318.

Nettle, D., Frankenhuis, W.E., and Rickard, I.J. (2013). The evolution of predictive adaptive responses in human life history. *Proceedings of the Royal Society. Series B: Biological Sciences*, *280*(1766), 20131343. doi:10.1098/rspb.2013.1343

Nettle, D., Monaghan, P., Gillespie, R., Brilot, B., Bedford, T., and Bateson, M. (2015). An experimental demonstration that early-life competitive disadvantage accelerates telomere loss. *Proceedings of the Royal Society. Series B: Biological Sciences*, *282*(1798), 20141610.

Newell, D.G., Clifton-Hadley, R.S., and Cheeseman, C.L. (1997). The kinetics of serum antibody responses to natural infections with *Mycobacterium bovis* in one badger social group. *Epidemiology & Infection*, *118*(2), 173–180.

Newman, C. (2000). The demography and parasitology of the Wytham Woods badger population. DPhil thesis, University of Oxford.

Newman, C. and Buesching, C.D. (2018). Mustelidae cognition. In *Encyclopedia of Animal Cognition and Behavior* (J. Vonk and T.K. Shackelford, eds). Springer International Publishing AG. doi:10.1007/978-3-319-47829-6_1193-1

Newman, C. and Buesching, C.D. (2019). Detecting the smell of disease and injury: scoping evolutionary and ecological implications. In *Chemical Signals in Vertebrates 14* (C.D. Buesching, ed., pp. 238–250). Springer, Cham.

Newman, C. and Byrne, A. (2017). Musteloid diseases: Implications for conservation and species management. In *Biology and Conservation of Musteloids* Newman, C. and Byrne, A. (2017). Musteloid diseases: implications for conservation and species management. In *Biology and Conservation of Musteloids* (D.W. Macdonald, C. Newman, and L.A. Harrington, eds, pp. 231–255). Oxford University Press, Oxford.

Newman, C. and Macdonald, D.W. (2015). The implications of climate change for terrestrial UK Mammals. Terrestrial biodiversity Climate change impacts report card Technical paper 2. Natural Environment Research Council (NERC), Swindon.

Newman, C., Macdonald, D.W., and Anwar, M.A. (2001). Coccidiosis in the European badger, *Meles meles* in Wytham Woods: infection and consequences for growth and survival. *Parasitology*, *123*(2), 133.

Newman, C., Buesching, C.D., and Macdonald, D.W. (2003). Validating mammal monitoring methods and assessing the performance of volunteers in wildlife conservation—'Sed quis custodiet ipsos custodies?' *Biological Conservation*, *113*(2), 189–197.

Newman, C., Buesching, C.D., and Wolff, J.O. (2005). The function of facial masks in 'midguild' carnivores. *Oikos*, *108*(3), 623–633.

Newman, C., Zhou, Y.B., Buesching, C.D., Kaneko, Y., and Macdonald, D.W. (2011). Contrasting sociality in two widespread, generalist, mustelid genera, *Meles* and *Martes*. *Mammal Study*, *36*(4), 169–188.

Newman, C., Noonan, M., and Buesching, C. (2016). Fair-weather badgers: how appearances can be deceptive in climate change ecology. https://www.ox.ac.uk/news/science-blog/fair-weather-badgers-how-appearances-can-be-deceptive-climate-change-ecology (accessed 6 May 2022).

Newman, C., Buesching, C.D., and Macdonald, D.W. (2017). Meline mastery of meteorological mayhem: the effects of climate changeability on European badger population dynamics. In *Biology and Conservation of Musteloids* (D.W. Macdonald, C. Newman, and L.A. Harrington, eds, pp. 420–433). Oxford University Press, Oxford .

Newsome, T.M., Ballard, G.A., Dickman, C.R., Fleming, P.J., and van de Ven, R. (2013). Home range, activity and sociality of a top predator, the dingo: a test of the Resource Dispersion Hypothesis. *Ecography*, *36*(8), 914–925.

Newton-Fisher, N., Harris, S., White, P., and Jones, G. (1993). Structure and function of red fox *Vulpes vulpes* vocalisations. *Bioacoustics*, *5*(1–2), 1–31.

Nickerson, R.S. (1998). Confirmation bias: a ubiquitous phenomenon in many guises. *Review of General Psychology*, *2*(2), 175.

Nicolas de Francisco, O., Esperón, F., Juan-Sallés, C., Ewbank, A.C., Das Neves, C.G., Marco, A., Neves, E., Anderson, N., and Sacristán, C. (2020). Neoplasms and novel gamma herpesviruses in critically endangered captive European minks (*Mustela lutreola*). Transboundary and Emerging Diseases, *68*(2), 552–564.

Nielsen, C.L.R. and Nielsen, C.K. (2007). Multiple paternity and relatedness in southern Illinois raccoons (*Procyon lotor*). *Journal of Mammalogy*, *88*(2), 441–447.

Nietzsche, F.W. (1990). *The Anti-Christ*, translated by R.J. Hollingdale. Penguin Books, London.

Niiniluoto, I. (2014). Scientific progress as increasing verisimilitude. *Studies in History and Philosophy of Science Part A*, *46*, 73–77.

Nilsson, A.L., Nilsson, J.Å., and Mettke-Hofmann, C. (2016). Energy reserves, information need and a pinch of personality determine decision-making on route in partially migratory blue tits. *PLoS ONE*, *11*(10), e0163213.

Nilsson, J.F. and Nilsson, J.Å. (2016). Fluctuating selection on basal metabolic rate. *Ecology and Evolution*, *6*(4), 1197–1202.

Nippert, J.B., Knapp, A.K., and Briggs, J.M. (2006). Intra-annual rainfall variability and grassland productivity: can the past predict the future? *Plant Ecology*, *184*, 65–74.

Noë, R. and Hammerstein, P. (1994). Biological markets: supply and demand determine the effect of partner choice in cooperation, mutualism and mating. *Behavioral Ecology and Sociobiology*, *35*(1), 1–11.

Noë, R. and Hammerstein, P. (1995). Biological markets. *Trends in Ecology & Evolution*, 10(8), 336–339. doi:10.1016/S0169-5347(00)89123-89125

Noë, R., Van Schaik, C.P., and Van Hooff, J. (1991). The market effect: an explanation for pay-off asymmetries among collaborating animals. *Ethology*, 87(1), 97–118.

Noguera, J.C., Metcalfe, N.B., and Monaghan, P. (2018). Experimental demonstration that offspring fathered by old males have shorter telomeres and reduced lifespans. *Proceedings of the Royal Society. Series B: Biological Sciences*, 285(1874), 20180268.

Nohl, H. and Gille, L. (2005). Lysosomal ROS formation. *Redox Report*, 10(4), 199–205.

Nol, E. and Smith, J.N. (1987). Effects of age and breeding experience on seasonal reproductive success in the song sparrow. *The Journal of Animal Ecology*, 56(1), 301–313.

Noonan, M.J., Markham, A., Newman, C., Trigoni, N., Buesching, C.D., Ellwood, S.A., and Macdonald, D.W. (2014). Climate and the individual: inter-annual variation in the autumnal activity of the European badger (*Meles meles*). *PLoS ONE*, 9(1), e83156.

Noonan, M.J., Newman, C., Buesching, C.D., and Macdonald, D.W. (2015a). Evolution and function of fossoriality in the Carnivora: implications for group-living. *Frontiers in Ecology and Evolution*, 3, 116.

Noonan, M.J., Markham, A., Newman, C., Trigoni, N., Buesching, C.D., Ellwood, S.A., and Macdonald, D.W. (2015b). A new magneto-inductive tracking technique to uncover subterranean activity: what do animals do underground? *Methods in Ecology and Evolution*, 6(5), 510–520.

Noonan, M.J., Rahman, M.A., Newman, C., Buesching, C.D., and Macdonald, D.W. (2015d). Avoiding verisimilitude when modelling ecological responses to climate change: the influence of weather conditions on trapping efficiency in European badgers (*Meles meles*). *Global Change Biology*, 21(10), 3575–3585.

Noonan, M.J., Newman, C., Markham, A., Bilham, K., Buesching, C.D., and Macdonald, D.W. (2018a). In situ behavioral plasticity as compensation for weather variability: implications for future climate change. *Climatic Change*, 149(3–4), 457–471.

Noonan, M.J., Tinnesand, H.V., and Buesching, C.D. (2018b). Normalizing gas-chromatography–mass spectrometry data: method choice can alter biological inference. *BioEssays*, 40(6), 1700210.

Norris, K. and Evans, M.R. (2000). Ecological immunology: life history trade-offs and immune defense in birds. *Behavioral Ecology*, 11(1), 19–26.

Nouvellet, P., Buesching, C.D., Dugdale, H.L., Newman, C., and Macdonald, D.W. (2011). Mouthing off about developmental stress: individuality of palate marking in the European badger and its relationship with juvenile parasitoses. *Journal of Zoology*, 283(1), 52–62.

Nouvellet, P., Newman, C., Buesching, C.D., and Macdonald, D.W. (2013). A multi-metric approach to investigate the effects of weather conditions on the demographic of a terrestrial mammal, the European badger (*Meles meles*). *PLoS ONE*, 8(7), e68116.

Novak, M. and Highfield, R. (2011). *Supercooperators*. Simon & Schuster, London.

Nowack, J., Tarmann, I., Hoelzl, F., Smith, S., Giroud, S., and Ruf, T. (2019). Always a price to pay: hibernation at low temperatures comes with a trade-off between energy savings and telomere damage. *Biology Letters*, 15(10), 20190466.

Nowak, M.A., Tarczy-Hornoch, K., and Austyn, J.M. (1992). The optimal number of major histocompatibility complex molecules in an individual. *Proceedings of the National Academy of Sciences*, 89(22), 10896–10899.

Nowak, M.A., Tarnita, C.E., and Wilson, E.O. (2010). The evolution of eusociality. *Nature*, 466(7310), 1057–1062.

Nunez-Garcia, J., Downs, S.H., Parry, J.E., Abernethy, D.A., Broughan, J.M., Cameron, A.R., Cook, A.J., de La Rua-domenech, R., Goodchild, A.V., Gunn, J., and More, S.J. (2018). Meta-analyses of the sensitivity and specificity of ante-mortem and post-mortem diagnostic tests for bovine tuberculosis in the UK and Ireland. *Preventive Veterinary Medicine*, 153, 94–107.

Nussey, D.H., Coulson, T., Festa-Bianchet, M., and Gaillard, J.M. (2008). Measuring senescence in wild animal populations: towards a longitudinal approach. *Functional Ecology*, 22(3), 393–406.

Nussey, D.H., Kruuk, L.E., Morris, A., Clements, M.N., Pemberton, J.M., and Clutton-Brock, T.H. (2009). Inter-and intrasexual variation in aging patterns across reproductive traits in a wild red deer population. *The American Naturalist*, 174(3), 342–357.

Nyakatura, K. and Bininda-Emonds, O.R. (2012). Updating the evolutionary history of Carnivora (Mammalia): a new species-level supertree complete with divergence time estimates. *BMC Biology*, 10(1), 1–31.

Nybo, L., Rasmussen, P., and Sawka, M.N. (2011). Performance in the heat—physiological factors of importance for hyperthermia-induced fatigue. *Comprehensive Physiology*, 4, 657–689.

Nybo, L., Schmidt, J.F., Fritzdorf, S., and Nordsborg, N.B. (2014). Physiological characteristics of an aging Olympic athlete. *Medicine and Science in Sports and Exercise*, 46(11), 2132–2138.

Nylin, S. and Gotthard, K. (1998). Plasticity in life-history traits. *Annual Review of Entomology*, 43(1), 63–83.

O'Hare, A., Balaz, D., Wright, D.M., McCormick, C., Skuce, R.A., and Kao, R.R. (2021). A new phylodynamic model of *Mycobacterium bovis* transmission in a multi-host system uncovers the role of the unobserved reservoir. *PLOS Computational Biology*, 17(6), e1009005. https://doi.org/10.1371/journal.pcbi.1009005.

O'Mahony, D.T. (2015). Badger (*Meles meles*) contact metrics in a medium-density population. *Mammalian Biology*, 80(6), 484–490.

O'Meara, D.B., Edwards, C.J., Sleeman, D.P., Cross, T.F., Statham, M.J., McDowell, J.R., Dillane, E., Coughlan, J.P., O'Leary, D., O'Reilly, C., and Bradley, D.G. (2012). Genetic structure of Eurasian badgers *Meles meles* (Carnivora: Mustelidae) and the colonization history of Ireland. *Biological Journal of the Linnean Society*, 106(4), 893–909.

Ober, C., Steck, T., van der Ven, K., Billstrand, C., Messer, L., Kwak, J., Beaman, K., and Beer, A. (1993). MHC class II compatibility in aborted fetuses and term infants of couples with recurrent spontaneous abortion. *Journal of Reproductive Immunology*, 25(3), 195–207.

Oelkrug, R., Polymeropoulos, E.T., and Jastroch, M. (2015). Brown adipose tissue: physiological function and evolutionary significance. *Journal of Comparative Physiology B*, 185(6), 587–606.

Ogura, M., Wakaiki, M., Rubin, H., and Preciado, V.M. (2017). Delayed bet-hedging resilience strategies under environmental fluctuations. *Physical Review E*, 95, 052404.

Oikawa, S. and Kawanishi, S. (1999). Site-specific DNA damage at GGG sequence by oxidative stress may accelerate telomere shortening. *FEBS Letters*, 453(3), 365–368.

Oli, M.K. (2004). The fast–slow continuum and mammalian life-history patterns: an empirical evaluation. *Basic and Applied Ecology*, 5(5), 449–463.

Oli, M.K. and Dobson, F.S. (2003). The relative importance of life-history variables to population growth rate in mammals: Cole's prediction revisited. *The American Naturalist*, 161(3), 422–440.

Oli, M.K. and Zinner, B. (2001). Partial life-cycle analysis: a model for birth-pulse populations. *Ecology*, 82(4), 1180–1190.

Oliver, T.H. and Morecroft, M.D. (2014). Interactions between climate change and land use change on biodiversity: attribution problems, risks, and opportunities. *Wiley Interdisciplinary Reviews: Climate Change*, 5(3), 317–335.

Ollivier, F.J., Samuelson, D.A., Brooks, D.E., Lewis, P.A., Kallberg, M.E., and Komáromy, A.M. (2004). Comparative morphology of the tapetum lucidum (among selected species). *Veterinary Ophthalmology*, 7(1), 11–22.

Olsson, M.J., Lundström, J.N., Kimball, B.A., Gordon, A.R., Karshikoff, B., Hosseini, N., Sorjonen, K., Olgart Höglund, C., Solares, C., Soop, A., and Axelsson, J. (2014). The scent of disease: human body odor contains an early chemosensory cue of sickness. *Psychological Science*, 25(3), 817–823.

Orr, T.J. (2012). The biology of reproductive delays in mammals: reproductive decisions, energetics, and evolutionary ecology. PhD thesis, UC Riverside.

Orr, T.J. and Zuk, M. (2014). Reproductive delays in mammals: an unexplored avenue for post-copulatory sexual selection. *Biological Reviews*, 89(4), 889–912.

Overy, C. and Tansey, E.M. (eds) (2015) *A History of Bovine TB c.1965–c.2000. Wellcome Witnesses to Contemporary Medicine*, vol. 55. Queen Mary University of London, London.

Owen-Smith, N. (1977). On territoriality in ungulates and an evolutionary model. *The Quarterly Review of Biology*, 52(1), 1–38.

Ozturk, S. (2015). Telomerase activity and telomere length in male germ cells. *Biology of Reproduction*, 92(2), 53. doi:10.1095/biolreprod.114.124008.

Pacifici, M., Foden, W.B., Visconti, P., Watson, J.E., Butchart, S.H., Kovacs, K. M. et al. (2015). Assessing species vulnerability to climate change. *Nature Climate Change*, 5(3), 215–224.

Packer, C. and Pusey, A.E. (1993). Should a lion change its spots? *Nature*, 362(6421), 595.

Packer, C., Gilbert, D.A., Pusey, A.E., and O'Brieni, S.J. (1991). A molecular genetic analysis of kinship and cooperation in African lions. *Nature*, 351(6327), 562–565.

Packer, C., Tatar, M., and Collins, A. (1998). Reproductive cessation in female mammals. *Nature*, 392(6678), 807–811.

Padodara, R.J. and Jacob, N. (2014). Olfactory sense in different animals. *Indian Journal of Veterinary Science*, 2(1), 1–14.

Page, R.E. and Erber, J. (2002). Levels of behavioral organization and the evolution of division of labor. *Naturwissenschaften*, 89(3), 91–106.

Page, R.J.C., Ross, J., and Langton, S.D. (1994). Seasonality of reproduction in the European badger *Meles meles* in southwest England. *Journal of Zoology*, 233(1), 69–91.

Paget, R.J. and Middleton, A.L.V. (1974a). *Badgers of Yorkshire and Humberside*. Ebor Press, York.

Paget, R.J. and Middleton, A.L.V. (1974b). Some observations on the sexual activities of badgers (*Meles meles*) in Yorkshire in the months December to April. *Journal of Zoology*, 173(2), 256–260.

Paiva, C.N. and Bozza, M.T. (2014). Are reactive oxygen species always detrimental to pathogens? *Antioxidants & Redox Signaling*, 20(6), 1000–1037. https://doi.org/10.1089/ars.2013.5447

Palme, R. (1997). Measurement of cortisol metabolites in faeces of sheep as a parameter of cortisol concentration in blood. *Mammalian Biology*, 62, 192–197.

Palme, R., Rettenbacher, S., Touma, C., El-Bahr, S.M., and Möstl, E. (2005). Stress hormones in mammals and birds: comparative aspects regarding metabolism, excretion, and noninvasive measurement in fecal samples. *Annals of the New York Academy of Sciences*, 1040(1), 162–171.

Palomares, F. and Caro, T.M. (1999). Interspecific killing among mammalian carnivores. *The American Naturalist*, 153(5), 492–508.

Palombit, R.A., Seyfarth, R.M., and Cheney, D.L. (1997). The adaptive value of 'friendships' to female baboons: experimental and observational evidence. *Animal Behaviour*, 54(3), 599–614.

Palphramand, K.L. and White, P.C.L. (2007). Badgers, *Meles meles*, discriminate between neighbour, alien and self scent. *Animal Behaviour*, 74, 429–436.

Palphramand, K.L., Newton-Cross, G., and White, P.C. (2007). Spatial organization and behaviour of badgers (*Meles meles*) in a moderate-density population. *Behavioral Ecology and Sociobiology*, 61(3), 401–413.

Papageorgiou, D. and Farine, D.R. (2020). Group size and composition influence collective movement in a highly social terrestrial bird. *Elife*, 9, e59902.

Papageorgiou, D., Christensen, C., Gall, G.E.C., Nyaguthii, B., Couzin, I.D., and Farine, D.R. (2019). The multi-level society of a small-brained bird. *Current Biology*, *29*(21), PR1120–PR1121. https://doi.org/10.1016/j.cub.2019.09.072

Parker, B.B. and Duszynski, D.W. (1986). Coccidiosis of sandhill cranes (*Grus canadensis*) wintering in New Mexico. *Journal of Wildlife Diseases*, *22*(1), 25–35.

Parker, G.A. (1974). Assessment strategy and the evolution of fighting behaviour. *Journal of Theoretical Biology*, *47*(1), 223–243.

Parker, G.A. (1990). Sperm competition games: raffles and roles . *Proceedings of the Royal Society of London. Series B: Biological Sciences*, *242*(1304), 120–126.

Parker, G.A. and Maynard Smith, J. (1990). Optimality theory in evolutionary biology. *Nature*, *348*(6296), 27–33.

Parmesan, C. (2006). Ecological and evolutionary responses to recent climate change. *Annual Review of Ecology, Evolution, and Systematics*, *37*, 637–669.

Parmesan, C., Root, T.L., and Willig, M.R. (2000). Impacts of extreme weather and climate on terrestrial biota. *Bulletin of the American Meteorological Society*, *81*(3), 443–450.

Parr, L.A., Matheson, M.D., Bernstein, I.S., and De Waal, F.B. (1997). Grooming down the hierarchy: allogrooming in captive brown capuchin monkeys, *Cebus apella* . *Animal Behaviour*, *54*(2), 361–367.

Parry, M.L., Canziani, O., Palutikof, J., Van der Linden, P., and Hanson, C. (eds) (2007). Climate Change 2007—Impacts, Adaptation and Vulnerability: Working Group II Contribution to the Fourth Assessment Report of the IPCC (Vol. 4). Cambridge University Press, Cambridge.

Paterson, D. and Ryder, R.D. (1977). *Animals' Rights: A Symposium*. Centaur Press Ltd, New York.

Patterson, M. (1998). Commensuration and theories of value in ecological economics. *Ecological Economics*, *25*, 105–125.

Patton, J.L. and Feder, J.H. (1981). Microspatial genetic heterogeneity in pocket gophers: non-random breeding and drift. *Evolution*, *35*(5), 912–920.

Pays, O., Ekori, A., and Fritz, H. (2014). On the advantages of mixed-species groups: impalas adjust their vigilance when associated with larger prey herbivores. *Ethology*, *120*(12), 1207–1216.

PBD Biotech (2021). Actiphage to offer greater decision support in the fight against bovine TB. https://www.pbdbio.com/news/actiphage-to-offer-greater-decision-support-in-the-fight-against-bovine-tb/ (accessed 10 March 2022).

Pearce, G.E. (2011). *Badger Behaviour, Conservation and Rehabilitation: 70 Years of Getting to Know Badgers*. Pelagic Publishing. Exeter.

Pearl, R. and Parker, S.L. (1921). Experimental studies on the duration of life. I. Introductory discussion of the duration of life in *Drosophila*. *The American Naturalist*, *55*(641), 481–509.

Pearson, R.G. and Dawson, T.P. (2003). Predicting the impacts of climate change on the distribution of species: are bioclimate envelope models useful? *Global Ecology and Biogeography*, *12*(5), 361–371.

Peccei, J.S. (2001). A critique of the grandmother hypotheses: old and new. *American Journal of Human Biology*, *13*(4), 434–452.

Pei, K. and Wang, Y. (1995). Some observations on the reproduction of the Taiwan ferret badger (*Melogale moschata subaurantiaca*) in southern Taiwan. *Zoological Studies*, *34*(2), 88–95.

Pellérdy, L.P. (1974). *Coccidia and Coccidiosis*, second edition. Verlag Paul Parey, Berlin and Hamburg.

Penn, D. and Potts, W. (1998a). How do major histocompatibility complex genes influence odor and mating preferences. *Advances in Immunology*, *69*(41), 1–436.

Penn, D. and Potts, W. (1998b). MHC-disassortative mating preferences reversed by cross-fostering. *Proceedings of the Royal Society of London. Series B: Biological Sciences*, *265*(1403), 1299–1306.

Penn, D.J. and Potts, W.K. (1999). The evolution of mating preferences and major histocompatibility complex genes. *The American Naturalist*, *153*(2), 145–164.

Penn, D.J. and Smith, K.R. (2007). Differential fitness costs of reproduction between the sexes. *Proceedings of the National Academy of Sciences*, *104*(2), 553–558.

Penn, D.J., Damjanovich, K., and Potts, W.K. (2002). MHC heterozygosity confers a selective advantage against multiple-strain infections. *Proceedings of the National Academy of Sciences*, *99*(17), 11260–11264.

Periquet, S., Mapendere, C., Revilla, E., Banda, J., Macdonald, D.W., Loveridge, A.J., and Fritz, H. (2016). A potential role for interference competition with lions in den selection and attendance by spotted hyenas. *Mammalian Biology*, *81*(3), 227–234.

Perrin, N. and Mazalov, V. (1999). Dispersal and inbreeding avoidance. *The American Naturalist*, *154*(3), 282–292.

Perrin, N. and Mazalov, V. (2000). Local competition, inbreeding, and the evolution of sex-biased dispersal. *The American Naturalist*, *155*(1), 116–127.

Pettersen, A.K., White, C.R., and Marshall, D.J. (2016). Metabolic rate covaries with fitness and the pace of the life history in the field. *Proceedings of the Royal Society. Series B: Biological Sciences*, *283*(1831), 20160323.

Pettersen, A.K., Marshall, D.J., and White, C.R. (2018). Understanding variation in metabolic rate. *Journal of Experimental Biology*, *221*(1), jeb166876.

Pettett, C.E., Johnson, P.J., Moorhouse, T.P., Hambly, C., Speakman, J.R., and Macdonald, D.W. (2017a). Daily energy expenditure in the face of predation: hedgehog energetics in rural landscapes. *The Journal of Experimental Biology*, *220*(3), 460–468. doi:10.1242/jeb.150359

Pettett, C.E., Moorhouse, T.P., Johnson, P.J., and Macdonald, D.W. (2017b). Factors affecting hedgehog (*Erinaceus europaeus*) attraction to rural villages in arable landscapes. *European Journal of Wildlife Research*, *63*(3), 54.

Pettett, C.E., Al-Hajri, A., Al-Jabiry, H., Macdonald, D.W., and Yamaguchi, N. (2018). A comparison of the ranging behaviour and habitat use of the Ethiopian hedgehog (*Paraechinus aethiopicus*) in Qatar with hedgehog taxa from temperate environments. *Scientific Reports*, 8(1), 1–10.

Philcox, C.K., Grogan, A.L., and Macdonald, D.W. (1999). Patterns of otter *Lutra lutra* road mortality in Britain. *Journal of Applied Ecology*, 36(5), 748–761.

Pianka, E.R. (1976). Natural selection of optimal reproductive tactics. *American Zoologist*, 16(4), 775–784.

Pigeon, G., Loe, L.E., Bischof, R., Bonenfant, C., Forchhammer, M., Irvine, R.J., Ropstad, E., Stien, A., Veiberg, V., and Albon, S. (2019). Silver spoon effects are constrained under extreme adult environmental conditions. *Ecology*, 100(12), e02886.

Pigozzi, G. (1992a). Behavioural ecology of the European badger (*Meles meles*): diet, food availability and use of space in the Maremma Natural Park, Central Italy. PhD thesis, University of Aberdeen.

Pigozzi, G. (1992b). Frugivory and seed dispersal by the European badger in a Mediterranean habitat. *Journal of Mammalogy*, 73(3), 630–639.

Pilley, J.W. and Reid, A.K. (2010). Border collie comprehends object names as verbal referents. *Behavioural Processes*, 86(2), 184–195. doi:10.1016/j.beproc.2010.11.007

Pitt, J.A., Larivière, S., and Messier, F. (2008). Social organization and group formation of raccoons at the edge of their distribution. *Journal of Mammalogy*, 89(3), 646–653.

Plaistow, S.J., Lapsley, C.T., Beckerman, A.P., and Benton, T.G. (2004). Age and size at maturity: sex, environmental variability and developmental thresholds. *Proceedings of the Royal Society of London. Series B: Biological Sciences*, 271(1542), 919–924.

Plant, T.M. (2006). The role of KiSS-1 in the regulation of puberty in higher primates. *European Journal of Endocrinology*, 155(1), S11.

Plard, F., Schindler, S., Arlettaz, R., and Schaub, M. (2018). Sex-specific heterogeneity in fixed morphological traits influences individual fitness in a monogamous bird population. *The American Naturalist*, 191(1), 106–119.

Pocock, M.J., Frantz, A.C., Cowan, D.P., White, P.C., and Searle, J.B. (2004). Tapering bias inherent in minimum number alive (MNA) population indices. *Journal of Mammalogy*, 85(5), 959–962.

Pocock, R.I. (1908). Warning coloration in the musteline Carnivora. *Proceedings of the Zoological Society of London*, 78(4), 944–959.

Pocock, R.I. (1911). LXXXVII—Some probable and possible instances of warning characteristics amongst insectivorous and carnivorous mammals. *Journal of Natural History*, 8(48), 750–757.

Pocook, R.L. (1920). On the external and cranial characters of the European badger (Melei) and of the American badger (Taxidea). *Proceedings of the Zoological Society of London*, 90(3), 423–436).

Podani, J., Kun, Á., and Szilágyi, A. (2018). How fast does Darwin's elephant population grow? *Journal of the History of Biology*, 51(2), 259–281.

Poessel, S.A. and Gese, E.M. (2013). Den attendance patterns in swift foxes during pup rearing: varying degrees of parental investment within the breeding pair. *Journal of Ethology*, 31(2), 193–201.

Poglayen-Neuwall, I. (1989). Procyonids. In *Grzimek's Encyclopedia of Mammals*, Volume 3 (S. Parker, ed., pp. 450–468). McGraw-Hill, New York.

Poirotte, C., Massol, F., Herbert, A., Willaume, E., Bomo, P.M., Kappeler, P.M., and Charpentier, M.J. (2017). Mandrills use olfaction to socially avoid parasitized conspecifics. *Science Advances*, 3(4), e1601721.

Polly, S. and Britton, K.H. (2015). *Character Strengths Matter: How to Live a Full Life*. Positive Psychology News, Greenwich, CT.

Pond, C. (1978). Morphological aspects and the ecological and mechanical consequences of fat deposition in wild vertebrates. *Annual Review of Ecology and Systematics*, 9, 519–570.

Pond, C. (2001). Ecology of storage and allocation of resources: animals. In *Encyclopedia of Life Sciences*. John Wiley & Sons, Ltd, Chichester.

Pontzer, H. (2015). Energy expenditure in humans and other primates: a new synthesis. *Annual Review of Anthropology*, 44, 169–187.

Pontzer, H., Yamada, Y., Sagayama, H., Ainslie, P.N., Andersen, L.F., Anderson, L.J., Arab, L., Baddou, I., Bedu-Addo, K., Blaak, E.E., and Blanc, S. (2021). Daily energy expenditure through the human life course. *Science*, 373(6556), 808–812.

Pope, L.C., Domingo-Roura, X., Erven, K., and Burke, T. (2006). Isolation by distance and gene flow in the Eurasian badger (*Meles meles*) at both a local and broad scale. *Molecular Ecology*, 15(2), 371–386.

Pope, L.C., Butlin, R.K., Wilson, G.J., Woodroffe, R., Erven, K., Conyers, C.M., Franklin, T., Delahay, R.J., Cheeseman, C.L., and Burke, T. (2007). Genetic evidence that culling increases badger movement: implications for the spread of bovine tuberculosis. *Molecular Ecology*, 16(23), 4919–4929.

Popper, K. (2012). *Objective Knowledge: an Evolutionary Approach*. Clarendon Press, Oxford.

Porter, R.D. and Wiemeyer, S.N. (1969). Dieldrin and DDT: effects on sparrow hawk eggshells and reproduction. *Science*, 165(3889), 199–200.

Portier, C., Festa-Bianchet, M., Gaillard, J.M., Jorgenson, J.T., and Yoccoz, N.G. (1998). Effects of density and weather on survival of bighorn sheep lambs (*Ovis canadensis*). *Journal of Zoology*, 245(3), 271–278.

Pörtl, D. and Jung, C. (2017). Is dog domestication due to epigenetic modulation in brain? *Dog Behavior*, 3(2), 21–32.

Post, E.S., Peterson, R.O., Stenseth, N.C., and McClaren, B. (1999). Ecosystem consequences of wolf behavioural response to climate. *Nature*, 401, 905–907.

Post, J.R., Parkinson, E.A., and Johnston, N.T. (1999). Density-dependent processes in structured fish populations: interaction strengths in whole-lake experiments. *Ecological Monographs, 69*(2), 155–175.

Powell, R.A. (1979). Mustelid spacing patterns: variations on a theme by Mustela. *Zeitschrift für Tierpsychologie, 50*(2), 153–165.

Powell, R.A. (1987). Black bear home range overlap in North Carolina and the concept of home range applied to black bears. Bears: Their Biology and Management, 7, 235–242.

Powell, R.A. and Zielinski, W.J. (1989). Mink response to ultrasound in the range emitted by prey. *Journal of Mammalogy, 70*(3), 637–638.

Prange, S., Gehrt, S.D., and Hauver, S. (2011). Frequency and duration of contacts between free-ranging raccoons: uncovering a hidden social system. *Journal of Mammalogy, 92*(6), 1331–1342.

Preston, B.T., Stevenson, I.R., Pemberton, J.M., Coltman, D.W., and Wilson, K. (2003). Overt and covert competition in a promiscuous mammal: the importance of weaponry and testes size to male reproductive success. *Proceedings of the Royal Society of London. Series B: Biological Sciences, 270*(1515), 633–640.

Preston, E.F., Thompson, F.J., Ellis, S., Kyambulima, S., Croft, D.P., and Cant, M.A. (2021). Network-level consequences of outgroup threats in banded mongooses: Grooming and aggression between the sexes. *Journal of Animal Ecology, 90*(1), 153–167.

Pretzlaff, I., and Dausmann, K.H. (2012). Impact of climatic variation on the hibernation physiology of *Muscardinus avellanarius*. In *Living in a seasonal world* (T. Ruf, C. Bieb, W. Arnold, and E. Millesi, eds, pp. 85–97). Springer, Berlin, Heidelberg.

Prinsloo, D. (2016). Parental care by a honey badger *Mellivora capensis* in Kruger National Park, South Africa. *Biodiversity Observations, 7*, 1–7.

Promislow, D.E. and Harvey, P.H. (1990). Living fast and dying young: a comparative analysis of life-history variation among mammals. *Journal of Zoology, 220*(3), 417–437.

Public Health England (2013). Tuberculosis in the UK. https://assets.publishing.service.gov.uk/government/uploads/system/uploads/attachment_data/file/332560/TB_Annual_Report_2012.pdf (accessed 11 March 2022).

Pusey, A. (1987). Sex-biased dispersal and inbreeding avoidance in birds and mammals. *Trends in Ecology & Evolution, 2*(10), 295–299.

Pusey A. (2012). Magnitude and sources of variation in female reproductive performance. In *The Evolution of Primate Societies* (J.C. Mitani, J. Call, P.M. Kappeler, R.A. Palombit, and J.B. Silk, eds, pp. 343–366). University of Chicago Press, Chicago, IL.

Pusey, A. and Wolf, M. (1996). Inbreeding avoidance in animals. *Trends in Ecology & Evolution, 11*(5), 201–206.

Putman, R.J. (1984). Facts from faeces. *Mammal Review, 14*(2), 79–97.

Puurtinen, M., Ketola, T., and Kotiaho, J.S. (2005). Genetic compatibility and sexual selection. *Trends in Ecology and Evolution, 20*(4), 157–158.

Qian, G., Sheng, H., and Wang, P. (1976). The winter diet of ferret badger. *Chinese Journal Zoology, 20*, 37.

Queller, D.C. and Goodnight, K.F. (1989). Estimating relatedness using genetic markers. *Evolution, 43*(2), 258–275.

Queller, D.C. and Strassmann, J.E. (2013). The veil of ignorance can favour biological cooperation. *Biology Letters, 9*(6), 20130365.

Racicot, K., Cardenas, I., Wünsche, V., Aldo, P., Guller, S., Means, R.E., and Mor, G. (2013). Viral infection of the pregnant cervix predisposes to ascending bacterial infection. *The Journal of Immunology, 191*(2), 934–941.

Radak, Z., Chung, H.Y., and Goto, S. (2008). Systemic adaptation to oxidative challenge induced by regular exercise. *Free Radical Biology and Medicine, 44*(2), 153–159.

Ragab, D., Salah Eldin, H., Taeimah, M., Khattab, R., and Salem, R. (2020). The COVID-19 cytokine storm; what we know so far. *Frontiers in Immunology, 11*, 1446.

Rainey, E., Badgers, S., Rainey, E., Butler, A., Bierman, S., and Roberts, A.M.I. (2009). Scottish Badger Distribution Survey 2006–2009: estimating the distribution and density of badger main sets in Scotland. Report prepared by Scottish Badgers and Biomathematics and Statistics Scotland.

Rajon, E., Desouhant, E., Chevalier, M., Débias, F., and Menu, F. (2014). The evolution of bet hedging in response to local ecological conditions. *The American Naturalist, 184*, E1–E15.

Randall, D.A., Marino, J., Haydon, D.T., Sillero-Zubiri, C., Knobel, D.L., Tallents, L.A., and Laurenson, M.K. (2006). An integrated disease management strategy for the control of rabies in Ethiopian wolves. *Biological Conservation, 131*(2), 151–162.

Rantala, M.J., Coetzee, V., Moore, F.R., Skrinda, I., Kecko, S., Krama, T., Kivleniece, I., and Krams, I. (2013). Adiposity, compared with masculinity, serves as a more valid cue to immunocompetence in human mate choice. *Proceedings of the Royal Society. Series B: Biological Sciences, 280*(1751), 20122495.

Rantanen, E., Macdonald, D.W., Sotherton, N.W., and Buner, F. (2015). Improving reintroduction success of the grey partridge using behavioural studies. In *Wildlife Conservation on Farmland Volume 1: Managing for Nature on Lowland Farms* (D.W. Macdonald and R.E. Feber, eds, pp. 241–254). Oxford University Press, Oxford.

Rasa, O.A.E. (1977). The ethology and sociology of the dwarf mongoose (*Helogale undulata rufula*). *Zeitschrift für Tierpsychologie, 43*(4), 337–406.

Rasa, O.A.E. (1987). The dwarf mongoose: a study of behavior and social structure in relation to ecology in a small, social carnivore. *Advances in the Study of Behavior, 17*, 121–163.

Rasmussen, G.S., Gusset, M., Courchamp, F., and Macdonald, D.W. (2008). Achilles' heel of sociality revealed by energetic poverty trap in cursorial hunters. *The American Naturalist, 172*(4), 508–518.

Rasphone, A., Bousa, A., Vongkhamheng, C., Kamler, J.F., Johnson, A., and Macdonald, D.W. (2022). Diet and prey selection of clouded leopards and tigers in Laos. *Ecology and Evolution*, 12:e9067.

Ratcliffe, E.J. (1974). *Through the Badger Gate*. Bell, London.

Raveh, A., Kotler, B.P., Abramsky, Z., and Krasnov, B.R. (2011). Driven to distraction: detecting the hidden costs of flea parasitism through foraging behaviour in gerbils. *Ecology Letters*, 14(1), 47–51.

Rawls, J. (1971). *A Theory of Justice*. Belknap Press of Harvard University Press, Cambridge, MA.

Réale, D., McAdam, A.G., Boutin, S., and Berteaux, D. (2003). Genetic and plastic responses of a northern mammal to climate change. *Proceedings of the Royal Society of London. Series B: Biological Sciences*, 270(1515), 591–596.

Réale, D., Dingemanse, N.J., Kazem, A.J., and Wright, J. (2010). Evolutionary and ecological approaches to the study of personality. *Philosophical Transactions of the Royal Society. Series B: Biological Sciences*, 365(1560), 3937–3946.

Reed, T.E., Kruuk, L.E., Wanless, S., Frederiksen, M., Cunningham, E.J., and Harris, M.P. (2008). Reproductive senescence in a long-lived seabird: rates of decline in late-life performance are associated with varying costs of early reproduction. *The American Naturalist*, 171(2), E89–E101.

Reed, T.E., Waples, R.S., Schindler, D.E., Hard, J.J. and Kinnison, M.T. (2010). Phenotypic plasticity and population viability: the importance of environmental predictability. *Proceedings of the Royal Society of London. Series B: Biological Sciences*, 277(1699), 3391–3400.

Reeg, K.J., Gauly, M., Bauer, C., Mertens, C., Erhardt, G., and Zahner, H. (2005). Coccidial infections in housed lambs: oocyst excretion, antibody levels and genetic influences on the infection. *Veterinary Parasitology*, 127(3), 209–219.

Reichert, S. and Stier, A. (2017). Does oxidative stress shorten telomeres in vivo? A review. *Biology Letters*, 13(12), 20170463.

Reichman, O.J. and Smith, S.C. (1990). Burrows and burrowing behavior by mammals. *Current Mammalogy*, 2, 197–244.

Reid, N., Etherington, T.R., Wilson, G.J., Montgomery, W.I., and McDonald, R.A. (2012). Monitoring and population estimation of the European badger *Meles meles* in Northern Ireland. *Wildlife Biology*, 18(1), 46–57.

Reid, A. T., Redgrove, K., Aitken, R. J., and Nixon, B. (2011). Cellular mechanisms regulating sperm–zona pellucida interaction. *Asian journal of andrology*, 13(1), 88.

Reiss, M. (1988). Scaling of home range size: body size, metabolic needs and ecology. *Trends in Ecology & Evolution*, 3(3), 85–86.

Remonti, L., Balestrieri, A., and Prigioni, C. (2006). Factors determining badger *Meles meles* sett location in agricultural ecosystems of NW Italy. *Folia Zoologica*, 55(1), 19.

Remonti, L., Balestrieri, A., Smiroldo, G., and Prigioni, C. (2011). Scent marking of key food sources in the Eurasian otter. *Annales Zoologici Fennici*, 48, 287–294.

Revilla, E. (2003a). What does the resource dispersion hypothesis explain, if anything? *Oikos*, 101(2), 428–432.

Revilla, E. (2003b). Moving beyond the resource dispersion hypothesis. *Trends in Ecology & Evolution*, 18(8), 380.

Revilla, E. and Palomares, F. (2002). Does local feeding specialization exist in Eurasian badgers? *Canadian Journal of Zoology*, 80(1), 83–93.

Revilla, E., Delibes, M., Travaini, A., and Palomares, F. (1999). Physical and population parameters of Eurasian badgers (*Meles meles* L.) from Mediterranean Spain. *International Journal of Mammalian Biology*, 64, 269–276.

Rewell, R.E. (1948). Diseases of tropical origin in captive wild animals. *Transactions of the Royal Society of Tropical Medicine and Hygiene*, 42(1), 17–36.

Rey-Rassat, C., Irigoien, X., Harris, R., and Carlotti, F. (2002). Energetic cost of gonad development in *Calanus finmarchicus* and *C. helgolandicus*. *Marine Ecology Progress Series*, 238, 301–306.

Reznick, D., Bryant, M.J., and Bashey, F. (2002). r- and K-selection revisited: the role of population regulation in life-history evolution. *Ecology*, 83(6), 1509–1520.

Rhodes, C.J., Atkinson, R.P.D., Anderson, R.M., and Macdonald, D.W. (1998). Rabies in Zimbabwe: reservoir dogs and the implications for disease control. *Philosophical Transactions of the Royal Society of London. Series B: Biological Sciences*, 353(1371), 999–1010.

Riahi, K., Rao, S., Krey, V., Cho, C., Chirkov, V., Fischer, G., Kindermann, G., Nakicenovic, N., and Rafaj, P. (2011). RCP 8.5—a scenario of comparatively high greenhouse gas emissions. *Climatic Change*, 109(1), 33–57.

Ribas, C., Damasceno, G., Magnusson, W., Leuchtenberger, C., and Mourão, G. (2012). Giant otters feeding on caiman: evidence for an expanded trophic niche of recovering populations. *Studies on Neotropical Fauna and Environment*, 47(1), 19–23.

Ribas, C., Cunha, H.A., Damasceno, G., Magnusson, W.E., Solé-Cava, A., and Mourão, G. (2016). More than meets the eye: kinship and social organization in giant otters (*Pteronura brasiliensis*). *Behavioral Ecology and Sociobiology*, 70(1), 61–72.

Richardson, P.R.K. (1987). Aardwolf mating system: overt cuckoldry in an apparently monogamous mammal. *South African Journal of Science*, 83(7), 405.

Rickard, I.J. and Lummaa, V. (2007). The predictive adaptive response and metabolic syndrome: challenges for the hypothesis. *Trends in Endocrinology and Metabolism*, 18(3), 94–99. https://doi.org/10.1016/j.tem.2007.02.004

Ricklefs, R.E. and Scheuerlein, A. (2001). Comparison of aging-related mortality among birds and mammals. *Experimental Gerontology*, 36(4–6), 845–857.

Ricklefs, R.E., Konarzewski, M., and Daan, S. (1996). The relationship between basal metabolic rate and daily energy expenditure in birds and mammals. *The American Naturalist*, 147(6), 1047–1071.

Riddell, E.A., Iknayan, K.J., Hargrove, L., Tremor, S., Patton, J.L., Ramirez, R., Wolf, B.O., and Beissinger, S.R. (2021). Exposure to climate change drives stability or collapse of desert mammal and bird communities. *Science*, 371(6529), 633–636.

Riordan, P., Delahay, R.J., Cheeseman, C., Johnson, P.J., and Macdonald, D.W. (2011). Culling-induced changes in badger (*Meles meles*) behaviour, social organisation and the epidemiology of bovine tuberculosis. *PLoS ONE*, 6(12), e28904.

Ripple, W.J., Newsome, T., Wolf, C., Dirzo, R., Everatt, K.T., Galetti, M. et al. (2015). Collapse of the world's largest herbivores. *Science Advances*, 1(4), e1400103. doi:10.1126/sciadv.1400103

Ripple, W.J., Chapron, G., Lopez-Bao, J.V., Durant, S.M., Macdonald, D.W., Lindsey, P. A. et al. (2017). Conserving the world's megafauna and biodiversity: the fierce urgency of now. *BioScience*, 67(3), 197–200.

Risques, R.A. and Promislow, D.E. (2018). All's well that ends well: why large species have short telomeres. *Philosophical Transactions of the Royal Society. Series B: Biological Sciences*, 373(1741), 20160448.

Robert-Gangneux, F. (2014). It is not only the cat that did it: how to prevent and treat congenital toxoplasmosis. *Journal of Infection*, 68, S125–S133.

Roberts, M.L., Buchanan, K.L., and Evans, M.R. (2004). Testing the immunocompetence handicap hypothesis: a review of the evidence. *Animal Behaviour*, 68(2), 227–239.

Roberts, M.S. and Kessler, D.S. (1979). Reproduction in red pandas, *Ailurus fulgens* (Carnivora: Ailuropodidae). *Journal of Zoology*, 188(2), 235–249.

Roberts, S.C., Little, A.C., Gosling, L.M., Jones, B.C., Perrett, D.I., Carter, V., and Petrie, M. (2005). MHC-assortative facial preferences in humans. *Biology Letters*, 1(4), 400–403.

Robertshaw, D. (2006). Mechanisms for the control of respiratory evaporative heat loss in panting animals. *Journal of Applied Physiology*, 101(2), 664–668.

Robertson, A., Palphramand, K.L., Carter, S.P., and Delahay, R.J. (2015). Group size correlates with territory size in European badgers: implications for the resource dispersion hypothesis? *Oikos*, 124(4), 507–514.

Robertson, L.J., Campbell, A.T., and Smith, H.V. (1992). Survival of *Cryptosporidium parvum* oocysts under various environmental pressures. *Applied and Environmental Microbiology*, 58(11), 3494–3500.

Robin, J.D., Ludlow, A.T., Batten, K., Magdinier, F., Stadler, G., Wagner, K.R., Shay, J.W., and Wright, W.E. (2014). Telomere position effect: regulation of gene expression with progressive telomere shortening over long distances. *Genes & Development*, 28(22), 2464–2476.

Robinson, P.A., Corner, L.A., Courcier, E.A., McNair, J., Artois, M., Menzies, F.D., and Abernethy, D.A. (2012). BCG vaccination against tuberculosis in European badgers (*Meles meles*): a review. *Comparative Immunology, Microbiology and Infectious Diseases*, 35(4), 277–287.

Rodgers, T.W., Giacalone, J., Heske, E.J., Pawlikowski, N.C., and Schooley, R.L. (2015). Communal latrines act as potentially important communication centers in ocelots *Leopardus pardalis* . *Mammalian Biology*, 80(5), 380–384.

Rodríguez, A., Martin, R., and Delibes, M. (1996). Space use and activity in a Mediterranean population of badgers *Meles meles* . *Acta Theriologica*, 41(1), 59–72.

Roellig, K., Menzies, B.R., Hildebrandt, T.B., and Goeritz, F. (2011). The concept of superfetation: a critical review on a 'myth' in mammalian reproduction. *Biological Reviews*, 86(1), 77–95.

Rogers, A.R. (1993). Why menopause? *Evolutionary Ecology*, 7, 406–420.

Rogers, L.M., Cheeseman, C.L., Mallinson, P.J., and Clifton-Hadley, R. (1997). The demography of a high-density badger (*Meles meles*) population in the west of England. *Journal of Zoology*, 242(4), 705–728.

Rogers, L.M., Forrester, G.J., Wilson, G.J., Yarnell, R.W., and Cheeseman, C.L. (2003). The role of setts in badger (*Meles meles*) group size, breeding success and status of TB (*Mycobacterium bovis*). *Journal of Zoology*, 260(2), 209–215.

Rolland, C., Macdonald, D.W., De Fraipont, M., and Berdoy, M. (2003). Free female choice in house mice: leaving best for last. *Behaviour*, 140(11–12), 1371–1388.

Rolland, Y., Czerwinski, S., Van Kan, G.A., Morley, J.E., Cesari, M., Onder, G., Woo, J., Baumgartner, R., Pillard, F., Boirie, Y., and Chumlea, W.M.C. (2008). Sarcopenia: its assessment, etiology, pathogenesis, consequences and future perspectives. *The Journal of Nutrition Health and Aging*, 12(7), 433–450.

Rood, J.P. (1983). The social system of the dwarf mongoose. In *Recent Advances in the Study of Mammalian Behavior* (J.F. Eisenberg and D.G. Kleiman, eds, pp. 454–488). American Society of Mammalogists, Shippensburg, PA.

Rood, J.P. (1990). Group size, survival, reproduction, and routes to breeding in dwarf mongooses. *Animal Behaviour*, 39(3), 566–572.

Roper, T.J. (1992). Badger *Meles meles* setts—architecture, internal environment and function. *Mammal Review*, 22(1), 43–53.

Roper, T.J. (1994). The European badger *Meles meles*: food specialist or generalist?. *Journal of Zoology*, 234(3), 437–452.

Roper, T.J. (2010). *Badger (Collins New Naturalist Library, Book 114)* (Vol. 114). HarperCollins, London.

Roper, T.J. and Mickevicius, E. (1995). Badger *Meles meles* diet: a review of literature from the former Soviet Union. *Mammal Review*, 25(3), 117–129.

Roper, T.J., Shepherdson, D.J., and Davies, J.M. (1986). Scent marking with faeces and anal secretion in the European badger (*Meles meles*): seasonal and spatial characteristics of latrine use in relation to territoriality. *Behaviour*, 97(1/2), 94–117.

Roper, T.J., Conradt, L., Butler, J., Christian, S.E., Ostler, J., and Schmid, T.K. (1993). Territorial marking with faeces in badgers (*Meles meles*): a comparison of boundary and hinterland latrine use. *Behaviour*, 127(3–4), 289–307.

Roper, T.J., Ostler, J.R., Schmid, T.K., and Christian, S.F. (2001). Sett use in European badgers Meles meles . *Behaviour*, 138, 173–187.

Rosalino, L.M., Macdonald, D.W., and Santos-Reis, M. (2004). Spatial structure and land-cover use in a low-density Mediterranean population of Eurasian badgers. *Canadian Journal of Zoology*, 82(9), 1493–1502.

Rosalino, L.M., Loureiro, F., Macdonald, D.W., and Santon-Reis, M. (2005a). Dietary shifts of the badger (*Meles meles*) in Mediterranean woodlands: an opportunistic forager with seasonal specialisms. *Mammalian Biology*, 70(1), 12–23.

Rosalino, L.M., Macdonald, D.W., and Santos-Reis, M. (2005b). Activity rhythms, movements and patterns of sett use by badgers, *Meles meles*, in a Mediterranean woodland. *Mammalia*, 69(3–4), 395–408.

Rosalino, L.M., Macdonald, D.W., and Santos-Reis, M. (2005c). Resource dispersion and badger population density in Mediterranean woodlands: is food, water or geology the limiting factor? *Oikos*, 110(3), 441–452.

Rosalino, L.M., Guedes, D., Cabecinha, D., Serronha, A., Grilo, C., Santos-Reis, M., and Hipólito, D. (2019). Climate and landscape changes as driving forces for future range shift in southern populations of the European badger. *Scientific Reports*, 9(1), 1–15.

Rosell, F., Gundersen, G., and Le Galliard, J.F. (2008). Territory ownership and familiarity status affect how much male root voles (*Microtus oeconomus*) invest in territory defence. *Behavioral Ecology and Sociobiology*, 62(10), 1559–1568.

Rosell, F., Jojola, S.M., Ingdal, K., Lassen, B.A., Swenson, J.E., Arnemo, J.M., and Zedrosser, A. (2011). Brown bears possess anal sacs and secretions may code for sex. *Journal of Zoology*, 283(2), 143–152.

Rosenheim, J.A. and Tabashnik, B.E. (1991). Influence of generation time on the rate of response to selection. *The American Naturalist*, 137(4), 527–541.

Ross, J., Hearn, A.J., and Macdonald, D.W. (2017). The Bornean carnivore community: lessons from a little-known guild. In *Biology and Conservation of Musteloids* (D.W. Macdonald, C. Newman, and L.A. Harrington, eds, pp. 326–339). Oxford University Press, Oxford.

Rossi, G., Crispell, J., Balaz, D., Lycett, S.J., Benton, C.H., Delahay, R.J., and Kao, R.R. (2020a). Identifying likely transmissions in *Mycobacterium bovis* infected populations of cattle and badgers using the Kolmogorov Forward Equations. *Scientific Reports*, 10(1), 1–13.

Rossi, G., Crispell, J., Brough, T., Lycett, S.J., White, P.C.L., Allen, A., Ellis, R.J., Gordon, S.V., Harwood, R., Palkopoulou, E., Presho, E.L., Skuce, R., Smith, G.C., and Kao, R.R. (2020b). Phylodynamic analysis of an emergent Mycobacterium bovis outbreak in an area with no previously known wildlife infections. *Journal of Applied Ecology*, 59, 210–222.

Rotenko, I. and Sidorovich, V. (2017). *Badger and Raccoon Dog in Belarus: Population Studies with Implication for the Decline in Badgers*. Chatyry Chverci, Minsk.

Rothschild, M.L., Schlein, J., Parker, K., Neville, C., and Sternberg, S. (1975). The jumping mechanism of *Xenopsylla cheopis* III. Execution of the jump and activity. *Philosophical Transactions of the Royal Society of London. Series B: Biological Sciences*, 271(914), 499–515.

Rousseau, K., Atcha, Z., and Loudon, A.S.I. (2003). Leptin and seasonal mammals. *Journal of Neuroendocrinology*, 15(4), 409–414.

Rowe, J.W. and Moll, E.O. (1991). A radiotelemetric study of activity and movements of the Blanding's turtle (*Emydoidea blandingi*) in northeastern Illinois. *Journal of Herpetology*, 25(2), 178–185.

Rowe, L. (1992). Convenience polyandry in a water strider: foraging conflicts and female control of copulation frequency and guarding duration. *Animal Behaviour*, 44, 189–202.

Rowell, T.E. (1974). The concept of social dominance. *Behavioral Biology*, 11(2), 131–154.

Rowe-Rowe, D.T. (1977). Food ecology of otters in Natal, South Africa. *Oikos*, 28(2/3), 210–219.

Royauté, R., Berdal, M.A., Garrison, C.R., and Dochtermann, N.A. (2018). Paceless life? A meta-analysis of the pace-of-life syndrome hypothesis. *Behavioral Ecology and Sociobiology*, 72(3), 1–10.

Rubenstein, D.I. and Hack, M.A.C.E. (2004). Natural and sexual selection and the evolution of multi-level societies: insights from zebras with comparisons to primates. In Sexual Selection in Primates: New and Comparative Perspectives (P.M. Kappeler and C.P. Van Schaik, eds, pp. 266–279). Cambridge University Press, Cambridge.

Rubin, L.H., Carter, C.S., Bishop, J.R., Pournajafi-Nazarloo, H., Drogos, L.L., Hill, S.K., Ruocco, A.C., Keedy, S.K., Reilly, J.L., Keshavan, M.S., and Pearlson, G.D. (2014). Reduced levels of vasopressin and reduced behavioral modulation of oxytocin in psychotic disorders. *Schizophrenia Bulletin*, 40(6), 1374–1384.

Rudolf, V.H. and Antonovics, J. (2005). Species coexistence and pathogens with frequency-dependent transmission. *The American Naturalist*, 166(1), 112–118.

Ruel, J.J. and Ayres, M.P. (1999). Jensen's inequality predicts effects of environmental variation. *Trends in Ecology & Evolution*, 14(9), 361–366.

Rufer, N., Brümmendorf, T.H., Kolvraa, S., Bischoff, C., Christensen, K., Wadsworth, L., Schulzer, M., and Lansdorp, P.M. (1999). Telomere fluorescence measurements in granulocytes and T lymphocyte subsets point to a high turnover of hematopoietic stem cells and memory T cells in early childhood. *The Journal of Experimental Medicine*, 190(2), 157–167. https://doi.org/10.1084/jem.190.2.157

Ruprecht, J.S., Ausband, D.E., Mitchell, M.S., Garton, E.O., and Zager, P. (2012). Homesite attendance based on sex, breeding status, and number of helpers in gray wolf packs. *Journal of Mammalogy*, 93(4), 1001–1005.

Rutte, C., Taborsky, M., and Brinkhof, M.W.G. (2006). What sets the odds of winning and losing? *Trends in Ecology & Evolution*, 21, 16–21. doi:10.1016/j.tree.2005.10.014

Rymešová, D., Králová, T., Promerová, M., Bryja, J., Tomášek, O., Svobodová, J., Šmilauer, P., Šálek, M., and

Albrecht, T. (2017). Mate choice for major histocompatibility complex complementarity in a strictly monogamous bird, the grey partridge (*Perdix perdix*). *Frontiers in Zoology*, 14, 9.

Ryszkowski, L., Wagner, C.K., Goszczyński, J., and Truszkowski, J. (1971, January). Operation of predators in a forest and cultivated fields. *Annales Zoologici Fennici*, 8(1), 160–168.

Saeki, M. and Macdonald, D.W. (2004). The effects of traffic on the raccoon dog (*Nyctereutes procyonoides viverrinus*) and other mammals in Japan. *Biological Conservation*, 118(5), 559–571.

Salim, S. (2014). Oxidative stress and psychological disorders. *Current Neuropharmacology*, 12(2), 140–147.

Salvador, A., Veiga, J.P., Martin, J., Lopez, P., Abelenda, M., and Puertac, M. (1996). The cost of producing a sexual signal: testosterone increases the susceptibility of male lizards to ectoparasitic infestation. *Behavioral Ecology*, 7(2), 145–150.

Sanchez-Hidalgo, A., Obregón-Henao, A., Wheat, W.H., Jackson, M., Gonzalez-Juarrero, M. (2017). Mycobacterium bovis hosted by free-living-amoebae permits their long-term persistence survival outside of host mammalian cells and remain capable of transmitting disease to mice. *Environmental Microbiology*, 19(10), 4010–4021. doi:10.1111/1462-2920.13810

Sánchez-Tójar, A., Schroeder, J., and Farine, D.R. (2017). A practical guide for inferring reliable dominance hierarchies and estimating their uncertainty. *Journal of Animal Ecology*, 87(3), 594–608. doi:10.1111/1365-2656.12776

Sandell, M. (1984). To have or not to have delayed implantation: the example of the weasel and the stoat. *Oikos*, 42(1), 123–126.

Sandell, M. (1989a). Ecological energetics, optimal body size and sexual size dimorphism: a model applied to the stoat, *Mustela erminea* L. *Functional Ecology*, 33(3), 15–324.

Sandell, M. (1989b). The mating tactics and spacing patterns of solitary carnivores. In *Carnivore Behavior, Ecology, and Evolution* (J.L. Gittleman, ed., pp. 164–182). Springer, Boston, MA.

Sandell, M. (1990). The evolution of seasonal delayed implantation. *Quarterly Review of Biology*, 65(1), 23–42.

Sandell, M. (2019). The mating tactics and spacing patterns of solitary carnivores. In *Carnivore Behavior, Ecology, and Evolution* (J.L. Gittleman, ed. pp. 164–182). Cornell University Press, Ithaca, NY.

Sandom, C.J., Faurby, S., Svenning, J.C., Burnham, D., Dickman, A., Hinks, A.E., Macdonald, E.A., Ripple, W.J., Williams, J., and Macdonald, D.W. (2017a). Learning from the past to prepare for the future: felids face continued threat from declining prey. *Ecography*, 41, 140–152. doi:10.1111/ecog.03303

Sandom, C.J., Williams, J., Burnham, D., Dickman, A.J., Hinks, A.E., Macdonald, E.A., and Macdonald, D.W. (2017b). Deconstructed cat communities: quantifying the threat to felids from prey defaunation. *Diversity and Distributions*, 2017(23), 667–679. doi:10.1111/ddi.12558

Sapolsky, R.M. (2005). The influence of social hierarchy on primate health. *Science*, 308, 648–652.

Sardell, R.J., Arcese, P., Keller, L.F., and Reid, J.M. (2011). Sex-specific differential survival of extra-pair and within-pair offspring in song sparrows, *Melospiza melodia*. *Proceedings of the Royal Society. Series B: Biological Sciences*, 278(1722), 3251–3259.

Sardell, R.J., Arcese, P., Keller, L.F., and Reid, J.M. (2012). Are there indirect fitness benefits of female extra-pair reproduction? Lifetime reproductive success of within-pair and extra-pair offspring. *The American Naturalist*, 179(6), 779–793.

Šárová, R., Gutmann, A.K., Špinka, M., Stěhulová, I., and Winckler, C. (2016). Important role of dominance in allogrooming behaviour in beef cattle. *Applied Animal Behaviour Science*, 181, 41–48.

Satchell, J.E. (1967). Lubricidae. In *Soil Biology* (A. Burges and F. Raw, eds, pp. 259–322). Academic Press, London.

Sato, J.J., Hosoda, T., Wolsan, M., and Suzuki, H. (2004). Molecular phylogeny of arctoids (Mammalia: Carnivora) with emphasis on phylogenetic and taxonomic positions of the ferret-badgers and skunks. *Zoological Science*, 21(1), 111–118.

Sato, J.J., Wolsan, M., Prevosti, F.J., D'Elía, G., Begg, C., Begg, K., Hosoda, T., Campbell, K.L., and Suzuki, H. (2012). Evolutionary and biogeographic history of weasel-like carnivorans (Musteloidea). *Molecular Phylogenetics and Evolution*, 63(3), 745–757.

Savill, P., Perrins, C., Kirby, K., and Fisher, N. (2010). *Wytham Woods: Oxford's Ecological Laboratory*. Oxford University Press, Oxford.

Sawyer, J., Mealing, D., Dalley, D., Dave, D., Lesellier, S., Palmer, S., Bowen-Davies, J., Crawshaw, T.R., and Chambers, M.A. (2007). Development and evaluation of a test for tuberculosis in live European badgers (*Meles meles*) based on measurement of gamma interferon mRNA by real-time PCR. *Journal of Clinical Microbiology*, 45(8), 2398–2403.

Scally, S.W., Petersen, J., Law, S.C., Dudek, N.L., Nel, H.J., Loh, K.L., and McCluskey, J. (2013). A molecular basis for the association of the HLA-DRB1 locus, citrullination, and rheumatoid arthritis. *Journal of Experimental Medicine*, 210(12), 2569–2582.

Scharlemann, J.P.W., Johnson, J.P., Smith, A.A., Macdonald, D.W., and Randolph, S.E. (2008). Trends in ixodid tick abundance and distribution in Great Britain. *Medical and Veterinary Entomology*, 22(3), 238–247.

Schassburger, R.M. (1993). *Vocal Communication in the Timber Wolf, Canis lupus, Linnaeus: Structure, Motivation, and Ontogeny; with 6 Tables*. Parey Scientific Publ., New York.

Scheffers, B.R., De Meester, L., Bridge, T.C., Hoffmann, A.A., Pandolfi, J.M., Corlett, R.T., Butchart, S.H., Pearce-Kelly, P., Kovacs, K.M., Dudgeon, D., and Pacifici, M. (2016). The broad footprint of climate change from genes to biomes to people. *Science*, 354(6313), aaf7671.

Schmitt, M.H., Stears, K., Wilmers, C.C., and Shrader, A.M. (2014). Determining the relative importance of dilution and detection for zebra foraging in mixed-species herds. *Animal Behaviour, 96*, 151–158.

Schindler, D.E., Armstrong, J.B., and Reed, T.E. (2015). The portfolio concept in ecology and evolution. *Frontiers in Ecology and the Environment, 13*, 257–263.

Schirmacher, R. (2002). Active design of automotive engine sound. *Audio Engineering Society Convention, 112*, 5544.

Schmid, V.S. and de Vries, H. (2013). Finding a dominance order most consistent with a linear hierarchy: an improved algorithm for the I&SI method. *Animal Behaviour, 86*(5), 1097–1105. doi:10.1016/j.anbehav.2013.08.019

Schoener, T.W. (2011). The newest synthesis: understanding the interplay of evolutionary and ecological dynamics. *Science, 331*(6016), 426–429.

Schreier, A.L. and Swedell, L. (2009). The fourth level of social structure in a multi-level society: ecological and social functions of clans in hamadryas baboons. *American Journal of Primatology: Official Journal of the American Society of Primatologists, 71*(11), 948–955.

Schreier, A.L. and Swedell, L. (2012). Ecology and sociality in a multilevel society: ecological determinants of spatial cohesion in hamadryas baboons. *American Journal of Physical Anthropology, 148*(4), 580–588.

Schulte, B.A., Müller-Schwarze, D., and Sun, L. (1995). Using anal gland secretion to determine sex in beaver. *The Journal of Wildlife Management, 59*(3), 614–618.

Schwartz, O.A. and Armitage, K.B. (1980). Genetic variation in social mammals: the marmot model. *Science, 207*(4431), 665–667.

Schweinfurth, M.K., Stieger, B., and Taborsky, M. (2017). Experimental evidence for reciprocity in allogrooming among wild-type Norway rats. *Scientific Reports, 7*(1), 4010. doi:10.1038/s41598-017-03841-3

Schwensow, N., Eberle, M., and Sommer, S. (2008). Compatibility counts: MHC-associated mate choice in a wild promiscuous primate. *Proceedings of the Royal Society Series B: Biological Sciences, 275*(1634), 555–564.

Scribner, K.T., Smith, M.H., and Johns, P.E. (1989). Environmental and genetic components of antler growth in white-tailed deer. *Journal of Mammalogy, 70*(2), 284–291.

Segerstrom, S.C. (2007). Stress, energy, and immunity: an ecological view. *Current Directions in Psychological Science, 16*(6), 326–330.

Selman, C., Blount, J.D., Nussey, D.H., and Speakman, J.R. (2012). Oxidative damage, ageing, and life-history evolution: where now? *Trends in Ecology & Evolution, 27*(10), 570–577.

Selman, C., Blount, J.D., Nussey, D.H., and Speakman, J.R. (2012). Oxidative damage, ageing, and life-history evolution: where now? *Trends in Ecology & Evolution, 27*(10), 570–577.

Senn, H., Barclay, D., Harrower, B., Ghazali, M., Kaden, J., Campbell, R. D., Kitchener, A.C., and Macdonald, D.W. (2019). Distinguishing the victim from the threat: SNP-based methods reveal the extent of introgressive hybridisation between wildcats and domestic cats in Scotland and inform future ex-situ management options for species restoration. *Evolutionary Applications, 12*(3), 399–414.

Serpell, J.A. and Jagoe, A. (1995). Early experience and the development of behaviour. In *The Domestic Dog: Its Evolution, Behaviour and Interactions with People* (J Serpell, ed., pp. 80–102). Cambridge University Press, Cambridge.

Service, K.M. (1997). Properties of badger urine as a substance used in scent marking. PhD thesis, University of Bristol.

Service, K.M. and Harris, S. (2001). Remote monitoring of badgers (*Meles meles*) for testing discrimination between urine samples from donors of different age and sex categories. In *Chemical Signals in Vertebrates 9* (A. Marchlewska-Koj, J.J. Lepri, and D. Muller-Schwarze, eds, pp. 451–457). Plenum Press, London.

Sestoft, L. (1980). Metabolic aspects of the calorigenic effect of thyroid hormone in mammals. *Clinical Endocrinology, 13*(5), 489–506.

Setchell, J.M., Vaglio, S., Moggi-Cecchi, J., Boscaro, F., Calamai, L., and Knapp, L.A. (2010). Chemical composition of scent-gland secretions in an Old World monkey (*Mandrillus sphinx*): influence of sex, male status, and individual identity. *Chemical Senses, 35*(3), 205–220.

Seyfarth, R.M. and Cheney, D.L. (1984). Grooming, alliances and reciprocal altruism in vervet monkeys. *Nature, 308*(5959), 541–543. doi:10.1038/308541a0

Shanahan, T. (2011). Phylogenetic inertia and Darwin's higher law. *Studies in History and Philosophy of Science Part C: Studies in History and Philosophy of Biological and Biomedical Sciences, 42*(1), 60–68.

Shankland, D.L. (1979). Action of dieldrin and related compounds on synaptic transmission. In *Neurotoxicology of Insecticides and Pheromones* (T. Narahashi, ed., pp. 139–153). Springer, Boston, MA.

Shao, W. (2016). Are actual weather and perceived weather the same? Understanding perceptions of local weather and their effects on risk perceptions of global warming. *Journal of Risk Research, 19*(6), 722–742.

Sharp, S.P. and Clutton-Brock, T.H. (2011). Reluctant challengers: why do subordinate female meerkats rarely displace their dominant mothers? *Behavioral Ecology, 22*(6), 1337–1343.

Sheldon, B.C. (1994). Male phenotype, fertility, and the pursuit of extra-pair copulations by female birds. *Proceedings of the Royal Society of London. Series B: Biological Sciences, 257*(1348), 25–30.

Shelley, W.B., Hurley, H.J., and Nichols, A.C. (1953). Axillary odor: experimental study of the role of bacteria, apocrine sweat, and deodorants. *AMA Archives of Dermatology and Syphilology, 68*(4), 430–446.

Shelton-Rayner, G.K., Macdonald, D.W., Chandler, S., Robertson, D., and Mian, R. (2010). Leukocyte reactivity as an objective means of quantifying mental loading during ergonomic evaluation. *Cellular Immunology, 263*(1), 22–30.

Shelton-Rayner, G.K., Mian, R., Chandler, S., Robertson, D., and Macdonald, D.W. (2012). Leukocyte responsiveness, a quantitative assay for subjective mental workload. *International Journal of Industrial Ergonomics, 42*(1), 25–33. doi:10.1016/j.ergon.2011.11.004

Sheppard, C.E., Inger, R., McDonald, R.A., Barker, S., Jackson, A.L., Thompson, F.J., Vitikainen, E.I., Cant, M.A., and Marshall, H.H. (2018). Intragroup competition predicts individual foraging specialisation in a group-living mammal. *Ecology Letters, 21*(5), 665–673.

Sherman, L. (2012). Eco-labeling: an argument for regulation and reform. *Pomona Senior Theses* 49. http://scholarship.claremont.edu/pomona_theses/49 (accessed 4 March 2022).

Shille, V.M., Munrot, C., Farmer, S.W., Papkoff, H., and Stabenfeld, G.H. (1983). Ovarian and endocrine responses in the cat after coitus. *Journal of Reproduction and Fertility, 69*(1), 29–39.

Shimmin, G.A., Skinner, J., and Baudinette, R.V. (2002). The warren architecture and environment of the southern hairy-nosed wombat (*Lasiorhinus latifrons*). *Journal of Zoology, 258*(4), 469–477.

Shkolnikov, V.M., Andreev, E.E., and Begun, A.Z. (2003). Gini coefficient as a life table function: computation from discrete data, decomposition of differences and empirical examples. *Demographic Research, 8*, 305–358.

Sibley, R.J. (2018). The investigation of a persistent outbreak of bovine tuberculosis using a novel enhanced cattle testing programme and evaluation of environmental contamination. *Cattle Practice, 26*(2), 148–153.

Sibly, R.M., Barker, D., Denham, M. C., Hone, J., and Pagel, M. (2005). On the regulation of populations of mammals, birds, fish, and insects. *Science, 309*(5734), 607–610.

Sibly, R.M., Collett, D., Promislow, D.E.L., Peacock, D.J., and Harvey, P.H. (1997). Mortality rates of mammals. *Journal of Zoology, 243*(1), 1–12.

Sidorovich, V.E., Macdonald, D.W., Pikulik, M.M., and Kruuk, H. (2001). Individual feeding specialization in the European mink, *Mustela lutreola* and the American mink, *M. vison* in north-eastern Belarus. *Folia Zoologica, 50*(1), 27–42.

Sidorovich, V.E., Rotenko, I.I., and Krasko, D.A. (2011). Badger *Meles meles* spatial structure and diet in an area of low earthworm biomass and high predation risk. *Annales Zoologici Fennici, 48*(1), 1–16.

Sies, H. (1997). Oxidative stress: oxidants and antioxidants. *Experimental Physiology: Translation and Integration, 82*(2), 291–295.

Sih, A., Hanser, S.F., and McHugh, K.A. (2009). Social network theory: new insights and issues for behavioral ecologists. *Behavioral Ecology and Sociobiology, 63*(7), 975–988.

Sih, A. (2013). Understanding variation in behavioural responses to human-induced rapid environmental change: a conceptual overview. *Animal Behaviour, 85*(5), 1077–1088.

Silberman, D.M., Wald, M., and Genaro, A.M. (2002). Effects of chronic mild stress on lymphocyte proliferative response. Participation of serum thyroid hormones and corticosterone. *International Immunopharmacology, 2*(4), 487–497.

Silk, J.B. (1983). Local resource competition and facultative adjustment of sex ratios in relation to competitive abilities. *The American Naturalist, 121*(1), 56–66.

Silk, J.B. (2007). The adaptive value of sociality in mammalian groups. *Philosophical Transactions of the Royal Society of London. Series B: Biological Sciences, 362*(1480), 539–559.

Sillero-Zubiri, C. and Macdonald, D.W. (1998). Scent-marking and territorial behaviour of Ethiopian wolves *Canis simensis* . *Journal of Zoology, 245*(3), 351–361.

Sillero-Zubiri, C., Gottelli, D., and Macdonald, D.W. (1996a). Male philopatry, extra-pack copulations and inbreeding avoidance in Ethiopian wolves (*Canis simensis*). *Behavioral Ecology and Sociobiology, 38*(5), 331–340.

Sillero-Zubiri, C., King, A.A., and Macdonald, D.W. (1996b). Rabies and mortality in Ethiopian wolves (*Canis simensis*). *Journal of Wildlife Diseases, 32*(1), 80–86.

Sillero-Zubiri, C., Marino, J., Gottelli, D., and Macdonald, D.W. (2004). Ethiopian wolves. In *The Biology and Conservation of Wild Canids* (D.W. Macdonald and C. Sillero-Zubiri, eds, pp. 311–322). Oxford University Press, Oxford.

Silva, A.P., Kilshaw, K., Johnson, P.J., Macdonald, D.W., and Rosalino, L.M. (2013). Wildcat occurrence in Scotland: food really matters. *Diversity and Distributions, 19*(2), 232–243.

Silva, A.P., Curveira-Santos, G., Kilshaw, K., Newman, C., Macdonald, D.W., Simões, L.G., and Rosalino, L.M. (2017). Climate and anthropogenic factors determine site occupancy in Scotland's northern-range badger population: implications of context-dependent responses under environmental change. *Diversity and Distributions, 23*(6), 627–639.

Silva, J.E. (1995). Thyroid hormone control of thermogenesis and energy balance. *Thyroid, 5*(6), 481–492.

Silvertown, J., Buesching, C.D., Jacobson, S.K., and Rebelo, T. (2013). Citizen science and nature conservation. *Key Topics in Conservation Biology, 2*(1), 127–142.

Simms, E. (1957). *Voices of the Wild*. Putnam Press, London.

Simpson, S.J. and Raubenheimer, D. (1995). The geometric analysis of feeding and nutrition: a user's guide. *Journal of Insect Physiology, 41*(7), 545–553.

Sin, Y.W., Buesching, C.D., Burke, T., and Macdonald, D.W. (2012a). Molecular characterization of the microbial communities in the subcaudal gland secretion of the European badger (*Meles meles*). *FEMS Microbiology Ecology, 81*(3), 648–659.

Sin, Y.W., Dugdale, H.L., Newman, C., Macdonald, D.W., and Burke, T. (2012b). Evolution of MHC class I genes in the European badger (*Meles meles*). *Ecology and Evolution, 2*(7), 1644–1662.

Sin, Y.W., Dugdale, H.L., Newman, C., Macdonald, D.W., and Burke, T. (2012c). MHC class II genes in the European badger (*Meles meles*): characterization, patterns of variation, and transcription analysis. *Immunogenetics, 64*(4), 313–327.

Sin, Y.W., Annavi, G., Dugdale, H.L., Newman, C., Burke, T., and MacDonald, D.W. (2014). Pathogen burden, co-infection and major histocompatibility complex variability in the European badger (*Meles meles*). *Molecular Ecology*, 23(20), 5072–5088.

Sin, Y.W., Annavi, G., Newman, C., Buesching, C., Burke, T., Macdonald, D.W., and Dugdale, H.L. (2015). MHC class II-assortative mate choice in European badgers (*Meles meles*). *Molecular Ecology*, 24, 3138–3150.

Sin, Y.W., Newman, C., Dugdale, H.L., Buesching, C., Mannarelli, M.E., Annavi, G., Burke, T., and Macdonald, D.W. (2016). No compensatory relationship between the innate and adaptive immune system in wild-living European badgers. *PLoS ONE*, 11(10), e0163773.

Sinclair, A.R.E. (1977). *The African Buffalo: a Study of Resource Limitation of Populations*. University of Chicago Press, Chicago, IL.

Skinner, C., Skinner, P., and Harris, S. (1991). The past history and recent decline of badgers *Meles meles* in Essex: an analysis of some of the contributory factors. *Mammal Review*, 21(2), 67–80.

Skogland, T. (1985). The effects of density-dependent resource limitations on the demography of wild reindeer. *The Journal of Animal Ecology*, 54(2), 359–374.

Skoog, P. (1970). The food of the Swedish badger, *Meles meles* L. PhD thesis, Stockholm University.

Slate, J., Kruuk, L.E.B., Marshall, T.C., Pemberton, J.M., and Clutton-Brock, T.H. (2000). Inbreeding depression influences lifetime breeding success in a wild population of red deer (*Cervus elaphus*). *Proceedings of the Royal Society of London. Series B: Biological Sciences*, 267(1453), 1657–1662.

Slawik, M. and Vidal-Puig, A.J. (2006). Lipotoxicity, overnutrition and energy metabolism in aging. *Ageing Research Reviews*, 5(2), 144–164.

Sleeman, D.P., Davenport, J., More, S.J., Clegg, T.A., Collins, J.D., Martin, S.W., Williams, D.H., Griffin, J.M., and O'Boyle, I. (2009). How many Eurasian badgers *Meles meles* L. are there in the Republic of Ireland? *European Journal of Wildlife Research*, 55(4), 333–344.

Sliwa, A. (1996). A functional analysis of scent marking and mating behaviour in the aardwolf, Proteles cristatus. PhD thesis, University of Pretoria.

Sliwa, A. and Richardson, P.R. (1998). Responses of aardwolves, *Proteles cristatus*, to translocated scent marks. *Animal Behaviour*, 56(1), 137–146.

Smadja, C. and Butlin, R.K. (2009). On the scent of speciation: the chemosensory system and its role in premating isolation. *Heredity*, 102(1), 77–97.

Smit, B. and Wandel, J. (2006). Adaptation, adaptive capacity and vulnerability. *Global Environmental Change*, 16(3), 282–292.

Smit, B., Burton, I., Klein, R.J., and Wandel, J. (2000). An anatomy of adaptation to climate change and variability. In *Societal Adaptation to Climate Variability and Change* (S.M. Kane and G.W. Yohe, eds, pp. 223–251). Springer, Dordrecht.

Smith, C.C. and Reichman, O.J. (1984). The evolution of food caching by birds and mammals. *Annual Review of Ecology and Systematics*, 15(1), 329–351.

Smith, G.C., Cheeseman, C.L., Clifton-Hadley, R.S., and Wilkinson, D. (2001). A model of bovine tuberculosis in the badger *Meles meles*: an evaluation of control strategies. *Journal of Applied Ecology*, 38, 509–519.

Smith, J.M. and Parker, G.A. (1976). The logic of asymmetric contests. *Animal Behaviour*, 24(1), 159–175.

Smith, J.N. (1974). The food searching behaviour of two European thrushes. II: the adaptiveness of the search patterns. *Behaviour*, 49(1–2), 1–60.

Smith, M.J. and Harper, D.G. (1995). Animal signals: models and terminology. *Journal of Theoretical Biology*, 177(3), 305–311.

Smith, N.H., Dale, J., Inwald, J., Palmer, S., Gordon, S.V., Hewinson, R.G., and Smith, J.M. (2003). The population structure of *Mycobacterium bovis* in Great Britain: clonal expansion. *Proceedings of the National Academy of Sciences*, 100(25), 15271–15275.

Smuts, B.B. and Smuts, R.W. (1993). Male aggression and sexual coercion of females in nonhuman primates and other mammals: evidence and theoretical implications. *Advances in the Study of Behavior*, 22(22), 1–63.

Snell-Rood, E.C. (2013). An overview of the evolutionary causes and consequences of behavioural plasticity. *Animal Behaviour*, 85(5), 1004–1011.

Snyder, R.E. and Ellner, S.P. (2018). Pluck or luck: does trait variation or chance drive variation in lifetime reproductive success? *The American Naturalist*, 191(4), E90–E107.

Sober, E. (1981). The principle of parsimony. *The British Journal for the Philosophy of Science*, 32(2), 145–156.

Somers, M.J. and Nel, J.A. (2004). Habitat selection by the Cape clawless otter (*Aonyx capensis*) in rivers in the Western Cape Province, South Africa. *African Journal of Ecology*, 42(4), 298–305.

Soulsbury, C.D. (2010). Genetic patterns of paternity and testes size in mammals. *PLoS ONE*, 5(3), e9581.

Soulsbury, C.D., Kervinen, M., and Lebigre, C. (2014). Sexual size dimorphism and the strength of sexual selection in mammals and birds. *Evolutionary Ecology Research*, 16(1), 63–76.

Spano, L.C., Gatti, J., Nascimento, J.P., and Leite, J.P.G. (2004). Prevalence of human cytomegalovirus infection in pregnant and non-pregnant women. *Journal of Infection*, 48(3), 213–220.

Sparks J. (1967). Allogrooming in primates: a review. In *Primate Ethology* (J. Sparks, ed., pp. 148–175). Routledge, Abingdon.

Speakman, J. (1997). Factors influencing the daily energy expenditure of small mammals. *Proceedings of the Nutrition Society*, 56(3), 1119–1136.

Speakman, J.R. and Król, E. (2010). Maximal heat dissipation capacity and hyperthermia risk: neglected key factors in the ecology of endotherms. *Journal of Animal Ecology*, 79(4), 726–746.

Speedy, A.W. (1980). *Sheep Production, Science into Practice*. Longman, Harlow.

Spencer, H. (1864). *The Principles of Biology* (2 vols). *System of Synthetic Philosophy*, 2. Williams and Norgate, London.

Spielman, A., Wilson, M.L., Levine, J.F., and Piesman, J. (1985). Ecology of *Ixodes dammini*-borne human babesiosis and Lyme disease. *Annual Review of Entomology*, 30(1), 439–460.

Spong, G.F., Hodge, S.J., Young, A.J., and Clutton-Brock, T.H. (2008). Factors affecting the reproductive success of dominant male meerkats. *Molecular Ecology*, 17(9), 2287–2299.

Spooner, W.A. (1914). The Golden Rule. In *Encyclopedia of Religion and Ethics*, Vol. 6 (J. Hastings, ed., pp. 310–312). Charles Scribner's Sons, New York.

Springate, L. and Frasier, T.R. (2017). Gamete compatibility genes in mammals: candidates, applications and a potential path forward. *Royal Society Open Science*, 4(8), 170577.

Srinivasan, S., Conlan, A.J.K., Easterling, L.A., Herrera, C., Dandapat, P., Veerasami, M., Ameni, G., Jindal, N., Raj, G.D., Wood, J., Juleff, N., Bakker, D., Vordermeier, M., and Kapur, V. (2021). A meta-analysis of the effect of Bacillus Calmette-Guérin vaccination against bovine tuberculosis: is perfect the enemy of good? *Frontiers in Veterinary Science*, 8, 637580. doi:10.3389/fvets.2021.637580

Stacey, P.B. and Ligon, J.D. (1991). The benefits-of-philopatry hypothesis for the evolution of cooperative breeding: variation in territory quality and group size effects. *The American Naturalist*, 137(6), 831–846.

Stamps, J.A. (2006). The silver spoon effect and habitat selection by natal dispersers. *Ecology Letters*, 9(11), 1179–1185.

Stamps, J.A. and Buechner, M. (1985). The territorial defense hypothesis and the ecology of insular vertebrates. *The Quarterly Review of Biology*, 60(2), 155–181.

Starrfelt, J. and Kokko, H. (2012). Bet-hedging a triple tradeoff between means, variances and correlations. *Biological Reviews*, 87, 742–755.

Stearns, S.C. (1977). The evolution of life history traits: a critique of the theory and a review of the data. *Annual Review of Ecology and Systematics*, 8(1), 145–171.

Stearns, S.C. (1992). *The Evolution of Life Histories*. Oxford University Press, Oxford.

Steele, V.R. and Temple-Smith, P.D. (1998). The northern hairy-nosed wombat recovery program: trials and triumphs. In *Wombats* (R.T. Wells and P.A. Pridmore, eds, pp. 113–124). Surrey Beatty & Sons, Chipping Norton (Australia).

Stenseth, N.C. and Ims, R.A. (eds) (1993). *The Biology of Lemmings*. Academic Press, London.

Stenseth, N.C. and Mysterud, A. (2002). Climate, changing phenology, and other life history traits: nonlinearity and match–mismatch to the environment. *Proceedings of the National Academy of Sciences*, 99(21), 13379–13381.

Stenseth, N.C. and Smith, J. M. (1984). Coevolution in ecosystems: red queen evolution or stasis? *Evolution*, 38(4), 870–880.

Stenvinkel, P. and Shiels, P.G. (2019). Long-lived animals with negligible senescence: clues for ageing research. *Biochemical Society Transactions*, 47(4), 1157–1164.

Stephens, D.W. and Krebs, J.R. (2019). *Foraging Theory*. Princeton University Press, Princeton, NJ.

Sterling, P. and Eyer, J. (1988). Allostasis: a new paradigm to explain arousal pathology. In *Handbook of Life Stress,Cognition and Health* (S. Fisher and J. Reason, eds, pp. 631–651). Wiley, New York.

Stewart, P.D. (1997). The Social Behaviour of the European Badger *Meles meles*. PhD thesis, University of Oxford.

Stewart, P.D. and Macdonald, D.W. (1997). Age, sex, and condition as predictors of moult and the efficacy of a novel fur-clip technique for individual marking of the European badger (*Meles meles*). *Journal of Zoology*, 241(3), 543–550.

Stewart, P.D. and Macdonald, D.W. (2003). Badgers and badger fleas: strategies and counter-strategies. *Ethology*, 109(9), 751–764.

Stewart, P.D. and Macdonald, D.W. (2003). Badgers and badger fleas: strategies and counter-strategies. *Ethology*, 109(9), 751–764.

Stewart, P.D., Anderson, C., and Macdonald, D.W. (1997). A mechanism for passive range exclusion: evidence from the European badger (*Meles meles*). *Journal of Theoretical*

Stewart, P.D., Bonesi, L., and Macdonald, D.W. (1999). Individual differences in den maintenance effort in a communally dwelling mammal: the Eurasian badger. *Animal Behaviour*, 57(1), 153–161.

Stewart, P.D., Macdonald, D.W., Newman, C., and Cheeseman, C.L. (2001). Boundary faeces and matched advertisement in the European badger (*Meles meles*): a potential role in range exclusion. *Journal of Zoology*, 255(2), 191–198.

Stewart, P.D., MacDonald, D.W., Newman, C., and Tattersall, F.H. (2002). Behavioural mechanisms of information transmission and reception by badgers, *Meles meles*, at latrines. *Animal Behaviour*, 63(5), 999–1007.

Steyaert, S.M., Reusch, C., Brunberg, S., Swenson, J.E., Hackländer, K., and Zedrosser, A. (2013). Infanticide as a male reproductive strategy has a nutritive risk effect in brown bears. *Biology Letters*, 9(5), 20130624.

Stockley, P., Searle, J.B., Macdonald, D.W., and Jones, C.S. (1993). Female multiple mating behaviour in the common shrew as a strategy to reduce inbreeding. *Proceedings of the Royal Society of London. Series B: Biological Sciences*, 254, 173–179.

Stockley, P., Searle, J.B., Macdonald, D.W., and Jones, C.S. (1994). Alternative reproductive tactics in male common shrews: relationships between mate-searching behaviour, sperm production and reproductive success as revealed by DNA fingerprinting. *Behavioural Ecology and Sociobiology*, 3, 71–78.

Stockley, P., Searle, J.B., Macdonald, D.W., and Jones, C.S. (1996). Correlates of reproductive success within alternative mating tactics of the common shrew. *Behavioural Ecology* 7(3), 334–340.

Stockmaier, S., Bolnick, D.I., Page, R.A., and Carter, G.G. (2018). An immune challenge reduces social grooming in vampire bats. *Animal Behaviour*, 140, 141–149. doi:10.1016/j.anbehav.2018.04.021

Stoeger, T. and Adler, H. (2019). 'Novel' triggers of herpes virus reactivation and their potential health relevance. *Frontiers in Microbiology*, 9, 3207. https://doi.org/10.3389/fmicb.2018.03207

Stonorov, D. and Stokes, A.W. (1972). Social behavior of the Alaska brown bear. *Bears: Their Biology and Management*, 2, 232–242.

Stopka, P. and Graciasová, R. (2001). Conditional allogrooming in the herb-field mouse. *Behavioral Ecology*, 12(5), 584–589.

Stopka, P. and Macdonald, D.W. (1999). The market effect in the wood mouse, *Apodemus sylvaticus*: selling information on reproductive status. *Ethology*, 105(11), 969–982.

Stopka, P. and Macdonald, D.W. (2003). Way-marking behaviour: an aid to spatial navigation in the wood mouse (*Apodemus sylvaticus*). *BMC Ecology*, 3(3).

Strain, S.A., McNair, J., McDowell, S.W., and Branch, B. (2011). Bovine tuberculosis: a review of diagnostic tests for *M. bovis* infection in badgers. Report: Agri-Food and Biosciences Institute. http://www.dardni.gov.uk/afbi-literature-review-tb-reviewdiagnostic-tests-badgers.pdf (accessed 22 March 2016).

Straub, R.H. (2017). The brain and immune system prompt energy shortage in chronic inflammation and ageing. *Nature Reviews Rheumatology*, 13(12), 743.

Strohacker, K., Carpenter, K.C., and Mcfarlin, B.K. (2009). Consequences of weight cycling: an increase in disease risk? *International Journal of Exercise Science*, 2(3), 191.

Strotz, L.C., Simoes, M., Girard, M.G., Breitkreuz, L., Kimmig, J., and Lieberman, B.S. (2018). Getting somewhere with the Red Queen: chasing a biologically modern definition of the hypothesis. *Biology Letters*, 14(5), 20170734.

Stubbe, M. (1971). Die analen Markierungsorgane des Dachses (*Meles meles*). *Der Zoolische Garten*, 40, 125–135.

Sugianto, N.A. (2018). Reproductive biology of the European badger (Meles meles): endocrinological insights into lifetime reproductive events, strategies and cub development in response to ecological factors. PhD thesis, University of Oxford.

Sugianto, N.A., Buesching, C.D., Heistermann, M., Newman, C., and Macdonald, D.W. (2018). Linking plasma sex steroid hormone levels to the condition of external genitalia in European badgers (*Meles meles*): a critical evaluation of traditional field methodology. *Mammalian Biology*, 93(1), 97–108.

Sugianto, N.A., Buesching, C.D., Macdonald, D.W. and Newman, C. (2019a). The importance of refining anaesthetic regimes to mitigate adverse effects in very young and very old wild animals: the European badger (*Meles meles*). *Journal of Zoological Research*, 3(3), 10–17.

Sugianto, N.A., Newman, C., Macdonald, D.W., and Buesching, C.D. (2019b). Extrinsic factors affecting cub development contribute to sexual size dimorphism in the European badger (*Meles meles*). *Zoology*, 135, 125688.

Sugianto, N.A., Newman, C., Macdonald, D.W., and Buesching, C.D. (2019c). Heterochrony of puberty in the European badger (*Meles meles*) can be explained by growth rate and group-size: evidence for two endocrinological phenotypes. *PLoS ONE*, 14(3), e0203910.

Sugianto, N.A., Newman, C., Macdonald, D.W., and Buesching, C.D. (2020). Reproductive and somatic senescence in the European badger (*Meles meles*): evidence from lifetime sex-steroid profiles. *Zoology*, 141, 125803.

Sugianto, N.A., Dehnhard, M., Newman, C., Macdonald, D.W., and Buesching, C.D. (2021a). A non-invasive method to assess the reproductive status of the European badger (*Meles meles*) from urinary sex-steroid metabolites. *General and Comparative Endocrinology*, 301, 113655.

Sugianto, N.A., Heistermann, M., Newman, C., Macdonald, D.W., and Buesching, C.D. (2021b). Alternative reproductive strategies provide a flexible mechanism for assuring mating success in the European badgers (*Meles meles*): an investigation from hormonal measures. *General and Comparative Endocrinology*, 310, 113823.

Sun, L. and Müller-Schwarze, D. (1998). Beaver response to recurrent alien scents: scent fence or scent match? *Animal Behaviour*, 55(6), 1529–1536.

Sun, Q., Stevens, C., Newman, C., Buesching, C.D., and Macdonald, D.W. (2015). Cumulative experience, age-class, sex and season affect the behavioural responses of European badgers (*Meles meles*) to handling and sedation. *Animal Welfare*, 24(4), 373–385.

Svanbäck, R. and Persson, L. (2004). Individual diet specialization, niche width and population dynamics: implications for trophic polymorphisms. *Journal of Animal Ecology*, 73(5), 973–982.

Svensson, E.I., Abbott, J., and Härdling, R. (2005). Female polymorphism, frequency dependence, and rapid evolutionary dynamics in natural populations. *The American Naturalist*, 165(5), 567–576.

Swain, S.L. (1983). T cell subsets and the recognition of MHC class. *Immunological Reviews*, 74, 129–142.

Swedell, L. and Plummer, T. (2012). A papionin multilevel society as a model for hominin social evolution. *International Journal of Primatology*, 33(5), 1165–1193.

Sweeney, J., Albanito, F., Brereton, A., Caffarra, A., Charlton, R., Donnelly, A., Fealy, R., Fitzgerald, J., Holden, N., Jones, M., and Murphy, C. (2008). *Climate Change—Refining the Impacts for Ireland: STRIVE Report (2001-CD-C3-M1)*. Environmental Protection Agency, Johnstown Castle.

Swift, B.M.C., Barron, E.S., Christley, R., Corbetta, D., Grau-Roma, L., Jewell, C., O'Cathail, C., Mitchell, A., Phoenix, J., Prosser, A., Rees, C., Sorley, M., Verin, R., and M. Bennet (in press). Tuberculosis in badgers where the bovine tuberculosis epidemic is expanding in cattle in England.

Swinton, J., Tuyttens, F., Macdonald, D., Nokes, D.J., Cheeseman, C.L., and Clifton-Hadley, R. (1997). A comparison of fertility control and lethal control of bovine tuberculosis in badgers: the impact of perturbation induced transmission. *Philosophical Transactions of the Royal Society of London. Series B: Biological Sciences*, 352(1353), 619–631.

Sykes, B. and Nicholson, J. (2000). The genetic structure of a Highland clan. https://electricscotland.com/history/articles/geneticstructureofahighlandclan.pdf (accessed 25 March 2022).

Symes, R.G. (1989). Badger damage: fact or fiction. In *Mammals as Pests* (R. Puman, ed., pp. 196–206). Springer Science & Business Media, Berlin, Germany.

Szemán, K., Liker, A., and Székely, T. (2021). Social organization in ungulates: revisiting Jarman's hypotheses. *Journal of Evolutionary Biology*, 34(4), 604–613.

Szulkin, M., Garant, D., McCleery, R.H., and Sheldon, B.C. (2007). Inbreeding depression along a life-history continuum in the great tit. *Journal of Evolutionary Biology*, 20(4), 1531–1543.

Taber, A.B. and Macdonald, D.W. (1992a). Spatial organization and monogamy in the mara *Dolichotis patagonum*. *Journal of Zoology*, 227, 417–438.

Taber, A.B. and Macdonald, D.W. (1992b). Communal breeding in the mara, *Dolichotis patagonum*. *Journal of Zoology*, 227, 439–452.

Taborsky, B. (2006). The influence of juvenile and adult environments on life-history trajectories. *Proceedings of the Royal Society. Series B: Biological Sciences*, 273(1587), 741–750.

Taborsky, M. and Brockmann, H.J. (2010). Alternative reproductive tactics and life history phenotypes. In *Animal Behaviour: Evolution and Mechanisms* (N. Anthes, R. Bergmüller, W. Blanckenhorn, H.J. Brockmann, C. Fichtel, L. Fromhage, and S. Zhang, eds, pp. 537–586). Springer Science & Business Media, Berlin, Heidelberg, Germany.

Taborsky, M., Oliveira, R.F., and Brockmann, H.J. (2008). The evolution of alternative reproductive tactics: concepts and questions. *Alternative Reproductive Tactics: An Integrative Approach*, 1, 21.

Taggart, D.A. and Temple-Smith, P.D.M. (2008). Southern hairy-nosed wombat. In *The Mammals of Australia* (pp. 204–206). New Holland Publishers (Australia) Pty. Ltd, Sydney.

Takahata, N. and Nei, M. (1990). Allelic genealogy under overdominant and frequency-dependent selection and polymorphism of major histocompatibility complex loci. *Genetics*, 124(4), 967–978.

Talwalkar, P. (2008). Understanding the stag hunt game: how deer hunting explains why people are socially late. https://mindyourdecisions.com/blog/2008/06/03/understanding-the-stag-hunt-game-how-deer-hunting-explains-why-people-are-socially-late/ (accessed 5 April 2021).

Tan, C.K.W., Moore, J., Bin Saaban, S., Campos-Arceiz, A., and Macdonald, D.W. (2015). The discovery of two spotted leopards (*Panthera pardus*) in Peninsular Malaysia. *Tropical Conservation Science*, 8(3), 732–737.

Tanaka, H. (2006). Winter hibernation and body temperature fluctuation in the Japanese badger, *Meles meles anakuma*. *Zoological Science*, 23(11), 991–997.

Tanaka, H., Yamanaka, A., and Endo, K. (2002). Spatial distribution and sett use by the Japanese badger, *Meles meles anakuma*. *Mammal Study*, 27(1), 15–22.

Tannerfeldt, M. and Angerbjörn, A. (1998). Fluctuating resources and the evolution of litter size in the arctic fox. *Oikos*, 83(3), 545–559.

Tatarenkov, A., Healey, C.I., Grether, G.F., and Avise, J.C. (2008). Pronounced reproductive skew in a natural population of green swordtails, *Xiphophorus helleri*. *Molecular Ecology*, 17(20), 4522–4534.

Tay, M.Z., Poh, C.M., Rénia, L., MacAry, P.A., and Ng, L.F. (2020). The trinity of COVID-19: immunity, inflammation and intervention. *Nature Reviews Immunology*, 20(6), 363–374.

Taylor, P.D. (1992). Altruism in viscous populations—an inclusive fitness model. *Evolutionary Ecology*, 6(4), 352–356.

Taylor, S.J., Ahonen, L.J., de Leij, F.A., and Dale, J.W. (2003). Infection of *Acanthamoeba castellanii* with *Mycobacterium bovis* and M. bovis BCG and survival of M. bovis within the amoebae. *Applied Environmental Microbiology*, 69(7), 4316–4319. doi:10.1128/AEM.69.7.4316-4319.2003

TBhub (n.d.a). Tuberculin skin testing. https://tbhub.co.uk/tb-testing-cattle/skin-testing/tuberculin-skin-testing/ (accessed 10 March 2022).

TBhub (n.d.b). The interferon gamma test (IFNγ). https://tbhub.co.uk/wp-content/uploads/2020/01/Factsheet_gamma_test_TB_hub.pdf (accessed 14 March 2022).

Temeles, E.J. (1994). The role of neighbors in territorial systems—when are they dear enemies. *Animal Behaviour*, 47, 339–350.

Ten Hwang, Y. and Larivière, S. (2003). *Mydaus javanensis*. *Mammalian species*, 2003(723), 1–3.

Ten Hwang, Y. and Larivière, S. (2004). *Mydaus marchei*. *Mammalian Species*, 2004(757), 1–3.

Terrien, J., Perret, M., and Aujard, F. (2011). Behavioral thermoregulation in mammals: a review. *Frontiers in Bioscience*, 16(4), 1428–1444.

Theis, K.R., Schmidt, T.M., and Holekamp, K.E. (2012). Evidence for a bacterial mechanism for group-specific social odors among hyenas. *Scientific Reports*, 2(1), 1–8.

Theis, K.R., Venkataraman, A., Dycus, J.A., Koonter, K.D., Schmitt-Matzen, E.N., Wagner, A.P., Holekamp, K.E., and Schmidt, T.M. (2013). Symbiotic bacteria appear to mediate hyena social odors. *Proceedings of the National Academy of Sciences*, 110(49), 19832–19837.

Thelen, E. and Farish, D.J. (1977). An analysis of the grooming behavior of wild and mutant strains of *Bracon hebetor* (Braconidae: Hymenoptera). *Behaviour*, 62, 699–715.

Thom, M.D., Johnson, D.D., and Macdonald, D.W. (2004a). The evolution and maintenance of delayed implantation in the Mustelidae (Mammalia: Carnivora). *Evolution*, 58(1), 175–183.

Thom, M.D., Macdonald, D.W., Mason, G.J., Pedersen, V. and Johnson, P.J. (2004b). Female American mink (*Mustela vison*) mate multiply in a free-choice environment. *Animal Behaviour*, 67, 975–984.

Thomas, G. (1974). The influences of encountering a food object on subsequent searching behaviour in *Gasterosteus aculeatus* L. *Animal Behaviour*, 22, 941–952.

Thompson, H.V. (1978). Wildlife as vectors in diseases: approaches to solving these problems in the United Kingdom. https://core.ac.uk/download/pdf/188050788.pdf (accessed 10 March 2022).

Thornhill, R. and Alcock, J. (1983). *The Evolution of Insect Mating Systems*. Harvard University Press, Cambridge, MA.

Thornhill, R., Gangestad, S.W., Miller, R., Scheyd, G., McCollough, J.K., and Franklin, M. (2003). Major histocompatibility complex genes, symmetry, and body scent attractiveness in men and women. *Behavioral Ecology*, 14(5), 668–678.

Thornton, P.D., Newman, C., Johnson, P.J., Buesching, C.D., Baker, S.E., Slater, D., Johnson, D.D., and Macdonald, D.W. (2005). Preliminary comparison of four anaesthetic techniques in badgers (*Meles meles*). *Veterinary Anaesthesia and Analgesia*, 32(1), 40–47.

Thornton, P.S. (1988). Density and distribution of badgers in south-west England—a predictive model. *Mammal Review*, 18(1), 11–23.

Thrall, P.H., Antonovics, J., and Dobson, A.P. (2000). Sexually transmitted diseases in polygynous mating systems: prevalence and impact on reproductive success. *Proceedings of the Royal Society of London. Series B: Biological Sciences*, 267(1452), 1555–1563.

Tian, X., Doerig, K., Park, R., Can Ran Qin, A., Hwang, C., Neary, A., and Gorbunova, V. (2018). Evolution of telomere maintenance and tumour suppressor mechanisms across mammals. *Philosophical Transactions of the Royal Society. Series B: Biological Sciences*, 373(1741), 20160443.

Tiddi, B., Aureli, F., and Schino, G. (2012). Grooming up the hierarchy: the exchange of grooming and rank-related benefits in a new world primate. *PLoS ONE*, 7(5), e36641.

Timmermann, A.G. (1993). How learning in financial markets generates excess volatility and predictability in stock prices. *The Quarterly Journal of Economics*, 108(4), 1135–1145.

Tinbergen, N. (1963). On aims and methods in ethology. *Zeitschrift für Tierpsychologie*, 20, 410–433.

Tinbergen, N., Impekoven, M., and Frank, D. (1967). An experiment on spacing out as a defence against predation. *Behaviour*, 28, 307–321.

Tinnesand, H.V., Buesching, C.D., Noonan, M.J., Newman, C., Zedrosser, A., Rosell, F., and Macdonald, D.W. (2015). Will trespassers be prosecuted or assessed according to their merits? A consilient interpretation of territoriality in a group-living carnivore, the European badger (*Meles meles*). *PLoS ONE*, 10(7), e0132432.

Toïgo, C., Gaillard, J.M., Van Laere, G., Hewison, M., and Morellet, N. (2006). How does environmental variation influence body mass, body size, and body condition? Roe deer as a case study. *Ecography*, 29(3), 301–308.

Tolhurst, B.A., Baker, R.J., Cagnacci, F., and Scott, D.M. (2020). Spatial aspects of gardens drive ranging in urban foxes (*Vulpes vulpes*): the resource dispersion hypothesis revisited. *Animals*, 10(7), 1167.

Tompkins, D.M., Buddle, B.M., Whitford, J., Cross, M.L., Yates, G.F., Lambeth, M.R., and Nugent, G. (2013). Sustained protection against tuberculosis conferred to a wildlife host by single dose oral vaccination. *Vaccine*, 31(6), 893–899.

Tonzetich, J. and Carpenter, P.A.W. (1971). Production of volatile sulphur compounds from cysteine, cystine and methionine by human dental plaque. *Archives of Oral Biology*, 16(6), 599–607.

Tonzetich, J., Eigen, E., King, W.J., and Weiss, S. (1967). Volatility as a factor in the inability of certain amines and indole to increase the odour of saliva. *Archives of Oral Biology*, 12(10), 1167–1175.

Toth, M.J. and Tchernof, A. (2000). Lipid metabolism in the elderly. *European Journal of Clinical Nutrition*, 54(3), S121–S125.

Tregenza, T. and Wedell, N. (2000). Genetic compatibility, mate choice and patterns of parentage: invited review. *Molecular Ecology*, 9(8), 1013–1027.

Trethowan, P., Fuller, A., Haw, A., Hart, T., Markham, A., Loveridge, A., and Macdonald, D.W. (2017). Getting to the core: internal body temperatures help reveal the ecological function and thermal implications of the lions' mane. *Ecology and Evolution*, 7(1), 253–262.

Triggs, B. (1996). *The Wombat: Common Wombats in Australia*. University of New South Wales Press, Sydney.

Trivers, R.L. (1971). The evolution of reciprocal altruism. *The Quarterly Review of Biology*, 46(1), 35–57.

Trivers, R.L. (1972). Parental investment and sexual selection. In *Sexual Selection and the Descent of Man* (B. Campbell, ed., pp. 136–179). Aldine, London.

Trivers, R.L. and Willard, D.E. (1973). Natural selection of parental ability to vary the sex ratio of offspring. *Science*, 179(4068), 90–92.

Tsai, M-S., Fogarty, U., Byrne, A.W., O'Keeffe, J., Newman, C., Macdonald, D.W., and Buesching, C.D. (2020). Effects of Mustelid gamma herpesvirus 1 (MusGHV-1) reactivation in European badger (*Meles meles*) genital tracts on reproductive fitness. *Pathogens*, 9(9), 769–786.

Tsai, M-S., François, S., Newman, C., Macdonald, D.W., and Buesching, C.D. (2021). Patterns of genital tract mustelid gammaherpesvirus 1 (MusGHV-1) reactivation are linked to stressors in European badgers (*Meles meles*). *Biomolecules*, 11(5), 716.

Tsai, M-S., François, S., Newman, C., Macdonald, D.W., and Buesching, C.D. (submitted). Infection with a recently discovered gammaherpesvirus variant in European badgers, *Meles meles*, is associated with higher relative viral loads in blood, *Pathogens*.

Tsai, M-S., Newman, C., Macdonald, D.W., and Buesching, C.D. (2022). Adverse weather during in utero development is linked to higher rates of later-life herpesvirus reactivation in adult European badgers, *Meles meles*. *Royal Society Open Science*, 9(5), 211749.

Tschirren, B., Fitze, P.S., and Richner, H. (2003). Sexual dimorphism in susceptibility to parasites and cell-mediated immunity in great tit nestlings. *Journal of Animal Ecology*, 72(5), 839–845.

Tsubota, T., Takahashi, Y., and Kanagawa, H. (1987). Changes in serum progesterone levels and growth of fetuses in Hokkaido brown bears. *Bears: Their Biology and Management*. Volume 7, A Selection of Papers from the Seventh International Conference on Bear Research and Management, Williamsburg, Virginia, USA, and Plitvice Lakes, Yugoslavia, February and March 1986 (1987), pp. 355–358.

Tsunoda, M., Newman, C., Buesching, C.D., Macdonald, D.W., and Kaneko, Y. (2018). Badger setts provide thermal refugia, buffering changeable surface weather conditions.

Tsunoda, H., Newman, C., Peeva, S., Raichev, E., Buesching, C.D., and Kaneko, Y. (2020). Spatio-temporal partitioning facilitates mesocarnivore sympatry in the Stara Planina Mountains, Bulgaria. *Zoology*, 141, 125801.

Tuljapurkar, S., Gaillard, J.-M., and Coulson, T. (2009a). From stochastic environments to life histories and back. *Philosophical Transactions of the Royal Society. Series B: Biological Sciences*, 364, 1499–1509.

Tuljapurkar, S., Steiner, U.K., and Orzack, S.H. (2009b). Dynamic heterogeneity in life histories. *Ecology Letters*, 12, 93–106.

Tuljapurkar, S., Zuo, W., Coulson, T., Horvitz, C., and Gaillard, J.M. (2020). Skewed distributions of lifetime reproductive success: beyond mean and variance. *Ecology Letters*, 23(4), 748–756.

Tully, T. and Lambert, A. (2011). The evolution of postreproductive life span as an insurance against indeterminacy. *Evolution: International Journal of Organic Evolution*, 65(10), 3013–3020.

Tuomi, J., Hakala, T., and Haukioja, E. (1983). Alternative concepts of reproductive effort, costs of reproduction, and selection in life-history evolution. *American Zoologist*, 23(1), 25–34.

Turbill, C., Ruf, T., Smith, S., and Bieber, C. (2013). Seasonal variation in telomere length of a hibernating rodent. *Biology Letters*, 9(2), 20121095.

Turner, D.C., Bateson, P., and Bateson, P.P.G. (eds) (2000). *The Domestic Cat: The Biology of Its Behaviour*. Cambridge University Press, Cambridge.

Tuyttens, F.A.M. (2000). The closed-subpopulation model and estimation of population size from mark-recapture and ancillary data. *Canadian Journal of Zoology*, 78, 320–326.

Tuyttens, F.A.M. and Macdonald, D.W. (2000). Consequences of social perturbation for wildlife management and conservation. *Proceedings of the Royal Society. Series B: Biological Sciences*, 274(1626), 2769–2777.

Tuyttens, F.A.M., Macdonald, D.W., Rogers, L.M., Mallinson, P.J., Delahay, R., Donnelly, C.A., and Newman, C. (1999). Differences in trappability of European badgers *Meles meles* in three populations in England. *Journal of Applied Ecology*, 36, 1051–1062.

Tuyttens, F.A.M., Barron, L., Mallinson, P.J., Rogers, L.M., and Macdonald, D.W. (2000a). Wildlife management and scientific research: a retrospective evaluation of two badger removal operations for the control of bovine tuberculosis. In *Mustelids in a Modern World: Management and Conservation Aspects of Small Carnivore: Human Interactions* (H.I. Griffiths, ed., pp. 247–265). Backhuys Publishers, Leiden.

Tuyttens, F.A.M., Delahay, R.J., Macdonald, D.W., Cheeseman, C.L., Long, B., and Donnelly, C.A. (2000b). Spatial perturbation caused by a badger (*Meles meles*) culling operation: implications for the function of territoriality and the control of bovine tuberculosis. *Journal of Animal Ecology*, 69, 815–828.

Tuyttens, F.A.M., Macdonald, D.W., Rogers, L.M., Cheeseman, C.L., and Roddam A.W. (2000c). Comparative study on the consequences of culling badgers (*Meles meles*) on biometrics, population dynamics and movement. *Journal of Animal Ecology*, 69, 567–580.

Tuyttens, F.A.M., Stapley, N., Stewart, P.D., and Macdonald, D.W. (2001). Vigilance in badgers *Meles meles*: the effects of group size and human persecution. *Acta Theriologica*, 46, 79–86.

UCD (2016). Cattle movements in Ireland—2016. https://www.youtube.com/watch?v=PTCdPMnenBw (accessed 14 March 2022).

Ulrich, Y., Saragosti, J., Tokita, C.K., Tarnita, C.E., and Kronauer, D.J.C. (2018). Fitness benefits and emergent division of labour at the onset of group living. *Nature*, 560(7720), 635–638.

Uttara, B., Singh, A.V., Zamboni, P., and Mahajan, R.T. (2009). Oxidative stress and neurodegenerative diseases: a review of upstream and downstream antioxidant therapeutic options. *Current Neuropharmacology*, 7(1), 65–74.

Vaiserman, A.M. (2015). Epigenetic programming by early-life stress: evidence from human populations. *Developmental Dynamics*, 244, 254–265, doi:10.1002/dvdy.24211

Valavanis-Vail, S. (1954). Leontief's scarce factor paradox. *Journal of Political Economy*, 62(6), 523–528.

Valeix, M., Loveridge, A.J., and Macdonald, D.W. (2012). Influence of prey dispersion on territory and group size of African lions: a test of the resource dispersion hypothesis. *Ecology*, 93(11), 2490–2496.

Valenti, G. and Schwartz, R.S. (2008). Anabolic decline in the aging male: a situation of unbalanced syncrinology. *The Aging Male*, 11(4), 153–156.

Valenzuela, D. and Macdonald, D.W. (2002). Home-range use by white-nosed coatis (*Nasua narica*). *Journal of Zoology*, 258, 247–256.

Van Buskirk, J. and Steiner, U.K. (2009). The fitness costs of developmental canalization and plasticity. *Journal of Evolutionary Biology*, 22(4), 852–860.

Van de Pol, M. and Cockburn, A. (2011). Identifying the critical climatic time window that affects trait expression. *The American Naturalist*, 177(5), 698–707.

Van Der Wal, J., Murphy, H.T., Kutt, A.S., Perkins, G.C., Bateman, B.L., Perry, J.J., and Reside, A.E. (2013). Focus on poleward shifts in species' distribution underestimates the fingerprint of climate change. *Nature Climate Change*, 3(3), 239–243.

Van Dongen, S., Sprengers, E., Löfstedt, C., and Matthysen, E. (1999). Fitness components of male and female winter moths (*Operophtera brumata* L.) (Lepidoptera, Geometridae) relative to measures of body size and asymmetry. *Behavioral Ecology*, 10(6), 659–665.

Van Helvoort, B.E., Melisch, R., Lubis, I.R., and O'Callaghan, B. (1996). Aspects of preying behaviour of smooth-coated otters Lutrogale perspicillata from southeast Asia. *IUCN Otter Specialist Group Bulletin*, 13(1), 3–6.

van Kesteren, F., Paris, M., Macdonald, D.W., Millar, R., Argaw, K., Johnson, P.J., Farstad, W., and Sillero-Zubiri, C. (2013). The physiology of cooperative breeding in a rare social canid: sex, suppression and pseudopregnancy in female Ethiopian wolves. *Physiology & Behaviour*, 122, 39–45. http://dx.doi.org/10.1016/j.physbeh.2013.08.016

van Lieshout, S.H., Bretman, A., Newman, C., Buesching, C.D., Macdonald, D.W., and Dugdale, H.L. (2019). Individual variation in early-life telomere length and survival in a wild mammal. *Molecular Ecology*, 28(18), 4152–4165.

van Lieshout, S.H., Sparks, A.M., Bretman, A., Newman, C., Buesching, C.D., Burke, T., and Dugdale, H.L. (2020a). Estimation of environmental, genetic and parental age at conception effects on telomere length in a wild mammal. *Journal of Evolutionary Biology*, 34(2), 296–308.

van Lieshout, S.H., Badás, E.P., Mason, M.W., Newman, C., Buesching, C.D., Macdonald, D.W., and Dugdale, H.L. (2020b). Social effects on age-related and sex-specific immune cell profiles in a wild mammal. *Biology Letters*, 16(7), 20200234.

van Lieshout, S.H.J., Badás, E.P., Bright Ross, J.G., Bretman, A., Newman, C., Buesching, C.D., Burke, T., Macdonald, D.W., and Dugdale, H.L. (2021). Early-life seasonal, weather and social effects on telomere length in a wild mammal. *Molecular Ecology*.

van Noordwijk, A. and J.de Jong, G. (1986). Acquisition and allocation of resources: their influence on variation in life history tactics. *The American Naturalist*, 128(1), 137–142.

Van Valen, L. (1962). A study of fluctuating asymmetry. *Evolution*, 16(2), 125–142.

Van Valen, L. (1965). Morphological variation and width of ecological niche. *The American Naturalist*, 99(908), 377–390.

Van Valen, L. (1973). A new evolutionary law. *Evolutionary Theory*, 1, 1–30.

Van Valen, L. (1977). The red queen. *The American Naturalist*, 111(980), 809–810.

Van Valkenburgh, B. (2001). The dog-eat-dog world. A review of past and present carnivore community dynamics. In *Meat-Eating and Human Evolution* (C.B. Stanford and H.T. Bunn, eds, pp. 101–120). Oxford University Press, Oxford.

Van Valkenburgh, B., Curtis, A., Samuels, J.X., Bird, D., Fulkerson, B., Meachen-Samuels, J., and Slater, G.J. (2011).

Aquatic adaptations in the nose of carnivorans: evidence from the turbinates. *Journal of Anatomy*, 218(3), 298–310.

Van Vuren, D. and Armitage, K.B. (1994). Survival of dispersing and philopatric yellow-bellied marmots: what is the cost of dispersal? Oikos, 69(2), 179–181.

Vanderhaar, J.M. and Hwang, Y.T. (2003). *Mellivora capensis . Mammalian Species*, 721, 1–8.

Vanderhaar, J.M. and Ten Hwang, Y. (2003). *Mellivora capensis. Mammalian Species*, 2003(721), 1–8.

Varadhan, R., Walston, J.D., and Bandeen-Roche, K. (2018). Can physical resilience and frailty in older adults be linked by the study of dynamical systems? *Journal of the American Geriatrics Society*, 66(8), 1455.

Varley, G.C. and Gradwell, G.R. (1960). Key factors in population studies. *Journal of Animal Ecology*, 29(2), 399–401.

Vedder, O., Komdeur, J., van der Velde, M., Schut, E., and Magrath, M.J. (2011). Polygyny and extra-pair paternity enhance the opportunity for sexual selection in blue tits. *Behavioral Ecology and Sociobiology*, 65(4), 741–752.

Vedder, O., Verhulst, S., Bauch, C., and Bouwhuis, S. (2017). Telomere attrition and growth: a life-history framework and case study in common terns. *Journal of Evolutionary Biology*, 30(7), 1409–1419.

Verdolin, J.L. (2009). Gunnison's prairie dog (*Cynomys gunnisoni*): testing the resource dispersion hypothesis. *Behavioral Ecology and Sociobiology*, 63(6), 789.

Verner, J. (1977). On the adaptive significance of territoriality. *The American Naturalist*, 111(980), 769–775.

Vervaecke, H., Stevens, J. M. G., Vandemoortele, H., Sigurjonsdottir, H., and de Vries, H. (2007). Aggression and dominance in matched groups of subadult Icelandic horses (*Equus caballus*). *Journal of Ethology*, 25, 239–248.

Vesey-Fitzgerald, B. (1942). *A Country Chronicle*. Chapman & Hall, London.

Vesterdorf, K., Blache, D., Harrison, A., Matthiesen, C.F., and Tauson, A.H. (2014). Low protein provision during the first year of life, but not during foetal life, affects metabolic traits, organ mass development and growth in male mink (*Neovison vison*). *Journal of Animal Physiology and Animal Nutrition*, 98(2), 357–372.

Vial, F. and Donnelly, C.A. (2012). Localized reactive badger culling increases risk of bovine tuberculosis in nearby cattle herds. *Biology Letters*, 8(1), 50–53.

Vial, F., Cleaveland, S., Rasmussen, G., and Haydon, D.T. (2006). Development of vaccination strategies for the management of rabies in African wild dogs. *Biological Conservation*, 131(2), 180–192.

Virgós, E. and Casanovas, J.G. (1999). Environmental constraints at the edge of a species distribution, the Eurasian badger (*Meles meles* L.): a biogeographic approach. *Journal of Biogeography*, 26(3), 559–564.

Virgós, E., Mangas, J.G., Blanco-Aguiar, J.A., Garrote, G., Almagro, N., and Viso, R.P. (2004). Food habits of European badgers (*Meles meles*) along an altitudinal gradient of Mediterranean environments: a field test of the earthworm specialization hypothesis. *Canadian Journal of Zoology*, 82(1), 41–51.

Vivier, E., Raulet, D.H., Moretta, A., Caligiuri, M.A., Zitvogel, L., Lanier, L.L., Yokoyama, W.M., and Ugolini, S. (2011). Innate or adaptive immunity? The example of natural killer cells. *Science*, *331*(6013), 44–49.

Vleck, D. (1981). Burrow structure and foraging costs in the fossorial rodent, *Thomomys bottae* . *Oecologia*, *49*, 391–396.

vom Saal, F.S., Finch, C.E., and Nelson, J.F. (1994). Natural history and mechanisms of reproductive aging in humans, laboratory rodents, and other selected vertebrates. *The Physiology of Reproduction*, *2*, 1213–1314.

von Schantz, T. (1984). Spacing strategies, kin selection, and population regulation in altricial vertebrates. *Oikos*, *42*(1), 48–58.

Vordermeier, H.M., Jones, G.J., Buddle, B.M., Hewinson, R.G., and Villarreal-Ramos, B. (2016). Bovine tuberculosis in cattle: vaccines, DIVA tests, and host biomarker discovery. *Annual Review of Animal Biosciences*, *4*, 87–109.

Vrba, E.S. (1985). Environment and evolution: alternative causes of temporal distribution of evolutionary events. *South African Journal of Science*, *81*, 229–236.

Vrba, E.S. (1992). Mammals as a key to evolutionary theory. *Journal of Mammalogy*, *73*(1), 1–28.

Vrba, E.S. (1995). On the connections between paleoclimate and evolution. In *Paleoclimate and Evolution, with Emphasis on Human Origins* (E.S. Vrba, G.H. Denton, T.C. Partridge, and L.H. Burckle, eds, pp. 24–45). Yale University Press, New Haven, CT.

Vucetich, J.A., Peterson, R.O., and Waite, T.A. (2004). Raven scavenging favours group foraging in wolves. *Animal Behaviour*, *67*(6), 1117–1126.

Vucetich, J.A. (2021). *Restoring the Balance: What Wolves Tell Us About Our Relationship with Nature*. JHU Press, Baltimore, MD.

Vucetich, J.A., Burnham, D., Macdonald, E.A., Bruskotter, J.T., Marchini, S., Zimmermann, A. and Macdonald, D.W. (2018). Just conservation: what is it and should we pursue it? *Biological Conservation*, *221*, 23–33.

Vucetich, J.A., Burnham, D., Johnson, P.J., Loveridge, A.J., Nelson, M.P., Bruskotter, J.T., and Macdonald, D.W. (2019). The value of argument analysis for understanding ethical considerations pertaining to trophy hunting and lion conservation. *Biological Conservation*, *235*, 260–272.

Vucetich, J.A., Macdonald, E.A., Burnham, D., Bruskotter, J.T., Johnson, D.D., and Macdonald, D.W. (2021). Finding purpose in the conservation of biodiversity by the commingling of science and ethics. *Animals*, *11*(3), 837.

Wade, M.G., Desaulniers, D., and Leingartner, K. (1997). Interactions between endosulfan and dieldrin on estrogen-mediated processes in vitro and in vivo. *Reproductive Toxicology*, *11*(6), 791–798.

Wade, M.J. and Kalisz, S. (1990). The causes of natural selection. *Evolution*, *44*(8), 1947–1955.

Wagner G.P. and Altenberg L. (1996). Perspective: complex adaptations and the evolution of evolvability. *Evolution*, *50*, 967–976.

Wagstaff, A. (2000). Socioeconomic inequalities in child mortality: comparisons across nine developing countries. *Bulletin of the World Health Organization*, *78*, 19–29.

Walker, F.M., Sunnucks, P., and Taylor, A.C. (2006). Genotyping of 'captured' hairs reveals burrow-use and ranging behavior of southern hairy-nosed wombats. *Journal of Mammalogy*, *87*(4), 690–699.

Walker, F.M., Sunnucks, P., and Taylor, A. C. (2008). Evidence for habitat fragmentation altering within-population processes in wombats. *Molecular Ecology*, *17*(7), 1674–1684.

Walker, J.M. and Berger, R.J. (1980). Sleep as an adaptation for energy conservation functionally related to hibernation and shallow torpor. *Progress in Brain Research*, *53*, 255–278.

Walker, W.F. (1980). Sperm utilization strategies in nonsocial insects. *The American Naturalist*, *115*(6), 780–799.

Wallach, E.E., Shoham, Z., Schachter, M., Loumaye, E., Weissman, A., MacNamee, M., and Insler, V. (1995). The luteinizing hormone surge—the final stage in ovulation induction: modern aspects of ovulation triggering. *Fertility and Sterility*, *64*(2), 237–251.

Walling, C.A., Pemberton, J.M., Hadfield, J.D., and Kruuk, L.E. (2010). Comparing parentage inference software: reanalysis of a red deer pedigree. *Molecular Ecology*, *19*(9), 1914–1928.

Walsh, D.M., Lewens, T., and Ariew, A. (2002). The trials of life: natural selection and random drift. *Philosophy of Science*, *69*(3), 429–446.

Wan, G.H. (2001). Changes in regional inequality in rural China: decomposing the Gini index by income sources. *Australian Journal of Agricultural and Resource Economics*, *45*(3), 361–381.

Wandeler, A., Müller, J., Wachendörfer, G., Schale, W., Förster, U., and Steck, F. (1974). Rabies in wild carnivores in central Europe III. Ecology and biology of the fox in relation to control operations. *Zentralblatt fuer Veterinaermedizin Reihe B*, *21*(10), 765–773.

Wang, J. (2013). A simulation module in the computer program COLONY for sibship and parentage analysis. *Molecular Ecology Resources*, *13*(4), 734–739.

Wang, J., Li, F., Xiao, L., Peng, F., Sun, W., Li, M., Liu, D., Jiang, Y., Guo, R., Li, H., and Zhu, W. (2018). Depressed TSH level as a predictor of poststroke fatigue in patients with acute ischemic stroke. *Neurology*, *91*(21), e1971–e1978.

Ward, C. (1966). Anarchism as a theory of organisation. www.panarchy.org (accessed 17 February 2022).

Ward, J.F., Macdonald, D.W., Doncaster, C.P., and Mauget, C. (1996). Physiological response of the European hedgehog to predator and nonpredator odour. *Physiology & Behavior*, *60*(6), 1469–1472.

Ward, J.F., Macdonald, D.W., and Doncaster, C.P. (1997). Responses of foraging hedgehogs to badger odour. *Animal Behaviour*, 53(4), 709–720.

Ward, J.F., Austin, R.M., and Macdonald, D.W. (2000). A simulation model of foraging behaviour and the effect of predation risk. *Journal of Animal Ecology*, 69(1), 16–30.

Ward, E.J., Parsons, K., Holmes, E.E., Balcomb, K.C., and Ford, J.K. (2009). The role of menopause and reproductive senescence in a long-lived social mammal. *Frontiers in Zoology*, 6(1), 1–10.

Ward, P. and Zahavi, A. (1973). The importance of certain assemblages of birds as 'information- centres' for food-finding. *Ibis*, 115, 517–534.

Warren, R., Price, J., VanDerWal, J., Cornelius, S., and Sohl, H. (2018). The implications of the United Nations Paris Agreement on climate change for globally significant biodiversity areas. *Climatic Change*, 147, 395–409. https://doi.org/10.1007/s10584-018-2158-6

Waser, P.M. (1981). Sociality or territorial defense? The influence of resource renewal. *Behavioral Ecology and Sociobiology*, 8(3), 231–237.

Waser, P.M. (1988). Resources, philopatry, and social interactions among mammals. *The Ecology of Social Behavior*, 109, 130.

Waser, P.M. and Jones, W.T. (1983). Natal philopatry among solitary mammals. *The Quarterly Review of Virology*, 58(3), 355–390.

Wasser, S.K. and Barash, D.P. (1983). Reproductive suppression among female mammals: implications for biomedicine and sexual selection theory. *The Quarterly Review of Biology*, 58(4), 513–538.

Wasser, S.K. and Starling, A.K. (1988). Proximate and ultimate causes of reproductive suppression among female yellow baboons at Mikumi National Park, Tanzania. *American Journal of Primatology*, 16(2), 97–121.

Waters, W.R., Palmer, M.V., Buddle, B.M. and Vordermeier, H.M. (2012). Bovine tuberculosis vaccine research: historical perspectives and recent advances. *Vaccine*, 30(16), 2611–2622.

Watt, N.J., Harkiss, G.D., Hayton, A., Cutler, K., Clarke, J., and O'Brien, A. (2020). Validation of Enferplex bTB antibody test. *Veterinary Record*, 187(11), 451.

Webb, D.R. and King, J.R. (1984). Effects of wetting on insulation of bird and mammal coats. *Journal of Thermal Biology*, 9(3), 189–191.

Webber, Q.M.R., Laforge, M.P., Bonar, M., Robitaille, A.L., Hart, C., Zabihi-Seissan, S., and Vander, W.E. (2020). The ecology of individual differences empirically applied to space-use and movement tactics. *The American Naturalist*, 196(1), E1–E15. doi:10.1086/708721

Weber, N., Bearhop, S., Dall, S.R.X., Delahay, R.J., McDonald, R.A., and Carter, S.P. (2013). Denning behaviour of the European badger (*Meles meles*) correlates with bovine tuberculosis infection status. *Behavioral Ecology and Sociobiology*, 67, 471–479.

Webster, B. (2013). Badger cull gets extra time despite advice from expert. https://www.thetimes.co.uk/article/badger-cull-gets-extra-time-despite-advice-from-expert-bdwc9vvd7xw (accessed 12 May 2022).

Webster, J.P., Brunton, C.F.A., and Macdonald, D.W. (1994). Effect of *Toxoplasma gondii* upon neophobic behaviour in wild brown rats, *Rattus norvegicus*. *Parasitology*, 109(1), 37–43.

Wedekind, C., Seebeck, T., Bettens, F., and Paepke, A.J. (1995). MHC-dependent preferences in humans. *Proceedings of the Royal Society of London*, 260(1359), 245–249. doi:10.1098/rspb.1995.0087

Wedlock, D.N., Denis, M., Vordermeier, H.M., Hewinson, R.G., and Buddle, B.M. (2007). Vaccination of cattle with Danish and Pasteur strains of *Mycobacterium bovis* BCG induce different levels of IFNγ post-vaccination, but induce similar levels of protection against bovine tuberculosis. *Veterinary Immunology and Immunopathology*, 118 (1–2), 50–58.

Wegner, K.M., Kalbe, M., Kurtz, J., Reusch, T.B., and Milinski, M. (2003). Parasite selection for immunogenetic optimality. *Science*, 301(5638), 1343–1343.

Wehtje, M. and Gompper, M.E. (2011). Effects of an experimentally clumped food resource on raccoon *Procyon lotor* home-range use. *Wildlife Biology*, 17(1), 25–32.

Weijs, P.J., Kruizenga, H.M., van Dijk, A.E., van der Meij, B.S., Langius, J.A., Knol, D.L., and van Schijndel, R.J.S. (2008). Validation of predictive equations for resting energy expenditure in adult outpatients and inpatients. *Clinical Nutrition*, 27(1), 150–157.

Weindruch, R. (1992). Effect of caloric restriction on age-associated cancers. *Experimental Gerontology*, 27(5–6), 575–581.

Weiner, J. (1992). Physiological limits to sustainable energy budgets in birds and mammals: ecological implications. *Trends in Ecology & Evolution*, 7, 384–388.

Weinstein, S.B., Moura, C.W., Mendez, J.F., and Lafferty, K.D. (2017). Fear of feces? Tradeoffs between disease risk and foraging drive animal activity around raccoon latrines. *Oikos*, 127(7), 927–934. doi:10.1111/oik.04866

Weinstein, S.B., Buck, J.C., and Young, H.S. (2018). A landscape of disgust. *Science*, 359, 1213–1215.

Weir, B.J. (1973). Another hystricomorph rodent: keeping casiragua (*Proechimys guairae*) in captivity. *Laboratory Animals*, 7(2), 125–134.

Weir, R.D., Kinley, T.A., Klafki, D.W., and Apps, C.D. (2017). Ecotypic variation affects the conservation of North American badgers endangered along their northern range extent. In *Biology and Conservation of Musteloids* (D.W. Macdonald, C. Newman, and L. Harrington, eds, pp. 410–419). Oxford University Press, Oxford.

Weiskopf, D., Weinberger, B., and Grubeck-Loebenstein, B. (2009). The aging of the immune system. *Transplant International*, 22(11), 1041–1050.

Weladji, R.B., Holand, Ø., Steinheim, G., Colman, J.E., Gjøstein, H., and Kosmo, A. (2005). Sexual dimorphism and intercohort variation in reindeer calf antler length is associated with density and weather. *Oecologia*, *145*(4), 549–555.

Welsh Government (2013). Bovine TB Eradication Programme IAA Badger Vaccination Project Year 1 report. http://www.bovinetb.info/docs/bovine-tb-eradication-programme-iaa-badger-vaccination-project-year-1-report.pdf (accessed 12 May 2022).

Welsh Government (2013). Bovine TB Eradication Programme IAA Badger Vaccination Project. Year 2 report. http://www.bovinetb.info/docs/bovine-tb-eradication-programme-iaa-badger-vaccination-project-year-2-report.pdf (accessed 12 May 2022).

Weng, N.P. (2001). Interplay between telomere length and telomerase in human leukocyte differentiation and aging. *Journal of Leukocyte Biology*, *70*(6), 861–867.

Werhahn, G., Senn, H., Ghazali, M., Karmacharya, D., Sherchan, A.M., Joshi, J., and Sillero-Zubiri, C. (2018). The unique genetic adaptation of the Himalayan wolf to high-altitudes and consequences for conservation. *Global Ecology and Conservation*, *16*, e00455.

Werhahn, G., Liu, Y., Meng, Y., Cheng, C., Lu, Z., Atzeni, L., and Joshi, J. (2020). Himalayan wolf distribution and admixture based on multiple genetic markers. *Journal of Biogeography*, *47*(6), 1272–1285.

West, G.B., Brown, J.H., and Enquist, B.J. (1997). A general model for the origin of allometric scaling laws in biology. *Science*, *276*(5309), 122–126.

Westerdahl, H., Asghar, M., Hasselquist, D., and Bensch, S. (2012). Quantitative disease resistance: to better understand parasite-mediated selection on major histocompatibility complex. *Proceedings of the Royal Society. Series B: Biological Sciences*, *279*(1728), 577–584.

Westermarck, E.A. (1921). *The History of Human Marriage*, 5th edn. Macmillan, London.

Westneat, D.F. and Stewart, I.R. (2003). Extra-pair paternity in birds: causes, correlates, and conflict. *Annual Review of Ecology, Evolution, and Systematics*, *34*(1), 365–396.

Weyrich, A., Lenz, D., Jeschek, M., Chung, T.H., Rübensam, K., Göritz, F., Jewgenow, K., and Fickel, J. (2016). Paternal intergenerational epigenetic response to heat exposure in male wild guinea pigs. *Molecular Ecology*, *25*(8), 1729–1740.

Whelan, A.O., Villarreal-Ramos, B., Vordermeier, H.M., and Hogarth, P.J. (2011). Development of an antibody to bovine IL-2 reveals multifunctional CD4 TEM cells in cattle naturally infected with bovine tuberculosis. *PLoS ONE*, *6*(12), e29194.

White, C.R. (2005). The allometry of burrow geometry. *Journal of Zoology*, *265*(4), 395–403.

White, P.R. and Chambers, J. (1989). Saw-toothed grain beetle *Oryzaephilus surinamensis* (L.) (Coleoptera: Silvanidae): antennal and behavioral responses to individual components and blends of aggregation pheromone. *Journal of Chemical Ecology*, *15*(3), 1015–1031.

Whiteman, N.K., Matson, K.D., Bollmer, J.L., and Parker, P.G. (2006). Disease ecology in the Galapagos hawk (*Buteo galapagoensis*): host genetic diversity, parasite load and natural antibodies. *Proceedings of the Royal Society. Series B: Biological Sciences*, *273*(1588), 797–804.

WHO/FAO Expert Committee on Zoonoses (1959). Second report. World Health Organization Technical Report Series No 169. WHO, Geneva.

Why Evolution is True (2011). Dawkins on Nowak et al. and kin selection. https://whyevolutionistrue.com/2011/03/24/dawkins-on-nowak-et-al-and-kin-selection/ (accessed 5 April 2021).

Wickens, A.P. (2001). Ageing and the free radical theory. *Respiration Physiology*, *128*(3), 379–391.

Wielebnowski, N. and Watters, J. (2007). Applying faecal endocrine monitoring to conservation and behavior studies of wild mammals: important considerations and preliminary tests. *Israel Journal of Ecology and Evolution*, *53*(3–4), 439–460.

Wiens, J.J. (2016) Climate-related local extinctions are already widespread among plant and animal species. *PLoS Biology*, *14*(12), e2001104. https://doi.org/10.1371/journal.pbio.2001104

Wijers, M., Trethowan, P., Markham, A., Du Preez, B., Chamaillé-Jammes, S., Loveridge, A., and Macdonald, D. (2018). Listening to lions: Animal-borne acoustic sensors improve bio-logger calibration and behaviour classification performance. *Frontiers in Ecology and Evolution*, *6*, 171.

Wijers, M., Trethowan, P., Du Preez, B., Chamaille-James, S., Loveridge, A.J., Macdonald, D.W., and Markham, A. (2020). Vocal discrimination of African lions and its potential for collar-free tracking. *Bioacoustics*, *30*(5), 575–593. https://doi.org/10.1080/09524622.2020.1829050

Wilbourn, R.V., Froy, H., McManus, M.C., Cheynel, L., Gaillard, J.M., Gilot-Fromont, E., and Nussey, D.H. (2017). Age-dependent associations between telomere length and environmental conditions in roe deer. *Biology Letters*, *13*(9), 20170434.

Wilbur, H.M. and Rudolf, V.H. (2006). Life-history evolution in uncertain environments: bet hedging in time. *The American Naturalist*, *168*(3), 398–411.

Wilbur, S.M., Barnes, B.M., Kitaysky, A.S., and Williams, C.T. (2019). Tissue-specific telomere dynamics in hibernating arctic ground squirrels (*Urocitellus parryii*). *Journal of Experimental Biology*, *222*(18), jeb204925

Wilesmith, J.W. (1983). Epidemiological features of bovine tuberculosis in cattle herds in Great Britain. *Epidemiology & Infection*, *90*(2), 159–176.

Wilkinson, D., Smith, G.C., Delahay, R.J., Rogers, L.M., Cheeseman, C.L., and Clifton-Hadley, R.S. (2000). The effects of bovine tuberculosis (*Mycobacterium bovis*) on mortality in a badger (*Meles meles*) population in England. *Journal of Zoology*, *250*(3), 389–395.

Wilkinson, D., Smith, G.C., Delahay, R.J., and Cheeseman, C.L. (2004). A model of bovine tuberculosis in the badger *Meles meles*: an evaluation of different vaccination strategies. *Journal of Applied Ecology*, 41(3), 492–501.

Willadsen P. (1980). Immunity to ticks. *Advances in Parasitology*, 18, 293. doi:10.1016/S0065-308X(08)60402-60409

Williams, B.M., Baker, P.J., Thomas, E., Wilson, G., Judge, J., and Yarnell, R.W. (2018). Reduced occupancy of hedgehogs (*Erinaceus europaeus*) in rural England and Wales: the influence of habitat and an asymmetric intra-guild predator. *Scientific Reports*, 8(1), 1–10.

Williams, G.C. (1957). Pleiotropy, natural selection, and the evolution of senescence. *Evolution*, 11, 398–411.

Williams, G.C. (1966). Natural selection, the costs of reproduction, and a refinement of Lack's principle. *The American Naturalist*, 100(916), 687–690.

Williams, G.C. (1975). *Sex and Evolution*. Princeton University Press, Princeton, NJ.

Wilmers, C.C., Stahler, D.R., Crabtree, R.L., Smith, D.W., and Getz, W.M. (2003). Resource dispersion and consumer dominance: scavenging at wolf-and hunter-killed carcasses in Greater Yellowstone, USA. *Ecology Letters*, 6(11), 996–1003.

Wilson, D.E. and Reeder, D.M. (eds) (2005). *Mammal Species of the World: A Taxonomic and Geographic Reference* (Vol. 1). Johns Hopkins University Press, Baltimore, MD.

Wilson, E.O. (2000). *Sociobiology: The New Synthesis*. Harvard University Press, Cambridge, MA.

Wilson, G., Harris, S., and McLaren, G. (1997). *Changes in the British Badger Population, 1988 to 1997*. People's Trust for Endangered Species, London.

Wilson, G.J. and Delahay, R.J. (2001). A review of methods to estimate the abundance of terrestrial carnivores using field signs and observation. *Wildlife Research*, 28(2), 151–164.

Wilson, G.S. (1943). The pasteurization of milk. *British Medical Journal*, 1(4286), 261.

Wilson, K., Knell, R., Boots, M., and Koch-Osborne, J. (2003). Group living and investment in immune defence: an interspecific analysis. *Journal of Animal Ecology*, 72(1), 133–143.

Wilson, R.P., White, C.R., Quintana, F., Halsey, L.G., Liebsch, N., Martin, G.R., and Butler, P.J. (2006). Moving towards acceleration for estimates of activity-specific metabolic rate in free-living animals: the case of the cormorant. *Journal of Animal Ecology*, 75(5), 1081–1090.

Wimsatt, W.A. (1945). Notes on breeding behavior, pregnancy, and parturition in some vespertilionid bats of the eastern United States. *Journal of Mammalogy*, 26(1), 23–33.

Wittemyer, G. and Getz, W.M. (2007). Hierarchical dominance structure and social organization in African elephants, *Loxodonta africana*. *Animal Behaviour*, 73(4), 671–681.

Wittig, R.M. and Boesch, C. (2003). Food competition and linear dominance hierarchy among female chimpanzees of the Tai National Park. *International Journal of Primatology*, 24(4), 847–867.

Wolf, M. and Weissing, F.J. (2012). Animal personalities: consequences for ecology and evolution. *Trends in Ecology & Evolution*, 27(8), 452–461.

Wolf, M., Van Doorn, G.S., Leimar, O., and Weissing, F.J. (2007). Life-history trade-offs favour the evolution of animal personalities. *Nature*, 447(7144), 581–584.

Wolff, J.O. (1993). Why are female small mammals territorial? *Oikos*, 68(2), 364–370.

Wolff, J.O. (1994). More on juvenile dispersal in mammals. *Oikos*, 71(2), 349–352.

Wolff, J.O. (1997). Population regulation in mammals: an evolutionary perspective. *Journal of Animal Ecology*, 66(1), 1–13.

Wolff, J.O. and Macdonald, D.W. (2004). Promiscuous females protect their offspring. *Trends in Ecology & Evolution*, 19(3), 127–134.

Wolff, J.O. and Peterson, J.A. (1998). An offspring-defense hypothesis for territoriality in female mammals. *Ethology Ecology & Evolution*, 10(3), 227–239.

Wolsan, M. and Sato, J.J. (2010). Effects of data incompleteness on the relative performance of parsimony and Bayesian approaches in a supermatrix phylogenetic reconstruction of Mustelidae and Procyonidae (Carnivora). *Cladistics*, 26(2), 168–194.

Wolsan, M. and Sotnikova, M. (2013). Systematics, evolution, and biogeography of the Pliocene stem meline badger *Ferinestrix* (Carnivora: Mustelidae). *Zoological Journal of the Linnean Society*, 167(1), 208–226.

Wong, B.B. and Candolin, U. (2005). How is female mate choice affected by male competition? *Biological Reviews*, 80(4), 559–571.

Wong, B. and Candolin, U. (2015). Behavioral responses to changing environments. *Behavioral Ecology*, 26(3), 665–673.

Wong, J., Stewart, P.D., and Macdonald, D.W. (1999). Vocal repertoire in the European badger (*Meles meles*): structure, context, and function. *Journal of Mammalogy*, 80(2), 570–588.

Woodroffe, R. & Macdonald, D.W. (1993). Badger sociality—models of spatial grouping. *Symposia of the Zoological Society of London*, 65, 145–169.

Woodroffe, R. (1995). Body condition affects implantation date in the European badger, *Meles meles*. *Journal of Zoology*, 236(2), 183–188.

Woodroffe, R. and Macdonald, D.W. (1995a). Costs of breeding status in the European badger, *Meles meles*. *Journal of Zoology*, 235(2), 237–245.

Woodroffe, R. and Macdonald, D.W. (1995b). Female/female competition in European badgers *Meles meles*: effects on breeding success. *Journal of Animal Ecology*, 64(1), 12–20.

Woodroffe, R. and Macdonald, D.W. (2000). Helpers provide no detectable benefits in the European badger (*Meles meles*). *Journal of Zoology*, 250(1), 113–119.

Woodroffe, R. and Vincent, A. (1994). Mother's little helpers: patterns of male care in mammals. *Trends in Ecology & Evolution*, 9(8), 294–297.

Woodroffe, R., Frost, S.D., and Clifton-Hadley, R.S. (1999). Attempts to control tuberculosis in cattle by removing infected badgers: constraints imposed by live test sensitivity. *Journal of Applied Ecology, 36*(4), 494–501.

Woodroffe, R., Donnelly, C.A., Johnston, W.T., Bourne, F.J., Cheeseman, C.L., Clifton-Hadley, R.S., Cox, D.R., Gettinby, G., Hewinson, R.G., Le Fevre, A.M., and McInerney, J.P. (2005). Spatial association of *Mycobacterium bovis* infection in cattle and badgers *Meles meles. Journal of Applied Ecology, 42*(5), 852–862.

Woodroffe, R., Donnelly, C.A., Cox, D.R., Bourne, F.J., Cheeseman, C.L., Delahay, R.J., Gettinby, G., McInerney, J.P., and Morrison, W.I. (2006a). Effects of culling on badger *Meles meles* spatial organization: implications for the control of bovine tuberculosis. *Journal of Applied Ecology, 43*, 1–10.

Woodroffe, R., Donnelly, C.A., Jenkins, H.E., Johnston, W.T., Cox, D.R., Bourne, F.J., Cheeseman, C.L., Delahay, R.J., Clifton-Hadley, R.S., Gettinby, G., and Gilks, P. (2006b). Culling and cattle controls influence tuberculosis risk for badgers. *Proceedings of the National Academy of Sciences, 103*(40), 14713–14717.

Woodroffe, R., Davies-Mostert, H., Ginsberg, J., Graf, J., Leigh, K., McCreery, K., and Somers, M. (2007). Rates and causes of mortality in endangered African wild dogs *Lycaon pictus*: lessons for management and monitoring. *Oryx, 41*(2), 215–223.

Woodroffe, R., Donnelly, C.A., Cox, D.R., Gilks, P., Jenkins, H.E., Johnston, W.T., Le Fevre, A.M., Bourne, F.J., Cheeseman, C.L., Clifton-Hadley, R.S., and Gettinby, G. (2009a). Bovine tuberculosis in cattle and badgers in localized culling areas. *Journal of Wildlife Diseases, 45*(1), 128–143.

Woodroffe, R., Donnelly, C.A., Wei, G., Cox, D.R., Bourne, F.J., Burke, T., Butlin, R.K., Cheeseman, C.L., Gettinby, G., Gilks, P., Hedges, S., Jenkins, H.E., Johnston, W.T., McInerney, J.P., Morrison, W.I., and Pope, L.C. (2009b). Social group size affects Mycobacterium bovis infection in European badgers (*Meles meles*). *Journal of Animal Ecology, 78*(4), 818–827. http://www.jstor.org/stable/27696431

Worldwide cost-effectiveness of infant BCG vaccination. *Archives of Disease in Childhood, 91*(8) (2006), 641.

Wrangham, R.W. (1993). The evolution of sexuality in chimpanzees and bonobos. *Human Nature, 4*(1), 47–79.

Wrangham, R.W., Gittleman, J.L., and Chapman, C.A. (1993). Constraints on group size in primates and carnivores: population density and day-range as assays of exploitation competition. *Behavioral Ecology and Sociobiology, 32*(3), 199–209.

Wright, E., Galbany, J., McFarlin, S.C., Ndayishimiye, E., Stoinski, T.S., and Robbins, M.M. (2019). Male body size, dominance rank and strategic use of aggression in a group-living mammal. *Animal Behaviour, 151*, 87–102.

Wright, J., Bolstad, G.H., Araya-Ajoy, Y.G., and Dingemanse, N.J. (2019). Life-history evolution under fluctuating density-dependent selection and the adaptive alignment of pace-of-life syndromes. *Biological Reviews, 94*, 230–247.

Wroe, S., McHenry, C., and Thomason, J. (2005). Bite club: comparative bite force in big biting mammals and the prediction of predatory behaviour in fossil taxa. *Proceedings of the Royal Society, Series B: Biological Sciences, 272*(1563), 619–625.

Wu, H.Y. (1999). Is there current competition between sympatric Siberian weasels (*Mustela sibirica*) and ferret badgers (*Melogale moschata*) in a subtropical forest ecosystem of Taiwan? *Zoological Studies, 38*(4), 443–451.

Wyatt, T.D. (2010). Pheromones and behavior. In *Chemical Communication in Crustaceans* (T. Breithaupt and M. Thiel, eds, pp. 23–38). Springer, New York.

Xiao, X., Newman, C., Buesching, C.D., Macdonald, D.W., and Zhou, Z.M. (2021). Animal sales from Wuhan wet markets immediately prior to the COVID-19 pandemic. *Scientific Reports, 11*(1), 1–7.

Yamaguchi, N. and Macdonald, D.W. (2001). Detection of Aleutian disease antibodies in feral American mink in southern England. *Veterinary Record, 149*(16), 485–488.

Yamaguchi, N. and Macdonald, D.W. (2003). The burden of co-occupancy: intraspecific resource competition and spacing patterns in American mink, *Mustela vison. Journal of Mammalogy, 84*(4), 1341–1355.

Yamaguchi, N., Gazzard, D., Scholey, G., and Macdonald, D.W. (2003). Concentrations and hazard assessment of PCBs, organochlorine pesticides and mercury in fish species from the upper Thames: river pollution and its potential effects on top predators. *Chemosphere, 50*(3), 265–273.

Yamaguchi, N., Sarno, R.J., Johnson, W.E., O'Brien, S.J., and Macdonald, D.W. (2004). Multiple paternity and reproductive tactics of free-ranging American minks, *Mustela vison. Journal of Mammalogy, 85*(3), 432–439.

Yamaguchi, N., Dugdale, H.L., and Macdonald, D.W. (2006). Female receptivity, embryonic diapause, and superfetation in the European badger (*Meles meles*): implications for the reproductive tactics of males and females. *The Quarterly Review of Biology, 81*(1), 33–48.

Yamazaki, K., Boyse, E.A., Mike, V., Thaler, H.T., Mathieson, B.J., Abbott, J., Boyse, J., Zayas, Z.A., and Thomas, L. (1976). Control of mating preferences in mice by genes in the major histocompatibility complex. *The Journal of Experimental Medicine, 144*(5), 1324–1335.

Yamazaki, K., Yamaguchi, M., Andrews, P.W., Peake, B., and Boyse, E.A. (1978). Mating preferences of F2 segregants of crosses between MHC-congenic mouse strains. *Immunogenetics, 6*(1), 253–259.

Yamazaki, K., Beauchamp, G.K., Imai, Y., Bard, J., Phelan, S.P., Thomas, L., and Boyse, E.A. (1990). Odor types determined by the major histocompatibility complex in germfree mice. *Proceedings of the National Academy of Sciences, 87*(21), 8413–8416.

Yanagimachi, R. (1978). Sperm-egg association in mammals. *Current Topics in Developmental Biology, 12*, 83–105.

Yasukawa, K. (1981a). Song and territory defense in the redwinged blackbird. *The Auk, 98*(1), 185–187.

Yasukawa, K. (1981b). Song repertoires in the red-winged blackbird (*Agelaius phoeniceus*): a test of the Beau Geste hypothesis. *Animal Behaviour, 29*(1), 114–125.

Yin, J., Schlesinger, M.E., and Stouffer, R.J. (2009). Model projections of rapid sea-level rise on the northeast coast of the United States. *Nature Geoscience, 2*(4), 262–266.

Yip, R. (2000). Significance of an abnormally low or high hemoglobin concentration during pregnancy: special consideration of iron nutrition. *The American Journal of Clinical Nutrition, 72*(1), 272S–279S.

Young, A.J., Spong, G., and Clutton-Brock, T. (2007). Subordinate male meerkats prospect for extra-group paternity: alternative reproductive tactics in a cooperative mammal. *Proceedings of the Royal Society. Series B: Biological Sciences, 274*(1618), 1603–1609.

Yu, B.P. (1994). Cellular defenses against damage from reactive oxygen species. *Physiological Reviews, 74*(1), 139–162.

Yun, C.H., Lillehoj, H.S., and Lillehoj, E.P. (2000). Intestinal immune responses to coccidiosis. *Developmental & Comparative Immunology, 24*(2–3), 303–324.

Zahavi, A. (1975). Mate selection—a selection for a handicap. *Journal of Theoretical Biology, 53*(1), 205–214.

Zahavi, A. (1981). Natural selection, sexual selection and the selection of signals. In *Evolution Today, Proceedings of the Second International Congress of Systematic and Evolution Biology* (G.G.E. Scudder and J.L. Reveal, eds, pp. 133–138).

Zeh, J.A. and Zeh, D.W. (1996). The evolution of polyandry I: intragenomic conflict and genetic incompatibility. *Proceedings of the Royal Society of London. Series B: Biological Sciences, 263*(1377), 1711–1717.

Zeh, J.A. and Zeh, D.W. (1997). The evolution of polyandry II: post-copulatory defenses against genetic incompatibility. *Proceedings of the Royal Society of London. Series B: Biological Sciences, 264*(1378), 69–75.

Zelano, B. and Edwards, S.V. (2002). An MHC component to kin recognition and mate choice in birds: predictions, progress, and prospects. *The American Naturalist, 160*(S6), S225–S237.

Zeveloff, S.I. (2002). *Raccoons: a Natural History*. UBC Press, Vancouver.

Zhang, J.X., Ni, J., Ren, X.J., Sun, L., Zhang, Z.B., and Wang, Z.W. (2003). Possible coding for recognition of sexes, individuals and species in anal gland volatiles of *Mustela eversmanni* and *M. sibirica*. *Chemical Senses, 28*(5), 381–388. PMID: 12826534

Zhang, L. Zhou, Y.B., Newman, C., Kaneko, Y., Macdonald, D.W., Jiang, P.P., and Ding, P. (2009). Niche overlap and sett-site resource partitioning for two sympatric species of badger. *Ethology Ecology & Evolution, 21*, 89–100.

Zhang, L., Zhou, Y.B., Newman, C., Kaneko, Y., Macdonald, D.W., Jiang, P.P., and Ding, P. (2009). Niche overlap and sett-site resource partitioning for two sympatric species of badger. *Ethology Ecology & Evolution, 21*(2), 89–100.

Zhang, L., Wang, Y., Zhou, Y., Newman, C., Kaneko, Y., Macdonald, D.W., and Ding, P. (2010). Ranging and activity patterns of the group-living ferret badger *Melogale moschata* in central China. *Journal of Mammalogy, 91*(1), 101–108.

Zhang, X., Lohmann, G., Knorr, G., and Purcell, C. (2014). Abrupt glacial climate shifts controlled by ice sheet changes. *Nature, 512*(7514), 290–294.

Zheng, S., Li, G., Song, S., Han, Y., and Ma, Z. (1988). Study on the ecology of sand badger. *Acta Theriologica Sinica, 3*, 65–72.

Zhou, Y.B., Zhang, L., Kaneko, Y., Newman, C., and Wang, X.M. (2008). Frugivory and seed dispersal by a small carnivore, the Chinese ferret-badger, *Melogale moschata*, in a fragmented subtropical forest of central China. *Forest Ecology and Management, 255*(5–6), 1595–1603.

Zhou, Y.B., Newman, C., Xu, W.T., Buesching, C.D., Zalewski, A., Kaneko, Y., Macdonald, D.W., and Xie, Z.Q. (2011). Biogeographical variation in the diet of Holarctic martens (genus *Martes*, Mammalia: Carnivora: Mustelidae): adaptive foraging in generalists. *Journal of Biogeography, 38*(1), 137–147.

Zhou, Y., Buesching, C.D., Newman, C., Kaneko, Y., Xie, Z., and Macdonald, D.W. (2013a). Balancing the benefits of ecotourism and development: the effects of visitor trail-use on mammals in a protected area in rapidly developing China. *Biological Conservation, 165*, 18–24.

Zhou, Y., Newman, C., Chen, J., Xie, Z., and Macdonald, D.W. (2013b). Anomalous, extreme weather disrupts obligate seed dispersal mutualism: snow in a subtropical forest ecosystem. *Global Change Biology, 19*(9), 2867–2877.

Zhou, Y., Chen, W., Buesching, C.D., Newman, C., Kaneko, Y., Xiang, M., Nie, C., Macdonald, D.W., and Xie, Z. (2015a). Hog badger (*Arctonyx collaris*) latrine use in relation to food abundance: evidence of the scarce factor paradox. *Ecosphere, 6*(1), 1–12.

Zhou, Y., Chen, W., Kaneko, Y., Newman, C., Liao, Z., Zhu, X., Buesching, C.D., Xie, Z., and Macdonald, D.W. (2015b). Seasonal dietary shifts and food resource exploitation by the hog badger (*Arctonyx collaris*) in a Chinese subtropical forest. *European Journal of Wildlife Research, 61*(1), 125–133.

Zhou, Y., Newman, C., Kaneko, Y., Buesching, C.D., Chen, W., Zhou, Z.M., Xie, Z., and Macdonald, D.W. (2017). Asian badgers—the same, only different: how diversity among badger societies informs socio-ecological theory and challenges conservation. In *Biology and Conservation of Musteloids* (D.W. Macdonald, C. Newman, and L.A. Harrington, eds, pp. 304–325). Oxford University Press, Oxford.

Zhou, Z.M., Buesching, C.D., Macdonald, D.W., and Newman, C. (2020). China: clamp down on violations of wildlife trade ban. *Nature, 578*(7794), 217–218.

Zhu, W., Mu, Y., Liu, J., and Wang, Z. (2015). Energy requirements during lactation in female *Apodemus chevrieri* (Mammalia: Rodentia: Muridae) in the Hengduan Mountain region. *Italian Journal of Zoology*, *82*(2), 165–171.

Zicsi, A. (1958). Einfluss der Trockenheit und der Bodenbearbeitung auf das Leben der Regenwürmer in Ackerböden. *Pedobiologia*, *9*, 141–146.

Ziegler, A., Kentenich, H., and Uchanska-Ziegler, B. (2005). Female choice and the MHC. *Trends in Immunology*, *26*(9), 496–502.

Zvereva, M.I., Shcherbakova, D.M., and Dontsova, O.A. (2010). Telomerase: structure, functions, and activity regulation. *Biochemistry (Mosc)*, *75*(13), 1563–1583.

Index

Note: *f*, *t*, and *b* following page numbers refer to figures, tables, and boxes. *Vs.* (versus) indicates a comparison or relationship. Footnotes are indicated by 'n' and the note number, e.g. 426n16.

Numbers

7/11 rule 44
7-year rule (dogs) 439n32

A

aardwolf 474
accelerometry 260
acoustic signals
    characteristics 93*t*
    see also auditory communication
acariasis ('mite rash') 318
Acrocephalus arundinaceus see great reed
        warbler
Actiphage test, bTB 340, 367
active radiofrequency identification
        (aRFID) 46, 260
    base station locations 61*f*
    border latrine visiting studies 48–9
    range overlap studies 62
active social domestication (ASD)
        hypothesis 443*b*
Activity
    vs. body condition 264, 265*f*
    climate change predictions 264–6,
        265*f*
    inter-individual variability 262
    mean *per noctem* durations 262*f*
    temperature effects 262, 263*f*
Arctonyx species *see* hog badgers
adaptation 218–20, 395–6, 440–1
    behavioural plasticity 259–60
    to climate change 249*b*, 257–8
    mechanisms 441, 442–4*b*
    predictions of activity
        changes 264–6, 265*f*
    rare-allele advantage hypothesis 331
adaptive cascade 259–60
adaptive immunity 306*b*
adaptive kinship hypothesis 423
additive genetic effects 426n16
adiponectin 289*b*
adult survival
    vs. age at first reproduction 296–7*f*
    vs. BCI 279–80, 282*f*, 290, 291*f*

breeding females 280, 292
    vs. rainfall 254
    role in population growth 244
    RTA mortality, weather
        influences 269–71, 270*f*
    vs. weather variability 254–5*f*
African wild dogs 144
    pack sizes 459
age at first reproduction 36
    vs. adult survival 296–7*f*
    vs. BCI 302
    females 298
    influence on population growth 245*b*
    prenatal influences 303*f*
    vs. reproductive success 416, 417,
        417*f*
    temperature effects 301–2
age at last reproduction 416, 417*f*
    females 422
    males 421
age differences
    in BCI 290
    MHV prevalence 310–11*f*
age estimation 228, 229*b*
ageing 285–6, 413–14
    vs. energy expenditure 278, 279*f*
    free radical theory of 319
    reproductive performance 278
    vs. senescence 414
    see also senescence; telomere attrition
aggression 74*b*
    vs. allogrooming 78*t*
    changes over time 81, 86–7, 88
    directed 77n4
    hierarchy linearity and steepness
        analyses 76*t*
    and male mounting frequency 126
    rarity, role of epigenetic
        switching 443*b*
    rarity during mating 120
aggression avoidance 142, 168*b*
    convenience polyandry 138–9
    role of allogrooming 162
aggression event matrices 75*f*

aggressive calls 83*b*
agranulocytes, telomere lengths 431
agranulocyte-to-granulocyte
        ratio 431–2
Albery, Greg 154–5
Albone, Eric 106, 108
Aleutian mink disease virus
        (AMDV) 307
alleles 333n43
Allen, Tanesha 103, 110–11, 186
Allen's rule 277*b*
Allobophora worms
    foraging tactic 200
    see also earthworms
allogrooming 73, 74*b*
    attention to body regions 151*f*
    biological trade model 77*b*
    cooperative flea hunting 149*b*, 150–2
    vs. dominance 77*b*
    duration of 151
    'Esther grooming' 158n12
    and game theory 160*b*
    hierarchy studies 77*b*, 79*t*
    initiation and termination 158–9
    ivermectin experiment 159–60
    and male mounting frequency 126
    mating behaviour 120
    motivations 78*b*
    origins of 162
    reciprocity 77–80, 159
    role in aggression avoidance 162
    social functions 153
allomarking 73, 74*b*, 76*b*, 94, 95–100,
        114
    bacterial exchange 107
    farm cats 99
    meerkats (*Suricata suricatta*) 100
    and reproductive status 96, 97, 98*f*
    seasonal variations 96*f*, 97*f*
    sett observation details 95*t*
    sex differences 96, 97
    sex preferences 96, 97
allomarking cliques 96
allometric size constraints 458*b*

allometry (allometric scaling) 13n11,
    277*b*
alloparental care 31, 146–7, 386–7,
    466–7
    vs. physique 147
allostatic load 307
allosuckling 146–7
allozyme studies 388–9
alpaca, induced ovulation 121
alternative reproductive tactics
    (ART) 138
altriciality 6
altruism 145
American badger (*Taxidea taxus*) 15,
    18, 459
    cooperative hunting 466n20
    distribution 16*f*
    social structure 462
American black bear (*Ursus amer-
    icanus*), population density,
    effects on puberty 38
American mink (*Neovision vision*) 307
    audition 93*b*
    cohabitation 144–5
    delayed implantation 123
    mating behaviour 140
amplitude of an ecosystem 274
Anakuma (Japanese badger) *see*
    Japanese badger (*Meles
    anakuma*)
anal glands 182
anal gland secretion (AGS) 173, 182
    behavioural trial methodology
        182–3, 183*t*
    collection technique 182*f*
    dear enemy phenomenon 183
    donor categories 183n10
    familiarity hypothesis 183
    information conveyed at
        latrines 185–6
    threat-level hypothesis 184
anal gland secretion responses
    sex differences 185
    subcaudal secretion
        over-marking 185
    variation with familiarity 184*f*, 185*f*
    variation with reproductive
        status 185, 186*t*
anarchic null hypothesis 80, 94
Anderson, Roy 153
andropause 417
anergy 353
Annavi, Geetha 387, 392, 400
antagonistic pleiotropy
    hypothesis 414, 433
antelopes, group size 192
antibiotics 337n3
antibodies 306*b*, 321n30, 332*b*
antigen-presenting cells 306*b*

antigen recognition 332*b*
antioxidant capacity (AOX) 9, 322*b*
    cubs 322–4
    vs. food shortage 322, 323*f*
    seasonal variations 322
antioxidant response 319, 321*b*
Anwar, Ali 25
appetite regulation 288–9*b*
apprenticeships 30
Apps, Peter 99
Arabian babblers (*Argya
    squamiceps*) 398
Arctic fox (*Vulpes lagopus*), litter
    sizes 244*b*
Arctic ground squirrel (*Urocitel-
    lus parryii*), telomere
    shortening 434
Arctomeles 18, 94
Ardrey, Robert 166
area-restricted searching 197n5, 198
Argya squamiceps see Arabian babblers
Aristotle 246
Asian badger (*Meles leucurus*) 18, 453
    distribution 16*f*, 20*f*
    facial stripes 458*b*
    human persecution 447
    rabies 316
assortative mating 409n25, 410
ATP (adenosine triphosphate) 320*b*
attention calls 85*b*
Attested Herd Scheme 341
auditory communication 93*b*, 94*b*
    characteristics of acoustic signals 93*t*
    see also vocalizations
August rainfall, *vs.* cub survival 251–2
autumn, energy expenditure 284–5
autumn weather, influence on
    BCI 292–3
averages, interpretation of 254
Axelrod, Robert 160*b*

B

Babesia missiroli 317–18
baboons (*Papio cynocephalus*),
    population density effects 38
bacterial populations, effect on
    subcaudal secretion 107–9
baculum 34, 120
badger BCG vaccination 33–4, 348*b*,
    353–4, 370, 375, 383
    costs 365–6
    efficacy 365, 366
    oral delivery 366
badger burial 438*b*
badger culls 267–9, 375
    cascading ecosystem effects 367n78
    'clean ring' strategy 346
    continuation after pilots 376, 382
    cull rates 371n87

current situation 368–71, 383
Cymag gassing 343, 345
effect on bTB prevalence 359, 361,
    362*f*
efficacy 383
efficacy studies 371, 373–4
efficiency 381–2
extrapolations to 2025 372*f*
financial costs 378, 382
future prospects 374
'interim' strategy 346
Irish 255
ISG conclusions 363
Krebs report 347
licencing
    licence extension, 2013 380
    original licence conditions 379
    plans to 2025/2026 371
live test trial 346
perturbation hypothesis 346, 355–6
perturbation studies 356–63
phasing out 383
pilot results 380, 381–2
political story 376–81, 383
proactive 362
proportion testing positive for
    bTB 382–3, 383n109
reactive 346, 361, 374
roll-out of cull zones 374*b*
shooting 346, 377n99, 378, 381–2
significant dates 347*b*
uninterpretable impact of 368n80
Badger Day 390
'badger debate,' bovine TB 346
Badger Edge Vaccination Scheme
    (BEVS-I) 353
Badger Found Dead Study 349
Badger 'H' (F298) 22, 23*f*, 30
badger-handling challenges 276
badger paradox 161, 168, 188, 191
badger persecution 345
badger protection 345, 346–7
    Wildlife and Countryside Act
        1981 346
badger species
    distribution map 16*f*
    evolution 18–21
    phylogenetic relationships 15–18,
        16*f*, 17*f*
    phylogeography 19*f*, 20*b*
Badger Trust 346–7
Bahaba taipingensis see giant yellow
    croaker
bait 223n6, 224
bait-marked latrine sites
    evidence for visiting behaviour 65–6
    kernel utilization distributions 63*f*,
        66

bait-marked latrine sites (*Continued*)
  minimum-area convex polygons 63*f*,
    65–6
  patterns of latrine use 180
bait marking 45, 46, 48, 52n6, 172*f*
Baker, Sandra 177n8
banded mongooses 474n34
bark 85*b*
  structural characteristics 84*t*
basal metabolic rate (BMR) 287
bat-eared foxes, group size 459
Bateson, Patrick 295
bats
  delayed implantation 403
  diet, *vs.* social organization 192
  extra-group paternity 405–6
  lifespans 414
BCG (Bacillus Calmette–Guérin) 338
  badger vaccination 353–4, 365–6,
    370, 375, 383
  cattle vaccination 366–8, 370, 375
Beacon Hill 53*f*
bears, induced ovulation 121
Beau Geste hypothesis 189
beavers, heterochrony 418
beetle consumption 194*f*, 195
behavioural adaptation 442*b*
behavioural oestrus 116, 132
behavioural plasticity 259–60, 273–4,
    442n3
Belarus badgers 447
Belyaev, Dmitri 443*b*
Bennett, Malcolm 349, 350, 355
Berdoy, Manuel 125
Bergmann's rule 277*b*
Bern Convention (Convention of
    European Wildlife and Natural
    Habitats) 381
'Beta Badger' 22, 30
beta-endorphin, Little B
bet-hedging 274, 295, 399, 411
Bialowieża Primeval Forest 447, 448–9
Bilham, Kirsten 322–5, 354
biogeographical distribution 446
biological trade model 77*b*
bird song, bluffing strategies 189
birth dates 5
births, season of 4
Biston betularia see peppered moth
bite wounds 81–2, 82*f*
  vs. body condition 87
  changes over time 86–7, 87*f*
  on cubs 86
  definitions 86
  effect of culling 359, 364
  vs. group size 87
  observations, 1990–99 86*t*
  seasonal variation 87

black and white colobus monkey
    (*Colobus guerezaullensis*) 192
blackberry consumption 194*f*
black swans 216n17
Bloat (M814) 22, 23*f*, 24, 30
blood composition, changes with
    age 431
bluffing, signalling strength of
    numbers 189
B lymphocytes 306*b*, 332*b*
body condition index (BCI)
  vs. Activity 264, 265*f*
  vs. age 290, 420*f*
  vs. age at first reproduction 302
  vs. antioxidant capacity 322, 323*f*
  BCI$_r$ 289
  vs. bite wounding 87
  and climate change 293–4
  vs. energy expenditure 285, 286*f*
  individual variation 293
  vs. male mounting frequency 124–5,
    126
  and oxidative damage 324–5
  vs. reproductive success 132, 290,
    292
  seasonal variations 289*f*–90
  vs. survival probability 279–80, 282*f*
  vs. survival rates 290, 291*f*
  vs. trapping efficiency 267, 268*f*
  trend (1990–1997) 246*f*
  vs. visiting behaviour 62, 65, 286
  vs. waistline assessment 289
  vs. weather 292–3*f*
  vs. winter conditions 252
body size
  vs. energy expenditure 277–8, 277*b*,
    279*f*
  ferret badgers 459
  and group living 461, 464, 474
  vs. home range size 464*f*
  vs. subcaudal secretion colour 103
  trade-offs 323–4, 325
Bonacic, Cristian 330n40, 331
bonobos, non-conceptive sex 143
border latrines *see* perimeter latrines
Bornean ferret badger (*Melogale
    everetti*) 16*f*
Bourne, John 363
bovine tuberculosis (bTB) 267, 307
  Area Eradication Plan, 1950 341
  Attested Herd Scheme 341
  badger culls *see* badger culls
  'badger debate' 346
  badger reservoir area
    identification 374
  badger susceptibility 354–5
  badger testing 352*b*
  badger vaccination *see* badger BCG
    vaccination

cattle diagnosis 368
cattle-oriented solutions 337
cattle vaccination *see* cattle BCG
    vaccination
cattle *vs.* badger paradigm 341–2
'clean ring' strategy 346
Consultative Panel 342
current situation 368, 383
diagnostic challenges 352*f*, 375
distribution, 1938 341*f*
financial costs 370
first discovery in a UK badger 342,
    344*f*
geographical distribution of
    breakdowns, 1992, 2004,
    2015 343*f*
hotspots, definition 375n93
human infection risk 338, 370
immune response
    development 338n6
'interim' strategy 346, 347*b*
ISG conclusions on badger
    culling 363
Krebs report 347
latent badger infections 346
live test trial 346
measures of herd bTB
    incidence 362n65
moral responsibilities 367–8, 375
OTF status 349n34, 351n43
perturbation hypothesis 346, 355–6,
    364–5
  see also perturbation effect studies
prevalence
  effect of badger culls 359, 361,
    362*f*
  rise since 1970s 341
reasons for test failures 340–1
role of cattle movements 342, 375
Royal Society review of evidence,
    2012 339
significant dates 347*b*
skin tests 338–40
spoligotyping 349
transmission 343, 349–50, 350, 375
treatment of infected cattle 367n77
WildCRU study on other
    mammals 354
Wytham badgers 351, 353–4
see also badger culls; SICCT (single
    intradermal comparative cervi-
    cal tuberculin) test; tuberculin
    skin test; tuberculosis
Bradbury, Jack 192
Briar Badger (F695) 1–2*f*, 3, 4*f*
  maternal care 5–6
Bright Ross, Julius 194, 227, 275, 296
Brisson, Mathurin Jaccques 15

British badgers, population
    estimates 445*b*
Brock TB ELISA test 352*b*
Brogdens Belt sett 53*f*, 57
brown hyaena 474
Buesching, Christina 33, 94, 95*f*, 105,
    165, 172, 182
Bunyan, James 1
Burketts Plantation 53*f*
Burmese ferret badger (*Melogale
    personata*) 16*f*
bush dog 473n31
bushmeat trade 455
butterfly effect 301n25
Byrne, Andrew 255–6

C

cacomistle 463
Callorhinus ursinus see northern fur seal
calorific content, subcaudal
    secretions 110–12, 111*f*
calorific requirements 195
camels, induced ovulation 121
camera-trap studies
    Scottish badgers 256–7
    weather influences 266
Campbell, Liz 476
canids
    female reproductive
        suppression 473
    genealogy 469–70
    hunting 472
    omnivorous diet 472
    paternal care 472, 473
    sexual size dimorphism 467
    social structures 467, 471–2, 474–5
    vulpine *vs.* lupine 472–3
    see also African wild dogs; dogs;
        hyaenas; wolves
canine distemper virus (CDV) 8, 307
canine infectious hepatitis 8
canine parvovirus enteritis 307
Canis latrans see coyotes
Capreolus capreolus see roe deer
capture–remark–capture (CMR)
    technique 227
capybaras
    sex ratio manipulation 11
    testis size 141
Carbone, Chris 459
Carnivora genealogy 469–70
carnivore societies 467–8
    female reproductive
        suppression 473
    fossoriality 475–8
    group living 471
    social organizations 470–1
Carr, Geoff 203

Carrington, David (Carrington
    leak) 380
carrying capacity (K) 239*t*, 302
    changes in 234, 246–7, 248
    identification of 238, 240*f*
cascading ecosystem effects
    of badger culls 367n78
    of fox rabies vaccination 365n67
Casanova B (M90) 214, 393, 395*b*,
    396–7, 396*b*
Castor canadensis see North American
    beaver
cats
    disease transmission 152
    hierarchies 74*b*
    infanticide 142
    rubbing behaviour 74*b*, 99
    social structure 99
cattle
    bTB-related slaughter, Defra data,
        2020 370
    see also bovine tuberculosis
cattle BCG vaccination 370, 375
    efficacy 366
    legal issues 350*b*
    and sensitive bTB diagnosis 367
    and tuberculin testing 366–7
cattle movements, role in bTB
    transmission 342, 375
causation *vs.* correlation 259, 266
causes of death 232–3
    human persecution 447–8
    road traffic accidents 269–71
cave dwelling 458*b*
CD4 and CD8 cells 332*b*
Cecil the lion 423*b*
cellular senescence 425
    see also telomere attrition
central place theory 55
centripetal model, cat behaviour 99
Cervus analysis 390n12, 391*f*
Cervus elaphus see red deer
chaetae 192
Chalet Outlier 53*b*
Chalet setts, cub integration 32
chambers, use of 89–90, 89*f*, 90*f*
chaos theory 301n25
Cheeseman, Chris 346, 361
cheetahs, social structure 471
chemokines 306*b*
Chesterton, G.K. 92
Chinese ferret badger (*Melogale
    moschata*) 457*f*
    distribution 15, 16*f*
chirp 85*b*
    structural characteristics 84*t*
chitter 85*b*
    structural characteristics 84*t*
Christaller, Walter 55

Christensen, Poul 378, 380, 381
churr 85*b*, 119
    structural characteristics 84*t*
Cimolestes 469
Cinclus cinclus see white-throated
    dippers
cingulum 15n14
clay setts 56
'clean ring' strategy, badger culls 346,
    347*b*
Clearing Outlier 1, 3n5, 53*b*
climate change 249*b*, 440
    adaptation to 257–8, 259, 271
    extinction predictions 441n2
    implications for BCI 293–4
    implications for oxidative stress 326
    implications for telomere
        attrition 435
    IPCC projections 256, 257
    predictions of activity
        changes 264–6, 265*f*
    role in population growth 251–3
Clinchy, Mike 446
Cluck call 85*b*
    structural characteristics 84*t*
Clutton-Brock, Tim 192
coatis (*Nasua* species) 468
    cooperative hunting 466
    cooperative vigilance and
        defence 466
    group size 459
    social structure 463
coccidiosis 305, 333
    Toxoplasma gondii 316–17
    see also Eimeria melis; Isospora melis
co-evolution, MHV 315
cognitive abilities 105–6
cohabitation 144–5
    costs of 166–7
    interspecific 447
Cole, Lionel 177
colobus monkeys, diet, *vs.* social
    organization 192
co-locations, aRFID studies 48–50, 49*f*,
    50*f*, 51*f*
Colony inference 390n12, 391*f*
Colony software, parental assignment
    rules 400*f*
Common Agricultural Policy
    (CAP) 208n13, 213–14
common shrew (*Sorex araneus*), mating
    behaviour 140
common wombat (*Vombatus
    ursinus*) 477*b*
competition 73
    disadvantages of 91
    vs. group size 459
    intra-guild 166
complement system 306*b*

Connelly, Erin 194
conservation of energy 260–4
conservation science, realities of 375–6, 378, 383–4
conservative bet-hedging 274
constraint hypothesis 416
contact and contact-seeking calls 85b
convenience polyandry 138–9, 142
Convention of European Wildlife and Natural Habitats (Bern Convention) 381
coo 85b
  structural characteristics 84t
cooperation 145, 465
  absence of 204
  alloparental care 466–7
  flea management 149–52, 149b
  and fossorial lifestyle 467–8
  hunting 465–6, 472
  inter-generational 478
  vigilance and defensive behaviour 466
cooperative care 146–7
cooperative digging 148–9
corpus luteum 37f, 136n18, 137n19
correlation vs. causation 259, 266
cortisol 306b
  stress response 330n40
Cota Doñana 448
Court Jester hypothesis 163, 274
Covid-19 306b, 355n53
  cytokine storm 321n33
Cox, Ruth 150
coyotes (Canis latrans), trappability 266
creodonts 469
Cresswell, Warren 87
Crook, John 191–2
crop protection, food aversion studies 177n8
Cruelty to Animals Act 1835 345
cryptic female choice 138, 141, 301, 387, 396, 403
cub cohort sizes 9, 10f
  POL effects 297–8
  and sex ratio 10
  vs. telomere length 433, 437
  weather effects 253
cub growth
  influencing factors 13–14
  sex differences 13–14
cub integration 32–3, 32t
  scent marking 33
  visual signalling 33
cub MHV susceptibility, prenatal influences 313f–14
cub mortality 245b
  role of coccidiosis 27–8
  role of MHV 308
  vs. weather 251–2, 253

see also cub survival
cub recruitment 12–13
cub relative leukocyte telomere length
  vs. spring weather 435–7, 435f, 436f
  vs. survival 428–9, 430f
  vs. weather 433
cubs
  adult indifference to 31, 146
  antioxidant capacity 322–4
  behavioural changes with increasing age 32t
  bite wounds 86
  capture and handling of 24–5f
  care from non-breeding adults 6
  cooperative care 146–7
  dispersal and philopatry 39–43
  Eimeria melis infection 9, 25–9
  growth curves 13, 14f
  growth vs. immune defences 323–4, 325
  maternal assignment 22–3
  maternal care 5–6
  measurement of 13f
  MHV infection prevalence 308
  opening of eyes 6
  oxidative stress, weather effects 325
  pre-emergence mortality 8
  puberty 34–9
  scent theft 33, 94, 108–9
  sex ratio effects 297–8
  socialization 6, 31f
cub survival 23–4f, 231–2, 246n27
  EGO vs. WGO 407
  vs. heterozygosity 401, 402f
  vs. rainfall 254
  vs. RLTL 428–9, 430f
  role in population growth 245b, 246
  vs. sett type 11–12f
  vs. weather 251–2, 253, 401
  vs. weather variability 254–5f
  see also cub mortality
Cuvier, Georges 15
Cymag 345
cytokines 306b
cytokine storm 321n33

D
daily energy expenditure (DEE)
  vs. age 278, 279f
  vs. body size 277, 279f
  individual variation 276–7
Daniels, Gilbert 271, 272
Dansgaard–Oeschger (D-O) events 249b
Darwin, Charles 60, 140, 222b, 266, 385, 441
da Silva, Jack 223, 389
Dawkins, Richard 441, 442–3b
DDT 248

dear enemy phenomenon (DEP) 183
death, causes of 232–3
  human persecution 447–8
  road traffic accidents 269–71
deciduous woodland
  importance of 208–9
  vs. mean adult male weight 209f
deer (Cervus species)
  bovine tuberculosis 354
  fallow deer population 208n14, 214
  see also red deer; roe deer
defence, cooperative 466
defence calls 83b
defence costs hypothesis 167
Defra
  bTB data, 2020 368–9
  bTB data by region, 1996–2020 369f
  bTB data by risk, 1996–2020 370f
  cattle slaughter data 370
  compromises on cull licence conditions 379
  conclusions on badger culling 368n80
  extrapolations from RBCT 376–7
Delahay, Dez 354
delayed dispersal, male offspring 452
delayed-dispersal threshold model 145
delayed implantation (DI) 3, 123–4, 252, 396, 403, 459b
  Chinese ferret badger (Melogale moschata) 457
  cryptic female choice 141
  interaction with superfoetation 133–8
  oestrone levels 134–5, 134f, 135n17
  roe deer 11
demographic forces of changes 222f
demography 222b
density dependence 238
  population growth curve 239t
  theta (θ) statistic 239
dentition 15
age estimation 229f
despotism 91
detection likelihood 266
devaluation hypothesis, multiple male mating 142
Developmental Origins of Health and Disease (DOHaD) paradigm 314
dice game, RDH 204b
dieldrin 248
diet 464
  calorific requirements 195
  Chinese ferret badger (Melogale moschata) 457
  diversity 193–5, 193f, 194f, 445–6
  earthworms 192–3
    abundance 195

alternatives to 198
availability 195–200
vs. fossoriality 476
foxes 199*f*
frugivorous groups 468
and group living 461, 468
hog badgers (*Arctonyx* species) 454,
455*f*
intra-specific variation 445–6, 449
Japanese badger (*Meles anakuma*) 452
omnivory 472
piscivorous groups 469
seasonal variations 194*f*
vs. social organization 192
stink badgers (*Mydaus* species) 460
dietary switching 452
digging behaviour 148–9
directed aggression 77n4
disassortative mating 410n27
diseases 305
coping abilities 319–26
protozoa 316–19
see also *Eimeria melis*; *Isospora melis*,
Toxoplasma
viruses 307
see also Mustelid herpesvirus;
rabies
see also bovine tuberculosis;
tuberculosis
disease susceptibility 305–6
and fossorial lifestyle 112
disease transmission 152
dispersal 30, 39–40, 72, 148, 167
age at 41–3
and gene frequency
heterogeneity 389
and habitat saturation 145
inter-specific variation 474n33
Japanese badger 452
see also Resource Dispersion
Hypothesis
dispersal distances 46
dispersal events 47*f*
dispersal rates 41, 42, 45–6
disposable soma hypothesis 414
DIT tests, bovine TB 367
diversified bet-hedging 274n4
DNA studies 387–8, 389–90
dogs
reactions to flea bites 152
sense of smell 93*b*
sperm 140n20
Dolichotis patagonum see maras
domestication
active social domestication
hypothesis 443*b*
silver foxes experiment 443*b*
dominance
definition 80–1

see also hierarchies
Domingo-Roura, Xavier 389–90
Doñana National Park 448
Doncaster, Patrick 58, 445–6
Donnelly, Christl 361
dormice 434
telomere shortening 434
doubly labelled water (DLW), energy
dynamics study 275–6
Dowager B (F227) 72, 393, 396*b*
Dugatkin, Lee Alan 145
Dugdale, Hannah 77*b*, 116, 146, 387,
390, 416
Dunbar, Robin 153

E
Earthwatch 31n2
earthworm abundance 195
vs. group size 207
increase, effect on badger
population 208
temporal changes in three habitat
types 212*f*
earthworm availability 195n2
vs. search strategies and
tactics 197–200
spatio-temporal heterogeneity 206
vs. weather 195–6, 198*f*, 251
earthworm consumption
vs. groups size 210*l*
intra-specific variation 445
earthworms 192–3
in badgers' diets 193–5, 193*f*, 194*f*
biomass estimation 210*b*
in foxes' diets 199*f*
hunting and capture techniques 196,
197*f*
ecological patterns 272
ecological regime shifts 234, 241
ecology, definitions 222*b*
ectoparasites 154–6
infection burden *vs.* population
density 155*f*
see also fleas; lice; ticks
Edleston, R.S. 219
Eimeria melis 9, 25–6, 25*f*, 154, 333
cub *vs.* adult oocyst counts 26
effect on adult size 28
effect on cubs 26, 28
effect on sett location 156
infection intensity 27*f*–8
infection prevalence 27*f*
life cycle 26n4
and palate marking asymmetry 28–9
vs. population density 155*f*
transmission to cubs 27
electro-ejaculation technique 139
electron transfer 320*b*
Ellwood, Stephen 46

Elton, Charles 222n2, 376n96
embryonic diapause 123–4
see also delayed implantation
emergence from sett 100–1
endosymbiosis 320*b*
energy conservation 260–4
energy dynamics 274–5
DLW study 275–6
energy expenditure
vs. age 278
vs. body condition 285, 286*f*
vs. body size 277–8, 277*b*
and climate change 293–4
immune defences 307
individual variation 278–9, 280*f*,
283, 286, 287
lactation 279
management tactics, implications for
resilience 287
measurement 260–1
vs. rainfall 284*f*
responses to environmental
changes 283
seasonal variations 283–5
task allocation 286
vs. temperature 283, 284*f*, 286*f*
and thyroid hormones 288*b*
vs. wind 284
Enferplex test, bTB 339n14, 340,
340n16, 341
enhanced MNA (eMNA) 227–8
environmental changes
responses to 283, 287
see also climate change; rainfall;
resilience; temperature; weather
environmental shocks 274
enzootic disease 305
epigenetics 302, 303, 442–4*b*
epizootic disease 305
erythema 339n8
EST-1 alleles 388
'Esther grooming' 158n12
Eurasian otters (*Luta lutra*), social
structure 469
European badger (*Meles meles*),
distribution 16*f*, 20*f*
European mink, feeding
specializations 286
eusocial species 144n1
Eustice, George 371
Evans, Peter 52, 388
evolution 18–21, 115
role of climate change 249*b*
of subcaudal gland 112–13
evolutionary ecology 145
extinctions 440, 441n2, 469n26
extra-group offspring (EGO)
performance *vs.* WGO 407–9

extra-group offspring (EGO)
(*Continued*)
  sociological disadvantage
    hypothesis 408
extra-group paternity (EGP) 391, 392*f*,
    401, 404
  advantages to males 401
  fitness hypothesis 407
  vs. infanticide 402
  vs. number of neighbouring-group
    candidate fathers 405, 406*f*
  vs. number of within-group
    females 405
  prevalence 404–5*f*
  random encounter hypothesis 405
  vs. within-group relatedness 405,
    406*f*
eyes, opening of, cubs 6

F

F134 128*b*, 129*f*, 130*b*, 131*b*
F140 68, 130*b*
F217 68, 69*b*, 70*b*, 73, 76, 214
F221 23, 68, 69*b*, 70*b*, 72*b*, 73, 76, 91,
    129*f*, 214
F227 (Dowager B) 72, 393, 395*b*, 396*b*
F240 69*b*, 76
F277 129*f*, 130*b*
F284 130*b*, 131*b*
F297 69*b*, 70*b*, 72*b*, 129*f*
F298 (Badger 'H') 22, 23*f*, 30
F301 70*b*, 128*b*, 131*b*
F305 69*b*, 70*b*, 128*b*, 129*f*
F323 23, 69*b*, 70*b*, 128*b*
F324 69*b*, 70*b*, 128*b*
F327 69*b*, 70*b*, 81, 128*b*
F351 130*b*, 396*b*
F353 130*b*, 395*b*
F380 70*b*, 72–3, 76, 129*f*, 214
F399 130*b*, 131*b*, 148
F400 22, 23*f*, 70*b*, 72*b*, 128*b*, 129*f*
F417 22, 70*b*
F461 128*b*, 129*f*
F507 22, 23*f*
F526 128*b*, 131*b*
F558 130*b*, 148
F591 70*b*, 128*b*
F594 70*b*, 128*b*
F638 421*f*, 422
F686 421*f*, 422
F695 (Briar Badger) 1–2*f*, 3, 4*f*
  maternal care 5–6
F699 232*b*, 393
F705 70*b*, 72*b*
F713 23, 129*f*, 421*f*, 422
F730 3, 4*f*
F765 129*f*, 421*f*, 422
F773 1, 3, 4*f*
F776 3, 4*f*

F779 22, 23*f*, 70*b*, 128*b*
F798 3, 4*f*
F816 22, 23*f*
F819 22, 23*f*, 51–2, 70*b*, 72*b*, 128*b*, 146–7
F869 6, 395*b*
F927 3, 4*f*
F945 2*f*, 3, 4*f*
  maternal care of 5–6
  raiding of rodent traps 7
  socialization 6–7
F959 3, 421*f*, 422
F960 2*f*, 3, 4*f*
  life history 7
  maternal care of 5–6
  raiding of rodent traps 7
  socialization 6–7
F1091 421*f*, 422
F1311 421*f*, 422
F1509 421*f*, 422
F/α (fertility to age at fist
    reproduction) ratio 245*b*
facial stripes 33, 458*b*
  warning function 82
facultative (lactational) delayed
    implantation 123n7
faeces
  diet assessment 193–5, 193*f*, 194*f*
  parasite examination 25
  see also bait marking
faeces matching 178, 189–90
FAI bTB outbreak 351, 353
fallow deer (*Cervus dama*)
    population 208n14, 214
familiarity hypothesis 183
farm cats *see* cats
Fasciola hepatis, effect on SICCT test 340
fat reserves 203, 205
  vs. Activity 264, 265*f*
  costs and benefits 285, 287–8
  importance in winter 252
  individual variation 287, 290
  inter-specific comparisons 460–1
  see also body condition index
fecundity 244*b*
feeding specializations 286–7
felids
  genealogy 469–70
  sexual size dimorphism 467
  social structures 467
  see also cats; lions; ocelot
Fell, Rebecca 32, 146
female dominance 74*b*, 76, 81
female mate choice 125–6, 396–9,
    405–6
  mice 404
  and MHC 409–11*f*
  mink 403–4
  summary 411–12
  see also extra-group paternity

female mating behaviour 120
female mounting success, predictors
    of 126–8, 290
female reproductive skew 282–3,
    299–300, 390, 393
female reproductive success
  vs. age 416
  elderly females 422–3
  number of mothers per social
    group-year 392*f*
  vs. survival probability 280, 292
female reproductive suppression 282,
    298, 423–4
  carnivores 473
  role of MHV 308–10
ferret badgers (*Melogale* species) 456–7,
    457*f*
  Chinese ferret badger 457*f*
  distribution 15, 16*f*
  facial stripes 458*b*
  fat reserves 460
  hunting of 455
ferrets, induced ovulation 121
fertility, F/α ratio 245*b*
fertility assurance, and oestrus
    schedules 132–3, 139
Firebreak 53*b*
  cub integration 32
fission–fusion societies 203n7
fissioning 53*b*, 54–5
  Jew's Harp sett 52
  Pasticks sett 52
fitness 385–6
  and senescence 414*f*
five Bs vignette 116, 118–19, 386
flea counts 150, 154
  vs. population density 155*f*
fleas (*Paraceras melis*) 149
  adverse impact 150
  avoidance strategies 148
  badgers' sensitivities to 160
  behaviour 150
  distribution of 152–3
  influence on sett location 156
  as motivation for allogrooming 78*b*
  mutual grooming 149*b*, 150–2
  Trypanosoma pestanai
    transmission 317
  variance in reactions to 152
flea treatment experiment 159–60
foetal programming 313–14
follicle-stimulating hormone
    (FSH) 36*b*
food aversion studies 177n8
food-related encounters 80–1
  Manna From Heaven experiment 81
food resource bottlenecks 209
food resources
  changes over time 214

spatio-temporal heterogeneity 206, 209, 210*b*
see also diet
food security 205, 216, 218, 248
  role in population growth 249
  winter 252
food shortage, *vs.* antioxidant capacity 322, 323*f*
food sources, *vs.* social structure 99
foraging 197
  energy conservation 260–4
  optimal foraging theory 199–200
  search strategies 198–9
Ford (E.B.(Henry)) 219
forelimb anatomy 458*b*
fossil species 223n2
Fossorial Benefits Hypothesis (FBH) 6
fossoriality 467–8
  allometric size constraints 458*b*
  disease hazards 112
  forelimb anatomy 458*b*
  and group living 461
  inter-specific comparisons 475–8
  maras 478*b*
  wombats 477*b*
  see also setts
foxes
  Arctic (*Vulpes lagopus*), litter sizes 244*b*
  audition 94*b*
  bat-eared, group size 459
  diet 199*f*
  earthworm consumption 193, 198
  earthworm hunting 196
  foraging strategies 198–200
  interaction with badgers 447*f*
  rabies vaccination 365
  silver, domestication experiment 443*b*
  spatial groups 203
  territory size 207
freeloading 161n19
free radicals 319n21
free radical theory of ageing 319
Freud, Sigmund 40n9
frost 251, 253, 285
frugivorous groups 468
fruit consumption 194*f*

G
game theory 160*b*
gamma herpes virus (MusGHV-1) 310
Gasterosteus aculeatus see three-spined stickleback
gatherings, distinction from groupings 202–3*f*
Gelling, Merryl 330–1, 354
genealogies 4*f*, 23*f*, 128*b*, 129*f*, 130*b*, 130*f*, 421*f*

Casanova B (M90) 395*b*
Dowager B (F227) 396*b*
Lecher B (M10) 69*f*
F699 232*b*
generation time (G) 237*b*, 242n25
genetic compatibility hypothesis 399
genetic diversity 21*b*
genetic relatedness
  vs. allogrooming 77*b*, 78*t*
  and hierarchies 74*b*, 76
genetic studies
  allozyme studies 388–9
  DNA studies 387–8
  microsatellite studies 389–90
genotypic adaptation 442*b*
Gentle, Louise 285
geology
  and formation of super-groups 56–7
  influence on sett location 56, 57*f*
gestation periods 4, 123
Ghiselin–Reiss small-male hypothesis 15
ghrelin 289*b*
giant otter (*Pteronura brasiliensis*) 418
  alloparental care 466
  cooperative hunting 465
  female reproductive suppression 473
  group living 462–3
  prey availability 469n25
  social structure 469, 473n32
giant yellow croaker (*Bahaba taipingensis*) 454n14
Gini index (GI) 299, 300*f*
Gladwell, Malcolm 116
Godfray, Charles 339, 383
Godfray Report, 2018 348*b*, 370–1
golden jackals (*Canis aureus*), territoriality 171
Golden Rule 145n2
Goldilocks zone 254
gonadotropin-releasing hormones (GnRH) 36*b*
good genes hypothesis 399
Gorman, Martyn 105, 107
Gosler, Andy 285
Gosling, Morris 168*b*
Gradwell, George 177
Grahame, Kenneth 92
grandmother hypothesis 424
grand-mothering 418
granulocytes, telomere lengths 431
Graunt, John 222*b*, 222n1
Great Ash Hill 3, 7, 53*f*
  fissioning 53*b*
greater hog badger (*Arctonyx collaris*) 448, 453
Great Oak sett 6, 53*f*
  group size 68

yearling breeders 36–7
great reed warbler (*Acrocephalus arundinaceus*) 437*b*
great tit (*Parus major*) 285
Great Wood sett 53*f*, 55
Greek (M243) 42, 69*b*, 70*b*, 128
grooming 73
  attention to body regions 151*f*
  control mechanisms 153
  cubs 31, 32*f*
  maternal care 5
  mating behaviour 116, 118–19, 120
  see also allogrooming
group Allee effect 144
group composition 67–72
  sex ratios 68
groupings
  distinction from gatherings 202–3*f*
  spatial groups 203, 204
group living
  associated factors 461
  badgers' adaptation to 162–3
  benefits for badgers 145–6, 160–3
  and body size 464
  canids 471–2
  carnivores 471
  costs *vs.* benefits 203–4
  and fat reserves 460
  fitness benefits 144
  flea distribution 152
  hypotheses and conclusions 158*t*
  otters 462–3
  RDH 201–6
  scent matching 169*b*
  selective pressures 191
  see also social structure
group mass, *vs.* home range size 465*f*
group partitioning
  mechanisms 217
  problems with RDH 213
  vs. range size reduction 214
  RDH predictions 211–13
  role of social cliques 216
group size
  vs. bite wounding 87
  calculation of 67, 68
  changes over time 215
  Chinese ferret badger (*Melogale moschata*) 457
  and defence 459
  determination of 204, 205, 206, 209, 214, 218
  vs. earthworm abundance 207
  vs. earthworm consumption 210*I*
  faeces matching 178, 189
  vs. hierarchies 76, 80
  hypothetical amalgamation of Wytham badgers 215
  influence of parasites 156

group size (*Continued*)
  influence on timing of puberty 38–9
  inter-specific variation 459
  intra-specific variation 448
  sociological factors 215
  spatio-temporal variation 67–8
  vs. territory size 207
growls 83*b*
  structural characteristics 84*t*
growth, trade-off with antioxidant
    capacity 323–4, 325
growth curves 14*f*
  vs. endocrinological maturity 38, 42*f*
  vs. testes conditions 43*f*
growth hormone 288*b*
grunt 85*b*
  structural characteristics 84*t*
guinea pigs, epigenetic
    adaptation 444*b*

H

habitat saturation 145
Hambly, Catherine 275
Hamilton, Bill 145, 149, 386, 398
Hamilton–Zuk hypothesis 398
Hancox, Martin 193
handicap principle 398
Hardy–Weinberg equilibrium 388n6
harvest mice (*Micromys minutus*)
    female mate choice 404
Harvey, Paul 414
hazelnut consumption 194*f*
Heackel, Ernst 222*b*
health
  effect on body odours 112
  see also diseases
hearing, sense of 93*b*
heat stress 285
  physiological response 278
hedgehogs, avoidance of badger
    odour 445–6
height determination (humans) 242n24
Heinrich events 249*b*
Heraclitus 250n4
Herodotus 142
Hersteinsson, Pall 201
Hesse's (heart–weight) rule 277*b*
heterochrony 418
heterogeneity *see* individual variation
heterozygosity 410
  vs. cub survival 401, 402*f*
  role in mate selection 399–401
heterozygosity–fitness correlations
    (HFCs) 400–1
Hewitt, Stacey 74*b*, 80*b*
hierarchies 67, 72, 73, 114
  aggression studies 74*b*, 76*t*
  allogrooming studies 77*b*
  determinants of 72–3

and food-related encounters 80–1
  sex differences 74*b*, 76
hierarchy linearity 74*b*
  analysis based on flow of
      aggression 76*t*
  analysis based on unreciprocated
      allogrooming 79*f*, 79*t*
hierarchy steepness 74*b*
  analysis based on flow of
      aggression 76*t*
  analysis based on unreciprocated
      allogrooming 79*f*, 79*t*
  and group size 76, 80
Hill Copse sett 3
  cub survival 12*f*
Hill End sett 53*f*, 58n11
hissing 83*b*
  structural characteristics 84*t*
histamine 306*b*
HLA-DR antigen-binding
    heterodimer 333n45
hoary badger (*Meles canescens*) 453
  distribution 18, 20*f*, 21
  facial stripes 458*b*
Hofer, Heribert 193, 210*b*, 212*b*
hog badgers (*Arctonyx* species) 18, 448,
    453–4, 454*f*, 457
  diet 454, 455*f*
  facial stripes 458*b*
  hunting of 455
  latrines 454
  scent marking 454n13
  social structure 454–5
Holly Hill 53*f*
  spatial arrangements 213*b*
home ranges 165–6
  super-groups 174*f*
  Wytham badgers 173*f*
  see also range sizes
honey badger (*Mellivora
    capensis*) 15–16, 459–60
  cooperative hunting 466n20
  distribution 16*f*
  social structure 462n19
hormones
  puberty 35*f*, 36
  see also oestradiol levels; oestrone
      levels; oxytocin; progesterone;
      testosterone titres
horse chestnut consumption 194*f*
house mice
  female mate choice 404
  reproductive senescence 419n9
huddling 476
human infertility 397
human leukocyte antigen (HLA)
    compatibilities 410
human persecution 447–8
  hog badgers 455

humans, badgers' fear of 446
humidity, *vs.* earthworm
    availability 195–6, 198*f*
hunting, cooperative 465–6, 472
hunting of badgers 447–8
  hog badgers 455
Hwange lion study 423*b*
hyaenas
  social structure 473–4
  see also spotted hyaena (*Crocuta
      Crocuta*)
hygiene hypothesis 432
hypersensitivity reactions 332*b*
hyperthermia 285
  physiological response 278
hypothalamic–pituitary–gonadal
    (HPG) axis 36*b*
hypoxia adaptation 442*b*

I

Ictitherium viverrinum 473
Ictonychinae 462
ideal free distribution (IFD) 152n9
identification of individuals 1, 224,
    226*f*
IDEXX antibody test, bTB 368n81
Ijzerman, Hans 148
immune system 306*b*
  agranulocyte-to-granulocyte
      ratio 431–2
  defences against mycobacteria 354–5
  energy expenditure 307
  hygiene hypothesis 432
  MHC 332*b*
  trade-off with growth 323–4, 325
immunocompetence handicap
    hypothesis 398–9
immunosenescence 431
implantation, triggering of 5
implantation dates 5, 253
  and female body condition 10*f*
  and sex ratio 10*f*
inbreeding 40, 388–9
  Casanova B (M90) 395*b*
  common shrew 140
inbreeding depression 400–1
incest taboo 40
inclusive fitness 386
Independent Scientific Group on Cattle
    TB (ISG) 363
individual variation 272–3, 303–4
  BCI 293
  BMR 287
  energy dynamics 274–5
  energy expenditure 276–7, 278–9,
      280*f*, 283, 286, 287
  female reproductive skew 282–3
  life history strategies 294–7
  as source of resilience 273–4

thyroid hormones 288*b*
induced ovulation 120–1, 405
  evolution 121–2
  interpretative implications 122–3
  triggers 121
inducible Nitric Oxide (iNOS) 354
inductive bias 259, 266, 269n8
induration 339n9
infant diseases 24
  Eimeria melis 25–9
infanticide 9*f*, 86
  and extra-group paternity 402
  and multiple male mating 141–2
infectious diseases 305
  vs. MHC profile 333–5, 334*f*
  protozoa 316
    see also Eimeria melis; Isospora melis
  viruses 307
    see also Mustelid herpesvirus;
      rabies
    see also bovine tuberculosis;
      tuberculosis
infertility, and ABO(H) blood type 397
infilling 61, 450–1, 468, 475
  carnivore societies 471
inflammatory response 306*b*, 321
information sharing 146
infrared (hot-eye) studies 196
innate immune response 306*b*
  and Reactive Oxygen Species
    (ROS) 321
insectivory 194*f*
  association with fossoriality 476
interferon gamma (IFN-γ) test,
    bTB 339, 351*b*, 353*b*
inter-generational cooperation 478
inter-group contacts 47*f*, 49–50
  itinerant individuals 51–2
  network patterns 51*f*
  see also visiting behaviour
interim strategy, badger culls 346,
    347*b*
inter-specific cohabitation 447
inter-specific comparisons
  American badger 459
  carnivore societies 467–8, 471–4
  Chinese ferret badger (*Melogale
    moschata*) 457*f*
  facial masks 458*b*
  fat reserves 460–1
  fossoriality 475–8
  frugivorous groups 468
  group size 459
  hog badger (*Arctonyx collaris*) 453–5
  honey badger (Mellivora
    capensis) 459–60
  martens 460
  Meles species 451–3
  piscivorous groups 469

stink badgers (*Mydaus* species) 460
inter-specific interactions 447*f*
intra-guild competition 166
intra-sexual territoriality 30n1, 40–1,
    60, 217, 450
  and induced ovulation 121–2
  inter-specific comparisons 462
  Japanese badger 452
intra-specific variation 219–20, 272
  competition and predators 446–7
  diet 445–6, 449
  population density 444–5
  and Resource Dispersion
    Hypothesis 449–51
  social organization 448–9
IPCC (Intergovernmental Panel
    on Climate Change)
    projections 294
  for Ireland 256
  for Scotland 257
Irish badgers
  body weights 256
  culls 255
  influences of weather
    variability 255–6
  MHV infection 309–10
  population estimates 445*b*
Isospora melis 29*f*, 154
Italian badgers 447, 448
itch perception 153

J

Japanese badger (*Meles anakuma*) 18,
    451–2*f*
  competition 459
  diet 452
  dispersal 452
  distribution 16*f*, 20*f*
  facial stripes 458*b*
  human persecution 447–8
  reproduction 452
  social structure 452, 453
Jarman, Peter 192
Javan ferret badger (*Melogale
    orientalis*) 16*f*
Jeffreys, Alec 387
Jensen's Inequality 254, 274n5
Jew's Harp 53*f*
  ontogeny 52
  range studies, 1970s 59–60, 60*f*
Johnson, Dominic 161*b*, 211*b*, 212*b*
Johnson, Paul 461, 467
Jordan, Bill 345
Jordan, Neil 165
juveniles
  allomarking behaviour 96, 97, 99
  substrate marking 101
  see also cubs

K

K = 3 system 55, 58
Kamler, Jan 459
Kaneko, Yayoi 172*f*
keckering 22n1, 83*b*
  structural characteristics 84*t*
Kent, Alice 308
kernel utilization distributions,
    bait-marked latrine sites 63*f*, 66
Kettlewell, Bernard 219
Keverne, Barry 153
Kilshaw, Kerry 180
kinkajou (*Potos flavus*) 468
  social structure 463
kin selection 386
Koch, Robert 337
Krebs, Charlie 222*b*
Krebs, Hans 320n28
Krebs, John 189, 347
Krebs Trial *see* Randomised Badger
    Culling Trial
Kruuk, Hans 44, 45, 46, 56, 58, 105,
    167, 171, 190, 193, 201, 444
  range studies 59–60, 60*f*

L

Lack, David 222n2
Lacks, Henrietta 427*b*
lactation
  assessment of 7, 22
  energy expenditure 279
  hormone levels 137–8
lactational (facultative) delayed
    implantation 123n7
'landscape of disgust' 154–6
land use changes 208n14, 213–14, 219,
    248–9
  vs. population growth 208
land use map xv*f*
Langton, Tom 371, 373–4
Lansing, Albert 437*b*
Lansing effect 426
Lasiorhinus krefftii see southern
    hairy-nosed wombat
latrines 181*f*
  characteristics 173
  Chinese ferret badger (*Melogale
    moschata*) 457
  contents 173
  Eimeria melis studies 28
  Facebook page analogy 176, 178,
    188, 189, 190
  faeces matching 189–90
  functions 188–90, 207–8, 208n12
  hog badgers (*Arctonyx* species) 454
  individuals' use of 176–8
  information conveyed by AGS 185–6
  information signalled 181
  Japanese badger (*Meles anakuma*) 452

latrines (*Continued*)
number *vs.* group range area 190*f*
patterns of use 179–80, 180*f*
positioning of 173–5
spatial distribution 175–6
temporal deployment 176
see also perimeter latrines
Leake, Jonathan 381
Lecher B (M10) 69*b*, 72, 91, 128, 214
reproductive success 68
Lefkovitch matrices 242*f*
lemming, reproductive strategy 294
Leopardus pardalis see ocelot
leptin 289*b*
Leptocyon 471–2
Leslie matrices 242
leukocyte composition, *vs.* RLTL 431–2
leukocyte coping capacity (LCC)
applications 330–1
and MHC system 334–5
rationale 327–8*b*
seasonal variations 334n48
technicalities 328n38
transport stress trial 328–30*f*, 329*f*
leukocytes
anti-bacterial actions 321n32
use of ROS 321
see also lymphocytes; macrophages;
neutrophils
Leyhausen, Paul 115
lice infestation 154
influence on sett location 156
vs. population density 155*f*
licking behaviour, cats 99
life history strategies
individual variation 294–7
pace of life 296–7
selection, risk-sensitivity 298
see also pace of life
lifespan 3, 231*f*, 232, 393–4, 439
vs cohort lifetime reproductive
success 397*f*
vs. cub RLTL 428–9, 430*f*
EGO *vs.* WGO 408*f*
Hwange lion study 423*b*
life tables 235*b*, 236*t*
life-time reproductive success
(LRS) 391, 393–4
vs. age attained 397*f*
by cohort 397*f*, 398*f*
effect on inbreeding 400–1
EGO *vs.* WGO 407
and extra-group paternity 408–9
linearity of hierarchies 74*b*
Linnaeus, Carl 15
lions (*Panthera leo*)
breeding behaviour 391
lifespan 423*b*
social structure 99, 471

litter sizes 7, 244*b*, 390–1
Little B (M478) 415*f*
Little Ice Age 250*b*
Little L 80, 447
local resource competition
hypothesis 10
long-distance feeding 198
Lorenz, Edward 301n25
Lotka, Alfred 235*b*
Loureiro, Filipa 449
Loveridge, Andrew 423*b*
LTRE analyses 242, 243*b*
Lumbricus terrestris see earthworms
lupine canids 472–3
luteinizing hormone (LH) 36*b*
ovulation induction 121
lymphocytes 306*b*, 431
lysosomes 321n29

M

M1 (Slash) 41, 128, 129*f*, 130*b*
M10 (Lecher B) 69*b*, 72, 91, 128, 214
reproductive success 68
M90 (Casanova B) 214, 393, 395*b*,
396–7, 396*b*
M222 128*b*, 129*f*
M232 69*b*, 70*b*, 130*b*, 131*b*
M238 41, 70*b*, 395*b*
M242 69*b*, 70*b*
M243 (Greek) 42, 69*b*, 70*b*, 128
M300 (Sandy) 22, 23*f*, 42, 69*f*, 70*b*, 76,
126, 128*b*, 129*f*, 131*b*
M302 69*b*, 70*b*
M317 128*b*, 129*f*
M328 70*b*, 128*b*
M340 (Submarine) 81, 128
M343 130*b*, 131*b*
M367 22, 23*f*, 70*b*, 72*b*, 126, 128*b*, 129*f*,
131*b*
M371 128*b*, 131*b*
M411 130*b*, 131*b*, 148
M478 (Little B) 415*f*
M479 28, 69*b*, 421*f*, 422
M494 70*b*, 72*b*
M497 128*b*, 129*f*
M507 3, 4*f*, 6–7
M531 395*b*, 396*b*
M553 22, 23*f*
M665 70*b*, 72*b*
M814 (Bloat) 22, 23*f*, 24, 30
M862 52, 232*b*
M873 128*b*, 129*f*
M893 3, 4*f*
M999 421*f*, 422
M1012 150, 421*f*, 422
M1020 421*f*, 422
M1021 421*f*, 422
M1019 52, 146
M1059 421–2, 421*f*

M1089 52, 421*f*, 422
M1090 421*f*, 422
M1141 3, 4*f*
M1208 421*f*, 422
M1357 421*f*, 422
M1365 421*f*, 422
M1459 421*f*, 422
macrophages 321n32, 431
defence against mycobacteria 354
deficiencies in badgers 355
magnetic inductance tracking 61, 88*f*
range studies 64–5
underground studies 89–91
main setts 3n5
major histocompatibility complex
(MHC) genes 332*b*
diversity 331
and female mate choice 409–11*f*
vs. infectious disease 333–5, 334*f*
and mate selection 331, 333, 399,
399n15
MHC-assortative mating 409, 410
olfactory communication 112
male body condition
vs. reproductive success 132
see also body condition index
male dominance 300
male mating behaviour 119–20
male mounting success, predictors
of 124–6
male offspring delayed dispersal 452
male reproductive inequality 299–301,
393
Casanova B (M90) 395*b*
number of fathers per social
group-year 392*f*
male reproductive success
vs. age 416
EGO *vs.* WGO 408
elderly males 421–2
Mallinson, Peter 346
Malthus, Thomas 222*b*, 246
maned wolf 473n31
mange 318–19
Manna From Heaven experiment 81
Mantoux tuberculin skin test 338
maps of setts xiii–xv
maras (*Dolichotis patagonum*) 396, 476,
477*b*, 478*b*
paternal care 396
Maremma National Park 448
Markham, Andrew 46, 88, 275
marking 73
Marley # 2 sett 52
group size 68
Marley Main 53*f*, 56
martelism 177–8
martens 460

yellow-throated (*Martes flavigula*) 468
MasterBayes 392n14
  parental assignment rules 400*f*
maternal care 2*f*, 5
  movement of cubs 5–6
maternal deaths 9
maternal diet, prenatal influences 314–15
maternal investment, cub sex differences 10, 13
maternal–offspring associations 448–9
mate selection 120, 125–7, 386, 394, 396–9
  for heterozygosity 399–401
  and MHC profiles 331, 333
  multiple matings 138–9
  non-conceptive sex 142–3
  visitors 132
Mathews, Fiona 354
mating behaviour 139–42, 387
  female 120
  female receptivity period 119
  five Bs vignette 116, 118–19, 386
  induced ovulation 120–3
  male 119–20
  mount durations 119
  mounting 116
  multiple mountings 119
  polyandry 119
  predictors of female mounting success 126–8
  predictors of male mounting success 124–6
  video data 116
mating call, males 119
Matthew Effect of Accumulated Advantage 12n10
May, Bob 153
Maynard-Smith, John 160*b*, 386
May rainfall, *vs.* cub survival 252*f*
Macdonalds, genealogy 470
McLaren, Graeme 328
Medawar, Peter 414
Medway, Lord Gathorne 345
Meerkats (*Suricata suricatta*) 218, 474n34
  allomarking 100
  cooperative digging 148
  telomere attrition 428
  territoriality 170, 177
Megalictis 18
Mehler reaction 320n27
Meles anakuma see Japanese badger
Meles canescens see hoary badger
Meles genus 451
  inter-specific variation 451–3
Meles leucurus see Asian badger
Meles maraghaus 18, 223n2

Meles meles see European badger
Meles meles polyomavirus 1 (MmelPyV1) 307
Meles polaki 18, 223n2
Meles thorali 18n20
Mellivora capensis see honey badger
Melodon 18
Melogale everetti see Bornean ferret badger
Melogale moschata see Chinese ferret-badger
Melogale orientalis see Javan ferret badger
Melogale personata see Burmese ferret badger
Melogale species see ferret badgers
menopause 417, 422
Mephitinae see skunks
metabolic rates
  BMR 287
    vs. body size 277*b*
    cubs 323
miacids 469–70
Mian, Rubina 326–7
mice
  olfactory communication 112
  see also harvest mice; house mice; wood mouse
microflora hypothesis 399
microsatellites 389n9
microsatellite sequences 392n13
microsatellite studies 389–92, 410
Milankovitch events 249n2
milkmaid's disease 338
minimum-area convex polygons (MCPs), bait-marked latrine sites 63*f*, 65–6
minimum defensible territory 207
minimum number alive (MNA) estimator 223n5, 226–7
  enhanced 227–8
  population graph, 1987–2015 230*f*
mink
  European, feeding specializations 286
  female mate choice 403–4
  maternal diet, prenatal influences 314–15
mites 318–19
mongooses 474n34
  breeding behaviour 391
  group size 459
  see also meerkats
monocytes 321n32
monoestrous, definition 132n13
Montes, Inigo 328–9, 330n40
Moorhouse, Tom 415
moose, tick infestations 154
Moreiro, Jose Roberto 11

Morris, Desmond 166
mother hypothesis 424
Mount sett 53*f*
mount durations 119, 122, 126–8
mounting behaviour 116
  females 120
  predictors of male success 124–6
  vs. reproductive success 128
  summary data 117*t*, 118*f*
  variance in 124, 127*f*
Muirhead, Roger 342, 344
multiple male mating (MMM) 142
multiple mountings 140
multiple paternity 140, 142, 391
multi-tasking 261
musclimorph morphology 113, 147, 478n37
Mustela nivalis see weasels
Mustelidae, phylogenetic tree 17*f*, 456*f*
mustelid ancestry 115
Mustelid herpesvirus (MHV) 307–8
  age differences in prevalence 310–11*f*
  co-evolution with badgers 315
  consequences for pregnant females 308, 309–10
  genetic studies 315
  genital infection prevalence 308–9, 310
  genotype distributions 316*f*
  inflammatory lymphoid hyperplasia 309
  latency 308
  lifetime susceptibility, prenatal influences 313*f*–14
  and MHC profile 333
  reactivation 310–14, 312*f*
  sex differences in prevalence 311
  transmission 308
  vertical transmission 308, 310
  viral load assessment 315n17
  weather influences 311–14, 312*f*, 313*f*
mustelid litter sizes 244*b*
Mustelinae see weasels
musteloid niches 467–8
mutation accumulation hypothesis 414
mutual allomarking ('bum kissing') 94, 95
  bacterial exchange 107
  seasonal variations 97*f*
  sex differences 96
  sex preferences 96
Mycobacterium bovis 338
  hosts 354n49
  immune defences 354–5
  see also bovine tuberculosis
Mycobacterium tuberculosis 337
  see also tuberculosis

Mydaus species *see* stink badgers
myeloperoxidase, anti-bacterial actions 321n32

N

Nadin, Catherine 77*b*, 80*b*
NADP+ (nicotinamide adenine dinucleotide phosphate) 320*b*
Nasua nasua see ring-tailed coati
National Federation of Badger Groups 346–7
  see also Badger Trust
Natural England (formerly Nature Conservancy) 384
concern over badger culling 345
  role in badger culling debate 377–8, 379–81
natural selection 218–19, 260, 385
Neal, Ernest 45, 345
Nealings Copse sett 53*f*
neighbourly relations 188
  anal gland secretion responses 184–5, 184*t*
  Paving Slab Experiment 177–8
  problems with RDH 213
  see also visiting behaviour
Neovison vison see American mink (also *Neogale vison*)
neutrophil activation, LCC test 328–9
neutrophils 306*b*, 321n32, 431
new herd incidents, bTB
  1996–2020, by region 369*f*
  1996–2020, by risk 370*f*
  2020 368–9
next-season BCI, *vs.* energy expenditure 279, 281*f*
niche variation hypothesis (NVH) 286–7
nightingales, Lodge Hill population 383–4
nitric oxide (NO), inducible 354
non-breeding adults 391, 392
  advantages 386
  interactions with cubs 6
non-communicable diseases (NCDs) 305
non-conceptive sex 142–3, 387
Noonan, Mike 6, 62, 64, 88, 105, 260–1, 275, 461, 475
Norreys, Sir Henry 55n8
North American beaver (*Castor canadensis*) 110n8
North American river otters (*Lontra canadensis*) 469
northern fur seal (*Callorhinus ursinus*) 38
northern hairy-nosed wombat (*Lasiorhinus krefftii*) 477*b*

northern hog badger (*Arctonyx albogularis*) 453
  diet 454, 455*f*
  latrines 454
  scent marking 454n13
  social structure 454–5
North Nibley perturbation effect study 356–7
  badger densities 1995–1999 360*f*
  evidence for perturbation 358–9
  numbers of badgers culled 1975–1994 364*f*
  post-cull bTB prevalence 359, 361
  previous population disturbance 363
  social interactions 364
Nouvellet, Pierre 238, 254
Novak, Martin 160*b*
nursery chambers 2*f*, 5
nut consumption 194*f*
Nyctereutus procyonoides see raccoon dog

O

occupancy analysis, Scottish badgers 256–7
ocelot (*Leopardus pardalis*), latrines 170
OD measure (ODBA/DEE) 276
  individual variation 278–9, 280*f*
  vs. next-season BCI 279, 281*f*
  vs. survival rates 279–80
Odocoileus virginianus see white-tailed deer
oestradiol levels 134
  urine 187
  vs. seasons and reproductive stage 135*f*, 137*f*
  vs. vulval condition 34, 39*f*
oestrone levels 134, 135n17
  vs. age 419*f*
  seasonal variation 42*f*
  vs. seasons and reproductive stage 134*f*, 137*f*
  threshold for ovulation 422
  vs. vulval condition 34, 39*f*
oestrus 116, 132
  muti-annual 132–3
  single, Wytham population 132, 137n19, 138–9
  terminology 132n13
  timing of 139
  urine hormone study 132
Old Common Piece sett 53*f*
olfactory communication 92, 94
  allomarking 95–100
  anal gland secretion trials 182–6, 184*t*
  latency 106
  latrine functions 170

  role in female mate choice 399
  shared group odours 106–8, 107*f*
  and social structure 114
  subcaudal gland 94–5*f*
    colour of secretions 103–6
    evolution 112–13
  substrate marking 100–2
  territory marking 168*b*
olfactory signals, characteristics 93*t*
olingos (*Bassaricyon* species), social structure 463
omnivorous diet 193
  canids 472
  see also diet, diversity
oogenesis 37*f*
optimal foraging theory 199–200
orcas, feeding specializations 259–60
Otodectes cynotis mite 319
otters 468
  alloparental care 466
  cooperative hunting 465
  cooperative vigilance and defence 466
  social structure 462–3, 469
  see also giant otter
Outlier 1, spatial arrangements 213*b*
outlier setts 3n5
  adult badger characteristics 11–12
  architecture 11
  construction of 216
  cub survival 11–12*f*
ovaries
  hormonal regulation 36*b*
  oogenesis 37*f*
overall dynamic body acceleration (ODBA) 261
  Activity studies 262–4
  vs. age 278, 279*f*
  vs. body condition 285, 286*f*
  vs. body size 278
  DLW (doubly labelled water) study 275–6
  individual variation 276, 286
  vs. rainfall 284*f*
  vs. soil temperature 284
  vs. temperature 283, 284*f*, 286*f*
  vs. wind 284
Overend, Eunice 345–6
over-marking 101–2
  decrease with age of existing mark 102*f*
ovulation 37*f*, 422
  endocrinology 36*b*
  induced *see* induced ovulation
  spontaneous 121n5
Oxford Foxlot 58n10
oxidative burst 321
oxidative phosphorylation 321*b*
oxidative stress 319, 320*b*

biomarkers 323n35
and body condition 324–5
implications of climate change 326
and reproduction 324
vs. telomere attrition 425
vs. weather 325–6, 327*f*
oxytocin 459*b*
release during grooming 153

P

pace of life (POL) 296–7, 304, 394,
413–14, 414
risk sensitivity 298
social influences 297–301
weather influences 301–2
pair living 448–9
canids 471
Palaemeles 18
palate markings 28–9, 29*f*
Papio cynocephalus see baboons
Parable of the Two Potters 295n21
Paraceras melis see fleas
parasite infestation
vs. population density 157*f*
vs. survival 158*f*
see also fleas; lice; ticks
parental age, and offspring telomere
length 437*b*, 438*f*
parental assignment rules 400*f*
Parker, Geoff 169*b*
Park View sett 52
partitioning of setts 52
parvovirus 8
Parvus major see great tit
Passive Range Exclusion Hypothesis
(PREH) 179
pasteurization of milk 338
Pasticks sett 53*f*
2001 cohort 22–3, 30
fissioning 215
reproductive pairings 69*b*, 71*f*
reproductive success 68
Pasticks super-group 52, 215
geology 56–7
patch feeding 198, 199
patch richness 204
paternal care 396, 471, 472, 473
and badgers 472n28
paternal heterozygosity *vs.* cub
survival 401, 402*f*
paternity, multiple 140, 142, 391
paternity confusion 141–2, 143, 409
paternity studies 387, 392
extra-group 391, 392*f*
number of cubs sired 391
pathogen resistance, and MHC
profiles 409–11
patriarch hypothesis 423
Paving Slab Experiment 177–8

Pearl, Raymond 235*b*
pedigree construction 387–92
computer analysis 390, 391*f*
early attempts 387–9
extra-group paternity 391, 392*f*
microsatellite studies 389–92
statistics 390–1
penis bone (baculum) 34, 120
peppered moth (*Biston betularia*) 219
perception, powers of 93*b*
perimeter latrines 46, 48*f*, 164–5, 165*f*,
169–70, 171*f*
aFRID studies 48–9
co-locations 49
evidence for visiting behaviour 65–6
Facebook page analogy 176, 178,
188, 189, 190
faeces matching 178, 189–90
functions 170, 188–90, 207–8, 208n12
golden jackals (*Canis aureus*) 171
individuals' use of 176–8
information signalled 181, 185–6
patterns of use 179–80, 180*f*
positioning of 174–5, 179
possible functions 165
spatial distribution 175–6
spotted hyaena (*Crocuta
Crocuta*) 167, 171
temporal deployment 176
time spent at 48
Perimyotis subflavus (tricolored
bat),delayed implantation 403
personality 273
perturbation effect studies 356–8
evidence for perturbation 358–9
policy implication 363
previous population
disturbance 363
social interactions 364
perturbation hypothesis 346, 355–6
relevance to bTB policy 364–5
Peto's paradox 427n18, 428
phagocytic leukocytes
use of ROS 321
see also macrophages; neutrophils
phenotypic plasticity 442*b*
philopatry 30, 39, 45, 46, 70*b*, 386
costs and benefits 145, 205
and gene frequency
heterogeneity 389
inbreeding 40
and inter-sexual territoriality 40–2
photosynthesis 320*b*
Phyllostomus hastatus (greater
spear-nosed bat), extra-group
paternity 405–6
phylodynamic modelling 349
phylogenetic constraints 40, 441, 458*b*

phylogenetic relationships 15–18, 16*f*,
17*f*, 456*f*
phylogeny, impact on social
structure 217
phylogeography 19*f*, 20*b*
physical adaptation 442*b*
physique, *vs.* alloparenting 147
picric acid 192
pine martens (*Martes martes*) 460
piroplasmosis 317
piscivorous groups 469
polar bear (*Ursus maritimus*)
population density, effects on
puberty 38
sense of smell 93*b*
Polish badgers 447, 448–9
activity studies 264
polyandry 119
convenience 138–9
polyoestrous, definition 132n13
population, MNA graph,
1987–2015 230*f*
population density 219, 221, 231, 247
effects on puberty timing 38
estimates from 1987–2015 231*f*
historical changes 44–5, 190
and inbreeding avoidance 388–9
influencing factors 444
intra-specific variation 444–5, 448–9
vs. oestrus schedule 138–9
vs. parasite burden 154–6, 155*f*
vs. parasitism 157*f*
same-sex, POL effects 298
population dynamics, parasite
effects 153–4
population estimates 445*b*
1987–2013 269*f*
capture–remark–capture
technique 227
enhanced MNA 227–8
importance of accuracy 267–9
minimum number alive
estimator 226–7
trapping efficiency 228
population growth 87, 208, 221, 228,
234
vs. body condition 246*f*
constraints 237*b*
influence of cub mortality 245*b*
and litter sizes 244*b*
LTRE analyses 242, 243*b*
minimum number alive model 240*f*
national statistics 250–1
possible constraints 248
role of adult survival changes 244
role of food availability 249
role of land use changes 248–9
role of weather 251–3
sensitivity/elasticity analyses 243*b*

population growth (*Continued*)
  social adaptations to 247
  UK badgers 445*b*
population growth curves, unconstrained *vs.* density-dependent growth 239*t*
population growth rate, *vs.* population density 240*f*, 241*f*
population growth rate quantification 235*b*
  projected growth 238*t*
population size
  limiting factors 221
  vs. mean air temperature 251*f*
  weather-induced variability 270*f*
population structures 67
  see also setts; social structure; super-groups
portfolio effect 273
Portuguese badgers 449
post-reproductive lifespan (PRLS) 422–3, 424*b*
  human 417–18
  theories of 423–4
predation risk, *vs.* group size 459
predators 446–7, 446*f*
predictive adaptive responses 241, 295, 314
pregnant females, ultrasound scanning 5, 8*f*, 252–3
prenatal influences
  on age at first reproduction 302–3*f*
  on cub MHV susceptibility 313*f*–314
  maternal diet 314–15
prey size, influence on social organization 472, 474
prior attributes hypothesis 73
Prisoner's Dilemma 160*b*
progenitor groups, fissioning 54–5
progesterone, variation with seasons and reproductive stage 135, 136*f*, 137*f*
Promeles 18
Promislow, Dan 414
propinquity 126n11
Protection of Badgers Act 1992 347
protozoan pathogens
  Babesia missiroli 317–18
  Toxoplasma gondii 316–17
  Trypanosoma pestanai 317, 318*f*, 319*f*
  see also Eimeria melis; Isospora melis
Pteronura brasiliensis see giant otter
puberty
  definition 34n7
  endocrinology 36*b*
  external signs 34
  hormonal changes 34, 35*f*, 36
  social influences 38–9
  testicular descent 35*f*
testicular volume 35*f*
testosterone titres 35*f*, 41*f*
variation in timing 37–8
public image of badgers 345
purifying selection 414n3
purr 85*b*
  structural characteristics 84*t*
pygmy dormouse (*Typhlomys cinereus*) 277*b*

Q

Quinn, Sharon 159

R

R_0 (net reproductive rate per generation) 237*b*
rabbit (Oryctolagus cuniculus), induced ovulation 121
rabies 316
  fox vaccination 365
raccoon dog (*Nyctereutus procyonoides*)
  cohabitation with badgers 447
  litter sizes 244*b*
raccoons (*Procyon lotor*) 218
  social structure 463
Radcliffe, Jane 345
rainfall
  vs. BCI 292, 293*f*
  vs. cub RLTL 433, 435*f*, 436*f*
  vs. cub survival 251–2*f*, 253
  vs. energy expenditure 284*f*
  influences on Irish badgers 256
  vs. MHV reactivation 312*f*
  vs. oxidative stress 327*f*
rainfall variation, *vs.* survival rates 254
Randomised Badger Culling Trial (RBCT) 357–8, 371
  changes in herd breakdowns, proactive vs, surrounding areas 377*t*
  evidence for perturbation 358–9
  outcome 376
  post-cull bTB prevalence 359, 361, 362*f*
  previous population disturbance 363
  social interactions 364
range overlaps
  1970s 59–60, 60*f*
  1990s 61
  sex differences 62, 65
ranges 59*f*
  effect of culling 358–9*f*
range sizes
  vs. body mass 464*f*
  Chinese ferret badger (*Melogale moschata*) 457
  determination of 210–11*b*, 214, 218
  vs. group mass 465*f*
Japanese badger 452
  vs. number of latrines 190*f*
range studies
  2010s 61–5, 62*b*
  aRFID 61–2
  magnetic inductance tracking 64
rare-allele advantage hypothesis 331, 333
rarity value 454n14
rats, mating behaviour 125–6
reactive oxygen species (ROS) 306*b*, 319, 321*b*
  collateral damage 321–2
  free radical theory of ageing 319
  oxidative bursts 321
reciprocity 145
recruitment 244*b*
  vs. rainfall 254
  vs. weather variability 254–5*f*
red colobus monkey (*Colobus badiustephrosceles*) 192
red deer (*Cervus elaphus*) 419n9, 420
red fox (*Vulpes vulpes*)
  audition 94*b*
  earthworm consumption 193
  spatial groups 203
  territory size 207
  see also foxes
Red Queen Hypothesis 241, 395, 414, 440
red-winged blackbird (Agelaius phoeniceus), diet *vs.* social organization 192
relative leukocyte telomere length (RLTL)
  cubs, *vs.* lifespan 428–30, 430*f*
  within-individual variation 429*f*
  lengthening of 431–2
  variation with age 427–31, 428*f*, 429*f*
Rensch's rule 13n12, 461, 467
reproduction
  costs of 301, 302
  delayed implantation *see* delayed implantation
  females, relationship to survival probability 280, 292
  first age at 36
  induced ovulation 120–3, 405
  Japanese badger 452
  non-conceptive sex 142–3
  oestrus 132–9
  oxidative damage 324
  sperm 139
  temperature effects 301–2
  see also mating behaviour
reproductive cessation, definition 418
reproductive inequality 72, 282–3, 298–301
  females 282–3, 390, 393

males 393
　Casanova B (M90) 395*b*
　number of fathers per social
　　group-year 392*f*
reproductive phenotypes
　female 422
　male 421
reproductive rate (R_0) 237*b*
reproductive senescence 415–18
　females 422–5
　heterochrony 418
　males 421–2
　post-reproductive lifespan 417–18,
　　422–3
　sex steroid analyses 418–21, 419*f*,
　　420*f*
reproductive status
　vs. allomarking 96, 97, 98*f*
　vs. anal gland secretion
　　responses 185, 186*t*
　vs. subcaudal secretion colour 104–5
　vs. substrate marking 101
　vs. urinary hormone levels 187
reproductive strategy, individual
　　variation 294–7
reproductive success 68, 91, 128, 391
　vs. age 416, 417*f*
　vs. age and sex 233*f*
　vs. age at first reproduction 416
　vs. age at last breeding 416–17
　vs. age attained 397*f*
　vs. BCI 291*f*, 292
　Casanova B (M90) 395*b*
　by cohort 394*f*, 397*f*, 398*f*
　vs. deciduous woodland
　　access 208–9
　distribution between breeders 393–4
　Dowager B (F227) 396*b*
　effect on inbreeding 400–1
　and extra-group paternity 408–9
　and group living 162
　vs. male body condition 132
　number of fathers per social
　　group-year 392*f*
　number of mothers per social
　　group-year 392*f*
　older females 278
　Pasticks sett 69*b*, 71*f*
　vs. winter conditions 252–3
residence, definition 45n4
resilience 234n14
　role of energy management
　　tactics 287
　role of individual variation 273–4
resistance 234n14
resource depletion 178–9*f*
Resource Dispersion Hypothesis
　　(RDH) 58, 145, 191, 202*f*, 203–6,
　　203n7, 449–50, 461–2, 466

applications and development
　of 201
application to badgers in
　general 206–7
application to Wytham
　badgers 208–11
and carnivore societies 467–8
and dice 204*b*
frugivorous groups 468
inter-specific comparisons 463–5
and intra-specific variation 450–1
origins of 201
piscivorous groups 469
predictions 205–6, 211–13, 214
primary and secondary
　occupants 205
problems with 213
Resource Holding Potential 169*b*
resource holding power (RHP)
　signals 109–10
resource hypothesis 167
respiration 320*b*
restraint hypothesis 417
Revilla, Emilio 205–6
Rhodanictis 18
ringtail, social structure 463
ring-tailed coati (*Nasua nasua*) 126
Riordan, Philip 358
r_m 237*b*
road traffic accidents (RTAs) 233
　vs. weather 269–71, 270*f*
Roberts, Stuart 369–70, 382
roe deer (*Capreolus capreolus*), sex
　ratio 10–11
Roland, Cecile 404
Roper, Tim 207, 438*b*
Rosalina, Miguel 449
Rothschild, Miriam 149
Rousseau, Jean-Jacques 160*b*
Royal Society, bovine TB review,
　2012 339
RSPCA Wild Animals Committee 345
rubbing behaviour, cats 99
Ryder, Charlotte 186
Ryder, Richard 345

S

Sambita, Sushruta 112
same-sex densities, POL effects 298
Sandy (M300) 22, 23*f*, 42, 70*b*, 76, 126,
　128*b*, 129*f*, 131*b*
Santos Reis, Margarida 449
Sarcoptes scabiei canis mites 318–19
Satyr (M4) 91
Saxton, Julie 5
scale allometries 13
scars 82*f*
　associated factors 86
　definitions 86

vs. movement history 45
observations, 1990–99 86*t*
scent marking 33
　attributes of territorial function 165
　hog badgers 454n13
　sources of scent 168*b*
　see also allomarking; latrines;
　　olfactory communication;
　　perimeter latrines
scent matching 169*b*
scent theft 33
Scottish badger studies
　influences of weather
　　variability 256–7
　population estimates 445*b*
search strategies 197, 198
　vs. earthworm availability 197–200
　encounter rate before changing
　　foraging location 200*f*
search tactics 197
　vs. earthworm availability 197–200
seasonally polyoestrous,
　　definition 132n13
seasonal variations
　allomarking behaviour 96*f*, 97*f*
　antioxidant capacity 322
　body condition 289*f*–90, 292–3
　early-life RLTL 434*f*
　energy expenditure 283–5
　fat reserves 285
　LCC 334n48
　subcaudal pouch energy
　　content 110, 111*f*
　subcaudal secretion colour 103–4,
　　104*f*
　substrate marking 100, 101*t*
　trapping efficiency 230*f*, 267, 268*f*
sedation 224
selection hypothesis 416
selection strength, *vs.* age 432–3
selective mortality 9
self-grooming
　attention to body regions 151*f*
　see also grooming
'selfish herd' effect 149
semi-detached relationship *see*
　　intra-sexual territoriality
senescence 414*f*
　vs. ageing 414
　cellular 425
　　see also telomere attrition
　Little B (M478) 415*f*
　reproductive *see* reproductive
　　senescence
　theories of 414
sense of smell 93*b*, 106
sensory bias hypothesis 399
sequential allomarking 94, 95
　of adult females 96, 97, 98*f*, 99

sequential allomarking (*Continued*)
  seasonal variations 96f, 97f
  sex differences 96, 97
  sex preferences 96, 97
  social cliques 96
serotonin, active social domestication
    hypothesis 443b
Service, Katrina 132
sett dispersion hypothesis 58–9
sett excavation 56
sett location
  geological factors 56, 57f
  influence of parasites 156
sett maps xiii–xv
setts
  Fossorial Benefits Hypothesis 6
  historical continuity 55–6
  increasing numbers 58
  main and outlier 3n5
  sites and ranges 59f
  spatial distribution 164
  tunnel mapping 88
  use of chambers 89–90, 89f, 90f
sett temperatures 5, 476
sett type
  vs. adult badger weight 11–12
  vs. cub survival 11–12f
seven ages (Shakespeare) 413n1
sex, *vs.* dominance 74b, 76
sex ratio 10
  maternal manipulation of 11
  POL effects 297–8
  reptiles, thermal determination 443b
  roe deer 10–11
sexual conflict 403
sexually-transmitted infections *see*
    Mustelid herpesvirus
sexual ornaments 398–9
sexual selection 14, 385, 387, 398, 403
  and MHC 409
  and sexual size dimorphism 461
sexual size dimorphism (SSD) 13–15,
    28, 41, 461–2
  carnivores 467
  relationship to female mass 462f
sexually transmitted infections 141
  see also Mustelid herpesvirus
sexy son hypothesis 397–8, 408
sharks, lifespan 414n2
Shaw, Brian 287
short tandem repeats (STRs) 389n9
shrews, lifespans 414
Sibly, Richard 238
SICCT (single intradermal comparative
    cervical tuberculin) test 338–9
  frequency of testing 342
  reasons for test failures 340–1
  sensitivity and specificity 339–40
  vs. SICT 338n5

Sidorovich, Vadim 286, 446n6
signalling pathway
  up-regulation 355n52
silver spoon hypothesis 295, 302, 314
Sin, Yung Wa (Simon) 108, 307, 331,
    387, 410
Singing Way sett 52
Skoog, Peravid 46
skunks (*Mephitinae*) 16
  social structure 462
Slash (M1) 41, 128, 129f, 130b
sleeping associations 91
sliding window bioinformatic
    method 311n14
Smith, Adrian 355
snail consumption 194f, 195
snarls 83b
  structural characteristics 84t
sniffing
  anal gland secretion responses 183,
    184–6, 184t, 185t
  cubs 31
  urine responses 187
snort 85b
social acceptance, role of shared group
    odours 108
social behaviour, population
    differences 272
social bonding, scent marking 33
social climate
social cliques 216
  allomarking behaviour 96
  and mating behaviour 126
social contact network (SCN)
    analysis 49–50
social dynamic hypothesis 73
social factors, influence on timing of
    puberty 38–9
social immune response 154
sociality, starting points 144
socialization, cubs 6
social plasticity 240–1, 247
social reconfigurations 214, 217
Lecher B dynasty 69–72b
social structure 67, 72
  allogrooming 153
  allomarking behaviour 95–100, 114
  American badger 459
  carnivore societies 467–8
  carnivore taxonomic families 470–1
  Chinese ferret badger 457
  vs. competition and predation 459
  ecological drivers 191–2
  vs. food sources 99
  vs. fossoriality 475–8
  frugivorous groups 468
  group composition 67–72
  hierarchies 73–81
  hog badgers 454–5

honey badger 459–60
hyaenas 473–4
  inter-specific variation 453, 462–3
  intra-specific variation 448–9
  Japanese badger 452, 453
  martens 460
  otters 462–3
  vs. pace of life 297–301
  phylogenetic influences 217
  piscivorous groups 469
  stink badgers 460
  super-groups 388
  underground behaviour 89–91
social thermoregulation 148
sociological disadvantage
    hypothesis 408
soil temperature, *vs.* energy
    expenditure 284
soil type map xvf
solitary behaviour, definition 161
Somerled (Highland chief) 470
Sorex areneus see common shrew
southern hairy-nosed wombat
    (*Lasiorhinus krefftii*) 477b
spacing out, as essence of
    territoriality 166
Spanish badgers 448
spatial groups 203, 204
spatial proximity, cats 99
spatio-temporal patchiness 58
Speakman, John 275
Spencer, Herbert 385
sperm
  dog 140n20
  electro-ejaculation technique 139
spermatogenesis 37f
sperm competition 34n5, 140–1, 142,
    143, 387, 403
spoligotyping, Wytham bTB 353
spontaneous ovulation 121n5
spotted hyaena (*Crocuta Crocuta*) 474
  intra-specific variation 272
  latrines 167, 171
spring
  energy expenditure 283–4, 285
  rainfall variability *vs.* cub RLTL 436f
  rainfall *vs.* cub RLTL 435f, 437
  temperature *vs.* early-life RLTL 435f,
    437
  weather, influence on BCI 292
squeak 85b
Stafford, Sophie 116
state-dependent responses 261
state-shifting 234
Statpak blood test 351, 352b
steepness of hierarchies 74b
Stewart, Paul 81, 83b, 116, 127, 159n14,
    160b, 171, 178–9, 389

stink badgers (*Mydaus* species) 16*f*, 17, 460
  fat reserves 460
Stockley, Paula 140
Stoddart, Mike 345
Stopka, Pavel 77*b*, 161*b*
stress 305
  leukocyte coping capacity test 327–31
  measurement techniques 326
  MHV reactivation 311, 314
  see also oxidative stress
stress response 321, 326
striped hyaena (*Hyeana hyaeana*) 474
Strix aluco see tawny owl
Strongyloides sp. 29
subcaudal gland 94–5*f*
  evolution 112–13
subcaudal secretion colour 103
  vs. body size 103
  and reproductive status 104–5
  seasonal variations 103–4, 104*f*
  sex differences 103–4, 104*f*
subcaudal secretions 33, 95
  bacterial populations 107–9
  changes with age 106, 420, 421*f*
  chemical analysis 105–6
  correlation with scarring 86
  costs and benefits 109–12
  energy content 110–12, 111*f*
  heavy molecules 109
  individual signatures 109
  information encoded by 106, 112
  overmarking of anal gland secretions 185
  role in mate selection 399
  sampling technique 103*f*
  sex differences 103, 104*f*
  shared group odours 106–8, 107*f*
  specimen preservation 103
  TRF analysis 108–9
  volume 103
Submarine (M340) 81, 128
substrate marking 94, 95, 100–2
  choice of object 100
  distinction from urination 100
  over-marking 101–2
  and reproductive status 101
  seasonal variations 100, 101*t*
  sex differences 100, 102
  timing of 100–1, 102
suckling 5
Sugianto, Nadine Adrianna 13, 34, 132, 186, 418
Sumatran hog badger (*Arctonyx hoevenii*) 453
summer
  energy expenditure 284, 285

weather, influence on BCI 292, 293, 294
Sun Tzu 182
super-breeders 393, 396–7
  Casanova B (M90) 395*b*
  Dowager B (F227) 396*b*
superfoetation (SF) 123, 136n18, 301
  interaction with delayed implantation 133–8
super-groups 52, 215
  allozyme studies 388
  geological factors 56–7
  ontogeny 52–5
  ranges 174*f*
  and resource availability 58
superior natal fit hypothesis 407
Supplementary Badger Control (SBC) 371
surface area *vs.* heat loss 277*b*
survival
  age-specific, changes over time (1987–1996) 239*t*
  role in population growth 244
  see also cub survival; lifespan
survival dynamics 231–3, 231*f*
  sex differences 232
  temporal trends 233*f*
survival tables 235*b*
sweat odour, *vs.* health 112
sweet chestnut consumption 194*f*
Swiss badgers 448

T

Taber, Andrew 476
tapetum lucidum 92
tattoos 1, 224, 226*f*
tawny owl (*Strix aluco*), diet 193
Taxidea taxus see American badger
Taxodon 18
teat distension 22
teat measurements 7
telomerase 426, 427*b*
  transcriptional repression 431
telomere attrition 425, 428–30
  RLTL (Relative Leukocyte Telomere Length), variation with age 427–31, 428*f*, 429*f*
  during winter 433–5
telomere length 439
  vs. cub cohort size 433, 437
  vs. cubs' weather experiences 433
  heritability 437*b*, 438*f*
  influencing factors 432*f*
telomere lengthening 431–2, 435
telomeres 425
  starting lengths 425–6
  structure 426*f*
Tembrock, Günter 478
temperature

vs. Activity 262, 263*f*
vs. BCI 292, 293*f*, 294
vs. cub RLTL 433, 435*f*
vs. cub survival 11
vs. earthworm availability 195
vs. energy expenditure 283, 284*f*, 286*f*
influence on reproduction 301–2
influences on Irish badgers 256
inside setts 5, 476
vs. MHV reactivation 312*f*
vs. oxidative stress 327*f*
vs. population size 251*f*
Scottish badger study, winter activity 256–7*f*
temperature variation, *vs.* survival rates 254
terminal investment (TI) 418
territorial boundaries 48*f*
territorial inheritance 476
territoriality 165–6
  anal gland secretion responses 182–6
  attributes of scent marks 165
  defence costs hypothesis 167
  early observations 45
  golden jackals (*Canis aureus*) 171
  planning ahead 166–7
  resource hypothesis 167
  scent marking 168–9
  spotted hyaena (*Crocuta Crocuta*) 171
  Tidy Territorial Model 45, 60, 171
  variation between populations 187–8, 190
territories 3
territory size
  determination of 204, 205, 206
  vs. group size 207
  influencing factors 444
  use of term 207n10
testes
  descent at puberty 34, 35*f*
  hormonal regulation 36*b*
  spermatogenesis 37*f*
testes conditions
  vs. growth curve 43*f*
  vs. subcaudal secretion colour 104–5
  vs. substrate marking 101
  vs. testosterone titres 40*f*
testis size 140–1, 142–3
  puberty 35*f*
testosterone titres
  elderly males 421–2
  vs. growth 42*f*
  immunocompetence handicap hypothesis 398–9
  puberty 35*f*, 41*f*
  vs. testes conditions 40*f*

testosterone titres (*Continued*)
  urine 187
  variation with age 419*f*
Tew, Tom 363
Thames Outlier 53*b*
thermoregulation 285
  age-related problems 278
  and fossoriality 476, 477*b*
  huddling 476
theta (θ) statistic 239
Thom, Mike 403–4
Thomas, Adrian 355
Thornbury Trial, badger culling 347*b*
Thorny Croft sett 53*f*
threat-level hypothesis 184
three-spined stickleback (*Gasterosteus
    aculeatus*) 334
thyroid hormones 288*b*
Tibbles, Maurice 99
ticks 318
  influence on sett location 156
  prevalence *vs.* population
      density 155*f*
Tidy Territorial Model 45, 60, 171
Tinbergen, Niko 45, 124n9, 161*b*, 196,
    197
Tinnesand, Veronica 182
'tipping day' 246n27
T lymphocytes 306*b*
  cytotoxic 332*b*
Toll-like receptors (TLRs) 354
  badger TLR 9 response 355
tooth wear, age estimation 229*f*
torpor 286, 294, 326
  fat reserves 460
  and telomere shortening 434–5
Toxoplasma gondii 316–17
tracking
  aRFID 46
  early techniques 44
trade-offs 275
  allogrooming 153
  extra-group paternity 402
  immune defences vs, growth 323–4,
      325
  pace of life 296–7
  subcaudal secretion production 110
  telomere maintenance vs, tumour
      suppression 427
transcriptional repressors 426n17,
    431n23
transport stress
  cortisol response 330n40
  LCC trial 328–30*f*, 329*f*
trappability 266–9
trapping 223–4, 225*f*, 267*f*
  and *Babesia missiroli* infection 317–18
  bait 223n6
  licencing 224n9

processing procedure 224, 226*f*
  release 224*f*, 225*f*, 227*f*
  trapping efficiency 225, 227, 228
    influencing factors 266–7, 268*f*
    seasonal variation 230*f*, 268*f*
  traps 223, 225*f*
  tree associations, latrines 174–5*t*
Trichodectes melis mite 318
Trichuris sp. (nematode) 29
Triplet E *see* Randomised Badger
    Culling Trial
Trivers, Bob 145
trophic niche breadth 455n11
Trypanosoma pestanai 317, 318*f*, 319*f*
Tsai, Ming-shan 302–3, 309–11, 315
T/S units 428
Tsunoda, Marie 476
tuberculin skin tests 337n1, 338–9, 375
  implications for cattle
      vaccination 350*b*
  reasons for failures 340–1
  regulations 351*b*
  sensitivity and specificity 339–40
  see also SICCT
tuberculosis (TB) 337, 337–8
  antibiotic treatment 337n3
  incidence in England 338n4
  skin tests 338–9
  see also bovine tuberculosis
Tuberculosis (England) Order 2007 (TB
    Order) 351*b*
tunnel mapping 88
Typhlomys cinereus see pygmy
    dormouse

U
ultrasound scanning, pregnant
    females 4, 8*f*
underground behaviour 88–91
  daytime 90
  nocturnal 89
  observation of 1, 2*f*
  sleeping associations 91
  transition between chambers 89–90,
      89*f*, 90*f*
Upper Follies sett 53*f*
urban badgers 444
urination, distinction from substrate
    marking 100
urination behaviour 186
urine
  hormone assays 34, 186
  information conveyed by 186
urine response study 186–7
Urocitellus parryii see arctic ground
    squirrel
Ursus americanus see American black
    bear
Ursus maritimus see polar bear

V
van Lieshout, Henricus Johannes
    (Sil) 427
variation, measures of 5n6
Varley, George 177
Vehrencamp, Sandy 192
'veil of ignorance' 402
vigilance 145–6, 459*b*, 466
viral infections 8
  rabies 316
  see also Mustelid herpesvirus (MHV)
vision 92, 94
visiting behaviour 45, 49–50, 164, 188
  allozyme study evidence 389
  bait-marking evidence 65–6
  vs. body condition index 62, 65, 286
  extra-group paternity 391, 392*f*
  itinerant individuals 51–2
  mountings 132
  network patterns 51*f*
  problems with RDH 213
  sex differences 62
visual pigments 92n3
visual signalling 33
  characteristics 93*t*
vitamins, antioxidant 322*b*
vocalizations 83*b*
  cluster analysis 83*f*
  male churr 119
  mating behaviour 118, 119
  structural characteristics 84*t*
Vombatus ursinus see common wombat
vomeronasal organ (VNO) 93*b*, 109
von Schantz, Torbjorn 207
Vucetich, John 459
vulnerable ape hypothesis 443*b*
Vulpes vulpes see red fox
vulpine canids 472–3
vulva 38*f*
  condition *vs.* hormone levels 34, 39*f*,
      116
  puberty 34

W
wail 85*b*
Waksman, Sellman 337n3
Warburg effect 320n27
Ward, Jane 445–6
Ware, Georges 106
warning calls 83*b*, 145
weasels (*Mustelinae*)
  audition 93*b*
  social structure 462
weather
  vs. BCI 292–3*f*
  vs. cub RLTL 433
  vs. cub survival 251–2, 401
  as a driver of population
      dynamics 251–3

early-life influences 302–3*f*
vs. earthworm availability 195–6, 198*f*, 251
effects on oxidative stress 325–6, 327*f*
vs. energy expenditure 283–4
impact on animals 249*b*
influence on trapping success 266
Irish badger studies 255–6
vs. MHV reactivation 311, 312*f*, 313*f*
POL effects 301–2
relationship RTA mortality 269–71, 270*f*
vs. trapping efficiency 267
see also rainfall; temperature; wind
weather variability 253–4
population impacts 254–5*f*
Scottish badger study 256–7*f*
vs. survival rates 254–5*f*
weaver birds (Ploceidae spp.) 192
weight
correlation with scarring 86
see also body condition index; body size
Werhahn, Geraldine 442*b*
white-nosed coatis 468
white-tailed deer (*Odocoileus virginianus*)
grandmother effect 325
reproductive senescence 419n9, 420

white-throated dippers (*Cinclus cinclus*) 437*b*
wicked problems 384n111
Wide Eyes (fox) 80
Wild Animals Committee, RSPCA 345
WildCRU (Wildlife Conservation Research Unit) vi, 58
Hwange lion study 423*b*
mission 375, 378
Wildlife and Countryside Act 1981 346
Wilson, E.O. 166
wind, *vs.* energy expenditure 284
winter conditions
vs. cub cohort size 253
effect on body condition 252
influence on BCI 292, 293
vs. reproductive success 252–3
vs. RTA mortality 269–71, 270*f*
Scottish badger study 256–7*f*
winter telomere shortening 433–5
within-group offspring (WGO)
performance *vs.* EGO 407–9
superior natal fit hypothesis 407
Wolff, Jerry 241
wolverines (*Gulo gulo*) 462
paternal interaction with cubs 466
wolves 473n31
breeding behaviour 391
hypoxia adaptation 442*b*

predation risk 446
territoriality 166
wombats 477*b*
Wong, Josephine 83*b*
Wood, Andrew 377, 378, 379, 380, 381
Woodchester Park badgers 350
allozyme studies 388–9
wood mouse (*Apodemus sylvaticus*), reproductive strategy 294
Woodroffe, Rosie 58, 61, 120, 337, 361, 389
wounds 85
see also bite wounds; scars
Wrangham, Richard 143

Y
yearling breeders 36–7, 38
yellow-throated martens (*Martes flavigula*) 468
yelp 85*b*
Younger Dryas stadial period 250*b*

Z
Zahavi, Amotz 398
Zanette, Lianne 446
zoonotic disease 307
zorillas (*Ictonyx striatus*) 462
Zuk, Marlene 398